ABC of
# Emergency Radiology

**Third Edition**

# ABC series

An outstanding collection of resources for everyone in primary care

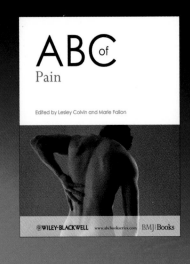

**ABC of Pain**

Edited by Lesley Colvin and Marie Fallon

WILEY-BLACKWELL   www.abcbookseries.com   BMJ Books

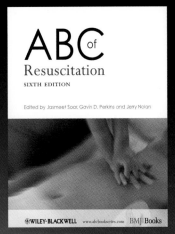

**ABC of Resuscitation**

SIXTH EDITION

Edited by Jasmeet Soar, Gavin D. Perkins and Jerry Nolan

WILEY-BLACKWELL   www.abcbookseries.com   BMJ Books

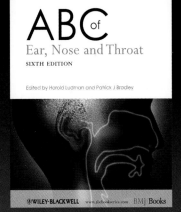

**ABC of Ear, Nose and Throat**

SIXTH EDITION

Edited by Harold Ludman and Patrick J Bradley

WILEY-BLACKWELL   www.abcbookseries.com   BMJ Books

**ABC of Occupational and Environmental Medicine**

THIRD EDITION

Edited by David Snashall and Dipti Patel

WILEY-BLACKWELL   www.abcbookseries.com   BMJ Books

The *ABC* series contains a wealth of indispensable resources for GPs, GP registrars, junior doctors, doctors in training and all those in primary care

▶ **Highly illustrated, informative and a practical source of knowledge**

▶ **An easy-to-use resource, covering the symptoms, investigations, treatment and management of conditions presenting in day-to-day practice and patient support**

▶ **Full colour photographs and illustrations aid diagnosis and patient understanding of a condition**

For more information on all books in the *ABC* series, including links to further information, references and links to the latest official guidelines, please visit:

## www.abcbookseries.com

**BMJ|Books**

# Emergency Radiology

## Third Edition

EDITED BY

*Otto Chan*

Consultant Radiologist
The London Independent Hospital
London, UK

**⟨W⟩WILEY-BLACKWELL**

A John Wiley & Sons, Ltd., Publication

BMJ|Books

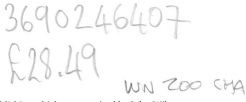

This edition first published 2013, © 2013 by Blackwell Publishing Ltd.

Previous editions 1995, 2007

BMJ Books is an imprint of BMJ Publishing Group Limited, used under licence by Blackwell Publishing which was acquired by John Wiley & Sons in February 2007. Blackwell's publishing programme has been merged with Wiley's global Scientific, Technical and Medical business to form Wiley-Blackwell.

*Registered office:* John Wiley & Sons, Ltd, The Atrium, Southern Gate, Chichester, West Sussex, PO19 8SQ, UK

*Editorial offices:* 9600 Garsington Road, Oxford, OX4 2DQ, UK

The Atrium, Southern Gate, Chichester, West Sussex, PO19 8SQ, UK

111 River Street, Hoboken, NJ 07030-5774, USA

For details of our global editorial offices, for customer services and for information about how to apply for permission to reuse the copyright material in this book please see our website at www.wiley.com/wiley-blackwell

The right of the author to be identified as the author of this work has been asserted in accordance with the UK Copyright, Designs and Patents Act 1988.

*Library of Congress Cataloging-in-Publication Data*

ABC of emergency radiology. – 3rd ed. / edited by Otto Chan.
    p. ; cm.
  Includes bibliographical references and index.
  ISBN 978-0-470-67093-4 (pbk. : alk. paper)
  I. Chan, Otto.
  [DNLM: 1. Radiography. 2. Emergencies. WN 200]
  616.07′572–dc23

2012032717

A catalogue record for this book is available from the British Library.

Wiley also publishes its books in a variety of electronic formats. Some content that appears in print may not be available in electronic books.

Cover images: Courtesy of the editor
Cover design by Meaden Creative

Set in 9.25/12 Minion by Laserwords Private Limited, Chennai, India
Printed in Singapore by Ho Printing Singapore Pte Ltd

1   2013

# Contents

# Contributors

**Muaaze Ahmad**
Consultant Musculoskeletal Radiologist, Barts Health NHS Trust, The Royal London Hospital, London, UK

**Syed Babar**
Consultant Radiologist and Honorary Senior Lecturer, Hammersmith & Charing Cross Hospitals, Imperial College, London, UK

**Dominic Barron**
Consultant Musculoskeletal and Trauma Radiologist, Leeds Teaching Hospitals, Leeds, UK

**Otto Chan**
Consultant Radiologist, The London Independent Hospital, London, UK

**Joe Coyle**
Fellow, Joint Department of Medical Imaging, University of Toronto, Toronto, ON, Canada

**Ahmed Daghir**
Clinical Fellow in Musculoskeletal Radiology, Oxford University Hospitals NHS Trust, Nuffield Orthopaedic Centre, Oxford, UK

**Marina J. Easty**
Consultant Radiologist, Great Ormond Street Hospital, London, UK

**David A. Elias**
Consultant Radiologist, King's College Hospital, London, UK

**Tim Fotheringham**
Consultant Interventional Radiologist, Barts Health NHS Trust, The Royal London Hospital, London, UK

**Simon Holmes**
Consultant Oral and Maxillofacial Surgeon, Bart's Health NHS Trust, London, UK

**Tudor Hughes**
Associate Professor of Clinical Radiology, Department of Radiology, University of California, San Diego, CA, USA

**Rosy Jalan**
Consultant Musculoskeletal Radiologist, Barts Health NHS Trust, The Royal London Hospital, London, UK

**Leonard J. King**
Consultant Musculoskeletal Radiologist, Southampton University Hospitals, Southampton, UK

**Andreas Koureas**
Associate Professor of Radiology, University of Athens, Athens, Greece

**Jimmy Makdissi**
Senior Lecturer/ Honorary Consultant, Barts Health NHS Trust, Institute of Dentistry, The London Hospital School of Medicine and Dentistry, London, UK

**Anmol Malhotra**
Consultant Radiologist, Radiology Department, Royal Free London NHS Foundation Trust, London, UK

**Lisa Meacock**
Consultant Radiologist, King's College Hospital, London, UK

**Amrish Mehta**
Honorary Senior Lecturer in Neuroradiology, Imperial College London, London, UK

**Arjun Nair**
Specialist Registrar, Radiology, St Georges Hospital, London, UK

**Ali Naraghi**
Staff Radiologist, Joint Department of Medical Imaging, University of Toronto, Toronto, ON, Canada

**Ravikiran Pawar**
Clinical Lecturer in Dental and Maxillofacial Radiology, Barts Health NHS Trust, The London Hospital School of Medicine and Dentistry, London, UK

**Katie Planche**
Consultant Radiologist, Royal Free London NHS Foundation Trust, London, UK

**Niall Power**
Consultant Radiologist, Upper GI Cancer, Royal Free London NHS Foundation Trust, London, UK

**Jeremy Rabouhans**

Consultant Radiologist (Locum), Royal Free London NHS Foundation Trust, London, UK

**Ian Renfrew**

Consultant Interventional Radiologist, Barts Health NHS Trust, The Royal London Hospital, London, UK

**R.J. Paul Smith**

Radiology Specialist Registrar, Barts Health NHS Trust, The Royal London Hospital, London, UK

**James Teh**

Consultant Radiologist, Oxford University Hospitals NHS Trust, Nuffield Orthopaedic Centre, Oxford, UK

**Suki Thomson**

Clinical Fellow in Neuroradiology, Lysholm Department of Radiology, National Hospital for Neurology and Neurosurgery, London, UK

**Sujit Vaidya**

Consultant Musculoskeletal Radiologist, Barts Health NHS Trust, The Royal London Hospital, London, UK

**Ioannis Vlahos**

Honorary Senior Lecturer, St George's, University of London, London, UK

**James A.S. Young**

Specialist Registrar in Trauma & Orthopaedics, St. Georges Hospital, London, UK

**Jeremy W.R. Young**

The Regional Medical Center, Orangeburg, SC, USA

# Preface

Emergency medicine is under scrutiny as never before, with daily newspaper reports highlighting the inadequacies for the provision of a 24/7 medical service. Reduced 'out-of-hours' survival rates for virtually all forms of acute medicine – not least trauma, strokes and heart attacks – have led to a rethink in strategy for emergency care in the UK.

Specific recommendations to address recognized deficiencies such as reorganisation of trauma care into regional systems has been shown to improve outcome and this relies on optimising all aspects of pre-hospital and hospital care and using a multidisciplinary approach to optimise patient care in polytrauma.

There have been dramatic technological advances in the past decade in diagnostic radiology that are central to the provision of a 24/7 service, in particular digital radiography (DR), picture archiving and communication systems (PACS), portable ultrasound (US), interventional radiology, magnetic resonance imaging (MRI) and multidetector computed tomography (MDCT).

Rapid acquisition and interpretation of DR, portable US and immediate availability of MDCT are now the mainstay of initial successful management of sick and traumatised patients admitted to Accident and Emergency departments (A&E).

Virtually any condition can present to A&E and so the volume of medical knowledge needed to manage these patients satisfactorily is enormous. Despite the reorganisation of acute medical services, unfortunately these patients are still initially seen and often treated by relatively inexperienced staff, most if not all with little or no training in radiology, in particular the interpretation of DR and CT. Although safety nets exist – in particular now that PACS is widely available in the UK – specialist radiological advice is still often not available at the time of presentation, when it is most needed.

Therefore, it is essential that all staff should be able to interpret DR and basic CT for fast, accurate and effective initial treatment, in order to avoid errors in interpretation, inappropriate treatment and the medicolegal consequences that may result from these errors.

This new edition of the *ABC of Emergency Radiology* has incorporated the latest technological advances (in particular replacing plain radiographs with digital radiographs) and changes in imaging protocols (in particular the use of MDCT).

Myself and the contributors have produced a simple and logical step-by-step approach on how to interpret DR and basic CT. The book is divided into anatomical chapters, followed by chapters in US, CT, paediatrics and major trauma. Each chapter starts with radiological anatomy, then recommended standard views, then a systematic approach to basic interpretation followed by a review of common abnormalities and finally a summary chart.

This book provides an up-to-date, simple, concise and systematic approach to the interpretation of DR and CT that should be very helpful to medical students, young trainee doctors, consultants in all specialties including radiologists and other health professionals working in A&E, not least radiographers and nurses.

Otto Chan

# CHAPTER 1

# Introduction: ABCs and Rules of Two

*Otto Chan*

The London Independent Hospital, London, UK

---

**OVERVIEW**

- Request the correct investigation
- Use a systematic approach to interpretation – ABCs
- Fundamental principles to avoid errors – Rules of two
- Always ASK for help – if in doubt!

---

Emergency medicine often brings together critically ill patients and inexperienced and tired doctors – a dangerous combination at the best of times with potentially serious clinical and medicolegal consequences. Virtually any medical condition can present in the emergency department (ED) and so the volume of medical knowledge needed to manage these patients satisfactorily is enormous.

There have been major technological advances in the past decade which have had a major impact on the management of patients in the ED, not least picture archiving and communication systems (PACS), digital radiography (replacing conventional plain to X rays), portable ultrasound (US; which is now readily available and often, but not often enough, performed by clinicians in the ED) and multidetector computed tomography (MDCT) in the ED. Despite all these advances, plain to X rays (whether conventional or digital) remain the mainstay of initial and successful management of most sick and traumatised patients in the ED.

---

**Radiological investigations**

- Plain to X rays (conventional or digital)
- Portable US
- MDCT
- MRI

---

The correct selection of imaging modality, rapid acquisition and the accurate interpretation of these investigations is often the key to quick and successful management of patients in the ED. Unfortunately these investigations are often done and interpreted

---

*ABC of Emergency Radiology*, Third Edition. Edited by Otto Chan.
© 2013 John Wiley & Sons, Ltd. Published 2013 by John Wiley & Sons, Ltd.

by medical staff who have little, if any, training in radiology and the usual safety net of a specialist radiological service is not available at the time of presentation, when it is most needed. This leads to delays and invariably results in increased morbidity and mortality! The selection of the correct imaging modality on admission saves time and saving time, saves lives! Ideally there should be a seamless 24/7 service.

## MDCT – initial imaging modality of choice in the ED

| | |
|---|---|
| Head injuries/headaches or epilepsy | Skull X-ray (SXR) no longer done. CT head ± contrast |
| Facial injuries | MDCT with multiplanar reconstructions (MPR) and 3D are essential |
| Chest pain (suspected aortic aneurysm (AA), myocardial infarction (MI), pulmonary embolus (PE) or pneumothorax (Px)) | Triple rule out CT scan |
| Severe abdominal pain (obstruction) | CT has replaced abdominal X-ray (AXR) |
| Renal/ureteric colic | CT kidneys, ureters and bladder (KUB) has replaced Intravenous urogram (IVU) in ED |
| Suspected leaking abdominal aortic aneurysm (AAA) | CT has replaced US and AXR |
| Suspected gastrointestinal (GI) bleeding | Initially CT angiography instead of angiography |
| Major trauma (adults) | Whole body CT instead of chest X-ray (CXR), AXR and US |

The Rules of Two (Ro2) is a helpful, simple set of guidelines, which relate to who, what, when and how to radiograph and how to get help or get out of trouble and therefore minimise the chances and the consequences of errors.

The ABCs systematic assessment is a simple systematic approach, which starts with basic essential normal radiographic anatomy, common normal variants, which may mimic pathology and in particular how to interpret imaging using a systematic approach, which is logical and easy to remember and therefore hopefully helps to minimise interpretive errors.

## Rules of Two

These rules represent a simple set of guidelines, most are obvious, some relate to specific clinical problems, but most are common sense useful general principles which should help in avoiding errors in interpretation and management of patients in the ED.

---

### Rules of Twos

- Two views – one view is always one view too few
- Two abnormalities – if you see one abnormality, always look for a second
- Two joints – image the joint above
- Two sides – if not sure or difficult X ray, compare with other side
- Two views too many – CT (and rarely US) has replaced plain X rays in many clinical situations
- Two occasions – always compare with old films IF available
- Two visits – bring patient back for repeat examination
- Two opinions and two records – always ask a colleague if not sure and record findings
- Two specialists – always get your ED specialist and also a radiologist's opinion
- Two investigations – always consider whether US, CT or MRI would help in diagnosis

---

### Rule 1 – two views ('One view is always one view too few')

Two views should be taken, preferably perpendicular to each other (Figure 1.1). This applies to all radiographs except the chest, abdomen and pelvis. It is not uncommon for a fracture or an abnormality to be visible only on one view (Figure 1.2).

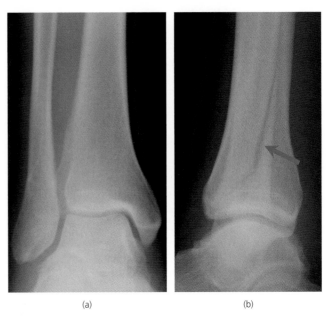

(a)                              (b)

**Figure 1.2** (a) Anteroposterior shows no obvious abnormality; (b) lateral shows oblique fracture of fibula.

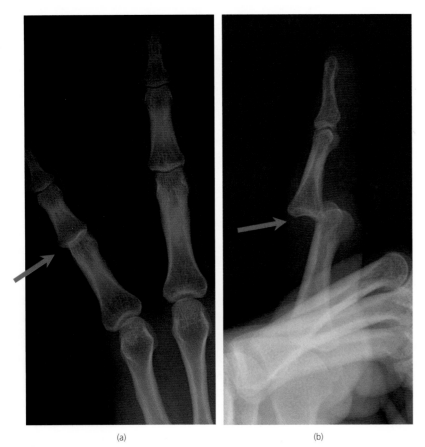

(a)                              (b)

**Figure 1.1** (a) Anteroposterior shows minimal overlap of proximal interphalangeal joint (PIPJ) of little finger; (b) lateral shows obvious dislocation of PIPJ.

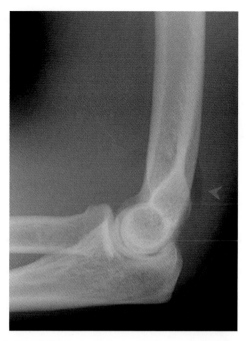

**Figure 1.3** Lateral elbow: anterior (arrow) and posterior fat pads (arrowhead). Anteroposterior view was normal so additional views of radial head were requested.

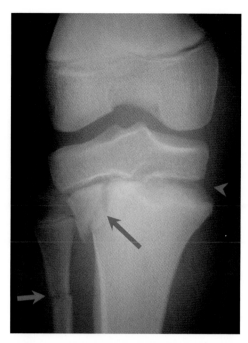

**Figure 1.5** Salter–Harris II of proximal tibia (large arrow for fracture of proximal tibial metaphysis, arrowhead for separation of epiphyseal growth plate) and fracture of shaft of proximal fibula (small arrow).

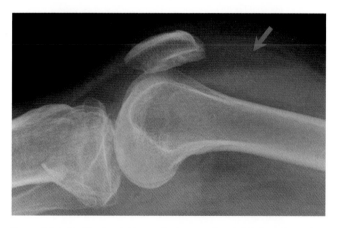

**Figure 1.4** Lateral horizontal beam: lipohaemoarthrosis fat above (arrow); blood below (arrowhead).

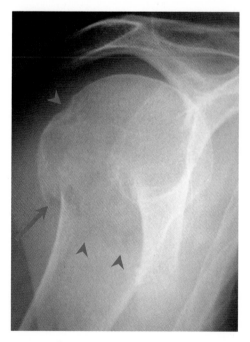

**Figure 1.6** Fracture of proximal humerus (arrow) with subtle lytic metastases (arrowheads).

In addition, if the two views fail to show an injury when there is a radiological suspicion of an injury (such as the presence of a fat pad sign (Figure 1.3) in the elbow or a lipo-haemoarthrosis (Figure 1.4) of the knee or if the findings don't fit in with the clinical presentation, then further views are warranted.

### Rule 2 – two abnormalities

Do not stop looking after detecting one abnormality; always keep looking for a second abnormality. There may be more than one fracture (Figure 1.5) or there may be an underlying predisposing abnormality, such as metastases (Figure 1.6). In addition, if there is a fracture in a ring-like structure such as the pelvis (Figure 1.7), mandible (Figure 1.8), radius/ulna or tibia/fibula, there will usually be a second fracture (a polo mint will always break in two places; see Figure 1.9).

### Rule 3 – two joints

In the forearm (Figure 1.10) and in the lower leg (Figure 1.11), always image the joint above and below the injury.

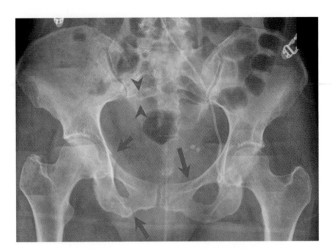

Figure 1.7 Numerous pelvic fractures (arrows), several in each ring with subtle right sacral foramina fractures (arrowheads – compare with intact left side).

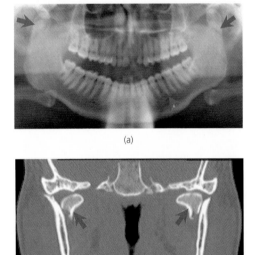

(a)

(b)

Figure 1.8 (a) Left mandibular fracture with bilateral displaced condylar neck fractures, better seen on CT (b).

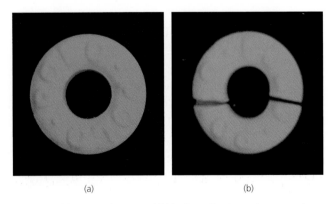

(a)

(b)

Figure 1.9 (a) Intact Polo mint and (b) broken mint shows there are at least two fractures in a ring.

Figure 1.10 Monteggia fracture-dislocation: fractured proximal ulnar shaft (arrow) with dislocated radial head (arrowhead).

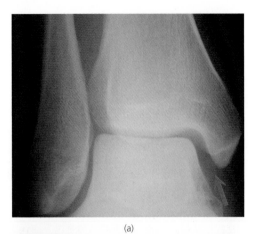

(a)

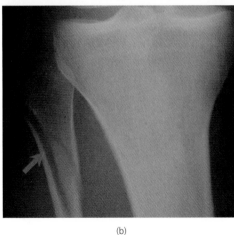

(b)

Figure 1.11 Maisonneuve fracture: (a) right ankle anteroposterior shows minimal widening of the medial joint margin (arrow); (b) this is associated with a proximal fibula fracture (arrow).

## Rule 4 – two sides

There are certain circumstances, where it is difficult to ascertain whether something is normal (Figures 1.12 and 1.13), or abnormal (Figures 1.14–1.16), in particular in children and on these occasions it is worth imaging the asymptomatic side for comparison.

## Rule 5 – two views too many

The advent of MDCT and the siting of MDCT scanners in the ED, has made MDCT the INITIAL imaging modality of choice in many

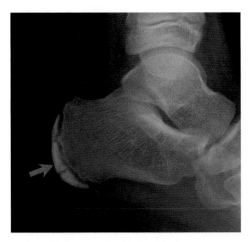

**Figure 1.12** Normal dense calcaneal apophyses (arrow) – same on both sides.

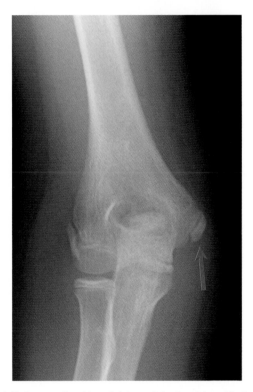

**Figure 1.13** Suspected avulsion of right medial epicondyle ossification centre, but identical on the left elbow (arrow).

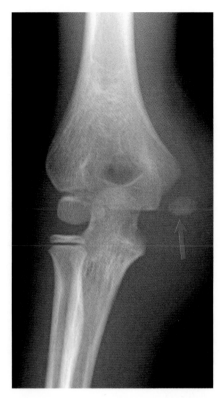

**Figure 1.14** Mild avulsion of right medial epicondyle ossification centre (arrow) when compared with left elbow.

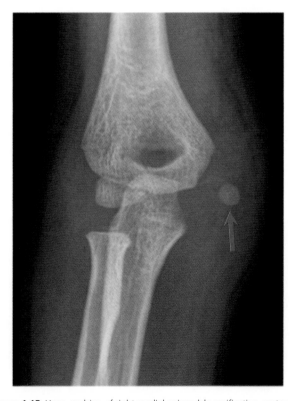

**Figure 1.15** Huge avulsion of right medial epicondyle ossification centre (arrow).

clinical presentations (Figures 1.17 and 1.18). This is emphasised in each relevant chapter, but it is critical that the correct imaging modality is chosen to save time and to avoid the increased morbidity and mortality of delayed management.

*Save time, saves lives!*
Furthermore, there are clinical situations which mandate immediate treatment of the patient to 'save life or limb' and clearly in these circumstances, imaging is inappropriate!

The extreme example is a patient with a penetrating injury to the chest, who clinically has all the signs of a cardiac tamponade and

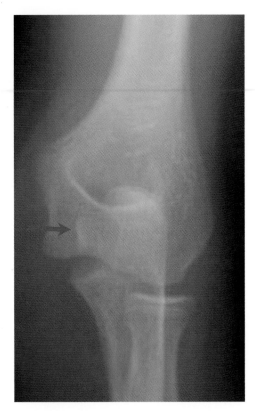

**Figure 1.16** Completely displaced avulsion of left medial epicondyle ossification centre (arrow).

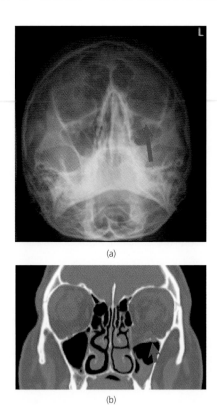

**Figure 1.18** (a) OM view with left tear drop – classic blow-out fracture; (b) Cor CT shows orbital floor fracture with inferior rectus muscle displaced.

who clearly needs an immediate open thorocotomy and instead the clinical team request a multiple X-rays! The other extreme is when a patient has an open bleeding wound, which just needs dressings and compression initially and instead the clinical team request plain X-rays to look for a fracture! Sadly, both of these examples occur commonly!

### Rule 6 – two occasions

ALWAYS look for old films for comparison, in particular if there is an abnormality. Old films are the cheapest and best investigation, in particular CXRs (Figure 1.19). Similarly, if there is a bony abnormality, an old film will not only confirm whether this is an old or new finding, but also the speed of change will help in the diagnosis (Figure 1.20).

### Rule 7 – two visits

ALWAYS repeat the X ray after a patient has undergone an intervention, in particular after insertion of a line or tube (central lines, endotracheal (ET) tubes, chest drains, etc.), reduction of a dislocation, putting a plaster of Paris (POP) (Figure 1.21) or cast or removal of a foreign body (FB).

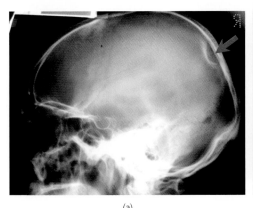

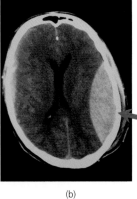

(a)                    (b)

**Figure 1.17** (a) Depressed fracture seen on SXR (arrow); (b) Large left extradural haematoma (arrow) and subtle right subdural collection (arrowhead) and numerous right intraparenchymal contusions seen on CT.

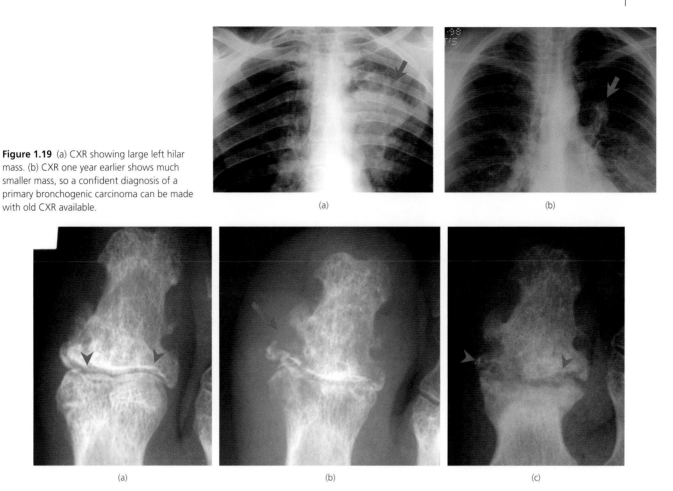

**Figure 1.19** (a) CXR showing large left hilar mass. (b) CXR one year earlier shows much smaller mass, so a confident diagnosis of a primary bronchogenic carcinoma can be made with old CXR available.

(a)

(b)

(a)

(b)

(c)

**Figure 1.20** (b) Initial film showed loss of joint space a large erosion (arrow) and a huge amount of soft tissue swelling, which was not present on X ray taken (arrow) 1 week earlier (a) or after treatment 1 month later (c). Confident diagnosis of a septic arthritis was made with old film available showing rapid progression.

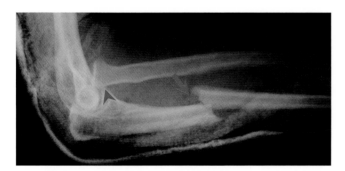

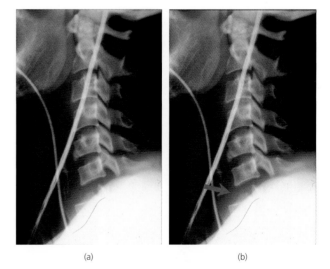

(a)

(b)

**Figure 1.21** Patient with fractured ulna (arrow) put in plaster of Paris, but follow-up check X-ray reveals a dislocated radial head (arrowhead) – a Monteggia fracture dislocation.

**Figure 1.22** (a) Lateral CS reported normal initially! (b) Lat CS clearly shows a hugely displaced C6 on C7 with marked separation.

## Rule 8 – two opinions and two records

If you are not sure, get a second opinion from a colleague or the radiographers (who are extremely knowledgeable). The red dot system is a form of a second opinion.

There are particular X-rays which are more challenging (children with unfused epiphyses and AXRs) or where missing an abnormality may have serious consequences (cervical spine X-rays (Figure 1.22) and suspected non-accidental injury (NAI) (Figure 1.23)). Under these circumstances, you should ALWAYS get a second opinion and record both your interpretations, both in the notes and also in the radiology information system (RIS). If at a later date, a major abnormality has been missed or overlooked, it means that you

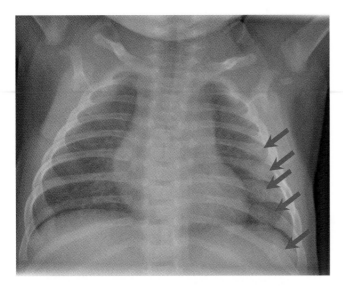

**Figure 1.23** CXR thought initially to be 'normal'; review shows clearly at least 5 left rib fractures consistent with NAI – referred immediately to paediatricians as suspected NAI.

can share the blame! The mere fact that you sought out a second opinion and recorded it, shows common sense and insight (even if it's not the case!).

### Rule 9 – two specialists

The report and the X-rays should be checked (Figure 1.23) both by your seniors (first specialist) and also a formal report should ALWAYS be sought from the radiologists (second specialist). If two specialists have missed the abnormality, the likelihood of the case of ending up in court in the UK is virtually zero!

### Rule 10 – two investigations

There are times when the initial X-ray has helped to exclude an obvious abnormality, but the history or clinical findings warrant further investigations (Figure 1.24). These investigations should be done as soon as possible but often are done the following day on a routine list. The problem is that delaying these investigations often leads to the investigation not being done or the team that looks after the patient forgetting about it!

Patients with head injuries and with suspected facial injuries often have a CT scan of the head but the facial injuries are not scanned at the time. Similarly, major trauma patients often have multiple intra-articular fractures detected on the plain X-rays and then have whole body CT, but the intra-articular fractures are not included in the CT scanning protocol. In both these groups of patients, the patient has to be brought back to the CT room. This is clearly not efficient use of CT and it is dangerous to move the 'stable' polytraumatised patient!

### ABCs systematic assessment

The ABCs systematic approach is without doubt the simplest and worldwide the most popular and widely used system for the assessment of plain X-rays. However, in the context of patients in the ED, the ABCs systematic approach is equally effective for the interpretation and assessment of whole body CT and seamlessly incorporates an advanced trauma life support (ATLS) approach in trauma.

Its as simple as ABC!

### Anatomy

The ABCs starts in ALL the chapters with a short section on 'basic' radiological anatomy, which is the 'bread and butter' of radiology!

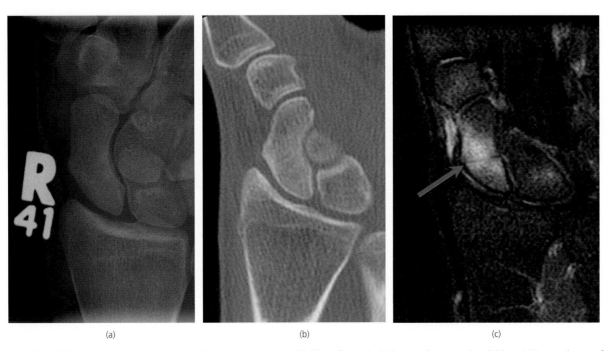

(a)                                            (b)                                            (c)

**Figure 1.24** (a) Initial X ray reported normal, but patient symptomatic, so (b) CT performed which was also normal and (c) an MRI was then performed confirming fracture though the waist of the scaphoid.

The images are colour coded, so that ALL "normal" anatomy diagrams, lines and annotations are shown in white or shades of blue.

| Know your anatomy and you know radiology! |

### Adequacy

Anatomy is followed in ALL the chapters by adequacy. Under adequacy, we have included the imaging investigation of choice and either basic knowledge necessary to understand the modality and the 'standard views' and also the other views available.

## Peripheral and axial skeleton

The ABCs for the bony skeleton are the same, although from an interpretive point of view, the order should be changed so that soft tissues are looked at first in the elbow (looking for the displaced fat pad sign) and the knee (looking for fat-fluid levels indicating a lipohaemoarthosis). Both these signs are easy to detect and immediately alert the clinician to the presence of a probable fracture.

### Alignment

- Exclude subluxations and dislocations

### Bone

- Follow the bony contour very carefully and exclude a fracture
- Check for disruption of the trabecular pattern, linear sclerosis or lucency indicating a fracture

### Cartilage and joints

- Check for even joint spaces and uneven loss of joint width

### Soft tissues

- Bright light X-rays or change 'window setting and contrast on The PACS system
- Look for FBs

## Head CT scan

### Airspaces

- Sphenoid sinus – look for fluid level
- Frontal sinus/mastoids/middle ear – look for fracture or infection
- Pneumocephalus – look for sinus or vault fracture

### Bones

- Look on bone windows and thin sections
- Look carefully over areas with soft tissue swelling

### Brain parenchyma

- Look for low density lesions
- Look for high density lesions

- Blood density and implications
- Look for signs of brain swelling

### CSF spaces

- Look for blood
- Look for mass effect
- Look for hydrocephalus

### Dura

- Look for subdural and extradural collections

### Eyes

- Globe injuries
- Optic nerve injuries
- Fractures
- Extraconal spaces

### Face

- Fractures
- Foreign body

### Survey (review areas)

- Scout
- Symmetry
- SAH
- Subtle
- Skull vault

### Face
### Alignment

- Check Dolan's, McGrigor's and Campbell's lines

### Bones

- Check all bones in the mid face and upper and lower third of the face.

### Cartilage and joints

- Check the ZF sutures and TMJs

### Sinuses and soft tissues

- Check for local swelling of the soft tissue
- Surgical/orbital emphysema
- Air-fluid levels and opaque sinuses
- Teardrop injuries
- FBs such as glass and metal

### Chest – CXR
### Airways

- Check endotracheal tube (ETT)
- Exclude FB in airway
- Position of trachea

## All lines

- ETT position
- Nasogastric (NG) tube position
- Venous catheters
- Chest drains

## Breathing

- Check lungs are clear

## Circulation

- Cardiac silhouette size
- Mediastinal position and contour
- Widening of mediastinum
- Hila: evaluate size, shape, and position

## Diaphragm

- Position
- Below the diaphragm

## Edges (pleura)

- Pneumothorax
- Effusion
- Empyema

## Skeleton

- Fractures
- Paraspinal lines
- Spine

## Soft tissues

- FB
- Emphysema
- Swelling/asymmetry

## Abdomen
### Air

- Exclude free intraperitoneal or abnormally sited air

## Bowel gas

- Check size, distribution and pattern

## Calcification

- Check for normal and abnormal calcification

## Densities

- Check for inserted or ingested foreign bodies

## Edges

- Check the hernial orifices
- Check the lung bases and pleural spaces

## Fat planes

- Check presence and symmetry of psoas shadows
- Check presence of perivesical fat plane
- Check that properitoneal fat planes are present

## Soft tissues

- Check for enlarged or absent organs. Confirm with US

## Skeleton

- In trauma, check that there are no obvious fractures
- If malignancy is suspected, exclude bony metastases

## Further reading

Krishnam MS, *Curtis J (eds). Emergency Radiology*. Cambridge University Press, 2009.

Marincek B, Dondelinger RF (eds). *Emergency Radiology: Imaging and Intervention*. Springer, 2006.

Mirvis SE, Shanmuganathan K, Miller LA. *Emergency Radiology: Case Review*. Mosby, 2009.

Raby N, Berman L, De Lacey G. *Accident and Emergency Radiology: A Survival Guide*, 2nd edn. Saunders, 2005.

Rogers LF. *Radiology of Skeletal Trauma*. Churchill Livingstone, 2001.

# CHAPTER 2

# Hand and Wrist

*Joe Coyle[1], Ali Naraghi[1] and Otto Chan[2]*

[1]University of Toronto, Toronto, ON, Canada
[2]The London Independent Hospital, London, UK

---

**OVERVIEW**

- Hands and wrist fractures account for 20% of acute fractures
- Age alone can accurately predict most injuries
- Plain radiographs remain the mainstay of imaging
- MRI (and CT) are developing increasing roles
- Aim is to restore function and avoid chronic disability

---

Injuries to the hand and wrist are very common, accounting for 20% of acute fractures presenting to emergency departments. The hand is the most active part of the body, is the least well protected and thus is often injured.

Most injuries to the wrist occur following a fall onto an outstretched hand (FOOSH). Mechanism of injury in these patients can accurately predict injury pattern. Age alone also can accurately predict likely fracture pattern (Table 2.1).

Clinical exam is usually accurate in this scenario and strong clinical suspicion for fracture can often direct close radiologic evaluation for subtle abnormalities.

The goal of treatment is rapid restoration of function with attention given to the prevention of chronic disability. Plain radiographs are the mainstay of imaging. Computed tomography (CT) and magnetic resonance imaging (MRI) are developing increasing roles, particularly as their availability increases.

## Anatomy

### Hand

Each ray, apart from the thumb, consists of a metacarpal and proximal, middle and distal phalanges. The thumb has a metacarpal

**Table 2.1** Age as a predictor of distal radial fractures following FOOSH.

| Age | Fracture pattern |
| --- | --- |
| <10 | Transverse metaphyseal (often incomplete) |
| 10–16 | Epiphyseal plate (Salter–Harris type injury) |
| 17–40 | Scaphoid and triquetral fractures |
| >40 | Transverse distal radial fractures |

*ABC of Emergency Radiology*, Third Edition. Edited by Otto Chan.
© 2013 John Wiley & Sons, Ltd. Published 2013 by John Wiley & Sons, Ltd.

and proximal and distal phalanges. At each metacarpophalangeal (MCP) joint and interphalangeal (IP) joint, lateral stability is provided by the collateral ligaments. The joint capsule at the MCP and IP joints also demonstrate on the volar aspect areas of dense fibrous thickening, known as the volar plate, which provide further strength. Each finger has two flexor tendons on the volar (palmar) surface and an extensor tendon complex on the dorsal surface.

### Wrist

The wrist (Figure 2.1a–c, e and f) consists of eight carpal bones arranged in two rows. The proximal row (scaphoid, lunate, triquetrum and pisiform) articulates wth the radius and ulna and the distal row (trapezium, trapezoid, capitate and hamate) articulates with the bases of the metacarpals. The distal row is more rigid and stable than the proximal. These bones are held together by a complex arrangement of strong ligaments. The radiocarpal joint has a 4–15° volar tilt and the hand is usually held in slight flexion and ulnar deviation. The radial styloid is distal to the ulnar styloid. Radial inclination to the ulna is assessed on the PA view and should be 20–25°.

In children the carpal bones first appear at the age of 3 months and all of the carpal bones are visible by 12 years. The age of a child can be estimated by counting the number of epiphyses minus one (see Chapter 16).

## ABCs systematic assessment

- **A**dequacy – check correct views have been obtained
- **A**lignment – check the relationship of the individual bones to each other
- **B**one – trace the contours of all the bones
- **C**artilage and joints – joint spaces should be uniform in width
- **S**oft tissues – change windows to look for soft tissue swelling and foreign bodies (FBs)

---

**Recommended radiological views**

- Hands – anteroposterior (AP), lateral or oblique
- Fingers – AP, lateral or oblique
- Wrist – AP, lateral ± oblique
- Scaphoid – coned scaphoid series (× 4)

---

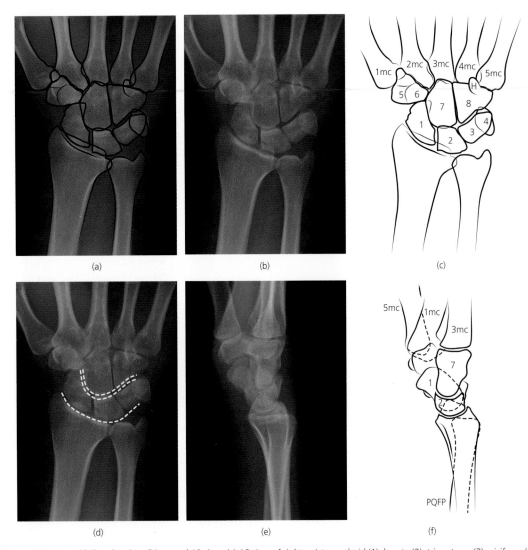

**Figure 2.1** (a) Normal AP view with line drawing; (b) normal AP view; (c) AP view of right wrist: scaphoid (1), lunate (2), triquetrum (3), pisiform (4), trapezium (5), trapezoid (6), capitate (7), hamate (8), hook of hamate (H), metacarpal (mc); (d) Gilula's three carpal arcs; (e) normal lateral view; (f) line drawing lateral view.

### Adequacy

The clinical findings should guide the radiologic views to be obtained. At least two views are mandatory and additional views may be necessary for specific injuries, such as scaphoid injuries, where coned views with an AP, lateral, oblique and a dedicated scaphoid view are indicated. MRI and CT are becoming increasingly available and are being used even in the acute setting.

### Hand

AP, lateral and oblique views are recommended for finger and hand injuries.

On the lateral views the fingers should be flexed to varying degrees to avoid overlap and confusing composite shadows.

### Wrist

In general a minimum of posteroanterior (PA) and lateral views are recommended, but in addition some centres advocate external

oblique views whereby the radial side of the wrist is elevated. If a scaphoid fracture is suspected, a PA view with ulnar deviation and also a dedicated scaphoid view with 20–30° of tube angulation are recommended in addition to a lateral and an oblique view (Figure 2.2).

On a true lateral wrist view, the palmar surface of the pisiform bone should overly between the palmar surfaces of the distal scaphoid pole and the capitate head.

CT is typically reserved for suspected fractures with negative initial and follow-up radiographs or for preoperative planning in cases with significant comminution and intra-articular extension. MRI is rarely indicated acutely, although some centres are doing MRIs on patient's with anatomical snuff box tenderness who have normal plain X rays (XRs) to exclude scaphoid fractures.

### Alignment

In alignment we look at bones and their relationship to each other.

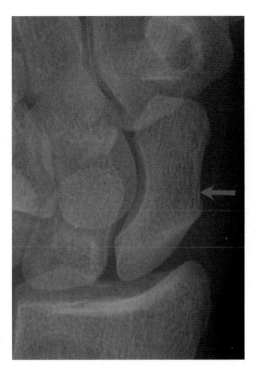

**Figure 2.2** Scaphoid coned PA view in ulnar deviation allows for visualisation of the full length of the scaphoid.

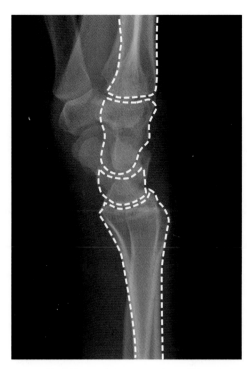

**Figure 2.3** Normal lateral view of the wrist outlining the normal relationship of the lunate and capitate bones and normal anatomy.

## Hand

Bony surfaces should be congruent along each ray from the metacarpals to distal phalanges. Alignment should always be assessed on at least two views. On AP views, overlap of joint margins may be the only indication of subluxation/dislocation. The carpometacarpal articulations in particular, where some degree of overlap is unavoidable, should be carefully scrutinised. Dislocations here may be overlooked.

## Wrist – AP view (see Figure 2.1a )

The intercarpal joint spaces should be uniform and <2 mm wide. Widening following injury, seen most commonly at the scapholunate articulation resulting in the Terry Thomas or Madonna sign, may be indicative of joint dissociation and ligamentous injury. The proximal and distal carpal rows form three arcs (Gilula's three carpal arcs). Arc 1 outlines the proximal surface of the scaphoid, lunate and triquetrium. Arc 2 outlines the distal surface of these same bones. Arc 3 outlines the proximal surface of the capitate and hamate bones. Disruption of one of these arcs suggests pathology at that site (Figure 2.1c).

The lunate should have a square (quadrilangular) shape. A 'pie-shaped' (triangular) lunate indicates a perilunate or lunate dislocation.

## Wrist – lateral view (Figure 2.3)

This can be daunting as there is significant overlap of many of the carpal bones! It is crucial to assess the alignment of the distal radius, lunate, capitate and 3rd metacarpal. The lunate is moon shaped (lunar = moon) and lies on its back on the distal radius (saucer on table). The proximal pole of the capitate sits into the concave distal surface of the lunate (cup in the saucer) and the third metacarpal should line up with the distal pole of the capitate. Interruption of this alignment is usually secondary to a perilunate or lunate dislocation (and should result in careful assessment for associated carpal and distal radial fractures).

## Bone

Trace the cortical contour of each bone on each projection. Fractures typically consist of a cortical step deformity, which may be visible on only one view. In more subtle cases there may be a subtle intramedullary lucency without significant visible cortical breach at initial presentation. Impacted or healing fractures may be manifested as an ill-defined sclerotic or dense band. As with all fractures, the location, direction, displacement, angulation and comminution of the fracture as well as the involvement of the articular surfaces should be assessed.

On the AP view, the normal fused distal radial epiphysis may present a slight irregularity on the radial aspect and may mimic a fracture. The dorsal surface of the distal radius typically shows a small area of irregularity representing Lister's tubercle, a normal anatomic landmark. Similarly vascular grooves in the mid shaft of the phalanges may mimic a fracture. Remember to use the digital 'windows' to look for soft tissue swelling associated with subtle fractures and to help with tricky calls.

## Cartilage

Joint spaces should be uniform in width. Narrowing may be due to technical factors (rotation, flexion, tilting) or disease (arthritis).

## Soft tissues

Careful attention should be paid to the cortical margins in regions of soft tissue swelling. Digital windowing of a radiograph may be required to adequately assess for soft tissue swelling particularly if the radiograph is overexposed. There is a fat plane volar to the distal radial metaphysis, along the volar aspect of the pronator quadratus muscle. This may be displaced (convex anterior surface) or obliterated in distal radial fractures. In the hand and wrist, soft tissue swelling often also spreads distal to the point of injury.

## Injuries

### Hand
### Phalanges
#### Shaft fractures

These may be transverse or spiral and may show angulation, shortening or rotational deformity. Angulation is particularly common in fractures through the proximal metaphysis of the proximal phalanx and should be carefully evaluated on the lateral view as there is often some overlap. The PA view often underestimates the degree of angulation. Rotation is often difficult to assess on plain radiographs and clinical examination, whereby rotation is more apparent on flexion, is important.

#### Crush fractures

Crush injuries to the terminal tuft are common and most often occur in the thumb or middle finger. These are generally stable but if associated with a significant soft tissue or nail bed injury may be treated as compound or open fractures (Figure 2.4).

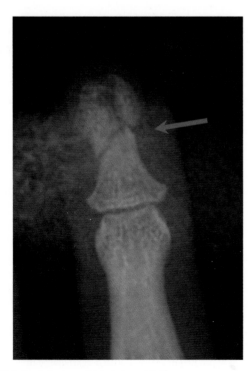

**Figure 2.4** Crush fracture distal phalanx, AP view.

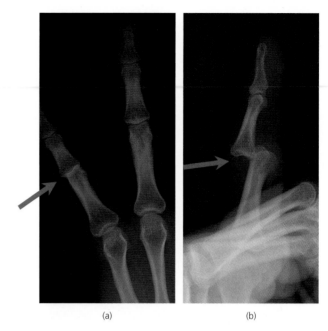

(a)                                             (b)

**Figure 2.5** Proximal interphalangeal joint dislocation. This injury is subtle on the AP view (a) and becomes obvious on the lateral view (b).

#### Dislocations/coach's finger

Interphalangeal joint dislocations are common. Most are dorsal and as well as significant ligamentous and soft tissue injury, there can be associated fractures. In sports, rapid reduction usually occurs at the pitchside, hence the term 'coach's finger' (Figure 2.5).

#### Boutonniere deformity

This is a less common acute injury, seen more often in the setting of rheumatoid arthritis, where there is rupture of the central slip of the extensor tendon at its insertion on the base of the middle phalanx. This produces a characteristic deformity with the proximal IP (PIP) held in flexion, and the distal IP (DIP) in extension. There is usually no other radiographic abnormality.

### Tendon and capsular avulsions
#### Mallet finger

This is caused by an acute flexion force on the distal phalanx of the finger such as by a direct blow from a ball in sport on the tip of the extended finger. The force results in avulsion of the extensor tendon at its insertion at the dorsal lip of the base of the distal phalanx. Clinically active extension is lost and the DIP is held in partial flexion. There may or may not be a small avulsed bony fragment. If greater than one-third of the bony articular surface of the DIP joint is avulsed, operative fixation may be indicated (Figure 2.6).

#### Jersey finger/volar plate avulsion

Both these injuries involve the volar aspect of the digits and occur as a result of a hyperextension injury. Jersey finger involves the base of the distal phalanx at the attachment of the flexor digitorum profundus tendon and the classic volar plate avulsion occurs at the base of the middle phalanx at the site of attachment of the

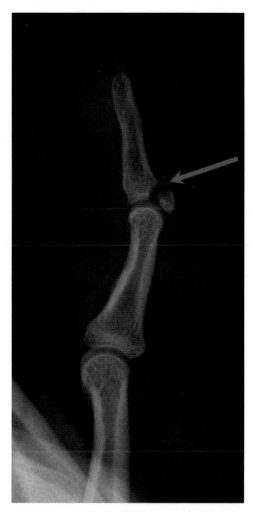

**Figure 2.6** Mallet finger, lateral view.

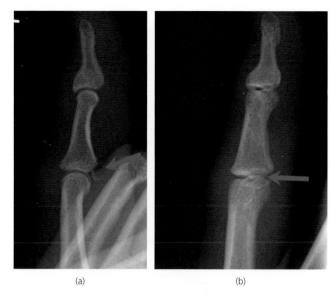

**Figure 2.7** Volar plate avulsion: (a) lateral and (b) oblique views. Occasionally these fractures are only visible on the oblique view.

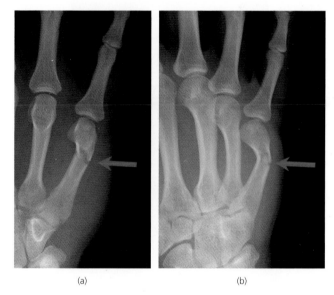

**Figure 2.8** Boxer's fracture: (a) AP and (b) oblique views showing the typical radial and volar angulation.

volar plate, an area of fibrous thickening of the volar capsule. Radiographically there is a small avulsed bony fragment. This can be very subtle and is often best seen on the lateral view. In subtle cases it may be missed on the lateral view but may be seen on the oblique view (Figure 2.7).

## Metacarpals

Metacarpal fractures account for at least a third of all hand fractures. The fifth metacarpal is the most commonly injured.

### Boxer's

This is a fracture of the neck of the fifth metacarpal. This results from a direct axial force as sustained in punching a solid object. The fracture typically occurs at the neck and angulates in a volar direction. Some degree of volar angulation is typically well tolerated. As with the phalangeal fractures, rotational abnormality is also of concern.

An axial load may also result in fracture dislocations through the bases of the metacrapals and at the carpometacarpal joints. This is most commonly seen through the base of the fourth and fifth metacarpals and may be associated with hamate fractures.

Following a punch injury occasionally a fracture is also seen through the base of the fifth metacarpal. Localised soft tissue swelling is often a clue to the site of injury (Figures 2.8 and 2.9).

## Carpometacarpal joint dislocations

An axial load may result in fractures or dislocations through the bases of the metacarpals. These are most commonly seen at the first and fifth carpometacarpal joints (Figures 2.10 and 2.11).

### Bennett's fracture

This is an intra-articular fracture dislocation involving the base of the first metacarpal along its palmar and ulnar aspect. The shaft

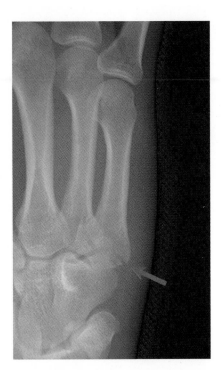

**Figure 2.9** Fracture of base of fifth metacarpal, oblique view.

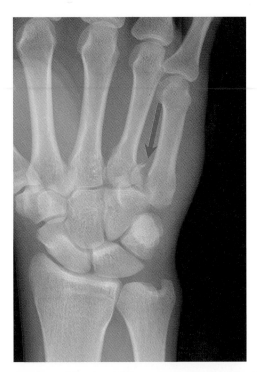

**Figure 2.11** Fracture dislocation of fifth carpometacarpal joint, PA view. Note the ulnar subluxation of the shaft of the fifth metacarpal with a longitudinal intra-articular fracture.

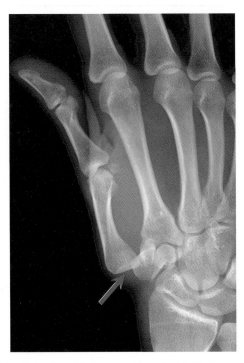

**Figure 2.10** Dislocation of first carpometacarpal joint, PA view.

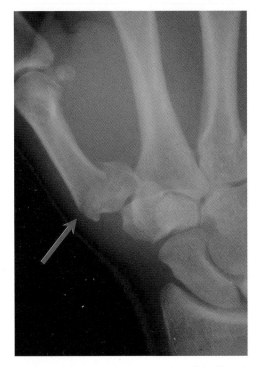

**Figure 2.12** Bennett's fracture, AP view. Note the radial subluxation of the distal fragment.

(distal fracture fragment) is typically displaced radially and dorsally by abductor pollicus longus. This an unstable injury at a very important articulation often requiring surgical fixation. Closed reduction is often complicated by redisplacement of the fracture fragments (Figure 2.12).

*Rolandos*

This is a comminuted intra-articular fracture through the base of the first metacarpal, often with Y or T configuration. It has a worse prognosis than a Bennett's fracture.

*Skier's (gamekeeper's) thumb*

This involves disruption of the ulnar collateral ligament at the first MCP joint from an acute stress such a fall with a ski pole causing forced abduction and extension. Injury typically occurs at the distal attachment of the ligament. A small avulsion fracture may be seen adjacent to the base of the proximal phalanx. In the absence of an osseous avulsion, integrity of the ligament is best assessed by MRI or ultrasound (Figure 2.13).

## Wrist
### Distal radius and ulna
Fractures here can be divided by age group and mechanism.

*Children/skeletaly immature (see Chapter 16)*
Childrens' bones are more pliable than those of adults and thus can bend a little. Fracture patterns reflect applied bending forces. Treatment of all these fractures is cast immobilisation.

*Torus*—This is a buckle fracture of the cortex from a longitudinal compression injury in a child. The cortex fractures on the compressive side. These occur most typically on the dorsal distal radius and may be seen on only one view (Figure 2.14).

*Greenstick*—This an incomplete transverse fracture from a longitudinal distraction injury in a child. The bone bends away from the injury and the distraction force causes a break in the cortex, like in immature fresh wood, 'greenstick'. This extends a variable distance across the bone but the opposite cortex is normal.

*Salter–Harris*—This is an injury to the physeal growth plate in an immature skeleton typically aged between 11 and 17 years. At this stage the bones are more rigid than those of the child and the point of inherent weakness is at the cartilaginous growth plate. There

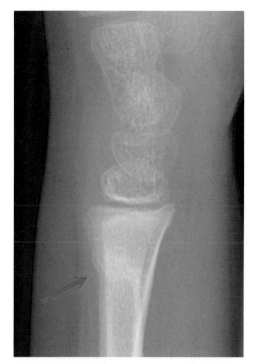

**Figure 2.14** Torus fracture, lateral view.

**Table 2.2** Salter–Harris type fractures.[a]

| Type | Pneumonic SALTE. . .R | Description |
| --- | --- | --- |
| 1 | S | Slip of the physis |
| 2 | A | Above the physis (spares the epiphysis) |
| 3 | L | BeLow the physis (involves the epiphysis) |
| 4 | T | Through the physis (involves both epiphysis and metaphysis) |
| 5 | EveRything | Everything (crush injury to the physis) |

[a] See Chapter 16.

are five fracture patterns (Table 2.2). At the distal radius, type 2 injuries are most common. This type involves the growth plate and the metaphysis, sparing the epiphysis. Treatment is usually cast immobilisation. If the epiphysis is fractured (type 3 and 4), surgery may be indicated (Figure 2.15).

*Adults/skeletally mature*
*Colles' fracture*—This is the most common fracture of the wrist. This is a transverse fracture of the distal radius with dorsal angulation and displacement of the fracture fragment. These typically occur in middle-aged and elderly patients following a FOOSH. Important to assess on the radiograph are radial length, dorsal tilt, radial inclination and possible intra-articular extension to the radiocarpal and distal radioulnar joints. Intra-articular extension is typically seen along the lunate fossa of the distal radius and is particularly well appreciated on the external oblique view. CT may be of value for evaluation of the degree of step deformity at the articular surface. Associated ulnar styloid fractures are common. Absence of an ulnar styloid fracture may suggest injury to the triangular

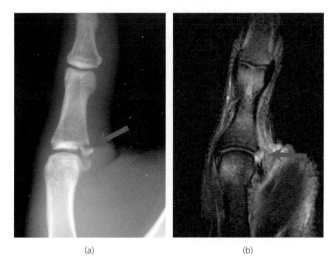

(a)          (b)

**Figure 2.13** Skier's thumb (gamekeeper). UCL tear with avulsion fracture: (a) AP oblique radiograph demonstrating an avulsed fragment adjacent to the sesamoid bone; (b) coronal STIR MRI of same injury. The MRI shows the tear gap but the osseous fragment is not well seen.

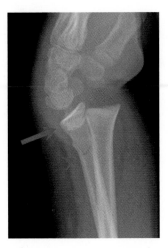

**Figure 2.15** Salter–Harris type 1 injury (arrow), with associated distal ulnar fracture (arrowhead). Lateral shoot through view. The entire distal radial epiphysis has slipped dorsally.

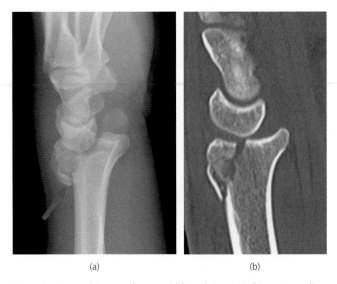

(a)　　　　　　　　　　　　(b)

**Figure 2.17** Dorsal Barton's fracture: (a) lateral view with (b) corresponding sagittal CT bony reconstruction.

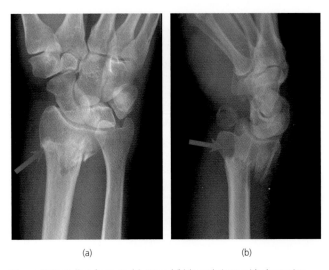

(a)　　　　　　　　　　　　(b)

**Figure 2.16** Colles' fracture: (a) AP and (b) lateral views with shortening, dorsal displacement and dorsal angulation.

fibrocartilage complex (TFCC). Associated scaphoid fractures are less common but of major importance. Treatment is generally closed reduction and cast immobilisation. Comminution and intra-articular involvement is common and may complicate attempts at achieving adequate closed reduction. In these cases open reduction and internal fixation may be warranted (Figure 2.16).

*Smith's fracture*—This is a reversed Colles' fracture where there is volar angulation and displacement of the distal fracture fragment. Volar comminution is commonly seen with this fracture. This is a less common injury than the Colles' fracture. This pattern typically occurs in younger patients with higher energy injuries.

*Barton's fracture*—This is secondary to a shear injury to the radiocarpal articulation. In the original description there is an intra-articular fracture of the dorsal rim of the radius with dorsal

subluxation of the carpus (dorsal Barton's fracture). Subsequently an intra-articular fracture of the volar rim of the radius with volar subluxation of the carpus was also described (volar Barton's fracture). Closed reduction is rarely successful and these injuries are treated with open reduction and internal fixation (IF) (Figure 2.17).

*Die punch*—This is an injury where an axial load is transmitted through the lunate onto the distal radius resulting in the 'punching out' of a depressed fracture of the articular surface of the distal radius. Closed reduction may be attempted; however, any intra-articular incongruity post-reduction may require CT evaluation and percutaneous pinning.

*Chauffeur (Hutchinson) fracture*—This is an oblique intra-articular fracture of the distal radius involving the base of the radial styloid. The fracture fragment may vary considerably in size (Figure 2.18).

*Galeazzi and Monteggia fractures*—A Galeazzi fracture is a fracture of the shaft of the radius, usually at the junction of middle and distal thirds, with associated dislocation of the distal radio-ulnar joint (Figure 2.19). The distal ulna typically dislocates dorsally.

A Monteggia fracture is a fracture of the ulna, often at the proximal or mid shaft with disruption of the radiocapitellar joint and dislocation of the radial head at the elbow. The radial head typically dislocates anteriorly. Both of these injuries disrupt the radioulnar fibro-osseous ring. Treatment of both these fractures is with open reduction and IF.

## Carpal bones
### Scaphoid
Scaphoid fractures comprise 80% of all carpal bone fractures. The majority tend to be subtle on initial radiographic examination.

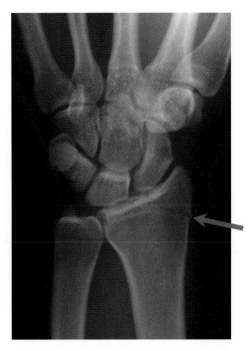

**Figure 2.18** Chauffeur's (radial styloid) fracture, AP oblique view.

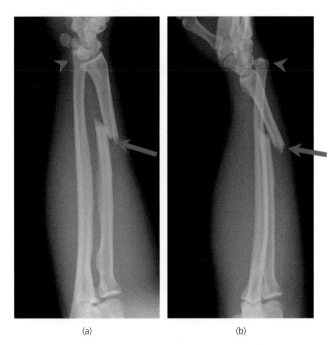

(a)                  (b)

**Figure 2.19** Galeazzi fracture: (a) AP and (b) lateral views. Fracture of the distal radius (arrow) and dorsal dislocation of the distal radioulnar joint (arrowhead).

Despite adequate radiographic projections, 30% of these fractures remain occult initially. Most (70%) fractures occur at the waist. A second set of radiographs 10 days after an initial negative assessment are often necessary if there is clinical concern regarding a scaphoid fracture with snuffbox tenderness and negative initial radiographs. CT or MRI are very sensitive for detecting subtle fractures.

The distal to proximal blood supply of the scaphoid, leaves the proximal pole at risk of avascular necrosis in fractures of the waist or proximal pole. Diagnosis is therefore crucial to avoid this outcome. A more common complication is fracture non-union. This is more common with displaced fractures (>1 mm) or fractures demonstrating a hump back deformity with palmar flexion of the distal fracture fragment. Immobilization alone may suffice for treatment unless there is displacement in which case surgical fixation may be warranted to avoid non-union (Figure 2.20 and 2.21).

*Triquetral fractures*

This is the second most common carpal bone fracture. This results from a ligamentous avulsion or a dorsal impaction from the ulnar styloid. Any small flake of bone on the dorsal aspect of the carpus seen on the lateral view following an acute injury likely represents a triquetral fracture. Treatment is cast immobilisation (Figure 2.22).

## Others

*Carpal dislocations*

These are a spectrum of injuries comprising ligamentous injuries with or without associated fractures. The mechanism is often high-energy trauma in young males. These injuries occur in a sequence as ligaments around the lunate fail in a radial to ulnar direction. The least severe type of injury consists of a tear of the scapholunate ligament with scapholunate dissociation. Progressive injury results in perilunate and lunate dislocations.

On the AP view, interruption of Gilula's arcs, widening of the scapholunate distance (Figure 2.23) or a pie-shaped lunate may be seen. On the lateral view, the position of the lunate relative to radius and capitate is crucial. In perilunate dislocations (Figure 2.24a and b), the lunate articulates normally with the radius but the capitate is displaced, typically in a dorsal direction, resulting in an empty distal articular surface of the lunate. In lunate dislocations (Figure 2.25), the lunate articulation with distal

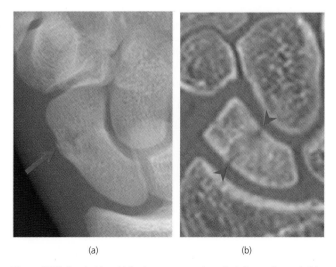

(a)                  (b)

**Figure 2.20** Scaphoid waist fracture seen on ulnar deviation radiograph (a) and on corresponding CT bony reconstruction (b) performed 2 weeks later.

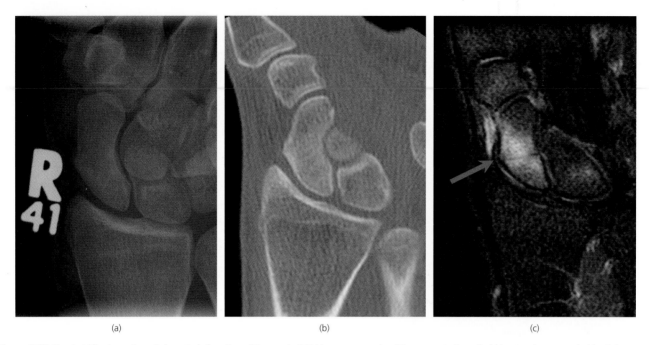

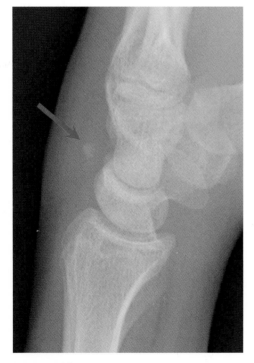

(a)                                    (b)                                    (c)

**Figure 2.21** Scaphoid fracture. Coned ulnar deviation view of the scaphoid (a) in a young male with a suspected scaphoid fracture is unremarkable. Subsequent CT bony reconstruction (b) performed the same day, did not demonstrate a fracture. Coronal STIR MRI (c) performed 4 days later due to ongoing pain reveals extensive bone oedema and a subtle linear fracture.

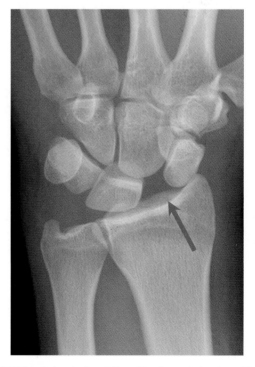

**Figure 2.22** Triquetral fracture.

**Figure 2.23** Scapholunate dissociation. AP radiograph showing widening of the scapholunate distance in comparison with the other intercarpal spaces.

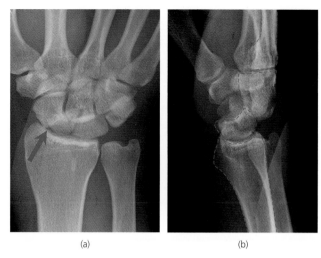

(a)                           (b)

**Figure 2.24** Transscaphoid perilunate dislocation. (a) AP view. Note the disruption of Gilula's arcs. (b) Lateral view. Note the loss of normal lunocapitate alignment, with the capitate displaced dorsally.

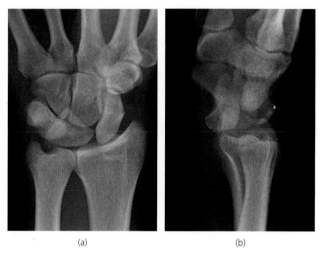

(a)                           (b)

**Figure 2.25** Lunate dislocation. (a) AP view showing pie-shaped lunate and (b) lateral view showing a volarly displaced and tilted lunate.

radius is lost and the lunate is displaced volarly. Often in lunate dislocation the capitate will maintain an alignment with the distal radius. Carpal dislocations may be associated with carpal fractures as well as fractures of the distal radius and ulnar styloid.

### ABCs systematic assessment

### Alignment

- There should be a uniform carpal joint space measuring 2 mm
- Check Gilula's three arcs
- The base of third MC, capitate, lunate and distal radius should lie in a line

### Bone

- Check each bone separately
- Check shape of the lunate and exclude an 'empty' lunate
- Check contour of distal radius
- Check there is no flake fracture on the dorsal aspect of the carpal bones
- Check scaphoid views VERY carefully

### Cartilage and joints

- Joint spaces should be uniform
- If joint spaces are wide, it suggests carpal instability

### Soft tissues

- Swelling indicates site of injury
- A displaced pronator quadratus fat pad may indicate a subtle distal radius fracture

## Further reading

Goldfarb CA, Yin Y, Gilula LA, Boyer M. Wrist fractures: What the clinician wants to know. *Radiology* April 2001;219:11–28.

Kaewlai R, Avery L, Novelline RA. Multidetector CT of carpal injuries: Anatomy, fractures, and dislocations. *Radiographics* October 2008;28: 1771–1784.

Peterson JJ, Bancroft LW. Injuries of the fingers and thumb in the athlete. *Clin Sports Med* 2006;25:527–542.

Rogers LF. *Radiology of Skeletal Trauma*, 3rd edn. Churchill Livingstone/ Harcourt Health Sciences, 2001. ISBN 0–443–06563–2.

# CHAPTER 3

# Elbow

*Muaaze Ahmad*

Barts Health NHS Trust, The Royal London Hospital, London, UK

---

**OVERVIEW**

- Commonly present following a FOOSH
- Injury patterns in children differ from adults due to the epiphyseal and apophyseal centres
- An understanding of age of ossification of these centres is important
- A methodical review, including lines and congruity of the joint spaces, is important to avoid missing injuries

---

Elbow injuries are common and usually result from a fall onto an outstretched hand (FOOSH). Detecting and interpreting abnormal features on radiographs can be difficult and challenging – particularly in children because they have multiple epiphyseal and apophyseal growth centres. An understanding of normal anatomy and adoption of a systematic approach when assessing the radiographs is essential. MRI, CT and US are not necessary in the acute presentation.

## Anatomy

The elbow (Figures 3.1a–c and 3.2a,b) is a hinge joint that consists of three articulations within a single synovial space. The lower end of the humerus is composed of two different shapes. On the lateral side, a partly spherical contour (capitellum) articulates with the concave articular surface of the head of the radius. On the medial side, a notched medial contour (trochlea) articulates with the ulna.

The stability of the radioulnar articulation is maintained by the annular ligament. This is a sling that holds the head of the radius against the ulna. The radial head is free to rotate within this sling.

The joint capsule comprises an inner layer of synovium, a layer of fat and an outer layer of fibrous tissue. The layer of fat results in anterior and posterior fat pads. These lie outside the synovial lining of the joint, but within the joint capsule. The anterior fat pad is radiographically visible in almost all normal elbows. The posterior fat pad lies deep in the olecranon fossa and is never visible in the flexed position unless a large effusion or haemarthrosis displaces it out of the fossa.

**Table 3.1** Particular problems in children.

| Potential traps | Helpful hints |
|---|---|
| Ossifying secondary centres can cause confusion, particularly when a centre shows multicentric ossification<br>An incompletely fused growth plate can mimic a fracture<br>A fracture involving the lateral condyle can be dismissed erroneously as the normal apophyseal growth plate | Doubt or confusion can usually be resolved with the help of a comparison radiograph of the opposite uninjured elbow |
| A pulled off medial epicondyle may be trapped in the joint. It can be mistaken for the normal trochlea ossification centre. This lesion is rare but is a recognised complication of a dislocated elbow – even one that has reduced spontaneously | Suspect this injury if there is slight widening of the medial joint space on the anteroposterior projection<br>Remember CRITOE... and the ossified trochlea never appears before the ossified internal epicondyle |

## Children

Unfused ossification centres in children (Figures 3.3a,b and 3.4a,b) lead to a particular set of injuries (Table 3.1). Children have three epiphyseal ossification centres (capitellum, trochlea and radius) and three apophyseal centres (internal epicondyle, external epicondyle and olecranon). These appear (begin to ossify) at different ages. The trochlear and olecranon centres are often multicentric and should not be mistaken for fracture fragments. If the appearance is confusing, refer to a textbook of normal variants or seek a specialist opinion. If necessary, radiograph the opposite uninjured elbow for comparison.

---

The acronym CRITOE (or CRITOL) (capitellum, radial head, internal epicondyle, trochlea, olecranon, external or lateral epicondyle) lists the most common sequence in which the secondary ossification centres appear on the radiograph (Table 3.2). Although the CRITOE order is the most common sequence, individual variation does occur. Nevertheless, one part of the sequence never varies: the internal epicondyle always ossifies before the trochlea. This has particular diagnostic relevance to an uncommon, but clinically important, injury involving major displacement of the internal epicondyle ossification centre.

---

*ABC of Emergency Radiology*, Third Edition. Edited by Otto Chan.
© 2013 John Wiley & Sons, Ltd. Published 2013 by John Wiley & Sons, Ltd.

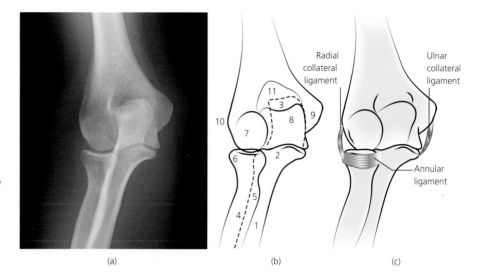

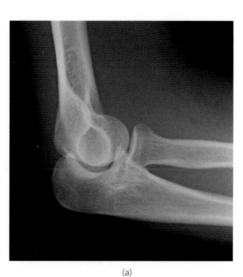

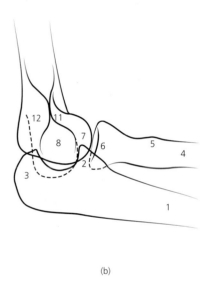

**Figure 3.1** (a)–(c) Normal anteroposterior view of right elbow and ligaments. 1, Ulna; 2, coronoid process; 3, olecranon; 4, proximal radius; 5, radial tuberosity; 6, radial head; 7, capitellum; 8, trochlea; 9, medial epicondyle; 10, lateral epicondyle; 11, coronoid fossa; 12, olecranon fossa.

(a)　　　　　　　　　　(b)　　　　　　　　　　(c)

**Figure 3.2** (a),(b) Normal lateral elbow (for explanation of numbers see legend for Figure 3.1).

(a)　　　　　　　　　　(b)

**Table 3.2** Approximate age at which secondary ossification centres appear on radiographs (CRITOE).

| Centre | Appears[a] | Age (years) |
| --- | --- | --- |
| Capitellum | First | 1 |
| Radial head | Second | 3–6 |
| Internal epicondyle *always before the...* | Third | 4–7 |
| Trochlea | Fourth | 7–10 |
| Olecranon | Fifth | 6–10 |
| External epicondyle | Sixth | 11–14 |

[a]NB The sequence in which the secondary centres appear (above) is the most common sequence. There are occasional individual variations.

## ABCs systematic assessment

- Adequacy
- Alignment
- Bones
- Congruity
- Soft tissues

### Recommended radiological views

- AP and lateral
- Additional radial head views rarely required
- CT necessary only for preoperative assessment
- US and MRI not necessary in acute setting

## Adequacy (Figures 3.1 and 3.2)

The standard radiographic projections are anteroposterior (AP) and lateral (lat). A severe injury will often make perfect positioning impossible. Other additional projections to show the head and neck of the radius are only necessary if the routine views are normal and there is a strong suspicion of a fracture (e.g. positive 'fat pad sign').

## Alignment
### Lateral radiograph

The olecranon articulates with the trochlea, and the radial head articulates with the capitellum. Note that the trochlea and

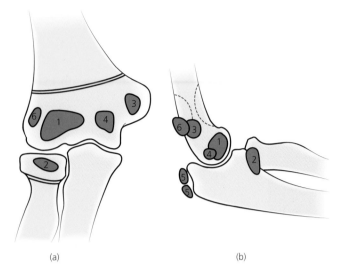

**Figure 3.3** (a),(b) Diagram of ossification centres. 1, Capitellum; 2, radial head; 3, medial or internal epicondyle; 4, trochlea; 5, olecranon; 6, lateral or external epicondyle.

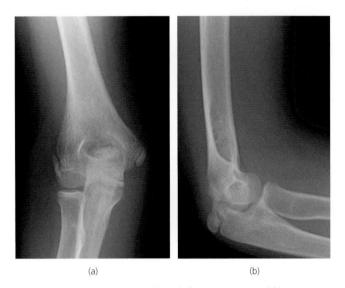

**Figure 3.4** (a),(b) Normal AP and lateral elbow in a 12-year-old boy.

capitellum are superimposed on each other on the lateral view. Two lines need to be assessed on the lateral view:

*Radiocapitellar line (RCL)* (Figure 3.5a,b) – The shaft of a normal radius, as seen on the AP and lat projections, is not always a perfectly straight line. Its proximal 2–4 cm may be set at an angle to the long axis of the rest of the bone. A line drawn along the centre of the long axis of this proximal 2–4 cm of the radius should pass through the capitellum. If it doesn't on either view, then a dislocated head of the radius is present (Figures 3.6 and 3.7).

*Anterior humeral line (AHL)* (Figure 3.5b) – There is a range of normal condylar shapes. Nevertheless, all normal elbows show a hockey stick or J-shaped contour. Loss of the hockey stick contour suggests a displaced supracondylar fracture.

A line drawn along the anterior cortex of the humerus should have a third or more of the blade of the hockey stick (the capitellum)

laying anterior to it. If this rule is broken, there will probably be a supracondylar fracture with posterior displacement of the distal fragment. The AHL is also useful to assess the degree of posterior displacement when a fracture is obvious.

### Anteroposterior radiograph

In normal adults, and in children, check the RCL to exclude a radial head dislocation.

### Bones

*Lateral radiograph (Figure 3.5b)*

Examine the cortical surfaces of the humerus, radius and ulna. A subtle break in the anterior humerus from a supracondylar fracture can be hard to detect, especially in a child. Check the X appearance (X-sign or hour glass sign) made up of the deep bone margins of the olecranon and coronoid fossae). A disrupted X appearance is indicative of a supracondylar fracture.

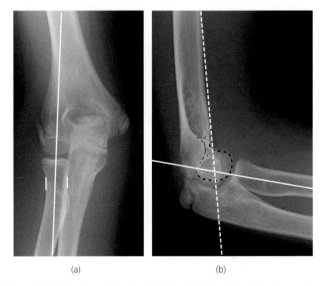

**Figure 3.5** (a) Radiocapitellar line (RCL) – white line; (b) anterior humeral line (AHL) – white dotted line.

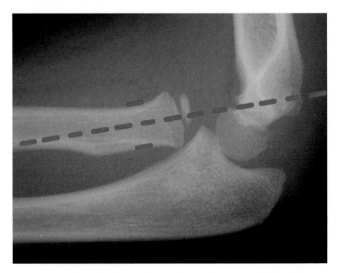

**Figure 3.6** Dislocated radial head with abnormal RCL as red dashed line.

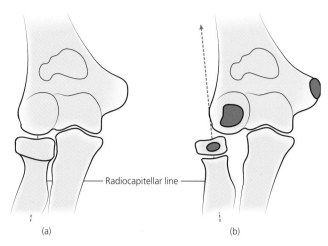

Radiocapitellar line

(a)                    (b)

**Figure 3.7** (a) Normal and (b) abnormal RCL.

Examine the internal trabecular pattern of the bones for bands of increased density. An impacted radial neck fracture may be visible only as a faint transverse band of increased density at the junction of the head and the neck. Bruising or damage to cartilage does occur, but it will not show on a radiograph.

### Anteroposterior radiograph

About half of all radial head fractures are undisplaced, and a radiographic abnormality can be subtle. Slight cortical disruption, faint depression, and/or slight angulation should be looked for.

In addition, in children:

- Check for a faint lucent line crossing the distal humerus – this is often the only evidence of either an undisplaced supracondylar fracture or a fracture of the lateral condyle of the humerus.
- Check the medical epicondyle is in a normal position. Specifically, make sure that it is not trapped in the joint and masquerading as a trochlear ossification centre. This is a rare injury. Children

with a dislocation of the elbow joint that reduces spontaneously are the group most at risk.

If the medial epicondyle is trapped within the joint, minor but detectable widening of the medial part of the joint will occur. Consequently, the joint's normal congruity is altered. The trapped epicondyle is rarely seen on an AP projection – it will be seen more clearly on a lateral radiograph.

### Cartilage and joint

The radiocapitellar and coronoidtrochlear joint spaces should be parallel and spaced equally.

Congruity of articular surfaces should be confirmed:

- The trochlea is congruous with the ulna.
- The capitellum is congruous with or parallels the articular surface of the radial head.

Loss of congruity or parallelism will be seen with some radial head fractures.

### Soft tissues
### Lateral radiograph

The normal anterior fat pad appears as a thin elongated radiolucency lying parallel and adjacent to the distal cortex of the humerus. A posterior fat pad is not identified in a normal elbow held in flexion. Displacement of these fat pads occurs when there is an intraarticular effusion (for example, a haemarthrosis) displacing the synovial lining (Figure 3.8).

- A displaced anterior fat pad appears as a triangular shaped black lucency anterior to the cortex of the humerus – but elevated off the bone. Sometimes this displacement is referred to as the '*sail sign*'.
- The posterior fat pad requires a large effusion to push it out of the deep olecranon fossa. It is then visualised as a black line just posterior to the cortex of the humerus.

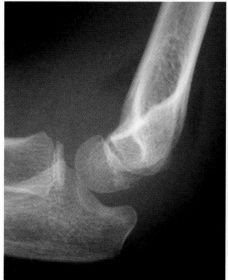
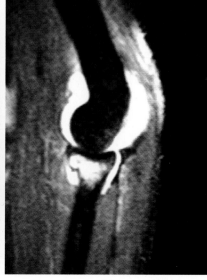

**Figure 3.8** (a) Lateral elbow with prominent anterior and posterior fat pads. (b) Sagittal fat-suppressed MRI showing haemoarthrosis causing fat pad displacement. The fracture of the radial head is seen on the MRI, but not on the X ray.

(a)                    (b)

Some authors refer to a supinator fat stripe. Claims that the appearance of this lucent line is helpful in diagnosing a bone injury have been shown to be over-optimistic. The supinator fat stripe can be ignored.

### Anteroposterior radiograph

If an injury has occurred to the medial or lateral epicondyles, adjacent soft tissue swelling will be present.

## Injuries

### Fractures

#### Supracondylar fracture

Supracondylar fractures (Figure 3.9) are commonly seen in children accounting for the majority (60%) of injuries around the elbow. The fracture line extends transversely across the condyles and through the coronoid and olecranon fossae. The majority (75%) of injuries are complete fractures with posterior displacement and occasionally anterior. However up to 25% are incomplete fractures without displacement that can be subtle.

#### Lateral condylar fracture

Lateral condylar fractures (Figure 3.10) are the second most common injuries in children. Injuries of the lateral condyle are either incomplete or complete. Complete injuries are type 4 Salter Harris injuries.

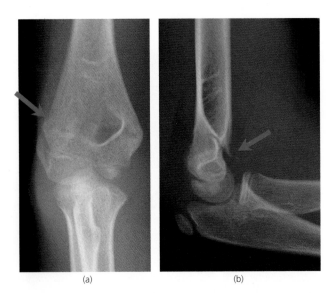

(a)       (b)

**Figure 3.9** (a),(b) Supracondylar fracture with disrupted X-sign.

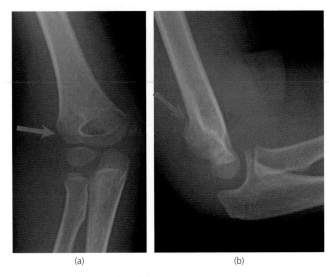

(a)       (b)

**Figure 3.10** (a),(b) Lateral condylar fracture.

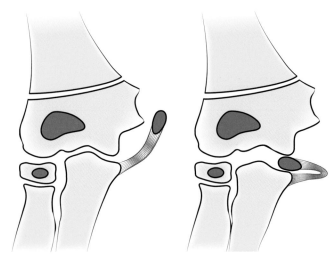

**Figure 3.11** Avulsion and displaced medial epicondyle.

### Medial epicondylar avulsion

This injury is only seen in children. If there is major separation the apophysis is displaced in to the medial joint space (Figures 3.11 and 3.12). This injury is often overlooked. It is important to identify the different ossification centres depending on age to ensure the medial epicondyle is correctly located.

### Olecranon fracture

Olecranon fractures (Figures 3.13) are usually well identified, however it is important not interpret an epiphyseal growth plate as a fracture in children.

### Radial head fracture

This is a common injury in adults accounting for 50% of injuries around the elbow (Figure 3.14). Often difficult to detect fracture

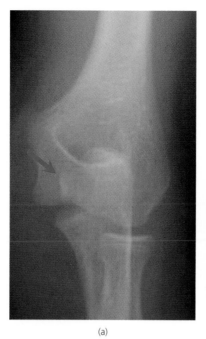

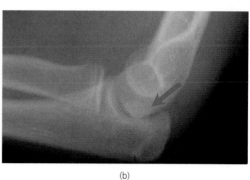

**Figure 3.12** (a) AP and lateral showing avulsed and displaced medial epicondyle; (b) lateral of a different patient.

(a)

(b)

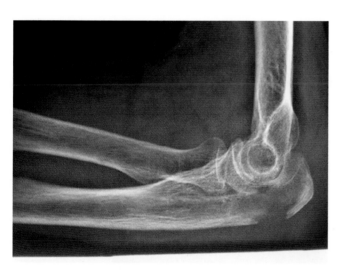

**Figure 3.13** Fracture of olecranon.

line however diagnosis often inferred in the presence of injury and a joint effusion.

## Dislocations

Elbow dislocation are seen commonly in adults and clearly identified on the lat view. It can be difficult to appreciate on the AP view (Figure 3.15).

### *Radial head*

Anterior radial head dislocations are occasionally seen in isolation.

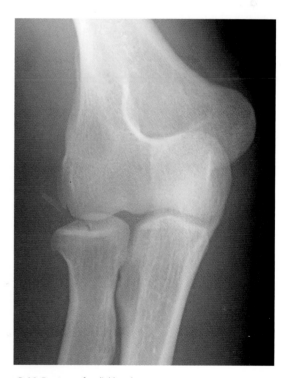

**Figure 3.14** Fracture of radial head.

### *Fracture dislocation of the capitellum*

A capitellar fracture may be associated with anterior displacement of the capitellum (Figure 3.16). This may present as an additional bony fragment seen only on the lateral view.

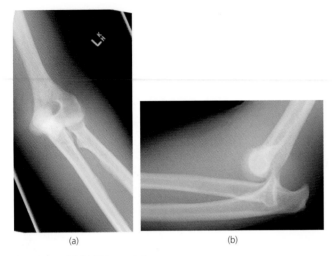

(a)            (b)

**Figure 3.15** (a),(b) Dislocated elbow.

*Pulled elbow*

Common injury in 2–6 year olds. Rare injury, with the diagnosis made on clinical examination rather than radiography which can often be normal. The annular ligament is stretched and there is slight radial head subluxation (Figure 3.17).

## Fracture-dislocations

A Monteggia fracture (Figure 3.18) is a fracture of the ulna, often at the proximal or mid shaft with disruption of the radiocapitellar joint and dislocation of the radial head at the elbow. The radial head typically dislocates anteriorly. Both of these injuries disrupt the radioulnar fibroosseous ring. Treatment of both these fractures is with open reduction and IF.

A Galeazzi fracture (Figure 3.19) is a fracture of the shaft of the radius, usually at the junction of middle and distal thirds, with associated dislocation of the distal radio-ulnar joint. The distal ulna typically dislocates dorsally.

---

**ABCs systematic assessment**

**Alignment**

- Radio-capitellar line – dislocated radial head
- Anterior humeral line – displaced supracondylar fracture

**Bones**

- Wrinkles of the cortex – fracture
- Faint depression of the cortex – fracture
- Slight angulation of the cortex – fracture
- Disrupted X sign – supracondylar fracture
- Additional fragment on lateral view. In adults suspect a radio-capitellar fracture or dislocation. In children suspect an avulsed medical epicondyle

**Cartilage and joints**

- Joint spaces not equidistant – dislocated elbow

**Soft tissues**

- Positive anterior or posterior fat pads – search for a fracture
- No visible fracture but both fat pads displaced. In adults – fracture of the radial head. In children – undisplaced supracondylar fracture

---

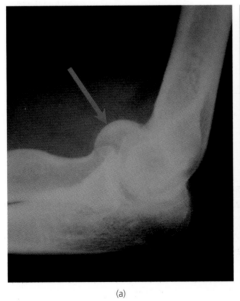
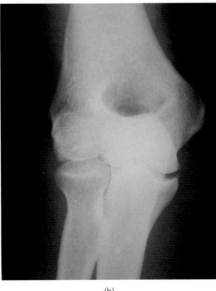

(a)            (b)

**Figure 3.16** (a),(b) Fracture-dislocation of the capitellum. There is an additional rounded bone fragment on the lateral, which should not be there (arrow) and represents the fracture-dislocation of the capitellum. AP looks almost normal but careful inspection confirms a subtle fracture.

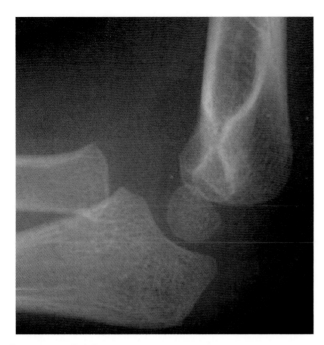

**Figure 3.17** Pulled elbow.

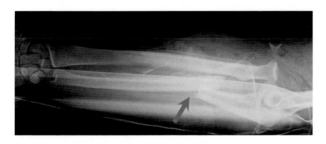

**Figure 3.18** Monteggia fracture dislocation: fractured ulna (arrow) with dislocated radial head (arrowhead).

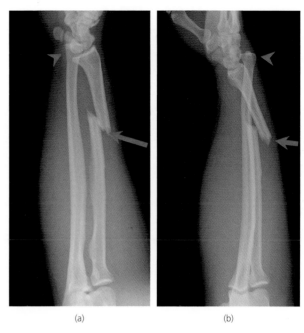

(a)  (b)

**Figure 3.19** Galeazzi fracture: (a) AP and (b) lateral fracture of the distal radius (arrow) and dorsal dislocation of the distal radioulnar joint (arrowhead).

## Further reading

Chessare JW, Rogers LF, White H. Injuries of the medial epicondylar ossification center of the humerus. *Am J Roentgenol* 1977;129:49–55.

Donnelly LF, Klostermeier TT, Klosterman LA. Traumatic elbow effusions in pediatric patients: are occult fractures the rule? *Am J Roentgenol* 1998;171:243–5.

Raby N, Berman L, de Lacey G. *Accident and Emergency Radiology: A Survival Guide*, 2nd edn. Philadelphia: Saunders, 2005.

Rogers LF. *Radiology of Skeletal Trauma*, 3rd edn. London: Churchill Livingstone, 2004.

Rogers LF, Malave S, White H, Tachdijan MO. Plastic bowing, torus and greenstick supracondylar fracture of the humerus: radiographic clues to obscure fractures of the elbow in children. *Radiology* 1978;128:145–50.

# CHAPTER 4

# Shoulder

*Ahmed Daghir and James Teh*

Oxford University Hospitals NHS Trust, Nuffield Orthopaedic Centre, Oxford, UK

### OVERVIEW

- The shoulder is very mobile and prone to dislocation
- Different patterns of injury in different age groups
- Plain radiographs remain the mainstay of imaging
- Anterior dislocations are obvious but posterior dislocations are subtle
- MRI, US, and CT are rarely necessary in the acute setting

Traumatic injury to the shoulder is a common presenting complaint to the emergency department. The shoulder girdle is highly mobile and it is particularly prone to dislocation (Box 4.1). There are different patterns of injury in different age groups. Plain radiographs are the initial investigation of choice for suspected fractures and dislocations. A variety of radiographic views of the shoulder may be obtained. The anatomy shown on each of these will be described. The radiological signs of pathology may be subtle so it is important to be familiar with the specific findings associated with certain injuries.

Box 4.1 **Shoulder girdle**

**Three joints**

- Glenohumeral
- Acromioclavicular
- Sternoclavicular

**Three bones**

- Scapula
- Humerus
- Clavicle

## Anatomy

The shoulder girdle (Figures 4.1–4.3) is made up of three bones – the scapula, clavicle, and proximal humerus – and three joints – the glenohumeral (GHJ), acromioclavicular (ACJ) and sternoclavicular

*ABC of Emergency Radiology*, Third Edition. Edited by Otto Chan.
© 2013 John Wiley & Sons, Ltd. Published 2013 by John Wiley & Sons, Ltd.

(SCJ) joints. The highly mobile GHJ is formed by the articular surfaces of the humeral head and the glenoid fossa. The glenoid cavity is deepened by a fibrocartilaginous ring – the glenoid labrum. The humeral head also includes the greater and lesser tuberosities, the sites of attachment of the rotator cuff tendons. The rotator cuff muscles and tendons are important dynamic stabilisers of the joint; the glenohumeral and coracohumeral ligaments also contribute to joint stability. The bicipital groove lies between the lesser and greater tuberosities and accommodates the long head of the biceps tendon. The ACJ is stabilised by ligaments around the joint itself as well as the strong coracoclavicular ligament, which anchors the clavicle to the scapula.

Important related neurovascular structures include the subclavian vessels and brachial plexus, which lie posterior to the clavicle, and the axillary neurovascular bundle passing inferior to the glenoid.

## ABCs systematic assessment

- Adequacy – check correct views have been obtained
- Alignment – check joint spaces are the same
- Bone – trace the contours of all the bones
- Cartilage and joints – joint spaces should be uniform in width
- Soft tissues – change windows to look for soft tissue swelling and FB.

**Recommended radiological views**

- Shoulder (GHJ) – AP and either a Y view or an axial view
- ACJ – AP and weight-bearing views

### Adequacy

Two projections should always be performed. The AP view is routinely obtained and is the most useful for identifying pathology. There are three alternative second views. The axial view is taken with the arm abducted; this view will show dislocation clearly and is particularly useful for demonstrating small, avulsed fracture fragments. When it is painful to abduct the arm, the 'Y' view is a useful alternative, as it requires no shoulder movement. A less commonly used second view is the axial oblique; this also requires

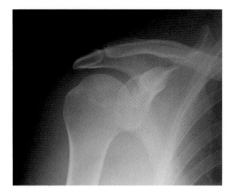 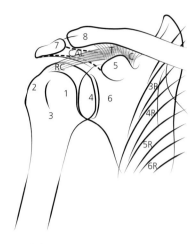

**Figure 4.1** Normal AP right shoulder. 1, Humeral head; 2, greater tuberosity; 3, lesser tuberosity; 4, glenoid fossa; 5, coracoid process; 6, neck of scapula; 7, acromion; 8, lateral end of clavicle.

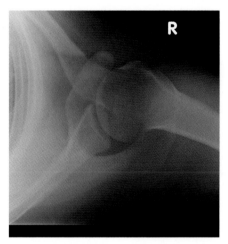 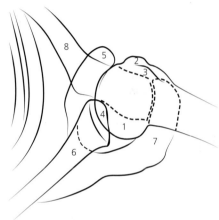

**Figure 4.2** Normal axial shoulder. Note the coracoid process and acromion both project anteriorly. 1, Humeral head; 2, greater tuberosity; 3, lesser tuberosity; 4, glenoid fossa; 5, coracoid process; 6, neck of scapula; 7, acromion; 8, lateral end of clavicle.

no shoulder movement and demonstrates the relation of the humeral head to the glenoid clearly. The AP view shows the ACJ well but an additional weight-bearing view may be requested if subluxation or dislocation is suspected.

## Alignment

Box 4.2 shows alignment and normal measurements.

> ### Box 4.2 **Alignment and normal measurements (Figure 4.4)**
>
> - GHJ space less than 6 mm
> - Inferior margin of the clavicle and acromion should be level
> - ACJ should be no greater than 7 mm
> - Coracoclavicular distance no greater than 13 mm
> - AHD of <7 mm is highly suggestive of a large rotator cuff tear

Assess the GHJ alignment on the AP view by checking there is an even joint space between the humeral head and glenoid. There should be an equal distance between the margins of their articular surfaces. The normal GHJ space is no greater than 6 mm. The axial view shows the humeral head normally aligned with the glenoid

fossa like 'a golf ball sitting on a tee'. On the 'Y' view check that the humeral head is positioned over the junction of the 'Y' shape, which indicates the position of the glenoid fossa.

The ACJ alignment is assessed on the AP view. The inferior margins of the acromion and lateral clavicle should be level with each other. This rule holds true in most instances but due to normal variation between individuals there is slight misalignment at this inferior margin in up to 20%. ACJ injury may result in widening of the joint space (normally no greater than 7 mm) or the coracoclavicular distance (normally no greater than 13 mm).

Check the space between the superior margin of the humeral head and the undersurface of the acromion. This acromiohumeral space accommodates part of the rotator cuff and if it is narrowed to less than 7 mm, it indicates the presence of a large rotator cuff tear.

## Bone

The contour of each bone should be carefully assessed to ensure it is smooth. Check there is no step or buckle of the cortex that may indicate a fracture. Other subtle signs of a fracture include disruption of the trabecular pattern and linear sclerosis that may indicate impaction. Each view needs to be systematically evaluated. The AP and axial views are particularly useful for identifying small

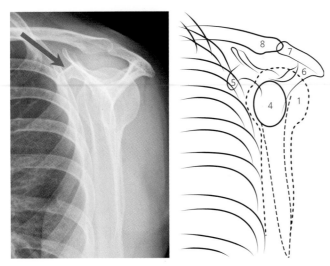

**Figure 4.3** Normal Y view of shoulder. The coracoid process projects anteriorly and may be used as a landmark to determine the direction of humeral head dislocation. The glenoid fossa has been outlined.

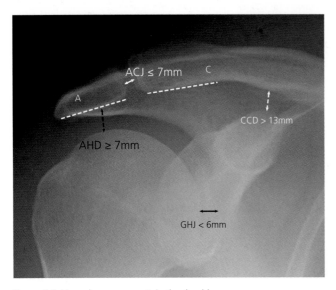

**Figure 4.4** Normal measurements in the shoulder.

fracture fragments. Anterior dislocation is commonly associated with a Hill–Sachs fracture in the posterosuperior humeral head and/or a Bankart fracture of the anteroinferior bony glenoid. Ribs shown on the AP view also need to be checked to exclude a fracture.

## Cartilage and joints

Ensure the joint spaces are preserved. The borders of the humeral head and glenoid should appear as two parallel lines. The GHJ space may appear reduced for technical reasons as well as true cartilage loss. Where there is true cartilage loss, there may be secondary findings including subarticular sclerosis and osteophytes. As primary degeneration of the GHJ is uncommon, cartilage loss due to another condition such as rheumatoid arthritis, haemophilia or, rarely, infection should be considered.

## Soft tissues

In the presence of ACJ disruption there is often marked overlying soft tissue swelling which may be apparent on the AP view. With intra-articular fractures of the humeral head there may be a lipohaemarthrosis visible on the AP view, seen as a sharp horizontal line. If there is a large haemarthrosis of the GHJ, this may displace the humeral head laterally and inferiorly giving the appearance of a 'pseudo-dislocation'. Calcification of the rotator cuff tendons is a common finding, which indicates calcific tendonitis. This condition is often painful in the acute phase and may become asymptomatic in the long term. The visible lung on the AP view needs to be carefully checked for pathology including a pneumothorax or unsuspected lung cancer.

## Injuries

### Dislocations

The GHJ is the most commonly dislocated joint in the body. The dislocation is described by the position of the humeral head with respect to the glenoid.

### Anterior dislocation

Anterior dislocations account for over 90% of shoulder dislocations. This occurs when a posteriorly directed force is applied to the arm held in abduction and external rotation. Anterior dislocations are usually clinically apparent as there is a squared appearance of the shoulder with a prominent lateral tip of the acromion. On an AP view, the diagnosis is usually obvious. The humeral head is shown displaced inferiorly and medially, frequently lying below the coracoid process. A second view confirms the humeral head is displaced anterior to the glenoid fossa, and more importantly helps to identify associated fractures. A Hill–Sachs fracture may be found in up to 50% of cases. This follows impaction of the posterosuperior humeral head on the anterior rim of the glenoid, and it usually appears as a hatchet-shaped indentation of the cortex. A Hill–Sachs fracture may be difficult to detect on the AP view unless the arm is held in internal rotation thereby showing the posterior humeral head in profile. The fracture may be better shown on the axial or 'Y' views (Figure 4.5). Greater tuberosity fractures are found in approximately 15% of anterior dislocations. Bankart fractures of the ferior antero-inferior glenoid occur in up to 10% (Figures 4.6 and 4.7).

Anterior dislocations may be further classified according to the humeral head position as subcoracoid, subclavicular, subglenoid and intrathoracic. A repeat radiograph following manipulation should be obtained to show successful reduction of the dislocation. Approximately 40% of anterior dislocations will recur, especially in younger age groups. Soft tissue injuries of the capsulolabral complex and associated ligaments – not shown on radiographs – may lead to chronic pain and instability and are best evaluated with dedicated MR arthrography.

### Posterior dislocation

Posterior dislocations (Figures 4.8 and 4.9) are uncommon and represent less than 10% of all shoulder dislocations. The clinical

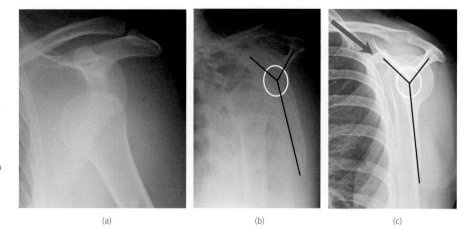

(a)  (b)  (c)

**Figure 4.5** (a) Anterior dislocation. On the AP view the humeral head is shown lying inferior to the coracoid process. (b) Y view: the humeral head is dislocated anterior to the glenoid fossa, which is sited at the centre of the Y shape. (c) Y view: normal for comparison.

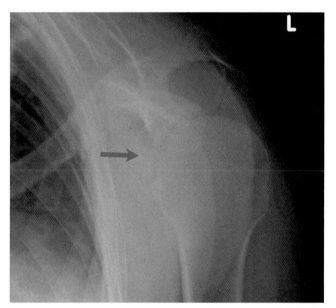

**Figure 4.6** Bankart fracture. A small fracture fragment (arrow) has become displaced from the anteroinferior glenoid.

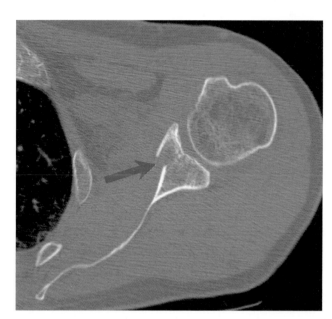

**Figure 4.7** The same fracture as in Figure 4.6 (arrow) is demonstrated more clearly on this axial CT image.

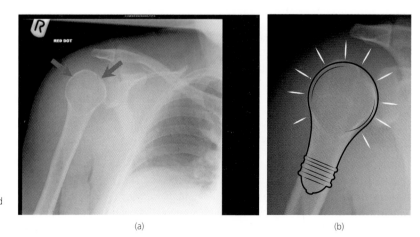

(a)  (b)

**Figure 4.8** (a) Posterior dislocation. (b) The humeral head is internally rotated resulting in a symmetric 'lightbulb' appearance. The glenohumeral joint space is increased, and there is loss of the normal parallelism of the articular surfaces.

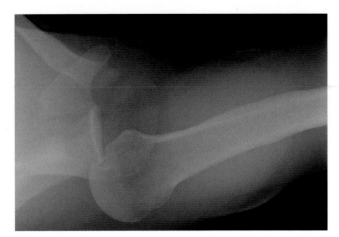

**Figure 4.9** Posterior dislocation on an axial radiograph. The humeral head is clearly displaced posteriorly and there is a hatchet-type impaction fraction of the humeral head ('reversed' Hill–Sachs fracture).

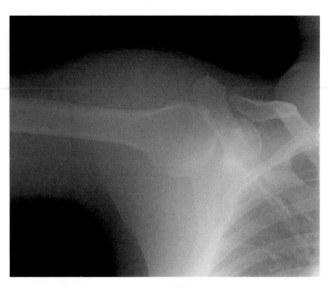

**Figure 4.10** Luxatio erecta. The arm is held in abduction. Note the articular surfaces of the glenohumeral joint do not overlap indicating dislocation.

and radiological signs may be subtle such that up to half of cases may be missed at the time of presentation. Posterior dislocation occurs when a posteriorly directed force is exerted on the humeral head with the arm held in internal rotation. Severe muscle spasm resulting from seizures or electrocution is a typical cause, and dislocation may be bilateral. Associated fractures are common, including a 'reversed' Hill–Sachs impaction of the anterior humeral head. There are three signs to look for on the AP view (Box 4.3): the 'lightbulb' sign describes a symmetrical appearance of the humeral head and neck due to internal rotation of the arm. This sign is not specific for posterior dislocation as internal rotation for any reason (e.g. pain) will give rise to this appearance. Secondly an increase in the normal glenohumeral joint space (greater than 6 mm) may be seen with associated loss of parallelism of the articular surfaces. Thirdly there may be a 'trough line', which is seen as a vertically orientated sclerotic line just lateral to the humeral head articular surface; this results from an impaction of the humeral head onto the glenoid. Posterior dislocations are more easily diagnosed on the second view, which shows the humeral head displaced posteriorly relative to the glenoid fossa. The coracoid process points anteriorly and is a useful landmark when evaluating the second view.

---

Box 4.3 **Signs of a posterior dislocation on an AP view**

- 'Lightbulb' sign
- Increase in GHJ space >6 mm
- 'Trough' line

---

## Luxatio erecta

Luxatio erecta is a very rare inferior dislocation, which classically results from a fall down an open manhole cover. The patient presents with the affected arm in the air (Figure 4.10).

**Table 4.1** Rockwood classification.

| Classification | Injury | Radiological findings |
|---|---|---|
| I | ACJ injury without complete tear of AC or CC ligaments | Normal appearance |
| II | Complete tears of AC ligaments | Widening of AC joint |
| III | Complete tears of AC and CC ligaments | Widening of AC joint and CC distance |
| IV | As in III but with posterior displacement of lateral clavicle through trapezius | Posterior displacement of lateral clavicle on axial view |
| V | As in III but with upward displacement of the lateral clavicle | Widening of AC joint and CC distance with elevation of lateral clavicle |
| VI | As in III but with inferior displacement of the lateral clavicle | Inferior displacement of lateral clavicle below coracoid process |

## Acromioclavicular joint injury

Injury of the ACJ is common in young adults. The degree of injury is quite variable and may be described using the Rockwood classification (Table 4.1). On the AP view the important features to look for are loss of alignment of the inferior margins of the acromion and clavicle, an increase in the ACJ space >7 mm and an increased coracoclavicluar distance >13 mm (Figure 4.11). If there is uncertainty, a weight-bearing AP view may be obtained to exaggerate any displacement resulting from injury. Also, when comparing with the contralateral side, there should be no more than 5 mm difference in the coracoclavicular distance.

## Sternoclavicular joint injury

Dislocation of the SCJ may be difficult to assess radiographically due to overlapping structures. The medial end of the clavicle usually

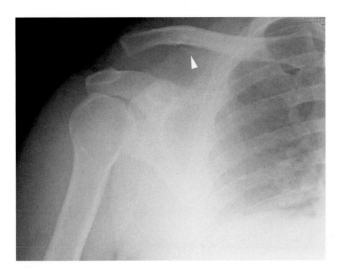

**Figure 4.11** Type III ACJ injury with widening of the ACJ distance and coracoclavicular distance. Note the fracture fragment (arrow) avulsed off the coracoid process by the coracoclavicular ligament.

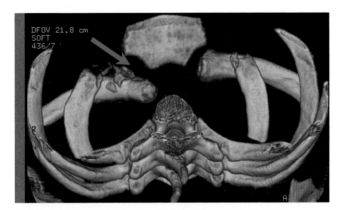

**Figure 4.12** Dislocation of the right sternoclavicular joint (arrow). Three-dimension axial CT reconstruction shows posterior displacement of the medial end of the clavicle.

dislocates anterosuperiorly. On a frontal view of the chest, the medial ends of the clavicles are shown lying at different levels. An oblique AP view may show the dislocation more clearly. Rarely the medial end of the clavicle dislocates posteriorly, thus potentially injuring adjacent mediastinal structures. A CT scan should be considered if there is clinical uncertainty about the diagnosis or for evaluation of suspected intrathoracic injury (Figure 4.12).

### Fractures
#### Clavicle
Clavicular fractures are common and usually result from falls onto the shoulder or outstretched hand. The AP view is usually sufficient to make the diagnosis, but if a second view is required an AP view with cranial angulation may be obtained. The middle third of the clavicle is involved in 80% of fractures, the lateral third in 15% and the medial third in 5%. Middle third fractures often result in overriding of the two fragments with inferior displacement of the distal end. The subclavian vessels or brachial plexus may be

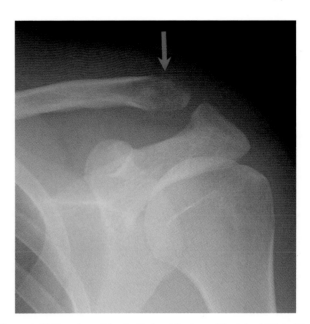

**Figure 4.13** There is a minimally displaced fracture of the lateral third of the clavicle (arrow).

injured with posterior displacement of the fractured bone. Lateral clavicular fractures are usually more stable due to ligamentous support (Figure 4.13). Medial third fractures are easy to miss due to overlapping structures on the AP radiograph.

### Scapula
Scapula fractures account for 5% of shoulder girdle injuries. Most scapular fractures are associated with injury to the head, thorax or spine. Fractures of the body of the scapula (Figures 4.14 and 4.15)

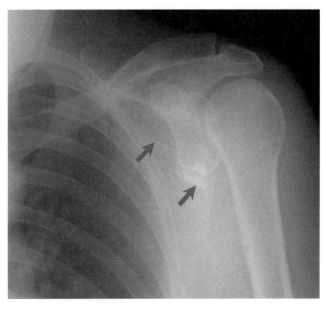

**Figure 4.14** A fracture of the body of the scapula. There is linear sclerosis (arrows) indicating the site of fracture.

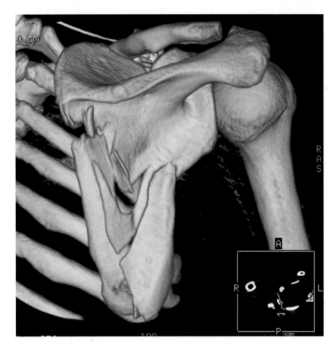

**Figure 4.15** A 3D CT reconstruction shows the extent of the scapular body fracture which is comminuted.

usually occur with severe trauma whereas fractures of the acromion typically result from a direct blow, for example a fall. A fracture of the coracoid process may occur with shoulder dislocations. If a scapular fracture is identified on a screening chest radiograph, dedicated shoulder views should also be obtained for better evaluation. Most scapula fractures can be managed conservatively but those involving the glenoid fossa or neck or the superior shoulder suspensory complex (SSSC) may require surgery.

The SSSC is a bone and soft-tissue ring attached to the trunk by a superior strut (middle third of the clavicle) and inferior strut (lateral scapular body and spine) from which the upper limb is suspended. The ring is comprised of the glenoid, coracoid process, coracoclavicular ligament, distal clavicle, ACJ and acromion.

Traumatic disruptions of a single component of the SSSC are common (e.g. simple clavicle fracture). If the ring is injured in two or more places (double disruption), this may result in altered shoulder biomechanics and instability, which may necessitate surgery.

The floating shoulder is an important injury to recognise consisting of ipsilateral fractures of the clavicle and scapular neck (Figure 4.16). Ligament disruption associated with isolated scapular neck fractures may result in the functional equivalent of this injury.

Scapulothoracic dissociation is a rare and potentially life-threatening injury. The scapula is distracted from the thoracic cage resulting in the equivalent of a closed forequarter amputation. Associated rib fractures are common and there is often neurovascular injury necessitating angiography. Plain radiographs demonstrate lateral displacement of the scapula with marked soft tissue swelling and a clavicle fracture or ACJ separation. On a well-centred chest radiograph the distance from the midline of the spine to the tips of both scapulae is unequal.

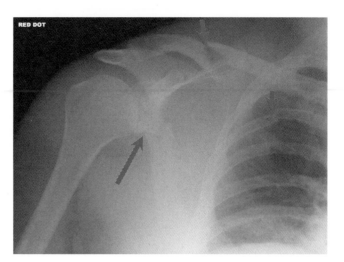

**Figure 4.16** Floating shoulder. The arrows show a scapular neck fracture, clavicle fracture and rib fracture.

## Proximal humerus fractures

Proximal humeral fractures are common following falls in the elderly, who may have coexisting osteoporosis. Adolescents with injuries of the proximal humerus may present with epiphyseal separation. The commonest site of fracture in adults is the surgical neck. Comminution of the fracture and involvement of the tuberosities is common (Figure 4.17). The Neer classification describes the displacement or angulation related to four parts of the proximal humerus – the articular surface, greater tuberosity, lesser tuberosity, and shaft. More than 80% are one-part fractures (without substantial displacement). Three-part and four-part fractures often require surgery. A CT scan is frequently necessary for

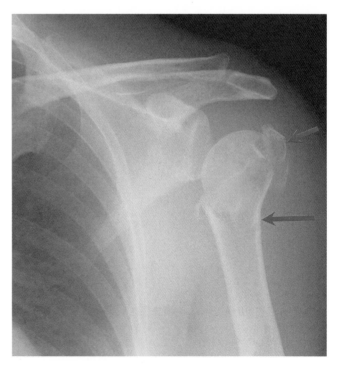

**Figure 4.17** A three-part fracture of the proximal humerus involving the surgical neck, greater and lesser tuberosities with displacement.

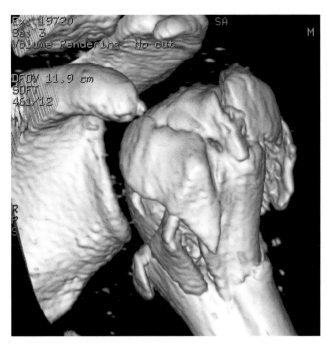

**Figure 4.18** A 3D CT reconstruction of the same injury AS IN Figure 4.17 demonstrates the relations of the fracture fragments and aids surgical planning.

surgical planning (Figure 4.18). Complications include neurovascular injury – including radial nerve damage – which is found in up to 17% of fractures involving the humeral shaft.

---

**ABCs systematic assessment**

**Alignment**

- GHJ space should be even and no greater than 6 mm
- On the axial view the GHJ appears as a 'golf ball on a tee'
- On the 'Y' view the humeral head should be centred over the junction of the Y shape
- ACJ alignment – the inferior margins of the acromion and clavicle should be level
- Check for widening of the ACJ and coracoclavicular distance

---

**Bone**

- The contours of the humeral head and scapula should be smooth
- Check for disruption of the trabecular pattern or linear sclerosis indicating impaction
- Small fracture fragments should be identified
- Anterior dislocation is frequently associated with a Hill–Sachs and/or Bankart fracture

**Cartilage and joints**

- Check for even joint spaces and loss of joint width
- Loss of GHJ space may be degenerative secondary to other causes

**Soft tissues**

- Check for a horizontal line indicating lipohaemarthrosis of the glenohumeral joint. There may associated lateral and inferior 'pseudo-dislocation' of the humeral head
- Calcification of the rotator cuff tendons is often painful in the acute phase
- Marked loss of acromiohumeral space indicates a large rotator cuff tear

## Further reading

Brucker PU, Gruen GS, Kaufmann RA. Scapulothoracic dissociation: evaluation and management. *Injury* 2005;36(10):1147–55.

Edelson G, Saffuri H, Obid E, Vigder F. The three-dimensional anatomy of proximal humeral fractures. *J Shoulder Elbow Surg* 2009;18(4):535–44.

Melenevsky Y, Yablon CM, Ramappa A, Hochman MG. Clavicle and acromio-clavicular joint injuries: a review of imaging, treatment, and complications. *Skeletal Radiol* 2011;40(7):831–42.

Owens BD, Goss TP. The floating shoulder. *J Bone Joint Surg Br* 2006; 88(11):1419–24.

Robinson CM, Shur N, Sharpe T, Ray A, Murray IR. Injuries associated with traumatic anterior glenohumeral dislocations. *J Bone Joint Surg Am* 2012;94(1):18–26.

# CHAPTER 5

# Pelvis and Hip

*Syed Babar[1], James A. S. Young[2], Jeremy W. R. Young[3] and Otto Chan[4]*

[1]Hammersmith & Charing Cross Hospitals, Imperial College, London, UK
[2]St Georges Hospital, London, UK
[3]The Regional Medical Center, Orangeburg, SC, USA
[4]The London Independent Hospital, London, UK

---

> **OVERVIEW**
>
> - Pelvic fractures in major trauma may be life-threatening (NB suspect vascular and pelvic organ injuries in these patients)
> - If one fracture is detected, always look for a second one
> - Hip fractures may occur after minor trauma in elderly patients
> - Plain radiographs are difficult to interpret and provide limited information. There should be a low threshold for use of CT and MRI

Pelvic and hip fractures are seen in the elderly population with trivial trauma whilst the mechanism in young patients generally involves high-impact injuries including road traffic accidents (RTAs). There is high morbidity and mortality associated with pelvic fractures. This results from internal visceral injuries (commonly bladder and urethra and rarely uterus, cervix, vagina and rectum) and bleeding due to high impact in RTAs, falls in young patients and associated underlying co-morbidities in elderly population. Prognosis is poor if the injuries are not detected and treated promptly. Pelvic fractures can be open or closed. The mortality rate for closed pelvic fractures is 27% and that for open fractures is 55%.

In contrast, hip fractures may occur after relatively minor trauma in elderly patients and are suspected from the clinical history and examination. The fractures may be subtle on plain radiographs and may be overlooked in particular in obese and elderly osteopenic patients.

## ANATOMY

### Bony anatomy
### Pelvis
The pelvis is the connection between lower limb and trunk and hence it is inherently unstable. It comprises three separate bones (the sacrum and two iliac/innominate bones) which are held together by a series of strong ligaments. The integrity of this pelvic bony ring can be compromised by disruption of these ligaments (Figures 5.1 and 5.2).

The ligaments are the anterior and posterior sacroiliac ligaments, the sacrotuberous ligament, sacrospinous ligaments and the ligaments of the symphysis pubis. The posterior group is strong and complex and attaches the spine to the pelvis. They resist posterior deformation. The anterior group in contrast is weak and prevents distraction and anteroposterior displacement. The anterior ligaments are the first to disrupt.

### Hip
The femoral head and the acetabulum form the hip joint. The acetabulum is formed by the anterior and posterior columns and connected by the supra-acetabular region. The anterior and posterior columns are connected to the axial skeleton through the sciatic buttress. Hip fractures are not commonly associated with dislocations because of the strong joint capsule. Hip fractures may be associated with avascular necrosis of the femoral head. This is more commonly seen in intracapsular rather than extracapsular fractures. There are various muscle attachments around the pelvis and hip region, which may be avulsed in traction injuries.

In children, the proximal capital femoral epiphysis is present from the age of 3 months until 18–20 years. There can be asymmetry of the epiphysis with irregularity and notching, which can be a normal finding. However, flattening is generally considered abnormal.

## ABCs systematic assessment

- Adequacy
- Alignment
- Bone
- Cartilage and joints
- Soft tissues

---

> **Radiographic projections of pelvis and hip**
>
> **Pelvis**
>
> **Standard**
> - AP view of hips
> - Lateral

---

*ABC of Emergency Radiology*, Third Edition. Edited by Otto Chan.

**Additional**

- Oblique view
- Inlet view
- Outlet view

**Hip**

**Standard**

- AP view of both hips

**Additional**

- Frog leg lateral view

# Pelvis
## Plain X-rays
### Adequacy

The routine view is a single AP view of the pelvis. The x-ray is generally done along with the AP CXR in the trauma protocol series. It is difficult to assess the stability of the pelvis on the AP view. In a comparative study between MDCT and plain X ray of the patients with blunt trauma, CT demonstrated 629 fractures in contrast to 405 fractures in a total of 226 patients. This gives an overall sensitivity of only 55% stressing the importance of pelvic CT in evaluation of patients with pelvic trauma.

Ideally the AP view has to cover the pelvis from the level of the iliac crests to the ischial tuberosity and laterally to include both greater trochanters.

Penetration should be adequate and is assessed by looking at the soft tissue structures. The soft tissue shadows which should be seen include the bladder with its perivesical fat, iliopsoas shadows and the rectum and bowel gas. It is important to note that the adequacy criteria may be difficult to meet due to reasons mentioned earlier. In addition elderly patients may have a large belly with thin proximal thighs and hence the exposure may vary significantly. The pelvic AP view is generally considered sensitive for the anteroinferior part of the pelvis, reasonable in the region of the acetabulum and Ilium and poor in the region of the posterior ring.

Pelvic centring also needs to be assessed. This can be done by aligning the symphysis with the sacrum and checking for the symmetry of the obturator foramina.

The AP view of the pelvis gives important information about the initial assessment of the traumatised patient so as to assess for other more significant underlying injuries. These significant injuries may be fractures of the acetabulum, obturator ring, injury to the bladder and urethra.

The AP view can be supplemented by inlet and outlet views of the pelvis in addition to the oblique (Judet) views especially in cases of suspected acetabular injury. However, nowadays these views have been replaced by CT with MPR and 3D reconstructions.

### Alignment and bones

The bony alignment is assessed by dividing the pelvis into three circles, one large circle and two smaller circles (Figure 5.3). The larger circle is the pelvic brim whilst the smaller circles are made by the obturator rings. The larger circle of the pelvic brim should be visualised as a continuous line around the margins of the brim

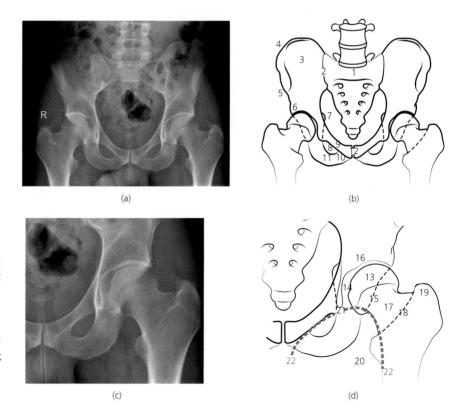

(a)

(b)

(c)

(d)

**Figure 5.1** (a)–(d) Normal pelvic and hip anatomy: 1, sacrum; 2, sacro-iliac joint; 3, ilium; 4, iliac crest; 5, anterior superior iliac spine; 6, anterior inferior iliac spine; 7, ischial spine; 8, obturator foramen; 9, superior pubic ramus; 10, inferior pubic ramus; 11, ischial tuberosity; 12, symphysis pubis; 13, femoral head; 14, fovea centralis; 15, posterior acetabular rim; 16, acetabulum; 17, neck of femur; 18, inter-trochanteric line; 19, greater trochanter; 20, lesser trochanter; 21, Kohler's tear drop; 22, Shenton's line.

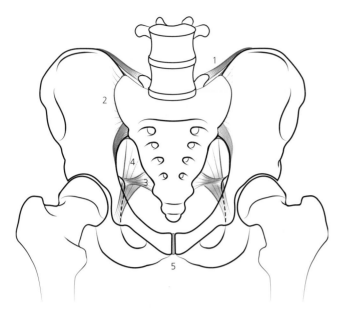

**Figure 5.2** Normal pelvic ligamentous anatomy: 1, left posterior sacro-iliac ligament; 2, right anterior sacro-iliac ligament; 3, right sacrospinous ligament; 4, sacrotuberus ligament; 5, symphysis pubis.

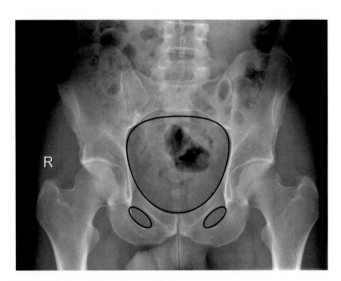

**Figure 5.3** Normal three bony pelvic rings.

There are certain other important lines, which can be assessed. These include the iliopectineal line, which assesses the anterior column, the ilioischial line, which assesses the posterior column, the anterior acetabular line and the posterior acetabular line. The latter two form the anterior and posterior margins of the acetabulum respectively. The teardrop line is seen at the medial margin of the acetabulum and hip joint, which indicates damage to the medial wall of the acetabulum.

In addition it is vital to follow the lines of the sacral foramina as a discontinuity of the sacral foramina line indicates a fracture, which can involve the sacral nerve roots.

Avulsion injuries tend to occur at the site of the attachment of the strong anterior and posterior hip and thigh muscles. The sites to look for avulsion injuries are the ischial tuberosity for hamstring tendons, anterior inferior iliac spine for rectus femoris muscle, anterior superior iliac crest for the sartorial muscle and also the sacral spines for sacrospinous and sacrotuberous ligaments.

In children, the pubis, ischium and ilium remain separated by a Y-shaped cartilage (called the triradiate cartilage) which fuses in puberty to form the acetabulum. In addition, there are other small accessory ossification centres, which should not be mistaken for fractures.

*Cartilage and joints*
The pubic symphysis is well visualised and should be checked for alignment, widening, asymmetry or overlapping bones. The symphysis pubis shows widening of the joint space in AP compression injuries. In vertical shear or injury to the pelvis from a fall, superior displacement of the symphysis can be seen.

The anterior margins of the SIJs are well visualised and should be checked for alignment, widening, asymmetry and overlapping. These changes tend to be more subtle and can be easily overlooked, in particular as there is often overlying bowel gas.

*Soft tissues*
Check for loss or displacement of soft tissue outlines, in particular displacement or asymmetry of the perivesical fat plane (surrounding the bladder) or the obturator internus fat plane (medial edge of the obturator internus muscle), which are strongly suggestive of a pelvic sidewall haematoma secondary to a fracture.

**Computed tomography (CT)**
In major trauma, the pelvic CT is covered as part of the whole body CT protocol. However, if a CT has not been performed, then pelvic CT should be carried out with intravenous contrast enhancement and is viewed on soft tissue and bone algorithm windows, in addition to MPR and 3D. CT angiography (CTA) should also be performed to look for vascular injuries. CT cystography (contrast is instilled into the bladder) can be performed to look for and assess bladder injuries.

MPRs and 3D reconstructions give detailed characterisation of simple and complex pelvic fractures and are now considered essential for preoperative planning for major pelvic reconstructions.

CT is also used to exclude and to assess injuries to the pelvic organs including the bladder, urethra, rectum, uterus and the

and continuing across the sacroiliac joints (SIJs) posteriorly and symphysis pubis anteriorly.

If a fracture is detected, always check for a second fracture or disruption/diastasis in the same ring (imagine trying to break a polo mint in one place!).

The smaller rings are formed by the two obturator foramina. (The same rule applies to the obturator foramina, if a fracture is detected, there is almost always a second fracture in the same ring or diastasis of the pubic symphysis.) There is a smooth uniform arc formed by drawing a line along the inner margin of the femoral neck and extending continuously to the superior margin of the obturator foramen. This continuous line is called Shenton's line. Disruption of this line indicates a femoral neck fracture. The only exception to this is if the fracture is non-displaced.

cervix and vagina. Pelvic haematomas can be detected and active contrast extravasation at the time of the CT, indicates active ongoing bleeding.

## Patterns of injury to the pelvis

There are numerous classifications for pelvic fractures. However, the two most commonly used are the Tile classification based on the integrity of the posterior sacroiliac complex and the Young classification based on the mechanism of injury. Young's classification divides the pelvic fractures into three main categories. These are AP compression, lateral compression and vertical shear and combinations of the three.

*Lateral compression (LC)*

This is the most common form. The direction of force is side to side. As a result the typical fractures are seen in the pubic rami in the horizontal direction (Figure 5.4). This common form of injury results from broadside traffic accidents or from fall on to the side. This is also associated with sacral impaction injury, fractures through the sacral foramina and rotational instability (Figure 5.5). Lateral compression causes an effect opposite to the AP directed force and causes a closed book phenomenon with internal rotation of the hemipelvis (Figure 5.6). There may also be an avulsion fracture from the posterior iliac bone. Further increase in the lateral compression force can cause the external rotation of the contralateral hemipelvis resulting in a windswept pelvis. This will cause both rotational and vertical instability. Lateral force can cause central dislocation of the femoral head and crush fracture of the femoral neck.

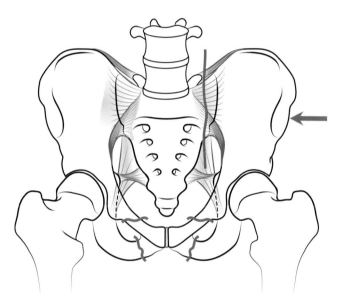

**Figure 5.4** Lateral compression injury with horizontal fracture of the pubic rami and through the neural foramina.

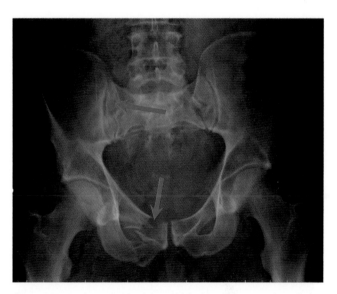

**Figure 5.5** Lateral compression injury with horizontal fractures through the pubic rami (arrowed).

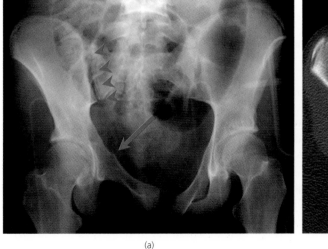

(a)

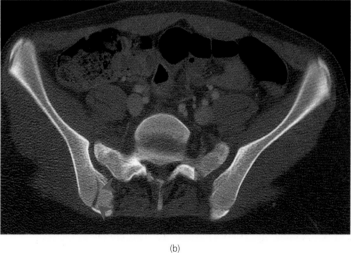

(b)

**Figure 5.6** (a) Lateral compression injury with an impacted horizontal fracture of the right superior pubic ramus (arrow) and diastases of the right sacro-iliac joint (arrowheads). (b) CT pelvis showing the vertical fracture (arrowheads) through the Ilium at the level of the right SIJ.

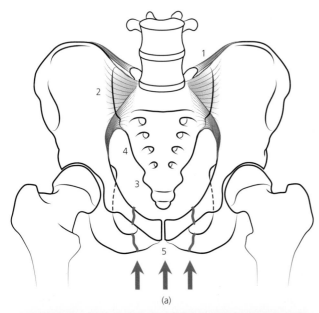

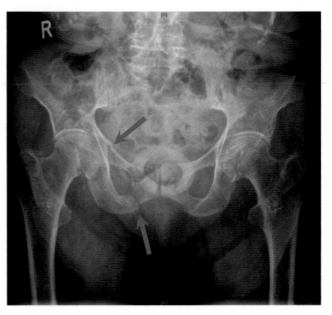

**Figure 5.8** AP compression injury with vertical fractures of the right superior and inferior pubic rami (arrowed).

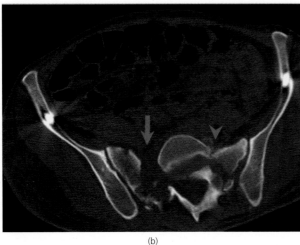

**Figure 5.7** (a) AP compression pelvic injury; (b) CT in pelvic trauma.

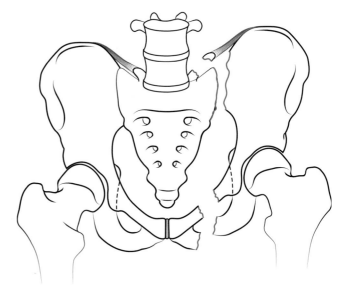

**Figure 5.9** Left vertical compression injury with vertical fractures through the pubic rami and the Ilium close to the SIJ.

*AP compression (APC)*

The direction of the force is in the anteroposterior direction usually from the front (or behind). As a result the typical fractures are seen in the pubic rami in a vertical orientation (Figure 5.7). This causes the ligaments of the symphysis pubis to disrupt. The distance between the symphysis pubis is normally less than 5 mm. A measurement of greater than 1 cm is considered abnormal. Disruption of the symphyseal ligaments does not lead to instability. However, excessive force in the AP direction results in opening up of the hemipelvis like a book if the anterior sacroiliac ligaments are also torn. This results in rotational instability and is generally seen if the symphyseal diastasis is greater than 2.5 cm. Further increase in the force vector will result in disruption of the posterior sacroiliac ligaments and hence cause not only rotational instability but also vertical instability (Figure 5.8.). This results in the loss of the tamponade effect on any underlying pelvic haematoma. In addition this sort of force vector may cause posterior dislocation of the femoral head with resultant posterior acetabular rim fracture.

*Vertical shear (VS)*

The direction of the force is in the vertical direction (up and down), such as a fall from a height. The typical fractures of the pubic rami and the iliac bones are in the vertical plane with disruption of the anterior and posterior sacroiliac ligaments and resultant superior displacement of the hemipelvis (Figure 5.9). The symphyseal ligaments are also disrupted. There is severe vertical pelvic instability and soft tissue injuries and haematoma formation (Figure 5.10). The normal distance between the SI joint is about 2–4 mm. If there is only disruption of the anterior SI ligaments the vertical stability of the pelvis is preserved. If however the posterior SI ligaments are disrupted as well, then the pelvis becomes vertically unstable (Figure 5.10). Displaced vertical fractures through the

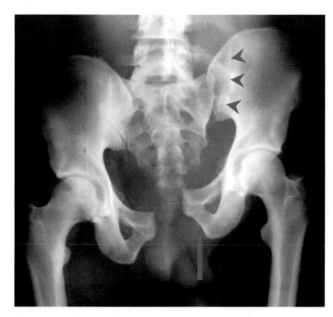

**Figure 5.10** Left vertical fracture with cranial displacement of the hemi-pelvis, pubic symphysis diastases (arrow) and SIJ displacement (arrowhead).

sacrum and the ilium adjacent to the SIJs have the same implications as fractures through the joint. The iliolumbar ligament is attached to the tip of the L5 transverse process. Fracture of the tip of the L5 may cause disruption of the posterior SI ligament and hence may cause vertical instability. So a fracture of the transverse process of L5 should not be dismissed lightly as this may be the only sign of pelvic instability.

*Complex combined injury*

This is the result of a combination of force vectors ending in a pattern of injuries with complex orientation (Figure 5.11). The

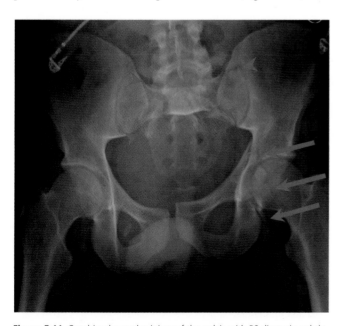

**Figure 5.11** Combined complex injury of the pelvis with PS diastasis, subtle left acetabular fracture (arrows) and multiple sacral fractures (arrowheads).

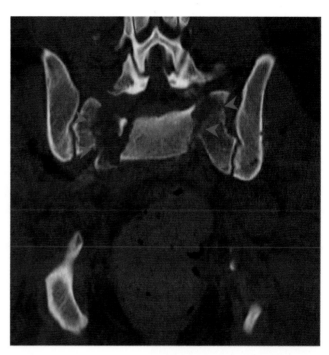

**Figure 5.12** CT with coronal reformats showing the vertical orientation of the sacral fractures.

basic principles of force vector however still hold true and the same mechanistic approach should be used to deal with these fractures, which are inherently unstable (Figure 5.12).

*Bleeding*

The bony pelvis is extremely vascular and hence it can bleed from the fractures. The mortality from pelvic haemorrhage is about 10–20%. Venous bleeding accounts for 90% of the bleeds. Superior gluteal artery is a large vessel and is commonly injured in posterior pelvic fractures. The obturator and pudendal arteries are injured in fractures involving the pubic rami. The retroperitoneum is a large space and can hold up to about 4 litres of blood. Diastasis of the pubic bone by more than 3 cm can double the pelvic volume.

There is morbidity and mortality associated with all the above mentioned injuries. The mortality rate is about 70% for VS, 20% for APC and about 7% for LC injuries. Haematomas are generally more commonly seen in VS and APC injuries.

*Soft tissue injuries*

The bony pelvis protects the internal pelvic organs and therefore if there is a fracture,then internal pelvic viscera may be damaged. Pelvic arch fractures have an incidence of about 20% for bladder and urethral injuries. Urethral injuries are more common in males and are seen in about 5–10% of all pelvic fractures. These injuries are seen with diastasis of the symphysis pubis and result from fractures of the inferomedial pubic bone. This may manifest as blood at the tip of the penis.

Extraperitoneal bladder rupture accounts for 80% of bladder injuries and is seen to involve the anterolateral aspect of the bladder

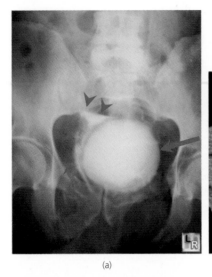

(a)

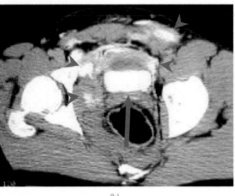

(b)

**Figure 5.13** (a, b) Plain X ray with intravesical contrast and corresponding CT showing extraperitoneal bladder rupture.

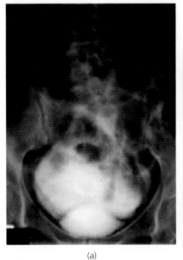

(a)

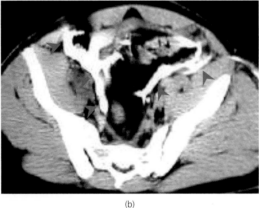

(b)

**Figure 5.14** (a, b) Plain abdominal X ray with intravesical contrast and corresponding CT demonstrates intraperitoneal bladder rupture.

base and is treated conservatively (Figure 5.13). The intraperitoneal rupture (20%) is seen more commonly in children and involves the weaker bladder dome (Figure 5.14). This results in contrast leak into the paracolic regions and is treated surgically.

## Hip
### AP view of the hip
*Adequacy*

AP and lateral are the standard views, which are requested when evaluating a patient with a suspected hip fracture. The AP view of the hip is very similar to the AP view of the pelvis except that it is centred lower and includes both hip joints. The reason for this is that similar symptoms may be caused by an injury to the pubic ring on the same or opposite side.

In children additional views like the frog leg lateral view may also be performed as a routine. This is useful for assessing the femoral capital epiphysis on both sides and will show the femoral head and neck in a position in between the AP and lateral views.

*Alignment and bones*

Check Shenton's line initially. AP view should be analysed in detail for both obvious and more subtle signs of fracture. If no obvious cortical breach is seen, then careful inspection of the bony trabecular pattern should be done. Disruption of bony trabeculae may be the only sign of fracture. Impacted fractures may present as sclerotic lines in the hip and hence should not be disregarded. Rarely, an undisplaced fracture can look completely normal and if the patient is in severe pain and cannot weight bear, then initially a CT or preferably an MRI is indicated.

In early childhood, the trabecula of the femoral neck may produce a striated pattern or unusual lucencies that simulate osteoporosis. If the plain film shows no abnormality in children who present with an irritable hip, other imaging is indicated, and orthopaedic referral is mandatory.

*Cartilage and joints*

In children, widening of the joint space between the teardrop and the cortex of the femoral head may be seen in joint effusions. A difference of more than 2 mm between the two sides is important clinically. Check that the physis (growth plate) is symmetrical in appearance and not widened or compressed (Salter-Harris types I and V fractures). Clearly ultrasound is advantageous in children to detect joint effusion, which in a trauma setting would suggest an underlying injury.

*Soft tissues*

Plain X-rays are usually not very good in assessing the soft tissues around the hip joint. This is due to the large muscle mass around the hip. However, a displaced gluteus medius fat plane is a reliably indicator of an effusion.

### Lateral view of the hip

*Adequacy*

The cross table lateral film should include the acetabulum, ischial spine and tuberosity, and proximal femur. The trochanters should overlap. The lateral view may be difficult to evaluate and is of limited value.

In the frog leg lateral film, the greater trochanter should project over the neck of the femur.

Computed tomography is often the preferred choice if the clinical diagnosis is difficult and initial imaging is equivocal.

Anteroposterior and lateral views of the femoral shaft and knee are indicated in patients with a history of severe trauma or when the clinical findings suggest more than one site of fracture. Gonad protection should always be used in children and adults of reproductive age, as long as it will not obscure a fracture.

*Alignment*

Femoral neck lies anteverted about 30° to the femoral shaft. Check that the entire metaphysis is covered by the epiphysis in children and adolescents. In a slipped upper femoral epiphysis, the centre of the femoral metaphysis lies anterior to its normal position over the central epiphysis. In patients with dislocated hips, the cross table lateral film will define whether dislocation is anterior or posterior.

*Bones*

Trace around the margins of the femur and then the acetabulum and ischium. If a dislocation is present, look for acetabular fragments. These are usually displaced in the same direction as the femoral head.

*Cartilage and joints*

Accessory ossification centres, recognised by their corticated margins, are commonly seen around the acetabular margins and may simulate fractures when partially fused in adolescence. They may persist into adult life. Acetabular roof notches and roof asymmetry are recognised normal variants. Symmetrical protrusion of the acetabular roofs medially is common in children aged 4–12 years.

Hypertrophic changes of the femoral head or inferior aspect of the neck may simulate fractures.

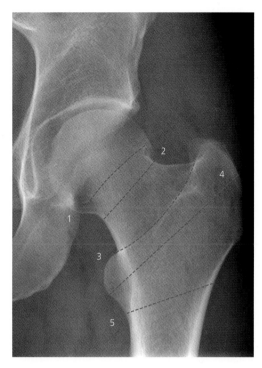

**Figure 5.15** Classification of hip fractures. Intracapsular: 1, subcapital; 2, transcervical; 3, basal or basicervical. Extracapsular: 4, transtrochanteric; 5, subtrochanteric.

*Soft tissues*

Skin folds superimposed over the intertrochanteric region extend past the outer cortical margins, and this differentiates them from fractures.

### Femoral neck fractures

Hip fractures are associated with a high morbidity and mortality especially in the elderly population. Mortality in the 1st year after a fracture is 25% and only 25% return to preinjury level of activity. Femoral neck fractures are most often seen after a fall in older women with osteopenia, although they also are seen in younger patients who have sustained major pelvic trauma.

Femoral neck fractures occur at four sites. Fractures may be intracapsular or extracapsular (Box 5.1; Figure 5.15).

Box 5.1 **Hip fractures**

**Intracapsular – based on level of neck fracture**

- Subcapital
- Transcervical
- Basicervical

**Extracapsular – trochanteric fractures**

- Intertrochanteric
- Subtrochanteric

Intracapsular fractures are subdivided depending on the level of the fracture in the neck of femur. Extracapsular fractures are also

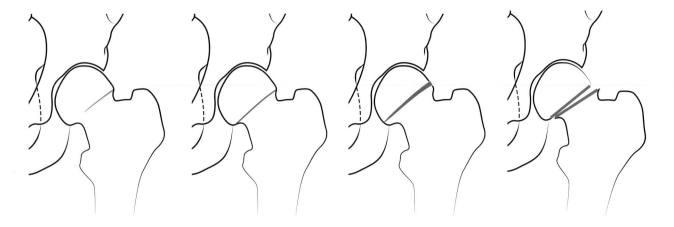

**Figure 5.16** Garden classification.

subdivided depending on whether they go through both trochanters (transtrochanteric, intertrochanteric or pertrochanteric ) and below the trochanters (subtrochanteric). Transtrochanteric fractures are often comminuted and the lesser trochanter is often displaced.

The fractures are usually visible in the anteroposterior view as a lucent line. The fracture line may be sclerotic if some impaction of the trabeculae has occurred.

The subcapital fractures, which are the most common type of intracapsular fractures are also classified by the severity and extent of the injury (Figure 5.16). This is important because the classification will decide the management of the fracture (Box 5.2).

---

Box 5.2 **Garden's classification of subcapital fractures**

- Grade I – incomplete fracture (Figure 5.17)
- Grade II – complete fracture but no displacement (Figure 5.18)
- Grade III – some separation of fracture (Figure 5.19)
- Grade IV – complete separation of fracture (Figure 5.20)

---

In children, considerable violence is needed to fracture the neck of the femur. In transepiphyseal fractures, the femoral capital epiphysis is separated from the metaphysis and dislocated out of the acetabulum; this often results in avascular necrosis (Box 5.3).

---

Box 5.3 **Delbet classification of femoral neck fractures in children**

- Type 1 – transepiphyseal (avascular necrosis usually follows)
- Type 2 – transcervical (avascular necrosis common if displaced)
- Type 3 – cervicotrochanteric
- Type 4 – pertrochanteric

---

## Acetabular fractures

Acetabular fractures may occur because of injury to the pelvic ring or they may occur separately. Fractures of the posterior rim are usually caused by posterior dislocation of the femur and are

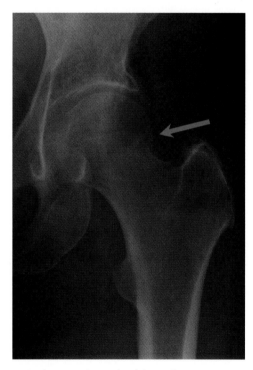

**Figure 5.17** Garden type 1 incomplete left NOF fracture.

therefore often seen in APC fractures of the pelvis (Figure 5.21). Less common anterior pillar fractures are seen in APC fractures of the pelvis, usually as a result of direct trauma to the anterior pelvis. Fractures of the quadrilateral plate, however, generally occur after LC fractures and are often part of a more complex pattern of acetabular injury that usually involves the posterior pillar or anterior pillar, or both. Acetabular fractures may be complicated by sciatic nerve palsy and by severe intrapelvic haemorrhage.

It is important to accurately classify the acetabular fractures as it has a direct bearing on the type of surgery to be performed (Box 5.4). Although there are various classification schemes for acetabular fractures, the most commonly used is that by Judet–Letournel. CT is the imaging modality of choice in classifying the acetabular

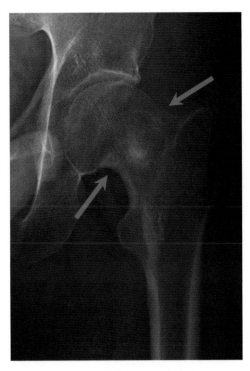

**Figure 5.18** Garden type 2 complete fracture of left NOF.

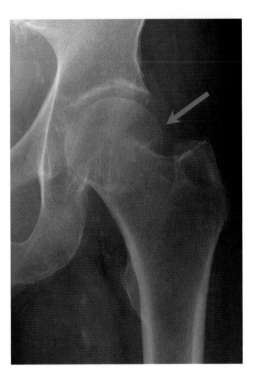

**Figure 5.19** Garden type 3 complete fracture of left NOF with some displacement.

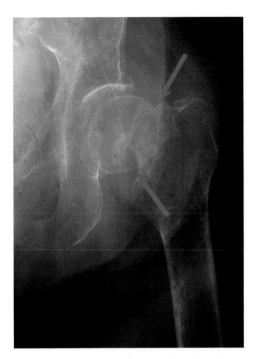

**Figure 5.20** Garden type 4 complete fracture of left NOF with complete displacement.

---

Box 5.4 **Acetabular fractures**

**Judet–Letournel Classification of the five common acetabular fractures**

- Both column
- T-shaped fracture
- Transverse fracture
- Transverse with posterior wall fracture
- Isolated posterior wall fracture

---

## Dislocation

The femoral head can dislocate anteriorly, posteriorly or centrally (Box 5.5). Central dislocation occurs when the femoral head impacts through the acetabulum because of lateral compression injury from a sideways fall or a blow to the greater trochanter. Falling onto the feet is often associated with a vertical fracture of the anterior or posterior pelvic columns. Posterior dislocation may result from a blow to the lumbar spine with the hip flexed – for example, from falling masonry. Dashboard injury in motor vehicle accidents

---

Box 5.5 **Complications of hip dislocation**

- Slipped femoral epiphysis (unfused skeleton)
- Sciatic nerve palsy
- Femoral nerve or artery compression (anterior dislocation)
- Failed reduction and recurrent dislocation
- Avascular necrosis of the femoral head
- Osteoarthritis
- Myositis ossificans
- Femoral head, neck or shaft fractures in major trauma

---

injuries. Although the Judet and Letournel classification includes 10 different types of fractures, there are five fractures which account for 90% of these injuries. These fractures may or may not involve the obturator rings. The detailed explanation of these fractures is beyond the scope of this chapter.

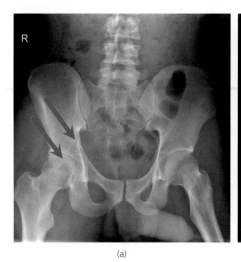
(a)

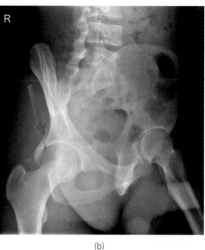
(b)

**Figure 5.21** (a, b) AP and oblique view demonstrating a right acetabular fracture with involvement of roof posteriorly.

results in posterior dislocation of the hip and is often associated with fracture of the femoral shaft or patella.

## Congenital dislocation of the hip

Successful treatment depends on correct and early recognition of congenital dislocation of the hip (CDH). Ultrasonography is the imaging modality of choice in diagnosis of CDH; however, diagnosis can be made from plain radiographs. At birth, the femoral epiphysis is not ossified, but the acetabular roof is often abnormal, with notching laterally and an increased acetabular angle. Once the epiphysis is ossified, the disorder becomes obvious from radiography.

## Idiopathic coxa vara

Idiopathic coxa vara is part of a spectrum of conditions known as proximal femoral focal deficiency (PFFD) of which two types exist: congenital and infantile. Lesions are usually bilateral and present with coxa vara with epiphysis that are low lying and look 'woolly', and there is epiphyseal or metaphyseal lucency.

## Slipped capital femoral epiphysis (adolescent coxa vara)

In this condition, the femoral neck moves proximally and externally rotates on the unfused epiphysis. In 20% of cases the condition is bilateral, and it occurs in overweight, hypogonadal, or tall but thin adolescents. Pain sometimes referred to the knee, or limp is a common presenting symptom. Both hips should be evaluated. Early slip is best assessed in the frog leg lateral film. The plain X-ray findings of this condition include indistinct epiphysis with associated widening; a line drawn along the lateral margin of the neck of the femur does not intersect the femoral head epiphysis. These patients are prone to avascular necrosis and degenerative arthritis.

---

**ABCs systematic assessment**

**Adequacy**

- Use one view for the pelvis and two views for the hips
- Ensure that all the pelvis and hips are visible

- Use a bright light to look at radiographs or
- Adjust windows on PACS

**Alignment**

- Check three rings
- Check SIJs and pubis
- Check Shenton's line
- Bone
- Check each bone carefully
- Check the trabecular lines
- If there is one pelvic ring fracture, look for a second

**Cartilage and joints**

- Check the pubic symphysis
- Check the sacroiliac joints
- Check the hip joints and acetabulum

**Soft tissues**

- Check the obturator internus fat planes inside the pelvis
- Check the perivesical fat plane
- Check the gluteus medius fat plane
- Check the femoral pulses and sciatic nerve

## Further reading

Judet R, Judet J, Letournel E. Fractures of the acetabulum: classification and surgical approaches for open reduction – preliminary report. *J Bone Joint Surg Am* 1964;46:1615–1646.

Letournel E, Judet R. *Fractures of the Acetabulum*, 2nd edn. Heidelberg, Germany: Springer-Verlag, 1993.

Lloyd E, Stambaugh, CC Blackmore. Pelvic ring disruption in emergency radiology. *Eur J Radiol* October 2003;48: 1:71–87.

Lyons AR. Clinical outcomes and treatment of hip fractures. *Am J Med* 1997;103:S51–S63.

Their, M.E, Bensch, F.V, Koskien, S. K, Handolin, L, Kiuru, M.J. Diagnostic value of pelvic radiography in the initial trauma series in blunt trauma. *Eur Radiol* 2005 Aug;15(8):1533–1537.

# CHAPTER 6

# Knee

*Lisa Meacock and David A. Elias*

King's College Hospital, London, UK

---

### OVERVIEW

- Conventional radiographs are useful in the acute setting
- Significant internal derangement can occur with normal XRs
- A lipohaemarthrosis indicates an intra-articular fracture
- Popliteal artery injuries have to be excluded in supracondylar fractures and femorotibial dislocations
- CT should be performed for intra-articular fractures and in pre-operative planning
- If internal derangement is suspected, then MRI should be performed
- Ultrasound is rarely indicated in acute trauma except in suspected extensor injuries

---

Conventional radiographs (XRs) are initially performed for knee trauma but severe injuries may be present with little or no abnormality on XRs. If XRs show a lipohaemarthrosis, then CT or MRI is indicated to confirm the intra-articular fracture and potential associated injuries. If internal knee derangement is suspected, then MRI should be requested. MRI allows comprehensive assessment of soft tissue and bony injuries. CT is indicated for suspected intra-articular fractures and in pre-operative planning of complex bony injuries. In supracondylar fractures and knee dislocations, popliteal artery injuries need to be excluded, generally using CT angiography.

## Anatomy

The knee is a hinge-type synovial joint formed by the articulation between the femoral condyles and tibial plateau. The two bones are separated by two C-shaped fibrocartilaginous structures: the medial and lateral menisci. The patella is a large sesamoid bone in the knee extensor mechanism. The quadriceps muscles form the quadriceps tendon, which inserts into the superior pole of the patella. The patella ligament extends from the inferior patella pole to insert onto the tibial tuberosity. The posterior surface of the patella articulates with the trochlear groove on the anterior surface of the femoral condyles and forms the patellofemoral joint (PFJ). The fibular head and posterolateral proximal tibia articulate at the proximal tibiofibular joint (Table 6.1).

Ligaments and musculotendinous structures provide the knee with stability (Table 6.2; Figure 6.1). The knee is surrounded by multiple bursae. The suprapatellar bursa is continuous with the knee joint and distends in the presence of a joint effusion. It lies between the suprapatellar fat and prefemoral fat above the level of the patella (Figure 6.3).

**Table 6.1** Timing of appearance and fusion of the secondary ossification centres about the knee.

| Bone | Secondary ossification centre | Age (years) | |
|------|------|------|------|
| | | At formation | At fusion |
| Femur | Distal femoral epiphysis | Birth–2 months | 15–17 |
| Tibia | Proximal tibial epiphysis | Birth–2 months | 15–17 |
| | Tibial tuberosity apophysis | 8–14 | 15 (with metaphysis) |
| Patella | May have multiple centres | 3–6 | |

**Table 6.2** Main supporting ligaments and musculotendinous structures of the knee.

| Structure | Origin | Insertion | Primary function |
|------|------|------|------|
| Anterior cruciate ligament (ACL) | Posterolateral aspect of roof of intercondylar notch of femur | Anterior intercondylar eminence of tibia | Resists anterior translation and internal rotation of tibia |
| Posterior cruciate ligament (PCL) | Anteromedial intercondylar notch of femur | Posterior tibial eminence | Resists posterior translation and external rotation of tibia |
| Medial collateral ligament (MCL) | Medial epicondyle of femur | Medial proximal tibial metaphysis | Resists valgus stress |
| Lateral collateral ligament | Lateral epicondyle of femur | Fibular head | Resists varus stress |
| Quadriceps mechanism | Anterior pelvis and proximal femur | Tibial tuberosity (via patellar) | Knee extension and patellar stabilisation |

*ABC of Emergency Radiology*, Third Edition. Edited by Otto Chan.
© 2013 John Wiley & Sons, Ltd. Published 2013 by John Wiley & Sons, Ltd.

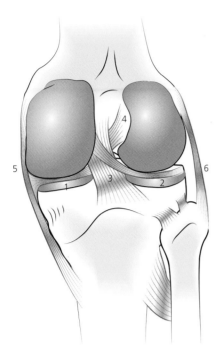

**Figure 6.1** The posterior knee demonstrating normal ligaments and menisci: 1, medial meniscus; 2, lateral meniscus; 3, posterior cruciate ligament; 4, anterior cruciate ligament; 5, medial collateral ligament; 6, lateral collateral ligament.

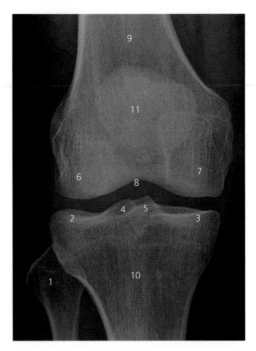

**Figure 6.2** Normal anteroposterior radiograph of the right knee in an adult: 1, fibular head; 2 lateral tibial plateau; 3, medial tibial plateau; 4 and 5, lateral and medial tibial spines; 6, lateral femoral condyle; 7, medial femoral condyle; 8, intercondylar notch; 9, femur; 10, tibia; 11, patella.

---

**Radiographic views**

- Anteroposterior (AP) (Figure 6.2) and lateral (lat) (Figure 6.3) projections are standard
- A skyline view (Figure 6.4) allows assessment of the patellofemoral articulation
- A tunnel or notch view is valuable to look at the intercondylar notch and to identify osteochondral fractures or intra-articular bodies
- Oblique views in internal and external rotation allow further evaluation of tibial plateau fractures and of the proximal tibiofibular joint

## ABCs systematic assessment

- Adequacy – check correct views have been obtained
- Alignment – check femorotibial alignment and patellar height
- Bones – trace the contours of all the bones
- Cartilage and joints – joint spaces should be uniform in width
- Soft tissues – change windows to look for soft tissue swelling, effusion, and fat/fluid level

### Adequacy

Routine AP and lat views should be obtained. Following acute trauma, the lat view should be performed with a horizontal beam so that a fat-fluid level can be identified. View the lat view first.

### Lateral view

*Soft tissues (on this occasion, S before ABC!)*

Separation of the suprapatellar and prefemoral fat pads (>5 mm) indicates an effusion (Figure 6.5). The presence of a fat-fluid level (lipohaemathrosis) within the joint is pathognomonic of an intra-articular fracture (causing bleeding and therefore a haemo-arthrosis), with leakage of marrow fat into the joint (hence the lipohaemoarthosis) (Figure 6.6).

### Alignment

On a lateral view, anterior tibial displacement indicates rupture of the anterior cruciate ligament, while posterior displacement indicates rupture of the posterior cruciate ligament.

On a lateral view with the knee in 20–30° flexion, the ratio of the patella tendon length to patella length should be in the range 0.8–1.2 (Figure 6.7). A high riding patella (patella alta) may be a congenital variant or the result of rupture of the patella tendon (Figure 6.8). A low riding patella (patella baja) may be a congenital variant or the result of rupture of the quadriceps tendon.

On the AP view, a line through the lateral edge of the lateral femoral condyle (lateral tibial line) should run to the lateral edge of the lateral tibial plateau. A significant step usually indicates a tibial plateau fracture (Figure 6.9).

### Bone

The cortices of the femur, tibia, patella, and fibula should be smooth, with no disruption of the trabecular pattern within the

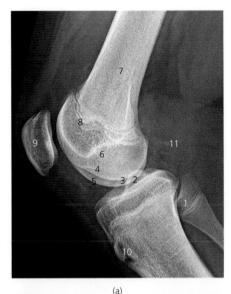

(a)

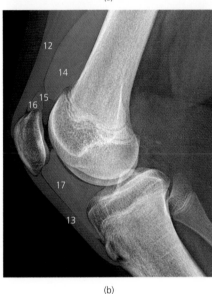

(b)

**Figure 6.3** Normal lateral knee radiograph of an adolescent. (a) 1, Fibula; 2 and 3, tibial spines; 4 and 5, lateral and medial femoral condyles (overlapping); 6, roof of intercondylar notch; 7, femur; 8, distal femoral growth plate; 9, patella; 10, tibial tuberosity; 11, fabella (sesamoid bone in the lateral head of gastrocnemius); (b) 12, quadriceps tendon (and distal muscle belly); 13, patellar ligament; 14, prefemoral fat pad; 15, suprapatella bursa; 16, suprapatella fat pad; 17, Hoffa's fat pad.

bones. Tibial plateau fractures may be identifiable as a subtle sclerotic line or subtle step defect only. A careful search should be made for intra-articular bodies, and for certain bony avulsions which may indicate significant ligamentous injuries.

### Cartilage and joints

The joint spaces in the medial and lateral compartments may be assessed on the AP view for height (reduced height may be the result of knee flexion or cartilage loss or arthritis) and for chondrocalcinosis (linear calcification within the cartilage, which may occur in numerous conditions).

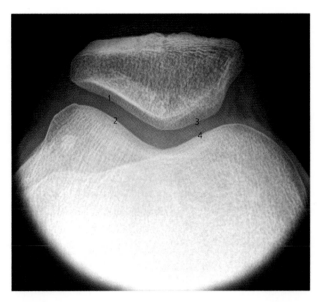

**Figure 6.4** Normal skyline view of the right knee in an adult: 1, lateral patellar facet; 2, lateral trochlear facet (anterior lateral femoral condyle; 3, medial patellar facet; 4, medial trochlear facet (anterior medial femoral condyle).

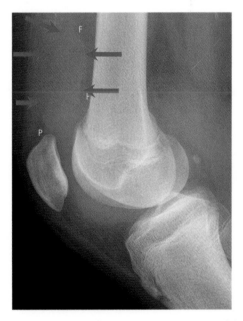

**Figure 6.5** Lateral radiograph of the knee in a 13-year-old boy following acute knee injury. There is a large effusion (arrows). Note how the effusion separates the suprapatellar (P) from the prefemoral fat pad (F).

## Injuries

### Distal femur

Femoral shaft fractures occur with considerable force. Anteroposterior and lateral views of the whole femur are essential to identify displacement and rotation.

### Supracondylar fractures

These fractures may have a variety of configurations. Evidence of intra-articular extension of the fracture line should be sought, as

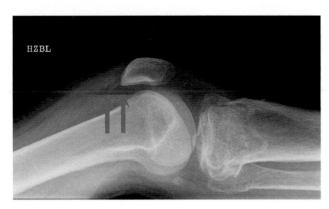

**Figure 6.6** Horizontal beam lateral radiograph of the knee in a patient with a tibial plateau fracture following a road traffic accident. Note the lipohaemarthrosis (fat-fluid level – arrows), which indicates the presence of an intra-articular fracture. The comminuted tibial plateau and associated fibular head fracture are also identified.

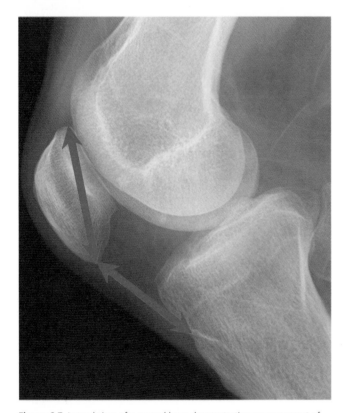

**Figure 6.7** Lateral view of a normal knee demonstrating measurement of the Insall Salvati ratio for patellar height. The ratio is measured as the patellar ligament length divided by the maximum patellar length and the ratio normally lies in the range 0.8–1.2.

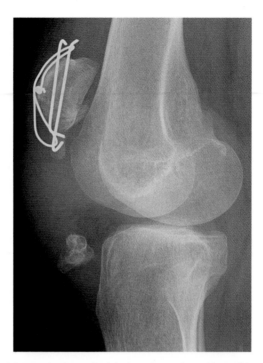

**Figure 6.8** Lateral radiograph of the knee in a patient with an old patellar fracture which was fixed with two K-wires and a tension band wire. Following a recurrent injury, the patellar ligament ruptured resulting in patellar alta with ossific fragments in the disrupted ligament.

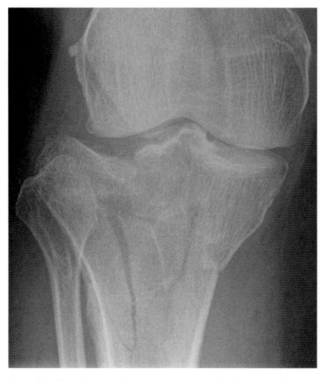

**Figure 6.9** AP radiograph of the knee in a patient showing a comminuted tibial plateau fracture following a road traffic accident. Note the malalignment with a large step between the lateral margin of the lateral femoral and the lateral margin of the tibial plateau.

this necessitates open reduction and internal fixation. The distal fragment may be angulated by the pull of gastrocnemius, and displacement can result in popliteal artery injury. Occasionally, supracondylar fractures are associated with fracture dislocations of the hip or tibial shaft (Figure 6.10).

## Femoral condylar fractures

These fractures may show displacement or comminution and are often seen best with computed tomography. Shearing or rotatory

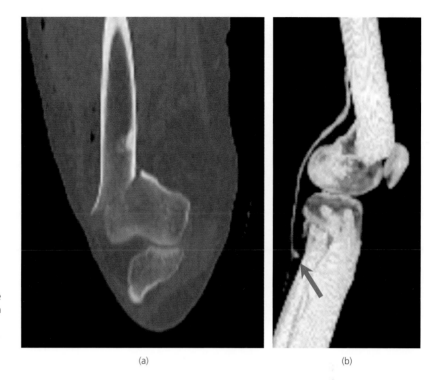

**Figure 6.10** Coronal plane reformatted CT image of the knee (a) and 3D volume rendered CT arteriogram (b) in a patient with a supracondylar fracture following a motorcycle accident. There is a traumatic avulsion of the origin of the anterior tibial artery (arrow) in association with the fracture.

(a)                              (b)

forces directed at the articular surface of a femoral condyle may produce fractures, known as osteochondral fractures, through the cartilage and subchondral bone. These are often occult on conventional radiographs, but irregularity in the articular surface may be present and intra-articular fragments of bone may be seen (Figure 6.11).

Acute osteochondral fractures should be distinguished from osteochondritis dissecans.

### Osteochondritis dissecans

This occurs in adolescents at the lateral margin of the medial femoral condyle, and there may be a separated osteochondral fragment with sclerotic margins. The aetiology of this condition is controversial, but it may involve trauma (Figure 6.12).

### Proximal tibia and fibula
### Tibial plateau fractures

These fractures occur most often in women >50 years, usually after twisting falls (Figure 6.13). Typically, valgus force is encountered, with impaction of the femoral condyle on the plateau, and involvement is confined to the lateral plateau in 75–80% of cases. Less than 25% of cases are the result of RTA and these typically result from the bumper of a car striking the knee. These fractures are often subtle, and AP and lat radiographs need careful examination.

**Figure 6.11** (a) Lateral knee radiograph and (b) coronal intermediate density fat saturated MR image of a patient with an acute osteochondral fracture. Note the ossific body in the knee joint on the lateral radiograph with a joint small effusion. On the MRI the corresponding osteochondral defect is seen in the lateral femoral condyle (arrows) with underlying marrow oedema.

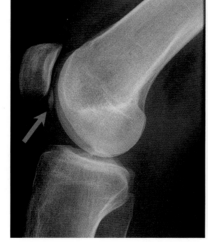

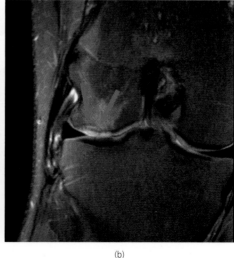

(a)                              (b)

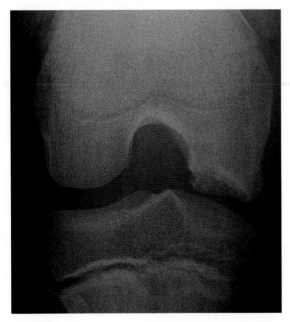

**Figure 6.12** Notch view of a 13-year-old boy with osteochondritis dissecans. Note the typical osteochondral defect at the lateral aspect of the medial femoral condyle.

Computed tomography is valuable, as the degree of depression of the fracture determines the need for surgery, and this is difficult to assess on conventional radiographs. Alternatively, magnetic resonance imaging may be used to assess these injuries; it also shows associated ligamentous and meniscal injuries, which are reported in 68–97% of cases.

### Tibial stress fractures

These fractures may be seen as transversely orientated sclerotic or lucent lines, with adjacent sclerosis in the medulla of the medial proximal tibial metaphysis extending to the cortex (Figure 6.14). The patient may show evidence of osteoporosis.

### Proximal fibular fractures

*Fibula head fractures* may be isolated injuries that result from a direct blow, but they are associated most often with tibial plateau fractures. A fracture of the fibula neck or proximal shaft may suggest the presence of an associated ankle joint injury (Maisonneuve). Proximal fibular fractures may be associated with common peroneal nerve injury.

## Patella
### Fractures

These fractures may be caused by a direct blow or by excessive quadriceps contraction during forced knee flexion. Fractures may be transverse, vertical, or comminuted, with or without displacement. A bipartite patella is a normal variant, which lies at the superolateral aspect of the patella, and which can be confused with a fracture (Figure 6.15).

### Patella dislocation

Dislocation of the patella is almost always in the lateral direction and classically occurs in teenage girls. Dislocation is usually transient

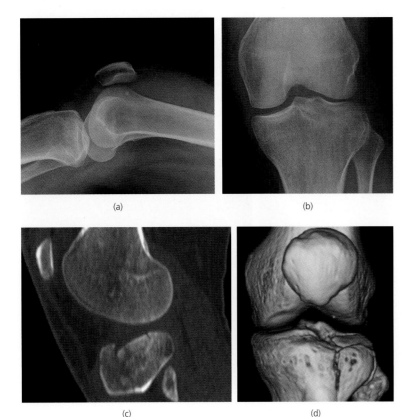

(a)

(b)

(c)

(d)

**Figure 6.13** Lateral tibial plateau fracture following a fall. The lateral radiograph demonstrates a large effusion with a lipohaemarthrosis but the fracture is relatively occult on both the lateral and AP radiographs (a and b). Subtle lucency is present on the AP radiograph at the site of fracture. The sagittal CT reformat (c) and 3D volume rendered CT image (d) demonstrate the lateral tibial plateau fracture and the extent of its depression.

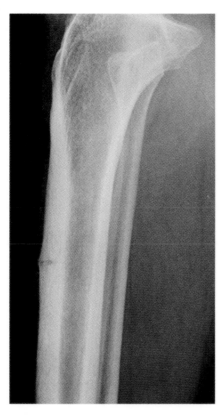

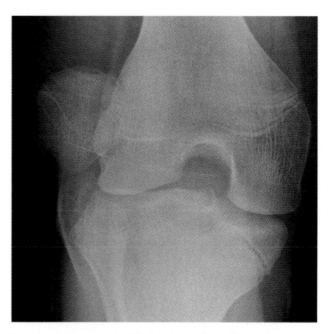

**Figure 6.16** AP radiograph of the knee in a patient with acute lateral patellar dislocation. Generally dislocations are transient and it is unusual to present with a persistent dislocation in this way.

**Figure 6.14** Lateral radiograph of the tibia in a runner with a chronic tibial stress fracture. A typical horizontal lucency is seen in the anterior cortex.

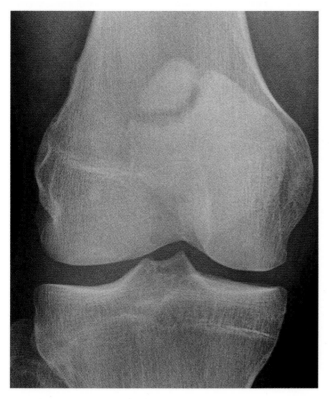

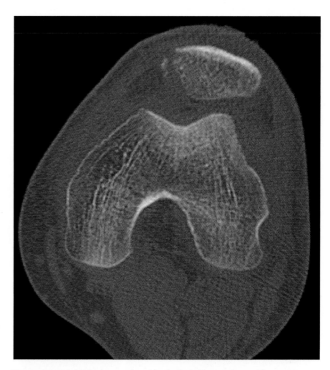

**Figure 6.17** Axial CT image through the knee in a patient with previous transient lateral patellar dislocation. The patellar is laterally subluxed and there is a separated avulsion fracture off the medial patellar margin.

**Figure 6.15** AP radiograph of a normal knee which shows a normal variant of a bipartite patella. There is a separated ossific fragment at the superolateral aspect of the patella and this should not be confused with a fracture.

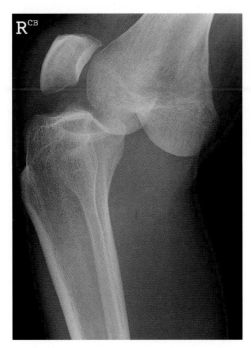

**Figure 6.18** AP radiograph of a knee dislocation following a road traffic accident. Such injuries have a significant association with popliteal arterial injury.

**Table 6.3** Some signs of ligamentous injury visible on conventional radiographs of knee.

| Ligament | Sign of injury | Cause |
|---|---|---|
| ACL | Avulsion of anterior tibial eminence | Avulsion of anterior cruciate ligament |
| | Segond fracture | Lateral capsular ligament avulsion producing a small avulsion fragment at the lateral tibial plateau (>75% associated with injuries of anterior cruciate ligament) |
| | Deep notch | Impaction of notch of lateral femoral condyle against posterior tibial plateau during injury of anterior cruciate ligament |
| PCL | Avulsion of posterior tibial eminence | Avulsion of tibial attachment of PCL |
| MCL | Pellegrini-Stieda lesion | Chronic recurrent injury of medial collateral ligament: shows linear calcification over the medial supracondylar edge |
| LCL | Avulsion of lateral epicondyle | Avulsion of lateral collateral ligament |
| QT | Avulsion of superior or inferior patella pole | Quadriceps contraction causing avulsion of distal quadriceps or proximal patella tendon |
| | Avulsion of tibial tuberosity[a] | Quadriceps contraction causing avulsion of distal patella tendon |

[a]This should be distinguished from Osgood-Schlatter's disease, a chronic condition occurring in adolescents characterised by anterior knee pain and fragmentation of the tibial tuberosity, with overlying soft tissue swelling.

and presents with non-specific acute haemarthrosis (Figure 6.16). In many patients dislocation becomes recurrent. Radiographs after injury rarely show a dislocated patella, as the patella has usually been reduced by the time the radiograph has been taken. Usually, plain radiographs are normal apart from a joint effusion. A skyline view occasionally shows osteochondral injury, with fragments separated from the medial aspect of the patella or the anterior tip of the lateral

femoral condyle, or both, as a result of impaction of the cortices at the time of transient dislocation (Figure 6.17).

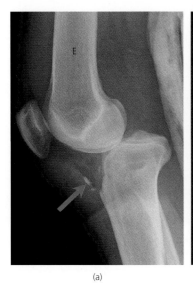

(a)

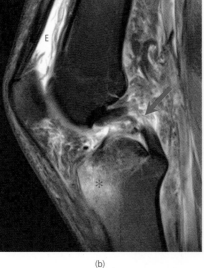

(b)

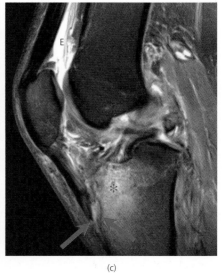

(c)

**Figure 6.19** Lateral radiograph of the knee (a) of a patient with knee dislocation following a road traffic accident. The tibia is posteriorly dislocated. Note the bony fragments (arrow) which indicate bony avulsion of the patellar ligament from the tibial tuberosity. There is a joint effusion ('E'). The corresponding sagittal intermediate density fat saturated MR images (b, c) demonstrate a mid substance tear of the posterior cruciate ligament (arrows) and bony avulsion of the distal patellar ligament (arrowheads), as well as extensive marrow contusion at the tibial tuberosity (asterisk).

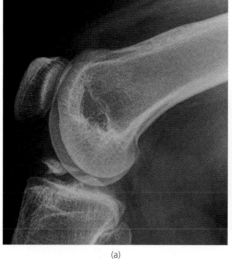
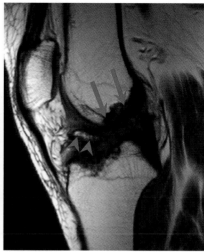

**Figure 6.20** Lateral radiograph of the knee (a) in a patient with a chronic anterior cruciate ligament (ACL) avulsion. The tibial bony footplate is avulsed and lies in the anterior intercondylar notch. The corresponding sagittal proton density MR image (b) demonstrates the bony avulsion fragment (arrowheads) with the attached ACL fibres (arrows).

(a)

(b)

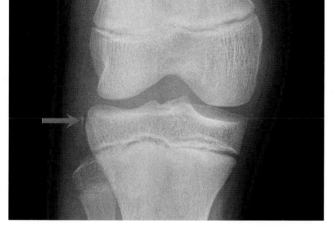

**Figure 6.21** AP radiograph of the knee in a patient with an acute anterior cruciate ligament (ACL) tear demonstrating a Segond fracture (arrow) of the lateral tibial plateau which is typically associated with ACL injury.

## Knee dislocation

Dislocation of the knee is rare. Most commonly, anterior translation of the tibia on the femur is present, and the risk of popliteal artery and peroneal nerve injury is considerable (Figure 6.18). True knee dislocation is invariably associated with rupture of multiple ligaments, but plain radiographs are often misleading, with little or no abnormality visible, apart from a joint effusion (Figure 6.19). Magnetic resonance imaging should be performed to evaluate associated injuries.

## Avulsion fractures and signs of ligamentous injury

Injury to ligamentous structures about the knee usually shows clinical signs of instability and a knee effusion. Avulsion fractures occasionally provide relatively specific evidence of particular ligamentous injuries (Figures 6.19–6.23; Table 6.3).

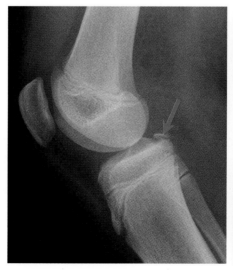
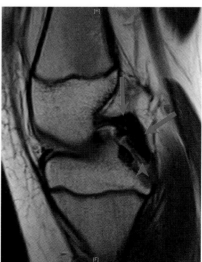

**Figure 6.22** Lateral radiograph of a knee (a) with bony avulsion of the posterior cruciate ligament (PCL). Note the typical elevated bony fragment (arrow) from the posterior margin of the intercondylar region of the tibia. The corresponding sagittal proton density MR image (b) demonstrates the bony fragment (arrowhead) with the attached PCL fibres (arrows).

(a)

(b)

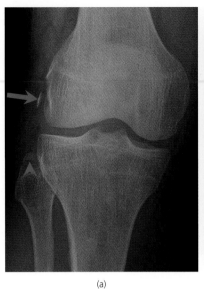
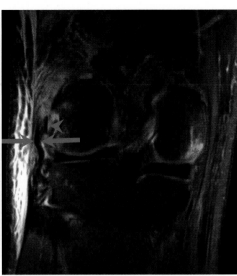

(a)                                                     (b)

**Figure 6.23** AP radiograph of the knee (a) in a patient with lateral collateral ligament avulsion following a fall from a height. Note the bony avulsion fragment from the lateral epicondyle (arrow) there is also a fibular head fracture due to posterolateral corner injury (arrowhead). The corresponding coronal intermediate density fat saturated MR image (b) shows the bony fragment (arrowhead) with the attached retracted lateral collateral ligament (arrows). Medial femoral condylar marrow contusion (asterisk) and lateral soft tissue bruising is also seen.

### ABCs systematic assessment

#### Easy 'AS ABC'!
*Adequacy*

- Minimum of lateral and AP views; lateral should be horizontal beam in acute trauma

*Soft tissues*

- Lipohaemarthrosis (fat-fluid level) indicates intra-articular fracture

*Alignment*

- Tibiofemoral alignment – anterior tibial displacement indicates ACL disruption; posterior tibial displacement indicates PCL disruption
- Patellar height – high riding patella ('alta') may indicate a congenital variant or patellar ligament disruption; low riding patella ('baja') may be a congenital variant or indicate quadriceps disruption

*Bones*

- Tibial plateau fractures may appear as subtle linear lucency or sclerosis
- Small bony fragments may represent significant ligamentous avulsions

*Cartilage*

- Joint space loss may be due to meniscal or hyaline cartilage injury.

## Further reading

Gottsegen CJ, Eyer BA, White EA, Learch TJ, Forrester D. Avulsion fractures of the knee: imaging findings and clinical significance. *Radiographics* 2008;28:1755–70.

Kapur S, Wissman RD, Robertson M, Verma S, Kreeger MC, Oostveen RJ. Acute knee dislocation: review of an elusive entity. *Curr Probl Diagn Radiol* 2009;38:237–50.

Miller LS, Yu JS. Radiographic indicators of acute ligament injuries of the knee: a mechanistic approach. *Emerg Radiol* 2010 Nov;17:435–44.

Rogers LF (Ed.). The Knee in: *Radiology of Skeletal Trauma*, 3rd edn. New York: Churchill Livingstone, 2002.

# CHAPTER 7

# Ankle and Foot

*Tudor Hughes*

University of California, San Diego, CA, USA

**OVERVIEW**

- Ankle and foot injuries are very common – use OTTAWA rules
- Meticulous assessment essential to avoid disastrous consequences
- Remember 7 review areas in a 'normal' radiograph
- CT, MRI or US should be requested where appropriate

Trauma to the ankle and foot is one of the most common reasons that people attend emergency departments. A spectrum of injuries can occur; from sprains through fractures to dislocations. The decision to obtain radiographs is made by careful clinical examination and knowledge of the mechanism of injury. The use of certain guidelines (such as the Ottawa rules, Box 7.1) will exclude serious injuries while reducing unnecessary exposure of the patient to radiation. A higher index of suspicion is needed in elderly.

Box 7.1 **Ottawa rules**

- Bone tenderness along the distal 6 cm of the posterior edge of the tibia or fibula
- Bone tenderness of the medial or lateral malleoli
- Bone tenderness at the base of the fifth metatarsal or the navicular
- An inability to bear weight immediately or in the emergency department

Injuries of the feet often result in a request for radiographs of the ankle and foot. Clinically, it should be possible to distinguish which area has been injured, and imaging of both is rarely needed. Injuries to the feet, however, often masquerade as ankle injuries.

## Anatomy

The ankle (Figure 7.1a–f) is a virtual hinge joint shaped as a mortice. The tibia and fibula form a ring with the proximal and distal tibiofibular joints.

The bony structure of the ankle is stabilised by three main groups of ligaments:

- Medial collateral ligament complex (deltoid ligament)
- Lateral collateral ligament, which includes anterior talofibular, posterior talofibular and calcaneofibular ligaments
- Tibiofibular syndesmotic complex.

The talus articulates inferiorly with the calcaneus and anteriorly with the navicular. It is made up of a body, neck, and anterior process and has a fragile blood supply that extends through the ankle joint capsule, which means that fractures of the talar neck may result in avascular necrosis of the body.

The foot (Figure 7.2a,b) is a complex structure of interdependent bones designed for weight bearing and movement. It can be divided into the forefoot, midfoot, and hindfoot. The joints are complex, but the articular surfaces are parallel and the joint spaces equidistant and symmetrical. Loss of articular parallelism and alteration of joint space width is always abnormal.

## Normal variants

Radiographs may show variation from the normal anatomy because of the presence of sesamoids, fused or partly fused bones, or accessory ossification centres (Figure 7.3a,b). Although these are normal variants, they can also be the sites of pathology. Commonly occurring ossicles are the os tibiale externum (medial to the navicular), os trigonum (posterior to the talus), and os peroneum (adjacent to the cuboid).

## ABCs systematic assessment

- Adequacy
- Alignment
- Bone
- Cartilage and joints
- Soft tissues

## Adequacy

Standard imaging of the ankle in the emergency department should include anteroposterior (AP), mortice and lateral projections and for the foot AP, oblique and lateral. Weight-bearing views are

*ABC of Emergency Radiology*, Third Edition. Edited by Otto Chan.
© 2013 John Wiley & Sons, Ltd. Published 2013 by John Wiley & Sons, Ltd.

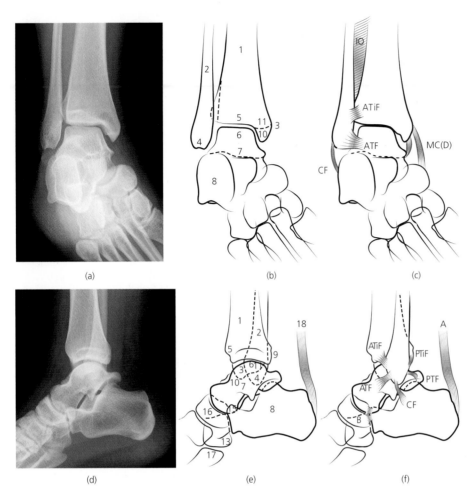

(a)  (b)  (c)

(d)  (e)  (f)

**Figure 7.1** (a),(b) AP view of ankle and drawing: (1), Tibia; (2), fibula; 3, medial malleolus; 4, lateral malleolus; 5, plafond; 6, dome; 7, talus; 8, calcaneum; 9, posterior malleolus; 10, anterior colliculus; 11, posterior colliculus. (c) AP ankle ligaments: A, Achilles tendon; ATiF, anterior tibiofibular; ATF, anterior talofibular; B, bifurcate; CF, calcaneofibular; D, deltoid; IO, interosseous; MC, medial collateral; PTiF, posterior tibiofibular; PTF, posterior talofibular. (d),(e) Lat view of ankle and drawing: 1, tibia; 2, fibula; 3, medial malleolus; 4, lateral malleolus; 5, plafond; 6, dome; 7, talus; 8, calcaneum; 9, posterior malleolus; 10, anterior colliculus; 11, posterior colliculus; 12, anterior tubercle; 13, peroneal groove; 14, cuboid; 15, anterior process; 16, navicular; 17, base of fifth metatarsal; 18, Achilles tendon. (f) Lat ankle ligaments.

Not all bones and joints will be seen clearly on one view, so multiple views will be needed (Figure 7.4a,b). Some injuries are visualised poorly on radiography, and computed tomography or other types of imaging may be needed (Box 7.2).

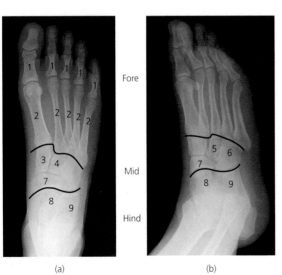

(a)  (b)

**Figure 7.2** (a) AP view and (b) oblique view of foot. Forefoot: 1, phalanges; 2, Metatarsals. Midfoot: cuneiforms – 3, medial; 4, middle; 5, lateral; 6, cuboid; 7, navicula. Hindfoot: 8, talus; 9, calcaneum.

Box 7.2

- US – Used to assess soft tissues, muscles, tendons and ligaments and for intervention; very operator dependent
- CT – Axial slices obtained with multiplanar reconstruction; good for looking at bones, bone bars and fractures
- MRI – Highly sensitive and specific; shows pathology in bones, joints, and soft tissues; multiplanar imaging, usually axial, sagittal and coronal
- Isotopes – Increase of isotope uptake in bones is a non-specific, highly sensitive indicator of disease

better when tolerable. If the pain is in the heel, lateral and axial (Harris) views of the heel are indicated. Advanced imaging should be performed only after consultation with a radiologist.

## Alignment

On the anteroposterior view, the uniform distance between the tibiotalar and fibulotalar joints should be <4 mm in adults. On the lateral view, the long axis of the tibia and fibula should overlap.

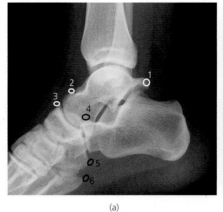

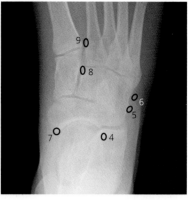

**Figure 7.3** (a),(b) Common accessory ossicles: 1, os trigonum; 2, os supratalare; 3, os supranaviculare; 4, os calcaneum secundarius; 5, os perineum; 6, os vesalaneum; 7, os tibiale externum; 8, os intercuneiforme; 9, os intermetatarseum.

(a)

(b)

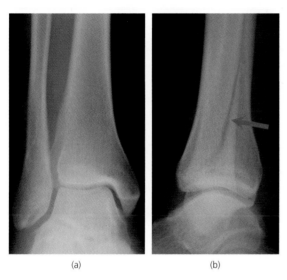

(a)                                      (b)

**Figure 7.4** (a) AP view (right) 'normal'; (b) lat view – there is clearly an oblique fracture of the fibular shaft only seen in this view (arrow).

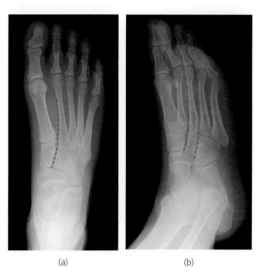

(a)                                      (b)

**Figure 7.5** The alignment of the midfoot is best assessed on standing views. On the AP (dorsoplantar) view (a) the medial side of the 2nd metatarsal should always align with the medial side of the middle cuneiform. On the AP pronation oblique view (b) the medial side of the 3rd and 4th metatarsal should align with the medial side of the lateral cuneiform and cuboid respectively.

On weight-bearing lateral view radiographs, the superior surfaces of the talus, navicular, medial cuneiform, and first metatarsal lie in a straight line. Böhler's angle (see Figure 7.18a) lies between the plane of the posterosuperior and anterosuperior surfaces of the calcaneus and measures 28–40° in normal feet. Flattening of the angle (<28°) follows calcaneal compression with trauma.

The midtarsal joint separates the talus and calcaneus from the navicular and cuboid and resembles a wave (cyma) on all 3 views of the foot (Figure 7.3a,b). An intact cyma line shows integrity of the midtarsal joints.

The midfoot articulates with the forefoot at the tarsometatarsal joint and must be evaluated carefully, as subluxations of this region can be subtle, and lead to disastrous consequences. In the anteroposterior view of the foot (Figure 7.5a), the medial aspect of the second metatarsal should align with the medial aspect of the middle cuneiform and on the oblique view (Figure 7.5b) the medial aspect of the third metatarsal should align with the medial aspect of the lateral cuneiform and the medial aspect of the fourth metatarsal should align with the medial aspect of the cuboid.

## Bone

Trace the cortical margins of the bones. Abnormal steps in the cortex, lucent, or sclerotic lines sometimes indicate the presence of fractures. If no abnormality is seen, check these review areas:

*Talar dome* – Look at the corners on the AP and mortice ankle views.

*Lateral process of talus* – Look carefully on the AP ankle view.

*Lateral malleolus* – Oblique fractures can look normal on the antero-posterior view, so look through the tibia on the lateral.

*Anterior margin of the tibial plafond* – Small flake fractures can be easily overlooked on the lateral view.

*Posterior malleolus* – Small and even large fractures can be difficult to detect and represent an unstable ankle injury.

*Superior surface of the talus or navicula* – Small capsular avulsion flake fractures should be sought.

*Calcaneus* – fractures *usually* occur as a result of a fall from a height. Compression fractures of the body of the calcaneus may extend to the subtalar joint. Stress fractures of the calcaneus usually occur in runners and may appear as a sclerotic linear band. Also anterolateral flake avulsion injuries by extensor digitorum brevis are common.

*Anterior process of the calcaneus* – These small avulsion or shear fractures may be seen on the lateral ankle view, but are best seen on the oblique view of the foot.

*Base of fifth metatarsal* – look carefully on the ankle films, unless foot views are available.

## Cartilage and joints

The joint surface should be smooth, with no discontinuity. Small fractures, particularly of the talar dome, can have important consequences for a patient if missed.

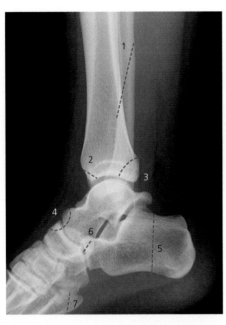

**Figure 7.6** The normal ankle X-ray (*see box* Review areas).

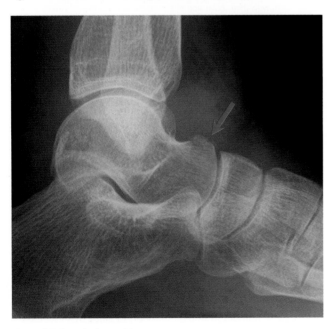

**Figure 7.7** Dorsal talar avulsion.

In the feet, subluxations can be very subtle on radiographs, and it is critical that they are not missed. Displacement can be made apparent with weight-bearing views.

## Soft tissues

The absence of soft tissue swelling usually rules out underlying pathology; conversely, soft tissue swelling may indicate local underlying disease. On the lateral view, look anteriorly for the tear-shaped sign of an ankle joint effusion and posteriorly for Achilles tendinosis as a fusiform swelling.

## Mechanisms of injury

### Ankle sprains

Sprains usually occur after supination or inversion injuries. They most commonly involve the sinus tarsi and the anterior talofibular ligament. Sinus tarsi syndrome occurs usually due to inversion injuries and the patient complains of very localised pain just inferior and anterior to the tip of the lateral malleolus, with instability in particular on uneven surfaces. Symptoms resolve rapidly with an US or CT guided injection with local anaesthetic and steroid. Severe ligamentous injuries may clinically simulate fractures, which radiographs exclude, or show joint incongruity. Ultrasonography or magnetic resonance imaging may be useful for further assessment of ligaments in selected patients.

### Ankle fractures

Fractures of the ankle follow distinct patterns that depend on the mechanism of injury. This helps in the planning of surgical management. The Weber classification of ankle fractures is based on the position of the fibular fracture in relation to the inferior tibiofibular joint. The classification helps identify injuries that are most likely to involve the syndesmosis, cause instability of the ankle and are more likely to require surgical intervention. Type A injuries are below the syndesmosis and stable. Type C injuries are above the syndesmosis and unstable. Type B injuries are at the level of the syndesmosis and stability is variable.

### Maisonneuve fracture

This is a variant of the pronation-external rotation fracture with a transverse medial malleolus fracture seen in the anteroposterior view; however, the fibular fracture is proximal and if is suspected, a full length radiograph of the fibula should be obtained (Figure 7.9a,b).

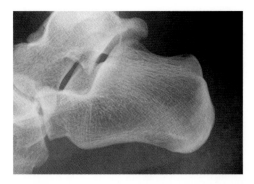

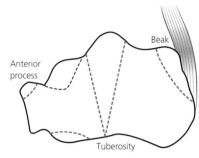

**Figure 7.8** Site of calcaneal fractures.

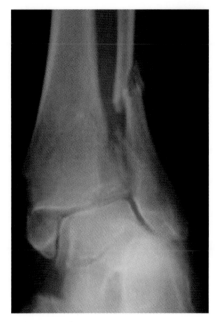

**Figure 7.9** Trimalleolar fracture.

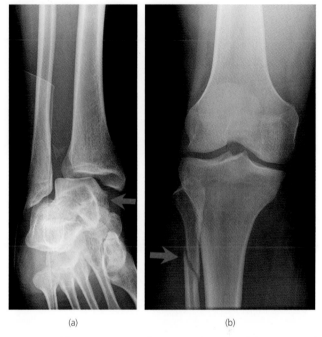

(a)  (b)

**Figure 7.10** (a) Maisonneuve: AP shows gross disruption of the mortice and distal tibiofibular syndesmosis; (b) Maisonneuve, AP of the knee shows the proximal fibula spiral fracture.

## Trimalleolar

These fractures are caused by severe forces of abduction or external rotation and involve all three malleoli. The fracture is often associated with an unstable ankle and may need surgical fixation, particularly if more than one-third of the articular surface of the tibia is involved by the posterior fracture (Figure 7.10a,b).

## Pilon – French for 'pestle'

These fractures occur after axial compression injuries (Figure 7.11a,b). The talus is driven upwards into the mortice with enough force to cause comminution of the tibial plafond and a fracture of the distal tibia above the ankle joint. Computed tomography is often used to assess the degree of comminution and the relation of fragments to the articulating surface for surgical planning.

## Ankle dislocation

Ankle dislocation (Figure 7.12) occurs as a result of substantial force and invariably there are associated fractures. The most common pattern of injury is posterior talar dislocation, often accompanied by a disruption of the tibiofibular syndesmosis or a fracture of the lateral malleolus. Lateral dislocations may occur in association with fractures of the malleoli. As with any dislocation, neurovascular injury is the main concern. Avascular necrosis of the talus may occur if joint reduction is delayed.

## Fractures involving the growth plate

*Juvenile Tillaux* – This fracture is characterised by avulsion of the medial distal epiphysis of the tibia by the anterior inferior tibiofibular ligament and constitutes a Salter–Harris III fracture.

*Triplane* (Figure 7.13a–d) – This complex fracture is similar to a juvenile Tillaux, but with an additional coronal plane fracture of the distal tibial metaphysis. These fractures can be subtle, but suspicion should be raised if the anteroposterior view shows a vertical fracture of the epiphysis. Computed tomography imaging of these fractures is invaluable in surgical planning.

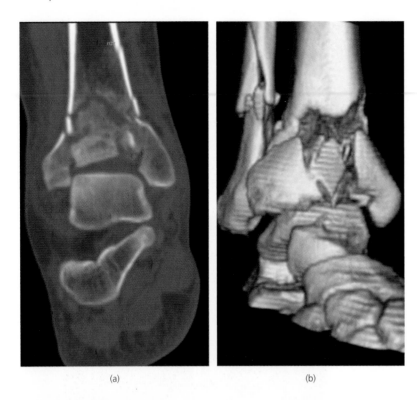

(a)                                    (b)

**Figure 7.11** (a),(b) Cor and 3D images (right) of a comminuted displaced tibial plafond pilon fracture.

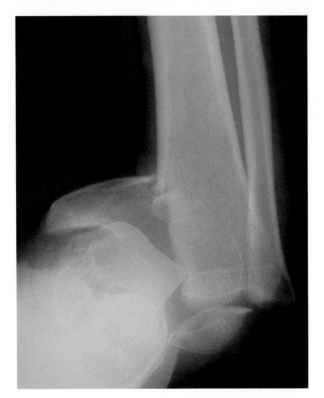

**Figure 7.12** AP view shows medial fracture dislocation of the left ankle, which needs to be reduced to prevent avascular necrosis.

## Talar fractures

Fractures of the talus mainly occur in adults and are uncommon. Chip or avulsion fractures are the most frequent fractures (50%) and fractures of the talar neck are the next most common (30%).

*Osteochondral fractures of the talus* (Figure 7.14) occur as the result of impaction of the talar dome against the tibial plafond. They typically occur on the anterolateral and posteromedial aspects of the talar dome. Usually seen on radiographs but best visualised with computed tomography or magnetic resonance imaging. If they are unstable, they may result in an intra-articular body.

*Talar neck fractures* (Figure 7.15a,b) are important because of the serious risk of ischaemic necrosis of the body of talus. The more displaced, the more likely this serious complication. Hawkins classification, relates displacement of the fracture with risk of ischaemic necrosis (Box 7.3). Radiographs and computed tomographic images show changes of avascular necrosis (talar dome flattening and sclerosis) at 6–8 weeks. Magnetic resonance imaging is much more sensitive at detecting such fractures.

> Box 7.3 **Hawkins classification of talar neck fractures and risk of AVN**
>
> - I – undisplaced fracture (0–10% risk)
> - II – displaced with subtalar joint subluxation (about a 30% risk)
> - III – as II with tibiotalar disruption (>90%)
> - IV – as III with talonavicular disruption (very high risk)

## Talar and subtalar dislocations

Most talar dislocations (Figure 7.16a,b) are associated with talar fractures. Pure talar dislocation is seen rarely but is very disabling and involves the ankle, subtalar, and talonavicular joints. Most subtalar dislocations are reduced easily and should be reduced if detected clinically. Subluxations (Figure 7.17a-b)can be very subtle

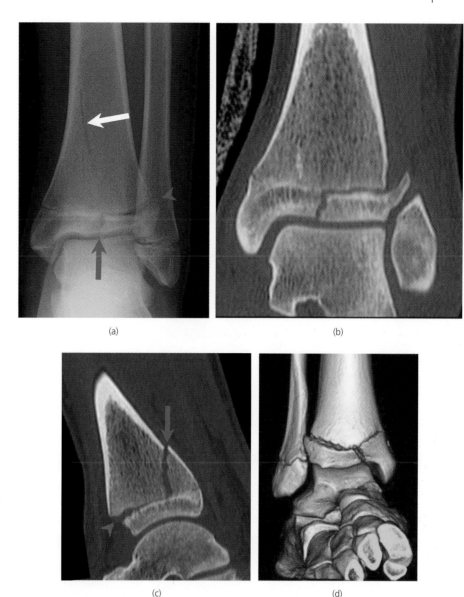

(a)

(b)

(c)

(d)

**Figure 7.13** (a)–(d) Triplane Fracture: AP view shows a sagittal plane fracture through the epiphysis, widening of the lateral physis, with an already partly fused medial physis, and a third fracture line extending up through the metaphysis in a different plane (Triplane fracture). The coronal, sagittal and 3D images of the right ankle confirm findings.

and unless there is meticulous evaluation of the subtalar joint, these are often missed. The key is to evaluate the alignment of the talonavicular and calcaneocuboid joints by confirming an intact cyma line.

Computed tomographic scanning should always be done as osteochondral fractures are common. These fractures may prevent reduction and may lie within the joints, resulting in premature osteoarthritis.

## Calcaneal injuries

Fractures in adults usually result from falls from a height and may be associated with lumbar vertebral body compression fractures (Figure 7.18a–c). In such patients the threshold for taking radiographs of the spine, pelvis, whole limb and the opposite calcaneum should be low.

Seventy-five per cent of adult calcaneal fractures involve the main part of the body and the subtalar joints and are often comminuted.

About half show serious displacement. In children, most calcaneal fractures are peripheral or extra-articular.

Multiplanar and 3D CT scanning should always be done to assess the true extent of injury and to plan surgery. CT images show the direction of fracture lines and the involvement of subtalar joint facets. Vertical compression is associated with mediolateral bursting of the fracture parts.

Fractures of the subtalar joints may result in articular incongruity and subsequent osteoarthritis. Compression of the superior surface of the calcaneus leads to a diminution of Böhler's angle.

Twenty-five per cent of calcaneal fractures in adults are extra-articular, involving the anterior process (Figure 7.19), tuberosity, or posterosuperior aspect (Figure 7.20) of the bone ('beak fracture'). The Achilles tendon may avulse the bone at its insertion.

Fractures of the apophysis occur in children, but the normal apophysis may have a fragmented or sclerotic appearance before fusion. Trauma will be associated with local soft tissue swelling and pain.

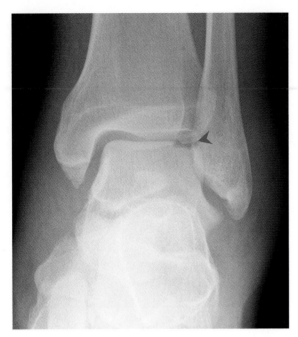

**Figure 7.14** AP view shows a rotated superolateral osteochondral fracture of the talar dome.

## Stress fractures of the calcaneus and foot

These fractures (Figure 7.21) appear as subtle (MRI may be required) sclerotic lines that usually parallel the posterior subtalar joint on radiographs. They are relatively common and may be due to fatigue in the young or insufficiency in the sick or elderly

(Figure 7.22). Patients have pain and tenderness in the foot, which is worse on exercise and relieved with rest.

## Navicular, cuboid and cuneiform fractures

Isolated fractures of these bones are relatively rare. Avulsion fractures of the navicular, at the talonavicular joint superiorly or the medial aspect by the posterior tibialis tendon, are the most common.

## Tarsometatarsal joints, Lisfranc fracture-dislocation

This is a not uncommon and important injury, in which the foot is forced into plantar flexion alone or is also rotated (Figure 7.23 and 7.24a,b). Different patterns of dislocation are seen at the tarsometatarsal joints, and these are best seen with radiographs. See alignment above. CT is useful to show the full extent of fractures and to plan surgery.

## Metatarsals

Fractures may result from crush injuries (Figure 7.25), which produce a comminuted or transverse fracture, while indirect trauma from a twist can result in spiral fractures. Stress fractures (Figure 7.26) often occur at the second and third metatarsal shafts and are seen in new army recruits, joggers, and dancers. The normal sagittal plane well corticated apophysis at the base of the fifth metatarsal may remain unfused and should not be confused with a fracture. Transverse fractures of the proximal few centimetres, especially in young athletic people, have a high incidence of non-union.

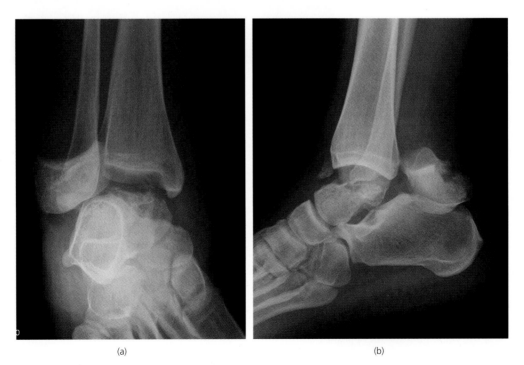

(a)                                         (b)

**Figure 7.15** (a), (b) AP and lat views of the right ankle show a displaced fracture of the talar neck (aviator's astragalus) with disruption of the tibiotalar and subtalar joints (Hawkins III). Note the talonavicular joint is still articulating.

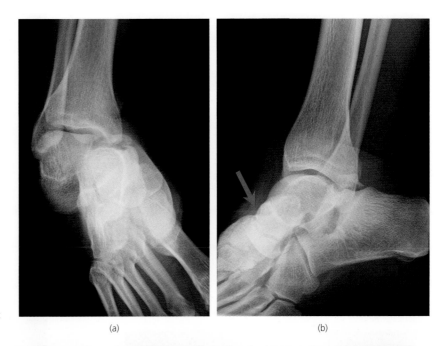

(a)  (b)

**Figure 7.16** (a), (b) The oblique and lateral radiographs of the ankle show the subtalar joints; talonavicular and posterior subtalar are dislocated, but the calcaneus maintains alignment with the midfoot.

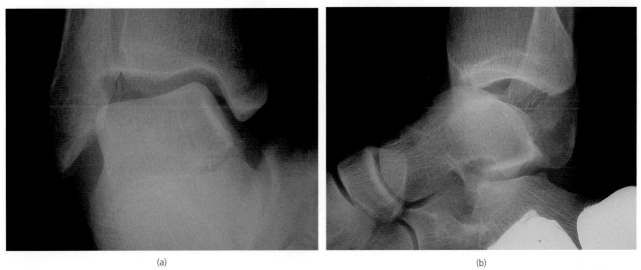

(a)  (b)

**Figure 7.17** (a), (b) AP and lat views show the value of stress views which can show ligamentous insufficiency with varus and anterior draw stress, showing subluxation.

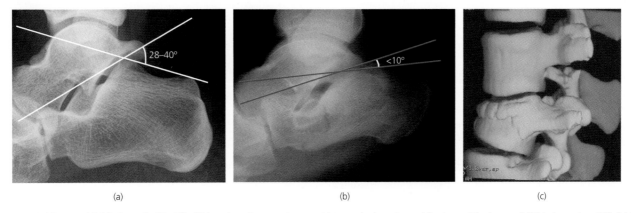

(a)  (b)  (c)

**Figure 7.18** (a) Normal Böhler's angle 28–40°. (b) Lat view shows a depressed intra-articular calcaneal fracture with abnormal Böhler's angle <10°. (c) 3D surface CT of the lumbar spine shows a L2–3 flexion distraction fracture.

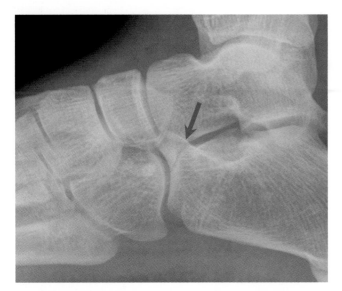

**Figure 7.19** Lat view shows an anterior process calcaneal fracture.

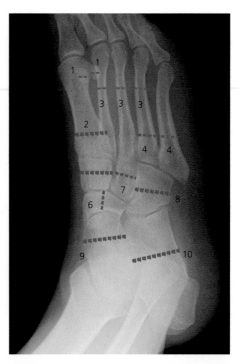

**Figure 7.21** Sites of stress fractures of the foot. 1, Sesamoids; 2, first metatarsal; 3, neck of second to fourth metatarsals; 4, base of fourth and fifth metatarsal; 5, medial cuneiform; 6, navicula; 7, lateral cuneiform; 8, cuboid; 9, talus; 10, calcaneum.

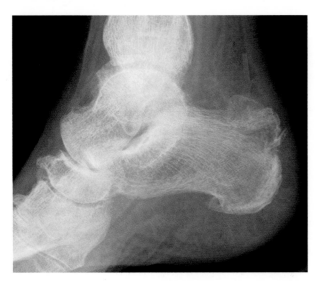

**Figure 7.20** Lat view shows a calcaneal beak fracture.

## Phalanges

Fractures to the phalanges usually result from direct trauma – for example, from crush injuries. Always look at adjacent phalanges and toes.

## Catches to avoid

Accessory ossicles can often be mistaken for avulsion injuries. They are usually rounded, with a well-corticated margin, whereas avulsion injuries have a sharp margin, and fit like pieces in a jigsaw puzzle to the adjacent bone.

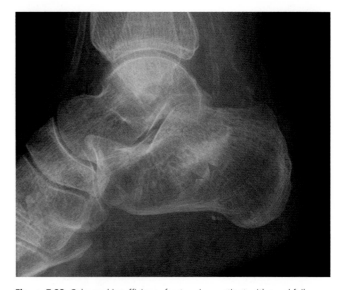

**Figure 7.22** Calcaneal insufficiency fracture in a patient with renal failure.

**Normal ankle movements**

- Dorsiflexion or plantar flexion – ankle joint
- Inversion or eversion – subtalar join

## Radiography

- Hindfoot – lateral, axial (Harris)
- Forefoot and midfoot – anteroposterior, oblique, lateral
- Phalanges – anteroposterior, oblique, elevated lateral of toe

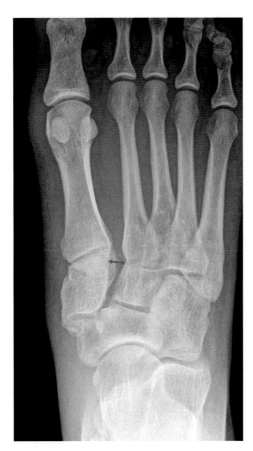

**Figure 7.23** AP view shows a first ray separation medially; isolated type of Lisfranc injury.

## Mechanism of injury

- Objects falling onto feet – forefoot injuries
- Kicking objects – big toe injuries
- Hitting objects barefoot – little toe fractures and dislocations
- Landing from a height (falls) – calcaneal fractures
- Twisting ankle – base of fifth metatarsal avulsion fractures
- Overuse – stress fractures

## Calcaneal fractures

- Calcaneus (60% of fractures of the foot)
- Intra-articular – Body and subtalar joint
- Extra-articular – Anterior process, tuberosity, posterior aspect (beak)

## Patterns of Lisfranc fracture-dislocation

- Hallux metatarsal may dislocate medially (divergent)
- Four lateral metatarsals may dislocate laterally (divergent)
- Associated with fracture of base of second metatarsal, which is proximal to the other metatarsal bases
- Laterally subluxed fourth and fifth metatarsal bases lie lateral to cuboid
- Dorsal and lateral dislocation of the lateral four, or all, metatarsals (homolateral)
- Proximal separation of the medial and middle cuneiforms

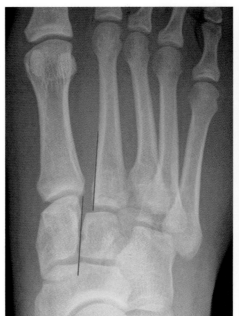
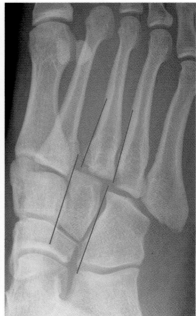

**Figure 7.24** AP (a) and oblique (b) views show a divergent Lisfranc fracture dislocation. Compare to image 7.5.

(a)     (b)

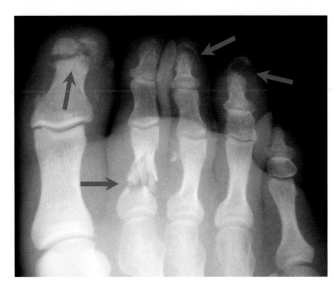

**Figure 7.25** Multiple crush fractures of toes (arrows).

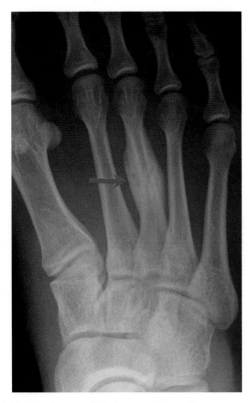

**Figure 7.26** Stress fracture of third metatarsal (arrow).

### ABCs systematic assessment

#### Alignment

- There should be a uniform ankle joint.
- The fibula should overlap the tibia on the lateral view
- Check talonavicular joint is intact to exclude a pure talus dislocation
- Make sure the cyma line is intact to exclude a midtarsal dislocation
- Carefully check tarsometatarsal alignment to exclude a Lisfranc injury

#### Bone

- Check distal fibula for an avulsion
- Check the cortical outline of the tibia and fibula
- Check the fibular shaft is intact on the lateral view
- Check the talar dome for fractures
- Check that the posterior malleolus is intact
- Follow contour of talar dome on mortice view to exclude an osteochondral fracture
- Fractures of the calcaneum can be subtle. Look for a disruption of the trabeculae and subtle sclerotic or lucent lines
- Check the anterior and posterior processes of the calcaneus
- Check for flake fractures of the superior surface of the navicular and talus on the lateral ankle view
- Isolated fractures of the cuboid and cuneiform are uncommon
- Do not confuse apophysis with fracture of fifth metatarsal

#### Cartilage and joints

- Joint space should be uniform
- Check cyma sign and Lisfranc joint

#### Soft tissues

- Swelling indicates site of injury
- Displacement of the fat pad at the anterior joint line on the lateral view may indicate an effusion which should prompt a search for a fracture

### Further reading

Fox JC. *Clinical Emergency Radiology*. Cambridge University Press, 2008.

Kelikan AS, Sarrafian S (eds). *Sarrafian's Anatomy of the Foot and Ankle: Descriptive, Topographic, Functional*, 3rd edn. Lippincott, Williams & Wilkin, 2011.

Rogers LF. Radiology of Skeletal Trauma, 3rd edn. Churchill Livingstone, 2002.

# CHAPTER 8

# Head

*Suki Thomson[1] and Amrish Mehta[2]*

[1]National Hospital for Neurology and Neurosurgery, London, UK
[2]Imperial College London, London, UK

---

**OVERVIEW**

- CT is the investigation of choice in the ER (NB Skull XRs are NOT indicated in head injuries)
- Delayed management of head injuries is a major cause of preventable death
- Have a low threshold to CT or repeat a CT
- MRI is rarely necessary in the acute setting

---

Computed tomography (CT) is the investigation of choice for cranial imaging in the emergency setting. The main reasons for CT scanning in emergency setting are for trauma, stroke or TIA, suspected subarachnoid haemorrhage or meningitis. Intravenous (IV) contrast is used in perfusion imaging following stroke and if space occupying lesions are detected. Other than in stroke, magnetic resonance imaging does not feature in the acute setting.

Plain skull radiographs (SXRs) have no role in the emergency setting in adults.

## CT technique

MDCT allows the spiral acquisition of data through the head with multiplanar reconstruction (MPR). Using different algorithms and altering window settings and MPR software has allowed more detailed analysis depending on which structures are being analysed. Most systems will have automatic selection for window levels to optimise viewing of different structures (Figure 8.1a–c)

Initially a scout image (scanogram) is obtained and then axial images are obtained without IV contrast (non-enhanced CT). These images are then reviewed and a decision is made whether to give intravenous contrast (contrast enhanced CT) (Box 8.1).

---

Box 8.1 **Indications for CT of the head**

- Non-contrast CT
- Trauma
- Stroke
- Subarachnoid haemorrhage

---

- Other intracranial haemorrhage
- Hypoxia
- Anoxia

**Consider** contrast if suspect:

- Meningitis
- Encephalitis (acute confusion)
- Raised intracranial pressure
- Obstructive hydrocephalus
- Following neurosurgery

**Give** contrast if suspect:

- Focal mass lesion
- Encephalitis
- Subdural empyema
- Venous sinus thrombosis
- Known primary malignancy
- HIV infection

## ABCs systematic assessment

- Adequacy
- Airspaces
- Bones
- Brain
- CSF
- Dura
- Eyes
- Face/ Foreign body
- Survey (review areas)

The ABC assessment is a simple systematic approach, which is easy to remember. It is important to review key areas, analogous to the secondary survey.

The key to diagnosis is knowledge of basic anatomy. Understanding and determining if a lesion is within the brain parenchyma (intra-axial) or in the dural or cerebrospinal fluid spaces (extra-axial) is crucial.

It is important to be able to recognise typical common imaging artefacts such as 'beam hardening' in the posterior cranial fossa and 'partial voluming' in the frontal lobes. Beam hardening can

*ABC of Emergency Radiology*, Third Edition. Edited by Otto Chan.
© 2013 John Wiley & Sons, Ltd. Published 2013 by John Wiley & Sons, Ltd.

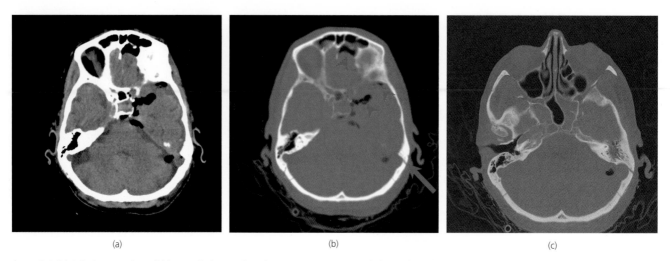

**Figure 8.1** (a) Soft-tissue settings; (b) bone window settings demonstrates pneumocephalus with pockets of air (arrow head) and a fracture through the left petrous temporal bone (arrow); (c) bony algorithm and bone windows makes images sharper for bony structures.

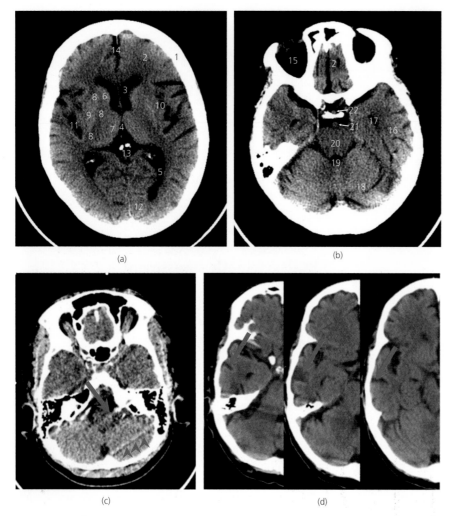

**Figure 8.2** (a),(b) 1, Skull left; 2, frontal lobe; 3, anterior horn of left lateral ventricle; 4, third ventricle; 5, occipital horn of the left lateral ventricle; 6, head of caudate nucleus; 7, thalamus; 8, internal capsule; 9, lentiform nuclei; 10, insular; 11, sylvian fissure; 12, occipital lobe; 13, pineal gland – typically calcified; 14, interhemispheric fissure; 15, orbit; 16, temporal lobe; 17, temporal horn of the left lateral ventricle; 18, left cerebellar hemisphere; 19, fourth ventricle; 20, brain stem; 21, basilar artery; 22, pituitary stalk. (c) Beam hardening artefacts – linear black streaks (arrow heads). (d) Partial voluming – the brain parenchyma appears of lower density in the middle scan.

be caused by a lot of bone, as present in the base of the skull vault, or by metal, where the XRs do not penetrate as well. The data manipulation of these areas can be spurious and cause straight high density lines to run across parts of the image. (Figure 8.2c arrow)

Conversely in areas of low density, such as CSF within sulci, if the data is not acquired in thin enough slices, part of the low density of the CSF can be included in the data manipulation, giving a false lower density in the brain parenchyma. (Figure 8.2d)

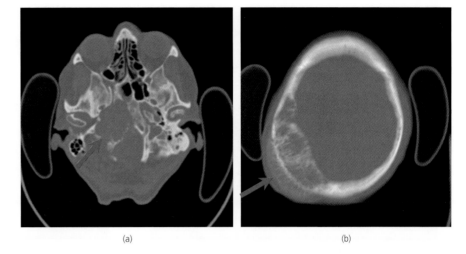

**Figure 8.3** (a) Large lytic metastasis of the right skull base. (b) Haemangioma of the right skull vault with coarse trabeculae and hair-on-end appearance.

(a)

(b)

## Adequacy

- Has the patient moved excessively?
- Is repeating the scan going to improve the image?
- Has the top of the skull vault been included?
- Has the foramen magnum and all the intracranial contents been included?
- Would contrast be helpful?

## Airspaces/bones (Figure 8.3a,b)

Look for:

- Sphenoid sinuses – an air-fluid level may indicate a base of skull fracture.
- Opacification of the paranasal sinuses – may be due to sinusitis, which may be the cause of a subdural empyema (infected subdural collection).
- Maxillary and ethmoid opacification – may be seen in facial fractures.
- Mastoid air cells – Fracture or infection leading to venous sinus thrombosis or abscess collection.
- Skull vault fractures (depressed or complex) – check for scalp soft tissue swelling as a marker of an acute injury.
- Focal bone lesions- primary bone tumour or metastases, myeloma
- Erosions at the skull base – foramina or pituitary fossa can be eroded in mass lesions
- Generalised skull vault thickening – Paget's, sickle cell, anticonvulsant therapy, long term ventricular shunting, diffuse metastatic disease.

## Brain

Examination of the brain parenchyma on every slice is necessary for evidence of low or high density lesions or for mass effect (Figure 8.4).

## Low density lesions

In the trauma setting, low density lesions such as non-haemorrhagic contusions or 'shear' injuries are typical. The latter include diffuse

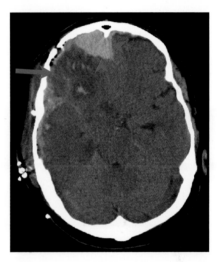

**Figure 8.4** Head trauma with mixed density in the right frontal and temporal lobes with non-haemorrhagic contusions (low density areas) and haemorrhagic contusions (high density areas). Note also the right frontal extra-axial haematoma underlying the right frontal fracture (arrow heads).

axonal injury which often occurs at the grey-white matter junction in the frontal and temporal lobes and in the corpus callosum, usually posteriorly.

In non-trauma cases, low density lesions may involve grey and white matter (such as infarcts, encephalitis and low grade gliomas) or white matter only. Demyelination and small vessel ischaemic disease in hypertension are common causes of white matter low density lesions.

Vasogenic oedema (Figure 8.5) is an area of low density around an abscess or tumour, which has finger like projections in to the white matter.

Diffuse low attenuation changes with reduced grey-white matter differentiation and associated generalised swelling usually indicates hypoxic brain injury – for example after cardiac arrest (Figure 8.6).

Other causes of focal low attenuation may be due to cystic lesions, for example, cystic tumours, such as pilocytic astrocytoma, or to

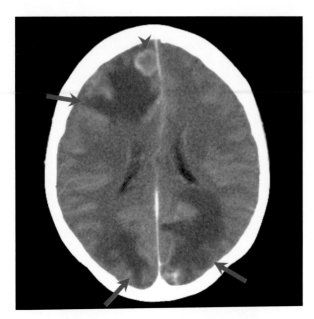

**Figure 8.5** Vasogenic oedema. Finger-like areas of low attenuation involving white matter is typical in the presence of peripherally enhancing (arrow head) metastases.

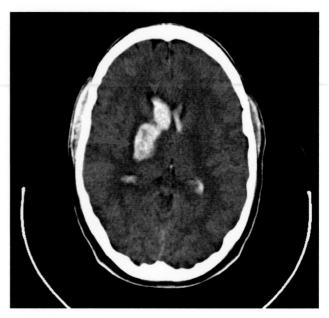

**Figure 8.7** Acute parenchymal haemorrhage. High density acute haematoma in the right basal ganglia (arrow) but with intraventricular extension.

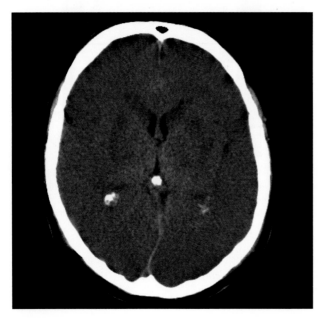

**Figure 8.6** Severe hypoxic brain injury with generalised cerebral oedema. Symmetric low attenuation involving the basal ganglia, associated with cerebral swelling and global loss of the differentiation of grey and white matter.

areas of necrosis in more aggressive tumours such as glioblastoma multiforme. Similarly, an abscess has a low density core with a thin enhancing wall.

## High density lesions

High density lesions usually represent acute haemorrhage (Figure 8.7). Certainly from day 1 to 7, a haemorrhage will be of high attenuation. In the acute setting, there may be surrounding low density, for example, non-haemorrhagic contusion, venous infarction or haemorrhagic tumour. The density of the haemorrhage progressively reduces with time, such that it may not be conspicuous at three- four weeks. In cases of ongoing parenchymal bleeding or in the presence of a coagulopathy, hyperacute haemorrhage can be of low density and may form fluid levels and 'swirls'. An acute haemorrhage may also be less dense than expected in the context of anaemia.

Alternatively, very high density areas may reflect calcification, possibly from congenital infection, or they could be related to tumour.

Lesions with high cellular density (Figure 8.8) (for example lymphoma) are often mildly hyperdense, and they usually have mass effect.

### Mass effect

Deciding if a lesion has mass effect or if there is evidence of mass effect (without necessarily appreciating the causative lesion) is crucial. If a lesion has mass effect it causes distortion of adjacent structures, initially locally, but when more severe there is a more widespread effect.

**Signs of mass effect**

- Gyral expansion
- Cerebrospinal fluid space effacement (local or generalised effect on sulci, basal cistern and ventricles)
- Subfalcine herniation
- Crowding of the foramen magnum
- Temporal uncal herniation, which may result in a posterior cerebral artery infarct.

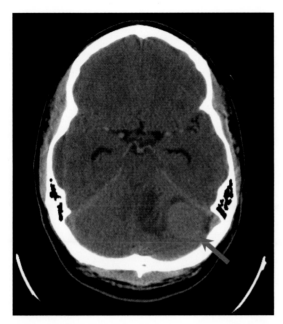

**Figure 8.8** Mildly hyperdense mass (arrow) in the left cerebellar hemisphere with surrounding vasogenic oedema and mass effect (with effacement of the fourth ventricle) consistent with a highly cellular haemangioblastoma.

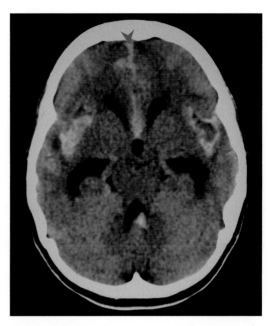

**Figure 8.9** Communicating hydrocephalus with dilatation of all the ventricles in subarachnoid haemorrhage (extensive high density acute blood in the subarachnoid spaces).

## CSF

The ventricles, basal cisterns and sulci contain cerebrospinal fluid. All components should be assessed for signs of acute haemorrhage (high density) or mass effect. Acute haemorrhage in the basal cisterns, sulci with or without ventricles is a subarachnoid haemorrhage, the most common causes of which are trauma and cerebral aneurysm rupture (Figure 8.9). Take care to look in the dependent locations, particularly in the occipital horns of the lateral ventricles and the interpeduncular fossa.

Haemorrhage in the sulci results in high density linear lesions between the cortical gyri. Mass effect on the cerebrospinal fluid spaces usually results in the loss of symmetry, for example of the ventricular system or sulci. The spaces are relatively small in normal young patients, so changes may be subtle.

### Hydrocephalus

In general, enlargement of the ventricular system may be a compensatory effect related to reduced volume of brain parenchyma seen in old age and certain degenerative conditions, or hydrocephalus. Acute hydrocephalus is usually associated with signs of mass effect such as effacement of sulci and periventricular oedema (low attenuation), and may be caused by obstruction to the flow of CSF somewhere along its pathway, for example at the foramen of Monro by a colloid cyst (Figure 8.10). In this case only the ventricles upstream from the obstruction will be dilated, and in this example only the lateral ventricles.

Obstruction to the absorption of CSF at the arachnoid granulation, mainly on the cerebral convexity surfaces can result in dilatation of the ventricles. This is known as communicating hydrocephalus. Typical causes of communicating hydrocephalus include subarachnoid haemorrhage and meningitis. It can be confused with hydrocephalus which is caused by obstruction to the outflow from the fourth ventricle.

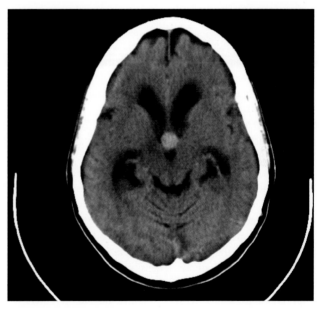

**Figure 8.10** Third ventricular colloid cyst (high density round mass) with an obstructive hydrocephalus.

## Dural spaces

**Extra-axial compartment**

1 Subarachnoid space (between the arachnoid/pia mater & the basal cisterns/sulci)
2 Subdural space (between the dura & arachnoid layer
3 Extradural space (between the dura & inner table of the skull)

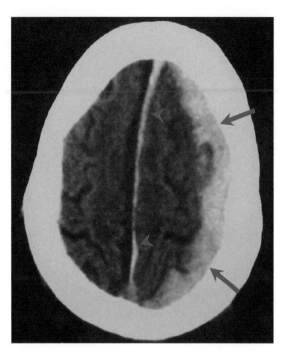

**Figure 8.11** Subdural haematoma: large acute high-density crescent-shaped collection extending into the interhemispheric fissure (arrow heads).

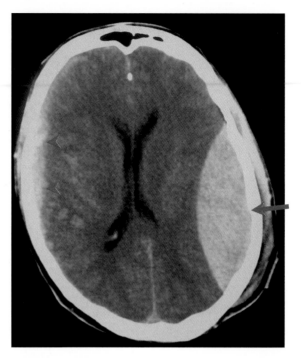

**Figure 8.12** Extradural haemorrhage: large left acute lentiform shaped collection with a shallow crescentic right sided subdural haematoma and associated right cerebral contusions. Note mild left to right subfalcine shift of the midline.

The principal abnormality to be identified is an extra-axial haemorrhage.

Subdural haematoma (Figure 8.11) is usually over the cerebral convexity but may be located in the interhemispheric fissure or related to the tentorium. Subdural haematomas are usually crescentic and often distant to the site of impact ('non-direct' or 'contre-coup').

Extradural haematomas (Figure 8.12) are typically related to the site of impact, often a skull vault fracture. They are biconvex and limited by the vault bone sutures.

The density of and extra-axial haemorrhage decreases over time. After about 12–15 days it may become isodense to brain and difficult to identify. Beyond three weeks, subdural haematomas are predominantly low density, occurring most often in elderly patients after minor trauma. In hyperacute haemorrhage or in patients with coagulopathy the subdural may be low density, but it usually contains swirls of mixed density.

Non-haemorrhagic, low density collections of subdural fluid are important. If present acutely after trauma, they may represent CSF effusion where there is a local arachnoid tear. They may have mass effect but are usually self limiting.

In the unwell, feverish patient, identifying a subdural empyema (Figure 8.13) (infected subdural effusion) is critical. Its margins are usually thickened and enhance after administration of intravenous contrast. Often they may present with sinusitis.

It is advisable to examine the dural venous sinuses for signs of thrombosis. In the acute setting a thrombosed venous sinus will be expanded and hyperdense on a plain CT scan. After contrast, a filing defect denoting the thrombus may be visible.

Deciding if a lesion is intra-axial or extra-axial is an important step in the interpretation of cranial CT. The differential diagnosis is quite different for the two compartments. An extra-axial lesion will displace inwardly the underlying cerebral cortex and pial blood vessels. A CSF cleft may be produced at the margin of the lesion. An extra-axial lesion will often have a broad based dural base or attachment and may be associated with overlying bony changes. Meningioma, which is an extradural tumour (benign tumour of the meninges) is typically well defined, hyperdense, enhancing and may cause calcification. Skull vault metastases may have an extradural soft tissue component.

## Eyes

- Orbital fracture (exclude muscle entrapment in blow out fracture)
- Orbital haematoma
- Globe injury
- Optic nerve injury
- Face (see Chapter 9)
- Check maxilla, mandible, zygoma or pterygoid plates for fractures.
- Foreign bodies

## Survey (key areas for review)

If the CT scan appears normal initially, survey key areas before calling it normal.
   Check:

- Scout (scanogram) – it may show an area that has not been scanned but has an abnormality
- Sides (extra-axial haemorrhage) – check the periphery of the brain and along the falx and tentorium

- Sulci and ventricular – effacement and symmetry
- Sinuses – examine the dural venous sinus for hyperdensity and expansion on the non-enhanced CT to indicate acute thrombosis
- Skull vault – bones again

## Head injury

Blunt trauma (non-penetrating) remains the most common cause of traumatic brain injury in the United Kingdom. Delay in diagnosis and treatment of head injuries is recognised as a major cause of preventable morbidity and mortality. Every effort should be made to accelerate the pathway to the management of these patients. Early recognition and suspicion of an intracranial injury and definitive surgery is essential and therefore there should be a low threshold for requesting CT head scans.

Trauma accounts for most cranial CTs performed as an emergency. In general, MRI is not used acutely although it does have long term applications in the investigation of cognitive or neurological deficits after head trauma.

Early CT, while maintaining a low threshold for re-imaging, is critical in the management of head injury patients to prevent or limit the extent of secondary brain injury.

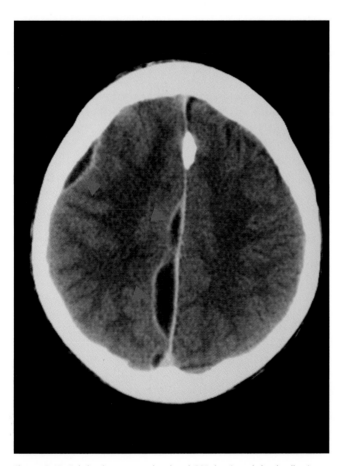

**Figure 8.13** Subdural empyema: loculated CSF density subdural collections over the right cerebral surface and in the interhemispheric fissure with enhancement of their dural margins.

## Primary brain injury

Primary brain injury results directly from the traumatic event, the mechanism of which may be penetrating or non-penetrating (blunt) (Box 8.2). In the UK, blunt trauma to the head is far more common and, depending on the mechanism, it can injure the brain at the point of impact (direct local coup) or distant from the point of impact (non-direct or contre-coup). Non-direct injuries are produced by shear-strain forced, which are mechanical stresses on brain tissue generated by sudden deceleration or angular rotation. Indeed, such injuries can also be produced with no direct cranial impact (Figure 8.14 and 8.15).

---

Box 8.2 **Primary brain injury – non-penetrating (blunt)**

**Direct (local impact = coup)**

- Skull fracture/scalp laceration
- Contusion
- Parenchymal haemorrhage
- Extradural haematoma
- Subdural; haematoma less commonly
- Traumatic subarachnoid haemorrhage

**Non-direct – Extra-axial**

- Traumatic subarachnoid haemorrhage, subdural haematoma

**Non-direct – Intra-axial**

- Cortical contusions – often called 'contre coup' injuries. They are caused by the impact of brain against rough bone or dura. The anterior temporal lobes are involved in 50% of cases with cortical contusions and the anteroinferior frontal lobes are involved in 35%. They are multiple and bilateral in 90% of cases. Parasagittal and dorsolateral brainstem lesions are less common
- Diffuse axonal injury: second most common lesion in closed head injury – 45% of cases. Typically patients present with immediate loss of consciousness. Diffuse axonal injury is a common cause of post-traumatic persistent coma. Multiple, usually small, white matter lesions are seen at the grey–white matter interface (mainly frontotemporal), corpus callosum (mainly posteriorly), internal capsule and brainstem. The initial CT is normal in 50–80% of patients. Later petechial haemorrhage may develop at these sites
- Deep cerebral and brain stem injury: associated with severe injury and a poor prognosis. Shearing of perforating arteries can lead to haemorrhaging into the basal ganglia or brainstem. CT shows hyperdense lesions of varying size with oedema

---

The manifestations of the primary brain injury are extra-axial haemorrhage and a range of intrinsic lesions. Typically, multiple coexisting lesions are induced by direct and non-direct means (Figure 8.16).

In general, the management of head injuries aims to prevent or limit the degree of secondary brain injury. These injuries include ischaemia, infarction, diffuse cerebral oedema, brain herniation and vascular complications and contribute to and result from cyclical deterioration in local or generalised cerebral perfusion and intracranial pressure (Figure 8.17).

NICE publish guidelines for the indications for non-enhanced CT in trauma which include Glasgow coma score (GCS) less than 15 at 2 hours after the injury, GCS less than 13 on initial assessment,

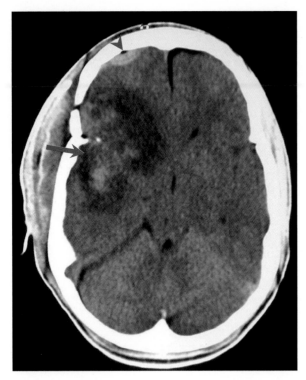

**Figure 8.14** Haemorrhagic contusions in the right frontal and temporal lobes directly beneath a right frontal fracture at the site of impact. Note also a small and shallow right frontal extradural haematoma containing a tiny locule of air (arrow head).

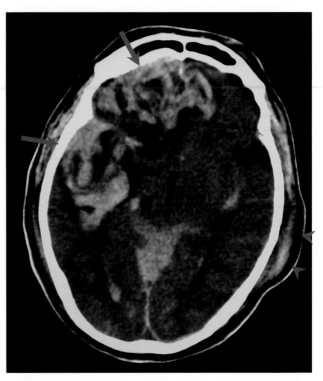

**Figure 8.15** Contrecoup haemorrhagic contusions in the right frontal and temporal lobes following impact to the left parietal region as shown by soft tissue swelling. Associated subdural haemorrhage layering on the tentorium and some intraventricular blood.

focal neurological deficit and post-traumatic seizure. Full guidance is available online.

## Stroke

The term stroke refers to a cerebrovascular event which may be ischaemic or haemorrhagic in aetiology (Box 8.3).

Box 8.3 **Causes of ischaemic stroke**

- Embolic phenomena from cardiac source – for example, atrial fibrillation and extracranial arterial source (typically atheroma in carotid bulb)
- Narrowing of extracranial carotid arteries leading to hypoperfusion and watershed or borderzone ischaemia
- Thrombosis or narrowing of major intracranial arteries (for example acute middle cerebral artery thrombosis
- Small vessel vasculopathy (secondary to ageing, diabetes, hypertension or vasculitis

In the UK, stroke is the third largest cause of death and the largest single cause of severe disability. The estimated cost to the NHS is over £2.8 billion per year. In 2008 the national stroke strategy was formed to try and improve the outcome for patients with stroke and a public awareness campaign was launched in 2009. Hyperacute stroke units across the UK were set up to provide rapid treatment to these patients to salvage brain at risk from infarction.

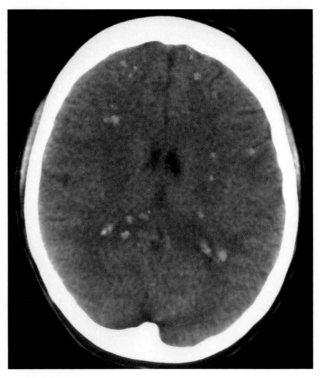

**Figure 8.16** Typical multiple foci of petechial haemorrhage at the grey-white interface and also in the posterior corpus callosum, in diffuse axonal injury (DAI).

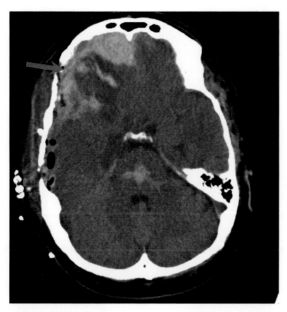

**Figure 8.17** Severe right frontal and temporal traumatic injuries. There is focal haemorrhage within the pons which is a poor prognostic sign.

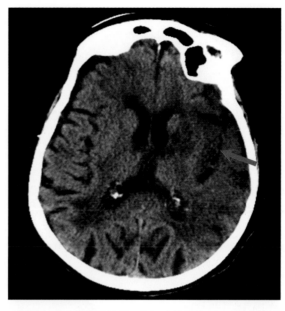

**Figure 8.19** Acute left MCA territory infarct. Note the decreased density of the left insula cortex compared to the other side.

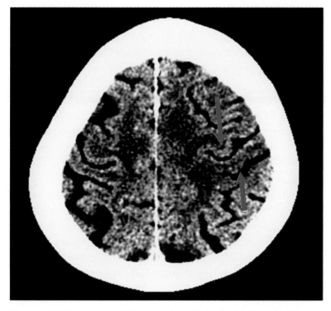

**Figure 8.18** There is loss of the cortical density in the left frontal lobe, up-pointing arrow, compared to the down-pointing arrow. This is seen in hyper-acute stroke.

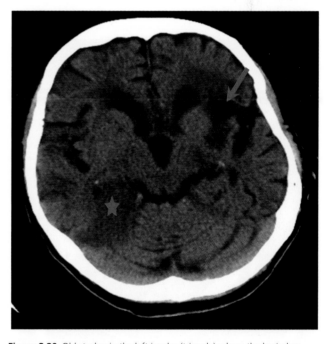

**Figure 8.20** Old stroke: in the left insular (triangle) where the brain has become gliotic and now looks like CSF. A recent stroke is seen in the right posterior temporal lobe (star) where there is loss of the normal grey-white differentiation and swelling. Note there are no sulci when compared to the same position on the other side.

A non-contrast CT is the first investigation to exclude acute intracranial haemorrhage or any other contraindication for thrombolysis. At present in the UK, emergency access to MRI for the assessment of stroke is generally not available. CT is also more sensitive at detecting acute haemorrhage.

Once acute haemorrhage has been excluded, the CT scan should be assessed for signs of acute stroke. A decrease in cortical density similar to white matter, is what is often seen in acute infarct (Figures 8.18 and 8.19). Once this has occurred the brain is not salvageable. As time passes the density will decrease further until the

point where it reaches that of CSF and is gliotic/encephalomalacia (Figure 8.20).

The vessels should also be analysed on thin slices, for increased density which may represent acute thrombus or occasionally calcification.

Swelling is also present in acute and subacute infarcts (Figure 8.21). Loss of sulci, in addition to the loss of cortical

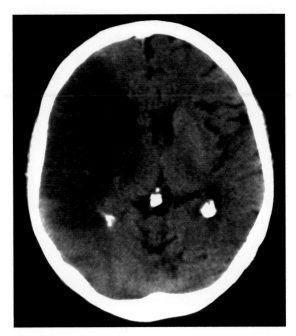

**Figure 8.21** Subacute right middle cerebral artery territory infarct with low attenuation involving grey and white matter and associated with local mass effect.

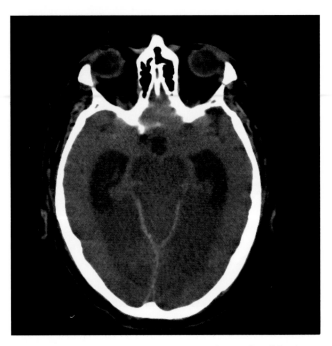

**Figure 8.23** Bilateral and complete posterior circulation infarct following hanging and bilateral vertebral artery occlusion.

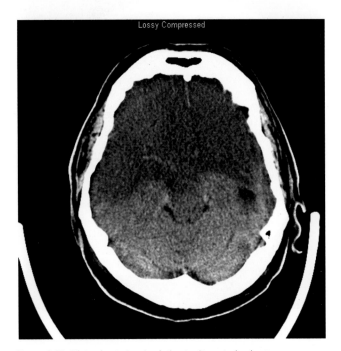

**Figure 8.22** Bilateral anterior circulation territory stroke due to strangulation.

density is a good indicator that there is an ongoing underlying process. In cases of large territory infarct this can cause significant mass effect and lead to midline shift, effacement of ventricles and herniation. In these cases the patients should be urgently referred to neurosurgery for consideration of a decompressive craniectomy (Figures 8.22 and 8.23).

In cases where it is unclear from the history when the onset of symptoms started, a CT perfusion scan can help to determine if there any salvageable brain. By injecting a contrast agent into the vein and then measuring the changes in the intracranial blood vessel opacification, a series of measurements and calculations can be made to determine the blood flow and blood volume. The time it takes to reach peak opacification, also known as time to peak (TTP), and the mean time taken for the contrast to enter and leave the various areas of the brain, also known as mean transit time (MTT), can also be measured. Using maps of these data it is possible to determine if an area has infarcted or is oligaemic and potentially salvageable if the blood flow to that area is improved.

A matched decrease in blood flow and blood volume is seen in areas of infarction, with prolonged MTT and decrease TTP. In areas of oligaemia, the blood flow is reduced, but the compensatory mechanisms mean the blood volume is maintained, thereby creating a mismatch. If the mismatch is moderate to large, these patients are more likely to benefit from thrombolysis (Figures 8.24 and 8.25).

### Intraparenchymal haemorrhage

It is important to differentiate a primary haemorrhage from a haemorrhage that is caused by an underlying lesion, in particular a tumour or an underlying vascular malformation. Trauma remains the most common cause of parenchymal haemorrhage, and in non-traumatic cases, often the patient has a history of hypertension. Haemorrhage within a primary or secondary brain tumour usually evolves in a different pattern where there is persistent mass effect and surrounding oedema, possibly recurrent or multifocal haemorrhage and areas of contrast enhancement after contrast administration. A delayed CT before and after contrast or MRI is often necessary after 2 months when the acute haemorrhage has resolved. CT or catheter angiography may be indicated acutely if an underlying vascular malformation is suspected and surgery is planned immediately (Figure 8.26).

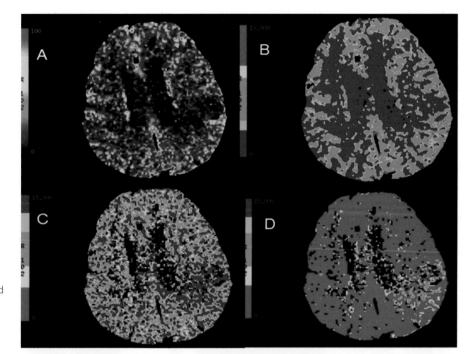

**Figure 8.24** Match perfusion deficit: A, blood volume; B, blood flow; C, mean transit time; D, time to peak. There is decreased blood flow and volume with increased mean transit time and time to peak which match in territory and size. This is consistent with an area of infarcted, non-salvageable brain.

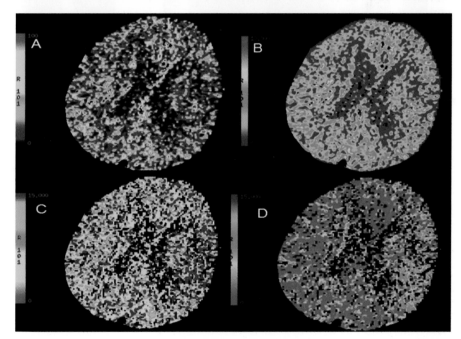

**Figure 8.25** Mismatch perfusion deficit. There is decreased blood flow and increased mean transit time and time to peak but the blood volume is virtually normal. This is consistent with a small area of infarction with a larger area of oligaemia or 'penumbra' which is the potentially salvageable brain.

- Younger adults: aneurysms and atrioventricular malformations
- Venous sinus thrombosis, vasculitis, haemorrhagic encephalitis, cavernoma
- Tumours
- Infection

Box 8.4 **Causes of intraparenchymal haemorrhage**

- Trauma
- Non-trauma
- Elderly patient: hypertension, amyloid angiopathy, haemorrhagic transformation in an infarct, haemorrhagic tumour, coagulopathy

On non-contrast enhanced CT, the typical appearance is of a high density mass in the striatocapsular region (in 65% of presentations). Haemorrhage in the thalamus occurs in 20%, although the brain stem, cerebellum and periphery of the cerebral hemispheres are other potential sites. There is often an intraventricular extension

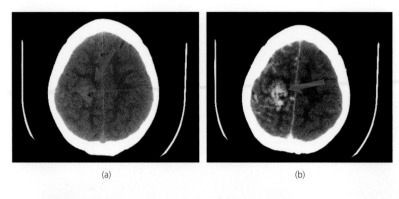

(a)                                  (b)

**Figure 8.26** Arteriovenous malformation in the right fronto-parietal region: (a) ill-defined high-density lesion with little mass effect on unenhanced images; (b) enhances avidly following contrast with serpiginous tubular structures in keeping with abnormal vessels.

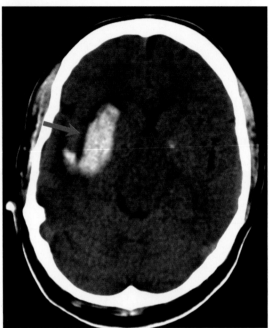

**Figure 8.27** Acute hypertensive haemorrhage in the right basal ganglia.

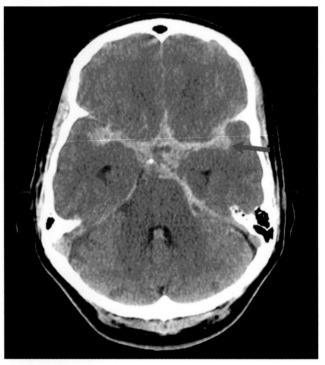

**Figure 8.28** Acute subarachnoid haemorrhage secondary to aneurysmal rupture with blood in the basal cisterns.

and evidence of a background of hypertension. In particular there may be evidence of small vessel disease with low attenuation lesions in the cerebral white matter, brainstem, basal ganglia, thalami and cerebellum.

---

**Imaging signs of hypertensive haemorrhage**

- Non-enhanced CT – high density mass
- Typically striato capsular (60–65% of hypertensive patients)
- Thalamus (20% of hypertensive patients), lobar (5–10% of hypertensive patients)
- With or without intraventricular extension
- Mass effect
- Hydrocephalus due to mass effect
- Evidence of hypertensive small vessel disease with low attenuation areas

---

## Subarachnoid haemorrhage (SAH)

SAH is defined as haemorrhage into the subarachnoid space. This space includes the basal cisterns, Sylvian fissures and cerebral sulci (Figure 8.28). It can involve the ventricular system and (rarely) the interhemispheric fissure. Usually SAH is caused by trauma. The most common non-traumatic cause in adults is a ruptured cerebral aneurysm.

Non-enhanced CT is the imaging modality of choice showing high density (acute haemorrhage) in the subarachnoid space. In aneurysmal subarachnoid haemorrhage, the distribution of blood may indicate the site of aneurysm (Figure 8.29).

---

**Other causes of subarachnoid haemorrhage**

- Aneurysm
- Atrioventricular malformations
- Venous thrombosis
- Cavernoma

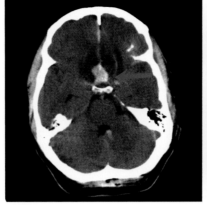 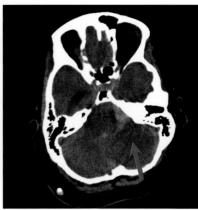

(a)                    (b)

**Figure 8.29** Acute subarachnoid haemorrhage: (a) acute inferior anterior interhemispheric clot from an anterior communicating artery aneurysm; (b) acute haemorrhage in the left cerebello-pontine angle, fourth ventricle and lateral aspect of the medulla from an aneurysm at the origin of the posterior inferior cerebellar artery.

Non-enhanced CT is highly sensitive in the detection of acute haemorrhage in the first 24 hours (approximately 98% and this may be higher if the scan is performed within 6 hours of the first presentation), but decreases with time to around 50% 1 week after the event.

### Complications of subarachnoid haemorrhage

- Early communicating hydrocephalus is typical
- Low attenuation areas in a vascular distribution indicates ischaemia related to vasospasm, especially at days 4 to 10
- Late hydrocephalus (after discharge from hospital)
- Re-bleed

## Meningitis

Suspected meningitis does not typically require CT before a lumbar puncture. A CT is usually normal in uncomplicated meningitis. Moreover, a normal study does not exclude the possibility of future brain herniation. A CT study is performed to exclude complications.

### Meningitis complications

- Obstructive hydrocephalus
- Intraparenchymal abscess
- Subdural effusion or empyema
- Vasculitis with ischaemia/infarction
- Venous sinus thrombosis

### Imaging signs in meningitis

#### General

- Often non-specific
- May be normal
- Signs of complications
- Subdural effusions (CSF density or intensity)

#### Non-enhanced CT

- Mild ventricular enlargement
- Effacement of basal cisterns
- High density subarachnoid space exudates.

#### Contrast Enhanced CT

- Enhancing exudates
- Prominent pial enhancement

#### Airspaces

- Sphenoid sinus – look for fluid level
- Frontal sinus/mastoids/middle ear – look for fracture or infection
- Pneumocephalus – look for sinus or vault fracture

#### Bones

- Look on bone windows and thin sections
- Look carefully over areas with soft tissue swelling

#### Brain Parenchyma

- Look for low density lesions
- Look for high density lesions
- Blood density and implications
- Look for signs of brain swelling

#### CSF spaces

- Look for blood
- Look for mass effect
- Look for hydrocephalus

#### Dura

- Look for subdural and extradural collections

#### Eyes

- Globe injuries
- Optic nerve injuries
- Fractures
- Extraconal

**Face**

- Fractures
- Foreign body

**Survey (review areas)**

- Scout
- Symmetry
- SAH
- Subtle
- Skull vault

## Further reading

Barber PA, Demchuk AM, Zhang J, Buchan AM. Validity and reliability of a quantitative computed tomography score in predicting outcome of hyperacute stroke before thrombolytic therapy. ASPECTS Study Group. Alberta Stroke Programme Early CT Score. *Lancet* 2000;355:1670–1674.

Brown SC, Brew S, Madigan J. Investigating suspected subarachnoid haemorrhage in adults. *BMJ* 2011 May 6;342.

Lucas EM, Sánchez E, Gutiérrez A, Mandly AG, Ruiz E, Flórez AF, Izquierdo J, Arnáiz J, Piedra T, Valle N, Bañales I, Quintana F de. CT protocol for acute stroke: tips and tricks for general radiologists. *Radiographics* 2008 Oct;28(6):1673–87.

NICE guidelines for emergency head imaging. Available online at http://www .nice.org.uk.

Wardlaw JM, Mielke O. Early signs of brain infarction at CT: observer reliability and outcome after thrombolytic treatment – systematic review. *Radiology* 2005;235:444–453.

# CHAPTER 9

# Face

*Simon Holmes[1], Ravikiran Pawar[2], Jimmy Makdissi[2] and Otto Chan[3]*

[1]Barts Health NHS Trust, London, UK
[2]Barts Health NHS Trust, The London Hospital School of Medicine and Dentistry, London, UK
[3]The London Independent Hospital, London, UK

---

## OVERVIEW

- Up to 70% of RTAs sustain facial injuries, mainly soft tissues. Head and cervical spine injuries must be ruled out before imaging the face
- Understand anatomy and classification of injuries
- Injuries occur mainly to the midface and mandible
- Always look for a second fracture in the mandible
- Plain radiographs are difficult to interpret and provide limited information. There should be a low threshold for use of CT and coned beam CT

---

The face is often injured in road traffic accidents, fights and assaults. Up to 70% of people who are in road traffic accidents sustain facial injuries, and most of these are soft tissue injuries. Associated injuries occur in up to half of patients with facial fractures, but, surprisingly, only 2% of these associated injuries occur to the cervical spine. The distribution of injuries to the face varies with patient population, but injuries usually occur to the midface or mandible.

Head and cervical spine injury should be excluded before patients are positioned for plain radiographs, in particular to avoid causing secondary neurological injury. Accurate radiological diagnosis is central to the management of maxillofacial trauma and other medical and surgical conditions that affect the facial bones. Although computed tomography (CT), and the more recently introduced cone beam CT (CBCT), is used increasingly in emergency assessment of such patients, plain radiographs play a central role in the initial management.

Plain radiographs are difficult to interpret, findings are subtle and there should be a low threshold for further imaging, in particular thin section MDCT with multiplanar reconstructions (MPRs) and 3D and CBCT. Using a systematic approach with a meticulous technique is essential, in particular studying the air-bone interfaces, cortical continuity, and symmetry helps the non-specialist to assess the facial bones.

## Recommended radiological views

- Frontal sinuses – posteroanterior (PA) and lateral
- Orbits including detection of foreign bodies (FBs) – PA20 and occipitomental (OM)
- Midface – OM and OM30
- Nasal bone – none usually or a coned lateral
- Mandible – OPG, PA mandible and lateral oblique

## Anatomy

The key to interpretation of imaging of maxillofacial injuries is to understand the basic anatomy and the radiological appearances of the face. The face can be divided into three areas (Figures 9.1 and 9.2):

- Midface (maxilla, zygoma, and nasal bones)
- Upper face (orbit and frontal sinuses)
- Lower face (mandible)

## ABCs systematic assessment

- Adequacy: choose the correct view and ensure correct positioning and exposure
- Alignment: check Dolan's lines, McGrigor's lines, and Campbell's lines
- Bone: check all the bones in the midface and above and below the midface
- Cartilage and joints: check the zygomaticofrontal sutures and temporomandibular joints
- Sinuses: check for opacification and polyps, and check for the presence of an air-fluid level
- Soft tissues: look for soft tissue swelling and surgical/orbital emphysema. Check for FBs

## Adequacy

The standard views are the occipitomental 0° (OM0) and occipitomental 30° (OM30) views. Lateral radiograph for facial injuries is rarely helpful and has been discarded as a routine view. It is important to obtain good quality radiographs in an emergency situation, although this is not always possible because patients can

*ABC of Emergency Radiology*, Third Edition. Edited by Otto Chan.
© 2013 John Wiley & Sons, Ltd. Published 2013 by John Wiley & Sons, Ltd.

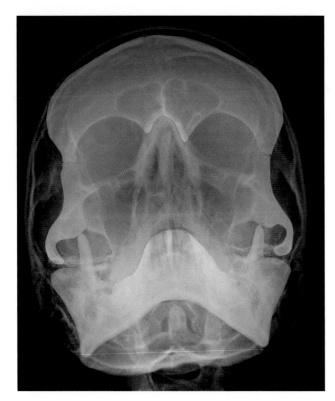

**Figure 9.1** Anatomical division of the face into upper, mid and lower regions.

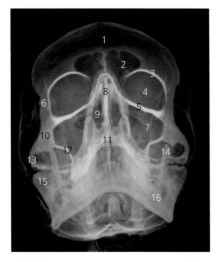

**Figure 9.2** Anatomy of the face. 1, frontal bone; 2, frontal sinus; 3, supra orbital rim; 4, orbit; 5, infra-orbital rim; 6, zygomatic frontal suture; 7, maxillary sinus; 8, nasal septum; 9, nasal cavity; 10, zygomatic bone; 11, maxilla; 12, postero-lateral wall of maxillary; 13, zygomatic arch; 14, coronoid process; 15, condylar process; 16, mandible.

be uncooperative, intoxicated, or unconscious. Additional views may be requested such as a posteroanterior (PA) views to look for injuries of the upper face or to look for radiopaque FBs (in particular glass, debris or metal) and submentovertical (SMV) to look for fractures of the zygomatic arch).

Thin-section MDCT provides exquisite detail of bony and soft tissue injuries and is able to add significant valuable information.

MPR and three-dimensional (3D) reconstructions can be done in seconds using relatively low doses.

## ABC assessment

The remainder of the ABCs systematic assessment is dealt with under each individual view.

## Occipitomental view

This view is simple to acquire and is usually obtained at 0° or 30°. It is able to demonstrate clearly most of the midface. Dolan described three lines that resemble an elephant's head and trunk.

Several other lines can be used to trace and look for step deformities, radiolucent lines, radio opaque areas and changes of contour. The left and the right sides of each image should be used to compare for symmetry (Figure 9.3).

- McGrigor's line 1 – Starts lateral to the right zygomaticofrontal suture. Check that the zygomaticofrontal suture is not wider than the opposite side, and then that the line goes up the superior orbital ridge, across the opposite side, and out through the contralateral zygomaticofrontal suture.
- McGrigor's line 2 – Starts at the superior and lateral aspect of the right zygomatic arch, runs medially to the infraorbital margin, over the contour of the nose, and across the other side.
- McGrigor's line 3 – Begins on the inferior and lateral surface of the zygomatic arch and moves medially. At the floor of the maxillary antrum, it goes across the alveolar process and repeats the movement across the other side.
- Campbell's line 4 – Runs along the medial aspect of the coronoid process inferomedially and across the superior surface of the body and ramus of the mandible. Finally, it crosses the midline over the superior surface of the symphysis menti and goes across the other side.
- Campbell's line 5 – Follows the inferior surface of the mandible from the lateral aspect of the right lateral condylar process to the angle of the mandible and along the inferior surface of the mandible, across the midline to the left condylar process.

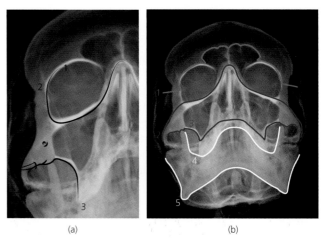

(a)  (b)

**Figure 9.3** (a) Dolan's lines representing elephants head with its tusk and trunk; (b) McGrigor's (1–3) and Campbell's (4–5) lines.

When drawing these lines across the entire face, look for lucent or sclerotic areas of bones that cross or breach the cortex. These are likely to be fractures. Remember to check the zygomaticofrontal suture which should be of equal size. Widening of the sutures could well be suggestive of a fracture. The super imposition of different structures on plain film radiography makes it significantly more difficult to detect fractures of the midface.

Always ensure, that two views with two different angles are acquired and used to assess facial fractures. In the case of mid-face fractures these could be a combination of the OM, OM30 and SMV.

## The panoramic radiograph – orthopantomogram (OPG)

This complex but very useful radiograph is able to demonstrate the full extent of the mandible, maxilla, arches and teeth. It is able to detect dental disease at the time of assessment of fractures. Some of these include teeth in the line of the fracture, retained roots and cystic areas that might weaken the mandible.

Check:

- Air–bone interface – Look for cortical discontinuity of the lower border of the mandible from the left to right condyle.
- Broken or missing teeth. Any missing teeth raise the possibility of aspiration by the patient.
- Condylar heads should lie centrally within the glenoid fossa with a smooth contour. Deviation of the condylar head from this position with a radiolucent line or increased radiopaque pattern (due to superimposition of fragments) could suggest an intra capsular condylar head/neck.
- Canal – Injuries with step deformities of the inferior alveolar canal often result in cortical disruption and might clinically present as paraesthesia.
- Dental occlusion – Spacing between the upper and lower teeth should be equal. Often a condylar neck fracture with shortening of the posterior face can be picked up in this way.
- Dento-alveolar injuries should be obtained.

- Diagnostic pitfalls of the OPG – Simulated fractures of the angle of the mandible (from air in the oropharynx) and symphysis (superimposition of the cervical spine).

## Fractures of the mandible

These injuries are common and are the result of moderate to severe energy transfer to the lower face (Figure 9.4). The clinical symptoms for mandibular fractures can include a combination of but not limited to subjective alteration in bite, pain, swelling, paraesthesia or anaesthesia of inferior alveolar nerve, or inability to open the mouth following trauma. The clinical signs of a fractured mandible include deranged occlusion, bony pain, paraesthesia, fragments mobility, bleeding from the ear, and sublingual haematoma.

The mandible is a rigid ring that is similar to the pelvis, therefore if you see one fracture, always look for a second or a dislocation. The panoramic radiograph may not clearly show the midline fractures, so a PA mandible view is always performed as well.

Condylar neck and head fractures are difficult to identify unless more dedicated views are used (Table 9.1). A combination of an OPG and PA mandible is often sufficient; however, a reverse Towne's view may be necessary (Figures 9.5–9.7).

| Middle third facial fractures |
| --- |
| **Central** |
| • Nasal |
| • Nasoethnoid |
| • Maxilliary – LeFort I, II, and III |
| **Lateral** |
| • Zygomatic |

The symptoms and signs of middle third fractures are epistaxis, diplopia, swelling, infraorbital nerve anaesthesia or paraesthesia, deranged occlusion, subconjunctival haematoma, facial asymmetry, and mobile or missing teeth.

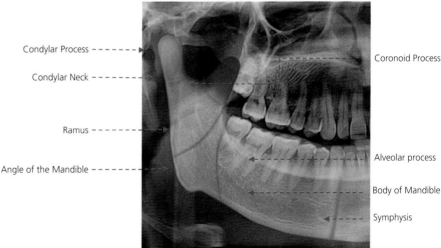

**Figure 9.4** Anatomy of the mandible:
1, condylar process; 2, condylar neck;
3, coronoid process; 4, ramus; 5, alveolar process; 6, angle of the mandible; 7, body of mandible; 8, symphysis.

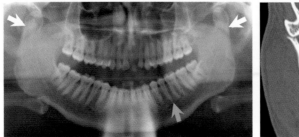

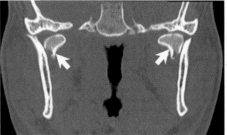

**Figure 9.5** OPG and coronal CT shows bilateral condylar fractures (white arrows) with a dentoalveolar fracture (orange arrow).

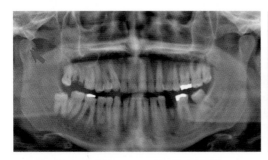

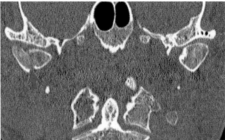

**Figure 9.6** OPG and CT show a right intracapsular unilateral condylar fracture.

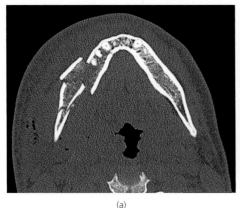

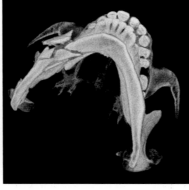

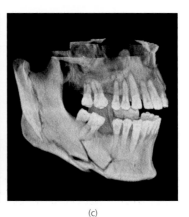

(a)          (b)          (c)

**Figure 9.7** Axial CT (a) and CT 3D reconstruction (b, c) shows comminuted fracture of the right mandible.

**Table 9.1** Fracture of the mandible – views.

| Site | Cause | View |
|------|-------|------|
| Angle | Unerupted third molar | OPG and PA |
| Body | Trauma | OPG and PA |
| Condyle | Anatomically thin | OPG, lat oblique, PA, Towne or CBCT |
| Symphysis | Trauma | OPG, PA or occlusal |

Nasal fractures are diagnosed clinically and usually do not need imaging. Maxillary fractures are classified using LeFort lines. In practice, the precise characterisation of these injuries is difficult and not of major importance in the emergency room. Mixed fracture configuration pattern is more common.

## LeFort injuries

Rene LeFort was a French doctor who studied facial injuries in cadavers. He dropped heads from a height and then described patterns of facial injuries. All LeFort injuries require plain imaging and computed tomography to determine the extent, distribution and pattern of the injuries before further management is attempted

- LeFort I: a fracture above the alveolar process, which leads to the alveolar process being separated from the rest of the maxilla. Clinically, the upper teeth can be moved away from the nose.
- LeFort II: the fracture line extends above the nose. In theory, the upper teeth and nose can be moved en bloc away from the rest of the face.
- LeFort III: the face is separated from the rest of the head (craniofacial dissociation).

Most of these injuries are not usually straightforward and not likely to be symmetrical either. Combination injuries are common and the final classification, however, is made on the basis of the highest or most severe injury.

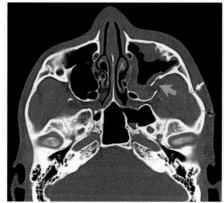
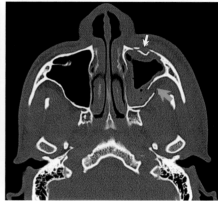

**Figure 9.8** Axial CTs shows fractures of the lateral wall (orange arrow) and anterior wall (white arrow) of the left maxillary sinus and ZA (white and black arrow).

(a)          (b)

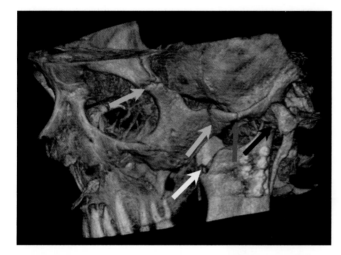

**Figure 9.9** 3D reconstruction of zygomatic complex fractures: ZF (green arrow), ZA (orange arrow), ZT (blue arrow), coronoid fracture (white arrow), condylar fracture (black arrow).

## Zygomatic fractures

These are the most common facial fractures and usually result from a blow to the cheek and direct injury to the zygoma. They are labelled as tripod fractures although they involve the frontal process of the zygoma at the level of the zygomaticofrontal (ZF) suture, the zygomatic arch (ZA), the posterolateral wall of the maxillary sinus and the floor of the orbit/infraorbital margin. The OM0 and OM30 views are commonly used to assess these fractures. Look for soft tissue swelling and air-fluid levels. Check also for step deformities at the level of the infraorbital margin, posterolateral wall of the maxillary sinus and rotation of the zygomatic bones (look for asymmetry) (Figures 9.8–9.10).

## Orbital injuries

These injuries are classified by the site of involvement.

- Orbital rim
- Orbital floor
- Orbital floor
- Medial wall
- Lateral wall

The symptoms and signs of orbital injuries include ecchymosis, subconjunctival haemorrhage, diplopia, enophthalmos, and infraorbital anaesthesia or paraesthesia (Figure 9.11).

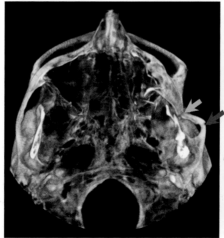
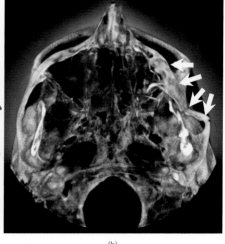

**Figure 9.10** CT 3D reconstruction showing a left ZA bowing fracture (orange arrow) and ZT fracture (blue arrow). Shaded area, outline of the left zygoma prior to fracture; white arrows, amount of displacement.

(a)          (b)

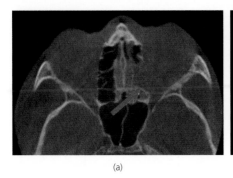 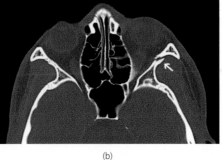

(a)  (b)

**Figure 9.11** (a) Axial CBCT shows fracture of the medial wall of the orbit. (b) Axial CT shows fracture of the lateral wall of orbit.

### Blowout fractures

These are injuries sustained by a direct blow to the eyeball, which then breaks either the floor and/or medial wall of the orbit, but not the orbital rim. Herniation of some contents may occur, particularly of orbital fat into the roof of the maxillary sinus in the case of orbital floor fracture or into the ethmoid air cells in the case of medial wall fracture. The extraocular muscles can become involved. The inferior rectus muscle can be trapped or herniate into the orbital floor fracture. The medial rectus muscle can be involved in medial wall fractures (Figures 9.12–9.14).

This may lead to enophthalmos, diplopia and numbness over the region of the inferior orbital margin.

Radiologically, the classic appearances are those of a teardrop on the roof of the maxillary sinus in the case of orbital floor fractures. Medial wall fractures are more difficult to see on plain film radiography. Plain radiography is the mainstay of imaging although CBCT should be performed in cases where plain radiographs fail to demonstrate a fracture in the presence of strong symptoms and signs of an orbital injury.

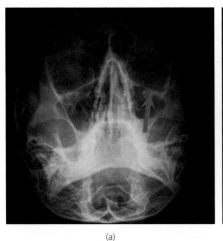 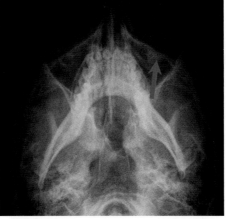

(a)  (b)

**Figure 9.12** OM0 (a) and OM30 (b) shows left blow ut fracture.

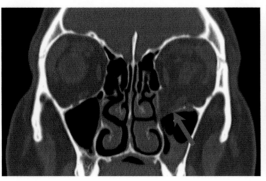 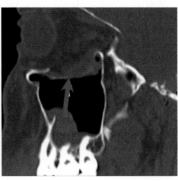

(a)  (b)

**Figure 9.13** Coronal (a) and sagittal (b) CT shows a very large blowout defect of the floor of the left orbit.

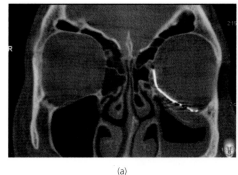
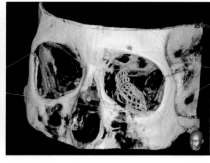

**Figure 9.14** Coronal CBCT (a) and 3D (b) of postoperative surgical repair of the floor of the orbit.

(a)

(b)

## Upper third facial fractures

These are unusual injuries and include fractures of the frontal bone, extended nasoethmoidal fractures and supraorbital injuries.

Epistaxis, soft tissue swelling, deformity, cerebrospinal fluid rhinorrhoea, anaesthesia of infraorbital nerve and pain on upward gaze.

Accurate diagnosis is essential for upper facial third fractures. Precise information about fractures of the frontal sinus with respect to anterior and posterior walls is key to management. For this reason, CT (with MPR and 3D) are invariably required in these cases, particularly if any signs of retrobulbar haemorrhage are present clinically (proptosis, pain, ophthalmoplegia and diminishing visual acuity) (Figure 9.15).

## Sinuses and soft tissues

Asymmetry of the face is an extremely helpful finding, particularly asymmetrical swelling of the soft tissues, air-fluid levels and/or mucosal thickening in the sinuses (Figures 9.16 and 9.17). Also, look for radio-opaque FBs, particularly glass and metal.

**Indirect signs of orbital trauma**

- Soft tissue swelling
- Surgical or orbital emphysema
- Opaque sinuses
- Air-fluid levels in the sinuses
- Teardrop sign

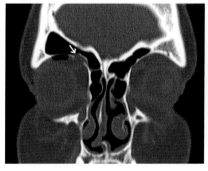
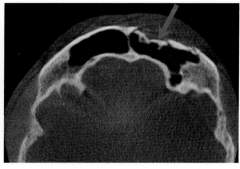

**Figure 9.15** (a) Coronal CT shows right supra orbital fracture. (b) Axial CBCT shows fracture of the left frontal sinus.

(a)

(b)

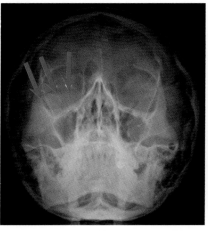
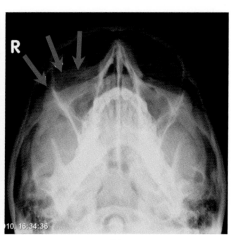

**Figure 9.16** OM0 (a) and OM30 (b) shows right infraorbital soft tissue swelling.

(a)

(b)

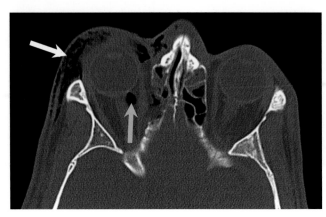

**Figure 9.17** Axial CT shows intraconal (orange arrow) and extraconal air (white arrow).

---

**ABCs systematic assessment**

**Adequacy**

- Select correct views
- Request CT when necessary

**Alignment**

- Check Dolan's, McGrigor's, and Campbell's lines

**Bones**

- Check all bones in the midface and upper and lower third of the face.

---

**Cartilage and Joints**

- Check the ZF sutures and TMJs

**Sinuses and Soft Tissues**

- Check for local swelling of the soft tissue
- Surgical/orbital emphysema
- Air-fluid levels and opaque sinuses
- Teardrop injuries
- FBs such as glass and metal

---

## Further reading

Holmes S. Reoperative orbital trauma: management of posttraumatic enophthalmos and aberrant eye position. *Oral Maxillofac Surg Clin North Am* 2011 Feb;23(1):17–29. Epub 2010 Dec 17.

Makdissi J. 3D imaging: the role of cone-beam computed tomography in dentistry: special reference to current guidelines *Faculty Dent J*, 2012;3(3):152–157.

Michael J. Gleeson (Ed.). Section editors: NS Jones, MJ Burton, R Clarke, G Browning, L Luxon, V Lund, J Hibbert, J Watkinson. *Scott-Brown's Otorhinolaryngology: Head and Neck Surgery*, 7th edn. Hodder Arnold Publishers, 2008.

Ramli R, Holmes S, Rahman RA. *Atlas Of Craniomaxillofacial Trauma*, 1st edn. Imperial College Press, 2011.

Shintaku WH, Venturin JS, Azevedo B, Noujeim M. Applications of cone-beam computed tomography in fractures of the maxillofacial complex. *Dent Traumatol* 2009 Aug;25(4):358–366.

# CHAPTER 10

# Cervical Spine

*Leonard J. King*

Southampton University Hospitals, Southampton, UK

**OVERVIEW**

- Most cervical spine injuries are relatively minor, but they can be potentially devastating, requiring prompt diagnosis and stabilisation
- Cervical spine radiographs are difficult to interpret and 20% of fractures may not be visible. Therefore, there should be a low threshold to proceed to CT
- CT is now the imaging modality of choice in major trauma
- MRI is indicated if there is suspicion of neurological, ligamentous or disc injury

Injuries to the cervical spine can occur either in isolation or in association with head injury or multisystem injury following major trauma. Most of these injuries are relatively minor, but they can be potentially devastating, requiring prompt diagnosis and stabilisation to minimise the risk and severity of associated neurological injury. The pattern, frequency and distribution of these injuries vary between different populations. In adults, the C1–2 and C5–6 levels are most typically affected. In children, injuries are less common and usually involve the upper cervical spine.

Up to 40% of cervical spine injuries are associated with neurological injury, with 5–10% reported as the result of missed injury and consequent lack of cervical stabilisation. About 0.1% of cervical spinal cord injuries do not present with a radiographic abnormality. These spinal cord injuries without radiographic abnormality (SCIWORA) most typically affect children and young adults but also occur in older patients, often with associated cervical spine degenerative disease.

## Anatomy

The cervical spine (Figure 10.1) is comprised of seven bony segments, separated by intervertebral fibrocartilaginous discs and supporting ligaments. The third to seventh vertebrae are morphologically similar, each with a vertebral body and a posterior neural arch that is comprised of bilateral pedicles, facets and laminae with a single posterior spinous process. They form a protective bony

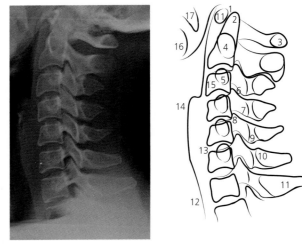

**Figure 10.1** Anatomy of cervical spine with line drawing.

spinal canal, around the spinal cord. Small transverse processes transmit the vertebral arteries via the foramina transversaria, usually from C2 to C6.

The ring-like C1 vertebra (atlas) has anterior and posterior arches but no vertebral body, articulating with C2 and the base of the skull via bilateral lateral masses and the anterior atlantodental joint. There is no C1–C2 intervertebral disc. The C2 vertebra (axis) is distinguished from the adjacent levels by a superior bony projection, the odontoid process or dens, which articulates with C1.

Stability of the cervical spine depends mainly on the integrity of soft tissue structures, in particular the spinal ligaments. The major longitudinal ligaments are the anterior and posterior longitudinal ligaments and the ligamentum flavum. The supraspinous and interspinous ligaments connect the spinous processes. Tough capsules support the facet joints. The apical ligament, alar ligaments and the tectorial membrane (continuation of the posterior longitudinal ligament) give support to the craniocervical junction, and the transverse ligament spans the interval between C1 lateral masses posterior to the dens supporting the C1–C2 articulation.

## Imaging of the cervical spine

Patients who are fully conscious with no history of alcohol consumption or drug intoxication, no head injury, no distracting

*ABC of Emergency Radiology*, Third Edition. Edited by Otto Chan.
© 2013 John Wiley & Sons, Ltd. Published 2013 by John Wiley & Sons, Ltd.

injuries, no symptoms and no clinical signs do not require any imaging. A three-view plain film series (lateral, anteroposterior and open mouth odontoid radiographs) is often performed as the first line investigation for patients whose cervical spine cannot be cleared by clinical assessment alone. Unfortunately, plain radiographs are often suboptimal and even good quality radiographs can fail to demonstrate up to 20% of fractures, so there should be a low threshold to proceed to CT.

CT is a useful adjunct to plain films and is now the first line imaging modality for the spine in major trauma patients, where there is a high risk mechanism as part of whole body CT.

---

### High risk parameters for cervical spine injury

- High velocity motor vehicle collision (>35 m.p.h.)
- Closed head injury
- Fall >10 ft
- Fractures (pelvic, multiple limbs)
- Spinal neurological symptoms
- Death at scene (motor vehicle collision)
- Neck pain/tenderness

---

### Mechanisms of injury to the cervical spine

- Hyperflexion
- Hyperextension
- Rotation
- Axial compression
- Distraction
- Lateral bending/shearing
- Complex or combined vectors

---

MRI is also indicated where there is neurological injury or possible major ligamentous or intervertebral disc damage (Figure 10.2). Lateral flexion and extension radiographs are no longer recommended.

## ABCs systematic assessment

- Adequacy
- Alignment
- Bone
- Cartilage and joints
- Soft tissues

---

### Radiological projections and adequate anatomical coverage

- Lateral: base of skull to T1 superior endplate
- AP: C3 to T1 vertebrae and C2 spinous process
- Open mouth odontoid: C1 and C2 margins should be visible

---

## Interpretation of lateral radiographs

---

Always review the lateral view as if the patient had turned to the right. This will orientate the lateral view in the same direction as the sagittal CT reconstructions and the sagittal MRI

---

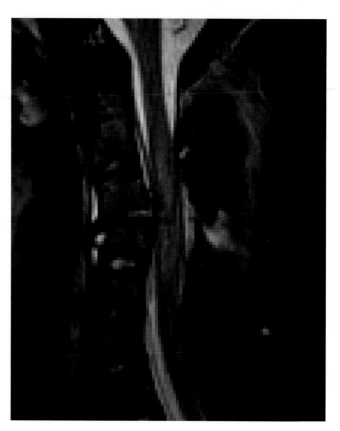

**Figure 10.2** Sagittal STIR MR image following a hyperflexion image demonstrating disruption of the C3/4 disc, ligamentous injury and cord contusion.

## Adequacy (Figure 10.3)

The base of skull to the superior endplate of the T1 vertebral body should be demonstrated with clear cortical and trabecular detail. The outline of the pre-vertebral soft tissues and tip of the spinous processes should also be shown. The lower cervical spine should not be obscured by overlying anatomy (shoulders) or extrinsic structures such as jewellery or monitoring devices. If the C7–T1 junction cannot be seen, a swimmer's view should be performed, progressing to CT if still inadequate.

## Alignment

The cervical spine normally forms a smooth lordotic curve, which may be flattened (Figure 10.4) or slightly reversed due to pain, muscle spasm, immobilisation or supine position. Pre-existing congenital anomalies or pathology such as degenerative change can also result in abnormal alignment. The anterior longitudinal line, posterior longitudinal line and spinolaminar line should all be traced looking for any abrupt alterations in alignment, which may infer ligamentous or disc injury and instability. The spinolaminar line may deviate slightly anteriorly between C1 and C3 on flexion and slightly posteriorly on extension.

The posterior margins of the facet joints should also form a smooth lordotic curve but this line can be difficult to assess due to rotation. The spinous processes form a tighter lordotic curve from C2 to C7 with no sudden widening of the interspinous distance although the curve may not be perfect due to anatomical variation.

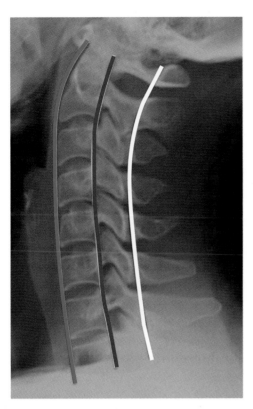

**Figure 10.3** Lateral radiograph demonstrating the normal anterior longitudinal line (light blue), posterior longitudinal line (blue) and the spinolaminar line (white).

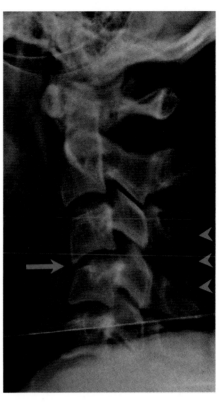

**Figure 10.5** Hyperflexion injury with acute angulation at C3/4, facet joint subluxation and interspinous widening.

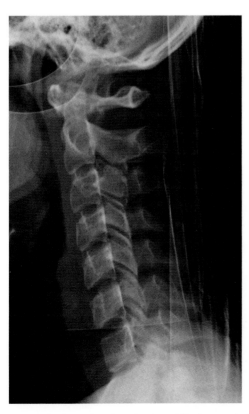

**Figure 10.4** Lateral radiograph with patient immobilised on a spinal board demonstrating flattening of the normal lordosis.

Hyperflexion injuries (Figure 10.5) result in a focal anterior angulation, which may be accompanied by abrupt widening of the interlaminar and interspinous distances, indicating injury to the posterior ligament complex and subluxation or dislocation of the facet joints.

Facet subluxation (Figure 10.6) is manifest by focal uncovering of the articular surfaces and anterior displacement of the more superior facet. Dislocation of a facet joint produces the 'naked facet sign' and when unilateral usually results in anterior displacement of the more superior vertebral body by less than one half of the vertebral body's width. The dislocated facet is displaced anteriorly, and the facet joints above the site of the injury may take on a 'bow tie' configuration as a result of associated rotation. Bilateral facet dislocation is characterised by anterior subluxation of the vertebral body by more than half of the vertebral body's width. The intervertebral disc space is usually narrowed, with little or no rotation.

Hyperextension injuries may be extremely subtle with widening of the anterior disc space and facet joints plus posterior displacement of the vertebrae above the injury.

Atlanto-occipital alignment (Figure 10.7) can be assessed by measuring the Powers ratio, which should lie within the range of 0.6–1.0. A ratio >1.0 indicates anterior subluxation/dislocation. The bony landmarks can however be difficult to visualise on plain radiographs and several alternative lines and measurements can be used (Figure 10.8 & 10.9).

Atlantoaxial alignment should be routinely evaluated on lateral films by measuring the atlantodental (predental) space, which

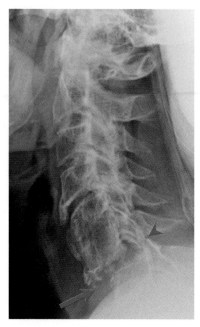

**Figure 10.6** Bilateral facet joint dislocation with anterior subluxation of C6 on C7.

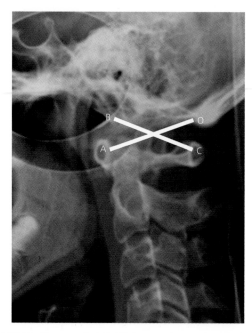

**Figure 10.7** The Powers ratio (BC AO) for assessing craniocervical alignment: 0.6–1.0, normal; >1.0, anterior dislocation.

should be less than 3 mm in adults and 5 mm in children. Widening indicates anterior subluxation of C1 on C2 with associated transverse ligament injury.

---

**Atlanto-occipital alignment**

- Basiodental interval – distance from basion to tip of dens
- Posterior axial interval – perpendicular distance between the basion and posterior axial line (cranial extension of posterior longitudinal line)
- Line drawn along clivus intercepts tip of dens
- C1 spinolaminar line intercepts posterior margin of foramen magnum

---

**Normal measurements**

- Basiodental interval: ≤12 mm
- Posterior axial interval: <12 mm when basion anterior to posterior axial line; <4 mm when posterior to posterior axial line

---

## Bone

The cortical outline and trabecular pattern of each vertebra should be examined looking for discontinuity, angulation, step off, bowing, or an abrupt alteration in density. No bony fragments should project over the spinal canal. Slight anterior wedging of the C3-C7 vertebral bodies by up to 3 mm can be a normal variant.

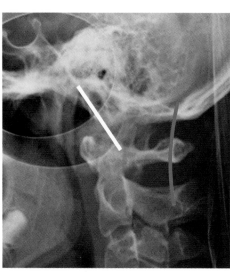

(a)

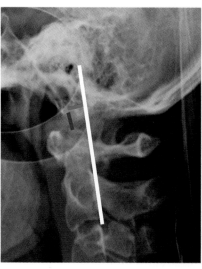

(b)

**Figure 10.8** (a). Line along clivus (white) intercepts odontoid. C1 spinolaminar line (light blue) intercepts opisthion. (b). Basion-dental line (dark blue) <12 mm, Line along posterior axis (white) within 12 mm of basion (light blue).

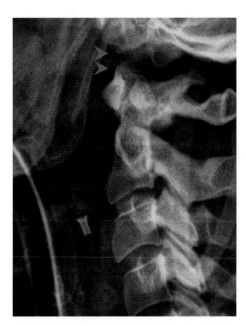

**Figure 10.9** Lateral radiograph demonstrating disruption of the craniocervical junction.

Where there is loss of anterior vertebral body height by more than 3 mm a compression fracture should be suspected and may be associated with focal disruption of the anterior wall cortex or the anterosuperior rim. If the posterior wall of the vertebral body has also lost height it is considered a 'burst fracture' (Figure 10.10) and may be associated with retropulsion of a posterior wall fragment into the spinal canal, which can compress the spinal cord.

A large inferior 'teardrop' fragment with posterior displacement of the vertebral body indicates a hyperflexion teardrop fracture (Figure 10.11), which frequently occurs at the C5 level with associated posterior ligament injury and spinal canal compromise. By comparison a hyperextension teardrop fracture (Figure 10.12) typically affects the C2 vertebral body. It is distinguished by a small, triangular fragment at the anteroinferior rim of the vertebral body and may be associated with anterior disc space widening but usually no posterior displacement.

The odontoid peg should be examined for a cortical break, displacement or increased angulation. Up to 36° of posterior angulation relative to the C2 body can be seen with normal variation.

The C2 Harris ring (Figure 10.13) is a composite ovoid shaped ring projected over the body of C2, which may be incomplete inferiorly. Disruption of the ring or widening of the C2 vertebral body (the 'fat C2' sign) (Figure 10.14) suggests a C2 injury, typically a low dens fracture or an atypical 'hangman's fracture' involving the C2 body (Figure 10.15). The typical hangman's fracture is a traumatic disruption of the C2 pars interarticularis, which may be associated with anterior displacement or angulation of C2 on C3.

Posterior element fractures occur in isolation or in association with vertebral body, disc or ligamentous injury. The spinous processes may fracture at any level and when occurring at the C6-T1 levels, are also known as a 'clay shoveler's fracture' (Figure 10.16). Fractures of the transverse processes may breach the foramina transversarium and can be associated with a vertebral artery injury.

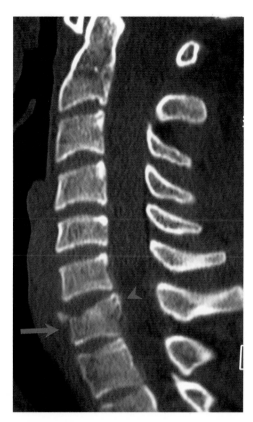

**Figure 10.10** Sagittal CT reconstruction demonstrating a burst fracture of C7 with minor retropulsion of the posterior wall.

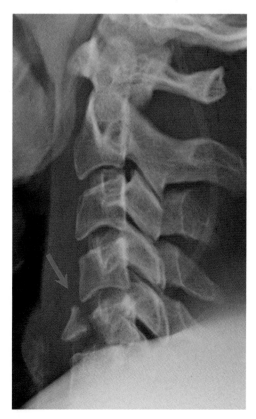

**Figure 10.11** Flexion teardrop fracture of C6.

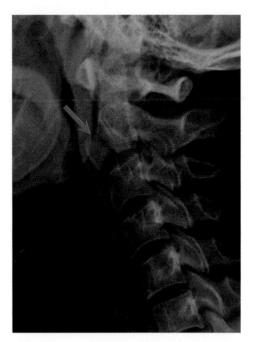

**Figure 10.12** Extension teardrop fracture of C2.

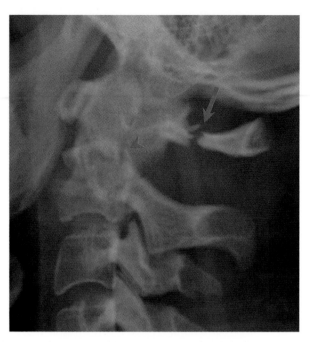

**Figure 10.14** Disruption of the C2 ring and the 'fat C2' sign due to a complex C2 body fracture. A posterior arch fracture of C1 is also present.

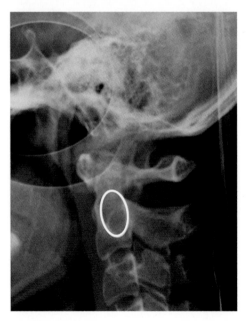

**Figure 10.13** Normal C2 Harris ring.

## Cartilage and joints

The intervertebral discs cannot be directly visualised unless calcified but may be involved in a variety of injury patterns. Injury can be inferred by widening, narrowing or asymmetry of the disc space or by angulation or translation between adjacent vertebrae. Degenerative narrowing is common in the adult population and may cause confusion but is often symmetric and associated with other features of degenerative change such as endplate sclerosis and osteophyte formation. The facet margins should be parallel, and incongruity should prompt scrutiny of the affected segment for other signs of injury, such as interspinous widening.

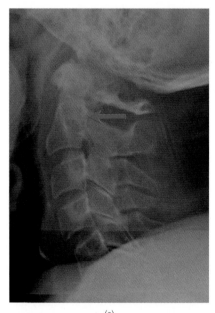

(a)

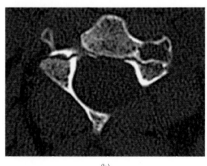

(b)

**Figure 10.15** (a) Lateral radiograph and (b) axial CT image demonstrating a hangman's fracture of C2.

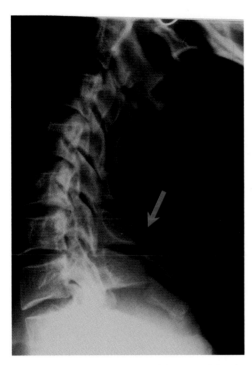

**Figure 10.16** Spinous process (clay shoveler's) fractures of the C6 and C7 spinous processes.

## Soft tissues

Cervical spine injuries may cause anterior soft tissue swelling (Figure 10.17) due to haemorrhage and oedema. In the upper cervical spine from C2 to C4 the soft tissues should not exceed 7 mm or one third of a vertebral body width, and below C4

should not exceed 21 mm or one vertebral body width. Diffuse or localised prevertebral soft tissue swelling may be the only indicator of an injury, although the finding is not specific for injury and a normal prevertebral soft tissue shadow does not exclude an injury. The airway and pharynx should be checked for swallowed foreign bodies, teeth, and malpositioned life support devices. Foreign bodies may also be projected over the soft tissues.

---

**Normal prevertebral soft tissue shadow (3–7–21 rule)**

- <7 mm or one-third of width of vertebral body at C3 (C2 to C4)
- <21 mm or width of vertebral body at C7 (below C4)

---

**Causes of prevertebral soft tissue widening (other than cervical spine injury)**

- Haemorrhage from face or skull base fracture
- Blood pooling in pharynx
- Prominent lymphoid tissue
- Endotracheal or orogastric tube
- Hypoaeration (child crying)
- Pus

---

## Interpretation of anteroposterior radiographs

### Adequacy

A true AP projection should be obtained which clearly demonstrates C3-T1. Overlying devices such as endotracheal tubes should not obscure vertebrae.

**Figure 10.17** Lateral radiograph (a) demonstrating upper cervical soft tissue swelling. Subsequent CT scan (b) demonstrates a type 2 odontoid peg fracture.

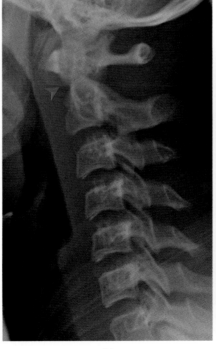

(a)

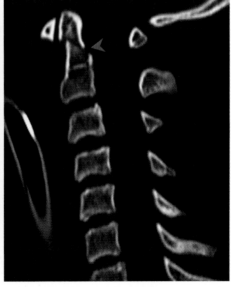

(b)

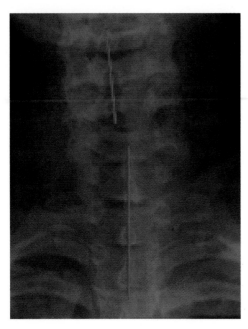

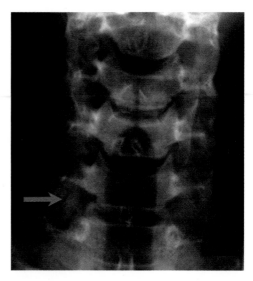

**Figure 10.20** AP radiograph demonstrating an isolated fracture of the right C7 articular pillar (arrow).

**Figure 10.18** AP radiograph demonstrating abrupt rotation at C5/6 due to a unifacet dislocation.

### Alignment

The vertebral body margins, articular pillars and spinous processes should be in line. If the neck is rotated, the spinous processes lose their vertical orientation with a gradual offset to the right or left. Abrupt alteration in alignment indicates injury, such as unilateral facet dislocation (Figure 10.18).

### Bone

The vertebral bodies (Figure 10.19) should have a symmetrical rectangular shape. The facets should also be rectangular: height

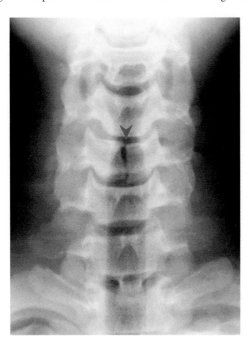

**Figure 10.19** Sagittal split fracture of the C5 vertebral body.

should be greater than width. An asymmetrically shortened facet with prominent joint spaces typically indicates an articular pillar fracture (Figure 10.20).

### Cartilage and joints

Each intervertebral disc space should be symmetric with parallel vertebral endplates. The facet joints should be symmetric.

### Soft tissues

Displacement or obscuration of the laryngotracheal air column may indicate a haematoma due to a spinal or soft tissue injury. Support tubes and lines should also be assessed and foreign bodies excluded.

## Interpretation of open mouth odontoid radiographs

### Adequacy

A true AP projection is required without rotation. The odontoid process should be visualised clearly (Figure 10.21) and not obscured by the occiput or overlying incisor teeth. The C1 and C2 lateral articulations should also be discernible.

### Alignment

The lateral margins of C1 and C2 should line up with no more than 1–2 mm of overlap, and the lateral atlantodental intervals should be symmetric. Asymmetry is present with rotation (which may be positional or due to injury) or in association with C1 or C2 fractures (Figure 10.23). Widening of C1 with overlap of one or both of the articular facets relative to the lateral masses of C2 indicates a Jefferson burst fracture of the atlas (Figure 10.24) which commonly splits the ring into four pieces with bilateral anterior and posterior arch fractures.

The dens should also lie vertically with respect to the body of C2 and lateral angulation is indicative of an odontoid fracture.

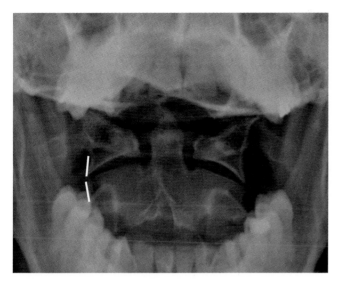

**Figure 10.21** Open mouth AP odontoid radiograph demonstrating normal alignment (no overlap – white lines).

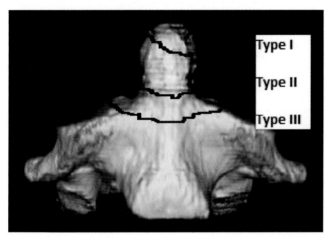

**Figure 10.22** 3D CT reformat illustrating Type I–III odontoid peg fractures.

**Types of dens fractures (Figure 10.22)**

- Type I: Tip of dens
- Type II: Base of dens
- Type III: Low dens fracture extending into C2 body

### Bones

The bone margins should be traced looking for any cortical break or distortion of the normal contour. Small cortical notches are commonly seen at the base of the odontoid peg on each side and are a normal variant.

### Cartilage and joints

The joint spaces between the inferior articular surface of C1 and the superior surface of C2 lateral should be parallel and symmetric.

### Soft tissues

Soft tissue swelling is rarely apparent on this projection.

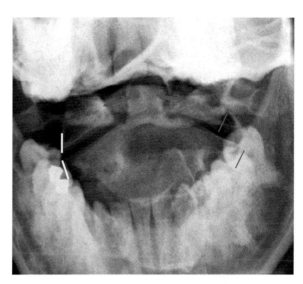

**Figure 10.23** Asymmetry of the lateral atlantodental spaces with a C1 left lateral mass fracture.

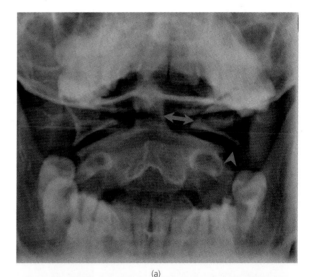

(a)

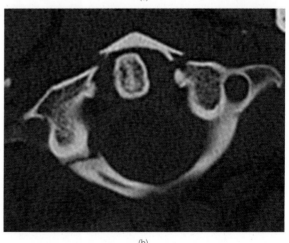

(b)

**Figure 10.24** Jefferson fracture of C1. (a) Open mouth odontoid radiograph demonstrating asymmetry of the lateral atlanto-odontoid spaces and overlap of the left C1 lateral mass. (b) Axial CT image confirms a three part fracture.

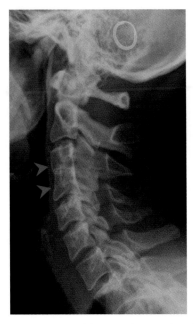

**Figure 10.25** Congenital fusion of the C3 and C4 vertebrae (Klippel–Feil).

## Pitfalls

All the cervical vertebrae from C1 to T1 may not be clearly demonstrated on plain radiographs, particularly the craniocervical and cervicothoracic junctions, both of which are prone to injury.

The presence of congenital variants (Figure 10.25), advanced degenerative changes or other spinal disorders such as ankylosing spondylitis (Figure 10.26) can make assessment of plain radiographs difficult. There should be a low threshold for progressing to CT in all these circumstances.

The upper cervical soft tissues may be artificially altered or widened by causes other than trauma.

Overlapping structures may mimic a fracture particularly at the C1 and C2 levels.

In children there are normal variants that can mimic injury such as pseudosubluxation where there is slight anterior translation of C2 on C3, or C3 on C4 in children up to eight years of age with normal alignment of the spinolaminar line.

---

**ABCs systematic assessment**

**Adequacy**

- Superior endplate of T1 is included on lateral view
- Open mouth odontoid view not rotated

**Alignment**

- Check the spinal lines (anterior longitudinal, posterior longitudinal line, spinolaminar line)
- Atlanto-occipital alignment
- Atlanto-dental space <3 mm (adults)

**Bone**

- Vertebral body height
- Bony contours
- Posterior neural arch

**Cartilage and Joints**

- Intervertebral disc space
- Facet joints
- Atlantodental joint and atlantooccipital joint

**Soft Tissues**

- Prevertebral soft tissues
- Exclude foreign bodies and teeth
- Check airway and life support devices

---

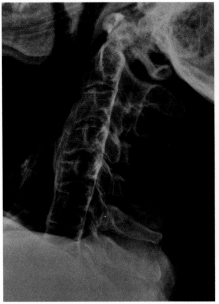

(a)

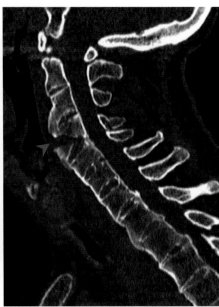

(b)

**Figure 10.26** Male patient with ankylosing spondylitis post trauma. (a) Lateral radiograph demonstrates cervical spine ankylosis but no clear fracture. (b) Sagittal reformat from a subsequent CT shows a C4 vertebral body fracture.

**Injuries – a summary**

- Atlanto-occipital dissociation – head and neck are separated
- Jefferson's fracture – burst fracture of C1 ring
- Odontoid fractures – types I – III
- Hangman's fracture – pars interarticularis fracture of C2
- Hyperflexion teardrop fracture – commonly C5
- Hyperextension teardrop fracture – commonly C2
- Unilateral facet dislocation – <50% subluxation of vertebral body. Abrupt rotation on AP film
- Bilateral facet dislocation – >50% anterior subluxation of vertebra body
- Clay shoveler's fracture – spinous process fracture C6 or C7

# Further reading

Cassar-Pullicino VN, Imhof H (Eds). *Spinal Trauma – An Imaging Approach*. Thieme, 2006.

Hoffman JR, Mower WR, Wolfson AB et al. Validity of a set of clinical criteria to rule out injury to the cervical spine in patients with blunt trauma. *N Engl J Med* 2000;343:94–99.

Hogan GJ, Mirvis SE, Shanmuganathan K, Scalea TM. Exclusion of cervical spine injury in obtunded patients with blunt trauma: is MR imaging needed when multidetector row CT findings are normal? *Radiology* 2005;37:106–113.

Stiell IG, Wells GA, Vandemheen KL, et al. The Canadian C-spine rule for radiography in alert and stable trauma patients. *JAMA* 2001;286(15): 1841–1848.

# CHAPTER 11

# Thoracic and Lumbar Spine

*Leonard J. King[1], Andreas Koureas[2] and Otto Chan[3]*

[1] Southampton University Hospitals, Southampton, UK
[2] University of Athens, Athens, Greece
[3] The London Independent Hospital, London, UK

---

**OVERVIEW**

- The TS and LS are better protected than the CS
- Larger forces are necessary to cause serious injuries
- Spinal injuries are often associated with injuries elsewhere
- In major trauma, TS and LS CT is covered as part of the whole body CT protocol
- MRI is indicated if there is suspicion of neurological, ligamentous or disc injury

---

Injuries to the thoracic and lumbar spine (TS and LS) commonly occur as a result of high-energy trauma such as falls from a height or motor vehicle collisions with axial loading, hyperflexion, extension, distraction, rotation or shearing forces. Lower energy trauma such as simple falls can result in significant injury, particularly where there is a predisposing condition such as osteoporosis, ankylosing spondylitis or spinal metastases. Patients may present with specific signs or symptoms suggesting injury to the spine such as back pain or a cauda equina syndrome. Most significant injuries are encountered in victims of major trauma who may have other life-threatening injuries. Clinical examination of the thoracic and lumbar spine is part of the secondary survey and is deferred until life-threatening conditions have been stabilised. Clinical evaluation can be misleading, especially in patients with other distracting injuries and is of limited value in unconscious patients. In these patients, CT of the TS and LS is covered as part of the whole body CT protocol.

Spinal injuries may be either stable or unstable with risk of mechanical or neurological deterioration. Injuries should be assumed to be unstable with appropriate immobilization and log rolling, until fully assessed by an experienced clinician and appropriate imaging has been performed.

## Anatomy (Figure 11.1)

There are normally 12 thoracic (Figure 11.2) and 5 lumbar vertebrae (Figure 11.3); however, variations are relatively common and it can

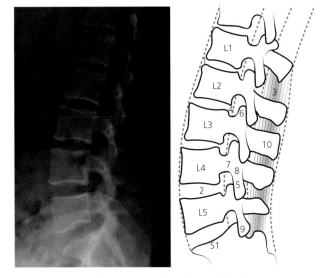

**Figure 11.1** Anatomy of lumbar spine: lateral LS and drawing. 1, ALL; 2, IVDS; 3, interspinous ligament; 4, PLL; 5, inferior articular facet of L4; 6, superior articular facet of L3; 7, pedicle of L4; 8, pars interarticularis of L4; 9, ligamentum flavum; 10, spinous process of L3.

be difficult on imaging to determine which are the T12/L1 or L5/S1 vertebrae. This is due partly to variation of the 12th ribs, but mainly to the presence of transitional vertebrae (Figure 11.4) at the lumbosacral junction with partial sacralisation of L5 or partial lumbarisation of S1. Six lumbar vertebrae may also occasionally be present. If images of the TS and LS are available then levels can be determined by counting down from T1; however, this may be further complicated by the presence of cervical ribs. The longest and most horizontal transverse processes are usually at L3.

Each vertebrae comprises a body and spinous process, plus two paired pedicles, transverse processes, superior and inferior articular facets, pars interarticularis and laminae. In the TS, there are articular facets on the lateral aspect of the vertebral bodies for articulation with the ribs. The lumbar vertebral bodies are larger and have a more horizontal spinous process.

Numerous strong ligaments support the spine, including the anterior and posterior longitudinal ligaments, the ligamentum flavum, and the interspinous and the supraspinous ligaments. The thoracic column is also stabilised by the upper ribs which form the

*ABC of Emergency Radiology*, Third Edition. Edited by Otto Chan.
© 2013 John Wiley & Sons, Ltd. Published 2013 by John Wiley & Sons, Ltd.

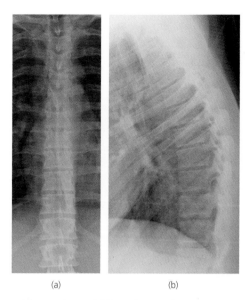

(a)                    (b)

**Figure 11.2** (a) Normal AP and (b) lateral views of the thoracic spine.

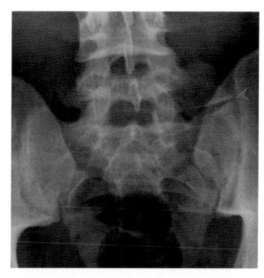

**Figure 11.4** AP view of the lumbar spine demonstrating a transitional lumbosacral vertebra with articulation on the left side (Bertolotti's syndrome if patient is symptomatic).

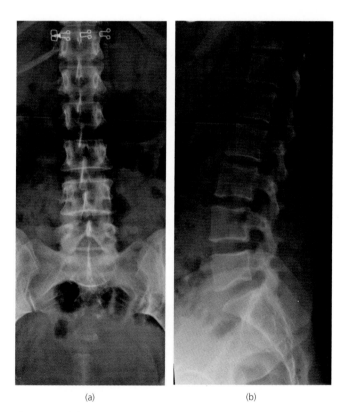

(a)                    (b)

**Figure 11.3** (a) Normal AP and (b) lateral views of the lumbar spine.

are most common at T4/5 and L2. The injuries are often multiple and contiguous.

Congenital vertebral anomalies such as hemivertebrae and butterfly vertebrae, or limbus vertebrae (Figure 11.5) due to non-fusion of the ring apophyses, can be confusing and may be mistaken for acute fractures.

The spinal cord terminates at the conus medullaris, which usually lies between T11 and L2. Below this level, neurological injuries are lower motor neurone due to nerve root injury, whereas more

thoracic cage. The lower ribs and the LS have strong surrounding muscles which augment the intrinsic stability provided by the discs, ligaments and facet joints.

The transition between the relatively immobile TS and the more flexible LS renders the thoracolumbar junction relatively vulnerable to injury, and around 60% of all injuries occur between T11 and L2. Upper and mid TS injuries are relatively uncommon in adults. By contrast, children have a relatively more mobile spine and injuries

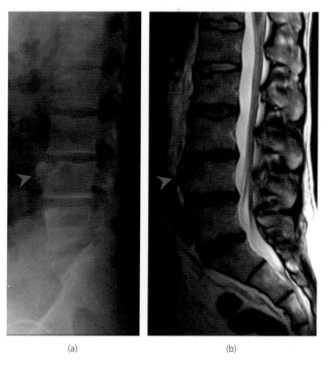

(a)                    (b)

**Figure 11.5** (a) Lateral radiograph and (b) sagittal T2 weighted MR image demonstrating an L4 limbus vertebra.

**Figure 11.6** Lateral radiograph demonstrating minor anterior wedge compression fractures of L1 and L2.

proximal spinal cord injuries are upper motor neurone resulting in hyperreflexia, spinal shock and spasticity.

## Mechanism of injury

There are several definable force vectors, which can result in spinal trauma, producing predictable patterns of injury (Box 11.1). Injuries are frequently the result of hyperflexion with forced bending around a fulcrum centered on the posterior third of the vertebral body. This produces anterior vertebral compression fractures (Figure 11.6) and when severe is associated with posterior element distraction (Figure 11.7). Axial loading for example due to a fall from a height produces compressive forces which result in burst fractures of the vertebral body (Figure 11.8). Hyperextension, rotation and shearing forces also produce defined patterns of injury although in many cases injuries are due to a combination of forces such as hyperflexion and axial loading, or flexion and rotation.

> Box 11.1 **Mechanisms of thoracic and lumbar spine injury**
>
> - Hyperflexion – usually at thoracolumbar junction (T11–L3) with a wedge fracture
> - Hyperextension – tears the anterior longitudinal ligament and widens the disc space
> - Axial compression – discs and vertebral bodies explode (burst injuries)
> - Distraction – rare but may cause ligament damage without bony injury
> - Shearing – slip in any direction, causing disruption of ligaments
> - Rotation – leads to facet joint and combination injuries
> - Complex or combined vectors

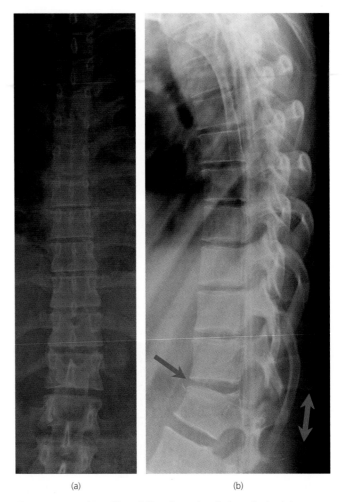

(a)            (b)

**Figure 11.7** AP (a) and lateral (b) radiographs of a hyperflexion injury at T12/L1 with interspinous widening, perched facets and disc disruption.

In motor vehicle collisions, seat belts act as a fulcrum around which the spine can move. Three-point fixation belts with a lap belt and sash combination may result in upper or mid TS injuries due to flexion and rotation. Lap belts however are associated

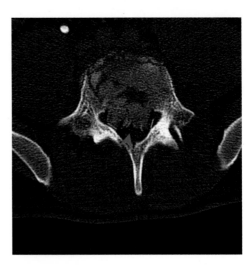

**Figure 11.8** Axial CT image demonstrating a burst fracture of L4 with retropulsed bony fragments.

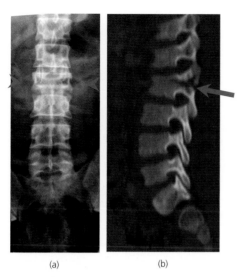

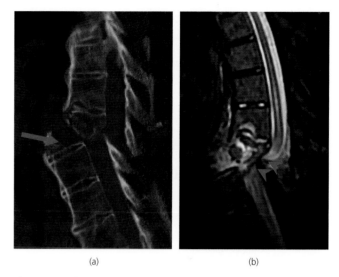

(a)　　　　　　　(b)

**Figure 11.9** (a) AP radiograph and (b) sagittal CT reformat image of an L2 Chance fracture in a child.

(a)　　　　　　　(b)

**Figure 11.11** (a) CT sagittal reformat and (b) short TI inversion recovery (STIR) MR image of a thoracic spine translation injury with spinal cord transection in a patient with ankylosing spondylitis.

with flexion distraction injuries where the spine is flexed around an anterior fulcrum (the lap belt) producing anterior as well as posterior distraction with little or no anterior compression. The classic 'Chance fracture' type of flexion distraction injury occurs at the thoracolumbar junction or the mid LS, horizontally splitting the posterior elements including the spinous process as well as the posterior portion of the vertebral body (Figure 11.9). There are several variants however which can disrupt the interspinous ligaments or the intervertebral disc.

Underlying bone disease such as osteoporosis, metastases (Figure 11.10), myeloma, Paget's disease or ankylosing spondylitis (Figure 11.11), predispose to spinal fractures and should be suspected if an injury appears out of proportion to the mechanism and degree of trauma (Box 11.2). Expansion, destruction, osteopenia, lucencies or sclerosis of a vertebrae, should also raise the suspicion of underlying pathology.

Box 11.2 **Conditions predisposing to spinal injury**

- Degenerative disease
- Malignancy (e.g. metastases or myeloma)
- Osteoporosis / osteomalacia
- Infection
- Paget's disease
- Haemangioma
- Ankylosing spondylitis
- Developmental or congenital anomalies

## Imaging of the thoracic and lumbar spine

In general, two orthogonal views are taken as the first line of investigation (Box 11.3)

Box 11.3

- Thoracic spine – anteroposterior and lateral views (Figure 11.2)
- Lumbar spine – anteroposterior and lateral views (Figure 11.3)
- Additional views of LS – coned LS view or oblique LS (suspected pars defect)

CT is a useful adjunct to plain films and is now the first line imaging modality for the whole spine in major trauma patients, where there is a high-risk mechanism as part of the whole body CT protocol.

MRI is also indicated where there is neurological injury or possible major ligamentous or intervertebral disc damage. This allows assessment of any cord injury and demonstrates ongoing neurological compression by disc, bone fragments, subluxation or epidural haematoma. It is also useful to assess ligamentous or disc disruption and can demonstrate radiographically occult vertebral body fractures.

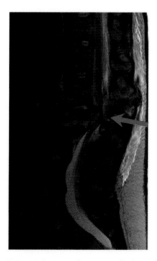

**Figure 11.10** Sagittal T2 MR image of an L2 pathological fracture through a vertebral body metastasis.

## ABCs systematic assessment

The thoracic and lumbar spine should be assessed in a systematic fashion using the ABCs technique.

- **A**dequacy
- **A**lignment
- **B**one
- **C**artilage and joint
- **S**oft tissue

## Interpretation of lateral radiographs

> Always review the lateral view as if the patient had turned to the right. This will orientate the lateral view in the same direction as the sagittal CT reconstructions and the sagittal MRI

### Adequacy

The whole of the TS and LS and the TL junction should be demonstrated. The upper TS can be difficult to demonstrate on lateral radiographs due to the overlying shoulders. Artefactual lines from clothing, sheets, drips, and ECG wires are common and should not obscure the spinal anatomy.

### Alignment

On lateral radiographs, the TS should have a gentle midthoracic kyphosis and the LS a slightly more pronounced lordosis. The anterior and posterior longitudinal lines (ALL and PLL), the spinolaminar line (SLL) and the facet joints should form smooth curves with no abrupt steps.

### Bones

The bony outlines of each vertebra should be sharply defined with no loss of continuity of the cortex other than the posterosuperior cortical margin which is difficult to define. The superior and inferior endplates should be minimally concave and the anterior cortex should be flat or minimally concave. Schmorl's nodes (Figure 11.12) are small areas of disc herniation though the vertebral endplates causing scalloping of the cortex, which are developmental or degenerate and may be associated with Scheuermann's disease. They are commonly seen in the TL spine and should not be confused with acute trauma. No bony fragments should be projected over the spinal canal. The vertebral bodies should increase slightly in height extending caudally from T1 to L5. The thoracic vertebrae often demonstrate slight anterior wedging as a normal variant particularly at T11-L1 which can be mistaken for a compression fracture.

### Cartilage

The intervertebral disc spaces should be relatively uniform increasing slightly in height extending caudally down to the L4-L5 level. The L5/S1 disc is usually slightly narrower than the L4/L5 disc. The lumbar discs are usually slightly wider anteriorly than posteriorly. Diffuse widening or posterior widening of an intervertebral disc

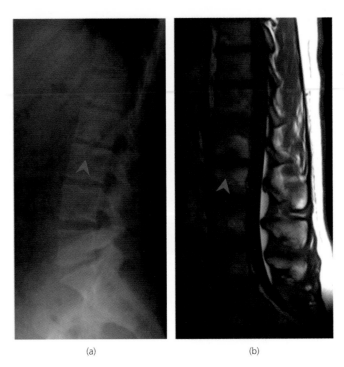

(a)                              (b)

**Figure 11.12** (a) Lateral radiograph and (b) sagittal T1 weighted MRI image demonstrating multiple Schmorl's nodes with prominent marrow reaction at L2/3.

space suggests a flexion distraction injury with disc disruption. Anterior widening can indicate a hyperextension injury. Disc narrowing is usually degenerative but can be the result of an acute disc injury or herniation.

### Soft tissues

The soft tissue outlines are of limited value on the lateral view but a large prevertebral haematoma may occasionally be seen in the thoracic region. Widening of the facet joints or interspinous distances indicates ligamentous injury. Foreign bodies such as bullets or glass fragments may also be demonstrated.

## Interpretation of anteroposterior radiographs

### Adequacy

The whole of the TS and LS and the thoracolumbar junction should be demonstrated without rotation. The bony structures and the paraspinal soft tissues should be visualised without overlying artefact.

### Alignment

The lateral margins of the vertebral bodies, the facet joints and the spinous processes should be vertically aligned with no sudden lateral deviation or angulation (Figure 11.13). The pedicles should form two slightly diverging columns with gradual widening of the interpedicular distance from T1 to L5. Abrupt widening can be a feature of burst fractures (Figure 11.14).

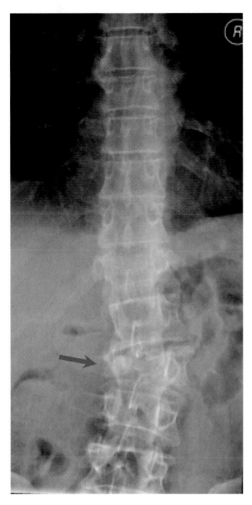

**Figure 11.13** AP radiograph demonstrating loss of mid lumbar coronal alignment due to an L2 fracture.

## Bones

As on the lateral view the cortical outlines should be smooth and sharp with minimally concave endplates. The height of the vertebral bodies should be similar. Loss of height is indicative of a wedge or burst fracture. The pedicles should be oval, distinct and symmetrical.

The spinous processes should lie in the midline, between the pedicles. Widening of the interspinous spaces indicates disruption of the supraspinous and interspinous ligaments.

## Cartilage

The disc spaces should be symmetric and parallel, increasing slightly in height from T1 to L4. The L5/S1 disc space is poorly visualised on the anteroposterior (AP) view.

## Soft tissues

The soft tissue contours and fat planes should be smooth and regular. In the TS, the left paraspinal line is usually wider and easier to see than the right paraspinal line, but focal or diffuse widening of the paraspinal lines indicates a posterior mediastinal

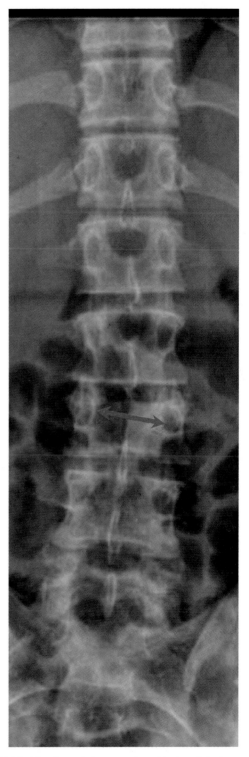

**Figure 11.14** AP radiograph demonstrating widening of the interpedicular distance due to an L3 burst fracture.

haematoma (Figure 11.15) due to a spinal or mediastinal injury. Airway deviation and apical pleural capping may also be demonstrated.

In the LS, loss, asymmetry or blurring of the psoas shadow may indicate a retroperitoneal haematoma or collection.

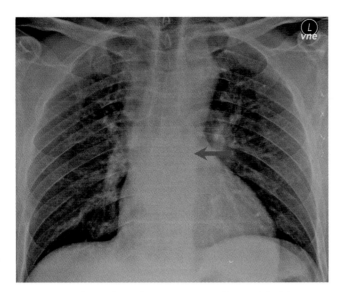

**Figure 11.15** Chest radiograph demonstrating widening of the left paraspinal line due to a mid thoracic fracture.

## Classification of injuries

Injuries are usually classified according to the pattern of damage or the mechanism by which the injury occurred (Box 11.4). Classification of injuries can help to determine their stability and the likely requirement for surgical stabilization. Numerous classification systems have been described for injuries to the thoracic and lumbar spine. The more complex systems classify injuries in detail but can be difficult to use in practice and are prone to lack of inter observer agreement. Simple classification systems such as the three column model described by Denis (Figure 11.16) are therefore commonly referred to although this system does not necessarily accurately predict which patients will require surgery.

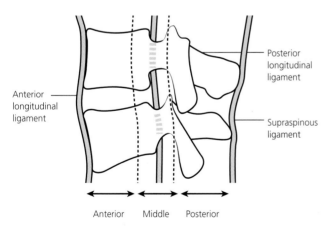

**Figure 11.16** Denis's three column model of spinal stability divides the spine into three columns – anterior, middle and posterior. Disruption of two columns indicates that an injury is unstable. Definite disruption of the middle column indicates that an injury is unstable. Assume instability until it is proved otherwise.

---

Box 11.4 **Injuries – a summary**

- Wedge fracture – an isolated anterior body compression fracture
- Hyperflexion injuries – compress the anterior column with distraction of the middle column and posterior columns. Usually unstable
- Burst fracture – common and usually unstable with a body fracture, widening of the pedicles and disruption of the posterior bony ring
- Chance fracture – flexion distraction injury with a horizontal body fracture, extending into the pedicles and posterior elements
- Translation injury – alignment disrupted by shear forces

---

**ABCs systematic assessment**

**Alignment**

- Normal kyphosis and lordosis on lateral view
- Check 3 lines (ALL, PLL and SLL) on lateral view
- Check lateral margins of vertebral body on AP view

**Bones**

- Check each vertebra for fractures
- Check height of each vertebra body on both AP and lat views
- Assess the anterior, middle, and posterior columns

**Cartilage**

- Check disc spaces
- Exclude focal and diffuse narrowing of disc spaces
- Exclude focal and diffuse widening of disc spaces

**Soft Tissues**

- Check the paraspinal lines
- Check for symmetrical psoas shadows
- Check interspinous distance – exclude widening

## Further reading

Cassar-Pullicino VN, Imhof H (Eds). *Spinal Trauma – An Imaging Approach.* Thieme, 2006.

Mirvis S, Shanmuganathan K (Eds). *Imaging in Trauma and Critical Care,* 2nd Edn). Saunders, 2003.

Wintermark M, Mouhsine E, Theumann N, Mordasini P. Thoracolumbar spine fractures in patients who have sustained severe trauma: depiction with multi-detector CT. *Radiology* 2003; 227:681–689.

# CHAPTER 12

# Chest

*Arjun Nair[1] and Ioannis Vlahos[2]*

[1]St George's Hospital, London, UK
[2]St George's, University of London, London, UK

---

**OVERVIEW**

- Plain radiographs remain the mainstay of imaging
- It is important to understand the basic technical concepts underpinning CXRs and thoracic CT to appreciate their benefits and limitations
- A systematic approach is crucial to evaluating the imaging appearances of various traumatic and non-traumatic conditions of the chest

---

The chest radiograph (CXR) remains the initial method of assessing the thorax in most patients presenting to the emergency department. At the same time, limitations of the CXR may mean that further imaging with other modalities, usually computed tomography (CT) is needed. This chapter provides a basis for understanding the normal appearances of the thorax on imaging, a structured evaluation approach, and descriptions of the common traumatic and non-traumatic conditions relevant to the emergency department.

## Imaging techniques

### Chest radiograph (CXR)

The posteroanterior (PA) view is the best CXR technique, taken with the patient in an erect position, facing the film, and with the X-ray tube behind the patient. The lungs are clearly viewed due to the full inspiration and the projection of scapulae outside the lungs (Figure 12.1).

However, the anteroposterior (AP) view is the most commonly performed view in the emergency department, as patients often cannot be positioned erect or require a portable film (Figure 12.2).

Both views have advantages and limitations (Table 12.1) (Figure 12.3).

The lateral view is not normally performed in the emergency setting but can be useful in evaluating suspected sternal fractures or manubriosternal dislocation.

### Computed tomography (CT)

CT scans can provide more cross-sectional detail of the thorax compared to standard CXR (Box 12.1). This is especially because

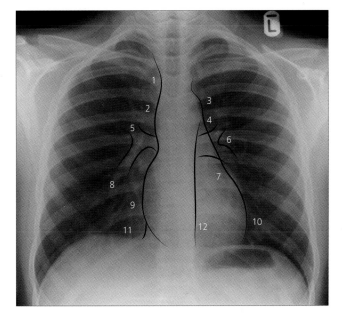

**Figure 12.1** Normal PA CXR, with some of the mediastinal and hilar structures outlined. Note the position of the scapulae, which do not overlap the lungs. 1, right brachiocephalic vein; 2, superior vena cava; 3, aortic arch; 4, main pulmonary artery; 5, right upper lobe pulmonary artery; 6, left pulmonary artery; 7, Left atrial appendage; 8, right lower lobe pulmonary artery; 9, right atrium; 10, left ventricle; 11, inferior vena cava; 12, descending thoracic aorta.

CT scans do not suffer from the superimposition of structures (e.g. the mediastinal structures over one another and over the vertebrae; or the ribs, clavicle and scapula over the upper lungs).

---

Box 12.1 **Uses of CT**

- To further characterise abnormalities detected on CXR
- To exclude pathology when clinical suspicion is high, but the CXR is normal (e.g. vascular injury, acute aortic syndromes and pulmonary embolism)
- In major trauma

---

The patient is placed on a table that is moved along its axis through a rotating gantry with an X-ray source and detectors positioned at

*ABC of Emergency Radiology*, Third Edition. Edited by Otto Chan.
© 2013 John Wiley & Sons, Ltd. Published 2013 by John Wiley & Sons, Ltd.

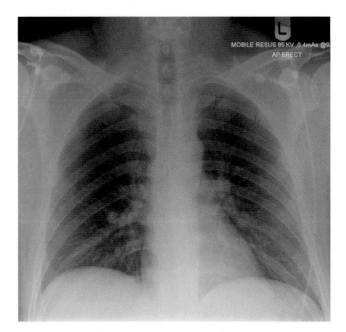

**Figure 12.2** Normal AP CXR. The scapulae partially overlap the lungs.

**Table 12.1** Advantages and limitations of PA and AP CXRs.

| Posteroanterior (PA) | Anteroposterior (AP) |
| --- | --- |
| Excellent visualisation of lungs and mediastinum | Reasonable for lungs, poor for mediastinum |
| Poor for evaluating the skeleton | Good for evaluating the skeleton |
| Cannot be performed supine or portable | Can be erect or supine, and portable if necessary |
| Patient can be well-centred | Patient may be rotated |
| Can ensure better inspiration | May have a poor inspiration |

opposite ends. Images reconstructed from the signals transmitted by the detectors provide detail regarding location (anatomy) as well as the density.

Scans are often performed with intravenous iodinated contrast, unless there are contraindications such as severe asthma, allergy or renal impairment. The timing of scanning with respect to contrast injection can be optimised to visualise the structures of most interest, such as the aorta during a CT aortogram.

Modern multidetector CT (MDCT) scanners allow quick scanning in a single breath-hold, with little distortion, and also allow synchronisation of scanning with the cardiac cycle (cardiac gating), as long as there is no limiting arrhythmia or tachycardia. This can provide good visualisation of the heart if required. MDCT scanners can provide 2D reconstructions in multiple planes, and 3D reconstructions to highlight specific structures.

On axial (transverse) CT images, the body is viewed from a 'bottom looking up' perspective, such that the patient's right will be located on the left of the image, and vice versa. The images can be viewed to show different structures, such as the lung, soft tissue or skeleton (greyscale windows) and in different planes (multiplanar reconstructions – MPR) or in 3D.

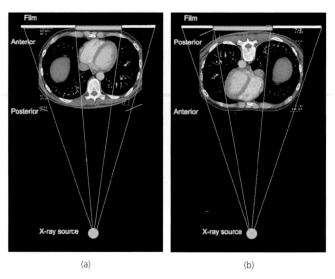

(a)   (b)

**Figure 12.3** Differences in magnification of the cardiac contour on PA and AP projections, illustrated using an axial CT. The cardiac contour is represented by the yellow region on the film. (a) On the PA image, the heart is closest to the film, and so the magnification is kept to a minimum, compared to (b) the AP image.

## Ultrasound (US)

US assessment of the thorax is usually helpful in confirming or excluding pleural fluid, and in guiding thoracocentesis (Figure 12.4). Thoracic US is usually performed with the patient sitting upright if possible.

## CXR anatomy

The lungs above the diaphragm are conventionally divided into upper, mid and lower 'zones' to facilitate description, but these

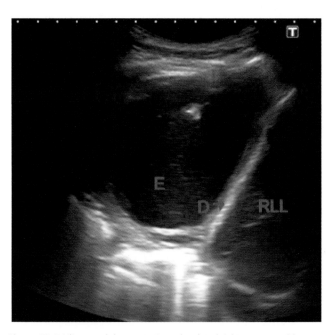

**Figure 12.4** Ultrasound demonstrating a loculated right empyema (E), above the right hemidiaphragm (D). RLL, right lobe of liver.

do not bear any correlation to normal anatomical landmarks. Each of these zones occupies arbitrarily one-third of the height of the lungs.

Only the horizontal fissure (which divides the right upper lobe from the middle lobe) is usually seen on a normal PA CXR.

A significant volume of lung parenchyma lies posterior to the heart and below the level of the diaphragm. The diaphragm is attached onto the 12th rib inferiorly. The posterior ribs lie horizontally and arise from the spine. The anterior ribs tend to slope inferiorly and anteriorly, and they fade medially (due to the orientation of the anterior aspects of the ribs as they join the sternum).

The right side of the mediastinum (superior to inferior) consists of the right brachiocephalic vein, superior vena cava and right atrium. The left mediastinal border (superior to inferior) comprises the left subclavian artery, aortic arch, main pulmonary artery, left atrial appendage and left ventricle.

The hila are usually smooth and taper laterally. The right hilum is usually lower than the left, but may be at the same level in about 3% of cases. The right hilum should never be higher than the left (Figure 12.1).

## ABCs structured assessment

An ABCs system of assessment can be applied to CXRs and adapted to thoracic CT scans. An ABC method specifically for reviewing the lungs may also be used. Once such approaches have been completed, however, one should always ask if the specific clinical question has been satisfactorily answered.

CXR
  ○ Adequacy, **a**irways, **a**ll lines
  ○ Breathing
  ○ Circulation
  ○ Diaphragm
  ○ Edges
  ○ Skeleton, **s**oft tissues
CT
  ○ Adequacy, **a**irways, **a**ll lines
  ○ Breathing
  ○ Circulation, **c**entral structures
  ○ Diaphragm
  ○ Edges
  ○ Skeleton, **s**oft tissues
Lungs
  ○ Apical zone
  ○ Basal zone
  ○ Central (middle) zone
  ○ Density
  ○ Extra signs

### Adequacy
Check:

- Demographic details
- Side markers
- Type of film

- Coverage: entire lung should be covered, from apices to costophrenic angles
- Inspiration: at least five anterior ribs should be seen above the midpoint of the hemidiaphragms
- Rotation: medial ends of the clavicles should be equidistant from the spinous process of the vertebra at that level. Some degree of rotation may be acceptable, as long as the film is still interpretable
- Exposure: on a correctly exposed film, the lower thoracic vertebrae (T8/T9 disc) and the left lower lobe pulmonary vessels should be visible through the cardiac silhouette

### Airways
The trachea should be central above the manubrium but deviates slightly to the right below this. The right mainstem bronchus should have a more vertical alignment than the left main bronchus. Check the position of the endotracheal tube, and exclude foreign bodies while assessing the airways.

Any deviation towards a particular side that is not accounted for by rotation suggests either:

- Ipsilateral volume loss: collapse (Figure 12.5), scarring, fibrosis, or surgery; or
- Contralateral mass effect: tension pneumothorax, haemothorax, or large effusion.

### All lines
### Endotracheal tube
The tip of the endotracheal tube should be at the level of the aortic arch or at least 3.5–5.5 cm above the carina, if the neck

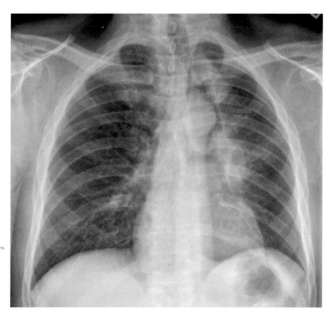

**Figure 12.5** Left upper lobe collapse with tracheal deviation to the left. The left upper lobe collapses anteriorly, resulting in a decreased size of the left lung, with a veil-like opacity. A lucency is noted adjacent to the aortic knuckle (arrow), due to the compensatory expansion of the apical segment of the left lower lobe between the collapsed left upper lobe and the aortic arch. This is known as the 'Luftsichel' sign. The collapse is due to a left hilar mass (arrowhead).

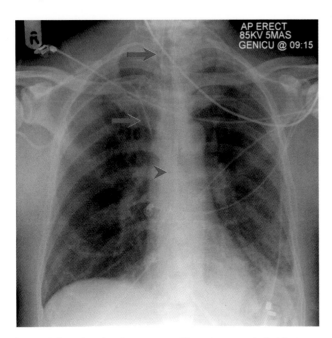

**Figure 12.6** Intubated patient on general intensive care unit. Satisfactory position of the endotracheal tube (arrow), right subclavian vein central venous catheter (block arrow), and nasogastric tube passing below the diaphragm (arrowhead).

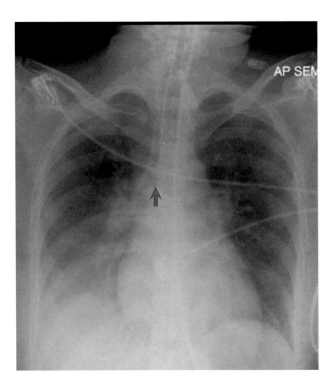

**Figure 12.7** Suboptimal endotracheal tube position, approximately 1.5cm proximal to the carina (arrow). Note the right lower lobe consolidation- this had developed secondary to collapse of the right lower lobe due to previous positioning of the endotracheal tube in the right lower lobe bronchus.

is in extension (Figures 12.6 and 12.7). Having the tip of the endotracheal tube at least 3.5 cm above the carina ensures that the tube does not extend past the carina if the neck is flexed, and so is less likely to obstruct a main bronchus.

## Venous catheters

An intravenous central venous catheter (CVC) should not be kinked and should follow the expected smooth curve of the vein into which it has been placed (Figure 12.6). The ideal position for the tip of a CVC is at the junction of the superior vena cava and right atrium. The CVC may occasionally be malpositioned in another vein, such as the azygos vein.

## Chest drains

Chest drains have a radio-opaque line within their wall. A break in this radio-opaque line indicates the site of the last hole in the tube, and it is essential to ensure that this hole lies within the thoracic cavity (Figure 12.8).

If there is still significant air leak and surgical emphysema despite adequate intrathoracic depth of the drain, its tip may be situated in the lung parenchyma itself (Figure 12.9).

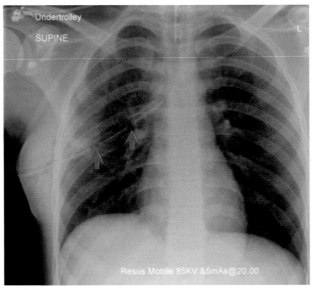

(a)

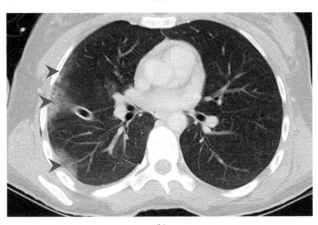

(b)

**Figure 12.8** (a) Good position of right intercostal drain, with the holes (arrows) well within the thoracic cavity. (b) CT more accurately demonstrates the position of the right-sided intercostals drain, with its tip in the right oblique fissure. A small right haemopneumothorax (arrow) and multiple pulmonary contusions (arrowheads) are present.

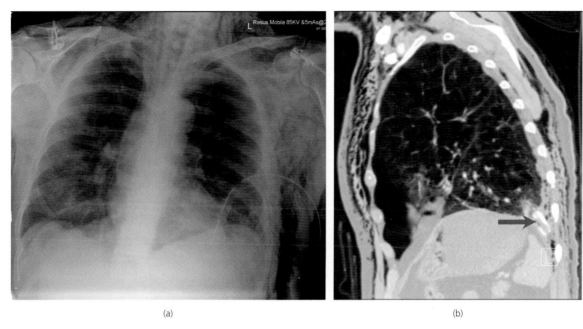

(a)                                                                     (b)

**Figure 12.9** (a) Suboptimal left intercostal drain. Although the drain is situated within the thoracic cavity, there is a large amount of surgical emphysema, suspicious for an intrapulmonary location of the drain tip. (b) Sagittal CT demonstrates the tip of the drain in the posterior left lower lobe (arrow).

### Nasogastric tube

If a nasogastric tube has been placed, ensure that its tip passes below the diaphragm into the stomach, and has not been inadvertently passed into an airway.

### Breathing

Systematically assess the lungs and pleural spaces, looking for any parenchymal or pleural abnormalities (see later).

### Circulation (mediastinum, hila and pulmonary vasculature)

Check:

- Cardiac silhouette size (remember: this will appear larger on an AP film). If enlarged, it could indicate cardiomegaly or a pericardial effusion
- Mediastinal position: one-third to the right, two-thirds to the left of the midline
- Mediastinal contour: clearly defined margins, especially aortic contour
- Widening of mediastinum: mediastinal widening, especially if > 8 cm may indicate mediastinal haematoma and traumatic aortic injury which should be excluded with computed tomography, in the trauma setting. In the absence of trauma, it usually indicates a mass or lymphadenopathy (Figure 12.10)
- Hila: evaluate size, shape, and position. Enlargement of one hilum usually indicates a mass or lymphadenopathy. Enlargement of both hila is either the result of lymphadenopathy or is vascular in nature (Figure 12.10)

### Diaphragm

Hemidiaphragms should have a sharp outline throughout their course. Loss of this outline indicates a pathological process in the lower lobe, or possible diaphragmatic rupture in the context

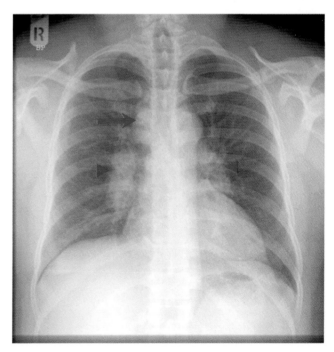

**Figure 12.10** Widened mediastinum due to right paratracheal lymphadenopathy (arrow), with coexistent bilateral hilar lymphadenopathy (arrowheads) in a young patient with sarcoidosis.

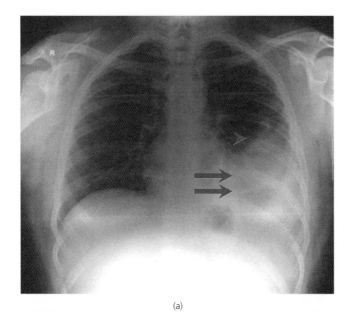

(a)

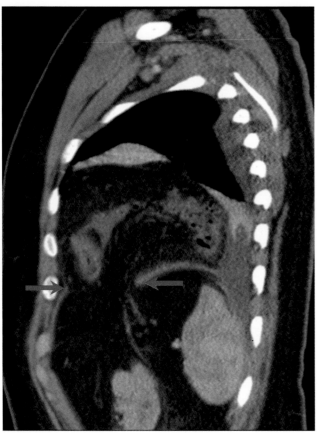

(b)

**Figure 12.11** (a) Obscuration of the left hemidiaphragm in a patient stabbed in the left chest. Increased abnormal lucencies are seen in the left lower zone (arrows) in association with some consolidation (arrowhead). (b) Sagittal CT demonstrates a rupture of the left hemidiaphragm with the contracted edges of the hemidiaphragm visible (arrows). There has been herniation of the transverse colon and stomach. The herniated transverse colon has caused the lucencies on the chest radiograph.

of penetrating trauma (Figure 12.11). The right hemidiaphragm normally lies 1.5–2.5 cm above the left hemidiaphragm. The area below the diaphragms should be assessed for free intraperitoneal air or abnormal areas of calcification.

## Edges (pleura)

The costophrenic angles normally form an acute angle. Obliteration of this angle is seen in the presence of pleural fluid or thickening. Pleura are not normally seen on a chest radiograph, except at the fissures. Visualisation of the pleura indicates the presence of air in the pleural cavity (pneumothorax) (Figure 12.12).

## Skeleton

Bones should be assessed for fractures or focal lesions such as metastatic deposits. In addition, the vertebrae can be assessed through the cardiac silhouette on a correctly exposed film. Bilateral paravertebral stripes are present: the left paravertebral stripe should measure <1 cm and the right paravertebral stripe should be <3 mm. These may become displaced by paravertebral haematomas secondary to the vertebral fractures. Assess the ribs for fractures, and flail segments in particular.

## Soft tissues

Soft tissues should be examined with regard to the presence of air (surgical emphysema) and foreign bodies. Check that the breasts are bilaterally present with symmetrical contours to exclude chest wall discrepancy as a cause for lung density differences.

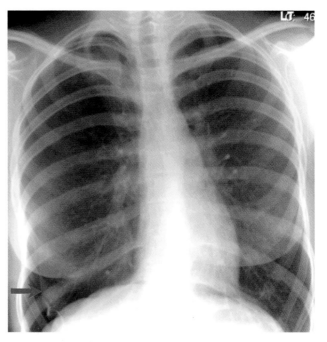

**Figure 12.12** Right-sided pneumothorax with visible visceral pleural line (arrow) and no markings lateral to this.

In addition to assessing the above criteria, the ABCs assessment of CT scans may involve a few additional points.

### Adequacy

Check that the scan covers the entire area of interest, and has been performed in the correct phase, with the correct reconstructions available. Note if there is a large amount of respiratory motion that may impair interpretation.

### Airways

Look for any abnormal contour of the tracheobronchial tree, or any abnormal extraluminal air, that could suggest tracheobronchial disruption.

### All lines

*Chest drains*

The tips of chest drains are more easily localised on CT as compared to CXR. Check if the tip is located within an intraparenchymal rather than intrapleural location (Figures 12.8b and 12.9b).

### Circulation

Assess:

- The thoracic aorta: its contour and main branches
- Cardiac chambers: thrombus or ventricular aneurysms may be visible (Figure 12.13)
- The pulmonary arteries: especially if pulmonary embolism is suspected (Figure 12.13)
- Relevant vessels in the setting of penetrating injuries

### Central structures

Pneumomediastinum can be more readily detected on CT, especially if small. This in turn can suggest oesophageal injury. A markedly distended and fluid-filled oesophagus can also identify a patient who is at potential risk of aspiration.

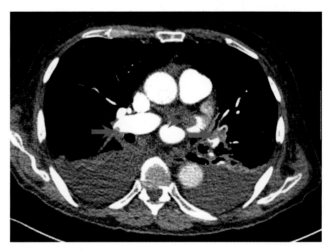

**Figure 12.13** Filling defects in the pulmonary arteries bilaterally (arrows), consistent with bilateral pulmonary emboli, as well as thrombus in the left atrium (arrowhead). Bilateral pleural effusions are present.

### Diaphragm

On CT, in addition to the loss of contour and change in position, the diaphragm itself can be evaluated for rupture or herniation (Figure 12.11b). Coronal or sagittal reconstructions can aid in this regard. Also, if the CT has been limited to the thorax and not included the abdomen, look at the available views of the upper abdomen to identify any overt abnormality, such as a subphrenic haematoma or free intraperitoneal air.

### Skeleton

Assess the bones not as readily visible on a frontal CXR, such as the scapula, sternum and the anterior and extreme lateral aspects of the ribs (Figure 12.14).

### ABCs approach to lung review

The following is only one of many methods to evaluate the lungs. This method employs a density-based approach to assess each zone.

**Figure 12.14** (a) Axial CT demonstrating manubrial fracture following a motor vehicle collision. (b) 3D oblique reconstruction from the same CT scan can demonstrate the fracture and its relationship to the sternoclavicular joints more clearly.

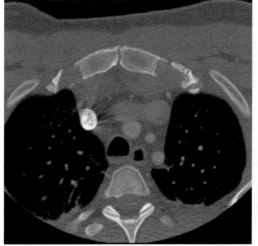

(a)

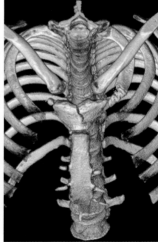

(b)

1 First divide the lungs into Apical, Basal and Central zones.
2 Next, assess a particular zone only for asymmetry of Density, while mentally ignoring ('masking') the other two zones.
3 If there is an asymmetry, decide which side is abnormal.
4 A more hyperlucent lung may still be normal, but be the result of mastectomy, atrophy of pectoral muscles, or rotation. In such cases it should still be possible to see pulmonary vessels extending to the periphery.
5 Abnormally hyperlucent lung may be due to pneumothorax, air-trapping, e.g. due to bronchial obstruction, or bullae.
6 Increased lung density may be due to collapse, air-space shadowing, linear opacities, nodules or masses (Table 12.2).
7 Look for Extra signs e.g. mediastinal deviation that support the likely diagnosis. Bear in mind that both sides could be abnormal.

## Common traumatic conditions in the emergency department

### Blunt chest trauma

The thorax is the third most commonly injured region in blunt trauma, after the head and extremities, most often as a result of motor vehicle accidents. CXR helps exclude life-threatening emergencies, but CT has an integral role to play, particularly in the exclusion of vascular and co-existent injuries.

**Table 12.2** Types and causes of increased lung density on chest radiograph.

| Type of abnormality | | Examples |
| --- | --- | --- |
| Airspace shadowing | Due to: | |
| | Water | Pulmonary oedema |
| | Pus | Infection |
| | Blood | Trauma, vasculitis |
| | Eosinophils | Pulmonary eosinophilia |
| | Tumour | Adenocarcinoma, lymphoma |
| Collapse | Passive: due to compression | Pleural effusion |
| | Obstructive | Tumour |
| | Cicatrical: contraction | Fibrotic lung disease |
| Lines | Septal lines | Pulmonary odema, lymphangitis carcinomatosis, interstitial lung disease |
| | Reticulation | Fibrotic lung disease |
| | Rings with tram-lines | Bronchiectasis |
| | Rings with no tram-lines | Cystic lung disease |
| | Tubular | AV malformations, mucus plugging ('finger-in-glove') |
| Nodule (defined as a rounded opacity ≤3 cm diameter) | ≤3 mm: miliary nodule | Miliary TB, metastases |
| | 3 mm–3 cm | Primary tumour, metastases |
| Mass (defined as opacity>3 cm diameter) | | Tumour, abscess, haematoma |

### Pneumothorax

Pneumothoraces may occur due to the rupture of alveoli, resulting from either a sudden rise in intrathoracic pressure, a crushing force, or sudden deceleration. There may be associated rib fractures.

Features on erect CXR include:

- Visible visceral pleural line (Figure 12.12)
- Absence of vessels lateral to the visceral pleural line
- Shift to the contralateral side, together with cardiovascular compromise, suggesting tension
- Beware contours or lines not paralleling the chest wall, with vessels or increased density seen peripherally; these likely reflect skin folds and will be absent on repeat films

However, in the setting of significant trauma and suspected spinal injury, patients often have to be kept immobilised and supine. In such cases, a pneumothorax can be harder to diagnose, although the following signs may help:

- Increased transradiancy of the affected hemithorax
- Increased sharpness of the adjacent hemidiaphragm, mediastinum or cardiac border
- A deep anterior costophrenic sulcus
  CT is invaluable in these situations (Figure 12.15).

### Pneumomediastinum

Pneumomediastinum in blunt trauma may be the result of alveolar rupture from compressive force, coexistent pneumoperitoneum, or less commonly from tracheobronchial disruption or oesophageal rupture. Features on CXR include:

- Air outlining the mediastinal vessels or the cardiac border
- Air dissecting the mediastinal pleura, sometimes outlining the thymus
- Air along the superior aspects of the diaphragm, tracking between the diaphragm and heart (the 'continuous diaphragm sign') (Figure 12.16)

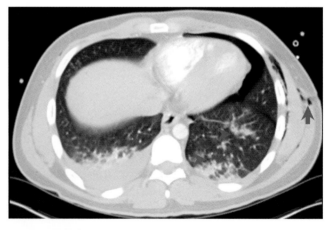

**Figure 12.15** Left pneumothorax from multiple stab wounds. CT shows the stab wound tract (arrow), as well as other associated injuries such as pulmonary contusions in the left lower lobe (arrowheads), and a right pleural effusion. The pneumothorax and contusions were not visible on the patient's supine anteroposterior radiograph.

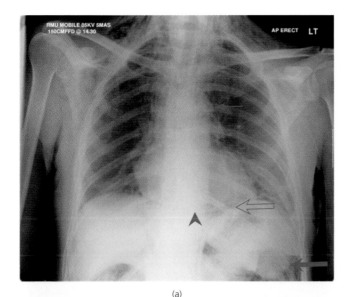

(a)

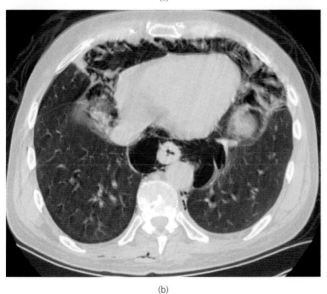

(b)

**Figure 12.16** (a) Pneumomediastinum and pneumoperitoneum. Lucency adjacent to the aortic arch is seen (arrow), and there is a continuous diaphragm sign (arrowhead). In addition, there is air below the left hemidiaphragm which outlines the spleen (transparent block arrow) as well as highlights the bowel wall of the splenic flexure inferiorly (solid block arrow). (b) Axial CT more clearly demonstrates the pneumomediastinum surrounding the oesophagus and within the pericardial fat.

CT is invaluable in confirming the above pathologies, suggesting the cause of pneumomediastinum (intra-or extrathoracic) (Figure 12.16b), ensuring correct chest drain placement, and identifying co-existent injuries such as haemopneumothorax and rib fractures.

## Pulmonary contusions and lacerations

Contusions are the most common form of lung injury in blunt trauma, manifesting as non-segmental patchy air-space shadowing

on CXR. CT is more sensitive in detecting contusions as they are usually seen immediately and may occur at the site of impact or in contrecoup locations (Figure 12.10b).

Pulmonary lacerations are less common, and heal more slowly as compared to contusions. They result in cavities that may be air-filled or blood-filled at CT.

## Traumatic aortic injury

Injury to the aorta most commonly occurs at the aortic attachments, most frequently at the ligamentum arteriosum due to a rapid deceleration force. Mediastinal or periaortic haematoma consequently develop. Such injuries are usually fatal.

Aortic injury may be suggested by the following signs on CXR (Figure 12.17):

Direct signs of mediastinal haematoma:

- Mediastinal widening >8 cm at the level of the aortic arch
- Deviation of the nasogastric tube and trachea to the right
- Depression of the left main bronchus
- Left apical pleural shadowing

Indirect signs (indicating significant force):

- First to third rib fractures
- Scapular, vertebral or sternal fracture
- Left haemothorax or pneumothorax
- Pulmonary contusion

The CXR is thought to be quite sensitive, but poorly specific for aortic injury. Overall, aortic injury cannot reliably be excluded by a normal CXR. In contrast, imaging with CT angiography can have a sensitivity and negative predictive value of 100%. CT angiography can characterise intimal tears, pseudoaneurysms, and periaortic haematoma, as well as illustrate other coexistent injuries, and help in planning therapeutic strategies (Figure 12.18).

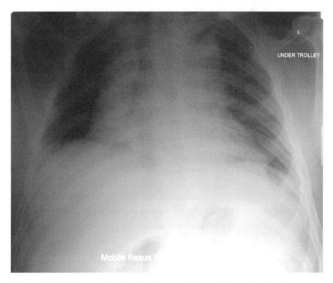

**Figure 12.17** Motorcyclist involved in a collision, with markedly widened mediastinum, and tracheal deviation to the right even allowing for film rotation.

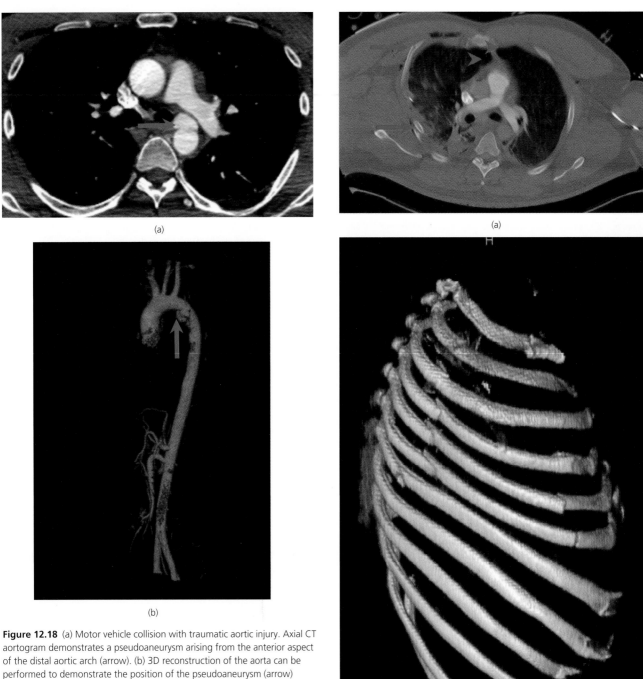

**Figure 12.18** (a) Motor vehicle collision with traumatic aortic injury. Axial CT aortogram demonstrates a pseudoaneurysm arising from the anterior aspect of the distal aortic arch (arrow). (b) 3D reconstruction of the aorta can be performed to demonstrate the position of the pseudoaneurysm (arrow) relative to the great vessels.

## Skeletal injuries

In the setting of significant blunt chest trauma, CXR can detect multiple or bilateral rib fractures which could suggest high energy impact.

Furthermore, a flail segment can be revealed on both CXR and CT (Figure 12.19). A flail segment occurs when at least three contiguous ribs have each been fractured in at least two locations, leading to paradoxical chest wall motion on spontaneous ventilation.

**Figure 12.19** (a) Patient who had a flail segment clinically as a result of being trapped under a train. Axial CT demonstrates fractures of the right 3rd and 4th ribs at this level (arrows), as well as bilateral pneumothoraces (arrowheads). (b) 3D reconstruction of the right thoracic cage, with the spine and left rib cage removed to obtain a clearer view, demonstrates a flail segment of the right fourth to sixth ribs.

Scapular fractures are more easily identified on CT, and should prompt suspicion of other injuries, such as spinal injuries.

## Penetrating Injury

The pulmonary, pleural and vascular injuries that can occur in blunt chest trauma can also occur in penetrating trauma. It is useful to place markers at the entry and exit (if present) sites on CXR, to more closely assess these regions. Any vessels close to the trajectory of penetration should be carefully assessed on CT.

Pneumothorax may also develop belatedly in such patients, and so it is sometimes useful to repeat the CXR 3–6 hours after injury.

Ten to thirty percent of patients with penetrating chest injury may have abdominal involvement.

## Common non-traumatic conditions in the emergency department

### Airspace shadowing

This is characterised by an increase in lung density. The causes of airspace shadows are outlined in Table 12.2.

Airspace shadowing encompasses a range of CXR descriptions, such as:

- Nodular opacities
- Ground-glass opacity:
  - on CXR- a homogeneous veil-like opacity which obscures the underlying vessels
  - on CT- a homogeneous opacity which does not obscure the underlying vessels
- Consolidation: increased opacity with air bronchograms (Figure 12.20)

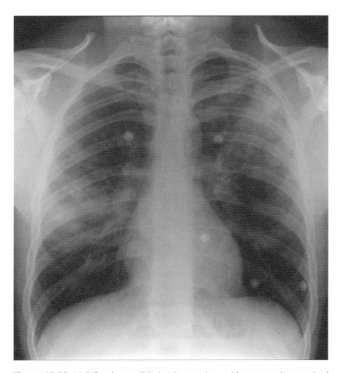

**Figure 12.20** Multifocal consolidation in a patient with community-acquired pneumonia.

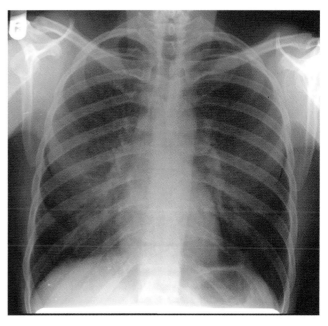

**Figure 12.21** HIV patient with severe shortness of breath. There are bilateral ill-defined ground-glass opacities in a perihilar distribution, with fine reticulation seen.

Infection in the immunocompromised host may have specific features. For example, Pneumocystis pneumonia may manifest as perihilar or sometimes diffuse ill-defined ground-glass opacities on CXR (Figure 12.21). On CT, this can appear as ground-glass opacities with or without interlobular septal thickening.

### Pulmonary oedema

This may be the result of cardiogenic (e.g. myocardial infarction) or non-cardiogenic (e.g. inhalation of noxious gases, raised intracranial pressure) insults.

CXR appearances include:

- Cardiac enlargement (if cardiogenic)
- Basal vasoconstriction
- Upper lobe blood diversion
- Loss of clarity of pulmonary vessels (perivascular haziness)
- Thickening of bronchi (peribronchial cuffing)
- Septal lines: thin linear opacities caused by interstitial fluid that are either located centrally, radiating from the hila, or peripherally, perpendicular to pleural surface
- Associated pleural effusions or airspace shadowing

CT is not normally necessary in the assessment of pulmonary oedema.

### Pulmonary embolism

The CXR signs of pulmonary embolism (PE), such as wedge-shaped consolidation and oligaemia, are neither specific nor sensitive. CT pulmonary angiography (CTPA) has become the most widely accepted method of detecting pulmonary embolism. However, it is

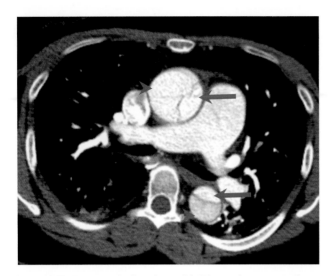

**Figure 12.22** Type A aortic dissection. Axial CT image demonstrates the true (arrows) and false (arrowheads) lumens. In the aortic arch, the false lumen lies close the aortic root, lying anterolateral and to the right of the true lumen, while in the descending thoracic aorta it lies posterior or posterolateral, and to the left, as the dissection usually has a spiral configuration.

important to determine the clinical pre-test probability of PE, and interpret the CTPA result in conjunction with this probability.

Signs of a PE on CTPA are:

- Intraluminal filling defect (Figure 12.15)
- In complete occlusion, the artery may be enlarged in comparison with adjacent patent vessels
- Parenchymal changes, e.g. haemorrhage, atelectasis
- Features of right ventricular strain if present, e.g. bowing of the interventricular septum to the left

If CTPA is unavailable or contraindicated, ventilation–perfusion scintigraphy (V/Q scan) is an alternative consideration, and Doppler ultrasound of the lower limbs can exclude coexistent DVT.

## Acute aortic syndromes

These encompass the acute non-traumatic presentation of aortic dissection, intramural haematoma and penetrating aortic ulcers. A normal CXR cannot exclude these pathologies in patients with appropriate presentation symptoms or signs. This is because even significant dissection of the aorta may result in no distortion of the mediastinal planes but is readily apparent on CT aortography (Figure 12.22).

## Acknowledgements

Dr Ali Naraghi and Dr Otto Chan, authors of the chest section of the second edition of *ABC of Emergency Radiology*.

## Further reading

Berger FH, van Lienden KP, Smithuis R, Nicolaou S, van Delden OM. Acute aortic syndrome and blunt traumatic aortic injury: pictorial review of MDCT imaging. *Eur J Radiol.* 2010;74(1):24–39.

Hansell DM, Lynch DA, McAdams H, Page, Bankier AA. *Imaging of Diseases of the Chest*, 5th edn. Chapter 2: The normal chest, and Chapter 17: Chest trauma. Elsevier, 2010.

Kaewlai R, Avery LL, Asrani AV, Novelline RA. Multidetector CT of blunt thoracic trauma. *Radiographics* 2008;28(6):1555–1570.

LeBlang, Suzanne D; Dolich, *Matthew O. Imaging of blunt penetrating thoracic trauma. J Thor Imag* 2000;15(2):128–135.

Schnyder P, *Wintermark M. Radiology of Blunt Trauma of the Chest.* Springer, 2000.

# CHAPTER 13

# Abdomen

*Katie Planche and Niall Power*

Royal Free London NHS Foundation Trust, London, UK

---

**OVERVIEW**

- The patient with acute abdominal pain requires rapid assessment to exclude life threatening pathology (e.g. ruptured aortic aneurysm)
- The AXR is quick and easy to obtain, but is difficult to interpret and provides limited information
- CT (rarely US) is required in patients with acute abdominal pain

---

**Table 13.1** Positions of organs.

| Organ | Position | Appearance on radiograph |
|---|---|---|
| Liver | RUQ | Subhepatic edge visible |
| Spleen | LUQ | Rarely seen below twelfth rib |
| Kidneys | Flanks | Outlined by fat and psoas (L1–4), left kidney higher. |
| Pancreas | Central | Not visible and retroperitoneal |
| Bladder | Pelvis | May be visible if full |

**Table 13.2** Position of bowel.

| Organ | Position | Appearance on radiograph |
|---|---|---|
| Stomach | Upper | Lies transverse across upper left and central abdomen |
| Small bowel | Central | <3 cm and contains air |
| Large Bowel | Peripheral | Contains air and faeces |

The patient presenting with abdominal pain may be suffering from a wide variety of conditions ranging from common, benign self-limited pathologies such as gastroenteritis to life-threatening emergencies such as a ruptured aortic aneurysm. It is vital to be able to differentiate these conditions using the most appropriate imaging modality (Box 13.1). These include the plain abdominal radiograph (AXR), ultrasound (US) and computed tomography (CT). Magnetic resonance imaging (MRI) does not currently have a role in the initial management of abdominal emergencies. There are only a few common causes of acute abdominal pain and these will be highlighted in this chapter. AXRs are difficult to interpret and often non-diagnostic and therefore in these patients, CT (or sometimes US) should be performed initially.

---

Box 13.1 **Indications for AXR**

- Suspected bowel obstruction
- Suspected perforation
- Renal colic
- Foreign body ingestion

---

## Anatomy

Normal appearances of the AXR vary (Figures 13.1 and 13.2), but Tables 13.1 and 13.2 show some useful general points to remember when assessing the radiographs.

The solid organs are outlined by fat planes and can therefore be identified on the AXR. The bowel normally contains a variable amount of gas and the different segments of bowel can usually be differentiated by their site and morphology. Any gas seen outside the bowel is abnormal and is highly suggestive of perforation. The AXR is difficult to interpret and gives limited information and further investigations such as US or CT are often needed. In patients with severe abdominal pain, CT (or sometimes US) is advised as the first line of investigation.

## ABCs systematic assessment

- **A**dequacy – must include the pubic symphysis
- **A**ir – exclude free intraperitoneal and retroperitoneal air
- **B**owel – check gas pattern, size, and distribution. Exclude small bowel (SB) and large bowel (LB) obstruction
- **C**alcifications – check renal tract, vascular and other structures
- **D**ensities – look for tablets and foreign bodies
- **E**dges – check hernial orifices and lung bases
- **F**at planes – check psoas, properitoneal and perivesical fat planes
- **S**olid organs – look for liver, spleen and kidneys
- **S**keleton – look at the bones for fractures

## Adequacy

The standard AXR (Figure 13.1) is an anteroposterior view taken with the patient supine, not rotated and should include the pubic symphysis and show the hernial orifices and the properitoneal fat

*ABC of Emergency Radiology*, Third Edition. Edited by Otto Chan.
© 2013 John Wiley & Sons, Ltd. Published 2013 by John Wiley & Sons, Ltd.

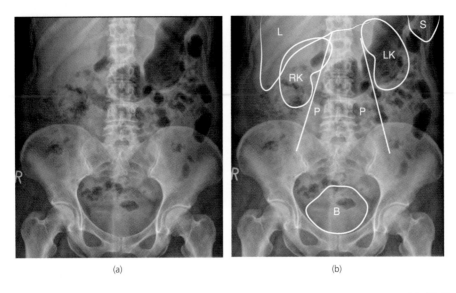

(a)                                        (b)

**Figure 13.1** (a),(b) Anteroposterior abdominal radiograph illustrating the position of the organs: L, liver; S, spleen; RK, right kidney; LK, left kidney; P, psoas; B, bladder.

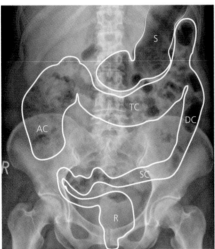

**Figure 13.2** Abdominal radiograph showing the position of the bowel. S, stomach; AC, ascending colon; TC, transverse colon; DC, descending colon; SC, sigmoid colon; R, rectum.

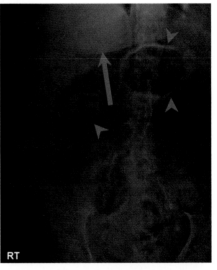

(a)

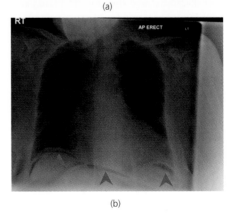

(b)

**Figure 13.3** Abdominal radiograph and erect chest radiograph showing free gas (arrows) in a patient with perforated diverticulitis.

planes. These fat planes are thin layers of fat between the parietal peritoneum and the lateral wall muscles. An additional view of the upper abdomen may be necessary in tall patients.

If a perforation is suspected an erect CXR should be performed (Figure 13.3). Small amounts of free air can be seen under the hemidiaphragm on the CXR, if enough time (at least 10 minutes) is allowed for the free air to rise. However, if perforation is suspected, a CT scan is indicated, as it is more sensitive at detecting free air and the underlying cause. The erect CXR may sometimes show chest pathology simulating an acute abdomen, such as pneumonia or aortic dissection.

## Air

Free intraperitoneal air rises to the front of the abdomen on a supine AXR (Figure 13.3). This free air can be very subtle and hard to detect, so an erect chest radiograph is mandatory if perforation is suspected. Look for extraluminal air (Figure 13.4) – any gas outside the bowel wall is abnormal.

## Bowel

When the patient is supine, bowel gas rises to the parts of the gastrointestinal tract that are most anterior, in particular the stomach, transverse colon and sigmoid colon. The stomach is located above

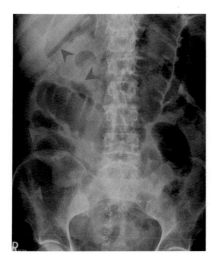

**Figure 13.4** Abdominal radiograph showing free gas (arrows) in a patient with a postoperative perforation.

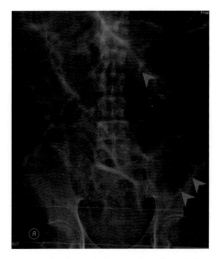

**Figure 13.6** Abdominal radiograph showing large bowel obstruction in a patient with sigmoid cancer. Note the haustra (arrowhead) extend part of the way across the bowel.

the transverse colon. The SB can be differentiated from the LB by the following features:

- Position: the SB lies in the central abdomen; the LB lies peripherally.
- Size: the SB should measure no more than 3 cm in diameter and is smaller than the LB. The large bowel has no definite measurement, but if the caecum dilates to more than 9 cm diameter in the presence of a suspected bowel obstruction, it infers impending perforation. In the presence of colitis, any part of the LB that measures more than 5.5 cm in diameter indicates a megacolon, but normal colon can easily be larger than this.
- Pattern: the SB has characteristic thin folds (Figure 13.5) (valvulae conniventes), which are close together and run across the whole bowel. The LB has thicker folds (Figure 13.6) (haustra) that do not run across the whole dilated LB. The distal ileum and sigmoid colon are relatively featureless and contain no folds.

- Content: the SB contains fluid and air, whereas the colon contains faeces, which have a characteristic mottled or solid appearance.

## Calcification

Normal calcifications:

- Costal cartilage (Figure 13.7) can sometimes be seen in the upper abdomen as an incidental finding
- Phleboliths (Figure 13.7) are small calcified veins in the pelvis. They can be confused with ureteric stones
- Coarse, nodular calcification in mesenteric lymph nodes is an incidental finding lying between the left L2 transverse process and the lower right sacroiliac joint
- Vascular calcification is often seen in the aorta

Abnormal calcifications:

- Around 45% of renal (Figure 13.8) and ureteric calculi are visible on an AXR

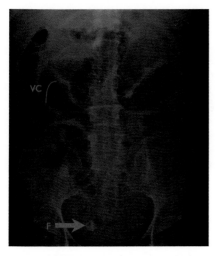

**Figure 13.5** Abdominal radiograph in a patient with small bowel obstruction. Note the valvulae conniventes (VC) extend all the way across the bowel. A calcified fibroid (F) is noted in the pelvis.

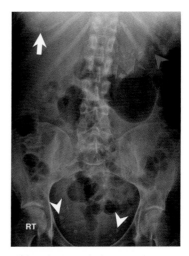

**Figure 13.7** Abdominal radiograph showing pancreatic calcification (red arrowhead), phleboliths (white arrowhead) and costal cartilage calcification (white arrow).

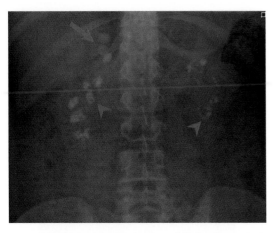

**Figure 13.8** Abdominal radiograph showing calcified gallstones (red arrow) and renal calcification (arrowhead).

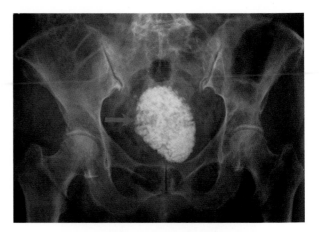

**Figure 13.10** Abdominal radiograph showing a calcified fibroid (arrow).

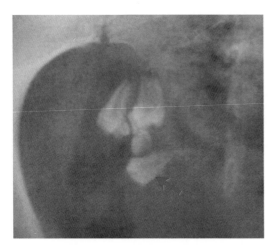

**Figure 13.9** AP magnified view of the pelvis showing a Dermoid with teeth.

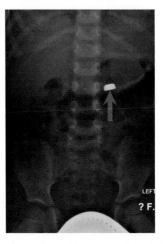

**Figure 13.11** Abdominal radiograph in a 4 year old showing a button battery (arrow) in the stomach.

- Abnormal vascular curvilinear calcification may outline aneurysms or thrombosed veins
- Only 10–15% of gallstones (Figure 13.8) are radio-opaque. Rarely, the gallbladder wall can calcify (porcelain gallbladder)
- Punctate fine stippled nodular calcification in the pancreas indicates chronic pancreatitis (Figure 13.7)
- Calcification in the spleen is usually caused by previous trauma or granulomatous or parasitic infections
- Appendicoliths and faecoliths may occasionally be seen
- Tuberculosis and schistosomiasis infections can cause calcification of the bladder wall
- Teeth can sometimes be seen in benign teratomas of the ovary (dermoid cysts) (Figure 13.9), and rarely, ovarian tumours can calcify.
- Calcification of the seminal vesicles is often seen in patients with diabetes or renal failure.
- Uterus: it is common for fibroids to calcify as large masses in the lower pelvis (Figures 13.5 and 13.10).

## Densities

These include foreign bodies (Figure 13.11 and 13.13), tablets and tampons. Foreign bodies of any type can be ingested or inserted

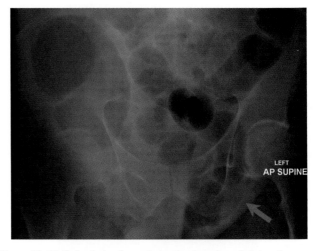

**Figure 13.12** Abdominal radiograph showing large bowel obstruction secondary to incarcerated left inguinal hernia (arrow).

via any orifice. Common sites of obstruction for foreign bodies are the distal oesophagus, pylorus of the stomach and terminal ileum. Tablets often contain calcium. Tampons can be seen as air-filled tubular structures in the lower pelvis (Figure 13.13).

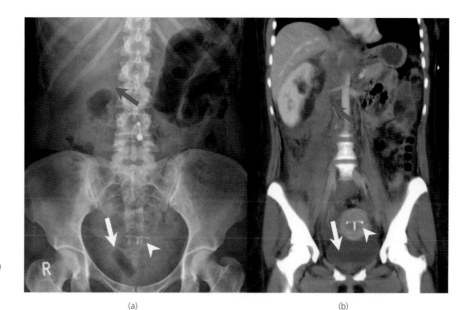

**Figure 13.13** (a) Abdominal radiograph and (b) coronal CT image showing displaced right kidney (red arrow) following haemorrhage from a renal mass. Also note piercing (Red arrowhead), intrauterine contraceptive device (IUCD, white arrowhead) and tampon (white arrow).

(a)　　　　　　　　　　　(b)

### Edges

In particular, look at the hernial orifices (Figure 13.12) for bowel in the hernia (this is a common cause of SB obstruction). Central lines are sometimes inserted via the groin. Pathology may be seen in the lung bases, such as pneumonia and pleural effusions.

### Fat planes

Identification of normal fat planes is important because absence, distortion or displacement (Figure 13.13) can indicate pathology. Visible fat planes include:

- Psoas – loss of the psoas fat plane may indicate a retroperitoneal mass or collection or haemorrhage
- Perirenal fat plane – outlines kidneys
- Perivesical fat plane – outlines bladder and is often displaced in the presence of pelvic fractures
- Properitoneal fat planes – outlines the lateral margin of the ascending and descending colon

### Skeleton

The visualised skeleton, in particular the lower ribs, spine, pelvis and hips should be assessed for fractures and bony metastases (Figure 13.14) .

## Important plain film findings

### Small bowel dilatation

The most common causes of dilated SB loops are mechanical obstruction and paralytic ileus and rarely infarction.

Mechanical obstruction of the small bowel may be caused by:

- Adhesions (75%) due to previous surgery (Figure 13.15)
- Obstructed hernia – commonest cause in the absence of previous surgery
- Inflammation
- SB volvulus
- Intussusception

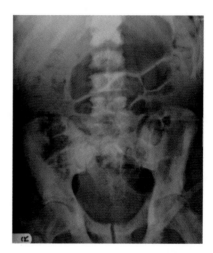

**Figure 13.14** Abdominal radiograph showing extensive sclerotic bony metastases in a patient with prostate cancer.

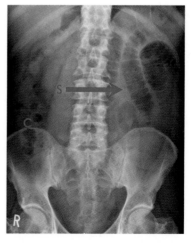

**Figure 13.15** Abdominal radiograph showing small bowel obstruction (S) secondary to adhesions. Note the colon (C) is collapsed.

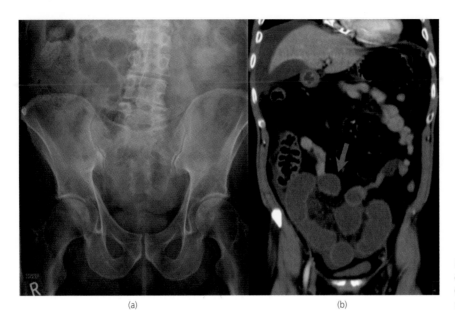

**Figure 13.16** (a) Abdominal radiograph and (b) coronal CT scan showing small bowel obstruction (red arrow), which cannot be seen on the abdominal radiograph.

- Malignancy
- Gallstone ileus – may be seen in elderly patients (Figure 13.26)

The cardinal features of SB obstruction are dilated loops of SB (usually >3 cm in diameter) containing variable amounts of air and fluid with collapse of the LB. SB obstruction can be difficult to differentiate from paralytic ileus. In the latter, the LB is also often dilated, and bowel sounds are absent. CT should be performed in all patients with suspected SB obstruction (Figure 13.16).

## Large bowel dilatation

LB can be dilated due to paralytic ileus, colonic pseudo-obstruction (Figure 13.17) or mechanical obstruction. Pseudo-obstruction is a form of ileus and is characterised by dilatation of the LB without an obstructing lesion. This is often seen in elderly patients who are clinically relatively well and comparison with previous radiographs is useful.

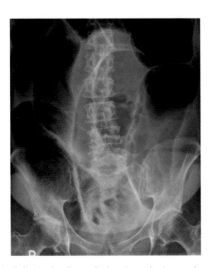

**Figure 13.17** Abdominal radiograph showing colonic pseudo-obstruction. The radiograph was unchanged over 2 years.

## Commonest causes of mechanical large bowel obstruction

- Malignancy (Figure 13.6)
- Diverticular disease
- Volvulus
- Adhesions

Volvulus is the twisting of a bowel loop, and it affects loops with redundant long mesentery. The most common sites are the sigmoid colon and rarely, the caecum and transverse colon.

In sigmoid volvulus (Figure 13.18), the redundant loop of sigmoid classically rotates towards the right upper quadrant to give an inverted U appearance devoid of haustra (coffee bean sign) and a characteristic central stripe with the apex above T10 and the liver (liver overlap sign) and descending colon (bowel overlap sign) can be seen through the dilated loop of LB.

Caecal volvulus usually rotates to the left upper quadrant with associated SB obstruction.

## Extraluminal air (Figure 13.4 and 13.19)

Air outside the lumen of the bowel has an array of causes of varying clinical significance. These include:

- Free intraperitoneal air (pneumoperitoneum) – In the absence of recent surgery, it implies a perforation, usually of a peptic ulcer or diverticulitis. Free air can be difficult to see on a supine radiograph and if clinically suspected, an erect CXR or CT should always be obtained.
- Air may be seen on the supine AXR in the hepatorenal recess (Morrison's pouch), subhepatic space, under the diaphragm (unicopula sign or visualisation of the falciform ligament), central abdomen (football sign), or between bowel loops. Rigler's sign is the visualisation of both sides of the bowel wall caused by the presence of air on both sides (normally bowel or fat abuts the outside of the wall).

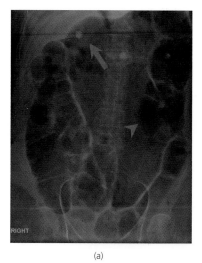
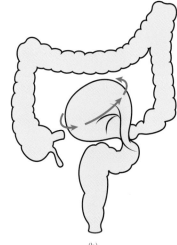

(a)

(b)

**Figure 13.18** (a) Abdominal radiograph (showing liver edge – 'liver overlap' sign and decending colon – 'bowel overlap' sign) and (b) diagram showing a sigmoid volvulus and illustrating how the sigmoid twists on its mesentery.

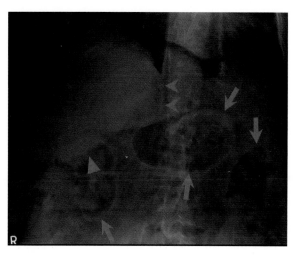

**Figure 13.19** Abdominal radiograph showing multiple sites of free air. Rigler's sign (small arrow), Unicupola sign (large arrow), subhepatic air (large arrowhead) and the falciform ligament (small arrowhead).

### Signs of free intraperitoneal air on supine AXRs (Figure 13.19)

- Rigler's sign (small arrow) – both sides of the bowel wall can be seen
- Unicopula sign (large arrow) – air in the central leaf of the diaphragm
- Subhepatic air (large arrowhead) – free air under the inferior margin of the liver
- Falciform ligament (small arrowhead) – air outlines the falciform ligament over the liver

*Air in the bowel wall – pneumatosis intestinalis*
Linear streaks of intramural air are important as they may indicate infarction of the bowel wall, which may be a sequelae of severe inflammation or ischaemia (Figure 13.20). In elderly patients air in the bowel wall, usually colon (Figure 13.21), may be secondary to benign causes such as longstanding obstruction or chronic lung disease (pneumatosis coli).

Other causes – Gas in the portal venous system should be differentiated from air in the biliary system (aerobilia) by its peripheral

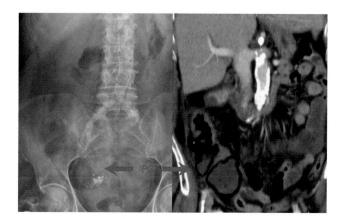

**Figure 13.20** Abdominal radiograph and coronal CT showing gas in the bowel wall – pneumatosis coli (arrow) in a patient with acute ischaemia of the ascending colon.

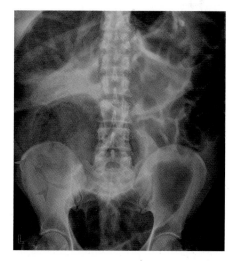

**Figure 13.21** Abdominal radiograph showing gas in the bowel wall (red arrowheads) – *Pneumatosis coli.*

location in the liver. This is a sinister sign in the adult patient that indicates transmural bowel infarction with tracking of air into the mesenteric veins.

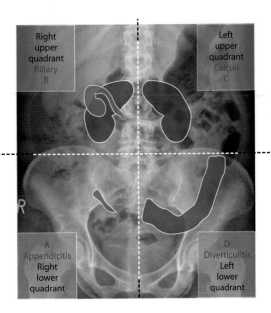

**Figure 13.22** Abdominal radiograph showing the most common sites of pain and causes of the acute abdomen.

Aerobilia and air in the gallbladder can be iatrogenic, caused by endoscopic procedures (ERCP), percutaneous transhepatic cholangiograms (PTC), biliary stents or erosion of a gallstone through the gallbladder wall into the SB. This may cause SB obstruction (gallstone ileus), usually in elderly patients (Figure 13.26).

Air may also be seen in an abscess, where it may have a mottled appearance or produce an abnormal air-fluid level. An abscess may have mass effect and displace adjacent structures. CT or US are the investigations of choice if an abscess is clinically suspected (Figure 13.23).

## Common abdominal emergencies

### Sites of pain (Figure 13.22)

- RUQ – Biliary – gallstones/cholecystitis
- RLQ – Appendicitis
- LLQ – Diverticulitis
- Central – SBO, pancreatitis, peptic ulcer, aneurysm

Box 13.2 gives the common causes of abdominal pain and Box 13.3 indicates which type of investigation to use.

---

Box 13.2 **Common causes of acute abdominal pain**

A – Appendicitis/aneurysm/acute pancreatitis
B – Bowel obstruction/perforation
C – Cholecystitis/calculi
D – Diverticulitis
E – Ectopic pregnancy

---

### Appendicitis (Figure 13.23)

This is the commonest abdominal surgical emergency. Peak incidence is in late teenage years, but also common in children and adults. The AXR will usually be normal, unless there is associated small bowel obstruction or if an appendicolith is present (30% but often not visible on AXR). For children or young women, the

---

Box 13.3 **Investigating acute abdominal pain**

If you suspect... – request

| | |
|---|---|
| Appendicitis | – US in children/young women, otherwise CT |
| Aneurysm | – CT |
| Acute pancreatitis | – US to exclude gallstones, CT after 48 hours if not settling |
| Bowel obstruction | – AXR then CT |
| Bowel perforation | – AXR + CXR then CT |
| Cholecystitis | – US |
| Calculi | – CT KUB |
| Diverticulitis | – CT |
| Ectopic pregnancy | – US |

---

investigation of choice is US. Adult patients should have a CT to exclude other causes and also to confirm the diagnosis. CT has a high negative predictive value for acute appendicitis, therefore this algorithm avoids unnecessary surgery.

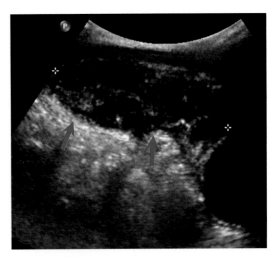

**Figure 13.23** Ultrasound showing appendix abscess in an 8 year old boy.

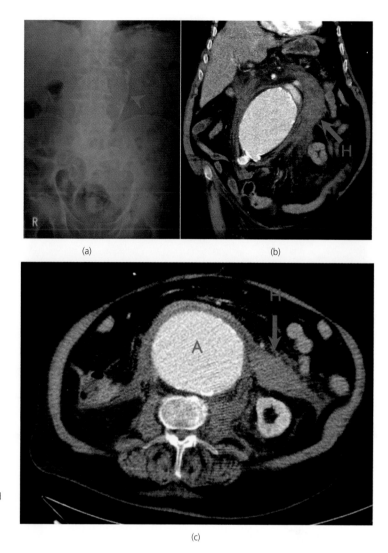

**Figure 13.24** (a) Abdominal radiograph showing calcification (red arrowhead) in the wall of a large abdominal aneurysm with corresponding (b) coronal and (c) axial CT scans showing large aneurysm (A) which has ruptured showing haemorrhage (H).

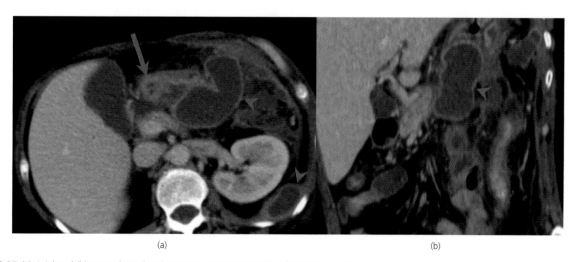

**Figure 13.25** (a) Axial and (b) coronal CT showing severe pancreatitis with inflammation of the pancreas (red arrow) and extensive peripancreatic collections (red arrowheads) within the abdomen.

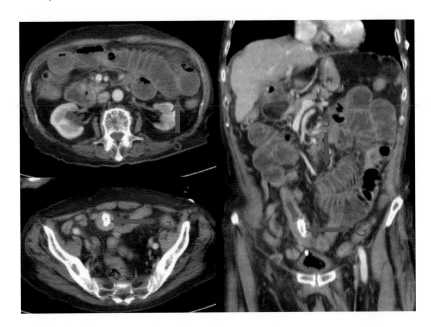

**Figure 13.26** Axial and coronal CT showing small bowel obstruction (S) secondary to an obstructing gallstone (G) in a patient with gallstone ileus.

## Abdominal aortic aneurysm (Figure 13.24)

An aneurysm may be suggested on the plain film by vertical curvilinear calcification, whilst a leak or rupture will present with signs of a retroperitoneal haematoma, with displacement or loss of the psoas fat plane or planes, displacement of the kidney and displaced bowel loops. Leak or rupture on CT is confirmed by extra luminal extravasation of contrast and the presence of a retroperitoneal haematoma.

## Acute pancreatitis (Figure 13.25)

If the clinical diagnosis is confirmed with an elevated amylase, no imaging is required, although ultrasound may be performed to look for gallstones. CT is useful after 48hours in patients who are not improving to look for complications such as pancreatic necrosis, abscess, pseudoaneurysms, and collections.

## Bowel obstruction and perforation (Figure 13.26, 13.27)

A normal AXR does not exclude either diagnosis and if clinical concern persists CT is indicated. If the AXR is abnormal, then CT is indicated.

In mechanical obstruction a transition zone from dilated proximal bowel to collapsed distal bowel should be sought (Figure 13.26). CT in perforation can confirm the diagnosis and the cause, usually peptic ulcer disease or diverticulitis (Figure 13.3, 13.4, 13.19).

## Cholecystitis (Figure 13.28)

There is no role for an AXR if this is suspected. Ultrasound typically shows a thick walled gallbladder with pericholecystic fluid and gallstones. Right upper quadrant tenderness on scanning over the gallbladder is known as an ultrasound positive Murphy's sign. Ultrasound is more sensitive than CT for gallstones.

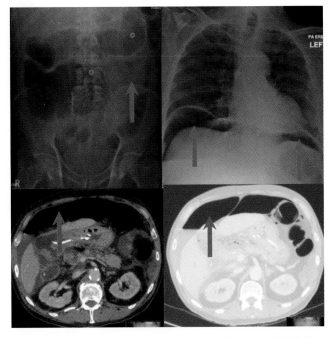

**Figure 13.27** Abdominal radiograph, erect chest radiograph and CT (soft tissue and lung windows) showing free gas in a patient with perforation – an anastamotic breakdown following surgery.

## Renal tract calculi (Figure 13.29)

Ultrasound and limited IVU have been superseded by low dose CT KUB. CT KUB can show calculi more accurately, as well as secondary signs of obstruction such as hydronephrosis, hydroureter and perinephric fat stranding, a urinary leak or an abscess and other pathologies such as a leaking abdominal aneurysm. Therefore unlike a limited IVU, CT KUB is quick, very sensitive and specific and can diagnose other causes of abdominal pain.

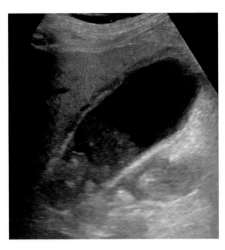

**Figure 13.28** Ultrasound scan showing thick-walled abnormal gallbladder (G) containing calculi in a patient with acute cholecystitis.

## Diverticulitis (Figure 13.30)

This is a common cause of an acute abdomen in elderly patients with pain classically localised to the left lower quadrant. CT is highly sensitive and specific and can show the typical findings of inflammation around a diverticulum, usually in the sigmoid colon, and can show complications such as perforation and abscess formation.

## Ectopic pregnancy (Figure 13.31)

Pregnancy should be excluded in all women of childbearing age with an acute abdomen. An ectopic pregnancy is a surgical emergency due to the risk of tubal rupture and associated haemorrhage. Transvaginal US is the investigation of choice.

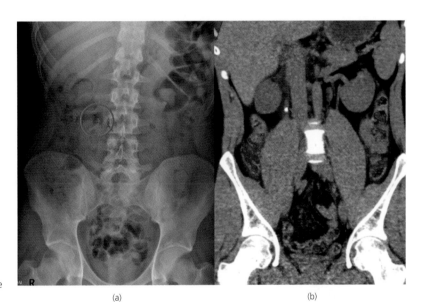

(a)　　　　　　　　　(b)

**Figure 13.29** (a) Abdominal radiograph and (b) coronal CT showing renal calculus in right proximal ureter (circled). Note that this is difficult to see on the abdominal radiograph.

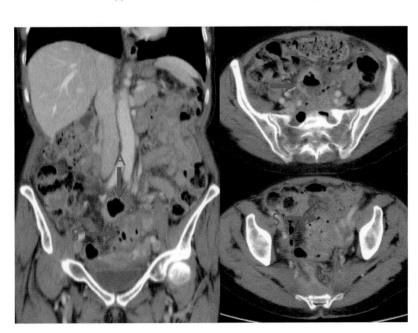

**Figure 13.30** Coronal and axial CT scans in a patient with perforated diverticulitis showing an abscess (A) containing gas and fluid. Extensive inflammatory change is seen around the sigmoid colon (S).

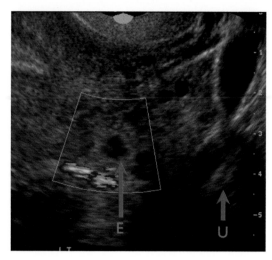

**Figure 13.31** Transvaginal ultrasound showing ectopic pregnancy (E), separate to the uterus (U).

### ABCs systematic assessment

**Air**

- Exclude free intraperitoneal or abnormally sited air

**Bowel gas**

- Check size, distribution, and pattern

**Calsification**

- Check for normal and abnormal calcification

**Densities**

- Check for inserted or ingested foreign bodies

**Edges**

- Check the hernial orifices
- Check the lung bases and pleural spaces

**Fat planes**

- Check presence and symmetry of psoas shadows
- Check presence of perivesical fat plane
- Check that properitoneal fat planes are present

**Soft tissues**

- Check for enlarged or absent organs. Confirm with US

**Skeleton**

- In trauma, check that there are no obvious fractures
- If malignancy is suspected, exclude bony metastases

## Further reading

http://www.learningradiology.com/toc/tocorgansystems/tocgi.htm

Stoker J, van Randen A, Laméris W, Boermeester MA. Imaging patients with acute abdominal pain. *Radiology* October 2009;253:31–46.

The Radiology Assistant – Abdomen. http://www.radiologyassistant.nl/en/ LearningRadiology.com

# CHAPTER 14

# Computed Tomography in Emergency Radiology

*Anmol Malhotra and Jeremy Rabouhans*

Royal Free London NHS Foundation Trust, London, UK

---

## OVERVIEW

- Introduction of how a CT scanner works
- The dramatic increase in the use of CT
- MDCT provides rapid and accurate diagnosis
- Ionising radiation, doses and the ALARA principle
- Dangers of ionising radiation in children and pregnant women

---

**Table 14.1** Typical window levels and widths.

|  | Window level | Window width |
|---|---|---|
| Abdomen | 60 | 360 |
| Liver | 100 | 200 |
| Lungs | 600 | 1600 |
| Bone | 800 | 2000 |
| Supratentorial | 40 | 80 |
| Infratentorial | 35 | 150 |
| Acute haemorrhage | 100 | 200 |
| Mediastium | 35 | 350 |
| Colon | 0 | 2000 |
| Cardiac | 90 | 750 |

There have been huge technological advances in the past decade, not least in diagnostic radiology, with the advent of multidetector CT (MDCT), US, MRI, digital radiography and picture archiving and communication systems (PACS). The role of the radiologist in trauma and emergency medicine has also dramatically changed, not least with almost an exponential increase in the use of CT. In 2007, there were over 60 million CT scans performed in the USA and it is estimated that over 100 million scans will be performed this year in the USA alone.

The new generation of MDCT scanners have over 256 detectors, rotation times of less than 0.3 s, 0.4 mm resolution and they can do whole body scans in less than 10 s and acquiring isotropic voxels. The continuingly evolving software packages process the continuously changing cross sections as the table (gantry) moves through the X-ray circle producing multiplanar reconstructions to allow viewing in any plane with similar resolution and 3D images at the touch of a button (almost real time).

The images are 'windowed' in order to be able to demonstrate the information based on the ability of the body structures to block (attenuate) the X-ray beam.

Typical windows are shown in Table 14.1.

As the number of applications for CT expand with these newer generation MDCT scanners, so the radiologist's role becomes central in the management of patients, not least in ER. Radiologists are involved not only in interpretation, but also in developing and implementing new protocols that take advantage of the new advances in CT technology. The increase in use has led to an increase in exposure to ionising radiation and although CT only represents about 20% of the general workload in radiology, it is responsible for over 90% of the radiation. Recent advances in dose reduction mean that whole body CT scans can now be done with a dose of under 5 mSv (Figure 14.1a,b). Despite this, it is essential that everyone adopts the ALARA (as low as reasonably achievable) principle when using CT and restricts the use in children and pregnant women. This chapter will concentrate on the general principles of CT and the role of CT in the acute abdomen.

## Definition of the acute abdomen

A clinical syndrome of sudden onset of severe abdominal pain requiring emergency medical or surgical treatment.

Imaging of the acute abdomen has been revolutionised by multi-detector computed tomography (MDCT) and every emergency department should now have access to one. Previously a supine AXR and an erect CXR were the first line of investigations, but the AXR is now almost obsolete. Kellow et al. (2008) stated 'when imaging is needed, the emergency physician should be encouraged to immediately request more definitive imaging modalities' (e.g. MDCT and ultrasound (US)).

There has been a dramatic increase in the use of MDCT in the past decade, in particular in the ER.

## Indications for acute abdominal CT

MDCT is accurate and cost effective for the evaluation of acute abdominal pain, and provides an earlier diagnosis. When surgical intervention may be required, CT provides additional information to facilitate and limit it to the minimum necessary, reducing patient hospital stay and peri-operative morbidity.

*ABC of Emergency Radiology*, Third Edition. Edited by Otto Chan.
© 2013 John Wiley & Sons, Ltd. Published 2013 by John Wiley & Sons, Ltd.

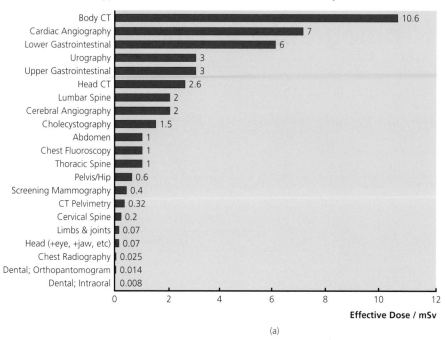

**Typical values of effective dose for various medical X-rays**

| Examination | Effective Dose / mSv |
|---|---|
| Body CT | 10.6 |
| Cardiac Angiography | 7 |
| Lower Gastrointestinal | 6 |
| Urography | 3 |
| Upper Gastrointestinal | 3 |
| Head CT | 2.6 |
| Lumbar Spine | 2 |
| Cerebral Angiography | 2 |
| Cholecystography | 1.5 |
| Abdomen | 1 |
| Chest Fluoroscopy | 1 |
| Thoracic Spine | 1 |
| Pelvis/Hip | 0.6 |
| Screening Mammography | 0.4 |
| CT Pelvimetry | 0.32 |
| Cervical Spine | 0.2 |
| Limbs & joints | 0.07 |
| Head (+eye, +jaw, etc) | 0.07 |
| Chest Radiography | 0.025 |
| Dental; Orthopantomogram | 0.014 |
| Dental; Intraoral | 0.008 |

(a)

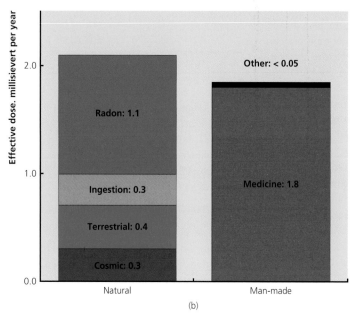

(b)

**Figure 14.1** (a) Graph illustrating the rapid increase in the number of CT scans per year in the USA. (b) Note the average number of CT scans per person per year is now more than 1 in 5.

US and CT complement each other as US has greater accuracy than CT for some conditions and involves no ionising radiation dose. This is the major limiting factor with CT, particularly with multiple phases or repeated examinations. The average dose of an adult abdominal and pelvic MDCT is 8.4 mSv; *equivalent to 3.4 years of UK background radiation or 400 chest radiographs.* Young patients are more sensitive to radiation-induced carcinogenesis, so the dose from CT should be considered when requesting imaging. Magnetic resonance imaging (MRI) also has no radiation dose but does not have an important role in the initial management of the acute abdomen. Your radiologist will advise on the most appropriate investigation given the relevant clinical information.

This chapter will present a systematic approach to the findings on MDCT of the most common causes of acute abdominal conditions.

## Relevant abdominal anatomy

Normal positions of the upper abdominal organs are shown in Figures 14.2 and 14.3.

## Causes of acute abdominal pain by quadrant

Causes of acute abdominal pain are shown in Figure 14.4 .

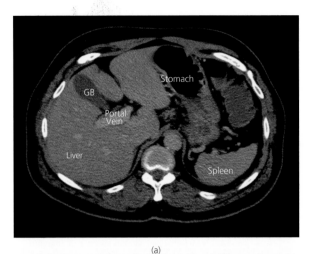

(a)

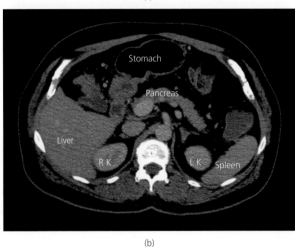

(b)

**Figure 14.2** (a) Level of the hepatic hilum; (b) level of the pancreas (© Commonwealth of Australia 2012, as represented by the Australian Radiation Protection and Nuclear Safety Agency (ARPANSA)).

## Abdominal CT protocols

The MDCT examination will vary according to the clinical question.

Most patients with suspected acute abdominal inflammation or infection require only a single phase intravenous iodinated

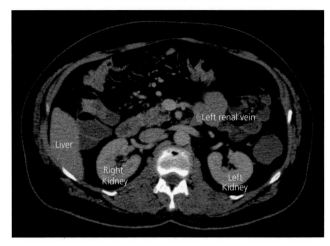

**Figure 14.3** Level of the renal hilum (Informationskreis KernEnergie, Berlin).

contrast-enhanced CT (CECT) from the diaphragm to the pubic symphysis. The portal venous phase (60–70 s delay after injection) provides optimal organ enhancement and is considered the standard protocol.

Certain indications will require non-enhanced (NECT), arterial phase imaging (30–40 s delay), or delayed imaging after several minutes, either alone or with a portal venous phase, and these will be discussed.

A CT KUB (kidneys, ureters and bladder) is a non-enhanced examination that is extremely sensitive and specific for the presence of urolithiasis (renal tract stones). The radiation dose of a CT KUB approaches that of a plain AXR, but it provides far more information. Although some extra-renal diagnoses can be made with NECT, a CT KUB should only be requested when the diagnosis of ureteric colic is suspected.

In most acute abdominal settings, there is no requirement or time to give oral contrast (drinking takes up to one hour). However, oral contrast is sometimes administered (either as water or diluted water-soluble iodinated contrast), in specific clinical scenarios.

If NECT is performed because of renal impairment or allergy, the images can be difficult to interpret; particularly in slender patients without intra-abdominal fat, as the organs and bowel are poorly

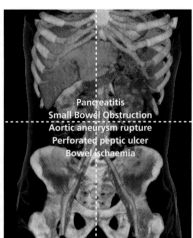

Cholecystitis
Liver abscess
Chest/cardiac pathology
Subphrenic collection

Gastric pathology
Chest/cardiac pathology
Subphrenic collection

Pancreatitis
Small Bowel Obstruction
Aortic aneurysm rupture
Perforated peptic ulcer
Bowel ischaemia

Appendicitis
Caecal Carcinoma
Diverticulitis
Crohn's disease
Urolithiasis
Pelvic inflammatory disease

Diverticulitis
Colonic carcinoma
Abscess Colitis
Large Bowel Obstruction
Urolithiasis
Appendagitis
Tuberculosis
Pelvic inflammatory disease

**Figure 14.4** Causes of acute abdominal pain by quadrant.

defined. In such cases we recommend the use of oral positive contrast or consider using US instead.

## Interpretation of abdominal CT

A systematic review is required because of the number of structures and pathologies that may be encountered.

An ABC approach is recommended (Table 14.2).

## CT features of conditions causing acute abdominal pain

### Signs of inflammation (non-specific)

- Normal fat looks 'clean' – dark grey/black on abdominal CT window
- 'Dirty' oedematous and inflamed fat is misty/streaky and lighter grey – a clue to local pathology
- Low density tissue oedema
- Mural thickening
- Inflammatory soft tissue density, enhancing mass (phlegmon, an abscess in the making)
- Free or localised fluid (Table 14.3; Figure 14.5)
- Poor or increased contrast enhancement
- Localised dilated loops of bowel (ileus)

**Table 14.2** ABCs approach to abdominal CT – $A^3B^3C^2DEFGHS^2$.

| | |
|---|---|
| A – All the organs | Each organ must be examined in turn |
| A – Air | Search for extra-luminal air (on lung CT window) |
| A – Appendicitis | If the appendix is not seen, it is normal or not present |
| B – Bowel | Trace bowel lumen carefully<br>Bowel dilatation, wall thickening and masses |
| B – Bleeding | Haematoma (previous haemorrhage)<br>Contrast extravasation (on-going haemorrhage) |
| B – Bases of lungs | Lung or pleural pathology can cause abdominal pain |
| C – Calcification | Urological tract (obstructing calculus)<br>Vascular (aneurysm and ischaemia) |
| C – Circulation | Vessel opacification and calibre |
| D – Dirty fat | Examine the organ-fat interface carefully, especially around appendix, colon, gallbladder and pancreas |
| E – Enhancement | Increased vascularity and enhancement with inflammation<br>Decreased perfusion and enhancement with tissue oedema, ischaemia or necrosis |
| F – Fluid | Free fluid and collections<br>Location may point to the site of pathology<br>Infected or not?<br>Take care to distinguish collections from bowel |
| G – Gynaecological | Pelvic inflammatory disease, ovarian torsion and ectopic pregnancy |
| H – Hernias | Examine hernial orifices and abdominal wall |
| S – Skeleton | Examine skeleton (on bone CT window, and in sagittal/coronal planes) |
| S – Summary | Check PACS for previous imaging and reports; often helpful information is available |

**Table 14.3** Fluid collections – CT characteristics.

| | |
|---|---|
| **Simple** | Homogeneous fluid of lower density than soft tissue<br>No wall<br>No surrounding fat stranding |
| **Infected** | Usually homogeneous or mixed low to medium density fluid<br>Thin enhancing wall<br>Surrounding fat stranding |
| **Abscess** | Heterogeneous variable density fluid<br>Less dense centrally, sometimes with gas locules<br>Thick irregular, enhancing wall<br>Marked surrounding fat stranding<br>Fistula tracts |
| **Haematoma** | Initially higher density than soft tissues on NECT<br>Over time, less dense from the periphery inwards<br>May have enhancing wall in absence of infection<br>Variable fat stranding depending on amount of blood, age and presence of infection |
| **Faeculant** | Heterogeneous solid material with mixed fat, fluid or soft tissue density with tiny specks of gas<br>Usually with associated signs of infection |

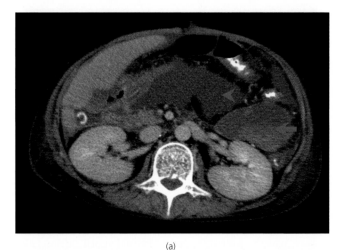

(a)

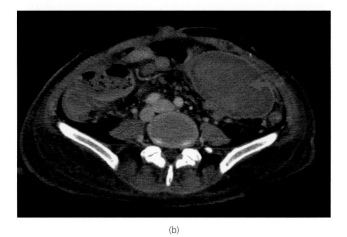

(b)

**Figure 14.5** (a) Central fluid collection (arrowhead) and (b) adjacent haematoma with evidence of layering (arrow).

## Signs of gastrointestinal tract perforation

Free intraperitoneal or retroperitoneal gas implies perforation (most commonly from a peptic ulcer). Pneumoperitoneum can also be caused by diverticulitis and can be present after a recent laparotomy. MDCT has a very high sensitivity and specificity for even small amounts (<2 ml) of free gas (Figures 14.6 and 14.7).

Extraluminal gas may also form in a walled off collection anywhere in the abdominal cavity, forming a gas-fluid level (as in the diverticular abscess described later) or as small bubbles of gas trapped in solid material or sited in anti-dependent locations when the patient is supine for the CT examination (as in the example of necrotising pancreatitis seen later).

## Tips

- Carefully look under the anterior abdominal wall and around the liver for tiny locules of gas that may be the only sign of a perforation.
- Lung windows should always be used to search for free gas.

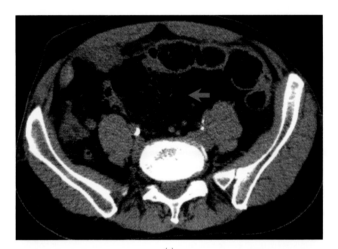

(a)

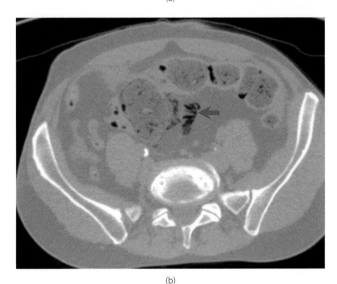

(b)

**Figure 14.6** Pneumoperitoneum (arrow). On soft tissue windows (a) this localised sigmoid colon perforation may be missed, whereas on lung windows (b) it is clearly seen to be extraluminal.

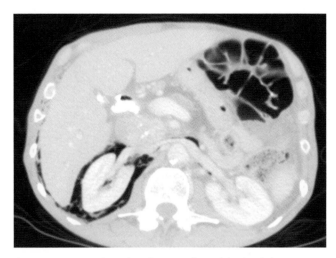

**Figure 14.7** Retroperitoneal gas from a perforated duodenal ulcer surrounds the right kidney and left renal vein.

## Acute appendicitis

### Role of CT

Ultrasound is the investigation of choice but CECT has a role if US is non-diagnostic or inconclusive; if the history is atypical (in 25–33%), or if complications or other pathologies are suspected such as caecal carcinoma in elderly patients. CT has increased the positive laparotomy rate, and reduced the surgical misdiagnosis rate to 5–10%.

### CT features

The normal appendix appears as a tubular serpiginous structure arising from the caecal pole. It is thin walled, <6 mm in diameter, and fluid or gas filled. Box 14.1 shows the signs of acute appendicitis.

Box 14.1 **Signs of acute appendicitis**

- Circumferential symmetric mural thickening
- Appendiceal enlargement
- Dilatation >7 mm with fluid
- Homogenous enhancement
- Peri-appendiceal fat haziness and stranding, or fluid (Figure 14.8)
- Calcified appendicolith (Figure 14.9)
- Mild caecal or terminal ileal thickening
- Right iliac fossa lymph node enlargement

### Complications

- Perforation
- Appendiceal mass (phlegmon)
- Abscess formation

### Consider the differential diagnosis

It is important to consider diagnoses for which the correct surgical approach is not appendicectomy or those which are managed conservatively.

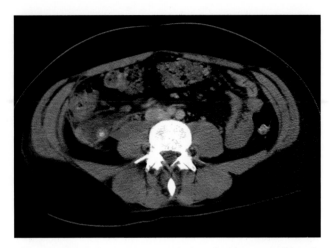

**Figure 14.8** Enlarged enhancing appendix (arrow) with appendicolith and peri-appendiceal fat stranding in acute appendicitis.

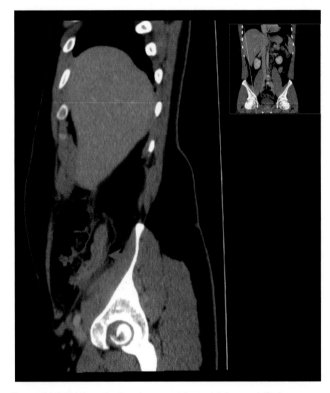

**Figure 14.9** Thickened enhancing appendix containing a calcified appendicolith (arrowhead). Sagittal reconstruction shows the retrocaecal location of the appendix (arrow).

- Mesenteric adenitis; enlarged mesenteric nodes
- Caecal carcinoma or right-sided diverticulitis
- Crohn's terminal ileitis/abscess

## Tips

- Coronal or sagittal reformats can aid identification of the appendix
- Important features which increase the difficulty of appendicectomy: a retrocaecal location, presence of perforation with fragments or appendiceal mass

## Diverticulitis

### Role of CT

Left lower quadrant pain in a patient over the age of 40 years should be investigated with CECT. The most likely diagnosis of colonic diverticulitis can be confirmed, complications discovered, and other pathology such as unsuspected colon carcinoma can be diagnosed.

### CT features

Uncomplicated diverticulosis is frequently an incidental finding; the prevalence increases with patient age and involves the sigmoid colon in 95%. It manifests as a diffuse symmetrical colonic mural thickening (>4mm), and multiple saccular out-pouchings. The peri-colonic fat should have normal density.

Diverticulitis occurs in 10–25% of patients with diverticulosis, resulting from local perforation of a diverticulum into the peri-colonic fat. The perforation is usually contained, and free pneumoperitoneum is uncommon. Box 14.2 shows the signs of diverticulitis.

> Box 14.2 **Signs of diverticulitis**
>
> - Presence of diverticulosis – suspect other pathology if this is absent elsewhere
> - Peri-colonic fat inflammation and increased vascularity in the mesocolon
> - Symmetrical mural thickening (usually <1 cm) over a long segment
> - Phlegmon formation
> - Free fluid

### Complications

- Abscess formation (Figures 14.10–14.12)
- Large or small bowel obstruction
- Intraperitoneal perforation
- Fistula formation with adjacent organs
- Portal vein gas, liver abscess
- Diverticular haemorrhage

### Consider the differential diagnosis

- The most important diagnosis to exclude is colonic carcinoma (eccentric wall thickening >1 cm over a short segment, enlarged lymph nodes, liver metastases)
- Ischaemic or infective colitis
- Appendagitis (inflammation of epiploic fat)

### Tips

- Look for other causes of acute abdomen with coexisting diverticulosis; fat inflammation is the clue
- Initial treatment is conservative, with percutaneous drainage of collections or bowel resection for severe complications

## Pancreatitis

### Role of CT

If the clinical diagnosis of acute pancreatitis is obvious, then there is no need for CT within the first 48 hours. Non-improving patients

If the diagnosis is not clear clinically, early CT can exclude other causes such as perforated peptic ulcer, but normal pancreatic appearances do not exclude pancreatitis. In this situation, a single portal venous abdominal CT is sufficient.

However, in severe pancreatitis, unenhanced, arterial and portal venous phases are required for assessment of haemorrhage, necrosis and pseudoaneurysm formation.

## CT features

The normal pancreas is a slender retroperitoneal organ with a smooth or slightly lobulated contour. The pancreatic duct can normally be seen as a thin fluid-filled tube, 2–3 mm in diameter. Box 14.3 shows the signs of mild and severe pancreatitis (Figures 14.13–14.16).

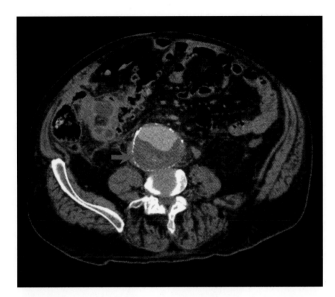

**Figure 14.10** Localised diverticular perforation with peri-colic abscess (arrowhead) and incidental non-ruptured aortic aneurysm (arrow).

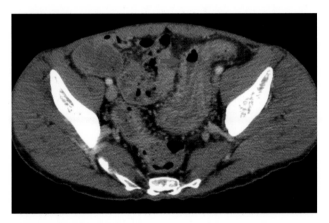

**Figure 14.11** Intramural diverticular abscess in the sigmoid colon.

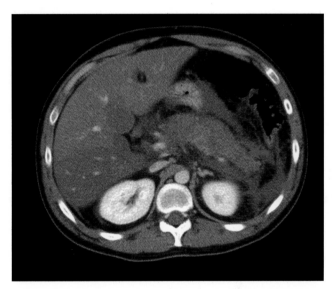

**Figure 14.13** In mild pancreatitis the pancreas is swollen, with an indistinct outline (arrow) and with inflammation of the surrounding fat (arrowhead).

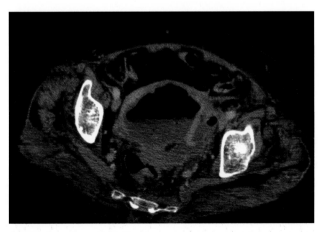

**Figure 14.12** Large sigmoid diverticular abscess with gas-fluid level.

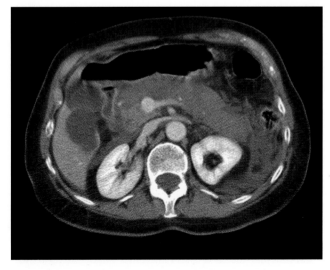

**Figure 14.14** In severe pancreatitis there is globally poor pancreatic enhancement (arrow) when compared to the liver parenchyma.

benefit from CT to stage the disease severity and prognosis, identify complications and to guide intervention.

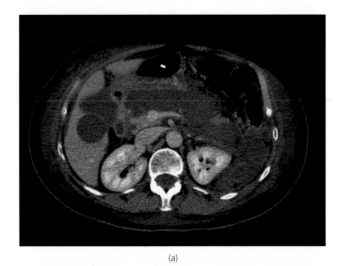

(a)

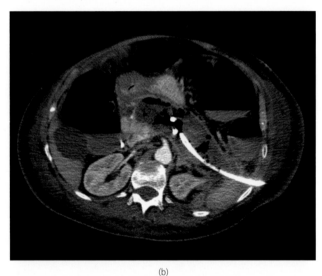

(b)

**Figure 14.15** Necrotising pancreatitis . Same patient as Figure 14. (a) One week later the pancreas is necrotic (arrowhead). It no longer enhances and has the same density as fluid in a liver cyst (b) One month later, after insertion of a percutaneous drain the pancreas is replaced by necrotic debris containing gas locules and an enhancing abscess wall. The splenic vein is also thrombosed.

## Complications

- **Abscess** – an infected collection of fluid without necrosis, usually >3 weeks after initial attack
- **Infected pancreatic necrosis** – partially or totally liquefied tissue with high mortality rate
- **Pseudoaneurysms** – usually splenic, gastroduodenal or pancreaticoduodenal arteries. **Life threatening haemorrhage** can occur, and therefore it is imperative to detect these so they can be treated by embolisation. Portal venous imaging alone may miss pseudoaneurysms, hence the need for an arterial phase
- **Pseudocyst** – localised peri-pancreatic fluid collection with a thick fibrous wall, occurring 4–6 weeks after an acute attack

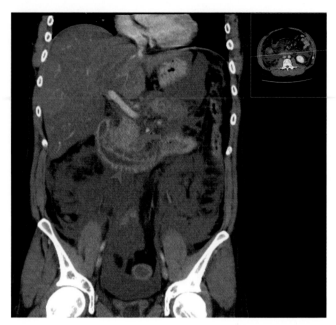

**Figure 14.16** Retroperitoneal fat stranding in acute pancreatitis. Coronal section through the retroperitoneum shows inflamed pancreatic head (arrow), with hyperenhancement of the adjacent duodenum (arrowhead), and widespread inflammation across the fascial planes of the retroperitoneum (arrowhead) with pelvic ascites.

Box 14.3 **Signs of pancreatitis**

**Signs of mild pancreatitis**

- Normal pancreas (28%)
- Oedematous swollen/enlarged gland
- Peri-pancreatic fat stranding and thickened fascial planes
- Usually homogenous enhancement
- Cause may be identified – gallstones (US more accurate)
- Parenchymal calcification – previous or chronic pancreatitis

**Signs of severe pancreatitis**

- Enlarged gland
- Haemorrhage: high density on NECT
- Necrosis: poorly enhancing parenchyma
- Obliterated peri-pancreatic fat
- Inflammation of fat penetrates across fascial and peritoneal boundaries
- Peri-pancreatic collections and free fluid
- Splenic vein thrombosis

## Cholecystitis

### Role of CT

US is the investigation of choice as the detection of gallstones is more sensitive. CT has a limited role, for example in suspected gallbladder perforation.

### CT features

The gall bladder is dilated, with an oedematous thickened wall (>3 mm), although this is non-specific, and there is surrounding fat surrounding or fluid (Figure 14.17).

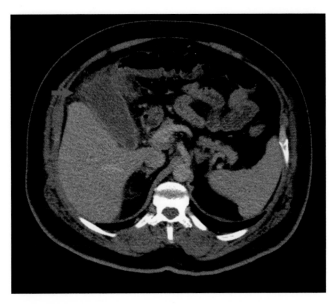

**Figure 14.17** Acute cholecystitis. Oedematous inflamed gall bladder wall with non-calcified gall stone.

Calcified gallstones may be seen on CT, but often stones appear isodense to bile, and sometimes contain gas (Figure 14.18).

### Tips

- Check carefully for biliary dilatation, as an obstructing stone in the common bile duct (CBD) warrants urgent ERCP.
- Coronal reformats can help show the length of the CBD.
- Gall bladder perforation may produce a liver abscess, but carcinoma can present similarly.

## Colitis/enteritis

### Role of CT

Infection is the commonest cause of enterocolitis and usually does not require imaging. CT is helpful in acute exacerbations or the initial presentation of inflammatory bowel disease (IBD).

### CT features

The gut responds to insult with non-specific mural thickening and surrounding fat stranding. Mucosal and serosal enhancement with intervening oedematous submucosa gives a target sign appearance. This may be focal (single or multiple 'skip' lesions) or diffuse, and involve small and/or large bowel (Table 14.4; Figures 14.19 and 14.20).

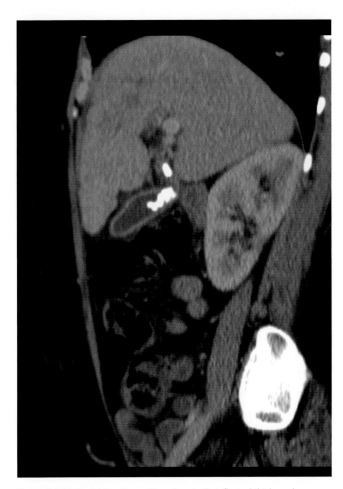

**Figure 14.18** Sagittal section with a chronically inflamed thickened enhancing gall bladder wall with calcified stones in the gall bladder and cystic duct.

**Table 14.4** Distinguishing CT features of bowel pathology.

| | |
|---|---|
| **All conditions** | Gut wall thickening |
| | Fat stranding |
| **Crohn's disease** | Terminal ileum (or anywhere along GI tract) |
| | Focal/skip lesions |
| | Acute/active disease: stratified wall enhancement (target sign) |
| | Chronic/inactive disease: homogenous wall enhancement |
| | Mesenteric fat proliferation and hypervascularity |
| | Lymph node enlargement |
| | Abscess or fistulae |
| **Ulcerative colitis** | Extends proximally from rectum |
| | Diffuse symmetrical thickening |
| | Target sign |
| | Lack of haustra, toxic megacolon (inflamed dilated colon) |
| | Ascites unusual |
| **Pseudomembranous colitis (*C. difficile* infection)** | Predisposing antibiotic or chemotherapy use |
| | Rectum/sigmoid/pan-colitis |
| | Irregular thickening with marked oedema |
| | Limited fat stranding |
| | Ascites |
| **Ischaemia** | Predisposing hypotension, vasculitis, obstruction, or embolus |
| | Arterial/venous origin |
| | Circumferential thickening ≫ fat stranding |
| | High density haemorrhage |
| | Poor enhancement or target sign |
| | Gas in bowel wall (pneumatosis) or portal vein |

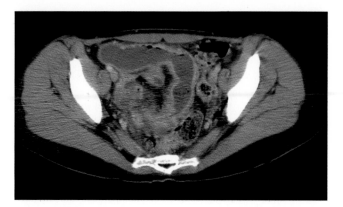

**Figure 14.19** Terminal ileitis in Crohn's disease. A long segment of ileum is dilated, with a thickened enhancing wall and surrounding mesenteric fat inflammation.

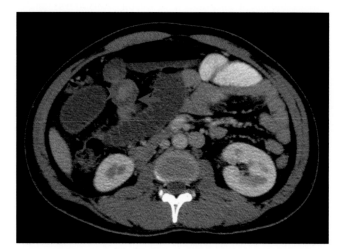

**Figure 14.20** Target sign in Crohn's disease. The bowel wall is thickened with three layers visible – inner mucosal (arrow) and outer serosal (arrowhead) enhancement with intervening low density oedema of the submucosa. Proximal small bowel is dilated.

## Tips

- If fat stranding is much greater than bowel wall thickening – consider external pathology: epiploic appendagitis or omental infarction (both treated conservatively).
- If bowel wall thickening is much greater than fat stranding – infection or ischaemia are more likely (Figures 14.21 and 14.22).

## Tuberculosis (TB)

TB is an important differential diagnosis for inflammation within the abdomen. The findings can be non-specific but in the right clinical context there are important signs on CT which can aid in its diagnosis.

The most common finding in the abdomen is lymph node enlargement, predominantly involving the lesser omental, mesenteric, and upper para-aortic lymph nodes although any group can be involved. They can demonstrate a variety of patterns of enhancement but the most characteristic is central low density (Table 14.5; Figure 14.23).

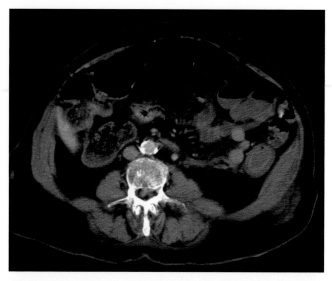

**Figure 14.21** Target sign in infective colitis with little fat stranding.

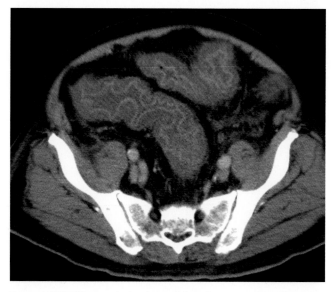

**Figure 14.22** Diffuse marked oedema of the colon in pseudomembranous colitis.

## Small bowel and large bowel obstruction (SBO/LBO)

### Role of CT

MDCT can often localise the level and cause of obstruction, differentiate between high and low grades of SBO, and assess the presence of complications. In patients with paralytic ileus, pseudo-obstruction can be seen (dilated large bowel with no obstructing cause).

### CT features

A mechanical blockage of the bowel causes dilatation of proximal loops. The key to diagnosis is the identification of a transition point from dilated to collapsed bowel. This is often best visualised by starting at the rectum and working backwards. Signs of small bowel obstruction are shown in Box 14.4 and Figures 14.24 and 14.25.

**Table 14.5** Signs of abdominal tuberculosis.

**Gastrointestinal tract**

| | |
|---|---|
| Bowel | Bowel wall thickening (usually terminal ileum/caecum) Often adjacent mesenteric lymphadenopathy |
| Lymph Nodes | Enlarged enhancing lymph nodes Often with central low attenuation |
| Peritoneum | Often present with extensive disease Wet type: viscous ascites (free or loculated) Fibrotic fixed type: omental masses, adherent loops of bowel/mesentery, and loculated ascites Dry type: caseous nodules, fibrous peritoneal reaction, and dense adhesions |
| Liver and Spleen | Miliary TB: tiny low density foci scattered in liver/spleen Macronodular form is rare: diffuse liver and splenic enlargement with multiple low attenuating lesions or a large tumour like mass |

**Genitourinary tract**

| | |
|---|---|
| Renal | Parenchymal calcifications Hydronephrosis: focal or generalised |
| Adrenal | Unilateral or bilateral adrenal mass with areas of low attenuation centrally |
| Ureters | Wall thickening and peri-ureteral inflammatory changes |
| Bladder | Shrunken thick walled bladder |

**Spine**

| | |
|---|---|
| Lumbar Spine | Affects the anterior part of the vertebral body adjacent to superior and inferior end plates Involves disc space leading to collapse of intervertebral disc space Paravertebral abscess Psoas abscess |

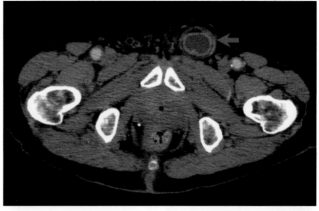

(a)

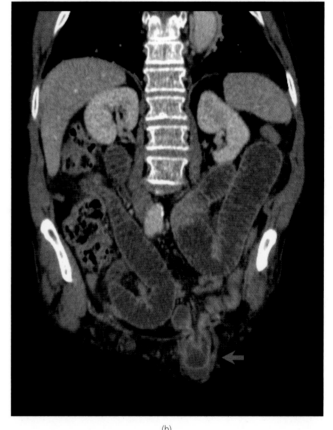

(b)

**Figure 14.24** Small bowel obstruction from an incarcerated left inguinal hernia (a) axial and (b) coronal.

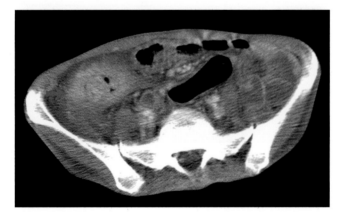

**Figure 14.23** Large inflammatory ileo-caecal mass and ascites in tuberculosis.

---

Box 14.4 **Signs of small bowel obstruction**

- Loops of small bowel >2.5 cm on MDCT; gas or fluid filled
- Transition point with distal collapsed small bowel loops
- Solid small bowel faeces proximal to obstruction
- Large bowel usually collapsed

---

Box 14.5 **Signs of large bowel obstruction**

- Dilated loops of large bowel; gas, fluid, or stool filled
- Distal colon collapsed to the obstruction
- If transverse colon is >5.5 cm and or the caecum is >9 cm there is a significant chance of perforation
- Look for underlying cause and complications

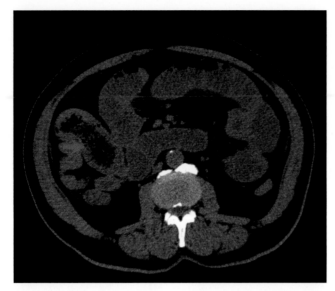

**Figure 14.25** The presence of solid faeces in the small bowel usually occurs just proximal to the obstruction point.

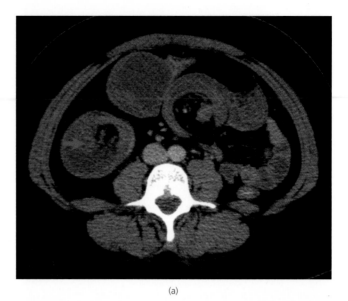

(a)

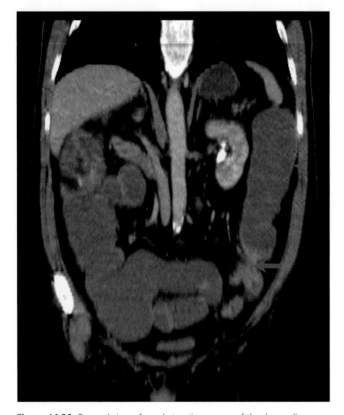

**Figure 14.26** Coronal view of an obstructing cancer of the descending colon (arrow), with fluid filled dilated colon proximally (arrowhead).

Signs of large bowel obstruction are shown in Box 14.5 and Figure 14.26.

## Consider the underlying cause

Table 14.6 gives the causes of bowel obstruction.

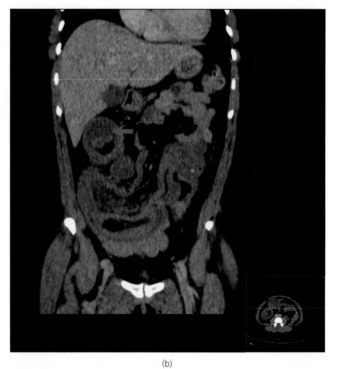

(b)

**Figure 14.27** A large ileo-colic intussusception. The fat of the small bowel mesentery is seen within the lumen of the colon as the ileum (the *intussusceptum*) (arrow) invaginates into the caecum (the *intussuscipiens*) (a) axial (b) coronal.

**Table 14.6** Causes of bowel obstruction.

| | |
|---|---|
| **Small bowel** | Commonest (80%): adhesions, hernias, and neoplasia (including non-gut tumours) |
| | Less common causes include volvulus, intussusceptions (Figure 14.27) and gallstone ileus |
| | Appendix abscess |
| **Large bowel** | Commonest: colorectal carcinoma |
| | Other causes: faecal impaction, volvulus |
| **Either** | Inflammation: Crohn's disease and diverticulitis |

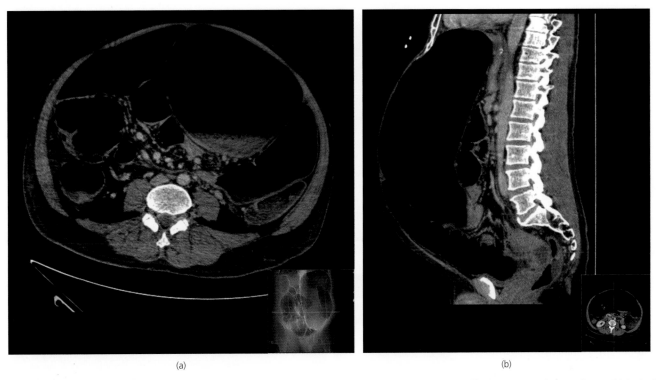

(a)  (b)

**Figure 14.28** Hugely distended loops of colon in sigmoid volvulus. (a) Axial view shows the typical radiographic 'coffee bean' sign in the scout image. (b) Sagittal view shows the extent of distension to the diaphragm.

- Colonic carcinoma appears as an enhancing mural mass that may cause obstruction. Search for secondary features of malignancy; lymphadenopathy and metastases (liver/lung/bone).
- Volvulus is a twisting of part of the gut on its mesenteric axis, resulting in a closed loop obstruction and can occur with stomach, small bowel, caecum and sigmoid colon. Colonic volvulus has characteristic AXR appearances.
- Sigmoid volvulus is commonest (50–75% of all colonic volvulus) presenting with a very large dilated loop arising from the left lower quadrant, and dilated proximal colon and small bowel (Figure 14.28). It is more commonly seen in institutionalised patients.
- Caecal volvulus presents with a dilated caecum in an ectopic location usually in the left upper quadrant. It is associated with markedly distended small bowel loops and a collapsed distal colon.

### Complications of bowel obstruction

- Ischaemia (poor enhancement, and adjacent free fluid)
- Infarction/necrosis (no enhancement of bowel wall, intramural gas)
- Gas in the mesenteric and/or portal veins (usually a pre-morbid sign; Figure 14.29)
- Perforation (pneumoperitoneum, pneumoretroperitoneum from colon)

### Tips

- SBO with a transition point but no visible cause is usually secondary to adhesions
- Gas in intrahepatic portal veins is distributed more peripherally than the common finding of gas in the biliary tree (aerobilia)

## Urolithiasis (renal tract calcification)

The formation of renal tract calculi is multifactorial in origin and heterogeneous in demographics. It commonly manifests as acute colicky flank pain radiating to the groin. Urolithiasis is divided into upper tract (calyceal, renal pelvis and pelvi-ureteric junction), ureteric and lower tract (bladder, urethral, prostatic, preputial).

### Role of CT

CT can assess the presence, number, location, and complications of calculi, and exclude other causes of abdominal pain, especially in the absence of any obstructing calculi.

### CT features

When assessing for urolithiasis, a non-enhanced examination is required (CT KUB) (Figure 14.30), as contrast may obscure the stone. Calculi are usually homogenous with calcium density on CT (check using bony windows) and most stones can be identified using NECT with the exception of indinavir (protease

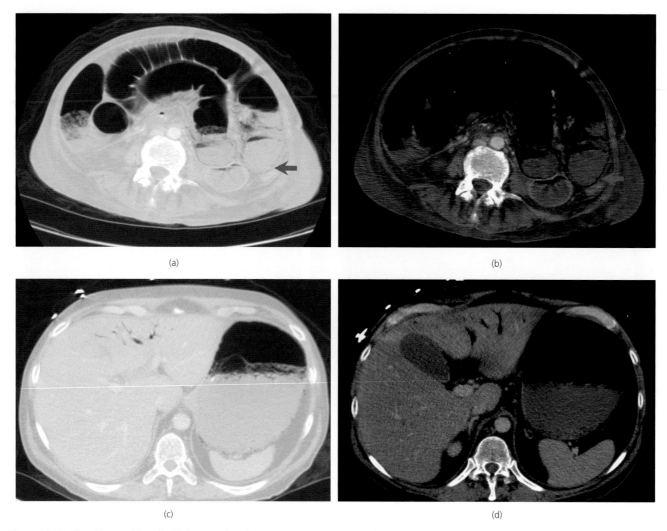

(a)        (b)

(c)        (d)

**Figure 14.29** Dilated loops of bowel with intramural and mesenteric venous gas in small bowel infarction on (a) lung and (b) soft tissue windows. Branching gas pattern in the intra hepatic portal vein on (c) lung and (d) soft tissue windows.

inhibitor used in HIV) stones which have low density. 93% of ureteric calculi <6 mm will pass spontaneously (Box 14.6).

> Box 14.6 **Secondary signs of obstructing renal tract calculus**
>
> - Perinephric/periureteric fat stranding
> - Dilated pelvicalyceal system (hydronephrosis)
> - Dilated ureter above calculus (hydroureter)
> - Ureteric rim sign – ureteric oedema around an impacted calculus
> - Pseudoureterocele: ureterovesical oedema around a calculus
> - If there is the possibility of a low density stone (indinavir) then the excretory phase of a MDCT urogram may localise the site of obstruction and stone

### Tips

Distinguishing between venous phleboliths and distal ureteric stones can be difficult; phleboliths usually are ring shaped rather than solid and lie lateral to the ureters (Figure 14.31).

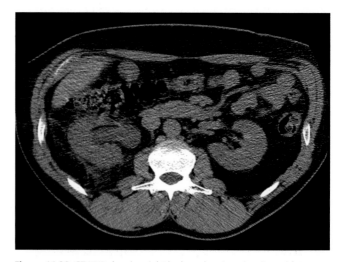

**Figure 14.30** CT KUB showing right hydonephrosis and peri-renal fat stranding from a distal ureteric calculus (not shown).

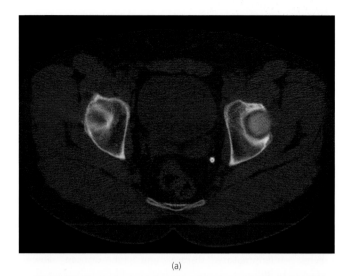

(a)

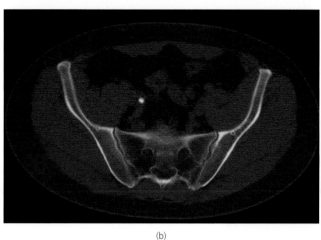

(b)

**Figure 14.31** (a) Phlebolith with ring like calcification, usually low in the pelvis. (b) Right ureteric calculus with solid calcification.

# Abdominal aortic aneurysm (AAA) rupture

## Role of CT

This is a life-threatening emergency that should be excluded in patients over 75 years with acute severe abdominal or back pain. The rupture rate with AAA diameter <4.9 cm is 1% per year, but 25% per year if diameter >6.0 cm. Patients that survive to hospital usually have a contained retroperitoneal rupture, and are usually

**Table 14.7** Signs of abdominal aortic aneurysm.

| | |
|---|---|
| **Aneurysm** | Aortic diameter >3 cm<br>Non-enhancing thrombus<br>Calcified wall |
| **Impending rupture** | Extravasation of contrast into thrombus or wall<br>Interruption of circumferential intimal calcification |
| **Contained rupture** | Bulging or indistinct aortic margin |
| **Frank rupture** | Large retroperitoneal haematoma<br>Extraluminal contrast |

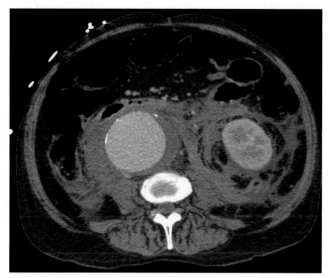

**Figure 14.32** Large abdominal aortic aneurysm rupture with extensive retroperitoneal haemorrhage (arrow), but no extravasation of contrast.

stable enough for emergency CT (Figure 14.32). An aneurysm or rupture can be confirmed with NECT (Table 14.7), but an arterial phase is paramount in determining management: either open or endovascular repair.

## Tips

AAA are often tortuous; measurements should be made perpendicular to the longitudinal axis, which may not be in the axial plane.

# Gastrointestinal haemorrhage

## Role of CT

If endoscopic measures are unsuccessful in identifying the source of upper or lower gastrointestinal tract bleeding, CT may localise the site prior to embolisation or surgery. A triple phase protocol is required (NECT, arterial and portal venous). Oral contrast should not be used as it will obscure luminal contrast extravasation.

## CT features

NECT demonstrates high density sentinel clot in the vicinity of the bleeding point, and confirms that any high density blush present on CECT truly represents intravenous contrast. If bleeding is not on-going at the time of imaging, this may be the only sign (Figure 14.33).

On-going haemorrhage is typified by a small jet of contrast in the arterial phase, (as dense as blood in the aorta) which dissipates and becomes less dense in the venous phase, however only the venous blush may be seen (Figure 14.34).

## Tips

- Advanced workstation techniques such as comparing the three phases together slice by slice, MIP (maximal intensity projection)

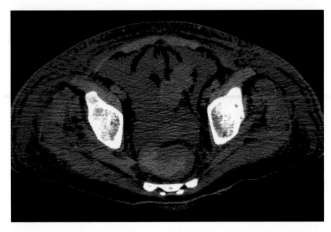

**Figure 14.33** Sentinel high density clot. On this unenhanced image there is high density blood within the rectum; the 'sentinel clot'. Blood may travel a long way proximally or distally in the bowel lumen.

slabs and multi-planar reformats aid localisation of the bleeding source.

• Intraluminal blood may pass proximally or distally away from the site of bleeding

## Other causes of acute abdominal pain

There are many causes of acute abdominal pain, and it is beyond the remit of this chapter to describe them all. Those that require urgent intervention have been discussed, but others which require only conservative treatment include:

• Mesenteric adenitis (Figure 14.35)
• Epiploic appendagitis (Figure 14.36)
• Omental infarction (Figure 14.37)
• Rectus sheath haematoma (Figure 14.38)

In women, consider gynaecological causes such as:

• Pelvic inflammatory disease and tubo-ovarian abscess (Figure 14.39)
• Torsion or rupture of ovarian cyst
• Ectopic pregnancy

Ultrasound is usually the best imaging modality to start with when assessing gynaecological disorders.

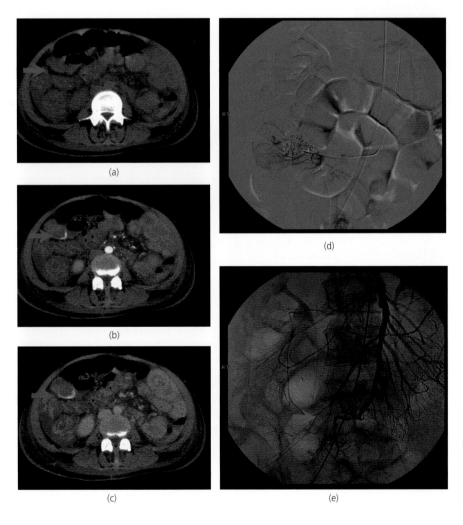

(a)

(b)

(c)

(d)

(e)

**Figure 14.34** Small bowel haemorrhage. Pre (a), arterial (b) and venous (c) phase CT showing arterial extravasation of contrast in the ileum, which dissipates. Selective mesenteric angiogram (d) locates the bleeding vessel and (e) a superior mesenteric arteriogram post embolisation with coils shows cessation of haemorrhage.

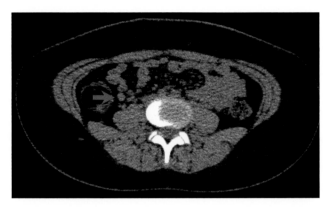

**Figure 14.35** Enlarged right iliac fossa lymph nodes in mesenteric adenitis.

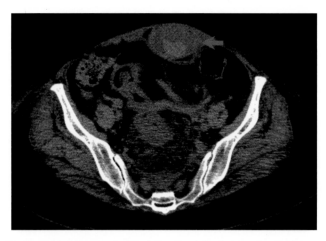

**Figure 14.38** Enlargement of the left rectus abdominal muscle with high density haematoma (arrow).

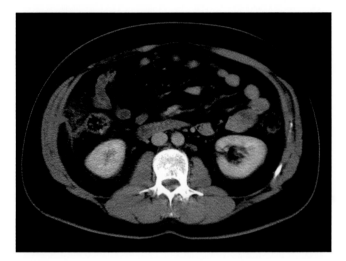

**Figure 14.36** Peri-colic inflammation in epiploic appendagitis.

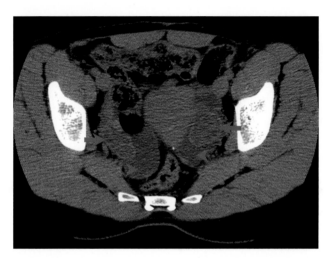

**Figure 14.39** Severe pelvic inflammatory disease with bilateral dilated fallopian tubes (pyosalpinx) (arrows) and tubo-ovarian abscesses (arrowheads).

## Further Readings

Federle M, Jeffrey RB, Woodward PJ, Borhani A. *Diagnostic Imaging: Abdomen*, 2nd edn. Amirsys 2010. ISBN 978-1-1931884-71-6.

Kellow ZS, MacInnes M, Kurzencwyg D, et al. The role of abdominal radiography in the evaluation of the non-trauma emergency patient. *Radiology* 2008;248(3):887–893.

Lameris W, van Randen A, van Es H, van Heesewijk JP, van Ramshorst B, Bouma WH, et al. Imaging strategies for detection of urgent conditions in patients with acute abdominal pain: diagnostic accuracy study. *BMJ* 2009;26:338. doi: 10.1136/bmj.b2431.

Marincek B, Dondelinger RF (Eds). *Emergency Radiology – Imaging and Intervention*, Springer, 2007. ISBN 978-3-540-26227-5.

Stoker J, van Randen A, Laméris W, Boermeester MA. Imaging patients with acute abdominal pain. *Radiology* 2009;253(1):31–46.

The Royal College of Radiologists. iRefer Guidelines: Making the best use of clinical radiology – Version 7.0.1. Available online at www.irefer.org.uk

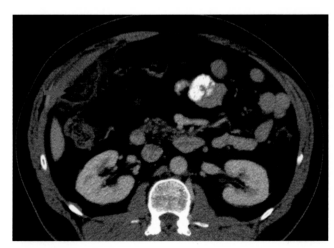

**Figure 14.37** Omental infarction.

# CHAPTER 15

# Emergency Ultrasound

*Tim Fotheringham[1], Otto Chan[2] and Ian Renfrew[1]*

[1]Barts Health NHS Trust, The Royal London Hospital, London, UK
[2]The London Independent Hospital, London, UK

---

### OVERVIEW

- The only imaging modality that is portable and real-time
- Can be used for diagnosis and treatment/intervention
- Quick, easy and safe – no radiation, so therefore safe for children and pregnancy
- MUST BE ADEQUATELY TRAINED BOTH FOR DIAGNOSIS AND INTERVENTION
  - NB Clinicians must therefore be aware of their limitations

---

Ultrasound (US) is the only imaging modality that is portable and real-time (Figure 15.1). US is now readily available to all clinicians and used widely in all clinical settings. The absence of ionising radiation makes it safe for all patient groups, in particular in children and pregnancy. US offers a quick and easy evaluation to diagnosis and to assist interventions.

Due to its portability, US is the only imaging modality that can be used at the emergency scene and – in addition to plain films – can be used in the emergency room (ER) without needing to move the patient.

US can be used in virtually any clinical setting, BUT operators need to be suitably trained and in particular need to understand their limitations and the limitations of US.

Short, focused training is sufficient for a novice to do a focused assessment with sonography for trauma (FAST), further directed training is necessary to detect aortic abdominal aneurysms (AAA), but specialist training is necessary for its general use in the abdomen, chest and in particular for intervention.

All patient groups presenting to ER can benefit from ultrasound, from the diagnosis of a deep vein thrombosis (DVT) to checking viability of a foetus, to aiding intervention in major trauma (insertion of arterial lines, central lines or chest drains).

Knowledge of the basic principles of US, artefacts and 'knobology' (where the buttons are and what they do!) is invaluable, but beyond the scope of this chapter (Box 15.1).

---

### Box 15.1 **Essential knobs**

- Selection of probes
- Selection of setting (abdomen, vascular, soft tissues . . .)
- Depth
- Focal zone
- Freeze/cine review
- Imaging – print or video

---

**Probes**

- Low frequency – abdomen, pelvis and chest
- Medium/high frequency – children, vascular, MSK, neck and scrotum

---

## ABCs systematic assessment – emergency uses

- A – appendicitis, AAA, ascites
- B – bladder
- C – calculi (gallbladder and kidneys), cardiac assessment (asystole), cysts
- D – DVT
- E – effusions/fluid (pleural, pericardial, peritoneal collections),
- F – FAST, foreign body (FB), foetus
- G – gallbladder
- S – soft tissues, skeleton

### Appendicits

The classical appearances of appendicitis (Figure 15.2) are a blind ending, dilated thick-walled structure/viscus with an appendicolith. Ancillary features may suggest the diagnosis – aperistaltic small bowel loops, localised fluid and a positive provocation test (pain on direct palpation with the probe). Unfortunately, the normal appendix is rarely visualised in adults and even in children and therefore US cannot reliably exclude the diagnosis.

---

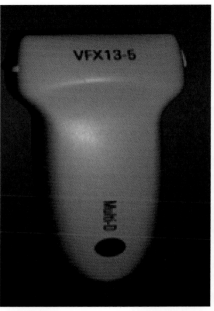
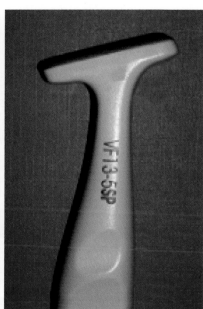

**Figure 15.1** Ultrasound machine and probes.

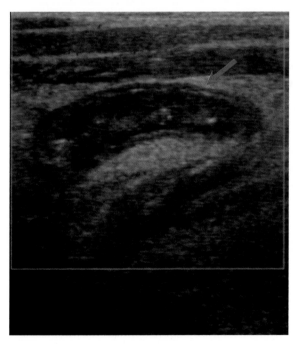

**Figure 15.2** Appendicitis. Dilated blind-ending non-compressible thick-walled structure.

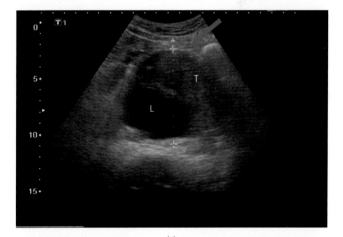

(a)

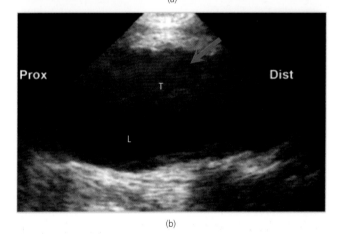

(b)

**Figure 15.3** (a) TS and (b) LS ultrasound of abdominal aortic aneurysm. L, lumen; T, thrombus.

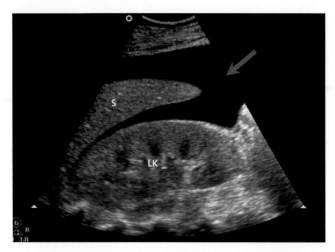

**Figure 15.4** Left upper quadrant: ascites with normal spleen (S) and left kidney (LK).

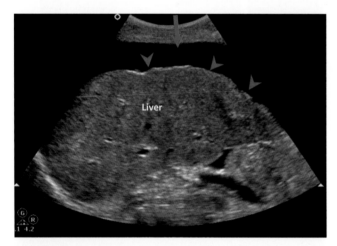

**Figure 15.5** Right upper quadrant: ascites with a nodular cirrhotic liver (arrowheads).

## Abdominal aortic aneurysm

US reliably detects AAA (Figure 15.3), but cannot be reliably used to exclude a leak. Therefore in the correct clinical setting, CT should be performed if the patient is haemodynamically stable.

## Ascites

US can detect even small volumes of intraperitoneal fluid, but is poor at determining the nature of the fluid (blood, pus, fluid) (Figures 15.4 and 15.5).

## Bladder

In anuric patients, it can be used to confirm bladder outflow obstruction or renal failure.

## Calculi

The presence of a mobile echogenic intraluminal structure with posterior acoustic shadowing is pathognomonic of a gallstone (Figure 15.6).

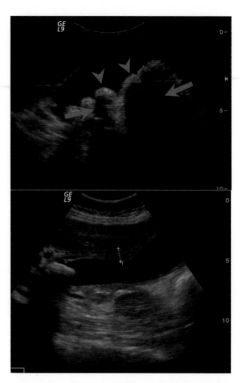

**Figure 15.6** Gallstones: echogenic well-defined lesions (arrowheads) with posterior acoustic shadowing (arrows).

## Renal calculi and hydronephrosis

The renal calculi can be parenchymal or pelvicalyceal or ureteric; the latter two lead to renal colic and may cause obstruction and therefore a hydronephrosis (Figure 15.7).

## Cardiac assessment

In a cardiac arrest situation, US is invaluable to confirm cardiac motion or asystole or a large pericardial effusion.

## Cysts – ovarian

Low abdominal pain can be caused by physiological or pathological ovarian cysts (Figure 15.8). Pathological cysts can bleed, undergo torsion or obstruct.

## Deep vein thrombosis

All veins should be anechoeic, fully compressible, and show variance with respiration and calf compression that can be seen and detected with Doppler. Clots can be seen as echogenic intraluminal material, which may be obstructive or non-obstructive and not compressible, typically causing distention of the vein (Figure 15.9).

## Effusions

Pleural and peritoneal – reliable identification of the diaphragm is essential to determine whether its pleural or inraperitoneal. A pleural effusion or ascites is usually seen as anechoeic fluid, with posterior acoustic enhancement but occasionally may contain echogenic debris (Figures 15.10 and 15.11).

Pericardial (Figure 15.12) – fluid is restricted between the heart and pericardium. A trace of fluid can be a normal finding.

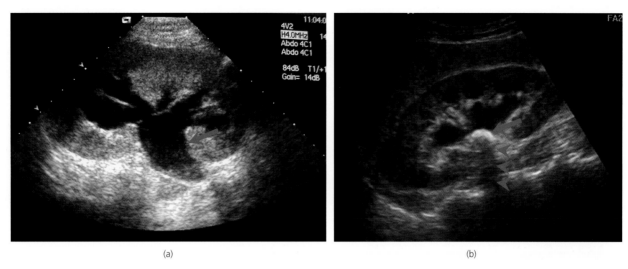

(a)                    (b)

**Figure 15.7** (a) Hydronephrosis: dilated PC system with normal kidney. (b) Renal calculus: PUJ calculus (arrow) with posterior acoustic shadowing (arrowheads) causing proximal dilated PC system.

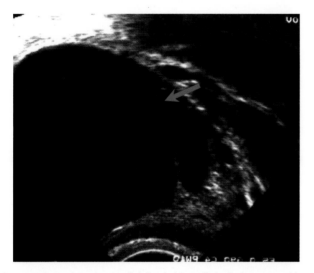

**Figure 15.8** Ovarian cyst: a well-defined, thin-walled echofree mass with posterior acoustic enhancement.

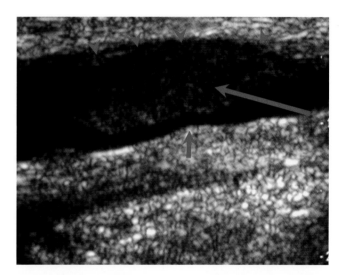

**Figure 15.9** DVT: non-compressible echogenic material (red arrow) almost occluding lumen (yellow arrow).

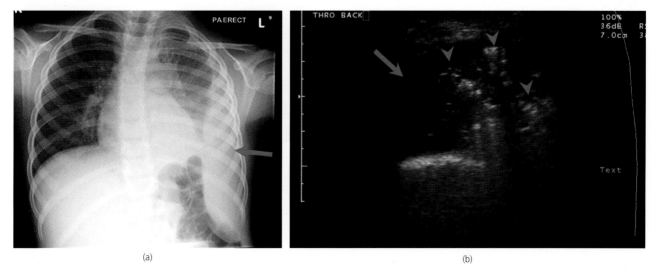

(a)                    (b)

**Figure 15.10** (a) Left pleural empyema on CXR (child). (b) US confirms fluid with echogenic material (arrowheads) consistent with pus.

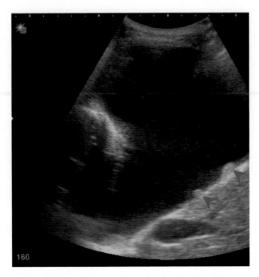

**Figure 15.11** Malignant effusion: pleural fluid with pleural metastases (arrowheads).

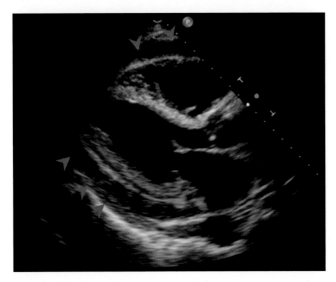

**Figure 15.12** Pericardial effusion (arrowheads).

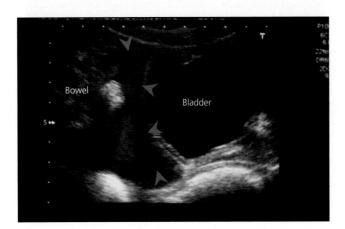

**Figure 15.13** FAST: free fluid (arrowheads) in the lower pelvis between bladder and bowel.

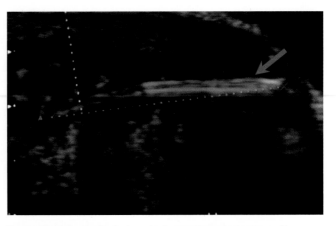

**Figure 15.14** Foreign body: irregular linear echogenic structure with no posterior acoustic shadowing consistent with wood splinter FB

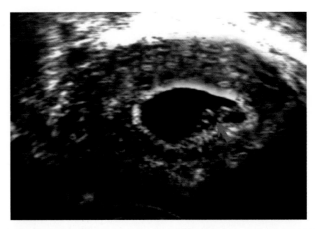

**Figure 15.15** Intra-uterine cystic structure with a foetus (arrow) and a visible foetal heart seen on real-time US (gestation 6 weeks).

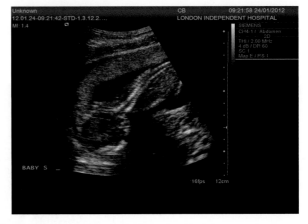

**Figure 15.16** Intra-uterine foetus (gestation 12 wks).

## FAST

Limited training has been shown to be sufficient for accurate detection of intraperitoneal fluid implying haemorrhage in trauma (Figure 15.13). FAST is now being used in pre-hospital assessment (at scene, ambulance or helicopter) of trauma patients.

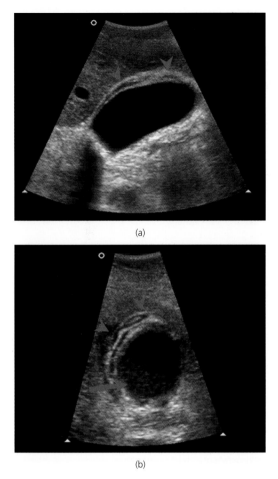

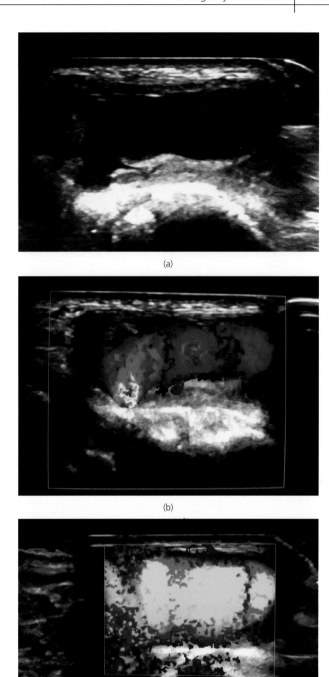

**Figure 15.17** (a) LS and (b) TS of acalculous cholecystitis: thick-walled structure with echopoor rim (arrowheads) and echogenic material (arrow) with no gallstones.

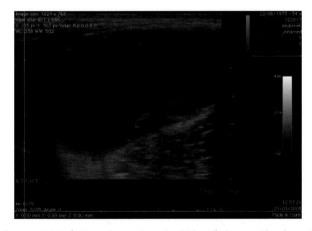

**Figure 15.18** Soft tissue abscess: irregular thick-walled mass with echogenic material.

**Figure 15.19** Pseudoaneurysm: US of hand following dog bite, with a well-defined thin-walled structure: (a) with a huge amount of colour; (b) power Doppler; (c) signal seen and swirling flow seen on real-time US.

## Foreign body

Most FBs can be seen as echogenic material with posterior acoustic shadowing (Figure 15.14). However, if the wound is open or there is gas within the wound, US may be of limited value.

## Foetus – viability and ectopic pregnancy

The normal foetus can be seen from about 4 weeks and the foetal heart at 6 weeks. Confirmation of an intrauterine pregnancy is possible with a transabdominal approach with a full bladder (Figures 15.15 and 15.16).

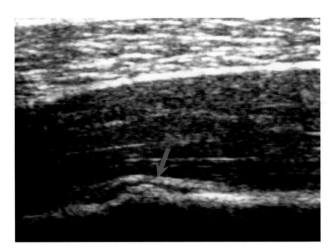

**Figure 15.20** Normal CXR with a rib fracture seen on US.

With a positive pregnancy test and an absent intrauterine pregnancy, an ectopic pregnancy should be excluded with a transvaginal US.

### Gallbladder – biliary colic, acute cholecystitis and empyema

US can be used to differentiate between biliary colic, cholecystitis and an empyema.

The features depend on seeing a thickened wall (3mms or more) and pericholecystic fluid (acute cholecystitis; Figure 15.17) and debris within the GB (empyema) and a positive provocation test (tenderness directly over the GB).

### Soft tissues

Induration, oedema and collections (including haematomas) can be differentiated (Figure 15.18).

In addition, US is widely used in MSK to assess joints, muscles, tendons and ligaments, but not in the emergency room.

Pseudoaneurysm (Figure 15.19) post dog bite seen in the right hand with a well-defined, echofree thick-walled mass with swirling flow seen real-time and confirmed with colour Doppler.

### Skeleton

US can show fractures and callus of superficial bones with normal XRs, in particular ribs (Figure 15.20). In children, it may show a periosteal reaction or a fracture not visible on an XR.

### Clinical scenarios

**Abdominal Pain**
- Ascites
- Appendicitis
- Abscess
- Bladder outflow obstruction
- Calculi – gallbladder and kidneys
- Cysts and pregnancy

**Breathlessness or Chest Pain**
- Pleural effusion
- Pericardial effusion
- Paralysis of diaphragm
- DVTs

**Cardiac Arrest**
- Cardiac motion – systole or asystole
- Size of IVC – hypovolaemia
- Pneumothorax
- Pleural effusion
- Pericardial effusion
- AAA

**Interventions**
- Arterial access
- Ascites,
- Abscess
- Bladder outflow obstruction
- Balloon puncture in foleys catheter
- Chest drains for pleural and pericardial collections *
- Central venous access **

*British Thoracic Society recommends.
**NICE guidelines.

## Further reading

Brooks A, Connolly J, Chan O. *Ultrasound in Emergency Care*. Blackwell Scientific 2004. ISBN 0-7279-1731-5.

http://guidance.nice.org.uk/TA49

http://www.brit-thoracic.org.uk/clinical-information/pleural-disease/pleural-disease-guidelines-2010.aspx

# CHAPTER 16

# Emergency Paediatric Radiology

*R. J. Paul Smith[1], Rosy Jalan[1] and Marina J. Easty[2]*

[1]Barts Health NHS Trust, The Royal London Hospital, London, UK
[2]Great Ormond Street Hospital, London, UK

---

**OVERVIEW**

- The common types of paediatric emergency differ with the age of the presenting child
- Paediatric fractures exhibit different types and patterns to adults
- Ultrasound is often the first line investigation of paediatric abdominal emergencies
- Be vigilant for possible non-accidental injury
- Minimise radiation exposure wherever possible

---

Children make up about one-third of all patients who attend emergency departments (Box 16.1). Skeletal injuries in infants (<1 year), children (>1 year), adolescents and adults all differ greatly.

---

Box 16.1 **Definitions used in paediatric medicine**

- Premature infant: ≤37 weeks' gestation
- Neonate: birth to 28 days old
- Infant: 1 month–1 year old
- Toddler: 1–3 years old
- Child: ≥3 years old

---

Children are more agile, more flexible and lighter than adults. When they fall, the forces generated are smaller. Boys sustain more injuries than girls, and most injuries occur at home or playing sport at school (Box 16.2).

---

Box 16.2 **Emergency paediatrics**

- Children often present at accident and emergency departments
- Boys present more commonly than girls
- Seasonal variations in the type of injury occur
- Children have different problems to adults

---

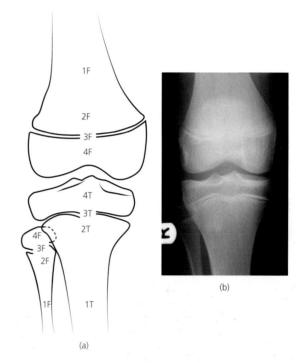

**Figure 16.1** Anteroposterior view of the right knee in a child.

Fractures in infants are rare. Toddlers tend to sustain skull and tibial fractures, whereas distal forearm, ankle and foot injuries are often seen in schoolchildren.

Minimising radiation exposure is an important principle in medical imaging. Children have an increased sensitivity to the risk posed from X-ray radiography. As such, any request for radiographs should give a clear indication and ask a relevant question.

This chapter gives an overview of common paediatric fractures and other radiological emergencies (Figure 16.1). It outlines a systematic approach of how to interpret the relevant radiographs. Common fractures in the paediatric population will be discussed, followed by a list of common abdominal and other childhood emergencies.

## Fractures

Fractures in children are different to those in adults because of anatomical, biomechanical, and physiological differences.

*ABC of Emergency Radiology*, Third Edition. Edited by Otto Chan.
© 2013 John Wiley & Sons, Ltd. Published 2013 by John Wiley & Sons, Ltd.

In addition, a range of fractures is caused in children, because paediatric bones are softer and more pliable than adult bones.

## Anatomy

At birth, many bones or the ends of bones are not visible. In time, ossification centres appear (sometimes several); they enlarge and coalesce, eventually fusing to the adjacent bone. The time interval between these changes varies from bone to bone. Predictable timescales vary slightly between boys and girls, and between different ethnic origins (Figure 16.2).

The physis (growth plate) is avascular after infancy. Damage to this area is shown radiographically by changes in its width or changes in adjacent bone. Damage to the epiphyseal vessels leads to death of the physeal chondrocytes and growth arrest.

The periosteum in the paediatric population is thick and strong. The attachment of the periosteum to the shaft of the bone is loose in children, and so periosteal reactions and subperiosteal collections are common.

## Biomechanical differences

Paediatric bones are more porous than adult bones, and they can bend more without breaking. The weak point lies at the physis, and so physeal fractures are common before bony fusion. The thick,

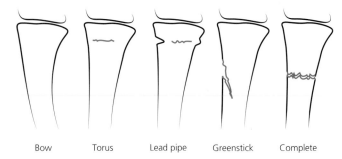

**Figure 16.3** Types of fracture.

strong periosteum resists displacement of the fracture (unless it is torn).

## Physiological differences

Fractures heal faster in children than in adults, and remodelling is quicker because children have rapid bone turnover. Normal alignment occurs in the plane of motion of the adjacent joint. Fracture healing results in longitudinal overgrowth, therefore in long bone diaphyseal fractures, overlap (up to 2 cm of bone) is accepted. The fracture should be described like an adult fracture – for example, transverse, oblique, comminuted, or compound. After the age of 12 years, fractures are treated more like those in the adult population due to slower remodelling.

## Types of fracture

Types of fracture are shown in Figure 16.3.

Complete diaphyseal fractures – The fracture site should be examined for an underlying bony abnormality, such as a bone cyst or generalised demineralisation. For infants and the non-ambulant, a careful history should be taken to exclude non-accidental injury.

Torus or buckle fractures – Failure on the compression side of a bending bone causes a torus or buckle fracture – an outward buckling of the cortex margin (torus is a latin term for bulge or swelling). Torus fractures usually occur near the metaphysis, where the cortex is thinnest. They often occur at the distal radius due to a fall on an outstretched hand (Figure 16.4).

Greenstick fracture – When a bone is bent beyond its limits, a greenstick fracture is produced. It is caused by the bone bending on the compression side, with complete failure on the tension side of the bone. The fracture may later hinge open because of muscle pull (Figure 16.5).

Bowing injuries – caused by acute plastic deformation of the bone secondary to longitudinal stress. An increase in longitudinal compression leads to bowing, buckle fractures, lead pipe fractures (Figure 16.6), greenstick fractures, and complete fractures.

## Physeal fractures
### Salter–Harris classification

The standard classification for physeal injuries is that of Salter and Harris (Box 16.3; Figure 16.7). This classification divides the

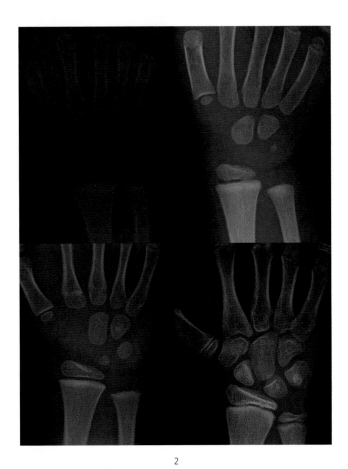

2

**Figure 16.2** Radiographs of the wrist in children at 18 months (top left), 3 years (top right), 6 years (bottom left), and 12 years (bottom right).

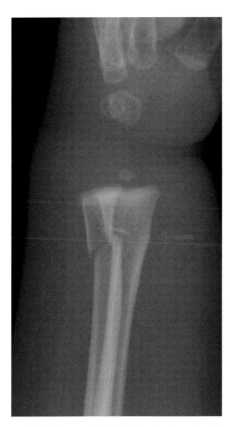

**Figure 16.4** Lead pipe fracture of the distal radius (arrow) and complete fracture of the distal ulna.

**Figure 16.5** Greenstick fracture of the mid radius (arrow).

common types (I–IV) according to the course of the fracture through the physis and the adjacent epiphyseal and metaphyseal bone. Type V injuries are rare, may be occult radiographically, and are caused by compression of the physeal cartilage (Box 16.4).

---

**Box 16.3 Salter–Harris classification**

- Type I – **S**lipped or **s**eparated
- Type II – **A**bove
- Type II – **L**ower
- Type IV – **T**hrough
- Type V – **E**venly **r**ammed

---

**Box 16.4 Common sites of physeal injuries**

**Type I (incidence 6% of Salter–Harris fractures)**

- Common locations – proximal humerus, distal humerus, proximal femur, distal tibia, and distal fibula

**Type II (incidence 75% of Salter–Harris fractures)**

- Common locations – distal radius, distal tibia, distal fibula, distal femur, distal ulna, and phalanges

**Type III (incidence 8% of Salter–Harris fractures)**

- Common locations – distal tibia, proximal tibia, and distal femur

---

**Type IV (incidence 10% of Salter–Harris fractures)**

- Common locations – distal humerus, and distal tibia

**Type V (incidence 1% of Salter–Harris fractures)**

- Common locations – ankle and knee

---

*Type I*

These are caused by a shearing stress through the physis (Figure 16.8). Most apophyseal injuries and slipped upper femoral epiphyses are type I fractures. Neonates may sustain these fractures at the proximal humerus. In the prepubertal child, a supination-inversion injury of the ankle may result in a type I fracture through the distal fibula. Invariably, this fracture will be reduced at the time of presentation and radiographical assessment. Type I fractures have a good prognosis.

*Type II*

These are the most common physeal fractures (75% of physeal fractures). The avulsion or shearing force fractures the physis and extends into the metaphysis (Figure 16.9). Radiographs show a triangular metaphyseal fracture known as the Thurston–Holland fragment. Between one-third and one-half of all type II injuries involve the distal radius. Reduction of the fracture is usually uncomplicated, and these injuries have a good prognosis.

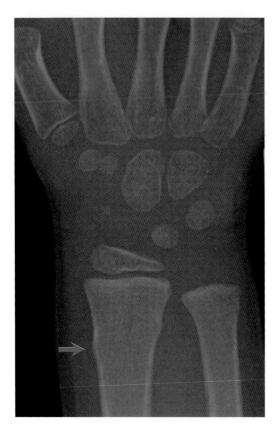

**Figure 16.6** Torus fracture of the distal radius (arrow).

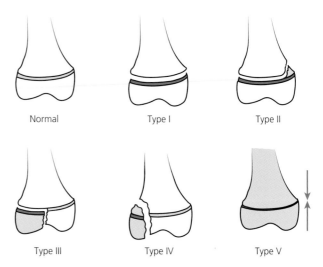

**Figure 16.7** Salter–Harris classification for physical injuries.

*Type III*

Type III injuries are partly intra-articular, with splitting of the epi-physis and a transverse fracture through the physis (Figure 16.10).

As they involve all layers of the physis, they cause growth arrest. Type III injuries occur in adolescents about the time of closure of the physis and they often require operative reduction to prevent displacement.

*Type IV*

These injuries cross the epiphysis, physis and metaphysis. They are caused by a longitudinally orientated splitting force, and they usually arise in the distal humerus or distal tibia. These injuries generally require open reduction to oppose the fracture frag-ments. Consequent angulation and leg length abnormalities may occur.

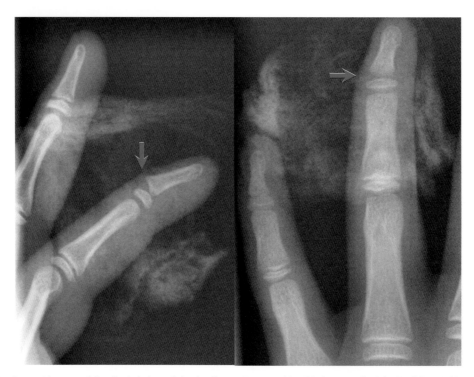

**Figure 16.8** Salter–Harris type I fracture of the distal phalanx of the ring finger. Compare this with the normal epiphysis of the neighbouring digit.

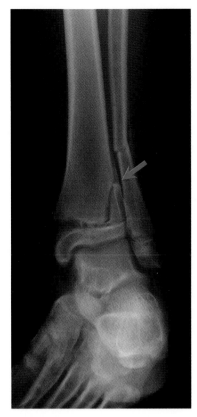

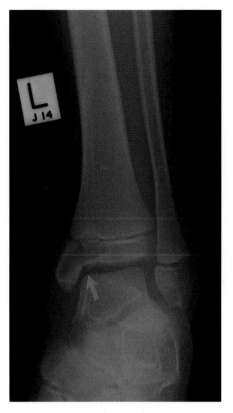

**Figure 16.9** Salter–Harris type II fracture of the distal tibial metaphysis (arrow). There is also a transverse fracture of distal fibula.

**Figure 16.10** Salter–Harris type III fracture of distal tibial epiphyses (arrow).

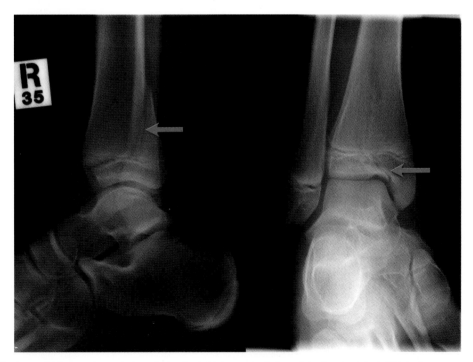

**Figure 16.11** Salter–Harris type IV fracture (arrows) through the distal tibial metaphysis (lateral view) and epiphysis (AP view).

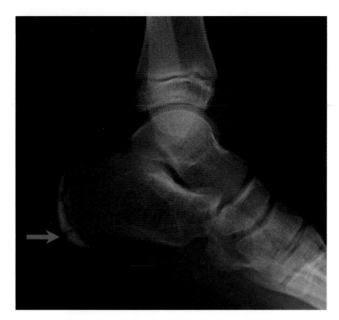

**Figure 16.12** The calcaneal apophysis is often fragmented as part of normal variation and should not be mistaken for a fracture.

*Type V*

Type V injuries are often diagnosed in retrospect, when growth arrest occurs after the injury. They are caused by a substantial loading or compressive force that damages the vascular supply and germinal cells of the growth plate (Figure 16.11). Most isolated type V injuries occur in the ankle or knee.

## Common pitfalls

Apophyses may be irregular and fragmented as part of natural variation, leading to a mistaken diagnosis of a fracture (Figure 16.12).

## Elbow injuries

Injuries of the three bones around the elbow are the most common fractures seen in infancy and childhood. There are six secondary ossification centres around the elbow (CRITOL, or CRITOE).

Fracture types include supracondylar fractures, epicondyle fractures and fractures and dislocations of the radial head. The frequency of the fracture types vary with the age of the child and some have specific radiographic signs (Box 16.5).

---

Box 16.5 **Average age of appearance of ossification centres (CRITOL)***

- **C**apitellum – 12 months
- **R**adial head – 3–6 years
- **I**nternal epicondyle – 4–7 years
- **T**rochlear – 7–10 years
- **O**lecranon – 6–10 years
- **L**ateral or external epicondyle – 11–14 years

*Appear earlier in girls

---

Evaluation of the elbow radiograph, and description of the paediatric ossification centres is dealt with in detail in Chapter 4 'Elbow' and will not be reiterated here.

## Forearm fractures

These include Monteggia fractures and Galeazzi fractures. A Monteggia fracture is a fracture of the proximal to the middle third of the shaft of the ulna with a dislocated radial head. Care must be taken when assessing apparent isolated ulna fractures, so that a radial head dislocation is excluded. Galeazzi fractures are rare in children. The radius is fractured and dislocation of the distal radio-ulnar joint is present.

Monteggia and Galeazzi's fractures are described in chapter 3 'Wrist'

## Painful hips

Perthes disease (Legg-Calve-Perthes disease) occurs commonly in Caucasian boys, with a male: female sex ratio of 4:1. Bilateral disease is present in up to 13% of patients who present with Perthes disease. The age of presentation ranges from 3–12 years, with children typically presenting at 5–8 years. Girls present at a younger age. Children with Perthes disease invariably have delayed bone age. They have pain in the hip, groin, thigh, or knee, and they have limited internal rotation.

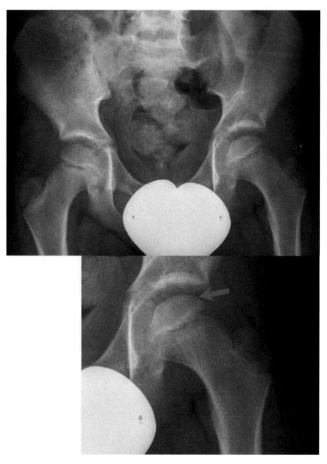

**Figure 16.13** Early Perthes disease of the left femoral head – note the subchondral lucency of the epiphysis (arrow)

Perthes disease is idiopathic avascular necrosis of the femoral capital epiphysis. The disease sometimes occurs after trauma or an effusion. The cause of Perthes disease, however, is not known.

The typical radiological findings depend on the stage of the disease. Early findings may show a small femoral capital epiphysis with a subchondral lucency (Figure 16.13). Magnetic resonance imaging and scintigraphy of the hip may pick up early changes better than plain radiography. Marrow oedema may be seen on a magnetic resonance image and absent radionuclide uptake in the affected epiphysis in the bone scan (Figure 16.14). Findings on plain radiography that occur later are fragmentation, flattening,

**Figure 16.14** Coronal magnetic resonance T2 weighted fat suppressed reconstruction of the hips – note the increase in signal in the left femoral epiphysis (arrow) indicating marrow oedema consistent with early Perthes disease.

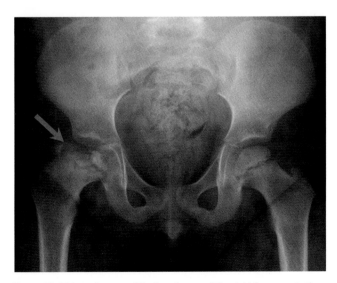

**Figure 16.15** Late changes of Perthes disease of the right femur- note the flattening, fragmentation and sclerosis of the capital epiphysis (arrow).

and sclerosis of the femoral capital epiphysis (Figure 16.15). Coxa magna may also develop during the reparative stage. Treatment may be minimal or simply rest. The aim is to prevent the hip subluxing, prevent pain, and minimise degenerative disease.

## Slipped upper femoral epiphysis

The male:female ratio is 2.5:1. The age of presentation is 12–15 years in boys and 10–13 years in girls. There may be familial cases of slipped upper femoral epiphysis (SUFE).

Children with SUFE are usually overweight or tall for their age and have some delay in skeletal maturation. Half of affected patients give a history of serious trauma. The most common presentation is hip pain and limp, but 25% of patients complain of knee pain. Bilateral slip occurs in 20–32% of patients, and more commonly in girls.

Slipped upper femoral epiphysis is a Salter–Harris type I injury. The initial imaging is an anteroposterior pelvic radiograph and a frog's leg lateral view of both hips (Figure 16.16).

The imaging findings may be subtle on the anteroposterior projection. The slip is initially posterior and therefore the frog's leg lateral view is essential, as only 75% of patients have a significant medial component to the slip.

The frontal view shows osteopenia of the affected femur. The physis may be wide. The metaphyseal margin of the physis is usually blurred. A line drawn tangential to the lateral femoral neck should bisect the femoral capital epiphysis, so that about one sixth of the diameter of the femoral capital epiphysis is lateral to this line. The epiphyseal height is reduced because of the posterior slip. In chronic slip, callus formation may be seen.

Treatment is to fix the hip to prevent further slip. The hip is pinned in situ, because realignment may lead to avascular necrosis.

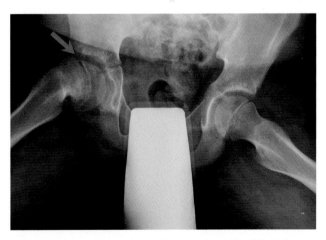

**Figure 16.16** Frog's leg lateral view of the hips shows slip of the right femoral epiphysis inferomedially (arrow) and blurring of the metaphyseal margin.

## Septic arthritis

Fever, pain, a raised white cell count, and raised levels of acute phase proteins may increase suspicion of septic arthritis. An ultrasound scan of the hip follows the pelvic radiograph (Figure 16.17). The findings on the plain film may be subtle and include osteopenia

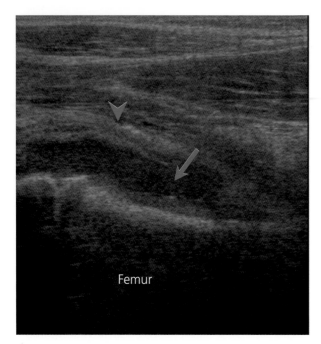

**Figure 16.17** An ultrasound image of septic arthritis in a 9-year-old girl taken in the sagittal plane along the long axis of the femur (superior to the right). The thickened joint capsule (arrowhead) is bowed by a joint effusion (arrow).

of the femur and bowing of the gluteus fat pad. The femoral head may be displaced laterally, with widening of the medial joint space caused by accumulated fluid. The hip joint is then imaged in the sagittal plane by ultrasonography, with comparison views of the unaffected side (Figure 16.18). Pus in the joint will cause bowing

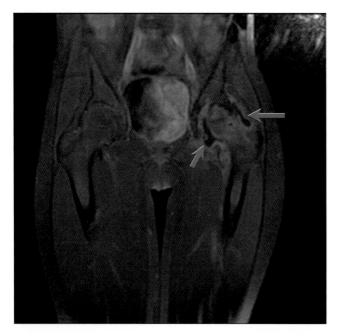

**Figure 16.18** A follow-up gadolinium-enhanced magnetic resonance scan 2 weeks later for the same child as in Figure 16.17 demonstrates avid enhancement of the thickened joint capsule on this T1 fat suppressed coronal image. Note patchy enhancement of the underlying bone. The child developed osteomyelitis of the femoral head.

of the capsule, and debris may be seen in the fluid. Treatment is surgical washout of the affected joint.

## Avulsion fractures around the hips

Avulsion fractures around the pelvic bones are common in children. Common sites are the anterior superior iliac spine (ASIS), the anterior inferior iliac spine (AIIS), and the ischial tuberosity. Plain films, and careful clinical examination, will usually identify these injuries (Figures 16.19 and 16.20).

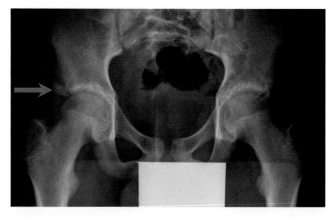

**Figure 16.19** Right anterior inferior iliac spine avulsion injury (arrow).

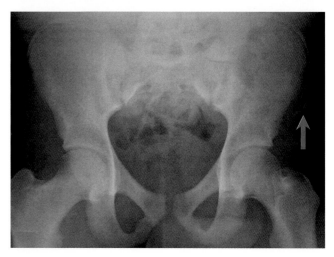

**Figure 16.20** Left anterior superior iliac spine avulsion injury (arrow).

## Toddler's fracture

This is a non-displaced oblique or spiral fracture of the midshaft of the tibia often sustained as toddler's begin to walk (Figure 16.21). Presentation may be with failure to bear weight on the leg or failure to continue to walk.

## Non-accidental injury

Skeletal presentations of non-accidental injury tend to occur in children who cannot talk, hence 50% occur before the age of one year and 80% before the age of two years. In 50% of proven cases of non-accidental injury, the skeletal survey is normal.

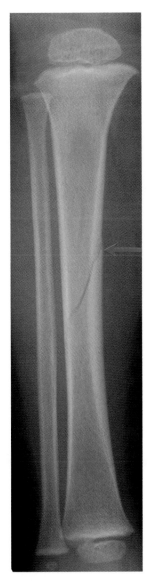

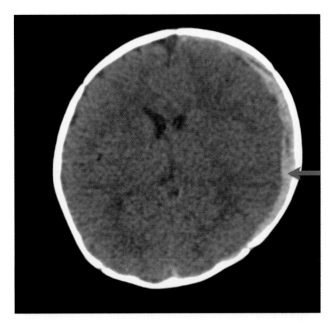

**Figure 16.22** CT axial reconstruction of the head in an 11-year-old child demonstrating an acute subdural haematoma (arrow).

**Figure 16.21** Minimally displaced spiral fracture of the midshaft of the right femur (arrow) – a toddler's fracture.

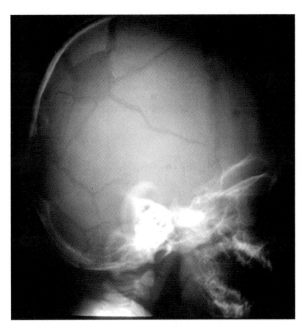

**Figure 16.23** Complex skull fracture in non-accidental injury.

Injuries with a delayed presentation, an unlikely explanation, a changing history, an unusual mechanism, or multiple fractures of differing ages should raise the suspicion of non-accidental injury. The importance of taking an accurate history in these cases cannot be overstated. The height of the fall, mechanism of the fall, the leading part involved and the surface onto which the child fell all need to be ascertained, as does the exact timing of the injury. Remember that a fall from a bed onto a carpeted floor usually will not lead to a fracture in a child with normal bones. Retinal haemorrhages caused by shaking are an important sign in non-accidental injury, as are marks on the skin, such as bruises and burns.

Many battered children present with skull fractures and underlying subdural haematomas. An unexplained skull fracture, particularly one that crosses sutures, or a diastased fracture are suggestive of non-accidental injury. A computed tomography

scan of the brain will show intracerebral injuries (Figure 16.22 and 16.23).

The classic fractures in non-accidental injury are metaphyseal corner fractures, caused by violent shaking or twisting of the baby; however, less than 50% of babies present with these fractures. They may present with other injuries, such as spiral or oblique fractures of the long bones, soft tissue injuries, or abdominal injuries (pancreatic and duodenal injuries). In babies younger than one year, an isolated long bone fracture should be considered suggestive of non-accidental injury, and a skeletal survey may be warranted (Box 16.6; Figure 16.24).

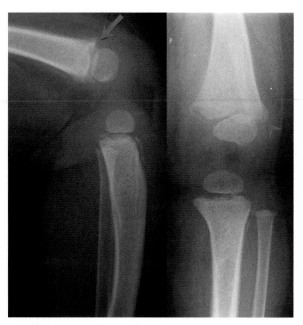

**Figure 16.24** Left image shows a metaphyseal corner fractures of the distal femur (arrow) and proximal tibia. Right image demonstrates a bucket handle fracture of the distal femur (arrowhead).

---

Box 16.6 **Skeletal survey for non-accidental injury**

- Skull radiograph – anteroposterior and lateral views, and Townes if an occipital fracture is suspected
- Lateral view of the whole spine
- Chest radiograph – repeat in 10 days
- Oblique views of the ribs
- Anteroposterior view of abdomen and pelvis
- Frontal views of arms and hands
- Frontal views of legs and feet
- Computed tomography scan of the brain

Note that lateral views of fractured bones are also advised

---

Rib, clavicular, spine, pelvis, and scapular fractures may all be seen in non-accidental injury. Rib fractures, particularly posterior fractures, are highly suggestive of non-accidental injury, although lateral and anterior fractures are also common in non-accidental injury (Figure 16.25). They result from squeezing the child. Anterior fractures occur by squeezing and shaking the child and are caused by costochondral separation. The fractures are seen best when callus forms, so a repeat chest radiograph is advised 10 days after presentation. Anterior rib fractures may be associated with intra-abdominal injury.

## Chest emergencies

### Airway obstruction
#### Inhaled foreign body
An expiratory radiograph is advised to reveal air trapping. The chest radiograph may show an area of consolidation, collapse, or air trapping, with a transradiant lung. The foreign body may be visible.

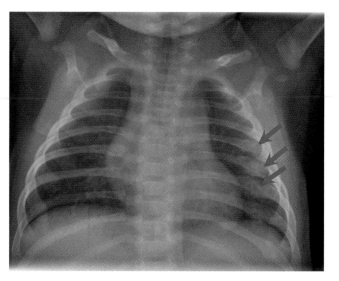

**Figure 16.25** Chest radiograph demonstrating multiple, healing, left posterior rib fractures (arrows).

The most common cause of partial upper airway obstruction in young children is acute laryngotracheobronchitis or croup. With a peak incidence in 1 year olds, this is often a clinical diagnosis, but upper airway radiographs may occasionally show subglottic narrowing (not routinely performed).

### Other chest emergencies
*Asthma* – No radiological abnormality is seen, except a degree of air trapping with large volume lungs. It is important to exclude a pneumothorax.

*Pneumonia* – May present as a focal area of consolidation (airspace shadowing).

*Viral lung infections* – may cause oedema of the bronchial walls, demonstrated on the chest radiograph as wall thickening radiating from the hila (Figure 16.26).

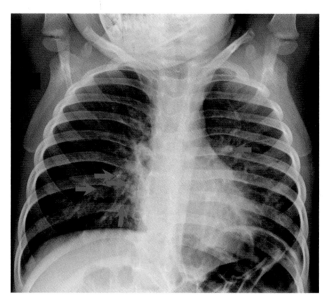

**Figure 16.26** Chest radiograph demonstrating bilateral bronchial wall thickening (arrows) consistent with a viral chest infection.

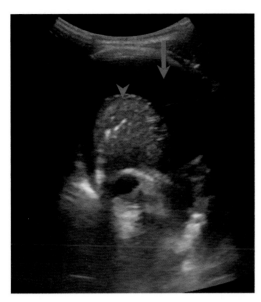

**Figure 16.27** This ultrasound scan taken in an axial plane between the ribs demonstrates a large pleural effusion (arrow), dark on the image, and underlying collapsed lung (arrowhead) bright on the image.

*Pleural empyema* – An ultrasound scan will show the pleural effusion and the presence of loculations (Figure 16.27). Drainage guided by ultrasonography is advocated.

*Pneumothorax* – May be difficult to diagnose if the chest radiograph has been taken supine.

*Cardiac abnormalities* – Chest radiograph may show cardiac enlargement, a right sided aortic arch, elevation of the cardiac apex, or selective chamber enlargement, but it is often unhelpful. Referral to specialists for echocardiography is advised for further management.

*Oesophageal atresia and tracheo-oesophageal atresia* – These children often present with respiratory symptoms for example choking on feeds. The chest radiograph may show atelectasis or airspace disease should aspiration have occurred.

> If the patient has a history of an inhaled foreign body, the chest film may be normal and a bronchoscopy may be required

## Acute abdominal emergencies

Clinical examination and ultrasonography are the initial investigations for children with acute abdominal emergencies (Figures 16.28 and 16.29). Unlike adults, computed tomography should be reserved for specific conditions in children and only requested by the paediatric specialists. A list of possible conditions is given in Box 16.7 without detailed description of the radiological abnormalities. Plain abdominal radiographs may be helpful in neonates but are less useful in older children.

Some of the more common abdominal emergencies are detailed below:

*Hypertrophic pyloric stenosis* – Presents with non bilious vomiting typically at 4–6 weeks. It can be diagnosed with ultrasound if the pylorus cannot be palpated.

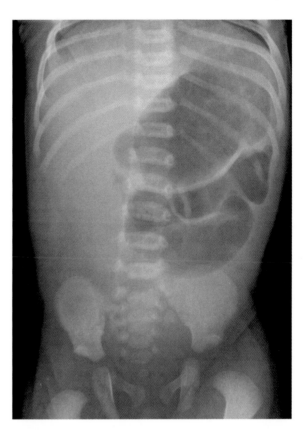

**Figure 16.28** High obstruction due to a malrotation. Distended bowel loops are evident in the upper abdomen with no gas below the level of obstruction.

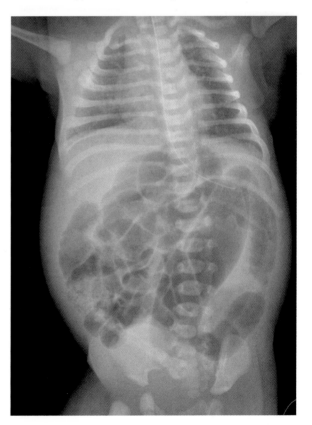

**Figure 16.29** Low obstruction. Note the multiple distended bowel loops throughout the abdomen.

Box 16.7 **Acute abdominal emergencies**

**Obstruction in the neonate**

*High obstruction*

- Malrotation and volvulus
- Duodenal atresia, duodenal web, duodenal stenosis, and annular pancreas all give a "double bubble" appearance
- Jejunal atresia
- Duodenal duplication cyst causing obstruction

*Low obstruction*

- Hirschsprung's disease
- Meconium ileus. May be diagnosed in utero. Note meconium peritonitis with calcification
- Meconium plug syndrome (left sided microcolon)
- Ileal atresia
- Ano-rectal malformation
- Anal stenosis
- Milk curd obstruction
- Obstructed hernia

**Obstruction in infants and older children**

- Intussusception, diagnosed with ultrasonography
- Appendicitis
- Adhesion obstruction
- Hernia
- Constipation
- Do not forget malrotation at all ages

**Abdominal masses**

- Duplication cysts
- Ovarian cysts
- Mesenteric cysts
- Tumours

*Intussusception* – Occurring in infants and young children often presenting with intermittent colicky pain, obstructive symptoms or 'redcurrent jelly' stools, one segment of bowel telescopes into another. It may be accurately diagnosed with ultrasound.

*Appendicitis* – The most common abdominal emergency in children, most do not require imaging. Ultrasound may help diagnose or exclude the condition in atypical presentations.

# Renal tract emergencies

If renal abnormalities are suspected, request an ultrasound scan.

## Urinary tract infection

Ultrasound imaging routinely in acute UTI is not recommended unless the child is seriously ill, has poor urine flow or abdominal mass, raised creatinine, septicaemia, fails to respond to antibiotics within 48hrs or has infection with non-ecoli organisms.

Outpatient ultrasound and fluoroscopy will aim to detect renal obstruction, ureteric reflux or renal parenchymal scarring.

## Other renal abnormalities

- Renal stones
- Haematuria, trauma, infection, renal vein thrombosis, glomerulonephritis, and nephrotic syndrome

## Renal masses

- Hydronephrosis
- Complicated duplex kidney
- Multicystic dysplastic kidney
- Renal tumour – usually a Wilms' tumour in children younger than 10 years with mesoblastic nephroma diagnosed at birth
- Autosomal recessive polycystic kidney disease

*Simple ovarian cysts* – These occur in pubertal girls when the follicle does not involute. Most are asymptomatic, but when complivated by haemorrhage, rupture or tortion the patient may present with lower abdominal pain. US is useful in the evaluation and differentiating the condition from other abdominal patholgy.

## Further reading

Borden S. Roentgen recognition of acute plastic bowing of the forearm in children. *AJR* 1975;125:524–30.

NICE clinical guideline 54 Urinary tract infection in children: diagnosis, treatment and long-term management

Rang M. *Children's Fractures*. Philadelphia: JP Lippincott, 1983

Salter RB, Harris WB. Injuries involving the epiphyseal plate. *J Bone Joint Surg Am* 1963;45:587–622.

# CHAPTER 17

# Major Trauma

*Dominic Barron[1], Sujit Vaidya[2] and Otto Chan[3]*

[1]Leeds Teaching Hospitals, Leeds, UK
[2]Barts Health NHS Trust, The Royal London Hospital, London, UK
[3]The London Independent Hospital, London, UK

---

**OVERVIEW**

- Trauma is common and a leading cause of morbidity and mortality
- Time is a critical factor in determining patient outcome
- MDCT is the single greatest recent advance in trauma care
- Advanced trauma life support (ATLS) principles for initial management – primary survey
- Primary CT survey versus primary clinical survey

---

Trauma has been and remains one of the leading causes of death, in particular in the 1- to 44-year-old age group. Despite this, trauma has been neglected and remained a subject unworthy of study, research and funding. There are an estimated 3.8 million deaths worldwide and 10 million people permanently disabled every year.

In 1966, a landmark white paper, 'Accidental death and disability, the neglected disease of modern society', highlighted the problems in the management of trauma, leading to guidelines to establish regionalised trauma care. The standard of trauma care and clinical outcomes varies hugely, not only between developed and third world countries, but also within developed countries. Severely injured patients in the UK have a 20% higher mortality than in the USA. It is estimated that 3000 of the 16,000 annual trauma deaths are preventable. Despite two highly critical reports of trauma care – 'Retrospective study of 1000 deaths from injury in England and Wales (1988)' and 'The management of patients with major injuries (1988)' – and the implementation of ATLS, there has been a 'Lack of change in trauma care in England and Wales since 1994 (2002)' and 'The National confidential enquiry of patient outcome and death (2007)' reported recently that 60% of trauma victims had substandard care in the UK. Specific recommendations were made to address these recognised deficiencies, firstly by organisation of trauma care into regional systems and secondly by a multidisciplinary approach.

There have been huge technological advances in radiology in the past decade, not least the advent of PACS (Picture Archiving and Communications systems) and multidetector computed tomography (MDCT). This has revolutionised the role of radiology and of the radiologist, requiring a 24/7 round-the-clock service in specialist trauma centres. It is widely accepted that MDCT is the single greatest recent advance in trauma care. A complete review of present practice is necessary in order to benefit from MDCT, not least the fundamental principle of ATLS that *definitive diagnosis is not necessary for the initial management of a traumatised patient*.

Nevertheless, ATLS remains the standard method of care for the initial management of severely injured patients. The principle is simple – treat the greatest threat to life first. Loss of airway will kill before inability to breathe, and inability to breathe will kill before bleeding and loss of circulation. Although a definitive diagnosis is not necessary for initial patient care, with the advent of MDCT, it is now recognised that a definitive diagnosis where possible is preferable and often guides the overall patient care. The most important principle to remember is *do no harm* to the patient during examination, investigation and treatment.

The notion that *Time is Golden* makes good common sense and two factors have been shown to correlate with improved patient outcome, namely *time from initial admission to CT* and *time from admission to definitive care*. Clinical examination of the obtunded traumatised patient has been shown to be unreliable at best and misleading at worst, therefore *don't examine patients to death* and taking unnecessary X rays or even taking X rays at all may delay definitive care, so *don't X-ray patients to death*.

The management of severely injured patients is divided into the primary and secondary survey. This chapter deals with the imaging during the primary survey.

## Primary survey – ATLS

The main goal of the pre-hospital phase of ATLS is to deliver a patient rapidly and safely to hospital (Box 17.1). Patients are transferred to the resuscitation room and the aim of the primary survey in ATLS is to do a rapid evaluation of the patient, resuscitate and stabilise the patient with a view to proceeding to definitive treatment. This process is called the ABCDE of trauma (Box 17.2). Adjuncts to the primary survey include relevant imaging during resuscitation and re-evaluation.

ATLS works on a step-by-step approach carried out by a single operator (vertical approach) with minimal assistance, treating the greatest threat to life first. Nowadays it is a concept or a common language and in practice, most of the steps of the ABCDE are carried out simultaneously by a trauma team (horizontal approach). Anaesthetists will usually deal with the airway and intravenous access, while the surgeon evaluates the chest, abdomen, and pelvis for potential life-threatening injuries.

Imaging is requested as part of the primary survey while the patient is assessed, life-threatening injuries are dealt with, and resuscitation procedures instituted. Imaging should not be performed if it interferes with the rest of the primary survey or definitive care, and only investigations that may have a direct effect on the patient's initial problems should be carried out.

Traditionally, imaging performed as part of the primary survey includes the supine chest and pelvis radiographs (Figures 17.1 and 17.2) and limited ultrasonography (FAST – focused assessment with sonography for trauma). The advent of MDCT scanners now means that in major trauma centres, CT can be incorporated into the primary survey, even in relatively unstable patients (Box 17.3).

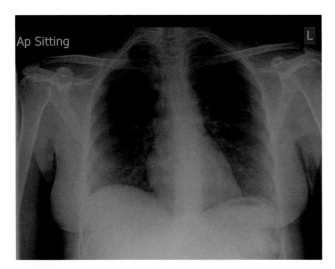

**Figure 17.1** Normal anteroposterior (AP) chest.

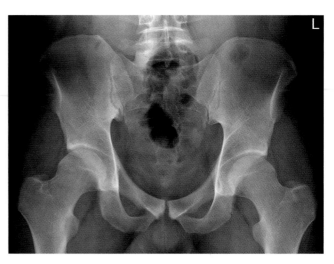

**Figure 17.2** Normal pelvis.

The main problem is delay of transfer from the emergency room to the CT room.

## Airway and cervical spine control

The airway should be assessed for patency. Foreign bodies and vomit should be removed and facial, mandibular, tracheal and laryngeal injuries should be excluded clinically.

If the patient is conscious and talking, there is usually no need for airway intervention. If the patient is unconscious and breathing spontaneously, an oropharyngeal airway may suffice as a temporary measure. Any patient who has a head injury and a score on the Glasgow coma scale of 8 or less should be intubated. However, intubation may be required for optimal control of airways in patients with higher scores.

If the patient has been intubated, a chest radiograph should be taken to check the position of the endotracheal tube. The tip of the tube should not lie below the level of the aortic arch in a supine chest radiograph and a minimum of 3.5 cm (and preferably 5 cm) above the carina.

Care should be taken to avoid worsening a potential cervical spine injury while establishing and safeguarding an airway. If the airway has been secured, the neck should be immobilised with a cervical collar, sandbag and tape. Should the collar need to be removed, an experienced member of the trauma team should carry out in-line manual immobilisation of the head and neck.

## Breathing and ventilation

A patent airway does not guarantee adequate ventilation. The lungs, chest wall and diaphragm must be assessed for potential injuries

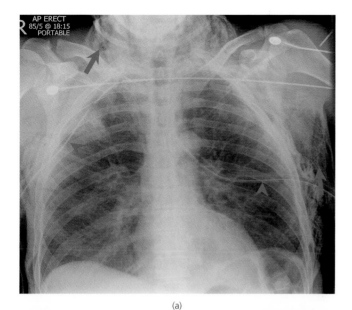

(a)

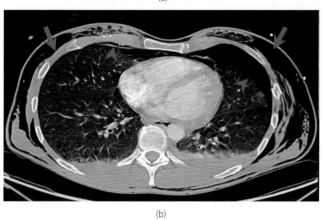

(b)

**Figure 17.3** (a) AP chest radiograph with bilateral chest drains. Extensive surgical emphysema. (b) CT chest in the same patient clearly demonstrating the left pneumothorax as well as the extensive surgical emphysema.

that could compromise ventilation acutely. These life-threatening injuries include tension pneumothorax, tension haemothorax, flail chest and open pneumothorax. It can be difficult to exclude these injuries in a patient with multiple trauma. A chest radiograph must be taken as soon as possible (Figure 17.3a,b). If the patient is subsequently intubated or ventilated, a second radiograph should be taken to confirm that the endotracheal tube is in a satisfactory position and that life-threatening injuries have not been made worse. Ventilation can cause a simple pneumothorax to become a tension pneumothorax.

## Circulation and haemorrhage control

The patient's haemodynamic state must be assessed quickly and accurately because bleeding is a major cause of preventable death. Clinical evaluation is essential, in particular the level of consciousness, skin colour and pulse. Any external source of bleeding should be identified and dealt with immediately using manual pressure. When the examination or history suggests internal injury a limited

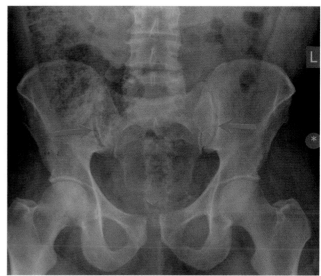

**Figure 17.4** Pubic symphyseal diastasis with bilateral widened SIJs, consistent with an AP compression injury.

ultrasonography (FAST) should be done to exclude hidden blood loss (Box 17.4).

---

Box 17.4 **Main causes of hidden blood loss**

- Chest, abdomen, and retroperitoneal injuries
- Pelvic fractures
- Multiple long bone fractures

---

FAST can be performed by a physician, surgeon, radiologist or paramedic and has been shown to be valuable in the assessment of blunt trauma patients in the emergency room, especially in unstable patients with multiple injuries. FAST performed in the pre-hospital setting is extremely useful, as this can then give the receiving unit advanced warning of potential injuries and help to triage patients.

Ultrasonography should be performed in five areas. These areas are the five Ps – perihepatic, peripluric, and pelvis in the abdomen, and pericardial (to exclude a pericardial tamponade) and pleural (to detect fluid or a pneumothorax) in the chest or consolidated lung.

The presence of a pelvic fracture (Figure 17.4) or free fluid on ultrasonography mandates a specialist opinion. In appropriate centres, CT is the investigation of choice in these patients as not only will this identify all bleeding sources, but it can differentiate between arterial bleeding and venous bleeding.

This is important as embolisation is now well recognised as a definitive treatment option for defined arterial bleeding. This particularly applies to the spleen, liver, kidneys and pelvis (Figures 17.5–17.7).

## Disability (neurological examination)

The patient's neurological state is assessed with the Glasgow coma scale (Box 17.5). It is easy and quick to use and is a determinant of patient outcome and possible further management.

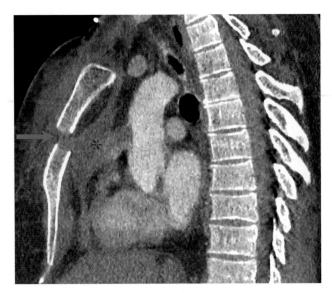

**Figure 17.5** Sagittal CT chest reformat showing sternal fracture (arrow) and retrosternal haematoma (asterisk).

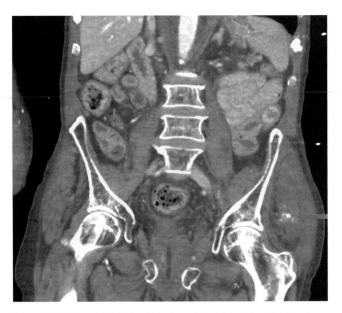

**Figure 17.7** Coronal CT abdominal reformat showing active left gluteal bleed.

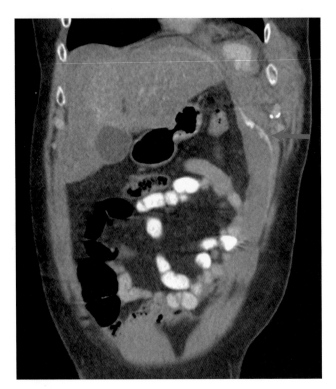

**Figure 17.6** Coronal CT abdominal reformat showing active arterial bleed (arrow).

Box 17.5 **Glasgow coma scale score**

**Eye opening (graded 1–4)**

- Spontaneous – 4
- To speech – 3
- To pain – 2
- None – 1

**Best motor response (graded 1–6)**

- Obeys command – 6
- Localises pain – 5
- Normal flexion – 4
- Abnormal flexion – 3
- Extension (decerebrate) – 2
- None – 1

**Verbal response (graded 1–5)**

- Orientated – 5
- Confused conversation – 4
- Inappropriate words – 3
- Incomprehensible sounds – 2
- None – 1

- Maximum score 15, minimum score 3
- Mild injury 14–15
- Moderate injury 9–13
- Severe injury 3–8
- Coma <8

All patients with a head injury or who are unconscious or ventilated should have CT of the head, especially if they have lost consciousness, have amnesia or have severe headaches. Up to 18% of patients with mild head injuries have abnormalities on CT, and 5% of these patients may require surgery.

If the patient has a head, scan it – missing a serious head injury may have catastrophic consequences

CT should be done as soon as possible because morbidity and mortality rises substantially if surgery is delayed. The intracranial findings of CT may include no abnormality, extradural haematoma, subdural haematoma, contusions and intracerebral haematomas,

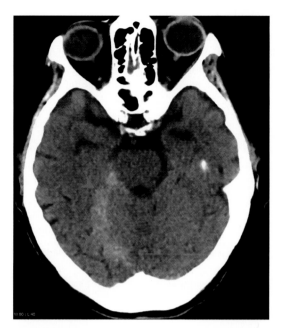

**Figure 17.8** CT head demonstrating right subdural haemorrhage.

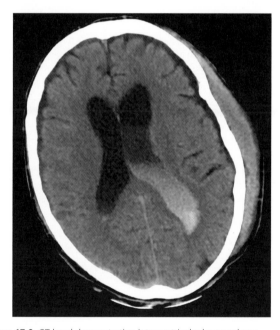

**Figure 17.9** CT head demonstrating intraventricular haemorrhage.

subarachnoid blood, diffuse axonal injury and combination injuries (Figures 17.8 and 17.9).

The National Institute of Clinical Excellence (NICE) introduced UK guidelines for management of head injury in 2003 that support the ATLS guidelines. They emphasise that CT must be done within an hour of the patient arriving at the hospital.

### Exposure and environment

The patient should be fully exposed (by cutting off all clothes) to allow a full examination. It is, however, critical to keep the patient warm with blankets and a heated emergency room. Large volumes of fluids may be infused, and these intravenous fluids should be warmed.

## Adjuncts to primary survey and resuscitation

As a minimum, patients should have electrocardiography, their blood pressure monitored, pulse oximetry, a nasogastric tube and a urinary catheter. Blood gases should also be monitored. If a fracture at the base of the skull is suspected, the nasogastric tube can be inserted after CT of the head or an orogastric tube placed.

## Interpreting primary survey images

All imaging must be supervised and done without fuss or undue delay and with meticulous technique. Attention to detail is essential. In particular, the film must be labelled (including the patient's name and a side marker) (Boxes 17.6 and 17.7).

---

**Box 17.6 ABCDEs interpretation of the supine chest radiograph (Figure 17.1)**

**Airways**

- Check trachea is clear and central
- Is airway patent?
- Check position of endotracheal tube
- Are there any teeth or foreign bodies?
- Check all lines and tubes

**Breathing**

- Exclude tension pneumothorax and haemothorax
- Check there is no radiological flail segment
- Exclude rib fractures
- Check lungs are clear

**Circulation**

- Check heart size and mediastinal contours are normal
- Make sure that the aortic arch is clearly seen
- Check the hila and vascular markings are normal

**Diaphragm**

- Check that diaphragms appear normal (size, shape, and position)
- Can both diaphragms be clearly seen?
- Check under each diaphragm

**Edges**

- Check the pleura and costophrenic recesses
- Exclude a subtle pneumothorax or effusion

**Soft Tissues and Skeleton**

- Look for surgical emphysema
- Check clavicles and shoulders and exclude rib fractures
- Look at the paraspinal lines and check the spine

---

The supine chest radiograph should be taken as soon as possible after the patient has been exposed and centred correctly. Attention must be paid to stop patients being rotated and to keep them in the middle of the trolley.

Box 17.7 **ABCs interpretation of pelvic radiographs (Figure 17.2)**

**Alignment**

- Check the pubic symphysis is symmetrical and not widened
- Carefully check that the sacroiliac joints are intact

**Bones**

- Check that all three pelvic rings are intact
- Use a bright light to check iliac crests and hips
- Look at the lumbar spine and hip joints separately

**Cartilage**

- Check the distance of the pubic symphysis
- Again check the sacroiliac joints
- Check both hips

**Soft Tissues**

- Check the soft tissue planes are symmetrical
- Look for obturator internus
- Carefully delineate the perivesical fat plane
- Make sure the gluteus medius and psoas fat planes are intact

A polytrauma CT should be dealt with in exactly the same way as all other trauma management. Therefore the initial readout should be the primary survey looking for major life-threatening injuries (see later). A secondary survey by a radiologist should then follow, which involves a detailed assessment of all the imaging.

All major trauma patients ideally should have a whole body CT
    Don't X-ray patients to death!

## Primary CT survey – ATLS and MDCT

The problem with the standard ATLS approach is that there are delays built into this step-wise approach, plain radiographs provide only limited information into the state of the patient and injuries sustained and in addition, almost all major trauma patients need further imaging with CT.

The need to save time and avoid unnecessary transfers to major trauma patients is critical and the most time-efficient method is to transfer patients from pre-hospital care directly to the CT scanner, performing the initial part of the primary survey (Airways and Breathing) on the CT scanner table and then doing a CT whole-body scan on a MDCT and then continuing with the primary survey (re-evaluation of A and B and then continuing with C).

The patient avoids transfer from the resuscitation room to the CT scanner room and back again to the resuscitation room, each transfer usually taking at least 45 minutes. Furthermore, the relatively 'stable patient' can potentially be made 'unstable' by the transfer, in addition to the other dangers involved in transferring

patients, such as tubes, drains and lines being moved and spinal injuries made worse.

Only in exceptional circumstances, where the patient is deemed extremely unstable and when the pre-hospital FAST scan shows free fluid in the abdomen or a pericardial collection in a penetrating injury, does the patient go directly to theatre for an emergency laparotomy or thoracotomy.

Dedicated resuscitation rooms are now available, where the patient can be resuscitated, have plain radiographs or MDCT, angiography and interventional procedures (such as embolisation) and the room is also an operating suite. Clearly these all in one rooms are expensive, need planning and will only be available in major trauma centres, but the advantages are clear, not least in saving time and transfers and, in effect, optimising treatment, improving outcome and avoiding preventable deaths.

**ABCs of CT interpretation**

**Primary Survey**

This should be done immediately as per ATLS protocol with the aim being to identify major life-threatening injuries.

**A – Airway**

- Check for airway obstruction
- Check for ET tube placement
- Look for obstructing foreign bodies

**B – Breathing**

- Exclude tension pneumothorax and haemothorax
- Check for pulmonary contusion
- Check for pulmaonary lacerations
- Assess chest drain placement

**C – Circulation**

- Asses for active bleeding in the Thorax / Abdomen / Pelvis and Soft Tissues

**D – Disability**

- Assess for major intra-cranial bleed / oedema
- Look for major spinal Injury

**Secondary Survey**

This should be done by a radiologist and involves a detailed review of everything. Ideally this is proforma driven to ensure nothing is missed.

## Further reading

ACS. *ATLS Student Course Manual*, 8th edn. Chicago: American College of Surgeons, 2008. ISBN 978-1-880696-31-6.

Mirvis SE, Shanmuganathan K. *Imaging in Trauma and Critical Care*, 2nd edn. Philadelphia: Saunders, Elsevier, 2003. ISBN 7216-9340-7.

Mirvis SE, Shanmuganathan K, Miller LA, Sliker CW. *Emergency Radiology: Case Review Series*. Mosby, Elsevier, 2009.

The Royal College of Radiologists. *Standards of Practice and Guidance for Trauma Radiology in Severely Injured Patients*. London: the Royal College of Radiologists, 2011. ISBN 987-4-905034-51-2.

# Index

# ABC of Sexually Transmitted Infections

## 6TH EDITION

### Karen E. Rogstad
Sheffield Teaching Hospitals NHS Foundation Trust, Sheffield

With sexually transmitted infections (STIs) a major cause of morbidity and mortality throughout the world, the new edition of *ABC of Sexually Transmitted Infections* is a much-needed introduction and reference guide. This sixth edition:

- Includes the latest guidance on the prevalence, prevention and treatment of STIs, screening programmes and new testing methods
- Features new chapters on service modernization and new care providers, high risk and special needs groups, systemic manifestations, and sexually transmitted infections in resource-poor settings
- Covers contraception in depth and reflects the increasing integration of STI and contraceptive services
- Is ideal for those providing community based STI diagnosis and management such as GPs, primary care physicians, pharmacists and contraceptive service providers. Those in the voluntary sector, junior doctors, medical students, and nurses working in community or specialist services will also find it a valuable resource

APRIL 2011 | 9781405198165 | 168 PAGES | £26.95/US$42.95/€34.90/AU$52.95

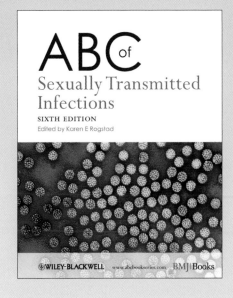

# ABC of Stroke

### Jonathan Mant & Marion F. Walker
UK Stroke Research Network and Addenbrooke's Hospital, University of Cambridge; UK Stroke Research Network and University of Nottingham

Stroke is the most common cause of adult disability and is of increasing importance within ageing populations. This practical guide to stroke:

- Covers the entire patient journey, from prevention through to long-term support
- Includes primary prevention and management of risk factors for stroke and secondary prevention including pharmaceutical, lifestyle and surgical intervention
- Addresses the general principles of stroke rehabilitation as well as mobility, communication and psychological problems, stroke in younger people and long-term support for stroke survivors and their carers
- Is invaluable to all aspects of stroke for health care professionals and is of particular relevance to GPs, junior doctors, nurses and therapists working with stroke patients and their careers

MARCH 2011 | 9781405167901 | 72 PAGES | £20.95/US$32.95/€26.90/AU$39.95

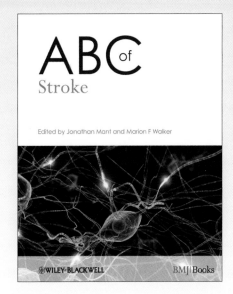

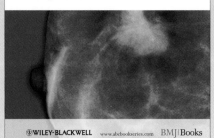

# ABC of Breast Diseases

## 4TH EDITION

**J. Michael Dixon**
Western General Hospital, Edinburgh, UK

Breast diseases are common and often encountered by health professionals in primary care. While the incidence of breast cancer is increasing, earlier detection and improved treatments are helping to reduce breast cancer mortality. The *ABC of Breast Diseases, 4th Edition*:

* Provides comprehensive guidance to the assessment of symptoms, how to manage common breast conditions and guidelines on referral
* Covers congenital problems, breast infection and mastalgia, before addressing the epidemiology, prevention, screening and diagnosis of breast cancer and outlines the treatment and management options for breast cancer within different groups
* Includes new chapters on the genetics, prevention, management of high risk women and the psychological aspects of breast diseases
* Is ideal for GPs, family physicians, practice nurses and breast care nurses as well as for surgeons and oncologists both in training and recently qualified as well as medical students

AUGUST 2012 | 9781444337969 | 168 PAGES | £27.99/US$46.95/€35.90/AU$52.95

# ABC of HIV and AIDS

## 6TH EDITION

**Michael W. Adler, Simon G. Edwards, Robert F. Miller,
Gulshan Sethi & Ian Williams**
University College London Medical School; Mortimer Market Centre, London; University College London; St Thomas' Hospital, London Medical School; University College London Medical School

Since the previous edition, big advances have been made in treatment, knowledge of the disease and epidemiology. The problem of AIDS in developing countries has become a major political and humanitarian issue.

* Edited by the Director of the Department for Sexually Transmitted Diseases, *ABC of HIV and AIDS, 6th Edition* is an authoritative guide to the epidemiology, incidence, and most up to date management of HIV and AIDS
* Reflects the constantly changing knowledge of the disease and its manifestations, new developments in drug and non-drug management, sociological and political issues
* Includes 6 new chapters on conditions associated with AIDS and further concentration on the community effects of the disease, and the situation of women with AIDS
* Ideal for all levels of health care workers caring for HIV and AIDS patients

JUNE 2012 | 9781405157001 | 144 PAGES | £24.99/US$49.95/€32.90/AU$47.95

# ABC of Pain

**Lesley A. Colvin & Marie Fallon**
Western General Hospital, Edinburgh; University of Edinburgh

Pain is a common presentation and this brand new title focuses on the pain management issues most often encountered in primary care. *ABC of Pain*:

- Covers all the chronic pain presentations in primary care right through to tertiary and palliative care and includes guidance on pain management in special groups such as pregnancy, children, the elderly and the terminally ill
- Includes new findings on the effectiveness of interventions and the progression to acute pain and appropriate pharmacological management
- Features pain assessment, epidemiology and the evidence base in a truly comprehensive reference
- Provides a global perspective with an international list of expert contributors

JUNE 2012 | 9781405176217 | 128 PAGES | £24.99/US$44.95/€32.90/AU$47.95

**ABC of Pain**
Edited by Lesley A. Colvin and Marie Fallon
WILEY-BLACKWELL   www.abcbookseries.com   BMJ|Books

# ABC of Urology

## 3RD EDITION

**Chris Dawson & Janine Nethercliffe**
Fitzwilliam Hospital, Peterborough; Edith Cavell Hospital, Peterborough

Urological conditions are common, accounting for up to one third of all surgical admissions to hospital. Outside of hospital care urological problems are a common reason for patients needing to see their GP.

- *ABC of Urology, 3rd Edition* provides a comprehensive overview of urology
- Focuses on the diagnosis and management of the most common urological conditions
- Features 4 additional chapters: improved coverage of renal and testis cancer in separate chapters and new chapters on management of haematuria, laparoscopy, trauma and new urological advances
- Ideal for GPs and trainee GPs, and is useful for junior doctors undergoing surgical training, while medical students and nurses undertaking a urological placement as part of their training programme will find this edition indispensable

MARCH 2012 | 9780470657171 | 88 PAGES | £23.99/US$37.95/€30.90/AU$47.95

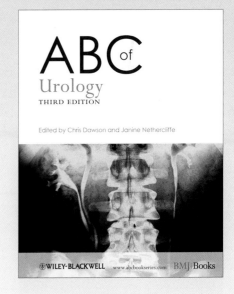

**ABC of Urology**
THIRD EDITION
Edited by Chris Dawson and Janine Nethercliffe
WILEY-BLACKWELL   www.abcbookseries.com   BMJ|Books

# ABC of Occupational and Environmental Medicine

## 3RD EDITION

**David Snashall & Dipti Patel**

Guy's & St. Thomas' Hospital, London; Medical Advisory Service for Travellers Abroad (MASTA)

Since the publication of last edition, there have been huge changes in the world of occupational health. It has become firmly a part of international public health, and in Britain there is now a National Director for Work and Health. This fully updated new edition embraces these changes and:

- Provides comprehensive guidance on current occupational and environmental health practice and legislation
- Concentrates on the newer kinds of occupational disease, for example 'RSI', pesticide poisoning and electromagnetic radiation, where exposure and effects are difficult to understand
- Places an emphasis on work, health and well-being, and the public health benefits of work, the value of work, disabled people at work, the aging workforce, and vocational rehabilitation
- Includes chapters on the health effects of climate change and of occupational health and safety in relation to migration and terrorism

NOVEMBER 2012 | 9781444338171 | 168 PAGES | £27.99/US$44.95/€38.90/AU$52.95

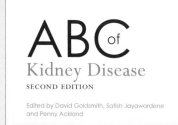

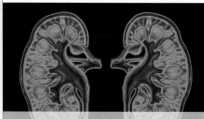

# ABC of Kidney Disease

## 2ND EDITION

**David Goldsmith, Satish Jayawardene & Penny Ackland**

Guy's & St. Thomas' Hospital, London; King's College Hospital, London; Melbourne Grove Medical Practice, London

Nephrology is sometimes considered a complicated and specialized topic and the illustrative ABC format will help GPs quickly and easily assimilate the information needed. *ABC of Kidney Disease, 2nd Edition*:

- Is a practical guide to the most common renal diseases to enable non-renal health care workers to screen, identify, treat and refer renal patients appropriately and to provide the best possible care
- Covers organizational aspects of renal disease management, dialysis and transplantation
- Provides an explanatory glossary of renal terms, guidance on anaemia management and information on drug prescribing and interactions
- Has been fully revised in accordance with new guidelines

OCTOBER 2012 | 9780470672044 | 112 PAGES | £27.99/US$44.95/€35.90/AU$52.95

# ALSO AVAILABLE

**ABC of Adolescence**
**Russell Viner**
2005 | 9780727915740 | 56 PAGES
£26.99 / US$41.95 / €34.90 / AU$52.95

**ABC of Antithrombotic Therapy**
**Gregory Y. H. Lip & Andrew D. Blann**
2003 | 9780727917713 | 67 PAGES
£26.50 / US$41.95 / €34.90 / AU$52.95

**ABC of Arterial and Venous Disease, 2nd Edition**
**Richard Donnelly & Nick J. M. London**
2009 | 9781405178891 | 120 PAGES
£31.50 / US$54.95 / €40.90 / AU$59.95

**ABC of Asthma, 6th Edition**
**John Rees, Dipak Kanabar & Shriti Pattani**
2009 | 9781405185967 | 104 PAGES
£26.99 / US$41.95 / €34.90 / AU$52.95

**ABC of Burns**
**Shehan Hettiaratchy, Remo Papini & Peter Dziewulski**
2004 | 9780727917874 | 56 PAGES
£26.50 / US$41.95 / €34.90 / AU$52.95

**ABC of Child Protection, 4th Edition**
**Roy Meadow, Jacqueline Mok & Donna Rosenberg**
2007 | 9780727918178 | 120 PAGES
£35.50 / US$59.95 / €45.90 / AU$67.95

**ABC of Clinical Electrocardiography, 2nd Edition**
**Francis Morris, William J. Brady & John Camm**
2008 | 9781405170642 | 112 PAGES
£34.50 / US$57.95 / €44.90 / AU$67.95

**ABC of Clinical Genetics, 3rd Edition**
**Helen M. Kingston**
2002 | 9780727916273 | 120 PAGES
£34.50 / US$57.95 / €44.90 / AU$67.95

**ABC of Clinical Haematology, 3rd Edition**
**Drew Provan**
2007 | 9781405153539 | 112 PAGES
£34.50 / US$59.95 / €44.90 / AU$67.95

**ABC of Clinical Leadership**
**Tim Swanwick & Judy McKimm**
2010 | 9781405198172 | 88 PAGES
£20.95 / US$32.95 / €26.90 / AU$39.95

**ABC of Complementary Medicine, 2nd Edition**
**Catherine Zollman, Andrew J. Vickers & Janet Richardson**
2008 | 9781405136570 | 64 PAGES
£28.95 / US$47.95 / €37.90 / AU$54.95

**ABC of COPD, 2nd Edition**
**Graeme P. Currie**
2010 | 9781444333886 | 88 PAGES
£23.95 / US$37.95 / €30.90 / AU$47.95

**ABC of Dermatology, 5th Edition**
**Paul K. Buxton & Rachael Morris-Jones**
2009 | 9781405170659 | 224 PAGES
£34.50 / US$58.95 / €44.90 / AU$67.95

**ABC of Diabetes, 6th Edition**
**Tim Holt & Sudhesh Kumar**
2007 | 9781405177849 | 112 PAGES
£31.50 / US$52.95 / €40.90 / AU$59.95

**ABC of Eating Disorders**
**Jane Morris**
2008 | 9780727918437 | 80 PAGES
£26.50 / US$41.95 / €34.90 / AU$52.95

**ABC of Emergency Differential Diagnosis**
**Francis Morris & Alan Fletcher**
2009 | 9781405170635 | 96 PAGES
£31.50 / US$55.95 / €40.90 / AU$59.95

**ABC of Geriatric Medicine**
**Nicola Cooper, Kirsty Forrest & Graham Mulley**
2009 | 9781405169424 | 88 PAGES
£26.50 / US$44.95 / €34.90 / AU$52.95

**ABC of Headache**
**Anne MacGregor & Alison Frith**
2008 | 9781405170666 | 88 PAGES
£23.95 / US$41.95 / €30.90 / AU$47.95

**ABC of Heart Failure, 2nd Edition**
**Russell C. Davis, Michael K. Davis & Gregory Y. H. Lip**
2006 | 9780727916440 | 72 PAGES
£26.50 / US$41.95 / €34.90 / AU$52.95

**ABC of Imaging in Trauma**
**Leonard J. King & David C. Wherry**
2008 | 9781405183321 | 144 PAGES
£31.50 / US$50.95 / €40.90 / AU$59.95

**ABC of Interventional Cardiology, 2nd Edition**
**Ever D. Grech**
2010 | 9781405170673 | 120 PAGES
£25.95 / US$40.95 / €33.90 / AU$49.95

**ABC of Learning and Teaching in Medicine, 2nd Edition**
**Peter Cantillon & Diana Wood**
2009 | 9781405185974 | 96 PAGES
£22.99 / US$35.95 / €29.90 / AU$44.95

**ABC of Liver, Pancreas and Gall Bladder**
**Ian Beckingham**
1905 | 9780727915313 | 64 PAGES
£24.95 / US$39.95 / €32.90 / AU$47.95

**ABC of Lung Cancer**
**Ian Hunt, Martin M. Muers & Tom Treasure**
2009 | 9781405146524 | 64 PAGES
£25.95 / US$41.95 / €33.90 / AU$49.95

**ABC of Medical Law**
**Lorraine Corfield, Ingrid Granne & William Latimer-Sayer**
2009 | 9781405176286 | 64 PAGES
£24.95 / US$39.95 / €32.90 / AU$47.95

**ABC of Mental Health, 2nd Edition**
**Teifion Davies & Tom Craig**
2009 | 9780727916396 | 128 PAGES
£32.50 / US$52.95 / €41.90 / AU$62.95

**ABC of Obesity**
**Naveed Sattar & Mike Lean**
2007 | 9781405136747 | 64 PAGES
£24.99 / US$39.99 / €32.90 / AU$47.95

**ABC of One to Seven, 5th Edition**
**Bernard Valman**
2009 | 9781405181051 | 168 PAGES
£32.50 / US$52.95 / €41.90 / AU$62.95

**ABC of Palliative Care, 2nd Edition**
**Marie Fallon & Geoffrey Hanks**
2006 | 9781405130790 | 96 PAGES
£30.50 / US$52.95 / €39.90 / AU$57.95

**ABC of Patient Safety**
**John Sandars & Gary Cook**
2007 | 9781405156929 | 64 PAGES
£28.50 / US$46.99 / €36.90 / AU$54.95

**ABC of Practical Procedures**
**Tim Nutbeam & Ron Daniels**
2009 | 9781405185950 | 144 PAGES
£31.50 / US$50.95 / €40.90 / AU$59.95

**ABC of Preterm Birth**
**William McGuire & Peter Fowlie**
2005 | 9780727917638 | 56 PAGES
£26.50 / US$41.95 / €34.90 / AU$52.95

**ABC of Psychological Medicine**
**Richard Mayou, Michael Sharpe & Alan Carson**
2003 | 9780727915566 | 72 PAGES
£26.99 / US$41.95 / €34.90 / AU$52.95

**ABC of Rheumatology, 4th Edition**
**Ade Adebajo**
2009 | 9781405170680 | 192 PAGES
£31.95 / US$50.95 / €41.90 / AU$62.95

**ABC of Sepsis**
**Ron Daniels & Tim Nutbeam**
2009 | 9781405181945 | 104 PAGES
£31.50 / US$52.95 / €40.90 / AU$59.95

**ABC of Sexual Health, 2nd Edition**
**John Tomlinson**
2004 | 9780727917591 | 96 PAGES
£31.50 / US$52.95 / €40.90 / AU$59.95

**ABC of Skin Cancer**
**Sajjad Rajpar & Jerry Marsden**
2008 | 9781405162197 | 80 PAGES
£26.50 / US$47.95 / €34.90 / AU$52.95

**ABC of Spinal Disorders**
**Andrew Clarke, Alwyn Jones & Michael O'Malley**
2009 | 9781405170697 | 72 PAGES
£24.95 / US$39.95 / €32.90 / AU$47.95

**ABC of Sports and Exercise Medicine, 3rd Edition**
**Gregory Whyte, Mark Harries & Clyde Williams**
2005 | 9780727918130 | 136 PAGES
£34.95 / US$62.95 / €44.90 / AU$67.95

**ABC of Subfertility**
**Peter Braude & Alison Taylor**
2005 | 9780727915344 | 64 PAGES
£24.95 / US$39.95 / €32.90 / AU$47.95

**ABC of the First Year, 6th Edition**
**Bernard Valman & Roslyn Thomas**
2009 | 9781405180375 | 136 PAGES
£31.50 / US$55.95 / €40.90 / AU$59.95

**ABC of the Upper Gastrointestinal Tract**
**Robert Logan, Adam Harris & J. J. Misiewicz**
2002 | 9780727912664 | 54 PAGES
£26.50 / US$41.95 / €34.90 / AU$52.95

**ABC of Transfusion, 4th Edition**
**Marcela Contreras**
2009 | 9781405156462 | 128 PAGES
£31.50 / US$55.95 / €40.90 / AU$59.95

**ABC of Tubes, Drains, Lines and Frames**
**Adam Brooks, Peter F. Mahoney & Brian Rowlands**
2008 | 9781405160148 | 88 PAGES
£26.50 / US$41.95 / €34.90 / AU$52.95

For more information on any of our medical books, please visit **www.wiley.com/go/medicine**

# Nursing

# Leadership & Management

## Third Edition

# Nursing Leadership & Management

## Third Edition

**Patricia Kelly,** RN, MSN
Professor Emerita
Purdue University Calumet
School of Nursing
Hammond, Indiana

and

Faculty
Evolve Testing & Remediation/
Health Education Systems, Inc.(HESI)
Houston, Texas

CENGAGE
Learning™

Australia • Brazil • Japan • Korea • Mexico • Singapore • Spain • United Kingdom • United States

**Nursing Leadership & Management,
Third Edition, International Edition**

Patricia Kelly

Vice President, Editorial: Dave Garza

Director of Learning Solutions: Matthew Kane

Executive Editor: Stephen Helba

Managing Editor: Marah Bellegarde

Senior Product Manager: Elisabeth F. Williams

Editorial Assistant: Jennifer Wheaton

Vice President, Marketing: Jennifer Baker

Marketing Director: Wendy E. Mapstone

Senior Marketing Manager: Michele McTighe

Marketing Coordinator: Scott A. Chrysler

Production Manager: Andrew Crouth

Content Project Manager: Allyson Bozeth

Senior Art Director: Jack Pendleton

For product information and technology assistance, contact us at
**Cengage Learning Customer & Sales Support, 1-800-354-9706**

For permission to use material from this text or product,
submit all requests online at **www.cengage.com/permissions.**
Further permissions questions can be e-mailed to
**permissionrequest@cengage.com**

International Edition:
ISBN-13: 978-1-111-30847-6
ISBN-10: 1-111-30847-0

Cengage Learning International Offices

**Asia**
www.cengageasia.com
tel: (65) 6410 1200

**Australia/New Zealand**
www.cengage.com.au
tel: (61) 3 9685 4111

**Brazil**
www.cengage.com.br
tel: (55) 11 3665 9900

**India**
www.cengage.co.in
tel: (91) 11 4364 1111

**Latin America**
www.cengage.com.mx
tel: (52) 55 1500 6000

**UK/Europe/Middle
East/Africa**
www.cengage.co.uk
tel: (44) 0 1264 332 424

**Represented in Canada by Nelson Education, Ltd.**
www. nelson.com
tel: (416) 752 9100 / (800) 668 0671

Cengage Learning is a leading provider of customized learning solutions with office locations around the globe, including Singapore, the United Kingdom, Australia, Mexico, Brazil and Japan. Locate your local office at **www.cengage.com/global**

For product information and free companion resources:
**www.cengage.com/international**
Visit your local office: **www.cengage.com/global**
Visit our corporate website: **www.cengage.com**

Printed in the United States of America
1 2 3 4 5 6 7 14 13 12 11

# CONTENTS

## CHAPTER 11:

### EFFECTIVE TEAM BUILDING / 267

## CHAPTER 12:

### POWER / 283

## CHAPTER 13:

### CHANGE, INNOVATION, AND CONFLICT MANAGEMENT / 297

## UNIT 3

# Leadership and Management of Patient-Centered Care / 322

## CHAPTER 14:

### BUDGET CONCEPTS FOR PATIENT CARE / 322

## CHAPTER 15:

### EFFECTIVE STAFFING / 343

## CHAPTER 16:

### DELEGATION OF PATIENT CARE / 368

## CHAPTER 17:

### ORGANIZATION OF PATIENT CARE / 400

## CHAPTER 18:

### TIME MANAGEMENT AND SETTING PATIENT CARE PRIORITIES / 424

# CONTRIBUTORS

**Rinda Alexander, PhD, RN, CS**
Professor Emeritus, Nursing
Purdue University Calumet
Hammond, Indiana
and
Nursing Consultant
Health Care and Administration
Schererville, Indiana
*Chapter 5: Evidence-Based Health Care*

**Kim Siarkowski Amer, PhD, RN**

Associate Professor
Department of Nursing
DePaul University
Chicago, Illinois
*Chapter 4: Basic Clinical Health Care
Economics*

**Margaret M. Anderson, EdD, RN**

Professor of Nursing and Past Chair
College of Health Professions
Department of Advanced Nursing Studies
Northern Kentucky University
Highland Heights, Kentucky
*Chapter 13: Change, Innovation,
and Conflict Management*

**Ida M. Androwich, PhD, RN, BC, FAAN**

Professor and Director,
Health Systems Management
Niehoff School of Nursing
Loyola University Chicago
Chicago, Illinois
*Chapter 4: Basic Clinical Health Care
Economics*
*Chapter 5: Evidence-Based Health Care*

*Chapter 10: Strategic Planning and
Organizing Patient Care*

**Crisamar J. Anunciado, RN, MSN, FNP-BC**

Inpatient Nurse Practitioner
Sharp Chula Vista Medical Center
Chula Vista, California
and
Adjunct Faculty
Southwestern Community College
Associate Degree Nursing Program
Chula Vista, California
*Chapter 11: Effective Team Building*

**Anne Bernat, RN, MSN, CNAA**
Vice President of Patient Care Services (Retired)
Arlington, Virginia
*Chapter 15: Effective Staffing*

**Nancy Braaten, RN, MS**

Adult Health Clinical Nurse Specialist
Clinical Analyst, Nursing Information
Systems
Northeast Health Acute Care Division
Troy, New York
*Chapter 19: Patient and Health Care
Education*

**Sister Kathleen Cain, OSF, JD**
Attorney
Franciscan Legal Services
Baton Rouge, Louisiana
*Chapter 23: Legal Aspects of Health Care*

**Carolyn Christie-McAuliffe, PhD, FNP**

Director of Research
Hematology Oncology Associates of CNY
East Syracuse, New York
and
Assistant Professor, SUNY Institute
of Technology
Utica, New York
*Chapter 5: Evidence-Based Health Care*
*Chapter 30: Healthy Living: Balancing*
*Personal and Professional Needs*

**Martha Desmond, RN, MS, Post Masters
Certificate in Nursing Education**
Clinical Nurse Specialist in Critical Care
Northeast Health Acute Care Division
Troy, New York
and
Adjunct Faculty
Excelsior College
Albany, New York
*Chapter 19: Patient and Health Care Education*

**Joan Dorman, RN, MS, CEN**
Clinical Assistant Professor
Purdue University Calumet
Hammond, Indiana
*Chapter 24: Ethical Aspects of Health Care*

**Deborah Erickson, PhD, RN**

Assistant Professor
Bradley University
Peoria, Illinois
*Chapter 8: Personal and Interdisciplinary*
*Communication*

**Barbara K. Fane, MS, RN, APRN-BC**

Clinical Nurse Specialist, Critical Care
Northeast Health Acute Care Division
Troy, New York
and
Adjunct Clinical Faculty
Southern Vermont College
Bennington, Vermont
*Chapter 22: Decision Making and*
*Critical Thinking*

**Mary L. Fisher, PhD, RN, CNAA, BC**

Professor and Department Chair
Environments for Health
Indiana University School of Nursing
Indianapolis, Indiana
*Chapter 15: Effective Staffing*

**Charlene C. Gyurko, PhD, RN, CNE**

Assistant Professor
Purdue University Calumet
School of Nursing
Hammond, Indiana
*Chapter 1: Nursing Leadership and*
*Management*

**Corinne Haviley, MS, RN**

Associate Chief Nursing Officer
Central DuPage Hospital
Winfield, Illinois
*Chapter 14: Budget Concepts for Patient*
*Care*

Photo by Allen Bourgeois Photography

**Paul Heidenthal, MS**
Senior Curriculum Developer
Texas Department of Health and Human Services
Austin, Texas
*Chapter 19: Patient and Health Care Education*

**Sara Anne Hook, JD, MLS, MBA**

Professor of Informatics, Indiana University
School of Informatics
Indianapolis, Indiana
and
Adjunct Professor of Law,
Indiana University School of Law
Indianapolis, Indiana
*Chapter 23: Legal Aspects of Health Care*

**Karen Houston, RN, MS**
Director of Quality and Continuum of Care
Albany Medical Center
Albany, New York
*Chapter 20: Managing Outcomes Using an*
*Organizational Quality Improvement Model*

**Ronda G. Hughes, PhD, MHS, RN**
Senior Health Scientist Administrator
Senior Advisor on End-of-Life Care
Center for Primary Care, Prevention, and Clinical Partnerships
Agency for Healthcare Research and Quality
Rockville, Maryland
*Chapter 2: The Health Care Environment*

**Mary Anne Jadlos, MS, ACNP-BC, CWOCN**

Coordinator—Wound, Skin and Ostomy
Nursing Service
Northeast Health Acute Care Division
Samaritan Hospital
Troy, New York
and
Albany Memorial Hospital
Albany, New York
*Chapter 21: Evidence-Based Strategies to*
*Improve Patient Care Outcomes*

**Josette Jones, PhD, RN**
Assistant Professor
Indiana University School of Informatics
School of Nursing
Indianapolis, Indiana
*Chapter 6: Nursing and Health Care Informatics*

**Stephen Jones, MS, RN, CPNP ET**
Pediatric Clinical Nurse Specialist/Nurse Practitioner
The Children's Hospital at Albany Medical Center
Albany, New York
and
Founder, Pediatric Concepts
Averill Park, New York
*Chapter 28: Emerging Opportunities*

**Patricia Kelly, RN, MSN**
Professor Emerita
Purdue University Calumet
Hammond, Indiana
and
Faculty
Health Education Systems, Inc. (HESI)
Houston, Texas
*Chapter 2: The Health Care Environment*
*Chapter 5: Evidence-Based Health Care*
*Chapter 16: Delegation of Patient Care*
*Chapter 31: NCLEX Preparation and Professionalism*

**Glenda B. Kelman, PhD, ACNP-BC**
Associate Professor and Chair
Nursing Department
The Sage Colleges
Troy, New York
and
Acute Care Nurse Practitioner
Wound, Skin, and Ostomy
Nursing Service
Northeast Health Acute Care Division
Samaritan Hospital
Troy, New York
and
Albany Memorial Hospital
Albany, New York
*Chapter 21: Evidence-Based Strategies to Improve Patient Care Outcomes*

**Mary Elaine Koren, RN, PhD**
Associate Professor of Nursing
Area Coordinator
Northern Illinois University
School of Nursing and Health Studies
DeKalb, Illinois
*Chapter 30: Healthy Living: Balancing Personal and Professional Needs*

**Lyn LaBarre, MS, RN**
Patient Care Service Director
Critical Care, Specialty and Emergency Services
Albany Medical Center
Albany, New York
*Chapter 29: Your First Job*

**Linda Searle Leach, PhD, RN, NEA-BC**
Assistant Professor
UCLA School of Nursing
Los Angeles, California
*Chapter 1: Nursing Leadership and Management*

**Camille B. Little, MS, RN, BSN**
Instructional Assistant Professor (Retired)
Mennonite College of Nursing
Illinois State University
Normal, Illinois
*Chapter 24: Ethical Aspects of Health Care*

**Sharon Little-Stoetzel, RN, MS, CNE**
Associate Professor of Nursing
MidAmerica Nazarene University
Independence, Missouri
*Chapter 22: Decision Making and Critical Thinking*

**Miki Magnino-Rabig, PhD, RN**
Assistant Professor
University of St. Francis
Joliet, Illinois
*Chapter 29: Your First Job*

**Patsy L. Maloney, EdD, MSN, MA, RN-BC, NEA-BC**
Professor and Director,
Continuing Nursing Education
School of Nursing
Pacific Lutheran University
Tacoma, Washington
*Chapter 9: Politics and Consumer Partnerships*
*Chapter 12: Power*
*Chapter 18: Time Management and Setting Patient Care Priorities*

**Richard J. Maloney, EdD, MA, MAHRM, BS**

Partner
Policy Governance Associates
Tacoma, Washington
*Chapter 9: Politics and Consumer
Partnerships*
*Chapter 12: Power*
*Chapter 18: Time Management and
Setting Patient Care Priorities*

**Maureen T. Marthaler, RN, MS**

Associate Professor
School of Nursing
Purdue University Calumet
Hammond, Indiana
*Chapter 16: Delegation of
Patient Care*

**Judith W. Martin, RN, JD**

Attorney
Franciscan Legal Services
Baton Rouge, Louisiana
*Chapter 23: Legal Aspects of Health Care*

**Edna Harder Mattson, RN, BN, BA(CRS), MDE**

Doctoral Student in Education
University of Phoenix
and
President
International Nursing Consultation
and Tutorial Services
Winnipeg, Manitoba, Canada
*Chapter 20: Managing Outcomes Using
an Organizational Quality Improvement
Model*
*Chapter 27: Career Planning*

**Mary McLaughlin, RN, MBA**

Assistant Director for Case Management and
Social Work
Albany Medical Center
Albany New York
*Chapter 20: Managing Outcomes Using an
Organizational Quality Improvement Model*

**Terry W. Miller, PhD, RN**

Dean and Professor
Pacific Lutheran University
School of Nursing
Tacoma, Washington
*Chapter 9: Politics and Consumer
Partnerships*
*Chapter 12: Power*

**Leslie H. Nicoll, PhD, MBA, RN, BC**

President and Owner
Maine Desk, LLC
Portland, Maine
*Chapter 6: Nursing and Health Care Informatics*

**Laura J. Nosek, PhD, RN**

Doctor of Nursing Practice Faculty
The Bolton School of Nursing
Case Western Reserve University
Cleveland, Ohio
and
Adjunct Associate Professor of Nursing
Marcella Niehoff School of Nursing
Loyola University Chicago
Chicago, Illinois
and
Course Facilitator
Excelsior College
Albany, New York
*Chapter 4: Basic Clinical Health Care Economics*

**Amy Androwich, O'Malley, MSN, RN**

Education and Program Manager
Medela, Inc.
McHenry, Illinois
*Chapter 5: Evidence-Based Health Care*
*Chapter 10: Strategic Planning and
Organizing Patient Care*

**Kristine E. Pfendt, RN, MSN**

Associate Professor
College of Health Professions
Department of Advanced Nursing Studies
Northern Kentucky University
Highland Heights, Kentucky
*Chapter 13: Change, Innovation,
and Conflict Management*

**Karin Polifko-Harris, PhD, RN, CNAA**

Vice President
Organization Development and Research
Naples Community Healthcare System
Naples, Florida
*Chapter 11: Effective Team Building*
*Chapter 25: Culture, Generational Differences, and
Spirituality*
*Chapter 27: Career Planning*

**Chad Priest, RN, MSN, JD**

Chief Executive Officer
MESH, Inc.
Indianapolis, Indiana
*Chapter 23: Legal Aspects of Health
Care*

**Jacklyn Ludwig Ruthman, PhD, RN**

Associate Professor, Retired
Bradley University
Peoria, Illinois
*Chapter 8: Personal and Interdisciplinary Communication*

**Patricia M. Lentsch Schoon, MPH, RN**

Adjunct Associate Professor
School of Graduate and Professional Programs
St. Mary's University,
Minneapolis, Minnesota
and
Clinical Instructor
School of Nursing
University of Wisconsin Oshkosh,
Oshkosh, Wisconsin
*Chapter 7: Population-Based Health Care Practice*

**Kathleen F. Sellers, PhD, RN**

Associate Professor
School of Nursing and Health Systems
SUNY Institute of Technology
Utica, New York
*Chapter 10: Strategic Planning and Organizing Patient Care*
*Chapter 17: Organization of Patient Care*

**Susan Abaffy Shah, MS, ACNP-BC**

Division of Cardiology
Albany Medical Center
Albany, New York
*Chapter 19: Patient and Health Care Education*

**Maria R. Shirey, PhD, MBA, RN, NEA-BC, FACHE, FAAN**

Associate Professor
University of Southern Indiana
College of Nursing and Health Professions
Evansville, Indiana
*Chapter 3: Organizational Behavior and Magnet Hospitals*

**Tanya L. Sleeper, GNP-BC, MSN, MSB**

Assistant Professor
Division of Nursing
University of Maine at Fort Kent
Fort Kent, Maine
*Chapter 2: The Health Care Environment*

**Nancy S. Sisson, MS, RN**

Adult Health Clinical Nurse Specialist
Clinical Nurse Specialist - Telemetry
Northeast Health Acute Care Division
Albany, New York
*Chapter 19: Patient and Health Care Education*

**Erin C. Soucy, MSN, RN**

Director, Division of Nursing
Assistant Professor of Nursing
University of Maine at Fort Kent
Fort Kent, Maine
*Chapter 29: Your First Job*

**Sara Swett, RN, BSN, MSN**

Clinical Instructor
Pacific Lutheran University
School of Nursing
Tacoma, Washington
*Chapter 25: Culture, Generational Differences, and Spirituality*

**Janice Tazbir, RN, MS, CCRN, CS**

Associate Professor of Nursing
Purdue University Calumet
School of Nursing
Hammond, Indiana
*Chapter 26: Collective Bargaining*

**Beth A. Vottero, PhD, RN, CNE**

Assistant Professor of Nursing
Purdue University Calumet
School of Nursing
Hammond, Indiana
*Chapter 15: Effective Staffing*

**Karen Luther Wikoff, RN, PhD**

Assistant Professor
California State University, Stanislaus
Turlock, California
*Chapter 25: Culture, Generational Differences, and Spirituality*

# REVIEWERS

**Karla Huntsman, RN, MSN/Ed**
Nursing Instructor
AmeriTech College
Draper, Utah

**Jaclynn A. Johnson, RNC-OB, MSN**
Otero Junior College
La Junta, Colorado

**Carolyn L. McKinney, RN, MSN/Ed, PHN**
Associate Professor of Nursing
California State University-Northridge

Northridge, California

**Cydney King Mullen, RN, PhD**
Professor of Nursing
Sandhills Community College
Pinehurst, North Carolina

**Tina Saunders, RN, MSN, CNE**
College of Nursing
Kent State University
Kent, Ohio

# PREFACE

Nurses play a crucial role in protecting patient safety and providing quality health care. A National Academy of Sciences, Institute of Medicine (IOM) report found that "how we are cared for by nurses affects our health, and sometimes can be a matter of life and death . . . nurses are indispensable to our safety" (Institute of Medicine, 2004). This finding was confirmed by an emerging body of research showing that nurses are much more likely than any other health professional to recognize, interrupt, and correct errors that are often life threatening (Rothschild et al., 2006), that higher staffing levels of nurses in hospital settings reduce mortality and failure to rescue rates (Aiken et al., 2010; Aiken, 2011, Rafferty et. al., 2007), and that inadequate nurse staffing levels may lead to a higher incidence of complications and inadequate care (Aiken, Clarke, Cheung, Sloane, & Silber, 2002; Joint Commission, 2002; Needleman & Buerhaus, 2003).

Another IOM report, *Health Professions Education: A Bridge to Quality* (2003), noted that nurses and other health professionals are currently not prepared to provide the highest quality and safest care possible. The IOM report concluded that education for the health professions is in need of a major overhaul and recommended that all programs that educate and train health professionals should adopt five core competencies. These core competencies are the ability to (1) provide patient-centered care, (2) work in interdisciplinary teams, (3) employ evidence-based practice, (4) apply quality improvement, and (5) utilize informatics (IOM, 2003).

The American Association of Colleges of Nursing (AACN) convened a task force to identify essential baccalaureate core competencies that should be achieved by professional nurses to assure high quality and safe patient care (AACN, 2006). These competencies include knowledge and skills related to critical thinking, including the application of evidence-based knowledge and quality improvement; health care systems and policies that contribute to safe and high-quality patient outcomes; communication, especially with the interdisciplinary team and shift handoff communication (Joint Commission, 2006); illness and disease management of individuals and communities; ethics; and information and health care technologies. These AACN competencies are essential for the nursing leader and manager of the future.

*Nursing Leadership & Management*, third edition, is designed to help beginning nurses address the IOM and AACN competencies. The text prepares beginning nurse leaders and managers for modern health care. It takes in to account the Patient Protection and Affordable Care Act of 2010 and the Health Care and Education Reconciliation Act of 2010. It also reviews information from the Agency for Healthcare Quality and Research, the Joint Commission, the Leapfrog Group, Thomson Medstat, the National Quality Forum, and the Institute for Healthcare Improvement (IHI), all focused on improvement of the quality and safety of patient care.

The chapter contributors to this third edition include nurse educators, nursing faculty, informatics faculty, clinical nurse specialists, lawyers, nurse practitioners, wound and ostomy care nurses, nurse entrepreneurs, and others. These contributors are from the United States and Canada, introducing a broad view of nursing leadership and management. There are contributors from California, Florida, Illinois, Indiana, Kentucky, Maine, Maryland, Minnesota, Missouri, New York, Ohio, Texas, Vermont, Washington, and Manitoba, Canada. Interviews with Dr. Loretta Ford, Dr. Susan Morrison, and many other nurse experts are also included.

## ORGANIZATION

*Nursing Leadership & Management*, third edition, consists of 31 chapters organized in a conceptual framework. This conceptual framework outlines the nurse's leadership and management responsibilities to the patient, to the community, to the health care team, to the institution, and to self. The five units of the framework provide beginning nurse leaders and managers with the knowledge needed in today's health care environment.

- Unit 1 introduces nursing leadership and management, including the health care environment, organizational behavior and magnet hospitals, basic clinical health care economics, evidence-based health care, nursing and health care informatics, and population-based health care practice.
- Unit 2 discusses leadership and management of the inter-disciplinary team, including personal and interdisciplinary communication; politics and consumer partnerships; strategic planning and organizing patient care; effective team building; power; and change, innovation, and conflict management.
- Unit 3 discusses leadership and management of patient-centered care, including budget concepts for patient care, effective staffing, delegation of patient care, organization of patient care, time management and setting patient care priorities, and patient and health care education.
- Unit 4 discusses quality improvement of patient outcomes, including managing outcomes utilizing an organizational quality improvement model; evidence-based strategies to improve patient care outcomes; decision making and critical thinking; legal aspects of health care; ethical aspects of health care; and culture, generational differences, and spirituality.
- Unit 5 discusses leadership and management of self and the future, including collective bargaining, career planning, nursing job opportunities, your first job, balancing personal and professional needs, and NCLEX preparation and professionalism.

Numerous additions and enhancements have been made to the text and supplements offerings to increase your understanding and broaden your learning experience. Selected enhancements are outlined below.

- A stronger foundation for evidence-based health care with attention to high quality, safe care, is emphasized throughout the text.
- Updated coverage of the role of the entry level staff nurse in nursing leadership and management gives a framework for understanding how the new nurse can become a leader and make a difference within the vast health care system.
- Chapters have been updated to include new information from national, federal, and state health care and nursing organizations. New coverage on transforming care at the bedside, based on the Institute for Healthcare Improvement (IHI) recommendations, is included, as is updated information from The Joint Commission (JC) on patient safety and legal aspects of nursing.

- A section on disaster planning has been added to the chapter on population-based health care.
- Content on time management, setting patient care priorities, budgeting, and health care economics, has been significantly reworked and expanded.
- Patient care assignment sheets have been added to facilitate hands-on practice with completing sheets correctly during nursing leadership experiences in the clinical area.
- Expanded test questions are included in all chapters to help in preparation for in-class exams as well as state and national board exams. Question formats include single choice and new NCLEX-style multiple response questions.
- The chapter on licensure preparation helps readers ensure that they are prepared for licensure by including a mini-nutrition chart, revised mini-medication chart, and tips on preparing for the licensure examination.
- A new companion **Premium Website**, accessed at www. CengageBrain.com, includes answers to text Review Questions and Review Activities, as well as a Glossary, and Chapter Summaries, Weblinks, and Objectives. Enter your passcode, found in the front of the book, and the Premium Website will be added to your bookshelf. **ISBN-13: 978-1-111-64105-4**

# CHAPTER FEATURES

Several standard chapter features are utilized throughout the text, which provide the reader with a consistent format for learning and an assortment of resources for understanding and applying the knowledge presented. Features include the following:

- Health care or nursing quote related to chapter content
- Objectives that state the chapter's learning goals
- Opening scenario, a mini entry-level nursing case study that relates to chapter, with two to three critical thinking questions
- Key Concepts, a listing of the primary understandings the reader is to take from the chapter
- Key Terms, a listing of important new terms presented in the chapter
- Review Questions, several NCLEX-style questions at the end of chapter content
- Review Activities to apply chapter content to entry-level nursing situations
- Exploring the Web research activities
- References
- Suggested Readings

Special elements are sprinkled throughout the chapters to enhance learning and encourage critical thinking and application:

- Evidence from the Literature with synopsis of key findings from nursing and health care literature

- Real World Interviews with nursing leaders and managers, including nursing staff, clinicians, administrators, risk managers, faculty, nursing and medical practitioners, patients, nursing assistive personnel (NAP), lawyers, and hospital administrators
- Critical Thinking exercises regarding an ethical, legal, cultural, spiritual, delegation, or quality improvement nursing or health care topic
- Case Studies to provide the entry-level nurse with a clinical nursing leadership/management situation calling for critical thinking to solve an open-ended problem

# INSTRUCTOR RESOURCES (IR)

**ISBN-13: 978-1-111-30669-4**

An *Instructor Resources (IR)* CD is available to adopters of the text. It is designed to assist faculty in presenting to nursing students the essential skills and information that are needed to help them secure a position as a beginning nursing manager and leader. The IR will assist faculty in planning and developing their programs and classes for the most efficient use of time and resources. The IR includes four must-have components:

1. An **Instructor Manual** offers practical resources for presenting material in the text and includes suggested answers to the text Critical Thinking exercises, Review Questions, Review Activities, and Case Studies; as well as a guided discussion of the chapter opening scenario.
2. **Lecture Slides in PowerPoint** ™ serve as guides for presentation in the classroom.
3. A **Test Bank** offers approximately 850 questions in multiple-choice and multiple response formats.
4. An **Image Library** includes photos and diagrams from the text, which can be imported into classroom presentations.

# REFERENCES

Aiken, L. H. (2011). Nurses for the future. *New England Journal of Medicine.* January 20;364(3):196–8. Epub 2010 Dec 15.

Aiken, L. H., Sloane, D. M., Cimiotti, J. P., Clarke, S. P., Flynn , L., Seago, J. A., Spetz, J., & Smith, H. L. (2010). Implications of the California nurse staffing mandate for other states. Health services research. August;45(4):904–21. Epub 2010 April 9.

Aiken, L. H., Clarke, S. P., Cheung, R. B., Sloane, D. M., Sochalski, J., & Silber, J. H. (2002, October 23). Hospital nurse staffing and patient mortality, nurse burnout, and job dissatisfaction. *Journal of the American Medical Association,* 288, 1987–1993.

American Association of Colleges of Nursing. (2006). *Hallmarks of quality and patient safety in baccalaureate nursing education.* Washington, DC: Author.

Institute of Medicine. (2003). *Health professions education: A bridge to quality.* Washington, DC: National Academies Press.

Institute of Medicine. (2004). *Keeping patients safe: Transforming the work environment of nurses.* Washington, DC: National Academies Press.

Joint Commission (JC). (2002). *Health care at the crossroads: Strategies for addressing the evolving nursing crisis.* Chicago: Author.

Joint Commission (JC). (2006). National Patient Safety Goals. Retrieved August 25, 2006, from http://www.jointcommission.org

Needleman, J., & Buerhaus, P. (2003). Nurse staffing and patient safety: current knowledge and implications for action. International *Journal of Quality Health Care.* August 15(4):275–7.

Rafferty, A. M., Clarke S. P., Coles J., Ball J., James, P., McKee, M., & Aiken , L.H. (2007). Outcomes of variation in hospital nurse staffing in English hospitals: cross-sectional analysis of survey data and discharge records. *International Journal of Nursing Studies.* February; 44(2):175–82. Epub 2006 Oct 24.

Rothschild, J. M., Hurley, A. C., Landrigan, C. P., Cronin, J. W., Martill-Waldrop, K., Foskitt, C. Burdick, E., Czeisler, C. A., & Bates, D. W. (2006, February). Recovering from medical errors: The critical care nursing safety net. *Joint Commission Journal on Quality and Patient Safety,* 32 (2), 63–72.

# ACKNOWLEDGMENTS

A book such as this requires great effort and the coordination of many people with various areas of expertise. I would like to thank all of the contributors for their time and effort in sharing their knowledge gained through years of experience in both the clinical and academic setting. All of the contributing authors worked within tight time frames to accomplish their work. Special thanks go to Robyn Pozza, Attorney, Austin, Texas, for her contributions to the legal chapter. Thanks also to Dr. Patricia Padjen, Madison, Wisconsin; and Dr. Susan Morrison, Houston, Texas, for their critical review and input into select chapters. I especially thank Jo Reidy and Corinne Haviley, Chicago, Illinois, for their help in arranging some of the photographs for the text.

I thank the reviewers for their time spent critically reviewing the manuscript and providing the valuable comments that have enhanced this text. Special thanks go to my Dad and Mom, Ed and Jean Kelly; my sisters, Tessie Dybel and Kathy Milch; my Aunt Pat and Uncle Bill Kelly (who convinced me to start writing); my Aunt Verna and Uncle Archie Payne; my nieces, Natalie Dybel Bevil, Melissa Milch Arredondo, and Stacey Milch; my nephew, John Milch; my grand nephew, Brock Bevil; my grand niece, Reese Bevil; my nephews-in-law, Tracy Bevil and Peter Arredondo; and my dear friends, Patricia Wojcik, Florence Lebryk, Lee McGuan, Dolores Wynen, and Joan Fox, who have supported me throughout my writing of this book. Thanks to Ron Vana, my friend, who makes me happy. Special thanks to my wonderful nursing friends, Zenaida Corpuz, Dr. Mary Elaine Koren, Dr. Barbara Mudloff, Dr. Patricia Padjen, Jane McKeon, Kerrie Ellingsen, and especially to Judy Ilijanich, Gerri Kane, Janice Klepitch, Sylvia Komyatte, and Julie Martini, as well as Anna Fizer, Trudy Keilman, Mary Kay Moredich, Judy Rau, Lillian Rau, and Ivy Schmude, who have supported me throughout this book and during our forty-nine years together as nurses. Special thanks to my faculty mentors, Dr. Imogene King, Dr. Joyce Ellis, and Nancy Weber.

I would like to acknowledge and sincerely thank the team at Delmar Cengage Learning who have worked to make this book a reality. Beth Williams, Senior Product Manager, is a great person who has worked tirelessly and brought knowledge, guidance, humor, and attention to help keep me motivated and on track throughout the project. Allyson Bozeth, Content Project Manager, skillfully managed the production process for the text. Steve Helba, Executive Editor at Delmar Cengage Learning, is a great support. Thanks also to Jane Woodruff for all the computer support.

# ABOUT THE AUTHOR

Patricia Kelly earned a Diploma in Nursing from St. Margaret Hospital School of Nursing, Hammond, Indiana; a Baccalaureate in Nursing from DePaul University in Chicago, Illinois; and a Master's Degree in Nursing from Loyola University in Chicago, Illinois. Pat is Professor Emerita, Purdue University Calumet, Hammond Indiana. She has worked as a staff nurse, school nurse, and nurse educator. Pat has traveled extensively in the United States, Canada, and Puerto Rico, teaching conferences for the Joint Commission, Resource Applications, Pediatric Concepts, and Kaplan, Inc. She currently teaches three day NCLEX-RN reviews for Evolve Testing & Remediation/Health Education Systems, Inc.(HESI), Houston, Texas, in various cities across the United States, and works part time as a volunteer nurse at Old Irving Park Community Clinic, Chicago, Illinois, a free clinic staffed by a team of volunteer nurses, physicians, and lay people for patients with no insurance, where she provides health care and teaches various health care topics to patients.

Pat was Director of Quality Improvement at the University of Chicago Hospitals and Clinics. She has taught at Wesley-Passavant School of Nursing, Chicago State University, and Purdue University Calumet, Hammond Indiana. She has taught Fundamentals of Nursing, Adult Nursing, Nursing Leadership and Management, Nursing Issues, Nursing Trends, Quality Improvement, and Legal Aspects of Nursing. Pat is a member of Sigma Theta Tau, the American Nurses Association, and the Emergency Nurses Association. She is listed in *Who's Who in American Nursing, 2000 Notable American Women,* and the *International Who's Who of Professional and Business Women.*

Pat has served on the Board of Directors of Tri City Mental Health Center, St. Anthony's Home, and the Quality Connection Journal. She is the author of *Nursing Leadership & Management,* Delmar Cengage Learning (2008, and 2003), *Essentials of Nursing Leadership & Management,* Delmar Cengage Learning (2010 and 2004). Pat is the co-author, with Maureen Marthaler, of *Nursing Delegation, Setting Priorities, and Making Patient Care Assignments,* Delmar Cengage Learning (2011), and *Delegation of Nursing Care,* Delmar Cengage Learning (2005). Pat contributed a chapter on " Obstructive Lung Disease " to Daniels, *Medical Surgical Nursing,* Delmar Cengage Learning (2010), and a chapter on "Preparing the Undergraduate Student and Faculty to Use Quality Improvement in Practice" to *Improving Quality,* second edition, by Meisenheimer. She has written several articles, including "Chest X-Ray Interpretation" and many articles on quality improvement. Pat has served as a Disaster Volunteer for the American Red Cross; as a volunteer nurse with health care teams of nurses, doctors, pharmacists, and lay people on health care visits to Nicaragua; as a volunteer at a church food pantry in Austin, Texas; and currently serves as a volunteer at a church food pantry in Chicago, Illinois. Throughout most of her career, she has taught nursing at the university level. Pat has been licensed and has worked in many states over her career, including Indiana, Illinois, Wisconsin, Oklahoma, New York, and Pennsylvania. Pat may be contacted at patkh1@aol.com. patkelly777@aol.com.

# HOW TO USE THIS BOOK

## QUOTE

*A nursing or health care theorist quote gives a professional's perspective regarding the topic at hand; read this as you begin each new chapter and see whether your opinion matches or differs, or whether you are in need of further information.*

## OBJECTIVES

*These goals indicate to you the performance-based, measurable objectives that are targeted for mastery upon completion of the chapter.*

## OPENING SCENARIO

*This mini case study with related critical thinking questions should be read prior to delving into the chapter; it sets the tone for the material to come and helps you identify your knowledge base and perspective.*

## CASE STUDY

*These short cases with related questions present a beginning clinical nursing management situation calling for judgment, decision making, or analysis in solving an open-ended problem. Familiarize yourself with the types of situations and settings you will later encounter in practice, and challenge yourself to devise solutions that will result in the best outcomes for all parties, within the boundaries of legal and ethical nursing practice.*

## EVIDENCE FROM THE LITERATURE

*Study these key findings from nursing and health care research, theory, and literature, and ask yourself how they will influence your practice. Do you see ways in which your nursing could be affected by these literature findings and research results? Do you agree with the conclusions drawn by the author?*

 **EVIDENCE** FROM THE LITERATURE

**Citation:** Buresh, B., & Gordon, S. (2006). *From Silence to Voice.* Ithaca, NY: Cornell University Press.

**Discussion:** This book discusses the fact that not enough nurses are willing to talk about their work. When nurses and nursing organizations do talk about their work, too often they unintentionally project an inaccurate picture of nurses using a virtue script instead of a knowledge script to highlight what nurses do. They discuss the virtuous and caring acts of nurses. They do not discuss the expert knowledge that nurses bring to patient care. When nursing groups give voice to nursing, they sometimes bypass, downplay, or even devalue the basic nursing work that occurs in direct care of the sick, while elevating the image of elite nurses in advance practice, administration, or academia. This contributes to social stereotypes that deride anyone who is "just a nurse."

**Implications for Practice:** If nursing is misunderstood by the public or by those with influence, it will be vulnerable to the budget axe, and new resources for nursing education and practice won't happen. A focus on the virtues of nurses is an invitation to seek not the best and brightest but rather the most virtuous, meekest, and self-sacrificing.

### Critical Thinking 8-5

Nurses need to be aware that different staff in the workforce may see things differently based on their age differences. For example, some staff often do only exactly what is asked. They work primarily to earn money to spend. Other staff may believe it is important to save, save, save. They work hard to eliminate a task. Some staff like to buy now, pay later, and work very efficiently. Some other staff work fast, sacrifice, and are very thrifty. These various staff members may have a hard time working together toward a goal. Do you see any of these age differences affecting the clinical area where you work? How can nurses bridge these differences?

## CRITICAL THINKING

*Ethical, cultural, spiritual, legal, delegation, and performance improvement considerations are highlighted in these boxes. Before beginning a new chapter, page through and read the Critical Thinking sections and jot down your comments or reactions, then see whether your perspective changes after you complete the chapter.*

## REAL WORLD INTERVIEWS

*Interviews with well-known nursing leaders are included as well as interviews with nursing and medical practitioners, hospital administrators, staff, patients, and family members. As you read these, ask yourself whether you had ever considered that individual's point of view on the given topic. How would knowing another person's perspective affect the care you deliver?*

 **REAL** WORLD INTERVIEW

A second shift occupational health nurse working in a factory setting was presented with a patient who entered the nursing office complaining that he didn't feel good. The nurse's initial assessment, including vital signs, revealed that the only abnormality was an elevated blood pressure. In this situation, like in any clinical situation, it is important to distinguish the urgent from the nonurgent. With hypertensive patients, it is important to realize that an urgent situation is suggested by evidence of acute end organ damage from the elevated BP. Specifically, in this situation, it was important to know whether the patient was experiencing altered sensorium, headache, visual disturbance, chest pain, or dyspnea. The presence of any of these findings should be communicated to the physician and would dictate urgent transport to the hospital. In their absence, the patient can be referred for more-elective blood pressure control.

In any clinical situation such as the one above, the nurse can check the patient for evidence of any signs or symptoms. The nurse can then facilitate communications by being organized and objective. Discuss the basics such as the patient's chief complaint, his vital signs, his medications, and any changes from baseline. Know why you are worried about observed changes, and communicate this to the practitioner.

**John C. Ruthman, MD**
Peoria, Illinois

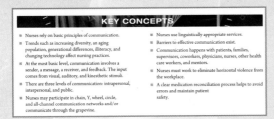

## KEY CONCEPTS

*This bulleted list serves as a review and study tool for you as you complete each chapter.*

## KEY TERMS

*Study this list prior to reading the chapter, and then again as you complete a chapter to test your true understanding of the terms and concepts covered. Make a study list of terms you need to focus on to thoroughly appreciate the material of the chapter.*

## REVIEW QUESTIONS

*These questions will challenge your comprehension of objectives and concepts presented in the chapter and will allow you to demonstrate content mastery, build critical thinking skills, and achieve integration of the concepts. Answers are found on the Premium Website.*

## REVIEW ACTIVITIES

*These thought-provoking activities at the close of a chapter invite you to approach a problem or scenario critically and apply the knowledge you have gained. You can find suggested responses to these on the Premium Website.*

## EXPLORING THE WEB

*Internet activities encourage you to use your computer and reasoning skills to search the Web for additional information on quality and nursing leadership and management.*

## REFERENCES

*Evidence-based research, theory, and general literature, as well as nursing, medical, and health care sources, are included in these lists; refer to them as you read the chapter and verify your research.*

## SUGGESTED READINGS

*These entries invite you to pursue additional topics, articles, and information in related resources.*

## PHOTOS, TABLES, AND FIGURES

*These items illustrate key concepts and offer visual reinforcement.*

# UNIT 1
# Nursing Leadership and Management

# CHAPTER 1

# Nursing Leadership
# and Management

LINDA SEARLE LEACH, PhD, RN, CNAA;
CHARLENE C. GYURKO, PhD, RN, CNE

*The future happens at the intersection of knowledge and service… if you have the knowledge you can provide the service… leadership requires a commitment to life-long learning and staying on the edge of the knowledge work that nurses do every day.*

(DAN PESUT, PhD, RN, PMHCNS-BC, FAAN)

## OBJECTIVES

Upon completion of this chapter, the reader should be able to:

1. Differentiate between leadership and management.
2. Distinguish characteristics of effective leaders.
3. Identify leadership theories.
4. Apply knowledge of leadership theory in carrying out the nurse's role as a leader.
5. Identify concepts of management.
6. Identify the management process.
7. Outline 10 roles that managers fulfill in an organization.
8. Relate management theories.
9. Summarize motivation theories.

Delmar/Cengage Learning

*Ed Harley was admitted to the cardiac observation unit earlier in the day. He had been diagnosed previously with heart disease and had experienced episodes of ventricular arrhythmias. His cardiologist had determined the need to change his antiarrhythmic medication to reduce the side effects Mr. Harley was experiencing. That evening, while Mr. Harley was talking to his wife on the phone and as his nurse Maria was walking to his bedside, he suddenly stopped talking and went into ventricular tachycardia and cardiac arrest. Maria reacted immediately and started cardiopulmonary resuscitation (CPR). Unable to use the phone to call for help, she gave a precordial thump to his chest. Normal sinus rhythm appeared on the monitor before anyone else could respond to the code. Mr. Harley was then transferred to the coronary care unit (CCU).*

*Maria had been a registered nurse (RN) less than one year at the time, and although she had participated in code arrests a few times, she had never witnessed one occur right before her eyes. Her knowledgeable action saved this patient's life. In nursing, CPR is a mandatory skill and considered part of a nurse's ordinary work. Yet it is quite extraordinary work.*

*Everything had happened so quickly that evening that Maria did not have a chance to talk to the patient before he was transferred. She entered his room the next morning in CCU, as the sun was just rising. As he awoke, Maria spent that quiet time with him. While he embraced the start of a new day, his thoughts were intense. What he chose to share was this acknowledgment: "You saved my life. Thank you." This precious moment was a celebration of both of their lives.*

*What leadership characteristics did Maria demonstrate in preventing a nurse-sensitive outcome of cardiac arrest?*

*Why is Maria considered a leader, even though she is not in a leadership or management position?*

*Why is leadership important at all levels throughout a health care organization?*

Professionals use their expertise and specialized knowledge to perform leadership roles. Many people think leaders are only top corporate executives and administrators, political representatives, military generals, or those who head organizations. This is because these leaders are highly visible and hold high-profile positions. We need leaders, though, at all levels of the organization. Leadership is a basic competency needed by all health professionals. Leadership development is a necessary part of the preparation of health care providers.

Nurses make a critical difference every day in the lives of their patients and patients' families, yet nurses believe those accomplishments are part of their ordinary work. Nurses are leaders, and by using their expert knowledge, they manage and meet patient care needs.

Leadership and management are different. Leadership influences or inspires the actions and goals of others. One does not have to be in a position of authority to demonstrate leadership. Not all leaders are managers. Bennis and Nanus (1985) popularized the phrase, "leaders are people who do the right thing; managers are people who do things right." Both roles are crucial but different. Management is a structural process, while leadership empowers followers and, by its inherent nature, is symbolic and political. Leadership and management are both human-resources oriented. Bolman and Deal (2003) address management as a structural process in that goals are achieved and problems are solved. As management and leadership are people oriented, people's skills, attitudes, energy, and commitment are people and human-resources oriented. People need to be managed as well as inspired and led. Leadership is part of management, not a substitute for it. We need both.

This chapter lays the groundwork for the development of knowledge about nursing leadership and management. Many concepts touched on in this chapter will be developed in depth in other chapters of the book. This chapter discusses motivation and leadership and provides a framework to differentiate leadership and management. Leadership characteristics, styles of leadership, and leadership theories are described. The chapter introduces the process of management and explains management theories and functions. Management is defined, and current trends are discussed.

## DEFINITION OF LEADERSHIP

**Leadership** is commonly defined as a process of influence in which the leader influences others toward goal achievement (Yukl, 1998). Influence is an instrumental part of leadership and means that leaders affect others, often by inspiring, enlivening, and engaging others to participate. The process of leadership involves the leader and the follower in interaction. This implies that leadership is a reciprocal relationship. Leadership can occur between the leader and another individual; between the leader and a group; or between a leader and an organization, a community, or a society. Defining leadership as a process helps us understand more about leadership than the traditional view of a leader being in a position of authority, exerting command, control, and power over subordinates. There are many more leaders in organizations than those who are in positions of authority. Each person has the potential to serve as a leader. What this means for nurses as professionals is that they function as leaders when they influence others toward goal achievement. Nurses are leaders.

Leadership can be **formal leadership**, as when a person is in a position of authority or in a sanctioned, assigned role within an organization that connotes influence, such as a clinical nurse specialist (Northouse, 2001). An **informal leader** is an individual who demonstrates leadership outside the scope of a formal leadership role or as a member of a group rather than as the head or leader of the group. The informal leader is considered to have emerged as a leader when she is accepted by others and is perceived to have influence.

## LEADERS AND FOLLOWERS

Leaders and followers are both necessary roles. Leaders need followers in order to lead. Followers need leaders in order to follow. Nurses are alternately leaders and followers when they work with other health care team members to achieve patient care goals, participate in meetings, and so forth. The most valuable followers are skilled, self-directed employees who participate actively in setting the group's direction and who invest time and energy in the work of the group, thinking critically and advocating for new ideas (Grossman & Valiga, 2008). Good followers communicate and work well with others, being supportive, yet thoughtful, in their approach to new ideas.

## LEADERS VERSUS MANAGERS

Kotter (1990a) describes the differences between leadership and management in the following way: Leadership is about creating change, and management is about controlling complexity in an effort to bring order and consistency. He says that leading change involves establishing a direction, aligning people through empowerment, and motivating and inspiring them toward producing useful change and achieving the vision, whereas management is defined as planning and budgeting, organizing and staffing, problem solving, and controlling complexity to produce predictability and order (Kotter, 1990b).

Nurses are leaders. Nurses function as leaders when they demonstrate leadership characteristics in their nursing roles and lead other nurses and their communities to achieve a vision of quality health care. See Table 1-1 for examples of nurses carrying out nursing leadership characteristics and role activities.

### Critical Thinking 1-1

Leaders create change. Managers control complexity. Note the selected components of leadership in Table 1-1 to help you determine your leadership and role activities. How could you apply one of the activities to your patient care?

### TABLE 1-1  Nursing Leadership Characteristics and Role Activities

| CHARACTERISTICS OF LEADERS | EXAMPLES OF NURSING ROLE ACTIVITIES |
| --- | --- |
| Leadership requires personal mastery. | Nurses demonstrate leadership when they show competence and mastery in the tasks they perform. Nurses are deemed competent by means of a license to practice nursing (NLN, 2010). |
| Leadership is about values. | Nurses exhibit leadership through their demonstration of cultural values that are embraced through individual belief systems. Nurses display their personal and professional values as they serve others. Values are often entwined with ethical conflicts (Dahnke, 2009). |
|  | The National League for Nursing implements its mission guided by four dynamic and integrated core values that permeate the organization and are reflected in its work:<br>• CARING: promoting health, healing, and hope in response to the human condition<br>• INTEGRITY: respecting the dignity and moral wholeness of every person without conditions or limitation<br>• DIVERSITY: affirming the uniqueness of and differences among persons, ideas, values, and ethnicities<br>• EXCELLENCE: creating and implementing transformative strategies with daring ingenuity (NLN, 2010) |

*(Continues)*

## TABLE 1-1 (Continued)

| CHARACTERISTICS OF LEADERS | EXAMPLES OF NURSING ROLE ACTIVITIES |
| --- | --- |
| Leadership is about service. | Service learning is a current buzz word in nursing. Service learning links information learned in the classroom or learning environment to the community and can enhance culturally congruent care (Amerson, 2010). |
| Leadership is about people and relationships. | Nurses demonstrate leadership and play roles in patient outcomes when they build relationships with patients and their significant others (Wong & Cummings, 2007). |
| Leadership is contextual. | Nurses demonstrate leadership when they adjust their leadership styles, depending on the context that surrounds a particular situation, to achieve nursing goals. A major context evolves around the interrelationships that nurses have with others (Spence Laschinger, Finegan, & Wilk, 2009). |
| Leadership is about the management of meaning. | Nurses demonstrate leadership when they monitor the meaning of what is being communicated, both verbally and nonverbally, and manage the situation to achieve goals for all involved. The communication must be clear and inspiring (Murphy, 2005). |
| Leadership is about balancing. | Nurses demonstrate leadership when they multitask and balance all that they do to achieve nursing goals. |
| Leadership is about continuous learning and improvement. | Nurses demonstrate leadership by continuing to increase and improve their knowledge and expertise. Nightingale argued, "…a nurse never stops learning" (Mills, 1964, p.35). |
| Leadership is about effective decision making. | Nurses demonstrate leadership when they make effective, evidence-based decisions. Nurses must be autonomous in their decision making and also work with other members of the health care team to assure the best care for their patients (Wong & Cummings, 2009). |
| Leadership is a political process. | Nurses demonstrate leadership when they participate in nursing organizations and various political processes in their states and nations. (Bishop, 2010). |
| Leadership is about modeling. | Nurses demonstrate leadership when they model learned beliefs and practices as they mentor other nurses. |
| Leadership is about integrity. | Nurses demonstrate leadership when they consistently model integrity, an expectation of a leader. |

**Source:** Compiled with information from Moore, J. (2004). Leadership: Lessons learned. PowerPoint presentation to Indiana State University PhD Educational Leadership and Foundation Students. Terre Haute, Indiana. Nonpublished PowerPoint presentation.

## LEADERSHIP CHARACTERISTICS

According to Bennis and Nanus (1985), there are three fundamental qualities that effective leaders share. The first quality is a guiding vision. Leaders focus on a professional and purposeful vision that provides direction toward the preferred future. As a nurse leader's actions are purposefully directed toward a preferred future, it seems logical that nurses increase their education as trends in patient care develop. The second quality is passion. Passion expressed by the leader involves the ability to inspire and align people toward the promises of life. Passion is an inherent quality of the nurse leader. This passion about assisting patients to recover helps the patient and significant others live life to the fullest extent possible. The third quality is integrity that is based on knowledge of self, honesty, and maturity that is developed through experience and growth. As a nurse leader, integrity is important and is enhanced throughout one's life span. McCall (1998) describes how self-awareness—knowing our strengths and weaknesses—can allow us to use feedback and learn from our mistakes. Daring and curiosity are also basic ingredients of leadership that leaders draw on to take risks, learning from what works as much as from what does not (Bennis & Nanus, 1985).

The American Association of Critical-Care Nurses in their landmark work, *AACN Standards for Establishing Healthy Work Environments: A Journey to Excellence,* cite authentic leadership as one of the key standards and assert that authentic leadership requires skill in the core competencies of self-knowledge, strategic vision, risk taking and creativity, interpersonal and communication effectiveness, and inspiration (AACN, 2004).

Certain characteristics are commonly attributed to leaders. These traits are considered desirable and seem to contribute to the perception of being a leader. They include intelligence, self-confidence, determination, integrity, and sociability (Stodgill, 1948, 1974). Research among 46 hospitals designated as magnet hospitals for their success in attracting and retaining registered nurses emphasized the value of leaders who are visionary and enthusiastic, are supportive and knowledgeable, have high standards and expectations, value education and professional development, demonstrate power and status in the organization, are visible and responsive, communicate openly, and are active in professional associations (McClure & Hinshaw, 2002; Scott et al., 1999; Kramer, 1990; McClure, Poulin, Sovie, & Wandelt, 1983; Kramer & Schmalenberg, 2005). Research findings from studies on nurses revealed that caring, respectability, trustworthiness, and flexibility were the leadership characteristics most valued. In one study, nurse leaders identified managing the dream, mastering change, designing organization structure, learning, and taking initiative as leadership characteristics (Murphy & DeBack, 1991). Research by Kirkpatrick and Locke (1991) concluded that leaders are different from nonleaders across six traits: drive, the desire to lead, honesty and integrity, self-confidence, cognitive ability, and knowledge of the business. Although no set of traits is definitive and reliable in determining who is a leader or who is effective as a leader, many people still rely on personality traits to describe and define leadership characteristics.

## LEADERSHIP THEORIES

Many believe that the critical factor needed to maximize human resources is leadership (Bennis & Nanus, 1985). A more in-depth understanding of leadership can be gleaned from a review of leadership theories. The major leadership theories can be classified according to the following approaches: behavioral, contingency, and contemporary.

### Behavioral Approach

Leadership studies from the 1930s by Kurt Lewin and colleagues at Iowa State University conveyed information about three leadership styles that are still widely recognized today: autocratic, democratic, and laissez-faire leadership (Lewin, 1939; Lewin & Lippitt, 1938; Lewin, Lippitt, & White, 1939). **Autocratic leadership** involves centralized decision making, with the leader making decisions and using power to command and control others. **Democratic leadership** is participatory, with authority delegated to others. To be influential, the democratic leader uses expert power and the power base afforded by having close, personal relationships. The third style, **laissez-faire leadership**, is passive and permissive, and the leader defers decision making. Lewin (1939) contrasted these styles and concluded that autocratic leaders were associated with high-performing groups, but that close supervision was necessary, and feelings of hostility were often present. Democratic leaders engendered positive feelings in their groups, and performance was strong whether or not the leader was present. Low productivity and feelings of frustration were associated with laissez-faire leaders.

Behavioral leadership studies from the University of Michigan and from Ohio State University led to the identification of two basic leader behaviors: job-centered behaviors and employee-centered behaviors. Effective leadership was described as having a focus on the human needs of subordinates and was called **employee-centered leadership** (Moorhead & Griffin, 2001). **Job-centered leaders** were seen as less effective because of their focus on schedules, costs, and efficiency, resulting in a lack of attention to developing work groups and high-performance goals (Moorhead & Griffin, 2001).

The researchers at Ohio State focused their efforts on two dimensions of leader behavior: initiating structure and consideration. **Initiating structure** involves an emphasis on the work to be done, a focus on the task, and production. Leaders who focus on initiating structure are concerned with how work is organized and on the achievement

**REAL** WORLD INTERVIEW

It is important that beginning nurses see themselves as leaders. Leadership doesn't mean you have to have an administrative title. The central task of leadership is catalyzing others to achieve shared values in a complex world that is constantly changing and requiring us to design new ways of achieving our goals. This is a description of what is required of nurses each and every day.

**Angela Barron McBride, PhD, RN, FAAN**
Distinguished Professor-Dean Emerita
Indiana University School of Nursing
Indianapolis, Indiana

# EVIDENCE FROM THE LITERATURE

**Citation:** Gordon, S. (2006). What do nurses really do? *Topics in Advanced Practice Nursing, 6*(1). Retrieved February 10, 2011 from http://www.medscape.com/viewarticle/520714.

**Discussion:** In this article by Suzanne Gordon, a journalist and author who writes about nursing, nurses are called upon to do a better job of accurately describing what nurses do and how they use expert knowledge acquired through scientific and technical mastery. She says, "What do nurses do? They save lives, prevent complications, prevent suffering, and save money." Her message to nurses is that there is a reason that the public has such little understanding of nursing and the importance of our work. The reason is twofold: traditional stereotypes about nursing cloud the reality of nursing as it is currently practiced, and nurses have been patterned to describe their contribution to health care in self-sacrificing and anonymous ways.

**Implications for Practice:** Nurses need to be clear about why it is important for the public to know what and how nurses contribute to health care. This article is an important vehicle from which nurses can begin to examine their own words and ways of discussing what nurses do and reflect upon the historical religious and societal practices that interfere with a clear, accurate, and realistic image of modern nursing. What nurses often think of as their ordinary work is really quite extraordinary. Nurses use scientific knowledge, expert judgment, and complex skills to make critical decisions that affect patient outcomes. Nurses need to be able to articulate how they do their work and the difference it makes.

## Critical Thinking 1-2

Among the individuals commonly identified as leaders (shown below), can you identify a set of leadership traits that they all possess, such as knowledge, self-confidence, determination, integrity, sociability, cognitive ability, caring, honesty, trustworthiness, flexibility, desire to lead, and drive? Now divide your class into groups. Have each group identify someone from this list whom they see as a leader and have each group describe that leader's traits. Then have the groups share with the class which leaders they chose. Do the leaders selected demonstrate any of the leadership traits identified above? When you work on the clinical unit, do you see any staff nurses displaying these leadership traits? How can you develop these traits in yourself?

**LEADERS AMONG US: PAST AND PRESENT**

| | |
|---|---|
| Mother Teresa | Martin Luther King |
| Fay Bower | Sister Rosemary Donley |
| Joyce Clifford | Jean Watson |
| Margaret Sovie | Roy Simpson |
| Virginia Henderson | Pam Cipriano |
| Linda Burnes Bolton | Florence Nightingale |
| Billye Brown | Gale Pollock |
| Imogene King | Leah Curtin |
| Muriel Poulin | Peter Buerhaus |
| Luther Christman | Linda Aiken |
| Rhonda Anderson | May Wykle |

of goals. Leader behavior includes planning, directing others, and establishing deadlines and details of how work is to be done. For example, a nurse demonstrating the leader behavior of initiating structure could be a charge nurse who, at the beginning of a shift, makes out a patient assignment.

The dimension of **consideration** involves activities that focus on the employee and emphasize relating and getting along with people. Leader behavior focuses on the well-being of others. The leader is involved in creating a relationship that fosters communication and trust as a basis for respecting other people and their potential contributions. A nurse demonstrating consideration behavior will take the time to talk with coworkers, be empathetic, and show an interest in them as people.

The leader behaviors of initiating structure and consideration define leadership style. The styles are as follows:

- Low initiating structure, low consideration
- High initiating structure, low consideration
- High initiating structure, high consideration
- Low initiating structure, high consideration

The Ohio State University studies associate the high initiating structure–high consideration leader behaviors with better performance and satisfaction outcomes than the other styles. This leadership style is considered effective, although it is not appropriate in every situation.

Another model based on these two dimensions is the managerial grid developed by Blake and Mouton (1985). Five styles identify the extent of structure, called *concern for production* and *concern for people*, demonstrated by the leader. The five leader styles are:

1. impoverished leader for low production concern and low people concern;
2. authority compliance leader for high production concern and low people concern;
3. country club leader for high people concern and low production concern;
4. middle-of-the-road leader for moderate concern in both dimensions; and
5. team leader for high production concern and high people concern.

## Critical Thinking 1-3

Identify the people who are leaders in your personal and professional life. In what way has their leadership affected you? As a person or as a nurse, identify the people you lead. In what way has your interaction with those people influenced them? Influenced you?

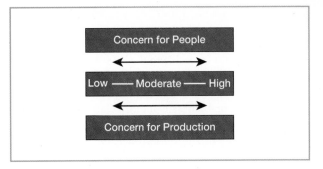

**FIGURE 1-1  Key leadership dimensions.** (*Source:* Compiled with information from Blake, Mouton, and McCanse *Leadership Grid from Leadership: Theory, Application, Skill Building* by R. N. Lussier and C. F. Achua, 2000, Cincinnatti, OH: South-Western College.)

Management is usually more effective when it does not overemphasize either concern for people or concern for production (Figure 1-1).

## Contingency Approaches

Another approach to leadership is **contingency theory**. Contingency theory acknowledges that other factors in the environment influence outcomes as much as leadership style and that leader effectiveness is contingent upon or depends upon something other than the leader's behavior. The premise is that different leader behavior patterns will be effective in different situations. Contingency approaches include Fielder's contingency theory, the situational theory of Hersey and Blanchard, path-goal theory, and the idea of substitutes for leadership.

**Fielder's Contingency Theory.**  Fielder (1967) is credited with the development of the contingency model of leadership effectiveness. Fielder's theory of leadership effectiveness views the pattern of leader behavior as dependent upon the interaction of the personality of the leader and the needs of the situation. The needs of the situation or how favorable the situation is toward the leader involves leader-member relationships, the degree of task structure, and the leader's position of power (Fielder, 1967). **Leader-member relations** are the feelings and attitudes of followers regarding acceptance, trust, and credibility of the leader. Good leader-member relations exist when followers respect, trust, and have confidence in the leader. Poor leader-member relations reflect distrust, a lack of confidence and respect, and dissatisfaction with the leader by the followers.

**Task structure** refers to the degree to which work is defined, with specific procedures, explicit directions, and goals. High task structure involves routine, predictable, clearly defined work tasks. Low task structure involves work that is not routine, predictable, or clearly defined, such as creative, artistic, or qualitative research activities.

**Position power** is the degree of formal authority and influence associated with the leader. High position power is favorable for the leader, and low position power is unfavorable. When all of these dimensions—leader-member relations, task structure, and position power—are high, the situation is favorable to the leader. When they are low, the situation is not favorable to the leader. In both of these circumstances, Fielder showed that a task-directed leader, concerned with task accomplishment, was effective. When the range of favorableness is intermediate or moderate, a human relations leader, concerned about people, was most effective. These situations need a leader with interpersonal and relationship skills to foster group achievement. Fielder's contingency theory is an approach that matches the organizational situation to the most favorable leadership style for that situation.

### Hersey and Blanchard's Situational Theory.

Situational leadership theory addresses follower characteristics in relation to effective leader behavior. Whereas Blake and Mouton focus on leader style and Fielder examines the situation, Hersey and Blanchard consider follower readiness as a factor in determining leadership style. Rather than using the words *initiating structure* and *contingency,* they use *task behavior* and *relationship behavior.*

High task behavior and low relationship behavior is called a *telling leadership* style. A high task, high relationship style is called a *selling leadership* style. A low task and high relationship style is called a *participating leadership* style. A low task and low relationship style is called a *delegating leadership* style.

Follower readiness, called maturity, is assessed in order to select one of the four leadership styles for a situation. For example, according to Hersey and Blanchard's situational leadership theory (2000), groups with low maturity, whose members are unable or unwilling to participate or are unsure, need a leader to use a telling leadership style to provide direction and close supervision. The selling leadership style is a match for groups with low to moderate maturity who are unable but willing and confident and need clear direction and supportive feedback to get the task done. Participating leadership style is recommended for groups with moderate to high maturity who are able but unwilling or are unsure and who need support and encouragement. The leader should use a delegating leadership style with groups of followers with high maturity who are able and ready to participate and can engage in the task without direction or support.

An additional aspect of this model is the idea that the leader not only changes leadership style according to followers' needs but also develops followers over time to increase their level of maturity (Lussier & Achua, 2000). Use of these four leadership styles helps a nurse manager assign work to others.

### Path-Goal Theory.

In this leadership approach, the leader works to motivate followers and influence goal accomplishment. The seminal author on path-goal theory is Robert House (1971). By using the appropriate style of leadership for the situation (i.e., directive, supportive, participative, or achievement oriented), the leader makes the path toward the goal easier for the follower. The *directive* style of leadership provides structure through direction and authority, with the leader focusing on the task and getting the job done. The *supportive* style of leadership is relationship oriented, with the leader providing encouragement, interest, and attention. *Participative* leadership means that the leader focuses on involving followers in the decision-making process. The *achievement-oriented* style provides high structure and direction as well as high support through consideration behavior. The leadership style is matched to the situational characteristics of the followers, such as the desire for authority, the extent to which the control of goal achievement is internal or external, and the ability of the follower to be involved. The leadership style is also matched to the situational factors in the environment, including the routine nature or complexity of the task, the power associated with the leader's position, and the work group relationship. This alignment of leadership style with the needs of followers is motivating and believed to enhance performance and satisfaction. The path-goal theory is based on expectancy theory, which holds that people are motivated when they believe they are able to carry out the work, they think their contribution will lead to the expected outcome, and they believe that the rewards for their efforts are valued and meaningful (Northouse, 2001).

### Substitutes for Leadership.

**Substitutes for leadership** are variables that may influence followers to the same extent as the leader's behavior. Kerr and Jermier (1978) investigated situational variables and identified some

### REAL WORLD INTERVIEW

Beginning nurses show leadership by bringing their understanding of evidence-based practices and by implementing them at the bedside. They are part of committees and have a voice on unit and hospital policies. Every nurse, no matter if brand new or seasoned, brings leadership qualities to the bedside. In the intensive care unit where I work, assessing changes in the patient's condition, notifying the residents, and making sure something gets done to help the patient in a matter of minutes requires leadership. It's what we do to make sure our patients get the best care.

**Johnny Tazbir, RN, BSN**
Crown Point, Indiana

## Critical Thinking 1-4

Historical leadership traits attributed to leaders are the following:

- Intelligence
- Self-confidence
- Determination
- Integrity
- Sociability
- Caring
- Respectability
- Trustworthiness
- Flexibility

Review each of these traits attributed to leaders. Are these the characteristics you think are important for nurse leaders? To what extent do you portray each of these leadership traits?

## Critical Thinking 1-5

In what way do professional nursing standards, such as standards for nursing practice, standards for nursing performance, and the Code of Ethics for nurses serve as a substitute for leadership?

aspects as substitutes that eliminate the need for leader behavior and other aspects as neutralizers that nullify the effects of the leader's behavior.

Some of these variables include follower characteristics, such as the presence of structured routine tasks, the amount of feedback provided by the task, and the presence of intrinsic satisfaction in the work; and organizational characteristics, such as the presence of a cohesive group, a formal organization, a rigid adherence to rules, and low position power. For example, an individual's experience substitutes for task-direction leader behavior (Kerr & Jermier, 1978). Nurses and other professionals with a great deal of experience already have knowledge and judgment and do not need direction and supervision to perform their work. Thus, their experience serves as a leadership substitute. Another substitute for leader behavior is intrinsic satisfaction that emerges from just doing the work. Intrinsic satisfaction occurs frequently among nurses when they provide care to patients and families. Intrinsic satisfaction substitutes for the support and encouragement of relationship-oriented leader behavior.

## CONTEMPORARY APPROACHES TO LEADERSHIP

Contemporary approaches to leadership address the leadership functions necessary to develop learning organizations. These approaches highlight charismatic theory, transformational leadership theory, servant leadership, and knowledge workers.

## Charismatic Theory

A charismatic leader has an inspirational quality that promotes an emotional connection from followers. House (1971) developed a theory of charismatic leadership that described how charismatic leaders behave as well as distinguishing characteristics and situations in which such leaders would be effective. Charismatic leaders display self-confidence, have strength in their convictions, and communicate high expectations and their confidence in others. They have been described as emerging during a crisis, communicating vision, and using personal power and unconventional strategies (Conger & Kanungo, 1987). One consequence of this type of leadership is a belief in the charismatic leader that is so strong that it takes on an almost supernatural purpose, and the leader is worshipped as if superhuman. Examples of charismatic leaders include Florence Nightingale and Martin Luther King.

Charismatic leaders can have a positive and powerful effect on people and organizations. Lee Iacocca, former chief executive officer (CEO) of Chrysler Corporation, and Herb Kelleher, former CEO of Southwest Airlines, are described as effective charismatic leaders. This type of leader can contribute significantly to an organization, even though all the leaders in an organization are not charismatic leaders. There are effective leaders who do not exhibit all the qualities associated with charismatic leadership. Charisma seems to be a special and valuable quality that some people have and some people do not.

## Transformational Leadership Theory

Burns defined transformational leadership as a process in which "leaders and followers raise one another to higher levels of motivation and morality" (Burns, 1978, p. 21). Transformational leadership theory is based on the idea of empowering others to engage in pursuing a collective purpose by working together to achieve a vision of a preferred future. This kind of leadership can influence both the leader and the follower to a higher level of conduct and achievement that transforms them both (Burns, 1978). Burns maintained that there are two types of leaders: the traditional manager concerned with day-to-day operations, called the **transactional leader**, and the leader who is committed to a vision that empowers others, called the **transformational leader**.

Transformational leaders motivate others by behaving in accordance with values, providing a vision that reflects

## REAL WORLD INTERVIEW

The development and mentoring of beginning nurses plays a critical role in delivering excellence across Community Healthcare Systems (CHS), its outpatient centers, physician offices, and three hospitals: Community Hospital in Munster; St. Catherine Hospital in East Chicago, and St. Mary Medical Center in Hobart. A strong focus on nursing education has facilitated the integration of best practices throughout the Northwest Indiana healthcare system, now ranked by Thompson Reuters as one of the Top 50 for quality and efficiency of care.

At CHS, our beginning nurses are supported in their profession by unit clinicians, educators, preceptors, and leaders committed to excellence. This affords an individualized orientation that builds on skill proficiency, incorporating evidenced-based practices. The new graduate nurse orientation includes classroom sessions, return demonstration of basic skills, equipment, and documentation along with online web-based programs to enhance clinical judgment. Education is further streamlined to focus on population specific care, and our bi-weekly evaluations set the path for continued learning relevant to each new nurse. CHS unit preceptors attend workshops designed to enhance mentoring and strong communication skills to provide accurate feedback of progress. The education team of unit area clinician, educators, and preceptors also plays an important role in the socialization of the new nurse to team members and unit routines.

CHS has taken another step in bringing state-of the-art technology to our nursing staff. With our addition of simulation lab programs, CHS educators will be able to provide realistic and challenging scenarios to integrate clinical decision making and critical thinking skills. This is an exciting path for the new nurse to be able to participate in problem-solving and augmentation of their professional development in the areas of assessment, monitoring, medication administration, and complex airway/ventilator management in a safe environment.

**Anthony Ferracone**
Vice President
Community Healthcare Systems
Munster, Indiana

---

mutual values, and empowering others to contribute. Bennis and Nanus (1985) describe this new leader as a leader who "commits people to action, who converts followers into leaders, and who converts leaders into agents of change" (p. 3). According to research by Tichy and Devanna (1986), effective transformational leaders identify themselves as change agents; are courageous; believe in people; are value driven; are lifelong learners; have the ability to deal with complexity, ambiguity, and uncertainty; and are visionaries. Yet transformational leadership may be demonstrated by anyone in an organization regardless of his position (Burns, 1978). The interaction that occurs between individuals can be transformational and motivate both to a higher level of performance (Bass, 1985).

Transformational leadership at the organizational level is about innovation and change. The transformational leader uses vision based on shared values to align people and inspire growth and advancement. It is both the inspiration and the empowerment aspects of transformational leadership that lead to commitment beyond self-interest, commitment to a vision, and commitment to action that

## CASE STUDY 1-1

A nurse is making rounds on her new postoperative laryngectomy patient. As the nurse enters the room, the patient begins to bleed from his neck incision. The nurse applies direct pressure to the patient's carotid artery with one hand and calls for assistance. Help arrives, and the patient is taken to surgery, with the nurse still maintaining pressure on the bleeding site. The patient lives and goes home a few days later.

How does nursing leadership and management on a patient care unit ensure good patient care in an emergency?

How can you develop your leadership and management skills to improve your ability to care for a group of patients?

creates change. Transformational leadership theory suggests that the relationship between the leader and the follower inspires and empowers an individual toward commitment to the organization.

Nurse researchers have described nurse executives according to transformational leadership theory and have used this theory to measure leadership behavior among nurse executives and nurse managers (Leach, 2005; Dunham-Taylor, 2000; Wolf, Boland, & Aukerman, 1994; McDaniel & Wolf, 1992; Young, 1992). Additionally, transformational leadership theory has been the basis for nursing administration curriculum (Searle, 1996) and for investigation of relationships such as between a nurse's commitment to an organization and productivity in a hospital setting (Leach, 2005; McNeese-Smith, 1997). Cassidy and Koroll (1998) explored the ethical aspects of transformational leadership, and Barker (1990) comprehensively discussed nursing in terms of transformational leadership theory. Of the contemporary theories of leadership, transformational leadership has been a popular approach in nursing.

Most recently, the Institute of Medicine identified transformational leadership theory as a precursor to any change initiative and stated that transformational leadership can be a crucial approach toward achieving work environments that optimize patient safety (IOM, 2003).

## Servant Leadership

In the early 1970s, Robert Greenleaf introduced another type of leadership, servant leadership. Greenleaf's definition of servant leadership focuses on wanting to put the needs of others above all else as the number-one priority. Only after this happens, Greenleaf contends, can a conscious choice to lead evolve (Greenleaf, 2002). Can a person be both a servant and a leader? A nurse is both. While serving and caring for patients, a nurse works in a leader role, making decisions based upon the best available evidence.

Characteristics of servant leadership include: listening, empathy, healing, awareness, persuasion, foresight, stewardship, growth, and building community (Jackson, 2007). For the nurse, each of these characteristics can stand alone or be entwined with each other. While the two words, servant leadership, seem to oppose each other, Greenleaf (2002) says they actually inspire one to collectively be more than the sum of individual parts.

Nurses listen to the patient and are empathetic. Nurses work to understand patients and help them heal, both physically and mentally. Nurses are aware of patient needs and may help to persuade patients of what is needed to make them whole. Nurses use their foresight, learn from past mistakes, and help patients learn from their past mistakes. Nurses are stewards. Stewardship includes the management of something entrusted to one's care (Merriam-Webster Dictionary, 2010). Nurses manage the care of the patients entrusted to their care.

Nurses care for the whole of the patient and their significant others, including their physical, mental, spiritual, and psychosocial being. Through a commitment to growth, nurses help patients grow as individuals. In helping patients become more independent in understanding how to manage the disease process, nurses help to build community by making the patients and their significant others aware of services beyond the walls of the hospital. Nurses are both servants and leaders.

## Knowledge Workers

As mentioned earlier, the organizations that nurses are a part of are changing. They reflect the advance and the promise of the technology that enables us to perform our work. Peter Drucker (1994) identifies the organization of the future as a knowledge organization composed of knowledge workers. Knowledge workers are those who bring specialized, expert knowledge to an organization. They are valued for what they know. The knowledge organization shares, provides, and grows the information necessary to work efficiently and effectively. Drucker says that knowledge organizations, in which the knowledge worker is at the front lines with the expertise and the information to act, will be the dominant organizational type (Drucker, 1994; Helgesen, 1995). In organizations such as these, the ideas of leadership at the top and leadership equated with the power of a position are obsolete notions. Knowledge workers with the expertise and information to act are the organization's leaders. They provide the service, interact with the customer, represent the organization, and accomplish its goals.

Knowledge workers work in the information age, where the rapid, instant access to information makes information the medium of exchange. Knowledge workers are valued for what they know. Knowledge workers with the expertise and information to act are valued for their human capital (Lawler, 2001).

In the information age, it is the development of new knowledge and innovation and its meaningful interpretation and application that becomes the source for transactions with patients and staff. Nursing's transition to the information age is occurring within the context of rapidly advancing technology and nanotechnology and is influenced by three key trends. These trends have been termed mobility, virtuality, and user-driven practices (Porter O'Grady, 2001).

*Mobility* refers to the ability to change skill sets as well as having the work dispersed among a variety of work locations, rather than work occurring at fixed sites (Bennis, Spreitzer, & Cummings, 2001). Nurses are working in many new settings today and are constantly adding to their knowledge as new technologies emerge.

*Virtuality* means working through virtual means using digital networks, where the worker may be far from the patient but present in a digital reality. Nurses are working in outpatient settings today where they carry a computer and are in instant communication with other practitioners and patients.

*User-driven practices* mean that the individual, at a time when digital mediums have given us more access to information and therefore more choices, acts more independently and is increasingly accountable for those choices and actions. Nurses are constantly assessing patients using traditional assessment methods as well as newer digital methods, for example, computerized vital sign monitors, and taking action to safeguard their patients and improve their care.

Nursing leadership practices are evolving to match nurses' work within this mobile, changing, environment with nurses who act without much supervision or guidance. This is facilitated by the growth and sophistication of nursing research, the application of nursing science, and the translation of available evidence into evidence-based nursing practice. The journey to the information age is continually fulfilling for nurses as they are able to display the rich and valuable contribution that knowledgeable nurses make to the quality of patient care and to quality health care outcomes.

## USING KNOWLEDGE AND EMOTIONAL INTELLIGENCE

Good nursing leadership and management of patients includes getting to know the patients; spending time assessing normal behavior, physiological and psychological responses to illness and hospitalization; and using knowledge to recognize even subtle changes in the patients' conditions and further evaluate them. Another key aspect of using knowledge is developing the ability to anticipate patient care problems. When a nurse intervenes with a postoperative patient who is bleeding from his incision by applying pressure at the site, the nurse minimizes the amount of bleeding. What assists a nurse to intervene correctly is using knowledge and anticipating such complications, thinking in advance of what should be done if a particular complication occurs, and then monitoring the patient to assess and identify complications early or, when possible, to prevent them. Another aspect of using knowledge for good leadership and management on the patient care unit is to have the right type of personnel and the right amount of personnel; in other words, having patient care unit staffing with enough registered nurses, licensed practical nurses, or nursing assistants on the unit, so that they all can fulfill their roles appropriately and with enough people to adequately care for the patients.

Nurses can develop their leadership and management skills with continuing education and by increasing their knowledge and expertise in caring for a group of patients by taking care of those patients regularly. The more experience a nurse gains in caring for particular patients, the more opportunity for learning to recognize patterns that occur with these patients. Patterns can include the type of symptoms that are common, possible complications or emergencies, and actions that can help prevent complications or

negative outcomes. Good planning on how to spend their time and determining what actions are nursing priorities and what can be delegated to others will also help a nurse manage a group of patients.

Knowledgeable nursing leadership and management on a nursing unit fosters good patient care by providing a supportive environment for nurses to deliver care. A supportive leadership and management environment does such things as providing a clear chain of command, clear job descriptions, patient care standards, good staffing ratios, good Internet and library resources, continuing education support, and so on. This allows the nurse to set goals, seek a mentor, and continue employment in a setting that is supportive of quality nursing care.

## Emotional Intelligence

**Emotional intelligence** is a component of leadership and refers to the capacity for recognizing your own feelings and those of others, for motivating yourself, and for managing emotions well in yourself and in your relationships. It describes abilities distinct from, but complementary to, academic intelligence. Many people who are *book smart* but lack emotional intelligence end up working for people who have lower IQs but excel in emotional intelligence skills (Goleman, 1998). Emotional intelligence includes these five basic emotional and social competencies:

1. *Self-awareness:* Knowing what you are feeling in the moment and using your preferences to guide your decision making; having a realistic assessment of your own abilities and a well-grounded sense of self-confidence
2. *Self-regulation:* Handling your emotions so that they facilitate rather than interfere with the task at hand; being conscientious and delaying gratification to pursue goals; and recovering well from emotional distress
3. *Motivation:* Using your deepest preferences to move and guide you toward your goals, to help you take initiative and strive to improve, and to persevere in the face of setbacks and frustrations
4. *Empathy:* Sensing what people are feeling; being able to take their perspective; and cultivating rapport and being in tune with a broad diversity of people
5. *Social skills:* Handling emotions in relationships well and accurately reading social situations and networks; interacting smoothly; using these skills to persuade and lead and negotiate and settle disputes; for cooperation and teamwork (Goleman, 1998)

## THE NEW LEADERSHIP

Margaret Wheatley, in *Leadership and the New Science* (1999), says, "There is a simpler way to lead organizations, one that requires less effort and produces less stress than the current practices." She presents a new view of leadership, one encompassing connectedness and self-organizing systems that follow a natural order of both chaos and uncertainty, which is different from a linear order in a hierarchy. The leader's function is to guide an organization using vision, to make choices based on mutual values, and to engage in the culture to provide meaning and coherence. This type of leadership fosters growth within each of us as individuals and as members of a group. The notion of connection within a self-organizing system optimizes autonomy at all levels because the relationships among the individual and the whole are strong. For nursing, such systems might be the infrastructure that will foster interdisciplinary decision making and strengthen the connection with nonprofessional workers.

In Wheatley's subsequent book, *Finding Our Way: Leadership for an Uncertain Time* (2005), she discusses how humans learn best when they are engaged in relationships with others and can exchange knowledge and expertise through informal, self-organized communities. Wheatley refers to these as communities of practice and encourages us to develop new leaders using communities of practice. Her notion of a community of practice represents several elements nurses are familiar with, that is, forming informal groups, using a group process of organizing, using principles of learning, and sharing information. What is unique in her description of these communities of practice is that they form via self-organization. They come together naturally. What makes these communities different from informal groups is Wheatley's

**REAL** WORLD
**INTERVIEW**

Transformational leadership encourages the retention of nurses through job satisfaction. Initially, it is important for a beginning nurse to have structure and supportive supervision. The nurse leader should be visible and approachable. The younger generation generally wants meaningful work, feedback, continuous learning, and flexibility for balancing activities. Transformational leadership fosters healthy staff-focused workplaces through such techniques as shared organizational goals, learning opportunities for career development, rewards, empowerment, autonomy, employee health programs, and shared governance. That facilitates quality of care, and consequently, job satisfaction and nurse retention.

**Anne Marriner-Tomey, PhD,RN**
Author, *Guide to Nursing Management and Leadership*

Professor Emerita
Indiana State University - College of Nursing
Terre Haute, Indiana

characterization of a community built from relationships and participation in a way that connects nurses and allows the creation of meaning from information or the exchange of knowledge. In work done for the Center of Creative Leadership, communities of practice are described as being different from the ideas or experiences we have had with groups, teams, and collective forming because communities of practice emerge from shared activity, shared knowledge, and ways of knowing that create meaning and thus a culture of engagement, participation, and relationships (Drath & Palus, 1994). Wheatley directs nurses to name these communities of practice that bring people together, support these connections, nourish the community, and illuminate their work. These exciting notions hold great promise for health professionals as we learn how to collaborate within and across disciplines and countries to advance health care practices.

## DEFINITION OF MANAGEMENT

**Management** is defined as a process of coordinating actions and allocating resources to achieve organizational goals. It is a process of planning, organizing and staffing, leading, and controlling actions to achieve goals. Planning involves setting goals and identifying ways to meet them.

### Critical Thinking 1-6

Becky is an Emergency Department (ED) nurse working with patients. Becky assesses her patients regularly and monitors their progress. She is certified in Advanced Cardiac Life Support and Trauma Nursing. One day, when she assesses her new patient, she notes that his pulse and respiration are increased and his blood pressure is decreased. She calls for the ED practitioner to see the patient immediately and simultaneously starts a large bore intravenous (IV) line and connects her patient to a cardiac monitor and the vital sign monitor. How has Becky demonstrated leadership? Do nurses regularly act to safeguard their patients in this way and prevent a negative nurse-patient outcome? Can the presence of safe staffing ratios affect a nurse's ability to deliver safe patient care and prevent negative nurse-sensitive patient care outcomes? Is Becky a knowledge worker?

Organizing and staffing is the process of ensuring that the necessary human and physical resources are available to achieve the planning goals. Organizing also involves assigning work to the right person or group and specifying who has the authority to accomplish certain tasks. Leading is influencing others to achieve the organization's goals and involves energizing, directing, and persuading others to achieve those goals. Finally, controlling is comparing actual performance to a standard and revising the original

## Critical Thinking 1-7

Note an administrative team during your clinical rotation. Who is part of the team? Does it include nurse managers, nursing staff, nurse clinicians, nurse practitioners, other health care practitioners, and staff from pharmacy, physical therapy, dietary, and so on? Does the team change depending on the problem it is working on?

plan as needed to achieve the goals (DuBrin, 2000). The daily activities of managers are diverse and fast paced, with regular interruptions. Priority activities are integrated among inconsequential ones. In the scope of one morning, a nurse manager may make serious decisions about a critically ill patient, a staff or patient complaint, a shortage of nurse staffing, and so forth. A nurse manager's work is driven by problems that emerge in random order and that have a range of importance and urgency. These circumstances create an image of the nurse manager as a *firefighter* involved in immediate and operational concerns. A significant proportion of a manager's time is spent in interaction with others, and more of the work is concerned with handling information than in making decisions (McCall et al., 1978). Nurse managers constantly interact with other members of a health care administrative team. This administrative team can include nurses, various health care practitioners, unit staff, and staff from other departments who share information and assure that quality patient outcomes are achieved.

## MANAGERIAL ROLES

One of the most frequently referenced taxonomies of managerial roles is from an in-depth, month-long study of five chief executives by Henry Mintzberg. A **taxonomy** is a system that orders principles into a grouping or classification. Mintzberg's observations led to the identification of three categories of managerial roles: (1) information-processing role, (2) interpersonal role, and (3) decision-making role (Mintzberg, 1973).

A role includes behaviors, expectations, and recurrent activities within a pattern that is part of the organization's structure (Katz & Kahn, 1978). Specific or distinct roles are part of each of the three categories of managerial roles. The information-processing roles are monitor, disseminator, and spokesperson, each of which is used to manage the information needs that people have. The interpersonal roles are figurehead, leader, and liaison, and each of these is used to manage relationships with people. The decisional roles are the entrepreneur, disturbance handler, allocator of resources, and negotiator roles that managers use to take action when making decisions.

## THE MANAGEMENT PROCESS

In the early 1900s, an emphasis on management as a discipline emerged with a focus on the science of management and a view that management is the art of accomplishing things through people (Follet, 1924). Henri Fayol, a manager, wrote a book in 1916 called *General and Industrial Management*. He described the functions of planning, organizing, coordinating, and controlling as the **management process** (Fayol, 1916/1949). His work has become a classic in the way that we define the process of managing.

Two other individuals, Gulick and Urwick, in some part as a result of their esteemed status as informal advisers to President Franklin D. Roosevelt, served to define the management process according to seven principles (Henry, 1992). Their principles form the acronym POSDCORB, which stands for planning, organizing, staffing, directing, coordinating (CO), reporting, and budgeting (Gulick & Urwick, 1937; Henry, 1992). Their work is also considered to be a classic description of management functions and is still a relevant description of how the management process is carried out today.

More recently, Yukl (1998) and colleagues (Kim & Yukl, 1995; Yukl, Wall, & Lepsinger, 1990) described 13 management functions that address two broad aspects of the management process: managing the work and managing relationships. The management functions for managing the work are planning and organizing, problem solving, clarifying roles and objectives, informing, monitoring, consulting, and delegating. The management functions for managing relationships are networking, supporting, developing and mentoring, managing conflict and team building, motivating and inspiring, and recognizing and rewarding.

The amount of time managers spend on particular roles or functions varies by the level of their positions in organizations, ranging from the first-level positions, to the middle-level positions, to the executive-level positions. A first-level managerial role or function in health care organizations is the nurse manager at the clinical bedside. First-level nurse managers spend the majority of time directly managing patient care and supervising others as they deliver care. The next highest percentage of their time is spent in planning. The rest of first-level nurse managers' functions take 10 percent or less of their time.

In contrast, middle-level nurse managers, often called nursing unit managers or nursing directors, spend less time in direct supervision and more time in the other managerial roles or functions, particularly, planning and coordinating. At the highest level of the organization, usually described as the executive level, planning and being a generalist are greatly expanded role functions. Direct supervision is not a major job assignment as it is in the other two levels. Nurses in executive-level roles in health care organizations usually have the title Chief Nurse Executive or, in acute care hospitals, the title may be Vice President of Patient Care Services.

## MANAGER RESOURCES

Nurse managers use four types of resources to accomplish their purpose (DuBrin, 2000). Nurse managers use human resources, such as the right staff on the health care team, to complete various assignments. They use financial resources wisely to help achieve organizational goals. Nurse managers also use physical resources, such as patient care equipment, to complete their work. Finally, nurse managers use information resources to stay up-to-date in delivering care to their patients.

## MANAGEMENT THEORIES

The current management practices have evolved from earlier theories. Management practices were actually a part of the governance in ancient Samaria and Egypt as far back as 3000 B.C. (Daft & Marcic, 2001). Most of our current understanding of management, however, is based on theories of management that were introduced in the 1800s during the industrial age as factories developed. Some of these theories are discussed here and in Table 1-2.

## SCIENTIFIC MANAGEMENT

While practicing managers, such as Fayol, who was mentioned earlier, were describing the functions of managers, a man named Frederick Taylor was focusing his attention on the operations within an organization by exploring production at the worker level. Taylor is acknowledged as the father of scientific management for his use of the scientific method and as the author of *Principles of Scientific Management* (1911). Productivity was the area of focus in scientific management. Taylor, an engineer, introduced

---

### Critical Thinking 1-8

Observe the patient unit during your next clinical rotation. Which of the following nurse management functions do you see the staff nurse performing?

____ Managing the work
    ____ Planning and organizing
    ____ Problem solving
    ____ Clarifying roles and objectives
    ____ Informing
    ____ Monitoring
    ____ Consulting
    ____ Delegating

____ Managing relationships
    ____ Networking
    ____ Supporting
    ____ Developing and mentoring
    ____ Managing conflict and team building
    ____ Motivating and inspiring
    ____ Recognizing and rewarding

## TABLE 1-2  Management Theories

| MANAGEMENT THEORY | KEY ASPECTS |
| --- | --- |
| **Scientific Management**<br>*Gulick & Urwick (1937),*<br>*Mooney (1947), Taylor (1947)* | Focuses on goals and productivity; organization is a machine to be run efficiently to increase production.<br><br>Selects the right person to do job; provides the proper tools, training, and equipment to work efficiently.<br><br>Uses time and motion studies to make the work efficient. |
| **Bureaucratic Management**<br>*Weber (1964)* | Focuses on hierarchical superior-subordinate communication transmitted from top to bottom via a clear chain of command.<br><br>Uses rational, impersonal management; distributes activities among personnel.<br><br>Uses merit and skill as basis for promotion and/or reward.<br><br>Uses rules and regulations; focuses on exacting work processes and technical competence.<br><br>Limits personal freedom.<br><br>Emphasizes career service, salaried managers. |
| **Human Relations**<br>*Argyris (1964), Barnard (1938),*<br>*Likert (1967), McGregor (1960)*<br>*Roethlisberger & Dickson (1939)* | Focuses on empowerment of the individual worker as the source of control, motivation, and productivity in meeting the organization's goals.<br><br>Hawthorne's studies at Western Electric plant in Chicago led to the belief that human relations between workers and managers and among workers are the main determinants of efficiency.<br><br>The Hawthorne effect refers to the phenomena of how being observed or studied results in a change in behavior.<br><br>Emphasizes that participatory decision making increases worker autonomy and provides training to improve work. |
| **Contingency**<br>*Burns & Stalker (1961),*<br>*Lawrence & Lorsch (1967),*<br>*Perrow (1967), Rundall, et al.*<br>*(1998), Thompson (1967)* | Highlights that organizational structure depends on the environment, task, technology, and the contingencies facing each unit.<br><br>Uses flexible approach; emphasizes that there is no one best way to manage work; encourages managers to study individuals and the situation before adapting efforts and deciding on a course of action to meet the requirements of the situation. |
| **Resource Dependence**<br>*Hickson, Hinings, Lee, Schneck,*<br>*& Pennings (1971), March &*<br>*Olsen (1976), Pfeffer & Salancik*<br>*(1978), Strasser (1983),*<br>*Williamson (1981)* | Emphasizes the need to secure necessary resources and provides reliable and valid data on patient care processes and outcomes. |
| **Strategic Management**<br>*Andrews (1971), Ansoff (1965),*<br>*Ouchi (1981), Porter (1980, 1985),*<br>*Schendel & Hofer (1979), Shortell &*<br>*Zajac (1990), Luke (2004)* | Emphasizes fit or alignment between the organization's strategy, external environment, and internal structure and capabilities.<br><br>Links quality improvement efforts to core strategies and capabilities of the organization to meet organizational needs. |
| **Population Ecology**<br>*Aldrich (1979), Delacroix &*<br>*Carroll (1983), Hannan &*<br>*Freeman (1985, 1989),*<br>*Kimberly & Zajac (1985)* | States that external environmental pressures are primary determinant of success; there is little managers and staff can do if action is not tolerable to the external environment.<br><br>Highlights powerful role played by the external environment; quality improvement efforts alone may not be sufficient if the organization is not well positioned for success in the environment. |

*(Continues)*

## TABLE 1-2 (Continued)

| MANAGEMENT THEORY | KEY ASPECTS |
|---|---|
| **Institutional**<br>*Alexander & Amburgey (1987), Meyer & Scott (1983), Scott (1987, 1995), Selznik (1966), Fligstein (1990), Powell & DiMaggio (1991), Scott, Ruef, et al. (2000)* | States that external norms, rules, and requirements cause organizations to conform in order to receive legitimacy; organizations in a similar institutional environment come to resemble each other.<br><br>Emphasizes that quality improvement efforts must take into account regulatory and accreditation pressures from local, state, and federal agencies and accrediting agencies, such as the Joint Commission (JC), as well as taking into account public expectations. |
| **Social Network**<br>*Uzzi (1997, 1999), Gulati (1995), Nohria & Berkley (1992), Ahuja (2000), Burt (1992)* | States that all behavior is social in nature, and that successful organizations will develop and use social networks to their advantage. |
| **Complex Adaptive Systems**<br>*Plsek (2001), Kauffman (1995), Anderson & McDaniel (2000), Begun, Zimmerman, & Dooley (2003), Berwick (1998)* | Emphasizes that an organization is a system of interrelated, unpredictable elements, and if a manager adjusts one part of a system, then other parts will be affected automatically; emphasizes that the organization is an open system that constantly interacts with its environment.<br><br>Encourages staff to look at their activities on one unit as part of a larger picture that may affect other units.<br><br>Emphasizes importance of innovation and rapid information sharing to improve performance. |

**Source:** Compiled with information from Shortell, S. M., & Kaluzny, A. D. (2006). *Health Care Management* (5th ed.). Clifton Park, NY: Delmar Cengage Learning.

precise procedures based on systematic investigation of specific situations. The underlying point of view is that the organization is a machine to be run efficiently to increase production.

Scientific management pioneered studies of time and motion that emphasized efficiency and culminated in "one best way" of carrying out work.

## BUREAUCRATIC MANAGEMENT

Max Weber is the German theorist recognized for the management theory of bureaucracy. Weber's beliefs were in stark contrast to the typical European organization that was based on a family-type structure in which employees were loyal to an individual, not to the organization. Weber, however, believed efficiency is achieved through impersonal relations within a formal structure, competence should be the basis for hiring and promoting an employee, and decisions should be made in an orderly and rational way based on rules and regulations. The **bureaucratic organization** was a hierarchy with clear superior-subordinate communication and relations, based on positional authority, in which orders from the top were transmitted down through the organization via a clear chain of command.

## HUMAN RELATIONS

The human relations movement started with the Hawthorne experiments. In contrast to the science of exact procedures, rules and regulations, and formal authority that characterized scientific management, the theories from the human relations school of thought espoused the individual worker as the source of control, motivation, and productivity in organizations. During the 1930s, labor unions became stronger and were instrumental in advocating for the human needs of employees. During this time, experiments were conducted at the Hawthorne plant of the Western Electric Company in Chicago that led to a greater understanding of the influence of human relations in organizations.

Electricity had become the preferred power source over gas; the Hawthorne plant experiments were run to show people that more light was necessary for greater productivity. This approach was designed to increase the use of electricity. Researchers Mayo (1933) and Roethlisberger and Dickson (1939) measured the effects on production of altering the intensity of lighting. They found that, with more and brighter light, production increased as expected. However, production also increased each time they reduced the light, even when the light was extremely dim. Their research findings led to the conclusion that something else besides the light was motivating these workers.

The notion of social facilitation, or the idea that people increase their work output in the presence of others, was a result of the Hawthorne experiments. They also concluded that the effect of being watched and receiving special attention could alter a person's behavior. The phenomena of being observed or studied, resulting in changes in behavior, is now called the **Hawthorne effect**. Emerging from this study was the concept that people benefit and are more productive and satisfied when they participate in decisions about their work environments.

# MOTIVATION THEORIES

The human relations perspective in management theory grew from the conclusion that worker output was greater when the worker was treated humanistically. This spawned a human relations point of view and a focus on the individual as a source of motivation. Motivation is not explicitly demonstrated by people but rather is interpreted from their behavior. **Motivation** is whatever influences our choices and creates direction, intensity, and persistence in our behavior (Hughes, Ginnett, & Curphy, 1999; Kanfer, 1990). Motivation is a process that occurs internally to influence and direct our behavior in order to satisfy needs (Lussier, 1999). Motivation theories are not management theories per se; however, they are frequently considered along with management theories.

There are content motivation theories and process motivation theories (Lussier, 1999). Content motivation theories define motivation in terms of satisfaction of needs. Process motivation theories define motivation in terms of rational cognitive processes. The process motivation theories are expectancy theory and equity theory. The content motivation theories include Maslow's needs hierarchy; Aldefer's expectancy-relatedness-growth (ERG) theory and model of growth needs, relatedness needs, and existence needs; Herzberg's two-factor theory; and McClelland's manifest needs theory and model of achievement, power, and affiliation. Maslow's hierarchy of needs and Herzberg's two-factor theory are presented here along with Theory X, Theory Y, and Theory Z (Table 1-3 and Figure 1-2).

## TABLE 1-3 Selected Motivation Theories

| MAIN CONTRIBUTORS | KEY ASPECTS |
| --- | --- |
| **Hierarchy of needs** <br> *Abraham Maslow* <br> *(1908–1970)* | Motivation occurs when needs are not met. Certain needs have to be satisfied first, beginning with physiological needs, then safety and security needs, then social needs, followed by self-esteem needs, and then self-actualization needs. Needs at one level must be satisfied before one is motivated by needs at the next higher level of needs. |
| **Two-factor theory: hygiene-maintenance factors and motivator factors** <br> *Frederick Herzberg* <br> *(1923–2000)* | Hygiene-maintenance factors include adequate salary status, job security, quality of supervision, safe and tolerable working conditions, and relationships with others. When these factors are absent, they can be sources of job dissatisfaction. When they are present, job dissatisfaction can be avoided. However, these factors alone will not lead to job satisfaction. |
| | Motivator factors include satisfying and meaningful work, development and advancement opportunities, and responsibility and recognition. When these factors are present, people are motivated and satisfied with their jobs. When these factors are absent, people have neutral attitudes about their jobs or organizations. |
| **Theory X** <br> *Douglas McGregor* <br> *(1906–1964)* | Leaders must direct and control because motivation results from reward and punishment. Employees prefer security, direction, and minimal responsibility, and they need coercion and threats to get the job done (Theory X). |
| **Theory Y** | Leaders must remove work obstacles because, under the right work conditions, workers have self-control and self-discipline. The workers' rewards are their involvement in work and in the opportunities to be creative (Theory Y). |

*(Continues)*

## TABLE 1-3 (Continued)

| MAIN CONTRIBUTORS | KEY ASPECTS |
|---|---|
| **Theory Z**<br>*William Ouchi (1981)* | Uses collective decision making, long-term employment, mentoring, holistic concern, and use of quality circles to manage service and quality. This is a humanistic style of motivation based on the study of Japanese organizations. |
| **Expectancy Theory**<br>*Victor Vroom (1964)* | Has three variables that are subdivided into three indicators: force, valence, and expectancy. *Force* describes the amount of effort one will exert to reach one's goal. *Valence* speaks to the level of attractiveness or unattractiveness of the goal. *Expectancy* is the perceived possibility that the goal will be achieved. Vroom's theory of motivation can be demonstrated in the form of an equation: *Force = Valence × Expectancy* (Vroom, 1964). The theory proposes that this equation can help to predict the motivation, or force, of an individual. |

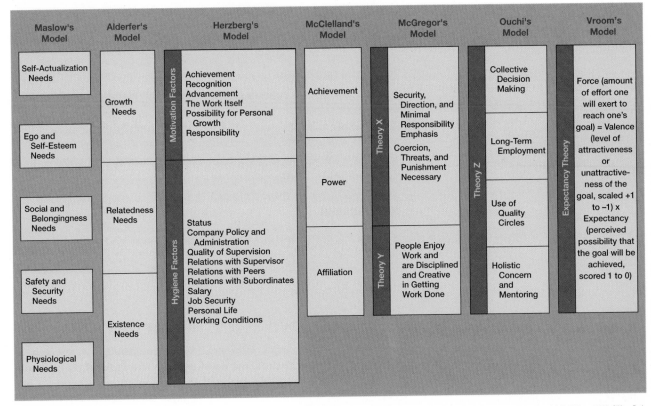

**FIGURE 1-2  A comparison of models of motivation.** (*Source:* Compiled with information from *Health Care Management* (5th ed.) by S. M. Shortell and A. D. Kaluzny, 2006, Clifton Park, NY: Delmar Cengage Learning; and Leadership and Management by L. S. Leach, from *Nursing Leadership & Management* (2nd ed.). by P. Kelly, 2008, Clifton Park, NY: Delmar Cengage Learning.)

Motivation theories are useful because they help explain why people act the way they do and how a manager can relate to individuals as human beings and workers. When you are interested in creating change, influencing others, and managing patient care outcomes, it is helpful to understand the motivation that is reflected in a person's behavior. Motivation is a critical part of leadership because we need to understand each other in order to lead effectively. See Table 1-4 for common motivation problems and potential solutions.

**TABLE 1-4  Common Employee Motivation Problems and Potential Solutions**

| MOTIVATION PROBLEMS | POTENTIAL SOLUTIONS |
|---|---|
| 1. Inadequate performance definition (i.e., lack of goals, inadequate job descriptions, inadequate performance standards, inadequate performance assessment) | 1. – Well-defined job descriptions<br>– Well-defined performance standards<br>– Goal setting<br>– Feedback on performance<br>– Improved employee selection<br>– Job redesign or enrichment |
| 2. Impediments to performance (i.e., bureaucratic or environmental obstacles, inadequate support or resources, poor employee-job matching, inadequate job information) | 2. – Enhanced hygiene factors (i.e., safe and clean environment, good salary and fringe benefits, job security, good staffing, time off job, good equipment)<br>– Behavior modification or positive reinforcement (individual or group)<br>– Pay for performance |
| 3. Inadequate performance-reward linkages (i.e., inappropriate or inadequate job rewards, poor timing of rewards, low probability of receiving rewards, inequity in distribution of rewards) | 3. – Enhanced job achievement or growth rewards (i.e., increased employee involvement and participation, job redesign or enrichment, career planning, professional development opportunities)<br>– Enhanced job esteem or power factors (i.e., job autonomy or personal control, self-management, modified work schedule, recognition, praise or awards, opportunity to display skills or talents, opportunity to mentor or train others, promotions in rank or position, improved information concerning organization or department, preferred work activities or projects, letters of recommendation, preferred work space)<br>– Enhanced affiliation or relatedness factors (i.e., recognized work teams and task groups; opportunities to attend conference, social activities, and professional and community group) |

**Source:** Compiled with information from Shortell, S. M., and Kaluzny, A. D. (2006). *Health Care Management* (5th ed.). Clifton Park, NY: Delmar Cengage Learning.

## MASLOW'S HIERARCHY OF NEEDS

One of the most well-known theories of motivation is Maslow's hierarchy of needs. Maslow (1970) developed a hierarchy of needs that shows how an individual is motivated. Motivation, according to Maslow, begins when a need is not met. For example, when a person has a physiological need, such as thirst, this unmet need has to be satisfied before a person is motivated to pursue higher-level needs. Certain needs have to be satisfied first, beginning with physiological needs, then safety and security needs, next social needs, followed by esteem needs, before an individual is motivated by the needs at the next level. The need for self-actualization drives people to the pinnacle of performance and achievement (Figure 1-3).

## TWO-FACTOR THEORY

Frederick Herzberg (1968) contributed to research on motivation and developed the two-factor theory of motivation. He analyzed the responses of accountants and engineers and concluded that there are two sets of factors associated with motivation. One set of motivation factors must be maintained to avoid job dissatisfaction. These factors include such items as salary, working conditions, status, quality of supervision, relationships with others, and so on. These factors have been labeled **maintenance or hygiene factors**. Factors such as achievement, recognition, responsibility, advancement, and so on also contribute to job satisfaction. These factors are intrinsic and serve to satisfy or motivate people. Herzberg proposed that, when these **motivation factors** are present, people are

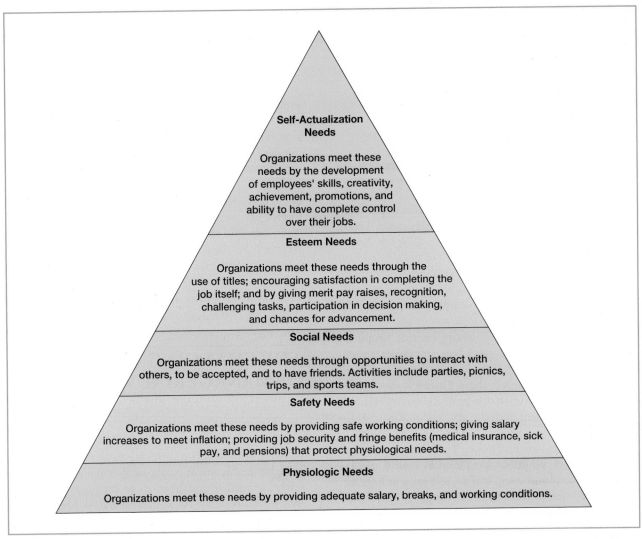

**FIGURE 1-3  How organizations motivate with hierarchy of needs theory.** (*Source:* From Lussier, R. N., & Achua, C. F. [2000]. *Leadership: Theory, Application, Skill Development* [p. 81]. Cincinnati, OH: South-Western College Publishing).

very motivated and satisfied with their jobs. When these factors are absent from a work setting, people have a neutral attitude about their organizations. In contrast, when the maintenance factors are absent, people are dissatisfied. Herzberg believed that, by providing the maintenance factors, job dissatisfaction could be avoided, but that these factors will not motivate people.

New graduate nurses can use Herzberg's theory by evaluating the maintenance factors present in health care organizations when they apply for jobs. The pay, working conditions, and the beginning relationship that has been established with the supervisor are aspects of the job that the nurse should consider. If these maintenance factors are not adequate to begin with, then the nurse may become easily dissatisfied with the job. The higher-level needs that Herzberg describes as motivation factors should also be evaluated by the nurse before joining an organization. Are

there opportunities for the nurse to achieve professional growth, to take on new responsibilities, to advance and be recognized for the contribution he has made?

It is important to recognize that not all employees respond to the same motivation factors. For example, a manager might be surprised to learn that some personnel will not respond to opportunities for autonomy and personal growth on the job. Rather, they may view their jobs as a means of providing income and seek various forms of personal fulfillment off-the-job through their families and leisure activities (Shortell & Kaluzny, 2006).

## THEORY X AND THEORY Y

Continuing the emphasis on factors that stimulate job satisfaction and what motivates people to be involved and contribute productively at work, McGregor capitalized on

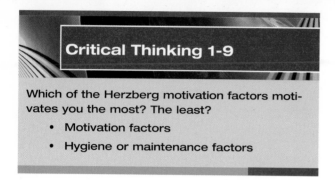

**Critical Thinking 1-9**

Which of the Herzberg motivation factors motivates you the most? The least?

- Motivation factors
- Hygiene or maintenance factors

his experience as a psychologist and university president to develop Theory X and Theory Y (McGregor, 1960). Theory X and Theory Y are about two different ways to motivate or influence others based on underlying attitudes about human nature. Each view reflects different attitudes about the nature of humans. The **Theory X** view is that, in bureaucratic organizations, employees prefer security, direction, and minimal responsibility. Coercion, threats, or punishment are necessary because people do not like the work to be done. These employees are not able to offer creative solutions to help the organizations advance. McGregor's beliefs about Theory X were related to the classical perspective of organizations that included scientific management, bureaucracy theory, and administrative principles.

The assumptions of **Theory Y** are that, in the context of the right conditions, people enjoy their work; can show self-control and discipline; are able to contribute creatively; and are motivated by ties to the group, the organization, and the work itself. In essence, this view espouses the belief that people are intrinsically motivated by their work. Theory Y was a guide for managers to take advantage of the potential of each person, which McGregor thought was being only partially utilized, and to provide support and encouragement to employees to do good work (McGregor, 1960).

## THEORY Z

**Theory Z** was developed by William Ouchi (1981) based on his years of study of organizations in Japan. He identified that Japanese organizations had better productivity than organizations in the United States and that they were managed differently with their use of quality circles to pursue better productivity and quality. Theory Z focuses on a better way of motivating people through their involvement. The organization invests in its employees and addresses both home and work issues creating a path for career development. Democratic leaders, who are skilled in interpersonal relations, foster employee involvement (Ouchi, 1981).

## VROOM'S EXPECTANCY THEORY

Vroom's Expectancy Theory of Motivation (1964) centers around what people want and their prospect of getting it (Marriner-Tomey, 2009). There are three variables in Vroom's motivation theory. They are force, valence, and expectancy. Force describes the amount of effort one will exert to reach one's goal. Valence speaks to the level of attractiveness or unattractiveness of the goal. Valence is graded between +1, an outcome that is highly attractive to the individual, and −1, a highly unattractive goal. A valence of 0 indicates the goal does not interest the individual. Expectancy is the perceived possibility that the goal will be achieved. Expectancy is evaluated between 1 and 0. A score of 1 indicates assurance that the goals will be achieved, and zero indicates that the individual sees the goal as impossible to achieve. Vroom's theory appears in the form of an equation as Force = Valence × Expectancy. Vroom proposes that this equation can help to predict the motivation of an individual to achieve a goal as negative, neutral, or positive.

## THE CHANGING NATURE OF MANAGERIAL WORK

Current trends indicate that the numbers of managers in an organization, particularly at the middle level, are being reduced and that downsizing of staff has been a common phenomenon in most health care institutions. Those left in higher management positions after reorganizations are fewer in number and have taken on more responsibility over more areas. As individual nurses become more involved in managing consumer relations and consumer care, managers will become managers of systems rather than managers of nurses per se. The systems they will manage include clinical systems, cost information systems, and data systems on consumer satisfaction and feedback.

## FEEDBACK

Because of their professional socialization and strong achievement needs, nurses want to deliver high-quality, excellent care to their patients. However, in order for them to know how well they are doing in this regard, they need to be able to assess how well they are doing with respect to their peers and their own past performance and to benchmark quality performance goals. One way of accomplishing this is to develop high-quality information systems that provide feedback on a frequent basis. Such a system can allow health care professionals to know not only how well they are doing but also can enhance their confidence that they are doing things right (Kongstvedt, 2002).

Feedback can be a powerful tool to assist managers in motivating behavior; however, there are several factors that should be considered to maximize feedback effectiveness. First, for feedback to have value, nurses must truly see that their behavior needs to change. Second, feedback needs to be frequent, timely, and given at precise time intervals to sustain new behaviors. Third, feedback must be usable, consistent, correct, and of sufficient diversity. Feedback should contain various important utilization, financial, and quality-related data that are valid representations of what is being measured. Otherwise, behavior problems can intensify as rewards flow to improvements based on flawed feedback data (Charns & Smith Tewksbury, 1993). Last, managers should not portray the feedback as *good* or *bad*. Professionals, such as nurses, know when they have missed the goal (Shortell & Kaluzny, 2006). In the future, leadership responsibilities will be integrated among all organizational participants who function as **knowledge workers** and provide their professional expertise. Knowledge workers are those involved in serving others through their specialized knowledge. Among nurses, specialized knowledge is the practice and science of nursing used to serve patients and families. So leadership responsibilities are dispersed among all nurses, who are knowledge workers by virtue of their professional nursing expertise.

## Critical Thinking 1-10

Look for opportunities to gather data about your clinical performance. As both a manager and leader of patient care, how can you use feedback about these quality measures to improve your patients' outcomes?

- *Patient Access:* Number of patients who are triaged within five minutes of arrival in the Emergency Department.
- *Utilization:* Number of patients who achieve quality outcomes under normal staffing ratios.
- *Financial:* Number of patients who meet patient care standards.
- *Quality:* Number of patients who state they are pleased with their nursing care.

## KEY CONCEPTS

- Nurses are leaders and make a difference through their contributions of expert knowledge and leadership to health care organizations. Leadership development is a necessary component of preparation as a health care provider.

- All nurses are leaders because they have expert knowledge that they contribute to coordinate and provide patient care. Building expertise by gaining experience, setting goals to direct experiential learning, and gaining clinical skills and judgment are some of the ways that new nurses develop as leaders.

- Leadership is a process of influence that involves the leader, the follower, and their interaction. Followers can be individuals, groups of people, communities, and members of society in general.

- Leadership can be formal or informal. It can occur by being in a position of leadership and authority in an organization, such as a manager. Leadership can also occur outside the scope of a formal role, such as when an individual or member of a group moves to assume leadership.

- Nurses are leaders. They lead nursing practice. Nurses lead other nurses, and they lead patients and communities toward improved health.

- Leadership styles are described as autocratic, democratic, and laissez-faire and have been studied by examining job-centered or task-oriented approaches versus employee-centered or relationship-oriented approaches.

- Blake and Mouton's leadership model has five styles to address high or low people concerns and high or low production concerns.

- Contingency theories of leadership acknowledge that other factors in the environment, in addition to the leader's behavior, affect the effectiveness of the leader. Fielder's contingency theory describes the leader's behavior/style as being dependent upon the nature of the task, leader-member relations, and the power associated with the leader's position. Hersey and Blanchard's situational theory matches task and relationship needs with maturity or readiness of the followers to participate by using leadership styles called telling, selling, participating, or delegating leadership styles. House's path-goal theory involves the leader using a directive style, a supportive style, a participative style, or an achievement-oriented leadership style to match the task.

- Substitutes for leadership are variables that eliminate the need for leadership or nullify the effect of the leader's behavior. These include work experience, professionalism, indifference to rewards, presence of routine tasks, feedback provided by the task, intrinsic satisfaction, cohesive groups, formal organizational structures, rigid adherence to rules, role distance, and low position power of the leader.

■ Charismatic leadership theory describes leader behavior that displays self-confidence, passion, and communication of high expectations and confidence in others. These types of leaders often emerge in a crisis with a vision, have an appeal based on their personal power, and often use unconventional strategies and their emotional connections to succeed.

■ Transformational leadership theory involves two styles of leadership: the transformational leader and the transactional leader. Transactional leaders focus on organizational operations and short-term goals. They use exchange and making trades as a way of accomplishing work. Transformational leaders inspire and motivate others to excel and participate in a vision that goes beyond self interests. Transformational leadership is believed to empower followers and contribute to their commitment to action and change.

■ "The servant-leader is servant first…It begins with the natural feeling that one wants to serve, to serve first…. The difference manifests itself in the care taken by the servant to make sure that other people's highest priority needs are being served" (Greenleaf, 2002). This leadership style is natural to nursing.

■ Management is a process used to achieve organizational goals. Management roles are classified as the information-processing role, the interpersonal role, and the decision-making role. Managers use these roles to manage the work and to manage relationships with people to accomplish the work.

■ The management process involves planning, coordinating, organizing, and controlling. The RN uses this process to manage patient care.

■ Scientific management, based on the work of Taylor and others, viewed industrial organizations as machines where work was to be carried out in the most scientifically exact and efficient way to increase production.

■ Weber is the theorist associated with bureaucratic theory. The idea of organizations as bureaucracies involves a formal organizational structure, impersonal relations, a hierarchy based on positional authority, and a clear chain of command.

■ Human relations management is also known as organizational behavior and led to participative management. The human needs of employees and the motivation and satisfaction of the individual and groups were the focus of efforts to increase production.

■ Motivation is an internal process that contributes to behavior in an effort to satisfy needs. Maslow's hierarchy of needs reflects the belief that the needs that motivate individuals have a priority order. Lower-level needs have to be satisfied first or individuals will not be motivated to address higher-level needs.

■ Herzberg's two-factor theory of motivation identifies maintenance factors, such as security and salary, that are needed to prevent job dissatisfaction, and motivator factors, such as job development and opportunities to advance, that contribute to job satisfaction.

■ Vroom's expectancy theory focuses very simply yet specifically on the assessment of the desires of what people want and their prospect of getting it.

■ Organizations need to be viewed as self-organizing systems where what initially looks like chaos and uncertainty is indeed part of a larger coherence and a natural order. Such a living, self-organizing system, when understood better by participants, will be a less stressful and more holistic environment in which to carry out work.

■ Future directions for nurses in organizations will continue to be influenced by technology and by the notions of mobility, virtuality, and user-driven practices. Knowledge workers, with specialized knowledge and expertise, are more self-directed. Future leadership practices need to adapt to them and to the changing work environments and circumstances.

■ Nursing is in transition to the knowledge age where nursing research advances nursing science, and nursing science is translated into evidence-based nursing practice.

## KEY TERMS

| | | | |
|---|---|---|---|
| autocratic leadership | formal leadership | leadership | substitutes for leadership |
| bureaucratic organization | Hawthorne effect | maintenance or hygiene factors | task structure |
| consideration | informal leader | | taxonomy |
| contingency theory | initiating structure | management | Theory X |
| democratic leadership | job-centered leaders | management process | Theory Y |
| emotional intelligence | knowledge workers | motivation | Theory Z |
| employee-centered leadership | laissez-faire leadership | motivation factors | transactional leader |
| | leader-member relations | position power | transformational leader |

# REVIEW QUESTIONS

1. Management is a process that is used today by nurses or nurse managers in health care organizations and is best described as
   A. scientific management.
   B. decision making.
   C. commanding and controlling others using hierarchical authority.
   D. planning, organizing, coordinating, and controlling.

2. A participative leadership style is appropriate for employees who
   A. are not able to get the task done and are less mature.
   B. are able to contribute to decisions about getting the work done.
   C. are unable and unwilling to participate.
   D. need direction, structure, and authority.

3. If you applied the concepts of Theory Y to describe nurses, which of the following statements would be the best description?
   A. Nurses prefer to be directed and want job security more than other things.
   B. Nurses use self-direction and self-control to achieve work objectives in which they believe.
   C. Nurses have a hard time accepting responsibility, but they learn to do this over time.
   D. Nurses don't really want to work and would quit if they could.

4. Nurse retention is an important focus for health care organizations as we face a growing shortage of health care professionals in the future. According to Herzberg's motivation factors, which of the following would most likely contribute to increased job satisfaction?
   A. The organization recognizes and rewards those nurses who advance their education and achieve certification, such as the CCRN certification for critical care RNs.
   B. Hiring bonuses of up to $5,000 are given to nurses to reduce the vacant positions and prevent short-staffing.
   C. Nurse managers place an emphasis on establishing effective relationships with the nurses who work for them.
   D. Salary is increased.

5. Consider your role as a staff nurse in a patient care unit of a hospital. What factors are present that may serve as a substitute for your need for leadership from your nurse manager?
   A. Your desire for a promotion and an increase in pay.
   B. Professional nursing standards, code of ethics, and the intrinsic reward you get from this important work.
   C. Your manager spends time telling you exactly what to do and how to do it.
   D. Your nurse manager is inspiring and highly motivating to work for.

6. Leadership is defined as
   A. being in a leadership position with authority to exert control and power over subordinates.
   B. a process of interaction in which the leader influences others toward goal achievement.
   C. managing complexity.
   D. being self-confident and democratic.

7. Why is leadership development important for nurses if they are not in a management position?
   A. It is not really important for nurses.
   B. Leadership is important at all levels in an organization because nurses have expert knowledge and are interacting with and influencing the patient.
   C. Nurse leaders leave their jobs sooner for other positions.
   D. Nurses who lead are less satisfied in their jobs.

8. Which motivation theory allows the student to determine the probability of achieving one's goal?
   A. Maslow's Hierarchy of Needs
   B. Two-Factor Theory
   C. Vroom's Expectancy Theory
   D. Theory Y

9. As both a manager and leader of patient care, how can you use feedback mechanisms to enhance your patient outcomes? Select all that apply.
   _____ A. Determine behavior needs of staff that need to change.
   _____ B. Give frequent feedback to the staff to sustain their new behaviors.
   _____ C. Portray feedback to staff as good or bad.
   _____ D. Give staff feedback that is understandable and usable.
   _____ E. Give staff feedback that is measurable.
   _____ F. Give staff feedback that is usable, consistent, and correct.

10. Which of the following leadership characteristics are important for the nurse to master to assure effective management of patients? Select all that apply.
   _____ A. Develop a professional and purposeful vision that provides direction toward the preferred future.
   _____ B. Portray a passion that involves the ability to inspire and align people toward the goal.
   _____ C. Display integrity that is based on knowledge of self, honesty, and maturity.
   _____ D. Know your strengths and weaknesses.
   _____ E. Be daring and curious when taking reasonable risks.

# REVIEW ACTIVITIES

1. Take the opportunity to learn about yourself by reflecting on five predominant factors identified as being influential in a nurse's leadership development: self-confidence, innate leader qualities or tendencies, progression of experiences and success, influence of significant others, and personal life factors. Consider what reinforces your confidence in yourself. What innate qualities or tendencies do you have that contribute to your development as a leader? Consider what professional experiences, mentors, and personal experiences or events can help you influence and change nursing practice.

2. Describe the type of leader you want to be as a nurse in a health care organization. Identify specific behaviors you plan to use as a leader. In what way are the transformational leadership and the charismatic leadership theories useful to your development as a leader?

3. Rate each of these 12 job factors that contribute to job satisfaction by placing a number from 1 to 5 on the line before each factor.

| Very important | | Somewhat important | | Not important |
|---|---|---|---|---|
| 5 | 4 | 3 | 2 | 1 |

_____ 1. An interesting job I enjoy doing

_____ 2. A good manager who treats people fairly

_____ 3. Getting praise and other recognition and appreciation for the work I do

_____ 4. A satisfying personal life at the job

_____ 5. The opportunity for advancement

_____ 6. A prestigious or status job

_____ 7. Job responsibility that gives me freedom to do things my way

_____ 8. Good working conditions (safe environment, nice office, cafeteria)

_____ 9. The opportunity to learn new things

_____ 10. Sensible company rules, regulations, procedures, and policies

_____ 11. A job I can do well and succeed at

_____ 12. Job security and benefits

Write the number from 1 to 5 that you selected for each factor. Total each column for a score between 6 and 30 points. The closer to 30 your score is, the more important these factors (motivating or maintenance) are to you.

| Motivating factors | Maintenance factors |
|---|---|
| 1. _____ | 2. _____ |
| 3. _____ | 4. _____ |
| 5. _____ | 6. _____ |
| 7. _____ | 8. _____ |
| 9. _____ | 10. _____ |
| 11. _____ | 12. _____ |
| Totals _____ | _____ |

From *Leadership: Theory, Application, Skill Development* (pp. 15–16), by R. N. Lussier and C. F. Achua, 2000, Cincinnati, OH: South-Western College Publishing.

4. What would a health care organization be like if it more closely resembled a self-organizing system and more holistic environment? How different would it be from a bureaucratic and more structured organization?

5. If nurses become more self-directed in the future, what kind of leadership practices will be the most effective?

6. What quality improvement projects are nurses in your work setting involved in? How is evidence being used to improve nursing practice? What sources of evidence from nursing science are you exploring through journals, attending educational conferences, the Internet, and participating in professional nursing associations?

# EXPLORING THE WEB

Search the Web, checking the following sites:

■ Emerging Leader: www.emergingleader.com

■ Leadership Skills Development: www.impactfactory.com Search for leadership.

■ LeaderValues: www.leader-values.com

■ Don Clark's Big Dog Leadership: www.nwlink.com/~donclark/leader/leader.html

■ American Association of Critical Care Nurse's Standards for Establishing and Sustaining Healthy Work Environments: www.aacn.org Under Priority Issues, click Healthy Work Environments.

■ American Organization of Nurse Executives Competencies: www.aone.org Click Resource Center and then click AONE Nurse Exec Competencies.

■ Analyze My Career: http://analyzemycareer.com

■ American Nurses Association Magnet Status Hospitals: www.nursecredentialing.org

■ This site on classic management functions can keep you busy all week: www.1000ventures.com Review your character and personality, and click on other areas of interest.

■ Joanna Briggs Institute for Evidence Based Nursing: www.joannabriggs.edu.au

# REFERENCES

Aiken, L. H., Clarke, S. P., Sloane, D. M., Sochalski, J., & Silber, J. H. (2002). Hospital nurse staffing, patient mortality, nurse burnout, and job dissatisfaction. *Journal of American Medical Association, 288*(16), 1987–1993.

American Association of Critical Care Nurses. (2003). Written testimony to the IOM Committee on work environment of nurses and patient safety. Retrieved February 15, 2007, from http://www.aacn.org.

American Association of Critical Care Nurses. (2004). *AACN standards for establishing and sustaining healthy work environments: A journey to excellence.* Retrieved September 26, 2010, from www.aacn.org/WD/HWE/Docs/HWEStandards.pdf.

Amerson, R. (2010). The impact of service-learning on cultural competence. Nursing Education Perspectives, 31(1), 18–22.

Argyris, C. (1964). *Integrating the individual and the organization.* Hoboken, NJ: Wiley.

Barker, A. (1990). *Transformational nursing leadership: A vision for the future.* Baltimore: Williams & Wilkins.

Barnard, C. (1938). *The functions of the executive.* Boston: Harvard University Press.

Bass, B. (1985). *Leadership and performance beyond expectations.* New York: Free Press.

Bennis, W., & Nanus, B. (1985). *Leaders: The strategies for taking charge.* New York: Harper & Row.

Bennis, W., Spreitzer, G. M., & Cummings, T. G. (2001). *The future of leadership.* San Francisco: Jossey-Bass.

Bidwell A., & Brasler, M. (1989). Role modeling versus mentoring in nursing education. Image: Journal of Nursing Scholarship, 21(1): 23–5.

Bishop, V. (2010). Coalition in leadership. Politics The big picture and the big game. Journal of Research in Nursing, 15(4), 291–293.

Blake, R. R., & Mouton, J. S. (1985). *The managerial grid III.* Houston, TX: Gulf.

Bolman, L., & Deal, T. (2003). *Reframing organizations: Artistry, choice, and leadership* (3rd ed.). San Francisco: Jossey-Bass.

Burns, J. M. (1978). *Leadership.* New York: Harper & Row.

Cassidy, V., & Koroll, C. (1998). Ethical aspects of transformational leadership. In E. Hein (Ed.), *Contemporary leadership behavior: Selected readings* (5th ed., pp. 79–82). Philadelphia: Lippincott.

Catalano, J. (2008). Nursing now! Today's issues, tomorrow's trends (5th. ed.). Philadelphia: F.A. Davis.

Charns, M., & Smith Tewksbury, L. (1993). *Collaborative management in health care.* San Francisco: Jossey-Bass.

Conger, J., & Kanungo, R. (1987). Toward a behavioral theory of charismatic leadership in organizational settings. *Academy of Management Review, 12,* 637–647.

Daft, R. L., & Marcic, D. (2001). *Understanding management* (3rd ed.). Philadelphia: Harcourt College.

Dahnke, M. (2009). The role of the American Nurses Association Code in ethical decision making. Holistic Nursing Practice. 23(2), 112–119.

Drath, W. H., & Palus, C. J. (1994). *Making common sense: Leadership as meaning-making in a community of practice.* Retrieved September 26, 2010, from http://www.ccl.org/leadership/pdf/publications/readers/reader156ccl.pdf.

Drucker, P. F. (1994). *The post-capitalist society.* New York: Harper & Row.

DuBrin, A. J. (2000). *Essentials of Management* (5th Edition). Cincinatti, OH: South-Western Educational Publishing.

Dunham-Taylor, J. (2000). Nurse executive transformational leadership found in participative organizations. *Journal of Nursing Administration, 30*(5), 241–250.

Fayol, H. (1916/1949). (C. Storrs, Trans.). *General and industrial management.* London: Pitman.

Fielder, F. (1967). *A theory of leadership effectiveness.* New York: McGraw-Hill.

Follet, M. (1924). *Creative experience.* London: Longmans, Green.

Gilbreth, F. (1912). *Primer of scientific management.* New York: Van Nostrand.

Goleman, D. (1998). Working with emotional intelligence. New York: Bantam Books.

Gordon, S. (2006). What do nurses really do? *Topics in Advanced Practice Nursing, 6*(1). Retrieved September 26, 2010, from http://www.medscape.com/viewarticle/520714_2.

Greenleaf, R. (1970). The Servant as Leader. Retrieved: 2010 from http://siliconvalley.alfnat.callahanpro.com/downloads/Greenleaf_servant_as_leader.pdf.

Greenleaf, R., & Spears, L. (2002). Servant leadership: A journey into the nature of legitimate power and greatness. Mahwah, NJ, US: Paulist Press.

Gulick, L., & Urwick, L. (Eds.). (1937). *Papers on the science of administration.* New York: Institute of Public Administration.

Hales, C. P. (1986). What managers do: A critical review of the evidence. *Journal of Management Studies, 23,* 88–115.

Heath, J., Johanson, W., & Blake, N. (2004). Healthy work environments: A validation of the literature. *Journal of Nursing Administration, 34*(11), 524–530.

Helgesen, S. (1995). *The web of inclusion: A new architecture for building organizations.* New York: Doubleday Currency.

Henry, N. (1992). *Public administration and public affairs* (5th ed.). Englewood Cliffs, NJ: Prentice Hall.

Hersey, P., & Blanchard, K. (2000). *Management of organizational behavior* (8th ed.). Englewood Cliffs, NJ: Prentice Hall.

Herzberg, F. (1968, January/February). One more time: How do you motivate employees? *Harvard Business Review, 46*(1), 53–62.

House, R. H. (1971). A path-goal theory of leader effectiveness. *Administrative Science Quarterly, 16,* 321–338.

Hughes, R. L., Ginnett, R. C., & Curphy, G. J. (1999). *Leadership: Enhancing the lessons of experience* (3rd ed.). San Francisco: Irwin McGraw-Hill.

Institute of Medicine (2003). *Health professions education: A bridge to quality.* Washington, DC: The National Academies Press.

Jackson, D. (2007). Servant Leadership in Nursing: A framework for developing sustainable research capacity in nursing, Collegian: Journal of the Royal College 15(1). 27–33.

Kanfer, R. (1990). Motivation theory in industrial and organizational psychology. In M. D. Dunnette & L. M. Hough (Eds.), *Handbook of industrial and organizational psychology: Vol. 1* (pp. 53–68). Palo Alto, CA: Consulting Psychologists Press.

Katz, D., & Kahn, R. L. (1978). *The social psychology of organizations* (2nd ed.). New York: John Wiley.

Kerr, S., & Jermier, J. (1978). Substitutes for leadership: Their meaning and measurement. *Organizational Behavior and Human Performance, 22,* 374–403.

Kim, H., & Yukl, G. (1995). Relationships of self-reported and subordinate-reported leadership behaviors to managerial effectiveness and advancement. *Leadership Quarterly, 6,* 361–377.

Kirkpatrick, S. A., & Locke, E. A. (1991). Leadership: Do traits matter? *The Executive, 5,* 48–60.

Kongstvedt, P. R. (2002). *Managed care: What it is and how it works.* Gaithersburg, MD: Aspen Publishers.

Kotter, J. (1990a). *A force for change: How leadership differs from management.* Glencoe, IL: Free Press.

Kotter, J. (1990b). What leaders really do. *Harvard Business Review, 68,* 104.

Kramer, M. (1990). The magnet hospitals: Excellence revisited. *Journal of Nursing Administration, 20*(9), 35–44.

Kramer, M., & Schmalenberg, C. E. (2005, July–Sep.). Best quality patient care: A historical perspective on Magnet hospitals. *Nursing Administration Quarterly, 29*(3), 275–287.

Laschinger, H. K. S., Almost, J., & Tuer-Hodes, D. (2003). Workplace empowerment and magnet hospital characteristics: Making the link. *Journal of Nursing Administration, 33*(7/8), 410–422.

Lawler, E. (2001). The era of human capital has finally arrived. In W. Bennis, G. Spreitzer, & T. Cummings (Eds.), *The future of leadership.* San Francisco: Jossey-Bass.

Leach, L. S. (2005). Nurse executive leadership and organizational commitment among nurses. *Journal of Nursing Administration, 35*(5), 228–237.

Lewin, K. (1939). Field theory and experiment in social psychology: Concepts and methods. *Journal of Sociology, 44,* 868–896.

Lewin, K., & Lippitt, R. (1938). An experimental approach to the study of autocracy and democracy: A preliminary note. *Sociometry, 1,* 292–300.

Lewin, K., Lippitt, R., & White, R. (1939). Patterns of aggressive behavior in experimentally created social climates. *Journal of Social Psychology, 10,* 271–299.

Likert, R. (1967). *The human organization: Its management and value.* New York: McGraw-Hill.

Lussier, R. N. (1999). *Human relations in organizations: Applications and skill building* (4th ed.). San Francisco: Irwin McGraw-Hill.

Lussier, R. N., & Achua, C. F. (2000). *Leadership: Theory, application, skill development.* Cincinnati, OH: South-Western College.

Manojlovich, M. (2006). The effect of nursing leadership on hospital nurses professional practice behaviors. *Journal of Nursing Administration, 35*(7/8), 366–374.

Marriner-Tomey, A. (2009). Guide to nursing management and leadership (8th ed). St. Louis: Mosby.

Maslow, A. (1970). *Motivation and personality* (2nd ed.). New York: Harper & Row.

Mayo, E. (1933). *The Human problems of an industrial civilization.* New York: Macmillan.

McCall, M. W., Jr. (1998). *High flyers: Developing the next generation of leaders.* Boston: Harvard Business School Press.

McCall, M. W., Jr., Morrison, A. M., & Hanman, R. L. (1978). *Studies of managerial work: Results and methods* (Tech. Rep.). Greensboro, NC: Center for Creative Leadership.

McClure, M., & Hinshaw, A. (Eds.). (2002). *Magnet hospitals revisited.* Washington, DC: American Nurses Publishing.

McClure, M., Poulin, M., Sovie, M., & Wandelt, M. (1983). *Magnet hospitals: Attraction and retention of professional nurses.* Kansas City, MO: American Nurses Association.

McDaniel, C., & Wolf, G. (1992). Transformational leadership in nursing service. *Journal of Nursing Administration,12*(4), 204–207.

McGregor, D. (1960). *The human side of enterprise.* New York: McGraw-Hill.

McNeese-Smith, D. (1997). The influences of manager behavior on nurses' job satisfaction, productivity, and commitment. *Journal of Nursing Administration, 27*(9), 47–55.

Merriam-Webster Online Dictionary. Retrieved July 29, 2010. http://mw2.merriam-webster.com/dictionary/stewardship.

Mills, E. (1964). Florence nightingale and state registration. International Nursing Review, 11, 31–36.

Mintzberg, H. (1973). *The nature of managerial work.* New York: Harper & Row.

Mooney, J. (1939). *Principles of Organization.* New York: Harper.

Moore, J. (2004). Leadership: Lessons learned. Power point presentation to Indiana State University PhD Educational Leadership and Foundation Students. Indianapolis, Indiana. non published PowerPoint presentation.

Moorhead, G., & Griffin, R. W. (2001). *Organizational behavior: Managing people in organizations* (6th ed.). Boston: Houghton Mifflin.

Murphy, L. (2005). Transformational leadership: A cascading chain reaction. Journal of Nursing Management, 13, 128–136.

Murphy, M., & DeBack, V. (1991). Today's nursing leaders: Creating the vision. *Nursing Administration Quarterly, 16*(1), 71–80.

National League for Nursing. *Core values* (2010). Retrieved September 26, 2010, from http://www.nln.org/aboutnln/corevalues.htm.

Neill, M., & Saunders, N. (2008, p. 398). Servant leadership: Enhancing quality of care and staff satisfaction. *Journal of Nursing Administration, 38*(9), 395–400.

Northouse, P. (2001). *Leadership: Theory and practice* (2nd ed.). Thousand Oaks, CA: Sage.

Ouchi, W. (1981). *Theory Z: How American business can meet the Japanese challenge.* Reading, MA: Addison-Wesley.

Porter O'Grady, T. (2001). Profound change: 21st century nursing. *Nursing Outlook, 41*(1), 182–186.

Pruitt, B., & Jacobs, M. (2006). Best practice interventions: Learn how you can prevent ventilator-associated pneumonia. *Nursing, 36*(2), 36–42.

Robbins, S. (2009). Organizational behavior (11th ed). Upper Saddle River, NJ: Prentice Hall.

Roethlisberger, J. F., & Dickson, W. J. (1939). *Management and the worker.* Cambridge, MA: Harvard University Press.

Scott, J. G., Sochalski, J., & Aiken, L. (1999). Review of magnet hospital research: Findings and implications for professional nursing practice. *Journal of Nursing Administration, 29*(1), 9–19.

Searle, L. (1996, January). 21st century leadership for nurse administrators. *Aspen's advisor for nurse executives, 11*(4), 1, 4–6.

Sheahan, M., Duke, M., Nugent, P. (2007). Leadership in Nursing. Australian Nursing Journal, 14 (7), 28.

Shortell, S. M., & Kaluzny, A. D. (2006). *Health Care Management* (5th ed.). Clifton Park, NY: Delmar Cengage Learning.

Spence Laschinger, H., Finegan, J., & Wilk, P. (2009). Context matters: The impact of unit leadership and empowerment on nurses' organizational commitment. The Journal of Nursing Administration, 39(5), 228–235.

Stodgill, R. M. (1948). Personal factors associated with leadership: A survey of the literature. *Journal of Psychology, 25,* 35–71.

Stodgill, R. M. (1974). *Handbook of leadership: A survey of theory and research.* New York: Free Press.

Taylor, F. (1911). *Principles of scientific management.* New York: Harper & Row.

Tichy, N., & Devanna, D. (1986). *Transformational leadership.* New York: Wiley.

Upenieks, V. (2003). Nurse leaders' perceptions of what comprises successful leadership in today's acute inpatient environment. *Nursing Administration Quarterly, 27*(2), 140–152.

Vroom, V. (1964). *Work and motivation.* New York: Wiley.

Weber, M. (1864), as reported in Mommsen, W. J. (1992). *The political and social theory of Max Weber: Collected essays.* Chicago: University of Chicago Press.

Wheatley, M. J. (1999). *Leadership and the new science: Learning about organization from an orderly universe.* San Francisco: Berrett-Koehler.

Wheatley, M. J. (2005). *Finding our way: Leadership for an uncertain time.* San Francisco, CA: Berrett-Koehler Publishers.

Wolf, G., Boland, S., & Aukerman, M. (1994). A transformational model for the practice of professional nursing. Part 1. *Journal of Nursing Administration, 24*(4), 51–57.

Wong, C. & Cummings, G., (2007). The relationship between nursing leadership and patient outcomes: a systematic review. Journal of Nursing Management. 15, 508–521.

Wong, C., & Cummings, G. (2009). Authentic leadership: A new theory for nursing or back to basics. Journal of Health Organization and Management, 23(5), 522–538.

Wren, D. (1979). *Evolution of management thought.* New York: Wiley.

Young, S. (1992). Educational experiences of transformational nurse leaders. *Nursing Administration Quarterly, 17*(1), 25–33.

Yukl, G. (1998). *Leadership in organizations* (4th ed.). Upper Saddle River, NJ: Prentice Hall.

Yukl, G., Wall, S., & Lepsinger, R. (1990). Preliminary report on validation of the managerial practices survey. In K. E. Clarke & M. B. Clark (Eds.), *Measures of leadership* (pp. 223–238). West Orange, NJ: Leadership Library of America.

# SUGGESTED READINGS

Akerjordet, K., & Severinsson, E. (2010). The state of the science of emotional intelligence related to nursing leadership: An integrative review. *Journal of Nursing Management, 18*(4), 363–382.

Baggett, M. M., & Baggett, F. B. (2005, July). Move from management to high-level leadership. *Nursing Management, 36*(7), 12.

Barker, A. M., Sullivan, D. T., & Emery, M. J. (2006). *Leadership competencies for clinical managers: The renaissance of transformational leadership.* Sudbury, MA: Jones and Bartlett.

Bass, B., & Riggio, R. (2005). *Transformational leadership* (2nd ed.). Mahwah, NJ: Erlbaum.

Bondas, T. (2010). Nursing leadership from the perspective of clinical group supervision: A paradoxical practice. *Journal of Nursing Management.* 18(4), 477–486.

Brady, G. & Cummings, G. (2010). The influence of nursing leadership on nurse performance: A systematic literature review. *Journal of Nursing Management.* 18(4), 425–439.

Coughlin, L., Wingard, E., & Holihan, K. (Eds.). (2005). *Enlightened power: How women are transforming the practice of leadership.* San Francisco: Jossey-Bass.

Davenport, T. H. (2005). *Thinking for a living: How to get better performances and results from knowledge workers.* Cambridge, MA: Harvard School Press.

Donley, R. (2005). *Reflecting on 30 years of nursing leadership: 1975–2005.* Indianapolis, IN: Sigma Theta Tau International.

Festa, M. S. (2005, July). Clinical leadership in hospital care: Leadership and teamwork skills are as important as clinical management skills. *British Medical Journal, 16,* 331(7509), 161–162.

Ganann, R., Underwood, J., Matthews, S., Goodyear, R., Stamier, L. L., Meagher-Stewart, D. M., & Munroe, V. (2010). Leadership attributes: A key to optimal utilization of the community health nursing workforce. *Nursing Leadership, 23*(2):60–71.

Gordon, S. (2005). *Nursing against the odds: How health care cost-cutting, media stereotypes, and medical hubris undermine nurses and patient care.* Ithaca, NY: Cornell University Press.

Jumaa, M. O. (2006, Nov.). Developing nursing management and leadership capability in the workplace: Does it work? (2005). *Journal of Nursing Management, 13*(6), 451–458.

Kelly, T. (2005). *Ten faces of innovation: IDEOs strategies for beating the devil's advocate and driving creativity throughout your organization.* New York: Doubleday.

Kerfoot, K. M. (2006, Oct.–Dec.). Nursing research in leadership/management and the workplace: Narrowing the divide. *Nursing Administration Quarterly, 30*(4), 373–374.

Lombardi, D. N. (2005, Nov.–Dec.). Preparing for the future: Management and leadership strategies. *Healthcare Executive, 20*(6), 8–12.

McBride, A. (2010). *The Growth and development of nurse leaders.* New York: Springer.

Richardson, A., & Storr, J. (2010). Patient safety: A literature review on the impact of nursing empowerment, leadership and collaboration. *International Nursing Review, 57*(1), 12–21.

Sellgren, S., Ekvall, G., & Tomson, G. (2006, July). Leadership styles in nursing management: Preferred and perceived. *Journal of Nursing Management, 14*(5), 348–355.

Senge, P. (1990). *The fifth discipline: The art and practice of the learning organization.* New York: Doubleday.

Tomey, A. M., Arvin, K., Brown, W., Eslinger, S., Hamilton, C., Lofton, S., et al. (2001, March–April). Review of leadership and management literature. *Nurse Educator, 26*(2), 53, 63.

# CHAPTER 2

# The Health Care Environment

Ronda G. Hughes, PhD, MHS, RN; Patricia Kelly, RN, MSN;
Tanya L. Sleeper, GNP-BC, MSN, MSB

*Health care in the twenty-first century will require a new kind of health professional: someone who is equipped to transcend the traditional doctor–patient relationship to reach a new level of partnership with patients; someone who can lead, manage, and work effectively in a team and organizational environment; someone who can practice safe, high-quality care but also constantly see and create the opportunities for improvement.*

(Donaldson, 2001)

## OBJECTIVES

Upon completion of this chapter, the reader should be able to:

1. Identify how health care is organized and financed in the United States.
2. Compare U.S. health care with that of other industrialized countries.
3. Identify the major issues facing health care.
4. Relate efforts for improving the quality, safety, and access to health care.

Delmar/Cengage Learning

*Your neighbor calls, asking you to come over and advise her as to what to do with her grandchild who is sick. Finding the three-year-old child with a runny nose, a slight fever, and a congested cough, you recommend that she take the child to her primary care clinician for an office visit, especially if the fever continues or rises. Your neighbor feels that there is no urgency because of the high cost of the office visit co-pay and her difficulty in getting the child to the clinician as the office hours coincide with her work schedule. Also, she is unable to get an appointment until two weeks later because her grandchild is covered by Medicaid. She opts to wait until after business hours on Friday and then take the child to the emergency department. By the time they arrive at the hospital five days later, the child is admitted to the pediatric intensive care unit with a temperature of 104°F (40°C) and is hospitalized for a week.*

*What do you think of the occurrence of this type of scenario in the United States?*

*How can access to health care be assured for all patients regardless of source of insurance?*

Currently, the U.S. health care system consists of a mix of health care providers from either nonprofit or for-profit organizations in both the public government and private sectors, organized to provide more than 300 million American citizens with access to cost-effective, quality health care. Reimbursement for health care services is paid in one or a combination of these four ways:

- Private insurers
- Public government-funded payers
- Charitable entities
- Direct payment by patients

Public government health care programs include Medicare and Medicaid, State Children's Health Insurance Program (SCHIP), Department of Veterans Affairs, and other public programs, such as the health care programs for the military, American Indians, and federal prisoners. Together, these public government health care programs represent 47% of all U.S. health care costs (Centers for Medicare & Medicaid Services, Office of the Actuary, National Health Statistics Group, 2008). Many Americans are surprised by this fact,

yet verbalize reluctance to move toward a government run health care program for all citizens.

Most Americans are in good health, but many citizens are children, elderly, sick, disabled, or otherwise in need of access to quality health care services at a reasonable cost. The need for access, quality, safety, and reasonable cost has driven various initiatives to improve health care in the past and present. Some of the major initiatives include the 1935 passage of the Social Security Act; the 1946 passage of the Hill-Burton Act; and the 1965 passage of Medicare and Medicaid. The 2010 passage of both the Patient Protection and Affordable Care Act (PPACA) and the Health Care and Education Reconciliation Act (HCERA) was a recent attempt to improve health care access for the millions of Americans without health care coverage.

The U.S. spends 15.3% ($6,402) of its GDP on health care, more than any other wealthy country, all of whom provide health care insurance for all their citizens (KFF & Health Research and Educational Trust, 2009). Many wealthy countries in the world spend less on health care, e.g., France ($3,414), Canada ($3,326), Germany ($3,286), the Netherlands ($3,146), and the United Kingdom ($2,462), yet provide Universal Health Care (UHC) to their citizens (Osborn, 2008). In these countries, per capita spending on health care is considerably less than in the U.S., yet health care outcomes for such things as infant mortality, immunization rates, and life expectancy in the U.S. are poorer by comparison (Nolte & McKee, 2008). Perhaps Americans need to study what other countries have done in health care and modify it to fit. Throughout the history of the U.S., efforts to implement a UHC program have been resisted, with costly social and economic consequences.

To avoid the financial burden of health care costs, many Americans delay obtaining care. Their contact with the health care system is episodic and usually in acute care settings. Even after their symptoms have progressed and are well-advanced, many Americans are likely to obtain only irregular, sporadic care. This means that they lack consistent care from a health care provider whom they see regularly, whether for health promotion, illness prevention, early detection, or health restoration.

Inability to pay for recommended treatments and medications also compromises adherence to health care recommendations, which in turn affects recovery. Medical debt is now the number one reason for personal bankruptcy in the U.S. The majority of people declaring bankruptcy because of medical debt are employed and have health insurance (Himmelstein, Warren, Thorne, & Woolhander, 2005). A cascading effect occurs for patients with soaring costs, high incidence of medical bankruptcy, progressively worsening health outcomes, and a high incidence of medical malpractice claims.

Spending on health care services is concentrated in disproportionate ways, which also adds to health care costs. For example, 10% of people account for 60% of spending on health care services. Twenty percent of health care expenditures are spent on 1% of the population, indicating that a small percentage of the population absorbs a tremendous amount of health care services and spending. From another perspective, 44% of health care expenses are concentrated in the treatment of five predominant health problems: heart conditions, cancer, trauma, mental health disorders, and pulmonary conditions. On the other side of the health care spectrum, 50% of the population contributes to 3% of health care expenditures (Kaiser Family Foundation, 2007a; Stanton, 2006).

The fight to achieve quality health care for all continues. Schulte (2011) recommends the development of evidence based guidelines by the nursing and medical professional associations as one way to tackle the health care problem. Adherence to these guidelines then could be regularly monitored on electronic health records to note the outcomes, improve both patient care and evidence based guidelines, and decrease the incidence of medical malpractice.

This chapter discusses a selected history of American health care. It discusses health care in various settings, disease management, and the influence of external forces on health care. The chapter reviews how health care is organized, funded, and accredited. It explores health care disparities and clinical variation, and reviews reports of the Institute of Medicine Committee on Health Care. Finally, the chapter discusses issues regarding quality health care and the education of health care professionals. Patient Protection and Affordable Care Act of 2010 and the Health Care and Education Reconciliation Act of 2010.

In 2010, 47 million uninsured people in the United States did not have health care insurance. This led to many discussions of rights to health care, access, fairness, sustainability, safety, quality, and discussions of the amount of money spent by the government on health care. The Patient Protection and Affordable Care Act (PPACA) (Public Law 111–148) (Kaiser Family Foundation, 2010a) was signed into law by President Barack Obama on March 23, 2010. Along with the Health Care and Education Reconciliation Act (HCERA) of 2010 (Public Law 111–152), signed March 30, the Acts are a product of the health care reform efforts of the 111th Congress and the Obama administration (Office of the Legislative Counsel. 111th Congress, 2d Session. May, 2010). PPACA and HCERA will require most U.S. citizens and legal residents to have health insurance and will introduce many insurance market reforms to take effect over the next four years. They prohibit insurers from establishing annual insurance coverage caps and provide funds for medical research. They create a new insurance marketplace with state-based American Health Benefit Insurance Exchanges through which individuals can purchase insurance coverage, with premium- and cost-sharing credits available to individuals and families with income between 133–400 percent of the federal poverty level (the poverty level is $18,310 for a family of three in 2009).

PPACA and HCERA also will create separate state Insurance Exchanges through which small businesses can purchase insurance coverage. They subsidize insurance premiums for people making up to 400 percent of the federal poverty level (FPL) ($88,000 for a family of four in 2010), so their maximum *out-of-pocket* payment for annual premiums will be from 2 percent to 9.8 percent of income, providing incentives for businesses to provide health care benefits, prohibiting denial of coverage and denial of claims based on pre-existing conditions, prohibiting insurers from establishing annual coverage caps, and giving support for medical research. The costs of these provisions are offset by a variety of taxes, fees, and cost-saving measures, such as new Medicare taxes for those in high-income brackets, taxes on indoor tanning, cuts to the Medicare Advantage program in favor of traditional Medicare, and fees on medical devices and pharmaceutical companies. There is also a tax penalty for those who do not obtain health insurance, unless they are exempt due to low income or other reasons. The Congressional Budget Office estimates that the net effect of both Acts will be a reduction in the federal deficit by $143 billion over the first decade.

PPACA will include fundamental changes to Medicare, expansion of the Medicaid program, and reforms to Part D, closing the Medicare donut hole by 2020. It includes initiatives to prevent fraud and abuse; includes more health information technology; and promotes disease prevention programs across the health care system.

HCERA makes a number of health-related financing and revenue changes to the PPACA of 2010. HCERA is divided into two titles, one addressing health care reform and the other addressing student loan reform. It is anticipated that PPACA and HCERA will impact health care significantly, though at the time of this writing in early 2011, political forces are gathering and the direction of health care changes is not clear. The Congressional Budget Office estimates that the net effect of both PPACA and HCERA will be a reduction in the federal deficit by $143 billion over the first decade.

## HISTORY OF HEALTH CARE

Florence Nightingale observed that noise, food, rest, light, fresh air, and cleanliness were instrumental in health and illness patterns, Thus, she maintained, the aim of nursing was to put the patient in the best condition for nature to act upon her or him (Nightingale, 1865/1970). Nightingale also discovered the link between adverse patient outcomes and a lack of cleanliness and hand washing. Yet generations after the insights of Florence Nightingale were first set forth, sporadic adherence by health care providers to hand washing accounts for 2 million hospital-acquired infections, is attributable for up to 90,000 deaths, and burdens the health care system with a cost of up to $29 billion annually (Jarvis, 2006).

One hundred years ago, illnesses such as tuberculosis or pneumonia required lengthy hospitalizations and were often catastrophic for individuals and families. Today, such illnesses are preventable and are often easily treated. Vaccination programs have been used extensively to prevent the spread of communicable diseases. Additionally, surgical interventions in hospitals (for example, tonsillectomies, appendectomies, and reproductive procedures) have improved to treat otherwise debilitating or mortal conditions (Mayo, 2007). Health care is delivered by professional nursing and medical practitioners who are science based and who use evidence-based practice. Health care is primarily directed at preventing and treating chronic and behavioral diseases. Health care advances have extended life expectancy, with the consequence of more elderly people requiring more health care for chronic and complex health problems. The majority of clinical care is still provided in hospitals, but length of stay is much shorter, and a variety of innovative models of care are now used to provide cost-effective care for people with acute, community, and long-term clinical needs (Health Workforce Solutions LLC, & Robert Wood Johnson Foundation, 2008).

Health care-associated infections currently result in increased length of stay, mortality, and health care costs. In addition, a Centers for Disease Control and Prevention (CDC) report estimates that the overall annual direct medical costs of health care-associated infections recently ranged from $28 to $45 billion (Scott, 2009). These infections are most often attributed to invasive supportive measures such as endotracheal intubation and the placement of intravascular lines and urinary catheters. Several studies have noted that health care associated infections can be prevented through a number of multidisciplinary, evidence-based interventions, reducing the incidence of infection by as much as 70 percent (Anderson et al., 2007; Harbarth, Sax, & Gastmeir, 2003; Muto et al., 2005).

## STRUCTURING HOSPITALS AROUND NURSING CARE

Nightingale also described the importance of structuring hospitals around nursing care. The initial design of hospitals followed that advice by building large wards where nurses could easily monitor and observe their patients. Later, hospital design evolved to placing patient rooms surrounding centrally located nursing stations. Then, as today, the physical environment of hospitals can create stress for patients, their families, and clinical staff. Research is finding links between the physical environment and patient outcomes, patient safety, and patient and staff satisfaction (Hamilton, 2003). Studies show that such elements of hospital design as exposure to natural light, private rooms, and facilities that are staff friendly and have less noise contribute to improved patient outcomes (Ulrich, Quan, Zimring, Joseph, & Choudhary, 2004).

Although little is known about how to best design the hospital environment to facilitate clinical advances and care delivery, an estimated $200 billion will be expended for new hospital construction across the United States during the next 10 years (Institute of Medicine [IOM], 2004a). The Robert Wood Johnson Foundation, the nation's largest philanthropy devoted exclusively to health and health care, has provided funding to the Center for Health Design, a nonprofit research organization, for the Designing the 21st Century Hospital Project, which is the most extensive review of the evidence-based approach to hospital design ever conducted. Launched in 2000, the Pebble Project is a joint research effort between the Center for Health Design and health care providers. The project engages health care providers that are building new health care facilities or renovating old ones using an evidence-based design. The project uses the latest available evidence to inform design innovations and then measure the outcomes of the innovations through carefully designed research projects. The results are shared with the larger health care community to promote change. The Pebbles Project is an example of facility design to improve quality of care (The Center for Health Design, 2011).

## COLLECTING DATA

Nightingale also astutely recognized the importance of collecting and using data to assess the quality of health care. She employed coxcomb diagrams to present visual images of the number of preventable deaths during the Crimean war (Figure 2-1) and then later in London hospitals.

Today, data is collected through patient records, surveys, and administrative systems. From these, reports are developed, such as To Err is Human (IOM, 1999); the Centers for Disease Control and Prevention (CDC) National Vital Statistics Reports (Martin, Hamilton, Sutton, & Ventura, 2006); and The National Healthcare Disparities Report (NHDR) (Agency for Healthcare Research and Quality [AHRQ], 2005). These reports provide invaluable information, and data is displayed with charts and pictures to emphasize the successes and failures of health care throughout our nation. Evidence of significant disparities and low quality continue to demonstrate the need for significant health care improvement.

## INFLUENCE OF EXTERNAL FORCES ON HEALTH CARE

Recognizing the influence of external forces on care delivery and scope of practice, Nightingale also kept informed of the activities of practitioners and government policy makers (Dossey, Selanders, & Beck, 2005). With health care being the largest sector of our economy, employers, clinicians, managers, and patients all have vested interests in proposed changes to health care financing, organization, and the responsibilities and scope of practice for clinicians.

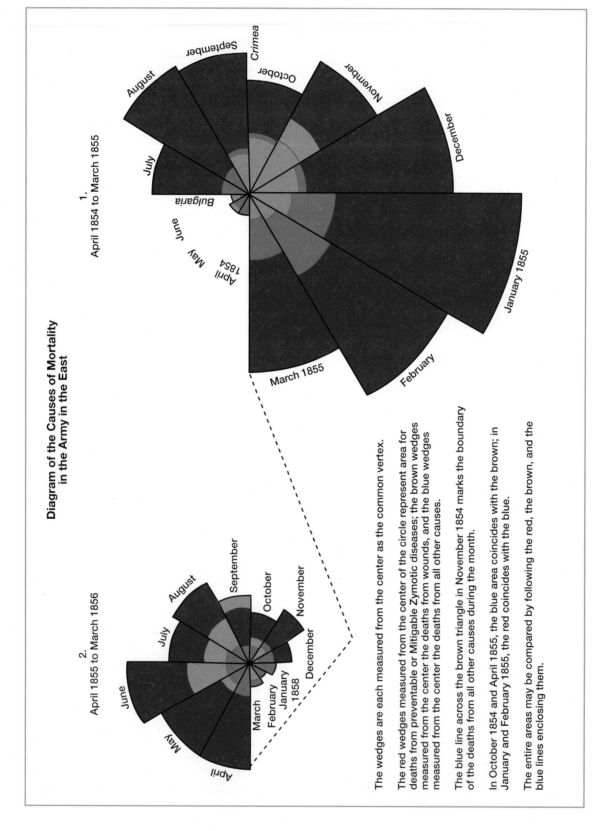

**Diagram of the Causes of Mortality
in the Army in the East**

1.
April 1854 to March 1855

2.
April 1855 to March 1856

The wedges are each measured from the center as the common vertex.

The red wedges measured from the center of the circle represent area for deaths from preventable or Mitigable Zymotic diseases; the brown wedges measured from the center the deaths from wounds, and the blue wedges measured from the center the deaths from all other causes.

The blue line across the brown triangle in November 1854 marks the boundary of the deaths from all other causes during the month.

In October 1854 and April 1855, the blue area coincides with the brown; in January and February 1855, the red coincides with the blue.

The entire areas may be compared by following the red, the brown, and the blue lines enclosing them.

**FIGURE 2-1** **Florence Nightingale's coxcomb diagram of the causes of mortality in the army in the east.** (*Source:* EC85.N5647.859n. Reprinted with permission of the Houghton Library, Harvard University, Cambridge, MA.)

Today, nursing leaders, managers, and staff need to be aware of and involved in the ongoing processes of making health policy.

# ORGANIZATION OF HEALTH CARE

Health care systems have three simple components: structure, process, and outcome. The **structure** component of health care includes resources or structures needed to deliver quality health care, for example, human and physical resources, such as nurses and nursing and medical practitioners, hospital buildings, medical records, and pharmaceuticals. The **process** component of health care includes the quality activities, procedures, tasks, and processes performed within the health care structures, such as hospital admissions, surgical operations, and nursing and medical care delivery following standards and guidelines to achieve quality outcomes. The **outcome** component of health care refers to the results of good care delivery achieved by using quality structures and quality processes and includes the achievement of outcomes such as patient satisfaction, good health and functional ability, and the absence of health care acquired infections and morbidity. See Table 2-1 for examples of structure, process, and outcome performance measures in clinical care, financial management, and human resources management. The American Nurses Association's Nursing Care Report Card for Acute Care (1995) also uses the structure, process, outcome framework for its indicators of quality.

It would be naïve to consider health care in the United States as it is currently being delivered as being an effective system of care. If that were true, it would imply that health care is based on shared values and goals; is organized around the patient; utilizes all pertinent information; ensures value-based and quality-based care; rewards quality care; is universally standardized and simplified; is available to everyone regardless of income, race, ethnicity, or education; is affordable; and reflects effective collaboration among clinicians and with patients (Davis, 2005; World Health Organization [WHO], 2000, p. 35). The World Health Organization (WHO) has put forth three primary goals for what good health care should do: (1) ensure that the health status of everyone is the best that is possible across the lifespan; (2) respond to patient's expectation of respectful treatment and include a focus on patients by health care clinicians; and (3) provide financial protection for everyone regardless of ability to pay (WHO, 2000). Consistent with these goals, Healthy People 2020 has also developed overarching goals to increase quality and years of healthy life and eliminate health disparities. These goals are:

Attain high-quality, longer lives, free of preventable disease, disability, injury, and premature death.
Achieve health equity, eliminate disparities, and improve the health of all groups.

Create social and physical environments that promote good health for all.
Promote quality of life, healthy development, and healthy behaviors across all life stages (Healthy People 2020, 2010).

## HEALTH CARE RANKINGS

Although state-of-the-art health care is available in the United States, access is limited to those who can afford the high costs associated with such care. The United States spends more money on health care than any other nation, yet health status and outcomes are significantly lower than in other industrialized or high-income countries. When compared to five other high-income nations—Australia, Canada, Germany, New Zealand, and the United Kingdom—the United States ranked last with respect to healthy lives, access, patient safety, efficiency, and equity (Davis et al., 2007). An overall score of 66 percent was recently given to the United States for its achievement across 37 core health indicators related to long, healthy, and productive lives; quality; access; efficiency; and equity of health care (Commonwealth Fund, 2008) (Figure 2-2).

Some major findings from the U.S. Scorecard include the following:

- The U.S. infant mortality rate is 7.0 deaths per 1,000 live births, compared with 2.7 deaths in the top three countries.
- Forty-nine percent of U.S. adults received preventive and screening tests according to guidelines for their ages and sex.
- Thirty-four percent of U.S. adults under age 65 have problems paying their medical bills or have medical debt they are paying off over time.
- U.S. insurance administrative costs were more than three times the rates of other countries.

## TABLE 2-1  Examples of Performance Measures by Category

| | CLINICAL CARE | FINANCIAL MANAGEMENT | HUMAN RESOURCES MANAGEMENT |
|---|---|---|---|
| Structure | *Effectiveness*<br>• Percent of nurses and physicians who are certified<br>• JC (formerly JCAHO) accreditation<br>• Presence of council for quality improvement planning<br>• Presence of magnet recognition | *Effectiveness*<br>• Qualifications of administrators in finance department<br>• Use of preadmission criteria<br>• Presence of an integrated financial and clinical information system and clinical decision-making technology | *Effectiveness*<br>• Ability to attract desired nursing and medical practitioners and other health professionals<br>• Size or growth of nursing and medical staff<br>• Salary and benefits competitive with competitors<br>• Quality of in-house staff education |
| Process | *Effectiveness*<br>• Ratio of medication errors<br>• Ratio of nurse-sensitive complications<br>• Ratio of health care-acquired infection<br>• Ratio of postsurgical wound infection<br>• Ratio of normal tissue removed during surgery | *Effectiveness*<br>• Days in accounts receivable<br>• Use of generic drugs and drug formulary<br>• Market share<br>• Size (or growth) of shared service arrangements | *Effectiveness*<br>• Number and type of staff grievances<br>• Number of promotions<br>• Organizational climate |
| | *Productivity*<br>• Ratio of total patient days to total full-time equivalent (FTE) nurses<br>• Ratio of total admissions to total FTE staff<br>• Ratio of patient visits to total FTE nursing and medical practitioners | *Productivity*<br>• Ratio of collections to FTE financial staff<br>• Ratio of total admissions to FTE in finance department<br>• Ratio of new capital acquisitions to fund-raising staff | *Productivity*<br>• Ratio of front-line staff to managers |
| | *Efficiency*<br>• Average cost per admission<br>• Average cost per surgery | *Efficiency*<br>• Average cost per debt collection<br>• Debt/equity ratio | *Efficiency*<br>• Recruitment costs |
| Outcome | *Effectiveness*<br>• Case-severity-adjusted mortality<br>• Patient satisfaction<br>• Patient functional health status<br>• Number of deaths from medical errors | *Effectiveness*<br>• Return on assets<br>• Operating margins<br>• Size and growth of federal, state, and local grants for teaching and research<br>• Bond rating | *Effectiveness*<br>• Staff turnover rate<br>• Number of absenteeism days<br>• Staff satisfaction |

**Source:** Compiled with information from Shortell, S. M., & Kaluzny, A. D. (2006). *Health care management* (5th ed.). Clifton Park, NY: Delmar Cengage Learning.

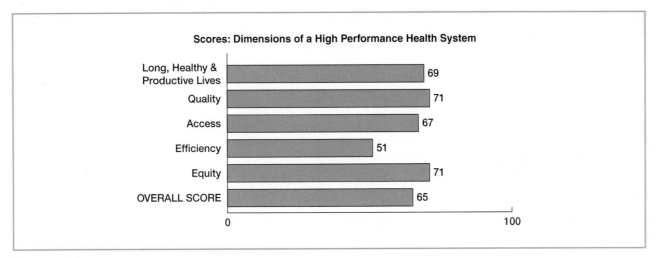

FIGURE 2-2  **U.S. scorecard on health system performance.** (*Source:* Commonwealth Fund National Scorecard on U.S. Health System Performance. [2008]. Retrieved September 11, 2008, from www.commonwealthfund.org/usr_doc/site_docs/slideshows/NatlScorecard/NatlScorecard.html).

- The United States lags well behind other nations in use of electronic medical records.
- There is a wide gap between low-income or uninsured populations and those with higher incomes and insurance.
- Hispanics are at particularly high risk of being uninsured, lacking a regular source of primary care, and not receiving essential preventive care.
- It would require a 24 percent or greater improvement in African American mortality, quality, access, and efficiency indicators to approach benchmark rates for Whites.
- Blacks are much more likely to die at birth or from conditions such as heart disease, diabetes, and cancer (Commonwealth Fund National Scorecard on U.S. Health System Performance, 2008).

## ACCESS TO HEALTH CARE FOR ALL

The United States is one of only a few large countries in the world without a universal system of health care. High health care costs in the United States in comparison to other countries has not resulted in increased access or utilization of services or more resources as the United States has fewer physicians, nurses, and hospital beds than other countries (Frogner & Anderson, 2006). In 2010, the Patient Protection and Affordable Care Act was signed into law. As mentioned earlier, the Act signifies a comprehensive health reform effort to expand health insurance coverage, control health care costs, and improve health care delivery in the United States. Through a number of provisions, the law, at the state and federal levels, will expand coverage through initiatives aimed at the individual employer. The Act also attempts to address rising health care costs through preventive strategies; increasing access to primary care; and targeting fraud, abuse, and waste (Kaiser Family Foundation, 2010a).

## HEALTH CARE PAYMENT IN OTHER COUNTRIES

Note that some countries, such as Britain, New Zealand, and Cuba, provide health care in government hospitals, with the government paying the bills. Others (e.g., Canada and Taiwan) rely on private-sector providers, paid for by government-run insurance. Many other wealthy countries (e.g., Germany, the Netherlands, Japan, and Switzerland) provide universal coverage using private doctors, private hospitals, and private insurance plans. In some ways, health care is less *socialized* overseas than in the United States, where almost all Americans sign up for government Medicare insurance at age 65. In Germany, Switzerland, and the Netherlands, seniors stick with private insurance plans for life. Meanwhile, the U.S. Department of Veterans Affairs is one of the world's purest examples of government-run health care (Reid, 2009).

## HOSPITAL CARE

The number of acute care hospital beds in the United States is 2.7 beds per 1,000 people. Since 1980, hospital bed use and length of stay have decreased in the United

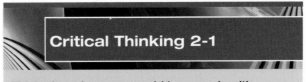

### Critical Thinking 2-1

Think about how you could improve health care delivery in the United States. If only 8 out of 1,000 people need hospitalization, and 800 have symptoms, how could we best deliver this care? Health care that focuses on more delivery of prevention and primary care would increase health and decrease disease throughout the United States. How can you work individually and with your community to begin to accomplish this?

## EVIDENCE FROM THE LITERATURE

**Citation:** Coddington, J. A., & Sands, L. P. (2008). Cost of health care and quality outcomes of patients at nurse-managed clinics. *Nursing Economics, 26*(2), 75–83.

**Discussion:** Lack of health insurance is a critical factor in access to appropriate health services and is directly associated with increased morbidity and mortality, lack of continuity of care, and rising health care costs. Nurse-managed clinics (NMCs) can serve as an important safety net in the health care delivery system by offering needed health services to populations of people affected by poverty and lack of insurance. NMCs remove barriers to care, improve health care access, and foster therapeutic relationships with nurse practitioners who provide primary care to vulnerable people. Much evidence also exists that nurse-managed clinics improve the use of preventive services, aid in the promotion of health, increase compliance with treatment, improve patient satisfaction, and reduce emergency room visits and re-hospitalizations.

**Implications for Practice:** The opportunity to provide quality care to vulnerable populations through NMCs is excellent. Overcoming the challenge of policies that restrict third-party reimbursement for nurse practitioners would allow an increased number of patients to be seen at NMCs.

In the March–April 2011 issue of *Consumer Digest,* author A. Christopher states that nurse practitioners are 95% of the health care providers in retail medical clinics in 42 states. The author reports on a September 2009 study in the Annals of Internal Medicine, stating that there is no evidence that the care that patients get at retail medical clinics is inferior to what they'd receive at a doctor's office and that that care is slightly superior to care in an emergency room.

## REAL WORLD INTERVIEW

As a nurse practitioner, there are many things about the American health care system that I really value. There are minimal wait times and there are lots of specialties.

I like that a patient whom I refer with a serious diagnosis can be seen by a specialist within two weeks. We develop and make top-notch technology available and we have great health care standards. For all of these reasons, we keep the world on track and it follows our lead. I'm also aware every day of the shortcomings of the health care system. It's far too expensive, there are too many special interest groups, and it's all going to collapse under its own weight, a classic example of capitalism gone amuck. For example, health insurance companies have too much power, and I hate how all the insurance hoops prevent me and my colleagues from giving the best care to our patients. We need a national health policy, and if the American people make enough noise, politicians will get behind it, too. I'm in favor of a national health plan, one that will ensure that all Americans have access to care. It needs to be one that incorporates what we already do best with what's useful from other countries such as Japan and Canada. We have a lot to learn from what they do well that we don't do.

**Nadine Lamoreau, RN, MSN, FNP, APRN-C**
Fort Fairfield, Maine

States, which corresponds to an increase in the use of outpatient and day-surgery facilities (Health System Change, 2006; Organisation for Economic Co-operation and Development, 2008). In the United States, the emphasis on acute care health care services has successfully driven health care costs higher but has not necessarily improved the quality of care or patient outcomes (Werner & Bradlow, 2006; Jeffrey & Newacheck, 2006).

When you look at a group of 1,000 people, it is estimated that 800 of them will experience symptoms of some disease or condition. Of this group of 800, 265 people will be seen in a practitioner's office or hospital outpatient department or emergency department, or they will use home health care. Only eight will eventually be hospitalized. The majority of the people don't need hospitalization and would benefit from more resources available for primary health care delivery outside the hospital (Green, Gryer, Yawn, Lanier, & Dovery, 2000). This finding seems odd when considering where the research dollars are targeted and where the majority of health care dollars are devoted, that is, acute care settings in hospitals. Note that a consistent focus on illness and injury, often referred to as a downstream focus, means fewer dollars are invested in upstream efforts. A focus on upstream efforts would be directed at keeping the population well through health promotion and illness prevention strategies and would be less costly.

## NEED FOR PRIMARY HEALTH CARE

Given that the majority of patient needs and patient care delivery occurs outside acute care settings, primary care "which provides integrated, accessible health care services by clinicians who are accountable for addressing a large majority of personal health care needs, developing a sustained partnership with patients, and practicing in the context of family and community" (IOM 1996, p. 1), should be better understood and appreciated for the role it has in improving patient's health status and health outcomes. The key foundations of primary care (Starfield, 1998) can be applied across the health care continuum and across organizational settings because **primary care** emphasizes seven important features: care that is continuous, comprehensive, coordinated, community oriented, family centered, culturally competent, and begun at first contact with the patient. According to Starfield (1998), patients and clinicians need to work together to appropriately utilize services, based on the following four foundations of primary care:

- *First Contact:* Conduct the initial evaluation and define the health dysfunction, treatment options, and health goals.
- *Longitudinality:* Sustain a patient–clinician relationship continuously over time, throughout the patient's illness, acute need, and disease management.

 **CASE STUDY** 2-1

You work in an Emergency Department (ED) that sees 6,000 patients a month. Patients are charged $200.00 per visit plus charges for tests and medications. Thus, these 6,000 patients can generate $1,200,000 in gross revenue for the hospital. Consider that there are 15 RNs making $30.00 per hour and 6 MDs, making $150.00 per hour working each shift. Salaries for the RNs total $324,000. Salaries for the MDs total $648,000. The total salary for these two groups is $972,000. Of the 6,000 patients, 50 percent have Medicare/Medicaid, 45 percent are covered by managed care or insurance, and 5 percent have no insurance. Thus, just 95 percent of patients can pay their bills. The other 5 percent of patient's bills are written off by the hospital as bad debt.

Medicare/Medicaid/Managed Care/Insurance companies often pay only 55 percent of the bills for these patients. They may deny payment for 45 percent of the bills. Thus, for the $1,140,000 billed (95 percent of $1,200,000), the hospital will receive approximately $627,000 (55 percent of the $1,140,000 billed). Approximately $513,000 of the bill will not be paid by Medicare/Medicaid/Managed Care/Insurance. Consider the following:

What other expenses besides salary must the hospital pay out of the $627,000 that it receives? Consider hospital space, liability insurance, technology costs, and so on.

Notice the effect that increasing the volume of patients has on your budget figures. What happens to your budget if the patient volume goes to 8,000 patient ED visits per month and staffing stays the same?

Are patients receiving useful information about future illness prevention and healthy living practices in the ED?

Is this a cost-effective way to deliver health care?

How could we better serve the health care needs of Americans?

- *Comprehensiveness:* Manage the wide range of health care needs, across health care settings and among different health care professionals.
- *Coordination:* Build upon longitudinality. Care received through referrals and other providers is followed and integrated, averting unnecessary services and duplication of services.

Primary care clinicians, which primarily include both medical and nursing practitioners, can be a patient's greatest asset in negotiating the health care system and improving patient outcomes. It is through understanding the patient's past and present that future health care needs can be anticipated. Primary care interventions, such as health promotion and timely preventive care and medication administration, can reduce the need for hospitalizations, improve the health of patients, and avert adverse morbidity and mortality outcomes. Patients and their families can communicate with clinicians to understand their health care needs, how to achieve the best possible health, and how to partner with clinicians to improve decision making. This is what patient-centered care is based on, both primary care and patient decision making. The World Health Report (2008b), *Primary Health Care: Now More than Ever*, underscores the need for primary health care. The report cites a disproportionate focus on specialist hospital care, fragmentation of health systems, and the proliferation of unregulated commercial care. The World Health Organization (2010) has also identified key elements in improving health status through primary care strategies aimed at reducing disparities through universal access, enhancing coordination and delivery of care, and increasing stakeholder participation at multiple levels.

## THE FEDERAL GOVERNMENT

The federal government is a major driver of health care organization and delivery. Distinct, major divisions of the U.S. Department of Health and Human Services (DHHS) include the following:

- *Agency for Healthcare Research and Quality (AHRQ):* Funds health services research on the effectiveness of health care services and outcomes of care.
- *Centers for Disease Control and Prevention (CDC):* Promotes health and quality of life by preventing and controlling disease, injury, and disability.
- *Centers for Medicare and Medicaid Services (CMS):* Administers the Medicare program and regulates the Medicaid program.
- *Food and Drug Administration (FDA):* Monitors the safety of food, the safety of cosmetic products, the safety and efficacy of drugs, and the safety and efficacy of medical devices.
- *Health Resources and Services Administration (HRSA):* Administers training programs for health care clinicians, funding for pregnant women and children, programs for persons with HIV/AIDS, and programs serving low-income, underserved, and rural populations.
- *Indian Health Service (IHS):* Maintains health services provided to American Indians and Alaska Natives.
- *National Institutes of Health (NIH):* Funds biomedical research through 18 research institutes primarily organized according to specific diseases.
- *Substance Abuse and Mental Health Services Administration (SAMHSA):* Provides leadership in services, policy, and information dissemination for mental health and substance abuse treatment and prevention.

## STATE AND LOCAL LEVELS

Public health services at the state and local levels include boards of health and state and local health departments. Even with the 1988 IOM Report on public health, the ability of public health departments to engage in improving the health of the public has become limited (Tilson & Berkowitz, 2006). In addition, efforts for bioterrorism and disaster preparedness have brought the nation's infrastructure desolation to light, causing increased funding for this nation's disaster preparedness efforts, with little money focused on public health care funding and infrastructure redevelopment.

## HOME HEALTH CARE

The location of care delivery is continually changing to adapt to technologies and patient needs. The use of home health care services continues to grow as more and more individuals access these services in the community setting, in lieu of institutional care. According to the National Health Care Expenditure Projections (Centers for Medicare & Medicaid Services [CMS], 2009), home health care expenditures have grown to 11.7 percent of national health care expenditures in 2009, with spending having reached $72.2 billion. This increase has been primarily driven by higher growth in Medicaid spending, partly due to Medicaid's continued shifting of long term care from institutional to home settings. In light of an aging population, home health care services will continue to serve an integral role in health care delivery.

## HEALTH CARE DISPARITIES

Inequalities in such things as gender, age, ethnicity, etc. (Figure 2-3) have been recognized as great influences on health outcomes. Socioeconomic status is the number one predictor of poor health. The 2008 National Healthcare Disparities Report found that, across the process of care

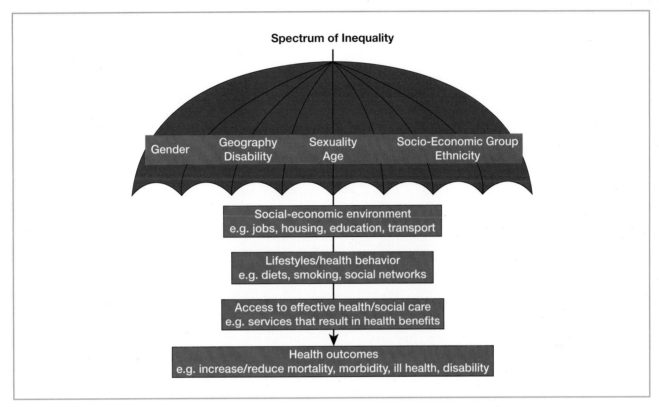

**FIGURE 2-3** **The spectrum of inequality.** (*Source:* BMJ Health Intelligence [2007]. Health Inequalities. Retrieved August 5, 2008, from healthintelligence.bmj.com/hi/do/public-health/topics/content/inequalities-in-health/definition.html).

measures tracked, patients received the recommended care less than 60 percent of the time. The report also noted, even when the overall quality of care improves, health care disparities often still persist across socioeconomic groups, racial and ethnic populations, and geographic areas. In addition, not enough health care delivery and attention is directed toward the top underlying causes of death in the United States (Table 2-2).

Analysis of large datasets illustrates that, as one ages, more health care services are utilized; women use health care services more frequently than men; and whites have greater health care access, and therefore higher utilization rates, than do patients of color (National Healthcare Disparities Report [NHDR], 2009); (National Healthcare Quality Report [NHQR], 2009). Both financial and nonfinancial barriers to care delivery result in lack of attention to health care disparities and factors contributing to the underlying causes of death, which affects health outcomes. In-depth information on national health care disparities is reported in the annual National Health Disparities Report

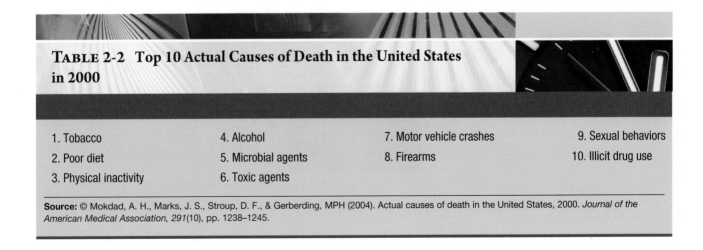

**TABLE 2-2  Top 10 Actual Causes of Death in the United States in 2000**

| | | | |
|---|---|---|---|
| 1. Tobacco | 4. Alcohol | 7. Motor vehicle crashes | 9. Sexual behaviors |
| 2. Poor diet | 5. Microbial agents | 8. Firearms | 10. Illicit drug use |
| 3. Physical inactivity | 6. Toxic agents | | |

**Source:** © Mokdad, A. H., Marks, J. S., Stroup, D. F., & Gerberding, MPH (2004). Actual causes of death in the United States, 2000. *Journal of the American Medical Association, 291*(10), pp. 1238–1245.

(NHDR, 2009), which now presents state variations (see http://statesnapshots.ahrq.gov/snaps09/index.jsp).

A disproportionate number of racial and ethnic minority groups are uninsured. Injury and illness are more common among Black, Asian, American Indian, Hispanic, and socioeconomically poor people, due in part to the lower level of care they receive as compared with Whites (AHRQ, 2008a). For example, disparities in infant morbidity and mortality, cardiovascular and pulmonary disease, diabetes, communicable disease, cancer, and disease prevention (i.e., immunization and health screening) are more likely to be experienced by people disadvantaged by poverty, age, skin color, or ability to speak English. Such differences are further aggravated by miscommunication and misunderstanding, stereotyping, discrimination, and prejudice between patients and providers. Lifestyle behaviors that contribute to illness are higher among vulnerable groups. Because of their financial difficulties and other difficulties in accessing the health care system, vulnerable people often postpone health care. They are more likely to use the acute care system when their illness symptoms are advanced. Use of emergency departments and other acute care facilities for treatment is the most expensive way to obtain health care. In countries with a national health care system, health disparities also exist, but virtually everyone in those countries regardless of socioeconomic background is assured of equal access to quality health care. In a recent study, health status, access to health care, and use of the health care system in the United States and Canada compared disparities such as race, income, and immigration status. Health care was determined to be less accessible in the United States, and disparities related to health care access were minimized in Canada because of universal health care (Lasser, Himmelstein, & Woolhandler, 2006).

## HEALTH CARE SPENDING

In the United States, health insurance has been generally employment-based, so long as it is affordable for the employer to offer this health care coverage for employees. The higher one's income in this country, the greater the likelihood of having health insurance coverage. The opposite is true for those with low incomes, especially those with poverty-level incomes. Patients who are at poverty levels often cannot afford insurance premiums nor can they afford, in the majority of instances, out-of-pocket health care costs. Since the inception of private health insurance in the late 1920s following the development of hospitals as the *center* of health care and subsequent rising health care costs (Starr, 1983), private health insurance from third-party payers such as insurance companies has been generally voluntarily offered as a benefit to employees and sometimes their families. Patients may make payments to providers of health care and third-party payers. Providers of health care deliver service to patients and bill third-party payers. Third-party payers may make payments to providers as direct payment fees for individuals, capitated payment for services for a group of patients, or as prospective payments for future patients. Health insurance distributes health care funds from the healthy to the sick (Figure 2-4).

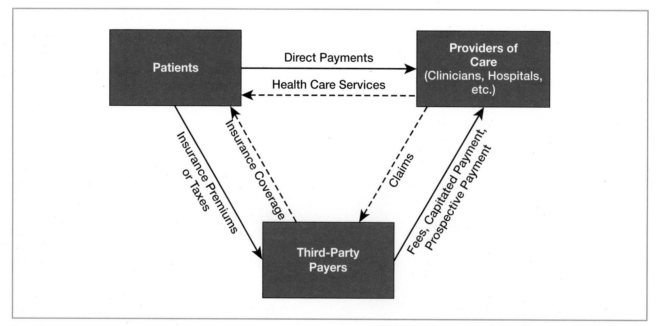

FIGURE 2-4 Economic relationships in the health care delivery system. (*Source:* Adapted from "What Can Americans Learn from Europeans?" by U. E. Reinhardt, 1989, *Health Care Financing Review* [Supplement], pp. 97–103.)

# MEDICARE AND OTHER HEALTH CARE COSTS

One reason proposed for the steady incline in health care costs is that the elderly have virtually universal health care coverage through Medicare. This universal health care coverage indicates that the United States will likely experience very rapid growth in overall health expenditures in coming years, as the population continues to age. Other sources of health care funding and where it went are shown in Figure 2-5 and Figure 2-6.

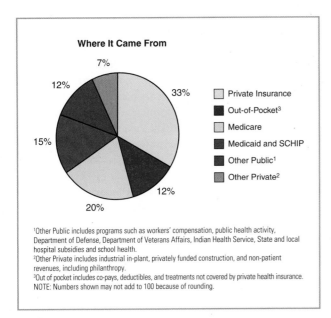

¹Other Public includes programs such as workers' compensation, public health activity, Department of Defense, Department of Veterans Affairs, Indian Health Service, State and local hospital subsidies and school health.
²Other Private includes industrial in-plant, privately funded construction, and non-patient revenues, including philanthropy.
³Out of pocket includes co-pays, deductibles, and treatments not covered by private health insurance.
NOTE: Numbers shown may not add to 100 because of rounding.

**FIGURE 2-5  The nation's health dollar, calendar year 2008: where it came from.** (*Source:* Centers for Medicaid & Medicare (CMS), Office of the Actuar, National Health Statistics Group)

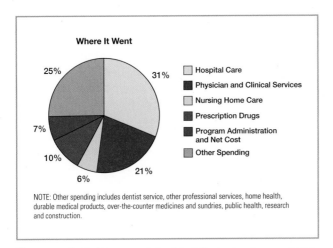

NOTE: Other spending includes dentist service, other professional services, home health, durable medical products, over-the-counter medicines and sundries, public health, research and construction.

**FIGURE 2-6  The nation's health dollar, calendar year 2008: where it went.** (*Source:* Centers for Medicaid & Medicare (CMS), Office of the Actuary, National Health Statistics Group)

## EVIDENCE FROM THE LITERATURE

**Citation:** Grose, T. (2007). How they do it better: Free health care for all. *U.S. News and World Report.* March 26, 2007, 65.

**Discussion:** Britain spends $2,546 annually per person for health care and offers all its citizens free care. The United States spends $6,102 annually per person for health care, and 47 million people are uninsured for health care. Britain's infant mortality rate is lower and its life expectancy rate is higher than the same rates in the United States. Patients in Britain wait for some elective surgeries, but they can buy private coverage for this, if desired. Appointments with general practitioners in Britain can be made quickly and Emergency Room treatment is good. Drugs are free to patients in Britain.

**Implications for Practice:** The United States would benefit from studies of other countries' health care systems. We can look to see what works well in other countries and what does not work so well and develop a system that is cost effective.

# HEALTH CARE INSURANCE

In the United States, health care insurance is one of the most significant factors in facilitating access to health care services. Recently, the number of people covered by insurance as well as the breadth and depth of health insurance coverage has decreased. According to data from the Kaiser Family Foundation-Health Research and Educational Trust Annual Employer Survey (2007c) and the U.S. Department of Labor, premiums for employment-based private insurance increased 114 percent from 1999 to 2007, while wage earnings increased only 27 percent. The 2010 passage of The Patient Protection and Affordable Care Act and the Health Care and Education Reconciliation Act is an attempt to improve health care coverage for all.

In the past, American health insurance companies have routinely rejected applicants with a *preexisting condition.* The insurance companies often denied claims. If a customer was hit by a truck and faced big medical bills, the insurance company dug through the records looking for grounds to cancel the policy, often while the victim was still in the hospital (Reid, 2009). Foreign health insurance companies, in

I was admitted to the hospital in December for what was later diagnosed as angina. After being stabilized in the emergency room, I was admitted to the cardiac care unit. A series of tests were ordered by my cardiologist to determine enzyme levels in my blood. This would show if I had suffered a heart attack. Blood was drawn every eight hours for twenty-four hours. The news was good—no heart attack—until I received the bill. My insurance company said the enzyme tests for angina were frivolous and not necessary, and they would not pay. I had my cardiologist write a letter of explanation, saying the tests he ordered were routine for determining a diagnosis. After this second appeal, the insurance company rejected my payment claim again. After several more appeals, my insurance company ultimately paid the laboratory bill for these tests, but would not pay the pathologist for his interpretation of the blood tests. It has now been six months since I received my initial bill, and the hospital and pathologist's office have turned my case over to a collection agency. They did not take into consideration the fact that I was appealing the bill. They wanted me to pay upfront and then appeal. That was not going to happen.

**Kathleen A. Milch**
Patient
Whiting, Indiana

contrast, must accept all applicants, and they can't cancel insurance as long as you pay your premiums. Everyone is mandated to buy insurance, to give the plans an adequate pool of rate payers. The key difference is that foreign health insurance plans exist only to pay people's medical bills, not to make a profit. The United States is the only developed country that lets insurance companies profit from basic health coverage. In many ways, foreign health care models are not really *foreign* to America, because our health care system uses elements of all of them. For Native Americans or veterans, we're Britain: The government provides health care, funding it through general taxes, and patients get no bills. For people

who get insurance through their jobs, we're Germany: Premiums are split between workers and employers, and private insurance plans pay private doctors and hospitals. For people over 65, we're Canada: Everyone pays premiums for an insurance plan run by the government, and the public plan pays private doctors and hospitals according to a set fee schedule. And up until the recent passage of health care legislation, for the tens of millions without insurance coverage, we're Burundi or Burma. In the world's poor nations, sick people pay out of pocket for medical care; those who can't pay stay sick or die. Seven hundred thousand Americans are forced into bankruptcy each year because of medical bills. In France, the number of medical bankruptcies is zero; Britain: zero; Japan: zero; Germany: zero (Reid, 2009).

Canadians have their choice of health care providers. In Austria and Germany, if a doctor diagnoses a person as *stressed*, medical insurance pays for weekends at a health spa. Canada does make patients wait weeks or months for nonemergency care, as a way to keep costs down. But studies by the Commonwealth Fund and others report that many nations—Germany, Britain, Austria—outperform the United States on measures such as waiting times for appointments and for elective surgeries. In Japan, waiting times are so short that most patients don't bother to make appointments. This includes waiting times for many surgical procedures (Reid, 2009).

Many gaps have existed in insurance programs in the United States, including incomplete health care coverage, need for copayments and deductibles, lack of provider choice, need for preauthorizations, and other difficulties in maneuvering through the insurance company requirements. **Copayments** are a fixed health care fee paid by the patient to the health care provider at the time of service; this amount is paid in addition to the money the health care provider will receive from the insurance company. **Deductibles** are a predetermined out-of-pocket fee paid by a patient for health care services before reimbursement through health insurance begins to be paid. For example, if an insurance plan has a $50 copayment and a $1,000 deductible, the patient is responsible for paying the $50 at each health care visit plus paying the first $1,000 of health care costs, after which costs will be reimbursed as allowed by the patients health insurance plan. Preauthorization requires that approval be obtained from the insurance company before care or treatment such as hospitalization or diagnostic testing is initiated if such services are to be reimbursed by the patients health insurance plan.

## MEDICARE AND MEDICAID

Because the United States does not have a national/universal health care program, public health care programs are intended to help fill the gap. Beginning in 1965 under

Titles XVIII (Medicare) and XIX (Medicaid) of the Social Security Act, eligibility for public insurance has been based on age (Medicare for those age 65 and older) or based on having a low income and/or having a disability (Medicaid). Medicare and Medicaid programs ensure access to many needed services for patients. Medicare is the largest federal program. Medicare Part A is an insurance plan for hospital, hospice, home health, and skilled nursing care that is paid through Social Security taxes. Nursing home care that is mainly custodial is not covered. Medicare Part B is an optional insurance that covers physician services, medical equipment, and diagnostic tests. Part B is funded through federal taxes and monthly premiums paid by the recipients. Part D is optional coverage for outpatient medications.

Medicaid, a state program financed by federal and state funds, pays for services provided to persons who are medically indigent, blind, or disabled and to children with disabilities. The federal government pays between 50 percent and 83 percent of total Medicaid costs based on the per capita income of the state. Services funded by Medicaid vary from state to state but must include services provided by hospitals, physicians, laboratories, radiology, prenatal and preventive care, and nursing home and home health care services.

## Other Public Programs

Other large public insurance programs are associated with American Indian and Alaska Native heritage, and the Military Services, that is, veteran's health and insurance for active military personnel and their families. The Indian Health Service (IHS), under the U.S. Department of Health and Human Services, provides health services to American Indians and Alaska natives enrolled in more than 500 tribes, villages, and pueblos (IHS, 2006). The Indian Self-Determination Act of 1975 gave many tribal organizations the responsibility for the provision of health care services. IHS maintains some hospitals and clinics, yet recent legislation has severely cut funding for these sites of care.

The Department of Defense (DOD) finances and manages TRICARE, the triple option benefit care plan available for military families, formerly called the Civilian Health and Medical Program of the Uniformed Services (CHAMPUS), for enlisted military personnel and their military dependents as well as for retirees and their dependents and survivors. The Department of Veterans Affairs, established in 1921, finances and manages the Veterans Health Administration (VHA). This program is available for U.S. veterans if they need any medical, prescription, surgical, and rehabilitation care services.

### Critical Thinking 2-2

You work in a large health care system. One of your patients is not following his health care regime. You wonder what thought processes this patient is using to justify continuing an activity that presents risk to his health.

How can you best assist the patient in your health care system and community? What kinds of structures, processes, and outcomes will your system want to develop to improve care to this patient and the population of patients that your system serves? How could you work in the community to enhance the population's choices for diet, exercise, or lifestyle?

## STATE REGULATION OF HEALTH INSURANCE

Three key pieces of federal legislation set forth national standards that the individual states use to regulate health insurance. First, the Employee Retirement Income Security Act (ERISA) of 1974 provides a framework for states to regulate health insurers. Second, the Consolidated Omnibus Budget Reconciliation Act (COBRA) of 1985 ensures that employees who resigned, were laid off, were terminated, or lost their jobs due to family-related reasons can retain their health insurance coverage for up to 18 months and, in some cases, up to a maximum of 36 months if they are deemed qualified and pay the full premiums. A third piece of legislation, the Health Insurance Portability and Accountability Act (HIPAA) of 1996 imposed restrictions on limitations and exclusions of insurance coverage for those with preexisting conditions and restricted other attempts to exclude employees from insurance coverage. It also provides protection of insurance coverage as employees change employers, and it provides tax exclusions for medical savings accounts (Table 2-3).

## INTERNATIONAL PERSPECTIVE

To the extent that the United States is similar economically and sociopolitically to countries such as France, Canada, and Japan, an examination of the health systems in those countries is useful. The differences in allocation of health care spending in the United States as compared to other countries are graphically displayed in Figure 2-7.

### TABLE 2-3   HIPAA Privacy Regulations

- Allows patient to review and request amendments to their medical records
- Gives consumers control over how their personal health information is used and limits the release of information without a patient's consent
- Restricts the amount of patient information shared between physicians and other caregivers to the *minimum necessary*
- Requires privacy-conscious business practices, such as hiring a privacy officer and training employees about patient confidentiality
- Requires that paper records and oral communications be protected from privacy breaches

**Source:** Compiled with information from U.S. Department of Health & Human Services. (1996). Summary of the Health Insurance Portability and Accountability Act (HIPAA) Privacy Rule, available at http://www.hhs.gov/ocr/privacy/.

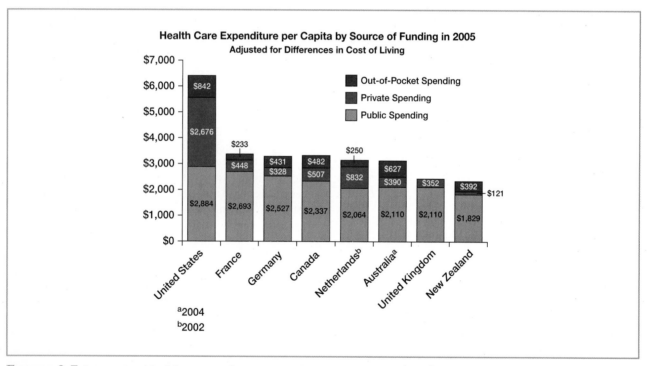

**FIGURE 2-7   International health care spending per capita by source of funding (2005).** (*Source:* Osborn, R. [2008]. Comparing health care systems performance: Opportunities for learning from abroad [slide #4]. Commonwealth Fund. Retrieved October 4, 2008, from www.allhealth.org/briefingmaterials/osborn-1189.ppt).

## France

France, ranked as having the best health care system in the world (Nolte & McKee, 2008), spends 11.1 percent of its GDP on health care (Organisation for Economic Co-operation and Development, 2008). This is approximately one-half of what the United States spends on health care per person. Comprehensive health care is guaranteed for all citizens and legal residents in France.

Similar to the United States, health care in France is provided through private and government insurance. Unlike Canada and Britain, there are no lengthy wait times in France. Unlike the United States, everyone is insured in France, and there are no additional patient charges for health insurance plan deductibles (Shapiro, 2008). Employed residents are covered by a national health insurance plan, referred to as *Sécurité sociale*, which includes spouses and children. Another plan, *couverture maladie universelle* (CMU), provides coverage for those people who do not qualify for the *sécurité sociale* program and is free to some people whose income is below a certain level. The national health insurance plan is funded through private and public means, with employees paying up to 21 percent of their incomes to the national health care system and employers making similar contributions. By comparison, Americans pay fewer taxes but pay more for health care (e.g., through paying health insurance premiums and other out-of-pocket expenses not covered by their insurance plans). In France, costs are dependent upon the type of provider seen; for example, a general practitioner is less expensive than a specialist. Likewise, it is more expensive to seek treatment at night, on the weekend, or on public holidays. Hospital care is reimbursed through the national health plan, and a percentage of the cost of prescription drugs is also reimbursed to the patient. Essentially, the sicker a person is, the more coverage is allowed, including for expensive drugs and experimental cancer treatments. Reducing cost and improving efficiency are the challenges for this system. Waste, such as *doctor shopping*, whereby a patient seeks treatment from more than one health care provider for the same ailment, and overuse of prescription drugs are partly responsible for high health care costs in France (National Coalition on Health Care, 2008c; Shapiro, 2008).

## Canada

Annually, the Canadian government spends 10 percent of its GDP on its national health care system. Canada has the eighth largest global health care budget in the world (Organisation for Economic Co-operation and Development, 2008). The Canadian health care system is administered by each Canadian provincial or territorial government. Seventy percent of health spending is publicly funded through federal and provincial taxation of individuals and corporations, and the remaining 30 percent is paid through private and out-of-pocket sources for additional services such as prescription medications or dental and vision care (Canadian Institute for Health Information, 2006). All Canadians have equal access to the same quality and quantity of health care. Under the Canada Health Act of 1984, comprehensive health care is publicly administered, portable between provinces, and accessible to all. Primary care is provided by physicians and nurse practitioners, who may work in private clinics or public institutions. These health care providers are reimbursed on a fee-for-service basis, which allows them to be reimbursed by each provincial or territorial health plan for each health care service rendered to a patient.

Unlike the privatized health care system in the United States, extra billing, deductibles, and copayments are not allowed. The health care provider bills the provincial or territorial health plan and is reimbursed with an agreed-upon amount for each health service given. No additional charges or costs can be billed to or recovered from the patient. With only one insurance payer, referred to as a single-payer system, many of the problems embedded within the American health care system are eliminated. The problem currently facing the Canadian health care system is the lengthy wait times to access family practitioners, specialists, emergency room services, diagnostic tests, and surgical procedures (National Coalition on Health Care, 2008c). Same-day access to the health care system is lowest in Canada (23 percent). The United States is slightly better (30 percent), but considerably lower than Germany (56 percent) (Commonwealth Fund, 2006). Wait times in Canada are circumvented based on the gravity of a patient's condition, but this solution aggravates the waiting time for less urgent cases. In a geographically vast country with a small population compared to the United States, some Canadians in rural or isolated locales often travel long distances to obtain specialized services.

As an example of the comprehensiveness and affordability of the Canadian health care service, a 51-year-old Canadian nurse working in the United States recently returned with her spouse and 15-year-old child to her home province of Alberta for one year. During that time, the family enrolled in the Alberta Health Care Insurance Plan for a cost of $88 per month (Alberta Health and Wellness, 2004). Health care for the nurse included an annual physical examination, lab tests, mammogram, bone density testing, and a routine colonoscopy. The colonoscopy required a five-month wait. Her husband underwent elective day surgery, which was booked seven months in advance. Her daughter was immunized at school, and she was once treated at a community health center for a persistent respiratory infection. Beyond payment of their monthly premium, the family was issued no bills for the health care they received.

## Japan

To pay for its national health care system, Japan spends 8.2 percent of its GDP on health care. This is less than France, Canada, or the United States, and Japan enjoys the longest life expectancy of all (i.e., 82.4 years) (Organisation for Economic Cooperation and Development, 2008). Membership in either of two broad health insurance programs is mandatory in Japan. The Employee Health Insurance Program provides health care coverage for people employed by medium to large companies, national or local government, or private schools. The National Health Insurance Program covers self-employed and unemployed persons, as well as those employed in agriculture, forestry, or the fisheries industries. Payment toward each health insurance plan varies, but in both programs, patients share in paying for their health care costs up to a certain level, after which the insurance plan provides full coverage (National Coalition on Health Care, 2008c).

The Japanese health care system provides comprehensive care and is one of the most advanced in the world. Physicians and hospitals are predominantly run as private businesses, and costs for care are reimbursed through the national health care system. Patients' health care costs are reimbursed at a fixed amount determined by the government. Refusal of coverage by insurers is impossible, and personal medical debt is nonexistent. Because health care providers are reimbursed for the number of services provided, Japan leads other countries in the number of drugs prescribed, the number of tests ordered per patient, and patient length of stay in a hospital. The Japanese average 16 consultations with a health care provider per year, three to four times the rate of consultations in the United States.

Factors contributing to lower health care costs in Japan include healthier diets and lifestyles, with a lower incidence of chronic diseases. Reimbursement rates for health care services are low, but reimbursement strategies for quantity rather than quality of services erodes the advantage a healthier lifestyle might have in terms of curbing costs. Despite cost containment efforts, health care expenses are increasing, especially with the earthquakes of 2011. A shortage of physicians also means the wait time to access the health system is an issue. With an aging population and shrinking workforce, concerns as to how access, affordability, and quality will continue in the future are being voiced (National Coalition Health Care, 2008c; Reid, 2008).

# THE RISING COST OF HEALTH CARE

Health care costs are measured as part of the U.S. Gross Domestic Product (GDP). The **Gross Domestic Product (GDP)** is an economic measure of a country's national income and output within a year and reflects the market value of goods and services produced within the country. The GDP is used as a barometer of the national economy Health care costs in the U.S. are increasing 2.5 percentage points faster than the annual U.S. GDP. Employer-based health care premiums have doubled since 2000, yet those who are insured incur greater financial burdens as they pay for more out-of-pocket expenses. Findings from the Employer Health Benefits Survey report increases in annual premiums, deductibles, co-insurances, co-payments and other out of pocket costs (KFF & Health Research and Educational Trust, 2009). These findings all influence health care utilization.

U.S. national health care expenditures were $1.9 trillion in 2004 (KFF & Health Research and Education Trust, 2006). From 1960 to 2000, the GDP for health care grew nearly 15-fold, from approximately $526 billion to the trillions of dollars spent today. It is projected that health care spending will rise to $4 trillion by 2015 (Borger et al., 2006). Health care spending continues to increase faster than the overall U.S. economy. This is 2.5 percentage points faster than the growth of the GDP (Centers for Medicare and Medicaid Services [CMS], 2006). In 2000, the percentage of GDP for health care was 15.3%, and analysts project this number to keep rising to 20% (Borger et al., 2006).

# FACTORS CONTRIBUTING TO RISING HEALTH CARE COSTS

There are many factors contributing to the rising costs of health care. The key factors include the aging of the population with growth in the demand for health care, increased utilization of pharmaceuticals, expensive new technologies, rising hospital care costs, practitioner availability and behavior, cost shifting, and administrative costs (Thorpe, Woodruff, & Ginsburg, 2005).

## Aging Population

Because more baby boomers are crossing the threshold age of 65, and the average life expectancy is increasing, the elderly are becoming the largest group in the population. According to the Administration on Aging (2010), by 2030, 72.1 million Americans will be age 65 and older and will comprise approximately 19 percent of the population. Although the aging population continues to grow, trends in nursing home utilization have declined in recent years as more and more seniors are receiving care within the community setting. In 2004, the number of nursing homes in the United States fell to 16,100 in comparison to 17,208 nursing homes in 1996. In addition, the number of available nursing home beds decreased to 1.7 million in 2004, with an 86 percent occupancy rate (CDC, 2010). Given the increased utilization of pharmaceuticals and the need to manage and treat chronic illnesses and long-term care needs with increasing age, those patients aged 75 and older incur per capita health expenditures five times higher than those of people between 25 and 34 years of age (Boddenheimer, 2005a). On average, annual per capita expenditures on those patients aged 65 and older is $11,089, significantly higher than the annual expenditure of $3,352 for those aged 19 to 64 (Keehan, 2004). Despite the projected increases in demand from both growth in the elderly population and rising health care costs, estimates are that, by the year 2018, federal funding will be able to cover only 80 percent of billings for inpatient care due to the projected nonsustainability of the long-run growth rate of Social Security or Medicare under current financing arrangements (CMS, 2006).

# Increased Utilization of Pharmaceuticals

Prescription drug spending is expected to continue to grow because of higher use of antiviral drugs as well as faster price growth for brand-name prescription drugs (CMS, 2009). Prescription medications are used by 92 percent of seniors and 61 percent of nonelderly adults (Woo, Ranji, Lundy, & Chen, 2007)(Table 2-4).

Prescription drug use increased among the elderly after Medicare Part D, a prescription drug benefit plan, was introduced in 2006 (Catlin, Cowan, Hartman, Heffler, & National Health Expenditure Accounts Team, 2008). Advertising directed to consumers has resulted in more prescriptions for the newer, costlier drugs, rather than prescriptions for the older drugs that are just as effective as the newer ones and have fewer side effects.

# Technological Advances

Although technological advances have facilitated earlier diagnoses and better treatment of disease, factors such as the greater availability of new technology drive per capita expenditures higher (Boddenheimer, 2005b). Treatment for the five most expensive health conditions, heart disease, cancer, trauma, mental disorders, and pulmonary conditions (Stanton, 2006), requires the use of expensive medications and technologies. While use of some technologies such as electronic record keeping may reduce costs, the presumed success of sophisticated drugs and technologies shapes consumer expectations of what the health care system can deliver. Medication costs add 10 percent to national health care expenditures (Woo, Ranji, Lundy, & Chen, 2007).

Health care consumers, including both patients and providers, also contribute to the cost of health care. These consumers' demands for intense services despite the lack of definite clinical need or evidence of efficacy strains health care spending. As an example, the United States performs more than three times (83) the number of magnetic resonance imaging (MRI) scans per person than does Canada (25.5) or Great Britain (19) (Canadian Institute for Health Information, 2006). Yet the United States lags behind these same countries with respect to patient care outcomes (Commonwealth Fund, 2006; Commonwealth Fund, 2008).

The United States is home to groundbreaking medical research but so are other countries with much lower cost structures. Any American who has had a hip or knee replacement is standing on French innovation. Deep-brain stimulation to treat depression is a Canadian breakthrough. Many of the wonder drugs promoted endlessly on American television, including Viagra, come from British, Swiss or Japanese labs. Overseas, strict cost controls actually drive innovation. In the United States, an MRI scan of the neck region costs about $1,500. In Japan, the identical scan costs $98. Under the pressure of cost controls, Japanese researchers found ways to perform the same diagnostic technique for one-fifteenth the American price and still make a profit (Reid, 2009).

# Rising Hospital Costs

A large proportion of health care dollars are devoted to hospital care. The Hill-Burton Act of 1946 funded the development of the nation's infrastructure of hospitals. Today, there are almost 6,000 hospitals nationwide. As technology and scientific knowledge grow and patients live longer, there has been a greater severity of illness in the hospitalized population. Hospital services contribute to the rise in health care costs due to the increased utilization of expensive technologies, high labor costs, the shortage of nurses, rising malpractice premiums, and increased costs

## TABLE 2-4  Cost of Selected Medications

| DRUG AND DOSES COST | |
|---|---|
| Acyclovir-90 | $43.49 |
| Advair Diskus-180 | $555.95 |
| Lipitor-90 | $237.99 |
| Lisinopril-90 | $31.99 |
| Lorazepam-90 | $43.99 |
| Nexium-90 | $429.97 |

**Source:** Compare drug prices. (2009). Available at www.drugstore.com/.

of hospitalization. Given the aging hospital infrastructure and increases in hospital reimbursements, future building of new hospital beds will also increase health care costs (Bazzoli, Gerland, & May, 2006).

## Practitioner Availability and Behavior

A shortage of frontline health care providers, most noticeably among registered nurses (RNs), also adds to the cost of health care. With 10.5 nurses per 1,000 members of the population, the number of RNs in the United States is slightly greater than among other wealthy countries, which average 9.7 nurses per 1,000 members of the population (Organisation for Economic Co-operation and Development, 2008). Regardless, the number of nurses is insufficient to meet the needs of the American population and will continue to be a problem with the increase in the number of patients under the Patient Protection and Affordable Care Act of 2010. See ftp://ftp.hrsa.gov/bhpr/workforce/behindshortage.pdf.

The number of physicians per person in the United States is less than in other wealthy countries such as France, Canada, or Japan (Organisation for Economic Co-operation and Development, 2008). This shortage is most pronounced in certain practice specialties, for example, family practice physicians and geriatric physicians. Physician shortages are also apparent in rural areas and will continue to be a problem with the increase in the number of patients under the Patient Protection and Affordable Care Act of 2010. In addition, American physicians earn more than their counterparts in other countries. A physician specialist who makes $300,000 in the United States or a family physician whose income is $175,000 per year in the United States would, on average, earn approximately 25 percent to 50 percent less in Canada (Eisenberg, 2006) (Table 2-5).

Defensive medicine and the high cost of medical malpractice insurance all add to increasing physician costs. Malpractice claims in the United States are filed 50 percent more often than in Great Britain and 350 percent more often than in Canada (Anderson,

## TABLE 2-5 Health Care Compensation

| | ANNUAL SALARY |
|---|---|
| Jeff Barbakow, Executive, Tenet Healthcare | $35 million |
| Miles White, Executive, Abbott Laboratories | $30.4 million |
| Richard Jay Kagan, Executive, Schering-Plough | $24.4 million |
| Chief Executive Officer (CEO)–Non-MD | $231,404 |
| Chief Executive Officer (CEO)–MD | $275,123 |
| Director of Human Resources | $82,690 |
| Medical Director | $241,670 |
| Director of Nursing | $82,138 |
| Nurse Practitioner | $82,513 |
| Physical Therapist | $63,117 |
| Cardiologist | $370,295 |
| Anesthesiologist | $344,691 |
| General Surgeon | $327,902 |
| Family Medicine MD | $185,740 |
| Emergency Care MD | $255,530 |
| Dentist | $165,599 |

**Source:** Useem, J. (2003, April 28). Have they no shame? *Fortune, 147*(8), 56–64, and 2007 American Medical Group Association Compensation Data available at www.cejkasearch.com/compensation/amga_administrative_compennsation_survey.htm.

Hussey, Frogner, & Waters, 2005). Two-thirds of claims in the United States are dropped, which results in a similar distribution of claim settlements among these countries. Regardless, the need for medical malpractice insurance to protect American physicians against such claims contributes to the cost of physician care. There are many calls for malpractice reform.

Physicians can be extremely persuasive in health care and hospitals as they advise patients where to go for treatment. The physician represents the primary source of patients for most hospitals. A physician affiliated with two or three hospitals can steer patients toward any of those hospitals. This power over admissions is one of the most important forces in the hospital. Physician control over admissions also affects nursing home admissions, home care agency patient referrals, and referrals to other physician specialists.

## Cost Shifting

The popular practice of **cost shifting**, whereby health care providers raise prices for the privately insured to offset the lower health care payments from both Medicare and Medicaid as well as the often nonpayment of health care premiums from the uninsured, continues to raise the cost of health care. Medicare and Medicaid payments are less than 50 percent of what private insurers pay. Health care providers shift charges for health care costs to the private insurance sector. Some estimates of the cost shift are being valued at $6 billion annually. Cost shifting increases the cost of all health care. The health care facility or the health care professional shifts the cost of health care to other patients with health insurance or to those patients who can afford to pay. Costs to the public for these programs continue to increase. While the intent of the Medicaid program has been to ensure access to health care for mainly low-income pregnant women and children, over 72 percent of Medicaid's $295.9 billion dollars in expenditures in 2004 went toward care of the disabled and dual-eligibles. Dual-eligibles are the elderly who are eligible for both Medicaid and Medicare, and they are the fastest growing proportion enrolled in Medicaid (Holahan & Cohen, 2006).

## Administrative Costs

According to Boddenheimer (2005a), the cost of administration of U.S. health care in 1999 was 24 percent of the nation's health expenditures. In an attempt to reduce these costs, providers such as practitioners' offices, clinics, hospitals, and so on, have invested in Information Technology (IT). IT has played a role in improving quality through availability of the electronic medical record, aiding with HIPAA compliance, and streamlining insurance coding and billing services. Because of the increased demand for such systems, the administrative cost of implementation has also gone up.

It may seem to Americans that U.S.-style free enterprise (i.e., private-sector, for-profit health insurance) is naturally the most cost-effective way to pay for health care. But in fact, many other countries' payment systems are more efficient than ours. U.S. health insurance companies have the highest administrative costs in the world; they spend roughly 20 cents of every dollar for nonmedical costs, such as paperwork, reviewing claims, and marketing. France's health insurance industry, in contrast, covers everybody and spends about 4 percent on administration. Canada's universal insurance system, run by government bureaucrats, spends 6 percent on administration. In Taiwan, a leaner version of the Canadian model has administrative costs of 1.5 percent; one year, this figure ballooned to 2 percent, and the opposition parties savaged the government for wasting money. The world champion at controlling medical costs is Japan, even though its aging population is a big consumer of medical care. On average, the Japanese go to the doctor 15 times a year, three times the U.S. rate. They have twice as many MRI scans and x-rays. Quality is high; life expectancy and recovery rates for major diseases are better than in the United States. And yet Japan spends about $3,400 per person annually on health care; the United States spends more than $7,000 (Reid, 2009).

### REAL WORLD INTERVIEW

Several years ago, I went to the hospital with excruciating pain. They admitted me to the hospital, and I had tests and x-rays for nine days before they found the cause. I had cancer in my left kidney, and I needed immediate surgery. I had the surgery and was discharged on my eighteenth hospital day. Thank heaven, I was now cancer free.

The hospital bill for this stay was $18,689.20. The radiologist and surgeon submitted additional bills. I was glad that I had Medicare and Blue Cross insurance, which paid it all. The only charge I had to pay was $25.00 per day for a private room. When I looked at the hospital bill, there were many charges for medications and treatments I never received. There were even charges for the day after I was discharged. I wonder how the hospital makes out the bill. I also wonder how people with no insurance pay these kinds of hospital bills.

**Leona McGuan**
Patient
Schererville, Indiana

## Other Factors Contributing to Rising Health Care Costs

Ongoing attention to healthy lifestyles, such as eating healthy foods, engaging in regular exercise, maintaining a healthy weight, living smoke free, and limiting alcohol intake, have not realized their potential to reduce the health care costs required to manage the chronic disease conditions associated with the absence of such behaviors (Knickman & Kovner, 2008).

Note other factors that can both increase and decrease utilization as listed in Table 2-6.

## COST CONTAINMENT STRATEGIES

Research has suggested the hazards and ethical problems in the overuse of services in fee-for-service settings (where payment is made based on service rendered to individuals for individual services) rather than service underuse in

### TABLE 2-6  Forces that Affect Overall Health Care Utilization

| FORCE | FACTORS THAT MAY DECREASE HEALTH SERVICES UTILIZATION | FACTORS THAT MAY INCREASE HEALTH SERVICES UTILIZATION |
|---|---|---|
| Financial incentives that reward practitioners and hospitals for performance (e.g., pay for performance [P4P] programs that reward quality practice) | • Changes in clinician practice patterns (e.g., encouraging patient self-care and healthy lifestyles; reduced length of hospital stay) | • Changes in clinician practice patterns (e.g., more aggressive treatment of the elderly) |
| Increased accountability for performance | • Consensus documents or guidelines that recommend decreases in utilization | • Consensus documents or guidelines that recommend increases in utilization |
| Technological advances in the biological and clinical sciences | • Better understanding of the risk factors of diseases and prevention initiatives (e.g., smoking-prevention programs, cholesterol-lowering drugs) | • New procedures and technologies (e.g., hip replacement, stent insertion, magnetic resonance imaging [MRI])<br>• New drugs, expanded use of existing drugs<br>• Increased supply of services (e.g., ambulatory surgery centers, assisted living residences) |
| Increase in chronic illness | • Aging of the population<br>• Discovery and implementation of treatments that cure or eliminate diseases<br>• Public health and sanitation advances (e.g., quality standards for food and water distribution) | • Growing elderly population:<br>  – more functional limitations associated with aging<br>  – more illness associated with aging<br>  – more deaths among the increased number of elderly (elderly are correlated with high utilization of services) |
| Increased ethnic and cultural diversity of the population | • Lack of insurance coverage<br>• Low income | • Growth in national population<br>• Efforts to eliminate disparities in access and outcomes |

*(Continues)*

## TABLE 2-6 (Continued)

| FORCE | FACTORS THAT MAY DECREASE HEALTH SERVICES UTILIZATION | FACTORS THAT MAY INCREASE HEALTH SERVICES UTILIZATION |
|---|---|---|
| Changes in the supply and education of health professionals | • Decreased supply (e.g., hospital closures, large numbers of nursing and medical practitioners and nurses retiring)<br><br>• Shifts to other sites of care may cause declines in utilization of staff at the original sites:<br>– as technology allows shifts (e.g., ambulatory surgery)<br>– as alternative sites of care become available (e.g., assisted living) | • Increase in chronic conditions<br><br>• Growth in national population |
| Social morbidity (e.g., increased AIDS, drugs, violence, disasters) | • Disparities in access to health services and outcomes | • New health problems (e.g., HIV/AIDS, bioterrorism, earthquakes) |
| Access to patient information | • Changes in consumer preferences (e.g., home birthing, more self-care, alternative medicine) | • Changes in consumer demand |
| Globalization and expansion of the world economy | • Growth in uninsured population | • Growth in national population |
| Cost control and competition for limited resources | • Insurance payer pressures to reduce costs | • Increased health insurance coverage<br><br>• Consumer and employee pressures for more comprehensive insurance coverage<br><br>• Changes in consumer preferences and demand (e.g., cosmetic surgery, hip and knee replacements, direct marketing of pharmaceuticals) |

**Source:** Adapted from Bernstein, A. B., Hing, E., Moss, A. J., Allen, K. F., Siller, A. B., Tiggle, R. B. (2003). *Health care in America: Trends in utilization.* Hyattsville, MD: National Center for Health Statistics; and Shortell, S. M., & Kaluzny, A. D. (2006). *Health care management* (5th ed.). Clifton Park, NY: Delmar Cengage Learning.

capitated care (where payment is made based on service rendered to a group of patients) (Berwick 1994; Leape et al., 1990). Over the years, cost containment strategies have targeted the financing and reimbursement sides of health care. Financing strategies have used health services regulation and limitation by means of taxes or insurance premiums and encouraged competition such as managed competition. Reimbursement containment strategies have used regulatory and competitive price controls and utilization controls, such as capitation, patient cost sharing, and utilization management. Capitation and prospective payment have had some of the most significant impact on cost containment.

## CAPITATION

Even though legislation creating managed care was passed in 1973 (the Health Maintenance Organization Act), managed care did not become a major player or driving force in health care until the late 1980s. In an effort to reduce the number of hospitalizations and control profit incentives for health care providers, managed care plans offered hospitals and practitioners a capitated set fee for office visits and hospitalizations for a group of patients. Capitation is the payment of a fixed dollar amount, per person, for the provision of health services to a patient population for a specified period of time, for example, one year. Under capitation, health care organizations benefit from using their financial resources to keep people well. Otherwise, health care providers bear the financial loss. Then in the mid-1990s, the U.S. population had quality concerns with this system and *backlashed* against managed care organizations. Research has shown that resource use is lower for managed care beneficiaries, but it is not clear whether this lowered resource use is appropriate or causes adverse patient outcomes (Murry, Greenfield, Kaplan, & Yano, 1992; Hohlen et al., 1990).

## PROSPECTIVE PAYMENT

Reacting to rapidly increasing costs to Medicare, the Tax Equity and Fiscal Responsibility Act (TEFRA) passed in 1982 mandated the Prospective Payment System (PPS) to control health care costs. For Medicare Part A services, PPS uses Medicare's administrative data to develop and continually refine PPS payments based on diagnosis-related

## EVIDENCE FROM THE LITERATURE

**Citation:** Reid, T. R. (2009). *The healing of America: A global quest for better, cheaper, and fairer health care.* NY: The Penguin Press.

**Discussion:** In his global quest to find a possible prescription for health care improvement in the United States, the author visited wealthy, free market, industrialized democracies like our own, including France, Germany, Japan, the United Kingdom, and Canada. He shares evidence from doctors, government officials, health care experts, and patients the world over. He discusses how other nations have higher performance health care systems that take care of everyone and at a much lower cost than in the United States. Other nations have adopted one of a few rational models of health care financing, though with variations. The basic models are Bismarck (Germany, France, Belgium, Switzerland, Japan), Beveridge (Great Britain, Italy, Spain, most of Scandinavia), and national health insurance (Canada, Taiwan, South Korea).

The United States has combined these three models into a patched-together system of financing health care. Our health care components are Bismarck (employer-sponsored plans), Beveridge (VA, Indian Health Service), and national health insurance (Medicare). But we've added one more model that the author discusses: out-of-pocket for the uninsured (Cambodia, Burkina Faso, rural India, rural China). The author makes a strong case that the dysfunctional, fragmented U.S. financing system is in a large part responsible for our very expensive health care system. He also found that the dreaded monster, *socialized medicine*, turns out to be a myth. Many developed countries provide universal coverage with private doctors, private hospitals, and private insurance. In addition to long-established health care systems, Reid also studies countries that have carried out major health care reform. The first question facing these countries and the United States is an ethical issue: Is health care a human right? Most countries have already answered with a resolute *yes*, leaving the United States in the murky moral backwater with nations we typically think of as far less just than our own. Reid sees problems elsewhere, too. He finds poorly paid doctors in Japan, endless lines in Canada, mistreated patients in Britain, and spartan facilities in France. Still, all the other rich countries operate at a lower cost, produce better health statistics, and cover everybody. In the end, this book finds models around the world that Americans can review and revise in order to guarantee health care for everybody who needs it. Finally, the author states that national health insurance, based on an improved model of Medicare, would be very popular once established and would enable us to reach our goal of affordable, high-quality care for everyone. He states that it is the least expensive, most equitable, most efficient, and most effective model of health care.

**Implications for Practice:** Nurses and all Americans must examine our current health care system and participate in the process of ensuring quality care for all.

groups (DRGs), that is, patients with similar diagnoses. The PPS is a method of reimbursement in which Medicare payment is made based on a predetermined, fixed amount for reimbursement to acute inpatient hospitals, home health agencies, hospices, hospital outpatient and inpatient psychiatric facilities, inpatient rehabilitation facilities, long-term care hospitals, and skilled nursing facilities. For Medicare Part B services, the Resource-Based Relative

Value Scale (RBRVS) is used to determine reimbursement amounts for practitioner services. The major problem that the CMS has encountered with funding prospective payment is DRG creep, in which health care providers *up code* or over bill a patient to indicate a need for financial reimbursement for more expensive health care services to recoup what the health care provider believes is a more equitable payment.

# EVIDENCE FROM THE LITERATURE

**Citation:** Lankshear, A. J., Sheldon, T. A., & Maynard, A. (2005). Nurse staffing and healthcare outcomes: A systematic review of the international research evidence. *Advances in Nursing, 28*(2), 163–174.

**Discussion:** The relationship between quality of care and the cost of the nursing workforce is of concern to policymakers. This study assesses the evidence for a relationship between the nursing workforce and patient outcomes in the acute sector through a systematic review of international research produced since 1990 involving acute hospitals and adjusting for case mix. Twenty-two large studies of variable quality were included. They strongly suggest that higher nurse staffing and richer skill mix (especially of RNs) áre associated with improved patient outcomes, such as improved failure to rescue rates and improved mortality rates, although the effect size cannot be estimated reliably.

Fundamental nursing care is often referred to as *basic*, for much of what nurses do appears deceptively simple. However, it is during these *basic* tasks that a complex interaction occurs—nurses assess patients' physical and psychological status, and patients talk to and receive information from nurses. This can be important in detecting early signs of clinical deterioration or complications. If the nursing resource is stretched because of contextual factors (geographical disposition, decreased skill mix, increased patient dependence, and unit activity), then the ability to provide proactive care, cope with the unpredictable, and maintain flexibility can be adversely affected. Where RN ratios are lower, much of the frontline care may be given by less qualified and less empowered staff. In addition, noting deterioration does not of itself improve outcomes, and, having decided that an intervention is needed, a nurse may need to persuade medical staff to attend to the patient. This requires nurses to be able to present the case logically and confidently. Prompt attendance by medical staff is more likely if the doctor called has respect for the nurse.

In the United States, the California Department of Health Services has set absolute minimum ratios for licensed nurses (RNs and licensed vocational nurses) at 1:6 (4 Care Hours Per Patient Day [CHPPD]), day and night for medical and surgical areas, although the introduction of the 1:5 (4.8 CHPPD) ratio was postponed. In Australia, the state of Victoria has recommended that RN ratios should be 1:4/5 (4.8-6 CHPPD) for day shifts in general medical and surgery units, depending on the type of hospital. However, the research evidence presented here does not support a precise recommendation on staffing levels, and evidence of diminishing returns implies that the cost effectiveness of using nurse staffing as a quality improvement lever must fall as levels increase. More research is needed to investigate the resource implications alongside the impact on patient outcomes.

**Implications for Practice:** Overall, there is accumulating evidence of a relationship between nurse staffing, especially higher nursing skill mix, and patient outcomes. However, the estimates of the nurse staffing effects are likely to be unreliable. There is emerging evidence of a curvilinear relationship that suggests that the cost effectiveness of using RN levels as a quality improvement tool will gradually become less cost effective. The research is not yet clear.

# HEALTH CARE QUALITY

The health care report, *To Err Is Human*, confronted health care clinicians and managers with concerns about the poor quality of health care attributable to misuse, overuse, and underuse of resources and procedures, which was responsible for thousands of deaths (IOM, 1999). The health care report, *Crossing the Quality Chasm* (IOM, 2001), and several large studies (McGlynn et al., 2003; Thomas et al., 2000) have shown that the quality of health care in the United States is at an unexpected low level and needs improvement in many dimensions given the amount of money the United States spends on health care (Table 2-7).

# HEALTH CARE VARIATION

Groundbreaking research beginning in the 1970s and continuing in the 1990s demonstrated that there was significant variation in utilization of specific health care services associated with geographic location, provider preferences and training, type of health insurance, and patient-specific factors such as age and gender (Wennberg & Gittelsohn, 1973; Leape, 1992; Adams, Fraser, & Abrams, 1973; Safran, Rogers, Tarlov, McHorney, & Ware, 1997; Greenfield et al., 1992). Associations between utilization rates of health care services have been found with availability of services and technologies, for example, MRIs, hospital beds, practitioners (Joines, Hertz-Picciotto, Carey, Gesler,

## TABLE 2-7    Health Care Dimensions Needing Improvement

| HEALTH CARE SHOULD BE | HEALTH CARE SHOULD |
|---|---|
| 1. Safe: Avoid injuries from care intended to help patients. | 1. Offer care based on continuous healing relationships: Make care available every day through face-to-face visits, telephone, Internet, and other means. |
| 2. Effective: Provide services based on scientific knowledge to all who could benefit, and refrain from providing services to those not likely to benefit (avoid overuse and underuse). | 2. Customize care based on patient needs and values: Provide care responsive to patient needs and preferences. |
| 3. Patient centered: Provide respectful and responsive care to individuals; patient preferences, needs, and values must guide clinical decision making. | 3. Have the patient as source of control: Foster patient empowerment and autonomy through information and shared decision making. |
| 4. Timely: Reduce wait time and harmful delays for those who receive and give care. | 4. Share knowledge and free flow of information: Facilitate patient access to his or her own medical information and to available clinical knowledge. |
| 5. Efficient: Avoid waste, for example, of equipment, supplies, ideas, energy, and other costly resources. | 5. Use evidence-based decision making: Provide consistent quality of care based on best available scientific knowledge. |
| 6. Equitable: Provide care consistent in quality irrespective of gender, ethnicity, geographical, and socioeconomic factors. | 6. Develop safety as a systems property: Develop systems of safety that mitigate error, promote patient safety, and reduce risk of injury. |
| | 7. Be transparent: Make information available to patients and families about health plans, hospitals, clinical practice, and alternative treatment options, including performance related to their safety, evidence-based practice, and patient satisfaction. |
| | 8. Anticipate needs: Anticipate patient needs rather than respond to events. |
| | 9. Continuously decrease waste: Use limited resources wisely. |
| | 10. Cooperate among clinicians: Collaborate and coordinate care between clinicians and institutions. |

**Source:** Compiled with information from the Institute of Medicine. (2001). *Crossing the quality chasm: A new health system for the 21st century.* Washington, DC: National Academy Press; and Berwick, D. M. (2002). A user's manual for the IOM's 'Quality Chasm' report. *Health Affairs, 2* (3), 80–90.

& Suchindran, 2003), prevalence and severity of morbidities (Dunn, Lyman, & Marx, 2005; National Healthcare Quality Report, 2005), race or ethnicity (National Healthcare Quality Report, 2005), patient adherence, health-seeking behaviors of patients (Calvocoressi et al., 2004), and many other factors. Variation in the delivery and quality of health services is also associated with socio-demographics, hospital types (e.g., urban and rural, teaching and nonteaching), and clinical areas (e.g., heart disease, diabetes, pneumonia, and clinical preventive services). According to Fisher, Goodman, and Chandra (2008), hospitalization for medical conditions such as worsening diabetes or heart failure increases the risk of medical error and complications, both of which entail more costs. Findings from a study by the Dartmouth Atlas Project revealed that the supply of services, not how sick people were, was more likely to determine the resources used (Fisher, Goodman, & Chandra, 2008). Regions of the country and health care providers with more resources had higher rates of use and cost. Efforts to decrease the variation of health care practices through standardization of care with quality, evidence-based guidelines are important to improve clinical decision making, care delivery, health outcomes, and cost efficiency.

Achieving **healthcare transparency** or truth in reporting is the ability to discover information about health care costs, medical errors, or practice preferences, preferably before receiving the service. Transparency is being encouraged by the Centers for Medicare and Medicaid (CMS), though transparency can be hampered by the fear of litigation or reprisal against the health care provider. The Patient Safety and Quality Improvement Act of 2005 addresses such concerns by encouraging health care providers to participate in developing and implementing evidence-based improvement initiatives. The Act also highlights the importance of recognizing and responding to the underlying hazards and risks to patient safety. Establishing national health benchmarks, such as those in Healthy People 2010 (USDHHS, 2000) is another strategy by which to achieve and measure quality improvement. Performance is monitored between states and reflects the health trends and improvements among groups demonstrated to be disadvantaged or vulnerable (AHRQ, 2008b).

## HIGHLY RELIABLE HEALTH CARE-IMPROVEMENTS TO STANDARDIZE CARE

Using the example of the management of heart disease, recent research findings illustrate the need for significant improvements to standardize the process of health care delivery. Each year, heart disease contributes to thousands of deaths. When evidence-based standards are used to guide the care of the patient with heart disease and aspirin and beta-blockers are given to patients who have had a myocardial infarction, it can lower health care dollars and save lives associated with heart disease (Schulte, 2011). This is true even if it is because aspirin use is being measured to assess provider performance (Williams, Schmaltz, Morton, Koss, & Loeb, 2005). Standardization of patient care can change the list of the top ten health care conditions, both in cost and mortality by making patient care delivery more reliable (Table 2-8).

### TABLE 2-8  Top 10 Conditions—Cost and Death

| COST | DEATH |
|------|-------|
| 1. Heart disease ($58B) | 1. Heart disease |
| 2. Cancer ($46B) | 2. Cancer |
| 3. Trauma ($44B) | 3. Cerebrovascular disease |
| 4. Mental disorders ($30B) | 4. Chronic lower respiratory disease |
| 5. Pulmonary conditions ($29B) | 5. Accidents (unintentional injuries) |
| 6. Diabetes ($20B) | 6. Diabetes mellitus |
| 7. Hypertension ($18B) | 7. Influenza and pneumonia |
| 8. Cerebrovascular disease ($16B) | 8. Alzheimer's disease |
| 9. Osteoarthritis ($16B) | 9. Kidney disease |
| 10. Pneumonia ($16B) | 10. Septicemia |

**Source:** Compiled with information from National Vital Statistics Reports. (2005). Vol. 53, No. 15.

# PERFORMANCE AND QUALITY MEASUREMENT

Performance and quality measurement is an essential component of health care improvement efforts. Performance and quality are measured to determine resource allocation, organize care delivery, assess clinician competency, and improve health care delivery processes. Hospitals and practitioners have been given past and present financial incentives to score well on measures of quality from both public and private health care payers. When the quality of care is measured, it improves (Brook, Kamberg, & McGlynn, 1996; Chassin & Galvin, 1998) possibly largely due to the Hawthorne effect, which has illustrated that observed activity shows improvement. The 2008 National Health Care Quality Report revealed areas in which health care performance has improved over time. It also found that, across the process of care measures tracked in the report, patients received recommended care less than 60 percent of the time. Nursing leaders have also recognized the need to establish classifications that can be used to measure nursing care. Selected classifications are listed in Table 2-9.

Note that setting standards for appropriate care and guideline development should have a basis in validated measures of quality, using reliable performance data, and making appropriate adjustments in care delivery. Reliable methods and measures need to be developed and tested. Some practitioners have been resistant to their care delivery being measured because they have believed that it would interfere with their professionalism and autonomy. If this belief persists, the majority of health care delivery will not be measured.

## Malcolm Baldridge National Quality Award

Health care organizations are eligible to consider another framework for health care quality and to apply for the Malcolm Baldridge National Quality Award. The Baldridge Award highlights the importance of leadership; strategic planning; and a customer focus in building a quality health care system. Baldridge also stresses the importance of measurement, analysis and knowledge management; workforce focus; operations focus and results (Baldridge Health Care Criteria for Program, 2011–2012: Available at http://www.nist.gov/baldrige/publications/hc_criteria.cfm.).

## Outcome Measurement

Outcome measurements can be done indicating an individual's clinical state, such as the severity of illness, course of illness, and the effect of interventions on the individual's clinical state. Outcome measures involving a patient's functional status evaluate a patient's ability to perform activities of daily living (ADLs). These can include measures of physical health in terms of function, mental and social health, cost of care, health care access, and general health perceptions. The measures can distinguish the

## TABLE 2-9   Selected Classification Systems

**North American Nursing Diagnosis Association (NANDA):** www.nanda.org

**Home Health Care Classification (HHCC):** www.sabacare.com

**PeriOperative Nursing Data Set:** www.aorn.org

**National Quality Forum-Endorsed Nursing-Sensitive Consensus Standards:** www.qualityforum.org

**Omaha System:** www.omahasystem.org

**ABC Codes:** www.alternativelink.com

**Logical Observation Identifiers Names and Codes:** www.loinc.org

**Nursing Interventions Classification:** www.nursing.uiowa.edu

**Nursing Outcomes Classification:** www.nursing.uiowa.edu

**National Database of Nursing Quality Indicators (NDNQI):** www.nursingworld.org. (Search for NDNQI.)

**SNOMED CT:** www.snomed.org

**International Classification of Nursing Practice:** www.icn.ch

concepts of physical and mental health and identify the five indicator categories of clinical status, functioning, physical symptoms, emotional status, and patient/family evaluation and perceptions about quality of life. Selected quality-of-life measures include quality-adjusted life years (QALY), quality-adjusted life expectancy (QALE), and quality-adjusted healthy life years (QUALY) (Drummond, Stoddart, & Torrance, 1994).

The Medical Outcomes Study (MOS) "Short Form 36" Health Survey is one of the many health indices that have been developed since 1950. The SF-36, as it is commonly known (Ware & Sherbourne, 1992), measures physical functioning, role limitations due to physical health, bodily pain, social functioning, general mental health, role limitations due to emotional problems, vitality, and general health perceptions.

## Other Health Assessment Tools

Other health status assessment surveys in use today include the Quality of Life Index (Spitzer, 1998), developed to measure the general health and well-being of terminally ill individuals; the COOP Charts for primary care practice patients; the functional status questionnaire (Jette & Cleary, 1987), a self-administered general health and social well-being survey for ambulatory patients; the Duke Health Profile (Parkerson, Broadhead, & Tse, 1990), which evaluates health status in primary care patients; the Sickness Impact Profile (Bergner, Bobbit, Carter, & Gilson, 1981), which was developed to measure changes in an individual's behavior as a result of illness; and the Nottingham Health Profile (Hunt, McKenna, McEwen, Williams, & Papp, 1981), developed as a measure of perceived general health status for primary care patients and general population health surveys.

## PUBLIC REPORTING OF PERFORMANCE

Public reporting of organizational performance and quality information is being driven by several forces. As more data about quality become available electronically, individuals reporting the data and those wanting to make comparisons among organizations want the data analyzed and the findings reported. This information can be used to determine where there are health care inefficiencies and poor quality of care. Public reporting of the information is also used to influence reimbursement policies where payment is linked to the ability to achieve standards and benchmarks, for example, pay-for-performance (P4P) (Dudley & Rosenthal, 2006). Performance reporting to demonstrate quality status on minimum reporting standards can also be used by major health care payers as a condition of doing business with an organization (Lansky, 2002). Performance reporting is also used to influence clinician and patient utilization behavior. It also

moves health care toward a population-based approach as opposed to focusing on individual patient care.

## INSTITUTE OF MEDICINE HEALTH CARE REPORTS

The Institute of Medicine (IOM), established in 1970 under the charter of the National Academy of Sciences, provides independent, objective, evidence-based advice to policymakers, health professionals, the private sector, and the public. In 1996, the Institute of Medicine (IOM) launched a concerted, ongoing effort focused on assessing and improving the nation's quality of care. The *Ensuring Quality Cancer Care Report* (1999) documented the wide gulf that exists between ideal cancer care and the reality many Americans with cancer experience.

Other reports released by the IOM on the quality of health care include:

> *To Err is Human; Building A Safer Health System* (1999);
> *Crossing the Quality Chasm: A New Health System for the 21st Century* (2001);
> *Strategies for Addressing the Evolving Nursing Crisis* (2002);
> *Patient Safety: Achieving a New Standard for Care* (2003);
> *Keeping Patients Safe: Transforming the Work Environment of Nurses* (2003);
> *Health Professions Education: A Bridge to Quality* (2003); and
> *Priority Areas for National Action: Transforming Health Care Quality* (2003);
> *Performance Measurement: Accelerating Improvement* (2005);
> *Preventing Medication Errors* (2006);
> *The Future of Nursing: Leading Change, Advancing Health* (2010).

A complete listing of IOM Reports is available at, www.iom.edu/Reports.aspx.

Eight principles are integral to health care reform as envisioned by the Institute of Medicine (2008). These eight principles are:

- Accountability
- Efficiency
- Objectivity
- Scientific rigor
- Consistency
- Feasibility
- Responsiveness
- Transparency

These principles are consistent with a professional nursing agenda, which states that all persons are entitled to affordable, quality health care services (American Nurses

Association [ANA], 2008). The *Quality Chasm* report described broader quality issues, defined six aims, and highlighted ten rules for care delivery redesign (Table 2-7) (IOM, 2001).

## OTHER NATIONAL PUBLIC QUALITY REPORTS

Several key national public quality sources of interest for health care and nursing leaders and managers for purposes of performance measurement and benchmarking or comparison are as follows:

- *AHRQ National Healthcare Quality Report 2009* Available at www.ahrq.gov/qual/nhqr09/nhqr09.htm.
- *AHRQ National Healthcare Disparities Report 2009* Available at www.ahrq.gov/qual/qrdr09.htm.
- *Healthy People 2010:* Accessible at www.healthypeople.gov.
- *Health Grades for Hospitals and Physicians:* Available at www.healthgrades.com.
- *Leapfrog:* Available at www.leapfroggroup.org.
- *The National Quality Forum:* Available at www.qualityforum.org/.
- *Health Plan and Employer Data and Information Set (HEDIS) & Quality Measurement, National Committee for Quality Assurance (NCQA):* Available at www.ncqa.org/.
- *Consumer Assessment of Healthcare Providers and Systems (CAHPS), Agency for Healthcare Research and Quality (AHRQ):* Available at www.cahps.ahrq.gov/default.asp.
- *Medicare Hospital Compare:* Available at www.hospitalcompare.hhs.gov/.
- *The Thomson Reuters 100 Top Hospitals*®: Available at www.100tophospitals.com/top-national-hospitals/.
- *U.S. News and World Report Best Hospitals, annual ranking:* Available at http://health.usnews.com/best-hospitals.

Public reporting of quality performance has been shown to influence declines in cardiac surgery mortality (Peterson, DeLong, Jollis, Muhlbaier, & Mark, 1998); improvements in the processes of obstetrics care (Bost, 2001); and employee enrollment and desire to switch health care providers (Beaulieu, 2002). While providers and policymakers do seek out these public quality reports, the general public does not search them out, does not understand them, distrusts them, and fails to make use of them (Marshall, Hiscock, & Sibbald, 2002).

In many respects, hospitals are providing quality care. Data to assess clinical performance from the Joint Commission core measures program, which uses standardized, evidence-based measures, and data from the Medicare program show improvements in the quality of care in hospitals (Williams et al., 2005). Yet at hospitals that do not meet the sample-size requirement for national comparisons, quality performance remains mediocre (Jha, Li, Orav, & Epstein, 2005).

## DISEASE MANAGEMENT

According to the Disease Management Association of America (DMAA) (2010), **disease management** is a system of coordinated health care interventions and communications for populations with conditions in which patient self-care is significant. What makes caring for patients with chronic diseases problematic is that the patients usually have multiple chronic conditions (e.g., the patient with congestive heart failure who also has hypertension, diabetes, emphysema, urinary incontinence, and chronic pain). Chronic diseases, such as heart disease, stroke, cancer, chronic respiratory diseases, and diabetes, are by far the leading cause of mortality in the world, representing 60 percent of all deaths (World Health Organization, 2010). Because of this, management of a single disease will not be successful given the likelihood of the presence of other diseases or co-morbidities. Documented widespread variation in disease management and treatment interventions has led health care payers to consider the option of paying-for-performance (P4P). Under this plan, disease management and treatment programs with good patient outcomes would be paid more. The efforts of many disease management programs have been successful in improving patient outcomes and providing high quality, cost-effective care (DePalma, 2006). In a recent study evaluating the impact of pay for performance and public reporting in hospitals, the results found that pay for performance and public reporting of performance by hospitals led to great improvements in all composite measures of quality (Lindenauer et al., 2007).

## BALANCED SCORECARDS

Another type of results reporting that is used by organizations is the balanced scorecard. **Balanced scorecards** are used to monitor customer perspective; financial perspective; internal processes and human resources; and learning and growth (Kaplan & Norton, 2004) for strategic management and as a way to examine performance throughout the organization. This examination allows the organization to review multiple key areas of performance, selected on the basis of their importance to the organization's strategic plan for quality.

## EVIDENCE-BASED PRACTICE

The body of evidence supporting clinical practice is steadily growing. However, even when evidence-based quality care guidelines are available for numerous conditions, for example, diabetes, congestive heart failure, and asthma, they have not been fully implemented in actual patient care, and variation in clinical practice is abundant (Timmermans & Mauck, 2005; IOM, 2001; McGlynn et al., 2003). Health care knowledge continues to expand. This requires practice guidelines and the measures of quality on which they are based to be continually updated. It also requires attention to continuing to develop health care quality.

## ACCREDITATION AND PATIENT SAFETY

Health care accreditation is a mechanism used to ensure that organizations meet certain national standards. Hospitals and other organizations seek accreditation to demonstrate their abilities to meet national quality standards. The Joint Commission (JC), formerly known as the Joint Commission on Accreditation of Healthcare Organizations (JCAHO), is the preeminent regulatory body overseeing health care quality. Its review processes are extensive, and payments to a hospital by government insurers of health care (Centers for Medicare & Medicaid Services) are dependent on the organization's ability to meet JC standards with a high degree of compliance (Table 2-10). In addition, other federal, state, local, and voluntary regulatory agencies oversee the quality of specific organizational components such as pharmacy, laboratory, long-term care, rehabilitative care, dietary, behavioral health, and fire safety. Accreditation, which signifies that the organization meets the standards for practice of these oversight agencies, influences market perception about the quality of health care that the organization provides and engenders trust and confidence in the organization. According to the Joint Commission's Annual Report on Quality and Safety (2008), "JC accredited hospitals have significantly improved the quality of care, saving lives and improving the health of thousands of patients" (p. 7).

Because of patterns of medical error that became apparent through JC and other national reporting mechanisms, National Patient Safety Goals (NPSG) were established by the JC in 2003. The 2009 National

---

### Critical Thinking 2-3

Review the case studies on reducing harm to patients at, www.commonwealthfund.org/Innovations/Case-Studies.aspx.

What can each staff member do to improve the quality of care, especially the safety of patient care?

How can we work toward a culture of continual improvement for those issues and situations that cause errors and almost lead to errors?

How much control do nurses have in identifying errors and reporting them?

What can you do to improve the quality of care afforded in your organization?

---

Patient Safety Goals for Hospitals, Ambulatory Health Care, Behavioral Health Care, Critical Access Hospitals, Disease-Specific Care, Home Care, Laboratories, Long Term Care, Office-Based Surgery, and others, are available at www.jointcommission.org/.

## IMPROVING QUALITY THROUGH HEALTH PROFESSIONS EDUCATION

There is a need to focus on retooling the health care workforce with new knowledge and requisite skills to function in better, redesigned health care systems. To begin to

---

### TABLE 2-10  Hospital Accreditation Standards Overview

- Environment of care
- Emergency management
- Human resources
- Infection prevention and control
- Information management
- Leadership
- Life safety
- Medication management
- Medical staff
- National patient safety goals
- Nursing
- Provision of care, treatment, and services
- Performance improvement
- Record of care, treatment, and services
- Rights and responsibilities of the individual
- Transplant safety
- Waived testing

**Source:** © Joint Commission: *CAMH: 2010 Comprehensive Accreditation Manual for Hospitals.* Oakbrook Terrace, IL: Joint Commission, 2011, available at www.jcrinc.com/Joint-Commission-Requirements/Hospitals/.

realize the quality agenda set forth by the IOM (IOM, 2001), a subsequent report, *Health Professions Education: A Bridge to Quality* (IOM, 2003), delineates a needed *overhaul* of the curriculum of health professionals' education to transform current skills and knowledge (IOM, 2003, p. 1). This curriculum includes training clinicians to develop:

1. *Ability to provide patient-centered care:* Patient-centered care emphasizes recognition of the patient or designee as the source of control and full partner in providing compassionate and coordinated care based on respect for the patient's preferences, values, and needs. It builds knowledge of effective communication approaches that allows patient access to information and achieves patient understanding. Patient-centered care respects patients' individuality, values, and needs, and uses related population-based strategies to improve appropriate utilization of health care services. Patient-centered care is important because research continues to find that involving patients in decision making about their care results in higher functional status, better outcomes, and lower costs.

2. *Ability to effectively work in interprofessional teams:* This competency calls for functioning effectively within nursing and interprofessional teams and fostering open communication, mutual respect, and shared decision making to achieve quality patient care. Interprofessional teams have been shown to enhance quality and lower costs, even though this training is challenged by differences in communication norms across disciplines and power and turf controversies among disciplines.

3. *Understanding of evidence-based practices:* Evidence-based practice integrates the best current research evidence with clinical expertise and patient and family preferences and values for delivery of optimal health care. To actively provide evidence-based care (EBC), clinicians need the following knowledge and skills: how to locate the best sources of evidence, how to formulate clear clinically-based questions, and how to determine when and how to translate new knowledge into practice.

4. *Ability to measure the quality of care:* Clinicians need to be able to use data to monitor the outcomes of care processes and use improvement methods to design and test changes to continuously improve the quality and safety of health care systems. Clinicians must use comparison benchmarks to identify opportunities for improvement; design, test, and assess quality improvement interventions; identify current and potential errors in care; and implement safety design

principles such as recognizing human factors and the need for standardization.

5. *Ability to use health information technology:* Health care informatics applications use information and technology to communicate, manage knowledge, mitigate errors, and support decision making. They enhance patient safety by driving standardization, as well as by facilitating knowledge management and communication. More technology becomes available, database systems are linked within and across health care settings. As our evidence-based measures and decision-making tools improve, clinicians will need to be able to fully utilize health information technology to improve the quality of health care delivery.

# QUALITY AND SAFETY EDUCATION FOR NURSES (QSEN)

The IOM's 2004 *Report on Patient Safety* was the first in a series of three reports published since the year 2000 to emphasize the connections among nursing, patient safety, and quality of care. *Keeping Patients Safe* sets forth the structures and processes health care workers use in the delivery of care and emphasizes the need to design the nurses' environments to promote the practice of safe nursing care (IOM, 2004b). The importance of organizational management practices, strong nursing leadership, and adequate nurse staffing for providing a safe care environment is critical (Laschinger & Leiter, 2006).

In 2008, the American Association of Colleges of Nursing (AACN) and the National League for Nursing (NLN) embraced the inclusion of quality improvement systems thinking, change strategies, and patient safety, etc., into undergraduate and graduate nursing education curricula. In 2005, the Robert Wood Johnson Foundation funded the University of North Carolina at Chapel Hill School of Nursing on a long-term project aimed at increasing the inclusion of quality in nursing education and the development of well-prepared faculty to teach the quality and safety competencies recommended by the Institute of Medicine (IOM) to make health care safe, effective, patient centered, timely, efficient, and equitable (IOM, 2001). In 2009, the AACN lent its support to this Quality and Safety Education for Nurses (QSEN) project (www.qsen.org) (Kovner et al., 2010).

The QSEN website is now a comprehensive resource for teaching strategies, etc., for the development of quality and safety competency in nursing. Faculty from 15 nursing education programs participated in the QSEN Learning Collaborative in Phase II.

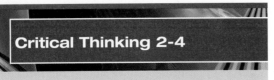

## Critical Thinking 2-4

Think about the competencies needed to improve health care education:

1. Patient-centered care delivery

2. Interdisciplinary teamwork

3. Evidence-based practice

4. Measurement of the quality of care

5. Health information technology skill

6. Culture of safety

How can you improve your education and experience in each of the competencies now and throughout your career?

## NEVER CONDITIONS

In 2008, the Centers for Medicare & Medicaid Services (CMS) identified hospital-acquired conditions (never conditions) for which they will not reimburse hospitals for a higher DRG payment (i.e., pressure ulcers; fractures,

dislocations, intercranial injury, crushing injury, and burns; catheter-associated urinary tract infections; vascular catheter-associated infections, object left in patient during surgery, air embolism, blood incompatibility, and mediastinitis after coronary artery bypass graft). Medicare no longer pays hospitals for these conditions (Waxman, 2008). Clearly the nurse of the future must be educated about quality.

## NURSE SHORTAGES

Until recently, shortages of nurses have been cyclical. These nurse shortages are associated with increased demand for patient care services at a time of falling nursing school enrollment, salary compression, and nominal increases in wages. With an aging nursing workforce (Norman et al., 2005), more nursing faculty is needed to train larger numbers of students. By 2020, the projected nursing shortage is expected to grow to an estimated 340,000. There will be an increased need for RNs in nursing homes by as much as 66 percent in comparison to 1991 data (KFF, 2008). Without significant changes in educational institutions, shortages of nursing and medical clinicians will have a devastating effect on the quality of health care and lead to changes in scope of practice (Cooper, Getzen, McKee, & Laud, 2002) and patient outcomes.

 **EVIDENCE** FROM THE LITERATURE

**Citation:** Soukup, Sr. M. (2000, June). Preface to section on Evidence-based Nursing Practice. *Nursing Clinics of North America, 35*(2), xvii–xviii.

**Discussion:** The author discusses a nurse's response to queries as to whether she has integrated evidence-based practice. The nurse responds, "Yes, I practice state-of-the-art nursing. My education and professional practice experiences have prepared me to care for more than 700 chronically ill patients annually, in the past five years. These patients have an average reported expected pain rating of 6.9 (using a scale of 1 to 10, with 10 being severe pain), and my pain management interventions have kept these patients, during my hours of care, at a reported actual pain rating of 4. Also, as a team member, these patients have not had any known pressure ulcers, skin tears, or catheter-related infections. On two occasions, for patients who were dying, I created a humanizing environment for the patients and their families when they were rapidly transferred from the critical care unit. My documentation has met organizational standards during monthly peer reviews; I have provided leadership for emergencies with positive outcomes; and physician and patient satisfaction ratings for clinical practice on our unit is 9.5 on a scale of 10, with 10 being the highest. Our unit-based team has not had a needle stick-related or back-related injury during the past two years. This has contributed to a significant cost avoidance and benefit to the organization."

**Implications for Practice:** Nurses practicing in the twenty-first century must embrace the principles of evidence-based practice, interdisciplinary teamwork, quality measurement, and use of health information technology as an approach to the provision of safe, patient-centered clinical care and professional accountability.

## CASE STUDY 2-2

Review the ratings of hospitals in your area of the country at www.healthgrades.com, www.hospitalcompare.hhs.gov/, www.100tophospitals.com/top-national-hospitals, and the U.S. News & World Report Best Hospitals, annual ranking, available at http://health.usnews.com/best-hospitals.

What kinds of ratings are given to hospitals in your area?

Review the criteria and evaluation system used to rate the hospitals. Is it valid and reliable?

Will you choose a hospital for your own family's care using a rating system like this?

## American Nurses Association

The American Nurses Association (ANA) is a full service, professional organization representing the nation's entire RN population. The ANA represents the 2.6 million RNs in the United States through its 54 constituent state and territorial associations. The ANA's mission is to work for the improvement of health standards and availability of health care services for all people, foster high standards for nursing, stimulate and promote the professional development of nurses, and advance their economic and general welfare (ANA, 2003).

The National Labor Relations Board (NLRB) recognizes the ANA as a collective bargaining agent. The fact that the ANA has a dual role of being a professional organization and a collective bargaining agent causes controversy. Some nurses believe that unionization is not professional and that the ANA cannot truly support nursing as a profession if it is also a collective bargaining agent. Nurse managers are excluded from union membership. The ANA lobbies Congress and regulatory agencies on health care issues affecting nurses and the general public. The ANA initiates many policies involving health care reform. It also publishes its position on issues ranging from whistle blowing to patients' rights. The American Nurses Credentialing Center (ANCC), a subsidiary of the ANA, created the Magnet Recognition Program to recognize health care organizations that provide the very best in nursing care. Since 1994, many institutions have received this award.

## KEY CONCEPTS

- Health care reports provide invaluable information that emphasizes the successes and failures of health care throughout our nation.

- Evidence of significant disparities and low quality continue to demonstrate the need for significant health care improvement.

- Today, leaders, managers, and staff need to be aware of and involved in the ongoing processes of the making of health policy.

- Health care systems have three simple components: structure, process, and outcome.

- The United States is one of only a few advanced countries in the world without a universal system of health care.

- In the United States, the emphasis on acute care health care services has successfully driven health care costs higher but has not necessarily improved the quality of care or patient outcomes.

- Primary care provides integrated, accessible health care services by clinicians who are accountable for addressing a large majority of personal health care needs, developing a sustained partnership with patients, and practicing in the context of family and community.

- Patients and clinicians need to work together to appropriately utilize services based on the following four foundations of primary care: First Contact, Longitudinality, Comprehensiveness, and Coordination.

- The federal government is a major driver of health care organization and delivery.

- Today, almost as many persons receive health care in the home as receive health care in acute-care settings.

- Enabling factors such as income, type of insurance coverage, gender, race or ethnicity, geographic proximity, and system characteristics affect a person's ability to have access to health care.

- The 2010 passage of The Patient Protection and Affordable Care Act and the Health Care and Education Reconciliation Act has the potential to make significant change in the U.S. health care system.

- The elderly have virtually universal health care coverage through Medicare.

- Because the United States does not have national/universal health care insurance, public health care programs are intended to fill the gap.

- Health care spending continues to increase faster than the overall U.S. economy.

- There are many contributing factors to the rising costs of health care. The key factors include the aging of the population with growth in the demand for health care, increased utilization of pharmaceuticals, expensive new technologies, rising hospital care costs, practitioner behavior, cost shifting, and administrative costs.

- Because rising health care costs are based on utilization, it is important to understand other factors that can both increase and decrease utilization.

- Capitation and prospective payment have had some of the most significant impact on cost containment.

- The health care report, *To Err Is Human,* confronted health care clinicians and managers with concerns about the poor quality of health care attributable to misuse, overuse, and underuse of resources and procedures, which was responsible for thousands of deaths (IOM, 1999).

- The health care report, *Crossing the Quality Chasm* (IOM, 2001) and several large studies (McGlynn et al., 2003; Thomas et al., 2000) have shown that the quality of health care in the United States is at an unexpected low level and needs improvement in many dimensions given the amount of money the United States spends on health care.

- Groundbreaking research beginning in the mid-1980s and continuing in the 1990s demonstrated that there was significant variation in utilization of specific health care services associated with geographic location, provider preferences and training, type of health insurance, and patient-specific factors such as age and gender.

- Recent research findings illustrate the need for significant improvements in the process of health care delivery.

- Health care performance and quality are measured to determine resource allocation, organize care delivery, assess clinician competency, and improve health care delivery processes.

- Public reporting of organizational performance and quality information is being driven by several forces.

- Several key national public quality reports of interest for health care and nursing leaders and managers for purposes of performance measurement and benchmarking are available.

- A key challenge for health care is the numerous deficiencies in the delivery of care of patients with chronic conditions.

- Evidence-based practice involves supplementing clinical expertise with the judicious and conscientious implementation of the most current and best evidence along with patient values and preferences to guide health care decision making.

- Health care accreditation is a mechanism used to ensure that organizations meet certain national standards.

- There is a need to focus on retooling the health care workforce with new knowledge and requisite skills to function in better, redesigned health care systems.

- Health professionals' education to transform current skills and knowledge includes training clinicians to effectively work in interdisciplinary teams; have an educational foundation in informatics; and deliver patient-centered care, fully exploiting evidence-based practice, quality improvement approaches, and informatics.

- The Institute of Medicine's 2004 *Report on Patient Safety* was the first in a series of reports published since the year 2000 to emphasize the connections among nursing, patient safety, and quality of care.

## KEY TERMS

| | | | |
|---|---|---|---|
| cost shifting | deductibles | Gross Domestic Product (GDP) | primary care |
| balanced scorecards | disease management | outcome | process |
| co-payments | healthcare transparency | | structure |

# REVIEW QUESTIONS

1. The national organization that accredits health care organizations is known as which of the following?
   A. American Nurses Association
   B. Health Professions Commission
   C. Agency for Health Care Research and Quality
   D. Joint Commission

2. The largest purchaser of health care in America is which of the following?
   A. Private individuals
   B. Private insurance companies
   C. Health Maintenance Organizations
   D. Medicare, Medicaid, and other governmental programs

3. Who identified a structure, process, and outcome framework for quality?
   A. Nightingale
   B. Donabedian
   C. Starr
   D. Lohr

4. What is the top underlying cause of health care disparity in the United States?
   A. Socioeconomic status
   B. Age
   C. Geographic location
   D. Chronic illness

5. Payment for health care services as a fixed dollar amount per member over a period of time is referred to as which of the following?
   A. Traditional fee for services
   B. Prospective payment system
   C. Capitation
   D. Diagnosis-related groupings

6. The Patient Protection and Affordable Care Act provides comprehensive health care reform through a number of initiatives including which of the following? Select all that apply.
   _____ A. Restricting in-patient hospital stays
   _____ B. Expanding health insurance coverage
   _____ C. Targeting fraud, abuse, and waste in health care
   _____ D. Increasing access to primary care services
   _____ E. Promoting preventive health strategies

7. Key factors contributing to rising health care costs in the United States include which of the following? Select all that apply.
   _____ A. An aging population
   _____ B. Advancements in technology
   _____ C. Increased utilization of pharmaceuticals
   _____ D. Rising costs of primary care
   _____ E. Administrative costs

8. The competencies for the education of health care professionals recommended by the Institute of Medicine to improve the quality of health care include all but which of the following?
   A. Primary care settings
   B. A patient-centered approach
   C. Use of health information technology
   D. Evidence-based practice

# REVIEW ACTIVITIES

1. Although it is difficult to modify the structure of health care, what could you do to implement a system to continually modify the process of health care delivery to improve health care quality in your organization?

2. What are strategies to ensure patient access to appropriate health care services in public and private health care agencies?

3. How can the five IOM health professions competencies and the need for safe, high-quality care be achieved in the current workplace?

# EXPLORING THE WEB

Federal Government:
- Agency for Healthcare Research and Quality (AHRQ): www.ahrq.gov
- Centers for Disease Control and Prevention (CDC): www.cdc.gov
- Centers for Medicare and Medicaid Services (CMS): www.cms.gov
- Department of Defense (DOD) TRICARE program: www.tricare.mil
- Food and Drug Administration (FDA): www.fda.gov

- Health Resources and Services Administration (HRSA): www.hrsa.gov
- Indian Health Service (IHS): www.ihs.gov
- National Center for Health Statistics: www.cdc.gov. Search for National Center for Health Statistics.
- National Guidelines Clearinghouse: www.guidelines.gov
- National Institutes of Health (NIH): www.nih.gov
- Substance Abuse and Mental Health Services Administration (SAMHSA): www.samhsa.gov
- Veterans Health Administration (VHA): www.va.gov

Private Foundations and Organizations:

- Commonwealth Fund: www.cmwf.org
- Henry J. Kaiser Family Foundation (KFF): www.kff.org
- Joint Commission (JC): www.jointcommission.org
- National Committee for Quality Assurance (NCQA): www.ncqa.org
- National Quality Forum (NQF): www.qualityforum.org
- Robert Wood Johnson Foundation (RWJF): www.rwjf.org
- Malcolm Baldridge Quality Award (MBQA): www.quality.nist.gov

# REFERENCES

Adams, D. F., Fraser, D. B., & Abrams, H. L. (1973). The complications of coronary arteriography. *Circulation, 48*(3), 609–618.

Administration on Aging. (2010). Aging statistics. Retrieved July, 2010, from http://www.aoa.gov/AoARoot/Aging_Statistics/index.aspx.

Agency for Healthcare Research and Quality (AHRQ). (2008a). National healthcare disparities report. Rockville, MD: Author. Retrieved August 4, 2008, from www.ahrq.gov/qual/qrdr07.htm.

Agency for Healthcare Research and Quality (AHRQ). (2008b). National healthcare quality report. Rockville, MD: Author. Retrieved August 4, 2008, from www.ahrq.gov/qual/qrdr07.htm#toc.

Alberta Health and Wellness. (2004). Health care insurance plan. Edmonton, Alberta. Retrieved August 2, 2008, from www.health.alberta.ca.

American Association for Accreditation of Ambulatory Surgery Facilities (AAAASF). (2001). About AAAASF. Retrieved October 4, 2006, from www.aaaasf.org/aboutAAAASt/about.cfm.

American Nurses Association. (1995). Nursing Care Report Card for Acute Care. Washington , DC. American Nurses Association.

Anderson, D. J., Kirkland, K. B., Kayes, K. S., Thacker, P. A., Kanafani, Z. A., & Sexton, D. J. (2007). Underresourced hospital infection control and prevention programs: penny wise, pound foolish? *Infection Control and Hospital Epidemiology, 28*(7), 767–773. doi: 10.1086/518518.

Anderson, G. F., Hussey, P. S., Frogner, B. K., & Waters, H. R. (2005). Health spending in the United States and the rest of the industrialized world. *Health Affairs, 24*(4), 903–914.

Antman, E. M., Lau, J., Kupelnick, B., Mosteller, F., & Chalmers, T. C. (1992). A comparison of results of meta-analyses of randomized control trials and recommendations of clinical experts: Treatments for myocardial infarction. *The Journal of the American Medical Association, 268*, 240–258.

Baldridge National Quality Program. (2010). *Health care criteria for performance excellence.* Gaithersburg, MD: Baldridge National Quality Program.

Bazzoli, G. J., Gerland, A., & May, J. (2006). Construction activity in U.S. hospitals. *Health Affairs, 25*(3), 783–791.

Beaulieu, N. D. (2002). Quality information and consumer health plan choices. *Journal of Health Economics, 21*(1), 43–63.

Bergner, M., Bobbit, R. A., Carter, W. B., & Gilson, B. S. (1981). The Sickness Impact Profile: Development and final revision of a health status measure. *Medical Care, 19*(8), 787–805.

Berwick, D. M. (1994). Eleven worthy aims for clinical leadership of health system reform. *Journal of the American Medical Association, 272,* 797–802.

Berwick, D. M. (2002). A user's manual for the IOM's "Quality Chasm" report. *Health Affairs, 21*(3), 80–90.

Bierman, A. S. (2004). Coexisting illness and heart disease among elderly Medicare managed care enrollees. *Health Care Financing Review, 25*(4), 485–488.

Boddenheimer, T. (2005a). High and rising health care costs. Part 1: Seeking an explanation. *Annals of Internal Medicine, 142*(10), 847–854.

Boddenheimer, T. (2005b). High and rising health care costs. Part 2: Technologic innovation. *Annals of Internal Medicine, 142*(11), 932–937.

Borger, C., Smith, S., Truffer, C., Keehan, S., Sisko, A., Poisal, J. (2006). Health spending projections through 2015: Changes on the horizon. *Health Affairs, 25*(2), 61–73.

Borger, C., Smith, S., Truffer, C., Keehan, S., Sisko, A., Poisal, J., et al. (2006). Health spending projections through 2015: Changes on the horizon. *Health Affairs, 25*(2), 61–73.

Bost, J. (2001). Managed care organization publicly reporting 3 years of HEDIS data. *Managed Care Interface, 14,* 50–54.

Buerhaus, P. I., & Staiger, D. O. (1999, Jan–Feb). Trouble in the nurse labor market? Recent trends and future outlook. *Health Affairs,* 214–222.

Calvocoressi, L., Kasl, S. V., Lee, C. H., Stolar, M., Claus, E. B., & Jones, B. A. (2004). A prospective study of perceived susceptibility to breast cancer and nonadherence to mammography screening guidelines in African American and white women ages 40 to 79 years. *Cancer Epidemiology Biomarkers Preview, 13*(12), 2096–2105.

Canadian Institute for Health Information. (2006). CIHI looks at how Canada measures up in health spending. Ottawa, Ontario: Author. Retrieved August 4, 2008, from www. cihi.ca/cihiweb/en/downloads/Dir_Wint06_ENG.pdf.

Centers for Disease Control and Prevention. (2010). Nursing home care. Retrieved July, 2010, from http://www.cdc.gov/nchs/fastats/nuringh.htm.

The Center for Health Design. 2011. Pebble Project. Available at, http://www.www.healthdesign.org/research/pebble/overview.php.

Centers for Medicare & Medicaid Services (CMS). (2006). Historical national health expenditure data. Retrieved June 4, 2006, from www.cms.hhs.gov/NationalHealthExpendData/02_ NationalHealthAccountsHistorical.asp#TopOfPage.

Centers for Medicare and Medicaid Services (CMS). (2010). National health expenditures projections 2009–2019. Retrieved July, 2010, from http://www.cms.gov/NationalHealthExpendData/downloads/proj2009.pdf.

Chassin, M. R., & Galvin, R. W. (1998). The urgent need to improve health care quality: Institute of Medicine National Roundtable on Health Care Quality. *The Journal of the American Medical Association, 280,* 1000–1005.

Christopher, A. (2011). Why drugstore clinics are under fire. Consumer Digest. March-April, 87–90.

Coddington, J. A., & Sands, L. P. (2008). Cost of health care and quality outcomes of patients at nurse-managed clinics. *Nursing Economics, 26*(2), 75–83.

Commonwealth Fund. (2006). International comparison: Access and timeliness. New York: Author. Retrieved July 23, 2008, from www. commonwealthfund. org/snapshotscharts/snapshotscharts_show. htm?doc_id=409110.

Commonwealth Fund. (2008). Why not the best? Results from the national scorecard on U.S. health system performance, 2008. The Commonwealth Fund Commission on a High Performance Health System. Retrieved July 30, 2008, from www.commonwealthfund.org/publications/.

Commonwealth Fund National Scorecard on U.S. Health System Performance. (2008). Available at www.commonwealthfund.org/Content/Publications/Fund-Reports/2008/Jul/Why-Not-the-Best—Resultsfrom-the-National-Scorecard-on-U-S-Health-System-Performance—2008.aspx.

Compare drug prices. (2009). Available at www.drugstore.com/

Cooper, R. A., Getzen, T. E., McKee, H. M., & Laud, P. (2002). Economic and demographic trends signal an impending physician shortage. *Health Affairs, 21*(1), 140–154.

Davis, K. (2005). Ten points for transforming the U.S. health care system. Retrieved October 4, 2006, from http://www.cmwf.org/aboutus/aboutus_show.htm?doc_id=259233.

DePalma, J. (2006). Disease management: Evidence support. *Home Health Care Management and Practice, 18*(3), 223–234.

Department of Defense (DOD). (2003). TRICARE: The Basics. Retrieved October 4, 2006, from http://www.tricare.mil/Factsheets/viewfactsheet.cfm?id=127.

Disease Management Association of America (DMAA). (2010). Population health. Retrieved July, 2010, from http://www.dmaa.org/dm_definition.asp.

Diagnosis-Related Group. (2009). Available at en.wikipedia. org/wiki/Diagnosis-related_group#References.

Donaldson, L. (2001). Safe high-quality health care: Investing in tomorrow's leaders. *Quality in Health Care, 10*(suppl II), ii8–ii12.

Dossey, B., Selanders, L., & Beck, D. (2005). *Florence Nightingale today: Healing, leadership, global action.* Washington, DC: American Nurses Publishing.

Drummond, M. F., Stoddart, F. L., & Torrance, G. W. (1994). *Methods for the economic evaluation of health care programmes.* Oxford, England: Oxford University Press.

Dudley, R. A., & Rosenthal, M. B. (2006). *Pay for performance: A decision guide for purchasers.* Rockville, MD: Agency for Healthcare Research and Quality. AHRQ Pub. No. 06-0047.

Dunn, W. R., Lyman, S., & Marx, R. G. (2005). Small area variation in orthopedics. *Journal of Knee Surgery, 18*(1), 51–56.

Eisenberg, M. J. (2006). An American physican in the Canadian health care system. *Archives of Internal Medicine, 166,* 281–282.

Fisher, E. S., Goodman, D. C., & Chandra, A. (2008). Disparities in health and health care among Medicare beneficiaries: A brief report of the Dartmouth Atlas Project. Dartmouth Institute for Health Policy and Clinical Practice/Robert Wood Johnson Foundation. Retrieved August 7, 2008, from www.dartmouthatlas.org/af4q/AF4Q_Disparities_Report.pdf.

Frogner, B. & Anderson, G. (2006). Multinational comparisons of health systems data, 2005. The Commonwealth Fund. Retrieved on June 12, 2010, from http://www.commonwealthfund.org/~/media/Files/Publications/Chartbook/2006/Apr/Multinational%20Comparisons%20of%20Health%20Systems%20Data%20%202005/825_Frogner_multinational_comphltsysdata%20pdf.pdf.

Gabel, J., Claxton, G., Holve, E., Pickreign, J., Whitmore, H., Dhont, K., et al. (2003). Health benefits in 2003: Premiums reach thirteen-year high as employers adopt new forms of cost sharing. *Health Affairs, 22*(5), 117–126.

Ginsburg, P. B. (2003). Can hospitals and physicians shift the effects of cuts in Medicare reimbursement to private payers? *Health Affairs, W3,* 472–479.

Goldstein, M. K., Lavori, P., Coleman, R., Advani, A., & Hoffman, B. B. (2005). Improving adherence to guidelines for hypertension drug prescribing: Cluster-randomized controlled trial of general verses patient-specific recommendation. *American Journal of Managed Care, 11*(11), 677–685.

Green, L. A., Gryer, G. E., Yawn, B. P., Lanier, D., & Dovery, S.M. (2000). The ecology of medical care revisited. *New England Journal of Medicine, 344,* 2021–2025.

Greenfield, S., Nelson, E. C., Zubkoff, M., Manning, W., Rogers, W., Kravitz, R. L., et al. (1992). Variations in resource utilization among medical specialties and systems of care. Results from the medical outcomes study. *Journal of the American Medical Association, 267*(12), 1624–1630.

Hamilton, K. (2003). The four levels of evidence based practice. *Healthcare Design, 3,* 18–26.

Harbarth, S., & Gastmeier, P. (2003). The preventable portion of nosocomial infections: an overview of published reports. *Journal of Hospital Infection, 53*(4), 258–266.

Healthy People 2020.(2010). U.S. Department of Health and Human Services. Available at http://www.healthypeople.gov/2020/about/default.aspx.

Health Care and Education Reconciliation Act of 2010, (March 30, 2010). GovTrack. Available at, http://www.govtrack.us/congress/bill.xpd?bill=h111-4872.

Health System Change. (2006). Tracking health care costs: Spending growth remains stable at high rate in 2005. *Data Bulletin, 33.* Retrieved August 3, 2008, from www.hschange.org.

Health Workforce Solutions LLC, & Robert Wood Johnson Foundation. (2008). Innovative care models. Retrieved July 25, 2008, from www.innovativecaremodels.com/about/about.

Himmelstein, D. U., Warren, E., Thorne, D., & Woolhandler, S. (2005, Feb. 2). Illness and injury as contributors to bankruptcy. Health Affairs Web Exclusive, pp. W5–63. Retrieved July 28, 2008, from content. healthaffairs.org/cgi/content/abstract/hlthaff.w5.63v1.

Hohlen, M. M., Manheim, L. M., Fleming, G. V., Davidson, S. M., Yadkowsky, B. K., Weiner, S. M., et al. (1990). Access to office-based physicians under capitation reimbursement and Medicaid case management: Findings from the Children's Medicaid Program. *Medical Care, 28,* 59–68.

Holahan, J., & Cohen, M. (2006). Understanding the recent changes in Medicaid spending and enrollment growth between 2000–2004. Issue Paper. Kaiser Commission on Medicaid and the Uninsured. Retrieved October 4, 2006, from http://www.kff.org/medicaid/upload/7499.pdf.

Hunt, S. M., McKenna, P., McEwen, J., Williams, J., & Papp, E. (1981). The Nottingham Health Profile: Subjective health status and medical consultations. *Social Science and Medicine, 15*(3, Pt. 1), 221–229.

Indian Health Service (IHS). (2006). Indian Health Service Fact Sheet. Retrieved October 4, 2006, from http://info.ihs.gov/Files/IHSFacts-June2006.pdf.

Institute of Medicine (IOM). (1996). *Primary care: America's health in a new era.* Washington, DC: National Academy Press.

Institute of Medicine (IOM). (1999). *To Err Is Human.* Washington, DC: National Academy Press.

Institute of Medicine (IOM). (2001). *Crossing the quality chasm: A new health system for the 21st century.* Washington, DC: National Academy Press.

Institute of Medicine (IOM). (2003, Jan. 7). Priority areas for national action: Transforming health care quality. Available at www.iom.edu/?id=35961.

Institute of Medicine (IOM). (2003, Apr. 8). Health professions education: A bridge to quality. Available at www.iom.edu/?id=35961.

Institute of Medicine (IOM). (2003, Nov. 4). Keeping patients safe: Transforming the work environment of nurses. Available at www.iom.edu/?id=35961.

Institute of Medicine (IOM). (2003, Nov. 20). Patient safety: Achieving a new standard for care. Available at www.iom.edu/?id=35961.

Institute of Medicine (IOM). (2004a). *Evidence-based hospital design improves healthcare outcomes for patients, families, and staff.* Washington, DC: National Academy Press.

Institute of Medicine (IOM). (2004b). *Insuring America's health: Principles and recommendations.* Washington, DC: National Academy Press.

Institute of Medicine (IOM). (2005, Dec. 1). Performance measurement: Accelerating improvement. Available at www.iom.edu/CMS/2955.aspx?show=0;3#LP3.

Institute of Medicine (IOM). (2006, July 20). Preventing medication errors: Quality chasm series. Available at www.iom.edu/?id=35961.

Jarvis, W. R. (2006). The state of the science of health care epidemiology, infection control, and patient safety. Infection Cont Retrieved July 25, 2008, from www.icas.org.sg/images/StateofScience.pdf.

Jeffrey, A. E., & Newacheck, P. W. (2006). Role of insurance for children with special health care needs: A synthesis of the evidence. *Pediatrics, 118*(4), 1027–1038.

Jennings, B. M., & Loan, L. A. (2001). Misconceptions among nurses about evidence-based practice. *Journal of Nursing Scholarship, 33*(2), 121–127.

Jette, A. M., & Cleary, P. D. (1987). Functional disability assessment. *Physical Therapy, 67,* 1854–1859.

Jha, A. K., Li, Z., Orav, E. J., & Epstein, A. M. (2005). Care in U.S. hospitals—The Hospital Quality Alliance program. *New England Journal of Medicine, 353*(3), 265–274.

Joines, J. D., Hertz-Picciotto, I., Carey, T. S., Gesler, W., & Suchindran, C. (2003). A spatial analysis of count-level variation in hospitalization rates for low back problems in North Carolina. *Social Science and Medicine, 56*(12), 2541–2553.

Joint Commission. (2008, November). Improving America's Hospitals: The Joint Commission's Annual report on Quality and Safety 2008. Retrieved July 2010 from http://www.jointcommission.org/NR/rdonlyres/833DB8A7-BF2E-48FB-8C30-668F79E11930/0/2008_Annual_Report.pdf.

Joseph, A., & Hamilton, D. (2008). The pebbles project: Coordinated evidenced based case studies. *Building, Research and Information, 36*(2), 129–145.

Kaiser Family Foundation (KFF). (2006). The uninsured: A primer. Key facts about Americans without health insurance. Publication number 7451-02. Retrieved August, 2010, from http://www.kff.org/.

Kaiser Family Foundation (KFF) and Health Research and Education Trust. (2007). Employer health benefits: 2007 annual survey. Publication number 7672. Retrieved August, 2010, from http://www.kff.org.

Kaiser Family Foundation (KFF). (2007). Health Care Spending in the United States and Organisation for Economic Co-operation and Development (OECD) Countries. Retrieved July, 2010, from www.kff.org/insurance/snapshot/chcm010307oth.cfm.

Kaiser Family Foundation (KFF). (2007a). Trends in health care costs and spending. Menlo Park: Kaiser Family Foundation. Retrieved September 28, 2008, from www.kff.org/insurance/upload/7692.pdf.

Kaiser Family Foundation (KFF). (2008). Addressing the nursing shortage. Retrieved August 10, 2010, from http://www.kaiseredu.org/topics_im.asp?imID=1&parentID=61&id=138.

Kaiser Family Foundation (KFF) and Health Research and Education Trust. (2009). Employer health benefits: 2009 annual survey. Publication number 7936. Retrieved August, 2010, from http://www.kff.org.

Kaiser Family Foundation (KFF). (2010). Summary of New Health Reform Law. Publication Number 8061. Retrieved July, 2010, from http://www.kff.org.

Kaiser Family Foundation. (2010a). Focus on Health Reform. Summary of New Health Reform law, Patient Protection and Affordable Care Act (P.L. 111-148). March 26, 2010. Available at http://www.kff.org/healthreform/8061.cfm. Accessed February 2009, 2011.

Kaplan, R. S. & Norton, D. P. (2004). Strategy Maps: Converting Intangible Assets into Tangible Outcomes. Harvard Business Review Press. Boston, MA.

Knickman, J. R., & Kovner, A. R. (2008). Overview: The state of health care delivery in the United States. In A. R. Kovner & J. R. Knickman (Eds.), *Jonas and Kovner's Health Care Delivery in the United States* (9th ed.). (pp. 3–11). New York: Springer.

Kovner, E.T., et al. (2010). New nurses: views of quality improvement education. Joint Commission Journal of Quality and Patient Safety. 26. Jan. 29-35.

Kurtzman, E., & Buerhaus, P. (2008). New Medicare payment rules: Danger or opportunity for nursing. American Journal of Nursing. 108. June. 30–35.

Lankshear, A. J., Sheldon, T. A., & Maynard, A. (2005). Nurse staffing and healthcare outcomes: A systematic review of the international research evidence. *Advances in Nursing, 28*(2), 163–174.

Lansky, D. (2002, July–Aug.). Improving quality through public disclosure of performance information. *Health Affairs,* 52–62.

Lapetina, E. M., & Armstrong, E. M. (2002, July–Aug.). Preventing errors in the outpatient setting: A tale of three states. *Health Affairs,* 26–39.

Laschinger, H. K. S., & Leiter, M. P. (2006). The impact of nursing work environments on patient safety outcomes: The mediating role of burnout/engagement. *Journal of Nursing Administration, 36*(5), 259–267.

Lasser, K. E., Himmelstein, D. U., & Woolhandler, S. (2006). Access to care, health status, and health disparities in the United States and Canada: Results of a cross-national population-based survey. *American Journal of Public Health, 96*(7), 1300–1307.

Leape, L. L. (1992). Unnecessary surgery. *Annual Review Public Health, 13,* 363–383.

Leape, L .L., Park, R. E., Solomon, D. H., Chassin, M. R., Kisecoff, J., & Brook, R. H. (1990). Does inappropriate use explain small area variations in the use of health care services? *Journal of the American Medical Association, 263,* 669–672.

Lindenauer, P., Ramus, D., Roman, S., Rothenberg, M., Benjamin, E., Ma, A., & Bratzler, D. (2007). Public reporting and pay for performance in hospital quality improvement. *The New England Journal of Medicine, 365*(5), 486–496.

Marshall, M. N., Hiscock, J., & Sibbald, B. (2002). Attitudes to the public release of comparative information on the quality of general practice care: A qualitative study. *British Medical Journal, 325*(7375), 1278.

Martin, J. A., Hamilton, B. E., Sutton, P. D., & Ventura, S. J. (2006). Births: Final data for 2004. *National vital statistics reports* (Vol. 55, No. 1). Hyattsville, MD: National Center for Health Statistics.

Mayo, T. W. (2007). U. S. Health Care Timelines. Southern Methodist University, Dedman School of Law. Retrieved July 25, 2008, from faculty.smu.edu/tmayo/health%20care%20timeline.htm.

McCanne, D. (2007, January). State plans miss the point. *USA Today,* 12A.

McGinnis, J. M., & Foege, W. H. (1993). Actual causes of death in the United States. *Journal of the American Medical Association, 270*(18), 2207–2212.

McGlynn, E. A., Asch, S. M., Adams, J., Keesey, J., Hicks, J., DeCristofaro, A., et al. (2003). The quality of health care delivered to adults in the United States. *New England Journal of Medicine, 348,* 2635–2645.

Mehrotra A, Liu H, Adams JL, Wang MC, Lave JR, Thygeson NM, Solberg LI, McGlynn EA.(2009). Comparing costs and quality of care at retail clinics with that of other medical settings for 3 common illnesses. Annals of Internal Medicine. September 1;151(5): 321–328.

Murry, J. P., Greenfield, S., Kaplan, S. H., & Yano, E. M. (1992). Ambulatory testing for capitation and fee-for-service patients in the same practice setting: Relationship to outcomes. *Medical Care, 30,* 252–261.

Muto, C., Harrison, E., Edward, J. R., Horan, T., Andus, M., Jernigan, J. A., Kutty, P. K. (2005). Reduction in central line-associated bloodstream infections among intensive care units - Pennsylvania, April 2001–March 2005. MMWR Weekly, 54(40), 1013–1016. Retrieved from http://www.cdc.gov/mmwr/preview/mmwrhtml/mm5440a2.htm.

National Healthcare Disparities Report. (2008). Agency for Healthcare Research and Quality, Rockville, MD. Retrieved August 2, 2010, from http://www.ahrq.gov/qual/nhdr08/nhdr08.htm.

National Healthcare Quality Report (NHQR). (2009). Agency for Healthcare Research and Quality, Rockville, MD. http://www.ahrq.gov/qual/nhqr09/nhqr09.htm.

National Coalition on Health Care (NCHC). (2008c). World health care data. Retrieved August 2, 2008, from www.nchc.org/facts/world/shtml.

Nightingale, F. (1865/1970). *Notes on nursing.* Princeton: Vertex.

Nolte, E., & McKee, C. M. (2008). Measuring the health of nations: Updating an earlier analysis. *Health Affairs, 27*(1), 58–71.

Norman, L. D., Donelan, K., Buerhaus, P. I., Willis, G., Williams, M., Ulrich, B., et al. (2005). The older nurse in the workplace: Does age matter? *Nursing Economics, 23*(6), 279, 282–289.

Organisation for Economic Co-operation and Development. (2008). OECD health data 2008: How does Japan compare. Paris: Author. Retrieved September 28, 2008, from www.oecd.org/document/46/0,3343,en_2649_33929_34971438_1_1_1_1,00.html.

Osborn, R. (2008). International health care spending per capita by source of funding (2005). Comparing health care systems performance: Opportunities for learning from abroad [slide #4]. Commonwealth Fund. Retrieved October 4, 2008, from www.allhealth. org/briefi ngmaterials/osborn-1189.ppt).

Office of the Legislative Counsel. 111th Congress, 2d Session (May, 2010). Compilation of the patient protection and affordable care act, as amended through May 1, 2010, including the patient protection and affordable care act and health related portions of the health care and education reconciliation act of 2010, Available at http://docs.house.gov/energycommerce/ppacacon.pdf, accessed February 20, 2011.

Parkerson, G. R., Jr., Broadhead, W. E., & Tse, C.-K. J. (1990). The Duke health profile: A 17 item measure of health and dysfunction. *Medical Care, 28,* 1056–1072.

The Patient Safety and Quality Improvement Act of 2005. U.S. Department of Health & Human Services. Agency for Healthcare Research and Quality. (2005). Available at www.ahrq.govqual/psoact.htm.

Peterson, E. D., DeLong, E. R., Jollis, J. G., Muhlbaier, L. H., & Mark, D. B. (1998). The effects of New York's bypass surgery provider profiling on access to care and patient outcomes in the elderly. *Journal of the American College of Cardiology, 32*(4), 993–999.

Reid, T. R. (2009). 5 myths about health care around the world. Sunday, August 23, 2009. The Washington Post Company. Retrieved January 5, 2011, from http://www.washingtonpost.com/wp-dyn/content/article/2009/08/21/AR2009082101778.html.

Robinson, J. C. (2001). Theory and practice in the design of physician payment incentives. *Milbank Quarterly, 79,* 149–177.

Safran, D. G., Rogers, W. H., Tarlov, A. R., McHorney, C. A., & Ware, J. E., Jr. (1997). Gender differences in medical treatment: The case of physician-prescribed activity restrictions. *Social Science Medicine, 45*(5), 711–722.

Schulte, M.F. (2009). Healthcare Delivery in the U.S.A.: An Introduction Productivity Press.

Scott, R. D. (2008). The direct medical costs of healthcare-associated infections in US hospitals and the benefits of prevention. CDC. Retrieved July 30, 2010 from http://www.cdc.gov/ncidod/dhap/pdf/Scott_CostPaper.pdf

Scott, R. D. (2009). The direct medical costs of healthcare-associated infections in US hospitals and the benefits of prevention. CDC. Retrieved July 30, 2010, from http://www.cdc.gov/ncidod/dhap/pdf/Scott_CostPaper.pdf.

Shapiro, J. (2008). Health care lessons from France. Washington: National Public Radio. Retrieved August 2, 2008, from www.npr.org/templates/story/story.php?storyId=91972152.

Shortell, S. M., & Kaluzny, A. D. (2006). *Health care management* (5th ed., p. 9). Clifton Park, NY: Delmar Cengage Learning.

Smith, M. A., Atherly, A. J., Kane, R. L., & Pacala, J. T. (1997). Peer review of the quality of care: Reliability and sources of variability for outcome and process assessment. *The Journal of the American Medical Association, 278*, 1573–1578.

Soukup, Sr. M. (2000, June). Preface to section on evidence-based nursing practice. *Nursing Clinics of North America, 35*(2), xvii–xviii.

Spitzer, W. O. (1998). Quality of life. In D. Burley & W. H. W. Inman (Eds.), *Therapeutic risk: Perception, measurement, and management.* New York: Wiley.

Stanton, M. W. (2006). The high concentration of U.S. health care expenditures. *Research in Action, 19*, 1–11. Retrieved July 31, 2008, from www.ahrq.gov/research/ria2019/expendria.pdf.

Starr, P. (1983). The social transformation of American medicine. Jackson, TN: Basic Books.

Starfield, B. (1998). *Primary care: Balancing health needs, services, and technology.* New York: Oxford University Press.

States take on national health insurance crisis, January 15, 2007, *USA Today,* 12A.

Thomas, E. J., Studdert, D. M., Burstin, H. R., Orav, E. J., Zeena, T., Williams, E. J., et al. (2000). Incidence and types of adverse events and negligent care in Utah and Colorado. *Medical Care, 38,* 261–271.

Thorpe, K., Woodruff, R., & Ginsburg, P. (2005). Factors driving cost increases. Retrieved on June 6, 2006, from http://www.ahrq.gov/news/ulp/costs/ulpcosts1.htm.

Tilson, H., & Berkowitz, B. (2006). The public health enterprise: Examining our twenty-first century policy challenges. *Health Affairs, 25*(4), 900–910.

Timmermans, S., & Mauck, A. (2005). The promises and pitfalls of evidence-based medicine. *Health Affairs, 24*(1), 18–28.

Ulrich, R., Quan, X., Zimring, C., Joseph, A., & Choudhary, R. (2004). *The role of the physical environment in the hospital of the 21st century.* Concord, CA: Center for Health Design.

Useem, J. (2003, April 28). Have they no shame? *Fortune, 147*(8), 56–64, and 2007 American Medical Group Association Compensation Data. Available at www.cejkasearch.com/compensation/amga_administrative_compennsation_survey.htm.

U.S. Department of Health & Human Services. (1996). Summary of the Health Insurance Portability and Accountability Act (HIPAA) Privacy Rule, available at http://www.hhs.gov/ocr/privacy/.

U.S. Department of Health and Human Services. (2000). *Healthy People 2010: Understanding and Improving Health* (2nd ed.). Washington, DC: U.S. Government Printing Office.

U.S. Department of Health and Human Services. (2000, November). *Healthy people 2010: Understanding and improving health* (2nd ed.). Washington, DC: U.S. Government Printing Office.

Ware, J. E., & Sherbourne, C. D. (1992). The MOS 36-item short form health survey I: Conceptual framework and item selection. *Medical Care, 30,* 473–478.

Waxman, K. T. (2008). *A practical guide to finance and budgeting: Skills for nurse managers* (2nd ed.). Marblehead, MA: HCPro.

Wennberg, J. E., & Gittelsohn, A. M. (1973). Small area variations in health care delivery. *Science, 182*(117), 1102–1108.

Werner, R. M., & Bradlow, E. T. (2006). Relationship between Medicare's hospital compare performance measures and mortality rates. *Journal of American Medical Association, 296*(22), 2694–2702.

Williams, S. C., Schmaltz, S. P., Morton, D. J., Koss, R. G., & Loeb, J. M. (2005). Quality of care in U.S. hospitals as reflected by standard measures, 2003–2004. *New England Journal of Medicine, 353*(3), 255–264.

Woo, A., Ranji, U., Lundy, J., & Chen, F. (2007). Prescription drug costs. Menlo Park: Kaiser Family Foundation. Retrieved August 4, 2008, from www.kaiseredu.org/topics_im.asp?id=352&parentID=68&imID=1.

World Health Organization (WHO). (2000). The World Health Report 2000—Health systems: Improving performance (pp. 27–35). Geneva: World Health Organization.

World Health Organization. (2008a). Cuba's primary care revolution: 30 years on. *Bulletin of the World Health Organization, 86*(5). Retrieved July, 2010, from http://www.who.int/bulletin/volumes/86/5/08-030508/en/index.html.

World Health Organization. (2008b). The world health report: 2008 primary health care - Now more than ever. Retrieved July, 2010, from http://www.who.int/whr/2008/whr08_en.pdf.

World Health Organization. (2010). Primary health care. Retrieved August, 2010, from http://www.who.int/topics/primary_health_care/en/.

World Health Organization. (2010). Chronic disease and health promotion. Retrieved July, 2010, from http://www.who.int/chp/en/.

## SUGGESTED READINGS

Anderson, R. M., Rice, T. H., & Kominski, G. F. (2007). *Changing the U.S. health care system: Key issues in health services policy and management* (3rd ed.). San Francisco: Jossey-Bass.

Brown, M. A., Draye, M. A., Zimmer, P. A., Magyary, D., Woods, S. L., Whitney, J., et al. (2006, May–Jun.). Developing a practice doctorate in nursing: University of Washington perspectives and experience. *Nursing Outlook, 54*(3), 130–138.

Centers for Medicare and Medicaid Services (CMS). (2006). 2006 Annual Report of the Boards of Trustees of the Federal Hospital Insurance and Federal Supplementary Medical Insurance Trust Funds. Retrieved October 4, 2006, from http://www.cms.hhs.gov/ReportsTrustFunds/downloads/tr2006.pdf.

Centers for Medicare & Medicaid Services, Office of the Actuary, National Health Statistics Group. (2006). Retrieved October 4, 2006, from http://www.cms.hhs.gov/NationalHealthExpendData/downloads/PieChartSourcesExpenditures2004.pdf.

Centers for Medicare and Medicaid Services, Office of the Actuary, National Health Statistics Group. (2006). Retrieved October 4, 2006, from http://www.cms.hhs.gov/NationalHealthExpendData.

Donabedian, A. (1966). Evaluating the quality of medical care. *Milbank Quarterly, 20*(1), 137–141.

Dracup, K., Cronenwett, L., Meleis, A. I., & Benner, P. E. (2005). Reflections on the doctorate of nursing practice. *Nursing Outlook, 53*(4), 177–182.

Draye, M. A., Acker, M., & Zimmer, P. A. (2006, May–Jun.). The practice doctorate in nursing: Approaches to transform nurse practitioner education and practice. *Nursing Outlook, 54*(3), 123–129.

Kahn, C. N., Ault, T., Isenstein, H., Potetz, L., & Van Gelder, S. (2006). Snapshot of hospital quality reporting and pay-for-performance under Medicare. *Health Affairs, 25*(1), 148–162.

Kaiser Family Foundation (KFF). (2005a). *Navigating Medicare and Medicaid, 2005: Medicaid.* Washington, DC: Kaiser Family Foundation. Accessible at http://www.kff.org/medicare/7240/medicaid.cfm.

Kaiser Family Foundation (KFF). (2006). *The uninsured: Key facts about Americans without health insurance.* Washington, DC: Kaiser Family Foundation. Accessible at http://www.kff.org/uninsured/upload/7451-021.pdf.

Mundinger, M. O. (2005). Who's who in nursing: Bringing clarity to the doctor of nursing practice. *Nursing Outlook, 53,* 173–176.

National Bureau of Economic Research. (2006). Healthcare expenditures in the OECD. Retrieved June 4, 2006, from http://www.nber.org/aginghealth/winter06/w11833.html.

U.S. Department of Health and Human Services. (2006). Annual Update of the HHS Poverty Guidelines. Accessible at http://aspe.hhs.gov/poverty/06fedreg.pdf.

# CHAPTER 3

# Organizational Behavior and Magnet Hospitals

## Maria R. Shirey, PhD, MBA, RN, NEA-BC, FACHE, FAAN

*Magnet hospitals are living evidence that creating professional nurse practice environments is the solution to the flight of nurses from hospital practice.*

(Aiken, 2002)

## OBJECTIVES

Upon completion of this chapter, the reader should be able to:

1. Relate organizational behavior.
2. Identify the evolution of organizational behavior and its impact on autocratic, custodial, supportive, and collegial organizational behavior.
3. Identify the characteristics of a high-performance organization.
4. Identify the organizational characteristics that define magnet nursing services.
5. Relate the historical evolution and significance of magnet hospitals.
6. Support the 14 Forces of Magnetism.
7. Describe elements of the Magnet Model.

Delmar/Cengage Learning

*Anne and Maria are new nurses who went to nursing school together. Now, one year after both became registered nurses, they still maintain their commitment to having dinner together at their favorite restaurant at least once a month. Both nurses started their careers with a great deal of excitement and anticipation. Anne is now concerned that Maria's "flame" is starting to lose its vibrancy.*

*When Anne and Maria get together, they invariably talk about their relatively new positions as staff nurses in two different local hospitals. Anne works on a medical-surgical unit at Midwest Community Hospital (MCH) that has been magnet designated for about five years. She joyously reports working on a cohesive unit adequately staffed with capable nurses who are true team players and contribute to outstanding patient outcomes. Anne raves about her unit-based clinical nurse specialist (her mentor) and about her nurse manager, who both seem very interested in Anne's personal and professional development. Maria, on the other hand, also works on a medical-surgical unit, but at a nonmagnet-designated facility. Maria reports that her nurse manager is not supportive of magnet designation. In fact, when Maria initiated a conversation to inquire about the possibility of her Good Spirit Hospital GSH pursuing such a journey, her manager's response was, "That magnet 'thing' is nothing more than a marketing ploy; it offers no real benefits." Fearing that her manager would see her as confrontational and not wanting to get on her manager's bad side, Maria dropped the conversation. Over the past few months, Maria has been working many mandatory overtime hours at her hospital. Every shift worked*

*on her unit feels like an exercise in "survival of the fittest." With more nurses resigning from her unit and patient care outcomes not being what they should be, Maria still loves nursing, yet she is beginning to wonder if she made the right choice in going to work at GSH.*

*Compare and contrast the characteristics of Anne's and Maria's professional nursing practice environments.*

*As a new nurse entering the workforce, what objective and subjective resources should you review to guide an assessment of the professional nursing practice environment at a potential place of employment?*

*From an organizational behavior standpoint, how can the practice environments at MCH and GSH be explained?*

The health care industry is in the midst of a workforce shortage that includes both nurses and other health care professionals. Facing an aging society and the rising need for health care services, the ability to attract and retain current and future health care professionals is of paramount concern. The literature suggests a link between desirable practice environments, such as those seen in magnet hospitals, and the ability to attract and retain health care professionals. A better appreciation of the links between employee attitudes and behaviors related to work environments, however, requires an understanding of organizational behavior.

This chapter introduces the concept of organizational behavior and explains how an understanding of organizational behavior may favorably shape the professional work environment to directly affect individuals, groups, and organizations. The chapter highlights **magnet hospitals** as high-quality health care organizations that 1) have met the rigorous nursing excellence requirements of the American Nurses Credentialing Center (ANCC), a division of the American Nurses Association (ANA), and 2) are supportive and collegial practice settings incorporating principles of organizational behavior to achieve positive individual, group, and organizational outcomes.

# ORGANIZATIONAL BEHAVIOR

**Organizational behavior** can be defined as the study of human behavior in organizations (Schermerhorn, Hunt, & Osborn, 2005). Organizational behavior is specifically concerned with work-related behavior and addresses individuals and groups, interpersonal processes, and organizational dynamics and systems. Organizational behavior draws from many disciplines, including psychology, sociology, social psychology, anthropology, and political science (Robbins, 2005).

An **organization** is a coordinated and deliberately structured social entity that consists of two or more individuals functioning on a relatively continuous basis to achieve a predetermined set of goals (Daft, 2006). Organizations are complex entities existing as **open systems**, that must interact with the environment in order to survive (Daft, 2006). An organization's long-term effectiveness may be determined by its capability to anticipate, manage, and respond to changes in its environment. These changes may result from **external forces**, influences originating outside the organization, such as the labor force and the economy. Or they may result from **stakeholders**, people or groups with an interest in the performance of the organization, such as customers, competitors, suppliers, government, and regulatory agencies.

The field of organizational behavior emphasizes people skills in addition to technical skills and involves the systematic study of the actions and attitudes people exhibit within organizations (Robbins, 2005). Attitudes of interest in organizational behavior include **job satisfaction**, how organizational members feel about their job, and **organizational commitment**, how committed or loyal employees feel to the goals of the organization. Actions or behaviors of interest in organizational behavior also include three important determinants of employee performance. The determinants are **productivity**, which is the quantity and quality of output an employee generates for an organization; **absenteeism**, which is the rate of employee absences from work; and **turnover**, which is the number of employees who have resigned divided by the total number of employees during the same time period. All of these determinants can be measured.

## IMPORTANCE OF ORGANIZATIONAL BEHAVIOR

Learning about organizational behavior enables organizational members to better understand their own behaviors as well as those of peers, superiors, and/or subordinates within an organization. This understanding helps individuals become more effective employees, team members, and managers within organizations. Research suggests that employees who demonstrate high levels of organizational commitment are generally more satisfied in their jobs and more likely to stay employed within their organizations (Lynn & Redman, 2005). Employees who experience empowering structures in the workplace are generally more engaged (Laschinger & Finegan, 2005) and may have lower absenteeism and lower turnover. Because excessive employee turnover is costly and detrimental to quality outcomes, organizations need to proactively concern themselves with the important issues related to organizational behavior.

Organizational behavior allows individuals to increase organizational effectiveness to ultimately meet the needs of the organization, its members, and society. **Organizational effectiveness** refers to an organization's sustainable high performance in accomplishing its mission and objectives (Schermerhorn et al., 2005). In the long run, the primary criterion to evaluate organizational effectiveness is the organization's capability to survive and thrive under conditions of uncertainty. Important contributors to the effectiveness of any organization are the quality of its workforce and their commitment to the goals and success of the organization. Maintaining a satisfied and committed workforce, however, is a planned effort that requires the contributions of many members within an organization. Organizational leaders play a pivotal role creating and sustaining desirable work environments that empower, engage, and retain employees.

## EVOLUTION OF ORGANIZATIONAL BEHAVIOR

Organizational behavior traces its roots to the original work of Frederick Taylor and to the advent of scientific management in the late 1800s and early 1900s. Proponents of scientific management emphasized the machine-like or assembly-line focus of work processes and the precise sets of instructions and time-motion studies assumed to enhance productivity. After World War I and the identification of the **Hawthorne effect**, which demonstrated that a change in employee behavior occurs as a result of being observed, the organizational behavior focus shifted to how human relations and psychology affected organizations. This era was followed by the introduction of the human motivation theories by Abraham Maslow (hierarchy of needs theory), Douglas McGregor (Theory X and Theory Y), and William Ouchi (Theory Z). See Chapter 1 for more on these theories.

Over the past century, the U.S. economy has shifted from an industrial focus and an assembly-line mentality in the 1900s to a knowledge economy in the 2000s. A knowledge economy requires highly educated employees for a more technologic information age and thus necessitates a new way of leading and developing future employees and organizations. Increasingly, today's health care professionals see themselves as **knowledge workers** who are well educated, technologically savvy, and who own their intellectual capital. This **intellectual capital** includes an individual's knowledge, skills, and abilities that have value and portability in a knowledge economy. This shift to a knowledge economy means that health care professionals today possess well-developed abilities and require supportive and collegial organizations to cultivate their much in-demand talents, or they will take their knowledge, skill, and ability to another organization. The supportive and collegial work environments that today's knowledge workers prefer are consistent with McGregor's Theory Y—that is, environments in which leaders remove obstacles for motivated and empowered individuals. These environments appear to be in sharp contrast with the autocratic and custodial organizational frameworks of earlier centuries.

**TABLE 3-1 Models of Organizational Behavior**

| TYPE OF MODEL | BASIS OF THE MODEL | MANAGERIAL ORIENTATION | EMPLOYEE ORIENTATION | EMPLOYEE NEED(S) MET | PERFORMANCE OUTCOMES |
|---|---|---|---|---|---|
| Autocratic | Power | Authority | Dependence on the boss | Subsistence | Minimal |
| Custodial | Economics | Money | Security<br>Benefits<br>Dependence on the organization | Security | Passive cooperation |
| Supportive | Leadership | Support | Good job performance<br>Participation | Status<br>Recognition | Awakened work drive and cooperation<br>Passion |
| Collegial | Partnership | Teamwork | Responsible behavior<br>Self-discipline | Self-actualization | Enthusiasm<br>Engagement |

Table 3-1 summarizes and compares the common organizational models of autocratic, custodial, supportive, and collegial organizations (Clark, 2000).

In addition to supportive and collegial organizations, today's health care industry places great emphasis on global diversity, technological intensity, change as a constant, superior quality and safety outcomes, and continuous learning and process improvements. Organizational behavior today continues to be influenced by the human relations movement and by the human motivation theories that capitalize on humanistic, motivational, team-based, and collaborative strategies. Business practice models emphasizing quality leadership and management (Drucker, 2006) and developing lifelong learning organizations (Senge, 1990) further contribute to contemporary organizational development and effectiveness.

# HIGH-PERFORMANCE ORGANIZATIONS

Health care organizations operate within a competitive environment of constant change and scarcity of human and financial resources. Given the rising shortages of qualified health care professionals, compounded by an aging population requiring greater access to health care services, many health care organizations have seen the need to reposition their organizations for the future. These changing forces have contributed to the reinvention of many health care institutions as high-performance organizations. A **high-performance organization** operates

in a way that brings out the best in people and produces sustainable high performance over time. High-performance organizations pay close attention to the dynamics of the workplace and are known for also having high quality-of-work-life environments (Schermerhorn et al., 2005). **High quality-of-work-life environments** are those work environments in which the quality of the human experience in the workplace meets and surpasses employee expectations. Employees in these environments are respected and treated well at work, thus keeping them motivated, engaged, continuously growing, and retained within their organizations.

Maintaining high quality-of-work-life environments requires the commitment of both leaders and employees in organizations. Leaders in high-performance organizations recognize that the single best predictor of an organization's success is its capability to attract, motivate, and retain talented people (Schermerhorn et al., 2005). So important is the quality of the work environment in health care that the Institute of Medicine (IOM), a major policy influencing organization, has specifically called for the transformation of the nurse's work environment in order to retain nurses in the profession and keep patients safe (IOM, 2004). Table 3-2 summarizes the five characteristics of high-performance organizations that contribute to a high quality-of-work-life environment and to quality patient care. A magnet hospital represents an example of a supportive and collegial work environment for nurses that may also be classified as a high-performance organization and a high quality-of-work-life environment.

**TABLE 3-2 Five Characteristics of High-Performance Organizations**

**HIGH-PERFORMANCE ORGANIZATIONS**

- Value people as human assets, respect diversity, and empower individuals to use their talents to advance personal and organizational performance

- Mobilize teams that build synergy from the talents of their members and are empowered to use self-direction and personal initiative to maximize performance

- Successfully bring people and technology together in a performance context

- Thrive on learning, encourage knowledge sharing, and enable members to continuously grow and develop

- Are achievement oriented, sensitive to the external environment, and focused on total quality management to deliver outstanding and sustainable results

# MAGNET HOSPITALS

As mentioned earlier, a magnet hospital is a health care organization that has met the rigorous nursing excellence requirement of the American Nurses Credentialing Center (ANCC), a division of the American Nurses Association (ANA). Magnet designation involves a voluntary credentialing process. Achieving magnet designation represents the highest level of recognition the ANCC accords to health care organizations that provide the services of registered professional nurses (ANCC, 2008a). As a testament to the increasing recognition given to magnet hospitals, *U.S. News and World Report* now includes magnet designation in its criteria for its annual "Best Hospitals of America" list (U.S. News & World Report, 2010).

Of 5,815 hospitals in the United States (AHA, 2009). 372 (or 6.4%) are magnet-designated facilities (ANCC, 2010a). This figure is rising daily. Community hospitals, teaching hospitals, and hospital systems—large (more than 1,000 beds) and small (less than 100 beds), in rural and urban settings—have achieved magnet recognition. Although pursuing magnet designation is an individual organizational decision, implementing the magnet standards in hospital settings can potentially benefit institutions, independent of whether or not they achieve magnet designation (Miller & Anderson, 2007).

## HISTORICAL OVERVIEW OF MAGNET HOSPITALS

In 1983, the American Academy of Nursing (AAN), an organization affiliated with the ANA, appointed a Task Force on Nursing Practice in Hospitals. The purpose of the task force was to identify workplace characteristics that were successful in recruiting and retaining hospital nurses. The task force studied 163 hospitals in the United States, based on their reputation for successfully attracting and retaining nurses and for delivering high-quality nursing care. Of the 163 hospitals studied, 41 (25%) were described as magnet hospitals (McClure, Poulin, Sovie, & Wandelt, 1983). A magnet designation was earned through demonstrated high nurse satisfaction, low nurse turnover, and low nurse vacancy rates. Interestingly, the 41 original magnet hospitals were able to recruit and retain nurses despite concurrent health care industry changes in the payment system; an unprecedented number of hospital mergers, acquisitions, and consolidations; and a major nursing shortage. The AAN's landmark study concluded that the 41 original magnet hospitals shared a set of core organizational attributes that were desirable. The study stimulated additional independent research that provided further evidence to highlight the achievement of superior outcomes in magnet hospitals.

By June 1990, the ANA established the ANCC as a separate, incorporated, nonprofit organization that was to serve as the credentialing arm for magnet hospitals. The initial proposal for the Magnet Recognition Program was approved by the ANA Board of Directors in December 1990. The MAGNET PROGRAM proposal indicated that it would be built upon the 1983 AAN magnet hospital study. Further, the MAGNET PROGRAM would use as a baseline for program development the 1999 *ANA Scope and Standards for Nurse Administrators*, now in its third revision (ANA, 2009).

## The ANCC Magnet Facilities

The University of Washington Medical Center in Seattle became ANCC's first magnet facility in 1994. By 1998, 13 hospitals achieved magnet designation. By the 2000s,

## REAL WORLD INTERVIEW

With magnet hospitals recognized for low RN turnover rates, higher nurse-patient staffing ratios, and greater autonomy and influence over practice decisions, it is anticipated that high nurse satisfaction will exist across nursing units. The professional nursing practice model that is needed in a healthy work environment, includes clinical practice development, positive nurse-physician relationships, supportive nurse-manager relationships, ongoing educational support, and adequate nurse staffing. It is the synergy of these factors within the nursing services department that will create high nurse satisfaction.

**Cherona Hajewski, RN, MSN, NEA-BC**
Vice President, Patient Care Services and Chief Nursing Officer
Deaconess Hospital
Evansville, Indiana

**TABLE 3-3   Nine Characteristics Defining Magnet Nursing Services**

- High-quality patient care
- Clinical autonomy and responsibility
- Participatory decision making
- Strong nurse leaders
- Two-way communication with staff
- Community involvement
- Opportunity and encouragement of professional development
- Effective use of staff resources
- High levels of job satisfaction

the growth of magnet hospitals was exponential, resulting in more than 300 magnet-designated facilities by 2010. Hundreds of facilities are continuously in the pipeline, seeking to become magnet hospitals. To date, the Magnet Recognition Program has been expanded to include both acute care hospitals and long-term facilities. The Magnet Recognition Program reviews applications from both U.S. and international health care organizations.

## GOALS OF THE MAGNET RECOGNITION PROGRAM

The Magnet Recognition Program (ANCC, 2008a) was created to achieve three major goals:

1. Promote quality in a milieu that supports professional nursing practice.
2. Identify excellence in the delivery of nursing services to patients.
3. Provide a mechanism for the dissemination of best practices in nursing services.

### Critical Thinking 3-1

To learn more about some of the most important questions that must be addressed at the beginning of the magnet application process, access and complete for practice, but do not submit the document entitled Nurse Opinion Questionnaire (ANCC, 2010b) at http://www.nursecredentialing.org/. Click on Magnet and choose Nurse Opinion Questionnaire. Click on the link to the Nurse Opinion Questionnaire. Note the questions that staff nurses can use to share their perspectives with the American Nurses Credentialing Center's Magnet Recognition Program®.

Nine characteristics define magnet nursing services (Table 3-3). These characteristics form the assessment framework for the Magnet Recognition Program's appraisal process.

# BENEFITS OF MAGNET RECOGNITION

Hospitals attaining magnet designation may achieve multiple benefits that include improved patient quality outcomes, enhanced organizational culture, improved nurse recruitment and retention, enhanced safety outcomes, and enhanced competitive advantage (Table 3-4). A major benefit of magnet designation is that it strengthens the image of nursing within the health care organization and the community. Magnet recognition raises the bar for nursing services and contributes to upgrading the quality of nursing services delivered at the local, national, and international levels.

## Improvement in Quality Patient Outcomes

Improved patient quality outcomes have been reported in magnet organizations. Much of the evidence documenting better patient outcomes in these hospitals supports the underlying assumption that work environments that are attractive to nurses yield better outcomes for patients (Aiken, 2002). Historical research by Aiken over the past two decades and more current studies conducted by her team and others continue to support the significant importance of the nurse's work environment (Lake & Friese, 2006).

In a study comparing patient quality outcomes between magnet and nonmagnet facilities, Aiken, Smith, and Lake (1994) demonstrated that magnet hospitals, after adjusting for differences in severity of patient illness, had a 4.6% lower Medicare mortality than did comparison hospitals. In another study comparing AIDS mortality in magnet versus nonmagnet facilities, researchers documented better outcomes in the magnet facilities (Aiken, Sloane, Lake, Sochalski, & Weber, 1999). The higher nurse-to-patient ratio in the magnet hospitals was found to be the major factor explaining the lower patient mortality.

Increased levels of patient satisfaction have also been documented in magnet hospitals. In a study comparing patients on dedicated AIDS units in magnet hospitals versus patients on medical units in conventionally organized hospitals, patient satisfaction was highest in units in the magnet hospitals (Aiken, Sloane, & Lake, 1997; Aiken et al., 1999). The researchers found that the single most important factor explaining differences in patient satisfaction was the superior nurse practice environment of the magnet hospitals and the dedicated AIDS units. Patients in the magnet hospitals and dedicated AIDS units reported better nurse accountability with their care, a factor presumed to enhance continuity of patient care. As news of outcomes in magnet hospitals continues to be disseminated in the lay literature, more patients are beginning to associate higher levels of perceived quality of

## TABLE 3-4   Benefits of Magnet Designation

Improved patient quality outcomes

- Lower patient morbidity and mortality
- Increased patient satisfaction

Enhanced organizational culture

- Greater nurse empowerment structures
- Improved nurse well-being
- Supportive, people-oriented, and visible nurse leaders and administrators
- Shared decision making
- Increased culture of respect for nurses
- Creation of culture of empowerment, respect, and integrity for all employees
- Increased employee morale

Improved nurse recruitment and retention

- Higher levels of nurse job satisfaction
- Increased perception by nurses of a work environment that allows them to give quality patient care

- Increased ability to attract high-quality nurses, physicians, and specialists
- Higher levels of nurse autonomy and control over practice
- More positive nurse-physician relationships
- Greater support for ongoing professional development

Enhanced safety outcomes

- Lower incidence of needle stick injury rates among nurses
- Lower incidence of near-miss patient injuries

Enhanced competitive advantage

- More validation of excellence in nursing services
- Enhanced public confidence with the facility
- Enhanced public perception of overall facility quality
- Lower nurse turnover
- Higher nurse job satisfaction
- Increased market share

care with magnet hospitals. Increasingly, patients are actively seeking magnet facilities to meet their health care needs.

## Enhanced Organizational Culture

Core values such as empowerment, pride, mentoring, nurturing, respect, integrity, and teamwork are reported in magnet facilities (ANCC, 2008a). There is evidence to suggest that nurses in magnet facilities work within greater empowerment structures that enhance the work environment, improve nurse well-being (Laschinger & Finegan, 2005), and make professional nursing practice more desirable in those settings (Laschinger, Almost, & Tuer-Hodes, 2003). Magnet workplace cultures report people-oriented, visible, and empowering nurse leaders (Upenieks, 2003; Steinbinder & Scherer, 2006) who contribute in a significant way toward building positive organizational cultures. Shared decision making is a hallmark of magnet cultures that allows nurses to practice in a workplace where professional autonomy is both valued and encouraged. Overall, magnet hospital nurses report a culture of respect for nurses. This culture of respect manifests itself in the form of more-supportive hospital administrators, more value attributed to nurses (Havens & Aiken, 1999), and an increase in nursing's contributions to the organization's mission and quality of care.

## Improved Nurse Recruitment and Retention

Magnet hospitals are considered to be good places to practice nursing (Scott, Sochalski, & Aiken, 1999). These facilities are named magnet hospitals due to their capability to attract and retain nurses and maintain high levels of job satisfaction (Brady-Schwartz, 2005). In fact, nurse turnover and vacancy rates are generally lower in magnet hospitals than they are in other facilities (Kramer, 1990; Kramer & Schmalenburg, 1991; Coile, 2001). It is not unusual to observe both new and experienced nurses making hospital employment decisions based on whether or not a facility is magnet designated. Organizations having magnet designation can showcase their advantage through multiple venues, including billboards (Figure 3-1), newspaper ads (Figure 3-2), word-of-mouth, and the Internet. Given today's ready electronic access, determination of an organization's magnet status requires a simple Web-based search (http://nursecredentialing.org/magnet) and clicking on Find a Magnet Facility. Interestingly, non-nursing health care professionals such as practitioners, pharmacists, physical therapists, social workers, and others (McClure & Hinshaw, 2002) also seem to benefit from magnet workplace cultures.

Additional research supports the essentials of "magnetism" and their role in nurse retention. For example, nurses in magnet facilities report higher levels of nurse autonomy and control over practice, with more-positive

**FIGURE 3-1** Billboard announcing hospital's Magnet designation, Stormont-Vail HealthCare Topeka, Kansas. (*Source:* Used with permission of Carol Perry, RN, BSN, MSM Vice-President, Patient Care Services and Chief Nursing Officer, Stormont-Vail HealthCare Topeka, Kansas).

**FIGURE 3-2** Newspaper ad created to promote Magnet designation, Stormont-Vail HealthCare Topeka, Kansas. (*Source:* Used with permission of Carol Perry, RN, BSN, MSM Vice-President, Patient Care Services and Chief Nursing Officer, Stormont-Vail HealthCare Topeka, Kansas).

## TABLE 3-5  Eight Essentials of Magnetism

- Opportunities to work with other nurses who are clinically competent

- Good nurse-physician relationships and communication

- Nurse autonomy and accountability

- Supportive nurse manager-supervisor

- Control over nursing practice and practice environment

- Support for education

- Adequate nurse staffing

- Concern for the patient is paramount

nurse-physician relationships (Aiken, Sloane, & Lake, 1997; Laschinger et al., 2003). These findings are significant given the important patient safety implications of cohesive interdisciplinary teams capable of initiating open communication and collaborative dialogue on behalf of patients. Nurses at magnet facilities also report strong organizational support for continuing professional development (Kramer & Schmalenberg, 2004). Such support is an important retention strategy that benefits patients and helps nurses meet their lifelong learning requirements.

Two decades of research by Kramer and Schmalenberg (2002) suggests that nurses in magnet facilities indicate high levels of job satisfaction because they perceive that magnet practice environments allow nurses the ability to give quality patient care. The eight essentials to giving quality care reported by staff nurses in magnet hospitals are also known as the Essentials of Magnetism (Table 3-5).

## Enhanced Safety Outcomes

Magnet hospitals are known for better patient and staff safety outcomes. Specifically, magnet hospitals have been found to have fewer needle stick injury rates among nurses (Aiken, Sloane, & Klocinski, 1997). Better nurse-to-patient ratios in magnet hospitals are also known to result in reduced near-miss patient injury (Clarke, Rockett, Sloane, & Aiken, 2002). The Agency for Health Care Research and Quality (AHRQ) defines a near miss as a close call or an event or situation that did not produce patient injury but could have done so (AHRQ, 2006). The literature clearly supports the value of nurses in establishing and maintaining the vigilant surveillance systems that are crucial for patient safety (Aiken, 2002; Aiken, Clarke, Cheung, Sloane, & Silber, 2003; Clarke & Aiken, 2006). Importantly, the empowerment structure evident in magnet organizations is consistent with fostering patient safety cultures (Armstrong & Laschinger, 2006).

## Enhanced Competitive Advantage

Magnet designation represents a gold seal of approval that validates excellence in nursing services. Achieving such distinction enhances the public's confidence with the facility and speaks to the organization's overall quality (ANCC, 2008a). The low turnover and high job satisfaction of nurses in magnet facilities also provides competitive advantage by increasing staff continuity, maintaining patient care quality, and minimizing the costs associated with employee turnover. Achieving the prestigious magnet hospital designation can be used powerfully in hospital-wide promotional materials, thus adding to the hospital's capability to gain marketing advantage. The business case for pursuing magnet designation is strong. For the typical 500-bed hospital, the direct costs associated with magnet designation range from $46,000 to $251,000, and the return on investment is approximately $2.3 million (Drenkard, 2010). These cost savings derive from the magnet benefits and improvements identified in Table 3-4.

## FORCES OF MAGNETISM

The 14 Forces of Magnetism represent the foundation of the Magnet Recognition Program (Urden & Monarch, 2002). These Forces incorporate the nine characteristics defining magnet nursing services and the eight essentials of magnetism discussed earlier in this chapter. The 14 Forces of Magnetism are the outcomes of innovative and dynamic implementation of the Scope and Standards for Nurse Administrators by visionary nurse leaders creating supportive and collegial environments for nursing practice (ANCC, 2008a) (Table 3-6).

Magnet designation requires the full expression of the 14 Forces of Magnetism (ANCC, 2008a). This means that facilities seeking magnet designation must show evidence to support the existence of all the Forces of Magnetism in the organization. An emerging body of literature exists to help navigate the magnet journey

## REAL WORLD INTERVIEW

We are a small group of seasoned nurses working on a busy outpatient infusion/procedure unit in a magnet hospital. Infusions include IV monoclonal antibody therapy such as Infliximab, daily long-term IV antibiotics, corticosteroid infusions, and hydration. The majority of infusions are blood and blood products. The unit can also accommodate staff surgeons when minor surgical procedures cannot be done in the office setting. Both the nurses and the technicians have well over 20 years of experience under their belts. We decided to suggest that we assume the responsibility of self-scheduling staff work time on our unit. Having control over work hours was a priority for us. We believed that this would satisfy our personal needs and make for happier staff to care for patients and their needs. A small team of our staff nurses reviewed the literature and interviewed department managers and staff within the hospital who currently do self-scheduling. Along with our unit manager, we examined the pros and cons of self-scheduling and requested to move forward with it. The department manager was instrumental in bringing our unit into the computerized scheduling system already used on several of the inpatient units at the hospital. This scheduling system allows our staff to access our unit work schedule from either work or home, a great staff incentive. Staff members can easily review their work schedule months in advance and cover shifts as needed. We can also make staffing requests with this system for such things as vacation time. Following our unit staffing guidelines, we can either work a set schedule or work a flexible schedule and modify the length of shifts or the number of days worked consecutively. With many staff members pursuing additional nursing degrees, this scheduling flexibility is valuable. The flexibility may also decrease staff fatigue and improve patient care. A magnet facility supports an environment of open communication and empowers nurses to make decisions that impact their practice. Self-scheduling was an idea started by the unit staff nurses, supported by unit management, and successfully implemented in our department. Self-scheduling has been a great decision for this unit and has worked very well.

**Beverly A. George, MS, RN, CRNI**
Central Dupage Hospital
Winfield, Illinois

(Goode, et al., 2005; Ellis & Gates, 2005; Havens & Johnston, 2004; Shirey, 2004). Articles that describe the magnet application process (Bumgarner & Beard, 2003; Bliss-Holtz, Winter, & Scherer, 2004) provide guidance in meeting difficult standards (Messmer, Jones, & Rosillo, 2002; Turkel, Reidinger, Ferket, & Reno, 2005; Turkel, Ferket, Reidinger, & Beatty, 2008). Help to document the Forces of Magnetism (Shirey, 2005; Drenkard, 2005; Poduska, 2005; Hitchings & Capuano, 2008) is available, as are sources assisting with the magnet journey (Horstman et al., 2006; Broom & Tilbury, 2007), the appraisal process (Lundmark & Hickey, 2006), the magnet site visit (Conerly & Thornhill, 2009), and the re-designation preparation (Upenieks & Sitterding, 2008).

## THE MAGNET MODEL

In 2004, the magnet program underwent a comprehensive evaluation using an independent consultant external to ANCC (Triolo, Scherer, & Floyd, 2006). The evaluation process ended in 2005, culminating with 22 evidence-based recommendations. In 2008, a new magnet model was developed that incorporated scholarly review and statistical analysis (Wolf, Triolo, & Ponte, 2008) and affirms what the magnet program represents(available at http://www.nursecredentialing.org/Magnet/NewMagnetModel.aspx).

The new magnet model emphasizes five new model components within global issues in nursing and health. Although the program continues to incorporate the Forces of Magnetism, the new model eliminates redundancy within the 14 Forces (ANCC, 2008c). The five components of the new magnet model include transformational leadership; structural empowerment; exemplary professional practice; new knowledge, innovation and improvements; and empirical outcomes.

Transformational leadership as defined by the new magnet model requires leaders to lead people where they need to be to meet future health care demands. According to ANCC (2008c), "such leadership requires vision, influence, clinical knowledge, and expertise" (p. 22). Forces of Magnetism 1 and 3 are incorporated within transformational leadership.

## TABLE 3-6   The 14 Forces of Magnetism

| FORCES OF MAGNETISM | DISCUSSION |
|---|---|
| 1. Quality nursing leadership | • Nurse leaders are perceived as knowledgeable, strong risk-takers who follow an articulated philosophy of nursing.<br>• Nurse leaders are strong staff advocates and supporters.<br>• The outcomes of quality nursing leadership are evident in nursing practice at the patient's bedside. |
| 2. Organizational structure | • Organizational structures are flat with decentralized, unit-based decision making.<br>• Strong nursing representation is evident in the organizational committee structure.<br>• Chief nursing officer reports to the organization's chief executive officer and is a member of the executive team. |
| 3. Management style | • Nursing and hospital administrators share a participative management style that incorporates feedback from staff at all levels of the organization. |
| 4. Personnel policies and programs | • Organization offers competitive salaries and benefits.<br>• Flexible staffing models are utilized.<br>• Personnel policies reflect staff involvement and clinical promotional opportunities. |
| 5. Professional models of care | • Nurses have accountability for their nursing practice.<br>• Practice model reflects nurses as coordinators of care. |
| 6. Quality of care | • Providing quality of care is seen as an organizational priority.<br>• Nurse leaders develop the work environment so that quality of care can be delivered, and nurses perceive that they are able to provide high-quality care. |
| 7. Quality improvement | • Staff nurses participate in quality improvement processes, and they perceive the process as educational and beneficial to quality patient care. |
| 8. Consultation and resources | • Knowledgeable experts, including advanced practice nurses, are available and utilized.<br>• Adequate consultation with other health care human resources is available within the organization. |
| 9. Autonomy | • Nurses engage in autonomous practice consistent with professional standards.<br>• Independent professional judgment is expected within the context of interdisciplinary collaboration in patient care. |
| 10. Community and the hospital | • The hospital maintains a strong community presence, with the community perceiving the hospital as a productive and positive corporate citizen. |
| 11. Nurses as teachers | • Nurses are expected to incorporate teaching in all aspects of their professional practice. |
| 12. Image of nursing | • Nurses are seen as crucial to the hospital's capability to provide patient care services, a perception also held by other members of the health care team. |
| 13. Interdisciplinary relationships | • Positive interdisciplinary relationships are present, with a sense of mutual respect exhibited among all disciplines. |
| 14. Professional development | • Significant emphasis is placed on the professional development of the staff, including orientation, in-service education, continuing education, formal education, and ongoing competency maintenance.<br>• Value is given to personal and professional growth, including emphasis on employee career development. |

**EVIDENCE** FROM THE LITERATURE

**Citation:** Havens, D. S., & Johnston, M. A. (2004). Achieving magnet hospital recognition: Chief nurse executives and magnet coordinators tell their stories. *Journal of Nursing Administration, 34*(12), 579–588.

**Discussion:** This research was designed to add to the understanding of how hospitals successfully pursue magnet recognition. Although extensive literature exists to demonstrate the benefits of magnet hospitals, few studies have been conducted to explain how to get there. A convenience sample of 24 participants (6 chief nurse executives and 18 magnet program coordinators) attending the October 2003 National Magnet Hospital Conference in Houston, Texas, was selected. Three two-hour focus groups were conducted, with questions guided by the literature on magnet hospitals. This article reports on the results of one question: How did you pursue ANCC magnet recognition?

Data analysis from the interviews revealed the following nine themes related to the research question. First, securing buy-in from key stakeholders is key. Having the support of hospital administrators, chief executive officers, the board of trustees, practitioners, nurses, and staff from other departments is important. The researchers found nurses to be the hardest sell, primarily because nurses were concerned with the perceived added workload that a magnet journey would entail. Second, the importance of celebrating throughout the process was highlighted. Because the magnet journey may be a long one, participants reported the importance of celebrating along the way with large, themed, kick-off events and functions after achieving specified milestones and formal galas after the attainment of magnet designation. Third, the use of external consultants was found to be mostly beneficial. The consultants were seen as expert guides who could serve as validators and encouragers. Satisfaction with the use of consultants was mixed. Fourth, putting the structure for the operation in place required time. Some facilities reported needing a lead time of about three years, whereas others appeared to have significant components of the magnet structure already in place earlier. The need to identify champions and cheerleaders early on to encourage the process was mentioned. Fifth, communicating frequently was crucial in helping to spread the magnet message. Newsletters, flyers, personal communications, and addressing an ongoing magnet agenda at multiple venues were strategies used to reach all levels of the organization. Sixth, educating nurses and others was identified as a key strategy for a successful magnet journey. Magnet education was included as part of an ongoing component of new employee orientation. Seventh, mentoring by ANCC magnet hospitals was reported to be most helpful early in the magnet journey. Mentoring took the form of on-site visits by teams at mentor hospitals, phone calls, and email exchanges. Eighth, telling the story while collecting the organization's magnet evidence was important. Although the hospitals reported a variety of methods for collecting magnet evidence, most agreed on the importance of gathering a variety of rich and robust examples from across the organization. Ninth, paying the costs of the magnet journey involved a significant time commitment, with the magnet coordinator assuming a great deal of the responsibility. Most organizations undertaking a magnet journey underestimate the time commitment it requires. Interestingly, there was no consensus among participants as to the monetary costs associated with the journey. Sources of funding for the magnet journey ranged from internal budgets to grants from hospital foundations. The participants agreed that a significant value of the magnet journey was in the magnet appraisal process.

**Implications for Practice:** The process of pursuing magnet designation is as meaningful as attaining the actual magnet designation. Pursuing a magnet journey is consistent with implementing a change process and should be approached from an organizational change perspective. The effort requires frequent communication and the involvement and support of multiple disciplines. The success of the magnet journey rests with the understanding that nursing exists within a larger culture, and thus the effort must be undertaken with the contributions of many within the organization. To this end, buy-in and support from multiple stakeholders is crucial to the success and sustainability of magnet hospitals.

Structural empowerment refers to structure and processes within organizations needed to support the magnet framework.

Influential and innovative leaders are essential for facilitating the mechanisms needed to allow professional practice to flourish and an organization's mission, vision, and values to achieve desired outcomes (ANCC, 2008c). A fundamental requirement of structural empowerment is that staff members must have freedom and resources to develop, direct, and find the best ways to accomplish organizational and desired outcomes. Forces of Magnetism 2, 4, 10, 12, and 14 are embedded within structural empowerment.

Exemplary professional practice is the essence of a magnet organization. Professional practice requires a comprehensive understanding of the autonomous role of nursing with patients and its interdisciplinary team approach needed to provide quality care and achieve desired outcomes. Forces of Magnetism 5, 6, 7, 8, 9, 11, and 13 are addressed within exemplary professional practice.

New knowledge, innovation, and improvements are the responsibility of a magnet organization. Excellence within this component of the model requires magnet-aspiring facilities to use new models of care, apply evidence to practice, make visible contributions to nursing science, and continuously improve quality. Forces of Magnetism 6 and 7 are addressed within new knowledge, innovations, and improvements.

Empirical outcomes refer to measurable results that derive from practice. These outcomes answer the question: "What difference did our organization make?" Over time, outcomes may be trended to determine different intervention effects and patient care quality changes. Regularly monitoring these empirical outcomes triggers identification of best practices and implementation of different solutions. Empirical quality outcomes may also be used as a means to provide an organizational report card and demonstrate excellence. Magnet Force 6 is reflected within empirical quality outcomes.

## MAGNET APPRAISAL PROCESS

The magnet appraisal process addresses specific requirements, processes, and activities necessary to achieve magnet designation. Magnet-aspiring organizations generally begin the process by purchasing the most current issue of the *Magnet Recognition Program Application Manual* (ANCC, 2008a). This manual guides the aspiring magnet organization's chief nursing officer (CNO), magnet program coordinator, and magnet steering team members in pursuing the magnet journey. The magnet appraisal process requires collecting detailed demographic information from applicant organizations and reviewing comprehensive documents and magnet evidence reflecting how applicants meet all program requirements. Additionally, the process considers feedback acquired from public comment opportunities. Magnet appraisers conduct a variety of site visits to verify and expand upon the submitted application materials.

The magnet appraisal process consists of four sequential phases: application, evaluation, site visit, and award decision. The application phase involves review of the application manual and the decision to apply for magnet designation. Early in the magnet journey, organizations will need to establish a database to collect data on **nursing-sensitive indicators**, that is, measures that reflect the outcome of nursing actions. Joining the National Database of Nursing Quality Indicators (NDNQI, 2006) is a means to achieve this data-collection requirement. Membership in the NDNQI is beneficial, because it provides organizations with the capability to benchmark data on nursing-sensitive indicators gathered at the unit level. Benchmarking is the process of comparing outcomes with those of similar organizations to identify and establish best practices.

Organizations will also conduct a gap analysis in the application process. A **gap analysis** is an assessment of the differences between the expected magnet requirements and the organization's current performance on those requirements. A **gap** is the space between where the organization is and where it wants to be. A gap analysis serves as a tool that provides direction in developing the necessary activities to bridge a gap.

The evaluation phase occurs following the written application and submission of the aspiring hospital's magnet evidence. These comprehensive documents and demographic data are reviewed by a team of ANCC magnet appraisers who independently score the evidence for each of the 14 Forces of Magnetism. Arrangements for a site visit will follow if the written documentation earns the necessary points to score at a level of excellence.

The site visit involves the magnet appraisal team making a planned site visit to the magnet-aspiring facility. While on site, the magnet appraisers visit the units of the organization where nurses work to verify the content of the written magnet evidence previously submitted and scored. Following the site visit, the appraisal team prepares a consensus report summarizing the written documentation review and the site visit findings.

The award decision involves review of the consensus report by ANCC's Commission on Magnet Recognition. Magnet awards are made when the Commission members agree that the evidence reflects magnet-defined excellence in an organization's nursing services (ANCC, 2008a). After magnet designation is conferred, facilities must maintain compliance with the magnet standards to sustain the magnet workplace culture and to position the organization for magnet re-designation. Magnet hospitals submit annual reports for interim reporting and repeat the original application, evaluation, and site visit activities every four years for the re-designation process.

# PROFESSIONAL NURSING PRACTICE

The quality of a professional nursing practice environment is crucial in attracting and retaining professional nurses. Nurse leaders play a key role in creating practice environments that are supportive of and conducive to professional nursing practice. Magnet hospitals represent one example of supportive and collegial work environments that are both high performance and high quality-of-work-life organizations for nurses and other health care professionals. These desired work environments do not happen overnight. They require significant investment of time, energy, and resources by individuals, groups, and organizations. Although the investment required to create magnet workplaces is significant, the rewards to individuals, groups, and organizations are even greater. Ultimately, the investment in building supportive and collegial work environments results in organizational effectiveness, a key desired outcome of organizational behavior.

## CASE STUDY 3-1

A new nurse, Latisha, has just joined the staff of a busy cardiovascular unit. The nurse manager has assigned a clinical nurse specialist and a senior staff nurse mentor to assist with Latisha's orientation and integration into the unit. As part of this orientation, Latisha has been asked to develop a personal career plan that addresses measurable goals at three months, six months, one year, and three years. The nurse manager also gives Latisha a list of the unit-based and hospital-based committees and asks her to think about which committees she would like to join. Latisha should pick committees that maximize her gifts and talents, to best contribute to her personal goals and to the hospital's goals. Involvement in committees and in quality improvement activities is very important. The organization is so committed to these activities that it not only regularly provides release time for employees to participate, but also funds attendance at outside educational programs to build the nursing knowledge base in these areas.

What model of organizational behavior from Table 3-1 does this case study depict?

What is your preliminary assessment regarding the 14 Forces of Magnetism and their presence in this unit and organization?

How can Latisha, a new staff nurse joining the staff on this cardiovascular unit, further contribute toward building and sustaining the workplace culture?

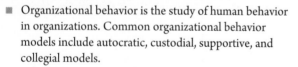

## KEY CONCEPTS

- Organizational behavior is the study of human behavior in organizations. Common organizational behavior models include autocratic, custodial, supportive, and collegial models.

- Creating desirable work environments is key to positively influencing employee attitudes, behaviors, and organizational performance. Failure to understand individual and group dynamics within organizations makes it difficult to create and sustain desirable work environments for professional practice.

- As organizations strive for sustainable high performance, they must consider material resources, such as technology, capital, quality improvement, and other information, yet they cannot lose sight of the human element, such as the people and teams that do the required work.

- Maintaining a satisfied and committed workforce requires a planned and dedicated effort by all members of the organization, to create and sustain desirable work environments for practice.

- Health care professionals in a knowledge economy possess valuable knowledge, skills, and abilities that individuals and organizations need to cultivate to retain employees, generate quality outcomes, and gain competitive advantage.

- High performance organizations that bring out the best in people generally focus on creating high quality-of-work-life environments. Nurses practicing in such organizations benefit from these organizational cultures and enjoy high levels of job satisfaction.

■ Magnet hospitals are known to have supportive and collegial work environments for nurses and may be classified as both high-performance and high quality-of-work-life environments. Nine characteristics define magnet nursing services.

■ A strong body of evidence exists to support the achievement of quality outcomes for nurses, patients, and organizations by magnet hospitals.

■ There are five major benefits of magnet designation.

■ The magnet process and magnet designation both create organizational value and meaning for individuals and groups within those organizations. Magnet designation should be seen as not merely a destination but rather an ongoing commitment to excellence in nursing services.

■ In the face of a highly competitive health care industry, health care leaders have a compelling obligation to create supportive and collegial work environments that are conducive to keeping patients safe and nurses retained within the profession.

■ The eight Essentials of Magnetism help a health care organization move toward achieving the 14 Forces of Magnetism.

## KEY TERMS

absenteeism
external forces
gap
gap analysis
Hawthorne
   effect

high quality-of-work-life
   environments
high-performance
   organizations
intellectual capital
job satisfaction

knowledge workers
magnet hospitals
nursing-sensitive indicators
open systems
organization
organizational behavior

organizational
   commitment
organizational effectiveness
productivity
stakeholders
turnover

## REVIEW QUESTIONS

1. As a new nurse attending the hospital's employee orientation program, you are told that the hospital recognizes its employees as its most valuable asset. Which of the following does not demonstrate congruence between this statement and what you observe in the workplace?
   A. Employees are treated with respect and fairness.
   B. The hospital supports employee attendance at continuing education programs.
   C. Employee suggestions are encouraged but are rarely acted upon.
   D. Employee job satisfaction is evaluated on a yearly basis.

2. Magnet hospitals foster the philosophy that nurse leaders are needed at all levels of the organization. As a staff nurse and clinical leader on your unit, which of the following activities would not be viewed as supportive of the hospital's philosophy in a magnet facility?
   A. Staff nurse serving as the chairperson of the nursing practice council.
   B. Medical practitioner serving in the role of chairperson of the nursing practice council.
   C. Staff nurses conducting breast cancer self-exam classes in the community.
   D. Medical practitioners collaborating with nursing personnel on an interdisciplinary task force.

3. A body of evidence spanning two decades of research exists to support the value of magnet hospitals. Which of the following statements offers the strongest support for the pursuit of magnet hospital designation?
   A. In a longitudinal, multisite study of more than 2,000 U.S. hospitals, researchers reported that one year and three years following implementation of a magnet workplace culture, employee turnover dropped by 50%.
   B. The nurses at Community Hospital stated that they liked working there because it was a magnet-designated hospital.
   C. The nurses at Harper Hospital stated that they should become a magnet hospital because others in the area also were becoming magnet designated.
   D. *People* magazine, one of the most popular and well-read magazines in one community, recently had an article supporting the benefits of magnet hospitals.

4. A nurse evaluating an organization's potential for long-term success knows that high-performance environments offer a competitive advantage. Which of the following examples does not exemplify a high-performance organization?
   A. The organization partners with a technology center that readily integrates new and emerging technologies into clinical practice.

B. The organization demonstrates a commitment to continuous quality improvement.

C. The organization controls information available to employees and follows the company policy that what employees don't know won't hurt them.

D. The organization developed a succession strategy that identifies and plans for the education of future leaders within the organization.

5. As a nurse, you are increasingly reading about magnet hospitals and noticing that these hospitals are also known as high-performance organizations. Which of the following most accurately characterizes why magnet hospitals may be classified as high-performance organizations?

A. High-performance organizations and magnet hospitals possess no similarities, and therefore this association in terminology is not well founded.

B. High-performance organizations such as magnet hospitals recognize the importance of attracting and retaining talented employees.

C. High-performance organizations such as magnet hospitals are known for high employee turnover.

D. High-performance organizations such as magnet hospitals focus on maintaining the status quo and benefit from the fact that they do not experience change or turmoil.

6. A nurse was visiting with a relative who is not a nurse. The relative had been reading about the benefits of magnet hospitals in a recent issue of *Reader's Digest*. She wanted to know more about these benefits. In responding to the relative's question, which of the following outcomes listed on a magnet hospital's Web site would the nurse not use to illustrate the benefits of magnet hospitals?

A. Patient satisfaction scores in the 99th percentile.

B. Nurse turnover rate of 2%.

C. Employee waiting lists for new hires in 50% of the hospital's departments.

D. Nurse-to-patient ratio of 1 nurse to 20 patients.

7. Part of the role of the staff nurse champion for a hospital's magnet journey is to prepare and present an in-service on the historical evolution of the Magnet Recognition Program. As part of this program's discussion, which primary goal should be presented to describe the main reason for creation of the Magnet Recognition Program?

A. To start the reengineering efforts in the health care industry.

B. To promote excellence in the delivery of nursing services to patients.

C. To increase the incidence of medication error reporting.

D. To promote staff and physician satisfaction.

8. Demonstrating that all the Forces of Magnetism exist in an organization is crucial to meeting the magnet requirements. As a staff nurse working on a magnet committee, you have been asked to help evaluate the magnet evidence to be submitted with the hospital's magnet application. Which of the following information pieces is not evidence to support professional development, the fourteenth Force of Magnetism?

A. Documentation of annual certification preparation courses offered in all the nursing specialty areas within the hospital.

B. Access to Web-based educational programs for nurses working at the unit level or nurses working from their own homes.

C. A 12-month financial report that details on a unit-by-unit basis the economic support provided to all staff nurses on that unit attending educational programs outside the health care organization.

D. A statement describing how decision making is done by the medical staff.

9. As a new nurse seeking a supportive and collegial work environment, which of the following workplace cultures would be most consistent with your aspirations? Select all that apply.

_____ A. The unit manager is known to hold staff meetings only when they are needed to address hospital financial concerns.

_____ B. The unit scheduling system is not flexible, and staff nurses work when they are told to work.

_____ C. The unit has a staff retention committee that hosts a quarterly luncheon and an employee-of-the-quarter recognition event.

_____ D. The nature of the nurse and practitioner relationship is paternalistic, and nurses seek practitioner orders for matters pertaining to independent nursing practice.

_____ E. The organization has a defined nursing career ladder in place.

10. A staff nurse just finished taking a course entitled "Becoming a Charge Nurse." Organizational behavior was part of the course content. Which of the following statements most readily explains the reason such a course would be useful? Select all that apply.

_____ A. The course may be beneficial to those interested in a charge nurse role, but it is not really beneficial for staff nurses at this time.

_____ B. The course helps employees better understand themselves and others as a basis for good contributions toward achieving top organizational performance.

_____ C. The course is not based on sound theoretical evidence and therefore its value is questionable for nurses.

_____ D. The course is helpful in assisting nurses to practice in static, noncomplex work environments.

_____ E. The course incorporates a personal strengths inventory assessment to ascertain the nurse's optimal career trajectory.

## REVIEW ACTIVITIES

1. To learn more about the magnet eligibility criteria, access and review the document entitled "Organization Eligibility Requirements" (2010c). To access the document, go to http://www.nursecredentialing.org/. Search for Organization Eligibility Requirements. After you have found the document, use this information to familiarize yourself with the requirements that must be in place to achieve magnet designation. Reflect on the extent to which you see your current health care work environment meeting the specified requirements.

2. To learn more about magnet re-designation self-assessment, access and review the document "Re-designation Self-Assessment for Magnet Excellence" (ANCC, 2008). To access the document, go to http://www.nursecredentialing.org/Documents/Magnet/Self-assessment.aspx.

## EXPLORING THE WEB

Search the Web, checking the following sites:

- Agency for Healthcare Research and Quality (AHRQ): www.ahrq.gov
  Explore the various elements of the Web site.

- Institute of Medicine: www.iom.edu
  Explore the various elements of the Web site.

- National Student Nurses' Association (NSNA) Career Center: www.nsna.org

## REFERENCES

Agency for Health Care Research and Quality. (2006). *AHRQ patient safety network: Glossary*. Retrieved May 6, 2006, from http://psnet.ahrq.gov/glossary.aspx.

Aiken, L. H. (2002). Superior outcomes for magnet hospitals: The evidence base. In M. L. McClure & A. S. Hinshaw (Eds.), *Magnet hospitals revisited: Attraction and retention of professional nurses* (pp. 61–81). Washington, DC: American Nurses Publishing.

Aiken, L. H., Clarke, S. P., Cheung, R. B., Sloane, D. M., & Silber, J. H. (2003). Educational levels of hospital nurses and surgical patient mortality. *Journal of the American Medical Association, 290*(12), 1617–1623.

Aiken, L. H., Sloane, D. M., & Klocinski, J. L. (1997). Hospital nurses' occupational exposure to blood: Prospective, retrospective, and institutional reports. *American Journal of Public Health, 87,* 103–107.

Aiken, L. H., Sloane, D. M., & Lake, E. T. (1997). Satisfaction with inpatient acquired immunodeficiency syndrome care: A national comparison of dedicated and scattered-bed units. *Medical Care, 36*(9), 948–962.

Aiken, L. H., Sloane, D. M., Lake, E. T., Sochalski, J., & Weber, A. L. (1999). Organization and outcomes of inpatient AIDS care. *Medical Care, 37*(8), 760–772.

Aiken, L. H., Smith, H., & Lake, E. (1994). Lower Medicare mortality among a set of hospitals known for good nursing care. *Medical Care, 32*(8), 771–785.

American Hospital Association. (2009). *Statistics and studies: Fast facts on U.S. hospitals*. Retrieved July 24, 2010, from http://www.aha.org/aha/resource-center.

American Nurses Association. (2009). *Scope and standards for nurse administrators* (3rd ed.). Washington, DC: American Nurses Publishing.

American Nurses Credentialing Center. (2008a). *Magnet recognition program: Application manual*. Washington, DC: American Nurses Credentialing Center.

American Nurses Credentialing Center. (2008b). Re-designation self-assessment for magnet excellence. Retrieved July 24, 2010, from http://www.nursecredentialing.org/Documents/Magnet/Self-assessment.aspx.

American Nurses Credentialing Center. (2008c). Modifying the magnet model: The shape of things to come. *American Nurse Today, 3*(7), 22.

American Nurses Credentialing Center. (2010a). Find a magnet facility. Retrieved July 24, 2010, from http://www.nursecredentialing.org/magnet/FindaMagnetFacility.aspx.

American Nurses Credentialing Center. (2010b). Nurse opinion questionnaire. Retrieved July 24, 2010, from http://www.nursecredentialing.org/magnet/nurseopinionsurvey.aspx.

American Nurses Credentialing Center. (2010c). Organization eligibility requirements. Retrieved July 24, 2010, from http://www.nursecredentialing.org/Magnet/ApplicationProcess/EligibilityRequirements/OrgEligibilityRequirements.aspx.

Armstrong, K. J., & Laschinger, H. (2006). Structural empowerment, magnet hospital characteristics, and patient safety culture: Making the link. *Journal of Nursing Care Quality, 21*(2), 124–132.

Bliss-Holtz, J., Winter, N., & Scherer, E. M. (2004). An invitation to magnet accreditation. *Nursing Management, 35*(9), 36–43.

Brady-Schwartz, D. C. (2005). Further evidence on the Magnet Recognition Program: Implications for nursing leaders. *Journal of Nursing Administration, 35*(9), 397–403.

Broom, C., & Tilbury. M. S. (2007). Magnet status: A journey, not a destination. *Journal of Nursing Care Quality, 22*(2), 113–118.

Bumgarner, S. D., & Beard, E. L. (2003). The magnet application. *Journal of Nursing Administration, 33*(11), 603–606.

Clark, D. (2000). *Big dog's leadership page: Organizational behavior.* Retrieved May 6, 2006, from http://www.nwlink.com/~donclark/leader/leadob.html

Clark, D. (2002). *Organizational behavior survey.* Retrieved May 15, 2006, from http://nwlink.com/~donclark/leader/obsurvey.html

Clarke, S. P., & Aiken, L. H. (2006). More nursing, fewer deaths. *Quality & Safety in Health Care, 15*(1), 2–3.

Clarke, S. P., Rockett, J. L., Sloane, D. M., & Aiken, L. H. (2002). Organizational climate, staffing and safety equipment as predictors of needlestick injuries and near-misses in hospital nurses. *American Journal of Infection Control, 30*(4), 207–216.

Coile, R. C. (2001). Magnet hospitals use culture, not wages, to solve nursing shortage. *Journal of Health Care Management, 46*(4), 224–227.

Conerly, C., & Thornhill, L. (2009). Magnet site visit preparation. *Nursing Management, 40*(7), 41, 42, 44, 46–48.

Daft, R. L. (2006). *Organization theory and design* (9th ed.). Mason, OH: South-Western College Publishing.

Drenkard, K. N. (2005). Sustaining Magnet: Keeping the forces alive. *Nursing Administration Quarterly, 29*(3), 214–222.

Drenkard, K. (2010). The business case for Magnet. *Journal of Nursing Administration, 40*(6), 263–271.

Drucker, P. F. (2006). *Classic Drucker: Wisdom from Peter Drucker from the pages of* Harvard Business Review. Boston: Harvard Business School Press.

Ellis, B., & Gates, J. (2005). Achieving Magnet status. *Nursing Administration Quarterly, 29*(3), 241–244.

Goode, C. J., Krugman, M. E., Smith, K., Diaz, J., Edmonds, S., & Mulder, J. (2005). The pull of magnetism: A look at the standards and the experience of a Western academic medical center hospital in achieving and sustaining magnet status. *Nursing Administration Quarterly, 29*(3), 202–213.

Havens, D. S., & Aiken, L. H. (1999). Shaping systems to promote desired outcomes: The magnet hospital model. *Journal of Nursing Administration, 29*(2), 14–20.

Havens, D. S., & Johnston, M. A. (2004). Achieving magnet hospital recognition: Chief nurse executives and magnet coordinators tell their stories. *Journal of Nursing Administration, 34*(12), 579–588.

Hitchings, K. S., & Capuano, T. A. (2008). The professional excellence council: Implications for all forces of magnetism. *Journal of Nursing Care Quality, 23*(2), 101–104.

Horstman, P. L., Bennett, C., Daniels, C., Fanning, M., Grimm, E., & Withrow, M. L. (2006). The road to Magnet: One magnificent journey. *Journal of Nursing Care Quality, 21*(3), 206–209.

Institute of Medicine. (2004). Keeping patients safe: Transforming the work environment of nurses. Washington, DC: The National Academies Press.

Kramer, M. (1990). The magnet hospitals: Excellence revisited. *Journal of Nursing Administration, 20*(9), 35–44.

Kramer, M., & Schmalenberg, C. (1991). Job satisfaction and retention insights for the 90's. *Nursing '91,* 50–55.

Kramer, M., & Schmalenberg, C. (2002). Staff nurses identify essentials of magnetism. In M. L. McClure & A. S. Hinshaw (Eds.), *Magnet hospitals revisited: Attraction and retention of professional nurses* (pp. 25–59). Washington, DC: American Nurses Publishing.

Kramer, M., & Schmalenberg, C. (2004). Magnet hospitals: What makes nurses stay? *Nursing 2004, 34*(6), 50–54.

Lake, E. T., & Friese, C. R. (2006). Variations in nursing practice environments: Relation to staffing and hospital characteristics. *Nursing Research, 55*(1), 1–9.

Laschinger, H. K. S., Almost, J., & Tuer-Hodes, D. (2003). Workplace empowerment and magnet hospital characteristics. *Journal of Nursing Administration, 33*(7/8), 410–422.

Laschinger, H. K. S., & Finegan, J. (2005). Empowering nurses for work engagement and health in hospital settings. *Journal of Nursing Administration, 35*(10), 438–449.

Lundmark, V. A., & Hickey, J. V. (2006). The magnet recognition program: Understanding the appraisal process. *Journal of Nursing Care Quality, 21*(4), 290–294.

Lynn, M. R., & Redman, R. W. (2005). Faces of the nursing shortage: Influences on staff nurses' intentions to leave their positions or nursing. *Journal of Nursing Administration, 35*(5), 264–270.

McClure, M., Poulin, M., Sovie, M., & Wandelt, M. (1983). *Magnet hospitals: Attraction and retention of professional nurses.* American Academy of Nursing Task Force on Nursing Practice in Hospitals. Kansas City, MO: American Academy of Nursing.

McClure, M. L., & Hinshaw, A. S. (2002). *Magnet hospitals revisited: Attraction and retention of professional nurses.* Washington, DC: American Nurses Publishing.

Messmer, P. R., Jones, S. G., & Rosillo, C. (2002). Using nursing research projects to meet magnet recognition program standards. *Journal of Nursing Administration, 32*(10), 538–543.

Miller, L., & Anderson, F. (2007). Lessons learned when Magnet designation is not received. *Journal of Nursing Administration, 37*(3), 131–134.

NDNQI. (2006). *National Database for Nursing Quality Indicators.* Retrieved May 6, 2006, from http://www.nursingquality.org.

NSNA. National Student Nurses' Association Career Center. Retrieved February 15, 2007, from http://www.nsna.org.

Poduska, D. D. (2005). Magnet designation in a community hospital. *Nursing Administration Quarterly, 29*(3), 223–227.

Robbins, S. P. (2005). Essentials of organizational behavior (8th ed.). Upper Saddle River, NJ: Pearson Prentice Hall.

Schermerhorn, J. R., Hunt, J. G., & Osborn, R. N. (2005). Organizational behavior (9th ed.). Hoboken, NJ: John Wiley & Sons.

Scott, J. G., Sochalski, J., & Aiken, L. H. (1999). Review of magnet hospital research: Findings and implications for professional nursing practice. *Journal of Nursing Administration, 29*(11), 1, 9–19.

Senge, P. (1990). *The fifth discipline: The art and practice of the learning organization.* New York: Doubleday.

Shirey, M. R. (2004). Preparing an organization for achieving magnet designation. (2004 Fellow project available from The American College of Health Care Executives, 1 N. Franklin Street, Suite 1700, Chicago, IL 60606).

Shirey, M. R. (2005). Celebrating certification in nursing: Forces of magnetism in action. *Nursing Administration Quarterly, 29*(3), 245–253.

Steinbinder, A., & Scherer, E. M. (2006). Creating nursing system excellence through the forces of magnetism. In K. Malloch & T. Porter O'Grady (Eds.), *Introduction to evidence-based practice in nursing and health care* (pp. 235–266). Boston: Jones & Bartlett Publishers.

Triolo, P. K., Scherer, E. M., & Floyd, J. M. (2006). Evaluation of the Magnet Recognition Program. *Journal of Nursing Administration, 36*(1), 42–48.

Turkel, M. C., Reidinger, G., Ferket, K., & Reno, K. (2005). An essential component of the Magnet journey: Fostering an environment for evidence-based practice and nursing research. *Nursing Administration Quarterly, 29*(3), 254–262.

Turkel, M. C., Ferket, K., Reidinger, G., & Beatty, D. E. (2008). Building a nursing research fellowship in a community hospital. *Nursing Economics, 26*(1), 26–34.

Upenieks, V. V. (2003). What constitutes effective leadership? Perceptions of magnet and nonmagnet nurse leaders. *Journal of Nursing Administration, 33*(9), 456–467.

Upenieks, V. V., & Sitterding, M. (2008). Achieving magnet redesignation: A framework for cultural change. *Journal of Nursing Administration, 38*(10), 419–428.

Urden, L. D., & Monarch, K. (2002). The ANCC Magnet recognition program: Converting research into action. In M. L. McClure & A. S. Hinshaw (Eds.), *Magnet hospitals revisited: Attraction and retention of professional nurses* (pp. 103–115). Washington, DC: American Nurses Publishing.

U.S. News & World Report. (2010). Best hospitals 2010-11: The honor roll. Retrieved July 28, 2010, from http://health.usnews.com/health-news/best-hospitals/articles/2010/07/14/best-hospitals-2010-11-the-honor-roll.html.

Wolf, G., Triolo, P., & Ponte, P. R. (2008). Magnet recognition program: The next generation. *Journal of Nursing Administration. 38*(4), 200–204.

## SUGGESTED READINGS

Aiken, L. H., Clarke, S., Sloane, D., Sochalski, J., & Silber, J. (2002). Hospital nurse staffing and patient mortality, nurse burnout, and job dissatisfaction. *Journal of the American Medical Association, 288*(16), 1987–1993.

Aiken, L. H., Havens, D. S., & Sloane, M. (2000). The magnet nursing services recognition program: A comparison of two groups of magnet hospitals. *American Journal of Nursing, 100*(3), 26–35.

American Nurses Credentialing Center. (2004). *Magnet: Best practices in today's challenging health care environment.* Washington, DC: American Nurses Credentialing Center.

Lundmark, V. A., & Hickey, J. V. (2007). The magnet recognition program: Developing a national magnet research agenda. *Journal of Nursing Care Quality, 22*(3), 195–198.

Maslow, A. (1970). Motivation and personality (2nd ed.). New York: Harper & Row.

McGregor, D. (1960). The human side of enterprise. New York: McGraw-Hill.

Ouchi, W. (1981). *Theory Z: How American business can meet the Japanese challenge.* Reading, MA: Addison-Wesley.

SearchSMB.com. (2006). *SMB definitions: Gap analysis.* Retrieved May 11, 2006, from http://searchsmb.techtarget.com/sDefinition/0,290660,sid44_gci831294,00.html.

Taylor, F. (1911). *Principles of scientific management.* New York: Harper & Row.

Urden, L. D. (2006). Transforming professional practice environments: The Magnet Recognition Program. In P. S. Yoder-Wise & K. E. Kowalski (Eds.), *Beyond leading and managing: Nursing administration for the future* (pp. 23–34). St. Louis, MO: Mosby Elsevier.

# CHAPTER 4

# Basic Clinical
# Health Care Economics

Laura J. Nosek, PhD, RN; Ida M. Androwich, PhD, RN, BC, FAAN;
Kim Amer, PhD, RN

*The purpose of creating and analyzing records of what transpires in hospitals is to know how the money is being spent; whether it is, in fact, doing good, or whether it is doing mischief.*

(Florence Nightingale, 1859)

## OBJECTIVES

Upon completion of this chapter, the reader should be able to:

1. Analyze why health care must be managed as a business.
2. Apply the cost equation to the mission statement of a health care enterprise to discover why the enterprise may be thriving or struggling.
3. Analyze the impact of health care reform legislation on the health care industry, including insurance companies, health care providers, and hospitals.
4. Apply the break-even formula to compute a break-even point for a piece of equipment your health care organization is planning to purchase.
5. Discover how a health care enterprise is balancing quality and profit by assigning its satisfaction rating and margin to an appropriate square on the Nosek-Androwich Profit: Quality (NAPQ) Matrix.

Delmar/Cengage Learning

*You and your spouse are vacationing in a foreign country. You rent small hand-held aqua scooters that pull you through the saltwater lagoon, and you gleefully romp in the quiet water. As you make a turn near the island of dead coral in the center of the lagoon, you lose your balance and are dragged along the sharp coral, lacerating both thighs. A lifeguard hears your screams and dashes to your aid. Hotel personnel put you in a taxi and send you to a nearby hospital emergency room. There, with your limited local language and the help of one person who speaks limited English, you discover that all health care is government run and provided free to all citizens. You are told that you cannot be treated, because there is no provision for accepting your American insurance and no provision for paying cash.*

*What do you think the hospital perceives the problem to be?*

*Do you see the problem the same way the hospital does?*

*When cost is removed from the equation, what drives the decision to provide or not provide health care?*

*If cost were not a driving consideration in providing health care in the United States, what would the health care system be like?*

Regardless of how expert, creative, collaborative, and altruistic a health care system may be, it cannot function without money. Over the ages, depending on the culture and country, the government or citizens have often provided the resources to keep the community healthy. The funding can vary, including philanthropy, volunteerism, fee for services, insurance, and government subsidies. Balancing cost with revenue is key to achieving the mission of providing health care and is now viewed as the shared responsibility of humans around the world.

Early nursing programs focused on nurses needing only to be educated in the art and the science of providing clinical care to patients. Today, nurses must be much more broadly educated. Nurses must be poised to work in a variety of roles, including hospitals, schools, home care, retail clinics, long-term health agencies, battlefields, and community health centers. The Institute of Medicine report on the Future of Nursing (2010) focuses on the importance of nurses advancing their education and expanding their practice breadth and depth to their full ability to diagnose, treat, and prescribe patient treatments for full reimbursement.

The 2010 Patient Protection and Affordable Care Act legislated broad health care overhauls in insurance company policy, funding of nursing education, and primary care provider support. And, for the first time, it rewards providers that do a good job providing quality care. Conversely, if mistakes are made or a hospital-acquired infection occurs, payment may be withheld (The Patient Protection and Affordable Care Act, 2010). Nightingale's early search for the "good" versus "mischievous" outcomes of the money spent on health care may have initiated an unspoken commitment to financial stewardship among nurses. All nurses need to be aware of new health care legislation so that confusion can be cleared up and patients and families can be educated about their insurance benefits and rights to health care coverage. Every nurse today needs to have a basic understanding of clinical health care **economics**—the study of how scarce resources are allocated among possible uses—in order to make appropriate choices among the increasingly limited resources of the future.

The study of economics is based on three general premises: (1) scarcity—resources exist in finite quantities, and consumption demand is typically greater than resource supply; (2) choice—decisions are made about which resources to produce and consume among many options; and (3) preference—individual and societal values and preferences influence the decisions that are made. Health care does not always fit well in this model, i.e.,

1. scarcity in nurses or supplies can affect the quality of care;
2. choice of providers can be limited by income or location; and
3. preference in health care providers may be available to only a limited portion of the population.

Even with limitations, health care costs may increase without a clear improvement of the "product" of health care. For example, consider the concept of price elasticity, which is related to the price that an individual is willing to pay for a given item. Normally, as the price goes up, the demand goes down. When the purchase is health care, however, the price may be viewed as irrelevant to the decision to purchase. Think of a wristwatch that you might readily purchase for $5, would likely not buy at $50, and would never consider at $500. Now, imagine that instead of a wristwatch the item in question is a medication or therapy needed to save your sick child. Now the consideration of price in the decision-making process is different. Thus, health care is much less "elastic" with reference to price than are many other consumer goods.

The health care system, however, is more complex than the seller of a product such as a watch or the traditional business model. In health care, the product is provided by hospitals, nurses, doctors, and health care team members and health care is often paid for by a government program such as Medicaid or Medicare or a health insurance company. The provider (nurse or physician) is not the payer, nor is the patient (buyer) who is using the hospital or treatment the payer. The actual **payer** is the third-party reimburser (insurance company or government). Consequently, the financial impact of the decision on the provider (buyer) and the patient user (buyer) is skewed. Neither of these buyers is the payer. Note that there are some health care settings such as clinics where patients who do not have health insurance pay the nurse or doctor directly for care. This is a more simple situation where the insurance company (third party payer) is left out of the equation.

This chapter presents basic clinical health care economic concepts that are important to the nurse entering clinical practice. Included are perspectives on the role of cost in directing health care delivery, the methods for determining the cost of delivering nursing care, and the effect of health care policy on the delivery of nursing care. Recognized nurse experts provide comments on the future impact of economics on clinical nursing.

# TRADITIONAL PERSPECTIVE ON THE COST OF HEALTH CARE: HEALTH CARE HELPS PEOPLE

The long-standing tradition of health care is to help people achieve their optimal level of health so that they can enjoy their maximum quality of life. Factors that are influencing recent policies on health reform, as described by Chen and Weir (2009), include political ideology, policy entrepreneurs, and interest groups.

Several early nursing leaders, including Florence Nightingale and Isabel Adams Hampton Robb, were members of socially prominent families instilled with the value that unselfish service to others was the expected role of the privileged. Such feelings of dedication to the less fortunate stemmed from medieval infirmaries established by convents and monasteries to care for the aged, orphaned, poor, and disabled. The first hospitals to care for the sick and injured were also charitable institutions established around the fourteenth century to provide illness care to those who did not have a home or who could not afford home care.

## NEED FOR HEALTH CARE DETERMINED BY PROVIDER

Prior to the 1980s, mainstream health care was delivered from a paternalistic model of governance and control. Health professionals, led by practitioners, controlled a vast body of scientific knowledge and skill rendered awesome and mystical by complex scientific language. Command of that scientific knowledge and skill required extensive and expensive education and was not shared with "outsiders." The practitioner determined what health care was needed, independent of the patient and even independent of professional colleagues. The practitioner also decided how much to charge for that care. Decision making about all aspects of health care was the exclusive domain of the professionals.

## RIGHT TO HEALTH CARE AT ANY COST

The cost of health care was not considered, let alone questioned, until the early 1960s. The American belief system firmly held that every individual was entitled to all the knowledge, skill, and technology related to health care at any cost. It was claimed to be a "right"; it was the "American way," although certain factors such as race, location, and income had some effect on the availability of health care. In an attempt to ease the burden of health care costs, the U.S. government stepped up in 1965 and enacted Titles XVIII and XIX, amendments to the Social Security Act. These are commonly referred to as the Medicare and Medicaid programs, which provide health care coverage for the elderly and the indigent, respectively. To be paid for health care services, documentation of the kind and amount of services provided was required. It was anticipated that by requiring health care providers to account for the cost of Medicare and Medicaid patients' care, spending would be curbed. Other insurers soon followed with their own documentation of care requirements, launching the overall budgeting of health care.

## COST PLUS

The cost of health care continued to spiral upward as hospitals became the preferred site for provision of intermediate care and the high technology necessary for state-of-the-art illness care. That cost was determined by the actual cost the provider incurred for the care plus a profit incentive for being in the business. The method was known as "cost plus," and clearly the incentive was "the more you spend, the more you get," not "how can this be accomplished more economically?"

# CONTEMPORARY PERSPECTIVE ON COST OF HEALTH CARE: HEALTH CARE AS A BUSINESS

Possibly the most common reason given for entering a health care profession is "to help people." Virginia Henderson, viewed by some as the contemporary Florence Nightingale, defined nursing as

*...primarily helping people (sick or well) in the performance of those activities contributing to health, or its recovery (or to peaceful death) that they would perform unaided if they had the necessary strength, will, or knowledge. It is likewise the unique contribution of nursing to help people to be independent of such assistance as soon as possible. (Henderson & Nite, 1978)*

Nurses fervently believe and state that this definition applies irrespective of the site where nursing care is given. Yet nurses also have begun to recognize that the cost of providing care in the traditional altruistic way was prohibitive and that achieving independence from nursing care as quickly as possible conserves scarce nursing resources.

Taxes to cover the ever-increasing costs of the government health care programs were climbing. In the late 1970s, Medicare insolvency was a threat. Again, the government, the major payer, stepped in. The Health Care Financing Administration (HCFA), the department responsible for the Medicare and Medicaid programs, was authorized to change the way it paid for health care. The Tax Equity and Fiscal Responsibility Act (TEFRA) of 1982 established new payment regulations. Instead of reimbursing the provider's cost, the government would henceforth pay a flat rate stated up front. The new system would therefore be called the prospective payment system.

The new payment system considerably changed institutions' incentives for spending. If the provider was able to provide the care for less than the prospective payment, the provider could make a profit. If the provider spent more than that payment, the provider lost money. Because length of hospital stay is a surrogate measure of hospital cost, reducing length of stay was seen as the easiest and most logical way to reduce the cost of care enough to ensure adequacy of the payment.

One of the first hospital lengths of stay to be shortened was that for obstetrical patients. It was an unforgettable October morning in 1983 when the headlines of the *Cleveland Plain Dealer* shocked the city with the news. Effective immediately, hospital care for those experiencing normal vaginal delivery would be three calendar days starting with the day of admission, the story read. At that time in that city, the usual stay for a normal vaginal delivery was five days; for a Cesarean delivery, it was seven days. If a woman were admitted shortly before midnight, that constituted day one. If she experienced a long labor, she could end up being discharged on her first postpartum day. Patients were crying in the halls. Nurses were outraged, trying to figure out how they could possibly teach breastfeeding when breast milk does not come in until postpartum day three. Practitioners were threatening a variety of actions, citing unsafe care.

Out with altruism. In with health care that clinicians had to recognize was truly a business. In with a whole new language, that of business and consumers and profit and margin and competition and cost, *cost*, COST. The bottom line (cost) became the focus, not only of managers and administrators but of all employees at all levels of all health care entities. Everyone needed to question what they did, how they did it, and how many it took to do it in order to determine whether it was required for safe care and quality outcomes and whether there was a less costly way of attaining a safe, quality outcome. In 1960, Abdellah had challenged health care providers to determine the care that was needed and to provide no more, no less (Abdellah, Beland, Martin, & Matheney, 1960). Nearly 50 years later, health care providers are still trying to come to grips with just that.

## NEED FOR HEALTH CARE DETERMINED BY THE CONSUMER

Attention has shifted toward safety and quality and the need for measurable outcomes. Total quality improvement (TQI) and continuous quality improvement (CQI) programs were initiated to assure society that cost management was not compromising safety or quality. These programs required all stakeholders, including patients, to work together to evaluate and improve outcomes. The expertise of allied health care providers was recognized, and through the growing access to information technology, consumers were empowered to better understand their own health, the complex technologies available, their options for choosing to manage the decisions about their care, and the cost implications of those decisions. No longer was health care the exclusive realm of the practitioner and other professionals. Despite such attempts at improving quality, there remains cause for concern. Hospital-acquired infections, uncontrolled bleeding during surgery, medication errors, and patient falls remain a primary risk for patients in hospitals. Landrigan (2010) identified 25.1 injuries per 100 admissions in his retrospective analysis of safety in North Carolina hospitals from 2002 to 2007. Most injuries were not life threatening, ranging from low blood sugar to urinary tract infections from catheters. But most importantly, 63% of the injuries were preventable.

Serb (2006) points out that, more than a decade ago, Harvard Business School Professor Regina Herzlinger predicted a revolution in health care toward a consumer-based model with greater choice, i.e., focused factories of provider teams, flexible insurance products, and widely available information on quality and cost. Many dismissed her ideas, and Herzlinger herself delayed publication of her award-winning book, *Market-Driven Health Care*, until after the heyday of HMOs—whose gatekeepers, top-down managements, and tight networks seemed the antithesis of consumerism. Since Herzlinger's predictions, access to reliable, extensive health care information, as well as cost and quality data, have become ever more available on the Internet. In addition, the 2010 Patient Protection and Affordable Care Act provides more-affordable health care insurance and accessibility through state-administered High Risk Pool plans. These High Risk Pool plans are for patients previously refused insurance due to preexisting health care conditions, and they provide affordable health care insurance for 45 million uninsured Americans. These uninsured persons can either choose a government-run insurance plan or they can choose from private insurance plans (Affordable Health Care for Americans, 2010).

## RIGHT TO HEALTH CARE AT A REASONABLE COST

The contemporary value system holds that individuals have the right to health care at a reasonable cost. Reasonable cost is currently determined by insurers. When it refers to fees charged for services, a reasonable cost is the usual and customary fee charged in the region. When referring to technology; complex and expensive procedures; or expensive, extensive pharmacologic therapies, there is no established standard for how much it should cost to provide someone enhanced quality of life over time. Clearly, the lack of consensus on what constitutes reasonable cost is at the heart of the controversies among insurers, patients, and their professional providers.

## MANAGED CARE

The method of paying doctors and hospitals 100% of their billing stopped in the 1980's. Managed Care was created to control health care costs and determine reasonable costs

for specific diagnoses and treatments, while factoring in geographical location variations. This effort to control cost extended through the Medicare and Medicaid programs and marked the beginning of health care reform. There was keen anticipation that if the care of the neediest—the elderly and the poor—was managed centrally, access, cost, and quality would be optimally controlled. When it became evident that the program costs had been woefully underestimated, health maintenance organizations (HMOs), or managed care, emerged as the answer to cost-efficient and quality care.

Managed care is not easily defined and categorized. It is the product of a series of efforts to establish an effective program for all **stakeholders** (providers, employers, customers, patients, and payers who may have an interest in, and seek to influence, the decisions and actions of an organization), and it has resulted in a complex, still-evolving array of structures and processes to deliver health care.

Managed care emphasizes delivery of a coordinated continuum of services across the care spectrum from wellness to death, using financial incentives to achieve cost efficiency. Figure 4-1 shows national health expenditures as a share of Gross Domestic Product, 1980-2008. The growth rate of expenditures is expected to result in a health share of GDP of 19.5% by 2017—nearly one-fifth of the economy, an estimated $4.3 trillion in health care spending (Keehan, et al., 2008).

Preapproval of care is required under managed care, and coverage is selective, effectively rationing care. Choice of practitioner or other provider and choice of site for care

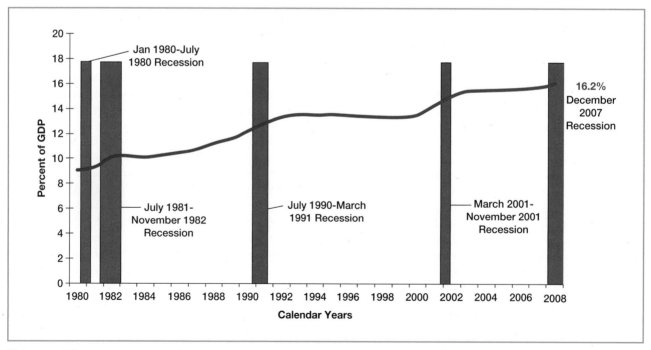

FIGURE 4-1 **National health expenditures as a share of gross domestic product, 1980–2008.** (*Source:* Centers for Medicare & Medicaid Services, Office of the Actuary, National Health Statistics Group, Department of Commerce, Bureau of Analysis and National Bureau of Ecvonomic Research, Inc.)

are restricted, which are additional methods of rationing care. An added incentive to rationing is a copayment for care that must be paid by the patient at the time that care is received. Despite industry assurances that care is not rationed, rationing is the undergirding concept of managed care. Managed care is not about providing health care; it is about being a for-profit brokerage business in which the managed care company acts as an agent who negotiates the contract about how the provision of health care will be accomplished.

There are a variety of models of managed care companies. Included are staff, group, network, **preferred provider organization (PPO)**, point of service (POS), and mixed, each having its own unique structure and risk arrangements. The most common form is the PPO. A PPO generally consists of a hospital and a number of practitioner providers. The PPO contracts with health care providers (both practitioners and hospitals) and payers (self-insured employers, insurance companies, or managed care organizations) to provide health care services to a defined population for predetermined fixed fees. Discount rates may be negotiated with the providers in return for expedited claims payment and a somewhat predictable market share. In the PPO model, patients have a choice of using PPO or non-PPO providers; however, financial incentives are built in to encourage utilization of PPO providers.

Nongovernmental health insurance has predominantly been accessed through employment. Employers provide coverage to employees as a benefit for working for their company. Therefore, the employer chooses the coverage with cost in mind and negotiates an acceptable package of benefits on behalf of the employees. If the employer offers a selection of benefit packages, the employee may choose the package that is most suitable. The range of available packages has narrowed to being nearly exclusively HMOs.

Managed care has become the focus for the anger of many health care providers and much of American society. Because care decisions are driven, in significant part, by the care options for which insurance coverage will pay rather than by the free choice of the patient in consultation with a professional provider, feelings of distrust and anger about necessary compromises in care are often strongly held by both patients and providers. As the public becomes more knowledgeable, it also becomes more demanding.

To ease the pressure from practitioners, patients, consumer advocates, and employers, many HMO programs recently dropped the requirement for managed care preapproval prior to hospitalization or consultation with a specialist. In response to the demand for greater choice of provider, care plans were adapted to permit those who can afford to pay higher premiums and copayments to have broader choices. Efforts to salvage the reputation of managed care have spawned another name—*coordinated care* is replacing the term *managed care* to better describe a system of mutual decision making among insurers, providers, and patients. In 2001, a patient's bill of rights was introduced into Congress to, among other things, allow patients to sue their coordinated care providers.

Such changes diminished the insurer's clout and with it the ability to contain costs. The news media reported that nationally renowned health care economist Uwe Reinhardt of Princeton University stated that "health plans can no longer bully and threaten the providers of care." Dr. David Lawrence, chief executive of the Kaiser Foundation Health Plan and Hospitals, noted, "When one uses financial tools to try to change the delivery of care, A, they are not very powerful, and B, they make people mad."

# UNIVERSAL HEALTH CARE

Universal health systems provide health care to all the citizens in a community, district, or nation, regardless of the person's ability to pay. Health care that uses a lot of high technology or that does a lot of elective surgery—e.g., plastic surgery—is usually not covered.

As noted in Chapter 2, many industrialized countries around the world provide a basic health care plan for all citizens. The differences in the health care plans depend on the type of priorities set by the citizens and the nation. Oregon has had a prioritized medical care services list for Medicaid recipients since 1989 (http://www.oregon.gov/DHS/healthplan/priorlist/main.shtml#overview, accessed Dec. 28, 2010).

Health care programs in Sweden and the United Kingdom are funded from income taxes and selected other taxes. Germany and the Netherlands rely on payroll taxes for funding. Canada's health care is funded through taxes and covers most patient needs, with a few exceptions—e.g., dental care, eye care, prescription drugs. Canada provides the same level of care to all of its citizens, has lower overall costs and much lower drug prices, and supplies education funds for health professionals. The U.S. system has shorter waits for seeing providers; allows greater choice in provider selection; provides higher wages and salaries for health professionals; and invests more in technology, research, and development.

Canada has better outcomes in infant mortality and life expectancy than does the United States but pays less than half the amount for health care. See Table 4-1. Direct comparisons of health statistics across nations are complex. The Commonwealth Fund, in its annual survey, "Mirror, Mirror on the Wall," compares the performance of the health care systems in Australia, New Zealand, the United Kingdom, Germany, Canada, and the United States. Its 2007 study found that although the U.S. system is the most expensive, it consistently underperforms compared to the other countries. A major difference between the United States and the other countries in the study is that the United States is the only country without universal

## TABLE 4-1

| COUNTRY | LIFE EXPECTANCY | INFANT MORTALITY RATE | PHYSICIANS PER 1,000 PEOPLE | NURSES PER 1,000 PEOPLE | PER CAPITA EXPENDITURE ON HEALTH (USD) | HEALTH CARE COSTS AS A % OF GDP | % OF GOVERNMENT REVENUE SPENT ON HEALTH | % OF HEALTH COSTS PAID BY GOVERNMENT |
|---|---|---|---|---|---|---|---|---|
| Australia | 81.4 | 4.2 | 2.8 | 9.7 | 3,137 | 8.7 | 17.7 | 67.7 |
| Canada | 80.7 | 5.0 | 2.2 | 9.0 | 3,895 | 10.1 | 16.7 | 69.8 |
| France | 81.0 | 4.0 | 3.4 | 7.7 | 3,601 | 11.0 | 14.2 | 79.0 |
| Germany | 79.8 | 3.8 | 3.5 | 9.9 | 3,588 | 10.4 | 17.6 | 76.9 |
| Japan | 82.6 | 2.6 | 2.1 | 9.4 | 2,581 | 8.1 | 16.8 | 81.3 |
| Norway | 80.0 | 3.0 | 3.8 | 16.2 | 5,910 | 9.0 | 17.9 | 83.6 |
| Sweden | 81.0 | 2.5 | 3.6 | 10.8 | 3,323 | 9.2 | 13.6 | 81.7 |
| UK | 79.1 | 4.8 | 2.5 | 10.0 | 2,992 | 8.4 | 15.8 | 81.7 |
| USA | 78.1 | 6.7 | 2.4 | 10.6 | 7,290 | 16.0 | 18.5 | 45.4 |

**Sources:** Life Expectancy vs. Health Care Spending in 2007 for OECD Countries, compiled with information from, http://www.oecd.org./home; OECD Health Data 2010—Frequently Requested Data, available at http://www.oecd.org/document/16/0,3343,en_2649_34631_2085200_1_1_1,00.html; OECD Health Data 2010—Country Notes and press releases, available at http://www.oecd.org/document/46/0,3343,en_2649_34631_34971438_1_1_1,00.html; Organisation for Economic Co-operation and Development available at http://en.wikipedia.org/wiki/Organisation_for_Economic_Co-operation_and_Development.

health care. Especially note the poor performance of the United States on infant mortality, per capita expenditure on health, and life expectancy. Many countries in the world face similar challenges of aging populations with chronic disease, the need to ration costly technology, and severe budget shortfalls.

With U.S. life expectancy at 78.1 years in 2007, the benefits reaped in the U.S. health care system seem to be those of enhanced quality of life rather than enhanced longevity. Quality of life must be defined within the specific cultures and value systems of each country. In the United States, qualities of life highly valued by Americans include prompt access to diagnostic and treatment services, even when health problems are not life threatening; ready availability of cutting-edge technology and pharmaceuticals; the ability to choose among health care practitioners and sites for care; and participation in health care decisions. All these contribute to the cost of health care. Is that enough to justify spending more than any other country in the world and nearly twice as much as European countries with similar health and financial circumstances?

## The Joint Commission (JC)

If the health care industry were an airline, Ann Scott Blouin, RN, PhD, wouldn't fly. In a 2010 article in the *Journal of Health Care Quality*, Blouin—executive vice president in the Division of Accreditation and Certification Operations at The Joint Commission, the highest-ranking nurse on the JC, and the first nurse to hold the position—says that she would like to see health care organizations have safety records similar to those of airlines, aircraft carriers, or nuclear plants. We have a lot more work to do to get to that standard. The Joint Commission evaluates and accredits more than 15,000 health care organizations and programs in the country. It is a private, nonprofit organization, although organizations must have the JC's accreditation to receive funding from the Centers for Medicare & Medicaid Services. Some states and insurers also require JC accreditation.

## THE INSTITUTE FOR HEALTHCARE IMPROVEMENT

The Institute for Healthcare Improvement (IHI) is a not-for-profit organization leading the effort for the improvement of health care throughout the world (www. ihi.org). IHI developed a campaign to reduce morbidity and mortality in the United States through the use of six health care interventions designed to save 100,000 lives. These six IHI interventions include the following: deploy rapid-response teams, improve care of patients with acute myocardial infarction, prevent adverse drug events through the use of medication reconciliation, and prevent central line infections, surgical site infections, and

ventilator-associated pneumonia. Several improvement stories report cost savings, decreased morbidity, better patient satisfaction, and several other areas of health care improvement in 2010 (http://www.ihi.org/IHI/Results/ImprovementStories/accessed February 27, 2011).

## Central Line Bundle

As part of the approaches for prevention of central line infections, several interventions have been developed (Institute for Healthcare Improvement, 2007). These interventions include:

1. Hand hygiene, including the following:
   - Before and after palpating catheter insertion site
   - Before and after inserting, replacing, accessing, repairing, or dressing an intravascular catheter
   - When hands are soiled
   - Before and after invasive procedures
   - Between patients
   - Before donning and after removing gloves
   - After using the bathroom
2. Maximal barrier precautions when inserting central lines, including the following:
   - Cap
   - Mask and sterile gown
   - Sterile gloves
   - Covering patient with large sterile drape
3. Chlorhexidine skin antisepsis
4. Optimal catheter site selection
   - Whenever possible and not contraindicated, the subclavian line site is preferred over the jugular and femoral sites for sterile catheters in adult patients
5. Daily review of central line necessity, with prompt removal of unnecessary lines

## Rapid-Response Team Deployment to Prevent Failure to Rescue

**Failure to Rescue** describes the clinician's inability to save a patient's life when the patient experiences complications. Rapid-response teams have been developed to rescue the patient by mobilizing hospital resources quickly, including bringing trained nursing and medical practitioners to the bedside when a patient's condition deteriorates. Failure to rescue often stems from having a lack of sufficiently trained clinicians who have too little time to maintain close surveillance. Two examples of failure to rescue include the following:

- Nursing or medical surveillance systems fail to act in a timely manner upon signs of a complication, resulting in a missed opportunity or seriously delayed rescue effort.
- Necessary supplies and equipment are not ready when a patient presents with a problem.

## PAY-FOR-PERFORMANCE PROJECTS

Medicare Pay-for-Performance (P4P) projects are an emerging movement in health care. Providers under this P4P arrangement are rewarded for meeting preestablished targets for delivery of health care services. This is a fundamental change from fee-for-service payment. P4P rewards physicians, hospitals, and other health care providers for meeting certain performance measures for quality and efficiency—e.g., clinical and patient satisfaction outcomes and overall costs. Disincentives, such as eliminating payments for negative consequences of care (e.g., medical errors, hospital readmissions, and increased costs), have also been proposed. Recently, as 10 medical groups concluded year four of a CMS physician group practice pay-for-performance demonstration project, CMS announced that the participants had saved $38.7 million in Medicare expenditures. As a result of this performance, CMS paid $31.7 million in performance payments to five physician groups, Geisinger, Marshfield, the Dartmouth-Hitchock Clinic, the University of Michigan Faculty Group Practice and St. John's Health System in Springfield, Missouri (Shinkman, 2010).

## REAL WORLD INTERVIEW

During my many experiences over the past years with the Ontario, Canada, health care system and the Ontario Health Insurance Program (OHIP), I have had annual visits to my general practitioner (GP) with referrals to specialists for various tests, including X-rays, blood work, and hospital stays as needed. All my medical expenses while in these doctors' care is covered completely by the OHIP funds. Medications, glasses, and dental work are not covered unless one's gross annual income is a very minimal amount, in which case some medications, dental, and hospitalizations are covered.

My husband was a diabetic, and he had a stroke 20 years ago at the age of 53. His GP met us at the hospital and assessed him through triage. My husband lost the use of his right arm and leg and his ability to comprehend language. The supreme care and compassion of his doctors and nurses helped my husband and family cope with the severity of this horrific disease. We paid for an upgrade to a semiprivate room. All other expenses while he was in the hospital were covered by OHIP. His rehabilitation began two weeks after his stroke, and he was transferred to the rehabilitation center, where he stayed for three months. All expenses while he was there were funded by OHIP.

After his acute illness, my husband saw a specialist every three months to monitor his blood sugar levels. He saw his GP every three months for his hypertension and saw his optometrist every year. He went for speech therapy once a week, to physical therapy three times a week, and to a nutritionist twice a year. This was all covered by OHIP.

He had several episodes of congestive heart failure through the years. Each time, I called 911, and the fire department came within minutes, administered oxygen, and inquired about his medical problems and medications while waiting for the ambulance medics to arrive. The medics took his vitals, gave him nitro, and called the admitting hospital.

We have four hospitals in Hamilton, Ontario, where I live. In the emergency department (ED), my husband was always processed immediately. When a heart specialist was called in, it took about an hour for him to arrive. Then the heart specialist would decide whether my husband was to be admitted to Intensive Care or a ward. There were times when a bed was not available until the next day, and we had to wait in the ED.

All emergencies in our ED are taken care of fairly promptly, the most severe first. Elective surgery such as knee or hip replacements, eye cataracts, etc., require a longer waiting period—sometimes up to six months. Hospital and emergency patients are on the priority list for magnetic resonance imaging (MRI) and CT scans. Other patients may have to wait for a couple of months for these tests. I also have the option, if I can afford it, of visiting the Buffalo, New York-area clinics for MRIs, etc., if it is not an emergency and I want to do it quicker. I would have to pay for these visits and tests out of pocket. All in all, I have been fairly pleased with our Canadian health system.

**Flo Paradisi**
Hamilton, Ontario, Canada

# FUTURE PERSPECTIVE ON COST OF HEALTH CARE

Futurists are in demand to guide health care to organize for success. Health care providers have been thrashing in chaos for many years, reinventing their structures and processes, right-sizing their enterprises, outsourcing to better focus on their core business, and merging to share scarce or expensive resources. Although there have been some short-term cost savings, and the rise in health care spending has slowed, several evolving trends keep the overall cost growing.

Highly complex and expensive technology, including microsurgery, continues to develop. Diseases that require expensive or long-term treatment, such as type 2 diabetes, obesity, AIDS, and other infectious diseases, continue to be a problem. With the eradication or successful management of selected diseases such as tuberculosis, populations are surviving longer. With that lengthened survival comes debilitating diseases of aging.

We look to futurists to help us make decisions about what services we ought to be prepared to provide. Will 45 million uninsured Americans find comprehensive primary care that is accessible? Is competition between health care providers a positive influence on health care delivery? Will a few strategically located hospitals provide only an intensive level of care, while acute care is managed on an ambulatory basis without invasive procedures? Will preventive, primary, and restorative care be the purview of advanced nurse practitioners practicing at community sites? Will unselfish concern for the welfare of others no longer be the basis for caring careers? Will nurses be "ordered to care" in a society that no longer values caring? Will the United States adopt a health care delivery system similar to those in European countries and Canada?

# THE COST EQUATION: MONEY = MISSION = MONEY

The mission statement of any health care business describes the purpose for existence of the business and the rationale that justifies that existence. The statement directs decision making about what is or is not within the purview of the business. The vision statement is a logical extension of the mission into the future, establishing long-range goals for the business. After the vision is established and the business can articulate where it wants to go, a strategic plan for how to achieve the vision, or how to reach the goals, is developed. There must be cohesion and consistency across the mission, vision, and strategic plan for the business to successfully achieve its mission. There must also be money, for without it no mission can be accomplished.

## COHESION AND CONSISTENCY OF THE BUSINESS

The question, then, is what is the cost of achieving the mission? Part of the cost may be in providing health care services that are not directly related to the mission, in the interest of political viability. Consider the Veterans Affairs (VA) health care system, established to provide intensive, acute, and rehabilitative care to military veterans (not active or reserve duty personnel or family members) who meet complex care-eligibility criteria for service-related illness and injury and/or indigency. Among the 72 VA medical centers in the United States and Puerto Rico are five Blind Centers that provide unique mechanical aids, training for activities of daily living, and job training. When a center still has beds available after admitting all veteran applicants who meet the VA eligibility criteria, it may admit others using more-lenient standards. This practice may result in inconsistency with the core mission, but it can maintain the program at capacity and better assure its ongoing viability. A great deal of soul-searching regularly occurs about whether the political obligation to provide this unique and valued health care service justifies the cost of existence—and its occasional extension beyond the VA mission—in a cost-sensitive health care environment.

A more familiar example of providing a costly service that is inconsistent with the mission of the health care business may be the small maternity service of a remote region that claims few women of childbearing age as inhabitants. Without the service, the women would need to relocate to a distant facility miles away for the duration of the pregnancy. A sense of commitment to the well-being of the surrounding community is viewed as justification for keeping the service open. Similarly, a commitment to charity as a component of their mission drives religious organizations that are otherwise astute businesses to provide free care to the poor. However, modern health care organizations have limited tolerance for diversification from their core business.

Refer to Chapter 10 in this text for more in-depth discussion of mission-vision-strategic plan cohesion.

## BUSINESS PROFIT

Revenue (income) minus cost (expense) equals profit. Profit is not restricted to for-profit businesses. Profit is not a dirty word. All businesses must realize a profit to remain in business. In for-profit businesses, a portion of the profit is distributed to stockholders in appreciation for their investing in the business, and the remainder is used to maintain and grow the organization. In nonprofit businesses, there are no stockholders to share the profit, so all of it is fed back into the business for maintenance and growth.

Not-for-profit organizations desiring a purer image than the term *profit* engenders refer to their profit as contribution to **margin**, with the rule of thumb being to secure

4% to 5% of the total budget as profit or margin. Mission and margin are strategically and operationally linked by the reality that resources are required to carry out the organization's strategic plan and achieve its mission. Without margin, or with limited margin, there would be a lack of money to replace worn-out equipment, to establish new services or enlarge existing services in response to changing community needs for health care, to purchase state-of-the-art technology, to maintain existing buildings or undertake new construction, and to replace heating and lighting systems. Failure to maintain such infrastructure can impair the organization's ability to be competitive, resulting in failure to meet its mission and eventual organizational failure. Profit is the elasticity that accommodates improvements in patient and staff education, recruitment and hiring of expert staff, and special programming that yields personal professional growth. Profit is a critical requirement for doing business. A truism of business: no margin, no mission.

Achieving and sustaining margin is a constant challenge for health care enterprises. The 2010 Patient Protection and Affordable Care Act will be implemented with unique pricing and budget structures in each state. Each state will have its own philosophy of how to generate health care dollars for citizens who don't have private insurance. New Jersey's governor once proposed a "sick tax" on acute care enterprises in his state at the rate of $1,400 per patient bed per month to help bolster the state's 2007 budget. In his letter to the editor in the June 2006 issue of *Hospitals and Health Networks*, Gary Carter, CEO and president of the New Jersey Hospital Association, walked through the ripple effect of such a heavy additional tax burden. Mr. Carter predicted that New Jersey's health care enterprises would be unable to maintain their buildings and infrastructure or to purchase the supplies, equipment, and personnel required to obtain expensive new technology required for exemplary care. He further predicted that the organizations would probably look first to cutting personnel costs by adjusting benefit packages, shifting more health care cost to employees and retirees. In addition, because hospitals in New Jersey are required by law to provide care to anyone who comes through their doors, some enterprises might not survive financially, depriving some populations of health care services (Carter, 2006).

## FUNDAMENTAL COSTS

There are many ways to examine or classify the costs of care. One fundamental method is to view costs as direct or indirect. A **direct cost** is directly related to patient care within a manager's unit, such as the cost of nurses' wages and the cost of patient care supplies. An **indirect cost** is not explicitly related to care within a manager's unit but is necessary to support care. The costs of electricity, heat, air conditioning, and maintenance of the facility are all considered indirect.

Those same costs may also be considered either fixed or variable. These distinctions are somewhat artificial and are related to the volume of services that are provided. A **fixed cost** is one that exists irrespective of the number of patients for whom care is provided, as shown in Figure 4-2. Examples of this are the cost of the rent or the monthly

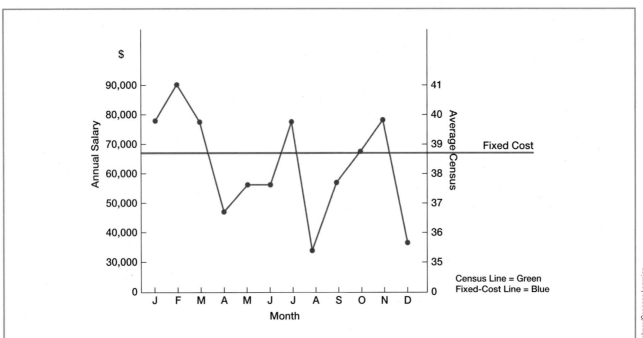

**FIGURE 4-2** Fixed salary cost of a salaried employee does not vary by patient volume. (Salary is about $70,000 annually regardless of census.)

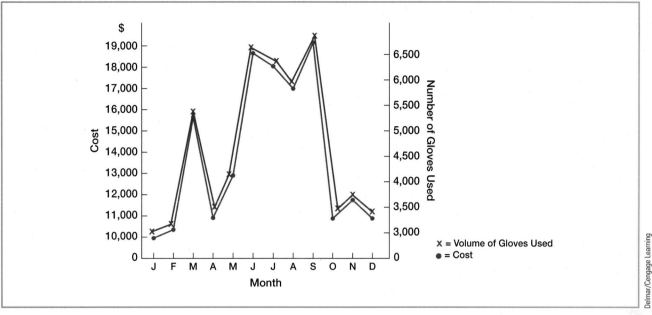

**FIGURE 4-3** Variable cost of latex-free gloves varies by volume used.

mortgage for the space in which the care is provided and the cost of salaried (but not hourly) wage earners such as the nurse manager or nurse administrator. These costs would be the same whether one or 1,000 patients were served. A **variable cost**, on the other hand, varies with volume and will increase or decrease depending on the number of patients, as shown in Figure 4-3. Medical supplies, laundry for the linens used in patient care, and patient meals are variable costs that increase or decrease in proportion to the number of patients served.

Some costs are step variable; that is, they vary with volume, but not smoothly. The key to step costs is that they are fixed over volume intervals but vary within the relevant range. For instance, a fixed number of nurses may be able to care for 11 to 21 patients. However, as depicted in Figure 4-4, when even one additional patient beyond 21 requires care, additional nurses are required.

## COST ANALYSIS

There is an old saying that numbers don't lie. There is also an old saying that statistics can be manipulated to show whatever is desired. Both sayings are true, and therein is the challenge, the frustration, and occasionally the glory of clinical cost management. The health care industry commonly embraces the position that the past is a prologue for forecasting its future. The ability to predict the behavior of cost in the future based on its past behavior, then, is considered requisite to successful cost management and thereby the achievement of mission and vision. A **budget** is a plan that provides formal quantitative expression for

acquiring and distributing funds over the ensuing time period (generally one year). The budget is based on what is known about how much was spent in the past and how that will inevitably change in the coming year. A cost prediction is simply a tool for developing a budget. The three most common methods of cost prediction are high-low cost estimation, regression analysis, and break-even analysis.

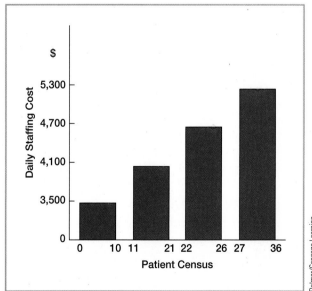

**FIGURE 4-4** Step variable costs are fixed within a range and then increase when volume exceeds the upper end of the range.

## High-Low Cost Estimation

This is not the tool to choose if a sophisticated, statistically rigorous prediction is needed, but it does surpass just guessing. Examining both fixed and variable cost information from the most recent five years for each category of expense provides a "good enough" cost projection for many items that remain relatively constant in volume of consumption and cost. Both fixed and variable dollars need to be adjusted upward to account for inflation, and, the total projected cost needs to be adjusted upward to cover anticipated or other wage increases and to cover bad debt when services are rendered but payment does not occur.

The following is an example that clarifies the math:

Highest wage cost = $500,000
Lowest wage cost = $300,500
Difference = $199,500

Highest number of patient days = 9,000
Lowest number of patient days = 7,500
Difference = 1,500

Difference in cost ($199,500)/Difference in volume (1,500) = $133 per patient day variable cost

Variable cost ($133) × Lowest number of -patient days (7,500) = $997,500 = Total annual variable cost

Total annual labor cost for the unit from this year's fiscal department records ($1,100,000) − Total annual variable cost ($997,500) = $102,500 Fixed cost

If 1,000 additional patient days are anticipated in the ensuing year, there would be an additional cost as follows:

$133 per patient day variable
cost × 1,000 additional days          = $133,000

+ Fixed cost (regardless of
the number of patient days)          = $102,500

Total additional cost                     = $235,500

Thus, the total cost for this unit next year, using high-low cost estimation, will be this year's cost of $1,100,000 plus $235,500 in new costs, or $1,335,500.

## Regression Analysis

A more precise prediction of cost can be realized using the statistical tool regression analysis. Whereas the high-low method relies on only two data points—the highest and the lowest—for historical cost behavior, regression analysis examines all available past cost information over a specific time period. It assumes that there is only one dependent variable (cost), with only one independent

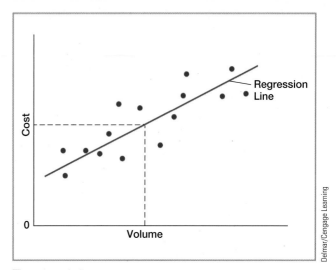

**FIGURE 4-5** Regression analysis.

variable (volume) causing change in that dependent variable. It also assumes that, mathematically, cost behavior can be shown in a linear fashion by drawing a straight line through a scatter diagram of all fixed and variable costs at all volumes of use. When all cost information is plotted on a vertical axis and all volume information is plotted on the horizontal axis, a scatter diagram results. The straight line through the scatter diagram that best approximates all the points is used to predict cost at a specific volume of use. Selecting a volume on the horizontal axis and examining where a vertical line from that point intersects the straight regression line, then moving horizontally to the vertical cost axis, provides the cost prediction for that specific volume as shown by the dotted lines in the scatter diagram in Figure 4-5. The analysis is carried out for each item for which cost needs to be predicted.

## Break-Even Analysis

Because accruing profit to enhance the quality of services provided and to achieve optimal competitive market position is the business goal of health care organizations, projecting whether and when profitability will be achieved is necessary for both proposed and well-established programs and services. The third basic cost analysis tool is break-even analysis. It assists the provider in predicting the volume of services that must be provided (and for which payment must be received) for the cost of providing the services to be equally matched by the payment received, yielding neither a profit nor a loss. The formula for computing a break-even analysis (Finkler & Kovner, 2000) is as follows:

Volume of procedures = Fixed cost/Payment − Variable cost

A common application of the break-even analysis is the determination of how many procedures must be completed using a new piece of equipment before the payments for the procedure cover the cost of the equipment and other resources consumed while doing the procedure (the **break-even point** at which income and expenses are equal), with all additional procedures generating profit. To make the use of the analysis more clear, consider that the purchase of a new piece of radiology equipment is proposed. The underlying question is, would the purchase generate a profit or a loss, or would the organization just break even?

Applying the break-even analysis formula, if the new procedure costs $50,000, the wages of the technician operating the equipment 1 hour for each use is $20 per use. Therefore, the payment for each procedure is $50, so it would take 1,667 procedures to pay for the equipment and technician wages before a profit would begin to accrue.

$$\text{Volume} = \$50,000 / (\$50 - \$20) = 1,667$$

If the payment for the procedure were $100, a profit would begin to accrue after only 625 procedures.

$$\text{Volume} = \$50,000 / (\$100 - \$20) = 625$$

Thus, a projection can be made about how long it will take before the new equipment would be a profitable venture, guided by the decision about how much cost the purchaser of the procedure is likely to tolerate.

## Critical Thinking 4-1

Reducing costs and improving quality at the same time is a very realistic goal. Frequently, quality problems are very costly. Examples of this are found in Kliger's (2010) article on Nurse-driven programs to improve patient outcomes in the Journal of Nursing Administration. The article describes a Transforming Care at the Bedside (TCAB) project funded by the Robert Wood Johnson Foundation and the Institute for Healthcare Improvement. The main focus of TCAB was to implement staff nurse identified solutions to quality or safety concerns. The identified solutions were adopted for two weeks and using a rapid cycle change process, the solutions were evaluated for effectiveness. If the staff nurse identified solution worked, it was adopted. If not, it was scrapped. One example given was the staff nurse identified the need for a shelf to place items outside the room in order to facilitate handwashing. With the simple added shelf, handwashing improved and the innovation was a success.

Identify some ways in which nurses can participate in reducing waste, influence best practices, and improve quality in these areas. Is keeping current with the literature in your specialty necessary to be effective in doing this?

## Critical Thinking 4-2

Nursing is a collaborative profession existing in a complex health care system. At one time it was influenced by three factors: cost, access, and quality. Today there are five influencing factors: cost, access, cost, quality, and cost (John M. Lantz, RN, PhD, Dean and Professor, School of Nursing, University of San Francisco).

Is this an accurate description of the things that influence contemporary nursing practice? What do these five influencing factors suggest may be an organization's focus as you interview for a staff nurse position?

You notice that a colleague frequently does not record patient charge items for elderly patients. When you inquire about it, you are told that your colleague feels sorry for those on fixed incomes and wants to save them money. Who pays when your colleague does this?

# DIAGNOSTIC, THERAPEUTIC, AND INFORMATION TECHNOLOGY COST

A common perception held by both society and the health care industry is that payroll costs constitute the largest expense item in organizational budgets and that the most expensive health care personnel are registered nurses (RNs). Therein lies the rush to downsize the number of RNs when a determination is made that costs need to be cut and better managed. Aiken et al. (2010) describe a clear link between nurse-patient ratios and the size of the nurse's patient assignment and how well the nurse can do the job. With more patients to care for and higher patient acuity, patient morbidity and mortality increases. With mandated ratios for nursing in California, morbidity and mortality has decreased.

Close examination of the entire hospital budget is likely to reveal that although the nursing payroll is the most expensive payroll item and the most expensive operating budget item, the most expensive item on the total budget is diagnostic, therapeutic, and information technology. This technology is required to meet society's demand for state-of-the-art care; professionals' demand for quicker, keener ways to work; and the organization's need to maintain a competitive business edge. Such items characteristically appear in the capital budget because of their considerable cost. With cost management generally focused on the operating budget, the cost of diagnostic, therapeutic, and information technology is often conveniently overlooked because it is deemed "strategic" and therefore untouchable during cost-cutting initiatives. Moreover, despite a rise in nursing payroll costs over the past 20 years, that rise has proportionately been considerably more gradual than that of diagnostic, therapeutic, and information technology as well as total hospital costs, resulting in a widening gap in costs that suggests that nursing is a cost bargain.

The high cost of technology in particular has been recognized by regulatory agencies for many years. In pursuit of cost control in hospitals, the states independently established laws more than 30 years ago creating Certificate of Need (CON) agencies to oversee, regulate, and approve major technology and construction expenditures. A secondary goal was to ensure equitable distribution of and access to high-end technology across the states. The CON approach was not successful, because it focused only on hospitals and provided no incentives to change either practitioner or patient behavior. Hospitals were given spending limits, but there was no incentive for practitioners to change their practice, so they didn't. Without incentives, patients' expectations and demands for care also remained unchanged.

Recently, managed care programs have exerted oversight of the use of complex, expensive technology by requiring justification and approval prior to its use for payment to occur. Only when less costly approaches had been exhausted would the possibility of using highly specialized technology be considered. Diagnostically, movement has been away from "fishing expeditions" such as ordering the comprehensive metabolic panel (CMP) and routine chest X-rays on all patients and toward completing only that which is minimally required to reach a reasonable diagnosis based on overt signs and symptoms. The question that arises is whether assurance of correct diagnosis has been critically compromised or whether sharper clinical skill has resulted.

## OREGON HEALTH PLAN

In 1993, the Oregon Health Plan, the state's Medicaid program, introduced rationing. It was intended to make health care more available to the working poor while rationing benefits. The system involves a treatment schedule that lists 649 potentially covered procedures. The state pegs the number of procedures the state will cover to the available funds. Patients requiring procedures above the cutoff line are out of luck. As of October 2010, only the first 502 treatments were covered The Oregon Health Plan also rations covered procedures under certain circumstances. Chemotherapy, for instance, is not provided if it is deemed to have a 5 percent or less chance of extending the patient's life for five years, meaning that a patient whose life might be extended a year or two with chemotherapy may not receive it (Oregonian PolitiFact Oregon, 2011). Neonates who are born with a weight of less than 500 grams are not provided intensive care, because the outcomes are very poor with such extreme prematurity. While controversial, many believe that citizens, patients, and health care providers must develop clear criteria for how health care dollars should be spent, as money for health care is not limitless. End-of-life care is the most expensive, and potentially the least effective if patients languish on long-term life support.

## MEDICARE MEDICATION COST

Medicare Part D was adopted in 2005 to provide a prescription drug benefit to enrollees with low incomes and high drug costs. Rules for membership and eligibility and a gap in coverage that has come to be called the "doughnut hole" have Americans confused, appalled, disappointed, and, in some cases, worse off than they were without the plan. The June 2006 AARP Bulletin (a publication of the former American Association of Retired People, now AARP) provided the following explanation of the initial doughnut hole:

Medicare drug coverage is generally divided into three phases:

- In the first phase, depending on your plan, you may pay a deductible and about 25% of your drug costs.
- In the second phase, there is a gap in coverage, commonly called the doughnut hole, when most people must pay 100% of their drug costs out of their own pocket.
- In the third phase, called catastrophic coverage, you pay about 5% of your drug costs.

You will be in the donut hole once you and your Medicare Part D drug plan have spent $2,840 for covered drugs.

The Affordable Care Act signed into law on March 23, 2010 (available at http://healthinsurance.about.com/od/reform/a/Affordable-Care-Act-What-You-Should-Know-About-The-Affordable-Care-Act.htm) makes several changes to Medicare Part D to reduce out-of-pocket costs when you reach the donut hole, including:

- Beginning in 2011, if you reach the donut hole, you will be given a 50% discount on the total cost of brand name drugs while in the donut hole gap. Medicare will phase in additional discounts on the cost of both brand name and generic drugs beginning in 2012. The donut hole continues until your total out-of-pocket cost reaches $4,550.
- By 2020, the above changes will effectively close the donut hole, and, rather than paying 100% of medication costs while in the donut hole, patient responsibility will be 25% of medication costs.

The annual out-of-pocket spending amount includes the yearly deductible, copayment, and coinsurance amounts. See Table 4-2.

## TABLE 4-2 Donut Hole Examples

### Charley Smith

Charley Smith takes three medications to treat his high blood pressure and high cholesterol. These medications will cost him about $1,200 in 2011. Charley is switching to a Medicare prescription drug plan that has a low premium and offers the standard Medicare drug benefit, including a deductible and no drug coverage in the donut hole. This is what his prescription medications will cost in the plan he has selected:

- Charley will pay a deductible of $310 ($1,200 − $310 = $890).
- He will then pay 25% out of pocket for the remaining $890 cost of his medications ($890 × 25% = $223).
- Because Charley did not reach the $2,840 initial coverage limit, he will not enter the donut hole.
- Charley's total estimated annual out-of-pocket prescription drug cost with his Medicare Part D plan will be $310 + $223 = $533 (plus his monthly premiums for the Medicare Part D plan).

### Mary Jones

Mary Jones takes three brand name medications to treat her type 2 diabetes, high blood pressure, and high cholesterol. These medications will cost her about $3,800 in 2011. Mary is joining a Medicare prescription drug plan that offers the standard Medicare drug benefit, including a deductible and no coverage for generic medications in the donut hole. This is what her prescription medications will cost in the plan she has selected:

- Mary will pay a deductible of $310 ($3,800 − $310 = $3,490).
- She will then pay 25% out of pocket for the cost of her medications for the next $2,530, until she reaches the donut hole coverage gap ($2,530 × 25% = $633).
- Mary reached $2,840 in drug spending ($310 + $2,530 = $2,840), and she will enter the donut hole.
- Prior to 2011, Mary would have been responsible for 100% of the remaining cost of $960 ($3,800 − $2,840 = $960). However, because all of Mary's medications are brand names, she will only have to pay about 50% of the drug costs while in the donut hole ($960/2 = $480).
- Mary's total estimated annual out-of-pocket prescription drug cost with her Medicare Part D plan will be $310 + $633 + $480 = $1,423, plus her monthly premiums for the Medicare Part D plan.

# COST OF MEDICATIONS AND HEALTH CARE EXECUTIVE SALARIES

Medications are a significant item in the overall budget. A recent health care Web site visit identified these drug charges. See Table 4-3. Also note the health care compensations shown in Table 4-4.

# NURSING COST

Hospital costs are different than patient charges in hospitals. Hospital costs are the "wholesale" version of how much hospitals pay for supplies, staffing, new equipment, and maintaining the physical plant of the hospital. Patient charges are what is billed to patients once they have received care—e.g., the hospital cost of acetaminophen may be 10 cents for one 500 mg tablet. The patient charge may be $20.00 for that one tablet because of the indirect cost of staff processing and administering the medication.

Presently, nursing in most hospitals is not a separate cost center for patient charges; nursing is included in overall hospital costs. Patient charges for one day in a hospital room may be $1,000, and a certain percentage of that patient charge is due to the cost of nursing care.

Fiscally, most organizations view nursing as a cost center that does not independently generate revenue. Although some deviation from that fiscal philosophy may occur when nurse practitioners are permitted by law to bill directly for their unique professional services, the cost of providing nursing care (wages, benefits, supplies

## TABLE 4-3  Cost of Selected Medications

| | |
|---|---|
| Acyclovir-90 $43.49 | Lisinopril-90 $31.99 |
| Advair Diskus-180 $555.95 | Lorazepam-90 $43.99 |
| Lipitor-90 $237.99 | Nexium-90 $429.97 |

**Source:** Compare drug prices. (2009). Available at www.drugstore.com/

## TABLE 4-4  Health Care Compensation

### ANNUAL SALARY

| | |
|---|---|
| Jeff Barbakow, Executive, Tenet Healthcare $35 million | Nurse Practitioner $82,513 |
| Miles White, Executive, Abbott Laboratories $30.4 million | Physical Therapist $63,117 |
| Richard Jay Kagan, Executive, Schering-Plough $24.4 million | Cardiologist $370,295 |
| Chief Executive Officer (CEO)–Non-MD $231,404 | Anesthesiologist $344,691 |
| Chief Executive Officer (CEO)–MD $275,123 | General Surgeon $327,902 |
| Director of Human Resources $82,690 | Family Medicine MD $185,740 |
| Medical Director $241,670 | Emergency Care MD $255,530 |
| Director of Nursing $82,138 | Dentist $165,599 |

**Source:** Useem, J. (2003, April 28). Have they no shame? *Fortune, 147*(8), 56–64; and 2007 American Medical Group Association. Compensation Data available at www.cejkasearch.com/compensation/amga_administrative_compennsation_survey.htm

and equipment, overhead) is commonly bundled into a catchall, room, or per diem cost that assumes that every patient consumes identical nursing resources each day. Such a view is not only antiquated, it is incorrect. Nursing care is not an identical product delivered in assembly-line fashion. It varies remarkably in intensity, in depth, and in breadth across patients, consistent with patients' unique, individual needs.

The cost of providing nursing care varies by patient acuity, type of surgical procedure, primary diagnosis, secondary diagnosis, etc. Depending on the type of patient, nurses and hospitals need to know the cost of providing nursing care per patient day; per acuity-adjusted patient day; per hour; per visit; per diagnosis-related group (DRG); or even per minute. Nursing costs can vary for the same patient during their hospital stay as the patient's needs change. A clear understanding of nursing costs is important so that nurses and hospitals have an improved understanding of the contribution that nursing makes to the organization as a whole. The information generated by improved cost information can be used to help nurses make effective decisions and to better control the cost of providing nursing services. Periodic nursing shortages make it even more important to understand the nursing resources needed by patients. Cost information is also useful for examining changes in the way nursing care is provided—e.g., the cost of primary nursing versus team nursing. When access to both nursing care and medical technology is needed, hospitalization is unquestionably appropriate. Consequently, the revenue generated from hospitalization is, in fact, payment, primarily for consumption of medical technology and nursing services, and should be recognized as such. The revenue generated from consumption of nursing resources should also be accurately assigned to each patient and charged at a rate consistent with the volume, acuity, and complexity of care consumed.

Ongoing efforts to measure and establish the cost of the diverse, yet related, components of nursing care are disappointing. That care includes direct hands-on care, teaching, and coordinating discharge, as well as documentation, consultation, critical problem solving and decision making, and supervision of multiple levels of workers. These same components contribute to the long-established cost of other practitioner and therapist services, yet the value, and thus the cost, of nursing care eludes measurement and agreement. If all nurses became a part of practice teams or groups as physical therapists do, nurses could contract out nursing services to hospitals and then charge per patient. This would prevent nursing charges from being bundled with the room charge and other hospital charges. Nurses would no longer be hospital employees; they would be contracted professionals just like physical therapists.

The efficiency and effectiveness of work flow, the way work is accomplished, affects nursing workload also. This may involve something as simple as the nurse making multiple trips back and forth to access supplies in an awkward physical configuration of the work environment. It may also involve something as complex as making innovative adjustments to a familiar work pattern to accommodate physical, cognitive, behavioral, or sociocultural challenges presented by a "difficult" patient.

## REAL WORLD INTERVIEW

The ongoing and escalating shortage of nurses and other health care professionals may be the major health care issue well into the twenty-first century and at least for the next 20 years. There must be valid and reliable mechanisms for calculating the requirement for nursing care time and intensity.

Models of care delivery that maximize the use of nursing assistive personnel (NAP) and provide for sufficient professional nursing expertise and time for both care delivery and supervision are a necessity. The number of nursing hours provided to each patient over a 24-hour period (nursing hours per patient day) that is commonly used to project staffing needs provides insufficient and inadequate information for planning and staffing decisions that can best assure quality clinical outcomes for patients. Mechanisms used to track or predict nursing workload must be able to differentiate between the care that requires professional nurses and the care that can safely and effectively be done by assistive personnel. Definitions and measurements of nursing workload must be standardized so that comparative data can be collected and analyzed.

**Sheila Haas, PhD, RN**
Dean and Professor
Loyola University, Chicago, Illinois

## EVIDENCE FROM THE LITERATURE

**Citation:** Aiken, L. H., Clarke, S. P., Cheung, R. B., Sloane, D. M., & Silber, J. H. (2003, Sept.). Educational levels of hospital nurses and surgical patient mortality. *Journal of American Medical Association, 290*(12), 1617–1623.

**Discussion:** Article discusses results of study done to determine the association between the educational levels of hospital RNs and the mortality of surgical patients. The study examined 168 adult acute care hospitals in Pennsylvania reporting a total of 232,342 surgical discharges to the Pennsylvania Health Care Cost Containment Council in 1999. The researchers also surveyed a random sample of 50% of hospital nurses who live in Pennsylvania and were registered with the Pennsylvania Board of Nursing. In all, 10,184 nurses (52% of nurses surveyed) responded.

According to the survey results, the average age of respondents was between 40 and 41 years, and between 30% and 31% of respondents had earned a BSN or higher degree. Hospital nurses who participated in the study had 14.2 years' nursing experience and a mean patient load of 5.7 per day. The researchers examined how the education of hospital nurses affected the death rates of surgical patients within 30 days of admission and those within 30 days of admission among patients who experienced complications. The study also took into consideration whether a board-certified surgeon performed the surgery. The types of surgeries examined included general surgery, orthopedic, and vascular procedures.

The study found that years of nursing experience don't predict a patient's outcome but that patients cared for in hospitals with a higher proportion of nurses holding a BSN degree or higher have a better chance of postsurgical survival. Specifically, the study stated that "a 10% increase in the proportion of nurses holding a bachelor's degree (in hospitals) was associated with a 5% decrease in both the likelihood of patients dying within 30 days of admission and the odds of failure to rescue." Failure to rescue was defined as "deaths in patients with serious complications." The researchers recognized two limitations to their study:

- The low (52%) response rate of the nurses surveyed
- The examination of hospitals from only one state

The researchers concluded that although these preliminary findings raise concerns over nurse education as it relates to patient outcomes, further study of nurses and hospitals nationwide would be required to make these results irrefutable.

**Implication for Practice:** This study offers powerful insights into the need for higher nursing education levels. The study bears repeating for validation.

## PATIENT CLASSIFICATION SYSTEM (PCS)

A tool used to identify nursing cost is the **patient classification system (PCS)**, a system for distinguishing among different patients based on their acuity, functional ability, or need for nursing care, in order to predict staffing needs and the cost of nursing care. Sicker patients requiring more nursing care are assigned higher acuity or PCS levels. Many hospitals have developed their own PCS system, and several commercial systems are widely used in hospitals throughout the United States. Patients with similar requirements for care are assigned to one of five progressively weighted categories of acuity from minimal (1) to maximal (5) acuity. The weights are the average number of hours of nursing care required by all of the

patients in each respective category; some patients classified as level 2 will require more or less care than level 2 calls for. However, if the system is functioning reasonably well, average patient resource consumption will match what is expected on the basis of the PCS. It would usually be expected that a level 2 patient will consume resources closer to the level 2 average than that of level 1 or level 3. (Finkler & McHugh, 2008).

Each work shift, nurses who are providing direct care and can therefore best judge the actual patient acuity, select the level of care each patient requires from a standard weighted menu of procedures. Patient acuity is often rated with a nursing-driven system yielding a number from 1 (low) to 5 (high) acuity. A formula (often institution specific) is then used to determine staffing. Besides the usual aspects of acuity, the formula may include considerations of psychosocial, education, family, and bereavement issues that are not included in usual staffing formulas. Other factors to consider in determining staffing level are the number of admissions, deaths, and discharges. When lengths of stay are short, more staffing is needed to manage the admissions. If there are many patients discharged alive, the discharge planning and family teaching burden is high. If there are many deaths in a given time period, this increases the staffing required.

Most units develop their own system of determining acuity. Frequently, this means a conversion factor is developed. Based on the kinds of factors described, an acuity level is ascribed to each patient (for example, 1–5). Then, a conversion factor for each acuity level is applied to determine the amount of staffing required. For example:

- (# patients) × (Acuity 1 conversion factor) = x hours nursing
- (# patients) × (Acuity 2 conversion factor) = x hours nursing
- (# patients) × (Acuity 3 conversion factor) = x hours nursing
- (# patients) × (Acuity 4 conversion factor) = x hours nursing
- (# patients) × (Acuity 5 conversion factor) = x hours nursing

The hours are totalled, and the number of FTE in nursing is determined.

Highest acuity usually means the patient requires 1:1 or 1:2 nursing care. Many patients today are acuity level 3 or 4. There are also other types of patient acuity scales. The higher the total weight of the required care, the higher the acuity. The higher the total acuity for all the patients assessed, the more nursing resources the PCS assigns. The more resources assigned, the higher the cost.

There are several shortcomings to a formula that measures only time. First, it cannot account for the presence (or absence) of knowledge, skill, or experience in the nurses providing the hours of care. Even the gross categories of RN and LPN cannot be taken into account. It cannot, therefore, account for the richness in clinical assessment and decision making that occurs when the same procedure or care is provided by an RN versus an LPN. Second, there is no ability to account for the atypical physical, cognitive, behavioral, or sociocultural challenges that consume above-average hours. Similarly, the way one patient manages pain, fear, or anxiety is seldom identical to the way another patient copes, and supporting one may require very different resources than supporting the other, identical acuity notwithstanding. In other words, a category 4 patient is not a category 4 patient is not a category 4 patient. Innumerable nursing hours are spent attempting to justify the actual nursing hours consumed when that number slides above the number of hours assigned by the PCS.

## EVIDENCE FROM THE LITERATURE

**Citation:** Hall, L. M., Pink, L., Lalonde, M., Murphy, G. T., O'Brien-Pallas, L., Laschinger, H. K., et al. (2006). Decision making for nurse staffing. *Policy Politics in Nursing Practice*, 7(4), 261–269.

**Discussion:** The effectiveness of methods for determining nurse staffing is unknown. Despite a great deal of interest, efforts conducted to date indicate that there is a lack of consensus on nurse staffing decision-making processes. This study explored nurse staffing decision-making processes, supports in place for nurses, nursing workload being experienced, and perceptions of nursing care and outcomes. Substantial information was provided from participants about the nurse staffing decision-making methods currently employed, including frameworks for nurse staffing, nurse-to-patient ratios, workload measurement systems, and "gut" instinct. A number of key themes emerged from the study that can form the basis for policy and practice changes related to determining appropriate workload for nursing. This includes the use of staffing principles and frameworks, nursing workload measurement systems, and nurse-to-patient ratios. There is a need for more evidence related to nurse staffing.

**Implications for Practice:** It appears that there is no one method to determine patient acuity and staffing requirements. While research must continue in this area, current systems use combinations of the above to determine safe staffing.

## RELATIVE VALUE UNIT (RVU)

A **relative value unit (RVU)** is an index number assigned to various health care services based on the relative amount of resources (labor and capital) used to produce the service. The actual consumption of nursing resources is not linear; that is, caring for an acuity level 2 patient does not consume twice as many nursing resources as caring for an acuity level 1 patient. RVUs provide a proportional comparison between the resources required by level 1 acuity (always a value of 1) and any other level. For example, only 20% (1.2 RVUs) more resources may be consumed by an acuity level 2 patient than a level 1, but twice as many resources may be consumed by a level 3 acuity (2.0 RVUs), and more than three times as much by a level 4 acuity (3.33 RVUs), and so forth, as shown in Table 4-5.

**TABLE 4-5  Comparative Values of Acuity, Care Hours, and Relative Value Units**

| ACUITY | CARE HOURS | RELATIVE VALUE WEIGHT |
|--------|-----------|----------------------|
| 1 | 3.00 | 1.00 |
| 2 | 3.60 | 1.20 |
| 3 | 6.00 | 2.00 |
| 4 | 9.99 | 3.33 |
| 5 | 15.00 | 5.00 |

The RVUs can be used to calculate the relative costs of nursing care, using the following reasoning:

For the time period of interest, the total cost of nursing wages = $1,250,000.

For the time period of interest, the total RVUs by acuity level and patient days are as follows:

| Acuity | Days | | RVU weight | | Total RVUs |
|--------|------|---|-----------|---|-----------|
| 1 | 125 | × | 1.00 | = | 125.00 |
| 2 | 200 | × | 1.20 | = | 240.00 |
| 3 | 500 | × | 2.00 | = | 1,000.00 |
| 4 | 550 | × | 3.33 | = | 1,831.50 |
| 5 | 400 | × | 5.00 | = | 2,000.00 |
| **Total** | 1775 | | | | 5,196.50 |

Dividing the total nursing costs by the total RVUs yields the cost per RVU.

$$\$1,250,000 / 5,196.50 = \$240.55 \text{ per RVU}$$

The cost for one level 5 acuity patient for one day, then, is $240.55 × 5 (the RVU weight for acuity level 5), or $1,202.75.

Patients' acuity varies, even within the same day, as patients experience invasive procedures, intensive treatments and medication, complications, or progress toward wellness. As acuity varies, consumption of nursing resources also varies. It follows that the cost for direct nursing care would be consistent with nursing resource consumption and could be determined using RVU calculations.

This approach provides a reasonably accurate per-patient costing approach. It does not, however, account for the difference in cost based on the category of worker providing the care. Indirect fixed costs that do not vary dependent on patient classification such as salaried wages, overhead, service contracts, and noncapital equipment purchases must be added to arrive at the full cost of nursing care. Both administrators and clinicians recognize the serious shortcomings of the current methods available to calculate the cost of nursing care and the critical need for something more accurate. But the current tools offer at least a rudimentary method for capturing nursing cost.

# USE OF THE COMPUTER TO CAPTURE NURSING COSTS

Many hospitals are beginning to examine the use of computerization to capture nursing costs. Computers will be much better able to assist in measuring the nursing resources expended on each patient when algorithms are developed that account for the cognitive and assessment work of nurses, as well as the physical care. At that point, the computer should be able to track sophisticated data on nursing resource usage by patient. With computers, nurses record not only when they are with each patient but also what they are doing for each patient. The specific cost of the nurse's care can be calculated by having the computer multiply the time spent with the patient by the salary of the nurse providing the care. When nurses are doing some indirect activities, such as documentation, the computer can also assign that cost to the appropriate patient, because the documentation is being done on the computer.

The system will need a fast way to record data so that it does not greatly increase the documentation work of the nurse. Radio frequency identification (RFID) chips may eventually assist with that task. An RFID chip embedded in the nurse's name tag can be detected by a sensor in the door frame. The sensor is a data receiver that transmits information to the computer. It records the times of the nurse's entry into and exit from the patient's room and links that information with the patient's hospital ID number. RFID technology is still under development (Finkler & McHugh, 2008).

# QUALITY MEASUREMENT

An evidence-based concept of quality is grounded on scientific evidence that a diagnostic or therapeutic approach improves patient outcomes. This concept exploded into the health care industry in the early 1990s with evidence-based clinical practices. The four core components are (1) a mechanism that establishes local or regional consensus about what constitutes the best practices based on scientific research findings, (2) strong feasible processes to accomplish such practices, (3) a deliberate program of outreach to the community on disease prevention and health promotion, and (4) a rigorous system to review actual performance and clinical outcomes, as well as identification and implementation of improvement methodologies that achieve a dynamic balance between economy and quality. The essence of that dynamic balance may be the "value" Dr. Tim Porter-O'Grady projects as the future focus of health care (see Real World Interview within this chapter).

Regardless of its particular size or mission, every contemporary health care **enterprise** (an organization of any size, established as a business venture) must commit to cost improvement and quality improvement as core goals for strategic business and clinical success. Clearly, this is not a new notion, but the recent intensification of the focus on cost and quality improvement and its expansion from hospitals to diverse ambulatory and home care sites is startling and is being referred to as the "Q-revolution" (Carpenter, 2006). The more critical cost containment and cost management become, the more critical attention to quality management becomes. Quality and cost are inextricably linked.

Logic suggests that higher quality could lead to a higher volume of use of the organization by patients and providers who have the flexibility to make choices about where they seek health care. Higher volume generally leads to higher profits, which, in turn, may be directed toward improving programs and services, thus achieving higher quality—a very positive spiral that can result in the organization's thriving.

Increased quality > Increased volume >
Increased profit > Enhanced programs/services >
Increased quality

An obverse spiral is more likely when quality is shoddy, a very negative and potentially fatal spiral.

Decreased quality > Decreased volume >
Decreased profit > Cutting corners >
Decreased quality

## Critical Thinking 4-3

Securing a profit might suggest that high quality is also secured, because it facilitates purchase of state-of-the-art equipment and expert practitioners. Quality and profit do not necessarily go together, however. The Nosek-Androwich Profit: Quality (NAPQ) Matrix shown here models four possible relationships between profit and quality that may exist in a health care organization. Any organization can fit into any quadrant, and a single organization may shift among quadrants from time to time in response to market forces. The challenge for the organization is to maintain its place in the high-profit, high-quality quadrant to be best positioned for both clinical success and business success, the mission of the organization. A common mission, consistent vision, collaboration, and constant vigilance to the elements of quality and profit by all employees and stakeholders together are keys to maintaining organizational positioning and achieving economic and quality success.

Access the annual report published by any health care enterprise. Based on the margin reported by the finance department, as compared to the rule of thumb of the 3% to 5% margin required for business success, would you rate that organization as high profit or as low profit? Based on the overall satisfaction score reported for all services of that organization, would you rate it as high quality or as low quality? (Hint: If the satisfaction score is not published, it may be because that score is relatively low.) In which quadrant of the NAPQ Matrix does the health care enterprise fit?

| | | |
|---|---|---|
| **Quality** ↑ | High Q<br>Low P | High Q<br>High P |
| | Low Q<br>Low P | Low Q<br>High P |
| | Profit → | |

Nosek-Androwich Profit: Quality (NAPQ) Matrix.

## CUSTOMER SATISFACTION

Regardless of how superior providers may perceive their own product or service to be, if customers fail to perceive it as a needed or wanted service provided conveniently by skilled and knowledgeable people in a caring manner at a reasonable cost and consistent with their own culture and value system, the organization may fail. Perception, accurate or not, is the key. Therefore, organizations use a variety of indices to measure both internal and external customer satisfaction.

Commercial surveys such as Press-Ganey measure how satisfied patients are with their care, food, physical environment, emotional ambiance, and interactions with various health care workers, and then provide comparison rankings across similar organizations and populations nationally. The Veterans Affairs system of hospitals conducts its own unique survey after patient discharge. Many organizations use instruments developed internally to measure attributes of unique interest to that organization. Research protocols use a variety of statistical methods to test satisfaction. Results of such surveys are analyzed as part of the organization's quality management program.

## HEALTH CARE SITE ECONOMICS

The study of economics focuses on how choices are made to overcome a scarcity of resources. Christensen, Bohmer, and Kenagy (2000) rate health care as possibly the most entrenched, change-aversive industry in the United States. Their thesis is that the three Rs—redesigning, restructuring, and reengineering—simply tweak the existing health care structure and processes (despite the latter's claim to turn organizations upside down and inside out through fundamental rethinking and radical redesign of processes to achieve dramatic improvements in critical performance). They believe that changes that offer cheaper, simpler, more convenient products or services aimed at the lower end of the market are key to the survival of the industry. They further believe that only a whole host of large and small changes—"disruptive innovations"—in technologies and business models that "sneak up from the bottom" can end the health care crisis. These disruptive innovations should come from workers in the trenches who know what would work better. Disruptive innovations in the usual setting for health care may already be underway as more care moves away from the hospital site.

## HEALTH CARE PROVIDER ECONOMICS

Economic risk is borne by individuals, as well as by organizations. Both individuals and organizations may experience actual or perceived pressure to provide less health care service than is optimal, in order to contain costs. Individual professionals risk what can be referred to as the "dumbing down" of their respective professions. Dumbing down occurs when cost-saving strategies include using individuals with less knowledge to perform health care services usually performed by people with advanced knowledge. When this happens, the quality of the services delivered may decrease without actual harm occurring to patients and without patient recognition that the services are not optimal. If the enterprise is willing to provide less-than-optimal services to save money, those with the advanced knowledge may face loss of job security. Eventually, if care is judged to be inadequate or inappropriate, there may be risk of expensive litigation that can threaten both individuals and the organization with loss of licensure and livelihood.

Organizations' attempts to provide more economical care by changing the mix of caregivers is ultimately regulated by the respective state's practice acts (laws), along with their accompanying rules. State nursing and medical practice acts define the scope of practitioner practice. Both types of acts regulate the practice components that can be delegated and the accountability that follows delegation of care. The law determines the extent to which an organization can stipulate that medical care or nursing care will be managed by delegation of that care in order to capture fiscal economies.

Individual providers such as practitioners and therapists receiving direct payment from insurers bear risk when they must lower their usual fees to a flat rate in order to be eligible for payment by the HMO. A catch-22 exists for individual providers. If the provider agrees to participate in an HMO as a preferred provider, a lower payment rate may be part of the agreement. If the provider chooses not to participate in an HMO as a preferred provider, there is risk of attracting only the limited number of patients willing and able to pay out-of-pocket fee-for-service rates. To minimize their own expenses for providing services, individual providers regulate the amount and selection of services they provide, as well as the time spent with each patient. As a result, patients may be dissatisfied with the services and choose an alternate provider.

Patients bear the risk of being unable to access services they regard as optimal or as representing the least possibility of harm. Clearly, frivolous services are not available under the managed care philosophy, but patients may find themselves co-opted from their preferences for health care services by incentives to accept less expensive levels of service. Their out-of-pocket expense, or copayment, is usually considerably lower if they accept care from a member of the PPO than if they choose to obtain care from a provider that is not a member of the PPO. Employers who provide health insurance as part of an employee's compensation package choose the insurance plan most economical for the employer, not necessarily the plan providing the best quality (Finkler, 2001).

## REAL WORLD INTERVIEW

The Joint Commission (JC) is a nonprofit organization that is committed to continuously improve the safety and quality of health care provided to the public. The JC accomplishes this mission primarily through regular inspection and survey of health care organizations by a multidisciplinary professional team using extensive standardized measures of facility, clinical care, and material standards that support improvement in quality. A health care organization is assigned a numeric score that must meet a minimum level for the enterprise to be accredited. The JC currently accredits nearly 19,000 health care organizations in the United States, including burn hospitals and home care organizations, and more than 8,000 organizations that provide long-term care or behavioral health care, as well as laboratory and ambulatory services. Extensive education about the nature of quality and how to best achieve it is also provided to health care enterprises by the JC.

Health care organizations commit a great deal of time and energy to preparing for both prescheduled and "surprise" on-site surveys, in order to successfully comply with the JC standards and receive a favorable accreditation status. The JC standards are developed through a definitive process of input from the field, expert consultation, and research to validate each standard as a measure of quality.

**Sally A. Sample, RN, MN, ND, DSc, FAAN**
Nurse at Large, Board of Governors, Joint Commission
Tucson, Arizona

## REAL WORLD INTERVIEW

One generally does not associate nurse practice acts with nursing economics. However, a connection can be formulated. Nurse practice acts delineate scope of practice and define the educational preparation necessary to perform nursing care. These laws restrict the health care agencies from unilaterally determining that an RN, practical nurse, or nursing assistant may work beyond their scope of practice. The agencies must conform to the law and employ qualified nurses, which invariably impacts health care costs.

**Anita Ristau, RN, MS**
Executive Director, Vermont State Board of Nursing
Montpelier, Vermont

Authors of the two seminal magnet hospital studies (McClure, Poulin, Sovie, & Wandelt, 1983; Kramer, 1990) eloquently concluded that a high level of performance by RNs is inseparable from high-quality patient care. Protagonists of disruptive innovation as the answer to the health care crisis believe that most ailments are relatively straightforward disorders whose diagnosis and treatment tap but a small fraction of what medical practitioners are educated to do and what nurse practitioners do so capably. They point out that most of the powerful innovations that have disrupted other industries have done so by enabling a larger population of less-skilled people to do in a more convenient and less expensive way things that historically were carried out only by a defined group of "experts." The findings of the cited studies are consistent with the magnet hospital findings and with disruptive innovation as an appropriate tool for change.

# EVIDENCE FROM THE LITERATURE

**Citation** Aiken, L. H., Sloan, D. M., Cimiotti, J. P., Clarke, S .P., Flynn, L., Seajo, J. A., Spetz, J., & Smith, H. L. (2010). Implications of the California Nurse Staffing Mandate for other states. *Health Research and Educational Trust.* doi:10.1111/j.1475-6773.2010.01114.x

**Discussion:** Article discusses survey data from hospital staff nurses in California, Pennsylvania, and New Jersey. Nursing workloads and patient outcomes including mortality, failure to rescue, and satisfaction were compared between the California nurses with mandated ratios and the Pennsylvania and New Jersey nurses. On average, the patient workload was one less patient for California nurses in intensive care units and two less patients on medical-surgical units. This workload appeared to influence patient mortality, and the nurses in California had lower burnout levels and lower job dissatisfaction. The California nurses also reported better quality of care for their patients.

**Implications for Practice:** This survey of RNs in California, New Jersey, and Pennsylvania explored self-reported patient workload, staffing ratios, and quality of care of patients. California nurses had lower patient workloads and less mortality for patients. Using a predictive model, the authors adjusted the lower risk of death that could have been experienced if lower patient workloads were implemented in Pennsylvania and New Jersey. The predictive model revealed 13.9% fewer deaths in New Jersey and 10.6% fewer deaths in Pennsylvania. The authors recommend multiple strategies for improving nurse staffing, such as mandatory reporting of nurse staffing or a mandated process for determination of staffing. Whether the exact number of patients per nurse is mandated or the hospital provides a reliable and safe staffing plan to provide the best quality and safest care, nurses must be aware of the research on this topic and be ready to discuss it with hospital administrators and decision makers.

# CASE STUDY 4-1

You have just been hired as a nurse in a home care agency. One of the first things you noticed when you took a tour of the facility during your interview process was the disarray of the supply area. In fact, as you peered into the area, you overheard a staff member letting off frustration at once again not being able to find the supplies he needed for his day's assignments. Although she looked a bit sheepish, your tour guide quickly dismissed the incident, stating that a STAT delivery of the supplies could easily and promptly be arranged. Now that you are an employee, you too are having difficulty locating things in that supply area. Broken and outdated supplies take up significant space, and new supplies sit unopened in stacks around the room. No one seems to have the time or interest to change the situation. You volunteer to lead a team effort to improve both the quality and the cost impact of the lack of organization in the supply area.

What method will you use to identify the issues?

What are some of the issues the staff might identify?

What are some of the options the staff might identify to address the issues?

How will you implement the options chosen?

## KEY CONCEPTS

- Health care economics is grounded in past values and culture. Nearly 150 years ago, Florence Nightingale recognized that the resources being used to care for sick people ought to be tracked and analyzed to improve clinical and business outcomes.

- Contemporary health care is characterized as a business struggling to balance cost and quality. Patients are fearful that health care is rapidly becoming unaffordable and are demanding care at a reasonable cost.

- In the United States, multiple programs exist to pay for health care. Managed care is a common nongovernmental structure for health care payment, with a variety of health maintenance organizations (HMOs) operating independently and offering a variety of health insurance packages.

- Government programs for eligible individuals are tax supported. Other industrialized countries around the world offer tax-supported socialized health care to every citizen, through centralized or decentralized programs, at about half of the U.S. per capita cost.

- Futurists agree that significant change will occur in the health care industry. They predict that the mechanism of change will be tumultuous. Futurists believe the focus on value received for dollars spent, the inextricable link of cost and quality, will grow.

- The ability to track and manage both cost and quality is critical. To achieve the organization's economic and quality goals, administrators and clinicians at all levels and in diverse health care organizations must focus on a common mission and consistent vision.

- Accounting for the cost of nursing care must be simplified and standardized. Regression analysis, patient classification systems (PCS), and relative value units (RVU) computations are complex, time consuming, and lack the support of both administrators and clinicians.

- Quality measurement in organizations is supplemented by external regulatory bodies that oversee safety and quality on behalf of society. Satisfaction indices may be measured both internally and externally and then compared to the performance of similar organizations.

- Nurses at all levels of all health care organizations are responsible for basic economic processes.

- An economic break-even formula can be used to compute the cost of health care equipment.

- The Nosek-Androwich Profit: Quality Matrix identifies the balance between quality and profit.

## KEY TERMS

break-even point
budget
direct cost
economics
enterprise

failure to rescue
fixed cost
indirect cost
margin

patient classification
system (PCS)
payer
preferred provider
organization (PPO)

reengineering
relative value unit (RVU)
stakeholder
variable cost

## REVIEW QUESTIONS

1. Identify the health care insurer that contributes the largest proportion of the health care dollar in the United States.
    A. U.S. Government
    B. Private health insurance
    C. Out-of-pocket payers
    D. Other private payers

2. Compare the life expectancy of women in Canada to that of women in the United States.
    A. Canadian women have the same life expectancy as U.S. women.
    B. Canadian women have a shorter life expectancy than U.S. women.
    C. Canadian women can expect to die about three years after U.S. women.
    D. Canadian women born in 2006 can expect to live six years longer than U.S. women born in 2006.

3. Select the most important reason for managing health care as a business.
    A. Patients need to know how the system works.
    B. Profit must be assured for the system to continue.
    C. Business ethics assure that patients are treated fairly.
    D. A competitive market keeps providers more responsive to patients.

4. Differentiate which of the following documents articulates the purpose for which a health care organization is in business?
   A. Strategic plan
   B. Mission statement
   C. Vision statement
   D. Corporate philosophy

5. Distinguish which of the following nurse leaders was not mentioned in this chapter as having a perspective on economics?
   A. Nightingale
   B. Sample
   C. Barton
   D. Henderson

6. Select the concept that is synonymous with profit.
   A. Dividends
   B. Billing privileges
   C. Contribution to margin
   D. Certificate of Need

7. For Medicare patients, the "donut hole" describes which of the following? Select all that apply.
   _____ A. Non-comprehensive prescription coverage by Medicare

   _____ B. A $20 co-payment for prescriptions
   _____ C. No insurance payment after one prescription is filled
   _____ D. A gap in coverage for prescriptions after the maximum is paid
   _____ E. A 20% decrease in reimbursement for obese patients
   _____ F. No payment for medications after 80 years of age

8. The cost of nursing practice in hospitals is typically paid for through which of the following? Select all that apply.
   _____ A. Nursing unions pay the salaries
   _____ B. Insurances companies bill for independent nurse practices
   _____ C. Nursing agencies make contracts with the chief nurse executive
   _____ D. A 25-30% cut of the physician billing to patients
   _____ E. Room charges that include nursing care, housekeeping and laundry services
   _____ F. Insurance companies that charge a daily rate for hospitalization

# REVIEW QUESTIONS

1. Identify the health care insurer that contributes the largest proportion of the health care dollar in the United States.
   A. U.S. Government
   B. Private health insurance
   C. Out-of-pocket payers
   D. Other private payers

2. Compare the life expectancy of women in Canada to that of women in the United States.
   A. Canadian women have the same life expectancy as U.S. women.
   B. Canadian women have a shorter life expectancy than U.S. women.
   C. Canadian women can expect to die about three years after U.S. women.
   D. Canadian women born in 2006 can expect to live six years longer than U.S. women born in 2006.

3. Select the most important reason for managing health care as a business.
   A. Patients need to know how the system works.
   B. Profit must be assured for the system to continue.
   C. Business ethics assure that patients are treated fairly.
   D. A competitive market keeps providers more responsive to patients.

4. Differentiate which of the following documents articulates the purpose for which a health care organization is in business?
   A. Strategic plan
   B. Mission statement

   C. Vision statement
   D. Corporate philosophy

5. Distinguish which of the following nurse leaders was not mentioned in this chapter as having a perspective on economics?
   A. Nightingale
   B. Sample
   C. Barton
   D. Henderson

6. Select the concept that is synonymous with profit.
   A. Dividends
   B. Billing privileges
   C. Contribution to margin
   D. Certificate of Need

7. For Medicare patients, the "donut hole" describes which of the following? Select all that apply.
   _____ A. Non-comprehensive prescription coverage by Medicare
   _____ B. A $20 co-payment for prescriptions
   _____ C. No insurance payment after one prescription is filled
   _____ D. A gap in coverage for prescriptions after the maximum is paid
   _____ E. A 20% decrease in reimbursement for obese patients
   _____ F. No payment for medications after 80 years of age

8. The cost of nursing practice in hospitals is typically paid for through which of the following? Select all that apply.

_____ A. Nursing unions pay the salaries

_____ B. Insurances companies bill for independent nurse practices

_____ C. Nursing agencies make contracts with the chief nurse executive

_____ D. A 25-30% cut of the physician billing to patients

_____ E. Room charges that include nursing care, housekeeping and laundry services

_____ F. Insurance companies that charge a daily rate for hospitalization

## REVIEW ACTIVITIES

1. Interview the chief financial officer of a health care organization to gain an understanding of how various costs are managed. Use the following questions to guide the interview:

   What method is used to measure nursing cost?

   What level of confidence does Fiscal Services have in its accuracy and why?

   How are contracts with various insurers such as Medicare, Medicaid, preferred provider organizations (PPOs), and Blue Cross discounted?

   What percentage of profit did the organization make last year, and how was it allocated? How typical is this?

   Which therapists' services are billed directly?

2. Consult a seasoned member of the medical staff for a personal perspective on the adjustments in practice that person made, if any, related to cost and quality issues in practice as his or her career unfolded. Compare and contrast what you discover to the experiences of a second person you consult who has been in practice for only the past 10 years. What level of passion about the discussion was demonstrated by each interviewee? What did they view as their greatest challenge?

3. Explore with the chief nursing officer what the most challenging clinical economic issue currently is for nursing in the organization and how it is being addressed.

4. Using the formulas provided in this chapter, determine the cost of nursing care for the past 24 hours for your work unit or for any manageable unit to which you have access.

## EXPLORING THE WEB

What sites could you recommend to a colleague interested in tracking health care cost trends over time?

- Centers for Medicare & Medicaid Services: www.cms.hhs.gov

- Search an alternate government bureau site. What does the Bureau of Labor Statistics offer at www.stats.bls.gov?

Is it relevant to cost or quality issues?

- Search the following Web sites for information of interest to nurses:

  www.florence-nightingale.co.uk

  www.aahn.org

  www.onhealth.com

  www.medexplorer.com

  www.medicarerights.org

## REFERENCES

Abdellah, F., Beland, I., Martin, A., & Matheney, R. (1960). *Patient centered approaches to nursing.* New York: Macmillan.

Affordable health care for Americans. (2010, March 20). The Health Insurance Exchanges, House Committees on Ways and Means, Energy and Commerce, and Education and Labor. Accessed at http://docs.house.gov/energycommerce/EXCHANGE.pdf. Retrieved January 4, 2011.

Aiken, L. H., Clarke, S. P., Cheung, R. B., Sloane, D. M., & Silber, J. H. (2003, September). Educational levels of hospital nurses and surgical patient mortality. *Journal of the American Medical Association, 290*(12), 1617–1623.

Aiken, L. H., Sloan, D. M., Cimiotti, J. P., Clark, S. P., Flynn, L., Seajo, J. A., Smith, H. L. (2010). Implications of the California nurse staffing mandate for other states. Health Research and Education Trust. doi:10.1111f. 1475–1773.

Blouin, A. S. (2010). Helping to solve healthcare's most critical safety and quality problems. Journal of Nursing Care Quality, 25(2), 95–99.

Carpenter, D. (2006, May). Attention, investors: The Q-revolution is spreading. *Hospitals & Health Networks,* 4–8.

Carter, G. (2006, June). "Sick tax" would hurt everyone. *Hospitals & Health Networks,* 6, 8.

Centers for Medicare & Medicaid Services. (2006). Changes in sources of funds for health care. Retrieved June 20, 2006, from http://www.cms.hhs.gov/statistics.

Centers for Medicare & Medicaid Services, Office of the Actuary, National Health Statistics Group. (2006). Retrieved March 31, 2006, from http://www.iii.org/media.

Chen, A. S., & Weir, M. (2009). The long shadow of the past: Risk pooling and the political development of health care reform in the States. *Journal of Health Politics, Policy and Law, 34*(5), 679–716.

Christensen, C., Bohmer, R., & Kenagy, J. (2000, September/October). Will disruptive innovations cure health care? *Harvard Business Review,* 102–112, 199.

Finkler, S., & McHugh, C. (2008). *Budgeting concepts for nurse managers* (4th ed.). St. Louis: Mosby.

Hall, L. M., Pink, L., Lalonde, M., Murphy, G. T., O'Brien-Pallas, L., Laschinger, H. K., et al. (2006). Decision making for nurse staffing. *Policy Politics in Nursing Practice, 7*(4), 261–269.

Henderson, V., & Nite, G. (1978). *Principles and practice of nursing* (6th ed.). New York: Macmillan.

Institute for Healthcare Improvement (IHI). Getting started kit: Prevent central line infections. Retrieved March 20, 2007, from http://www.ihi.org/NR/rdonlyres/of4cc102-CS64-4436-AC3A-OC57B1202872/o/Ccentrallineshowtoguidefind720.pdf.

Institute of Medicine of the National Academies. (2010). The future of nursing: leading change, advancing health. Report Brief. Available at http://www.iom.edu/nursing. Accessed February 28, 2011.

Keehan, S., Sisko, A., Truffer, C., Smith, S., Cowan, C., Poisal, J., Clemens, M.K., & the National Health Expenditure Accounts Projections Team.(2008). Health Spending Projections Through 2017: The Baby-Boom Generation Is Coming To Medicare. Health Affairs. 27, no. 2 (2008): w145–w155 (published online 26 February 2008; 10.1377/hlthaff.27.2.w145). Available at, http://www.cnbcasia.com/images/documents/CMC%20 Healthcare%20Study.PDF. Accessed February 26, 2011.

Kliger J., Lacey, S. R., Olney, A., Cox, K. S., & O'Neil, E. (2010). Nurse-driven programs to improve patient outcomes: Transforming care at the bedside, integrated nurse leadership program, and the clinical scene investigator academy. Journal of Nursing Administration, 40(3), 109–114.

Liberman, A., & Rotarius, T. (2001). Managed care evolution—Where did it come from and where is it going? In E. Hein (Ed.), *Nursing Issues in the 21st century: Perspectives from the literature.* Philadelphia: Lippincott.

Life expectancy. (2006). Retrieved June 20, 2006, from http://www.en.Wikipedia.org.

McClure, M., Poulin, M., Sovie, M., & Wandelt, M. (1983). *Magnet hospitals: Attraction and retention of professional nurses.* Kansas City, MO: American Academy of Nursing.

Mirror, mirror on the wall: An international update on the comparative performance of American health care. The Commonwealth Fund. (2007, May 15). Available at http://www.commonwealthfund.org/Content/Publications/Fund-Reports/2007/May/Mirror—Mirror-on-the-Wall—An-International-Update-on-the-Comparative-Performance-of-American-Healt.aspx. Retrieved January 4, 2011.

National Coalition on Health Care. (2006). Health care cost as percentage of gross domestic product. Retrieved June 20, 2006, from http://www.nchc.org.

Nightingale, F. (1859). *Notes on hospitals;* being two papers read before the National Association for the Promotion of Social Science, at Liverpool in October, 1858. With evidence given to the Royal Commissioners on the State of the Army in 1857 (2nd ed.). London: Parker.

Oregonian PolitiFact Oregon. (2011). Truth-o-Meter. UPDATE, Feb. 21, 2011. Available at, http://politifact.com/oregon/statements/2011/feb/19/wesley-smith/one-pundit-says-oregon-health-plan-rations-covered/.

The 2010 Patient Protection and Affordable Care Act. (2010). Available at, http://www.gpo.gov/fdsys/pkg/PLAW-111publ148/pdf/PLAW-111publ148.pdf. Accessed February 26, 2011.

Serb, C. (2006, June). Financial fitness test: 10 ways to get in shape for a new payment era. *Hospitals & Health Networks,* 34–36, 38, 40, 49.

Shinkman, R. (2010). CMS pay-for-performance project saves $38.7 Million. Healthcare Finance News. December 14, 2010. Available at, http://www.fiercehealthfinance.com/story/cms-pay-performance-project-saves-387m/2010-12-14. Accesssed February 27, 2011.

## SUGGESTED READINGS

Aging baby boomers will have less impact than other trends on inpatient needs. (2006, May). *Hospitals & Health Networks,* 73.

Borkowski, N. (Ed). (2005). *Organizational behavior in health care.* Sudbury, MA: Jones and Bartlett.

California Health Care Foundation. (2005, March). Health care costs 101. Retrieved March 19, 2007, from http://www.chcf.org.

Dunham-Taylor, J., & Pinczuk, J. Z. (2006). *Health care financial management for nurse managers: Merging the heart with the dollar.* Sudbury, MA: Jones and Bartlett.

Fortin, L., & Douglas, K. (2006). Shift bidding technology: A substantial return on investment. *Nurse Leader, 4*(1), 26–28.

Gullatte, M. (2005). *Nursing management: Principles and practice.* Pittsburgh, PA: Oncology Nursing Society.

Henry, J. Kaiser Family Foundation. (2005, Sept.). Employee health benefits: 2005. Annual Survey.

Herzlinger, R. (2006, May). Why innovation in healthcare is so hard. *Harvard Business Review, 84*(5), 58–66, 156.

Hospitals & Health Networks. (2006, May). U.S. ranks lowest of six affluent nations in patients' opinions, health care equity. Author, *80*(5), 74.

Malloch, K., & Porter-O'Grady, T. (2005). *The quantum leader: Applications for the new world of work.* Sudbury, MA: Jones and Bartlett.

National League for Nursing Public Policy Action Center. (2006). Health Resources and Services Administration 2004 National Sample Survey of Registered Nurses, p. 2. Retrieved February 28, 2006, from http://capwiz.com/nln/home.

Premier Hospital Quality Incentive Demonstration. (2006). Retrieved January 30, 2006, from http://www.cms.hhs.gov/Hospital-QualityInits/ 35_HospitalPremier.asp.

Sandrick, K. (2006, February). A tale of two turnarounds. *Hospitals & Health Networks,* 66–70.

Watson Wyatt Worldwide. (2006). Health care costs: Up, down and around the world. Retrieved June 20, 2006, from http://www.WatsonWyattWorldwide.com.

# CHAPTER 5

# Evidence-Based Health Care

Carolyn Christie-McAuliffe, PhD, FNP; Amy Androwich O'Malley, MSN, RN; Rinda Alexander, PhD, RN, CS; Ida M. Androwich, PhD, RN, BC, FAAN; Patricia Kelly, RN, MSN

*Not all "literature" is "evidence" and not all evidence is valid or relevant to the patient at hand.*

(Dr. Feldstein, Wisconsin Medical Journal, 2005)

## OBJECTIVES

Upon completion of this chapter, the reader should be able to:

1. Discuss the history of evidence-based practice (EBP) in nursing.

2. Develop an understanding of evidence and its use in decision making.

3. Assume responsibility for developing an evidence-based approach to patient care.

4. Understand selected terminology used to describe types of evidence and evidence-based care processes.

5. Develop an informed view of the current state of evidence-based care and an understanding of the role of nursing in evidence-based decision making.

6. Apply the steps needed to incorporate evidence-based care in practice.

7. Conduct a search for evidence on a given topic, using the PICO method.

Delmar/Cengage Learning

*You are at the annual Nurses' Day luncheon, and the vice president for nursing has just announced that your institution is applying for magnet status. The nursing manager of your unit and three other units stops to chat. She says that she is concerned because she has noticed variability in the use of nursing interventions on the units and that this variability may be negatively affecting patient outcomes in the institution. In light of the plan to achieve magnet designation, she would like to see all units incorporate evidence in patient care processes. The nursing manager wants to appoint you to a task force to consider using evidence-based practice (EBP) as a means of reducing variability and improving patient outcomes. This scenario is becoming more and more common in health care institutions as managers of care explore new ways to improve the quality of care.*

*What do you know about EBP?*

*What are some things you need to know before you can accept the appointment to the task force?*

*How will you prepare yourself for the task?*

Many professionals, and society in general, accept the situation in which nurses are the coordinators of care within health care institutions. However, as the health care system continues to evolve, nurses and other health professionals must consider new ways to deliver more effective and efficient interventions. In fact, there is probably no greater challenge for nursing than to ensure that we have the competencies needed in the twenty-first century for evidence-based health care delivery. Nursing leaders and managers are particularly well placed to see that health care institutions have work processes in place that provide professional nurses with support to meet new challenges in the clinical delivery of care. All nurses must work to remove barriers to evidence-based health care delivery in organizations, such as lack of staff authority to change practices, insufficient time to work on new practice development, and limited research knowledge and awareness (Strickland & O'Leary-Kelley, 2009).

**Evidence-based nursing (EBN)** is an integration of the best evidence available, nursing expertise, and the values and preferences of the individuals, families, and communities who are served (Sigma Theta Tau International, 2005). This assumes that optimal nursing care is provided when nurses and health care decision makers have access to a synthesis of the latest research, a consensus of expert opinion, and are thus able to exercise their judgment as they plan and provide care that takes into account cultural and personal values and preferences. This approach to nursing care bridges the gap between the best evidence available and the most appropriate nursing care of individuals, groups, and populations with varied needs.

Review of Rogers's theory of diffusion of innovations, 2003, is useful when introducing an EBN practice into a setting. According to Rogers, diffusion of innovations such as EBN practices occurs in stages: knowledge, persuasion, decision, implementation, and confirmation. Knowledge of an EBN practice could be gleaned from a literature review. The persuasion of self and others to adopt the practice might occur as the result of a patient's need for the innovation. A decision to adopt the EBN practice innovation could then be made and implementation occurs, perhaps through the intervention of change agents. During confirmation of the practice innovation, evaluation, possible integration into an organization, and final adoption would then occur.

The purpose of this chapter is to define what constitutes evidence for clinical practice, to discuss the importance of evidence to patients and nursing, and to develop individual strategies to provide EBN. Future trends in health care will require the use of sophisticated evidence-based tools to deliver care and measure outcomes of care. This suggests that nursing must become more comfortable using a scientific process driven by evidence-based standards and practice guidelines, while also emphasizing continuing quality improvement.

## HISTORY OF EBP

The term *evidence-based practice* (EBP) was coined at McMaster Medical School in Canada during the 1980s. D. L. Sackett, along with his Oxford colleagues, encouraged EBP as a way to integrate individual clinical medical experience with external clinical evidence, using a systematic research approach. Sackett, Rosenberg, Gray, Haynes, and Richardson, in their classic 1996 definition, define **evidence-based practice (EBP)** as "...the conscientious, explicit, and judicious use of current best evidence in making decisions about the care for individual patients". These decisions are based on information from clinical expertise, research evidence, and patient values and preferences. The search for best evidence should be systematic. A systematic critical appraisal of literature is an effective strategy for identifying

## REAL WORLD INTERVIEW

There are several key components of an effective strategy for the implementation of evidence-based practice. First, it is essential to find a source of evidence-based content developed using a strict and rigorous methodology that includes both review and classification of the sources of the evidence and routine updating. Second, the evidence-based content itself must be efficient for clinicians to use at the bedside. This efficiency factor can be accomplished through easy-to-read, succinct summaries of the evidence. And also, it can be achieved by integrating evidence-based content in order sets, plans of care, and documentation forms. This integration delivers evidence-based content and direction to the clinician at the bedside, at the point of decision making.

**Patricia S. Button, EdD, RN, Director**
Nursing Content, Zynx Health
Lyme, New Hampshire

recommendations for improving practice (Garrard, 2007; Melnyk & Fineout-Overholt, 2005). Then, practitioners use their clinical judgment to determine the relevance of evidence for changing practice.

Currently, there are a number of excellent sites for finding information about evidence-based nursing and evidence-based practice (EBP), which are listed at the end of this chapter. Since the early work of the McMaster's group, methods for review and summarization of evidence have undergone dramatic advances. A. Cochrane of the Cochrane Library group (http://www.cochrane.org/) was a pioneer in the movement and preparation of high-quality reviews. In 1978, Cochrane suggested that only 15% to 20% of practitioner interventions were supported by objective evidence. This led to much variation in patient care delivery and patient care outcomes. Since then, with increasing technology, there have been major improvements in our ability to access information.

## EVIDENCE-BASED PRACTICE MODELS

A variety of models for EBP now exist, e.g., the conduct and utilization of research in nursing (CURN) project (Horsley, 1983), the Stetler model (Stetler, 2001), the ACE

Star Model of Knowledge Transformation (Stevens, K. R., 2004), and the Iowa Model of Evidence-Based Practice to Promote Quality Care (Titler, et al., 2001).

The ACE Star Model is a frequently used model for evidence-based practice that provides a framework for nurses to transition through five stages of discovery of evidence, evidence summary, translation, integration, and evaluation. The Iowa Model of Evidence-Based Practice to Promote Quality Care (Titler et al., 2001, Figure 5-1) starts by considering a trigger that focuses the nurse on new knowledge and research or one that focuses the nurse on a problem or opportunity for patient care improvement. Other triggers for seeking new evidence for practice may also occur. If the topic is a priority for the organization, the nurse can help form a team and assemble relevant research and related literature on the topic in order to critique and synthesize research for use in practice. If adequate evidence or a sufficient research base exists to propose a change in practice, the nurse can pilot the proposed change in practice. Within the Iowa Model, lack of sufficient evidence then prompts the nurse to seek evidence for best practices from other sources of information and/or to initiate the research process. Nurses in practice can seek answers to questions through collaboration with clinical nurse specialists and/or nurse scientists. Many times, these expert nurses can be found in the staff development department within a hospital. Nurse managers can also serve to provide direction and guidance for this purpose. Nurse manager support is particularly important, because the last element within the Iowa Model directs the nurse to ask if the change proposed is appropriate for adoption into practice. If it is, the nurse can institute the change in practice and monitor and analyze structure, process, and outcome data and disseminate the results. If the change proposed is not appropriate for adoption in practice, the nurse will continue to evaluate the quality of care and review new knowledge until another trigger for seeking new evidence occurs.

## IMPORTANCE OF EBP

Why is EBP important? Despite the national and international EBP initiatives described, studies reveal that up to 40% of patients still receive health care that is not based on scientific evidence, and up to 25% receive unnecessary and/or potentially harmful care (Graham, Logan, Harrison, Straus, Tetroe, et al., 2006). Consequently, there is nothing more important to patients and professional nursing than evidence-based clinical interventions that can be linked to clinical outcomes and used as a basis for care within the institution. However, there has been a lack of generally agreed-upon standards or processes that are based on evidence. This lack of standards has been addressed of late with the development of EBP.

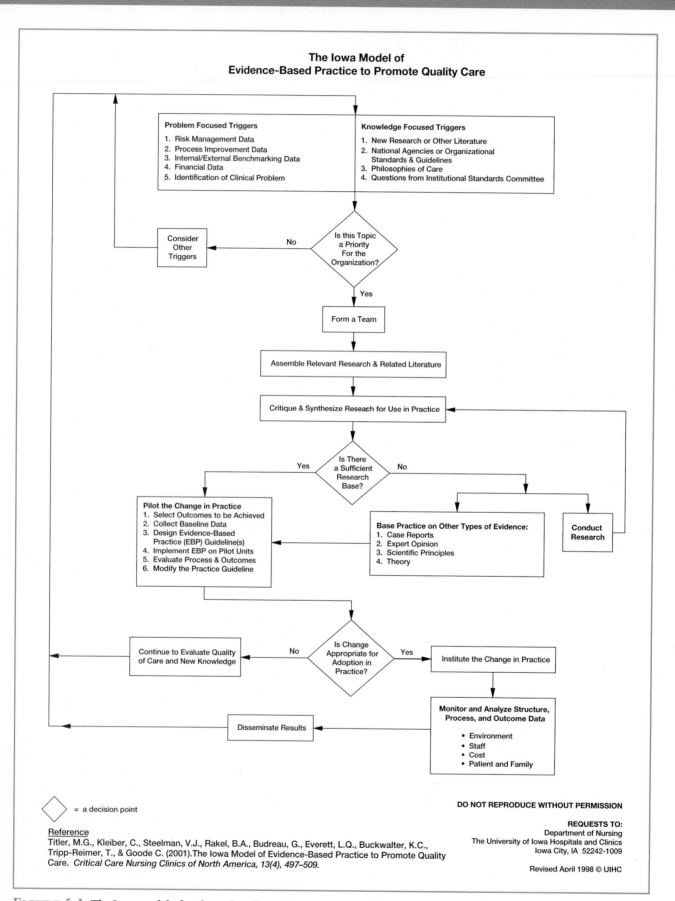

**The Iowa Model of Evidence-Based Practice to Promote Quality Care**

Problem Focused Triggers

1. Risk Management Data
2. Process Improvement Data
3. Internal/External Benchmarking Data
4. Financial Data
5. Identification of Clinical Problem

Knowledge Focused Triggers

1. New Research or Other Literature
2. National Agencies or Organizational Standards & Guidelines
3. Philosophies of Care
4. Questions from Institutional Standards Committee

Is this Topic a Priority For the Organization?

No → Consider Other Triggers

Yes

Form a Team

Assemble Relevant Research & Related Literature

Critique & Synthesize Research for Use in Practice

Is There a Sufficient Research Base?

Yes

No

Pilot the Change in Practice
1. Select Outcomes to be Achieved
2. Collect Baseline Data
3. Design Evidence-Based Practice (EBP) Guideline(s)
4. Implement EBP on Pilot Units
5. Evaluate Process & Outcomes
6. Modify the Practice Guideline

Base Practice on Other Types of Evidence:
1. Case Reports
2. Expert Opinion
3. Scientific Principles
4. Theory

Conduct Research

Continue to Evaluate Quality of Care and New Knowledge

No ← Is Change Appropriate for Adoption in Practice? → Yes → Institute the Change in Practice

Disseminate Results

Monitor and Analyze Structure, Process, and Outcome Data

• Environment
• Staff
• Cost
• Patient and Family

◇ = a decision point

DO NOT REPRODUCE WITHOUT PERMISSION

REQUESTS TO:
Department of Nursing
The University of Iowa Hospitals and Clinics
Iowa City, IA 52242-1009

Revised April 1998 © UIHC

Reference
Titler, M.G., Kleiber, C., Steelman, V.J., Rakel, B.A., Budreau, G., Everett, L.Q., Buckwalter, K.C., Tripp-Reimer, T., & Goode C. (2001).The Iowa Model of Evidence-Based Practice to Promote Quality Care. *Critical Care Nursing Clinics of North America, 13(4), 497–509.*

**FIGURE 5-1** **The Iowa model of evidence-based practice to promote quality care.** Reprinted with copyright permission from University of Iowa Hospitals and Clinics.

## REAL WORLD INTERVIEW

With over 20 years in nursing and hospice care, I have increasingly realized that a large part of the health care system, both doctors and nurses, do not understand the need for adequate pain control. One of my patients was a young man with cancer of the throat. He had received a radical laryngectomy and had had his tongue removed. He was left with a tracheotomy and was rendered speechless. His doctors at a top-rated university hospital seemed amazed that he was in as much pain as he was. It was not until he entered hospice care, where we emphasized an evidence-based holistic pain protocol, that he received some degree of comfort. This man was an exceptional person who never complained. He also had exceptionally strong support from his wife and family.

As nurses, please advocate for adequate pain relief for your patient. I have seen over and over that patients do not abuse the medications or the system. Our hospice saying is, "Meet the patients where they are, not where you think they should be!" When our patients are comfortable, there is peace for everyone.

**Sylvia Komyatte, RN, MPS**
Chaplain, Hospice
Munster, Indiana

## REAL WORLD INTERVIEW

The first and most important tool in searching the literature for peer-reviewed, evidence-based journal articles are your institution's librarians. Speak to one of them and ask them to tell you about your library's key journals and databases for nursing. While many databases exist for health care literature, MEDLINE (http://www.nlm.nih.gov/databases/databases_medline.html) and the Cumulative Index to Nursing and Allied Health Literature (CINAHL) (http://www.ebscohost.com/cinahl/) are two of the most essential for nursing.

MEDLINE is the National Library of Medicine's electronic journal database, indexing thousands of journal publications in the fields of medicine, nursing, dentistry, veterinary medicine, the health care system, and the preclinical sciences. MEDLINE provides a complete citation for each article within those journals, with an abstract of the article provided in most cases. PubMed (http://www.pubmed.gov) is the free search engine for MEDLINE, provided by the National Library of Medicine. Ovid Medline (http://www.ovid.com/site/catalog/DataBase/901.jsp) and its electronic access to full-text journal articles is available to you only if your parent institution's library subscribes to it.

CINAHL is an electronic database that indexes the contents of nursing and allied health publications, including journals, dissertations, and other materials. As with MEDLINE, whether or not the full text of an article found in CINAHL can be accessed electronically depends on whether your parent institution's library subscribes to it.

Other literature searching tools for nursing exist, such as the Cochrane Library (http://www.thecochranelibrary.com/view/0/index.html), PsycINFO (http://www.apa.org/pubs/databases/psycinfo/index.aspx), and Embase (www.elsevier.com/wps/product/cws_home/523328). These tools will get you started. When using electronic databases such as these, keep in mind that they often do not work like Internet search engines. There may be a few additional techniques to learn. Remember, if you have trouble, ask a librarian.

**Scott Thomson, MLIS**
Reference and Education Librarian
Health Learning Centers
Northwestern Memorial Hospital
Chicago, Illinois

Generally speaking, nursing, medicine, health care institutions, and health policy makers recognize EBP as care based on state-of-the-art science reports. It is a process approach to collecting, reviewing, interpreting, critiquing, and evaluating research and other relevant literature for direct application to patient care. EBP uses evidence from research; performance data; quality improvement studies such as hospital or nursing report cards, program evaluations, and surveys; national and local consensus recommendations of experts; and clinical experience.

The EBP process further involves the integrating of both clinician-observed evidence and research-directed evidence. This then leads to state-of-the-art integration of available knowledge and evidence in a particular area of clinical concern that can be evaluated and measured through outcomes of care.

Applying the best available evidence does not guarantee good decisions, yet it is one of the keys to improving outcomes affecting health. EBP should be viewed as the highest standard of care so long as critical thinking and sound clinical judgment support it. Nurses and practitioners will always need to search for the best evidence available to support their clinical decisions. Sometimes, there is little research backing for clinical actions. In that case, nurse and practitioners should use their critical thinking skills and apply the consensus of experts. Institutions of care have a responsibility to provide nurses and others in health care with an environment supportive of EBP.

Demonstrating that outcomes of health care are effective, efficient, and safe is a major responsibility for nursing. It is evident as we consider the art and science of nursing that recognizing the importance of EBP and stimulating an environment within institutions in which evidence-based models of care can flourish will result in improved outcomes of clinical care. Table 5-1 summarizes current issues and trends driving the development of EBP in nursing.

A common misconception about EBP is that it ignores patient values and preferences. This is not the case; rather, the nurse works with the patient in deciding treatment

## TABLE 5-1 Current Issues and Trends Driving Development of EBP in Nursing

- Significant contributions to the body of clinical research

- Need for decreased variability in implementation of nursing practice

- Need for implementation of research in practice to improve effectiveness and efficiency

- Societal demands for evidence-based, clinically competent care (Institute of Medicine, 1999 and 2001)

- Growth of advanced practice roles with development of prescriptive power and evidence-based diagnostic decision making

- Increased experience in clinical pathways, standards, protocols, and algorithms

- Increase in integrated systematic reviews of research studies found in the nursing, medical, and health care literature

- Need for outcome data to guide patient care

- Explosion in information technology with better-organized, rapidly retrievable information in the literature and on the Internet

- Improved knowledge base facilitating research capable of supporting EBP models

- Need to collaborate in complex decision making with patients and other members of the health care team

- Need to improve patient safety, e.g., search for "clinical guidelines" at each of these Web sites: http://www.guideline.gov/, http://www.ihi.org/ihi; http://www.nice.org.uk/, http://www.cdc.gov/; and http://www.g-i-n.net/

- Requirement for evidence-based standards of care implemented by the Joint Commission (JC) and National Database on Nursing Quality Indicators (NDNQI) (www.nursingquality.org)

options based on which trade-offs the patient is willing to accept. Studies conducted by Michaels, McEwen, and McArthur (2008) as well as Eddy (2005) confirm and reinforce the unique abilities of nurses to facilitate these sometimes difficult and confusing conversations.

## NURSING AND EBP

It was inevitable that nursing would move to EBP. One of the earlier proponents for EBP in nursing was the Joanna Briggs Institute for Evidence Based Nursing and Midwifery (http://www.joannabriggs.edu.au/, accessed February 27, 2011), established in 1996. Significant work has been done worldwide to implement EBP in Australian, Canadian, and UK institutions of care.

In the United States, the Agency for Healthcare Research and Quality (AHRQ) has provided stimulus for the EBP movement through recognition of a need for evidence to guide practice throughout the health care system. In 1997, the AHRQ launched its initiative establishing 12 evidence-based practice centers. This partnered AHRQ with other private and public organizations in an effort to improve the quality, effectiveness, and appropriateness of care. The initiative is discussed later in this chapter.

## ATTRIBUTES OF EBP

A new culture that can support EBP needs to evolve in institutions. There is a need to define evidence in each agency, a need to begin to use the term *evidence* in daily practice, and a need to look for the best evidence when evaluating new goals and programs. Where possible, the use of visible, formal supports for EBP should be encouraged, along with the development of systems in the health care agency that support EBP on an ongoing basis. Though written in 2000, a manual by Guyatt and colleagues remains a pivotal document that sets out the two fundamental principles of EBP:

- Evidence alone is never sufficient to make a clinical decision.
- EBP involves a hierarchy of evidence to guide decision making.

In a landmark study completed as recently as 2005, Pravikoff, Tanner, and Pierce surveyed more than 2,000 nurses and found that while 64.5% of nurses report needing information weekly or several times a week, only 26.7% have received training in using tools to access evidence. Furthermore, only 11% of nurses state that they search for information from the evidence weekly or more often, and nearly half (48.5%) were not familiar with the term *evidence-based practice.*

## LEVELS OF EVIDENCE

There is no standard formula for how much EBP factors, i.e., evidence; patient values and preferences; and a clinician's expertise should be weighed in the process of making decisions about the type of care to be delivered to patients. However, there are levels of evidence that identify the strength or quality of evidence generated from a study or report (Table 5-2). This is important so that nurses can be

## TABLE 5-2 Levels of Evidence

| LEVEL OF EVIDENCE | |
| --- | --- |
| Level I: | Evidence from a systematic review or meta-analysis of all relevant randomized controlled trials (RCTs), or evidence-based clinical practice guidelines based on systematic reviews of RCTs |
| Level II: | Evidence obtained from at least one well-designed RCT |
| Level III: | Evidence obtained from well-designed controlled trials without randomization |
| Level IV: | Evidence from well-designed case-control and cohort studies |
| Level V: | Evidence from systematic reviews of descriptive and qualitative studies |
| Level VI: | Evidence from a single descriptive or qualitative study |
| Level VII: | Evidence from the opinions of authorities and/or reports of expert committees |

**Source:** Compiled with information from Melnyk & Fineout-Overholt, 2005.

confident about how much emphasis they should place on a study when making decisions about patient care. Nurses will want to refer to Table 5-3 to become comfortable with the language of evidence and research.

Level I, systematic reviews or meta-analyses of randomized controlled trials (RCTs) and evidence-based clinical practice guidelines, is considered to be the strongest level of evidence upon which to guide practice decisions (Guyatt & Rennie, 2002). The Cochrane Review is one group that conducts systematic reviews for use by health care professionals in its Cochrane Library. Evidence-based clinical practice guidelines also provide the highest level of evidence to guide practice as they are often based on systematic reviews of RCTs of the best evidence on specific topic areas. For example, the Agency for Healthcare Research and Policy (AHRQ), U.S. Department of Health and Human Services, has established and maintains a database of the most currently evaluated evidence-based

## TABLE 5-3 Research Terminology

| | |
|---|---|
| Best practice | In application, best practice includes the use of rigorous scientific evidence to support the effectiveness of specific clinical interventions for explicit patients, groups, or populations; implementation monitoring to assure accurate application; and outcome measurement to validate effectiveness. |
| Case-control study | A case-control study compares certain characteristics of an individual, for example a child with asthma, with someone who does not have that characteristic, for instance a similar child without asthma. This type of study is conducted for the purpose of identifying variables that might predict the condition— e.g., exercise, environmental allergies. |
| Clinical practice guidelines (or practice guidelines) | Systematically developed statements or recommendations to assist practitioner and patient decisions about appropriate health care for specific clinical circumstances. Practice guidelines present indications for performing a test, procedure, or intervention, or the proper management for specific clinical problems. Guidelines may be developed by government agencies, institutions, organizations such as professional societies or governing boards, or by expert panels. Evidence-based clinical practice guidelines provide the strongest level of evidence to guide practice, as they are based on systematic reviews of RCTs of the best evidence on specific topic areas. |
| Cohort study | A longitudinal study that begins by gathering two groups of individuals (the cohorts), one group that received exposure to a disease or condition—e.g., a developmental disability—and one that did not. The groups are then followed prospectively over time to measure the outcomes. |
| Control group | Subjects in an experiment who do not receive the experimental treatment and whose performance provides a baseline against which the effects of the treatment can be measured. |
| Dependent variable | The outcome variable of interest; the variable that is hypothesized or thought to depend on or be caused by another variable, called the independent variable. |
| Descriptive research | Research studies that have as their main objective the accurate portrayal of the characteristics of people, situations, or groups, and the frequency with which certain phenomena occur. |
| Independent variable | The variable that is believed to cause or influence the dependent variable; in experimental research, the independent variable is the variable that is manipulated. |
| Integrative review | A type of evidence summary; concludes with implications from research for practice. |
| Meta-analysis | A systematic review that uses quantitative measurement methods such as surveys and questionnaires to summarize the results of multiple studies. It often produces a summary statistic that represents the effects of an intervention across multiple studies and, therefore, is more precise than individual findings from any one study used in the review. |

*(Continues)*

## TABLE 5-3 (Continued)

| | |
|---|---|
| Nonexperimental research | A study in which the researcher collects data without introducing an intervention. |
| Outcomes research | Research designed to document the effectiveness of health care services and the end results of patient care. |
| Prospective study | A study that begins with an examination of presumed causes (e.g., cigarette smoking) and then goes forward in time to observe presumed effects (e.g., lung cancer). |
| Qualitative analysis | The organization and interpretation of nonnumeric data for the purpose of discovering important underlying dimensions and patterns of relationships. |
| Quantitative analysis | The manipulation of numeric data through statistical procedures for the purpose of describing phenomena or assessing the magnitude and reliability of relationships among them. |
| Randomized clinical trial (RCT) | An experimental research study in which subjects are randomly assigned to experimental and control groups. The experimental group then receives an experimental preventive, therapeutic, or diagnostic intervention. This type of study is strong in internal validity, i.e., the ability to say that it was the experimental intervention that caused a change in the outcome variable of interest and not other extraneous variables. |
| Systematic review of the literature | Type of evidence summary that identifies all relevant research studies on a particular topic and uses a rigorous scientific process for retrieving, critically appraising, and synthesizing studies in order to answer a clinical question. |
| Variable | An attribute of a person or object that varies (i.e., takes on different values) within the population under study (e.g., body temperature, heart rate). |

**Source:** Compiled with information from Polit, D., & Beck, C. (2008). *Nursing research: Generating and assessing evidence for nursing practice* (8th ed.). Philadelphia: Lippincott; DiCenso, A., Guyatt, G., & Ciliska, D. (2005). *Evidence-based nursing: A guide to clinical practice.* St. Louis, MO: Elsevier Mosby; and Ciliska, D., Cullum, N., & Marks, S. (2001). Evaluation of systematic reviews of treatment or prevention interventions. *Evidence-Based Nursing, 4*,100–104.

practice guidelines. These guidelines are available online at the National Guideline Clearinghouse. This database is updated weekly and is free and accessible to health care professionals as well as to the public. One of the most useful and unique aspects of this online resource is its feature allowing visitors to conduct comparative analysis of guidelines on similar topics. In addition, each guideline clearly lists its process of evaluation with complete citations for reference. The Canadian Medical Association also maintains a database of clinical practice guidelines.

Level II in the evidence hierarchy is that evidence that is generated by at least one well-designed randomized controlled trial (RCT). With random assignment, a procedure in which subjects are placed in experimental or control groups by probability strategies—e.g., tossing a coin—there is a good probability that subjects in each of the experimental and control study groups will be equal on important demographic and clinical variables at the start of a study. As a result, if differences are found in the outcomes being measured in the study—e.g., weight loss—it can be assumed that these differences were the direct result of the weight loss intervention that was used in the RCT.

Level III evidence is obtained from well-designed, controlled trials without randomization. Because random assignment is not used to assign subjects to experimental and control groups, this type of evidence is less strong in internal validity—it cannot be assumed that the subjects in the study are equal on major demographic and clinical variables at the beginning of the trial. Therefore, less confidence can be placed in the outcomes of the study, because other extraneous variables could be the cause of the change in the outcome variables rather than the experimental intervention itself.

Level IV evidence is generated from well-designed case-control and cohort studies.

Evidence from Levels V, VI, and VII studies are also useful in developing guidelines for patient care. Note that regulations exist with the Centers for Medicare & Medicaid Services

## REAL WORLD INTERVIEW

I have been a nurse for 39 years and have seen many changes in health care during that time. About five years ago, I had a ruptured brain aneurysm, which occurred after having an angiogram. I was airlifted to a nearby university hospital. The knowledge and competence of the air transport team was impeccable. I was glad to be in the hands of people who were up to date on the best way to care for me. They were definitely in control—a nice thing to have when you're not.

Before the surgery, I was unconscious, and my family was informed of the dangers and the necessity to either coil or clip the aneurysm. During the 14-hour operation, the neurosurgeon's nurse updated them periodically on my progress. During my entire hospital stay, she visited me daily and was available by pager if my husband had any questions, doubts, or fears. While I was in the ICU, I was the only patient cared for by one nurse for three days. Although I had short-term memory loss at the time, I do remember that I did not have any concerns about my needs not being met. In fact, I felt the nurses predicted the problems I had before I even vocalized them. I was given quality patient care, and as an RN, I am very proud to say that. I am back working as a nurse now, and my experience has given me a whole new level of understanding for patients. I am glad that the people who cared for me were up to date on the latest!

**Janice Klepitch, RN**
Office Nurse
Schererville, Indiana

## EVIDENCE FROM THE LITERATURE

**Citation:** Moch, S. D., Cronje, R. J., & Branson, J. (2010). Part 1. Undergraduate nursing evidence-based practice education: Envisioning the role of students. *Journal of Professional Nursing, 26*(1): 5–13.

**Discussion:** Nursing educators have embraced the integration of EBP into the nursing education curriculum in numerous ways. As this review of nursing literature demonstrates, most of these approaches are built upon long-standing commitments to help students understand the scientific research process, think critically, and develop the information literacy skills that will enable them to find the evidence that can inform their practice. Many reports in the nursing literature recount various strategies used to teach EBP to nursing students. Another category of nursing articles discusses ways in which EBP education can be suffused throughout the nursing school curriculum. Few educators, however, have envisioned students as having a role beyond that of mere recipients of EBP education. Nonetheless, a small but growing number of nurse educators have begun to envision students as enablers of practice change in clinical settings. These innovators advocate a paradigm that places students into socially meaningful partnerships with practicing nurses as a means to promote the update of EBP in clinical settings. This article is followed by Part 2 and Part 3 articles.

**Implications for Practice:** There is growing knowledge of EBP in nursing. The implications of this for patient care and safety are impressive.

(CMS) requiring health care professionals to utilize and demonstrate that the care they deliver is based upon scientific evidence. It can be difficult, however, to evaluate the volumes of medical and scientific literature continuously published.

Malloch and Porter-O'Grady (2006) state that patient decision making may require a variety of measures to gauge the efficacy or appropriateness of a treatment. For example, if a patient is not responding to prescribed pain measures, the nurse might search out evidence from a Level I meta-analysis of treatments for pain in order to identify what might be best for the patient. The nurse would then evaluate the safety and efficacy of that evidence within the specific response of the patient and his or her unique circumstances. While guidelines are often based on Level I evidence, the nurse may need to adjust a guideline, based on an individual patient's pain. The assessment garnered from the patient might not be consistent with documented guidelines. In that case, the nurse might decide to base decision making more on the patient's perspective than on the Level I guidelines. The art and science of nursing comes from the intelligent evaluation of evidence in the context of sensitivity to intuition and response to individual and unique circumstances.

## PICO-BASED APPROACH TO GUIDING AN EVIDENCE SEARCH

A PICO-based approach may be used to guide an evidence search (Melnyk & Fineout-Overholt, 2005):

    P—Population
    I—Intervention
    C—Comparison
    O—Outcome

For example,

(P) In a group of patients over the age of 60,
(I) how does early ambulation on the day of surgery

### Critical Thinking 5-1

Go to this site at the University of Minnesota: http://www.biomed.lib.umn.edu/learn/ebp/mod01/index.html. Take the tutorial. Note that the practice of evidence-based practice includes five fundamental steps:

Step 1: Formulating a well-built question
Step 2: Identifying articles and other evidence-based resources that answer the question
Step 3: Critically appraising the evidence to assess its validity
Step 4: Applying the evidence
Step 5: Reevaluating the application of evidence and areas for improvement

What did you learn about EBP and PICO from the site?
Be sure to work through the case studies at the site.
Construct a PICO question when you are done, and do a search for the key PICO terms in http://www.ncbi.nlm.nih.gov/pubmed/.

(C) versus ambulation the next morning
(O) improve patient outcomes?

Key terms from the PICO-based approach can be entered in databases to identify key references. Review the steps of evidence-based practice at http://www.biomed.lib.umn.edu/learn/ebp/mod01/index.html. This PICO-based approach can be applied to both clinical and organizational situations and can lead to improvement of care and/or function. Some common examples of types of evidence are listed in Table 5-4. Several evidence-based resources are listed in Table 5-5.

## TABLE 5-4 Common Examples of Types of Evidence

- Published research

- Published quality improvement report

- Published meta-analysis

- Published systematic literature review

- Policies, procedures, protocols

- Published guidelines

- Conference proceedings, abstracts, presentations

## TABLE 5-5  Evidence-Based Resources

| | |
|---|---|
| National guideline clearinghouse, Agency for Health Care Quality and Research | www.guideline.gov |
| The CINAHL Rehabilitation Guide, 7 Step Evidence-Based Methodology and Protocol | www.ebscohost.com/rrc_editorial/ |
| Evidence-Based Nursing | http://ebn.bmj.com |
| Sigma Theta Tau International | www.nursingsociety.org |
| Cochrane collaboration | www.cochrane.org |
| Joanna Briggs Institute for Evidence Based Nursing and Midwifery | www.joannabriggs.edu.au |
| Sarah Cole Hirsh Institute for Best Nursing Practices Based on Evidence | http://fpb.case.edu |
| Centre for Evidence-Based Medicine–Toronto | http://ktclearinghouse.ca/cebm/ |
| Centre for Evidence-Based Medicine in Oxford, United Kingdom | http://www.cebm.net |
| The Canadian Medical Association | www.cma.ca |
| Centre for Evidence-Based Nursing | www.york.ac.uk |
| Royal College of Nursing | www.rcn.org.uk |
| Registered Nurses' Association of Ontario (guidelines available in English, French, and Italian) | www.rnao.org |
| Academic Center for Evidence-Based Nursing (ACE) | www.acestar.uthscsa.edu |
| Cumulative Index to Nursing and Allied Health (CINAHL) | http://www.ebscohost.com/cinahl/ |
| *Evidence-Based Nursing* Journal | www.evidencebasednursing.com |
| McGill University Health Centre, Clinical and Professional Staff Development, Research & Clinical Resources for Evidence-Based Nursing | www.muhc-ebn.mcgill.ca |
| MEDLINE via PubMed, free resource provided by the National Library of Medicine | www.ncbi.nlm.nih.gov |
| PubMed Tutorial (NLM), an in-depth tutorial from the National Library of Medicine | www.nlm.nih.gov |
| National Institute for Health and Clinical Excellence in the United Kingdom | http://www.nice.org.uk |
| Association of Women's Health Obstetrical and Neonatal Nurses | http://www.awhonn.org |
| Oncology Nursing Society | http://www.ons.org |

## CASE STUDY 5-1

You are a nurse working on a busy oncology floor. Mrs. Smith has been a patient on your unit for over a month and has recently been placed on the hospice service due to progression of her disease. She wishes to have a glass of wine with dinner now that she is waiting for placement. She remains on a palliative chemotherapy regimen to treat her metastatic bone cancer. As a nurse, you realize that there could be potential drug interactions with the wine that you must assess. What resources would you access to help determine this risk? How would you evaluate the strength and validity of what you find? As a patient advocate, you are committed to helping this patient's wish become a reality. However, no policy exists that allows for this wine option for patients in your facility. What sources of evidence would help make this decision?

## EBP CENTERS IN THE UNITED STATES

As mentioned earlier, clinicians have coped with accusations for many years that only a small percentage of treatments provided have scientific foundations. Many within the health care system have supported this accusation. In response to this ongoing lack of evidence to make clinical decisions, the Agency for Healthcare Research and Quality (AHRQ) has undertaken the development and funding of 12 evidence-based centers to carry out development and dissemination of best practice models based on available scientific information and data. Development of these special centers has been a driving force for state-of-the-art evaluations of current knowledge used in EBP in the United States. An example of an evidence-based center for nursing is the ACE Center at the University of Texas Health Science Center in San Antonio (www.acestar. uthscsa.edu).

## EVIDENCE FROM THE LITERATURE

**Citation:** Melnyk, B. M., & Fineout-Overholt, E. (2005). Evidence-based practice (EBP) in nursing and healthcare: A guide to best practice. Philadelphia: Lippincott Williams & Wilkins.

**Discussion:** A challenge in using evidence is translating the evidence found in the literature into the actual, real world environment. Melnyk and Fineout-Overholt set out to accomplish just that. They developed a five-step process of EBP:

1. Asking the relevant clinical questions
2. Searching for the best evidence
3. Appraising the evidence critically
4. Integrating evidence with clinical expertise, patient preferences, and values in making a practice decision or change
5. Evaluating the practice decision or change

**Implications for Practice:** Nurses need to be on the alert for opportunities to improve practice. Nurses can get started with EBP by: asking their librarians to give them an orientation to key EBP Web sites for best evidence and clinical guidelines; using the PICO search process to identify the best literature on a given topic; paying attention to the latest news, research, and standards; subscribing to one or more nursing and health care journals; starting a journal club for discussion of EBP articles; incorporating EBP guidelines into the revision of procedures and guidelines as they are reviewed; collaborating with researchers at their hospital or local university, as needed; and by partnering with faculty, other health care professionals, and students on EBP. The use of evidence-based practice can significantly improve patient care.

## Critical Thinking 5–2

You work on a unit delivering care to patients. You want to be sure your care delivery is state of the art.

How will you, as a new nurse, begin to provide EBP?

What contributions can you make to EBP?

How can we as nurses help overcome potential resistance to EBP?

# PROMOTING EVIDENCE-BASED BEST PRACTICES

The U.S. health care system is a $1 trillion industry, and yet it is difficult to get all clinical health care providers to carefully consider the findings of both nursing and medical research and then deliver quality outcomes. Clinicians, nursing leaders, and nursing managers must promote the use of EBP to develop best practices at all levels of care. Research can be facilitated within the institution, and then the findings can be reviewed and implemented. A change in practice can be facilitated through the collaboration of nursing and medicine, working closely with quality improvement teams to deliver quality outcomes. There are many potential outcomes of integrating evidence-based practice into an organization's culture (Table 5-6).

**TABLE 5-6   Potential Outcomes of Integrating Evidence-Based Practice into Organizational Culture**

| ORGANIZATION | PATIENTS |
| --- | --- |
| Improved recruitment of nurses | Reduced length of stay |
| Improved retention of nurses | Reduced readmissions |
| Improved employee satisfaction | Reduced mortality and morbidity |
| Higher percentage of nurses pursuing or attaining advanced degrees in nursing | Improved patient satisfaction |

**Source:** Compiled with information from LoBiondo-Wood, G. & Haber, J. (2010). *Nursing research: Methods and critical appraisal for evidence-based practice.* St. Louis, MO: Mosby.

## REAL WORLD INTERVIEW

I entered the hospital last year for a complete knee replacement. I was confident that I had the best orthopedic surgeon. He had told me of the pain that would follow the surgery and accompany the therapy. He told me that the hospital followed a clinical protocol for pain management that had been developed from the best research possible, so I approached surgery confidently. The anesthetist administered the epidural and enough anesthesia to keep me semi-awake during surgery. I relaxed and slept until I was moved to my room. When I was awake, I experienced the most excruciating pain I had ever had. When I asked for the nurse, she came in and told me I was going to have pain. She also said I would be able to press a button and receive measured doses of the epidural and be relieved. Each time she checked on me, she asked at what number between one and ten was my pain. I told her for seven hours that the pain was unbearable. My family pleaded

*(Continues)*

## REAL WORLD INTERVIEW (CONTINUED)

for the nurse to get help. She said she had called the anesthetist, who finally appeared and checked the epidural. He said it was not working, but he would be happy to insert a new tube, and if he did it, it would work. That means that for seven solid hours I had no working treatment to manage post-surgery bone pain. What's the use of having well-researched pain protocols if they are not used properly?

Instead of having another epidural inserted, I asked for morphine, which I was not able to tolerate. The next day, they gave me Demerol, which made my therapy bearable. I did not fill out the patient satisfaction survey they sent following my dismissal, because I was disgusted, and I did not feel they would ever take any action anyway. Something was definitely wrong with their pain-management system. So much for the best research possible! It only works if it is administered properly. I want to add that caring and compassion should be the most important qualities a health caregiver should be taught. I did have one nurse who was very considerate and informative. I wish she were working the day I went to surgery. Maybe she would have gotten help for my pain.

**Patricia A. Murry Kelly**
Patient
Munster, Indiana

## KEY CONCEPTS

- The focus on EBP can be expected to remain a driving force in the health care arena in the foreseeable future.
- Nursing can make significant contributions to the advancement of EBP.
- Nursing leaders and managers can promote a culture receptive to the practice of EBP, and all nurses can support this.

- Ultimately, EBP is the gold standard in clinical care.
- By accepting the challenge to provide EBP, nursing can pursue its future, confident of its ability to contribute to an increasingly complex health care system.

## KEY TERMS

evidence-based nursing (EBN)

evidence-based practice (EBP)

## REVIEW QUESTIONS

1. What is the major purpose of evidence-based practice (EBP)?
   A. To increase variability of care
   B. To cause a link to be missing in clinical care
   C. To determine what medical models can be applied by nursing
   D. To provide EBP in support of clinical competency

2. Which of the following is an accurate statement concerning EBP at this time?

A. EBP takes the place of continuous quality improvement.
B. Because we can already demonstrate effective and efficient care, EBP is redundant.
C. Leaders and managers in nursing are not clinicians, generally speaking, and so do not have a part in EBP processes.
D. Generally speaking, EBP is recognized by nursing, medicine, and health policy makers as state-of-the-art science reports.

3.  Which of the following organizations develops clinical practice guidelines on multiple clinical conditions?
    A.  American Heart Association
    B.  Agency for Healthcare Research and Quality (AHRQ)
    C.  Pew Health Professions Commission
    D.  Joint Commission (JC)

4.  Which of the following is true about evidence for use in clinical practice?
    A.  It is typically adopted quickly by clinicians.
    B.  It always needs to be interpreted for appropriateness for a specific patient.
    C.  It should never be challenged by nurses.
    D.  It provides unarguable direction for patient care.

5.  Which of the following is likely to provide the strongest evidence about diabetic patient care?
    A.  A study found in the literature about a patient with type 2 diabetes
    B.  An article about diabetes in a nursing research journal
    C.  An article about diabetes in a medical research journal
    D.  A systematic meta-analysis of the literature on diabetic nursing care.

6.  Assessment of evidence from what source must be sought first when a patient complains of pain?
    A.  The literature
    B.  Your nursing textbook
    C.  The patient
    D.  Your unit's policy manual

7.  Potential outcomes for organizations that adopt policies based on EBP include which of the following? Select all that apply.
    _____ A.  Improved quality of care
    _____ B.  Improved cost savings
    _____ C.  Decreased patient satisfaction
    _____ D.  Decreased lengths of stay
    _____ E.  Improved patient outcomes

8.  Nurses must consider which of the following when working with patients?
    A.  What the patient wants as well as what is medically best
    B.  What science decides is best for the patient
    C.  Only what the patient wants
    D.  Only what the patient's family wants

## REVIEW ACTIVITIES

1.  Review Table 5-2, Levels of Evidence. Are the evidence levels clear to you? Look at a patient's condition that you encounter in your clinical lab experience. Which level of evidence supports the care delivery approaches for this patient? Are any clinical pathways or standards in use in caring for this patient?

2.  Using the five-step process for EBP outlined in the Evidence from the Literature by Melnyk and Fineout-Overholt in this chapter, work to improve patient care on a health care condition of your choice. Use the PICO approach to focus your literature review.

## EXPLORING THE WEB

Go to the following sites to develop your knowledge of evidence-based health care:

■ American College of Physicians (ACP) ACP Journal Club. ACP Journal Club's general purpose is to select from the biomedical literature those articles reporting original studies and systematic reviews that warrant immediate attention by clinicians attempting to keep pace with important advances in internal medicine. www.acpjc.org

■ Agency for Healthcare Research and Quality (AHRQ): www.ahrq.gov

■ Center for Evidence-Based Medicine, University of Toronto. Note the evidence-based nursing syllabi and resources. www.cebm.utoronto.ca

■ Cumulative Index to Nursing and Allied Health (CINAHL). The CINAHL database covers nursing, allied health, biomedical and consumer health journals, and publications of the American Nursing Association and the National League for Nursing. More than 350,000 records and 900 journals are included. www.cinahl.com

■ Cochrane Library. The Cochrane Library contains four databases: the Cochrane Database of Systematic Reviews (CDSR), the Database of Abstracts of Reviews of Effectiveness (DARE), the Cochrane Controlled Trials Register (CCTR), and the Cochrane Review Methodology Database (CRMD). View introductory information for free: www.cochrane.org/index.htm

- Free access to Cochrane Reviewer's Handbook:
  www.cochrane.org/training/cochrane-handbook

- EBM Education Center of Excellence, North Carolina:
  http://library.ncahec.net

- Evaluating the Literature: Quality Filtering and Evidence-Based Medicine and Health. Available from the National Library of Medicine:
  www.nlm.nih.gov

- Evidence-Based Medicine:
  www.acponline.org

- Evidence-Based Medicine ToolKit:
  www.med.ualberta.ca

- Health Information Research Unit, McMaster University:
  http://hiru.mcmaster.ca

- Searching CINAHL Using the Ovid Web Gateway, Duke University Medical Center Library:
  www.mclibrary.duke.edu

- University of North Carolina Health Sciences Library:
  www.hsl.unc.edu

## REFERENCES

Bernadette Mazurek Melnyk, PhD, RN, CPNP/NPP, FAAN, FNAP, is Associate Dean for Research and Professor; Director, Center for Research & Evidence-Based Practice and Pediatric Nurse Practitioner (PNP) and Dual PNP/Psychiatric Mental Health Nurse Practitioner Programs, University of Rochester School of Nursing, Rochester, NY.

Ciliska, D., Cullum, N., & Marks, S. (2001). Evaluation of systematic reviews of treatment or prevention interventions. *Evidence-Based Nursing, 2*, 102–104.

DiCenso, A., Guyatt, G., & Ciliska, D. (2005). *Evidence-based nursing: A guide to clinical practice* (Glossary, pp. 547–573). St. Louis, MO: Elsevier Mosby.

Eddy, D. M. (2005). Evidence-based medicine: A unified approach. *Health Affairs, 14*(1), 9–17.

Fager, J., & Melnyk, B. M. (2004). The effectiveness of intervention studies to decrease alcohol use in college undergraduate students: An integrative analysis. *Worldviews on Evidence-Based Nursing, 1*(2), 102–119.

Garrard, J. (2007). Health sciences literature review made easy: The matrix method. Gaithersberg, MA: Aspen.

Glasziou, F. T., Vandenbrouke, J., & Chalmers, I. (2004). Assessing the quality of research. *British Medical Journal, 328*, 39–41.

Graham, I. D., Logan, J., Harrison, M. B., Straus, S. E., Tetroe, J., Caswell, W., & Robinson, N. (2006). Lost in knowledge translation: Time for a map? *Journal of Continuing Education in the Health Professions, 26*(1), 13–24.

Guyatt, F., Haynes, B., Jaeschke, R., Cook, D., Green, L., Naylor, C., Wilson, M., & Richardson, W. (2000). EBM: Principles of applying users' guides to patient care. *JAMA, 284*(10), 1290–1296.

Guyatt, G., & Renee, D. (2002). Users' guides to the medical literature. Washington, DC: American Medical Association (AMA) Press.

Haller, K. (2006, Summer). Safe staffing saves lives. *John Hopkins Nursing, 4*(2), 16.

Horsley, J. A., Crane, J., Crabtree, M. K., & Wood, D. J. *Using research to improve nursing practice: A guide.* New York: Grune & Stratton; 1983.

LoBiondo-Wood, G., & Haber, J. (2010). *Nursing research: Methods and critical appraisal for evidence-based practice.* St. Louis, MO: Mosby.

Malloch, K., & Porter-O'Grady, T. (2006). Evidence-based practice in nursing and health care. Boston: Jones and Bartlett Publishers.

Melnyk, B. M., & Fineout-Overholt, E. (2005). Evidence-based practice in nursing & healthcare. A guide to best practice. Philadelphia: Lippincott Williams & Wilkins.

Melnyk, B. M., & Fineout-Overholt, E. (2010). *Evidence-based practice in nursing and healthcare: A guide to best practice.* Philadelphia: Lippincott.

Michaels, C., McEwen, M. M., & McArthur, D. B. (2008). Saying "no" to professional recommendations: Client values, beliefs, and evidence-based care. *Journal of the Academy of Nurse Practitioners, 20*, 585–589.

Moch. S. D., Cronje, R. J., & Branson, J. (2010). Part 1. Undergraduate nursing evidence-based practice education: Envisioning the role of students. *Journal of Professional Nursing, 26*(1), 5–13.

Munten, G., van den Bogaard, J., Cox, K., Garretsen, H., & Bongers, I. (2010). Implementation of evidence-based practice in nursing using action research: A review. *Worldviews on Evidence-Based Nursing, 7*(3), 135–157.

Pravikoff, D. S., Tanner, A. B., & Pierce, S. T. (2005). Readiness of U.S. nurses for evidence-based practice. *American Journal of Nursing, 105*(9), 40–51.

Polit, D. F., & Beck, C. T. (2008). Nursing research: Generating and assessing evidence for nursing practice (8th ed.). Philadelphia: Lippincott.

Rogers, E. (2003). Diffusion of innovations (5th ed.). New York: Free Press.

Sackett, D. L., Rosenberg, W. M. C., Gray, J. A. M., Haynes, R. B, & Richardson, W. S. (1996). Evidence-based medicine: What it is and what it isn't. *British Medical Journal, 312*(7023), 71–72.

Sigma Theta Tau International. (2005). Evidence based nursing position statement. Retrieved December 26, 2010, from http://www.nursingsociety.org/aboutus/PositionPapers/Pages/EBN_positionpaper.aspx.

Stevens, K. R. (2004). *ACE Star Model of EBP: Knowledge Transformation.* Academic Center for Evidence-based Practice. The University of Texas Health Science Center at San Antonio. Retrieved April 5, 2011, from http://www.acestar.uthscsa.edu/acestar-model.asp.

Stetler, C.B. (2001). Updating the Stetler Model of Research Utilization to Facilitate Evidence-Based Practice. Nursing Outlook 2001;49:272–9. Retrieved April 5, 2011, from http://www.rqt.qc.ca/repDoc/Symposiums/2007/Lundi%2026%20novembre%20PM/Article_Stetler%20model%20%C3%A9valuation.pdf.

Strickland, R. J., & O'Leary-Kelley, C. (2009). Clinical Nurse Educators' Perceptions of Research Utilization. *Journal for Nurses in Staff Development, 25*(4), 164–171.

Titler, M. G., Kleiber, C., Steelman, V. J., Rakel, B. A., Budreau, G., Everett, L. Q., Buckwalter, K. C., Tripp-Reimer, T., & Goode, C. (2001). The Iowa model of evidence-based practice to promote quality care. *Critical Care Nursing Clinics of North America, 13*(4), 497–509.

Retrieved April 5, 2011, from http://www.mosescone.com/documents/public/Nursing%20Research/Iowa%20Model%201998.pdf.

Tompkins, E. S. (1985). The CURN project: Uniting research and practice. *New Jersey Nurse, 15*(8), 20.

Tuite, P. K., & George, E. L. (2010). The role of the clinical nurse specialist in facilitating evidence-based practice within a university setting. *Critical Care Nursing Quarterly 33*(2), 117–125.

Van Achterberg, T., Schoonhoven, L., & Grol, R. (2008). Nursing implementation science: How evidence-based nursing requires evidence-based implementation. *Journal of Nursing Scholarship, 40*(4), 302–310.

Varnell, G., Haas, B., Duke, G., & Hudson, K. (2008). Effect of an educational intervention on attitudes toward and implementation of evidence-based practice. *Worldviews on Evidence-Based Nursing, 5*(4), 172–181.

# SUGGESTED READINGS

Allen, D. E., Bockenhauer, B., Egan, C., & Kinnaird, L. S. (2006). Relating outcomes to excellent nursing practice. *Journal of Nursing Administration (JONA), 36*(3), 140–147.

Arnold, L., Campbell, A., Dubree, M., Fuchs, M. A., Davis, N., Hertzler, et al. (2006). Priorities and challenges of health system chief nursing executives: Insights for nursing educators. *Journal of Professional Nursing, 22*(4), 213–220.

Brewer, C. S. (2005). Health services research and the nursing workforce: Access and utilization issues. *Nursing Outlook, 53*(6), 281–290.

Clancy, C., Sharp, B. A., & Hubbard, H. B. (2005). Guest editorial: Intersections for mutual success in nursing and health services research. *Nursing Outlook, 53*(6), 263–265.

Engelke, M. K., & Marshburn, D. M. (2006). Collaborative strategies to enhance research and evidence-based practice. *The Journal of Nursing Administration (JONA), 36*(3), 131–135.

Ervin, N. E. (2006). Does patient satisfaction contribute to nursing care quality? *Journal of Nursing Administration (JONA), 36*(3), 126–130.

Fineout-Overholt, E., Mazurek Melynk , B. Stillwell, S.B. & Williamson, K.M. (2010). Evidence-Based Practice step by step: implementing an evidence-based practice change. *American Journal of Nursing.* March;111(3):54–60.

Goode, C. J. (2000). What constitutes the "evidence" in evidence-based practice? *Applied Nursing Research, 13,* 222–225.

Healy, B. (2006). Who says what's best? *U.S. News & World Report, 141*(9), 75.

Hutchinson, A. M., & Johnston, L. (2006). Beyond the BARRIERS scale: Commonly reported barriers to research use. *Journal of Nursing Administration (JONA), 36*(4), 189–199.

Jangold, K. L., Pearson, K. K., Schmitz, J. R., Scherb, C. A., Specht, J. P., & Loes, J. L. (2006). Perceptions and characteristics of registered nurses' involvement in decision making. *Nursing Administration Quarterly, 30*(3), 266–272.

Joanna Briggs Institute for Evidence Based Nursing & Midwifery. Retrieved March 9, 2007, from http://www.joannabriggs.edu.au/services/search.php.

Jones, C. B., & Mark, B. A. (2005). The intersection of nursing and health services research: An agenda to guide future research. *Nursing Outlook, 53*(6), 324–332.

Jones, C. B., & Mark, B. A. (2005). The intersection of nursing and health services research: Overview of an agenda setting conference. *Nursing Outlook, 53*(6), 270–273.

Meyer, T. C. (2005). Following up on evidence-based medicine. *Wisconsin Medical Journal, 104*(3), 1.

Newhouse, R. P., Pettit, J. D., Poe, S., & Rocco, L. (2006). The slippery slope: Differentiating between quality improvement and research. *Journal of Nursing Administration (JONA), 36*(4), 211–219.

Potylycki, M. J., Kimmel, S. R., Ritter, M., Capuano, T., Gross, L., Riegel-Gross, K. I., et al. (2006). Nonpunitive medication error reporting: 3-year findings from one hospital's primum non nocere initiative. *Journal of Nursing Administration (JONA), 36*(7/8), 370–376.

Pravikoff, D. S., Pierce, S. T., & Tanner, A. (2005, Jan.–Feb.). Evidence-based practice readiness study supported by academy nursing informatics expert panel. *Nursing Outlook, 53*(1), 49–50.

Preheim, G. J., Armstrong, G. E., & Barton, A. J. (2009). The new fundamentals in nursing: Introducing beginning quality and safety education for nurses' competencies. *Journal of Nursing education, 48*(12), 694–697.

Sackett, D. L., Rosenberg, W. M., Gray, J. A., Haynes, R. B., & Richardson, W. S. (2007). Evidence based medicine: What it is and what it isn't. *Clinical Orthopaedics and Related Research, 455,* 3–5.

Schmidt, N. A., & Brown, J. M. (2009). Evidence-based practice for nurses: appraisal and application of research. Jones and Bartlett Pub. LLC.

Simpson, R. L. (2006). Evidence-based practice: How nursing administration makes it happen. *Nursing Administration Quarterly, 30*(3), 291–294.

Stevens, K. R., & Staley, J. M. (2006). The Quality Chasm reports, evidence-based practice, and nursing's response to improve healthcare. *Nursing Outlook, 54*(2), 94–101.

# CHAPTER 6

# Nursing and Health Care Informatics

JOSETTE JONES, RN, PhD; LESLIE H. NICOLL, PhD, MBA, RN, BC

*I think it's fair to say that personal computers have become the most empowering tool we've ever created. They're tools of communication, they're tools of creativity, and they can be shaped by their user.*

(BILL GATES, N.D.)

## OBJECTIVES

Upon completion of this chapter, the reader should be able to:

1. List the components that define a nursing specialty and discuss how nursing informatics meets these requirements.

2. Relate educational opportunities for nurses interested in pursuing a career in nursing informatics.

3. Identify current challenges for health information technology applications.

4. Relate how ubiquitous computing and virtual reality have the potential to influence nursing education and practice.

5. Use established criteria to evaluate the content of health-related sites found on the Internet.

6. Identify the role of informatics in evidence-based practice (EBP) and patient safety.

Delmar/Cengage Learning

*Consider the health care consumer of the year 2015:*

*Everyone could carry around a Personal Health Record (PHR) on a type of pocket-sized electronic Smart card that stores individual health care data accessible anytime, anywhere by patients and health care providers. Patients could be diagnosed and treated with the aid of telemedicine at the point where they are living.*

*Clinicians are no longer the only sources of health care advice for consumers. Paging "Doctor Google" and using online patient communities are complementing traditional health care delivery. Electronic e-patients are placing themselves at the center of a network of electronic resources. They are also reading personal online journals (blogs) of patients like themselves, listening to podcasts, updating their social network profile, and posting comments. Many people, once they find health information online, talk with someone about it offline. Providers can be contacted by e-mail 24/7. Self-diagnosis, self-care, and self-medication are in!*

*How will all of this affect the health care consumer and the practice of nursing?*

Computers and the Internet have changed the way we communicate, obtain information, work, entertain, and make important health decisions. As documented in the Pew Report from April 19, 2006 (Horrigan & Rainie, 2006), some 21 million Americans rely on the Internet in a crucial or important way for career training. Another 17 million rely on it when choosing a school for themselves or for a child.

Computers and the Internet are changing health care delivery too. 61% of all adults get health info online (PEW Research Center 2010). A literature review on the growing application of clinical information technology in managing depression care highlights that computer reminders were shown to be superior to manual reminders in improving adherence to a clinical practice guideline (Kilbourne, McGinnis, Belnap, Klinkman, & Thomas, 2006). An integrated electronic reminder system also resulted in variable improvement in care for diabetes and coronary artery disease (Sequist, et al., 2005). Another study showed that onscreen computer-generated reminders sent to practitioners of patients lacking a recent potassium test increased potassium testing by 9.8% (p , 0.001) (Hoch, et al., 2003; Matheny, et al., 2008). On a regional and national level, health care communities are successfully moving forward with health information exchanges to create Local or Regional Health Information Infrastructures (LHII) and are improving the efficiency, quality, and safety of care by interconnecting local as well as national health information resources (Sicotte & Paré, 2010).

Nurses are not immune to the changes that computers are bringing to both everyday life and nursing practice. By 2015, use of a certified electronic health record (EHR) is mandated under the Health Information Technology for Economic and Clinical Health (HITECH) Act (Centers for Medicare and Medicaid Services, 2010). HITECH created new Medicare and Medicaid incentive payment programs totaling as much as $27 billion to help eligible physicians, other professionals, and hospitals as they transition from paper-based medical records to EHRs. The Medicare and Medicaid EHR Incentive Programs provide a financial incentive to health care providers for the "meaningful use" of certified EHR technology to achieve health and efficiency goals. By putting into action and meaningfully using an EHR system, providers must show they're using certified EHR technology in ways that can be measured significantly in quality and in quantity, e.g., to improve the quality of health care and/or to submit information about clinical quality and other measures. Providers will reap benefits beyond financial incentives–such as a reduction in errors, improved availability of records and data, increased clinical reminders and alerts, support for clinical decisions, and increased use of e-prescribing/refill automation. To demonstrate meaningful use successfully, eligible professionals and hospitals are required to report information about clinical quality measures specific to them (Centers for Medicare and Medicaid Services, 2010). Providers who do not meet the requirements and become "meaningful users" of EHR by 2015 will have their reimbursement reduced (Health law blog, 2010). EHR can provide many benefits for providers and patients with more complete and accurate health information and better access to it. As professionals, information technology can help achieve the goals of quality patient-centered care and increased patient safety. Whether you are a nursing student learning a clinical procedure using a computer-based instruction program; a nurse on the floor using electronic devices such as ventilators, intravenous pumps, telemetry, and the electronic health record; a nursing administrator using a spreadsheet and database to plan a budget; or a nursing researcher or clinician keeping updated with the latest evidence-based nursing care, it is evident that information technology has become an essential part of professional nursing practice, both on the individual and institutional level.

In this day and age, most everyone is involved with computers to some degree. In the past decade alone, there has been a significant increase in the demand for nurses whose knowledge of nursing and specialization in informatics contributes to nursing practice, leadership, education, and research throughout the United States, Canada, and other countries (Carroll, Bradford, Foster, Cato, & Jones, 2006).

In addition, many nurses are testing the limits of new patient care models using computers. Go to the American Academy of Colleges of Nursing Web site and review their Edge Runners initiative (http://www.aannet.org, clicking on Raise the Voice and then Edge Runners, accessed March 23, 2011). It features many innovative models of care developed by nurse entrepreneurs. In addition, health care home (HCH) projects, also known as patient-centered medical home demonstration projects, are being conducted nationwide to integrate patients as active participants in their own health and well-being, using the best available technology and evidence (Centers for Medicare and Medicaid Services, 2011).

This chapter will introduce you to the world of nursing informatics, computers, and information technology. The professional RN of the twenty-first century will not be effective in the role without a solid base of knowledge related to the impact of nursing informatics, computers, and information technology on nursing practice, patient care, and patient outcomes.

# NURSING INFORMATICS

The definition of **nursing informatics** agreed upon by the International Medical Informatics Association—Nursing Informatics (IMIA-NI), Special Interest Group, at their General Assembly in Stockholm in 1997 (amended for clarity in Seoul, 1998) is: the integration of nursing, its information, and information management with information processing and communication technology to support the health of people worldwide. The focus of IMIA is to foster collaboration among nurses and others who are interested in nursing informatics (www.imia.org).

## E-HEALTH AND TELEHEALTH

Related terms in the field of informatics are *e-health* and *telehealth*. E-health is an emerging field in the intersection of medical informatics, public health, and business, referring to health services and information delivered or enhanced through the Internet and related technologies. In a broader sense, the term characterizes not only a technical development but also a state of mind, a way of thinking, an attitude, and a commitment for networked, global thinking to improve health care locally, regionally, and worldwide by using information and communication technology (Oh, Rizo, Enkin, & Jadad, 2005).

Telehealth is the delivery of health-related services and information for health promotion, disease prevention, diagnosis, consultation, education, and therapy, via telecommunications technologies and computers to patients and providers at another location. It may be as simple as two health professionals discussing a case over the telephone or as sophisticated as using satellite technology to broadcast a consultation between providers at facilities in two countries, using videoconferencing equipment or robotic technology. It may include transmission of patient images or data, e.g., pulse oximetry, heart rate, blood pressure, etc., for diagnosis or disease management; groups or individuals exchanging health services, education, health care advice, or research via live videoconference; or home monitoring of dialysis or cardiac patients. Telehealth can improve health care access in vulnerable patient populations through the use of electronic devices in patients' homes that monitor and assess for early complications (Prinz, Cramer, & Englund, 2008).

## ELEMENTS OF NURSING INFORMATICS

Some common elements of nursing informatics (NI) include the following:

- Computerized order entry
- Electronic health records
- Patient decision support tools, clinical and business related
- Laboratory and X-ray results reporting as well as picture retrieval and viewing systems
- Electronic prescribing, order entry, and medication administration systems including barcoding
- Community and population health management and information
- Communication using Internet, intranet, and e-mail; staffing; and administrative systems, e.g., billing
- Evidence-based knowledge and information retrieval systems with access to remote library and Internet resources
- Quality improvement data collection/data summary systems/business intelligence
- Documentation and care planning
- Patient monitoring for vital signs and other measurements directly into the patient's record
- Problem alerts for vital signs and other measurements
- Electronic bed boards to review bed status and availability
- Data mining techniques for sifting through large amounts of data to discover knowledge
- Disease surveillance systems
- Web pages to personalize information
- Computer-generated nursing care plans, critical pathways, and patient documentation such as discharge instructions and medication information

- Staff reminders of planned nursing interventions and documentation prompts to ensure comprehensive charting
- Access to computer-archived patient data from previous patient encounters
- Collaboration with patients, other nurses, and health care providers (DeLaune, 2009)

## NEED FOR NURSING INFORMATICS

Few American hospitals and practitioners have implemented health information technology (Medical Records Institute, 2006), (Johnston, et al., 2003). Despite a consensus that the use of health information technology should lead to more efficient, safer, and higher-quality care, in 2009, the adoption rate of the EHR is only 43.9% (Ford, Menachemi, Peterson, & Huerta, 2009). The vast majority of health care transactions in the United States still take place on paper, a system that has remained unchanged since the 1950s. While other industries have reaped the benefits of their information technology (IT) investments, the adoption and diffusion of information systems in the health care arena has been growing more slowly (Menachemi, Randeree, Burke, & Ford, 2008).

The development of standards for EHRs is important to the national health care agenda. Without accessible, standardized EHRs, practicing clinicians, pharmacies, and hospitals cannot share patient information, which is necessary for timely, patient-centered, and portable care. There are currently multiple competing vendors of EHR systems, each selling a software suite that in many cases is not compatible with those of their competitors. Only counting the outpatient vendors, there are more than 25 major brands currently on the market. The position of National Coordinator for Health Information Technology was created through an Executive Order in 2004. In 2009, the Health Information Technology for Economic and Clinical Health Act (HITECH), which was part of the American Recovery and Reinvestment Act, was passed with the goals of addressing computer interoperability issues. Interoperability is the ability of a computer to connect with other computers in various settings in a secure, accurate and efficient way without special effort on the part of the user and without any restricted access or implementation. HITECH also established a National Health Information Network (NHIN). Under the Office of the National Coordinator for Health Information Technology (ONC), Regional Health Information Organizations (RHIOs) have been established in many states to promote the sharing of health information.

The Veterans Affairs (VA) hospital system has the largest enterprise-wide health information system that includes an electronic health record, known as Veterans Health Information Systems and Technology Architecture (VistA). Health care providers can review and update a patient's electronic health record at any of the VA's over 1,000 health care facilities. The EHR has the ability to place orders regarding medications, diet, labs, and, x-rays, etc. The U.S. Indian Health Service has an EHR that is similar to VistA.

As of 2005, one of the largest projects for a national EHR is by the National Health Service (NHS) in the United Kingdom. The goal of the NHS is to have 60,000,000 patients with a centralized electronic health record. The Canadian province of Alberta began an EHR project in 2005 that is expected to encompass all of Alberta (Electronic Health Record, 2008).

## JOINT COMMISSION NATIONAL PATIENT SAFETY GOALS

Several other forces have highlighted the need for increased patient technology and nursing informatics. The Joint Commission (JC), formerly the Joint Commission on Accreditation of Healthcare Organizations (JCAHO), has set the following National Patient Safety Goals for 2011, many of which require the use of technology:

- Use at least two ways to identify patients. For example, use the patient's name and date of birth. This is done to make sure that each patient gets the correct medicine and treatment.
- Make sure that the correct patient gets the correct blood when they get a blood transfusion.
- Get important test results to the right staff person on time.
- Before a procedure, label medicines that are not labeled. For example, medicines in syringes, cups and basins. Do this in the area where medicines and supplies are set up.
- Take extra care with patients who take medicines to thin their blood.
- Use the hand cleaning guidelines from the Centers for Disease Control and Prevention or the World Health Organization. Set goals for improving hand cleaning. Use the goals to improve hand cleaning.
- Use proven guidelines to prevent infections that are difficult to treat.
- Use proven guidelines to prevent infection of the blood from central lines.
- Use proven guidelines to prevent infection after surgery.
- Find out what medicines each patient is taking. Make sure that it is OK for the patient to take any new medicines with their current medicines.
- Give a list of the patient's medicines to their next caregiver. Give the list to the patient's regular doctor before the patient goes home.
- Give a list of the patient's medicines to the patient and their family before they go home. Explain the list.

- Some patients may get medicine in small amounts or for a short time. Make sure that it is OK for those patients to take those medicines with their current medicines.
- Find out which patients are most likely to try to commit suicide.
- Make sure that the correct surgery is done on the correct patient and at the correct place on the patient's body.
- Mark the correct place on the patient's body where the surgery is to be done.
- Pause before the surgery to make sure that a mistake is not being made ( Joint Commission, 2011).

Attainment of many of these Goals requires the judicious use of technology.

## LEAPFROG GROUP

The Leapfrog Group is another force advocating for technology. Leapfrog is a voluntary program aimed at using employer purchasing power to alert America's health industry that big leaps in health care safety, quality, and customer value will be recognized and rewarded. Among other initiatives, Leapfrog works with its employer members to encourage transparency and easy access to health care information as well as rewards for hospitals that have a proven record of high quality care (www.leapfroggroup.org). Data support services are offered to Leapfrog by Thomson Medstat (www.medstat.com).

## THE NATIONAL QUALITY FORUM

The National Quality Forum (NQF) is a not-for-profit membership organization created to develop and implement a national strategy for health care quality measurement and reporting (www.qualityforum.org). A shared sense of urgency about the impact of health care quality on patient outcomes, workforce productivity, and health care costs prompted leaders in the public and private sectors to create the NQF as a mechanism to bring about national change. The NQF measurement framework focuses on patient outcomes, processes of care, the cost of resources used for an episode of care, and the use of health information technology.

 **EVIDENCE** FROM THE LITERATURE

**Citation:** Stevens, K. R., & Staley, J. M. (2006). The Quality Chasm reports, evidence-based practice, and nursing response to improve healthcare. *Nursing Outlook, 54*(2), 94–101.

**Discussion:** In a growing set of landmark reports, the institute of Medicine (IOM) set in motion a sweeping quality initiative for reform of the health care system. Many of the IOM recommendations incorporate evidence-based practice and technology applications. New rules to redesign and improve care are highlighted in this article as follows. Patient care requires:

- *Care based on continuous healing relationships:* Patients should receive care whenever they need it and in many forms, not just face-to-face visits. This rule implies that the health care system should be responsive at all times (24 hours a day, every day) and that access to care should be provided over the Internet, by telephone, and by other means in addition to face-to-face visits.

- *Customization based on patient needs and values:* The system of care should be designed to meet the most common types of needs but have the capability to respond to individual patient choices and preferences.

- *The patient as the source of control:* Patients should be given the necessary information and the opportunity to exercise the degree of control they choose over health care decisions that affect them. The health care system should be able to accommodate differences in patient preferences and encourage shared decision making.

- *Shared knowledge and the free flow of information:* Patients should have unfettered access to their own medical information and to clinical knowledge. Clinicians and patients should communicate effectively and share information.

- *Evidence-based decision-making (EBDM):* Patients should receive care based on the best available scientific knowledge. Care should not vary illogically from clinician to clinician or from place to place.

*(Continues)*

# EVIDENCE FROM THE LITERATURE (Continued)

- *Safety as a system property:* Patients should be safe from injury caused by the care system. Reducing risk and ensuring safety require greater attention to systems that help prevent and mitigate errors.

- *The need for transparency:* The health care system should make information available to patients and their families that allows them to make informed decisions when selecting a health plan, hospital, or clinical practice, or when choosing among alternative treatments. This should include information describing the system's performance on safety, *EBDM*, and patient satisfaction.

- *Anticipation of needs:* The health system should anticipate patient needs, rather than simply react to events.

- *Continuous decrease in waste:* The health system should not waste resources or patient time.

- *Cooperation among clinicians:* Clinicians and institutions should actively collaborate and communicate to ensure an appropriate exchange of information and coordination of care (Institute of Medicine [IOM], 2001).

The IOM also offers analysis of the current health system and recommends changes in education of health professionals. These suggested changes include several core competencies requiring the use of technology. Health professionals must:

- *Provide patient-centered care:* Identify, respect, and care about patients' differences, values, preferences, and expressed needs; relieve pain and suffering; coordinate continuous care; listen to, clearly inform, communicate with, and educate patients; share decision making and management; and continuously advocate disease prevention, wellness, and promotion of healthy lifestyles, including a focus on population health.

- *Work in interdisciplinary teams:* Cooperate, communicate, and integrate care in teams to ensure that care is continuous and reliable.

- *Employ evidence-based practice (EBP):* Integrate best research with clinical expertise and patient values for optimum care; participate in learning and research activities to the extent feasible.

- *Apply quality improvement:* Identify errors and hazards in care; understand and implement basic safety design principles such as standardization and simplification; continually understand and measure quality of care in terms of structure, process, and outcome in relation to patient and community needs; and design and test interventions to change processes and systems of care, with the objective of improving quality.

- *Utilize informatics:* Communicate, manage knowledge, mitigate error, and support decision making using information technology (IOM, 2003).

**Implications for Practice:** Health care professionals will be expected to be familiar with technology, EBP, and EBDM as they deliver care to patients in the future. Schools of Nursing must move to prepare students and faculty to meet this demand.

## THE TIGER INITIATIVE AND QUALITY AND SAFETY EDUCATION FOR NURSES (QSEN)

The Technology Informatics Guiding Educational Reform (TIGER) Initiative aims to enable practicing nurses and nursing students to fully engage in the unfolding digital electronic era in health care. The TIGER Initiative is working to catalyze a dynamic, sustainable, and productive relationship between the Alliance for Nursing Informatics (ANI), with its 20 nursing informatics professional societies, and the major nursing organizations including the American Nurses Association (ANA), the Association of Nurse Executives (AONE), the American Association of Colleges of Nursing (AACN) and others which collectively represent over 2,000,000 nurses. The TIGER Initiative is focused on using informatics tools, principles, theories, and practices to enable nurses to make health care safer, more effective, efficient, patient centered, timely, and equitable (http://www.tigersummit.com/About_Us.html, accessed March 24, 2011). Collaborative teams have researched best practices from both nursing education and practice within

## EVIDENCE FROM THE LITERATURE

**Citation:** Cibulka N.J., Crane-Wider L. (2011). Introducing personal digital assistants to enhance nursing education in undergraduate and graduate nursing programs. Journal of Nursing Education. February; 50(2):115–8.

**Discussion:** This article describes how a school of nursing implemented an innovative program to introduce personal digital assistants (PDAs) to undergraduate and graduate nursing students. Undergraduate students studying pharmacology and nurse practitioner graduate students in an adult health course were asked to purchase a personal digital assistant privately or through the university bookstore. Faculty selected an appropriate software package. After students were oriented to the hardware and software package, innovative teaching strategies were implemented to help guide students to use their mobile devices to access clinically relevant information. Student feedback about this experience was positive. The most important elements for successful adoption of personal digital assistants are to provide training for both faculty and students, and to develop learning opportunities using the technology.

**Implications for Practice:** The use of technology such as PDAs and other technology such as the Blackberry and iPhone is improving care. Nursing students and all nurses will see more use of technology to improve patient care. The use of mobile technologies is an important competency that will improve the quality of nursing practice. Its use should be considered in all nursing curricula.

---

nine key topic areas so that this knowledge can be shared through information technology.

The nine topic areas are:

- Standards and interoperability
- National health information technology (IT) agenda
- Informatics competencies
- Education and faculty development
- Staff development
- Usability and clinical application design
- Virtual demonstration center
- Leadership development
- Consumer empowerment and personal health records

(Technology Informatics Guiding Education Reform [TIGER], 2009). NI competencies are available on the TIGER web site and the Quality and Safety Education for Nurses (QSEN) web site (http://www.qsen.org/, accessed April 3, 2011). QSEN was developed as part of a Robert Wood Johnson–funded project designed to facilitate reform in nursing education in the areas of quality and safety. QSEN is a comprehensive resource for quality and safety education for nurses. This website is a place to learn and share ideas about educational strategies that promote quality and safety competency development in nursing.

## THE SPECIALTY OF NURSING INFORMATICS

According to the American Nurses Association (ANA, 2008), nursing informatics (NI) is a discipline-specific practice within the broader perspective of health informatics.

NI is a specialty that integrates nursing science, computer science, and information science to manage and communicate data, information, knowledge and wisdom into nursing practice (ANA, 2008). NI includes, but is not limited to, the following:

- Use of decision-making systems or artificial intelligence to support the nursing process
- Use of software applications to support health care organization, for example, staffing, bed allocation, and so on
- Integration of information technology in patient education
- Use of computer-aided learning for nursing education
- Development and use of nursing databases and nursing information systems (NIS)
- Use of research related to nurses' information management and communication

The focus of NI is on representation of nursing data, information, and knowledge, as well as the management and communication of nursing information within the broader context of health informatics. The Informatics Nurse Specialist has at least a master's degree in nursing informatics and functions in the role of project manager, consultant, educator, researcher, development supporter, policy developer, or entrepreneur related to nursing information technology applications. The Informatics Nurse Specialist is expected to demonstrate the competencies enumerated in the Standard of Practice for Nursing Informatics (ANA, 2008). There are numerous universities nationwide

# EVIDENCE FROM THE LITERATURE

**Citation:** Fineout-Overholt, E., Mazurek Melynk, B. Stillwell, S.B. & Williamson, K.M. (2010). Evidence-Based Practice step by step: implementing an evidence-based practice change. American Journal of Nursing. March;111(3):54–60.

**Discussion:** This is the ninth article in a series of articles that began November, 2009, from the Arizona State University College of Nursing and Health Innovation's Center for the Advancement of Evidence-Based Practice. Evidence-based practice (EBP) is a problem-solving approach to the delivery of health care that integrates the best evidence from studies and patient care data with clinician expertise and patient preferences and values. When delivered in a context of caring and in a supportive organizational culture, the highest quality of care and best patient outcomes can be achieved. The purpose of this series is to give nurses the knowledge and skills they need to implement EBP consistently, one step at a time. Articles appear every other month to allow time to incorporate information as readers work toward implementing EBP at their institution. Also, the authors have scheduled "Chat with the Authors" calls every few months to provide a direct line to the experts to help readers resolve questions.

**Implications for Practice:** Practicing nurses, nursing faculty, and students will want to review this series of articles. The articles in this series review various elements of EBP, e.g., use of PICO, rapid critical appraisal of research, etc. The articles can assist in planning continuing education opportunities and designing nursing curricula that prepare nurses for use of EBP and twenty-first century professional practice.

offering master's degrees, doctoral degrees, and postgraduate certifications in Nursing Informatics.

## Formal Programs in Informatics

As the health care industry relies more on information technology for the delivery of care, it is imperative that basic computer skills and nursing informatics competencies are incorporated in all levels of professional nursing education programs.

It was as recent as 1989 that the first Masters program in Nursing Informatics was established at the University of Maryland, followed by a doctoral program in 1992. Now, there are numerous universities nationwide offering master's degrees, doctoral degrees, and postgraduate certifications in Nursing Informatics. For a comprehensive list of programs for nursing, medical, and health informatics, refer to the AMIA site at https://www.amia.org/informatics-academic-training-programs. The American Nurses Credentialing Center (ANCC) offers certification examinations for a variety of specialties in nursing, including informatics. The ANCC Web site (www.nursingworld.org/ancc) details the nursing candidates requirements for the Informatics Nurse Certification Exam.

## Informal Education

There are numerous venues for all nurses to stay abreast of emerging health information technologies and trends. A popular continuing education seminar is the Weekend Immersion in Nursing Informatics (WINI), which can be

accessed at www.winiconference.net. Additionally, groups such as the American Medical Informatics Association (AMIA) and the Healthcare Information and Management Systems Society (HIMSS) have annual conferences that attract thousands of health information technology professionals and vendors, offering educational and networking opportunities for all informatics appetites. Many regions have formed nursing informatics groups that function at the local, national, and even international levels. These groups, such as ANIA-CARING (formerly the American Nursing Informatics Association and the Capital Area Roundtable on Informatics in Nursing), offer vital networking and educational services to their members. In 2004, many regional nursing informatics groups, representing 2,000 nurses, formed the Alliance for Nursing Informatics (ANI) in collaboration with HIMSS, the American Nurses Association (ANA), and ANIA-CARING (http://www.allianceni.org/about.asp). There are several nursing and health informatics scholarly journals, such as *Computers, Informatics, and Nursing* (www.cinjournal.com) and *Journal of Medical Informatics Association* (www.jamia.org), that provide essential elements of continuing industry trends and nursing informatics education.

## Certification

Certification in a specialty is a formal, systematic mechanism whereby nurses can voluntarily seek a credential that recognizes their quality and excellence in professional practice and continuing education (American Nurses

**Critical Thinking 6-1**

Search for the American Nurses Credentialing Center at www.nursingworld.org. Click on Certification and choose another specialty in which a nurse can be certified, comparing it to the certification in informatics.

How are they similar? How are they different?

Credentialing Center, 1993). For many nurses, becoming certified is a professional milestone and validation of their qualifications, knowledge, and skills in a defined area of nursing practice. The American Nurses Credentialing Center (ANCC) offers certification examinations for a variety of specialties in nursing, including informatics. The ANCC Website (www.nursingworld.org/ancc) details the nursing candidate's requirements for the Informatics Nurse Certification Exam.

## Career Opportunities

Career opportunities in the fields of computer science and information technology are growing at an exponential rate, and nursing is no exception. Nurses working in informatics can look forward to multiple job opportunities, with new roles continuously being developed as technology changes and matures. Changes in health care delivery, particularly managed care, have caused shifts in computer systems in care management, clinical systems, clinical data repositories, care mapping, and outcomes measures (Oroviogoicoechea, Elliott, & Watson, 2008). The Nursing Informatics Working Group of the AMIA maintains a repository of NI role descriptions at www.amia.org (Carroll, et al., 2007).

## HEALTH INFORMATION SYSTEMS

Health information systems (HISs) systems are changing the way that health care is delivered, whether in the hospital, the clinic, the provider's office, or the patient's. A health information system (HIS) is a comprehensive and integrated information system designed to manage clinical and administrative aspects of health care. A **clinical information system (CIS)** is a computerized system which manages clinical data with subsystems, such as electronic health records, clinical decision support systems, bedside medication administration using positive patient identification, computerized prescribing practitioner order entry, patient surveillance, and clinical data warehouses (Hebda & Czar, 2009). Administrative information systems manage financial and demographic information in hospitals and other health care systems and provide reporting capabilities to support patient care. Increasingly, health information systems function as the mechanisms to improve outcomes and reduce errors, as well as to control costs by realizing a host of efficiencies in clinical data entry and patient care.

An HIS can be patient focused or departmental in focus. In patient-focused systems, automation supports patient care processes and identifies patient outcomes. Typical applications found in a patient-focused system include order entry, test results reporting, clinical documentation, care planning, and clinical pathways. As data are entered into the system, data repositories are established that can be accessed to look for trends and outcomes of patient care. Departmental health information systems have evolved to meet the operational needs of particular departments, such as laboratory, radiology, pharmacy, medical records, or billing.

## HEALTH INFORMATION SYSTEMS IMPORTANT TO NURSING

Health information systems important to nursing practice include the **electronic health record (EHR)** which may or may not include Computerized Provider Order Entry (CPOE), Clinical Decision support (CDS), documentation systems, e-prescribing, and the **personal health record (PHR)**.

## ELECTRONIC HEALTH RECORD (EHR)

An electronic health record is a digital record of an individual's health information across settings and over time.

The EHR is a replacement for the paper medical record as the primary source of information for health care, meeting all clinical, legal, and administrative requirements. However, the EHR is more than a paper-based health record. Information technology permits much more data to be captured, processed, and integrated, which results in information that is broader than that found in a linear paper record.

The EHR is not a record in the traditional sense of the term. *Record* connotes a repository with limitations of size, content, and location. The term has traditionally suggested that the sole purpose for maintaining health data is to document events. Although this is an important purpose, the EHR permits health information to be used to support the generation and communication of evidence-based knowledge for health care delivery.

Data may be entered in an EHR in free text (such as in progress notes); in a structured form via a drop-down pick list; as images; or as digitized signals with associated meta data (e.g., electrocardiograms). Even if the EHR system collects data via drop-down pick lists, there is no guarantee that the values in the pick lists will be compatible with those of other systems in use at the health care institution.

## REAL WORLD INTERVIEW

To me, the greatest contribution of health care information technology and systems is its power to support clinician decision making. Far too often we focus on the technical aspects of hardware and software design while minimizing the real intent of this powerful technology, which is to support clinicians in their practice endeavors. We need to make clinicians' knowledge paramount as we transform our clinical practice environments through the design and implementation of new health care information technology and systems.

**Rita Snyder-Halpern, PhD, RN, CNAA**
Associate Professor
San Diego, California

In order to resolve computer term confusion, a controlled vocabulary system such as NANDA, NIC, NOC (NNN), or SNOMED-CT (Anderson, Keenan, & Jones, 2009), must be used to capture the data. In addition, data must be captured in such a way that the system can recognize the appropriate terms and place them in the proper context. The more multidisciplinary the data coding by the system, the more knowledge and discipline are required from each provider entering the data, and the more efforts within the organization are required to manage the structure and link the vocabularies/nomenclatures being used. Structured data that uses concepts or vocabularies not appropriate for multidisciplinary use will not produce valid results and as thus compromise patient safety.

Most EHRs are designed to combine data from large ancillary services, such as pharmacy, laboratory, and radiology, with various clinical care components (such as nursing plans, medication administration records [MAR], and physician orders). The number of integrated components and features involved in any given EHR is dependent upon the data structures and systems implemented by the technical teams. The EHR may have a number of ancillary system vendors that are not necessarily integrated into the EHR. The EHR, therefore, may import data from the ancillary systems via a custom interface or may provide interfaces that allow clinicians to access the silo storehouse system through a portal.

By 2015, in addition to controlled vocabularies, certified EHR need to incorporate the following functionalities:

computerized physician order entry (CPOE),
e-prescribing,
clinical decision support (CDS), and
clinical e-documentation systems.

CPOE permits clinical providers to electronically order laboratory, pharmacy, and radiology services. CPOE systems offer a range of functionality, from e-prescribing capabilities alone to more sophisticated systems such as complete ancillary service ordering, alerting, customized order sets, and result reporting. CPOE and clinical decision support systems have been well demonstrated to reduce medication-related errors. (Poissant, Pereira, Tamblyn, & Kawasumi, 2005). Electronic clinical documentation systems enhance the value of EHRs by providing electronic capture of clinical notes; patient assessments; and clinical reports, such as medication administration records (MAR). As with CPOE components, successful implementation of a clinical documentation system must coincide with workflow redesign and buy-in from all the stakeholders in order to realize clinical benefits, which may be substantial—as much as 24 percent of a nurse's time can be saved.

Examples of clinical e-documentation include:

Physician, nurse, and other clinician notes
Flow sheets (vital signs, input and output, problem lists, MARs)
Peri-operative notes
Discharge summaries
Transcription document management
Medical records abstracts
Advance directives or living wills
Durable powers of attorney for health care decisions
Consents (procedural)
Medical record/chart tracking
Release of information (including authorizations)
Staff credentialing/staff qualification and appointments documentation
Chart deficiency tracking
Utilization management

Medical devices can also be integrated into the flow of clinical information and used to generate real time alerts as the patient's status changes. The major value of integrated clinical systems is that they enable the capture of clinical data as a part of the overall workflow. An EHR enables the administrator to obtain data for billing, the physician to see trends in the effectiveness of treatments, a nurse to report an adverse reaction, and a researcher to analyze the efficacy of medications in patients with co-morbidities.

# Personal Health Record (PHR)

Nurses want to provide the best care possible to every patient they encounter. But the quality of care is limited by the extent to which nurses have access to their patients' complete health care profile. In the past this was not an issue, because patients visited one primary care provider (PCP) who provided health care or coordinated meeting their health care needs for their entire life. Often that same PCP treated the patient's parents and siblings, thus PCPs had access to their patients' family health care history. Health information was centralized with the PCP and provided PCPs with their patients' entire health care profile. This is no longer the case in the United States, for three primary reasons. The first reason is that patients often see specialists, and therefore visit several providers, and even different facilities, to meet their health care needs. The second reason is that we are an increasingly mobile society, where people move around and don't see the same PCP for their entire life and they don't see the same PCP that treated their parents. The third reason is that patients might need to select a new provider in their insurance network when they change insurance companies or when their current provider leaves the insurance network. The result is decentralized patient health information and less continuity of care for patients.

One potential solution for the need to centralize patient information is a personal health record (PHR). According to the Health Information and Management Systems Society (HIMSS) the PHR is ". . . a universally accessible, layperson comprehensible, lifelong tool for managing relevant health information, promoting health maintenance and assisting with chronic disease management via an interactive, common data set of electronic health information and e-health tools (http://www.himss.org/ASP/topics_phr.asp). A PHR is typically a health record that is initiated and maintained by an individual. An ideal PHR would provide a complete and accurate summary of the health and medical history of an individual by gathering data from many sources and making this information accessible online to anyone who has the necessary electronic credentials to view the information. PHRs provide an avenue for patients to track their personal health information, such as doctor's visits, medications, allergies, lab results, surgeries, immunization records, chronic illnesses, hospitalizations, family history, insurance, medical directives, and vision and dental information.

Patients can use their PHR during follow-up visits to update their health care provider with any new information, such as new or discontinued medications or visits at other institutions or specialists. Not only does this provide health care providers with more accurate information and their patients with the potential for better care than is currently available, it also can get patients more involved in their own health care. Despite the mentioned benefits, adoption of PHRs is still its infancy. Several security, privacy, and interoperability issues with the PHR need to be resolved.

The most basic form of a PC-based PHR would be a health history created in a word-processing program. The health history created in this way can be printed, copied, and shared with anyone with a compatible word processor.

PHR software can provide more sophisticated features such as data encryption, data importation, and data sharing with health care providers. Some PHR products allow the copying of health records to a mass-storage device such as a CD-ROM, DVD, smart card, or USB flash drive. PC-based PHRs are subject to physical loss and damage of the personal computer and the data that it contains. Some other methods of device solution may entail cards with embedded chips containing health information that may or may not be linked to a personal computer application or a web solution.

## Information Processing

Computer functions provide for effective retrieval and processing of data into useful information. These include decision-support tools such as alerts and alarms for drug interactions, allergies, and abnormal laboratory results. Reminders can be provided for appointments, critical path actions, medication administration, and other activities. The systems may also provide access to consensus-driven and evidence-driven diagnostic and treatment guidelines and protocols (Weaver, Warren, & Delaney, 2005). The nurse could integrate a standard guideline, protocol, or critical path from a site such as, http://www.joannabriggs.edu.au/, or http://ebn.bmj.com/, into a specific individual's EHR, modify it to meet unique circumstances, and use it as a basis for managing and documenting care. Outcome data communicated by various caregivers and by health care recipients themselves also may be analyzed and used for continual improvement of the guidelines and protocols. The EHR must also provide access to point-of-care information databases and knowledge sources, such as pharmaceutical formularies, referral databases, and reference literature.

Electronic data sharing of any kind of data raises concerns of security and confidentiality. Existing challenges in protecting security while allowing for increased ease of data retrieval have been organized by the Health Insurance Portability and Accountability Act of 1996 (HIPAA) and revised by the Health Information Technology for Economic and Clinical Health (HITECH) Act (2009). Specific legal and ethical issues vary from state to state, from specialty to specialty, and from caregiver to caregiver. Incorporating these variations into a cohesive, comprehensive CIS presents a considerable challenge.

## Critical Thinking 6-2

Go to, http://www.qsen.org/. Click on Quality/ Safety Competencies. Then click on, Informatics. Scroll down and click on, Website Evaluation Exercise. Identify the learning needs of a patient you are assigned to and search for web sites that address these needs. Use either of the two resources listed and/or other similar resources and evaluate the quality of the web sites. Complete the Health Information on the Internet: Evaluation Criteria form, i.e., Web eval form pdf. Compile a list of high quality websites that can be shared with nurses on the unit. In discussion with the patient/family, present the information you have found and describe the evaluation criteria that should be used when searching for health information on the Internet.

## Security

Electronic health record systems may provide better protection of confidential health information than do paper-based systems, because they support controls that ensure that only authorized users with legitimate uses have access to health information. Security functions address the confidentiality of private health information and the integrity of the data. Security functions must be designed to ensure compliance with applicable laws, regulations, and standards. Security systems must ensure that access to data is provided only to those who are authorized and have a legitimate purpose for using the data and provide a means to audit for inappropriate access.

Three important terms are used when discussing security: privacy, confidentiality, and security. It is important to understand the differences among these concepts.

- *Privacy* refers to the right of individuals to keep information about themselves from being disclosed to anyone. If a patient had an abortion and chose not to tell a health care provider this fact, the patient would be keeping that information private.
- *Confidentiality* refers to the act of limiting disclosure of private matters. After a patient has disclosed private information to a health care provider, the provider has a responsibility to maintain the confidentiality of that information.
- *Security* refers to the means to controlling access and protecting information from accidental or intentional disclosure to unauthorized persons and from alteration, destruction, or loss. When private information is placed in a confidential EHR, the system must have

controls in place to maintain the security of the system and not allow unauthorized persons access to the data. Computer security mechanisms use a combination of two basic protection measures: logical and physical restrictions such as firewalls and the installation of antivirus and spyware detection software. Automatic sign-off is an example of a logical restriction; this mechanism logs off a user from the system after a specific period of inactivity on the computer. A firewall is a barrier between systems to protect those systems from unauthorized access. A firewall screens traffic, allowing only approved transactions to pass through them, and restricts access to other systems or sensitive areas such as patient information, payroll, or personal data (Hebda & Czar, 2009).

Antivirus software refers to a set of computer programs that can locate and eradicate viruses and other malicious programs from scanned memory sticks, storage devices, individual computers, and networks. These systems require updating to combat the constant creation of new viruses.

Another security concern is spyware. Spyware is a self-installing data collection mechanism that is installed without the user's permission. This may happen when the user is browsing the Web or downloading software. Spyware often includes cookies that track Web use as well as applications that capture credit card, bank, and PIN numbers or other health information stored on that computer

## EVIDENCE FROM THE LITERATURE

**Citation:** Levine, C. (2006). HIPAA and talking with family caregivers. *American Journal of Nursing, 106*(8), 51–53.

**Discussion:** HIPAA has alarmed and confused many conscientious health care providers. Providers can share needed information with family and friends or with anyone a patient identifies as involved in his care as long as the patient does not object and/ or the provider believes that doing so is in the best interests of the patient. The article shares the U.S. Department of Health & Human Services Web site for a list of frequently asked questions about HIPAA (available at www.hhs.gov/hipaafaq).

**Implications for Practice:** Health care providers concerned with quality patient care will want to review HIPAA and ensure that they do not use HIPAA to avoid difficult conversations.

## REAL WORLD INTERVIEW

Several thoughts immediately come to mind about the marriage between nursing and trends in computing. First, nursing leadership within organizational informatics is pivotal, especially within the next few years. Without strong nursing informatics leadership, nursing requirements stand to be overlooked or ignored during large systems installation. Nursing is at the heart of information integration at the patient level, and systems need to support this kind of patient-centered design, including functions needed by nurses. Having an educated informatics nurse specialist reporting directly to the executive staff is the beginning for solving this problem. Second, barriers must be addressed, such as chief financial officers (CFOs) balking at the cost of electronic patient records and the other resources required to install these large-scale informatics projects. Clinical care can no longer exist efficiently and effectively without electronic tools such as the electronic patient record, decision support, and Internet access. Additionally, the external impetus to install electronic patient records is great. For example, the Leapfrog Group has set standards for hospitals to meet in order to obtain business from large corporations such as General Motors, Ford Motor Company, IBM, and 75 other companies. One of those standards is computerized physician order entry, which is projected to reduce prescribing errors by more than 50%. CFOs need to realize that their institutions will not be competitive in the health industry without electronic access for patients, computerized orders, and protections for patients such as decision support. It is no longer a question of whether to install these computerized functions, but when.

**Nancy Staggers, PhD, RN, FAAN**
Professor, Informatics
College of Nursing
Salt Lake City, Utah

for illicit purposes (Hebda & Czar, 2009). Indicators of a computer spyware infection are the presence of pop-up ads, keys that do not work, random error messaging, and poor system performance. Spyware detection software should be installed on computers because of this security threat (DeLaune, 2009).

An aspect of security that many people fail to consider is disaster planning. Natural disasters, such as earthquakes and floods, as well as manmade disasters like broad attacks on an individual company, or on a network, or on the nation's Internet infrastructure may also impact the security of mobile health devices and the systems that run them. Individual health care devices and laptop computers may be particularly at risk, being small and easy to steal or lose. Also, since these are often wireless devices, there is increased risk that personal health information could be compromised. On the other hand, the trend is for increased virtualization of computer systems to third-party vendors, known as cloud computing. Not only does this pose additional security risks, but this adds another layer of questions on the extent to which personal health information is protected. Mobile health application devices, which have the advantages of portability and accessibility, can also present security threats, (Wright & Sittig, 2007b).

Fortunately, these devices also can also be equipped with encryption and security features, such as passwords, and are designed to be attached to a keychain or lanyard and carried with the user (Wright & Sittig, 2007a).

## Information Presentation

The wealth of information available through EHR systems must be managed to ensure that authorized caregivers, including nurses and others with legitimate uses, have the information they need in their preferred presentation form. A nurse, for example, may like to see data organized by source, caregiver, encounter, problem, or date. Data can be presented in detail or summary form. Tables, graphs, narratives, and other forms of information presentation must be accommodated. Some users may need to know only of the presence or absence of certain data, not the nature of the data itself. For example, blood donation centers test blood for HIV, hepatitis, and other conditions. If a donor has a positive test result, the center may not be given the specific information regarding the test, but just general information that a test result was abnormal and that the patient should be referred to an appropriate health care provider.

# INTERFACE BETWEEN THE INFORMATICS NURSE AND THE CLINICAL INFORMATION SYSTEM

Information demands in health care systems are pushing the development of EHRs. The ongoing development of computer technology—smaller, faster machines with extensive storage capabilities and the ability for cross-platform communication—is making the goal of an integrated electronic system a realistic option, not just a long-term dream. As these systems evolve, informatics nurses will play an important role in their development, implementation, and evaluation.

Informatics nurses, because of their expertise, are in an ideal position to assist with the development, implementation, and evaluation of CISs. Their knowledge of policies, procedures, and clinical care is essential as work-flow systems are redesigned within a CIS. It is not unusual for nurses within an institution to have more hands-on interaction with and knowledge of different departments than any other group of employees in an institution. Jenkins (2000) suggests that the process model of nursing (assessment, planning, implementation, and evaluation) works well during a CIS implementation; thus nurses have a familiar framework from which to understand the complexity of a major system change.

# TRENDS IN COMPUTING

The explosion in information—some estimate that all information is replaced every 9 to 12 months—requires nurses to be on the cutting edge of knowledge to practice ethically and every nurse must be computer literate. **Computer literacy** is defined as the knowledge and understanding of computers, combined with the ability to use them effectively (McDowell & Ma, 2007). For health care professionals, computer literacy requires having an understanding of systems used in clinical practice, education, and research settings. In clinical practice, for example, electronic patient records and clinical information systems are being used more often. The computer literate nurse must be able to use these systems effectively and address issues discussed earlier, such as confidentiality, security, and privacy. At the same time, the more advanced nurse must be able to effectively use applications typically found on PCs, such as word-processing software, spreadsheets, and PowerPoint. The nurse will want to develop the ability to use statistics for research. Finally, the computer literate nurse must know how to access information from a variety of electronic sources and how to evaluate the appropriateness of the information at both the professional and patient level. The computer literate nurse

has information literacy. **Information literacy** is defined as the understanding of the architecture of information; the ability to navigate among a variety of print and electronic tools to effectively access, search, and critically evaluate appropriate resources; and the ability to synthesize accumulated information into an existing body of knowledge and practice (Barnard, Nash, & O'Brien, 2010).

## DEVELOPMENT OF MODERN COMPUTING

Weiser and Brown (1996) have characterized the history and future of computing as having three phases. The first phase is known as the "mainframe era," in which many people shared one computer. Computers were found behind closed doors and run by experts with specialized knowledge and skills. Although we have mostly moved beyond the mainframe era, it still exists in large mainframe systems, such as banking, weather forecasting, and academic institutions.

The archetypal computer of the mainframe era must be the ENIAC, developed at the University of Pennsylvania in 1945. The Electronic Numerical Integrator and Computer (ENIAC) was conceived by John Mauchly, an American physicist, and built at the Moore School of Engineering by Mauchly and J. Presper Eckert, an engineer. It is regarded as the first successful digital computer. It weighed more than 60,000 pounds and contained more than 18,000 vacuum tubes. About 2,000 of the computer's vacuum tubes were replaced each month by a team of six technicians. Even though one vacuum tube blew approximately every 15 minutes, the functioning of the ENIAC was still considered to be reliable! Many of the ENIAC's first tasks were for military purposes such as calculating ballistic firing tables and designing atomic weapons. Because the ENIAC was initially not a stored program machine, it had to be reprogrammed for each task.

Phase two in modern computing is the "PC era," which is characterized by one person to one computer. The first harbinger of the PC era was in 1948 with the development of the transistor at Bell Telephone Laboratories. The transistor, which could act as an electric switch, replaced the costly, energy-inefficient, and unreliable vacuum tubes in computers and other devices, including televisions. By the late 1960s, integrated circuits, tiny transistors, and other electrical components arranged on a single chip of silicon replaced individual transistors in computers. Integrated circuits became miniaturized, enabling more components to be designed into a single computer circuit. In the 1970s, refinements in integrated circuit technology led to the development of the modern microprocessor, made up of integrated circuits that contained thousands of transistors. Weiser and Brown (1996) date the true start of the second phase as 1984, when the number of people using personal computers surpassed the number of people using shared computers.

Manufacturers used integrated circuit technology to build smaller and cheaper computers. The first PCs were sold by Instrumentation Telemetry Systems. The Altair 8800 appeared in 1975. Graphical user interfaces were first designed by the Xerox Corporation in a prototype computer, the Alto, developed in 1974. This prototype computer incorporated many of the features found on computers today, including a mouse, a graphical user interface, and a "user friendly" operating system. However, Xerox Corporation made a corporate decision to not pursue commercial development of the PC, the rationale being that the core business strategy of Xerox was copiers, not computers. One only has to look at how PCs have proliferated throughout the world to realize that this may not have been the smartest business decision ever made. In fact, this whole episode has become a bit of a computer history legend (Hiltzik, 2000; Smith & Alexander, 1988).

## UBIQUITOUS COMPUTING

Phase three has been dubbed the "era of ubiquitous computing (UC)," in which computers are no longer distinct objects but are integrated into the environment, thus enabling people to interact with information-processing devices more naturally and casually. Weiser and Brown (1996) estimated that the crossover with the personal computer (PC) era is between 2005 and 2020. In this phase, computers are everywhere—in walls, chairs, clothing, light switches, cars, appliances, and so on. Computers have become so fundamental to our human experience that they have "disappeared," and we have ceased to be aware of them. The result is "calm technology," in which computers do not cause stress and anxiety for the user but, rather, recede into the background of life. In 2011, our information and even news is accessed through computers or mobile devices; many books are read through wireless reading devices (Kindle, iPad); mobile phones are part of daily life; and Global Positioning Systems (GPS) are in many cars, phones, etc.

One only has to look around a typical house to see how UC is becoming part of our lives. Microprocessors exist in every room: appliances in the kitchen, remote controls for the TV and stereo in the den, and clock radios and cordless phones in the bedroom. And the bathroom? Matsushita of Japan has developed a prototype toilet (dubbed the "smart toilet") that includes an online, real-time health monitoring system. It measures the user's weight, fat content, and urine sugar level; plots the recorded data on a graph; and sends the data instantaneously to a health care provider for monitoring (Watts, 1999; Hadidi, 2010).

## MOBILE APPLICATIONS

The rapid uptake of the iPhone, iPad, BlackBerry, HP, Android phones, etc., by both consumers and providers for health care purposes is astounding compared to the slower adoption of health information technology (HIT). Mobile applications (commonly called apps) are software programs for smartphones and other handheld computing devices. The creation of health care apps for these mobile devices is also moving quickly. What is it that makes these mobile devices so attractive? Unlike any other HIT platform, they are basically inexpensive, attractive, and easy to use anytime they are needed. They are so intuitive and user-friendly that most people can download and use the many computer applications available for them quickly without much training or basic computer literacy (Sarasohn-Kahn, 2010). It is projected that by 2015, there will be 1.4 billion smartphone users and 500 million of them will use mobile health applications. According to recent surveys conducted by the Pew Internet & American Life Project, currently nine percent of all mobile phone users in the US have downloaded an app to track or manage their health (Dolan, 2011).

Some of the most widely used mobile applications by nurses are drug and clinical references and clinical tools such as drug calculators. Some free mobile apps are available (See Critical Thinking 6-3 in this chapter). Free references usually have limited features and may not include automatic updates. Updates usually are available with subscription programs or included in the purchase price of the app. Online sources for purchase of specific apps include the iTunes App Store (http://www.apple.com/iphone/apps-for-iphone/); Android Market (http://www.android.com/market); BlackBerry App World (http://appworld.blackberry.com/webstore); imedicalapps.com; and the Palm App Catalog (http://www.palm.com) (Innocent, 2010). A number of mobile apps that are combined with sensor technologies connect patients with chronic diseases on a continual basis with their clinicians and caregivers. Mobile applications are empowering patients and health care providers alike.

## SOCIAL NETWORK SITES

Social network sites are also invading health care. They have been growing in popularity across broad segments of Internet users and are a convenient means to exchange information and support. For example, Facebook groups (www.Facebook.com) have become a popular tool for awareness-raising, fundraising, and support-seeking related to breast cancer, attracting over one million users (Bender, et al, 2011). Patient communities, such as PatientsLikeMe (http://www.patientslikeme.com/) are flourishing. PatientsLikeMe is committed to providing patients with access to the tools, information, and experiences that they need to take control of their disease. Medical centers are diving into social media with caution because of potential confidentiality, privacy, and security risks associated with these new technologies.

# EVIDENCE FROM THE LITERATURE

**Citation:** Miller, J., Shaw-Kokot, J. R., Arnold, M. S., Boggin, T., Crowell, K. E., Allegri, F., et al. (2005). A study of personal digital assistants to enhance undergraduate clinical nursing. *Nursing Outlook, 44*(1), 19–24.

**Discussion:** This article reports on personal digital assistants (PDAs) as a means to prepare competent nurse professionals who value and seek current information. Through the incorporation of PDAs in nursing clinical courses, the value and skill of seeking current information will become something integral that nursing students take into their professional practice. PDA software is available for such things as DrugGuides, Medical Laboratory Reference, Medical Abbreviations, Medical Spanish, and so on.

**Implications for Practice:** PDAs bring evidence-based practice (EBP) to the bedside. The development of skill in using this technology is a basis for the increased use of EBP.

## VIRTUAL REALITY

Virtual reality (VR) puts people inside a computer-generated world. Virtual reality, while still somewhat limited in its development, does have enormous potential in health care applications, such as health games like Ruckus Nation: An Online Idea Competition to Get Kids Moving! (http://www.delicious.com/gamesforhealth, accessed, March 24, 2011). With VR, a person can see, move through, and react to computer-simulated items or environments. Using certain tools such as a head-mounted computer display and a handheld input device, the user feels immersed in and can interact with this world. The virtual world can represent the current world or a world that is difficult or impossible to experience firsthand—for example, the world of molecules, the interior of the human body, or the surface of Pluto. By putting the sensors on the person (head-mounted computer display, sensors in gloves, shoes, and glasses), the person can move and experience the world in a typical way—by walking, moving, and using the senses of touch, sight, smell, and hearing.

VR has allowed practitioners to develop minimally invasive surgical techniques. This allows direct viewing of internal body cavities through natural orifices or small incisions. The manipulation of instruments by the surgeon or assistants can be direct or via virtual environments. In the latter case, a robot reproduces the movements of humans, using virtual instruments. The precision of the operation may be augmented by data or images superimposed on the virtual patient (Nunes, 2008).

Applications in health care education also exist. Virtual reality allows information visualization through the display of enormous amounts of information contained in large databases. Through 3-D visualization, students can understand important physiological principles or basic anatomy. Students can go "inside" the body to visualize structures and see how they work. It is also possible to observe changes in physiologic functioning. For example, a student can visualize the vascular system of a patient going into shock (Kilmon, Brown, Ghosh, & Mikitiuk, 2010).

A popular use in psychology has been in exposure therapy for patients with specific phobias. VR simulation has been used for a decade to treat patients with a fear of flying (Brinkman, Mast, Sandino, Gunawan, & Emmelkamp, 2010). Other researchers have found significant improvements in patients with post-traumatic stress disorder (PTSD) by using exposure therapy and VR (Gerardi, Cukor, Difede, Rizzo, & Rothbaum, 2010).

VR may prove to be an exciting and safe tool for stroke rehabilitation but its evidence base is too limited to permit a definitive assessment of its value yet. Research in this area is still at an early stage and further study in the form of

## Critical Thinking 6-3

Some free mobile apps for health care providers include Archimedes (http://www.skyscape.com); CheckRx (http://www.skyscape.com); Diagnosaurus (http://www.skyscape.com or iTunes App Store ($0.99); Epocrates (http://www.epocrates.com or iTunes App Store); Epocrates Med Tools (add-on to Epocrates) http://www.epocrates.com; and Shots 2010 (http://www.immunizationed.org/ShotsOnline.aspx) (Innocent, 2010).

Visit some of these sites and take a look at some of the mobile apps there. What did you find?

# EVIDENCE FROM THE LITERATURE

**Citation:** Henneman, E. A., Cunningham, H., Roche, J. P., & Curnin, M. E. (2007). Human patient simulation: Teaching students to provide safe care. *Nurse Educator, 32*(5), 212–217.

**Discussion:** The use of human patient simulation as a teaching methodology for nursing students has become popular. It effectively demands paying careful attention to the details of the simulation, debriefing staff after use, and employing good evaluation processes. Our experience in designing simulation experiences and evaluating student behaviors confirms the resource-intensive nature of human patient simulation and the need for clear, measurable objectives. When used properly, human patient simulation offers a unique opportunity to teach nursing students important patient safety principles. For example, one simulator model allows the insertion of a chest tube or the application of a trauma or wound care kit. Such features support educators' abilities to create learning situations that address a variety of specific clinical problems or needs. Human patient simulation can also provide clinicians an opportunity to care for a simulated patient with other acute clinical problems, such as airway obstruction or cardiac arrest, hemorrhage, shock, and various other common emergencies.

**Implications for Practice:** Working with patient simulators allows students to solve problems, work as a team, and communicate effectively with their colleagues and other providers. Role-play provides an opportunity to improve communication and enhance patient safety. By integrating concepts related to patient safety—such as human factors engineering, staff management, and situational awareness—participants learn approaches and concepts related to patient safety and develop clinical skills that reduce the potential for errors. Patient safety and avoidable medical errors are a concern in any institution. The Institute of Medicine report *To Err Is Human: Building a Safer Health System* recommends simulation training as one strategy to prevent errors in the clinical setting.

rigorous controlled studies is warranted. (Crosbie, Lennon, Basford, & McDonough, 2007; Sandlund, McDonough, & Häger-Ross, 2009).

Applications in nursing are similar. For students learning clinical procedures, VR gives them the opportunity to practice invasive and less commonly occurring procedures in the lab so that they have both the skill and the confidence necessary when encountering a patient requiring the procedure for the first time. Likewise, VR enhances patient education materials. Diabetic patients needing to understand the physiologic processes of the pancreas may visualize the organ to more fully understand their disease and treatment. Patients requiring painful or unusual procedures may experience a VR simulation as a means of preparation. By providing an alternate environment, VR also has the potential to mitigate or minimize the side effects of certain procedures such as chemotherapy in patients with cancer.

## USING THE INTERNET FOR CLINICAL PRACTICE

It is important for nurses to develop information literacy to ensure accurate and up-to-date information for patient care. Information literacy is the ability to identify when and what information is needed, understand how the information is organized, identify the best sources of information for a given need, locate those information sources, evaluate the information sources critically, and share that information as appropriate (ANA, 2008). Information literacy is critically important for evidence based practice; especially since the amount and complexity of information is expanding daily. Not all information is created equal: some is authoritative, current, reliable, but some is biased, out of date, misleading, and false. Also, to find the "right" information, keep in mind that you must be persistent. No one search strategy is going to work all the time, nor is any one search engine more effective than any other. Here are some strategies and tactics to render Internet searches more efficient and to reduce search time (Anderson & Klemm, 2008):

- Use Web sites published by trusted governmental or professional organizations for the information they specialize in.
  - American Heart Association (www.americanheart.org) or the National Heart, Lung & Blood Institute (www.nhlbi.nih.gov), cardiovascular diseases
  - CancerNet (http://cancernet.nci.nih.gov), cancer
  - National Institute of Diabetes and Digestive and Kidney Diseases (http://digestive.niddk.nih.gov), diabetes and digestive disorders

## EVIDENCE FROM THE LITERATURE

**Citation:** Jha, A. K., Doolan, D., Grandt, D., Scott, T., Bates, D. W. (2008). The use of health information technology in seven nations. *International Journal of Medical Informatics*, (July 24), 254–262.

**Discussion:** The authors assessed the state of health information technology (HIT) adoption and use in seven industrialized nations. They used a combination of literature review, as well as interviews with experts in individual nations, to determine the use of key information technologies. They examined the rate of electronic health record (EHR) use in ambulatory care and some hospital settings, along with current activities in health information exchange (HIE) in seven countries: the United States (U.S.), Canada, United Kingdom (UK), Germany, the Netherlands, Australia, and New Zealand (NZ). Four nations (the UK, the Netherlands, Australia, and NZ) had nearly universal use of EHRs among general practitioners (each >90%), and Germany also exhibited widespread use (40%–80%). The U.S. and Canada had a minority of ambulatory care physicians who used EHRs consistently (10%–30%). Although there are no high-quality data for the hospital setting from any of the nations examined, the study concludes that HIT adoption in the seven nations was high in ambulatory settings but appears to lag in inpatient EHR and HIE.

**Implications for Practice:** Increased efforts will be needed if interoperable EHRs are soon to become universally available and used in these seven nations. Nurses and other practitioners in these nations must be part of the solution to make the EHR universally used and available.

- Center for Disease Control and Prevention (www.cdc.gov), a primary resource for developing and applying disease prevention and control, environmental health, and health promotion activities
- National Institute on Aging (www.nia.nih.gov), well-being and health of older adults
- American Dietetic Association (www.eatright.org), offering information related to food and diet
- American Medical Association's consumer health information site (www.ama-assn.org), resource for general health information
- Use reputable health care organizations' Web sites such as www.mayoclinic.com from the Mayo Clinic and www.intelihealth.com from Harvard Medical School.
- Use Consumer health sites organized by health care librarians. These offer a wealth of organized information on disease management. Examples are MEDLINE*plus* (www.medlineplus.gov), maintained by the National Library of Medicine, and the New York Online Access to Health (NOAH) (www.noah-health.org), a multilingual Web site organized and maintained by health care librarians in New York City.
- Use precise terms, such as "diabetes type 1" instead of just "diabetes," to reduce the number of hits when searching for very specific information. Use PICO to focus your search (see Chapter 5).
- Draw on search engines such as Mayo Clinic (www.mayoclinic.com), WebMD (www.webmd.com), and

others that collect information from reliable online health resources rather than relying on the "bots," or robots, typically used by search engines to "crawl" the Web, as with Google.

- Refine your Internet searches with filters. Filtering is mechanically blocking Internet content from being retrieved through the identification of key words and phrases. For example, you can narrow your search by the type of medical viewpoint (traditional or alternative), reading level (easy, moderate, or complex), and type of site (commercial, noncommercial, government, or nonprofit) that you use in your key words to filter your search.

The result of your searches after using these strategies will probably be a more focused and helpful list of links matching your specific request. You will then want to evaluate your search data.

## INTERNET RESOURCES FOR PATIENTS

The Internet can also be used by patients to manage their health or diseases. A major problem with using the Internet for this purpose is that information on the Internet lacks the conventional standards by which traditional published resources are evaluated. Though many general instruments can be used in evaluating health-related Websites, most of them are incomplete, and many do not measure what they claim to measure (Anderson & Klemm, 2008). In addition, they are geared toward professional and regulatory organizations, and some patients may lack the knowledge to understand the information provided. Yet, it is critical that non-clinicians be able to evaluate health information on the Internet. Helping patients determine this quality and relevance becomes a key responsibility for clinicians. Criteria to evaluate Internet health information have developed over the years. However, these criteria often assume knowledge of health care content and some familiarity with traditional standards for evaluating such resources. None of them focus on helping patients filter the information found on the Web. Based on published criteria and taking into account patient context and needs, the nurse can provide the patient with a simple set of criteria, subjective and objective, to evaluate Website content, design, navigation, and credibility (Alpay, Overberg, & Zwetsloot-Schonk, 2007; J. Jones, 2003). Health information Websites adhering to eight Health on the Net (HON) principles, i.e., authority, complementarity, confidentiality, attribution, justifiability, transparency, financial disclosure, and advertising, are accredited with the HON code and can be verified at http://www.healthonnet.org/.

## INTERNET AND SEARCHING FOR EVIDENCE

Information literacy is a necessary skill for extracting evidence from research and practice resulting in evidence-based-practice (EBP). EBP is the process of systematically finding, appraising, and using contemporaneous research findings as the basis for clinical decisions. Evidence-based health care asks questions, finds and appraises the relevance based on accurate analysis of current nursing knowledge and practice, and harnesses that information for everyday clinical practice. The primary sources for this type of information are web-based resources such as online databases (e.g., Medline, CINAHL) and also electronic practice documents such as EHRs.

Evidence-based health care follows four steps:

1. Formulate a clear clinical question from a patient's problem;
2. Search for clinical research articles or prior practice applications relevant to the intervention at hand;
3. Evaluate and critically appraise the evidence for its validity and usefulness;
4. Implement useful findings in clinical practice (Magrabi, Westbrook, & Coiera, 2007).

The skills that underpin evidence-based clinical practice and nursing are:

- Forming a good clinical question
- Sinding evidence using effective search techniques

- using resources for evidence based practice including databases, e-journals, e-textbooks and other Internet sites (Cushing/Whitney Medical Library, 2009).

For nurses, sources of evidence are studied and the wealth of information that exists is confronted. The challenging question is how to integrate a more evidence-based approach into the every-day practice of nurses. Although the nursing profession recognizes that research is the basis for knowledge development, lack of organizational support for providing nursing time to access, use, and conduct research on clinical units can limit the ability of nurses to practice EBP.

## CHALLENGES RELATED TO THE USE OF THE INTERNET IN HEALTH CARE

As mentioned earlier, The Health Information Technology for Economic and Clinical Health (HITECH) Act is part of the American Recovery and Reinvestment Act of 2009 (ARRA). ARRA contains incentives related to health care information technology in general, e.g. creation of a national health care infrastructure, and contains specific incentives designed to accelerate the adoption of electronic health record (EHR) systems among providers. Because this legislation anticipates a massive expansion in the exchange of electronic protected health information, HITECH also widens the scope of privacy and security protections available under The Health Insurance Portability and Accountability Act (HIPAA) of 1996. It increases the potential legal liability for non-compliance and it provides for more enforcement. HITECH addresses computer interoperability, privacy, and security concerns associated with the electronic transmission of health information over the Internet through funding assistance, incentives, disincentives, and education.

## INTEROPERABILITY

Interventions to address a person's health combine data from different sources, i.e., health care services, insurance payers, laboratories, dental services, nutritional services, thus all of these sources must be interoperable and able to contribute to and access appropriate information. The information must be understood by everyone involved in the care of the individual (Hufnagel, 2009). The National Broadband Plan (2010) and the push to adopt health information technology support interoperability priorities by dramatically improving the collection, presentation and exchange of health care information, and by providing clinicians and consumers the tools to transform care in a "meaningful way". Technology alone cannot heal, but when appropriately incorporated into care, technology

**EVIDENCE** FROM THE LITERATURE

**Citation:** Fox, S. (2005). Health information online [Electronic version]. *Pew Internet & American Life Project.* Retrieved March 1, 2006, from http://www.pewinternet.org/pdfs/PIP_Healthtopics_May05.pdf

**Discussion:** The Pew Internet & American Life Project's previous survey suggested that online health information seekers were often motivated to search out information that related to actions they might need to take for specific medical issues in their lives. For instance, they, or people they love, might have experienced health symptoms that worried them, or Internet users might search for information about whether they would be wise to visit a doctor. They might have just received a diagnosis and wanted to learn more about their medical condition. They might have had a new medical treatment or new medicine prescribed, and they wanted to learn more about it. In many cases, online health information seekers were action-oriented and highly purposeful, because there was a pressing medical issue for them to address.

In the current survey, those concerns remain reflected in many respondents' answers. At the same time, there are also notable changes that relate to specific kinds of health searches. Online investigations for information about diet, fitness, exercise, and over-the-counter drugs have grown. This suggests that online health information seekers are increasingly interested in wellness information and material that could be unconnected to worrisome symptoms, a doctor's diagnosis, or another kind of health crisis. Two other notable categories of growth were seen in searches related to health insurance and material about specific doctors and hospitals. This suggests that health seekers are doing more health homework online before they make big decisions about health care. The article asks, what is the best approach to online, information-seeking health care consumers?

Suggested answers are the following:

- Keep in mind that you can turn this behavior into an opportunity to teach. Encourage the patient to take the time to research and learn more about his or her health condition.

- React in a positive manner about information from the Internet, but remind the patient that its quality and reliability may be unknown.

- Inform patients that time constraints will not permit you to read the information on the spot but that you will gladly read it if they send it to you via e-mail, perhaps before a scheduled appointment.

- Accept patients' contributions and acknowledge that they may have valuable information that you may not have come across yet.

Things you should not do include the following:

- Be dismissive or paternalistic.
- Be derogatory about others' comments on the Internet.
- Refuse Internet material.
- Try to "one-up" the patient or family members regarding the information.
- Break normal rules of patient confidentiality.

**Implications for Practice:** Clinicians may improve their care delivery by acknowledging Internet information gathered by patients. It is also important to help patients evaluate the validity of any information they find on the Internet.

can assist health care professionals and consumers to make better decisions, become more efficient, engage in innovation, and understand both individual and public health more effectively (F. Jones, Hook, Park, & Scott, 2011 in press).

## PRIVACY

There are important issues regarding privacy – who controls sharing and accessing of the information in the health record, how to optimally design EHR in order to allow patients to maximize the security of their information, and how to develop authentication methods that ensure both privacy and security, yet do not present a major barrier to access (Kaelber, Jha, Johnston, Middleton, & Bates, 2008).

In terms of protecting patient privacy in the information age, Kendall (2008) notes that "the loss of privacy seems to be a foregone conclusion in the information age. Polls show that most Americans believe they have lost all control over the use of personal information by companies. Americans are also concerned about the threats posed by identity theft and fraudulent Internet deceptions like phishing. Phishing is a scam by which an e-mail user is duped into revealing personal or confidential information which the scammer can use illicitly (Merriam-Webster Online Dictionary 2011). People are learning the hard way to withhold personal information unless it is absolutely necessary to disclose it. A reluctance to provide personal health and financial information is not

difficult to understand, given the many high-profile reports of data security breaches from top companies and organizations. These breaches are not just caused by a lapse in computer security practices; they have often occurred because of stolen laptops and misplaced back-up tapes. With medical identity theft the fastest growing crime in American, followed closely by Medicare fraud, the heightened awareness about data security on the part of the public is not surprising, particularly where the individual consumer or patient has little control and can only take remedial action once a breach has occurred. Many people are now taking advantage of laws in various states that allow them to "freeze" access to their credit reports. Any health information technology device must be able to ensure the privacy of its information, especially the personal health information, while still being appealing, accessible, and easy to use (J. F. Jones, et al., 2011 in press).

## CASE STUDY 6-1

You are working at a women's health clinic with a number of nursing and medical practitioners. The clinic receives at least two to three telephone calls a day from women with a urinary tract infection (UTI). The question comes up: Do all these women need to be seen by a practitioner, or is there a way to manage some of the cases by telephone? You are asked to be on a committee to explore this issue and possibly come up with a protocol. Where do you begin? Is a protocol for telephone management of UTI realistic?

## KEY CONCEPTS

- Computers and the Internet are no longer a novelty, but a fact of life with exponential growth on a worldwide basis.

- Within nursing, the combination of computer science and information science with clinical expertise allows us to develop systems that have the ultimate goal of improving patient outcomes.

- Nursing informatics is the specialty that integrates nursing science, computer science, and information science in identifying, collecting, processing, and managing data and information to support and expand nursing practice, administration, education, research, and nursing knowledge.

- Nurses can pursue formal education in nursing informatics at both the graduate and undergraduate levels. Informal education in informatics can be pursued through self-study, attendance at conferences, and reading the informatics literature.

- Certification in nursing informatics is voluntary and recognizes superior achievement and excellence in the specialty.

- Virtual reality simulations have been used in education and patient care. Minimally invasive surgery has been developed in large part through the technology of virtual reality.

- Telehealth applications are developing rapidly in nursing.

- Effective searching for information on the Internet requires that you target your search.

- Information literacy is critically important for evidence based practice, especially since the amount and complexity of information is expanding daily.

- It is important to evaluate information found on the Internet and to educate your patients about the importance of evaluating the information found on the Internet.

- Several Institute of Medicine reports have highlighted the need for health care technology.

- Government recognizes the potential of information technology for quality care and cost containment.

# KEY TERMS

clinical information
    system (CIS)
computer literacy

electronic health record
    systems (EHR)
information literacy

nursing informatics
personal health record

# REVIEW QUESTIONS

1. E-health includes which of the following activities?
    A. Obtaining health information online
    B. Shopping for health products online
    C. Online case management
    D. All of the above

2. A person who is HIV positive and chooses not to reveal this information to a nurse during an admission assessment is keeping this information
    A. anonymous.
    B. secure.
    C. confidential.
    D. private.

3. If a patient came to you asking for an Internet site where he could learn more about diabetes, which of the following would be appropriate for you to suggest?
    A. MEDLINE at the National Library of Medicine
    B. OncoLink
    C. MEDLINEplus
    D. All of the above

4. You are interested in learning more about amyotrophic lateral sclerosis (ALS). You find a Web site where the author states that he is a worldwide leading researcher into causes for this disease. As part of your evaluation of the author's credentials, you will do which of the following?
    A. Take him at his word.
    B. E-mail him and ask for a list of references.
    C. Ask several colleagues whether they are familiar with his research.
    D. Do an author search on MEDLINE.

5. Decipher the following URL: www.noah-health.org. This is the URL for which of the following Web sites?

    A. The National Library of Medicine
    B. A local health organization for immigrants
    C. A consumer health Web site
    D. A Web site published by medical librarians.

6. Staff nurses at a hospital are extremely resistant to using the new automated medication administration system. This resistance most likely is due to the fact that the system did which of the following? Select all that apply.
    _____ A. Was not integrated into the network system.
    _____ B. Operated from the mainframe computer.
    _____ C. Did not consult end users during the design phase.
    _____ D. Was not part of the nursing process.
    _____ E. Used optical disk storage as a backup.

7. Maintaining confidentiality of the computer-based patient record is complex due to which of the following? Select all that apply.
    _____ A. Storage technology required for the electronic record.
    _____ B. Need to meet Joint Commission physical plant security requirements.
    _____ C. Complexity of patient and clinical data.
    _____ D. Access to medical records via telecommunications networks.
    _____ E. Confusing security and privacy rules regarding health care.

8. Informatics nursing is distinguished from other nursing specialties by its focus on which of the following?
    A. Computerized health care records.
    B. Data coding and the use of abbreviations.
    C. Content and representation of data and information.
    D. Training and education.

# REVIEW ACTIVITIES

1. The EHR and health information exchanges between health care agencies are supported by the government, insurers, and many health care organizations. Do you think consumers will encourage the development of such systems or not? Why?

2. A popular area of research in nursing informatics has been the integration of information technology in all areas of nursing practice. Another topic that has been studied widely is the development and implementation of terminologies to document and communicate nursing practice. As we move into the era of EHR, do you think that we need to refocus on more-generic health care terminologies such as SNOMED-CT, which serve the health care community rather than nursing alone? Why or why not?

# EXPLORING THE WEB

- Pick a specialty area in nursing that interests you. Visit the American Nurses Credentialing Center (www.ana.org) to determine whether the ANCC certifies nurses in this specialty. If yes, what are the requirements? If no, is there another organization that credentials nurses in this specialty? Do a Web search to find the organization and the requirements for certification.

- Search for "consumer empowerment" and then "meaningful use" at: http://healthit.hhs.gov

- The Unified Medical Language System (UMLS) (http://umlsks.nlm.nih.gov) is developed to compensate for differences in concepts in several biomedical terminologies.

  How many nursing languages are incorporated in the UMLS?

  How are the languages used in clinical practice?

  How would you obtain more information on each of the approved languages?

**Check these sites:**

Center for the Health Professions at the University of California, San Francisco:
www.futurehealth.ucsf.edu

TIGER Initiative:
www.tigersummit.com

Health Information Management Systems Society:
www.himss.org

ANI: Alliance for Nursing Information:
www.allianceni.org

Institute of Medicine:
www.iom.edu

Pepid Medical Information Resources
www.pepid.com

Argonne National Laboratory:
http://www.anl.gov

University of Southern California:
www.usc.edu

University of Virginia:
www.virginia.edu

University of Wisconsin-Madison:
www.wisc.edu

- The Nursing Information and Data Set Evaluation Center of the ANA was established "to review, evaluate against defined criteria, and recognize information systems from developers and manufacturers that support documentation of nursing care within automated Nursing Information Systems (NIS) or within the electronic health record (EHR). Visit the center at www.nursingworld.org to find answers to the following questions:

  How many nursing languages have been recognized by the ANA?

  Why have these languages been developed?

  What are they designed to do?

  How would you obtain more information on each of the approved languages?

# REFERENCES

Alpay, Laurence L., Overberg, Regina I., & Zwetsloot-Schonk, Bertie (2007). Empowering citizens in assessing health related websites&#58; a driving factor for healthcare governance. *International Journal of Healthcare Technology and Management, 8*(1–2), 141–160.

American Nurses Credentialing Center (ANCC). (1993). *Statement of philosophy.* Washington, DC: American Nurses Association.

American Nurses Association (ANA). *Standards of practice for nursing informatics.*

Anderson, A., & Klemm, P. (2008). The Internet: Friend or foe when providing patient education? [10.1188/08.CJON.55–63]. *Clinical Journal of Oncology Nursing, 12*(1), 55–63.

Ball, M. J., Hannah, K. J., Newbold, S. K., & Douglass, J. V. (2000). *Nursing informatics: Where caring and technology meet* (3rd ed.). New York: Springer Verlag.

Barnard, A. G., Nash, R. E., & O'Brien, M. (2010). Information literacy: Developing lifelong skills through nursing education. *Journal of Nursing Education, 44*(11), 505–510.

Bender, J.L., Jimenez-Marroquin, M., & Jadad, A.R. (2011). Seeking Support on Facebook: A Content Analysis of Breast Cancer Groups. *J Med Internet Res, 13*(1), e16.

Bill Gates. (n.d.). BrainyQuote.com. Retrieved March 24, 2011, from http://www.brainyquote.com/quotes/quotes/b/bill gates173261.html.

Brinkman, W.-P., Mast, C. V. D., Sandino, G., Gunawan, L. T., & Emmelkamp, P. M. G. (2010). The therapist user interface of a virtual reality exposure therapy system in the treatment of fear of flying. *Interacting with Computers, 22*(4), 299–310.

Carroll, K., Bradford, A., Foster, M., Cato, J., & Jones, J. (2007). An emerging giant: Nursing informatics. *Nursing Management, 38*(3), 38–42.

Centers for Medicare and Medicaid Services. (2010). Meaningful use. https://www.cms.gov/EHRIncentivePrograms/30_Meaningful_Use.asp#BOOKMARK1, accessed April 3, 2011. http://www.qsen.org/, accessed April 3, 2011.

Centers for Medicare and Medicaid Services. (2011). Medicare demonstrations. http://www.cms.gov/DemoProjectsEvalRpts/MD/list.asp?intNumPerPage=all, accessed April 3, 2011.

Cibulka N.J., Crane-Wider L. (2011). Introducing personal digital assistants to enhance nursing education in undergraduate and graduate nursing programs. Journal of Nursing Education. February; 50(2):115–8.

Crosbie, J. H., Lennon, S., Basford, J. R., & McDonough, S. M. (2007). Virtual reality in stroke rehabilitation: Still more virtual than real. *Disability & Rehabilitation, 29*(14), 1139–1146.

Cushing/Whitney Medical Library (2009). Evidence Based Practice. Accessed March 23, 2011 at http://info.med.yale.edu/library/nursing/education/ebhc.html.

DeLaune, S.C. & Ladner, P.K. (2009). *Fundamentals of Nursing: Standards and Practice (3rd ed.).* Delmar Cengage Learning.

Dolan, B. (2011). Smartphones to drive mHealth to 500 million users? mobihealthnews. Nov 16, 2010 . Retrieved April 6, 2011, at http://mobihealthnews.com/9523/smartphones-to-drive-mhealth-to-500-million-users/

Electronic health record. (2008). Available at http://en.wikipedia.org/wiki/Electronic_health_record. Accessed January 4, 2011.

Ford, E. W., Menachemi, N., Peterson, L. T., & Huerta, T. R. (2009). Resistance is futile: But it is slowing the pace of EHR adoption nonetheless. *Journal of the American Medical Informatics Association, 16*(3), 274–281.

Fox, S. (2005). Health information online. [Electronic version]. *Pew Internet & American Life Project.* Retrieved March 1, 2006 from http://www.pewinternet.org/pdfs/PIP_Healthtopics_May05.pdf

Gerardi, M., Cukor, J., Difede, J., Rizzo, A., & Rothbaum, B. (2010). Virtual reality exposure therapy for post-traumatic stress disorder and other anxiety disorders. *Current Psychiatry Reports, 12*(4), 298–305.

Hadidi, T. (2010). Model and simulator of the activity of the elderly person in a Health Smart Home. Paper presented at the e-Health Networking Applications and Services (Healthcom), 2010 12th IEEE International Conference.

Health law blog, (2010). EHR Incentive Program Registration Begins in 11 States. http://www.healthlaw-blog.com/2010/12/ehr-incentive-program-registration-begins-in-11-states/. Accessed April 4, 2011.

Hebda, T.L. & Czar, P. (2009). 4th Ed. Handbook of Informatics for Nurses and Healthcare Professionals. Prentice Hall. Upper Saddle River, NJ .

Henneman, E. A., Cunningham, H., Roche, J. P., & Curnin, M. E. (2007). Human patient simulation: Teaching students to provide safe care. *Nurse Educator, 32*(5), 212–217 210.1097/1001.NNE.0000289379.0000283512.fc.

Hiltzik, M. (2000). *Dealers of lightning: Xerox PARC and the dawn of the computer age.* New York: Harper Business.

Hoch, I., Heymann, A. D., Kurman, I., Valinsky, L. J., Chodick, G., & Shalev, V. (2003). Countrywide computer alerts to community physicians improve potassium testing in patients receiving diuretics. *Journal of American Medical Informatics Association, 10*(6), 541–546.

Horrigan, J., & Rainie, L. (2006). When facing a tough decision, 60 million Americans now seek the Internet's help. Observation Deck: Analysis of public opinion, demographic and policy trends.

Retrieved May 3, 2006, from http://pewresearch.org/obdeck/?ObDeckID=19

Hufnagel, Stephen P. (2009). Interoperability. *Military Medicine, 174,* 43–50.

imedicalapps.com. (2011). Apple's New iPad 2 Commercial Prominently Features Medical Apps, 2011, from http://www.imedicalapps.com/2011/04/apple%E2%80%99s-new-ipad-2-commercial-prominently-features-medical-apps/

Innocent, K. (2010). Mobile apps for nurses. September 1 Nursing2010Critical Care. pp. 45–47. Retrieved April 5, 2010 at http://www.nursingcenter.com/pdf.asp?AID=1059927.

Institute of Medicine (IOM). (2001). *Crossing the quality chasm: A new health system for the 21st century.* National Academies Press: Washington, DC.

Institute of Medicine (IOM). (2003). *Health professions education: A bridge to quality.* Washington, DC: National Academies Press.

International Medical Informatics Association (IMIA)—Special Interest Group, Nursing Informatics, Helsinki, Finland, 1997.

Jenkins, S. (2000). Nurses' responsibilities in the implementation of information systems. In M. Ball, K. Hannah, S. Newbold, & J. Douglas (Eds.), *Nursing informatics: Where caring and technology meet* (pp. 207–223). New York: Springer-Verlag.

Jha, A. K., Doolan, D., Grandt, D., Scott, T., Bates, D. W. (2008, July 24). The use of health information technology in seven nations. *International Journal of Medical Informatics*, 254–262.

Johnston, D., Pan, E., Walker, J., Bates, D. W., & Middleton, B. (2003). *The value of computerized provider order entry in ambulatory settings: Executive preview.* Wellesley, MA: Center for Information Technology Leadership.

Joint Commission. (2011). Hospital National Patient Safety Goals. Retrieved April 5, 2011 at http://www.jointcommission.org/assets/1/6/2011_NPSG_Hospital_3_17_11.pdf.

Jones, J. (2003). Patient education and the WWW. *Clinical Nurse Specialist, 17*(6), 281–283.

Jones, Josette F., Hook, Sara A., Park, Seong C., & Scott, LaSha M. (2011 in press). *Privacy, Security and Interoperability of Mobile Health Applications.* Paper presented at the HCII 2011, Orlando, FL.

Kaelber, David C, Jha, Ashish K, Johnston, Douglas, Middleton, Blackford, & Bates, David W. (2008). A Research Agenda for Personal Health Records (PHRs). *Journal of the American Medical Informatics Association, 15*(6), 729–736. doi: 10.1197/jamia.M2547

Kendall, David B. (2008). Improving Health Care in America: Protecting Patient Privacy in the Information Age. 2 *Harvard Law & Policy Review.*

Kilbourne, A., McGinnis, G., Belnap, B., Klinkman, M., & Thomas, M. (2006). The role of clinical information technology in depression care management. *Administration and Policy in Mental Health and Mental Health Services Research, 33*(1), 54–64.

Kilmon, C., Brown, L., Ghosh, S., & Mikitiuk, A. (2010). Immersive Virtual Reality Simulations in Nursing Education. *Nursing Education Perspectives, 31*(5), 314–317.

Levine, C. (2006). HIPAA and talking with family caregivers. *American Journal of Nursing, 106*(8), 51–53.

Magrabi, Farah, Westbrook, Johanna I., & Coiera, Enrico W. (2007). What factors are associated with the integration of evidence retrieval technology into routine general practice settings? *International Journal of Medical Informatics, 76*(10), 701–709.

Matheny, M. E., Sequist, T. D., Seger, A. C., Fiskio, J. M., Sperling, M., Bugbee, D., et al. (2008). A randomized trial of electronic clinical reminders to improve medication laboratory monitoring. *Journal of the American Medical Informatics Association, 15*(4), 424–429.

Medical Records Institute (2006). Retrieved July 25, 2006, from http://www.medrecinst.com

Menachemi, N., Randeree, E., Burke, D. E., & Ford, E. W. (2008). Planning for hospital IT implementation: A new look at the business case. *Biomedical Informatics Insights,* 2008(BII-1-Ford-et-al), 29.

Merriam-Webster Online Dictionary (2011). Accessed March 24, 2011, at http://www.merriam-webster.com/dictionary/phishing

Miller, J., Shaw-Kokot, J. R., Arnold, M. S., Boggin, T., Crowell, K. E., Allegri, et al. (2005). A study of personal digital assistants to enhance undergraduate clinical nursing. *Nursing Outlook, 44*(1), 19–24.

National Institute of Nursing Research (NINR). (1997). Nursing informatics: Enhancing patient care. Retrieved August 25, 1998, from http://www.nih.gov/ninr/research/vol4

Nunes, F. L. S. (2008). The virtual reality challenges in the health care area: A panoramic view. Paper presented at the Proceedings of the 2008 ACM symposium on Applied computing - SAC '08.

Oh, H., Rizo, C., Enkin, M., & Jadad, A. (2005). What Is eHealth (3): A systematic review of published definitions. *Journal of Medical Internet Research, 7*(1), e1.

Oroviogoicoechea, C., Elliott, B., & Watson, R. (2008). Review: Evaluating information systems in nursing. *Journal of Clinical Nursing, 17*(5), 567–575.

Pew Research Center (2010). Online health seeking, How Social Networks Can be Health Communities. Accessed March 22, 2011, at http://www.pewinternet.org/~/media/Files/Presentations/2010/Oct/2010%20-%2010.25.10%20-%20Online%20health%20seeking%20-%20pdf%20Newport.pdf.

Poissant, L., Pereira, J., Tamblyn, R., & Kawasumi, Y. (2005). The impact of electronic health records on time efficiency of physicians and nurses: A systematic review. *Journal of the American Medical Informatics Association, 12*(5), 505–516.

Prinz, L., Cramer, M. & Englund, , A. (2008). Telehealth: a policy analysis for quality, impact on patient outcomes, and political feasibility. Nursing Outlook. July-August;56(4):152–8.

Sandlund, M., McDonough, S., & Häger-Ross, C. (2009). Interactive computer play in rehabilitation of children with sensorimotor disorders: A systematic review. *Developmental Medicine & Child Neurology, 51*(3), 173–179.

Sarasohn-Kahn, Jane (2010). How Smartphones Are Changing Health Care for Consumers and Providers. In California HealthCare Foundation (Ed.).

Sequist, T. D., Gandhi, T. K., Karson, A. S., Fiskio, J. M., Bugbee, D., Sperling, M., et al. (2005). A randomized trial of electronic clinical reminders to improve quality of care for diabetes and coronary artery disease. *Journal of the American Medical Informatics Association, 12*(4), 431–437.

Sicotte, C., & Paré, G. (2010). Success in health information exchange projects: Solving the implementation puzzle. *Social Science & Medicine, 70*(8), 1159–1165.

Smith, D., & Alexander, R. (1988). *Fumbling the future: How Xerox invented, then ignored, the first personal computer.* New York: W. Morrow.

Stevens, K. R., & Staley, J. M. (2006). The Quality Chasm reports, evidence-based practice, and nursing response to improve healthcare. *Nursing Outlook, 54*(2) 94–101.

Technology Informatics Guiding Education Reform (TIGER). (2009). *Collaborating to integrate evidence and informatics into nursing practice and education: An executive summary.* Washington, DC: Alliance of Nursing Informatics (ANI).

The Internet: Friend or Foe When Providing Patient Education? [10.1188/08.CJON.55–63]. Clinical Journal of Oncology Nursing, 12(1), 55–63.

U.S. Department of Health and Human Services. (2011). The office of the coordinator for health information technology. http://healthit.hhs.gov/portal/server.pt/community/healthit_hhs_gov_home/1204, accessed April 3, 2011.

Watts, J. (1999). The healthy home of the future comes to Japan. *Lancet,* 353(9164), 1597–1600.

Weaver, C. A., Warren, J. J., & Delaney, C. (2005). Bedside, classroom and bench: Collaborative strategies to generate evidence-based knowledge for nursing practice. *International Journal of Medical Informatics,* 74(11–12), 989–999.

Weiser, M., & Brown, J. (1996, October 5). The coming age of calm technology. Retrieved July 19, 2001, from http://www.ubiq.com/hypertext/weiser/acmfuture2endnote.htm

Work Group on Confidentiality Privacy & Security. (1995). *Guidelines for establishing information security policies at organizations using computer-based patient records.* Schaumburg, IL: Computer-Based Patient Record Institute.

Wright, Adam, & Sittig, Dean F. (2007a). Encryption Characteristics of Two USB-based Personal Health Record Devices. *Journal of the American Medical Informatics Association,* 14(4), 397–399.

Wright, Adam, & Sittig, Dean F. (2007b). Security Threat Posed by USB-Based Personal Health Records. *Annals of Internal Medicine,* 146(4), 314–315.

## SUGGESTED READINGS

Artinian, N. T. (2007, January–February). Telehealth as a tool for enhancing care for patients with cardiovascular disease. *Journal of Cardiovascular Nursing, 22*(1), 25–31.

Bowles, K. H., & Baugh, A. C. (2007, January–February). Applying research evidence to optimize telehomecare. *Journal of Cardiovascular Nursing, 22*(1), 5–15.

Demiris, G. (2007). Interdisciplinary innovations in biomedical and health informatics graduate education. *Methods in Infectious Medicine, 46*(1), 63–66.

Doebbeling, B. N., Vaughn, T. E., McCoy, K. D., & Glassman, P. (2006). *Informatics implementation in the Veterans Health Administration (VHA) healthcare system to improve quality of care.* American Medical Informatics Association Annual Symposium Proceedings, 204–208.

Gurses, A. P., Hu, P., Gilger, S., Dutton, R. P., Trainum, T., Ross, K., et al. (2006). *A preliminary field study of patient flow management in a trauma center for designing information technology.* American Medical Informatics Association Annual Symposium Proceedings, 937.

Hao, A. T., Chang, H. K., & Chong, P. P. (2006). *Mobile learning for nursing education.* American Medical Informatics Association Annual Symposium Proceedings, 943.

Palm, J. M., Colombet, I., Sicotte, C., & Degoulet, P. (2006). *Determinants of user satisfaction with a clinical information system.* American Medical Informatics Association Annual Symposium Proceedings, 614–618.

Poon, E. G., Keohane, C., Featherstone, E., Hays, B., Dervan, A., Woolf, S., et al. (2006). *Impact of barcode medication administration technology on how nurses spend their time on clinical care.* American Medical Informatics Association Annual Symposium Proceedings, 1065.

# CHAPTER 7

# Population-Based
Health Care Practice

### Patricia M. Lentsch Schoon, MPH, RN

*Population health improvement involves more than providing health care to individuals and families who make their way to hospitals and clinics. Achieving health and preventing disease requires the collaborative efforts of health care providers, policy makers and enforcers, educators, employers, and individuals and families.*

(Zahner & Block, 2006)

## OBJECTIVES

Upon completion of this chapter, the reader should be able to:

1. Discuss the social mandate to provide population-based health care at the global, national, state, and local levels.

2. Describe how population-based nursing is practiced within the community and the health care system.

3. Identify vulnerable and high-risk population groups for whom specific health promotion and disease-prevention services are indicated.

4. Outline a multidisciplinary population-based planning and evaluation process that includes partnerships with the community and health care consumers.

5. Discuss the nurse's role in disaster preparedness and response.

Native American Community Clinic staff and community board members, Minneapolis, MN.
Photo by David Schoon

*You are completing a clinical experience in an elementary school located in a large metropolitan area. This school has 600 students from kindergarten through sixth grade; 95% of them are from families living at or below the poverty level. Seventy-five percent of the student body transfers in or out of the school during each school year. Of these children, 40% are Southeast Asian, 25% are Hispanic, 10% are African American, 10% are white, 10% are Somali refugees, and 5% are Native Americans. Sixty percent of the students speak English as a second language. Although 95% of the students are eligible for state-sponsored health insurance, only 30% of the children have health insurance, and only 10% have dental insurance. You observe that many children have significant dental problems and many are below the fifth percentile in height and weight on the standardized growth grid. The school nurse encourages you to think about what nursing actions you might take with these two groups of children.*

> *What health problems can you identify in this group of schoolchildren?*
>
> *What social and environmental factors might contribute to these health problems?*
>
> *What actions directed at groups of children, rather than individuals, could you take to help these children?*
>
> *How can this be addressed at the population level?*

Nurses have acted to improve the health of populations and communities since the time of Florence Nightingale. Nightingale's actions to improve the health care of soldiers on the battlefields of the Crimean War and of the poor and infirm in London were directed at vulnerable population groups (Falk-Rafael, 2005). Nightingale's actions were based on the recognition that vulnerable population groups were not able to advocate effectively for themselves. She became their advocate. Nightingale was able to intervene to improve the health status of disenfranchised groups of people by influencing the health policies of the English government and changing the health care delivery systems in London and on the battlefield. That same spirit of advocacy and call to action is alive today among nurses throughout the world.

The primary thrust of this advocacy is directed at population groups that are at greatest risk for a decrease in health status and those groups that are most vulnerable to the socioeconomic forces that interfere with access to affordable quality health care. Nurses are united in partnership with other health care disciplines, the community, and health care consumers to achieve population-based global, national, state, and local health care goals related to access, cost, and quality. This chapter provides readers with an understanding of how population-based health care is practiced in the public and private health care sectors.

Population-based care requires active partnership of providers and recipients of care. The phrase *health care consumers* includes current recipients of care, potential or past recipients of care, and other interested parties within the community.

This chapter introduces the reader to the application of the nursing process to population-based nursing practice. Collaboration with other health and social service providers and partnership with community members are key strategies to improve population-based health outcomes, particularly among disenfranchised and vulnerable population groups. The diverse health care needs of the twenty-first century will require innovation in how health care is delivered. Nurses can take the lead in creating culturally inclusive population-based health care services and interventions that meet the needs of diverse population groups in the community.

## POPULATION-BASED HEALTH CARE PRACTICE

**Population-based health care practice** is the development, provision, and evaluation of multidisciplinary health care services to population groups experiencing increased health risks or disparities, in partnership with health care consumers and the community, to improve the health of the community and its diverse population groups. Vulnerable population groups are subgroups of a community that are powerless, marginalized, or disenfranchised, and are experiencing health disparities.

**Health risk factors** are variables that increase or decrease the probability of illness or death. Health risk factors may be modifiable (that is, they can be changed), for example, health care prevention practices; or they may be nonmodifiable (cannot be changed), for example, age, sex, race, or other inherent physical characteristic.

**Health determinants** are variables that include biological, psychosocial, environmental (physical and social), and health systems factors or etiologies that may cause changes in the health status of individuals, families, groups, populations, and communities. Health determinants may be assets (positive factors) or risks (negative factors). **Health assets** are health-promoting attributes of individuals/families, population groups, and communities.

# HEALTH STATUS DISPARITIES

**Health status disparities** are differences in health risks and health outcome measures that reflect the poorer health status that is found disproportionately in certain population groups. Individuals, families, population groups, and communities may experience health status disparities. These disparities lead to unequal burdens in disease morbidity and mortality rates borne by racial and ethnic groups in comparison to the dominant racial or ethnic group in a society (Baldwin, 2003; Lillie-Blanton, Maleque, & Miller, 2008). The Office of Minority Health, Center for Disease Control (CDC), describes the increasing disease burden and risk factors experienced by people of color and Native Americans (www.cdc.gov/omh, click on, About Minority Health, and then click on Disease Burden & Risk Factors). Significant health status disparities persist in the United States among population groups representing racial, ethnic, socioeconomic, and geographic differences (American College of Physicians, 2004; Eberhardt & Parnuk, 2004; Hartley, 2004; Institute of Medicine [IOM], 2002; IOM, 2004).

The absolute rate of health status disparities has decreased in the last decade. For example, rates of many diseases have decreased among African Americans and white Americans. However, the relative rates of health status disparities have remained about the same. In other words, the gap between the health status of African Americans and white Americans has not changed. It is estimated that by the year 2045, half the people in the United States will be members of racial minority population groups. If there is no significant change in the quality and availability of health care, health status disparities will affect a greater proportion of the United States population at mid-century (Institute of Medicine [IOM], 2010).

Selected examples of health status disparities that exist in the United States include the following (Secretary's Advisory Committee on Health Promotion and Disease Prevention Objectives for 2020, 2008, p. 73):

- Black infants have higher mortality rates than white infants.
- Maternal mortality is higher among black women.
- Among the elderly, women's health and functional status are worse than men's.
- Black women are more likely than white women to die from breast cancer.
- Life expectancy at age 26 is shorter and rates of heart disease and diabetes are higher among people of lower incomes or educational levels and among blacks, Hispanics, and Native Americans.

- Poor or fair health (contrasted with good, very good, or excellent health) is more prevalent among children in low-income families.
- In elderly adults, disability rates are inversely related to income (Minkler et al., 2006).
- Obesity appears to be more prevalent in adults with sensory, physical, and mental health conditions (Weil et al., 2002).

**Health care disparities** are differences in health care coverage, health care system access and quality of care for different racial, ethnic, and socioeconomic population groups that persist across settings, clinical areas, age, gender, geography, and health needs and disabilities (Lillie-Blanton, Maleque, & Miller, 2008). These health care disparities result in poorer health care outcomes. Significant health care disparities exist in the U.S. health care system (Agency for Healthcare Research and Quality [AHRQ], 2005; IOM, 2010). Social and economic factors, including health care disparities, are the primary cause of health disparities among people of color and American Indians in the United States.

Health status disparities are the result of many societal causes, including health care disparities. *Unnatural Causes*, a seven-part documentary produced in 2008 by California Newsreel, documents the diverse social, economic, environmental, and health care variables that influence the course of individual and population health status across the lifespan. For example, white neighborhoods have four times as many supermarkets as black and Latino neighborhoods. In black neighborhoods, 80% of billboards advertise tobacco or alcohol (*Unnatural Causes*, 2010). Living under chronic inhospitable environmental conditions such as inadequate housing and unsafe neighborhoods, lack of employment, and absence of educational opportunities results over time in long-term chronic health conditions. As more people of color live in these environmental conditions, the incidence of physical health disparities is greater among people of color (Jackson et al., 2010).

Poverty has a significant impact on life span. People with incomes greater than 400% of the poverty level live an average of 6.5 years longer than those with incomes below the poverty level (U.S. Department of Health and Human Services [USHHS], 2007). Inadequacy of health care may be a major factor in this disparity. Trends of core measures of health care quality and access demonstrate that African Americans, Hispanics, and people living in poverty receive worse quality of care than do other comparable groups and those not living in poverty (USHHS, 2007). For most core health care measures, the health status disparities remained the same or got worse over time.

The level of uninsured people in a community also influences the availability of health resources in a community. Clinics and other health resources generally locate in neighborhoods with a large patient base, i.e., people who have health insurance. Thus, neighborhoods with higher levels of poverty, more people of color, and people who are less likely to be insured have fewer available health resources located in their communities. This results in less access to primary care providers and medical specialists (IOM, 2004). Both the insured and uninsured that live in such communities are adversely affected by the lack of available health resources.

Global measures of health status of population groups include morbidity (illness) and mortality (death) rates as well as measures of quality of life.

# HEALTHY PEOPLE 2020

"The context of people's lives determine their health, and so blaming individuals for having poor health or crediting them for good health is inappropriate. Individuals are unlikely to be able to directly control many of the determinants of health" (WHO, 2010).

Population health status is measured by health outcomes and health indicators (behaviors) at the national level. In 1979, the Surgeon General released a report on health promotion and disease prevention for the United States. Ten-year evidence-based health objectives for the

nation were established in 1980, 1990, 2000, 2010, and 2020 in a series of *Healthy People Reports.*

National health goals and plans for their achievement have been published every 10 years since 1980 in a series of *Healthy People Reports.* These reports are a statement of national values and the willingness and ability of the nation to secure better health and well-being for all (Secretary's Advisory Committee on National Health Promotion and Disease Prevention Objectives for 2020, 2008). See Table 7-1 for the *Healthy People 2020* mission and overarching goals. Key focus areas have been identified to carry out the mission, vision, and goals of *Healthy People 2020.* These focus areas cover four broad areas of health concerns:

1. *Population groups over the lifespan:* adolescent health, early and middle childhood; genomics; global health; lesbian, gay, bisexual, and transgender health issues; maternal, infant, and child health; and older adults
2. *Health behaviors, lifestyle, and health determinants:* access to health services; immunization and infectious diseases; family planning; health-related quality of life; injury and violence protection;

## Critical Thinking 7-1

Review the *Healthy People 2010* Focus Areas.

These major focus areas were tracked to determine progress toward *Healthy People 2010* goals. These focus areas and others will also be tracked for *Healthy People 2020.*

| | |
|---|---|
| Access to health care | Injury and Violence Prevention |
| Access to Quality Health Services | Maternal, Infant, and Child Health |
| Arthritis, Osteoporosis and Chronic Back Conditions | Medical Product Safety |
| Cancer | Mental Health and Mental Disorders |
| Chronic Kidney Disease | Nutrition and Overweight |
| Diabetes | Occupational Safety and Health |
| Disability and Secondary Conditions | Oral Health |
| Educational and Community-Based Programs | Physical Activity and Fitness |
| Environmental Health | Public Health Infrastructure |
| Family Planning | Respiratory Diseases |
| Food Safety | Sexually Transmitted Diseases |
| Health Communication | Substance Abuse |
| Heart Disease and Stroke | Tobacco Use |
| HIV | Vision and Hearing |
| Immunizations and Infectious Diseases | |

Find the health data for these major health focus areas for the United States and the state that you live in. Go to http://wonder.cdc.gov/data2010/focus.htm to find the Healthy United States 2010 Database. How does your state compare to other states or to the United States as a whole in achievement of the *Healthy People 2020* major health focus areas?

**TABLE 7-1  Healthy People 2020**

**Mission**

*Healthy People 2020* strives to:

- Identify nationwide health improvement priorities

- Increase public awareness and understanding of the determinants of health, disease, and disability and the opportunities for progress

- Provide measurable objectives and goals that are applicable at the national, state, and local levels

- Engage multiple sectors to take action to strengthen policies and improve practices that are driven by the best available evidence and knowledge

- Identify critical research, evaluation, and data collection needs

**Overarching Goals**

- Attain high-quality, longer lives free of preventable disease, disability, injury, and premature death

- Achieve health equity, eliminate disparities, and improve the health of all groups

- Create social and physical environments that promote good health for all

- Promote quality of life, healthy development, and healthy behaviors across all life stages

**Source:** Compiled from *Healthy People 2020* Framework. Accessed October 29, 2010, from http://www.healthypeople.gov/hp2020/Objectives/framework.aspx.

nutrition and weight Status; physical activity; sexually transmitted diseases; sleep health; substance abuse; tobacco use; and environmental health
3. *Health-related infrastructure:* blood safety; educational and community-based programs; food safety; health communication and health information technology; health care–associated infections; medical products safety; occupational safety and health; preparedness; and public health infrastructure
4. *Health conditions and chronic diseases:* arthritis, osteoporosis, and chronic Back conditions; blood disorders; cancer; chronic kidney diseases; dementias including Alzheimer's disease; diabetes; disability and health; hearing and other sensory or communication disorders; heart disease and stroke; HIV; mental health and mental disorders; oral health; respiratory diseases; and vision

It is imperative that health providers and consumers address the health priorities identified in *Healthy People 2020,* as population health indicators demonstrate that population health has been declining steadily (American Public Health Association, 2009). For example, there has been a significant reduction in smoking cessation and a significant increase in obesity throughout the U.S. population during the first decade of this century. It is estimated that if the current growth in chronic conditions continues through 2023, the incidence of the seven most common chronic diseases will increase by 42 percent. These diseases include cancer, diabetes, hypertension, stroke, heart disease, pulmonary conditions, and mental disorders.

Progress toward achievement of these objectives is measured by major health indicators. The Partners in Information Access for the Public Health Workforce (http://phpartners.org/hp) Web site helps you search for published research on *Healthy People 2010* and *Healthy People 2020,* as well as evidence-based strategies that help to improve the health of population groups.

The goals of population-based health care include: (1) improvement of access to health care services, (2) improvement of quality of health care services, (3) reduction of health status disparities among different population groups, and (4) reduction of health care delivery costs.

Population-based interventions are provided at three levels: (1) individual/family; (2) community; and (3) systems. A community may be a group of people with a shared characteristic that provides a cohesive bond (i.e., deaf community, cultural group) or a physical place or geographic area (i.e., city, county, region of the country). A system is an institution or organization that may exist within one or more communities (i.e., health care system, school, religious institution, legislature). Outcomes of these interventions are measured in three domains: population health status, quality of life, and functional health status.

**Health status** is the level of health of an individual, family, group, population, or community. It is the sum of existing health risk factors, level of wellness, existing diseases, functional health status, and quality of life. **Quality of life** is the level of satisfaction one has with the actual conditions of one's life, including satisfaction with socioeconomic status, education, occupation, home, family life, recreation, and the ability to enjoy life, freedom, and independence. Quality of life assessment reviews the perceived and actual ability to be autonomous and independent in making life choices; one's sense of happiness, satisfaction, and security; and the ongoing ability to strive to reach one's potential. **Health-related quality of life** refers to one's level of satisfaction with those aspects of life that are influenced either positively or negatively by one's health status and health risk factors.

**Functional health status** is the ability to care for oneself and meet one's human needs. Functional abilities are the combined abilities to be independent in both activities of daily life and in the instrumental activities of daily living. **Activities of daily living** are activities related to toileting, bathing, grooming, dressing, feeding, mobility, and verbal and written personal communication. **Instrumental activities of daily living** (IADLs) are activities related to food preparation and shopping; cleaning; laundry; home maintenance; verbal, written, and electronic communications; financial management; and transportation, as well as activities to meet social and support needs, manage health care needs, access community services and resources, and meet spiritual needs. Functional health status affects health-related quality of life.

Addressing the priority health needs of the most vulnerable population groups at the population level rather than the individual level challenges nurses to target finite health care resources more effectively. Vulnerable population groups are often underserved. These vulnerable groups may have decreased health status and increased risk for morbidity and mortality because of multiple and complex medical and social problems. They may be marginalized, meaning they exist at the margins of mainstream society, without access to the majority of community resources and networks. They may be disenfranchised, meaning they do not have the ability to participate in or influence decisions that affect their health care status.

# CULTURALLY INCLUSIVE HEALTH CARE

The U.S. population continues to become more diverse. Today about 34 percent of the U.S. population identify themselves as a member of a racial or ethnic minority. By 2050, 54 percent of the U.S. population will identify themselves as a member of a racial or ethnic minority. About 44 percent of U.S. children today are members of a racial or ethnic minority. This number will increase to 62 percent by 2050. Hispanics are the fastest growing ethnic group. The Census Bureau says that current U.S. minorities will be the majority by 2042 (America.gov,[2008]). If immigration and birth patterns continue, the number of Hispanics, Asians, and African Americans in the U.S. population will increase far more rapidly than the white population. Nurses graduating today will experience significantly more diversity in the workplace than older nurses experienced. Increased health care disparities among racial and ethnic minorities are primarily a result of social and economic structures rather than genetic racial and ethnic factors (Tashiro, 2005).

The proportion of ethnic minorities in the RN workforce in 2008 (HRSA, 2010) continues to lag behind the proportion of ethnic minorities in the U.S. population (Table 7-2). Thirty-four percent of the U.S. population report themselves as minority. By 2042, minorities are expected to make up more than 50% of the U.S. population (U.S. Census Bureau, 2010, accessed November 8, 2010, from http://www.uscensus.gov). Nurses need to play a key role in reshaping the health care system to be more culturally inclusive, increasing the ethnic diversity of the nursing profession, and becoming more culturally inclusive in their nursing practice. Barriers to cultural competence and cultural inclusivity in the health care workplace include lack of awareness of differences; lack of time; ethnocentrism, bias, and prejudice; lack of skills to address differences; and lack of organizational support (Taylor, 2005).

A culturally inclusive health care system is one in which health care is population based. This system includes multiple methods of providing cultural health care that reflect the diverse history and cultures of the population groups served. All communities have a culture. Culture refers to the learned patterns of behaviors and the range of beliefs attributed to a specific group that are passed on through generations. It includes ways of life, norms and values, social institutions, and a shared construction of the physical world (Dreher et al., 2006).

It is essential for nurses to consider the culture and ethnicity of the community and population groups when working to improve population health (Dreher et al., 2006). A culturally inclusive health care system promotes positive health outcomes using a variety of intervention strategies to achieve outcome measures tailored to the

## TABLE 7-2  Race/Ethnicity/Gender of Registered Nurse Workforce Compared to U.S. Population

| | REGISTERED NURSES | U.S. POPULATION | DIFFERENCE |
|---|---|---|---|
| White Non-Hispanic | 83.2% | 65.1% | +18.1% |
| Black, African American (Non-Hispanic) | 5.4% | 12.9% | −7.5% |
| Asian/Pacific Islander/Native Hawaiian (Non-Hispanic) | 5.8% | 4.8% | +1% |
| Hispanic/Latino | 3.6% | 15.8% | −12.2% |
| American Indian/Alaskan Native | 0.3% | 1.0% | −0.7% |
| Two or More Racial Backgrounds | 1.7% | 1.7% | 0.0% |
| Males | 6.2% | 47.3% | −41.1% |

**Source:** National sample survey of registered nurses—preliminary findings, 2008. Compiled from http://bhpr.hrsa.gov/healthworkforce/rnsurvey/initialfindings2008.pdf, accessed November 8, 2010; U.S. Census Quick Facts. (2010), accessed November 8, 2010, from http://quickfacts.census.gov/qfd/states/00000.html; America.gov. (2010), accessed November 8, 2010, from http://www.america.gov/index.html; and American Association of Critical Care Nurses. (2010), Enhancing diversity in the nursing workforce, accessed November 8, 2010, from http://www.aacn.nche.edu/media/factsheets/diversity.htm.

diversity of the population groups served. The U.S. health care system demonstrates significant disparities in the provision of just and equitable health care. It tends to be more culturally exclusive than inclusive (Gustafson, 2005). The development of a culturally inclusive health care system requires significant change in the current system. To start this change, nurses and other health care providers need to work together to create it. Increased awareness of the injustice of existing health care systems may lead to both individual and system change (Giddings, 2005). Individual and systems change will require increased diversity in the health care workforce. Increasing the proportion of bicultural and bilingual health care providers to fit the diversity of the U.S. population will increase the probability that health care will be delivered in a culturally inclusive manner that is truly population based. This is a start!

A community-based action research model in which the community or population group is an active part of the research process is another effective strategy in developing culturally inclusive health care programs (Garwick, Rhodes, Peterson-Hickey, Hellerstedt, 2008) Student nurses of similar racial and ethnic origins to the population to be served can play a significant role in contributing to a culturally inclusive health care system by participating in community-based action research. All health care providers must work together to develop this research and other strategies to achieve a culturally inclusive health care system.

# HEALTH DETERMINANT MODELS

Health determinant models provide conceptual tools to use in assessing and addressing the priority health needs of at-risk population groups. Assessment and intervention models take into account the importance of community systems such as health, social service, government, and economics in influencing the health outcomes in population groups as well as individuals (U.S. Department of Health and Human Services [USDHHS], 2000; Falk-Rafael, 2005). Health determinant models are holistic and reflect the multiple causes of health disparities in diverse population groups. Existing health determinants models identify income inequality in concert with social, cultural, and political conditions as a key health determinant influencing population health outcomes such as morbidity and mortality rates globally (Falk-Rafael, 2005). Some health determinants have a greater impact than others on health outcomes and health status. Pawlak (2005) proposes a health determinants model in which health literacy, the ability to find and use health information to make decisions, is a key

## REAL WORLD INTERVIEW

The Experience of Hmong Women Living with Diabetes Community-Based Collaborative Action Research (CBCAR) includes the community in the definition of the research question, data analysis, dissemination, and action planning (Pharris et al., 2002). One example of a culturally sensitive use of CBCAR occurred when the advisory council of a community clinic identified diabetes as a health disparity among Hmong women, which needed to be addressed. Female Hmong nursing researchers from the clinic's academic partner wondered if clinic health care providers were labeling Hmong women with diabetes as noncompliant without understanding the women or their culture.

CBCAR using Dr. Margaret Newman's nursing theory of health as expanding consciousness (Newman, 1999) was the framework used to engage Hmong women with diabetes in a dialogue to understand their life patterns and to envision health-promoting actions. Five Hmong women with Type 2 diabetes and HgbA1c levels over 7.0 were recruited from the clinic and interviewed in their homes by female Hmong nursing researchers to identify common patterns. Researchers worked with a female Hmong playwright to weave common patterns into a play. Hmong women were invited via Hmong radio and community advertisements to a dinner, performance, and dialogue. Two senior female Hmong nursing students performed the play for the Hmong women. After the play, the Hmong women were grouped into four small dialogue circles facilitated by female Hmong students and female Hmong community leaders. The dialogue focused on what needed to be added so that the play could better reflect the women's life experiences. It also focused on what actions needed to be taken so that Hmong women with diabetes could live healthy, happy lives in the United States.

**Avonne A. Yang, RN, BSN**
Faculty Assistant/Director of Community Health Nursing
Student Internship Program (CHNSIP)
Department of Nursing
College of St. Catherine
St. Paul, Minnesota

determinant of population health. The *Healthy People 2020* Framework (Figure 7-1) highlights the relationship between health determinants including the physical environment, the social environment, individual behavior, biology and genetics, and health services to achieve positive health outcomes. The social environment includes social factors or social determinants of health such as socioeconomic conditions, social norms, and social supports. Policy making at federal, state, and local levels and availability and quality of health services are also health determinants. This model depicts the constantly changing and cumulative effect of health determinants on the health outcomes of individuals, families, and population groups (U.S. DHHS. 2010. *Healthy People 2020*: Determinants of Health retrieved March 22, 2011 from http://healthypeople.gov/2020/about/DOHAbout.aspx#policymaking). This model will continue to be used over the next decade. Examples of selected community-level, health systems–level, and population-level health determinants are provided in Table 7-3. This schema can be used to identify those at greatest risk for specific health events and diseases.

Data sources used for population-based assessment are seen in Table 7-4.

# POPULATION-BASED NURSING PRACTICE

*Population-focused nursing practice* and *population-based nursing practice* are terms that are used interchangeably in contemporary community health nursing literature. Population-focused practice has been an integral part of nursing since the profession began. Florence Nightingale's use of aggregated statistics as population-based indices and outcome measures demonstrates her population-focused practice. Lillian Wald established population-focused nursing practice in the United States in the early 1900s. Her population-focused efforts included founding the Henry Street Nurses' Settlement; helping to establish the Children's Bureau; and advocating for the rights of children, the mentally ill, the indigent, and immigrants. Wald's vision of nurses was that they would be "carriers of health" (Peters, 1995). According to Wald (1915), nurses were

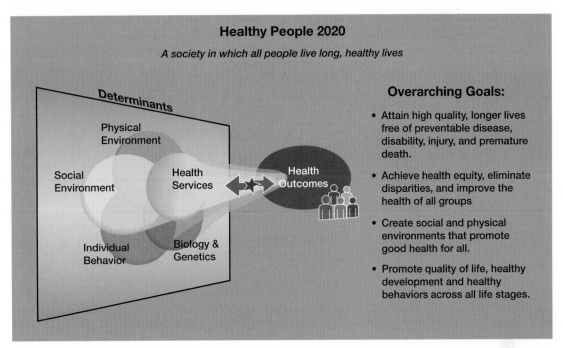

**FIGURE 7-1** U.S. Department of Health and Human Services (HHS). Office of Disease Prevention and Health Promotion. (2011) Framework, The Vision, Mission, and Goals of *Healthy People 2020.* (Accessed March 12, 2011, at http://healthypeople.gov/2020/consortium/HP2020Framework.pdf).

**TABLE 7-3 Population-Based Health Determinants Assessment Template (Excerpt)**

**Community Level**

*Physical Environment*

• Housing and geographic location, safety of neighborhoods, community, and school quality

• Environmental quality: air, water, ground, chemical, physical, and biological hazards

• Availability of transportation, communication systems, parks, and recreation facilities

*Social Environment*

• Community norms, values, and patterns of behavior; political structures within community

• Incidence of crime and violence within the population group and the larger community

• Employment opportunities within community, economic viability of community

*Policies and Interventions*

• National, state, county, and city health and social policies

• Policies of public, private, and voluntary organizations that provide health and social services

• Policies of health insurance companies, health maintenance organizations, health systems, and health care provider groups

*(Continues)*

## TABLE 7-3 (Continued)

### Health Systems Level

*Access to Quality Health Care*

- Appropriate primary, secondary, and tertiary health services and providers

- Health and social services workforce (numbers, diversity, interdisciplinary mix, deployment, sustainability); educational institutions offering health care provider education and training

- Availability of health and social services resources 24 hours a day, seven days per week

### Population Level

*Biological*

- Demographic data (age, gender, racial/ethnic patterns)

- Biological and genetic factors, patterns of health and disease (morbidity and mortality data)

*Behavioral*

- Education patterns and levels, cultural patterns (lifestyle, languages, religion)

- Socioeconomic status (employment patterns, housing, health and dental insurance)

- Cultural health patterns (health beliefs and self-care practices, nutrition, fitness, previous experiences with health care system, current health providers, family and intergenerational health patterns)

*Data Analysis*

- Prioritize the health needs of the community and prioritize the vulnerable and at-risk population groups based on need and health status.

- Identify the modifiable and nonmodifiable health risks of the at-risk population.

- Identify the biological, psychosocial, environmental, cultural, political, financial, and iatrogenic causes of the identified health risks.

"part of the community plan for the attainment of communal health" (p. 60). As nursing in the community became more organized, particularly in public health agencies, nursing leaders such as Ruth Freeman encouraged nurses to consider the health needs of the total community as their mission (Freeman, 1957).

In the future, both nursing education and nursing practice will place a greater emphasis on use of population-based mortality and morbidity statistics for assessment of community health needs. They will focus on maximizing health status, functional abilities, and improving the quality of life of groups of health care consumers.

Contemporary public health nursing leaders have consistently stressed the need for nurses in the community to be population focused in their practice (ANA, 2007). More-recent discussions have included the need for nurses across all health care settings to be population focused.

**Population-based nursing practice** is defined as the practice of nursing in which the focus of care is to improve the health status of vulnerable or at-risk population groups within the community by employing health promotion and

**TABLE 7-4 Data Sources for Population-Based Assessment**

**Common Sources of Primary Data**

- Key informant interviews and surveys of health and social services providers, community leaders, media, and governmental agency officials and personnel

- Key informant interviews, surveys, and observations (participant and nonparticipant) of members of at-risk or vulnerable population groups (may require formal approval process if research done on human subjects)

- Windshield and walk-through surveys of the community and organizations involved

**Common Sources of Secondary Data**

- Health data, vital statistics, and census data obtained from governmental sources

- Community planning documents

- Health reports on subpopulations from governmental, voluntary, private organizations, and consumer groups

- Scientific and professional literature and databases

- Proprietary patient/member population data from health care insurers and providers (may be available only to employees of the organization)

- The news and communications media, including newspapers, journals, television, radio, and the Internet

Data utilized may be primary (collected by group conducting assessment) or secondary (collected from other sources).

disease prevention interventions across the health continuum. Health care consumers are involved as full partners in the planning and evaluation of the nursing services provided. Population-based nursing practice is holistic in nature, taking into account cultural and ethnic diversity, religious and spiritual uniqueness, economic disparities, and geographic and regional differences. It seeks to empower population groups by enhancing their protective factors and resiliency. **Protective factors** are strengths and resources that patients can use to combat health threats that compromise core human functions. **Resilience** is the social and psychosocial capacity of individuals and groups to adapt, succeed, and persevere over time in the face of recurring threats to psychosocial and physiologic integrity. Population groups that experience greater threats to health based on biological, physical, or social environmental risk factors will experience poorer health and safety outcomes. The level of resilience in the population group or the ability of the population group to thrive despite the presence of risk factors will help to reduce the group's health and safety risks (Davis, Cook, & Cohen, 2005). Promoting resilience in marginalized and disenfranchised population

groups and simultaneously working to reduce the barriers to health care are dual responsibilities for nurses practicing population-based nursing.

The goals of population-based nursing practice are consistent with the health goals identified by the World Health Organization (WHO Task Force, 2005) and *Healthy People 2020*. Population-based nursing practice goals address the health needs of individuals, families, communities, and population groups, and focus on the goals of health care access, quality, cost, and equity.

## POPULATION-BASED NURSING PRACTICE INTERVENTIONS

A population-based public health interventions model for public health nursing practice, developed by the Minnesota Department of Health (2001), is in concert with the public mandate that directs public health agencies to protect the health of the public they serve. A primary principle of population-based nursing practice is that it is initiated with a community health assessment.

## REAL WORLD INTERVIEW

Last April, a 12-year-old boy from a nearby village was brought to our clinic at Nuestros Pequeños Hermanos (NPH). He had fallen out of a tree from about 30 feet. He was unresponsive and vomiting. His head was bleeding, and there was blood coming from both ears. On the 45-minute drive to the hospital in our clinic ambulance, I sat in the back of the pickup with the boy, trying to physically immobilize him with my own body until we got to the hospital. I left the hospital that night feeling sure the little boy wasn't going to make it.

Two weeks later, the boy was discharged to his home to die, so we brought him to our clinic. He could not speak, had casts on his left arm and leg, and couldn't sit on his own. Within two weeks, he was eating food and sitting up. After the boy went home, he came back to our clinic for treatments. Several months later, I saw the little boy in the clinic waiting room. When he saw me, his eyes lit up and he broke into a big smile. He hadn't remembered me until now and hadn't been able to speak more than a few words. Tears welled in my eyes when the little boy looked at me with eyes that had been so full of fear that April night, as he asked me for water to share with his little brother.

The children that live at NPH have lived through unimaginable horrors. Their parents have been killed in front of them. Some parents have died from AIDS or cancer. Many children have been abandoned and left to survive on their own, children caring for children. Some have been beaten, abused, or neglected. It means the world to me to see these kids enjoying a happy and loving life. Every step I took as a volunteer has brought me to a better place, with a greater outlook and understanding of life.

**Annie Kautza, BSN**
Volunteer
Nuestros Pequeños
Hermanos, Honduras

---

Multidisciplinary interventions are developed based on the community health priorities identified in the assessment process.

Population-based interventions encompass three levels of practice: (1) the community; (2) systems within the community; and (3) individuals, families, and groups. Population-based, community-focused practice changes community norms, attitudes, practices, and behaviors. Population-based, systems-focused practice changes laws, power structures, policies, and organizations. Population-based, individual-focused practice changes the knowledge, attitudes, beliefs, practices, and behaviors of individuals, families, and groups (Keller, Strohschein, Lia-Hoagberg, &

Schaffer, 2004a; Keller, Strohschein, Schaffer, & Lia-Hoagberg, 2004b). Figure 7-2 depicts the Minnesota Department of Health Public Health Interventions II Wheel developed by the Section of Public Health Nursing.

These interventions have a logical sequence. For example, to provide health services to underserved and vulnerable population groups, outreach (finding the people at risk in the community) and screening must precede referral, teaching, counseling, and consultation.

Nursing interventions need to be provided in a culturally sensitive and appropriate manner to be effective with culturally diverse population groups.

## TRADITIONAL VERSUS NONTRADITIONAL MODEL

The traditional model of population-based public health nursing practice starts with public health and community health agencies working in partnership with the community to carry out a community assessment. Priorities are established, and a plan is developed in partnership with community members. After the plan is implemented, evaluation is also conducted in partnership with the community. In recent years, private health care organizations

### Critical Thinking 7-2

Assess your community using the template in Table 7-3. How does your community score? Do you have any suggestions to improve your community's health, based on this assessment?

## REAL WORLD INTERVIEW

The Native American Community Clinic (NACC) is a nonprofit clinic established to provide health services to low-income Native Americans and others in this inner-city Minneapolis community. We have enrolled over 4,800 patients and have had over 30,000 patient visits since opening. Eighty-five percent of our patients are Native American. NACC health providers, including nursing and medical practitioners, have 66 years of combined experience in the community. These practitioners have established relationships with many families, groups, and organizations, and are well connected to services and organizations serving the Native American community. All of the NACC's Governing Board are Native American community members, and many of them are directors of Indian organizations and leaders in the community.

NACC's mission is to promote wellness and regular health maintenance in Native American families, decrease health disparities in Native Americans in the metropolitan area, and provide access to care regardless of ability to pay. We are committed to serving Native Americans because of their significant health disparities and because they do not have any clinic other than NACC in the metropolitan area that they can say is "theirs." Although all are welcome, we are committed to preserving NACC as a place where Native Americans feel welcome and respected. Our vision is centered on involving the Indian community in the work of disease prevention, health promotion, and the elimination of health disparities. We believe in spending time out in the community, involved in projects, activities, and other efforts to make the issue of health disparities more understandable and practical for individuals.

Despite years of funding and services from governmental sources such as the Indian Health Service, Indian people have many of the worst health disparities of any racial group in the state and nation. American Indians experience rates of disease and premature death significantly greater than those of whites and other racial and ethnic groups. Significant health disparities exist across the life span. The rate of inadequate or no prenatal care is almost six times higher than the rate for whites. American Indian babies die at a rate more than three times higher than that for white babies. The teen pregnancy rate is five times higher than the white rate. Disease rates for heart disease, diabetes, cervical cancer, HIV/AIDS, and sexually transmitted diseases are two to five times higher than for whites. Less than 50% of American Indian elders are vaccinated against pneumococcal disease, one of the leading causes of death in American Indians. American Indian males have a suicide rate 16 times higher than for whites. Child maltreatment is three times higher, and Native American people suffer from significant mental health disorders such as depression, post-traumatic stress disorder, and chemical dependency (MDH, 2005).

Many NACC patients suffer from poverty and chronic medical conditions. Many have a history of physical and/or sexual abuse during childhood. Domestic abuse is common among women of child-bearing age. Forty to sixty percent of pregnant women use alcohol during the first trimester of their pregnancy, and alcohol dependency is a common struggle in many families. Forty-seven percent of our diabetic patients had a positive depression screen. NACC provides comprehensive primary health care services to all, with major emphasis on health disparities across the life span.

**Lydia Caros, DO**
Executive Director
Native American Community Clinic
Minneapolis, Minnesota

have launched population-based health care initiatives to improve the health status of their members and to reduce the health systems' costs. The nontraditional initiatives generally start with an assessment of specific population groups within the health system's membership that have complex health care problems. These might include older adults with multiple chronic diseases who are experiencing significant health disparities, such as Native Americans with diabetes or patients with complex health and social problems, for example, pregnant adolescents. Health services such as outreach and case management are commonly used. For example, a hospital might undertake

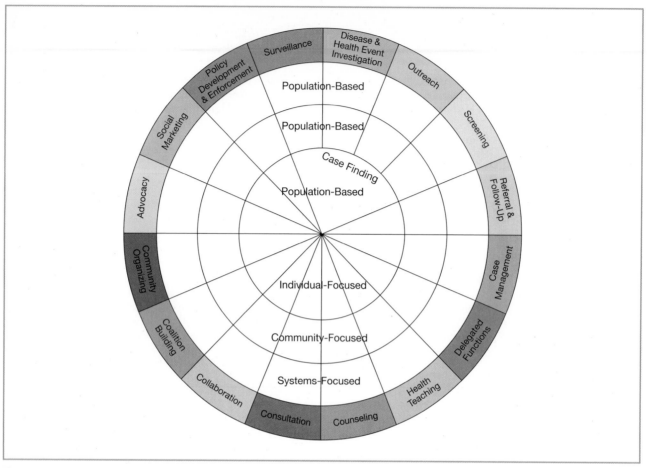

FIGURE 7-2 **Minnesota Department of Health Public Health Interventions II Wheel. Used with permission of the Section of Public Health Nursing, Minnesota Department of Health.** (Retrieved November 3, 2010, from http://www.health.state.mn.us/divs/cfh/ophp/resources/docs/wheelbook2006.pdf).

## CASE STUDY 7-1

You are a public health nurse working for a county public health department. An immigrant community in your county has a low rate of childhood immunizations compared to the rest of the county. The county health board has directed your agency to take actions to improve the childhood immunization rate in this immigrant community. In order to do this, you need to identify the causes of the low immunization rate. You also want to know how the community would be able to help itself and to participate in improving their childhood immunization rate. You decide to collect data about the health assets of the immigrant community and the health barriers related to childhood immunization rate. How might you go about doing this? Is there any other important information about this community that it would be helpful to know?

a survey of frequent nonacute visitors to its emergency department (ED) to determine if other more appropriate and less expensive health services could be developed to better meet the health needs of these frequent visitors.

Public Health Nursing (PHN) agencies and many Visiting Nurse associations provide the traditional model of population-based health care. These nursing agencies work with other health and social service organizations to provide comprehensive health services to population groups in their communities who are at risk for or experiencing health disparities. Public health nurses and community health nurses use a variety of independent nursing interventions in their population-based practice. Interventions most commonly used are displayed in Figure 7-2 in the Minnesota Public Health Intervention Wheel (Minnesota Department of Health, 2001). PHN agencies throughout the United States provide these interventions. A recent Public Health Workforce Study (HRSA, 2005) of four states documented the importance of these public health interventions to public health nursing agencies, as well as the percentage of agencies providing these interventions.

A nontraditional model of population-based nursing practice is also emerging. This model is being developed by private for-profit and nonprofit health care organizations. In this model, the vulnerable or at-risk population group is identified before community assessment occurs, and the subsequent community assessment process focuses on health determinants related to the at-risk group. The organizations generally focus on the population groups within their service areas, their market niche, and current membership. The primary goal of these organizations is usually to contain or reduce health care costs. The secondary goal is to improve the quality of care provided to their membership to improve health outcomes. Generally, the population groups who have the highest health risks, the poorest health outcomes, and the highest service costs are the groups targeted for additional services. Population groups with multiple diagnoses, called comorbidities, as well as complex therapeutic treatment regimes and a pattern of noncompliance and missed visits are also targeted for services.

## APPLICATION OF NURSING PROCESS TO POPULATION-BASED NURSING PRACTICE

Population-based nursing practice involves the application of the nursing process in working with communities, organizations, and population groups. Depending on the focus of a project, nurses may spend more of their time and effort on different components of the nursing process. Assessment, diagnosis, planning, implementation, and evaluation may take different forms depending on the nature of the project.

Assessment methods using the assets approach identified in Table 7-5 provide more opportunity for an egalitarian relationship between health providers and community members and are more consistent with the partnership model recommended by most community and public health professionals. The assessment phase may focus on the total health resources and needs of a community, or may be limited to one population group or one health concern. Use of community-based action research in which the community is an equal partner throughout the assessment, diagnosis planning, implementation, and an evaluation phase is becoming more common. This method enhances the ability of diverse communities to manage their own health care needs.

Population-based health care is interdisciplinary and involves participation of community members and population groups experiencing health disparities. Use of a

**TABLE 7-5  Nursing Process Applied To Population-Based Nursing—Assessment Examples**

**Assets Approach to Community Health Assessment:** Assessment process begins with an assessment of the assets—that is, strengths and resources—of a community or population group. An assets-based approach lends itself to identifying how the community can manage its own health needs by building on community and population group strengths and resources.

**Community Mapping:** Community mapping involves identifying the strengths and resources of the community, including its natural and built physical environment and its social environment: e.g., social systems, networks, formal and informal organizations, and cultural and lifestyle patterns.

**Focused Community Assessment:** An assessment of a specific at-risk population group with a specific health concern is carried out to identify population group assets, priorities, health needs, existing health services, and service gaps.

**Community-Based Action Research:** An assessment is completed in partnership with the community or specific population group. The assessment goals; data collection tools; and methods of data analysis, reporting, and documentation are determined and designed in partnership with the community.

nursing diagnosis may be a barrier to collaboration on an interdisciplinary team. One option would be a community-impact or goal statement that reflects the community's concerns about a specific health issue or population group and that reflects the culture and lifestyle of the community as understood by all members of the partnership. Another option is use of an interdisciplinary collaborative diagnosis (Table 7-6). Goal setting and implementation should be culturally and developmentally specific to the population group served (Table 7-7).

Community members should be involved in planning how, what, when, and where health care services will be

## TABLE 7-6  Nursing Process Applied to Population-Based Nursing—Diagnosis Examples

**Nursing Diagnosis:** Ineffective therapeutic management of mental health needs of community homeless related to lack of community awareness, lack of funding, and lack of community mental health workers as evidenced by a 20% increase in homeless people visiting the emergency departments of local hospitals for psychotropic medications, as well as an increase in police calls to attend to homeless people exhibiting psychotic behavior

**Community Impact Statement:** Community concern that homeless people have mental health needs that are not being addressed by community mental health professionals because of lack of funding and staff resulting in the community's fear of the homeless and increased calls to police to manage homeless people exhibiting frightening behaviors

**Collaborative Diagnosis:** Unmet mental health needs of community homeless due to lack of funding, staffing, and outreach as identified by a 20% increase in homeless people visiting the emergency department for psychotropic medications, an increase in police calls to attend to homeless people exhibiting psychotic behavior, and an increase in incarceration of homeless people with psychiatric diagnoses

## TABLE 7-7  Nursing Process Applied to Population-Based Nursing—Planning and Implementation Examples

**Planning:** Planning always involves a collaborative process. When working with the community, key community or population group representatives, members of the interdisciplinary health team, and representatives of community organizations and agencies all participate in the planning process. Nurses or other members of the health team may be functioning as consultants to the community, project managers, or coordinators for a specific community or population-based initiative, or be part of an interdisciplinary service team working with the community or local health care agency.

**Implementation:** Implementation is designed to fit the specific culture and lifestyle of the community or population group. Goals and outcomes demonstrate joint planning in partnership with the community. Interventions are interdisciplinary, build on community or population group assets, take into account existing resources and services, do not duplicate existing services and fill gaps in service needs, allocate resources based on need, and improve the health of the community or population group in a cost-effective manner. Interventions may be at the population group, system, or community levels of practice. Interventions are public health oriented, as described in the 17 public health interventions found in the Minnesota Public Health Intervention Wheel (MDH, 2001), (Figure 7-2). Interventions should be evidence based. This may be difficult to achieve if complementary therapies and culture-specific interventions that have not been studied are employed.

## REAL WORLD INTERVIEW

I work with pregnant and parenting teens at an alternative high school program that provides educational options for teens whose lives don't fit in with the traditional school day. Our program includes teens from a variety of cultures and backgrounds. The program currently has eight young women who will deliver their babies during the school year. This year, we also have four young fathers enrolled. The program has an on-site child care center, so these students bring their children to school with them and are able to visit their child during the school day.

Another public health nurse and I share this public health nursing assignment. We teach weekly prenatal classes in conjunction with the life skills class that all our students are required to take for graduation. Each student spends time working in the child care rooms, both in their own child's room as well as in the room of the next age group. This provides us with the opportunity to discuss growth and development, to discuss health care, and to role model **parent-child** interactions.

We also work with each student to look at family planning options for him or her and are very proud of a program we've started called The Pregnancy-Free Club. This is a voluntary "club" that allows each student to have private time with a public health nurse to talk about how to use birth control correctly and look at barriers that prevent the student from effectively using birth control.

Peer support has also become an unexpected part of this program, making it easier and more acceptable for the students to talk about their birth control choice or their choice of abstinence. Our program currently has a repeat pregnancy rate significantly lower than the national average.

**Barbara Reilly, PHN**
City of Bloomington Health Department
Bloomington, Minnesota

**Source:** From Reilly, B. (2001). "Health teaching" in Getting Behind the Wheel, *Public Health Nursing Online Newsletter, 13.* Section of Public Health Nursing, Minnesota Department of Health. Retrieved July 13, 2001, from http://www.health.state.mn.us/divs/chs/phn/phnnews13.html.

---

The Center for Disease Control and Prevention (CDC) has identified six public health emergency action-based preparedness disaster goals (Langan & James, 2005).

## CASE STUDY 7-2

Go to the State of Michigan Web site for emergency burn triage and management in a mass casualty disaster: http://www.michiganburn.org/index.shtml (accessed December 18, 2010). This Web site includes disaster plans; burn essentials such as fluid resuscitation protocols and nutrition algorithms; learning modules such as sedation, pain management, and debridement protocols; and training videos such as Silvadene burn applications and burn dressings. How might you use this Web site in a burn disaster?

- Detect—Identify the cause and distribution of potential threats to the public's health through epidemiologic, laboratory, and intelligence agency surveillance.
- Control—Provide medical countermeasures and health guidance to those affected by threats to the public's health.
- Maintain—Assure continuity of essential services during a public health emergency.
- Recover—Restore public health services and assure environmental safety following threats to the public's health.
- Plan—Complete and refine key public health response plans.
- Train and Exercise—Improve the ability of the public health workforce to respond to emergencies.

Two types of preparedness are critical to an effective disaster response: public health preparedness and medical preparedness. Nurses may be involved in both efforts.

- **Public health preparedness** is the ability of the public health system, community, and individuals to prevent, protect against, quickly respond to, and recover from health emergencies, particularly those in which scale, timing, or unpredictability threaten to overwhelm routine capabilities.

- **Medical preparedness** is the ability of the health care system to prevent, protect against, quickly respond to, and recover from health emergencies, particularly those whose scale, timing, or unpredictability threaten to overwhelm routine capabilities. Medical preparedness generally is the responsibility of agencies other than CDC (CDC, 2010).

Nine comprehensive disaster preparedness strategies have been identified by the International Federation of Red Cross and Red Crescent Societies (2000). These strategies include:

- Hazard, risk, and vulnerability assessments
- Response mechanisms and strategies
- Preparedness plans
- Coordination
- Information management
- Early-warning systems
- Resource mobilization
- Public education, training, and rehearsals
- Community-based disaster preparedness

Every local, state, or national community and health care facility can have a disaster plan. Schools and businesses also have disaster plans. The potential for a global avian flu pandemic was recognized around the year 2005. Colleges and universities across the United States updated their disaster preparedness plans. St. Catherine University baccalaureate nursing students participated in pandemic flu preparedness planning for all five stages of the disaster cycle as part of their public health nursing coursework (Table 7-10).

There is a current movement for families to have their own individual disaster plans. In order for nurses to be ready to respond to a disaster, they should first have a family disaster plan prepared. There are many Web sites that provide public information on how to develop a family disaster plan. Nurses should also know their designated disaster roles and responsibilities in their employment settings as well as their state law related to nurses as first responders. Nurses may be required by their employer or their government to take on a specific role in a disaster at their place of employment or in their community. It is also important for nurses to know their personal and professional rights in case of a disaster in their community. Actions nurses may want to take to prepare themselves for potential disasters include:

- Formation of a family disaster plan
- Continuing education in disaster preparedness
- Disaster preparedness training at work
- Obtaining information about rights and responsibilities in the employment setting and under state law
- Becoming a volunteer for local or national disaster relief organizations. Visit some of the following Web sites for information on how to develop a family disaster plan:
  - http://www.ready.gov/
  - http://emergency.cdc.gov/preparedness/
  - http://www.getreadyforflu.org/newsite.htm

You may also want to think about how you can prepare yourself to respond to a disaster as part of your professional nursing role. Visit the International Nursing Coalition for Mass Casualty Education, Nursing Curriculum for Emergency Preparedness at http://www.nursing.vanderbilt.edu/incmce/modules.html.

## COLLABORATION AND COMMUNITY ACTION

Nurses in the twenty-first century will be continuously challenged to look for new solutions to health care needs for populations at risk. The elderly, the poor, and marginalized population groups continue to grow nationally and globally. We must continue to find ways to partner with local communities in finding solutions and in providing services. Nurses must take the lead in working collaboratively within a variety of health care and social service disciplines and agencies within the community. There is growing evidence that interdisciplinary and community partnerships are effective in reducing health disparities in populations at risk.

### THE ACTION MODEL TO ACHIEVE HEALTHY PEOPLE 2020 OVERARCHING GOALS

The Secretary's Advisory Committee on National Health Promotion and Disease Prevention Objectives for 2020, U.S. Department of Health and Human Services, 2008, provides a graphic display of how nurses must be involved with community partners working to improve the health of diverse populations across the life span (Figure 7-6).

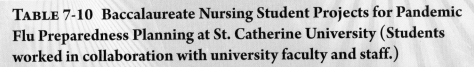

**TABLE 7-10  Baccalaureate Nursing Student Projects for Pandemic Flu Preparedness Planning at St. Catherine University (Students worked in collaboration with university faculty and staff.)**

**Non-Disaster and Pre-Disaster Stage Preparedness Projects**

- Online campus risk survey and key informant interviews

- Campus needs assessment

- Residence life needs assessment

- Access and success (organization for returning students with families) needs assessment

- Early Childhood Center needs assessment and program development for communicable disease prevention and control

- Campus security preparedness needs assessment

- Development of health education and social marketing posters, family emergency plan, and pandemic flu kit for Communication Department

**Impact and Emergency Stage Preparedness Planning Projects**

- Faculty and staff plan—developed with Human Resources Committee

- Student plan—developed with committees from Residence Life, Student Affairs, Counseling and Student Development, and Health and Wellness Center

- Systemwide plans—worked with Facilities Management, Public Safety, and Associate Dean for Health Programs

**Emergency and Reconstruction Stage Preparedness Projects**

- Managing pandemic flu outbreak and caring for ill and well students on campus

- Assessing level of needed resources and level of preparedness

- Assuring essential services

- Developing tertiary prevention plans with key departments

Nurses are being challenged to expand their professional boundaries by participating in the political process, working within their professional organizations, and working as volunteers with disenfranchised populations. Nurses will continue to participate in local, national, and international rescue and humanitarian efforts using population-based nursing practice interventions.

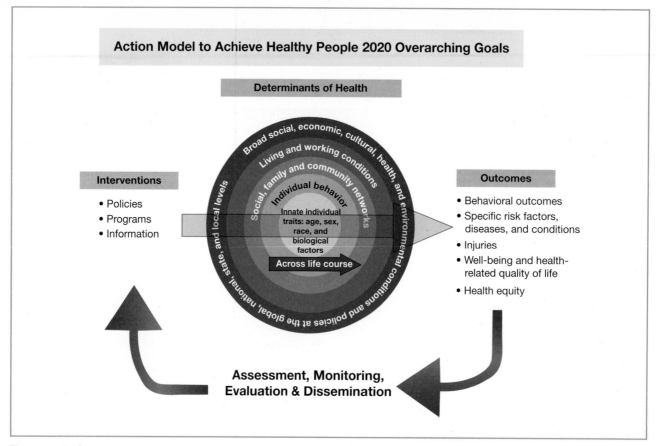

FIGURE 7-6  Action Model to Achieve *Healthy People 2020* Overarching Goals. Secretary's Advisory Committee on National Health Promotion and Disease Prevention Objectives for 2020. U.S. Department of Health and Human Services. (2008). Phase I Report—Recommendations for the Framework and Format of *Healthy People 2020.* (Accessed August 11, 2010, from http://www.healthypeople.gov/hp2020/advisory/PhaseI/PhaseI.pdf).

 **EVIDENCE** FROM THE LITERATURE

**Citation:** Fowler, B., Rodney, M., Roberts, S., & Broadus, L. (2005). Collaborative breast health intervention for African-American women of lower socioeconomic status. *Oncology Nursing Forum, 32*(6), 1207–1216.

**Discussion:** A collaborative community program using specially trained community health advisors was developed to increase the rate of mammography in African American women of low socioeconomic status. African American women leaders in the community were influential in encouraging community women to participate in the mammography screening project. Outreach, case finding, referral, monitoring, follow-up, and health education interventions were provided by the community health advisors. Of the women recruited, 90 (81%) met the study inclusion criteria, and 68 (76%) followed through and received mammography screening. Health education provided at the time of the mammogram resulted in the African American women demonstrating increased knowledge about breast health and increased sharing about their health concerns.

**Implications for Practice:** Collaboration with community members is an effective way to reach population groups at risk who may not be comfortable participating in health care programs provided by the established health care system. Nurses should consider collaboration with community members as well as other health care providers when trying to reach at-risk and marginalized population groups within the community.

# REAL WORLD INTERVIEW

On August 29, 2005, Hurricane Katrina slammed into the Gulf Coast of eastern Louisiana and Mississippi. Katrina, a Category 4 storm when it made landfall with wind speeds of 140 mph, was the largest storm ever to hit the United States. Hundreds of thousands of residents were evacuated before and after the storm, going to motels, homes of relatives, or shelters. On September 5, the American Refugee Committee (ARC) sent a seven-member advance team to Louisiana to determine how to deploy Minnesota Lifeline, a volunteer consortium of Minnesota health care professionals. The ARC advance team included the director of International operations, the international programs manager, a medical director, and a registered nurse. The team worked with the Office of Public Health (OPH) to develop a plan for providing evacuees with primary health care that would be sustainable after the volunteer consortium left.

Minnesota Lifeline was an interdisciplinary group of physicians, physician's assistants, advance practice nurses, RNs, psychologists, social workers, respiratory care therapists, phlebotomists, health information managers, and clinic assistants from St. Catherine University, the University of Minnesota Health Corps, and the Mayo Clinic. Minnesota Lifeline made a 60-day commitment to work in Region Four, which encompassed seven parishes near Lafayette—that is, Acadia, Evangeline, Iberia, Lafayette, St. Martin, St. Landry, and Vermillion parishes. The 60-day commitment was divided into four two-week waves of workers. The first wave included about 95 persons. The following groups were smaller. The final group focused on integrating the primary care services for the evacuees into the local health care delivery systems.

Interdisciplinary teams were dispatched to one of the approximately 68 sanctioned shelters in Region Four. The shelters primarily housed New Orleans area residents and were located in churches, campgrounds, American Legion halls, community centers, and indoor sports facilities. Truckloads of supplies arrived with the first wave of workers and included personal care kits, medications, nebulizers, blood glucose monitors, dressings, gloves, blood pressure cuffs, and everything else needed to run a primary care clinic. The Mayo Foundation donated most of the supplies. Clinic supplies and medications were packed into large totes and brought to whichever shelter was in need of care. A recreational vehicle provided by the Northwest Medical Team was first used as a mobile clinic and later as a mobile pharmacy.

Prior to participating in a clinic, permission had to be obtained from multiple levels of governmental and community organizations, including the Region Four OPH medical director, the public health nursing director, the parish emergency preparedness coordinator, the city emergency preparedness coordinator, and the shelter manager. Clinics were also set up in churches, motels, and parking lots to serve those who were living at those sites. Teams participated in community meetings to assess health needs, to gain community trust, and to help the team learn how to fit into the community culture.

The Minnesota Consortium provided primary health care for longstanding health needs as well as acute health care needs. The budget for providing public health in Louisiana had been severely cut many years ago, and little public health care was available. Most parishes had a clinic facility, which was used during weekday work hours to provide immunizations, reproductive health, and Women, Infant, and Children (WIC) clinics. Uninsured Louisiana residents received health care through the Charity Hospital system. Emergency care was provided at Charity Hospitals, but diagnostic workups and preventative care were provided on a sliding fee scale and required cash up front prior to receiving care. Only about 15% of children in Region Four were up to date on immunizations.

A three-page health form was developed by combining the CDC forms with OPH forms. Vaccines were administered to tens of thousands of evacuees and residents. Tetanus-diphtheria, hepatitis A, hepatitis B, and influenza vaccines were administered to persons who had been or would be in contact with the floodwaters or would be working on demolition of buildings. Most medications prescribed were for chronic illnesses such as diabetes and hypertension. Antibiotics, asthma and allergy medications, blood glucose monitors, and nebulizers were given out. Reproductive health care needs were common. Those who had contact with floodwaters often experienced skin rashes and a dry

*(Continues)*

## REAL WORLD INTERVIEW (Continued)

cough coined the "Katrina cough." Psychological trauma from separation of family members and flight from the hurricane were common. Those with mental health illnesses who were without their medications experienced symptoms of medication withdrawal and increased severity of symptoms.

The arrival of Hurricane Rita required evacuation of the first wave of relief workers into northern Mississippi to wait out the storm. Many New Orleans evacuees who had just gotten permanent housing and jobs and whose children were adjusting to new schools in their newly adopted towns had to again pack up belongings and evacuate. Some were later able to return. Many evacuees had to adjust to another new setting. The economically depressed area of the Gulf Coast has received a devastating blow. The development of a viable private-public health care infrastructure will be slow.

**Pam Hamre, RN, PhD, CNM, CNE, CNM**
Associate Professor, Department of Nursing, St. Catherine University
St. Paul, Minnesota

## KEY CONCEPTS

- The health status of population groups is measured by global measures such as morbidity, mortality, and quality of life. These are the vital signs of population groups.

- Health determinants give us a holistic picture of both the health assets and the health risks of population groups within the community. When nurses identify the health determinants of population groups, they can build on population strengths and target the most significant health risks to help improve the health status of population groups.

- There are significant health disparities within the U.S. population. The causes for these disparities are complex and involve societal factors such as governmental health policies, lack of insurance, poverty, lack of health literacy, barriers to health access, culture, and ethnicity. Societal interventions are necessary to reduce health disparities. Traditional nursing and medical care alone will not reduce health disparities significantly.

- The U.S. health care system provides unequal access, treatment, and quality to people of different ethnic and cultural groups. These disparities persist over time, geography, and different health care systems. Nurses need to take the lead in creating culturally inclusive health care systems that meet the needs of diverse population groups.

- The United States has been tracking health status and health disparities among population groups since the 1970s. National goals and objectives with an action plan and an evaluation plan have been developed and implemented every decade since 1980.

- Population-based health care starts with community assessment, identifies health disparities among population groups within the community, and establishes community health priorities. The unequal needs of diverse populations within the community require an unequal division of available resources. Nurses need to be able to use ethical principles in determining how to use resources in an ethical and equitable manner.

- Population-based nursing practice is the application of the nursing process to population groups and communities. It focuses on the priority health needs in a community or in specific population groups. Nurses practice population-based care at three levels: individual, system, and community, across the prevention continuum. This includes primary, secondary, and tertiary health care. A set of population-based public health nursing interventions has been identified as specific to this area of nursing practice. Almost all of the interventions are independent nursing functions within the scope of practice of the baccalaureate nursing graduate.

- Empowering communities and population groups to manage their own health care needs is an important aspect of population-based nursing practice. To empower these populations, nurses must develop egalitarian partnerships with the population groups they wish to help. Using an assets-based approach, in which community and population group strengths and resources are identified first before health needs, facilitates the development of partnerships.

- The health needs of vulnerable population groups are complex and require the combined resources of the interdisciplinary team, including paraprofessionals. Nurses need to be skilled in communications, group processes, care coordination, and collaboration to work effectively within the interdisciplinary team.

■ Nurses are involved in improving the health status of populations and communities as well as individuals.

These actions may be carried out as part of their professional roles or as part of their personal lives as concerned citizens.

■ Whether dealing with a randomly occurring natural disaster or an accidental or purposeful man-made disaster, nurses must be ready to participate in relief efforts.

## KEY TERMS

activities of daily living
disaster
functional
   health status
health assets
health care disparities

health care systems
   disparities
health determinants
health-related quality of life
health risk factors
health status

health status disparities
instrumental activities
   of daily living
medical preparedness
population-based health
   care practice

population-based nursing
   practice
protective factors
public health reparedness
quality of life
resilience

## REVIEW QUESTIONS

1. You have just attended a community hearing about the increasing rate of sexually transmitted infections (STIs) among adolescents in your community. The community is divided on what actions to take. You want to help the community reach some consensus on a plan of action. What intervention would be the most appropriate to do first?
   A. Develop a social marketing campaign to increase awareness of STIs.
   B. Talk to adolescents at a local high school about their concerns.
   C. Encourage the passage of a new law funding a free condom program.
   D. Form a coalition of community leaders and interest groups to take action.

2. The *best* indication that a community clinic is interested in providing culturally inclusive care is that the clinic does which of the following?
   A. Provides care for people from all cultures in the community
   B. Recruits board members that reflect community diversity
   C. Provides cultural diversity training for its staff annually
   D. Conducts a health survey of the community annually

3. Health status disparities among the poor, people of color, and American Indians persist in the United States. Which of the following changes in health determinants would help to reduce these disparities? Select all that apply.
   _____A. Decrease in the poverty level
   _____B. Decrease in the birth rate
   _____C. Increase in the average life span
   _____D. Increase in the availability of primary care
   _____E. Improvement of housing conditions in poor neighborhoods

4. The incidence of obesity among children and teenagers is increasing in the United States. In order to reduce the incidence of obesity, the risk factors that cause obesity need to be reduced. Elimination of candy, snack food, and high-sugar beverages from school vending machines is one action that many school districts have taken. What health determinants have been modified by changing the contents of the vending machine? Select all that apply.
   _____A. Body mass index of the students
   _____B. Physical environment of the school building
   _____C. Policies and interventions of the school district
   _____D. Peer group pressure within the social environment
   _____E. Individual choices available to students

5. You want to increase the resilience of the people living in an inner city neighborhood where there are high rates of poverty and unemployment. Which of the following actions would help you achieve this goal?
   A. Work with local businesses to increase employment opportunities.
   B. Convince the neighborhood health clinic to offer evening and weekend hours.
   C. Use community-action research to identify the health priorities of the people.
   D. Collaborate with local churches to expand food shelves and homeless shelters.

6. You have been given the charge to determine the health status of a group of children living in an inner city housing development. What health status data should you collect? Select all that apply.
   _____A. Mortality data
   _____B. Morbidity data

_____C.  Health behaviors data

_____D.  Health care access information

_____E.  Family support information

7. You are a county public health nurse. Many families in your community do not have health insurance, do not have a primary care provider, and their children are often not immunized against common childhood illnesses. You decide to hold a free immunization clinic at the local grade school during normal school hours. Which health system disparity are you trying to overcome?

   A.  Childhood immunization rate
   B.  Quality of the immunization process
   C.  Access to child health services
   D.  Parental resistance to immunizations

8. You are developing health teaching for fall prevention in the elderly. You plan to present this program to older adults at churches, community centers, and neighborhood clinics. This health teaching intervention is an example of nursing practice at which of the following levels?

   A.  Systems level
   B.  Individual level
   C.  Community level
   D.  Group level

9. Your community is under a hurricane alert. The hurricane is expected to hit your community in approximately

10 hours. You are participating in the Predisaster Stage of the community Disaster Preparedness Plan. Activities you might carry out include which of the following?

   A.  Setting up a shelter at a local school and evacuating a mobile home park
   B.  Planning how the public and private sectors of the community will work together
   C.  Reading to preschoolers at a local day care facility to keep them calm and safe
   D.  Creating family emergency plans and determining the availability of bottled water

10. You want to evaluate the effectiveness of a blood pressure (BP) screening clinic for the elderly. Which of the following measures would be best in determining the impact of the BP screening clinic on the health status of the people screened?

   A.  Number of people screened who were found to have elevated blood pressure and were not aware of it
   B.  Cost of the BP screening for each person screened
   C.  Number of people with previously diagnosed hypertension whose BP was within normal limits and who were taking their antihypertensive medications
   D.  Number of people with elevated BP who followed through with referral to their primary care provider and were placed on antihypertensive medications

## REVIEW ACTIVITIES

1. What are the health determinants for students on your college campus? Consider using an assets-based approach. Use the health determinants model to identify the health promoting social and physical environmental factors that influence the holistic health of students: body, mind, and spirit. Then identify the social and physical environmental barriers to holistic health. Consider college policies and access to health and social services on campus.

2. What are the major health risks for the people in your community? Take a walk in your community and see what physical and environmental health risks you can identify for one or more vulnerable groups in the community.

3. Think about a service project that your nursing class could do to improve the health of one vulnerable group within your community. What might you want to do before you launch any initiative?

## EXPLORING THE WEB

Where could you find information to help you search the health needs of immigrants and refugees?

- Center for Cross Cultural Health: www.crosshealth.com
- National Institute of Health: www.nih.gov
- Office of Minority Health: www.cdc.gov Search for Office of Minority Health.
- Office of U.S. Surgeon General: www.surgeongeneral.gov
- United Nations: www.un.org
- U.S. Department of Health and Human Services: www.dhhs.gov

- U.S. Committee for Refugees and Immigrants: www.refugees.org
- World Health Organization: http://who.org

What sites could you search for health information on your community and the nation?

- Centers for Disease Control and Prevention: www.cdc.gov
- DATA 2010: http://wonder.cdc.gov/data2010/
- FEDSTATS: www.fedstats.gov
- Governmental health initiatives: www.health.gov

- *Healthy People 2010*: www.healthypeople.gov
- *Healthy People 2020*: www.healthypeople.gov/hp2020/
- Leading Health Indicators (LHI): www.healthypeople.gov/LHI/lhiwhat.htm

- National Institute of Environmental Health Services: www.niehs.nih.gov
- Office of Disease Prevention and Health Promotion: http://odphp.osophs.dhhs.gov

## REFERENCES

Agency for Health Care Research and Quality (AHRQ). (2005). *National Healthcare Disparities Report 2005*. USDHHS. Retrieved June 18, 2006, from http://www.ahrq.gov/qual/nhdr05/nhdr05.pdf.

America.gov. (2008). U.S. Minorities Will Be the Majority by 2042, Census Bureau Says. Accessed March 21, 2011, at http://www.america.gov/st/peopleplace-english/2008/August/20080815140005xlrennef0.1078106.html.

American Association of Colleges of Nursing. (2010). Enhancing diversity in the nursing workforce. Accessed November 8, 2010, from http://www.aacn.nche.edu/media/factsheets/diversity.htm.

American College of Physicians. (2004). Racial and ethnic disparities in health care. *Annals of Internal Medicine, 141*(3), 221–225.

American Nurses Association. (2010). *Public health nursing: scope and standards of practice*. Silver Springs, MY: NursesBooks.org.

American Public Health Association, French. M. (2009). Issue brief: Our nation's health—Prevention and wellness as national policy. Washington, D.C.: Author. Accessed July 15, 2010, from http://www.apha.org.

American Red Cross. (1975). Disaster Relief Program, 2235. Washington, DC: Author.

Association of City and County Health Offices. (2010). Unnatural Causes Web site at http://www.unnaturalcauses.org/.

Baldwin, D. (2003). Disparities in health and health care: Focusing efforts to eliminate unequal burdens. *Online Journal of Issues in Nursing, 8*(1), 113–122. Retrieved June 1, 2006, from EBSCOhost at http://sas.epnet.com.pearl.

California Newsreel. (2008). Unnatural causes [videos]. San Francisco: Author. Accessed October 28, 2010, from http://www.unnatural-causes.org/.

CDC, Emergency Preparedness and Response, 10. Accessed September 25, 2010, from http://emergency.cdc.gov/cdc/.

Davis, R., Cook, D., & Cohen, L. (2005). A community resilience approach to reducing ethnic and racial disparities in health. *American Journal of Public Health, 95*(12), 2168–2173.

Dreher, M., Shapiro, D., & Asselin, M. (2006). *Healthy places, healthy people*. Indianapolis, IN: Sigma Theta Tau International.

Eberhardt, M., & Parnuk, E. (2004). The importance of place of residence: Examining health in rural and nonrural areas. *American Journal of Public Health, 94*(10), 1682–1686.

Falk-Rafael, A. (2005). Speaking truth to power—Nursing's legacy and moral imperative. *Advances in Nursing Science, 28*(3), 212–223.

Fowler, B., Rodney, M., Roberts, S., & Broadus, L. (2005). Collaborative breast health intervention for African-American women of lower socioeconomic status. *Oncology Nursing Forum, 32*(6), 1207–1216.

Freeman, R. B. (1957). *Public health nursing practice* (2nd ed.). Philadelphia: Saunders.

Garwick, A., Rhodes, K., Peterson-Hickey, M., & Hellerstedt, L. (2008). Native teen voices: Adolescent pregnancy prevention recommendations. *Journal of Adolescent Health, 42*(1), 81–88.

Giddings, L. (2005). A theoretical model of social consciousness. *Advances in Nursing Science, 28*(3), 224–239.

Gustafson, D. (2005). Transcultural nursing theory from a critical cultural perspective. *Advances in Nursing Science, 28*(1), 2–17.

Hartley, D. (2004). Rural health disparities, population health, and rural culture. *American Journal of Public Health, 94*(10), 1675–1678.

Health Resources and Services Administration (HRSA). (2010). *National sample survey of registered nurses—Preliminary findings, 2008*. Accessed November 8, 2010, from http://bhpr.hrsa.gov/healthworkforce/rnsurvey/initialfindings2008.pdf.

Her, M. (2001). Case study: Target population, Hmong elderly at risk for depression-intervention, outreach in Behind the Wheel. *Public Health Nursing Newsletter, 13*. Retrieved July 13, 2001, from http://health.state.mn.us/divs/chs/phn/phnnews15.html (no longer available).

Institute of Medicine (IOM). (2002). *Unequal treatment confronting racial and ethnic disparities in healthcare*. Institute of Medicine Report. Washington, DC: National Academy Press.

Institute of Medicine. (2004). *A shared destiny: Community effects of uninsurance*. Washington, D.C.: National Academies Press.

Institute of Medicine. (2010). Future directions for the National Healthcare Quality and Disparities Reports. Accessed July 25, 2010, from www.iom.edu/ahrqhealthcarereports.

International Federation of Red Cross and Red Crescent Societies. (2000). Disaster Preparedness Training Programme: Introduction to disaster preparedness.

Jackson, S., Knight, K., & Rafferty, J. (2010). Race and unhealthy behaviors: Chronic stress, the HPA axis, and physical and mental health disparities over the life course. *American Journal of Public Health, 100*(5), 933–939.

Keller, L., Strohschein, S., Lia-Hoagberg, B., & Schaffer, M. (2004a). Population-based public health interventions: Practice-based and evidence-supported. Part I. *Public Health Nursing, 21*(5), 453–468.

Keller, L., Strohschein, S., Schaffer, M., & Lia-Hoagberg, B. (2004b). Population-based public health interventions in practice, teaching, and management. Part II. *Public Health Nursing, 21*(5), 469–487.

Langan, J., & James, D. (2005). Preparing Nurses for Disaster Management. Upper Saddle River, NJ: Pearson Prentice Hall.

Lillie-Blanton, M., Maleque, S., & Miller, W. (2008). Reducing racial, ethnic, and socioeconomic disparities in health care: Opportunities in national health reform. *Journal of Law, Medicine & Ethics, 26*(4), 693–702.

Minkler, M., Fuller-Thompson, E., Gurdnik, J. (2006). Gradient of disability over the socioeconomic spectrum in the United States. *New England Journal of Medicine, (355)*7, 695–703.

Minnesota Department of Health (MDH). (2005). *Background infor-mation. Priority health areas for elimination of health disparities in Minnesota.* St. Paul, MN: Author. Retrieved May 10, 2006, from http://www.health.state.mn.us/divs/chs/pdf/gdlinebkgrd9.pdf.

Minnesota Department of Health, Section of Public Health Nursing. (2001). *Public health interventions: Application for public health nurs-ing practice.* St. Paul, MN: Minnesota Department of Health. Retrieved April 12, 2006, from http://www.health.state.mn.us/divs/cfh/ophp/resources/docs/ph-interventions_manual2001.pdf.

Newman, M. A. (1999). Health as expanding consciousness. New York: National League for Nursing Press.

Noji, E. K. (1997). The nature of disaster: General characteristics and public health effects. In E.K. Noji (Ed.), *The public health consequences of disasters* (pp. 3–20). New York: Oxford University Press.

Pawlak, R. (2005). Economic considerations of health literacy. *Nursing Economics, 23*(4), 173–180.

Peters, R. M. (1995). Teaching population-focused practice to baccalau-reate nursing students: A clinical model. *Journal of Nursing Education, 34*(8), 378–383.

Pharris, M. D., Sankofa, P., Amaikwu-Rushing, L., Fitzgerald, D., & Ollom, K. (2002). *Racism, health, and well being: The experience of women and girls of color in North Minneapolis, Centers of Excellence in Women's Health.* St. Paul, MN: College of Catherine.

Reilly, B. (2001). "Health teaching" in Getting Behind the Wheel, *Public Health Nursing Newsletter, 13.* Retrieved July 13, 2001, from http://health.state.mn.us/divs/chs/phn/phnnews13.html (no longer available).

Secretary's Advisory Committee on National Health Promotion and Disease Prevention Objectives for 2020. U.S. Department of Health and Human Services. (2008). Phase I report—Recommendations for the framework and format of Healthy People 2020. Accessed August 11, 2010, from http://www.healthypeople.gov/hp2020/advisory/PhaseI/PhaseI.pdf.

Tashiro, C. (2005). Health disparities in the context of mixed race—Challenging the ideology of race. *Advances in Nursing Science, 28*(3), 203–211.

Taylor, R. (2005). Addressing barriers to cultural competence. *Journal for Nurses in Staff Development, 21*(4), 135–142.

U.S. Census Bureau. (2005). Race and Hispanic origin in 2005. Population profile of the United States: Dynamic Version 1. Accessed August 25, 2010, from http://www.census.gov/population/pop-profile/dynamic/RACEHO.pdf.

U.S. Department of Health and Human Services (HHS). (2011) Healthy People 2020 Framework: Topics & Objectives Index - Healthy People. Accessed March 22, 2011, at http://healthypeople.gov/2020/topicsobjectives2020/default.aspx.

U.S. Department of Health and Human Services (HHS). Office of Disease Prevention and Health Promotion. (2011) Framework, The Vision, Mission, and Goals of Healthy People 2020. Accessed March 21, 2011, at http://healthypeople.gov/2020/consortium/HP2020Framework.pdf.

U.S. Department of Health and Human Services, Agency for Healthcare Research and Quality. (2007). *National health disparities report.*

U.S. Department of Health and Human Services. (2010). Healthy People 2020 framework. Accessed on November 1, 2010, from http://www.healthypeople.gov/hp2020/Objectives/framework.aspx.

U.S. Department of Health and Human Services. Healthy People 2010. 2nd ed. With Understanding and Improving Health and Objectives for Improving Health. 2 vols. Washington, DC: U.S. Government Printing Office, November 2000.

Wald, L. D. (1915). *The house on Henry Street.* New York: Dover Publications.

Weil, E., Wachterman, M., McCarthy, E., Davis, R., O'Day, B., Lezzoni, L., Wee, C. (2002). Obesity among adults with disabling conditions. *Journal of the American Medical Association, (288)*10, 1265–1268.

WHO Task Force on Research Priorities for Equity in Health & WHO Equity Team. (2005). Priorities for research to take forward the health equity policy agenda. *Bulletin of the World Health Organization, 83*(12), 948–953.

WHO. (2010). The determinants of health. Accessed August 27, 2010, from http://www.who.int/hia/evidence/doh/en/index8.html.

Zahner, S., & Block, D. (2006). The road to population health: Using Healthy People 2010 in nursing education. *Journal of Nursing Education, 45*(3), 105–108.

## SUGGESTED READINGS

Alexander, G., Wingate, M., & Boulet, S. (2008). Pregnancy outcomes of American Indians: Contrasts among regions and with other ethnic groups. *Maternal Child Health, (12),* S5–S11.

Anderson, J., Felton, G., Wewers, M., Waller, J., & Humbles, P. (2005). Sister to sister: A pilot study to assist African-American women in subsidized housing to quit smoking. *Southern Online Journal of Nursing Research, 1*(6), 1–23. Retrieved June 1, 2006, from http://www.snrs.org.

Baker, M. (2005). Creation of a model of independence for community-dwelling elders in the United States. *Nursing Research, 54*(5), 288–295.

Baltrus, P., Lynch, J., Everson-Rose, S., Raghunathan, T., & Kaplan, G. (2005). Race/ethnicity, life-course socioeconomic position, and body weight trajectories over 34 years: The Alameda County study. *American Journal of Public Health, 95*(9), 1595–1601.

Bent, K. (2003). Culturally interpreting environment as determinant and experience of health. *Journal of Transcultural Nursing, 14*(4), 305–312.

Brackbill, R., Thorpe, L., DiGrande, L., Perrin, M., Sapp, J., Wu, D., et al. (2006). Surveillance for world trade center disaster health effects among survivors of collapsed and damaged buildings. *Morbidity and Mortality Weekly Report, (55),* SS-2, 1-18. Retrieved May 1, 2006, from http://www.cdc.gov.

Collins, J., Wambach, J., David, R., & Rankin, K. (2009). Women's lifelong exposure to neighborhood poverty and low birth weight: A population-based study. *Maternal Child Health, (13),* 326–333.

Gruskin, S., Cottingham, J., Hilber, A., Kismodi, E., Lincetto, O., & Roseman, M.J. (2008). Using human rights to improve maternal and neonatal health: History, connections and a proposed practical approach. *Bulletin of the World Health Organization, (86),* 589–593.

Hinton, A., Downey, J., Lisovicz, N., Mayfield-Johnson, S., & White-Johnson, F. (2005). The community health advisor program and the Deep South Network for Cancer Control: Health promotion programs for volunteer community health advisors. *Family & Community Health, 28*(1), 20–27.

Kara, M. (2005). Preparing nurses for the global pandemic of chronic obstructive pulmonary disease. *Journal of Nursing Scholarship, 37*(2), 127–133.

Katz, D., O'Connell, M., Yeh, M., Nawaz, H., Njike, V., Anderson, L., et al. (2005). Public health strategies for preventing and controlling overweight and obesity in school and worksite settings. *Morbidity and Mortality Weekly Report*, (54), RR-10, 1–11. Retrieved May 1, 2006, from http://www.cde.gov.

Kelly, D. (2005). From the emergency department to home. *Journal of Clinical Nursing, 14*, 776–785.

Kim, J., Must, A., Fitzmaurice, G., Gillman, M., Chomitz, V., Kramer, E., et al. (2005). Incidence and remission rates of overweight among children aged 5 to 13 years in a district-wide school surveillance system. *American Journal of Public Health, 95*(9), 1588–1594.

Kim, S., Flaskerud, J., Koniak-Griffin, D., & Dixon, E. (2005). Using community-partnered participatory research to address health disparities in a Latino community. *Journal of Professional Nursing, 21*(4), 199–209.

Lavery, S., Smith, M., Esparza, A., Hrushow, A., Moore, M., & Reed, D. (2005). The community action model: A community-driven model designed to address disparities in health. *American Journal of Public Health, 95*(4), 611–616.

Lewis, L., Sloane, D., Nascimento, L., Diamart, A., Guinyard, J., Yancey, A., et al. (2005). African Americans' access to healthy food options in South Los Angeles restaurants. *American Journal of Public Health, 95*(4), 668–673.

Ludwig, T., Buchholz, C., & Clarke, S. (2005). Using social marketing to increase the use of helmets among bicyclists. *Journal of American College Health, 4*(1), 51–58.

Morello-Frosch, R., & Shenassa, D. (2006). The environmental "risk-scape" and social inequality: Implications for explaining maternal and child health disparities. *Environmental Health Perspectives, 114*(8), 1150–1153.

Pate, R., Ward, D., Saunders, R., Felton, G., Dishman, R., & Dowda, M. (2005). Promotion of physical activity among high-school girls: A randomized controlled trial. *American Journal of Public Health, 95*(9), 1582–1587.

Pavlish, C. (2005). Action responses of Conglolese refugee women. *Journal of Nursing Scholarship, 37*(1), 10–17.

Schraeder, C., Dworak, D., Stoll, J., Kucera, C., Waldschmidt, V., & Dworak, M. (2005). Managing elders with comorbidities. *Journal of Ambulatory Care Management, 28*(3), 201–209.

Schultz, A., Senk, S., Odoms-Young, A., Hollis-Neely, T., Nwankwo, R., Lockett, M., et al. (2005). Healthy eating and exercising to reduce diabetes: Exploring the potential of social determinants of health frameworks within the context of community-based participatory diabetes prevention. *American Journal of Public Health, 95*(4), 645–651.

Tembreull, C., & Schaffer, M. (2005). The intervention of outreach: Best practices. *Public Health Nursing, 22*(4), 347–353.

Welch, D., & Kneipp, S. (2005). Low-income housing policy and socioeconomic inequalities in women's health: The importance or nursing inquiry and intervention. *Policy, Politics, & Nursing Practice, 6*(4), 335–342.

Wu, P., Duarte, C., Mandell, D., Fan, B., Liu, X., Fuller, et al. (2006, May). Exposure to the World Trade Center attack and the use of cigarettes and alcohol among New York City public high-school students. *American Journal of Public Health, 96*(5), 804–807. Epub 2006, March 29.

# UNIT 2
## Leadership and Management of the Interdisciplinary Team

# CHAPTER 8

# Personal and Interdisciplinary Communication

JACKLYN RUTHMAN, PhD, RN; DEBORAH ERICKSON, PhD, RN

*Before we can change things, we must call them by their real names.*

(CONFUCIUS)

## OBJECTIVES

Upon completion of this chapter, the reader should be able to:

1. Analyze how current trends in society affect communication.
2. Focus on the elements of the communication process.
3. Describe the effects of the Health Insurance Portability and Accountability Act on communication.
4. Use various modes of communication.
5. Identify common communication networks.
6. Describe organizational communication and communication skills in the workplace.
7. Identify barriers to communication and strategies to overcome them.
8. Identify linguistically appropriate communication strategies.
9. Describe the SBARR communication tool.

Delmar/Cengage Learning

*As a newly licensed RN working on a medical unit in a community hospital, Steve begins the shift by making rounds on patients to perform initial assessments. He enters the room of Mr. Mason, who has a long history of chronic obstructive pulmonary disease. He was admitted yesterday with pneumonia. Mr. Mason is well known to the experienced staff. As Steve assesses the patient's breath sounds, Mr. Mason says, "I don't think I'm going to make it this time." His wife, who is at his bedside, replies, "Don't talk like that."*

*What nonverbal cues might be used to help Steve interpret the patient's message?*

*What communication skills will Steve use to respond appropriately?*

Today's nurses use basic principles of communication to facilitate interactions with patients, family members, peers, and colleagues from other disciplines. These principles allow nurses to adapt to trends that affect the profession of nursing and its practice. Nurses rely on communication skills to effectively promote patient care and professionalism in a variety of settings for an increasingly diverse society. These skills enable nurses to engage the complex, interactive process of communication, which uses both verbal and nonverbal modes. Nurses are aware of the context in which communication occurs. Nurses must be aware of potential barriers to communication to be able to overcome them. Awareness of principles and skills of communication empowers nurses to manage a variety of communication demands in the workplace.

## TRENDS IN SOCIETY THAT AFFECT COMMUNICATION

Good communication will grow in importance because of trends in our culture. Among the trends affecting nursing practice is the increasing diversity in society. The United States is made up of people with different ethnic, racial, cultural, and socioeconomic backgrounds. Increased diversity causes once-dominant values and beliefs to be replaced or diluted with diverse values and beliefs. These differences become a source of possible misunderstanding that can be bridged by effective communication.

Another trend is our aging population. It is estimated that 20% of the population will be 65 years of age or older by 2030. Our aging society will challenge nurses to maintain effective communication to compensate for the diminished sensory abilities that typically accompany aging. Multiple sensory deficits can occur simultaneously so that patients may experience losses in a variety of combinations that include hearing, seeing, smelling, tasting, and touching. The potential diminished input challenges nurse and patient alike to creatively compensate for these deficits.

At the same time as the population is aging, it is also shifting to an electronic mode, with computer technology playing an increasingly dominant role. As electronic communication assumes this role, the nurse's ability to effectively communicate in writing will grow in importance. Reliance on written communication using electronic input shifts the source of input away from traditional visual, auditory, and kinesthetic modes to the written word. To use electronic tools effectively, tomorrow's nurses will require keen writing skills. These trends have influenced nursing today.

## ELEMENTS OF THE COMMUNICATION PROCESS

**Communication** is an interactive process that occurs when a person (the sender) sends a verbal or nonverbal message to another person (the receiver) and receives feedback. The communication process is influenced by emotions, needs, perceptions, values, education, culture, goals, literacy, cognitive ability, and the communication mode (Figure 8-1). The input comes from visual, auditory, and kinesthetic stimuli.

Communication in health care is used to coordinate patient care. Several studies of ICUs indicate that effective communication and coordination among clinical staff results in more efficient and better quality care (Baggs, Ryan, Phelps, Richeson, & Johnson, 1992; Knaus, Draper, Wagner, & Zimmerman, 1986; Shortell et al., 1994). More recent studies of other health care delivery settings also indicate that effective coordination of staff leads to better clinical outcomes (Gittell et al., 2000; Young et al., 1997; Young et al., 1998).

On the other hand, ineffective coordination and communication among hospital staff contributes substantially to adverse events. Physicians, nurses, pharmacists, diagnostic laboratory staff, office staff, patients, and family members can correct errors before they affect patients, through vigilance, perseverance, and by using their power to change course, protocols, or systems (Parnes et al., 2007). Because no system can catch all errors, all hospital staff must take seriously their role in stopping the error cascade.

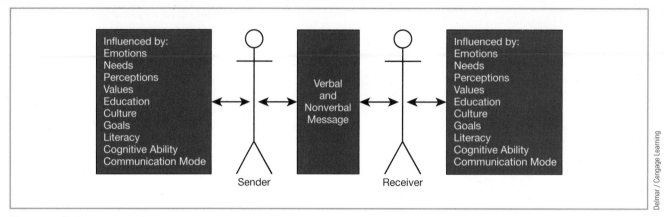

FIGURE 8-1 **Communication process.**

For many types of health care organizations, staff communication and coordination is also relevant to their ability to comply with the requirements of accrediting bodies. In particular, both the Joint Commission (JC) (www.jointcommission.org) and the National Committee on Quality Assurance (http://www.ncqa.org), two leading accrediting bodies in the health care industry, have adopted standards that address coordination among professional groups, patient care units, and service components within health care organizations. See the JC Web site for information about their National Patient Safety Goals. Included in these goals are recommendations for improving communication among staff. In addition, organizations such as the American Association of Critical Care Nurses and the American College of Chest Physicians have established complementary initiatives to impact patient-focused care and healthy work environments (McCauley & Irwin, 2006). While setting standards obviously cannot ensure good communication and coordination, it does symbolize the growing recognition among accrediting and other oversight bodies that communication and coordination are highly important to the performance of health care organizations. Another important area of communication involves following patient privacy guidelines.

## HEALTH INSURANCE PORTABILITY AND ACCOUNTABILITY ACT (HIPAA) OF 1996

As nurses communicate today, they must increasingly be aware of patient privacy. The Department of Health and Human Services (HHS) has issued regulations known as the HIPAA Privacy Rule that were required as of April 14, 2003. The **HIPAA Privacy Rule** protects all individually identifiable health information held or transmitted by a covered entity or its business associate, in any form or media, whether electronic, paper, or oral. It applies to all health care plans,

health care clearinghouses, and to any health care provider who transmits health information. The law introduced new standards for protecting the privacy of individuals' identifiable health information. There are 18 personal health identifiers. This law may also apply to health information that is shared for research purposes (www.hhs.gov); see Table 8-1. De-identifying data is one way to share data legally and ethically. De-identified data are data that cannot be used alone or in combination with information from other sources to identify an individual.

While the primary goal of the Privacy Rule is to protect individuals' identifiable health information, a secondary goal of the Privacy Rule is to permit the flow of information needed to provide quality health care (Anderson, 2007). Health care providers do not need patients' permission to share with other health care providers health information that is necessary to provide treatment. Health information may also be shared, when appropriate for treatment, with family members and friends who participate in the patient's care, unless the patient specifically objects. Tips for protecting patient privacy are seen in Table 8-2.

## MODES OF COMMUNICATION

Two traditional modes of communication, verbal and nonverbal, are exemplified in the nurse-patient scenario that follows in the Real World Interview. Because face-to-face encounters usually allow for verbal and nonverbal exchange, they have been regarded as the most effective modes of communication and hence have been preferred. Verbal communication is a conscious process, so the sender has the ability to control what is said. While it is generally accepted that tone of voice is more important than the words spoken, it has long been suggested that nonverbal facial expression is even more important than either tone of voice or the words used. Nonverbal communication tends to be unconscious and more difficult to control.

## TABLE 8-1  Elements That Are Considered Patient Identifiers under HIPAA

- Names
- All geographic subdivisions smaller than a state, including street address, city, county, precinct, zip code
- All elements of dates (except year) for dates directly related to an individual, including birth date, admission date, discharge date, date of death; all ages over 89
- Telephone numbers
- Fax numbers
- E-mail addresses
- Social Security numbers
- Medical record numbers

- Health plan beneficiary numbers
- Account numbers
- Certificate/license numbers
- Vehicle identifiers and serial numbers
- Device identifiers and serial numbers
- Web universal resource locators (URLs)
- Internet Protocol (IP) address numbers
- Biometric identifiers, including fingerprints and voiceprints
- Full-face photographic images and any comparable images
- Any other unique identifying number, characteristic, or code unless otherwise permitted

**Source:** Compiled with information from U.S. Department of Health and Human Services. (2003). *Protecting personal health information in research: Understanding the HIPAA Privacy Rule.* NIH Publication Number 03-5388.

## TABLE 8-2  Tips for Protecting Patient Privacy

1. When using an electronic documentation system
   - Close results on computer screens when not working on the screen.
   - Have screens point away from visitor traffic areas.
   - Change passwords often.
   - Do not share passwords with others.
2. Refrain from removing copies of personal health information (PHI) from the hospital or other work settings.
3. Dispose of PHI properly—shred or dispose of them in locked bins.
4. Discuss patients and their care in appropriate settings (not in the cafeteria, elevators, or other public areas).
5. Place hard copies of patient charts and health care provider's handwritten notes facedown.
6. When discussing PHI with other health care providers via the telephone, be cognizant of who is in the vicinity who may overhear you.
7. When discussing PHI with the patient
   - Ask visitors to leave.
   - Offer to move patient to a private area if in a semiprivate room.
8. Before looking at a patient's PHI, ask yourself, "Do I need to know this information to do my job?"
9. Refrain from taking pictures of patients without their written consent.

When face-to-face encounters are not possible or practical, other approaches are used. Historically, the next most effective approach is the telephone, followed by voice messages, electronic pages, e-mail messages, and written documents. These electronic methods comprise the third mode of communication, and they will grow in importance as nurses increasingly rely on technology, particularly computers, to communicate interpersonally. One example of a wireless e-mail device is the BlackBerry.

# ELECTRONIC COMMUNICATION

Electronic communication is playing an increasingly dominant role. Electronic health care communication is slowly inching its way into the twenty-first century. Patients are being monitored long distance and connecting to their health care providers using a variety of technologies, including telephones, voice mail, and e-mail. Devices such as the Apple iPhone, Blackberry or other Smart Phones allow for rapid connections. These methods, where caregiver and care receiver interact using technology rather than the traditional face-to-face or voice-to-voice encounter, require careful communication to maintain patient safety. For example, e-mail now allows almost instantaneous communication around the world, but it also accommodates individual preferences with respect to the timing of the response. This allows a patient to provide an update on a condition early in the day and affords the caregivers the opportunity to respond as their schedules permit. Using e-mail may save a patient and caregiver from travel or loss of work. However, using e-mail requires that nurses acting in such a caregiver role have keen writing skills. An explanation of all the considerations that are important to effective writing is beyond the scope of this text. However, a few tips are worth sharing. The speed with which exchanges can now be made using technology has reduced the acceptable response time. Therefore, the first tip when communicating using technology is that it is important that both parties have an understanding about the circumstances under which different modes of communication will be used. Although one practitioner who is "connected" may be comfortable receiving an electronic email message with urgent patient information such as an elevated potassium level, other practitioners expect a telephone call if the data being shared is potentially life threatening. Practitioners may be satisfied to receive a fax or access data electronically if the data are not urgent. Often, organizations have policies that guide under what circumstances a particular mode of communication is used, so be sure to understand your institution's policy for communicating urgent information.

Another tip is to respond in a timely manner. Timeliness is defined by what information is being shared and the route being used. A fax delivered to a practitioner's office over the weekend will likely not generate a reply before Monday. E-mail, in general, provides greater immediacy, but the telephone remains the primary tool for communicating urgent information. Other tips for communicating on e-mail include the following:

- NO CAPITAL LETTERS—this looks like you are shouting.
- Be brief and reply sparingly, as appropriate.
- Use clear subject lines.
- Cool off before responding to an angry message. Answer tomorrow.
- Spell out your message clearly; avoid abbreviations, as patients and others may not understand them.
- Forward e-mail messages from others only with their permission.
- Forward jokes selectively, if ever.
- Use good judgment; e-mail may not be private.

Keep in mind that accurate spelling, correct grammar, and organization of thought assume greater importance in the absence of verbal and nonverbal cues that are given in face-to-face encounters. Always proofread correspondences prior to sending them. Imagine yourself the recipient of the document. Look for complete sentences, logical development of thought and reasoning, accuracy, and appropriate use of grammar such as punctuation and capitalization.

Electronic health records (EHR) are increasingly being adopted by health care systems, particularly acute care settings. However, these EHR systems are expensive to implement, so there is a lag between what's available to improve record keeping and what is actually being used. The types and features of the EHR systems adopted are almost as numerous as the institutions using them, so specific details will not be elaborated here. Orientation to each institution likely includes an introduction to the system(s) in use.

## Critical Thinking 8-1

Upon entering a patient's room, you identify yourself as his nurse and greet the patient. You then ask him how he is feeling. He responds with a whisper and a grimace, "I've never felt better." You notice that the patient is slightly cyanotic. He is supporting himself with his elbows so that he is sitting upright over the bedside table. You note that he is using pursed lip breathing. Respirations are 36 per minute and shallow.

What kinds of problems occur when verbal and non verbal communications are incongruent? How would you handle the incongruent verbal and non verbal communications? Which message, the verbal or nonverbal, is easier to identify?

## REAL WORLD INTERVIEW

I view my primary responsibility as a nurse to be that of patient advocate. As a team leader, I am responsible for coordinating patient care for a group of patients. I am responsible for setting patient care goals and then directing my team to achieve the goals. I make those patient care goals the focus of my team's efforts. Patient care is rendered with the assistance of subordinates, including certified nurse assistants (CNAs), nursing student externs, and occasional high school student volunteers. Communication is the key to a successful team. A recent patient typifies how I interact with my team.

An elderly nonverbal patient with a history of schizophrenia was admitted to our surgical unit for dehydration. She was in need of total care, especially with respect to hygiene, which had been neglected. She was dependent on staff to turn and position her. Her level of awareness suggested that she was unable to use a call light for help.

This patient challenged staff for a variety of reasons. First, due to multiple other health problems, she was not a candidate for surgery. This placed her among the patients who don't really "fit" the surgical unit where she was admitted. Nonetheless, my goal was to advocate for comfort care with her physician while also encouraging subordinates to provide quality care even though the goal was not for cure with this particular patient. The patient's inability to communicate verbally added to the challenge. It was unclear how aware the patient was of the care she was receiving. Her nonverbal status blocked her ability to dialogue. This caused us to rely on nonverbal cues. Respect for patients with or without their verbal feedback is essential. The CNA and I tackled the needed bed bath together. Teamwork kept the focus on the goal for the patient, which was to optimize comfort and maintain skin integrity. It allowed me to complete a thorough assessment and to model desired communication with the patient, whom I addressed by name. I inquired whether she was in pain, to which she responded with twisting motions. I continued the one-way conversation, attempting to clarify what her nonverbal responses meant. She pointed to her shoulder, so we repositioned her and she settled down, resting quietly. As is often the case, the CNA willingly returned to reposition the patient with confidence the remainder of the shift. The patient's inability to verbalize needs was perceived as less of a barrier once we were successful in overcoming it together.

I find that CNAs will often volunteer to complete entire tasks they feel capable of performing independently. They also need to be assured that they will not be expected to handle clinical situations for which they do not feel qualified. This mutual respect for each other is essential to an ongoing working relationship. They honor my standard of care and will often complete tasks, going above and beyond what I ask. For example, later in the afternoon, the CNA returned to the patient and washed and braided her hair. Because this same patient would not likely use the call light, I also explained our goal to the high school student volunteer, and I asked her to check the patient's position whenever she went by the room. I instructed her to let me know if the patient appeared uncomfortable, assuring her that I would reposition the patient as needed. The student expressed that she thought it was cool how nurses communicate with patients who can't talk. I believe that, through effective communication, our team achieved the goal of optimizing this patient's comfort in spite of many potential barriers.

**Lari Summa, RN, BSN**
Team Leader
Peoria, Illinois

An EHR that can be used by all health care organizations has been advocated by many, but to date has not been widely adopted in the United States. Although there is limited high-quality data for EHRs in hospital settings in any of seven nations recently studied (i.e., the United States, Canada, the United Kingdom, Germany, the Netherlands, Australia, and New Zealand), evidence suggests that only a small fraction of hospitals (<10%) in any of the seven nations had the key components of an EHR (Jha, Doolen, Grandt, Scott, & Bates (2008). In general, as with all patient records, issues of confidentiality are of utmost importance, so nurses must be mindful of their important role in maintaining privacy and appropriately accessing and granting access to the system.

## EVIDENCE FROM THE LITERATURE

**Citation:** Cleary, M., & Freeman, A. (2005). Email etiquette guidelines for mental health nurses. *International Journal of Mental Health Nursing, 14,* 62–65.

**Discussion:** Most practitioners have e-mail addresses. E-mail provides greater immediacy than traditional modes of communication. Used correctly, e-mail is a fast and effective mode of communication.

1. Ask yourself, is e-mail the best route for communicating? If yes, does the subject line match the content?

2. While e-mail does not necessitate the same level of social interchange as a telephone call, nurses consider courtesy and respect for others to be important principles. Use an appropriate level of formality. Is it readable, concise, and professional?

3. E-mail is about as private as putting a message on the back of a postcard. Limit information and get patient's permission prior to using e-mail.

4. Avoid common pitfalls by keeping e-mails brief and arranging for how communication will be managed during periods of absences.

5. Use standard font. Avoid forwarding e-mail that could be construed as attacking. Use the high priority feature sparingly.

6. Copy information prudently so that only those in need of information receive it.

7. Open attachments only from known sources to avoid viruses. Forward attachments sparingly.

8. Acknowledge important e-mail when it is received.

**Implications for Practice:** Nurses can avoid communication problems by following these e-mail guidelines.

# LEVELS OF COMMUNICATION

The level of communication involves who the audience is at the time of communication. Communication can be thought of as having three levels: public, intrapersonal, and interpersonal.

## PUBLIC COMMUNICATION

Nurses rely primarily on interpersonal communication. However, they also use the other levels, so brief descriptions will be given for them as well. The nurse educator presenting a workshop on signs and symptoms of menopause to a room full of middle-aged women engages in public communication. Her audience is a group of people with a common interest. As presenter, she acts primarily as a sender of information. By design, feedback is typically limited in public speaking, though it does occur. Technology is changing public communication. Online courses often increase interactions between student and learner and increase the time needed for teaching and learning. Mastery of concepts taught may be communicated through several electronic modes other than testing, e.g., electronic chat rooms and discussions.

## INTRAPERSONAL COMMUNICATION

Another level is **intrapersonal communication**, which can be thought of as self-talk. As the name suggests, it is what individuals do within themselves, and it can present as doubts or affirmations. Complex mechanisms are involved in understanding how subconscious beliefs drive behavior, and what is needed to change them is based on social learning theory and social cognitive theory. A new nurse may engage in intrapersonal communication as he simultaneously doubts and affirms his ability to complete a procedure. For example, the first time the newly licensed RN has to catheterize a patient, he may simultaneously doubt his ability to insert a Foley catheter, with one message, "I have only done this on a mannequin," while affirming his ability to insert a Foley catheter with an "I can do this" message to himself. He is engaging in intrapersonal communication. The so-called competing voices within himself act as sender and receiver in this intrapersonal conversation whose outcome will be influenced by the feedback that follows.

Communication with self, or intrapersonal communication, is the first important element in developing a sphere of

## Critical Thinking 8-2

A diabetic patient who lives in the rural West is excited to have recently become "connected" to the World Wide Web with the acquisition of a computer and Internet services. His wife, who accompanies him to his office visit, wonders whether some of the monitoring that currently occurs during office visits might not be accomplished using e-mail. They reason that winters are hard and travel is difficult, the patient's circulation is poor, he fatigues easily, and it could save them the six-hour round-trip. Furthermore, they're hoping to spend several weeks in the South this winter. It seems to them that, using e-mail, they can maintain contact as needed without the drawbacks they identified.

What are the advantages of this suggestion for the patient? The caregiver? What safeguards can you think of to facilitate communicating using e-mail and to assure quality care?

communication. From self-awareness and understanding of oneself, a nurse can move confidently into one-to-one interactions with others, and then into interactions with smaller and larger groups.

## INTERPERSONAL COMMUNICATION

The last level, **interpersonal communication**, involves communication between individuals, person-to-person or in small groups. Not surprisingly, nurses engage in this level regularly. Interpersonal communication allows for a very effective level of communication to occur and incorporates all of the elements, channels, and modes previously discussed. The nurse, who observes a patient grimace when he moves, interprets the nonverbal cue as indicating that the patient is experiencing pain. Using verbal communication, she clarifies her perception by asking the patient to describe and rate his pain. He describes it as tolerable and states that he is expecting a visitor and he does not want to be drowsy. The communication goes back and forth until, ideally, both parties' understanding of the message match. This is the goal of communication. Note that good nursing care and effective nursing and medical practitioner communication has been linked to patient survival in ICUs (Arford, 2005).

## ORGANIZATIONAL COMMUNICATION

Networks of communication are often defined by an organization's formal structure. The formal structure establishes who is in charge and identifies how different levels of personnel and various departments relate within the organization. Some of these relationships are typically depicted by an organizational chart. When the chief executive officer of an organization announces that the company will adopt a new policy that all employees will follow, that is downward communication. The message starts at the top and is usually disseminated by levels through the chain of communication. Upward communication is the opposite of downward communication. The idea originates at some level below the top of the structure and moves upward. For example, when a nurse recommends a more efficient approach to organizing care to his nurse manager, who takes the recommendation to her superior, who uses the recommendation to develop a new policy, that is upward communication. Besides the upward and downward chain network of communication, other common networks of communication have been identified, e.g., a Y network, a wheel network, a circle network, and an all-channel network of communication (Longest & Darr, 2008). A Y pattern of communication would be two people reporting to another person who reports to others. An example is two staff nurses who report to the nursing unit director, who reports to the vice president for nursing, who reports to the president. A wheel network of communication looks like an *X in a wheel* and could be a situation in which four nurses report to one nurse manager. There is no interaction among the four nurses, and all communications are channeled through the nurse manager at the center of the wheel. This network is rare in health care organizations. Even though this type of network communication is not used routinely, it may be used in circumstances in which urgency or secrecy is required. For example, the president of an organization who has an emergency might communicate with the vice presidents in a wheel network because time does not permit using other modes.

A circle network of communication allows communicators in the network to communicate directly only with two others, next to them in the circle. Because each communicates with another communicator in the circle network, the effect is that everyone communicates with someone, and there is no central authority or leader. Finally, an all-channel network of communication is a circle network, except that each communicator may interact with every other communicator in the network, not just the communicator next to them (Longest & Darr, 2008).

Communication networks vary along several dimensions. The most appropriate network depends upon the situation in which it is used. The wheel and all-channel networks tend to be fast and accurate compared with the chain

## REAL WORLD INTERVIEW

In the future world of work and service, one of the most important things that you can do for yourself and others is to take responsibility for creating and sustaining high-quality connections. Develop a reputation as someone who energizes people at work rather than someone who de-energizes people at work. High-quality connections (HQCs) foster admiration, appreciation, support, challenge, and hope for the future. Low-quality connections (LQCs) foster distrust, alienation, and regret in work environments.

How do you tell the difference between a high- and low-quality connection? Think of the people who energize you. Use them as role models, and pay attention to how they act and what they say. Develop a plan to model their behavior and interactions with others.

At the same time, pay attention to those people who de-energize you. Think about what they say and do. Pay attention to how co-workers react when these de-energizers are around. Researchers Dutton and Heaphy (2003) believe HQCs are like strong, flexible, resilient, and healthy blood vessels that feed connective tissue to sustain health and nourish vitality.

In contrast, LQCs block the positive flow of life-giving energy in interpersonal relationships. Dutton and Heaphy observe, "with a low quality connection there is a little death in every interaction." Develop a reputation for being affirming and life giving rather than negative and de-energizing. People skilled at creating and maintaining HQCs create more-meaningful work for themselves and the organizations within which they work.

**Daniel J. Pesut, PhD, APRN, BC, FAAN**
Professor and Associate Dean for Graduate Programs
President of the Honor Society of Nursing, Sigma Theta Tau International, 2003–2005
Indianapolis, Indiana

and Y-pattern networks. The chain and Y-pattern networks promote clear-cut lines of authority and responsibility but may slow communications. The circle and all-channel networks enhance morale among those in the networks better than other patterns. Nurses must construct communication networks to fit the various communication situations they face. Note that e-mail has had the effect of flattening some organizational communications to sometimes allow more direct access between levels that were formerly controlled through middle managers.

A final avenue worth mentioning, which is not a formal avenue, is the grapevine. The grapevine is an informal avenue in which rumors circulate. It ignores the formal chain of command. The major benefit of the grapevine is the speed with which information is spread, but its major drawback is that it often lacks accuracy. For example, nurses who inform an oncoming shift about a rumor that layoffs or mandatory overtime is imminent, in the absence of any information from the hospital's administration, are participating in grapevine communication.

## COMMUNICATION SKILLS

Because nurses are often placed in positions of leadership and are responsible for representing nursing's concerns to others, including the patient, it is important that they have the requisite skills to be effective communicators. There is no one correct way to ensure effective communication. Rather, effective communication requires that both parties engaged in communicating use skills that enhance that particular interaction. In general, the most important considerations for facilitating communication are to be open, assertive, and willing to give and receive feedback. Some of the most important skills upon which nurses rely to do this are listed in Table 8-3.

### Critical Thinking 8-3

How does your organization communicate? How does the organization assure effective communication within itself? How does the organization communicate with other organizations in the community?

## TABLE 8-3 Communication Skills

| SKILL | DESCRIPTION |
|-------|-------------|
| Attending | Active listening for what is said and how it is said as well as noting nonverbal cues that support or negate congruence, for example, making eye contact and posturing. |
| Responding | Verbal and nonverbal acknowledgement of the sender's message, such as "I hear you." |
| Clarifying | Restating, questioning, and rephrasing to help the message become clear, for example, "I lost you there." |
| Confronting | Identifying the conflict, for example, "We have a problem here" and then clearly delineating the problem. Confronting uses knowledge and reason to resolve the problem. |
| Supporting | Siding with another person or backing up another person: "I can see that you would feel that way." |
| Focusing | Centers on the main point: "So your main concern is …" |
| Open-ended questioning | Allows for patient-directed responses: "How did that make you feel?" |
| Providing information | Supplies one with knowledge she did not previously have: "It's common for people with pneumonia to be tired." |
| Using silence | Allows for intrapersonal communication. |
| Reassuring | Restores confidence or removes fear: "I can assure you that tomorrow …" |
| Expressing appreciation | Shows gratitude: "Thank you" or "You are so thoughtful." |
| Using humor | Provides relief and gains perspective; may also cause harm, so use carefully. |
| Conveying acceptance | Makes known that one is capable or worthy: "It's okay to cry." |
| Asking related questions | Expands listener's understanding: "How painful was it?" |

# BARRIERS TO COMMUNICATION

Barriers are obstacles to effective communication. The nurse who can identify potential barriers to communication will be better equipped to avoid them or to compensate for them. Some barriers can be physical, such as trying to communicate with someone on a poor phone connection. Some of the other most common barriers are language, gender, culture, anger, generational differences, illiteracy, and conflict. Conflict management is so central to good patient care that it will be discussed in another chapter.

## USE OF LANGUAGE

Language is the primary means we use to communicate with each other. Humans have developed written, sign, and oral languages to share messages. We use language to express ideas, feelings, and emotions and to communicate information, reactions, and directions to each other and to negotiate with each other.

Oral language is a feature of every society. In a health care setting, oral language is used to verbally communicate with patients and other health care professionals. Individuals also communicate nonverbally using body language. It is important to remember that there are many cross-cultural similarities in body language, but there are also key differences. The meaning of different gestures varies from culture to culture. Never assume that a gesture holds the same meaning for you and your patient, especially if your patient is from another culture.

As you work with patients, you will encounter a great diversity in spoken languages. For instance, more than 6,900 languages and dialects are spoken today, with Mandarin Chinese topping the list, with 1 billion 213 million speakers. The other top five languages spoken are Spanish, numbering 329 million speakers; English, 328 million; Arabic, 221 million; and Hindi, 182 million (WhoYOUgle, 2009–2011).

## Linguistically Appropriate Services

Any individual who is seeking health care services and who has limited English proficiency (LEP) has the right, based on Title VI of the Civil Rights Act of 1964, to have an interpreter available to facilitate communication within the health care system. Open and clear communication is essential to develop an appropriate diagnosis and treatment. Potential errors in this area can occur when the health care provider is unable to obtain accurate information from the patient with LEP. These individuals must be able to access language services, such as using an interpreter and translated materials for information necessary to understand the health care services and benefits available. It is important to note that some patients may not be literate even in their native language; therefore, they may not benefit from translated material that requires them to read instructions or health information.

Recognizing the challenge of obtaining an interpreter 24 hours a day, the Office of Civil Rights suggests the need for the availability of interpreter services during the main hours of operation. Language assistance services need to be comprehensive in that the patient should receive these services from the initial point of contact with the provider, during the initial health interview, while receiving health care services, and during planning for discharge and home care. Furthermore, the health care provider must ensure that the trained interpreter follows ethical practices and guidelines. This means that strict confidentiality and privacy of information must be ensured and maintained at all times. It is the responsibility of the health care organization to ensure that the interpreters are competent and properly trained in the medical and health context. Unfortunately, use of family members or friends as interpreters is convenient and continues to be common practice in some health care settings. This practice is discouraged and can be acceptable only when the patient expresses the preference to have a family member or friend be the interpreter. Clearly, there may be situations in which a formally trained interpreter is unavailable and the use of telephone interpretation is not practical; family members may then be used with permission from the patient. Although telephone interpreter services are permissible, the use of a face-to-face, in-person interpretation is desirable, acceptable, and appropriate.

Health care organizations are expected to make known to all their patients and patients' families, the availability of interpreter services at no cost to the patient. Information about available bilingual staff can also help the patient access these services. Posting translated signs in the agency that are clearly visible will be very helpful to the patient.

Linguistically and culturally appropriate services may be posted in community newspapers and on radio stations to inform the community of the available service. To maintain the quality improvement process, health care organizations are also encouraged to share with the public and the recipients of their services how they have implemented culturally and linguistically appropriate services. An annual report may be one way to inform the public. Although these standards are directly addressed to the health care organizations, their application and relevance are clearly an expectation in all professional practice (Munoz & Luckmann, 2005).

## GENDER

Gender interferes with communication when men and women lack the understanding that they may process information differently. In general, some men are more interested in using reason, where as some women may want to be heard and validated through communication (Gray, 1992). Gender differences and patterns do not preclude working together. Rather, both sides must realize the other's preference and make accommodations so that effective communication results.

Gender differences have been attributed, in part, to gender socialization in which males are sometimes provided with more opportunities to develop confidence and assertiveness than are females. Fortunately, the feminist movement and increased sexual equality in Western society, in general, have lessened traditional sociological patterns of competitiveness and decisiveness in men and passivity and nurturing in women. However, remnants of the traditional model persist, particularly in health care settings. Nurses who lack assertiveness and confidence are encouraged to acquire the requisite skills to be assertive and confident so they be an effective patient advocate and also so they can communicate in a confident manner.

### Critical Thinking 8-4

Nurses who practice assertiveness are direct, honest, and appropriate. They say, "I need you to do ..." rather than, "You should do ..." The next time you work in the clinical area, note how nurses interact with other staff and nursing and medical practitioners. Do the nurses and other staff always speak assertively? Do they become aggressive or passive instead? Is assertiveness always easy? Is being assertive part of being a professional nurse? Is it possible to meet your patient's needs if you are not assertive?

## CULTURE

As was stated previously, our culture grows increasingly diverse. This diversity reduces the likelihood that patients, nurses, and other health care providers will share a common cultural background. In turn, the number of safe assumptions about beliefs and practices decreases, and the probability for misunderstanding increases. For example, shortly after delivering a baby, women are often hungry and thirsty. Some cultures believe that for women to restore their energies appropriately, they should eat hot foods and beverages, while other cultures believe cold foods and beverages are appropriate. A well-intentioned nurse who does not consult with the patient about her preferences may arrange culturally inappropriate nourishment. Broadly defined, culture encompasses a learned pattern of beliefs, customs, language, norms, and values. Sanders and Ewart (2005) stress that developing communication that is culturally competent requires respect for each individual's unique mix of values and beliefs. These values and beliefs can only be identified by carefully listening to each individual. Culture and communication are intrinsically intertwined. Culture is discussed more in Chapter 25. Nurses are responsible for bridging gaps between themselves and their patients through first being accepting of differences. They can also overcome cultural differences by learning about other cultures. Nurses bridge cultural differences by vigilantly using the skills previously described to facilitate communication.

## GENERATIONAL DIFFERENCES

Generational differences can create tensions among workers because of the divergent outlooks on life. Different generations can have different values about work, motivation, lifestyle, and communication. It is important to be aware that even though generalizations can be misleading and unfair, four generations make up the current workforce. Awareness of generational differences allows you as a nurse to use this information as a tool to maximize strengths and deal with conflicts (Greene, 2005). This is discussed more in Chapter 25.

## HEALTH LITERACY

Nurses need to be mindful that many patients and their families lack the health literacy skills to understand the health care system and function successfully as health care consumers. Health literacy is an individual's ability to read, understand and use health care information to make decisions and follow instructions for treatment. There are multiple definitions of health literacy, in part because health literacy involves both the context (or setting) in which health literacy demands are made (e.g., health care, media, Internet or fitness facility) and the skills that people bring to that situation (Rudd, Moeykens, & Colton, 1999).

This definition encompasses the elements of personal empowerment and action and views health literacy as an out-

## EVIDENCE FROM THE LITERATURE

**Citation:** Greene, J. (2005). What nurses want. *Hospitals and Health Networks*, 79(3), 34–42.

**Discussion:** The author discusses four distinct generations working together in the workforce today:

- *Matures, veterans (born 1922–1946):* They believe in hard work, paying dues, conformity, and long-term commitment to one employer.
- *Baby boomers (born 1946–1964):* They define selves through employment, that is, self worth = work. They are willing to work long hours and like to change things.
- *Generation X (born 1964–1980):* They are independent and may change places of employment often. They seek connection with managers on an equal footing and are very comfortable with technology.
- *Generation Y (born 1980–2000):* They are optimistic, are street smart, expect diversity, crave structure, and are technologically savvy.

**Implications for Practice:** Awareness of generational differences helps the nurse work well with coworkers. Learning styles, work habits, beliefs about family and work balance, computer literacy, and comfort with technology may all vary in the different generations.

come of health promotion and health education efforts that have both personal and social benefits. Patients, families, and even subordinates simply do not always understand what nurses and other health care providers are trying to communicate. For example, while most health care materials are written at the tenth-grade level or higher, most adults read between an eighth and ninth grade level (Safeer & Keenan, 2005). Almost a quarter of the adult population in the United States is functionally illiterate, and nearly half have limited literacy skills. Literacy challenges are increased for patients whose primary language is not English. It is also increased for patients from some racial or ethnic backgrounds, patients from certain areas of the country, patients with low-income levels, patients who have been in prison, and patients older than 60 years of age (Safeer & Keenan, 2005).

The Health Literacy of America's Adults is the first release of the National Assessment of Adult Literacy (NAAL) health literacy results (U.S. Department of Education Institute of Education Sciences National Center for Health Statistics. 2006). The results are based on assessment tasks designed specifically to measure the health literacy of adults living in the United States. Health literacy was reported using four performance levels: Below Basic, Basic, Intermediate, and Proficient. The majority of adults (53%) had Intermediate health literacy. About 22% had Basic and 14% had Below Basic health literacy. Relationships between health literacy and background variables (such as educational attainment, age, race/ethnicity, where adults get information about health issues, and health insurance coverage) were also examined and reported. For example, adults with Below Basic or Basic health literacy were less likely than adults with higher health literacy to get information about health issues from written sources (news papers, magazines, books, brochures, or the Internet) and more likely than adults with higher health literacy to get a lot of information about health issues from radio and television. Limited health literacy also leads to additional costs (Kutner et. al., 2006). It is important to recognize the health literacy communication barrier and to create a patient-centered and shame-free health care environment (Cornett, 2009). Clues indicating low health literacy skills include: (1) patients make excuses when asked to read or fill out forms; (2) poor readers often lift page closer to their eyes or point to text with finger while reading; (3) patients provide incomplete medical history or answer no to avoid follow-up questions; (4) poor readers often miss appointments and; (5) poor readers show signs of nervousness, providing incorrect answers when questioned about what they've read (Cornett, 2009).

Most health care materials are written at the tenth-grade level or higher, while most adults read between an eighth and ninth grade level (Safeer & Keenan, 2005). Literacy challenges are increased for patients whose primary language is not English. It is also increased for patients from some racial or ethnic backgrounds, patients from certain areas of the country, patients with low-income levels, patients who have been in prison, and patients older than 60 years of age (Schyve, 2007; Safeer & Keenan, 2005). These literacy challenges interact in different ways to different degrees for different patients (Singleton & Krause, 2009).

Low health literacy reduces the success of treatment and increases the risk of medical error. Various interventions, such as simplified information and illustrations, avoiding jargon, "teach back" methods and encouraging patients questions, have improved health behaviors in persons with low health literacy. Health literacy is of continued and increasing concern for health professionals as it is a primary factor behind health disparities. The government's Healthy People 2020 has included it as a pressing new topic, with objectives for addressing it in the decade to come (Healthy People 2020).

Assessing patients' literacy is one way that you as the nurse can decide what patient education material to use and how best to present it. It is a mistake to rely on patients' self-report of their reading skills, because the majority of patients who have low health literacy say they read "well." The Rapid Estimate of Adult Literacy in Medicine is a quick assessment that takes two to three minutes to complete (Bass, et al. 2003). Always try to communicate in short, clear, and simple ways with patients, using pictures if possible to aid understanding. Limit the amount of information to three to five "need to know" points and provide written material that is prepared using at least a size 12 print font. Be aware that broad strategies have been identified to improve health literacy and patient safety. They are: (1) make effective communication a priority; (2) address patients' communication needs across the continuum of care; and (3) pursue policy changes that promote improved provider-patient communications (Murphy-Knoll, 2007). Focus on the patient's experience of the condition rather than elaborating on the patient's pathophysiology (Safeer & Keenan, 2005; Murphy-Knoll, 2007; Cornett, 2009).

## ANGER

Anger is a universal, strong feeling of displeasure that is often precipitated by a situation that frustrates or prevents a person from attaining a goal or getting what is wanted from life. Anger is influenced by one's beliefs. Ellis (1997) describes anger as an irrational response that arises from one of four

### Critical Thinking 8-5

Nurses need to be aware that different staff in the workforce may see things differently based on their age differences. For example, some staff often do only exactly what is asked. They work primarily to earn money to spend. Other staff may believe it is important to save, save, save. They work hard to eliminate a task. Some staff like to buy now, pay later, and work very efficiently. Some other staff work fast, sacrifice, and are very thrifty. These various staff members may have a hard time working together toward a goal. Do you see any of these age differences affecting the clinical area where you work? How can nurses bridge these differences?

## TABLE 8-4  Additional Barriers to Communication

| BARRIER | DESCRIPTION |
| --- | --- |
| Offering false reassurance | Promising something that cannot be delivered. |
| Being defensive | Acting as though one has been attacked. |
| Stereotyping | Unfairly categorizing someone based on his or her traits. |
| Interrupting | Speaking before another has completed her message. |
| Inattention | Not paying attention. |
| Stress | A state of tension that gets in the way of reasoning. |
| Unclear expectations | Ill-defined direction to perform tasks or duties that make successful completion unlikely. |
| Incongruent responses | When words and actions in a communication don't match the inner experience of self and/or are inappropriate to the context. This response commonly presents itself as blaming, placating, being super-reasonable, or using irrelevant information when communicating with another person. |
| Giving advice | Assumes that the other person is unable to solve his or her own problems. It is better to share problem solving options when communicating. This allows the other person to choose one of the options for solving a problem. |

irrational ideas: (1) feeling that the treatment one received was awful (awfulizing); (2) feeling that one can't stand having been treated so irresponsibly and unfairly (can't-stand-it-itis); (3) believing that one should not, must not behave as he did (shoulding and musting); and (4) thinking that because a person acted in a terrible manner, he is a terrible person (undeservingness and damnation). Ellis maintains that beliefs remain rational as long as the evaluation of the action does not involve an evaluation of the person. Rational and appropriate responses are feelings of disappointment. Anger, on the other hand, can be unmanageable and self-defeating. Ellis believes that we all have the ability to choose our response to anger.

Anger can be dealt with in one of several ways. Three methods that may work from time to time but that may have serious and potentially destructive drawbacks are denying and repressing anger, which may lead to resentment; expressing anger, which may lead to defensiveness on the part of the respondent; and turning the other cheek, which may lead to continued mistreatment and lack of trust. Anger can stem from deep-seated feelings of unassertiveness. Assertiveness involves taking a stand, where as aggression involves putting another down. If unassertiveness is the source of anger, then a solution is to learn to act assertively.

Additional barriers to communication are seen in Table 8-4. Ways to overcome these communication barriers are seen in Table 8-5.

# WORKPLACE COMMUNICATION

It is probably clear by now that how individuals communicate depends, in part, on where communication occurs and in what relationship. Patterns of communication in the workplace are sensitive to organizational factors that define relationships. Nurses have diverse roles and relationships in the workplace that call for different communication patterns with supervisors, coworkers, subordinates, practitioners and other health care professionals, patients, families, and mentors. Nurses need to keep in mind what the educational levels are of the people with whom they are communicating. Using medical terminology is appropriate with another practitioner who shares a common understanding. Discussions with others such as LPNs, nursing assistive personnel (NAP), patients, and family members will more likely result in understanding when language is adjusted to their level of understanding. Always remember that the goal is to communicate effectively.

## TABLE 8-5 Overcoming Communication Barriers

| METHOD | ACTIONS |
| --- | --- |
| Understand the receiver | Ask yourself what's in it for the other person. |
| | Work to develop understanding of the other person's needs. |
| Communicate assertively | Be direct. |
| | Explain ideas clearly and with feeling. |
| | Repeat important messages. |
| | Use various communication channels, for example, written, e-mail, verbal, and so on. |
| Use two-way communication | Ask questions. |
| | Communicate face-to-face. |
| Unite with a common vocabulary | Define the meaning of important terms, such as *high quality*, so that everyone understands their meaning. |
| Elicit verbal and nonverbal feedback | Request and offer verbal feedback often. |
| | Document important agreements. |
| | Observe nonverbal feedback. |
| Enhance listening skills | Pay attention to what is said, what is not said, and to the nonverbal signals. |
| | Continue listening carefully even when you don't like the message. |
| | Give summary reflections to assure understanding; for example, "You say you are late giving medications because the pharmacy did not deliver meds on time." |
| | Engage in concluding discussions, such as, "Has your unit been late with medications due to problems with pharmacy deliveries before?" |
| | Ask questions to explore problems. |
| | Paraphrase a speaker's words to decrease miscommunication rather than blurting out questions as soon as the other person finishes speaking. |
| Be sensitive to cultural differences | Know what cultural communication barriers exist. |
| | Show respect for all workers. |
| | Minimize use of jargon specific to your culture. |
| | Be sensitive to cultural etiquette such as use of first names, eye contact, hand gestures, personal appearance. |
| Be sensitive to gender differences | Be aware that men and women may have some differences in communication style; for instance, men may call attention to their accomplishments, and women may tend to be more conciliatory when facing differences. |
| | Know that male-female stereotypes often don't fit the person you are working with. |
| | Avoid barriers by knowing that differences exist, and don't take things personally. |
| | Males can improve communication by showing more empathy and females by becoming more direct. |
| Engage in metacommunication | Communicate about your communication to resolve a problem; e.g., "I'm trying to get through to you, but either you don't react to me or you get angry. What can I do to improve our communication?" |

**Source:** Adapted from DuBrin, A. J. (2000). *The active manager.* United Kingdom: South-Western Pub.

## SUPERVISORS

Communicating with a supervisor can be intimidating, especially for a new nurse. Observing professional courtesies is an important first step. For instance, begin by requesting an appointment to discuss a problem when it arises. This demonstrates respect and allows for the conversation to occur at an appropriate time and place. Dress professionally. Arrive for the appointment on time, and be prepared to state the concern clearly and accurately. Provide supporting evidence, and anticipate resistance to any requests. Separate out your needs from your desires. State a willingness to cooperate in finding a solution and then match behaviors to words. Persist in the pursuit of a solution (Table 8-6).

## COWORKERS

Nurses depend on their coworkers in many ways to collectively provide quality patient care. Nowhere is this more important than in the acute care setting where nursing services are nonstop around the clock. Transfer of patient care from nurse to nurse is one of the most important and frequent communications between coworkers. Fluid communication in end of shift hand-off reports are crucial for achieving quality nursing care. However, time constraints demand that the change of shift handoff report be accurate, informative, and succinct. How the nursing care is organized influences who gets the report. Nursing hand-off reports are discussed in Chapter 18.

### TABLE 8-6  How to Improve Your Ability to Work with Your Boss

*Know your boss's:*

- Goals and objectives
- Pressures
- Strengths, weaknesses, and blind spots
- Working style

*Understand your own:*

- Objectives
- Pressures

- Strengths and weaknesses
- Working style
- Predisposition toward dependence on authority figures

*Develop a relationship that:*

- Meets both your objectives and styles
- Keeps your boss informed
- Is based on dependability and honesty
- Selectively uses your boss's time and resources

**Source:** Adapted from Gabarro, J., & Kotter, J. P. (1993, May–June). Managing your boss. *Harvard Business Review*, 150–157.

### Critical Thinking 8-6

You are having a coffee break with another nurse who mentions a problem she is having with care delivery. The nurse is not sure how to solve it. You want to be helpful and supportive and yet avoid giving advice. Ask the nurse if she can describe the problem fully for you. Do not interrupt. Then, ask the nurse some questions about the problem, and seek clarification until you are clear on the problem and the nurse has fully described it. Do not give advice. Use your communication skills, such as attending, clarifying, and responding, and ask the nurse such things as, tell me more about that, what did you think about it, and so forth, until you are both clear on the issue. At the end of this process, you can just finish by relaxing for the rest of your break or you can ask the nurse, "Do you want advice about your problem?"

Many times, this process will help the nurse solve the problem by himself or herself. If the nurse does want advice, you can give some suggestions if you are comfortable doing so. Do you think this process can strengthen people's ability to find answers to their problems? Do you think this process would be helpful to you in working with others on the unit? Do you think this approach honors people's integrity and ability to solve their own problems?

*Source:* Adapted from M. Parsons (personal communication, 2003). San Antonio, Texas.

Tips for communicating with coworkers include remembering professional courtesies and being mindful of an appropriate time and place to share your concerns. Stay focused on what is needed to get the job done, and seek a win-win solution to conflicts, where all parties are satisfied with the outcome.

An excellent guide for directing communication with coworkers is the golden rule: "Do unto others as you would have them do unto you." As a nurse who will be responsible for overseeing others' work, a valuable perspective for you to maintain is that all members of the team are important to successfully realize quality patient care. Communication between nurses and coworkers will most likely involve delegating. This important topic is covered in Chapter 16, and you are encouraged to review this material as it relates to coworkers. In addition to delegating, a few other communications skills are worth mentioning. Offering positive feedback such as "I appreciate the way you interacted with Mr. T to get him to ambulate twice this shift" goes a long way toward team building, and it improves coworkers' sense of worth. Nurses also have an opportunity to act as teachers to coworkers. Often in a hospital setting, nurses teach by example. Demonstrating the desired behavior allows the coworker the opportunity to copy the behavior. It is important to allow time for return demonstrations to evaluate that the coworker has learned the intended skill. Offer constructive feedback. Be patient. Remember your own learning curves when mastering new skills and behaviors, and allow those you supervise the opportunity to grow. Be open to the possibility that coworkers, particularly those with experience, may have a few pearls of wisdom to share with you as well.

# HORIZONTAL/LATERAL VIOLENCE

New nurses need to be aware that not all working environments are hospitable. According to Duffy (1995), horizontal or lateral violence is hostile or aggressive behavior by an individual or group member(s) towards another member or members of the group. This may also be called bullying. Manifestations of horizontal violence include overt or covert hostile behaviors that are a symptom of a hostile "system" or work group culture rather than individual pathology. See Table 8-7 for signs of horizontal/lateral violence.

## TABLE 8-7   Signs of Horizontal/Lateral Violence

Aggressive or mocking behavior

Verbal retorts, abrupt responses, vulgar language

Belittling gestures such as deliberate rolling of eyes, folding arms, staring into space

Undermining behavior such as constantly ignoring questions, devaluing comments

Criticizing or excluding an individual from discussion

Withholding needed information or advice

Sabotage, such as setting up a new-hire nurse for failure

Constant confrontation, demonstrating negativity

Infighting and bickering

Scapegoating

Blaming and gossiping behind a colleague's back

Humiliation and confrontations in public

Failure to respect privacy, broken confidences

Shouting, yelling, or other intimidating behaviors

Judging others based on age, gender, sexual orientation, etc.

Punishing activities by management, e.g., bad schedules, chronic understaffing

Physical violence

**Source:** Hawaii Nurses Association. (2008).

## REAL WORLD INTERVIEW

A second shift occupational health nurse working in a factory setting was presented with a patient who entered the nursing office complaining that he didn't feel good. The nurse's initial assessment, including vital signs, revealed that the only abnormality was an elevated blood pressure. In this situation, like in any clinical situation, it is important to distinguish the urgent from the nonurgent. With hypertensive patients, it is important to realize that an urgent situation is suggested by evidence of acute end organ damage from the elevated BP. Specifically, in this situation, it was important to know whether the patient was experiencing altered sensorium, headache, visual disturbance, chest pain, or dyspnea. The presence of any of these findings should be communicated to the physician and would dictate urgent transport to the hospital. In their absence, the patient can be referred for more-elective blood pressure control.

In any clinical situation such as the one above, the nurse can check the patient for evidence of any signs or symptoms. The nurse can then facilitate communications by being organized and objective. Discuss the basics such as the patient's chief complaint, his vital signs, his medications, and any changes from baseline. Know why you are worried about observed changes, and communicate this to the practitioner.

**John C. Ruthman, MD**

Peoria, Illinois

There are several explanations for horizontal/lateral violence in nursing, including role, gender, self-esteem, anger as an unacceptable emotion, and nursing as an oppressed group. Whatever its origins, violence in the nursing workplace is demoralizing for nurses and also negatively impacts patient and nursing safety. Several organizations have developed no-tolerance policies. Personal strategies to avoid horizontal violence and create a safe and happy workplace include: (1) name the problem using the term, horizontal violence, to label the situation; (2) break the silence about horizontal violence by raising the issue at staff meetings; (3) ask about the process for dealing with horizontal violence; (4) engage in reflective practices such as journaling to raise your self-awareness about your own behaviors, beliefs, values, and attitudes; (5) engage in self-care activities to maintain your own health and happiness; (6) be willing to speak up when you observe horizontal violence behaviors; and (7) discuss strategic options with your union representative or nursing or hospital administrators. (**Hawaii Nurses Association 2008**).

## PHYSICIANS, NURSE PRACTITIONERS, AND OTHER HEALTH CARE PROFESSIONALS

One of the most intimidating experiences for new nurses may be communicating with physicians, nurse practitioners (NP), or physician assistants. Despite gender and role challenges that have already been discussed, this need not be a stressful event. The nurse's goal is to strive for collaboration, keeping the patient goal central to the discussion. Collaboration allows all parties to be satisfied, improves quality, and is an attribute of magnet hospitals (Arford, 2005). It involves seeking creative, integrative solutions while also working through emotions. To communicate effectively with the practitioner, the nurse presents information in a straightforward manner, clearly delineating the problem, supported by pertinent evidence. This is especially important when reporting changes in patient

### Critical Thinking 8-7

Sometimes you may work with staff who are difficult. These staff can include those who usually work in another setting; staff who are more knowledgeable than you; staff who are older than you; staff who think they are better than they are; and staff who are defensive when you ask them to do something. With all these staff members, it can help to develop strategies that identify your performance expectations clearly and involve them in the work that needs to be done. This is useful with all types of staff—those who are easy to work with and those who are difficult. How can you work to improve your ability to communicate your performance expectations clearly to the staff you work with?

## EVIDENCE FROM THE LITERATURE

**Citation:** Haig, K. M., Sutton, S., & Whittington, J. (2006). The SBARR technique: Improves communication, enhances patient safety. *Joint Commission's Perspectives on Patient Safety, 5*(2), 1–2.

**Discussion:** Communication failures are the root cause of nearly two-thirds of sentinel events in hospitals. This is due, in part, because nursing and medical practitioners are trained to communicate differently. The SBARR technique (Situation, Background, Assessment, Recommendation, Response) is designed to improve communication among health care personnel and improve patient safety.

- *Situation:* What is going on with the patient? Identify self, unit, patient, room number. Briefly state the problem, when it started, and its severity.

- *Background:* Provide background information related to the situation, as needed. Be aware of patient's admitting diagnosis, date of admission, current medications, allergies, IV fluids, most recent vital signs, lab results with date and time each was performed, other clinical information, and patient code status. The practitioner may ask you for these when you call.

- *Assessment:* What is your assessment? Do you think the patient's condition is deteriorating? Do you think the patient needs medication?

- Recommendation: What is your recommendation or what do you want? Know what you want from the practitioner before you call. Don't hang up without communicating this to the practitioner and assuring that your patient's needs are met—e.g., patient needs to be admitted, patient needs to be seen, order needs to be changed, or medication needs to be added.

- *Response:* Document response of practitioner and any further actions you take to meet patient needs.

**Implications for Practice:** Nursing and medical practitioners who use the SBARR technique will improve their communications. Patients will be the beneficiaries.

conditions. Nurses are responsible for knowing classic symptoms of conditions, apprising the practitioner of changes, and recording all observations in the chart. It is important that the nurse remain calm and objective even if the practitioner does not cooperate. Calfee (1998) offers suggestions for handling telephone miscommunications. For example, if a practitioner hangs up, document that the call was terminated and fill out an incident report. If the practitioner gives an inappropriate answer or gives no orders, for example for a patient complaint of pain, document the call, the information relayed, and the fact that no orders were given. In addition, document any other steps that were taken to resolve the problem, for example notifying the nursing supervisor. If a practitioner cannot be reached, first follow the institution's procedure for getting the patient treated and then document the actions taken (Table 8-8). The SBARR tool can be very useful for all communication with other nursing or medical practitioners.

## PATIENTS AND FAMILIES

Communication with patients and families is optimized by the many skills previously described in this chapter. There are a few additional skills that have not yet been mentioned.

The first is touch. Nurses routinely use touch as a way to communicate caring and concern. Occasionally, language barriers will limit communication to the nonverbal mode. For instance, a stroke patient who cannot process words can still interpret a gentle hand on his shoulder.

Communication requires an openness and honesty with concurrent respect for patients and families. In addition, it is important to honor and protect patients' privacy with the nurse's actions and words. Information that patients share with nurses and other health care providers is to be held in confidence. Verbal exchanges regarding patient conditions are private matters that should not occur in the hallway or just outside a patient's room where others will hear them. Nurses are obligated to not discuss patient conditions with others, even family members, without patient permission.

## Medication Reconciliation

Medication reconciliation is a communication process in which a patient's complete list of home medications is compared with medications ordered at transitions of care to resolve any discrepancies (Hall, 2010). The reconciliation process involves communication between

**TABLE 8-8  SBARR Tool to Organize Information for Calling Another Nursing Or Medical Practitioner for Assistance**

*Situation:*
Identify date and time of call.
Identify self, unit, patient, room number, admitting diagnosis.
State the problem: What it is, when it started, and its severity.

*Background:*
Review background information related to the situation. What are current medications, allergies, IV fluids, vital signs, level of consciousness, urine output, status of airway, breathing, circulation, pain level, pulse oximeter, cardiac rhythm, lab results, patient code status, and other clinical information?

*Assessment:*
What is your assessment of the patient? Is the patient's problem severe? Is his condition deteriorating? Does the patient need medication?

*Recommendation:*
What is needed from the practitioner? Know what you want from the practitioner before you call. Don't hang up without communicating this to the practitioner and assuring that your patient's needs are met—for example, patient needs to be admitted, patient needs to be seen, patient needs medication, and so on.

*Response:*
Document response of practitioner. Document all calls to practitioner and messages left. Notify supervisor when needed for follow-up. Document all actions you take to assure patient needs are met and patient safety is assured.

**Source:** Adapted from Haig, K. M., Sutton, S., & Whittington, J. (2006). The SBARR Technique: Improves Communication, Enhances Patient Safety. Joint Commission's Perspectives on Patient Safety, 5 (2), 1–2.

members of an interdisciplinary team that includes nurses, physicians, and pharmacists. While the process may vary from one institution to another, members of the interdisciplinary team engage in checks and balances to prevent medication errors. The following is an example of one hospital's medication reconciliation process (Hall, 2010). At the time of the patient's admission to the hospital, the nurse obtains a list of the patient's home medications (including the time, dose, route, frequency, next dose, and reason for taking) from the patient or caregiver and provides that information to the pharmacist and physician. The admitting physician reviews the list of home medications and then writes orders for all medications, making note of whether a home medication is to be administered or held. The pharmacist contacts the physician if there are discrepancies between home medications and those ordered on admission. When a patient is discharged or transferred, the physician determines which medications will be continued. The nurse then provides a complete list of medications to the patient as well as to the next provider of care. Communication between members of the interdisciplinary team is important to resolve any medication discrepancies and prevent medication errors.

From the time a medication order is placed in the system to the time the medication is administered to the patient, there are numerous opportunities for medication errors to occur in health care settings today. Nurses must be aware that medication administration is the most interrupted nursing activity. Interruptions during medication administration can lead to medication errors (Biron, Loiselle, & Lavoie-Tremblay, 2009). Sources of interruptions include people (health care professionals, patients, and family members) and technical interruptions (missing equipment, IV infusion pump failure, etc.). Nurses and nursing assistive personnel are the primary source of interruptions, with discussions related to patient care issues or personal issues. While little research has been done to test interventions to help nurses create a culture of safety and cope with interruptions, some hospitals have attempted to reduce errors by creating interruption-free medication preparation zones.

## MENTOR AND PRODIGY

The final pattern of communication that occurs in the workplace that will be discussed is between mentor and prodigy. Mentoring may be an informal process that occurs between an expert nurse and a novice nurse, but it may also be an assigned role. This one-on-one relationship focuses on professional aspects and is mutually beneficial. The optimal novice is hardworking, willing to learn, and anxious to succeed. Communication entails using the skills previously described in this chapter to help the novice develop expert status and career direction. The novice accomplishes this by gleaning the mentor's wisdom. This wisdom is typically shared through listening, affirming, counseling, encouraging, and seeking input from the novice. A strategy that facilitates mentoring is to share the same work schedule so that the novice is exposed to the mentor. This allows for sharing and shadowing opportunities. The mentor can also anticipate added challenges that will likely occur with increasing responsibility. Outlining these challenges with suggestions for how to manage them prepares the novice for his expanding responsibilities (Ihlenfeld, 2005). Role-playing, in which the expert preceptor nurse describes a theoretical situation and allows the novice to practice his response to new and sometimes challenging situations, is another strategy that can be used.

## CASE STUDY 8-1

As a new graduate, you have received notice that you have passed your NCLEX exam and have finished orientation. The nursing care manager is relieved, because two of the other regular nurses are pregnant and will soon be off on maternity leave. One of them is your preceptor. These absences will create a staffing crunch. Therefore, the nursing care manager is anxious to acclimate you to the role of team leader, because you will soon be expected to assume those responsibilities.

How can your preceptor help you take on this additional responsibility?

What techniques will you use to enhance your communication with all staff to assure that the care you oversee is safe? How will you negotiate your workload with your nurse manager?

## KEY CONCEPTS

- Nurses rely on basic principles of communication.

- Trends such as increasing diversity, an aging population, generational differences, illiteracy, and changing technology affect nursing practices.

- At the most basic level, communication involves a sender, a message, a receiver, and feedback. The input comes from visual, auditory, and kinesthetic stimuli.

- There are three levels of communication: intrapersonal, interpersonal, and public.

- Nurses may participate in chain, Y, wheel, circle, and all-channel communication networks and/or communicate through the grapevine.

- Nurses use linguistically appropriate services.

- Barriers to effective communication exist.

- Communication happens with patients, families, supervisors, coworkers, physicians, nurses, other health care workers, and mentors.

- Nurses must work to eliminate horizontal violence from the workplace.

- A clear medication reconciliation process helps to avoid errors and maintain patient safety.

## KEY TERMS

communication
grapevine

interpersonal communication
intrapersonal communication

HIPAA Privacy Rule
SBARR Technique

## REVIEW QUESTIONS

1. An RN asks the nursing assistive personnel (NAP) to take a set of vital signs on a patient who has just had an arterial venous shunt placement. The RN reminds the NAP not to take the blood pressure (BP) on the operative side. An hour later, the RN finds the deflated blood pressure cuff on the operative arm of the patient. The NAP has done this before and has been counseled about it. What should the RN do first?
   A. Assess the patient's condition.
   B. Avoid asking the NAP to take BPs in the future.
   C. Discuss the situation with the NAP and the supervisor.
   D. Find the NAP and review the importance of taking the blood pressure on the nonoperative side.

2. A new graduate RN has received an unfamiliar treatment order from the physician. How should the nurse proceed?
   A. Refuse to do the treatment.
   B. Do the treatment to the best of the nurse's ability.
   C. Inform the physician and then proceed to do the treatment.
   D. Inform the physician and ask the physician or charge nurse for assistance in doing the procedure.

3. What part of the communication process returns input to the sender?
   A. Feedback
   B. Message
   C. Receiver
   D. Sender

4. Which of the following characteristics is most relevant to to verbal communication?
   A. Eye contact
   B. Nodding
   C. Smiling
   D. Tone of voice

5. All but which of the following are steps to improve communication using the SBARR technique?
   A. Share the situation.
   B. Provide background information.
   C. Assure patient safety.
   D. Ask for a recommendation from the practitioner.

6. Which of the following skills involves active listening and is a very important skill used by nurses to gain an understanding of the patient's message?
   A. Attending
   B. Clarifying
   C. Confronting
   D. Responding

7. The most prevalent language spoken in the world is which of the following?
   A. Arabic
   B. English
   C. Hindu
   D. Mandarin Chinese

8. All but which of the following is a personal health identifier?
   A. Age (under 89) in years
   B. Certificate/license number
   C. Fax number
   D. Name

9. When administering medications, which items completed by the nurse during the medication administration process would increase the nurse's risk of making a medication error? Select all that apply.
   _____ A. Troubleshooting an IV infusion pump alarm
   _____ B. Answering questions from nursing colleagues
   _____ C. Calling pharmacy to check on a missing medication
   _____ D. Checking the patient's armband to verify patient identity
   _____ E. Obtaining a handoff report on a patient returning from a diagnostic test

10. Nurses can proactively create a safe and happy workplace by using which techniques? Select all that apply.
   _____ A. Engage in self-care activities.
   _____ B. Exclude an individual from discussion.
   _____ C. Name the problem, labeling horizontal violence, as appropriate.
   _____ D. Use reflection to raise self-awareness.
   _____ E. Withhold needed information.

## REVIEW ACTIVITIES

1. Your nursing care manager has asked you to serve on a committeee to explore how your unit might communicate more effectively. What elements of communication might affect the group's plan?

2. The charge nurse apologizes as she informs you that your assignment includes the "problem patient" on the unit. What communication skills will you use to enhance communication with this patient? How will you avoid barriers to communication?

3. When you report for your evening shift one month after completing orientation, you discover that you are assigned to be the team leader. What communication skills will you use to communicate with coworkers?

## EXPLORING THE WEB

- What sites would you consider to improve your communication skills? Google identifies 351,000,000 sites for communication for nurses. Explore this at www.google.com, and look at the incredible possibilities online, recognizing that the quality of this online information varies.

- Would you like to keep up with what is happening in the field of nursing and informatics? Visit www.ania.org. What did you find?

- Search this site to order articles from over 2,900 journals indexed in the CINAHL database: www.cinahl.com

- Note this site for links to various cultural sites to improve your communication awareness: www.diversityrx.org

## REFERENCES

Anderson, R. (2007). Finding HIPAA in your soup: Decoding the Privacy Rule. *American Journal of Nursing, 107*(2), 66–71.

Arford, P. H. (2005). Nurse-physician communication: An organizational accountability. *Nursing Economics, 23*(2), 72–77.

Baggs, J. G., Ryan, S. A., Phelps, C. E., Richeson, J. F., & Johnson, J. E. (1992). The association between interdisciplinary collaboration and patient outcomes in a medical intensive care unit. *Heart and Lung, 21*(1), 18–24.

Bass, P.F. Wilson, J.F., & Griffith, C.H. (2003). A Shortened Instrument for Literacy Screening. Journal of General Internal Medicine. December; 18(12): 1036–1038, accessed March 4, 2011 from, http://www.ncbi.nlm.nih.gov/pmc/articles/PMC1494969/.

Biron, A. D., Loiselle, C. G., & Lavoie-Tremblay, M. (2009). Work interruptions and their contribution to medication administration errors: An evidence review. *Worldviews on Evidence-Based Nursing*, (2nd Qtr., 2009), 70–86.

Calfee, B. E. (1998). Making calls to the physician. *Nursing 1998* Oct; 28(10):17.

Cleary, M., & Freeman, A. (2005). Email etiquette guidelines for mental health nurses. *International Journal of Mental Health Nursing, 14*, 62–65.

Cornett, S. (2009). Assessing and addressing health literacy. *Journal of Issues in Nursing, 14*(3), 1–17.

Davis, T. C., Crouch, M. A., Long, S. W., Jackson, R. H., Bates, P., George, R. B., & Bairnsfather, L. E. (1991). Rapid assessment of literacy levels of adult primary care patients. *Family Medicine, 23*(6), 433–435.

DuBrin, A. J. (2000). The active manager. United Kingdom: South-Western Pub.

Duffy, E. (1995). Horizontal violence: a conundrum for nursing. Collegian: *Journal of Royal College of Nursing Australia, 2*(2), 5–9, 12–17.

Dutton, J., & Heaphy, E. (2003). The power of high quality connections. In K. Cameron, J. Dutton, & R. Quinn, (Eds.), *Positive organizational scholarship: Foundations of new discipline* (pp. 263–279). San Francisco: Berrett-Koehler.

Ellis, A. (1997). *Anger: How to live with it and without it.* New York: Citadel Press, Kensington.

Gittell, J. H., Fairfield, K. M., Bierbaum, G., Head, W., Jackson, R., Kelly, M., et al. (2000). Impact of relational coordination of quality of care, postoperative pain and functioning, and length of stay: A nine-hospital study of surgical patients. *Medical Care, 38*(8), 807–819.

Gray, J. (1992). *Men are from Mars, women are from Venus.* New York: HarperCollins.

Greene, J. (2005). What nurses want. *Hospitals and Health Networks, 79*(3), 34–42.

Haig, K. M., Sutton, S., & Whittington, J. (2006). The SBARR technique: Improves communication, enhances patient safety. *Joint Commission's Perspectives on Patient Safety, 5*(2), 1–2.

Hall, D. (2010). Med reconciliation: Do the right thing. *Nursing Management*, (February, 2010), 32–36.

Hawaii Nurses Association. (2008). Horizontal Violence or Beyond the Bully ... Accessed March 4, 2011 from, http://www.hawaiinurses.org/news-a-publications/hna-news-archive/44-general-news/187-horizontal-violence-or-beyond-the-bully.html.

Healthy People 2020. Health Communication and Health Information Technology. Accessed March 4, 2011, at http://www.healthypeople.gov/2020/topicsobjectives2020/overview.aspx?topicid=18.

Ihlenfeld, J. T. (2005). Hiring and mentoring graduate nurses in the intensive care unit. *Dimensions of Critical Care Nursing, 24*(4), 175–187.

Jha, A. K., Doolen, D., Grandt, D., Scott, T., & Bates, D. W. (2008). The use of health information technology in seven nations. *International Journal of Medical Informatics*, (July 24), 254–262.

Knaus, W. A., Draper, E. A., Wagner, D. P., & Zimmerman, J. E. (1986). An evaluation of outcome from intensive care in major medical centers. *Annals of Internal Medicine, 104*(3), 410–418.

Kutner, M., Greenberg, E., Yin, Y., & Paulsen, C. (2006). The health literacy of American adults: Results from the 2003 National Assessment of Adult Literacy (NCEA2006-483). Washington, DC; U.S. Department of Education, National Center for Education Statistics.

Longest, B. B., & Darr, K. (2008). *Managing health services organizations and systems* (5th ed.). Baltimore, MD: Health Professions Press.

McCauley, K., & Irwin, R. (2006). Changing the work environment in ICUs to achieve patient-focused care: The time has come. *CHEST, 130*(5), 1571–1578.

Microsoft. (1996). *Microsoft Encarta 97 Encyclopedia.* Microsoft Corporation.

Munoz, C., & Luckmann, J. (2004). *Transcultural Communication in Nursing* (2nd ed.). Clifton Park, NY: Delmar Learning.

Murphy-Knoll, L. (2007). Low health literacy puts patients at risk: The Joint Commission proposes solutions to national problem. *Journal of Nursing Care Quality, 22*(3), 2-5-209.

Parnes, B., Ferrell, D., Quintela, J., Araya-Guerra, R., Westfall, J., Harris, D. et al., (2007). Stopping the error cascade: A report on ameliorators from the ASIPS collaborative. *Quality and Safety in Health Care, 16*, 12–16.

Rudd, R., Moeykens, B. Colton, TC. (1999). Health and literacy: A review of medical and public health literature. In J. Comings, B. Garners, & C. Smith, eds. Annual Review of Adult Learning and Literacy, Volume I. New York, NY: Jossey-Bass.

Safeer, R. S., & Keenan, J. (2005). Health literacy: The gap between physicians and patients. *American Family Physician, 72*(3), 463–468.

Sanders, J. & Ewart, B. (2005). Developing cultural competency for healthcare professionals through work based learning. *Work Based Learning in Primary Care, 3*, 99–105.

Schyve, P. M. (2007). Language differences as a barrier to quality and safety in health care: The Joint Commission perspective. *Journal of General Internal Medicine*, November 22 (S2), 360–363.

Shortell, S. M., & Kaluzny, A, D. (2006). *Health Care Management: Organization design and behavior.* Clifton Park, NY: Delmar Cengage Learning.

Shortell, S. M., Zimmerman, J. E., Rousseau, D. M., Gillies, R. R., Wagner, D. P., Draper, E. A., et al. (1994). The performance of intensive care units: Does good management make a difference? *Medical Care, 32*(5), 508–525.

U.S. Department of Education Institute of Education Sciences National Center for Health Statistics. (2006). The Health Literacy of America's Adults: Results from the 2003 National Assessment of Adult Literacy. Accessed March 4, 2011 from, http://nces.ed.gov/pubsearch/pubsinfo.asp?pubid=2006483.

U.S. Department of Health and Human Services. (2000). *Healthy People 2010 Vol.1: Understanding and improving health and objective for improving health* (2nd ed., pp. 11–21). Washington, DC: U.S. Government Printing Office.

WhoYOUgle. (2009–2011). World's most popular languages. Accessed at http://whoyougle.com/texts/most-popular-languages. March 4, 2011.

World Health Organization (WHO). (1998). Division of Health Promotion, Education and Communication. Health Education and Health Promotion Unit. *Health promotion glossary.* Geneva: World Health Organization.

Young, G. J., Charns, M. P., Daley, J., Forbes, M. G., Henderson, W., & Khuri, S. F. (1997). Best practices for managing surgical services: The role of coordination. *Health Care Management Review, 22*(4), 72–81.

Young, G. J., Charns, M. P., Desai, K., Khuri, S. F., Forbes, M. G., Henderson, W., et al. (1998). Patterns of coordination and surgical outcomes: A study of surgical services. *Health Services Research, 33*(5).

## SUGGESTED READINGS

American Journal of Nursing. (2009). HIPAA: Not so bad after all? *American Journal of Nursing, 109*(7), 22–24.

Biron, A. D., Lavoie-Tremblay, M., & Loiselle, C. G. (2009). Characteristics of work interruptions during medication administration. *Journal of Nursing Scholarship, 41*(4), 330–336.

Calcagno, K. M. (2008). Listen up. Someone important is talking. *Home Healthcare Nurse, 26*(6), 333–336.

Cooper, J. R. M., Walker, J. T., Winters, K., Williams, P. R., Askew, R., Robinson, J. C. (2009). Nursing students' perceptions of bullying behaviours by classmates. *Issues In Educational Research, 19*(3), 212–226. http://www.iier.org.au/iier19/cooper.html.

Green, S. D., & Thomas, J. D. (2008). Interdisciplinary collaboration and the electronic medical record. *Pediatric Nursing, 34*(3), 225–227, 240.

Joint Commission. (2005). Focus on five strategies to improve hand-off communication. *Joint Commission on Perspectives on Patient Safety, 5*(7), 11.

Kalisch, B. J., & Aebersold, M. (2010). Interruptions and multitasking in nursing care. *The Joint Commission Journal on Quality and Patient Safety, 36*(3), 126–132.

Kleinpell, R., Thompson, D., Kelson, L., & Pronovost, P. (2006). Targeting errors in the ICU: Use of a national database. *Critical Care Nursing Clinics of North America, 18*, 509–514.

Mannahan, C. A. (2010). Different worlds: A cultural perspective on nurse-physician communication. *Nursing Clinics North America, 45*, 71–79.

Manojlovich, M., & DeCicco, B. (2007). Healthy work environments, nurse-physician communication, and patients' outcomes. *American Journal of Critical Care, 16*(6), 536–543.

Robles, J. (2009). The effect of the electronic medical record on nurses' work. *Creative Nursing, 15*, 31–35.

Speros, C. (2005). Health literacy: Concept analysis. *Journal of Nursing Administration, 50*(6), 633–640.

Trbovich, P., Prakash, V., & Stewart, J. (2010). Interruptions during the delivery of high-risk medications. *Journal of Nursing Administration, 40*(5), 211–218.

Weydt, A. (2010). Mary's story, relationship-based care delivery. *Nursing Administration Quarterly, 34*(2), 141–146.

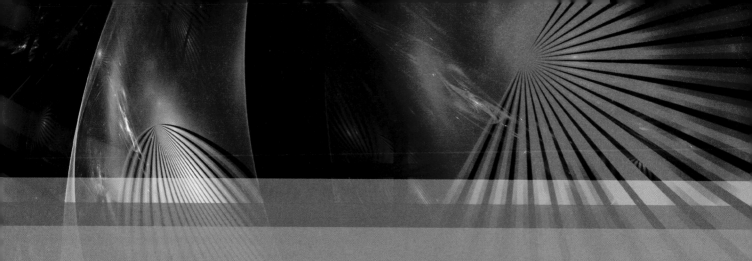

# CHAPTER 9

# Politics and Consumer Partnerships

Terry W. Miller, PhD, RN; Patsy Maloney, EdD, MSN, MA, RN-BC, NEA-BC; Richard J. Maloney, EdD, MA, MAHRM, BS

*Baby boomers will demand higher standards, more choice, and respect for their autonomy and dignity than any previous generation.*

(Becker, 2009)

## OBJECTIVES

Upon completion of this chapter, the reader should be able to:

1. Relate how politics defines health care services and affects nursing practice.
2. Explain the need for nurses to be politically involved with the consumer movement in health care.
3. Identify the role of a nurse as a consumer advocate and political force.
4. Plan a political strategy for strengthening nurse-consumer relationships.
5. Devise a service-oriented plan for providing nursing services to a selected consumer interest group.
6. Analyze the impact of demographic changes on nurses and nursing practice.

Delmar/Cengage Learning

*A 60-year-old man lying in a hospital bed, pushing a call button and waiting for a nurse, is not thinking about insurance, Medicare, Medicaid, or any other payer for the service. All he knows is that he is not getting the care that he expected, and what little care he is receiving is costing far more than he or his family can afford. He is imagining losing his home and his meager savings. Because he had lost his job and his health insurance, he ignored his headaches and dulled thinking as long as he could, hoping that his very treatable symptoms would resolve. He probably does not know that he is one of more than 45 million uninsured Americans (U.S. Census Bureau, 2009b).*

*How does a new graduate approach the care of adults such as these?*

*How do politics play a role in health care?*

*Should a new graduate work with a consumer group to improve the care of this man and others like him? If you answered yes, what consumer group or groups would you choose to work with?*

**P**olitics is predominantly a process by which people use a variety of methods to achieve their goals. These methods inherently involve some level of competition, negotiation, and collaboration for the power to achieve desired outcomes, as well as to protect and enhance the interests of groups or individuals. Nurses who can effectively compete, negotiate, and collaborate with others to get what they want or need develop strong political skills. They have the greatest ability to build strong bases of support for themselves, patients, and the nursing profession. Nurses consistently show up as rated number one in consumer opinion polls asking who are considered to be trusted professionals. Nurses can garner consumer support for professional nursing positions to help patients and help the profession of nursing by tapping into this strong support. Nursing is important as a profession only as it meets its societal mandate for professional nursing service. Nurses must gather political support to do this most effectively.

Politics exist because resources are limited, and some people control more resources than others. **Resources** include people, money, facilities, technology, and rights to properties, services, and technologies. Individuals, groups, or organizations that have the ability to provide or control the distribution of desirable resources are politically empowered. The consumer movement in health care is a political movement about health care resources. It reflects consumer perceptions and values and influences patient care delivery.

The purpose of this chapter is to support the need for nurses as the largest health care group to be politically active for the good of patients and the health care system. A major focus is how the consumer movement in health care creates new opportunities for nurses to advance nursing services by giving patients, including all people who receive health care, a stronger voice in their health care as consumers. Nurses are encouraged to develop strong political skills and partner with their professional nursing organization and with consumer groups to take the lead in improving health care.

## HEALTH CARE REFORM IN THE UNITED STATES IN 2010

After much political activity, the Patient Protection and Affordable Care Act became law on March 23, 2010. This Act provided for the phased introduction over four years of a comprehensive system of mandated health insurance, with reforms designed to eliminate some of the worst practices of the insurance companies: precondition screening and premium loading, policy rescinds on technicalities when illness seems imminent, and lifetime and annual coverage caps. The Act also sets a minimum ratio of direct health care spending to premium income, creates price competition bolstered by the creation of three standard insurance coverage levels to enable like-for-like comparisons by consumers, and provides a Web-based health insurance exchange where consumers can compare prices and purchase plans. The system preserves private insurance and private health care providers and provides more subsidies to enable the poor to buy insurance (Lochhead, 2010). The Act was shortly thereafter amended by the Health Care and Education Reconciliation Act of 2010 (H.R. 4872), which became law on March 30, 2010. Previous health care reform achievements at the national level include the following:

1965: Medicare, which covers both hospital and general medical insurance for senior citizens, paid for by a federal employment tax over the working life of a retiree; and Medicaid, which permits the federal government to partially fund a health care program for the poor, with the program managed and cofinanced by individual states.

1985: The Consolidated Omnibus Budget Reconciliation Act of 1985 (COBRA), which amended the Employee

Retirement Income Security Act of 1974 (ERISA) to give some employees the ability to continue health insurance coverage after leaving employment.

1997: The State Children's Health Insurance Program (SCHIP), which was established by the federal government in 1997 to provide health insurance to children in families at or below 200% of the federal poverty line.

The Patient Protection and Affordable Care Act mandates that by 2014 people must have adequate insurance coverage or else pay a fine. It seeks to improve the health of the nation by increasing insurance coverage. A study published in the *American Journal of Public Health* found that more than 44,800 deaths annually in the United States are associated with lack of insurance (Wilper et al. 2009). It was estimated that the number of people in the United States—insured and uninsured—who die per year because of lack of medical care was nearly 100,000 (Woolhandler et al., 1997). Health care expenditures are highly concentrated, with the most expensive 5% of the population accounting for half of aggregate health care spending and the bottom 50% of health care spenders accounting for only 3% of the population, which means that insurers' gains to be had from avoiding serving the sick greatly outweigh any possible gains from managing their care. As a consequence, insurers' resources have been devoted to such avoidance, at a direct cost to effective care management, which is against the interests of the insured (Blumberg & Holahan, 2009). According to a study from Cambridge Hospital, Harvard Law School, and Ohio University, 62% of all 2007 personal bankruptcies in the United States were due to an inability to pay medical costs (Himmelstein, 2009). Many of these people forced into bankruptcy had medical insurance, but the effect of insurance caps, exclusions, and inability to fund or continue COBRA coverage was behind the financial distress of many. Medical impoverishment is almost unheard of in wealthy countries other than the United States, either because the state covers everyone or everyone is obliged by law to have insurance. In the United States, many stakeholders play a role in the political process affecting health care delivery.

# STAKEHOLDERS AND HEALTH CARE

Control of health care resources is spread among a number of vested interest groups called **stakeholders**. Everyone is a stakeholder in health care at some level, but some people are far more politically active about their stake in health care than others. See Table 9-1 for a list

### TABLE 9-1 Washington's Top 20 Spenders on Lobbying for all years 1998–2010

| | |
|---|---|
| 1. U.S. Chamber of Commerce | 11. Exxon Mobil |
| 2. American Medical Association | 12. Verizon |
| 3. General Electric | 13. Edison Electric |
| 4. Pharmaceutical Research & Manufacturers of America | 14. Business Roundtable |
| 5. American Association of Retired Persons (AARP) | 15. Boeing |
| 6. American Hospital Association | 16. Lockheed Martin |
| 7. American Telephone and Telegraph (AT&T) | 17. Pacific Gas and Electric Corporation |
| 8. Northrop Grumman | 18. General Motors |
| 9. Blue Cross/Blue Shield | 19. Southern Company |
| 10. National Association of Realtors | 20. Pfizer |

**Note:** All lobbying expenditures on this page come from the Senate Office Records. Data for the most recent year was downloaded on July 26, 2010.

**Source:** From Center for Responsive Politics (2010). Retrieved from http://www.opensecrets.org/lobby/top.php?showYear=a&indexType=s.

## EVIDENCE FROM THE LITERATURE

**Citation:** McDonald, L. (2006). Florence Nightingale as a social reformer. *History Today,* *56*(1), 9–15.

**Discussion:** The purpose of this historical research on the work of Florence Nightingale is to illuminate her work as a social reformer of a public health system. She advocated health promotion and disease prevention based on evidence. Ms. Nightingale extended the use of nurses that she had trained in hospitals to the care of the poor in pauper houses. She not only placed nurses in these workhouses for the poor, but she lobbied the powerful for passage of laws for the poor. She persuaded key figures to support her ideas for reform. Due to her hard work, the Metropolitan Poor Bill was passed by Parliament in 1867 and was followed by other reforms that improved the lot of the poor and infirmed in Britain. Ms. Nightingale was able to obtain the support of such powerful and influential people due to her meticulous attention to detail and careful, methodical preparation. Florence Nightingale used what is now dubbed "Nightingale methodology." First, study the best information in print, especially government reports and statistics. Second, interview experts, and if the available information is inadequate, survey others with a questionnaire. When you have a proposed plan, test it at one institution, consult with the practitioners that implemented it, and send out the draft reports for comment before sending the final report out for publication and dissemination to the influential.

**Implications for Practice:** New nurses often believe that their responsibilities begin and end at the bedside. But historical research demonstrates that from the very beginning of modern nursing, nurses have not only given care but partnered with others to influence public opinion and to change legislation for the benefit of the consumer of health care, especially the vulnerable—the poor, the mentally ill, soldiers, and children. So, as a new nurse, view yourself as a patient advocate who is willing to join with others through professional associations or consumer groups to improve health care and the system within which it is delivered.

of Washington's most powerful lobbying groups. These lobbyists are stakeholder groups and include insurance companies; consumer groups; professional organizations such as the American Nurses Association; and health care groups such as nursing and medical practitioners, pharmacists, dieticians, physical therapists, administrators, and educational groups. Stakeholders exert political pressure on health policy makers—local, state, and federal legislative bodies—in an effort to make the health care system work to the economic advantage of the stakeholder.

Not all stakeholders in health care support the consumers' potentially dominant role in health care politics, for a variety of reasons. Some contend that consumers do not necessarily know what is best for them. Instead, they support the idea that health care experts, such as practitioners, are better able to direct health care policy. Others maintain that only those directly paying for the services should make policy decisions and that health care is not necessarily a right, because services should be based on ability to pay.

Becker notes that baby boomers "are the products of the higher education boom; they are assertive, media savvy, ecologically aware, increasingly health conscious, and well informed about life's choices" (2009, p. 108).

This is leading to increasing political activism by third-party payers. Third-party payers include the government, business, and health insurance companies. Exposure of Medicare/Medicaid fraud has led to the very nature and control of professional practice being questioned by government payers. Nurses have come to understand how the control and distribution of resources in health care can drastically affect their incomes, workloads, work environments, and patients. Nurses across the country have reported that the patient load per nurse provider has increased significantly. However, these nursing concerns without political influence do little to change health care at any level.

Nurses must work to strengthen the long tradition of pulling together the various stakeholders of health care, such as patients, practitioners, administrators, pharmacists, physical therapists, dieticians, and so on. These stakeholders are needed to coordinate health care services and ensure that patients obtain the health services they need (Middaugh & Robertson, 2005). Although unknown by many of today's practicing nurses, this tradition began with Florence Nightingale and her "Nightingale methodology" (McDonald, 2006). New nurses who recognize

## Critical Thinking 9-1

Consider these health care stakeholders:

- Medical practitioners
- Patients
- Pharmacists
- Physical therapists
- Nurse aides
- Hospital administrators
- Nurse practitioners
- Social workers
- Dieticians
- Insurance companies

How might each of these stakeholders see the importance of efficient, high-quality care to patients on a clinical unit? How might they define the need for patient access to services differently?

their critical role in addressing the major issues in health care delivery at the bedside will ensure that nursing enters into a partnership with agency executives who have control in the wider health care system of which the agency is a part. As partners, they will be able to compete, negotiate, and collaborate with other stakeholders at the system level to be politically effective. These nurses must also be concerned with the price of health care at the system level and understand that resources are controlled and distributed through health policy decisions.

Many authors of nursing articles, some books, and a few research studies support nurses' involvement in public policy and health care politics (Grindel, 2005; Hughes, Duke, Bamford, & Moss, 2006; Kelly, 2007; Mason, Leavitt, & Chaffee. 2006; Milstead, 2007). Several other authors promote greater inclusion of policy content and political process in nursing curricula (Hahn, 2010; Harrington, Crider, Benner, & Malone, 2005; Keepnews, 2005; Morris & Falk, 2007; Vesley, 2008). Nurses' involvement in policy arenas such as policy-making committees and institutional boards includes advocating for recipients of health care when those in need have little or no voice and advocating for those who need a stronger voice. Any professional nurse should understand and be able to articulate the relevance of politics to nursing practice. Making a difference in health care arenas is an outcome of involvement in policy making. As Margaret Mead said, "Never doubt that a small group of thoughtful committed citizens can change the world, indeed it's the only thing that ever has."

## REAL WORLD INTERVIEW

Nurses serve as advocates and allies for consumers by helping consumers obtain what they perceive they need. Nurses have the opportunity to help consumers better understand what is available to them, as well as what they can legitimately expect to get from both the provider and the system. Competent nurses need to work at understanding how the system works, because that is "where they live." Consumers move in and out of the system, so they are not acclimated to the limitations or pathways of the system. The consumer and the nurse become natural allies, whether in a patient care setting or a public policy setting. One of my biggest frustrations is when nurses fail to see themselves as connected to the patient and the whole health care system. Nurses become myopic in their approach to problems in direct patient care, because they do not see that they are a piece of something bigger. No nurse's practice occurs in isolation. We are all part of an interdependent, highly complex system with governing economic and political relationships. I found consumer partnerships to be most useful when working as a lobbyist for the Washington State Nurses' Association, because, as nurses, we were able to build politically powerful coalitions with consumer groups and subsequently define the direction of long-term health care policy—specifically, state policies governing the long-term care industry. Because we were successful in partnering with selected consumer interest groups, we were able to assure passage and funding of seven significant legislative bills. These bills included the AIDS Omnibus legislation, a long-term care reform act, and an act enabling nurses to declare a patient dead for the purpose of preventing unnecessary stress, care, and cost to consumers.

**Robert S. Ball, MSN, RN**
Nursing Care Manager
Tacoma, Washington

# THE POLITICS AND ECONOMICS OF HUMAN SERVICES

Many nurses want to avoid the political nature of their work, because they believe that human service should not be politically motivated. They may also ignore the business aspects of health care until they find themselves responsible for a budget. Yet all health care is inextricably linked to politics and economics as well as to the availability and services of providers. As a human service, health care has yielded remarkable returns in terms of improving overall quality of life as well as in extending life spans. As a business, health care has afforded millions of people, including nurses, with economic opportunities and lifelong careers.

Health care in the United States depends heavily on a continuous supply of resources from both public and private sectors. These resources include people, such as the providers of health care services, and the money to educate and pay these providers. Buildings, technology, administration, and equipment are just some of the other resources needed for the health care system to be serviceable. With health care requiring so many resources on such a large economic scale, thousands of people are directly involved in the allocation of those resources—hence, the politics of health care. Most of those people have good intentions, but they often disagree about how the resources can best be used to support health care.

Many consumers are aware that the ongoing redistribution of health care resources may not meet their health care needs, especially as they age and become more dependent upon related services. They are frightened by media reports of increasing national health care expenditures. Although most consumers do not directly pay for the majority of their health care, their individual portions of expenses incurred are rapidly increasing.

Health policy is formulated, enacted, and enforced through political processes at the local, state, or federal levels. For example, at the local level, policies are established and implemented by an individual hospital board or by directors of a total health care system regarding whether or when flu injections will be available to high-risk populations being served by that institution. At the state level, policies govern nurses within a state by defining nursing practice, nursing education, and nursing licensure. These policies are often governed by a state nurse practice act that designates a state nursing commission or health professions board as the authority for enforcing the policies. Federal policies are evident in the rules and regulations governing Medicare and Medicaid funding.

When bills affecting health care are being developed in state and federal legislative bodies, it is impor-

## CASE STUDY 9-1

Maria is a maternity support nurse for First Steps, a specific state-funded program designed to provide care to underserved and underinsured pregnant women. Maria is in her third year of professional practice and has become highly resourceful as well as able to work in new situations with minimal supervision. Many of her case referrals come through a partnership with the local hospital's Teen Parent Resource Center, which targets girls under 20 who have dropped out of school during their pregnancy. Many of Maria's patients are from families at high risk for domestic violence and substance abuse. Recently, she was informed that the Teen Parent Resource Center will be discontinuing its partnership with First Steps because of funding issues.

**What does Maria need to know to continue to serve her high-risk population?**

**What are her options?**

**How could you use the Nightingale methodology (see McDonald Evidence from the Literature earlier in this chapter) to study the best information in print, especially government reports and statistics; interview experts and survey others with a questionnaire if the information is inadequate; propose and test a plan; consult with the practitioners who proposed it; and send out draft reports for comment before sending out the final report for publication and dissemination to the influential?**

tant that nurses be aware of those actions and obtain copies of those bills, thus adding to their political knowledge base.

Ultimately, health care will be defined and controlled by those wielding the most political influence. If nurses fail to exert political pressure on the health policy makers, they will lose ground to others who are more politically active. It is unrealistic to believe that other stakeholders will take care of nursing while the competition for health care resources increases. Historically, some stakeholders in health care have never supported nursing as a profession or acknowledged professional roles for nurses. Nurses, like other health care providers, must compete, negotiate, and collaborate with others to ensure their future in health care.

# CULTURAL DIMENSIONS OF PARTNERSHIPS AND CONSUMERISM

By 2005, the U.S. minority population was one-third of the total population, that is, 98 million people out of the nation's 296.4 million people. These numbers identify the increasing racial and ethnic diversity of the U.S. population. The largest minority group continues to be Hispanics at 42.7 million. The Hispanic population increased 3.3% in one year, from July 2004 to July 2005, to clearly become the fastest-growing minority group. Black/African Americans are the second largest minority group with 39.7 million people, followed by Asians at 14.4 million, 4.5 million Alaska Natives and American Indians, and almost 1 million native Hawaiian and other Pacific Islanders. Non-Hispanic Whites totaled 198.4 million (Bernstein, 2006). Non- Hispanic Whites have decreased from 70% of the population in 2000 to 66.9% in 2005. During the same time, the minority population increased from 30% of the total population to 33.1%.

If nurses intend to form partnerships with consumer groups distinguished by cultural heritage, racial makeup, or ethnic background, they must understand and value diversity. Strong partnerships will frame nursing services in ways that respect cultural differences.

**EVIDENCE** FROM THE LITERATURE

**Citation:** Vesely, R. (2008). 'Unleash the energy.' Activists, professors try to use their own sphere of influence to affect U.S. health-care policy and improve patient care. *Modern Healthcare, 38*(7), 6–7, 16.

**Discussion:** Six professors who are health care policy experts were interviewed. They shared their passion for policy and the belief that they must use their own personal spheres of influence to impact health care policy and improve health care for all. Kathy Dracup, dean and professor of nursing at University of California, San Francisco was the only nurse out of the six professors. Dr. Dracup emphasizes to her students the importance of nurses serving as patient advocates and the need for nurses to speak out in support of improved access to health care.

**Implications for Practice:** As new nurses and other health care providers graduate, they face a complex health care system that needs to change to meet the needs of the ever-growing population of older adults, the increasing diversity of patients, an epidemic of patients without health care, and technology that can preserve life but not necessarily quality of life. New nurses will need to join with experienced nurses and other health care professionals to answer the call to both individually and collectively influence policy to improve health care to meet the needs of the changing population.

# POLITICS AND DEMOGRAPHIC CHANGES

Certainly not all consumers agree about what health care should be, who should provide it, or how it should be paid for. The social, cultural, economic, psychological, and demographic characteristics of consumers largely determine their attitudes and inclination toward the health care system, its providers, and its services. Consumers also recognize some level of personal risk when changes are made in the system, especially involving payment for services and providers. If consumers such as retired persons perceive that their out-of-pocket cost for health care extends beyond their capacity to pay or will increase in the future, they are highly motivated to exert political pressure on their legislators to reverse the perceived trend.

The fastest-growing consumer group now and for years to come is the elderly—persons 65 and older—(U.S. Census Bureau, 2009a). An estimated 76 million baby boomers turned 65 by 2011. The number of elderly people (85 years of age and older) is growing at an explosive rate and is expected to reach 27 million by 2050 (U.S. Census Bureau, 2009a). People over age 50 control approximately one-half of the country's disposable income as well as three-quarters of its financial assets and 80% of its savings. Yet most elderly Americans are not wealthy. Twenty percent of the elderly are projected to qualify for Medicare benefits by the year 2020 (Shi & Singh, 2010).

Without doubt, this aging of the U.S. population will profoundly affect health care at every level. The dramatic increase in the number of older adults in the United States means that 40% of employees nationwide were eligible for retirement in 2010 (King, 2008). Studies of voting behavior of U.S. citizens show that the elderly have no predictable political orientation on anything except obvious threats to perceived entitlements, the most widely recognized entitlement being Social Security benefits.

Many seniors are joining consumer groups to have a greater political voice, influence health policy decisions, and ensure that they receive the health care services they will need for years to come. A growing number are bridging the

## Critical Thinking 9-2

Demographic changes present multiple challenges to health care. Of the additional 185,000 nurses that joined the nursing workforce in 2002 and 2003, 70% were age 50 and over (Norman et al., 2005). By 2005, the nursing shortage had entered its eighth year, "making it the longest lasting nursing shortage in half a century" (Buerhaus, Donelan, Ulrich, Norman, & Dittus, 2005, p. 62). It is predicted that by 2016 the United States will need more than one million new and replacement nurses (Ferlise & Baggot, 2009). Sister Rosemary Donley (2005) believes that it is time to think outside the box for new solutions to the nursing shortage. Doing business as we have in the past will not work.

How do you anticipate that the aging workforce, coupled with the aging population, will affect your nursing practice?

What are the political implications?

gap of social isolation, prominent in the past, through the Internet as well as through involvement in consumer groups. They are establishing closer contact with the outside world and are managing to successfully strengthen their relationships with other stakeholders in health care.

The AARP (formerly known as the American Association of Retired Persons), with more than 15 million members, constitutes a growing political powerhouse and an ideal consumer partner for nursing in many ways. A large percentage of nurses are 50 years of age or older and qualify for membership in the AARP. Few other consumer groups appear to have the potential that the AARP has for defining the health care system of the future.

## NURSE AS POLITICAL ACTIVIST

Nurses are the largest health care group, and nurses who are politically active have a definitive voice in their work environments for patient welfare as well as for themselves. Nurses must set their political goals as individual nurses, nurse citizens, nurse activists, and nurse politicians for the future (Table 9-2). Nurses must study the issues, garner political support, and contact policy makers such as the chairperson of the hospital board or legislators

through phone calls, letters, and e-mail messages. Nurses join professional organizations and actively participate to ensure a more collective, unified voice supporting health care issues and policies that have value for consumers and nursing. Nurses who are most involved will be seen supporting political activities and candidates, assisting during campaigns, helping to draft legislation, and running for political office.

As nurses develop politically, they come to understand the need for political strategy. The purpose of developing political strategies is to understand different ways to achieve your goals, or the goals you are advocating for, while identifying the other stakeholders and their goals. Political strategy attempts to persuade those people supporting an issue, formulating a policy, or taking an action to take the position in support of those using the political strategy. To be feasible, a political strategy requires commitment by those using it, as well as their awareness of the other stakeholders. Effective political strategy implies considerable forethought and clarity of purpose in even the most ambiguous situations. Nurses who are most likely to wield political influence operate with strategy in mind before taking political action, voicing concerns, making demands, or even advocating for others. It is important to study the political issues and the major stakeholders' positions regarding the issue prior to becoming involved and to seek opportunities for collaboration.

Every nurse should be cognizant of what other involved groups think regarding any relevant political health issue. It is critical that nurses listen to other policy perspectives and understand as many facets of the issue as possible when making health policy proposals. Proposals need to include a rationale to neutralize opposing views. This ensures that unnecessary political fights can be avoided and that more collaboration will occur prior to any policy proposal being made to policy-making bodies such as hospital boards or state legislatures. The more support obtained from the various stakeholders in any policy arena, the better chance of a workable policy being developed and implemented.

To be most politically effective, nurses must be able to clearly articulate several dimensions of nursing to any audience or stakeholder: what nursing is; why nursing is important to society; what distinctive services nurses provide to consumers; how nursing benefits consumers, including the prevention of nurse-sensitive outcomes such as pneumonia, cardiac arrest, and so on; and what nursing services cost in relation to other health care services. Although anecdotal stories and emotional appeals may be effective with certain audiences, it is far more powerful to present research-based evidence to support the political position of the nursing profession. Table 9-3 details essential dimensions of nursing.

## TABLE 9-2  Political Roles for Nurses

| ROLE | ACTIVITIES | EXAMPLE |
|---|---|---|
| Nurse individual | • Highlights important role of nurse to prevent nursing-sensitive outcomes, for example, pneumonia, cardiac arrest, and so on.<br><br>• Sets goals to strengthen nursing as a profession (Chapter 31).<br><br>• Highlights the essential dimensions of nursing (Table 9-3).<br><br>• Participates as a member in health care consumer groups, for example, the AARP, and so on. | **You**—as a new graduate have a political role as a nurse individual. |
| Nurse citizen | • Votes and writes members of Congress and state legislators on issues of interest.<br><br>• Educates patients on how to evaluate Website sources of health care information.<br><br>• Works as a Nurse activist | **You**—as a new graduate have a political role as a nurse citizen. Be informed, vote, and participate. |
| Nurse activist | • Participates as a member of a professional organization that lobbies and influences state and federal legislation.<br><br>• Notifies hospital Board of Trustees of any quality issues. | **Mary Wakefield, RN, PhD**—Dr. Wakefield was an advocate for rural health care. Now she has been appointed as Administrator for Health Resources and Services Administration (HRSA) by President Barack Obama on February 20, 2009 (http://www.hrsa.gov/about/organization/biowakefield.html, accessed August 18, 2010). |
| Nurse politician | • Runs for a political office and serves society as a whole.<br><br>• Collaborates with other health care professionals to improve care at the local/state | **Lois Capps, RN, MA**—Congresswoman Capps was first sworn into Congress in 1998. She represents the 23rd Congressional District and has spent her life working to improve access to quality healthcare. Her experiences as a registered nurse have influenced her as she plays a key role in many of the major health policy debates in Congress (http://capps.house.gov/, accessed August 18, 2010). |

## POLITICS AND ADVOCACY

The concept of patient advocacy has been a fundamental aspect of nursing since nursing's beginning. The role of a nurse as a patient advocate has changed as nursing has evolved. Originally, nurses acted as an intermediary and pleader for the patient, and now they act to ensure that the patient's rights to self-determination and free choice are not violated (Rudolph, 2005). Advocacy can be seen as representing the patient to others in the health care organization, which has extended into what has been referred to as cultural brokering or interpreting the health care environment for the patient. There is a strong argument that advocacy actually stems from patient power and consumerism rather than a lack of power or a vulnerability, as some authors have implied (O'Connor & Kelly, 2005). If a nurse is to act as a patient advocate that interprets the health care environment to the patient and actually guides the patient through the maze, the nurse needs knowledge of the system. Sometimes a nurse's advocating for a patient's rights may be contrary to the wishes of another health care provider or to the organization as a whole. This can be risky for a nurse. Nurses from the beginning of the

**TABLE 9-3 Essential Dimensions of Nursing**

- Providing a caring relationship that enhances healing and health

- Focusing on the full range of experiences and human responses to illness and health within both the physical and social environments

- Appreciating the subjective experience and the integration of such experience with objective data

- Diagnosing and intervening in care by using scientific knowledge, judgment, and critical thinking

- Advancing nursing knowledge through scholarly inquiry

- Influencing social and public policy to promote social justice

- Assuring safe, quality, and evidence-based care.

**Source:** From American Nurses Association. (2010). *Nursing's social policy statement* (3rd ed.). Washington, DC: American Nurses Publishing.

profession have served as patient advocates. They have acquired the necessary knowledge, and they have taken on the risk. Some early nurses were even jailed for taking unpopular stands. Entry-level nurses need knowledge of the health care organization and the courage of their convictions to act as patient advocates.

Interestingly, some patients have begun advocating for nurses, a role reversal. Those patients have perceived nurses as overextended by the nature of the work environment and have contended that something must be done to improve the working conditions of nursing. Nurses in California used consumer support to build their case for counteracting significant staffing cuts affecting nursing practice in their state. With strong consumer support, California nurses developed a legislative proposal in 1999 that has been enacted into state law. This law mandates a minimal staffing level of registered nurses for patient care. The intent is to protect patients from unqualified or dangerous staffing levels while receiving nursing services.

## ADVOCACY AND CONSUMER PARTNERSHIPS

Nurses must understand the political forces that define their relationships with consumers. Consumers expect the best people to be health care providers, but they are confused about what the roles and responsibilities of professional nurses are. Informed consumers understand how health policy directly affects them but are less likely to recognize how health policy affects nurses. Consumers may

**Critical Thinking 9-3**

Political conflict occurs because people may hold significantly different or conflicting opinions about any given topic. Consumers may disagree about what health care should be, who should provide it, how much it should cost, and/or who should pay for it. As a new nurse, you have a professional responsibility to promote consumer dialogue and offer creative, thoughtful, and evidence-based solutions to health care problems. Yet nurses disagree with each other in regard to the same issues consumers may have about health care.

What do you think your responsibility is as a consumer of health care?

What is your responsibility as a health care provider?

Do you think your responsibility as a consumer could conflict with your responsibility as a provider?

expect nurses to be their advocates only in the context of providing direct patient care.

Working through their professional organizations, nurses can collaborate with consumer groups by creating formal partnerships, which serve to promote the role of nurses as consumer advocates in health policy arenas and strengthen the political position of both partners. These partnerships have a stronger political voice than either group has alone. The partners gain power when interacting with any policy-making body, because they represent (1) a larger **voting block** (a group that represents the same political position or perspective); (2) a broader funding base (a source of financial support); and (3) a stronger **political voice** (an increase in the number of voices supporting or opposing an issue) to any policy-making body. Increasingly, professional organizations in nursing recognize the value of partnering with consumers to build a better health care system. See Table 9-4 for steps in establishing a partnership with a consumer group.

## TABLE 9-4  Steps in Establishing a Partnership with a Consumer Group

| STEP | DESCRIPTION |
|---|---|
| 1. Listen | Become sensitized to the health care needs and political nature of the potential consumer partner. |
| 2. Study | Seek both representative and opposing perspectives from consumer group meetings, focus groups, relevant literature, and interviews. |
| 3. Assess | Determine the need, value, context, and boundaries for establishing the partnership. |
| 4. Focus | Mutually identify the purpose, and articulate the goals and specific, realistic objectives for the partnership. |
| 5. Compromise | Work through nonessential and noncritical points and issues. |
| 6. Negotiate | Agree on your position and responsibilities in the partnership. |
| 7. Plan | Develop a political strategy for achieving the goals and fulfilling the objectives. |
| 8. Test | Test the political waters. Gather feedback on the plan from key people before taking action. |
| 9. Model | Model the political work. Define the structure for working the political strategy with partners. |
| 10. Direct the political action | Understand the bigger picture, and concentrate on what can be changed. |
| 11. Implement | Line up political support and take action. |
| 12. Network | Be committed to the mutually recognized goal, and consistently work to have an adequate base of support in terms of people, money, and time. |
| 13. Build political credibility | Participate in local, state, and national policy-making efforts that support the partnership and its political agenda. |
| 14. Soothe and bargain | Downplay rivalry, and address conflict in a timely, constructive manner. |
| 15. Report, publicize, and lobby | Report, publicize, and lobby the group's political cause. Draw public attention to the needs of the consumer group. |
| 16. Reaffirm, redefine, or discontinue | Regularly evaluate work with the consumer group. |

## Critical Thinking 9-4

The American public has become increasingly aware of and interested in health promotion (Hood & Leddy, 2006). The relationship between personal lifestyle and the incidence of several diseases has been demonstrated through the mainstream media with public education campaigns. Many health promotion programs include the expectation that people invest in themselves, but people who live in poverty lead precarious lives. How do you think people in the lowest socioeconomic class perceive their health care? Do you assume that most people will invest in themselves by living a lifestyle that promotes higher education, planned savings, healthy eating, regular exercise, deferred gratification, avoidance of smoking and excessive alcohol consumption, planned birth control, and regular physical checkups? Do you know people who seem to live only from one day to the next because their perspective of time is in the immediate and they do not seem to recognize the benefits of long-term planning?

## CONSUMER DEMANDS

As recipients of health care are required to pay a larger portion of the cost for health care services, consumers are demanding to be treated as something more than passive recipients of health care. They are very vocal in their requests to providers, payers, and agencies that they be more consumer friendly and service oriented, and they are seriously requesting a voice in how health care is regulated.

Nurses, working through professional organizations such as the American Nurses Association (ANA), have been strong, early supporters for patients' rights, regardless of the patient's ability to pay. Other professional groups, such as the American Medical Association (AMA) and the American Hospital Association (AHA), have received far more media recognition for their support of patients' rights. This is an indication that the AMA and AHA are better funded, wield more political power, and may do a better job than the ANA in presenting their positions on consumer issues to the media.

Any political vision to make health care more consumer friendly and service oriented must address cost, access, choice, and quality. Perhaps the vision starts with this formula: the highest quality of care for all people at the least cost. Yet defining, much less evaluating, quality is culturally bound and very complicated. Many people cannot afford minimal health care services. Other people are increasingly unwilling to subsidize the care of other patients

by paying increased costs for their health care. A vision for high-quality, low-cost care will require multiple stakeholders collaborating with one another to develop a workable philosophy, to include a mechanism for checks and balances to minimize abuse and misuse and encourage intelligent, ethical decisions by those wielding the most political power.

## TURNING A CONSUMER-ORIENTED VISION INTO REALITY

Nurses have opportunities to be more than supporters of a consumer-oriented vision for health care; they can be co-creators. To make this vision real, nurses will need to be more educated and articulate about what value they add to the overall health care system. Getting other stakeholders and the policy makers to understand and promote the value of nursing to consumers will take considerable political work. This work will have to be more than anecdotal pleas, arguments in support of some consumer cause, or reactions to some particular health care issue or workplace injustice. Believing in a vision and working hard are not enough. Nurses must have a clear image of the vision; develop a sound philosophy; demonstrate intelligent, strategic thinking; and wield more political influence.

Although nurses may think they are the ones primarily affected by the changes in health care, it is more powerful and therefore strategic to understand that everyone, especially the consumer, is affected.

As patient advocates, nurses need to seize the opportunity to make a difference. New nurses can make a difference by implementing the strategies outlined in Table 9-5.

### THE CONSUMER DEMAND FOR ACCOUNTABILITY

The vigilance of government, payers, and even attorneys is understandable. Some providers focus on "What's good for the agency, my interest group, or me?" rather than on "What is going to work for the consumers and all the other stakeholders over the long run?"

When stakeholders are motivated and directed solely by their own perceived needs, competitive political strategies replace more-collaborative approaches to addressing the consumer's health care needs. Accountability becomes a serious issue, because the goal of overtaking the competition supersedes the goal of offering the highest-quality services. People who will own the future of health care must address this growing problem of accountability. They will have to establish and sustain their credibility during a time when more people are distrustful of the health care system, its providers, third-party payers, and legislators.

## REAL WORLD INTERVIEW

Gary, age 46, and Laurie, age 42, have been married seven years and have two daughters, ages 4 years and 6 years. Gary is a practicing commercial architect recently diagnosed with Ménière's disease. Laurie is a professional photographer working from home. She was recently diagnosed with sarcoidosis. Gary said the effectiveness of the U.S. health care system is "very restricted by the dictates of insurance companies and by what health care providers want. It's like doing the minimum for whatever reason. There's a reluctance to take a holistic approach." Laurie agreed, adding that, "as it stands, service providers are needlessly competing with each other. As a consumer, I do not feel comfortable with how they collaborate with each other, much less with the consumer. Nursing has been channeled into a corporate culture because so many nurses fail to think or practice as professionals." Gary said, "Nursing is taken for granted; we really do not see nurses as providers as much as technicians. There is a missed opportunity to meet the expanding needs of consumers. A system that is more accessible from a daily living or health maintenance standpoint is needed." Consumers need their "concerns addressed openly and should be able to get answers or at least options without going through so many gatekeepers. I have come to believe that health care is a very inexact science. They are just guessing, or don't seem concerned enough to do more than what the provider wants or will get paid for." Laurie said, "Information is lost between service providers, hospitalizations, and clinic visits. You provide the same information several times, and you wonder if anyone is really using or thinking about the information they are gathering. If nurses were better educated, they could be stronger advocates for the consumer." Gary adds, "Health care is becoming more inaccessible unless you are dying. Nurses certainly have the opportunity to do more and to be first-level providers. We think that nurses did more to shape our hospital experiences to be a positive experience than anyone. They are the worker bees of health care. They are vital, but they could do a better job of working with consumers of health care. We are willing to pay more if we trust the competence of the provider and can understand the need for the service. This includes nursing."

**Gary and Laurie Maples**
Consumers
Tacoma, Washington

## TABLE 9-5  Political Strategies for Mounting Consumer Campaigns

- Lobbying at state and federal levels for health care regulations and guidelines that serve a consumer group's interest

- Consulting with representatives from a consumer group when health care regulations and guidelines are being debated or written

- Monitoring the enforcement of health care regulations and exacting corrective or punitive action when noncompliance occurs

- Encouraging providers and payers to make changes in the delivery of services voluntarily to meet changing consumer demands

- Changing consumer perceptions and behaviors through the distribution of educational materials or other media

Most people comprehend that being accountable requires being held responsible for one's behavior, decisions, and affiliations with others. Notwithstanding, some nurses claim they are not culpable for their actions because they are merely doing what they must do as defined by their employer or some other larger entity. These nurses fail to understand that professional accountability goes beyond responsibility in a particular employment situation. The practice of nursing is based on a social contract with society that gives nurses certain rights and responsibilities and requires that nursing be accountable to the public (ANA, 2010). In addition, the individual nurse practice acts of each state address these concepts.

Increasingly, consumers are more educated and have more access to the Internet, for example, www.webMD.com. Consumers are demanding positive results and are holding those in the health care system accountable for better health care outcomes. If the trend to increased litigation related to professional negligence and medical malpractice is any indication, just having an ethical process for providing care will not satisfy the consumer who experiences negative health outcomes. The strongest potential for litigation in health care comes from too few health care professionals accepting personal responsibility for ensuring that health care services are provided in a safe, competent manner at a system level as well as at a personal level. Health care professionals, including nurses, depend upon each other to ensure the quality, consistency, and overall effectiveness of health care within their work environments.

## CASE STUDY 9-2

Juan and Casey are study partners in the final semester of a nursing program. Juan has served as a medical corpsman and worked as an emergency medical technician for several years, whereas Casey entered college right out of high school, starting as a business major. One of their class assignments is to develop a strategy for reducing malpractice risks in a hospital setting. Casey proposes redefining the patient as the customer of health care services, because adopting a more customer-oriented approach to health care services in hospitals would improve patient satisfaction and subsequently reduce malpractice claims. Juan opposes that strategy because he thinks that too many health care providers are adopting the culture of corporate America when they define patients as customers. He views patients as something more than customers but also thinks that patients do not know what is best for them in most health care situations.

Do you think patients should be defined as health care customers?

Would patients be less likely to sue health care providers if they were approached as customers instead of patients?

## PATIENT SAFETY: AN IMPORTANT PARTNERSHIP

Historically, patients have been passive recipients of health care, but with the consumer movement, this is changing. Readily available information through Internet resources—such as www.pubmed.org, www.webMD.com, and www.askanurse.com—has facilitated this move from patient to consumer of health. Patients who have chronic diseases may be active members in organizations that educate and advocate for those with the disease. No longer do the health care providers have all the answers. In addition to greater understanding of health care and its risks, consumers have expectations of greater involvement in health care decisions (Holme, 2009). Informed consumers are important partners in ensuring safe care. A patient-clinician partnership model that empowers the patient and uses clinician strengths should replace the old patriarchal model of health care (Wiggins, 2008). Consumers can demand competent, educated care providers as emphasized in ANA's campaign entitled "Every Patient Deserves a Nurse." Consumers can require information about statistical indicators of care quality, such as medical error rates, nosocomial infection rates, morbidity and mortality rates, lengths of stay for patients with certain conditions, and incidents of malpractice. Consumers can ask facilities about their performance on the Joint Commission's 2010 National Patient Safety Goals, available at http://www.jointcommission.org/patientsafety/nationalpatientsafetygoals/. Individual consumers can ask questions when given prescriptions, treatments, and diagnostic tests. Collectively, educated consumers can partner with nursing organizations or other health care organizations to support policies and legislation that facilitate safe care (Table 9-6).

Professional nursing organizations have a major role in promoting patient safety. The role includes but is not limited to developing policy statements, lobbying for regulations and legislation that serve and protect the consumers of nursing care, and advocating for patients. There are times when a professional organization's need to advocate for its members conflicts with its usual purpose of advocating for the welfare of the public. The organization must balance these conflicting roles and continue to promote patient safety.

Since the shift from a service focus to a business focus within health care, nurses have found themselves working to maintain quality care and patient safety when the powers that be are focused on efficiency (Wiggins, 2008). Where do nurses turn when they have concerns but no power to change the situation? Both new nurses and experienced nurses need to turn to their professional organizations. Alone, a nurse is only one voice, but together, nurses are powerful. There are many professional organizations that speak for nurses. Most are specialty organizations

## TABLE 9-6  How a Bill Becomes Law

### Step 1: A Bill Is Born

Anyone may draft a bill; however, only members of Congress can introduce legislation and, by doing so, become the sponsor(s). The president, a member of the cabinet, or the head of a federal agency can also *propose* legislation, but a member of Congress must introduce it.

### Step 2: Committee Action

As soon as a bill is introduced, it is referred to a committee. At this point, the bill is examined carefully and its chances for passage are first determined. If the committee does not act on a bill, the bill is effectively "dead."

### Step 3: Subcommittee Review

Often, bills are referred to a subcommittee for study and hearings. Hearings provide the opportunity to put on the record the views of the executive branch, other public officials and supporters, experts, and opponents of the legislation.

### Step 4: Markup

When the hearings are completed, the subcommittee may meet to "mark up" the bill, that is, make changes and amendments prior to recommending the bill to the full committee. If a subcommittee votes not to report legislation to the full committee, the bill dies. If the committee votes for the bill, it is sent to the floor.

### Step 5: Committee Action to Report a Bill

After receiving a subcommittee's report on a bill, the full committee votes on its recommendation to the House or Senate. This procedure is called "ordering a bill reported."

### Step 6: Voting

After the debate and the approval of any amendments, the bill is passed or defeated by the members voting.

### Step 7: Referral to Other Chamber

When the House or Senate passes a bill, it is referred to the other chamber, where it usually follows the same route through committee and floor action. This chamber may approve the bill as received, reject it, ignore it, or change it.

### Step 8: Conference Committee Action

When the actions of the other chamber significantly alter the bill, a conference committee is formed to reconcile the differences between the House and Senate versions. If the conferees are unable to reach agreement, the legislation dies. If agreement is reached, a conference report is prepared describing the committee members' recommendations for changes. Both the House and Senate must approve the conference report.

### Step 9: Final Action

After both the House and Senate have approved a bill in identical form, it is sent to the president. If the president approves of the legislation and signs it, it becomes law. Or, if the president takes no action for 10 days, while Congress is in session, it automatically becomes law. If the president opposes the bill, the president can veto it; or if the president takes no action after the Congress has adjourned its second session, it is a "pocket veto," and the legislation dies.

### Step 10: Overriding a Veto

If the president vetoes a bill, Congress may attempt to override the veto. If both the Senate and the House pass the bill by a two-thirds majority, the president's veto is overruled, and the bill becomes a law.

**Source:** www.genome.gov/pfv.cfm?pageID=12513982.

## Critical Thinking 9-5

Access, quality, and timing of information available to the public have greatly enhanced the consumer health care movement. Using the Internet, people can do customized searches on practically any health care concern; garner input from a wide audience; and do comparative shopping for services, providers, and products. Several uniform sources of information have been developed for the U.S. health care delivery system. These data sets offer information requested by the decision makers about some predetermined dimension of health care. They also establish standard definitions, classifications, and measurements for making evidence-based decisions. Nursing has been relatively absent from the data sought, collected, and disseminated to the decision makers.

Is there a need for nursing-sensitive consumer outcome measures that can be used by decision makers? What steps could be taken to make such information available to the public?

with small memberships. Some of the specialty organizations, such as Emergency Nurses Association and the Association of Critical Care Nurses, have large memberships (Table 9-7).

The American Nurses Association (ANA) is the largest representative of registered nurses in the United States. ANA is not specialty focused and it represents nurses from all specialties, all work sites, and all levels of education. ANA has many goals, but one of them is to assure safe, quality care (ANA, 2010). To achieve its patient safety goal, ANA develops and disseminates documents such as *Nursing's Social Policy Statement* and the *Code of Ethics*

*for Nurses.* These documents define nurses' contract with society and each nurse's obligation to act ethically. Patient safety is part of the contract as well as an ethical responsibility of nurses. ANA lobbies for regulations and legislation that ensure patient safety. This is done largely through state nursing associations, state nursing boards, and state legislatures, which define nursing practice in each state. ANA advocates for the consumers of health care by pushing for evidence-based patient outcome data and encouraging certification and credentialing. Educated consumers can choose facilities with improved patient outcomes and well-prepared nurses.

### TABLE 9-7 Nursing Organizations

| ORGANIZATION | WEB SITE |
|---|---|
| Academy of Medical-Surgical Nurses | *www.medsurgnurse.org* |
| Academy of Neonatal Nursing | *www.academyonline.org* |
| Air & Surface Transport Nurses Association | *www.astna.org* |
| American Academy of Ambulatory Care Nursing | *www.aaacn.org* |
| American Academy of Nurse Practitioners | *www.aanp.org* |
| American Assembly for Men in Nursing | *www.aamn.org* |
| American Association for the History of Nursing | *www.aahn.org* |
| American Association of Colleges of Nursing | *www.aacn.nche.edu* |
| American Association of Critical-Care Nurses | *www.aacn.org* |
| American Association of Heart Failure Nurses | *http://aahfn.org* *(Continues)* |

## TABLE 9-7 (Continued)

| ORGANIZATION | WEB SITE |
| --- | --- |
| American Association of Legal Nurse Consultants | www.aalnc.org |
| American Association of Managed Care Nurses | www.aamcn.org |
| American Association of Neuroscience Nurses | www.aann.org |
| American Association of Nurse Anesthetists | www.aana.com |
| American Association of Nurse Life Care Planners | aanlcp.org |
| American Association of Occupational Health Nurses | www.aaohn.org |
| American Association of Spinal Cord Injury Nurses | http://nurses.ascipro.org |
| American College of Nurse-Midwives | www.acnm.org |
| American College of Nurse Practitioners | www.acnpweb.org |
| American Forensic Nurses | www.amrn.com |
| American Holistic Nurses Association | www.ahna.org |
| American Nephrology Nurses' Association | www.annanurse.org |
| American Nurses Association | www.ana.org |
| American Nursing Informatics Association | www.ania.org |
| American Organization of Nurse Executives | www.aone.org |
| American Pediatric Surgical Nurses Association | www.apsna.org |
| American Psychiatric Nurses Association | www.apna.org |
| American Society for Pain Management Nursing | www.aspmn.org |
| American Society of Ophthalmic Registered Nurses | http://webeye.ophth.uiowa.edu/asorn |
| American Society of PeriAnesthesia Nurses | www.aspan.org |
| American Society of Plastic Surgical Nurses | www.aspsn.org |
| Association of Camp Nurses | www.campnurse.org |
| Association of Child Neurology Nurses | www.acnn.org |
| Association of Nurses in AIDS Care | www.nursesinaidscare.org |
| Association of Operating Room Nurses (same as Association of PeriOperative Registered Nurses) | www.aorn.org |
| Association of Pediatric Gastroenterology and Nutrition Nurses | http://apgnn.org |

*(Continues)*

## TABLE 9-7 (Continued)

| ORGANIZATION | WEB SITE |
| --- | --- |
| Association of Pediatric Hematology/Oncology Nurses | http://www.aphon.org |
| Association for Radiologic and Imaging Nursing | www.arinursing.org |
| Association of Rehabilitation Nurses | www.rehabnurse.org |
| Association of Women's Health, Obstetric and Neonatal Nurses | www.awhonn.org |
| Chi Eta Phi Sorority, Inc. | www.chietaphi.com |
| Dermatology Nurses' Association | www.dnanurse.org |
| Developmental Disabilities Nurses Association | www.ddna.org |
| Emergency Nurses Association | www.ena.org |
| Family Medicine Residency Nurses Association | www.fmrna.org |
| Gerontological Advanced Practice Nurses Association | www.gapna.org |
| Home Healthcare Nurses Association | www.hhna.org |
| Hospice and Palliative Nurses Association | www.hpna.org |
| Infusion Nurses Society | www.ins1.org |
| International Association of Forensic Nurses | www.iafn.org |
| International Organization of Multiple Sclerosis Nurses | www.iomsn.org |
| International Society of Nurses in Cancer Care | www.isncc.org |
| International Society of Nurses in Genetics | www.isong.org |
| National Association of Clinical Nurse Specialists | www.nacns.org |
| National Association of Hispanic Nurses | http://thehispanicnurses.org |
| National Association of Neonatal Nurses | www.nann.org |
| National Association of Nurse Practitioners in Women's Health | www.npwh.org |
| National Association of Orthopedic Nurses | www.orthonurse.org |
| National Association of Pediatric Nurse Practitioners | www.napnap.org |
| National Association of School Nurses | www.nasn.org |
| National Association of School Nurses for the Deaf | www.nasnd.net |
| National Black Nurses Association | www.nbna.org |
| National Council of State Boards of Nursing | www.ncsbn.org |

*(Continues)*

## TABLE 9-7 (Continued)

| ORGANIZATION | WEB SITE |
| --- | --- |
| National Gerontological Nursing Association | www.ngna.org |
| National League for Nursing | www.nln.org |
| National Nursing Staff Development Organization | www.nnsdo.org |
| National Organization of Nurse Practitioner Faculties | www.nonpf.com |
| Nurses Organization of Veterans Affairs | www.vanurse.org |
| Oncology Nursing Society | www.ons.org |
| Pediatric Endocrinology Nursing Society | www.pens.org |
| Preventive Cardiovascular Nurses Association | www.pcna.net |
| Sigma Theta Tau International | www.nursingsociety.org |
| Society of Gastroenterology Nurses and Associates, Inc. | www.sgna.org |
| Society of Otorhinolaryngology and Head-Neck Nurses, Inc. | www.sohnnurse.com |
| Society of Pediatric Nurses | www.pedsnurses.org |
| Society of Trauma Nurses | www.traumanurses.org |
| Society of Urologic Nurses and Associates | www.suna.org |
| Transcultural Nursing Organization | www.tcns.org |
| Wound, Ostomy and Continence Nurses Society | www.wocn.org |

# CREDIBILITY AND POLITICS

To have credibility, nurses must demonstrate professional competence and a degree of professional accountability that exceeds consumer expectations. Nurses who are most able to successfully overcome these challenges assert their professional credibility in several ways. They are lifelong learners and demonstrate professional growth throughout their careers in nursing. They approach their vocation as a service to the public and to the nursing profession and as an honorable way to make a living. They take ownership of the situations in which they find themselves and work to resolve problems and overcome obstacles to providing the best care possible. Nurses strengthen their political position by sharing accountability for health care problems with other health care providers. When nurses point fingers at others such as supervisors, administrators, politicians, or practitioners for health care problems or look to others for improvement, their power is weakened. Nursing's accountability to the public is part of its professional responsibility. This means that nurses must work cohesively with partners to assure quality care (Hood & Leddy, 2006).

Nursing ownership, however, is not enough to guarantee the political credibility of nursing in the future. This is because others see the political gains to be made from identifying themselves as consumer advocates. As other service providers board the consumer bandwagon, nurses will need to continuously demonstrate that they are more valuable to the consumer than providers of alternative services that cost less. Consumers will increasingly demand tangible evidence from nurses that their services are worth the price.

## KEY CONCEPTS

- Individuals or groups take political action to get what they want or to prevent others from getting something they do not want them to have.

- Entry-level nurses can join organizations that influence policy such as gun control, tobacco sales, and mandatory helmets for certain activities.

- Politics are inherent in any system in which resources are absolutely or relatively scarce and where there are competing interests for those resources.

- Nurses have a critical role in addressing the major system-level issues in health care delivery.

- Political, economic, and social changes such as aging, diversity, and the costs of technology in the United States are transforming the health care system.

- Nurses must articulate what nursing is, what distinctive services they provide, how these services benefit consumers, and how much these services cost in relation to other health care services.

- If nursing is defined through politics to be less than critical or professional, nurses will be less empowered and paid less.

- The aging of the U.S. population constitutes a growing political force and affords nursing a wonderful opportunity to become stronger in health policy arenas.

- Consumer partnerships will become more critical for all stakeholders in health care.

- When a consumer group forms a political coalition with other groups, such as nurses, in a given community, the political influence of both is strengthened.

## KEY TERMS

political voice
politics
resources

stakeholders
voting block

## REVIEW QUESTIONS

1. Political methods inherently involve which of the following? Select all that apply.
   - _____ A. Competition
   - _____ B. Naiveté
   - _____ C. Negotiation
   - _____ D. Collaboration
   - _____ E. Consultation
   - _____ F. Negativity

2. Politics exist because of which of the following statements?
   A. They are required by law.
   B. Resources cannot be limited by political process.
   C. Some people want to control more resources than others.
   D. Resources must be equally distributed among stakeholders.

3. The consumer movement in health care is characterized by which of the following statements?
   A. It is a socialist movement about health care resources.
   B. It is growing because of the Internet and organizations such as the AARP.
   C. It is supported by all stakeholders in the health care system.
   D. It is inclusive of only people who are not health care providers.

4. In that the national expenditures on health care for the uninsured exceeds $100 billion annually, which of the following is true?
   A. All Americans receive adequate health care.
   B. Forty-three million Americans are uninsured.
   C. Medicaid provides insurance for the working poor.
   D. Medicare ensures that all patients over 60 receive health care.

5. Many consumers are concerned about media reports of health care expenditures rising because of which of the following?
   A. They are uninsured.
   B. Their individual share of expenses is rising.
   C. They pay the entire cost of health care out of pocket.
   D. Rationing is inevitable.

6. Which of the following is true concerning the minority population in the United States?
   A. It includes less than 25 million racially and ethnically diverse people.
   B. It makes up less than one-tenth of the U.S. population.
   C. It has resisted the consumer movement.
   D. It is a well-organized, cohesive group of consumers.

7. Which of the following is true regarding the consumer movement in the United States?
   A. It is a passing fad.
   B. It challenges nurses to be more professionally accountable.
   C. It is sustainable only through partnerships with the nursing profession.
   D. It is encouraged by all health care systems and providers.

8. ANA describes the essential elements of nursing. Which of the following is not an essential element?
   A. Establishing a caring relationship
   B. Delivering problem-focused care
   C. Focusing on the full range of experiences and human responses
   D. Appreciating the patient's subjective experience and integrating such experience with objective data

9. As a new school nurse, you become aware of some state laws that actually get in the way of providing the best care for your students. What are the best steps in becoming politically active and changing this? Select all that apply.
   _____ A. Join the state school nurse organization, become active, and partner with parent groups.
   _____ B. Vote in the next election, and encourage all your friends to do so.
   _____ C. Study politics, and run for political office.
   _____ D. Write your state legislators, describe your problem, and ask them to initiate laws that will resolve the problem.
   _____ E. Bad-mouth the current legislators and laws.
   _____ F. Talk to anyone who will listen about the problem.

10. Nursing political influence is increased by:
    A. Pursuing lifelong learning and accountability to the public.
    B. Pointing out the physician monopoly on health care leadership.
    C. Blaming administration for the lack of safe staffing.
    D. Looking to politicians for solutions to health care challenges.

## REVIEW ACTIVITIES

1. Find out who your congressional representatives are. Write or e-mail them to find out what health care legislation they are supporting.

2. Notice who is supporting current health care legislation. Are consumer protections being emphasized in any proposed legislation?

3. Identify a consumer group in which you are interested. Use the steps identified in Table 9-4 to establish a partnership with the group. What did you learn?

## EXPLORING THE WEB

- Identify some Web sites for consumer groups:
  AARP (American Association of Retired Persons): www.aarp.org
  Citizen's Council on Health Care: www.cchc-mn.org and www.cchconline.org

- Go to the consumer site for combating health-related fraud: www.quackwatch.org

- Note the consumer tips at this site: www.consumertips.com

- Search for health care on this Consumers for Quality Care (CQC) site: www.consumerwatchdog.org

- Search for health on this site for consumer reports: www.consumersunion.org

- Identify some sites for government offices and health care agencies:
  U.S. Congress: www.congress.org
  U.S. Department of Health and Human Services: www.hhs.gov

Government consumer health Web site: www.healthfinder.gov

Medicare and Medicaid programs: www.cms.hhs.gov

- Read about the Milbank Memorial Fund, whose purpose is to improve health by helping the decision makers: www.milbank.org

- Identify some Web sites that new nurses should know about concerning policy:
  ANA keeps nurses abreast of legislative issues that involve them: www.nursingworld.org. Search for *government.*
  The Library of Congress Thomas site provides a searchable database of federal legislation: http://thomas.loc.gov
  The Library of Congress state government information provides state and local government links: www.loc.gov. Search for state and local government news.

# REFERENCES

American Nurses Association. (2009b). http://www.nursingworld.org/MainMenuCategories/ANAMarketplace/ANAPeriodicals/OJIN/TableofContents/Volume82003/No3Sept2003/AssociationsRole.aspx.

American Nurses Association. (2010). *Nursing's social policy statement: The essence of the profession* (3rd ed.). Washington, DC: American Nurses Publishing.

Becker, R. (2009). Can the palliative care services of today keep up and match the expectations of the 'baby boomer' generation? *International Journal of Palliative Nursing, 15*(3), 108–109.

Bernstein, R. (2006, May 10). Immediate news release: Nation's population one-third minority. U.S. Census Bureau. Retrieved June 29, 2006, from http://www.census.gov/Press-Release/www/releases/archives/population/006808.html.

Blumberg,, L. J., & Holahan, J. (2009). The individual mandate—An affordable and fair approach to achieving universal coverage. New England Journal of Medicine, 361, 6–7.

Buerhaus, P., Donelan, K., Ulrich, B., Norman, L., & Dittus, R. (2005). Part one: Is the shortage of hospital registered nurses getting better or worse? Findings from two recent national surveys of RNs. *Nursing Economics, 23*(2), 61–71, 96.

Centers for Disease Control and Prevention. State-by-state breakout of excess deaths from lack of insurance. Accessed December 4, 2010, from http://pnhp.org/excessdeaths/excess-deaths-state-by-state. pdf.

Center for Responsive Politics. (2010). Retrieved March 7, 2011, from http://www.opensecrets.org/lobby/top.php?showYear=a&indexType=s.

Donley, R. (2005). Challenges for nursing in the 21st century. *Nursing Economics, 23*(6), 312–318.

Ferlise, P., & Baggot, D. (2009). Improving staff nurse satisfaction and nurse turnover: Use of a closed-unit staffing mode. *Journal of Nursing Administration, 39*(7–8), 318–320.

Grindel, C. (2005). Influencing health care policy with our children in mind. *MedSurg Nursing, 14*(5), 277–278.

Hahn, J. (2010). Integrating professionalism and political awareness into the curriculum. *Nurse Educator, 35*(3), 110–113.

Harrington, C., Crider, M., Benner, P. E., & Malone, R. (2005). Advanced nursing training in health policy, designing, and implementing a new program. *Policy, Politics & Nursing Practice, 6*(2), 99–108.

Himmelstein, D. U., Thorne, D., Warren, E., Woolhandler, S. (2009, August). Medical bankruptcy in the United States, 2007: Results of a national study. American Journal of Medicine, 122(8), 741–746. Accessed March 7, 2011, from http://www.amjmed.com/article/S0002-9343(09)00404-5/abstract.

Holme, A. (2009). Exploring the role of patients in promoting safety: Policy to practice. *British Journal of Nursing, 18*(22), 1392–1395.

Hood, L. J., & Leddy, S. (2005). *Conceptual bases of professional nursing* (6th ed.). Philadelphia: Lippincott.

Hughes, F., Duke, J., Bamford, A., & Moses, C. (2006). Enhancing nursing leadership through policy, politics and strategic alliances. *Nurse Leader, 4*(2), 24–27.

Keepnews, D. M. (2005). Health policy—A nursing specialty? *Policy, Politics & Nursing Practice, 6*(4), 275–276.

Kelly, K. (2007). From apathy to political activism. *American Nurse Today, 2*(8), 55–56.

King, M. (2008, April 9). Companies find ways to retain expertise of older workers. *Seattle Times.* Retrieved March 7, 2011 from http://seattletimes.nwsource.com/html/localnews/2004336246_olderworkers09m.html.

Lochhead, Carolyn. (2010, March 22). Houses passes health care bill 219–212. Accessed December 5, 2010, from http://www.webcitation.org/5oS1boYV0.

Mason, D., Leavitt, J. K., & Chaffee, M. (2006). *Policy and politics in nursing and health care* (5th ed.). W. B. Saunders. Philadelphia, Pa.

McDonald, L. (2006). Florence Nightingale as a social reformer. *History Today, 56*(1), 9–15.

Middaugh, D. J., & Robertson, R. D. (2005). Politics in the workplace. *Nursing Management, 14*(6), 393–394.

Milstead, J. A. (2007). *Health policy & politics: A nurse's guide* (3rd ed.). Sudbury, MA: Jones and Bartlett Publishers.

Morris, A. H., & Faulk, D. J. (2007). Perspective transformation enhancing the development of professionalism in RN-to-BSN students. *Journal of Nursing Education, 46*(10), 445–491.

Norman, L., Donelan, K., Buerhaus, P., Willis, G., Williams, M., Ulrich, B., & Dittus, R. (2005). Part five: The older nurse in the workforce: Does age matter? *Nursing Economics, 23*(6), 282–289.

O'Connor, T., & Kelly, B. (2005). Bridging the gap: A study of general nurses' perceptions of patient advocacy in Ireland. *Nursing Ethics, 12*(5), 453–467.

Rudolph, B. J. (2005). *How nurses define the role of patient advocacy.* Unpublished master's thesis, Division of Nursing, Wilmington College, Wilmington, DE.

Shi, L., & Singh, D. A. (2008). *Delivering health care in America* (4th ed.). Gaithersburg, MD: Aspen.

U.S. Census Bureau. (2009a). An aging world: 2008. Washington, DC: U.S. Government Printing Office. Retrieved March 7, 2011, from http://www.census.gov/prod/2009pubs/p95-09-1.pdf.

U.S. Census Bureau. (2009b). Income, poverty, and health coverage in the United States: 2008. Washington, DC: U.S. Government Printing Office. Accessed March 7, 2011, from http://www.census.gov/prod/2009pubs/p60-236.pdf.

Vesely, R. (2008). 'Unleash the energy.' Activists, professors try to use their own sphere of influence to affect U.S. healthcare policy and improve patient care. *Modern Healthcare, 38*(17), 6–7.

Wiggins, M. S. (2008). The partnership care delivery model: An explanation of the core concept and the need for a new model of care. *Journal of Nursing Management, 16*(5), 629–638.

Wilper, A. P., Woolhandler, S., Lasser, K. E., McCormick, D., Bor, D. H., & Himmelstein, D. U. (2009). Health Insurance and Mortality in US Adults. American Journal of Public Health, 99(12), pp. 1–7. Accessed March 7, 2011, from http://pnhp.org/excessdeaths/health-insurance-and-mortality-in-US-adults.pdf.

Woolhandler, S., Himmelstein, D. U. (1997, March). Costs of care and administration at for-profit and other hospitals in the United States. New England Journal of Medicine, 336(11), 769–74. Accessed December 4, 2010, from http://www.nejm.org/doi/full/10.1056/NEJM199703133361106.

# SUGGESTED READINGS

Aaron, H. J. (2006). Longer life spans: Boon or burden? *Daedalus, 135*(1), 9–19.

American Nurses Association. (2009a). Hill basics: The legislative process. Retrieved March 7, 2011, from http://nursingworld.org/MainMenuCategories/ANAPoliticalPower/Federal/ToolKit/LegislativeProcess.aspx.

American Nurses Association. (2009b). Tips: Contacting Members of Congress. Retrieved March 7, 2011 from http://nursingworld.org/MainMenuCategories/ANAPoliticalPower/Federal/ToolKit/ContactCongress.aspx.

Buerhaus, P., Donelan, K., Ulrich, B., Norman, L., & Dittus, R. (2005). Six part series on the state of the registered nurse workforce in the United States. *Nursing Economic, 23*(2), 58–60.

Buerhaus, P., Donelan, K., Ulrich, B., Norman, L., & Dittus, R. (2005). State of the registered nurse workforce in the United States. *Nursing Economics, 23*(2), 6–12.

Buerhaus, P., Donelan, K., Ulrich, B., Norman, L., Williams, M., & Dittus, R. (2005). Hospital RNs' and CNOs' perceptions of the impact of the nursing shortage on the quality of care. *Nursing Economic, 23*(5), 214–221.

Buerhaus, P.I. (2009). Massachusetts health care reform: Lessons for the nation? *Nursing Economics 27*(3), 194–196.

Donlan, K., Buerhaus, P. I., Ulrich, B. T., Norman, L., & Dittus, R. (2005). Awareness and perceptions of the Johnson & Johnson campaign for nursing's future. *Nursing Economic, 23*(4), 150–156, 180.

Hersey, P., Blanchard, K.H., & Johnson, D. E. (2007). *Management of organizational behavior.* Upper Saddle River, NJ: Prentice Hall.

Siegel, E., & Bennett, P. (2006). Creating partnership through patient safety awareness week. *Nursing Economic, 24*(3), 162–165.

Vestal, K. (2007). The power and intrigue of workplace politics. *Nurse Leader, 5*(1), 6–7.

# CHAPTER 10

# Strategic Planning and Organizing Patient Care

**AMY ANDROWICH O'MALLEY, MSN, RN; IDA M. ANDROWICH, PhD, RN, BC, FAAN; KATHLEEN F. SELLERS, PhD, RN**

*Strategic planning is a process by which we can envision the future and develop the necessary procedures and operations to influence and achieve that future.*

(CLARK CROUCH, MANAGEMENT CONSULTANT AND PRINCIPAL, THE RESOURCE NETWORK)

## OBJECTIVES

Upon completion of this chapter, the reader should be able to:

1. Understand the importance of an organization's mission and philosophy and the impact of these on the structure and behavior of the organization.

2. Define the purpose and identify the steps in the strategic planning process.

3. Articulate the importance of aligning the organization's strategic vision with both its mission, philosophy, and values and with the goals and values of the communities and stakeholders served by the organization.

4. Apply a basic understanding of common organizational structures and the advantages and disadvantages of each in identifying which structures would be best suited to meeting differing organizational objectives.

5. Discuss high-reliability organizations and a culture of safety.

6. Discuss organizational culture.

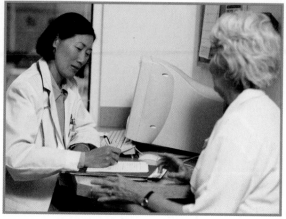

Delmar/Cengage Learning

*A friend is discussing her plans to step down as the assistant unit coordinator on 3 West. You recall how pleased and excited she was to be offered the position only a few months ago and wonder what has changed. She begins to describe her frustration with never feeling that she has sufficient information for decision making nor receiving the information in a timely manner. She feels that this puts her in a poor position to be a staff and patient advocate for her unit. She states, "I feel as if my manager has so many areas that are taking up all her time, and she really can't concentrate on our unit's needs. It seems as if we are always putting out fires and never have a chance to step back and actually plan programs and operational processes that could make a big improvement on the unit. The organization's mission statement says that we value education, but I continually have to turn down requests from staff to go to educational programs because of inadequate staffing. We are not recruiting and hiring nurses for all our budgeted positions. There are plenty of nurses routinely scheduled on days and not nearly enough on evenings when most patients are now discharged. Consequently, we have our major needs for discharge planning and patient education in the evening, and there is no plan for providing this necessary information in a consistent manner. The result is dissatisfied patients, staff, and physicians."*

*What are your thoughts about this situation?*

*What advice do you have for your friend?*

*Is this situation unusual? Can it be improved? How?*

There are increasing opportunities for nurses to become involved in strategic and tactical planning for the delivery of health care services in their organizations and communities. Yet, to be effective in leadership roles, nurses need a basic understanding of the way in which organizations are structured, how organizational systems function, and how to engage in the strategic planning process. In the past, many health care organizations were structured in a highly formal, top-down, militaristic manner. These bureaucratic organizations worked well in a relatively stable environment when communication channels could be hierarchical. They tend not to be as useful in a dynamic health care system in which information needs

and practice are rapidly changing. In addition, workers in today's health care systems are considered knowledge workers—professionals hired for their knowledge, skills, and expertise. They need a system that supports their ability to practice to the full extent of their professional accountability.

Leadership in the health care organizations of the twenty-first century demands competent nurses with different skill sets than in the past. That is because performance in a leadership role in today's highly complex health care environment requires a good understanding of how systems function and how to improve health care delivery. Yet health care providers, including professional nurses, have been slow to integrate this information into their clinical practice. Planning for continuous improvement of quality, service, and cost-effectiveness is a critical competency for nurses in twenty-first century health care organizations.

The seminal Institute of Medicine (IOM) report *To Err Is Human* (1999) states that preventable adverse events cause between 44,000 and 98,000 deaths each year, at an annual cost of between $37.6 billion and $50 billion. That report and the follow-up IOM report, *Crossing the Quality Chasm* (2001), have changed the way we view quality and patient safety. It is now generally understood that patient safety is dependent on the implementation of collaborative teams and interdisciplinary-focused care delivery systems that address the realities of practice and patient care. Recent research studies stress that the way a nurse's work is organized is a major determinant of patient welfare. Consequently, nurses in leadership positions must be educationally prepared to be able to develop and implement sound models for the effective delivery of patient care. Although many health care organizations collect large sets of data and are beginning to use scientific methods to improve the services they render, these activities are typically fragmented, isolated from day-to-day nursing management, and lack alignment with organizational strategy. The American Nurses Association (ANA) concurs with the IOM (1999) that errors occur as a result of system failure rather than human failure.

This chapter will discuss the strategic planning process and the importance of aligning the organization's strategic vision with the mission, philosophy, and values of the organization and the communities served by the organization. Also, because the manner in which organizations are structured has an impact on their ability to be effective in providing care, it is important for nurses to have a good understanding of the structures that support and promote health care delivery.

Generally, the organization starts with a defined vision, purpose, and mission. The strategic planning process allows the organization to scan the environment and develop strategic goals based on changes in the environment. This is a process that occurs within the defined vision, purpose, and mission of the organization. The March of Dimes is a good example of an organization that was able to change its strategic plan in response to an environmental change. In 1938, the March of Dimes was established as the National Foundation for Infantile Paralysis (polio). With funding provided by the organization, Dr. Jonas Salk was able to develop an experimental polio vaccine in 1952, and, in 1962, Dr. Sabin, also with funding from the March of Dimes, developed the oral polio vaccine. Today, virtually all American babies receive the vaccine to prevent polio, and there has not been a new case since 1991 in the United States. Consequently, since 1958, the March of Dimes has changed its strategic plan to funding cutting-edge research and innovative programs to save babies from birth defects, premature births, and low birth weight (March of Dimes, 2006).

Many organizations do not have such a dramatic need for a mission change. The strategic plan also identifies the direction for the budget of the organization.

# ORGANIZATIONAL VISION, MISSION, AND PHILOSOPHY

Every organization has a guiding vision and mission. Most often, the purpose and philosophy are explicitly stated and detailed in a formal mission statement. Typically, this mission statement reflects the organization's values and provides the reader with an indication of the behavior and strategic actions that can be expected from that organization. Most health care organizations have mission statements that speak to providing high quality or excellence in patient care. Some mission statements focus exclusively on providing care, whereas others assume a broader view and consider the education of health care professionals and the promotion of research as contributing to their broader mission. The mission of other organizations may be community based, and these organizations consequently will focus on providing community outreach and population-based services to a specific community or population within a community.

## MISSION STATEMENT

The **mission statement** is a formal expression of the purpose or reason for existence of the organization. It is the organization's declaration of its primary driving force or its vision of the manner in which it believes care should be delivered. An institution's mission statement defines how it is unique and different from other organizations that provide a similar service. (For examples of actual mission statements, refer to the Exploring the Web section at the end of the chapter.)

## PHILOSOPHY

The **philosophy of an organization** is typically embedded in the mission statement. It is, in essence, a value statement of the principles and beliefs that direct the organization's behavior. A careful reading of the mission statement will usually provide a good understanding of the institutional philosophy or value system. Mission statements with phrases such as "without consideration for ability to pay," "with respect for the dignity of each elderly resident," "a brighter future for all children," or "vigorous rehabilitation to maximize each individual's utmost potential" provide

---

### Critical Thinking 10-1

Examine these two mission statements and then respond to the questions that follow.

Hospital A: "Our mission is to ensure the highest quality of care for the patients in our community. We believe that each patient has the right to the most innovative care that current science and technology can provide. To that end, we have assembled a world-renowned medical staff who will strive to ensure that the latest developments in medical science are used to combat disease."

Hospital B: "Our mission is to provide excellence in care to all. Our health care staff, nurses, practitioners, and other professionals believe that care can best be provided in an atmosphere of collaboration and partnership with our patients and community. We believe in education—for our patients, for our staff, and for future health care providers. At all times, we strive for optimal health promotion and the prevention of disease and disability."

Which of these institutions do you think would be more likely to have a patient lecture series on "Living with Diabetes"? Value the contributions of nursing? Provide experimental therapy for cancer? Be open to scheduling routine patient care visits for uninsured patients?

clues to the type of service that you could expect from an organization. In the best of worlds, there is congruence among the stated mission, vision, and philosophy and the behavior of the organization.

Sometimes, this consequence is not the case. For example, does the mission statement read "Our patients are our highest priority" only to organize the environment and services in such a manner that there are inadequate directional signs and registration staff? The result is that these "highest priority" patients are often not able to find their way around the health care organization and have unduly long waits for registration. When a mission is formally stated, quality can be measured (Table 10-1).

---

## TABLE 10-1  Mission, Goals, and Quality Measures

### Mission

The Peoples Choice Healthcare Center provides excellent health care to all patients through partnerships with patients and the community and collaboration with nursing and medical practitioners as well as with other health care staff. We believe in continuous education for patients, health care staff, and future health care providers. We are committed to optimal health care promotion and the prevention of disease and disability.

### Goals

1. Collaborate with all health care staff to improve patient care
2. Increase customer satisfaction scores
3. Increase number of emergency room visits
4. Increase patient days
5. Increase use of computers by all staff
6. Increase funding for staff's continued education
7. Encourage all staff to attend one education program yearly
8. Increase number of specialty certifications of staff
9. Monitor nurse-sensitive patient outcomes such as incidence of cardiac arrest, Urinary Tract Infection (UTI), Gastrointestinal (GI) bleeding, thrombophlebitis, failure to rescue, etc.
10. Decrease medication errors

### Nursing Sensitive Measures. The National Quality Forum (NQF). 2004.

1. Falls prevalence
2. Falls with injury
3. Restraint prevalence (vest and limb only)
4. Skill mix (RN, LPN, NAP*, and contract)
5. Nursing care hours per patient day (RN, LPN, and NAP*)
6. Practice Environment Scale—Nursing Work Index (composite and 5 subscales)
7. Voluntary turnover

### Quality Measures for Emergency Department (ED)

*Customer/Patient*

1. Increase in patient satisfaction
2. Increase in customer return
3. Decrease in patient complaints
4. Increase in market share
5. Decrease in repeat asthma patient visits
6. Develop patient education materials explaining norms for ED stays, well and ill child care, and so on
7. Develop evidence-based standards for care of patients with cardiac arrest, UTI, upper GI bleeding, and thrombophlebitis
8. Review all emergency department deaths

*Financial*

1. Increase use of computers on all units
2. Monitor budget compliance
3. Improve nurse staffing ratios
4. Develop computerized order-entry system for medications.
5. Develop electronic health records

*Internal Processes*

1. Achieve 90% on key performance improvement measures
2. Decrease sick time and overtime by 10%
3. Increase number of nursing research projects
4. Achieve magnet status
5. Increase use of best practice educational materials for all patients and staff
6. Increase participation of nursing, medicine, and NAP as well as pharmacy staff in quality improvement activities
7. Arrange for all staff to attend one outside conference yearly
8. Set up interdisciplinary committee on medication administration safety

## TABLE 10-1 (Continued)

*Employee Growth and Learning*

1. 50% of the nursing department join a nursing professional association

2. All nurses working in the ED are certified in ACLS and PALS** trauma nursing

3. One-third of nurses are continuing their nursing education

4. 50% of all staff are cross-trained and can work in an ED

5. 90% of employees are very satisfied

6. 90% of staff are retained

7. All nurses are able to use the computer for patient information, to search literature, and so forth

8. 20% of nursing staff present a community program on topics such as stroke and other pertinent annually

---

*RN, LPN, and nursing assistive personnel (NAP)

**Advanced Cardiac Life Support (ACLS) and Pediatric Advanced Life Support (PALS)

***Intensive Care Unit (ICU)

**Source:** Compiled with information from Kelly, P. (2009). *Essentials of nursing leadership & management* (2nd ed.). Clifton Park, NY: Delmar Cengage Learning; and Kurtzman, E. T., & Jennings, B. M. (2008). Capturing the imagination of nurse executives in tracking the quality of nursing care. *Nursing Administration Quarterly, 32*(3), 235–246.

---

It is always important to assess an organization's mission, vision, and philosophy prior to considering employment because when important individual and organizational philosophies collide, it is likely to be a constant source of frustration for the employee and the employer.

## Examining a Mission Statement

In 1956, the United Mine Workers of America (UMWA) dedicated the Miners Memorial Hospital Association's (MMHA) facilities, which consisted of ten hospitals in Kentucky and West Virginia and included services to patients from Appalachia. This hospital system soon became a model for similar health care organizations across the country. The MMHA system was such a rarity that one publication described the newly opened hospital locations as "ten places where hospital and medical care history will be written." Today, the health care system, now known as Appalachian Regional Healthcare (ARH) to more accurately describe its far-ranging activities, continues its mission by offering residents of eastern Kentucky and southern West Virginia a local option for state-of-the-art

## REAL WORLD INTERVIEW

Strategic planning and organizing patient care are critical issues in the field of clinical information systems. The strategic planning process consists of several phases, each of which requires specific activities for designing and implementing an electronic health record. One important area to consider is the selection of a standardized coded terminology for documenting patient care. I recommend that the Clinical Care Classification (CCC) system or another similar system be selected as the terminology of choice. It is important to select a research-based and ANA-recognized documentation terminology such as CCC, Systematized Nomenclature of Medicine (SNOMED), North American Nursing Diagnosis Association (NANDA), Nursing Intervention Classification (NIC), Nursing Outcomes Classification (NOC), Omaha, or the International Classification for Nursing Practice. These allow nurses to document nursing diagnoses, nursing interventions and actions, and patient outcomes. The documentation is critical in the development of a computerized information system that supports and informs nursing practice.

**Virginia K. Saba, EdD, RN, FAAN**

CEO and president, Sabacare.com
and principal investigator of the
CCC System Arlington, Virginia

**TABLE 10-2  Mission, Vision, and Values of Appalachian Regional Healthcare (ARH)**

**ARH Mission:** To improve health and promote well-being of all the people in Central Appalachia in partnership with our communities.

**ARH Vision:** To earn the confidence and trust of the diverse communities we serve by offering healthcare excellence, delivered with compassion in a timely manner.

**ARH Values:** Excellence, compassion, safety, teamwork, inclusion, professionalism

**Source:** Mission, Vision and Values (1997–2011). Appalachian Regional Healthcare. Accessed March 8, 2011 at http://www.arh.org/AboutUs/mission.php.

technology and high quality health care services. Since its inception, ARH has been committed to growing and meeting the needs of its communities (Table 10-2).

## Vision

In addition to the Mission Statement, part of strategic planning is clarifying the vision of the organization. A **vision** statement tells us how the people of the organization plan to actualize the mission. It portrays what people would like to think of the organization and how it envisions itself to be in three to five years. Even though it is focused on the future, a vision statement is written in the present tense, using action words as if the vision had already been achieved. In health care, a vision statement describes a balance of addressing the needs of the providers, the patients, and the environment.

# STRATEGIC PLANNING

As Lewis Carroll observed in *Alice's Adventures in Wonderland*, "If you don't know where you are going, any road will do." A health care organization needs to have a good idea of where it fits into its environment and what types of programs and services are needed and demanded by its customers or stakeholders. This is true at a broad organizational level as well as at a unit level. It is important that a nurse manager have an understanding of which programs and services are valued by a patient population and plan for how the unit's ongoing activities fit in with the overall strategy of the larger organization.

## STRATEGIC PLANNING DEFINITION

A **strategic plan** can be defined as the sum total or outcome of the processes by which an organization engages in environmental analysis, goal formulation, and strategy development with the purpose of organizational growth and renewal. Drucker (1973) defines strategic planning as "a continuous, systematic process of making risk-taking decisions today with the greatest possible knowledge of their effects on the future" (p. 125). Strategic planning is ongoing and is especially needed whenever the organization is experiencing problems, including internal/external review problems.

## PURPOSE OF STRATEGIC PLANNING

The purpose of strategic planning is twofold. First, it is important that everyone have the same idea or vision for where the organization is headed, and second, a good plan can help to ensure that the needed resources and budget are available to carry out the initiatives that have been identified as important to the unit or agency. In addition, a clear plan allows the manager to select among seemingly equal alternatives based on the alternatives' potential to move the organization toward the desired end goal.

## STEPS IN STRATEGIC PLANNING PROCESS

In any strategic planning process, there are steps to be followed. This process is similar to the nursing process. You assess and plan before implementing a nursing treatment. It is equally important when developing an organizational, unit, or program plan to progress in a systematic manner. Table 10-3 lists the steps in the strategic planning process.

### Environmental Assessment

An environmental, or a situational, assessment requires a broad view of the organization's current environment.

## REAL WORLD INTERVIEW

The pivotal value of strategic planning is that it requires an organization to focus on its raison d'être, its mission, and to test how its operations are leading to accomplishment of that mission. Determined by the degree of dynamic change present in both its internal and external environments, an organization's strategic planning may extend years, or only months, into the future. However, at least annually, the strategic plan must be examined, reasserted as appropriate, and used as the standard against which short-term initiatives are measured for congruency with mission accomplishment.

**Laura J. Nosek, PhD, RN**
Health Care Consultant
Auburn, Ohio

### TABLE 10-3  Steps in Strategic Planning Process

1. Perform environmental assessment, SWOT analysis.

2. Conduct stakeholder analysis.

3. Review literature for evidence-based best practices.

4. Determine congruence with organizational mission.

5. Identify planning goals and objectives.

6. Estimate resources required for the plan.

7. Prioritize according to available resources.

8. Identify timelines and responsibilities.

9. Develop a marketing plan.

10. Write and communicate the business plan/strategic plan.

11. Evaluate.

For example, an environmental analysis of the type of undergraduate nursing education that would be needed for the twenty-first century professional nurse led one school of nursing to begin planning a curriculum revision that would incorporate the increasing emphasis on safety and patient centered care. In addition, this analysis of the environment led faculty to understand that new models for clinical education will be needed to promote improved and expanded linkages between education and practice.

It is important that the internal environment as well as the external environment be carefully appraised. Whereas the external environmental assessment is broad based and attempts to view trends and future issues and needs that could impact the organization, the internal

assessment seeks to inventory the organization's assets and liabilities.

## SWOT Analysis

A **SWOT analysis** is a tool that is frequently used to conduct these environmental assessments. SWOT stands for strengths, weaknesses, opportunities, and threats. A SWOT analysis identifies both strengths and weaknesses in the internal environment and opportunities and threats in the external environment. The SWOT analysis is useful both for initial brainstorming and for a more formal planning document. Figure 10-1 is an example of a SWOT analysis that could be conducted by a university health care center (Jones & Beck, 1996).

## External Environment

### Internal Environment

| Opportunities | Strengths | Weaknesses | Threats |
|---|---|---|---|
| New Programs and Services | Financial Status/Cash Flow | Union Demands/ Relations | Competition |
| Advanced Technology | Supporting Practitioners/Nurses Alumni | Lag Time for Management Information System | Staffing Shortages |
| Population Growth | Programs and Services | Escalating Costs for Benefits/ Salaries | Increased Legislative Regulations |
| Legislative Changes | Employee Suggestion Program | Unprofitable Services | Unionization |
| Practitioner/Nurse Recruitment/ Retention | Equipment/ Technology | | Cuts in Funding of Nursing/Medical Education |
| New Markets | Administrative Team | | Revenue Loss for Caring for the Uninsured |
| Managed Care | Expertise of Staff | | Decreased Reimbursement |
| Prestige in the Community | Staff Satisfaction | | |
| | Location | | |
| | Marketing | | |
| | Patient Satisfaction | | |

FIGURE 10-1  **Key to success in strategic planning: SWOT analysis.** (Compiled with information from Jones, R., & Beck, S. (1996). Decision making in nursing. Clifton Park, NY: Delmar Cengage Learning.)

## REAL WORLD INTERVIEW

The electronic health record (EHR) plays an essential role in achieving the quality, safety, and efficiency goals that are defined in our strategic plan. Automation of the EHR is an evolutionary process whereby specific application implementations are sequenced and timed in accordance with the priorities established within the strategic plan. For this reason, the plan for automating the EHR is closely connected to the strategic plan. The financial and clinical indicators used for evaluating success of the strategic plan are similar to the indicators used for evaluating the success of the EHR implementation plan. These indicators include improvement in quality patient outcomes, reduction in adverse events, and increases in efficiency, along with many other desired outcomes as defined by the organization. An effective model for deriving value from both the strategic plan and the use of information technology is one where nursing takes the lead in setting the critical foundation for success. Nursing has a critical role in leading the process and ultimately driving the creation of new paradigms in improving health care.

**Rosemary Kennedy, RN, MBA**

Chief nursing informatics officer, Siemens Medical
Solutions Philadelphia, Pennsylvania

## Community and Stakeholder Assessment

A frequently overlooked but highly important area for analysis is the stakeholder assessment. A **stakeholder** is any person, group, or organization that has a vested interest in the program or project under review. A **stakeholder assessment** is a systematic consideration of all potential stakeholders to ensure that the needs of each of these stakeholders are incorporated in the planning phase.

For a program to be successful, the involvement of those who will be affected is essential. This is true whether the stakeholders are in the community or they are the unit staff who will be affected by a proposed strategic plan. When stakeholders are not involved in the project planning, they do not gain a sense of ownership and may accept a program or strategic goals only with limited enthusiasm, or not at all.

## Other Methods of Assessment

A number of methods can be used to support involvement in the strategic planning process. Thoughtful planning is required to determine the method and when to use the method.

**Surveys and Questionnaires** Frequently, surveys or questionnaires are used when there is a large number of stakeholders and only a general idea of the options available. For example, staff might be polled to see whether they would attend continuing education and which days and times would be most desirable.

**Focus Groups and Interviews** **Focus groups** are small groups of individuals selected because of a common characteristic, such as a recent diagnosis of diabetes. The focus group is invited to meet in a group and respond to questions about a topic in which they are expected to have interest or expertise. An example of a focus group would be a group of patients who have recently had experiences with childbirth. They might be asked to come together to discuss their obstetric experiences at the institution in the hope that the discussion will lead to insights or information that could be used for improving care or marketing services in the future. Focus groups are usually more time consuming and expensive to conduct than questionnaires or surveys. They work best when the topic is broad and the options are not clear. For example, an organization might conduct patient focus groups to determine what programs would be of most interest to a community.

**Advisory Board** Large projects often benefit from the formation of an advisory board selected from various constituencies affected by a proposed program. The advisory board does not have formal authority over a program, but it is instrumental in reviewing the planned program and making recommendations and suggestions. Because the advisory board is deliberately selected to reflect representation from various stakeholders and areas of expertise, it is expected that the board will be able to identify potential concerns and provide sound guidance for the program.

**Review of Literature on Similar Programs** A review of the literature should be completed prior to strategic planning or beginning any new project or program. This allows the project team to identify similar programs, their structures and organization, potential problems and pitfalls, and successes. This is an ongoing process that includes tentatively identifying programs, searching the literature for successes and issues, and then refining the program ideas.

**Best Practices** Identifying best practices or evidence-based innovations that have been adopted with success by other organizations can facilitate strategic planning. Consequently, nurses planning to develop a new program need to carefully examine the existing evidence and practices prior to beginning planning.

### Critical Thinking 10-2

Think about a community organization with which you are familiar. What is the mission of that organization? Compare it to the mission in Table 10-1. Is it clearly communicated to the stakeholders? Are the activities of the organization you have chosen reflective of its mission? If not, why do you think this is so?

## RELATIONSHIP OF STRATEGIC PLANNING TO THE ORGANIZATION'S MISSION

All strategic planning, goals, and objectives must be examined with an eye to the purpose or mission of the organization. Sometimes, organizations get into trouble when they move too far afield of their core mission. Consequently, each new project needs to be evaluated in light of its congruence with the main mission that has been identified. It is fine for an organization to move to

## REAL WORLD INTERVIEW

Loyola University Chicago is an independent, urban University grounded in the Jesuit Catholic tradition of service to humanity through learning, justice and faith. Loyola is dedicated to higher education, health care and community service. The University enrolls more than 15,000 undergraduate students; offers master's and doctoral degrees in 31 departments and has professional schools of nursing, medicine, business, law, education, and social work. The Marcella Niehoff School of Nursing is located at the Lakeshore Campus in Chicago and a new School of Nursing building is being built at the Loyola Health System Campus in Maywood, Illinois.

Prior to the early 1990s, there was little engagement with the Maywood, Illinois, community where the Health System Campus is located. At that time, members of the professional schools and the Health System met with leaders of the Maywood community to determine major health issues and concerns. They identified teen pregnancy, hypertension, sexually transmitted diseases, HIV, substance abuse and access to information about careers in the health profession for minority youth as a major concern. Based on these identified needs, the Marcella Niehoff School of Nursing and Stritch School of Medicine applied for and received grant funding from the PEW Foundation's Health of the Public Initiative, a national award designed to support partnerships between academic health professionals and local communities. Since community concerns focused on youth, the concept of Healthy Teens 2000 was the focus of grant funding. The goal was to build an infrastructure that would provide widespread support for the project among the partnering University, Health System, and community. An advisory board was formed which included representatives from multiple stakeholders, including faith based organizations, schools, parents, the local Chamber of Commerce, the Marcella Niehoff School of Nursing, the Stritch School of Medicine, and youth from the community. Initial projects including a youth mentoring program in which high school students were trained by Loyola Health professionals to be peer mentors and engaged school aged children in a variety of programs focusing on healthy lifestyle choices. During this period, it became apparent that students in the local high school also had many unmet health related needs. Based on risk assessment surveys and needs assessment data, it was determined that there was a need to establish a School-based Health Center (SBHC) at Proviso East High School which serves the Maywood community. Faculty from the Marcella Niehoff School of Nursing applied for and received over two million dollars in grants from the Health Resources and Services Administration (HRSA) and the Illinois Department of Human Services (IDHS) to develop the SBHC. The SBHC is operated by Loyola School of Nursing and delivers health promotion programs, primary health care, nutrition counseling, and mental health services for the 1,800 young people served by the school. The SBHC is now in its tenth year of operation, is fully accredited by IDHS and has been continually funded by HRSA, IDHS, and numerous private foundations. Members of the Healthy Teen Advisory Board were the first to envision such a project and continued to meet as a Healthy Teens Board until they were incorporated into the SBHC's Advisory Board in the 2000's. Today the SBHC is a thriving clinical site for graduate and undergraduate nursing students, dietetic interns, medical students and residents, and is a focal point for the high schools health related activities. It is a model of partnership between Marcella Niehoff School of Nursing, Stritch School of Medicine, and the community.

**Diana Hackbarth, PhD, RN, FAAN**
Professor and Director, Proviso East School-based Health Center
Maywood, Illinois

another project, but only if the new project is in line with the mission. Otherwise, there is a risk that the new programs will drain energy from the main mission. Figure 10-2 depicts the relationship of strategic planning to the organization's vision and mission.

## Planning Goals and Prioritizing Objectives

After all strategic goals and objectives have been identified, they need to be prioritized according to strategic importance, resources required, and time and effort involved. A timeline should be set. This will allow a thoughtful evaluation of

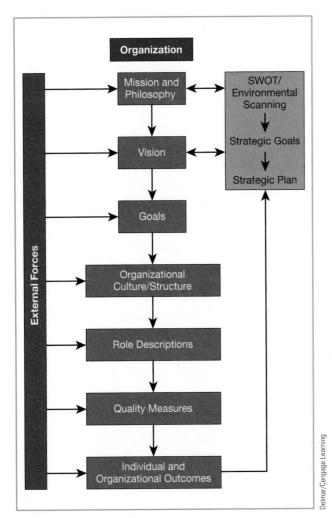

FIGURE 10-2 **Relationship of strategic planning to organizing care.**

each goal and objective and the degree to which each can be implemented in the specified time frame and with the available resources.

## Setting a Timeline with Responsibilities Identified

Setting a timeline for completion of a strategic plan is similar to the prioritization process in that the strategic importance, resources, and effort required are major considerations. Realistic timelines and individual responsibilities must be developed, specified, clarified, and communicated to all stakeholders. This will help to avoid misunderstandings and unmet expectations.

## Developing a Marketing Plan

If a part of strategic planning involves new programming for external audiences or if only internal redesign or restructuring is involved, the strategic plan, goals, and objectives will need to be communicated to all involved constituencies. Such communication will be needed, for example, when an institution is planning to implement a new information system to ensure that it remains competitive in the market. Designing, implementing, training, and evaluating this new system will require substantive changes in work flow and in the way that employees carry out their day-to-day work processes. If there has not been adequate thought to communication across the organization about the project, there is less chance of success and a greater risk of poor cooperation.

A marketing plan ensures that all stakeholders have the needed information. **Marketing** is the process of creating a product or health care service for patients, and it uses the four Ps of marketing—patient, product, price, and placement—to place desirable health care services or products in desirable locations at a price that benefits both patients and the health care facility. In this way, the health care facility, the patient, and the community benefit. Marketing of services does have a price tag, such as the cost of advertising campaigns on television and radio. Using printed materials, mailing information to patient residences, and advertising in journals, magazines, and newspapers are all examples of ways to educate and stimulate the public for future referrals for health care services. Once marketing strategies are implemented, most organizations attempt to measure their effectiveness, or return on investment.

# ORGANIZATIONAL CULTURE

**Organizational culture** consists of the deep underlying assumptions, beliefs, and values that are shared by members of the organization and typically operate unconsciously. Effective organizations that demonstrate high levels of four cultural traits can be measured with the 60-item Denison Organizational Culture Survey (DOCS). Each culture trait is measured by three indices. The traits include adaptability, with the indices of creating change, customer focus, and organizational learning; involvement, with the indices of empowerment, team orientation, and capability development; consistency, with the indices of core values, agreement, and coordination and integration; and mission, with the indices of strategic direction and intent, goals and objectives, and vision (available at http://www.denisonconsulting.com/products/cultureProducts/surveyOrgCulture.aspx, accessed December 28, 2010). The DOCS items were developed to address those aspects of culture that had demonstrated links to organizational effectiveness, such as having a shared sense of responsibility, possessing consistent systems and procedures, being responsive to the market, and having a clear purpose and direction for the organization.

# HIGH-RELIABILITY ORGANIZATIONS—A CULTURE OF SAFETY

High-reliability organizations have generated much interest from nurses in recent years. Nurses want to know how to halt the alarming rate of errors and preventable complications and develop health care systems that are safe for patients. High-reliability organization theory (HROT) is a framework describing characteristics of high-risk yet safe systems that were developed in non–health care systems such as nuclear submarines, aircraft carriers, and aviation. In these systems, a small slip or error could lead to a catastrophic

event. HROT describes five characteristics of high-reliability systems, including preoccupation with failure, reluctance to simplify, sensitivity to operations, commitment to resilience, and deference to expertise (Kemper & Boyle, 2009). Within highly reliable systems is a unique teamwork culture called a "culture of safety." An organization's **culture of safety** is the result of shared values and behaviors that demonstrate communications based on mutual trust, agreement on the importance of safety, and confidence in the ability to prevent errors through the use of known safety practices (Kemper & Boyle, 2009). Organizations that want to develop high reliability and a culture of safety should take these actions:

- Determine areas of high risk, e.g., monitor outcomes and use survey tools such as the DOCS and the National Database of Nursing Quality Indicators (NDNQI) RN Survey (available at http://www.nursingworld.org/MainMenuCategories/ANAMarketplace/ANAPeriodicals/OJIN/TableofContents/Volume122007/No3Sept07/NursingQualityIndicators.aspx).
- Develop action plans to improve performance in weak areas.
- Ask staff to speak up if they witness an unsafe practice.
- Learn from errors and near misses. Focus on analysis of system problems, not on blaming individuals.
- Evaluate the culture of safety. Consider use of the following:
  - Hospital Survey on Patient Safety Culture (available at http://www.ncbi.nlm.nih.gov/pmc/articles/PMC2912897/, accessed December 28, 2010)
  - Nursing Home Survey on Patient Safety Culture (available at http://www.ahrq.gov/qual/patientsafetyculture/nhsurvindex.htm, accessed December 28, 2010)
  - Safety Climate Survey (available at http://www.ahrq.gov/downloads/pub/advances/vol4/Connelly.pdf, accessed December 28, 2010)
  - Safety Attitude Questionnaire (available at http://psnet.ahrq.gov/resource.aspx?resourceID=3601, accessed December 28, 2010)
- Enhance teamwork skills. Consider use of TeamSTEPPS (available at TeamSTEPPS, accessed December 28, 2010), an evidence-based teamwork system available through the Agency for Healthcare Research and Quality.
- Safeguard patients (Kemper & Boyle, 2009). Develop a culture of safety; use standardized approaches to common procedures; develop decision aids, e.g., an OR checklist, standing orders, etc.; develop system redundancy, e.g., use of two nurses to confirm that the correct type of blood is being hung for a patient; develop care bundles that are a collection of best practices for a particular patient condition, e.g., a ventilator bundle (Sanford, 2010).

Finally, high-reliability organizations will encourage all staff to use their professional judgment and assess each patient individually. Not all patients fit the standard.

# ORGANIZATIONAL STRUCTURE

Organizations are structured or organized to facilitate the execution of their mission, goals, reporting lines, and communication within the organization. This is true of entire organizations as well as individual nursing units. There are a number of ways to describe organizational structures that involve classifying them by identifying selected characteristics. Each of these characteristics tends to exist on a continuum. For example, under the category of type or level of authority in an organizational structure, a highly bureaucratic, highly authoritarian structure is at one end of the continuum and a highly democratic, participative structure is at the other end. The highly authoritarian model is seen in the military and is well suited for the purpose of the military. When decisions need to be made quickly, with clarity and not with challenges or discussion, as in a battle situation, a highly authoritarian organizational structure works well.

An example at the other end of the continuum would be a multidisciplinary group of professionals meeting to determine the care management of a patient or patient population. An example of this group might be a hospice team task force made up of nurses, social workers, practitioners, home care aides, bereavement specialists, and chaplains, all meeting to discuss the care planning for a dying patient. In this situation, it will be important for team members of each discipline to freely contribute according to their particular area of knowledge and expertise. An organizational structure such as this can function successfully only in a participative, democratic manner.

## TYPES OF ORGANIZATIONAL STRUCTURES

Most often, the existing organizational structures are communicated by means of an organizational chart. Figure 10-3 is an example of an organizational chart for a typical acute care general hospital (Shortell & Kaluzny, 2006). This organization has a tall, bureaucratic structure with many layers in the hierarchy or chain of command and a centralized formal authority in the board of trustees. It represents a formal, top-down reporting structure.

## EVIDENCE FROM THE LITERATURE

**Citation:** Allison, M., & Kaye, J. (2005). Strategic planning for non-profit organizations. Hoboken, NJ: John Wiley & Sons, Inc.

**Discussion:** Why strategic planning? Because a well-wrought strategic plan helps you set priorities and acquire and allocate the resources needed to achieve your goals. It provides a framework for analyzing and quickly adapting to future challenges. And it helps all board and staff members focus more clearly on your organization's priorities, while building commitment and promoting cooperation and innovation. But to be effective, your plan will need to address the special needs of your institution and/or practice area. This workbook is packed with real world insights and practical pointers. It shows you how to: 1) Develop a clear mission, vision, and set of values; 2) conduct SWOT analyses and program evaluations; 3) assess patient needs and determine stakeholder concerns; 4) set priorities and develop core strategies, goals, and objectives; 5) balance the dual bottom lines of mission and money; 6) write and implement a solid strategic plan; 7) develop a user-friendly annual work plan; 8) establish planning cycles, gauge progress, and update strategies.

View Empire State College, State University of New York's strategic plan and the president's quarterly report that demonstrates the successful implementation of that plan. These can be found at www.esc.edu/presidentsoffice. They are titled 2006–2010 Strategic Plan and Vision 2015, respectively.

**Implications for Practice:** This book provides a clear guide and workbook for strategic planning tailored to the special circumstances of non-profit organizations such as many academic institutions, hospitals, and health care practices. The particular academic institution mentioned here provides a great example of a successful strategic plan that has been implemented.

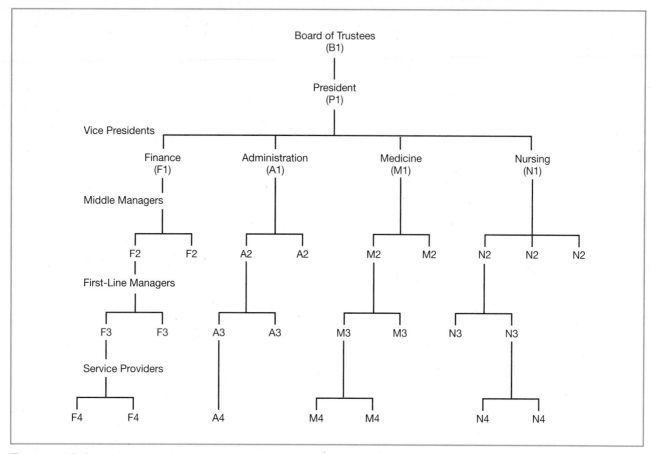

**FIGURE 10-3**  Organizational chart, formal tall bureaucratic, multilayered authority structure: Acute care general hospital.
(From Shortell, S., & Kaluzny, A. (2006). Health care management: Organization design and behavior [5th ed.] Clifton Park, NY: Delmar Cengage Learning.)

For example, three middle managers report to the nursing vice president, and two first-line unit nurse managers (NMs) report to each middle manager. Two service providers report to each first-line unit NM.

## Matrix Structure

Today, given the greater complexity of the health care system, more organizations are using matrix structures. Figure 10-4 shows a matrix design (Shortell & Kaluzny, 2006).

## Flat versus Tall Structure

Organizations are considered flat when there are few layers in the reporting structure. A tall organization would have many layers in the chain of command. An example of a flat organizational structure in a hospital would be one director of nursing with two head nurses reporting to her, one for maternal and child patient care units and one for medical-surgical patient care units (Figure 10-5). Contrast this flat type of structure in Figure 10-5 with Figure 10-6, which has many layers.

## Decentralized versus Centralized Structure

The terms *centralized* and *decentralized* refer to the degree to which an organization has spread its lines of authority, power, and communication. A tall, bureaucratic design such as that in Figure 10-6 would be considered highly centralized. A matrix design such as that in Figure 10-4 would be on the decentralized end of the continuum. As can be seen in Figure 10-4, the nursing manager can interface with the Alzheimer's disease program manager without going through a central, hierarchical core, as would happen in a bureaucratic structure such as that in Figure 10-3.

Other characteristics or attributes can be used to assess organizations. Many typologies exist that may be used for this purpose. For example, Shortell and Kaluzny (2006) suggest using external environment, mission/goals, work groups/work design, organizational design, interorganizational relationships, change/innovation, and strategic issues.

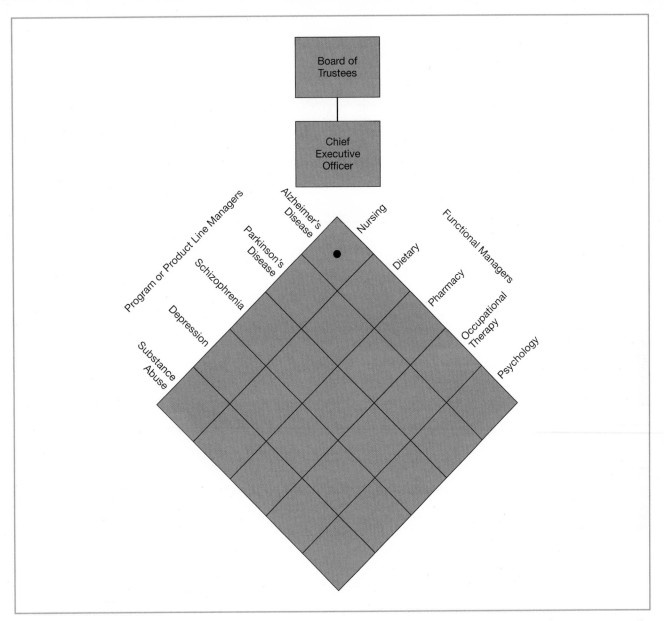

**FIGURE 10-4** Matrix design: A psychiatric center. An individual worker in this example is part of the Alzheimer program as well as a member of the nursing department. (From Shortell, S., & Kaluzny, A. (2006). Health care management: Organization design and behavior [5th ed.]. Clifton Park, NY: Delmar Cengage Learning.)

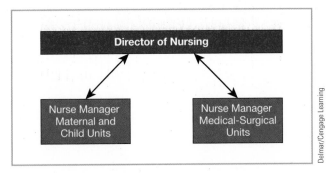

Delmar/Cengage Learning

**FIGURE 10-5** Example of a flat organizational structure.

Refer to Table 10-4 for an example of how some of these attributes can be assessed in four different health service organizations.

## DIVISION OF LABOR

The way that the labor force is divided or organized has an impact on how the mission is accomplished. The organizational chart in Figure 10-7 graphically depicts a functional design of how the formal authority in this organization is structured. At the highest level, the board of trustees delegates authority to the chief executive officer,

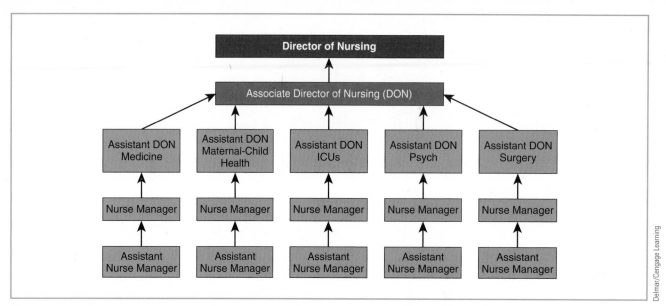

FIGURE 10-6  Example of a tall organizational structure.

## TABLE 10-4  Attributes of Four Health Services Organizations

| ATTRIBUTE | HEALTH MAINTENANCE ORGANIZATIONS (HMOS) | HOME HEALTH CARE AGENCIES | HOSPITALS | PHARMACEUTICAL COMPANIES |
|---|---|---|---|---|
| Mission/goals | Primary care emphasis; keep people well | Quality of life; maintaining functional status | Acute care emphasis; curing illness | Research and development (R&D) emphasis; new product development |
| Work group/ work design | Primary care teams; coordinated referrals | Simple design: primary nursing; one-on-one patient contact | Departmental and across-departmental teams; high need for coordination | Separation of functions possible; R&D vs. sales; relatively low need for coordination |
| Organizational design | Functional and divisional | Functional | Divisional, product line, or matrix | Divisional and strategic business units |
| Change/ innovation | Creates and responds to new patient care management approaches | Respond to demographic and social changes | Respond to the new paradigm; implement new role within vertically integrated systems | Respond to new product development demands |
| Strategic issues | Expand the concept of managed care | Demonstrate continuing value, and therefore, reimbursement for services | Fit into an expanded and changing delivery system | Decrease time to develop new drugs |

**Source:** Compiled with information from Shortell, S., & Kaluzny, A. (2006). *Health care management: Organization design and behavior* (5th ed.). Clifton Park, NY: Delmar Cengage Learning.

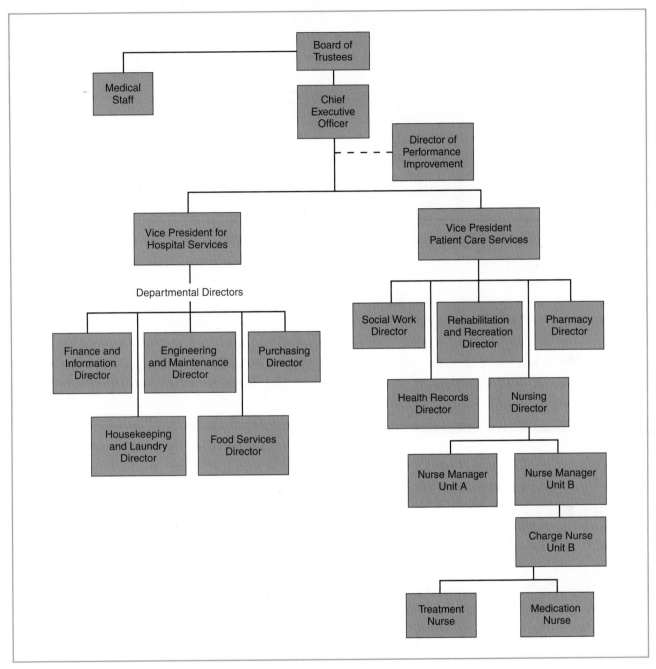

**FIGURE 10-7** **A functional design: Nursing home.** (From Shortell, S., & Kaluzny, A. (2006). Health care management: Organization design and behavior [5th ed.]. Clifton Park, NY: Delmar Cengage Learning.)

who delegates to the two vice presidents. The vice president of Patient Care Services has five directors, that is, directors of Social Work, Rehabilitation, Pharmacy, Nursing, and Health Records departments. The nurse managers (NMs) of Unit A and Unit B report to their department director of nursing. The charge nurse reports to the NM of Unit B. The treatment nurse and the medication nurse report to the charge nurse. In this functional design, the division of labor is efficient and specialized. A danger with functional division of labor is that each individual may be so focused

on a specific area that he or she has little perspective about the overall picture. For example, a treatment nurse may focus only on treatments and have little information about the total patient.

In the matrix structure shown in Figure 10-4, the structure is less important, and the workforce roles and reporting relationships are based on the project or task to be accomplished, rather than on a rigid hierarchy. An example of this is the planning involved in the preparation for a Joint Commission (JC)—formerly Joint Commission

## CASE STUDY 10-1

A patient developed a rash from a new medication, unbeknownst to the medication nurse, who never asked about any signs of problems. The treatment nurse noticed the rash during a routine dressing change but never thought to inquire about any new dietary or medication changes. It was not until the time of discharge when the patient read the drug information sheet advising that any skin changes be reported that the patient asked the discharge planning nurse if the week-old rash was significant.

What could have been done differently?

Was anyone at fault? Who?

Why is good communication especially important in a situation in which there is a functional division of labor?

What types of problems could you expect if staff members focused on their own tasks and failed to communicate with each other about the patient's emotional, psychosocial, educational, and discharge needs?

---

on Accreditation of Healthcare Organization (JCAHO)—review. The JC team could be composed of various individuals at varying levels of responsibility and from programs across the organization, but they could interact with staff at all levels and report as a task force at a high level in the organization.

## Span of Control

The term *span of control* is used to designate the number of individuals that report to one person. If the span of control is too narrow, an organization may become top heavy, and much time may be wasted in unnecessary communications up and down the chain of command, resulting in lost efficiency. On the other hand, if the span of control is too broad, it is difficult for one manager to give adequate attention to the support and development of all the individuals that report to him or her.

## Roles and Responsibilities

Note that exact roles and responsibilities within each level and division are not defined on the organizational charts beyond specifying the given division, for example, nursing. Scope of responsibilities, specific duties, and specific job requirements are found in documents such as individual job or position descriptions.

## Reporting Relationships

An organizational chart, such as the one in Figure 10-7, allows you to determine the formal reporting relationships, which are shown with a solid line. Sometimes dotted lines are used in an organizational chart to depict dual

or secondary reporting relationships. An example of this might be the role of the director of performance improvement. This individual might report directly to the chief executive officer but also have position accountabilities to the board of trustees. The formal reporting relationships may or may not reflect the actual communications that occur within the institution.

## Basis for Division of Labor

There are a number of ways to divide the workload in an organization. The important consideration is that the manner in which the work is distributed should match the goals of the organizational unit and should contribute maximally to the efficient, effective attainment of the desired outcomes.

**Functional Division of Labor**   In a functional division of labor, work is divided by job activity. Nursing care that would be distributed in this manner might consist of a medication nurse role, a treatment nurse role, and an education or a discharge planner role. Each nurse would be specialized and would potentially become highly proficient in a given functional role.

**Division of Labor by Geographic Area**   Care delivery divided according to geography or location can be efficient. It might consist of the hospital and ambulatory care or, at smaller unit levels, the North Team and the West Team. Frequently, care provided by home health agencies is divided by geographic district for efficiency in travel. At the health care system level, geographical division could mean that each major area, such as the hospital or the clinics, would have separate supporting services, such as two pharmacies, one in the outpatient clinic and one in the hospital. Both clinic and inpatient areas could, and often do, have separate medical records departments. An obvious concern in such arrangements is lack of coordination and duplication of services.

**Division of Labor by Product or Service** Sometimes, care delivery is organized around product lines or service lines. This is a type of functional division of work, but it is based on a patient's diagnosis or the specialty care required by a patient. For example, there might be a cardiology service line, a woman's health service line, and an oncology service line. This can lead to improved quality of care and decreased confusion for the patient, because the information and protocols used in the outpatient side would be consistent with the information and protocols used in the hospital and across the entire health care system. Figure 10-8 demonstrates a product line design (Shortell & Kaluzny, 2006).

# FACTORS INFLUENCING ORGANIZATIONAL STRUCTURES

Shortell and Kaluzny (2006) identify situations in which an organization's structure should be rethought. These situations include the experiencing of severe problems in the areas of either performance, customer satisfaction, or internal/external review. Other reasons to rethink an organization's structure include a significant change in the internal or external environments; new programs, services, or product lines; or a change in the leadership. It is also important to note that design changes can be made in many levels. These range from changes in individual positions, work groups, clusters of work groups, total organizations, to changes in networks and, finally, to entire systems.

## Technology

Changes in technology have wrought many design changes in health care organizations. The ability to provide safe same-day surgery has led to patients staying in hospitals for

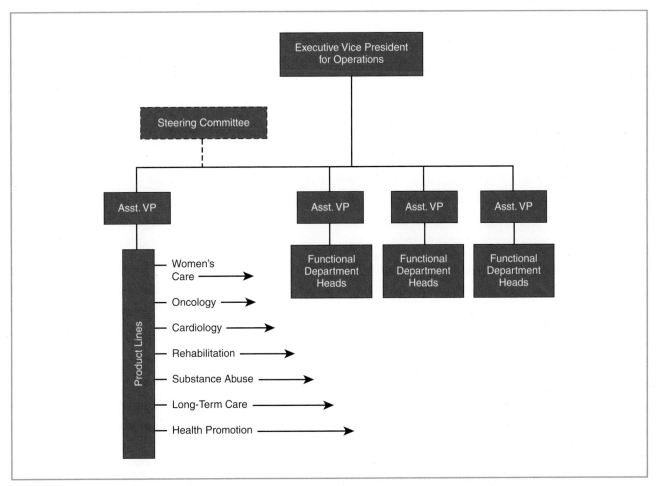

**FIGURE 10-8 Product line design.** (From Shortell, S., & Kaluzny, A. (2006). Health care management: Organization design and behavior [5th ed.]. Clifton Park, NY: Delmar Cengage Learning.)

less than 24 hours. This means that there is an increased need for nurses to provide discharge planning with patients that they may have just met and in a very rapid time frame. Information technology or medical information systems departments in most institutions were relatively small 10 to 15 years ago. Today, there may be numerous employees involved in information technology throughout the organization. Communication and information exchange is instant and broad based with e-mail and other technologic advances. This has had tremendous impact on the information flow in an organization. Information can rapidly be in more than one place at a time. Nurses and practitioners can update their nursing and medical care and medications quickly. Treatment plans may be revised instantaneously. In general, nurses and practitioners are less dependent on old communication patterns, which is changing the way we deliver care.

## Sociocultural Environment

We are working in an increasingly diverse workforce. Educational, language, ethnic, cultural, age, and values divisions are greater than ever before. Nurses entering the health care field will need to be increasingly sensitive and skilled in dealing successfully with the differences that exist between themselves, their coworkers, and patients.

## Size

Often, the size of an organization, system, or department will determine the organizational structure that is used. The larger an organization, the more complex structures are needed. In a smaller, rural hospital, one nursing department with a director and an assistant director of nursing may be adequate to provide leadership resources. However, in an urban, multihospital health care system, there may need to be a corporate nursing department, headed by a systemwide chief nurse executive, and separate nursing departments at each hospital in the system.

## Repetitiveness of Tasks

Another factor that determines organizational structure is the repetitive nature of the tasks to be accomplished. It is possible to manage a larger number of individuals, units, or departments if they are all engaged in similar activities and processes. If there is a great deal of differentiation among their tasks, more levels of management are usually needed.

## TRENDS IN ORGANIZATIONS

There is a need for a leadership culture that promotes sound ethical values and quality assurance. In turn, cultures with strong leadership promote productivity and performance. One nurse leader can help to make this difference. Consequently, transformational nurse leaders are needed who can create ethically sound environments and assist nurses to strive for quality outcomes and personal mastery. There is a need for concerted efforts to reconnect nursing education and nursing service and assure that nurses employed in leadership positions are educationally prepared to function in management and clinical decision-making roles.

## KEY CONCEPTS

- There are increasing opportunities for nurses to become involved in strategic planning for the delivery of health care services in their organizations and communities. To do so effectively, however, they will need a basic understanding of the way in which organizations are structured, how organizational systems function, and how to engage in the strategic planning process.

- The mission statement reflects the organization's values and provides the reader with an indication of the behavior and strategic actions that can be expected from that organization.

- A health care organization needs to have a good idea of where it fits into its environment and what types of programs and services are needed and demanded by its customers or stakeholders.

- The pivotal value of strategic planning is that it requires an organization to focus on its raison d'être and its mission and to test how its operations are leading to accomplishment of that mission.

- The purpose of strategic planning is twofold. First, it is important that everyone has the same idea or vision of where the organization is headed; second, a good plan can help to ensure that the needed resources are available to carry out the initiatives that have been identified as important to the unit or agency.

- A stakeholder assessment is a systematic consideration of all potential stakeholders to ensure that the needs of each of these stakeholders are incorporated in the planning phase. For a program to be successful, the involvement of those who will be affected is essential.

- Organizations are structured or organized in a manner that is designed to facilitate the execution of their mission and their strategic plans.

- There are a number of ways to describe organizational structures that involve classifying them by identifying selected characteristics.

# KEY TERMS

culture of safety
focus groups
marketing
mission statement

organizational
culture
philosophy of an
organization

stakeholder
stakeholder
assessment
strategic plan

SWOT analysis
vision

# REVIEW QUESTIONS

1. A document that describes the institution's purpose and philosophy is which of the following?
   A. the organizational chain of command
   B. the organizational chart
   C. the mission statement
   D. the strategic plan

2. Which of these statements is true about a strategic plan? Select all that apply.
   _____ A. Requires focus on its mission and vision
   _____ B. Should not be attempted during turbulent times
   _____ C. Requires an assessment of the environment
   _____ D. Needs to be revisited periodically
   _____ E. Needs to be preceded by a SWOT analysis

3. The most formal and hierarchical organizational structure would be expected to have an organizational chart with which of the following?
   A. A matrix design
   B. Many layers of command
   C. A product line design
   D. A number of dotted lines representing reporting relationships

4. SWOT means which of the following?
   A. Strengths, weaknesses, opportunities, threats
   B. Strengths, worries, outcomes, threats
   C. Strengths, weaknesses, opportunities, treatment
   D. Structures, worries, outcomes, threats

5.

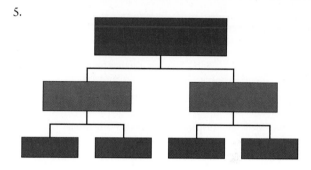

   The above figure is an example of what type of organizational structure?
   A. Tall
   B. Matrix
   C. Product line
   D. Flat

6. Which of the following are steps of the strategic planning process? Select all that apply.
   _____ A. Stakeholder analysis
   _____ B. Developing organizational charts
   _____ C. Identifying planning goals and objectives
   _____ D. Hiring well-prepared individuals
   _____ E. Developing mission and vision statements

7. A small hospital with 25 beds would probably be best served by which of the following organizational structures?
   A. Matrix
   B. Tall
   C. Flat
   D. Product line

# REVIEW ACTIVITIES

1. Identify a situation in which a strategic plan could guide an organization in its choices among alternative actions.

2. Write a beginning mission statement and strategic plan for your professional nursing career.

3. You are asked to plan for the advisory board for your institution's proposed hospice program. How would you go about determining who to include on the advisory board? What

groups of professionals and consumers would you want to see represented on a hospice advisory board? Identify several candidates and the stakeholder group they might represent.

4. Examine the organizational structure of an organization or institution with which you are familiar. How would you characterize it using the types of structures that were discussed in this chapter?

# EXPLORING THE WEB

Upon completion of your nursing degree, you are planning to interview for a position at an area hospital. In preparation for your interview, you want to understand the mission as well as other information about that institution. Today, that information is readily available on the Web. For example, if you were planning to apply at Loyola University Chicago, you would go to www.luhs.org. Another example is Children's Memorial Hospital in Chicago at www.childrensmemorial.org.

Look at these Web pages, paying particular attention to the descriptions they provide of the organizations' missions. What impressions do you form about these organizations and their missions? Does the stated mission seem to fit with the general "feel" that you get from the Web site? Could you easily find information about positions available? About the institution? Try this exercise with your local hospital or medical center.

# REFERENCES

Allison, M., & Kaye, J. (2005). Strategic planning for non-profit organizations. Hoboken, NJ: John Wiley & Sons, Inc.

Drucker, P. (1973). *Management tasks, responsibilities, and practices.* New York: Harper & Row.

Institute of Medicine. (1999). *To err is human.* Washington, DC: National Academies Press.

Institute of Medicine. (2001). *Crossing the quality chasm.* Washington, DC: National Academies Press.

Jones, R., & Beck, S. (1996). *Decision making in nursing.* Clifton Park, NY: Delmar Cengage Learning.

Kemper, C., & Boyle, D. K. (2009). Leading your organization to high reliability. *Nursing Management,* (April), 14–18.

March of Dimes. (2006). The March of Dimes story. Retrieved September 6, 2006, from http://www.marchofdimes.com/printableArticles/789_821.asp.

Mission, Vision and Values (1997–2011). Appalachian Regional Healthcare. Accessed March 8, 2011, at http://www.arh.org/AboutUs/mission.php.

National Quality Forum Announces Publication of New Report on Nursing-Sensitive Care Performance Measures. (2004). The National Quality Forum. Accessed March 7, 2011, at http://www.rwjf.org/files/research/prNursingReportAvailableFINAL100404.pdf.

Sanford, K. D. (2010). Designing more reliable nursing systems. Healthcare Financial Management Association. Available at http://www.hfma.org/Publications/E-Bulletins/Business-of-Caring/Archives/2010/Spring/Designing-More-Reliable-Nursing-Systems/. Accessed December 28, 2010.

Shortell, S., & Kaluzny, A. (2006). *Health care management: Organization design and behavior* (5th ed). Clifton Park, NY: Delmar Cengage Learning.

# SUGGESTED READINGS

Arnold, L., Campbell, A., Dubree, M., Fuchs, M. A., Davis, N., Hertzler, B., et al. (2006). Priorities and challenges of health system chief nursing executives: Insights for nursing educators. *Journal of Professional Nursing,* 22(4), 213–220.

Brewer, C. S. (2005). Health services research and the nursing workforce: Access and utilization issues. *Nursing Outlook,* 53(6), 281–290.

Buerhaus, P., Donelan, K., Ulrich, B., Norman, L., Williams, M., & Dittus, R. (2005). Hospital RNs' and CEOs' perceptions of the impact of the nursing shortage on the quality of care. *Nursing Economic$,* 23(5), 214–221.

Clancy, C., Sharp, B. A., & Hubbard, H. B. (2005). Guest editorial: Intersections for mutual success in nursing and health services research. *Nursing Outlook,* 53(6), 263–265.

Covaleski, M. A. (2005). The changing nature of the measurement of the economic impact of nursing care on health care organizations. *Nursing Outlook,* 53(6), 310–316.

Jones, C. B., & Mark, B. A. (2005). The intersection of nursing and health services research: Overview of an agenda setting conference. *Nursing Outlook,* 53(6), 270–273.

Jones, C. B., & Mark, B. A. (2005). The intersection of nursing and health services research: An agenda to guide future research. *Nursing Outlook,* 53(6), 324–332.

Mick, S. S., & Mark, B. A. (2005). The contribution of organization theory to nursing health services research. *Nursing Outlook,* 53(6), 317–323.

Ricketts, T. C., & Goldsmith, L. J. (2005). Access in health services research: The battle of the frameworks. *Nursing Outlook,* 53(6), 274–280.

Spetz, J. (2005). The cost and cost-effectiveness of nursing services in health care. *Nursing Outlook,* 53(6), 305–309.

# CHAPTER 11

# Effective Team Building

Karin Polifko-Harris, PhD, RN, CNAA;
Crisamar J. Anunciado, RN, MSN, FNP-BC

*The strength of the team is each individual member ...the strength of each member is the team.*

(Coach Phil Jackson, Chicago Bulls)

## OBJECTIVES

Upon completion of this chapter, the reader should be able to:

1. Identify advantages and disadvantages of teamwork.
2. Review key concepts of creating an effective team.
3. Relate the stages of a team process.
4. Relate ways to create a conducive environment for teamwork
5. Identify the qualities of an effective team leader.
6. Discuss Crew Resource Management.
7. Identify elements of groupthink.

Delmar/Cengage Learning

*You are a new graduate nurse who has six months of clinical experience working on an oncology unit. Your nurse manager has observed your skill and compassion for terminally ill patients and your warm interaction and interpersonal relationship with the other staff nurses on the unit. Complimenting you on your skills and personal work ethic, the nurse manager has also asked if you would be interested in joining the Interdisciplinary Cancer Support Committee to help address pain management and other issues to improve patient care. If you accept this responsibility, this will be your first committee membership.*

*How would you respond to this request?*

*Would you be ready at this point to accept the responsibility?*

*How would you prepare yourself for this unfamiliar role?*

*What qualities or skills do you need to possess to become a productive member of the team?*

In today's health care environment, great demands are placed on each health care professional to provide the best quality of care efficiently, safely, and cost-effectively to optimize patient care outcomes. Many administrators and nurse managers recognize that effective interprofessional communication and collaboration through teamwork is needed to do this. In the interest of creating a safe patient care environment, health care professionals recognize that each discipline can no longer work alone. However, collaboration among health care professionals with varying specialties and focuses is a complex process. It requires careful planning, time, and effort to achieve the common goal of positive patient care outcomes.

A popular trend in human resource development is team training. Team training is well developed by many in the airline industry, major businesses, large corporations, bank institutions, and health care organizations. The demand for team training is so great that in the last few decades, there has been an explosion of team training companies providing tool kits, seminars, and conferences to improve teamwork and collaboration among executives, administrators, and other employees. Collaboration and teamwork among staff nurses and other disciplines in the health care setting is so critical to optimizing patient care safety and outcomes that

it is a priority for most health care administrators, directors, and managers (Amos, Hu, & Herrick, 2005).

This chapter focuses on the different aspects of effective team building. It discusses the advantages and disadvantages of teams and the various types of teams and committees using the Tuckman and Jensen Conceptual Model of the team process (1977). The chapter also dissects the components of a successful team.

# DEFINITION OF A TEAM

According to Buchholz and Roth (1987), authors of *Creating a High Performance Team*, "Wearing the same T-shirt doesn't make you a team." If a collection of T-shirts does not make a team, what then makes a team? What characteristics of people make a team? What is a team?

An often-quoted definition of a team is by Katzenbach and Smith (2003), authors of *Creating a High-Performance Team*: "A **team** is a small number of people with complementary skills who are committed to a common purpose, performance goals, and approach for which they are mutually accountable." There are several types of teams, and these exist for specific purposes. Thomas Edison, the inventor, said, when asked why he had a team of 20 first assistants, "If I could solve all the problems myself, I would!"

## CHARACTERISTICS OF A SUCCESSFUL TEAM

A multidisciplinary or interdisciplinary team is an interprofessional collaborative team involved in sharing, partnership, working together, and power issues. Teamwork occurs when multiple health professionals with varied backgrounds work together with patients, their families or caregivers, and the community in order to optimize the delivery of the best possible care (McDonald & McCallin, 2010). Characteristics of a successful team are:

1. They have a definite purpose.
2. They are comprised of health care professionals with varied skills and backgrounds.
3. They communicate closely.
4. They understand each other's roles.

Teams are created for specific purposes. They can be permanent or temporary (come together for a specific purpose and then disband) depending on the needs and goals of the health care organization and an individual patient's care (Riley, 2009).

## COMMITTEES

In the health care setting, teams serve on standing committees, advisory committees, and ad hoc committees, which are created for specific goals or tasks. An example of a standing committee is the nursing policy and procedure committee, which meets routinely to review, revise, and approve nursing-related policies and procedures. Other types of standing committees may be created to help meet requirements of federal regulatory agencies such as the Department of Health and Human Services (DHHS) or national accreditation agencies such as the Joint Commission (JC), formerly the Joint Commission on Accreditation of Healthcare Organizations (JCAHO). Examples of these standing committees are the bioethics committee, medication safety committee, and the quality improvement council. An example of an ad hoc committee is a hypoglycemia treatment policy and procedure committee, which will meet only until the policy and procedure is developed, approved, and adopted by the institution. An example of an advisory committee is the nursing clinical leadership committee. This committee may be overseen by the chief nursing officer and include nurse managers and other members of the interdisciplinary team who meet to discuss and advise on concerns pertaining to the professional nursing staff.

Any type of team can be effective if the members possess clarity of purpose, a unified goal, and an understanding of each other's role in achieving team objectives. Central to any health care team is a general goal of improving various aspects of the patient's quality of care (Bennett, Perry, & Lapworth, 2010).

## ADVANTAGES OF TEAMWORK

Teams do not evolve by happenstance, nor is the path to effective teamwork easy. Developing an effective team requires ample planning, with conscious and deliberate intentions focused on building its foundation through an organized system. Teamwork has both advantages and disadvantages.

Communication, along with the other characteristics of successful teamwork discussed earlier, is central in effective interprofessional collaboration for efficient and safe delivery of patient care. Collaboration among health care professionals with increasingly diverse specialization optimizes safe patient care outcomes.

Teamwork equalizes power through shared governance. Porter-O'Grady (2005) advocates shared governance, where power is more evenly distributed among the nursing staff and leaders, as an effective tool to promote empowerment, ownership of one's own clinical practice, and responsibility and accountability among nursing staff members. Power comes from having knowledge of each team member's strengths and weakness, focusing on the strengths and working around the weaknesses; cultivating team trust; focusing on the positive; creating a common vision; making the vision a reality; fostering open and honest communication; encouraging creativity; and creating an environment of constant renewal, giving permission for personal downtime and promoting a work/life balance (DiMichele & Gaffney, 2005).

Shared governance also facilitates interprofessional collaboration through working towards common goals. This increases personal and professional growth and staff morale. When staff are valued and included in the decision-making process, there is a boost in professional autonomy, which in turn eventually improves the standard of care (Caples & March, 2009).

 **EVIDENCE** FROM THE LITERATURE

**Citation:** Miller, K., Riley, W., & Davis, S. (2009). Identifying key nursing and team behaviours to achieve high reliability. *Journal of Nursing Management, 17,* 247–255.

**Discussion:** The aim of this study was to identify key nursing and interdisciplinary team behaviors that promote high reliability during critical events. Technical and team competence are necessary to achieve high reliability in interdisciplinary teams to ensure optimum patient care safety. Technical competence is generally guaranteed because of professional training, licensure, and practice standards. During critical events, team competence is difficult to observe, measure, and evaluate in interdisciplinary teams. Markers of team competence include having clear situational awareness; using the Situation, Background, Assessment, Recommendation, and Response (SBARR) method of communication (Table 8-8); using closed-loop communication, i.e., repeating back information when one team member makes a request of another team member; and having a shared mental model of requirements, procedures, and role responsibilities of the team. These markers, which are necessary for nurses to contribute to highly reliable interdisciplinary teams, are not consistently observed during critical events. This constitutes a breach in defensive barriers for ensuring patient safety.

**Implications for Practice:** Nurses make an important contribution to assuring effective technical and team competence through appropriate and timely transfer of information during critical events. Nurses need to identify important clinical and environmental cues and act to ensure that the team progresses along the optimal course for patient safety.

Teamwork improves interpersonal relationships and job satisfaction. In a study conducted among medical-surgical nursing staff on the factors improving job satisfaction, the authors noted that team building decreased the employee turnover rate from 13.42% to 6.56%. It also improved productivity, decreased absenteeism, and stabilized the workforce. This stability in employee turnover allowed the hospital to improve quality of care and provided positive economic benefits for the hospital (Amos, et al., 2005).

## DISADVANTAGES OF TEAMWORK

In this era of multicultural nursing, Caples and March (2009) found several obstacles to working together as a team. These obstacles include discrimination, exploitation, exclusion by colleagues, conflicts with local nursing practices, and language difficulties. These are unique and difficult but not impossible challenges in teamwork. Although it is often true that "two heads are better than one," there are some disadvantages of teamwork. Teams may take longer to achieve a goal than would one individual. On a team composed of varied disciplines, or even on a team with all members from the same discipline but with various levels of experiences and backgrounds, one may expect that a single patient care situation may produce very diverse solutions. The team members may have disagreements on the best course of action to take for a specific situation. Another disadvantage is that teams develop through time, going through predictable stages of selecting the right members for the team, organizing team goals and roles, and collaborating as a team. This team process takes time, effort, and resources.

Some team members may lack interest, motivation, ability, or skill to participate in the team process. These members may have been appointed or self-appointed for whatever reason, but they may not do the work as expected. Factors such as personality differences, personal work ethics, and varied perceptions of team goals may impede effective team collaboration. This, in turn, may create tension among other team members, cause delay in achieving goals, and cause frustration (Polifko-Harris, 2003; Katzenbach & Smith, 2003).

## INFORMAL TEAMS

Anyone working in an organization needs to be aware of the informal teams that influence the organization. Shortell and Kaluzny (2006) state that the importance of informal work-group structure and group processes has been recognized for many years. The Hawthorne experiments firmly established the proposition that an individual's performance is determined in large part by informal relationship patterns that emerge within work groups (Roethlisberger & Dickson, 1939). The work group has an impact on individual behaviors and attitudes because it controls so many of the stimuli to which the individual is exposed in performing organizational tasks (Hasenfeld, 1983).

Informal groups are not directly established or sanctioned by the organization but are often formed naturally by individuals in the organization to fill a personal or social interest or need. Shortell and Kaluzny (2006) identify a number of circumstances under which informal groups can have a negative impact on an organization. Groups may become overly exclusionary and lead to interpersonal conflict. In other cases, informal groups can become so powerful that they undermine the formal authority structure of the organization.

Informal groups can assume a change agent role. They are often responsible for facilitating improvements in working conditions. Such informal groups sometimes evolve into formal groups. Informal groups may also emerge to deal with a particular organizational problem or to work toward changes in organizational policies and procedures. In sum, informal groups play a unique role in organizations. These roles may be positive or negative.

## STAGES OF A TEAM PROCESS

Teams evolve through a predictable development process. A widely used theory of the team development process was introduced by Tuckman (1965) and then modified by Tuckman and Jensen (1977). They identified five stages of team development: forming, storming, norming, performing, and adjourning. Teams develop at various paces, depending on the team's composition, experiences, relationships, and type of tasks. Understanding the phases of the team development process may help improve team development and participation (Amos, et al., 2005) (Table 11-1).

### FORMING STAGE

The first phase of the team process is the forming stage. This stage occurs when the group is created and they meet as a team for the first time. The team members come to the meeting with zest and a sense of curiosity, adventure, and even apprehension as they orient themselves to each other and get to know each other through personal interaction and perhaps team-building activities. With the help of the team leader or facilitator, they will explore the purpose of the team, why they are called to be a part of the team, and what contribution they can bring to the table. When the purpose of the team is clearly identified, they may proceed to establishing their team goals and expectations and setting boundaries for the teamwork.

### STORMING STAGE

The second phase of the team process is the storming stage. As the group relaxes into a more comfortable team setting, interpersonal issues or opposing opinions may arise that may cause conflict between members of the team and with the team leader. This may cause feelings of uneasiness in the

## TABLE 11-1  Tuckman and Jensen's Stages of Team Process

| STAGES | DESCRIPTION |
| --- | --- |
| Forming | *Relationship development:* Team orientation, identification of role expectations, beginning team interactions, explorations, and boundary setting occurs. |
| Storming | *Interpersonal interaction and reaction:* Dealing with tension, conflict, and confrontation occurs. |
| Norming | *Effective cooperation and collaboration:* Personal opinions are expressed and resolution of conflict with formation of solidified goals and increased group cohesiveness occurs. |
| Performing | *Group maturity and stable relationships:* Team roles become more functional and flexible and structural issues are resolved, leading to supportive task performance through group-directed collaboration and resource sharing. |
| Adjourning | *Termination and consolidation:* Team goals and activities are met, leading to closure, evaluation, and outcomes review. This may also lead to reforming when the need for improvement or further goal development is identified. |

**Source:** Compiled with information from Tuckman & Jensen, 1977; Hall & Weaver, 2001; Polifko-Harris, 2003; Amos et al., 2005.

group. It is important at this stage to understand that conflict is a healthy and natural process of team development. When members of the team come from various disciplines and specialties, there is always a tendency to approach an issue from several completely different standpoints. These differences need to be openly confronted and addressed so that effective resolution of the issue may occur in a timely manner. Real teams don't emerge unless individuals on them take risks involving conflict, trust, interdependence, and hard work (Katzenbach & Smith, 2003).

## NORMING STAGE

The third stage is called norming. After resistance is overcome in the storming stage, a feeling of group cohesion develops. Team members master the ability to resolve conflict. Although complete resolution and agreement may not be attained at all times, team members learn to respect differences of opinion and may work together through these obstacles to achieve team goals. Communication of ideas, opinions, and information occurs through effective cooperation among the team members. Overcoming barriers to performance is how groups become teams (Katzenbach & Smith, 2003).

## PERFORMING STAGE

The fourth phase of the team development process is the performing stage. In this stage, group cohesion, collaboration, and solidarity are evident. Personal opinions are set aside to achieve group goals. Team members are openly communicating, know each other's roles and responsibilities, are taking risks, and are trusting or relying on each other to complete

assigned tasks. The group reaches maturity at this stage. One of the biggest strengths of this stage is the emphasis on maintaining and improving interpersonal relationships within the team as members function as a whole. Kenneth Blanchard, author of the *One Minute Manager,* sums it up with his comment, "None of us is as smart as all of us."

## ADJOURNING STAGE

The fifth and final stage of team process development is the adjourning stage. Termination and consolidation occur in this stage. When the team has achieved their goals and assigned tasks, the team closure process begins. The team reviews their activities and evaluates their progress and outcomes by answering the questions: Were the team goals sufficiently met? Was there anything that could have been done differently? The team leader summarizes the group's accomplishments and the role played by each member in achieving the goals. It is important to provide closure or feedback regarding the team process to leave each team member with a sense of accomplishment.

## ANATOMY OF A WINNING TEAM

Henry Ford once said, "Coming together is a beginning. Keeping together is progress. Working together is success." Effective teamwork is essential in any setting, whether it be in a large business corporation, in a complex health care system, in dynamic social assemblies, and even within the close

## Critical Thinking 11-1

Teamwork on a patient care unit—day shift routine:

Throughout the day shift, nursing and unit staff communicate and work together to deliver quality patient care.

7:00 a.m.    Day shift takes hand-off patient shift report from night shift
7:15 a.m.    Charge nurse reviews patient care assignments with all nurses and unit staff; goals and priorities are set
7:20 a.m.    Patient assessment, vital sign assessment, lab work, IV assessment, etc.
7:30 a.m.    Breakfast served
7:45 a.m.    A.M. care begins
8:00 a.m.    Medications given, practitioners make rounds, patient sent for diagnostic tests, nursing care standards followed, documentation begun, hourly regular patient rounds and planning care with other disciplines, etc.
11:30 a.m.   Vital sign assessment
12:00 p.m.   Lunch served; noon medications given
2:00 p.m.    Intake and output reports completed; documentation completed
3:00 p.m.    Hand-off patient shift report from day shift to evening shift

How does teamwork get patient care completed? Does patient care on your clinical unit follow a time sequence similar to the routine above?

network of a family unit. People cannot avoid interacting with others. This is deeply ingrained in our culture and society. In today's health care setting, effective teamwork is not considered an option; it is a necessity. Patient welfare and safety depends on health care professionals collaborating together.

Effective teamwork is achieved when there is synergy. Mark Twain defines synergy as the bonus that is achieved when things work together harmoniously. Steven Covey says synergy means that the whole is greater than the sum of its parts (1989). The American Association of Critical Care Nurses has a Synergy Model of Professional Caring & Ethical

## CASE STUDY 11-1

You are a nurse in a busy 42-bed telemetry unit. Both the staff nurses and nursing assistive personnel (NAP) work 12-hour shifts. Change of shift hand-off report occurs at 0700 and 1900 daily. During the staff meeting, the nurse manager charges everyone to think of ways to improve patient satisfaction. You note that patients are dissatisfied during the change of shift hand-off report times when the unit hallways become crowded and noisy. Most nurses and NAP are not available to answer call lights and attend to patient needs during this time.

What are some suggestions you can make to improve patient care during the change of shift hand-off report?

If you are asked to lead a team to problem-solve and identify solutions to these issues, what qualities do you possess that will be essential in this team leadership role?

What qualities would you look for in selecting your team members?

How can you use the five stages of team development in Table 11-1—forming, storming, norming, performing, and adjourning—to develop your team?

Practice that guides the nurse and results in synergy, where the needs and characteristics of a patient, clinical unit, or system are matched with a nurse's competencies (Hardin & Kaplan, 2005). Effective nurses and teams achieve this synergy. They develop the ingredients for creating a winning team where people with different ideas, backgrounds, and beliefs work together synergistically and harmoniously.

In order to succeed as a team, first and foremost the team must have a clearly stated purpose: What are the goals? What are the objectives? What does the leader see the team accomplishing? Are any budget requirements, decision-making ability, and lines of authority for the team spelled out? An effective team keeps the larger organization's goals in mind as it progresses; otherwise, its goals will be inconsistent with those of the parent organization.

Second, an assessment of the team's composition is needed. What are the team members' personal strengths and weaknesses? How do the team members see themselves as individuals? Do they see themselves as part of a cohesive team? Are the contributions of all team members valued? Are all team members' opinions respected? Does the team have a plan to avoid groupthink? Are any additional members with special expertise needed? What is the role of each team member?

Third, the communication links must be clear: Are effective communication patterns in place? Is there a need to improve communication, either in written or verbal format? Does the team work well together, and is communication open, with minimal hidden agendas of the members? Can the truth be told in a compassionate and sympathetic manner in order to reach a difficult decision?

Active participation by all team members is a critical fourth item needed for a team to succeed: Does everyone have a designated responsibility? Do people listen to one another? Is "we versus they" thinking discouraged? Are all team members involved in shaping plans and decisions? Are they all carrying their weight on the team, or are some members not doing their part? What are the relationships of the team members? Is there mutual trust and respect for members and their decisions, however unpopular? Are there political turf issues that must be resolved before proceeding? The climate of the team should be relaxed but supportive.

Is there a clear plan as to how to proceed? Is there a way to acknowledge team accomplishments and positive change? This fifth element of a successful team leads to an action plan that everyone agrees with early on, and one that is revisited at certain designated times. Feedback by team members and others affected by the team's decisions is necessary to keep focus.

The sixth guideline for a successful team is actually ongoing, in that assessment and evaluation are continuous throughout the team's history. Outcomes have to be consistent and related to the expectations of the organization.

Creativity is also encouraged at the team level; perhaps a member has an idea to solve a problem that no one has ever tried. In a supportive environment, pros and cons of all reasonable ideas should be freely discussed.

## CONDUCIVE ENVIRONMENT FOR TEAMWORK

Developing a supportive and conducive environment for teamwork to succeed requires ongoing time and effort. Lindeke and Siekert (2006) point out that physical facility design can impact teamwork by influencing productivity, work attitudes, confidentiality, and the professional image of the health care personnel. Facility design that allows for adequate allotted space enhances team collaboration and interaction. Factors to consider in facility design include noise control, privacy, seating space, and convenience.

DiMichele & Gaffney (2005) identified these values essential to a successful team:

1. *Know the team's strengths and weaknesses:* Each team member has unique talents that can be maximized to benefit the team's goals.
2. *Build Trust:* Each team member has confidence in the good intentions of fellow team members.
3. *Focus on the positive:* Each team member views the experience as an opportunity to learn and grow instead of focusing on failures.
4. *Create a vision:* Each team member collaborates to create a vision or goal that helps facilitate professional and personal growth.
5. *Market the vision:* Each team member shares the vision with other team members and potential team members. This inspires a more desirable team spirit and team environment.
6. *Make the vision a reality:* Each team member assumes responsibility for keeping the vision alive.
7. *Use open, honest communication:* Each team member uses direct communication and opens discussion on tough issues to foster the team's goals. Team members focus on the positive and not the negative.
8. *Encourage creativity:* Encourage creativity to keep the team's vision alive. Creativity fuels positive achievements. With each achievement, celebrate. Celebration is a central part of a successful team.
9. *Provide an environment of constant renewal:* it is imperative that team members allow themselves personal downtime. There should be a good balance of work and personal life to allow team members to regroup, renew, and revitalize.

Following these values has been proven to successfully create a winning team.

The type of communication that occurs between a team and those outside the team is also important.

Ancona and Caldwell (1992) use the following classification to describe the range of communication activities seen within a team and observed in their research:

- *Ambassador activities:* Members carrying out these activities communicate frequently with those above them in the organizational hierarchy. This set of activities is used to protect the team from outside pressures, to persuade others to support the team, and to lobby for resources.
- *Task coordinator activities:* Members carrying out these activities communicate frequently with other groups and persons at lateral levels in the organization. These activities include discussing problems with others, obtaining feedback, and coordinating and negotiating with outsiders.
- *Scout activities:* Members carrying out these activities are involved in general scanning for ideas and information about the external environment. These activities differ from the other two in that they relate to general scanning instead of specific coordination issues.

Another key factor to team success is its political environment. Is there sufficient support and buy-in from the administrative level for the staff? Are there enough resources available to the team—that is, financial, support staff, time allotted, and so on? Administrative support means leadership support for team efforts. Does administration empower staff by encouraging decision making at the team level? Does administration allow for individual creativity and self-governance? Is the role of the team clear from the beginning of the team's work as to who has the final power to make a decision? (Table 11-2).

## TEAM SIZE

It is believed that team size affects performance in that too few or too many members reduce performance (Cohen & Bailey, 1997). As teams become larger, communication and coordination problems tend to increase, and a climate of fairness and cohesiveness decreases (Colquitt, Noe, & Jackson, 2002; Liberman, Hilty, Drake, & Tsang, 2001). However, teams have to be large enough or small enough to accomplish the task assigned. Probably, smaller groups are less cumbersome, with fewer social distractions. Smaller teams also have lower incidences of social loafing (Liden, Wayne, Jaworski, & Bennett, 2004). Individuals in large teams are able to maintain a sense

### TABLE 11-2    Team Evaluation Checklist

| | YES | NO |
|---|---|---|
| 1. Is the environment/climate conducive to team building? | ____ | ____ |
| 2. Do the team members have mutual respect and trust for one another? | ____ | ____ |
| 3. Are the team members honest with one another? | ____ | ____ |
| 4. Does everyone actively participate in the decision making and problem solving of the team? | ____ | ____ |
| 5. Are the purpose, goals, and objectives of the team obvious to all participants? | ____ | ____ |
| 6. Are the goals met? | ____ | ____ |
| 7. Are creativity and mutual support of new ideas encouraged by all team members? | ____ | ____ |
| 8. Does the team work to avoid groupthink? | ____ | ____ |
| 9. Is the team productive, and does it see actual progress toward goal attainment? | ____ | ____ |
| 10. Does the team begin and end its meetings on time? | ____ | ____ |
| 11. Does the team leader provide vision and energy to the team? | ____ | ____ |
| 12. Do any persons on the team serve as ambassador, task coordinator, or scout? | ____ | ____ |

**Source:** Compiled with information from Polifko-Harris (2003) and Ancona & Caldwell (1992).

of anonymity and gain from the work of the group without making a suitable contribution (Shortell & Kaluzny, 2006).

## STATUS DIFFERENCES

Status is the measure of worth conferred on an individual by a group. Status differences are seen throughout organizations and serve some useful purposes (Shortell & Kaluzny, 2006). Differences in status motivate people, provide them with a means of identification, and may be a force for stability in the organization (Scott, 1967).

Status differences have a profound effect on the functioning of teams. Research findings are fairly consistent in showing that high-status members initiate communication more often, are provided more opportunities to participate, and have more influence over the decision-making process (Owens, Mannic, & Neale, 1998). Thus, an individual from a lower-status professional group may be intimidated or ignored by higher-status team members. The group, as a result, may not benefit from this person's expertise. This situation is very likely in health care, where status differences among the professions are well entrenched (Topping, Norton, & Scafidi, 2003). Often, multidisciplinary teams are idealistically expected to operate as a company of equals, yet the reality of the situation makes this difficult (Shortell & Kaluzny, 2006). In a study of end-stage renal disease teams in which the equal participation ideology was accepted by most team participants, it was clear that the medical practitioners, who were perceived as having higher professional status than other individuals, had greater involvement in the actual decision-making process (Deber & Leatt, 1986). The mismatch between expectations and reality made many team members, particularly staff nurses, have a sense of role deprivation, with accompanying implications for morale and job satisfaction. This problem is exacerbated in teams characterized by sex diversity. In one study, men were more likely to want to exit teams that were female dominated for those that were male dominated or homogenous. Men have historically been perceived as having higher status in managerial roles in organizations, and men's satisfaction with a team (Chatman & O'Reilly, 2004).

Status differences may have very significant impacts on patient outcomes. According to the report *Keeping Patients Safe: Transforming the Work Environment of Nurses*, "counterproductive hierarchical communication patterns that derive from status differences" are partly responsible for many medical errors (Institute of Medicine, 2003, p. 361). The relationship between the nurse leadership group and the corresponding physicians is so important that the American Organization of Nurse Executives (AONE) devoted an entire section of the Nurse Executive Competencies on Communication and Relationship-Building 2005 to medical staff relationships (AONE, 2005).

A review of medical malpractice cases from across the country found that medical practitioners, perceived by some as the higher-status members of the team, often ignored important information communicated by nurses, perceived by some as the lower-status members of the team. Nurses in turn were found to withhold relevant information for diagnosis and treatment from medical practitioners (Schmitt, 2001). In this status-conscious environment, opportunities for learning and improvement can be missed because of unwillingness to engage in quality-improving communication.

Shortell and Kaluzny (2006) suggest that if status inequality exists, it is advisable to build a trust-sensitive environment in which members can disagree with the leader and others on the team without repercussions. The use of training with nonmember team facilitators early in the team development process may be able to help a team cope with any problems brought about by status differences. In well-managed multidisciplinary groups, all individuals are encouraged to contribute to the team goals. It is important to develop a team where all members are highly regarded and respected if the team's goals are to be fully achieved.

## CREW RESOURCE MANAGEMENT

Patient care, like other technically complex and high-risk fields, is an interdependent process carried out by teams of individuals with advanced technical training who have varying roles and decision-making responsibilities. While technical training assures proficiency in specific tasks, it does not address the potential for errors created by communication and decision making in dynamic environments. Experts in aviation have developed safety training focused on effective team management, known as Crew Resource Management (CRM). Improvements in the safety record of commercial aviation may be due, in part, to this training (Helmreich, et al., 1999). Over the past 20 years, lessons from aviation's approach to team training have been applied to medicine (Pizzi, et al., 2000), notably in the intensive care unit (ICU) (Shortell, et al., 1994) and anesthesia training (Howard, et al., 1992). CRM training encompasses a wide range of knowledge, skills, and attitudes, including communication, assertiveness, situational awareness, problem solving, decision making, and teamwork.

Assertiveness is the willingness to actively participate, state, and maintain a position until convinced by the facts that other options are better. It requires initiative and the courage to act. Assertiveness differs from passive behavior, which is often submissive to avoid conflict and demonstrates a lack of initiative. Assertiveness also differs from aggressive behavior, which can be dominating, hostile, belligerent, and argumentative.

Situational awareness refers to the degree of accuracy by which one's perception of his or her current environment mirrors reality. Factors that reduce situational awareness include insufficient communication, fatigue/stress, task overload, task underload, group mind-set, a "press on regardless" philosophy, and degraded operating conditions.

CRM fosters a climate or culture where the freedom to respectfully question authority is encouraged. However, the primary goal of CRM is not enhanced communication, but enhanced situational awareness. It recognizes that a discrepancy between what is happening and what should be happening is often the first indicator that an error is occurring. This is a delicate area for many organizations, especially ones with traditional hierarchies, such as health care. Appropriate communication techniques must be taught to all nursing and medical practitioners so that they understand that the questioning of authority need not be threatening and so that all understand the correct way to question orders.

Todd Bishop, a CRM expert, developed a five-step assertive statement process that encompasses inquiry and advocacy steps. The five steps are:

- Opening or attention getter—Address the individual as, for example, "Dr. Karen" or "Michelle" or whatever name or title will get the person's attention.
- State your concern—State what you see in a direct manner. "Mr. Jones has a pulse of 160."
- State the problem as you see it—"Mr. Jones is going into ventricular tachycardia."
- State a solution—"Mr. Jones needs an antiarrhythmic medication."
- Obtain agreement (or buy-in)—"Do you want to order an antiarrhythmic medication?"

The five steps are difficult but important skills to master, and they require a change in interpersonal dynamics and organizational culture.

## PSYCHOLOGICAL SAFETY

Psychological safety describes individuals' perceptions about the consequences of interpersonal risks in their work environment—that is, beliefs, largely taken for granted about how others will respond when one puts oneself on the line, such as by asking a question, seeking feedback, reporting a mistake, or proposing a new idea in the team context (Shortell & Kaluzny, 2006). In psychologically safe teams, people believe that if they make a mistake, other team members will not penalize them or think less of them for it. This belief fosters the confidence to experiment, discuss mistakes and problems, and ask others for help. Psychological safety is created by mutual respect and trust among team members, and leader behavior is a powerful influence on the level of psychological safety in teams (Edmondson, 1999, 2003).

It is important to note that psychological safety is distinct from group cohesiveness. As noted, team cohesiveness can lead to groupthink, a reduction in willingness to disagree and challenge others' views (Janis, 1972). Groupthink leads to a lack of interpersonal risk-taking and is discussed later in this chapter. Psychological safety describes instead a climate that fosters productive discussion enabling early prevention of problems and accomplishment of shared goals, because people feel less need to focus on self-protection (Shortell & Kaluzny, 2006).

## QUALITIES OF EFFECTIVE TEAM MEMBERS

To be an effective team member, you need to possess certain characteristics conducive to team collaboration. You must be proactive, motivated, have a certain personal sense of purpose or mission, and possess personal and time management skills (Covey, 1989).

Being proactive and motivated means taking charge of your life and the circumstances around you. Proactive, motivated people are not easily affected by situations in their surroundings, because they avoid being reactive. Being proactive and motivated is to take full responsibility for your own actions, decisions, and behavior. People who end up with the good jobs are the proactive and motivated ones who are the solutions to the problems, not the problems themselves. They seize the initiative to do whatever is necessary, consistent with correct principles, to get the job done (Covey, 1989).

To succeed in nursing as a new graduate and as a member of a team, you need to develop a personal sense of purpose, mission, and professional goals early on and work toward meeting them. Covey (1989) calls this "working toward goals and beginning with the end in mind." When joining a team, nurses need to examine their own skills, assess what contributions they can provide, and be confident in their role as team members. Knowing your priorities and managing personal and professional time wisely and efficiently will go a long way towards teamwork.

## QUALITIES OF EFFECTIVE TEAM LEADERS

A team leader will organize, facilitate, and manage the entire team. Nurse leaders need to examine their own leadership styles, strengths, and weaknesses, and learn to capitalize on their strengths. Effective team leaders must understand how various learning styles, cultural diversity, and personality differences play into the dynamics of teamwork. Qualities of a good team leader should include good communication skills, conflict resolution skills, and leadership skills (Hall & Weaver, 2001).

Open and honest communication is an essential skill to develop as a team leader. Respectful negotiations, clear and

**EVIDENCE** FROM THE LITERATURE

**Citation:** Sirota, T. (2008, July). Nurse/physician relationships: Survey report. 28–31.

**Discussion:** More than half of respondents (57%) to this survey of nurses say they're generally satisfied with their professional relations with physicians; a significant minority, 43%, report dissatisfaction. Although relationships have improved some since an earlier survey in 1991, respondents' comments indicate that they perceive several factors to be at play here:

- male physicians' perceptions of traditional gender and cultural roles
- physicians' arrogance and feelings of superiority
- nurses' feelings of inferiority
- hospital culture or policy reinforcing a subordinate role for nurses

Important ways to improve nurse-physician overall collaboration identified in the article include workplace empowerment for nurses, nurse-physician rounds, team meetings, collaborative educational programs, and collaborative membership on hospital committees. The article states that the bottom line is that nurses aren't expendable. In the current nursing shortage, the climate is ripe for nurses to speak up as a group and let facility administrators know that they need to pay attention to nurses' legitimate concerns about nurse-physician collaboration and then correct deficiencies. Perhaps most important, improving relationships between nurses and physicians will benefit both professional groups by improving job satisfaction and productivity and will also benefit patients by enhancing their overall safety, welfare, and clinical progress. This can be accomplished by promoting greater nurse-physician professional respect, improving communication and collaboration, educating physicians about nursing roles and skills, and addressing physician misconduct (Rosenstein, Russell, & Lauve, 2002).

**Implications for Practice:** Nursing and medical practitioner communication has improved since the 1991 survey. Good communication by all members of the health care team affects patient safety and must be facilitated by nursing, medical, and hospital team members to build an environment for patient safety.

tactful expression of ideas and messages, and empathetic listening must occur. The team leader must work to develop these skills. Empathetic listening allows a leader to seek first to understand the other person, become truly interested in and attentive to what the other person has to say, and then place himself or herself in the other person's shoes (Covey, 1989).

Conflict resolution skill is also an important quality of a good team leader. In interdisciplinary practice collaboration, people come to the table with varying ideas about the goal, as well as with different levels of commitment and diverse attitudes. These differences are the breeding ground for conflict. Conflict may be minimized if the team learns conflict resolution skills early on. The team leader can help by providing a clear statement of goals, setting team guidelines, and identifying role delineations to avoid or minimize conflict (Hall & Weaver, 2001).

## CLINICAL TEAMS

When problems occur on clinical teams with substandard clinical practice, immediate intervention is necessary. This usually starts with review by nursing and medical peers.

Failure to act puts patients and the institution at risk. If the problem is not resolved, administration gets involved and, if needed, the final level of intervention may need to come from the state licensing board and the Joint Commission.

When problems occur because of inappropriate behavior by a physician or a nurse, determination of a cause for the disruptive behavior begins with exploring the existence of a medical condition such as diabetes, depression, or dementia. If no medical condition exists, immediate action is needed. Document disruptive behavior completely, and refer it to the proper person or group in the organizational chain of command. Many of these situations can be resolved by productive, nonjudgmental confrontation by the appropriate person or group in possession the facts. If needed, programs to help physicians and nurses are often available through the hospital employee assistance program or the state licensing board, as appropriate.

All nurses and physicians must communicate clearly and support each other. Communication should be assertive, direct, open, and honest. Acting in an indecisive or uncertain way can lead to a lack of respect from other team members. Support means refusing to participate in or listen to abuse, gossip, blame, or judgment of others. If someone is rude or abusive, speak to them calmly and request that they do the same. Ask for clarification, if needed. Some organizations have instituted a "Code Pink" alert to deal with difficult physicians. When Code Pink is called, other nurses come to the aid of a coworker by surrounding the physician who is out of control and saying nothing (Durham, n.d.).

## GUIDELINES FOR MEETINGS

A fundamental skill in effectively working with a team is leadership. The leader must implement appropriate leadership or guidance through the stages of the team process. For example, when the team is just forming, members may need more guidance and highly directive leadership during meetings (Table 11-3).

As the team matures and begins working on the goals, the leader may need to take a step back from directive leadership and become a facilitator, mentor, and guide. The leader focuses on encouraging and guiding the team's sense of self-direction, ownership, and responsibility to accomplish desired goals (Hall & Weaver, 2001). The leader reinforces teamwork by focusing the team on outcome improvement, visually tracking results, and recognizing members who make significant contributions (Cox, 2003). In one study, leadership in ICUs was positively related to efficiency of operation, satisfaction, and lower turnover of nurses (Shortell, et al., 1994). Successful leaders adopted a supportive, formal or informal leadership style, emphasizing standards of excellence, encouraging interaction, communicating clear goals and expectations, responding to changing needs, and providing support resources when possible. Successful leaders communicate a compelling rationale for change, motivate others

to exert the necessary effort, and also minimize the status difference between themselves and other members of the team to facilitate others' ability to speak up with questions, observations, and concerns (Shortell & Kaluzny, 2006).

In deciding upon a leadership style, group leaders need to consider in realistic terms their formal and informal authority within the group. Use of a coercive or forceful style may backfire when the individual does not have the power to back up decisions. Such a leader may find that an informal leader is able to veto, modify, or sabotage demands. It is best for the formal leader not only to consider the views of informal leaders but also to collaborate with them, if possible (Shortell & Kaluzny, 2006).

## AVOIDING GROUPTHINK

Effective leaders work to avoid groupthink. The concept of groupthink emerged from Janis's studies of high-level policy decisions by government leaders, including decisions about Vietnam, the Bay of Pigs, and the Korean War. **Groupthink** occurs when the desire for harmony and consensus overrides members' rational efforts to appraise the situation. In other words, groupthink occurs when maintaining the pleasant atmosphere of the team implicitly becomes more important to members than reaching a good decision (Shortell & Kaluzny, 2006). There is a reduced willingness to disagree and challenge others' views in groupthink. Some of the following symptoms may indicate the presence of groupthink (Janis, 1972):

- *The illusion of invulnerability:* Team members may reassure themselves about obvious dangers and become overly optimistic and willing to take extraordinary risks.

---

### TABLE 11-3    Guidelines for Meetings

In managing meetings, leaders should be aware of the following principles:

- Set a time frame for meetings, and stick to it.
- At the beginning of the meeting, review the progress made to date and establish the task facing the group.
- Help group members feel comfortable with one another.
- Establish ground rules governing group discussions.
- As early in a meeting as possible, get a report from each member who has been preassigned a task.
- Sustain the flow of the meeting by using informational displays.
- Manage the discussion to achieve equitable participation.
- Work to avoid groupthink by using critical appraisal of all ideas.
- Close the meeting by summarizing what has been accomplished and reviewing assignments.
- Identify a time frame for future meetings.

- *Collective rationalization:* Teams may overlook blind spots in their plans. When confronted with conflicting information, the team may spend considerable time and energy refuting the information and rationalizing a decision.
- *Belief in the inherent morality of the team:* Highly cohesive teams may develop a sense of self-righteousness about their role that makes them insensitive to the consequences of decisions.
- *Stereotyping others:* Victims of groupthink hold biased, highly negative views of competing teams. They assume that they are unable to negotiate with other teams, and rule out compromise.
- *Pressures to conform:* Group members face severe pressures to conform to team norms and to team decisions. Dissent is considered abnormal and may lead to formal or informal punishment.
- *The use of mind guards:* Mind guards are used by members who withhold or discount dissonant information that interferes with the team's current view of a problem.
- *Self-censorship:* Teams subject to groupthink pressure members to remain silent about possible misgivings and to minimize self-doubts about a decision.
- *Illusion of unanimity:* A sense of unanimity emerges when members assume that silence and lack of protest signify agreement and consensus.

Shortell and Kaluzny (2006) state that the consequences of groupthink are that teams may limit themselves, often prematurely, to one possible solution and fail to conduct a comprehensive analysis of a problem. When groupthink is well entrenched, members may fail to review their decisions in light of new information or changing events. Teams may also fail to adequately consult with experts within or outside the organization and fail to develop contingency plans in the event that the decision turns out to be wrong.

Team leaders can help avoid groupthink. First, leaders can encourage members to critically evaluate proposals and solutions. Where a leader is particularly powerful and influential yet still wants to get unbiased views from team members, the leader may refrain from stating his or her own position until later in the decision-making process. Another strategy is to assign the same problem to two separate work teams. Most importantly, groupthink can be avoided by proactively engaging in a process of critical appraisal of ideas and solutions, and by understanding the warning signs of groupthink (Shortell & Kaluzny, 2006).

## REAL WORLD INTERVIEW

As an emergency medicine physician, I am frequently interfacing with nurses during life-threatening medical scenarios. Whether it is an acute myocardial infarction or respiratory failure or even a very sick child, the dialogue is standard. There is a set tone of urgency on the part of each team member, and we get straight to work with little discussion. I think when we work as a team, we are like a well-oiled machine. The nurse anticipates my needs and I hers or his, and we follow our protocols. The absolute focus is on the patient and getting him or her out of immediate danger. That is what the emergency department does best. We "stabilize" the patient's acute life-threatening event. The rapport between MD and RN is built from an understanding of mutual competency. I know the nurses I work with in the ED, and they know me. We could not save lives day in and day out without the team work mentality.

**Dr. Elizabeth Horvath, D.O.**
Crystal Lake, Illinois

## KEY CONCEPTS

- Effective teamwork and collaboration is essential to improving patient care outcomes.
- Teams and committees are formed for a variety of reasons depending on the level of collaboration required or the specific purpose desired.
- Each team goes through the stages of a team process.
- Successful teamwork requires a conducive physical, social, and political environment for it to succeed.
- An effective team member must be proactive, be motivated, have a certain personal sense of purpose or mission, and possess personal and time management skills.
- Teams are affected by team size, status differences, and psychological safety.
- Guidelines for conducting meetings are useful when managing meetings.

- Successful teams work to avoid groupthink, which can cause the group to fail to analyze a problem.
- The team leader must possess effective communication skills, conflict resolution skills, and leadership skills.
- Great teams have clear goals, well-defined role delineation, organized processes that are outcomes oriented, and open and honest interpersonal relationships.

- Crew Resource Management training encompasses a wide range of knowledge, skills, and attitudes including communication, assertiveness, situational awareness, problem solving, decision making, and teamwork.
- Groupthink occurs when the desire for harmony and consensus overrides members' rational efforts to appraise a situation.

## KEY TERMS

groupthink
team

## REVIEW QUESTIONS

1. Based on the Tuckman and Jensen team process, what stage of team development is it when team members work harmoniously together, have open communication, take risks, and trust each other to complete assigned tasks?
   A. Forming
   B. Storming
   C. Norming
   D. Performing

2. A good leader does all except which of the following?
   A. Encourages open communication and interpersonal relationship
   B. Uses only an authoritative leadership style
   C. Actively listens to team members' concerns and opinions
   D. Clearly identifies goals, roles, and the team process

3. Which of the following is not a characteristic of groupthink?
   A. Use of mind guards
   B. Illusion of unanimity
   C. Free discussion of ideas
   D. Pressure to conform

4. Which of the following communication roles is used to protect the team from outside pressures from those above them in the organizational hierarchy?
   A. Ambassador
   B. Task coordinator
   C. Scout
   D. Leader

5. Which of the following showed that an individual's performance is often determined in large part by the work group?
   A. Kaluzny study
   B. Dickson research
   C. Carnegie study
   D. Hawthorne experiments

6. Which of the following are characteristics of a successful team? Select all that apply.
   _____ A. They socialize outside of the work environment.
   _____ B. They understand the role of each team member.
   _____ C. They communicate effectively .
   _____ D. They possess identical expertise.
   _____ E. They have an explicit purpose.

7. According to Miller, et al., key nursing behavioral markers for interdisciplinary interaction include which of the following? Select all that apply.
   _____ A. Having clear situational awareness
   _____ B. Using a situation, background, assessment, recommendation, response (SBARR) communication method
   _____ C. Being prompt for shift duty
   _____ D. Using closed-loop communication
   _____ E. Having a shared mental model

## REVIEW ACTIVITIES

1. Ask to attend a team meeting in your clinical agency. Who is the formal leader of the group? What does the leader do to facilitate the attainment of team objectives?

2. Note a team of which you are a member. Have you ever seen groupthink operate in the team? What did you do?

3. Note the communication roles of ambassador, task coordinator, and scout. Have you seen these roles operate on any teams to which you belong?

## EXPLORING THE WEB

Check these sites for information on teams:

- American College of Health Care Administrators (ACHCA): www.achca.org
- Belbin Team-Role Theory, products, conferences, and free online newsletter: www.belbin.com
- Free online newsletter on leadership: www.injoy.com
- Health care teams information: www.learningcenter.net
- Tuckman and Jensen stages of the team process: www.infed.org

Search for Tuckman and Jensen.

- Like a Team www.likeateam.com. Click on, "Like a Team" You Tubes.
- Nursezone.com for work and for life www.nursezone.com
- AHRQ site on teamwork training http://psnet.ahrq.gov. Click on Patient safety primers. Click on Teamwork training.
- IHI site on teamwork and SBAR www.ihi.org Click on, Programs. Click on, Audio and Web Programs. Click on, On Demand: Effective Teamwork as a Care Strategy - SBAR and Other Tools for Improving Communication Between Caregivers.

## REFERENCES

American Organization of Nurse Executives (AONE). (2005) AONE Nurse Executive Competencies. Accessed March 17, 2011 at, http://www.aone.org/. Search for, Nurse executive competencies.

Amos, M., Hu, J., & Herrick, C. A. (2005). The impact of team building on communication and job satisfaction of nursing staff. *Journal for Nurses in Staff Development, 21*(1), 10–16.

Ancona, D. G., & Caldwell, D. F. (1992). Briding the boundary: External activity and performance in organizational teams. *Administrative Science Quarterly, 37*, 634–665.

Bennett, C., Perry, J., & Lapworth, T. (2010). Leadership skills for nurses working in the criminal justice system. *Nursing Standard, 24*(40), 35–40.

Buchholz, S., & Roth, T. (1987). *Creating the high performance team.* New York: John Wiley & Sons, Inc.

Caples, M., & March, L. (2009). Working together. *World of Irish Nursing, 11*, 42–43.

Caramanica, L. (2004). Shared governance: Hartford Hospital's experience. *Online Journal of Issues in Nursing, 9*(1), 2. Retrieved April 23, 2006, from http://www.nursinworld.org/ojin/topic23/ tpc23_2 htm.

Carnegie, D. (1984). *How to win friends and influence people.* New York: Galahad Books.

Chatman, J., & O'Reilly, C. (2004). Asymmetric effects of work group demographics on men's and women's responses to work group composition. *Academy of Management Journal, 47*(2), 193–208.

Cohen, S. G., & Bailey, D. E. (1997). What makes teams work: Group effectiveness research from the shop floor to the executive suite. *Journal of Management, 23*, 239–290.

Colquitt, J. A., Noe, R. A., & Jackson, C. L. (2002). Justice in teams: Antecedents and consequences of procedural justice climate. *Personnel Psychology, 55*, 83–100.

Covey, S. (1989). *The 7 habits of highly effective people.* New York: Fireside.

Cox, S. (2003). Nursing management: Building dream teams. Retrieved April 18, 2006, from http://www.findarticles.com/p/articles/mi_qa3619/is_200303/ai_n9221079/print.

Deber, R. B., & Leatt, P. (1986). The multidisciplinary renal team: Who makes the decisions? *Health Matrix, 4*(3), 3–9.

DiMichele, C., & Gaffney, L. (2005). Proactive teams yield exceptional care. *Nursing Management, 36*(5), 61–64.

Durham, L. (n.d.). Ideas on Lightening Up Workplaces. SpeakerNet News. Accessed March 17, 2011 at, http://www.speakernetnews.com/post/lightenwork.html.

Edmondson, A. C. (1999). Psychological safety and learning behavior in work teams. *Administrative Science Quarterly, 44*, 350–383.

Edmondson, A. C. (2003). Speaking up in the operating room: How team leaders promote learning in interdisciplinary action teams. *Journal of Management Studies, 40*(6), 1419–1452.

Hall, P., & Weaver, L. (2001). Interdisciplinary education and teamwork: A long and winding road. *Medical Education, 35*, 867–875.

Hardin, S. R., & Kaplan, R. (2005). *Synergy for clinical excellence, American Association of Critical Care Nurses.* Boston: Jones & Bartlett.

Hasenfeld, Y. (1983). *Human service organizations.* Englewood Cliffs, NJ: Prentice Hall.

Huber, G. (1980). *Managerial decision making.* Glenview, IL: Scott Foresman.

Institute of Medicine (2003). *Keeping patients safe: Transforming environment of nurses.* Washington, DC: National Academics Press.

Janis, L. (1972). *Victims of groupthink.* Boston: Houghton-Mifflin.

Katzenbach, J. R., & Smith, D. K. (2003). *The wisdom of teams: Creating the high-performance organization.* New York: HarperCollins Publishers Inc.

Liberman, R. P., Hilty, D. M., Drake, R. E., & Tsang, H. (2001). Requirements for multidisciplinary teamwork in psychiatric rehabilitation. *Psychiatric Services, 52*(10), 1331–1342.

Liden, R. C., Wayne, S. J., Jaworski, R. A., & Bennett, N. (2004). Social loafing: A field investigation. *Journal of Management, 30*, 285–305.

Lindeke, L. L., & Siekert, A. M. (2004). Nurse-physician workplace collaboration. Retrieved April 18, 2006, from http://www.nursingworld.org/mods/mod775/nrsdrfull.htm#team.

McDonald, C., & McCallin, A. (2010). Interprofessional collaboration in palliative nursing: What is the patient-family role? *International Journal of Palliative Nursing, 16*(6), 285–288.

Miller, K., Riley, W., & Davis, S. (2009). Identifying key nursing and team behaviours to achieve high reliability. *Journal of Nursing Management, 17,* 247–255.

Owens, D. A., Mannic, E. A., & Neale, M. A. (1998). Strategic formation of groups: Issues in task performance and team member selection. In D. H. Gruenfeld (Ed.), *Research on managing groups and teams* (pp. 149–165). Stamford, CT: MAI Press.

Polifko-Harris, K. (2003). Effective team building. In P. Kelly-Heidenthal (Ed.) *Nursing leadership and management.* Clifton Park NY: Delmar Cengage Learning.

Porter-O'Grady, T. (2005). Strong leaders or empowered staff: Where is real empowerment? Retrieved April 18, 2006, from http://www.tpogassociates.com/considerthis/fall2005.htm.

Riley, W. (2009). High reliability and implications for nursing leaders. *Journal of Nursing Management, 17,* 238–246.

Roethlisberger, F. J., & Dickson, W. J. (1939). *Management and the worker.* Cambridge, MA: Harvard University Press.

Rosenstein, A. H., Russell, H., & Lauve, R. (2002). *Disruptive physician behavior contributes to nursing shortage. Study links bad behavior by doctors to nurses leaving the profession.* Physician Executive. November-December; 28(6):8–11.

Schmitt, M. H. (2001). Collaboration improves the quality of care: Methodological challenges and evidence from health care research. *Journal of Interprofessional Care, 15,* 47–66.

Scott, W. G. (1967). *Organization theory.* Homewood, IL: Irwin.

Shortell, S. M., & Kaluzny, A. D. (2006) *Health care management* (5th ed.). Clifton Park, NY: Delmar Cengage Learning.

Shortell, S. M., Zimmerman, J. E., Rousseau, D. M., Gillies, R. R., Wagner, D. P., Draper, E. A., et al. (1994). The performance of intensive care units. Does good management make a difference? *Medical Care, 32,* 508–525.

Sirota, T. (2008, July). Nurse/physician relationships: Survey report. 28–31.

Topping, S., Norton, T., & Scafidi, B. (2003). Coordination of services: The use of multidisciplinary, interagency teams. In S. Dopson & A. L. Mark (Eds.), *Leading health care organizations* (pp. 100–112). New York: Palgrave Macmillan.

Tuckman, B. W. (1965). Developmental sequences in small groups. *Psychological Bulletin, 63,* 384–399.

Tuckman, B. W., & Jensen, M. A. C. (1977). Stages of small group development revisited. *Group and Organizational Studies, 2,* 419–427.

## SUGGESTED READINGS

Buchbinder, S. B. (2006). Building an effective team. Retrieved April 18, 2006, from http://community.nursingspectrum.com/MagazineArticles/article.cfm?AID=20377.

Hinkle, J. L., Steffen, K. A., Heck, C. E., McBride, J., & Wenograd, D. (2006). A team approach to neuroscience nursing critical care orientation. *Journal of Neuroscience Nursing, 38*(5), 390–394.

Lingard, L., Regehr, G., Espin, S., Devito, I., Whyte, S., Buller, D., et al. (2005). Perceptions of operating room tension across professions: Building generalizable evidence and educational resources. *Academic Medicine, 8*(10), S75–S79.

Murray, T., & Kleinpell, R. (2006). Implementing a rapid response team: Factors influencing success. *Critical Care Nursing Clinics of North America, 18*(4), 493–501.

Smith, M. K. (2005). Bruce W. Tuckman—Forming, storming, norming and performing in groups. *Encyclopedia of Informal Education.* Retrieved April 23, 2006, from http://www.infed.org/thinkers/tuckman.htm.

Williamson, S. (2006). Training the team: Talk is cheap—but vital in keeping patients safe. *Inside: Duke University Medical Center & Health System Newsletter, 15*(5).

# CHAPTER 12

# Power

**Terry W. Miller, PhD, RN; Richard J. Maloney, EdD, MA, MAHRM, BS;**
**Patsy L. Maloney, EdD, MSN, MA, RN-BC, NEA-BC**

*The sole advantage of power is that you can do more good.*

(Baltasar Gracian, 17th Century)

## OBJECTIVES

Upon completion of this chapter, the reader should be able to:

1. Define power and describe it at the personal, professional, and organizational levels.

2. Describe each of the following sources of power and analyze its likely relative strength for an entry-level nurse: expert, legitimate, referent, reward, coercive, connection, and information.

3. Apply an understanding of power to help nurses improve their effectiveness.

4. Analyze how new nurses can increase their power.

Delmar/Cengage Learning

*Nurse Pat, a new graduate who has just finished her medical-surgical nursing orientation, is working with a patient for whom a surgical consult has been written. The unit clerk and a long-time nurse on the unit remark that Dr. Killian, the practitioner doing the surgical consultation, should be named Dr. Killjoy because she humiliates new nurses to try to put them in their place. Based on previous reports by other nurses on the unit, Pat knows Dr. Killian has the reputation of being demeaning and inappropriately demanding when interacting with new nurses. Two hours later, Dr. Killian appears on the unit and asks to see the nurse who did the surgical admission sheet.*

*What would you do if you were Pat?*

*How would you approach Dr. Killian?*

Effective nurses are powerful. With objectivity, creativity, and knowledge, they influence others through their practice. They develop and exert power from multiple perspectives; they use an understanding of power to motivate others, accomplish organizational goals, and provide safe, competent care. The political process, discussed in Chapter 9, is another venue through which effective nurses achieve desired outcomes and protect their own as well as their patients' interests.

This chapter will discuss power and how nursing power affects patient care. It looks at sources of power, levels of power, and how nurses can increase their power.

## DEFINITIONS OF POWER

Power has been defined in multiple ways. Commonly, **power** is described as the ability to create, get, and use resources to achieve one's goals. Goals within an organization vary widely across departments, health care groups, and individuals. Nurses should strive to gain knowledge of the distribution of power in their organization, the circumstances under which power is used, and strategies associated with the use of power. Power can be defined at various levels: personal, professional, or organizational. Personal power derives from characteristics in the individual. For example, parents and teachers are often seen as personally powerful because of the trust or knowledge they possess. Professional power is conferred on members of a profession by others and by the larger society to which they belong. This power comes from offering a service that society values. Organizational power comes from one's position in an organizational hierarchy, as well as from understanding the organization's structures and functions and from being authorized to function powerfully within an organizational culture. Power, regardless of level, comes from the ability to influence others or affect others' thinking or behavior.

Power at the personal level is closely linked to how one perceives power, how others perceive the individual, and the extent to which the individual can influence events. Nurses who are empowered at a personal level are likely to manifest a high level of self-awareness and self-confidence. They are more likely to understand nursing as a profession because it represents a group to which they belong. They also understand the structure and operations of health care because it represents their work environment.

People who are perceived as experts in health care have a significant amount of authority and influence, which makes them more effective than those not perceived as experts. There are at least two ways to wield the influence of an expert. The first way is to be introduced and promoted to a group as an expert, which validates one's expertise; the second way is to actually become an expert based on knowledge, skills, and abilities that are consistently demonstrated in practice settings. Remarkably, nurses are sometimes reluctant to be identified as experts to patients, practitioners, administrators, other nurses, other health care workers, and the public in general. This lack of identification must be addressed if the nursing profession is to become more visible and achieve the status and degree of empowerment it seeks.

The personal power of nurses is evident in the decisions they make on a daily basis about how their lives and work are organized to accomplish what they want and obtain what they need for themselves and their patients. The more nurses believe they can influence events through personal effort, the greater their sense of power. Many nurses believe they can make a difference and influence events in their lives. These nurses are likely to participate actively in trying to get what they want. This participation will make them feel more powerful even if their efforts are not always successful.

## Critical Thinking 12-1

The work and contributions of some nurses are so significant that they change the world. To be effective, these nurses conceptualize themselves in terms far more powerful than society may expect. Such an empowering self-image begins for entry-level nurses on day one. When we consider the relevance and significance of nursing history, we realize the contributions made by nurses who saw themselves as powerful and acted on that empowered self-image to improve the lives of others. A lack of historical awareness within the nursing profession empowers others to discount nursing's importance to society. Ultimately, we limit our own future as nurses by hindering our potential for making even greater contributions to the future of health care.

Can you name two nurses who represent powerful figures in modern history and tell why their contributions are so significant? See Exploring the Web in this chapter.

Can you identify some obstacles they experienced because they were nurses?

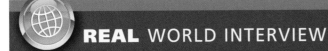

## REAL WORLD INTERVIEW

I try to use power in a positive manner in the pain management clinic that I manage. I attempt to get people to move without them necessarily knowing they are being helped. I think that power at the bedside is a personal thing. Confidence, authenticity, and genuineness are qualities I most value and see as empowering nurses in practice. I feel that the middle manager in nursing gets power by developing staff, building self-confidence on a personal level, and delegating authority in a way that supports others. In higher levels of management I found, through my military experience, that the best leaders almost give power away with the caveat that the persons are being delegated to represent the leader and are accountable for the decisions they make. I believe that the tighter a person holds onto power, the less powerful she or he becomes. I also believe that nurses run the risk of losing power when they look for power outside nursing to direct them. New graduates should seek experiences and input from a variety of people and glean what they can from each experience. It is somewhat like a smorgasbord. Ultimately, the new nurse has to decide what she or he wants because there is no one answer or approach to power. No one really has *the* answer; instead, they have *an* answer.

**Nancy Safranek, MSN, RN**
RN Director, OPS, PACU, SP, PMC
Puyallup, Washington

# POWER AND ACCOUNTABILITY

Effective nurses see power as positive and view their ability to understand and use power as a significant part of their responsibilities to patients, their coworkers, the nursing profession, and themselves. Nurses therefore have a professional obligation not to view power as a negative concept, thus avoiding power struggles and those who seem to savor power.

Traditionally, accountability has been considered one of the major hallmarks of the health care professions. Nursing is a profession and, as such, nurses have the primary responsibility for defining and providing nursing services. Yet some nurses appear to have a difficult time understanding the underlying accountability that comes with this powerful claim. Inherent in the role of the nurse are professional accountability and direct responsibility for decisions made and actions rendered.

## EVIDENCE FROM THE LITERATURE

**Citation:** Buresh, B., & Gordon, S. (2006). *From Silence to Voice*. Ithaca, NY: Cornell University Press.

**Discussion:** This book discusses the fact that not enough nurses are willing to talk about their work. When nurses and nursing organizations do talk about their work, too often they unintentionally project an inaccurate picture of nurses using a virtue script instead of a knowledge script to highlight what nurses do. They discuss the virtuous and caring acts of nurses. They do not discuss the expert knowledge that nurses bring to patient care. When nursing groups give voice to nursing, they sometimes bypass, downplay, or even devalue the basic nursing work that occurs in direct care of the sick, while elevating the image of elite nurses in advance practice, administration, or academia. This contributes to social stereotypes that deride anyone who is "just a nurse."

**Implications for Practice:** If nursing is misunderstood by the public or by those with influence, it will be vulnerable to the budget axe, and new resources for nursing education and practice won't happen. A focus on the virtues of nurses is an invitation to seek not the best and brightest but rather the most virtuous, meekest, and self-sacrificing.

## SOURCES OF POWER

Most researchers agree that the sources of power are diverse and vary from one situation to another. They also agree that these **sources of power** are a combination of conscious and unconscious factors that allow an individual to influence others to do as the individual wants (Fitzpatrick, Campo, Graham, & Lavandero, 2010). Articles and textbooks about nursing administration, educational leadership, and organizational management commonly include references to the work of Hersey, Blanchard, and Natemeyer (1979) as well as an expansion of the power typology originally developed by French and Raven in 1959 (cited in Hersey, Blanchard, & Natemeyer, 1979). The typology helps nurses understand how different people perceive power and subsequently relate to others in the work setting and in attempts to achieve their goals. Power is described as having a basis in expertise, legitimacy, reference, reward, coercion, or connection. More recently, another power source—information—has been added to the typology (Wells, 1998). Generally speaking, nurses exert influence derived from one or a combination of these power sources. Some of these power sources may be from the organization, and other power sources may be from the individual. Power that an organization confers on someone—for example, the ability to work as a staff RN—is not necessarily transportable to another organization unless the second organization also agrees to confer power.

Power derived from the knowledge and skills nurses possess is referred to as **expert power**. There are, however, special considerations to keep in mind about expertise and power. The geometric explosion of knowledge has made expertise more valuable, and technological advances for accessing information have enabled more people to acquire expertise on any given subject. Knowing more about a subject than do others, combined with the legitimacy of holding a position, gives an individual a decided advantage in any situation. But the less acknowledged that experts are in a group, the less effective their expert powers become. Visible reciprocal acknowledgment of expertise among group members balances power and enhances productivity, whereas lack of reciprocal acknowledgment has the opposite effect. Combining expertise with high position is most powerful if the person consistently demonstrates expertise. Entry-level nurses will enhance their expert power and their ability to get the patient care mission accomplished if they add to their current knowledge through professional reading and seek additional training to improve their clinical skills.

**Legitimate power** is power derived from the position a nurse holds in a group, and it indicates the nurse's degree of authority. The more comfortable nurses are with their legitimate power as nurses, the easier it is for them to fulfill their role. Nurses in legitimate positions are expected to use what authority they have and may be punished for not doing so. Sometimes, too little legitimacy or authority is delegated to nurses who are given the responsibility for leading. People generally follow legitimate leaders with whom they agree. Although legitimacy is a significant part of influence and control, it is not universally effective and is not sufficient as one's only source of power. New nurses may be placed in positions of legitimate power and authority relatively quickly following their orientation. For example, they may be assigned as charge nurses, supervising paraprofessionals on the shift. They must be aware of the limits of their legitimate authority so they don't overstep their bounds and should seek guidance in situations in which they must act. Appropriate and timely use of legitimate power assures positive patient care outcomes.

Power derived from how much others respect and like any individual, group, or organization is referred to as **referent power**. Nurses who are identified with respected, trustworthy individuals or groups will benefit from referent power by virtue of such identification. A nurse identified as a graduate of a respected university gains prestige and power from such identification. New nurses often start the job with a blank slate as far as referent power is concerned but can quickly gain influence among coworkers by identifying with other individuals or groups for which the coworkers have respect. Through identification with their university, church membership, or some other agent of influence, the entry-level nurse taps into referent power. Even dressing and looking like a nurse who has earned respect can confer referent power.

The ability to reward others to influence them to change their behavior is commonly termed **reward power**. Meaningful rewards exist other than money, such as formal recognition before one's nursing peers at an awards ceremony. The manner in which rewards are distributed is important. Rewards seldom motivate as effectively as a vision that unifies the members of a group, thus reward power is an uncertain instrument for long-term change. New nurses may lack a direct ability to dispense rewards, but they can often indirectly exercise reward power by giving appropriate positive feedback to supervisors about the performance of others with whom they work. Rewards are not likely to permanently change attitudes. Withholding rewards or achieving a goal by instilling fear in others often results in resentment.

People who have the ability to administer punishment or take disciplinary actions against others to influence them to change their behaviors have **coercive power**. This type of power is often considered the least desirable tactic to be used by people in positions of authority. Typically, people do not enjoy being coerced into doing something other than what they choose to do and often perceive punishment as humiliating. A new nurse can indirectly influence others' behavior by reporting incidents that require disciplinary action. This course of action should not be taken lightly, for it affects working relationships and should therefore be sparingly employed because of possible unintended consequences.

The extent to which nurses are connected with others having power is called **connection power**. Leaders can dramatically increase their influence by understanding that people are attracted to those with power and their associates. As a new nurse, when you go to the office of the director of nursing services or the vice president for patient care services, do not forget that the clerical workers in the outside office have relationships with their boss—thus they have connection power. If you try to go around them and take their power lightly, and insult or patronize them, you have risked your own power base in relation to the director or vice president. Similarly, if an entry-level nurse bypasses a person who is directly responsible for a situation, the attempted circumvention reflects negatively on the nurse. Nurses should work to resolve issues at the appropriate level before they take their concerns to a higher level of authority. Nurses are expected to understand the structure and policies of the organizations in which they provide services.

Nurses who influence others with the information they provide to the group are using **information power**. Regardless of a nurse's leadership style, information plays an increasingly critical role. Legitimate power, reward power, and coercive power tend to be bestowed on individuals by their organizations. They tend to be effective only for a short period of time unless they are accompanied by other forms of power, such as information power. Information power is especially important because, to be functional, health care teams and organizations require accurate and timely information that is shared. To be perceived as having information power, nurses must share knowledge that is both accurate and useful. Nurses must discuss the expert knowledge they bring to patient care to be seen as experts (Buresh & Gordon, 2006). Information sharing can improve patient care, increase collegiality, enhance organizational effectiveness, and strengthen one's professional role and connections. New nurses who value team membership will empower coworkers and themselves by giving and receiving information with which to do the job more effectively. See Table 12-1 for a summary of the different sources of power and examples of each for nursing.

Effective nurses use the sources of power covered thus far: expertise, legitimacy, reference, reward, coercion, connection, and information. They have the ability to combine referent (charismatic) power and expert power from a legitimate power base, adding carefully measured portions of reward power and little or preferably no coercive power (Fitzpatrick et al., 2010). These emerging leaders gather and use information in new and creative ways. They understand that power should be a means to accomplish a goal rather than a goal in itself.

## TABLE 12-1 Understanding and Using Sources of Power

| TYPE | SOURCE | EXAMPLES FOR NURSING |
|------|--------|----------------------|
| Expert | Power derived from the knowledge and skills nurses possess. The more proficiency the nurse has, the more the nurse is received as an expert. | Communicating information from current evidence-based journals and bringing expert knowledge to patient care. |
| Legitimate | Power derived from an academic degree, licensure, certification, experience in the role, and job title in the organization. | Wearing or displaying symbols of professional standing, including license and certification. |
| Referent | Power based on the trust and respect that people feel for an individual, group, or organization with which one is associated. | Gaining power by affiliating with nurses and others who have power in an organization. |
| Reward | Power that comes from the ability to reward others to influence them to change their behavior. | Using a hospital award to alter others' behavior. |
| Coercive | Power that comes from the ability to punish others or take disciplinary actions against others to influence them to change their behavior. | Using the hospital disciplinary evaluation system to alter another's behavior. |
| Connection | Power that comes from personal and professional relationships that enhance one's resources and the capacity for learning and information sharing. | Developing good working relationships and mentoring with your boss and other powerful people. |
| Information | Power based on information that someone can provide to the group. | Sharing useful knowledge gleaned from the Internet and other sources with coworkers. |

**Source:** Developed with information from Hersey, Blanchard, & Natemeyer (1979); French & Raven (1959); and Wells (1998).

# POSITIVE PERSONAL ORIENTATION TO POWER

A person's desire for power takes one of two forms. One form is an orientation toward achieving personal gain and self-glorification. Another form is an orientation toward achieving gain for others or the common good. Orientation to personal gain and power as a bad thing and therefore something to be avoided is reflected in a quotation from Lord Acton (Seldes, 1985, p. 234): "Power tends to corrupt, and absolute power corrupts absolutely." People having this orientation tend to believe that those wielding or afforded power ultimately should not have power because of their potential to misuse it, and that people desiring power should not be trusted because their motivation for acquiring power is inherently wrong—they want power for personal gain at any cost.

The other point of view, that power is a good thing, a force that is used for good purposes, is reflected in Gracian's (1892, p. 172) saying: "The sole advantage of power is that you can do more good." Nurses today are likely to see power as a positive thing and are more inclined to use this positive power to help others.

# EMPOWERMENT

Empowerment is a popular term in the nursing literature related to management, leadership, and politics. Authors describing empowerment usually view it as something positive or highly desirable to be aspired to, advocated for, or attained. Ahmad and Oranye (2010) described empowerment as a management strategy related to job satisfaction and commitment. Wagner, Cummings, Smith, Olson, Anderson, and Warren (2010) describe the relationship between structural and psychological empowerment. Using the basis of power of French and Raven (1959), we can conceive of empowerment as a form of capacity building, in which one's capacity to influence others is enhanced by an increase in any of the sources of power.

Nurses empower themselves and others in many ways. At the most basic level, they empower others because they perceive them to be powerful. If an individual, a group, or an organization is perceived as being powerful, that perception can empower that individual, group, or organization. Hahn (2010) describes how nurses, as individuals and through professional associations, can mobilize political power to influence local and national policy. Nurses disempower themselves if they see nurses or nursing as powerless.

## POWER AND THE MEDIA

People who work in the media recognize the relationship between power and perception. Those who work in advertising, marketing, and public relations understand how media can be used to create or change perceptions. They have long recognized that the public's perception can be created or changed through advertising and marketing campaigns, damage control, timely press releases, and well-orchestrated media events.

The way the media present nursing to the public will empower or disempower nursing. Nurses must work to consistently use the media as effectively as other more powerful occupational groups. To date, the media have failed to recognize nursing as one of the largest, most trusted groups in

> ### Critical Thinking 12-2
>
> Think about the challenges you will face as a new nurse on a busy medical-surgical unit. How will you fit into the group? Who can you rely on among your new coworkers? What is the status of nursing vis-à-vis the medical practitioners, pharmacy service, and food service? What sources of power do you bring into the work setting? How might you establish and enhance your knowledge and skills as a competent member of the team? Are you able to appropriately assert yourself in this new environment? What additional assets do you bring to the job?

health care. The media's presentation of the rapidly growing nursing shortage over the next decade can improve the public's perceptions of nursing as a career and human service. The media can show nurses as decision makers, coordinators of care, and primary care providers in health care. Too often, the media has presented a stereotypical, insignificant view of nurses (The Center for Nursing Advocacy, Inc.

## REAL WORLD INTERVIEW

Ms. Cox is 38 years old and has been hospitalized four times. She underwent surgery this past spring and has encountered nursing care and nurses in various roles throughout the health care system. Ms. Cox is articulate and reflective, having earned a degree in English and holding a position at a selective liberal arts university as an admissions counselor. She states, "I don't think nurses know they are powerful.... Nurses can take on more than they think they can. They have the power to change the system in which they work. Yet I also see nurses as the most overworked, underpaid, and underachieving professions. There is so much more they could be doing if they didn't spend so much time railing against the machine. They are telling the wrong people—each other—that they are frustrated. They should be telling the ones with real power, or, better yet, more of them should become the ones in power. Instead, they suffer with each other and stay angry. It appears almost passive-aggressive how nurses deal with power. My concern is that it can affect patient care in such a negative way. Believe me when I say patients value nurses, but the people writing the paycheck for nurses must value nurses. Patients need nursing far more than they need anything else. The better nurses that have cared for me have been instrumental in my healing. Beyond knowing when I need medication or performing some procedure, it is the smile, the touch, and the well-placed word of encouragement that has gotten me through. This is where nurses have power, because no one but a real nurse can provide it. It comes from the heart."

**Audrey Cox**
Patient
University Place, Washington

2003–2008). According to the Woodhull Study on Nursing and the Media (Sigma Theta Tau International, 1997), nurses are nearly invisible to the media. Sometimes, even nurses fail to view nursing as the honorable profession it is. One strategy for empowering nursing is to employ the media to create a stronger, more powerful image of nursing—for example, by writing opinion editorial (op-ed) pieces and letters to the editor for the local newspaper. Examples on a larger scale include a series of television spots promoting a positive nursing image, such as that recently sponsored by the Johnson & Johnson Corporation as part of its Campaign for Nursing's Future (Johnson & Johnson, 2006). Advocacy for nursing on a national scale requires more nurses to become active participants in some formal part of their profession—that is, the American Nurses Association (ANA), the National League for Nursing (NLN), or one of the nursing specialty organizations, for example, Emergency Nurses Association.

## PERSONAL POWER DEVELOPMENT

Understanding power helps the novice nurse become more effective, make better decisions, and help others better. Understanding power from a variety of perspectives is not just important for nurses professionally, it is important for them personally as well. It allows nurses to gain more control of their work lives and personal lives. There are three

### Critical Thinking 12-3

Personal empowerment for a beginning nurse requires imagining the future; setting concrete, achievable goals that represent the imagined future; and then creating that future by taking specific action steps that are likely to result in each desired goal. Think for a moment about the kind of nurse you might become. What kind of skills and expert knowledge are needed by someone who meets that description?

What are the possibilities? Are you really imagining all possible scenarios? What is probable at this time, knowing what you know now? What do you prefer? Finally, what would it take, in steps, for you to achieve your preference?

ways to imagine the future: (1) what is possible, for example, brainstorming various scenarios about what could be; (2) what is probable, for example, making a judgment about the likelihood of a given scenario; and (3) what is preferred, for example, choosing a desired scenario, even one that may be less likely to happen than the probable future. A nurse who wants to experience a preferred future should think about

## CASE STUDY 12-1

Maria and Haley work on the same nursing unit in a large, metropolitan hospital. Both predominantly work the evening shift and have less than one year's experience since graduating from nursing school. Maria has been offered increasingly difficult patient assignments, given charge duties, and recently was selected for a two-week leadership training program. Haley has not adapted as well and has withdrawn from what was once a close relationship with Maria. Haley seeks consolation with the nurses she and Maria claimed they would never emulate. Haley takes her breaks and eats dinner with two nurses who complain that the best nurses are undervalued in the organization, yet these same nurses were not supportive of Maria or Haley or any other new nurses oriented to their unit. Maria seeks out others she perceives to be knowledgeable and more satisfied in their professional roles. She strives to participate in nonmandatory meetings as well as clinical rounds, using them as an opportunity to ask questions; thus, she is beginning to increase her personal level of power by connecting with the other staff and gathering information. One night after a difficult shift, Haley accuses Maria of abandoning her and playing up to administration, saying that Maria is being used by the unit's nurse manager. Haley tells Maria that the other nurses are planning to file a complaint against the unit supervisor for selecting Maria, over the nurses that have been on the unit longer, to attend the leadership training program.

What does Haley's behavior tell you about her personal orientation toward power?

How should Maria react to Haley in this situation?

Apply your understanding of power to suggest how Maria can empower Haley.

## EVIDENCE FROM THE LITERATURE

**Citation:** DeCicco, J., Laschinger, H., & Kerr, M. (2006, May). Perceptions of empowerment and respect: Effect on nurses' organizational commitment in nursing homes. *Journal of Gerontological Nursing*, 49–56.

**Discussion:** This study supported Kanter's theory of structural empowerment by finding that access to empowering work structures—such as information support, opportunity, and resources—empowers individuals to get their work accomplished and thus contributes to overall organizational effectiveness.

**Implications for Practice:** Access to empowering work structures lead to empowerment and respect and are strong predictors of organizational commitment. Entry-level nurses should be able to recognize and value such conditions as access to information, opportunity, support, and resources as means that enable them and their coworkers to get the job done. They will be better prepared to act and thus to increase their success and that of the organization with self-reinforcing intermediate benefits such as feeling respected and increasing their commitment to the organization.

## REAL WORLD INTERVIEW

Power is the ultimate responsibility to care for the patient. Our assessment and subsequent actions can determine whether a patient will live or die. I feel confident in my abilities for the future, but as a new nurse, it is difficult to feel powerful because there is so much to learn. I believe that nurses tend to perceive power as being able to stand up to those above or higher up in the work setting. I think nursing is changing for the better because of the power gained with nurses having more autonomy. There is more trust put in the nurse's judgment because women's roles in society have changed. Equal opportunity programs have created the structure to protect women and others who have been vulnerable to those in positions of greater power. I have learned that you do not have to be afraid to advocate for your patient. If you need to call the attending physician during the night, then you do it. I have come to realize that power is not abusing the people working under you. Also, to understand power, it is important to understand people's roles and where they are coming from.

**Julie Bergman**
New Nursing Graduate
Tacoma, Washington

what is happening to him or her as a person and as a nurse, consider what possibilities he or she faces as a person and as a nurse, and then take action to create the future.

## POWER AND THE LIMITS OF INFORMATION

Even if nurses could fully trust the completeness and accuracy of information they have in their practice, they would have insufficient data. "There is no end to information just as there is no end to what we could know about

something" (Wells, 1998, p. 29). To make good decisions, nurses must be able to gather enough information and realistically interpret its value, as well as share and apply information in a safe, competent manner. Effective nurses understand time constraints and set priorities to ensure that what is most important receives the most attention. These nurses are willing to take the inherent risk of making a decision, while understanding that there will always be more information to gather and analyze. They recognize that choosing to make no decision is a decision in itself and that information is never complete. Table 12-2 presents a framework for becoming empowered.

## TABLE 12-2  A Framework for Becoming Empowered

| TYPE OF POWER | STEPS TO EMPOWERMENT |
| --- | --- |
| Personal power | Find a mentor. |
| | Notice who holds power in your personal, professional, and organizational life. Introduce yourself to them. |
| | Find and maintain good sources of evidence-based information. |
| | Seek answers to questions. |
| | Make a plan to develop all sources of personal, professional, and organizational power. |
| | Evaluate the plan. |
| Professional power | Assess patient's condition using relevant, objective measurements. |
| | Collaborate with administrators and other nursing and medical practitioners and health care workers involved in the care of your patients. |
| | Join your professional nursing organization. |
| | Consult with significant others, friends, and members of the patient's family. |
| | Monitor and improve patient care quality. |
| Organizational power | Get involved beyond direct patient care. |
| | Volunteer for committee assignments that will challenge you to learn and experience more than what is expected of you in a staff nurse role. |
| | Think about the following when involved with committees: |
| | 1. What is the committee trying to do? |
| | 2. What specific information does the committee use to operate and make decisions? |
| | 3. How does the committee apply to my practice, to my colleagues, to my patients, to my organizational unit, and to the organization as a whole? |
| | Continually improve and add to the knowledge you have in relation to your patients, your colleagues, your organizational unit, the organization as a whole, and the profession of nursing. |
| | Readily share appropriate knowledge with others who will value it and use it to a good end. |
| | Evaluate your plans. Did you achieve the expected outcomes? If not, why? Were there staffing problems or patient crises? Were the activities that were necessary for outcome achievement carried out? |
| | How can you apply lessons learned from this evaluation to the future? |
| | Periodically review Table 12-1. Continue to develop all your sources of power. |
| | Volunteer to be involved with health care at the local, state, and national levels. |

# MACHIAVELLI ON POWER

It would be naive to think that one can necessarily expect easy acceptance, understanding, or even support for what one is attempting to do. Machiavelli, an early authority on power, is reported to have said:

> *There is nothing more difficult to take in hand, more perilous to conduct than to take a lead in the introduction of a new order of things, because the innovation has for enemies all those who have done well under the old conditions, and lukewarm defenders in those who may do well under the new. (Machiavelli & Rebhorn, 2003)*

The entry-level nurse will do well to heed Machiavelli's warning. Machiavelli recognized that the power to innovate even small changes should be employed thoughtfully.

## Critical Thinking 12-4

As an entry-evel nurse, you empower yourself not only by using information for providing patient care, but also by using information in areas beyond direct patient care. Think about the following when involved with committees:

What is the committee trying to do?

What specific information does the committee use to make decisions?

How does the committee's work apply to your practice, your colleagues, your organizational unit, and the organization as a whole?

What is the strength of the information you have in relation to your patients, your colleagues, your organizational unit, and the organization as a whole?

Are you readily sharing information with others who will value it and use it to good end?

## KEY CONCEPTS

- Power can be described as the ability to get and use resources to achieve goals.
- Power can be defined at personal, professional, and organizational levels.
- Sources of power include expert, legitimate, referent, reward, coercive, connection, and information power.
- Effective nurses have a positive orientation toward power and feel comfortable discussing the expert knowledge that nurses bring to patient care.
- Effective nurses understand power from multiple perspectives.
- Effective nurses increase their own power sources and use power for safe, competent care.
- The personal power of effective nurses is evident in the decisions they make.

## KEY TERMS

coercive power
connection power
expert power

information power
legitimate power
power

referent power
reward power
sources of power

## REVIEW QUESTIONS

1. If, as a new nurse, you propose a change in a long-standing unit routine, you can expect which of the following?
   A. support from the head nurse.
   B. resistance from your coworkers.
   C. resistance from anyone who is comfortable with the status quo.
   D. support from your coworkers.

2. Which of the following can interfere with a new nurse's ability to influence his or her peers? Select all that apply.
   _____ A. Seeking knowledge of recent advances in nursing practice
   _____ B. Viewing the acquisition and use of power as a bad thing
   _____ C. Sharing helpful information about patients with coworkers
   _____ D. Interacting with nonnursing hospital staff
   _____ E. Being unwilling to acknowledge errors when they are made
   _____ F. Being a loner who is unwilling to share information with coworkers

3. Which of the following sources of power should a new nurse expect to increase during the first few months on the job?
   A. Expert power
   B. Legitimate power
   C. Referent power
   D. Coercive power

4. The most effective nurses use power is which of the following?
   A. in one primary way.
   B. to influence others or affect others' thinking or behavior.
   C. predominantly at an organizational level.
   D. only to gain the necessary resources to be a better nurse.

5. When a person fears another enough to act or behave differently than he or she would otherwise, the source of the other person's power is called which of the following?
   A. coercive power.
   B. reward power.
   C. expert power.
   D. connection power.

6. How can a nurse's personal power be enhanced? Select all that apply.
   _____ A. Collaborating with colleagues on special projects outside the work setting.
   _____ B. Taking part in gossip on the nursing unit.
   _____ C. Volunteering to serve on organizational committees led by nonnurses.
   _____ D. Getting to know one's coworkers off duty and sharing recreational activities together
   _____ E. Becoming more self-aware of personal nursing knowledge and skills
   _____ F. Identifying a competent coworker and learning from him or her

7. Three levels of power include all but which of the following?
   A. Personal
   B. Professional
   C. Organizational
   D. Unit

8. Power has been described in the literature
   A. consistently as a negative concept.
   B. most often as a manifestation of personal ambition.
   C. as maintained only through one's work.
   D. in multiple ways, some not so positive.

9. Which of the following can most interfere with a new nurse's base of personal power?
   A. Wearing a badge that indicates the nurse's name and professional license
   B. Hesitating when giving instructions to a nursing aide
   C. Introducing oneself to patients by giving first and last name
   D. Explaining a nursing procedure by citing the research that supports it

10. How can a new nurse gain influence on a nursing unit within a few weeks on the job? Select all that apply.
    _____ A. Identify a mentor on the unit and ask for guidance.
    _____ B. Associate with competent and respected coworkers.
    _____ C. Suggest improvements in unit routines that seem unclear.
    _____ D. Consult the Internet for more information about patients' diagnoses and treatment.
    _____ E. Steer clear of physicians and other nurses in daily nursing activities.
    _____ F. Seek information about patients from their family members and friends.

## REVIEW ACTIVITIES

1. Identify a nursing leader. Observe the nurse and note what type of power the nurse uses to meet objectives.

2. Watch a television show that portrays nurses. Note how nurses use or do not use the different types of power available to them. What do you observe?

3. Observe our national leaders. What examples of the use of power do you see? Is power used in helpful or unhelpful ways? Explain.

4. Observe a nursing unit during a shift. How do nursing coworkers use various types of power to influence one another?

## EXPLORING THE WEB

- This site critiques and advocates for accuracy in the media's portrayal of the role of nursing and of nurses. www.nursingadvocacy.org

- This site has a funny, not scholarly, synopsis of nursing power. www.NursingPower.net

- This site supports political power for patients. www.healthcarereform.net

- This site discusses a variety of nursing resources and issues, including collective power. www.nursingworld.org

- Go to the site for the Center for Health Policy, Research and Ethics, College of Health and Human Services, George Mason University. Note the seminars offered to nurses and other health care staff interested in health policy. http://www.gmu.edu. Search for, Center for Health Policy Research and Ethics. Click again on, Center for Health Policy Research and Ethics. Then click on, Institute.

- On the Google search engine, perform a search using the term "nursing leaders." www.google.com

- Find two nurses identified on the following site: www.distinguishedwomen.com

## REFERENCES

Ahmad, N., & Oranye, N. O. (2010). Empowerment, job satisfaction and organizational commitment: A comparative analysis of nurses working in Malaysia and England. *Journal of Nursing Management, 18*(5), 582–91.

Buresh, B., & Gordon, S. (2006). *From silence to voice.* Ithaca, NY: Cornell University Press.

DeCicco, J., Laschinger, H., & Kerr, M. (2006). Perceptions of empowerment and respect: Effect on nurses' organizational commitment in nursing homes. *Journal of Gerontological Nursing, 32*(5), 49–56.

Fitzpatrick, J. J., Campo, T. M., Graham, G., & Lavandero, R. (2010). Certification, empowerment, and intent to leave current position and the profession among critical care nurses. *American Journal of Critical Care, 19*(3), 218–226.

French, J. P. R., Jr., & Raven, B. (1959). The bases of social power. In D. Cartwright and A. Zander (Eds.), *Group dynamics* (pp. 607–623). New York: Harper and Row.

Gracian, B. (1892). *The art of worldly wisdom.* (J. Jacobs, Trans.) Boston: Dover Publications, 2005. (Original work published 1647).

Hahn, J. (2010). Integrating professionalism and political awareness into the curriculum. *Nurse Educator, 35*(3), 110–113.

Hersey, P., Blanchard, K., & Natemeyer, W. (1979). Situational leadership, perception and impact of power. *Group and Organizational Studies, 4,* 418–428.

Johnson & Johnson. (2006). Campaign for nursing's future. Retrieved July 3, 2006, from http://www.jnj.com/our_company/advertising/discover_nursing.

Machiavelli, N., & Rebhorn, W. A. (2003). *The prince and other writings.* New Providence, NJ: Barnes & Noble.

Seldes, G. (1985). *The great thoughts.* New York: Ballantine Books.

Sigma Theta Tau International. (1997). Woodhull study on nursing and the media. Retrieved August 18, 2010 from Increasing public understanding of Nursing. The Center for Nursing Advocacy. Accessed March 19, 2011, at www.nursingsociety.org/media/woodhullextract.html.

The Center for Nursing Advocacy, Inc. (2003–2008). Increasing public understanding of Nursing. Accessed March 18, 2011, at http://www.nursingadvocacy.org/faq/media_affects_nursing.html.

Wagner, J. I., Cummings, G., Smith, D. L., Olson, J., Anderson, L., & Warren, S. (2010). The relationship between structural empowerment and psychological empowerment for nurses: A systematic review. *Journal of Nursing Management, 18*(4), 448–62.

Wells, S. (1998). Choosing the future: *The power of strategic thinking.* Boston: Butterworth-Heinemann.

# SUGGESTED READINGS

American Nurses Association. (2010). *Nursing's social policy statement: The essence of the profession* (3rd ed.). Washington, DC: American Nurses Publishing.

Benner, P. E. (2000). *From novice to expert: Excellence and power in clinical nursing practice* (Commemorative ed.). Upper Saddle River, NJ: Prentice Hall.

Gordon, S. (2010). *When chicken soup isn't enough: Stories of nurses standing up for themselves, their patients, and their profession*. Ithaca, NY: ILR Press.

Government Technology. (any recent issue). (Available free online at www.govtech.net or by writing to Government Technology at 100 Blue Ravine Road, Folsom, CA 95630.)

Grindel, C. (2006). The power of nursing: Can it ever be mobilized? *MedSurg Nursing, 15*(1), 5–6.

Hughes, F., Duke, J., Bamford, A., & Moses, C. (2006). Enhancing nursing leadership through policy, politics and strategic alliances. *Nurse Leader, 4*(2), 24–27.

Kalisch, B., & Kalisch, P. (1986). A comparative analysis of nurse and physician characters in the entertainment media. *Journal of Advanced Nursing, 11*(2), 179–195.

Leddy, S., & Pepper, J. M. (2009). *Conceptual bases of professional nursing* (7th ed.). Philadelphia: Lippincott.

Manojlovich, M. (2007). Power and empowerment in nursing: Looking backward to inform the future. OJIN: *The Online Journal of Issues in Nursing, 12*(1). Retrieved at http://www.medscape.com/viewarticle/553403. Accessed March 18, 2011.

Middaugh, D. J., & Robertson, R. D. (2005). Politics in the workplace. *Nursing Management, 14*(6), 393–394.

Ponte, P. R., Glazer, G., Dann, E., McCollum, K., Gross, A., Tyrrell, R., Branowicki, P., Noga, P., Winfrey, M., Cooley, M., Saint-Eloi, S., Hayes, C., Nicolas, P., Washington, D. (2007). The power of professional nursing practice—An essential element of patient and family centered care. *OJIN: The Online Journal of Issues in Nursing, 12*(1). Retrieved at http://www.medscape.com/viewarticle/553405. Accessed March 18, 2011.

Vesely, R. (2008). 'Unleash the energy.' Activists, professors try to use their own sphere of influence to affect U.S. healthcare policy and improve patient care. *Modern Healthcare, 38*(17), 6–7

Vestal, K. (2007). The power and intrigue of workplace politics. *Nurse Leader, 5*(1), 6–7.

# CHAPTER 13

# Change, Innovation, and Conflict Management

## Kristine E. Pfendt, RN, MSN; Margaret M. Anderson, EdD, RN

*Change and innovation are closely related concepts. In our society, change is inevitable and innovation is a necessary component of our culture.*

(Lillian Sims, 2000)

## OBJECTIVES

Upon completion of this chapter, the reader should be able to:

1. Explain change from personal, professional, and organizational perspectives.
2. Identify the change theorists.
3. Explain the concept of the learning organization.
4. Identify driving and restraining forces of change within a structured setting context.
5. Relate change strategies.
6. Explain the role and characteristics of a change agent in the change process.
7. Plan, implement, and evaluate a change project using the change process.
8. Apply the concept of innovation to health care.
9. Identify conflict situations.
10. Identify steps in the conflict management process.

Delmar/Cengage Learning

*Dwayne is working on a patient care unit as a new computerized charting system is introduced. He notices that several of the staff start using the system quickly. Other staff are resistant and angrily say they want no part of the new system. Dwayne feels comfortable with the system and begins to use it right away.*

> *How can Dwayne contribute to the staff's smooth transition to the new charting system?*

> *How can Dwayne's nurse manager help staff with the transition to the new system?*

> *What knowledge and skills help a new graduate adapt to a new environment?*

Living organisms must constantly adapt to changes in their environment in order to thrive. In this same way, human beings must also successfully adapt and manage change in order to maintain homeostasis and equilibrium. Rapid changes and innovations occurring in the world today require continuous human responses. How these changes and innovations are perceived by the individual can mean the difference between successful adaptation and maladaptation, survival and extinction.

Today's health care environment is also constantly changing. Luthans and Jensen (2005) state that "the health care industry is suffering from an unprecedented shortage of qualified nurses as well as increasing demands and complexity due to technological innovations and changing consumer expectations" (p. 304). Access to information has transformed the relationship between the patient and health care provider. Individuals in many cases are better informed about their medical conditions and treatment options. These same individuals view themselves as partners in their health care rather than just consumers.

Evidence-based practice is changing the way decisions are made regarding health care treatment and how nursing care is delivered. Interventions such as surgery and diagnostic testing that once required lengthy and costly stays in hospitals are often provided now in outpatient settings. The length of stay in hospitals has been shortened both by the impact of enhanced technology and the requirements of managed care.

Changing demographics within the population has resulted in a diversity of cultures and languages as well. Challenges continue for nursing and others involved with health care delivery to understand these cultures and to communicate effectively. The aging of the baby boomers and the continuous need for nurses to care for them will also impact health care in the future.

The rising costs of health care services have had a major impact on society as well. Premiums are constantly rising for those consumers who are fortunate enough to have health care insurance. The cost of Medicare and Medicaid also continues to rise, and the nation must grapple with payment. For those without any access to health care, the costs continue to soar. Delays in seeking care because of an individual's inability to pay for services and medications often result in more seriously ill patients, with life-altering consequences.

Issues regarding patient safety can be translated into the cost of care. An Institute of Medicine (IOM) report states that there are staggering consequences to unsafe care and medication errors. "Medication errors alone, occurring either in or out of the hospital, are estimated to account for over 7,000 deaths annually...One recent study conducted at two prestigious teaching hospitals, found that about 2 out of every 100 admissions experienced a preventable adverse drug event, resulting in average increased hospital costs of $4,700 per admission or about $2.8 million annually for a 700-bed teaching hospital" (IOM, 2002).

All of these factors affect the ways that nurses function in the health care environment and how they care for their patients. How nurses manage these changes and innovations will impact quality of care, the health care system of the future, and the lives of all health care consumers. This chapter discusses the change process, innovation, and conflict management. This information will enhance the nurse's ability to manage future change and innovation successfully.

## CHANGE

There are many definitions of *change*, and there are many types of change. For simplicity, **change** can be defined as "making something different from what it was" (Sullivan & Decker, 2009, p. 67). The outcome may be the same as prior to the change, but the actions performed to get to the outcome may be different. For instance, because of road closures, how you get to work may have to change. The goal of getting to work remains the same, but the method may be different, perhaps by bus rather than automobile or by use of a different route. In the professional nursing setting, new patient admission forms may necessitate a different method of assessing the patient or change the number of people involved in the admission process. Rather than one registered nurse conducting the entire admission process, the process may be broken down so that individuals with different skill levels conduct different parts of the process. The goal is still the admission of the patient to a unit; how it is done may be different. Most change is implemented for a good or reasonable purpose. Beason (2005) states that to be effective, change should affect all levels of the health care organization and that communication throughout all levels is essential.

For purposes of discussion, **personal change** is a change made voluntarily for your own reasons, usually for self-improvement. This may include changing your diet for health reasons, taking classes for self-improvement, or removing yourself from a destructive or unhealthful environment or situation. **Professional change** may be a change in position or job, such as obtaining education or credentials that will benefit you in a current position or allow you to be prepared for a future position. Professional change is often planned and involves extensive alterations in both personal and professional lives. Although either personal or professional change may be stressful, if it is voluntary and carries intrinsic or extrinsic rewards, it is often considered important and worth the stress.

Organizational change often causes the most stress or concern. Generally speaking, **organizational change**, which is a planned change in an organization to improve efficiency, is thrust upon employees. Sometimes there is a lot of preparation and prior discussion. Change that is unexpected causes a great deal of consternation and stress. Organizational change that is planned and purposeful is generally better accepted. It is used to improve efficiency or improve financial standing or for some other organizational purpose. Change is planned to meet organizational goals. One of the goals most organizations have is to maintain a positive financial balance. Change to ensure a healthy financial standing maintains jobs and increases the organization's capability to meet its mission. Improved efficiency is good for the organization's capability to get goods and services to the customer. Improved efficiency in health care provides better-quality care for patients and improves the workload of the employees.

Organizations that adapt successfully to change are able to adapt quickly to new market conditions. These are the organizations that will survive in a competitive environment. Employees who are educated about the need for change and who embrace new ideas are key to the achievement of successful outcomes (Recklies, 2006). Organizational change must be embraced by all levels of employees, including managers, to be effective (Beason, 2005).

Budget changes may be necessary for the survival of a health care institution. Nurses caring for patients are often impacted by staff reductions due to budget cuts. Nursing staff may believe that the quality of patient care will diminish as a result. As staffing levels are cut, those nurses who are left to care for patients often face increased workloads. Staff burnout may result, and pressures may rise as turnover increases. In the larger picture, the increased costs of hiring and orienting new nurses to replace the ones who have left may exceed the savings from the initial staff layoffs Ervin, (2009).

Effective communication to nursing staff about why budget changes are necessary is essential to adaptation to changes that impact the quality of patient care. Education is needed about the current marketplace and how nursing units are impacted. For example, changes in third-party payment practices may influence the number of nurses that a patient care unit can have. Nurse leaders and managers often have the responsibilities for communicating and educating staff concerning why these changes are necessary (Tvedt, Saksvik, & Nytvek, 2009).

At times, when organizational change is planned, the employees may be the last to know what the anticipated change is but may be the ones most affected by it. The staff nurse is expected to implement the new care delivery system, but he or she may also be the last one to know about the change before it is to be implemented. For example, in an organization in which primary nursing has been the care delivery system for several years, the implementation of modified team nursing is a major change in philosophy and thinking. It is important that proper care and planning of the change process be used, so that the staff will not resist the change and make the implementation much more stressful than necessary. Table 13-1 summarizes the types of change.

## TRADITIONAL CHANGE THEORIES

The classic change theories discussed here are Lewin's force-field model (1951), Lippitt's phases of change (1958), Havelock's six-step change model (1973), and Rogers' diffusion of innovations theory (1983). These are classic change theories based on Lewin's original model.

Lewin's model has three simple steps. The steps are unfreezing, moving to a new level, and refreezing. See Figure 13-1. Unfreezing means that the current or old way of doing things is flawed. People begin to be aware of the need for doing things differently, that change is needed for a specific reason. In the next step, the intervention or change is introduced and explained. The benefits and disadvantages are discussed, and the change—the move to

**FIGURE 13-1 Lewin's force field model of change.** (Source: Lewin, K. (1951). *Field theory in social science*. New York: Harper & Row).

## TABLE 13-1 Types of Change

| TYPE OF CHANGE | DEFINITION |
| --- | --- |
| Personal | A change made voluntarily for one's own reasons, usually for self-improvement. |
| Professional | A change for a position or job, such as obtaining education or credentials that will benefit one in a current position or allow one to be prepared for a future position. |
| Organizational | A planned change in an organization to improve efficiency. |

## EVIDENCE FROM THE LITERATURE

**Citation:** Haase-Herrick, K. (2005). The opportunities of stewardship. *Nursing Administration Quarterly, 28*(2), 115–118.

**Discussion:** This article discusses goals for improving the current health care system that were first recommended by the Institute of Medicine. These were published in *Crossing the Quality Chasm: A New Health System for the 21st Century* in 2002. The recommendations were that health care should be safe, effective, patient-centered, efficient, and equitable. Haase-Herrick challenges nurses to strive to change the health care delivery system as it currently exists and to change the status quo. She recommends that nurses do this by increasing their knowledge level of health care economics, health care financing, and statistical and financial analysis.

Nurses are applying these strategies and are in fact changing the way that health care is delivered. Through the establishment of new nurse-run centers, care is being provided that is cost effective and patient centered. These centers are meeting the needs of many underserved populations in their respective communities. (See the article "Nurse-Managed Health Centers Are Changing the Face of Health Care" National Nursing Centers Consortium, 2011).

The author also advocates for nurses to become more actively involved in political processes, to ensure that the most current standards of practice are used by professional organizations and licensing and accrediting agencies. The profession of nursing can impact the quality of care through continuous involvement in these processes.

**Implications for Practice:** Nurses must ask themselves in what way can they become involved in the political processes that are changing health care continuously. Change is inevitable!

a new level—is implemented. In the third step, refreezing occurs. This means that the new way of doing things is incorporated into the routines or habits of the affected people. Although these steps sound simple, the process of change is, of course, more complex.

Lippitt's phases of change are built on Lewin's model. Lippitt defined seven stages in the change process. These steps are: (1) diagnosing the problem, (2) assessing the motivation and capacity for change, (3) assessing the change agent's motivation and resources, (4) selecting progressive change objectives, (5) choosing an appropriate role for the change agent, (6) maintaining the change after it has been started, and (7) terminating the helping relationship. Lippitt emphasized the participation of key personnel and the change agent in designing and planning the intended change project. Lippitt also emphasized communication during all phases of the process.

Havelock designed a six-step model of the change process. This model is based on Lewin's, but Havelock included more steps in each stage. The planning stage includes: (1) building a relationship, (2) diagnosing the problem, and (3) acquiring resources. This planning stage is followed by the moving stage, which includes (4) choosing the solution and (5) gaining acceptance. The last stage, the refreezing stage, includes (6) stabilizing and self-renewing. Havelock emphasized the planning stage. He believed that resistance to change can be overcome if there is careful planning and inclusion of the affected staff. Havelock also believed that the change agent, the person responsible for planning and implementing the change, encourages participation on the part of the affected people. The more the people affected by the change participate in the change, the more they are likely to make the change successful and to support the necessity for the change (Sullivan & Decker, 2009).

In 1983, Rogers published his diffusion of innovations theory. Though based on Lewin's model, this theory is much broader in scope and approach. He developed a five-step innovation/decision-making process. Rogers believes that the change can be rejected initially and then adopted at a later time. He believes that change is a reversible process and that initial rejection does not necessarily mean the change will never be adopted. This also works in reverse—the change may initially be adopted and then rejected at a later time. Rogers' approach emphasizes the capriciousness of change. Timing and format take on new meaning and importance. Rogers means that as the time involved in change implementation grows, the more the change process takes on a life of its own, and the original change and reasons for it may be lost. The change process must be carefully managed and planned to ensure that it survives mostly intact.

## Commonalties and Differences

These change theories, models, and phases are helpful in understanding change and the dynamics involved in change. They do have similarities and differences, as shown in Table 13-2.

All of these theorists relate to Lewin's three simple steps of unfreezing, moving, and refreezing. Table 13-2 also identifies the theories' uses in change projects. The theories described in Table 13-2 are linear in nature in

### TABLE 13-2  Comparison Chart of Change Theories and Their Uses

|  | LEWIN | LIPPITT | HAVELOCK | ROGERS |
|---|---|---|---|---|
| **Theorist and Year** | Lewin (1951) | Lippitt (1958) | Havelock (1973) | Rogers (1983) |
| **Title of Model** | Force-Field Model | Seven Phases of Change | Six-Step Change Model | Diffusion of Innovations Theory |
| **Steps in Model** | 1. Unfreeze.<br>2. Move.<br>3. Refreeze. | 1. Diagnose problem.<br>2. Assess motivation and capacity for change.<br>3. Assess change agent's motivation and resources.<br>4. Select progressive change objectives.<br>5. Choose appropriate role of change agent.<br>6. Maintain change.<br>7. Terminate helping relationship. | 1. Build relationship.<br>2. Diagnose problem.<br>3. Acquire resources.<br>4. Choose solution.<br>5. Gain acceptance.<br>6. Stabilize and self-renew. | 1. Awareness<br>2. Interest<br>3. Evaluation<br>4. Trial<br>5. Adoption |
| **Use in Change Projects** | General model for most situations and organizations. | Good for changing a process and general change. | Often used for educational change or cultural change. | Used in organizational change, individual change, and group change. |

**Source:** Compiled with information from Lewin, K. (1951). *Field history in social science.* New York: Harper & Row; Lippitt, R. (1958). *The dynamics of planned change.* New York: Harcourt Brace; Havelock, R. G. (1973). *The change agent's guide to innovation in education.* Englewood Cliffs, NJ: Educational Technology; Rogers, E. M. (1983). *Diffusion of innovations* (3rd ed.). New York: Free Press.

that they more or less proceed in an orderly manner from one step to the next. This linearity and the fact that they are all based on Lewin's theory make them similar in complexity and in use. These theories work well for low-level, uncomplicated change. They may not always work well in highly complex and nonlinear situations. Havelock's theory is often applied to educational change, whereas Rogers's is useful for individual change. Health care organizations are very complex and require more-sophisticated theories of change, which will be presented next.

# CHAOS THEORY

Chaos theory hypothesizes that chaos actually has an order. That is, although the potential for chaos appears to be random at first glance, further investigation reveals some order to the chaos. Health care organizations have experienced chaos repeatedly during the past twenty years. Chaos theory would say that this is normal. Most organizations go through periods of rapid change and innovation and then stabilize before chaos erupts again. Even though each chaotic occurrence is similar to the one that occurred before, each is different. The political, scientific, and behavioral components of the organization are different from before, so the chaos looks different. Order emerges through fluctuation and chaos. Thus, the potential for chaos means that nurses and the organization must be able to organize and implement change quickly and forcefully. There is little time for orderly linear change.

# THEORY OF LEARNING ORGANIZATIONS

Peter Senge (1990) first described his classic Learning Organization theory, which focused on ways organizations learn and adapt. Learning organizations demonstrate responsiveness and flexibility. Senge believed that because organizations are open systems, they could best respond to unpredictable changes in the environment by using a learning approach in their interactions and interdisciplinary workings with one another. The whole cannot function well without a part regardless of how small that part may seem. An example in health care is that the billing department cannot submit an accurate bill to the insurance company without the cooperation of the nursing staff. If the patient is not charged appropriately for items used in his care, then the biller cannot prepare an accurate invoice. Without this invoice, the organization cannot be paid for the actual services and supplies used. The learning organization understands these interrelationships and responds quickly to improve relationships. This may be through dialogue or team problem solving, but all parties must understand what is at stake for cooperation and working together to occur.

Senge, Kleiner, Roberts, Ross, and Smith (1999) emphasize that the core of the learning organization is based on five "learning disciplines"—lifelong programs of study and practice.

- **Personal Mastery.** Learning to expand our personal capacity to create the results we most desire, creating an organizational environment that encourages all members of the team to develop themselves toward the goals and purposes they choose.
- **Mental Models.** Reflecting on, continually clarifying, and improving our internal pictures of the world, and seeing how this vision shapes our actions and decisions.
- **Shared Vision.** Building a sense of commitment in a group by developing shared images of the future we seek to create, and developing the principles and guiding practices by which we hope to get there.
- **Team Learning.** Encouraging conversational and collective thinking skills, so groups of people can reliably develop intelligence and ability greater than the sum of individual members' talents.
- **Systems Thinking.** A way of thinking about and understanding the forces and interrelationships that shape the behavior of systems.

*Systems thinking* highlights the fact that change in one area will affect other areas of the system. This discipline helps us to see how we can act more in tune with the larger processes of the natural and economic world.

In organizations, Senge believes the individuals who contribute most to the enterprise are the ones who are committed to the practice of these five learning disciplines. They are developing personal mastery of their environment; continually clarifying and improving their internal mental model of the world and the future they hope to create; sharing their vision and building group commitment to the future; encouraging team learning; and working to understand the behavior of the systems they are developing.

# THE CHANGE PROCESS

Planned change in the work organization is similar to planned change on a personal level. The major difference is that more people are involved, the scale is larger, and more opinions must be considered. There are three basic reasons to introduce a change: (1) to solve a problem, (2) to improve efficiency, and (3) to reduce unnecessary workload for some group (Marquis & Huston, 2006). To plan change, one must know what has to be changed. Change for the sake of change is unnecessary and stressful (Bennis, Benne, & Chin, 1969).

## Steps in the Change Process

The change process can be related to the nursing process. Using the nursing process as a model, the first step in the change process is assessment.

**Assessment**   In assessment, you identify the problem or the opportunity for improvement through change by collecting and analyzing data. Sometimes the change is planned to meet a previously determined goal. The current methods to meet that goal may not be working, so a new plan needs to be developed. Often the change is needed to take advantage of an opportunity rather than to solve a problem. Assessment must be aimed at the perceived problem or opportunity, and then the plan is focused on that. For example, if the nurse manager observes that the evening shift is not able to finish its work on time or within a reasonable time, then what is the problem? It could be that staff are not very experienced, or maybe there are too many transient staff. Perhaps there is a new patient service offered and the care for these patients is taking more time than is anticipated. Is there a structural problem? For example, maybe the supplies or care products are stored too far away for the number of staff members working this shift. Maybe the supply stations should be less far apart. Are supplies or drugs not delivered in an appropriate time frame? Is new or old technology slowing down the staff response time? These are the issues to be assessed to identify the problem. A quick examination or a hasty decision will not accurately diagnose the problem.

After assessment has been completed to determine the problem, attention must be paid to data collection and data analysis. This may be different from data with which the nurse is familiar. The data collection and analysis should be from different sources: structure, technology, and people. A structural problem is one of physical space or the configuration of physical space. For instance, a medical-surgical unit in a hospital may move to the space vacated by the old obstetrics unit. The problem may be that the space is large enough, but it is not configured to be conducive to the care of medical-surgical patients. Assessment of structural components may include examination of the location of elevators, supply stations, patient charts, telephones, call lights, or any other physical or structural components. If nurses have not been included in the space configuration planning, then more-expensive remodeling may be necessary. Structural components often mandate how the work is done or the process of doing the work. Poor structural configuration may require extra steps or work to accomplish the goals of the work team.

Technological problems may include a lack of wall outlets for necessary equipment, poorly situated computer locations, and lack of systems capability to interface among computers. Often, technology lags behind the goals of a work team and therefore slows the team down. The team spends more time troubleshooting technology to meet their needs than in providing care.

People problems may include personnel with inadequate training, willingness, commitment, or understanding to change and meet any new goals.

A work process problem (how the work is done), an equipment problem, or a people problem may be the cause of the need to change. Data collection must focus on the work, equipment, or people problem, in order to reach desired goals.

Assessment data are collected from internal and external sources. Lewin identified forces that are supportive of as well as those that are barriers to change. He called these driving and restraining forces. If the restraining forces outweigh the driving forces, then the change must be abandoned because it cannot succeed. Driving and restraining forces include political issues, technology issues, cost and structural issues, and people issues. See Figure 6-2. The political issues include the power groups in favor of or against the proposed change. This may include practitioners, administrators, civic and community groups, or state or federal regulators. The technology issues include whether to update old equipment, computer systems, or methods for accounting for supply use. Structural issues include the costs, desirability, and feasibility of remodeling or building new construction for the change project. People issues include the commitment of the staff, their level of education and training, and their interest in the project. The most common people issue is fear of job loss or fear of not being valued. It bears repeating that if the restraining forces outweigh the driving forces, then the change will not succeed and it should be abandoned or rethought.

During data analysis, potential solutions may be identified, sources of resistance may come to light, determination of strategies may become apparent, and some areas of consensus may become evident. Statistical analysis is an important component of analysis. This should be carried out whenever possible to provide persuasive information in favor of the change, especially if meeting cost or mission objectives is the issue. The goal of data analysis is to support the need to change and offer data to support the potential solution selected. The people who are potentially going to be most affected by the change need to be involved in the assessment, data collection, and data analysis. They have a vested interest in the change and must not only support it but also be willing to implement change (Bennis, Benne, & Chin, 1969).

**FIGURE 13-2 Forces driving and restraining change.**

# Planning

The next step is the planning step. Here, the who, how, and when of the change is determined. All of the potential solutions are examined. The driving and resisting forces are again examined, and strategies are determined for implementing the change. The target date for implementation and the outcomes or goals are clearly determined and stated in measurable terms. Although these items were examined in the assessment step, they are now made more specific. The potential for error is reduced, and the forces for success are marshaled. Again, the individuals who will be most affected must be involved. The most successful plan for change is one in which the affected people are involved, satisfied, and committed. Change cannot be thrust upon people without expecting resistance. Most of the time, resistance can be overcome with planning and involvement of those most affected by the change.

As you plan the change, consider, for instance, how many work groups or units will implement the change at once. Will the change implementation be staggered from month to month or week to week? Will the supports necessary to manage the change go into effect first? Just how will this change be implemented? Finally, the overall plan includes plans for evaluation. It is crucial that evaluation be built in. Expected outcomes must be identified in measurable terms, and the plan to evaluate those outcomes and a timetable for evaluation must be evident. Unevaluated change will not succeed. The status quo will seep back into play, and all the efforts directed at change will be for naught. Planned evaluation keeps the project in the forefront of people's minds. In health care, there is no change that does not involve people. Patients and staff are affected by any changes.

**Implementation of Change Strategies**   The next step is the implementation step. Here, the plan actually goes live! There are several ways to implement change. The most common method to encourage change by individuals is to provide information. Information regarding the change and its anticipated advantages must be disseminated early and often. The change should be viewed by those who designed it and those who must implement it as a positive solution to a problem or a wonderful method for addressing an opportunity. Optimism is the key to a higher success rate.

Another method used to change individuals is competency-based education. The educator provides information and practice. This is especially useful for equipment or technology change. The individual is shown how to use the equipment and when to use it. The information provides useful data for incorporating the equipment into daily work life. Large groups of individuals can be taught together for the best results. They help each other achieve the goals of safe practice and appropriate use of the product. Bennis,

Benne, & Chin (1969) identified three strategies to promote change in groups or organizations. Different strategies work in different situations. The power or authority of the change agent has an impact on the strategy selected. Most change agents use a variety of strategies to promote successful change. The power-coercive strategy is very simple—"do it or get out." This is a strategy based on power, authority, and control. It is a strong indication of the political clout of the change agent. There is very little effort to encourage the participation of employees, and there is little concern about their acceptance of or resistance to the proposed change. Sullivan and Decker (2009) use the federal government's decision to impose prospective payment on Medicare hospital patients as an example of the power-coercive strategy—no discussion, simply this is what will be. This type of strategy is generally reserved for situations in which resistance is expected but not important to the power group.

The second group change strategy is normative-reeducative. This strategy is based on the assumption that group norms are used to socialize individuals. It assumes that because people are social beings, social relationships are important and people will go along with a change if the social group sanctions it. The change agent uses satisfactory interpersonal relationships rather than power to gain compliance with the desired change. This strategy uses noncognitive methods of inducing compliance by focusing on social relationships and perceptual orientations. This means that information and knowledge are not used to gain compliance, but rather the individual's need for satisfying social relationships in the workplace is used. Very few individuals can withstand social isolation or rejection by the work group. So compliance and support for a change are garnered by focusing on the perceived loss of social relationships in the workplace. Although some resistance to change may be expected, this strategy assumes that people are interested in preserving relationships and will go along with the majority.

The third group of strategies is rational-empirical. It assumes that humans are rational people and will use knowledge to embrace change. It is assumed that once the self-interests of a group are evident, members will see the merit in a change and embrace that change. Knowledge is the component used to encourage compliance with change. This is a very successful strategy when little resistance is anticipated. Table 13-3 summarizes the strategies for change.

These strategies are important to the success of the change and are often used together to effect a necessary change. The implementation phase is very important to success. When compliance is gained, implementation will be smoother. The project has a greater chance of being successful.

Two characteristics indicate successful implementation of a project. The most important is that the people affected by the change begin to own the change and speak

## TABLE 13-3  Strategies for Change

| STRATEGY | DESCRIPTION | EXAMPLE |
|---|---|---|
| Power-coercive approach | Used when resistance is expected but change acceptance is not important to the power group. Uses power, control, authority, and threat of job loss to gain compliance with change—"Do it or get out." | Student must achieve a passing grade in a class project to complete the course requirements satisfactorily. |
| Normative-reeducative approach | Uses the individuals need to have satisfactory relationships in the workplace as a method of inducing support for change. Focuses on the relationship needs of workers and stresses "going along with the majority." | A new RN who is working 8-hour shifts is encouraged by the other unit staff to embrace a new unit plan for 12-hour staffing. |
| Rational-empirical approach | Uses knowledge to encourage change. Once workers understand the merits of change for the organization or understand the meaning of the change to them as individuals and the organization as a whole, they will change. Stresses training and communication. Used when little resistance is anticipated. | Staff are educated regarding the scientific merits of a needed change. |

well of it. The benefits stated as positive outcomes of the change actually begin to materialize. The importance of the change is apparent, and those involved demonstrate interest in and enthusiasm for the change. The communication about the change supports its value, benefits, and usefulness. The second characteristic of successful implementation is that the change is perceived as an improvement and is received with positive regard and anticipation throughout the organization.

**Evaluation of Change**  In the evaluation step, the effectiveness of the change is evaluated according to the outcomes identified during the planning and implementation steps. This is the most overlooked step, and it is considered by some experts to be the most important. Usually, enough time is not allowed for the change to be effective or stable. This is especially true in health care. The learning curve is not identified and, therefore, the change is prematurely assumed to be ineffective. This is a terrible error. The time intervals for evaluation should be identified and allowed to elapse before modifications and declarations of failure are asserted. A certain period of confusion and turmoil accompanies all changes, whether large or small. If the outcomes are achieved, then the change is a success. If not, then some revision or modification may be necessary to achieve the anticipated outcomes. Note that although it is important to understand why a change was not successfully implemented, it is equally important to understand why a change is successful.

## REAL WORLD INTERVIEW

One of the things I have learned about change is that fear of the change is worse than the change itself. Now I concentrate on assisting my staff to focus on the reasons the change is necessary and involve them in the process of finding solutions to the problem.

**Joy Churchill, RN**
Team Leader
Highland Heights, Kentucky

**Stabilization of Change**  After the effectiveness of the change is determined, then stabilization of the change is completed. The project is no longer a pilot or experiment but is a part of the culture and function of the organization. This is when the change agent bows out and the affected employees own the change. Although there is no magical time frame for stabilization to occur, it should be encouraged as soon as possible, to make the change project complete. Often, reevaluation is planned after the first six months or year of implementation to assure that stabilization has occurred.

## Critical Thinking 13-1

Nirmala was nearing the end of her 12-hour day shift and looking forward to attending her son's final high school soccer game after work. Just when she was ready to leave the unit, a new patient unexpectedly arrived and needed to be admitted. Judy, the nurse on the next shift, asked Nirmala to admit the new patient before she left. Nirmala refused, and a heated exchange occurred. What were the conditions leading to this conflict? What was the core of the conflict? Who was right? How can conflicts like this be avoided in the future?

## RESPONSES TO CHANGE

People do have responses to change. In health care, there has been such chaos and confusion at times, that employees are often resistant to change simply because it is a change. The most typical response to change is resistance. Human beings like order and familiarity. Humans enjoy routine and the status quo, so it is common for humans to resist change. The more the relationships or social mores are challenged to change, the more resistance there is to change. Marquis and Huston (2006) point out that nurses are more likely to accept a change in an intravenous pump rather than a change in who can administer the intravenous fluid. This suggests that the social mores of a group are more important than technology in a change. The social mores dictate the roles and responsibilities of groups of workers such as registered nurses, licensed practical

nurses, nurse aides, and so on. Registered nurses are often less concerned with technology and more concerned with maintaining traditional roles and responsibilities. The trick to successful change is to manage the resistance rather than try to eliminate it.

Several factors affect resistance to change. The first is trust. The employee and employer must trust that each is doing the right thing and that each is capable of producing successful change. In addition to capability, predictability is important. The employee wants a predictable work environment and security. When change is introduced, predictability begins to come into question. Another factor is the individual's ability to cope with change. Jones (2007) points out four factors that affect an individual's ability to cope with change:

1. Flexibility for change, that is, the ability to adapt to change
2. Evaluation of the immediate situation, that is, if the current situation is unacceptable, then change will be more welcome
3. Anticipated consequences of change, that is, the impact change will have on one's current job
4. Individual's stake or what the individual has to win or lose in the change, that is, the more individuals perceive they have to lose, the more resistance they will offer

Change is a frightening prospect for those who have not had much experience with change or have had negative experiences with change. It is important to help individuals remember that change is inevitable and ever present. Developing an attitude of embracing and accepting change is desirable.

Bushy and Kamphius (1993) have identified six behavioral responses that individuals have to planned change. See Table 13-4.

## TABLE 13-4  Responses to Change

1. *Innovators:* Change embracers. Enjoy the challenge of change and often lead change.

2. *Early adopters:* Open and receptive to change but not obsessed with it.

3. *Early majority:* Enjoy and prefer the status quo but do not want to be left behind. They adopt change before the average person.

4. *Late majority:* Often known as the followers. They adopt change after expressing negative feelings and are often skeptics.

5. *Laggards:* Last group to adopt a change. They prefer tradition and stability to innovation. They are somewhat suspicious of change.

6. *Rejectors:* Openly oppose and reject change. May be surreptitious or covert in their opposition. They may hinder the change process to the point of sabotage.

Regardless of the importance and necessity of change, the human response is very important and cannot be dismissed. So often, in the zeal to respond to a need, the change agent forgets that the human side of change must be dealt with. People have a right to their feelings and a right to express them. The important point is that the change agent helps people respond and then move on to the goal of implementing the change. Gently but firmly, people must be guided toward acceptance.

## THE CHANGE AGENT

Throughout this discussion of change, the term **change agent** has been used instead of manager, leader, or administrator. The change agent leads and manages change. This person may be from inside or outside an organization. The change agent may be a leader or manager or may have become a leader or manager because he or she is an innovator and likely to enjoy change. The change agent is the person who is ultimately responsible for the success of the change project, large or small. The role of the change agent is to manage the dynamics of the change process. This role requires knowledge of the organization, knowledge of the change process, knowledge of the participants in the change process, and an understanding of the feelings of the group undergoing change. Probably the most important role of the change agent is to maintain communication, momentum, and enthusiasm for the project while still managing the process. Table 13-5 identifies change agent approaches.

The change agent should possess some important characteristics. These include trust and respect from the recipients of change as well as from the chief executives in the organization. The recipients of change must trust not only the change agent's interpersonal skills to provide information and manage change, but also the change agent's personal integrity and standing as an honest, principled individual.

### TABLE 13-5  Change Agent Approaches

1. Begin by articulating the vision clearly and concisely. Use the same words over and over. Constantly remind people of the goals and vision.

2. Map out a tentative timeline and sketch out the steps of the project. Have a good idea of how the project should go.

3. Plant seeds with, or mention some ideas or thoughts to, key individuals from the first step through the evaluation step so that some idea of what is expected is under consideration.

4. Select the change project team carefully. Make sure it is heavily loaded with those who will be affected and other experts as needed. Select a variety of people. For example, an innovator, someone from the late majority group, a laggard, and a rejector are probably good to include. These people provide insight into what others are thinking.

5. Set up consistent meeting dates and keep them. Have an agenda and constantly check the timeline for target activities.

6. For those not on the team but affected by the project, give constant and consistent updates on progress. If the change agent does not, someone on the project team will, and the change agent wants to control the messages.

7. Give regular updates and progress reports both verbally and in writing to the executives of the organization and those affected by the change.

8. Check out rumors and confront conflict head on. Do not look for conflict, but do not back away from it or ignore it.

9. Maintain a positive attitude, and do not get discouraged.

10. Stay alert to political forces both for and against the project. Get consensus on important issues as the project goes along, especially if policy, money, or philosophy issues are involved. Obtain consensus quickly on major issues or potential barriers to the project from both executives and staff.

11. Know the internal formal and informal leaders. Create a relationship with them. Consult them often.

12. Having self-confidence and trust in oneself and one's team will overcome a lot of obstacles (Lancaster, 2008).

Good communication is essential to "buy in"
of nursing staff to the change process. It is
very important for nurses to understand why
a change is necessary in order for them to
embrace change.

**Caron Martin, RN**
Staff Nurse
Highland Heights, Kentucky

The executives in the organization must trust that the change agent will accomplish the established goals given the proper support. Credibility and flexibility are also important characteristics for the change agent to possess. The change agent cannot be temperamental or rigid. The ability of the change agent to compromise and negotiate is also important.

An essential characteristic of the change agent is the ability to maintain and communicate the vision of the change. No matter how small or large the project is, the vision or picture of that change must be maintained and kept in the forefront of everyone's memory. The ability to articulate the vision and to mold disparate concepts into that vision is paramount to success. The astute change agent recognizes that those affected by the change may not have the same vision, so the change agent must paint a vivid mental picture of the change and how it will work during change implementation. The change agent will also have to recognize that those developing the project have some definite inclusion concepts that must be folded in to the vision for success. Inclusion concepts are those ideas or concepts that the affected parties believe are absolutely necessary for their peace of mind or for moral value. Including these concepts in the change process helps people feel ownership and value, that is, a piece of them or their idea is in the plan.

Perhaps the most outstanding skill the change agent needs is the ability to communicate and have good interpersonal skills. Communication throughout the change process cannot be emphasized enough. Each stage must be communicated to all who want or need to know about the project. The ability to establish trusting relationships, interact with others, and manage conflict is of utmost importance to the success of the project. Honesty is the key to truly excellent communication. People are better able to deal with the truth than with a half-truth or a lie. If a project team member's ideas are not acceptable for whatever reason, let them know immediately so there are no misunderstandings later. Communication includes using positive, concise, clear words that communicate accurately and responsibly. Ambiguity in communication is an error that leads to conflict and distrust later in the project.

Credible communication maintains open communication lines and helps to manage the process. The change agent must be sensitive to the project team members' feelings and involvement with the way things were. Do not denigrate or disrespect what was. Concentrate on how much better things can be.

Another important characteristic of the change agent is the ability to empower people to control the change project as it affects their lives. Those most affected by the change must be involved in assessing, planning, implementing, and evaluating the change. Without this support, the change cannot be successful. The change agent needs to be open and empowering about the selection of people to work on the project. Empowerment and participation in the project help people own the project and support its successful implementation. It is important for the change agent to respond appropriately to people's responses to change and not dismiss or be disrespectful of those reactions. Empowerment is a powerful tool in helping people realize that although something is changing, they have some control and input into that change. The change agent needs to use that empowerment to move the change along and to help people respond positively to the change.

The change agent must have some intuition during the evaluation steps so that he or she can bow out of the change and allow those affected to accept ownership of the change. This is a matter of timing and insight into when the staff are ready to accept and incorporate the change as its own. Bennis (1989) warns that to not step away from the project and cut the ownership bonds means that it remains the change agent's project for many years to come, even after the change agent might have left the organization. During evaluation, the change agent must support modifications and revisions that help transfer project ownership.

As has been reiterated previously, change is an inevitable part of life and will continue to affect the health care system for several years to come. It is important to maintain an attitude that change is preferable to stagnation. This will help leaders identify opportunities for change and embrace those changes for a better quality work life or for better care for the patient. Nothing can ever stay the same for long.

# INNOVATION

**Innovation** is defined as "a process for inventing something new or improving on that which already exists" (Blakeney, Carleton, & Coakley, p. 2). Tom Kelly, author of the *The Ten Faces of Innovation* (2005), stresses that the innovative process is now recognized as a pivotal management tool in all industries, including health care. Kelly emphasizes that innovation is a team event that is made up of individuals who possess different strengths and points of view. This team approach results in new, innovative ways to effectively solve problems using a variety of skills and abilities (Figure 13-3).

---

Nurse leaders rely on a variety of skills and abilities to facilitate innovation. The following are tools that can be used or analyzed to facilitate change:

Ideas
Creativity
Teamwork
Resources
Imagination
Collaboration
Opportunities
Interpersonal Skills
Effective Leadership
Communication Skills
Achievable Outcomes
Critical Thinking Skills
Problem Solving Ability
Evidence-Based Best Practices
Supportive Environment for Change

**FIGURE 13-3** Innovation tools for nurses.

Two of the biggest problems in health care today are related to patient safety and soaring costs. The nursing profession can find practical solutions in both of these areas. Through ongoing research, application of evidence-based practice, safe care initiatives, and innovation, nurses can improve the quality of patient care (Manojlovich, 2005).

An example of innovation in health care has been applied to the problem of medication errors. Injuries and death from medication errors have been identified by internal and external groups creating pressure for change in performance. Once the medication performance gap was recognized and identified by interdisciplinary health care groups, nurses (working in collaboration with practitioners, pharmacists, and other team members) analyzed why medication errors were occurring. Rather than blaming the person who administered the medication, an innovative "systems" analysis revealed why the errors were occurring. Systems errors included illegible handwriting, unfamiliar medications, dosage calculation errors, food/drug interactions, and lack of documentation of patient allergic reactions.

By applying a systems approach to problem solving, new safety structures and health care processes were implemented and institutionalized. Health care structures and processes were developed to include a computerized medication order-entry system and education of all personnel on the system. This system changed the process of how health care orders are written. Handwritten orders that are prone to interpretation errors are replaced by clear, concise, computer-generated orders. Multiple checks and balances that document patient allergies, health care conditions, and current height and weight were incorporated into the computer system to assist in appropriate medication ordering and dosing. Nurses and dietitians review the computerized patient information profiles for possible food/drug allergies and interactions. Pharmacists review orders using this computer system before dispensing medications, to analyze whether the medication dosage is indeed correct based on the patient's height and weight. Nurses review computer-generated medication administration records (MARS). Barcoding systems are used now to ensure that the right drug is being administered to the right patient at the right time.

Centralized computerized charting for nurses and other health care providers now aid in the accurate and timely flow of information. Patient histories and current laboratory results can be accessed quickly. Home care nurses access this same database from their portable laptop computers. This use of portable technology speeds the flow of information from the medical practitioner or nurse practitioner to the nurse caring for the patient in the community. By improving the flow of essential information, patient safety is enhanced. Hopefully, ongoing evaluations of these innovative measures will indicate that medication errors are occurring less frequently and that patient outcomes are improving.

Challenges for the future include encouraging continued innovative approaches to solve other problems in health care. Nurses play an important role in this process through their professional knowledge, clinical expertise, leadership, and conflict management skills.

## INNOVATION IN NURSING

Innovation in nursing is essential for maintaining and improving the quality of patient care. In the current economic downturn, cost-effectiveness drives innovation (Nursing Update, 2009.) The International Council of Nursing appropriately chose the theme "Delivering Quality/Serving Communities: Nurses Leading Care Innovations" in 2010 to celebrate the 100th anniversary of Florence Nightingale's death. Throughout nursing history, nursing practitioners have used innovation as a way to improve health care, but today there is an increasing emphasis on it (Manchester, 2009).

Florence Nightingale utilized innovation to change the way in which care was provided to wounded soldiers during the Crimean War. She collected data in order to identify reasons for the excessively high mortality rates of soldiers at the Scutari military barracks. She was able to demonstrate that the primary cause of death of these soldiers was from poor environmental conditions, not battle wounds. Her evidence demonstrated that excessively high mortality rates were related to inadequate nutrition, poor sanitation, lack of clean drinking water, and shortages of supplies such as clean clothing and warm blankets. Florence Nightingale was able to advocate for change to improve the way necessary supplies were distributed. She enhanced sanitation and nutrition standards in the hospital environment, improved health outcomes, and decreased the mortality rate (Alligood & Tomey, 2010). Florence Nightingale is also credited with developing documentation notes as a way to communicate more effectively with physicians (Manchester, 2009).

Innovation is not new to the nursing profession. Hughes (2006) states that "Nurses worldwide are engaged in innovative activities on a daily basis; activities motivated by the need to improve patient outcomes and reduce costs to the health system. Many of these activities by nurses have resulted in significant improvements in the health of patients, populations, and health systems. In most health systems, nurses are the main professional component of 'frontline' staff providing up to 80% of primary health care. As such, nurses are critically positioned to provide creative and innovative solutions for current and future global health challenges—challenges such as aging, HIV/AIDS, tuberculosis, malaria, increases in non-communicable diseases, poverty, inadequate resources, and workforce shortages. The need for innovative solutions has never been greater as health care environments globally struggle to provide equitable, safe and effective health services, while at the same time contain costs" (Hughes, p. 94).

Innovation creates change in systems. Nurse leaders are instrumental in pinpointing the need for change in systems and for guiding the change process. Student nurses who are learning about the change process and innovation can develop leadership projects during their educational experiences that enhance their abilities to become leaders of effective change in the future. These change projects can serve to demonstrate the leadership potential of students who are interviewing for a nursing position after graduation. Employers are very interested in candidates who have the ability to manage a project that brings about effective and innovative change to a health care setting.

# CONFLICT

An important part of the change process is the ability to resolve **conflict**. Conflict management skills are leadership and management tools that all registered nurses should have in their repertoire. Conflict itself is not bad. Conflict is healthy. It, like change, allows for creativity, innovation, new ideas, and new ways of doing things. It allows for the healthy discussion of different views and values and adds an important dimension to the provision of quality patient care. Without some conflict, groups or work teams tend to become stagnant and routinized. Nothing new is allowed to penetrate the "way we have always done it" mentality. Conflict can be stimulated by such things as scarce resources, invasion of personal space, safety or security issues, cultural differences, scarce nursing resources, increased workload, group competition, and various nursing demands and responsibilities.

# DEFINITION OF CONFLICT

There are a variety of definitions of conflict. Conflict is a disagreement about something of importance to the people involved. Conflict can also be defined as two or more parties holding differing views about a situation (Tappen, Weiss, & Whitehead, 2007), or as the consequence of real or perceived differences in mutually exclusive goals, values, ideas, and so on within one person or among groups of two or more (Sullivan & Decker, 2009). As can be surmised from these definitions, conflict can be defined as a disagreement about something of importance to each person involved. Not all disagreements become conflicts, but all disagreements have the potential for becoming conflicts, and all conflicts involve some level of disagreement. It is the astute manager who can determine which disagreements might become conflicts and which ones will not. This discussion of conflict management does not include professional communication skills. These are discussed elsewhere in this book.

# SOURCES OF CONFLICT

Whenever there is the opportunity for disagreement, there is a potential source of conflict. The common sources of conflict in the professional setting include disputes over resource allocation or availability, personality differences, differences in values, threats from inside or outside an organization, cultural differences, and competition. In recent years, organizational, professional, and unit goals have served as a major source of conflict. Nurses frequently see financial goals and patient care goals as being in direct conflict with one another.

Sources of conflict in personal arenas include differences in values, threats to security or well-being, financial problems, and cultural differences. Living conditions and social contacts can increase or decrease the sources of conflict in someone's personal life. When people believe they have control over their living conditions and their social contacts, they can work to minimize conflict. Family relationships may present sources of conflict because of the complexity of these relationships.

# TYPES OF CONFLICT

There are three broad types of conflict: intrapersonal, interpersonal, and organizational. Intrapersonal conflict occurs within the individual. When opposing values or differences in priority arise within an individual, they may suffer intrapersonal conflict. For example, if Marilyn is not granted her requested day off, she may have internal conflict about whether or not to call in sick or to take the day off without pay or to go to work. Or Marilyn may have a conflict about priorities. Should she attend her daughter's softball game or write her paper for school? Individuals often have internal conflict about values. For example, in Marilyn's case, the requested day off may have been to attend the softball game, so the values in conflict would be the values of family versus the values of the work ethic.

In interpersonal conflict, the source of disagreement may be between two people or groups or work teams. There may be disagreement in philosophy or values, or policy or procedure. It may be a personality conflict; for example, two people just rub each other the wrong way. This type of conflict is not unusual in the work situation. People new to a team may bring ideas with them that are not totally acceptable to the members in place. Individuals who transfer from one unit to another often stir up a certain amount of conflict over processes and procedures. For example, the nurse transferring from the intensive care unit (ICU) to the coronary care unit (CCU) may be comfortable with one way of making assignments and then try to encourage his new peers to adopt that methodology without sharing the rationale for why his way is better.

On occasion, there may be some interpersonal conflict between ICU nurses and medical-surgical nurses when they are required to work in one another's areas. Little regard is sometimes offered to each other about differences among patients, equipment, or required organizational skills. This lack of regard may sometimes lead to conflict based on preconceived notions rather than fact.

Organizational conflict is often referred to as intergroup conflict. It is at times a healthy way of introducing new ideas and encouraging creativity. Competition for resources, organizational cultural differences, and other sources of conflict help organizations identify areas for improvement. Conflict helps organizations identify legitimate differences among departments or work teams based on corporate need or responsibility. When organizational conflict is highlighted, corporate values and differences are aired and resolved.

In today's health care environment, conflict between the organizational goals of quality patient care and the need for a healthy financial bottom line can occur. These are both important values to the organization. Organizational conflict may help individual work groups or teams clarify goals and become more cohesive, hopefully with few interpersonal disagreements. This process may also work for organizations in direct competition with one another as they unite internally against an outside threat to their well-being.

## THE CONFLICT PROCESS

In 1975, Filley suggested a process for conflict management that is widely accepted. In this process, there are five stages of conflict:

1. antecedent conditions,
2. perceived and/or felt conflict,
3. manifest behavior,
4. conflict resolution or suppression, and
5. resolution aftermath.

In Filley's model, conflict and conflict management proceed along a specific process that begins with specific preexisting conditions called antecedent conditions. The situation develops so that there is perceived or felt conflict by the involved parties that initiates a behavioral response or manifest behavior. The conflict is either resolved or suppressed, leading to the development of new feelings and attitudes, which may create new conflicts. Conflict management is vital in change. The antecedent conditions that Filley suggests may or may not be the cause of the conflict, but they certainly move the disagreement to the conflict level. The sources of these conditions include those discussed earlier: disagreement in goals, values, or resource utilization. Other issues may also serve as antecedent conditions, such as the dependency of one group on another. For instance, the nursing department is dependent on the pharmacy department to provide drugs for the nursing unit in a timely fashion. The goals and priorities of pharmacy and nursing may be different at the time the nurse requests the drugs, so a source of disagreement arises. If the circumstances for disagreement continue, a conflict will develop. According to Sullivan and Decker (2009), goal incompatibility is the most important antecedent condition to conflict.

An example of this is the incompatibility of quality nursing care and financial goal setting. There is often suspicion on the part of nurses that cutting costs in any area of nursing will lead to poor quality nursing care, and the antecedent condition of goal disagreement is established. If the cost cutting occurs too often, patient care may or may not be negatively affected. There is no way of actually knowing the exact impact of financial cuts on quality, but there is some predictability that at some point continual financial cuts will negatively affect the quality of care provided for the patient. The point here is that the antecedent condition or preexisting condition for conflict is present with apparently incompatible goals; that is, financial goals and quality-care goals are not always compatible.

The antecedent conditions lead to frustration, and a conflict is born. The frustration is often described as a felt conflict—that is, when one party feels in conflict with another and perceives conflict and when each party believes he or she knows what the other party's position is in the conflict. Of course, in the case of groups, one group or person may not be aware of another person's or group's feelings of conflict. This adds to the frustration. In the case of intrapersonal conflict, the frustration surrounds the internal conflict and how to resolve it. Once frustration has been felt, then resolution of some kind is necessary. The emotional components of frustration are anger and resignation. Both of these are powerful emotions that require some kind of action. For example, if Julio is not aware of the conflict between him and Tamika, Tamika may feel frustrated and upset. She is then obligated to let Julio know of her feelings so the conflict can be managed. If Tamika does not let Julio know of her feelings, then the frustration builds to anger, and Tamika will not be able to resolve the issues with Julio until that frustration is alleviated. This often results when frustration erupts into anger over some little thing rather than over the actual conflict.

## Meaning of the Conflict

Conceptualization of the meaning of a conflict develops when individuals form an idea or concept of what the conflict is about, such as a conflict over control, professional standards, values, goals, and so on. Each party may or may not be aware of the other's conceptualization of the meaning of the conflict, but the parties do have what they believe is a clear concept of the conflict in their own minds. To determine the accuracy of the beliefs about the conflict, both parties need to sit down and determine the existence and nature of the conflict and the reasons it exists. People may disagree on the four aspects of a conflict (Brinkert, 2010). These aspects include facts, goals, methods of goal achievement, and the values or standards used to select the goals or methods. This means that the actual facts of the dispute may be in question, the goals each side wishes to achieve may not be the same, how to achieve the agreed-upon goal is not acceptable to one side or the other, or values are in dispute. After the nature of the conflict or the points of disagreement are known, then conflict management can begin.

The actions taken to resolve the conflict can take many forms:

- Discussion of the conflict may move it toward resolution.
- Someone in power may take steps to end the conflict or at least suppress it.
- One or both parties may decide to do nothing to resolve the conflict.

Successful conflict management will be discussed later. The important point here is that failure to successfully manage the conflict leads to more frustration and a further heightening of the conflict. Communication breaks down, and fighting spreads to the pettiest of issues. After it is apparent that a conflict is occurring, conflict management of some type must occur. People simply do not get over it and move on. The conflict is a source of friction and pain that must be resolved.

Outcomes are, of course, the result of any action taken. Positive resolution of conflict leads to positive outcomes; negative resolution or no resolution of conflict leads to negative or angry consequences. To determine if successful outcomes were achieved, you can ask whether or not important goals were achieved to some degree and what the relationships are between the parties at this point. Although the relationships may not be affectionate, it is hoped that the parties can at least be cordial and collegial

with one another. The hope is that each side has gained some measure of trust and respect, but if they are at least still talking, progress has been made. The best outcome is one in which both sides feel that something is won and their self-respect is intact.

## CONFLICT MANAGEMENT

There are essentially seven methods of conflict management. These methods dictate the outcomes of the conflict process. Although some methods are more desirable or produce more successful outcomes than do others, there may be a place in conflict management for all the methods, depending on the nature of the conflict and the desired outcomes. Table 13-6 is a summary of these methods, highlighting some of their advantages and disadvantages. The five techniques most commonly acknowledged are avoiding, accommodating, competing, compromising, and collaborating. Other techniques are negotiating and confronting.

*Avoiding* is a very common technique. The parties involved in the conflict ignore it, either consciously or unconsciously. Use of this is common where there is interpersonal conflict in a highly cohesive group. The threat to cohesiveness is greater than the potential gains from the conflict. There may be circumstances where avoidance is appropriate, such as when (1) one of the parties is leaving so the conflict will resolve itself; (2) the conflict is not solvable and not all that important; (3) there are other, more important issues at stake and conflict management is not worth the time and energy at this point. An example of avoiding is when two members of a highly cohesive nursing team disagree on some minor treatment modality or the amount of initial pain relief medication. The cohesiveness of the group is more important than the issue, so the two nurses agree to disagree, avoid discussing the issue, and monitor patient outcomes for more information. Each nurse uses the treatment or provides the pain relief according to his or her own values until outcomes are clear.

*Accommodating* is often called smoothing or cooperating. In this technique, one side of the disagreement decides or is encouraged to accommodate the other side by ignoring or sidestepping their own feelings about the issue.

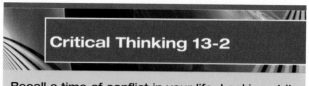

### Critical Thinking 13-2

Recall a time of conflict in your life. Looking at it with a different perspective, what conditions led to the conflict? What was the core of the conflict? How could you look at the conflict to prevent similar problems in the future?

## TABLE 13-6 Summary of Conflict Management Techniques

| CONFLICT MANAGEMENT TECHNIQUE | ADVANTAGES | DISADVANTAGES |
|---|---|---|
| Avoiding—ignoring the conflict | Does not make a big deal out of nothing; conflict may be minor in comparison to other priorities | Conflict can become bigger than anticipated; source of conflict might be more important to one person or group than to others |
| Accommodating—smoothing or cooperating; one side gives in to the other side | One side is more concerned with an issue than is the other side; stakes not high enough for one group, and that side is willing to give in | One side holds more power and can force the other side to give in; the importance of the stakes are not as apparent to one side as to the other |
| Competing—forcing; the two or three sides are forced to compete for the goal | Produces a winner; good when time is short and stakes are high | Produces a loser; leaves anger and resentment on losing side |
| Compromising—each side gives up something and gains something | No one should win or lose, but both should gain something; good for disagreements between individuals | May cause a return to the conflict if what is given up becomes more important than the original goal |
| Negotiating—high-level discussion that seeks agreement but not necessarily consensus | Stakes are very high, and solution is rather permanent; often involves powerful groups | Agreements are permanent, even though each side has gains and losses |
| Collaborating—both sides work together to develop the optimal outcome | Best solution for the conflict and encompasses all important goals of each side | Takes a lot of time; requires commitment to success |
| Confronting—immediate and obvious movement to stop conflict at the very start | Does not allow conflict to take root; very powerful | May leave impression that conflict is not tolerated; may make something big out of nothing |

This is often done when the stakes are not all that high and the need to move on is pressing. Frequent use of this method, however, can lead to feelings of frustration or being used—one person is "used" to get the cooperation of another. This can be an effective technique as long as the losing side is agreeable to the situation. This is clearly a technique where one party gains, or wins, and the other loses. The losing side often agrees to this method to gather "credits" that can be called in at another time. The leader/manager who uses this technique needs to do so prudently and sparingly. An example of this method is when one nurse will agree to last-minute schedule changes to accommodate other team members. This nurse will then call in these credits at a later date to use for an extended vacation or leave of absence.

*Competing* is a conflict management technique that produces a winner and a loser. The concept is that there is an all-out effort to win at all costs. This technique may be used when time is too short to allow other techniques to work or when a critical, though unpopular, decision has to be made quickly. This is often called forcing, because the winner forces the loser to accept his or her stance on the conflict.

Some authorities see negotiation and compromise as the same; others see them as separate activities. They will be discussed separately here. Compromising is a method used to achieve conflict management in situations where both sides can win and neither side should lose. Compromise is rampant in our society and is useful for goal achievement when the stakes are important but not necessarily critical. Compromise is often seen as appeasement; each side gives up something and each side gains something. Compromise is a good technique for minor conflicts or conflicts that cannot be resolved satisfactorily for both sides. Both parties win and lose. An example of compromise is when two nurses want the same shift off on the same day. One way to compromise is to split the shift so each nurse has part of the shift off.

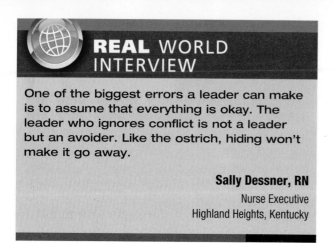

Negotiating is an advanced skill that requires careful communication techniques and highly developed skills.

The optimum solution for the conflict may not be reached, but each side has some wins and some losses. The term *negotiation* is often used in reference to collective bargaining or politics. It is, however, a very useful technique for conflict management at all levels. Negotiating is used when the stakes are high or when return to the conflict cannot occur. Return to the conflict may not be possible for a variety of reasons, such as a union contract, a permanent change in policy or governance, or career or life changes. The idea of negotiation is that each party will gain something, so general agreement is reached, but consensus is not necessarily the goal. Consensus means that the negotiating parties reach an agreement that all parties can support, even if it is not completely what they want. Consensus does not satisfy everyone completely, but reaching a consensus indicates that all parties to a decision will accept the conditions of the agreement. Negotiating a consensus among staff nurses may be accomplished during scheduling. Each nurse may negotiate his or her best, though not necessarily optimum, schedule.

According to Lewicki, Hiam, and Olander (1996), there are five basic approaches to negotiating: collaborative (win-win), competitive (win at all costs), avoiding (lose-lose), accommodating (lose to win), and compromise (split the difference). These five approaches to negotiation are influenced by the importance of maintaining the relationship relative to the importance of achieving desired outcomes (Coles, 2010). Figure 13-4 is a graph of negotiation strategies that indicates the importance of the relationship of groups or individuals and the importance of the outcome in the five approaches to managing conflict. The conflicting parties must decide if the individual or group relationships are more or less important than the desired outcome of the project. For example, if relationships are more important than achieving an outcome, then an accommodating or collaborative approach may be chosen to protect or continue the relationship. If outcomes are more important, than a competitive approach may be chosen to achieve the outcome.

*Collaborating* occurs in conflict management when both sides work together to develop the optimal outcome. It is a creative endeavor designed to find the best solution to the conflict so that all of the perceived important goals are achieved. Often, this involves the creation of new higher goals that encompass the goals of both sides. This is a very high-level technique that requires maturity and a spirit of cooperation to reach each other's goals. To achieve collaboration is a very time-consuming process. Of course, in most conflicts, collaboration is preferred. An example of collaboration is the merger of two nursing teams. Instead of one team's methods of operation being forced on the other team, the two teams work together to create an even better environment than each had separately. This is accomplished by collaborating to pick the best policies, methods of operation, and so on.

One other technique used in conflict resolution is *confronting*. This technique heads off conflict as soon as the first symptoms appear. Although most commonly used

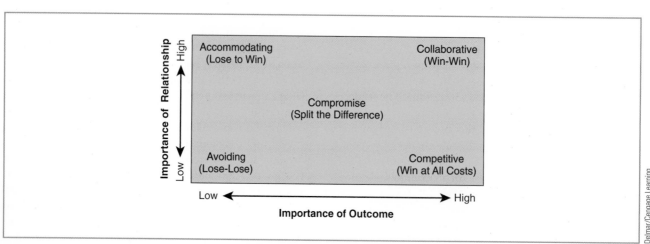

**FIGURE 13-4** Negotiation strategies.

between supervisor and employee, it can be used very successfully with conflict management. Both parties are brought together, the issues are clarified, and some outcome is achieved. Often, conflict is the result of people misunderstanding or testing if the new policy is really going to be enforced. Confrontation heads off both situations early and quickly. The danger to confrontation is that when used extensively, it would appear that there is little tolerance for conflict. As mentioned earlier, conflict is often healthy and encourages creativity and open discussion and debate. The technique of confrontation should be used judiciously to avoid the impression of intolerance.

Success of the techniques presented here depends on several factors. The importance to the various sides of the issue in the conflict has an enormous impact on the technique selected and the degree of success that will be achieved. Conflict management is never really permanent, because new issues will always arise. The trick for the nurse is to determine what conflicts require intervention and what techniques stand the best chance of success. If one technique does not work, try another. Conflicts should be suppressed or avoided only under special circumstances that are dependent on the issues involved and the importance of the issues to the parties. Keep in mind that little problems become big problems later if the stakes are high and the issue is important to someone.

## STRATEGIES TO FACILITATE CONFLICT MANAGEMENT

Open, honest, clear communication is the key to successful conflict management. The nurse and all parties to the conflict must agree to communicate with one another openly and honestly. Courtesy in communicating is to be encouraged. This includes listening actively to the other side. This does not include interrupting, being aggressive, or being overbearing in demeanor. Most importantly, use of derogatory language or gestures is not acceptable or tolerable. Voice level should be calm and at a normal tone. This sounds easy, but it may not always be easy to do in practice.

The setting for the discussions for conflict management should be private, relaxed, and comfortable. If possible, external interruptions from phones, pagers, overhead speakers, and personnel should be avoided or kept to a minimum. The setting should be on neutral territory so that no one feels overpowered. The ground rules, such as not interrupting, who should go first, time limits, and so on, should be agreed upon at the beginning. Adherence to ground rules should be expected.

The management of conflict should be entered into by both sides in the spirit of expectation of compliance to the results. Threats on either side of the conflict should not be tolerated. Conflict management cannot be achieved if either party is threatening one another. If one party cannot agree to comply with the decisions or outcomes, there

is no point to the conflict management process. A tool for assessing conflict is identified in Figure 13-5. This tool can be used to determine interpersonal or intergroup conflict within an organization and whether a given conflict is functional or dysfunctional.

## LEADERSHIP AND MANAGEMENT ROLES

Marquis and Huston (2006) have identified some leadership and management roles in conflict management. Leadership roles include the role modeling of conflict management methods as soon as the conflict is evident. This strategy demonstrates awareness of the intrapersonal or interpersonal conflict and works to resolve it and set the goal of conflict management so that both parties win. The leader also works to lessen the perceptual differences of the conflicting parties about the conflict and tries to encourage each side to see the other's view. The nurse assists conflicting parties to identify techniques that may resolve the conflict and accepts differences between the parties without judgment or accusation. The leader fosters open and honest communication.

The nursing manager role includes the creation of an environment conducive to conflict management. The manager uses his or her authority to solve conflicts, including the use of competition for immediate or unpopular decisions. The manager facilitates conflict resolution in a formal manner when necessary. The manager competes and negotiates for available resources for unit needs when necessary. The manager can compromise unit goals when necessary to achieve another, more important unit goal. The manager negotiates consensus or compliance to conflict resolution outcomes or goals. Although the roles of leader and manager often appear in the same person, the leadership roles in conflict management are often more important to resolution and compliance. The manager has formal power that can be used when necessary, but it should be reserved for truly unmanageable or important issues.

## CONFLICT MANAGEMENT AND CHANGE

Conflict management is an important part of the change process. Change can often threaten individuals and groups, so conflict is an inevitable part of the process. It is important to keep in mind that some conflicts resolve themselves, so the change agent should not be too quick to jump into an intervention mode. Figure 13-6 provides a guide for assessment of the level of conflict. A low conflict level supports change and encourages the healthy exchange of thoughts and ideas. This level of conflict is often resolved between individuals or between groups in an organization. Note that a conflict level that is too low may lead to complacency or a sense of dread

## Interpersonal or intergroup?

1. Who?

- Who are the primary individuals or groups involved? Characteristics (values, feelings, needs, perceptions, goals, hostility, strengths, past history of constructive conflict management, self-awareness)?
- Who, if anyone, are the individuals or groups that have an indirect investment in the result of the conflict?
- Who, if anyone, is assisting the parties to manage the conflict constructively?
- What is the history of the individuals' or groups' involvement in the conflict?
- What is the past and present interpersonal relationship between the parties involved in the conflict?
- How is power distributed among the parties?
- What are the major sources of power used?
- Does the potential for coalition exist among the parties?
- What is the nature of the current leadership affecting the conflicting parties?

2. What?

- What is (are) the issues(s) in the conflict?
- Are the issues based on facts? Based on values? Based on interests in resources?
- Are the issues realistic?
- What is the dominant issue in the conflict?
- What are the goals of each conflicting party?
- Is the current conflict functional? Dysfunctional?
- What conflict management strategies, if any, have been used to manage the conflict to date?
- What alternatives in managing the conflict exist?
- What are you doing to keep the conflict going?
- Is there a lack of stimulating work?

3. How?

- What is the origin of the conflict? Sources? Precipitating events?
- What are the major events in the evolution of the conflict?
- How have the issues emerged? Been transformed? Proliferated?
- What polarizations and coalitions have occurred?
- How have parties tried to damage each other? What stereotyping exists?

4. When/Where?

- When did the conflict originate?
- Where is the conflict taking place?
- What are the characteristics of the setting within which the conflict is occurring?
- What are the geographic boundaries? Political structures? Decision-making patterns? Communication networks? Subsystem boundaries?
- What environmental factors exist that influence the development of functional versus dysfunctional conflict?
- What resource persons are available to assist in constructive conflict management?

| Functional or dysfunctional? | Yes | No |
| --- | --- | --- |
| Does the conflict support the goals of the organization? | [] | [] |
| Does the conflict contribute to the overall goals of the organization? | [] | [] |
| Does the conflict stimulate improved job performance? | [] | [] |
| Does the conflict increase productivity among work group members? | [] | [] |
| Does the conflict stimulate creativity and innovation? | [] | [] |
| Does the conflict bring about constructive change? | [] | [] |
| Does the conflict contribute to the survival of the organization? | [] | [] |
| Does the conflict improve initiative? | [] | [] |
| Does job satisfaction remain high? | [] | [] |
| Does the conflict improve the morale of the work group? | [] | [] |

A yes response to the majority of the questions indicates that the conflict is probably functional. If the majority of responses are no, then the conflict is most likely a dysfunctional one.

FIGURE 13-5 **Guide for the assessment of conflict.** (*Source:* From McFarland, G., Skipton Leonard, H., & Morris. [1984]. *Nursing leadership & management.* Clifton Park, NY: Delmar Cengage Learning.)

| **Is conflict too low?** | **YES** | **NO** |
|---|---|---|
| Is the work group consistently satisfied with the status quo? | [] | [] |
| Are no or few opposing views expressed by work group members? | [] | [] |
| Is little concern expressed about doing things better? | [] | [] |
| Is little or no concern expressed about improving inadequacies? | [] | [] |
| Are the decisions made by the work group generally of low quality? | [] | [] |
| Are no or few innovative solutions or ideas expressed? | [] | [] |
| Are many work group members "yes-men"? | [] | [] |
| Are work group members reluctant to express ignorance or uncertainties? | [] | [] |
| Does the nurse manager seek to maintain peace and group cooperation regardless of whether this is the correct intervention? | [] | [] |
| Do the work group members demonstrate an extremely high level of resistance to change? | [] | [] |
| Does the nurse manager base the distribution of rewards on popularity as opposed to competence and high job performance? | [] | [] |
| Is the nurse manager excessively concerned about not hurting the feelings of the nursing staff? | [] | [] |
| Is the nurse manager excessively concerned with obtaining a consensus of opinion and reaching a compromise when decisions must be made? | [] | [] |

A yes response to the majority of these questions can be indicative of a work group conflict level that is too low.

| **Is conflict too high?** | **YES** | **NO** |
|---|---|---|
| Is there an upward and onward spiraling escalation of the conflict? | [] | [] |
| Are the conflicting parties stimulating the escalation of conflict without considering the consequences? | [] | [] |
| Is there a shift away from conciliation, minimizing differences, and enhancing goodwill? | [] | [] |
| Are the issues involved in the conflict being increasingly elaborated and expanded? | [] | [] |
| Are false issues being generated? | [] | [] |
| Are the issues vague or unclear? | [] | [] |
| Is job dissatisfaction increasing among work group members? | [] | [] |
| Is the work group productivity being adversely affected? | [] | [] |
| Is the energy being directed to activities that do not contribute to the achievement of organizational goals (e.g., destroying the opposing party)? | [] | [] |
| Is the morale of the nursing staff being adversely affected? | [] | [] |
| Are other parties getting dragged into the conflict? | [] | [] |
| Is a great deal of reliance on overt power manipulation noted (threats, coercion, deception)? | [] | [] |
| Is there a great deal of imbalance in power noted among the parties? | [] | [] |
| Are the individuals or groups involved in the conflict expressing dissatisfaction about the course of the conflict and feel that they are losing something? | [] | [] |
| Is absenteeism increasing among staff? | [] | [] |
| Is there a high rate of turnover among personnel? | [] | [] |
| Is communication dysfunctional, not open, mistrustful, and/or restrictive? | [] | [] |
| Is the focus being placed on non-conflict-relevant sensitive areas of the other party? | [] | [] |

A yes response to the majority of these questions can be indicative of a work group conflict level that is too high.

**FIGURE 13-6  Guide for the assessment of level of conflict.** (*Source:* From McFarland, G., Skipton Leonard, H., & Morris. [1984]. Nursing leadership & management. Clifton Park, NY: Delmar Cengage Learning.)

## CASE STUDY 13-1

LaTonya works the night shift on her unit. She is concerned that the medication carts stock high dosages of intravenous potassium chloride. LaTonya knows that high dosages of this medication are lethal. She believes that this medication should be made up by the pharmacy only on an "as needed" basis in response to a specific patient's needs. How can LaTonya work as a change agent, unfreeze the current situation, and move to a new way of preparing this medication? How can she then refreeze this needed change?

regarding change. On the other hand, conflict that is too high is not healthy, as it promotes anxiety and anger. It may dissolve into conflict that cannot be managed without drastic steps by the manager or change agent. These drastic steps may include a change in group composition or assignment of some members of the group to another type of project. Successful change may be difficult to achieve when groups or individuals have either too little or too high levels of conflict. It is the nurse manager's responsibility to assure that the conflict level is neither too low nor too high. Both of these levels of conflict can interfere with achievement of the objectives of the change project. Both change and conflict are positive processes that promote creativity, idea exchange, and innovation. Leaders, managers, and staff should be encouraged to embrace them and explore them as opportunities for positive growth development and professional expansion.

## KEY CONCEPTS

- Change is inevitable, exciting, and anxiety provoking.

- Change is defined as making something different from what it was.

- Major change theorists include Lewin, Lippitt, Havelock, and Rogers.

- Senge's model of five disciplines describes the learning organization. This model describes organizations undergoing continuous and unrelenting change.

- The change process is similar to the nursing process. The steps in the change process include assessment; data collection; and analysis, planning, implementation, and evaluation.

- Strategies for change include the power-coercive approach, the normative-reeducative approach, and the rational-empirical approach.

- Evaluation of change is a very important part of determining success and cannot be overlooked or skimmed.

- The change agent must be honest, open, optimistic, and, above all, trustworthy.

- The change agent is an important part of the change process. The change agent is responsible and accountable for the project.

- Chaos theory says that most organizations go through periods of rapid change and innovation and then stabilize before chaos erupts again.

- Innovation is the process of creating new services or products.

- Conflict is a normal part of any change project and is often healthy and positive.

- Conflict comes from many sources, including value differences, fear, goal disagreement, and cultural differences.

- The four steps in the conflict management process are frustration, conceptualization, action, and outcomes.

- There are several strategies for conflict management. Clear, open communication is key. There must be commitment to conflict management.

- Useful tools for conflict management include a guide for the assessment of conflict and a guide for the assessment of the level of conflict.

- The techniques for conflict management include avoiding, accommodating, compromising, competing, negotiating, confronting, and collaborating.

- Conflict can move the change process along if it is handled well. Conflict can stop the change process if it is handled poorly or allowed to get out of control.

## KEY TERMS

| | | | |
|---|---|---|---|
| change | innovation | organizational change | professional change |
| change agent | learning organization | personal change | |
| conflict | theory | | |

## REVIEW QUESTIONS

1. What is often the most desirable conflict resolution technique?
   A. Avoiding
   B. Competing
   C. Negotiating
   D. Collaborating

2. The change agent and the person responsible for conflict management have what characteristic in common?
   A. Secretive and willful
   B. Trustworthy and a good communicator
   C. Ambitious and avoiding
   D. Powerful and dictatorial

3. Select the best reason why change is necessary.
   A. To maintain the status quo
   B. To enhance the quality of health care
   C. To encourage staff turnover
   D. To increase the cost of patient care

4. Identify the theorist who first proposed the original change theory model.
   A. Rogers
   B. Havelock
   C. Lewin
   D. Lippitt

5. Identify the least reliable form of communication on a nursing unit.
   A. The grapevine
   B. Minutes from a staff meeting
   C. A memo from the unit director
   D. A medical center newsletter

6. Which of the following strategies will be least helpful to the nurse who is managing a change project?
   A. Articulating a vision of what the change will entail
   B. Mapping a timeline for how the project will evolve
   C. Selecting a team of personal friends to develop the project
   D. Choosing effective means of communication regarding project status

7. The nurse leader develops an educational program for the unit personnel on a new policy for determining patient-to-nurse staffing ratios. This is an example of what kind of change strategy?
   A. Rational-empirical
   B. Power-coercive
   C. Normative-reeducative
   D. Authoritarian

8. A nurse's conflict with a peer regarding who will admit a patient at the change of shift is an example of what type of conflict?
   A. Intrapersonal
   B. Interpersonal
   C. Organizational
   D. Multidisciplinary

9. Which of the following statements are true regarding innovation? Select all that apply.
   _____ A. Innovation is new to the profession of nursing.
   _____ B. Nursing practice innovation can improve patient outcomes.
   _____ C. Evidence-based practice is a valuable resource for the process of innovation.
   _____ D. Leadership projects can drive innovative change in nursing practice.
   _____ E. Cost-effectiveness is not a concern for nursing practice.

10. As a nurse, you are asked to lead a change project that impacts the unit where you work. With your knowledge of the change process, identify the order in which you will do the following. Select all that apply.
    _____ A. Set up a schedule of meeting dates and times.
    _____ B. Define the problem.
    _____ C. Select a qualified team of individuals to examine the issue.
    _____ D Identify key players who have knowledge about the problem.
    _____ E. Brainstorm possible solutions to the problem.
    _____ F. Select a problem solution that is supported by evidence-based practice.

## REVIEW ACTIVITIES

1. Select a change project that you have either personally achieved or have experienced in a clinical situation, and discuss with your classmates how you felt and how the change agent maintained momentum and enthusiasm for the project. If this is a personal change, how did you maintain enthusiasm?

2. Recall a conflict with which you have been involved in the clinical situation. Discuss each of the methods of conflict management identified in the chapter. Identify which ones would have worked. Did the conflict ever get resolved? How?

3. Discuss with a nurse manager how he or she determines whether a conflict is occurring and what steps the nurse manager takes to bring it out in the open. Share the information with your classmates.

4. Discuss with a nurse manager how he or she feels about constant change on a personal level. How did the nurse present an impending change to the staff? Did the nurse use any of the techniques discussed in this chapter? Was the change successful?

## EXPLORING THE WEB

- Look up the Journal of Conflict Resolution (http://jcr.sagepub.com) and describe its purpose. Would this journal be useful to the new nurse manager? A new nurse? Anyone else in health care?

- Explore the following Web sites to learn more about change, conflict management, and innovation. www.beginnersguide.com

Scroll down to A Beginner's Guide to Management. Note the discussion of change management.

- International Association of Conflict Management: www.iacm-conflict.org

- National League for Nursing: www.nln.org

Search for position statement and innovation.

## REFERENCES

Alligood, M., & Tomey, A. (2010). *Nursing theorists and their work* (7th ed.). Maryland Heights, MO: Mosby Elsevier.

Beason, C. F. (2005). The nurse as investor: Using the strategies of Sarbanes-Oxley corporate legislation to radically transform the work environment of nurses. *Nursing Administration Quarterly, 29*(2), 171–178.

Bennis, W. (1989). *On becoming a leader.* Reading, MA: Addison-Wesley.

Bennis, W., Benne, K., & Chin, R. (Eds.). (1969). *The planning of change* (2nd ed.). New York: Holt, Rinehart, Winston.

Blakeney, B., Carleton, P., & Coakley, E. (2009). Unlocking the power of innovation. *OJIN: Online Journal of Issues in Nursing, 14*(2), 1–12.

Brinkert, R. (2010). A literature review of conflict communication causes, costs, benefits and interventions in nursing. *Journal of Nursing Management, 18*(2), 145–156.

Coles, D. (2010). Because we can—Leadership responsibility and the moral distress dilemma. *Nursing Management. 41*(3), 26–30.

Filley, A. C. (1975). *Interpersonal conflict resolution.* Glenview, IL: Scott Foresman.

Haase-Herrick, K. (2005). The opportunities of stewardship. *Nursing Administration Quarterly, 29*(2), 115–118.

Havelock, R. G. (1973). *The change agent's guide to innovation in education.* Englewood Cliffs, NJ: Educational Technology.

Hughes, F. (2006). Nurses at the forefront of innovation. *International Nursing Review, 53,* 94–101.

Institute of Medicine. (2002). *To err is human: Building a safer health care system.* Washington, DC: National Academies Press.

Jones, D. B. (2007). *The merger process as experienced by nurse middle managers: A grounded theory study.* (Unpublished doctoral dissertation, Indiana University).

Kelly, T. (2005). *The ten faces of innovation.* New York: Doubleday.

Lancaster, J. (2008). Nursing issues in leading and managing change. St. Louis, MO: Mosby.

Lewin, K. (1951). *Field theory in social science.* New York: Harper & Row.

Lippitt, R. (1958). *The dynamics of planned change.* New York: Harcourt, Brace.

Luthans, K. W., & Jenson, S. M. (2005). The linkage between psychological capital and commitment to organizational mission: A study of nurses. *Journal of Nursing Administration, 35*(6), 304–310.

Manchester, A. (2009). Recognizing [sic] innovation. *Kai Tiaki Nursing New Zealand, 15*(3), 11.

Manojlovich, M. (2005). Promoting nurses' self-efficacy. *Journal of Nursing Administration, 35*(5), 271–278.

Marquis, B. L., & Huston, C. J. (2006). *Leadership roles and management functions in nursing: Theory applied* (5th ed.). Philadelphia: Lippincott.

McFarland, G., Skipton Leonard, H., & Morris, M. (1984). *Nursing leadership and management contemporary strategies.* Clifton Park, NY: Delmar Cengage Learning.

National Nursing Centers Consortium. (2011). Nurse-Managed Health Centers are Changing the Face of Health Care. Accessed March 25, 2011 at, http://www.nncc.us/about/nmhc.html.

Recklies, O. (2006). Managing change—Definitions and phases in change processes. Retrieved September, 2006, from http://www.themanager.org/Strategy/Change_Phases.htm.

Rogers, E. M. (1983). *Diffusion of innovations* (3rd ed.). New York: Free Press.

Senge, P. M. (1990). *The fifth discipline: The art and practice of the learning organization.* New York: Doubleday.

Senge, P., Kleiner, A., Roberts, C., Ross, R., Roth, G., & Smith, B. (1999). *The Dance of Change: The Challenges of Sustaining Momentum in Learning Organizations.* New York: Doubleday/Currency.

Sullivan, E. J., & Decker, P. J. (2009). *Effective leadership & management in nursing* (6th ed.). Upper Saddle River, N.J.: Pearson Education Inc.

Tappen, R. M., Weiss, S. A., & Whitehead, D. K. (2007). *Essentials of nursing leadership and management* (3rd ed.). Philadelphia: F. A. Davis.

## SUGGESTED READINGS

Ball, M. J. (2005, Feb.). Nursing informatics of tomorrow. One of nurses' new roles will be agents of change in the health care revolution. *Healthcare Informatics, 22*(2), 74, 76, 78.

Begun, J., Zimmerman, B., & Dooley, K. (2006). *Health care organizations as complex adaptive systems.* Retrieved September 8, 2006, from http://www.change-ability.ca/ComplexAdaptive.pdf.

Block, L. & LeGrazie, B. (2006). Don't get lost in translation: Successfully communicate evidence with fact and follow-up. *Nursing Management,* (May), 37–40.

Crow, G. (2006). Diffusion of innovation: The leaders' role in creating the organizational context for evidence-based practice. *Nursing Administration Quarterly, 30*(3), 236–242.

Erwin, D. (2009). Changing organizational performance: Examining the change process. *Hospital Topics, 87,* 28–40.

MacGuire, J. M. (2006, Jan.). Putting nursing research findings into practice: Research utilization as an aspect of the management of change. *Journal of Advanced Nursing, 53*(1), 65–71.

McCartney, P. (2006, Nov.–Dec.). The International Council of Nurses innovations database. *American Journal of Maternal Child Nursing, 31*(6), 389.

National Nursing Centers Consortium. (2006). Nurse-managed health centers are changing the face of health care. Retrieved February 16, 2006, from http://www.nncc.us/about/nmhc.html.

Pfeffer, J., & Sutton, R. (2006, January). Evidence-based management. *Harvard Business Review,* 63–74.

Porter-O'Grady, T., & Malloch, K. (2007). *Quantum leadership: A textbook of new leadership.* Sudbury, MA: Jones and Bartlett Publishers.

Senge, P., Flowers, B. S., & Scharmer, C. O. (2005). *Presence: An exploration of profound change in people, organizations, and society.* New York: Doubleday.

Thomas, L. M., Reynolds, T., & O'Brien, L. (2006). Innovation and change: Shaping district nursing services to meet the needs of primary health care. *Journal of Nursing Management, 14*(6), 447–454.

# UNIT 3
## Leadership and Management of Patient-Centered Care

# CHAPTER 14

# Budget Concepts for Patient Care

CORINNE HAVILEY, RN, MS

*A higher proportion of nursing care provided by registered nurses (RNs) and a greater number of hours of care by RNs per day are associated with better care for hospitalized patients.*

(JACK NEEDLEMAN, 2002)

## OBJECTIVES

Upon completion of this chapter, the reader should be able to:

1. Identify the budget preparation process for health care organizations.
2. Illustrate commonly used types of budgets for planning and management.
3. Select key elements that influence budget preparation and monitoring.
4. Illustrate a scope of service.
5. Identify expenses associated with the delivery of service.

Delmar/Cengage Learning

*You are assigned to a patient care unit for your clinical experience. You are wondering what types of services are provided to patients on this unit. You talk with your instructor and the nursing manager of the unit to review unit staffing. You also review the unit's scope of service and budget.*

*What kinds of patients are cared for on this unit?*

*What kinds of services are provided to patients on this unit?*

*How does the staffing model help ensure provision of care for patients on this unit?*

*How would a nurse shortage affect the model?*

A key factor that influences patient care is the cost involved in the delivery of service. Resources—people, equipment, and time—are required to support the services delivered by nurses. These resources cost money. The economic success of a health care organization depends on those who are involved with service delivery. The decline in health care reimbursement, as well as escalating costs and increasing competition, have required hospitals to improve operational efficiency and to make economically sound decisions. The challenge in health care is to ensure that the quality of care and the caliber of the staff are not compromised in this ever-changing, cost-controlled environment.

Nurses need to understand how to manage the cost of patient care as it relates to their own clinical practice. Nurses are accountable for the distribution and consumption of resources, whether that equates to time, supplies, drugs, or staff. It is essential that appropriate decisions be made regarding cost-effective practices. Cost containment affects the patient's bill and the financial viability of a nursing department or unit. Hence, nurses need to be informed and partner with the management team to generate revenue and control expenses in relation to patient care. According to regulatory and accrediting health care organizations, departmental budgets need to be developed in collaboration with staff from respective services involved in care.

The purpose of this chapter is to provide an overview of the operational budget process, including budget development, implementation, performance, and evaluation.

Common financial language and tools are discussed so nurses can understand the process involved in cost-effective care.

## TYPES OF BUDGETS

Hospitals use several types of budgets to help with future planning and management. These include operational, capital, and construction budgets.

### OPERATIONAL BUDGET

An **operational budget** accounts for the income and expenses associated with day-to-day activity within a department or organization. Revenue generation is based upon billable services and expenses associated with equipment, supplies, staffing, and other indirect costs. Revenue may be based on the number of days that a patient stays on an inpatient unit or the number of hours spent in a procedure room. Revenue may be also based on the types of procedures delivered to a patient. Depending on reimbursement rates and requirements, expenses are sometimes bundled or included in a procedure or room charge—for example, an admission packet that includes a washbasin, cup, soap holder, and so on. In other situations, supply items may be billed separately, such as IV start kits, leukocyte removal filters, and so on.

### CAPITAL BUDGET

A **capital budget** accounts for the purchase of major new or replacement equipment. Equipment is purchased when new technology becomes available or when older equipment becomes too expensive to maintain because of age-related problems such as inefficiencies resulting from the speed of the equipment or downtime (amount of time it is out of service for repairs). Sometimes the expense and lack of availability of replacement parts make it prohibitive to maintain equipment. Finally, equipment may become antiquated because of its inability to deliver service consistently, meet industry or regulatory standards, or provide high-quality outcomes.

Because a significant expense is associated with equipment acquisition, organizations want to make the best and most economical and informed decisions. Staff members from a variety of areas—including materials management, clinical experts, legal counsel, biomedical engineering, information technology, finance, and management—often participate in planning for equipment purchases because they may have important input. Substantial analysis is required, because equipment features, benefits, and limitations have to be understood as they relate to a department or institution's needs and goals. Often, multiple vendors or companies sell similar or varying products with different terms, conditions, and warranty and maintenance agreements that can have short- and long-term effects.

Some organizations may differ regarding the dollar amount that is considered a capital purchase. Capital purchases are based upon the equipment cost and the life expectancy (also known as shelf life), or how long the equipment is expected to perform over time. Generally, capital purchases cost more than $500 and last five years or longer. For example, in one organization, a stent costing $2,000 is a supply item used during a surgical procedure. This supply is considered an operational expense because its life expectancy is two years, whereas a CT scanner costing more than $500 with a life expectancy of seven years is considered a capital expense.

## CONSTRUCTION BUDGET

A **construction budget** is developed when renovation or new structures are planned; it typically includes labor, materials, building permits, inspections, equipment, etc. If it is anticipated that a department will need to close during construction, then projected lost revenue is accounted for in the budget. Revenue and expenses may also be shifted to another department that absorbs the services on a temporary basis.

## BUDGET OVERVIEW

An operational budget is a financial tool that outlines anticipated revenue and expenses over a specified period. A process called **accounting**, which is an activity that managers engage in to record and report financial transactions and data, assists with budget documentation. The budget translates operational plans into financial and statistical terms so that income can be projected with associated costs. Budgets serve as standards to plan, monitor, and evaluate the performance of a health care system. Details regarding a budget are specific to the area governed. Budgets account for the income generated as compared to the expenses needed to deliver the service. **Profit** is determined by the relationship of income to expenses. Profitability results when the income is higher than the expenses.

Budgets make the connection between operational planning and allocation of resources. This is especially important because health care organizations measure multiple key indicators of overall performance. For example, along with financial performance, organizations routinely evaluate quality patient outcomes and customer and staff satisfaction. All these indicators are intertwined and hold value in terms of patient care. Collectively, they reflect organizational success.

Figure 14-1 is a balanced scorecard, or dashboard, showing a variety of indicators that illustrate the connectivity between performance and quality outcomes. A **dashboard** is a documentation tool providing a snapshot image of pertinent information and activity at a particular point in time (Frith, Anderson, & Sewell, 2010). The purpose of the dashboard is to provide a visual pulse of how a unit or department is achieving its goals. The dashboard is considered a picture of a moment in time, because the unit can change within days or weeks, impacting the unit's outcomes displayed on the dashboard. A dashboard identifies any of four perspectives about an organization: finances, customer satisfaction and services, internal operating efficiency, and learning and growth (Advisory Board, 2009b). Figure 14-1 shows two separate units, the gastrointestinal laboratory (GI lab) and a medical nursing unit. Dashboards display measurable unit activity, e.g., the number of procedures delivered in the GI lab. Additionally, they illustrate the amount of revenue generated and the expenses incurred. **Variance**, or the difference between what was budgeted and the actual result, can be tracked. Key activities that affect the number of patients that can be cared for, such as the room turnaround time, are also monitored. Finally, specific patient satisfaction indicators are visualized to determine whether the goals are met in high-priority areas.

Similar to controlling personal funds, such as managing a checking or savings account, budgeting helps to define services by projecting how much cash is generated (revenue) and how much services will cost to operate (expenses). Budgeting requires forward thinking so that problems can be planned for and ways to work around any obstacles can be anticipated. Budgets also serve as a benchmark to measure whether the planning expectations are being met. Typically, budgets are monitored monthly, so that if deficiencies arise throughout the year, financial improvement plans can be instituted early. Drilling into core health care processes early is imperative so that organizations use labor, technology materials, and facilities to their fullest potential (Advisory Board, 2009a). Corrective action is often initiated to prevent long-term effects in a particular area, such as wastage or loss of supply items. This corrective action may entail scrutinizing and mapping every step in the process for care delivery while examining the value in everything that nurses and other professionals do on a daily basis. The budget functions as a tool to foster collaboration, because individuals within departments must work together to achieve its goals.

## BUDGET PREPARATION

Formulating a budget begins with preparation and planning. Budgets are generally developed for a 12-month period. The yearly cycle can be based on a fiscal year as determined by the organization (e.g., September 1 through August 31) or a calendar year (e.g., January 1 through December 31). Shorter- or longer-term budgets may also be developed depending upon the organizational planning process.

Prior to the beginning of the budget year, most organizations devote approximately six months to preparing and developing the operational budget. To prepare a budget, organizations gather fundamental information about a variety of elements that influence the organization, including demographic and marketing information, competitive analysis, regulatory influences, and strategic plans. Additionally, it is helpful to review the department's scope of service, goals, and history.

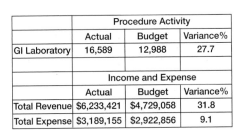

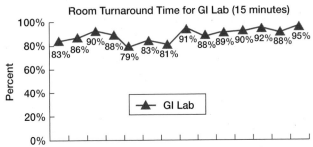

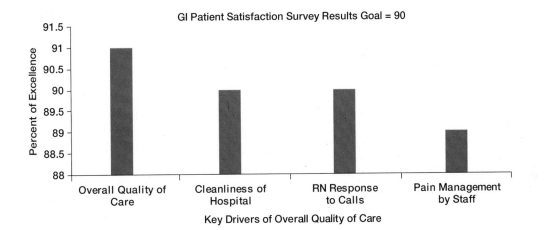

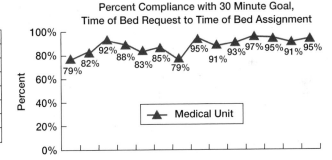

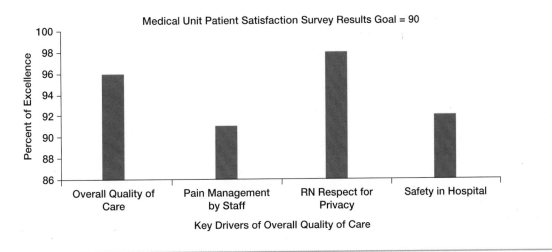

FIGURE 14-1 **Patient satisfaction, room turnaround time, and budget activity dashboard.** *Values are for instructional purposes only (*Source:* Adapted with permission of Central DuPage Hospital, Winfield, IL).

## DEMOGRAPHIC INFORMATION AND MARKETING

Pulling together demographic information relative to the population the organization serves is most helpful, because it identifies unique market characteristics such as age, race, sex, income, and so on that influence patient behavior. For example, an obstetrical practice would be expected to attract women of childbearing age rather than an older male population. Therefore, understanding the demographics and capture rate (the percent of the population that has been "captured" by the organization as a consumer) of the immediate or distant market region helps paint a clear picture regarding the patient population.

It is important to understand demographics as it relates to determining what type of patients are consumers that institutions target. Marketing strategies are built around the population types that an organization is attempting to attract. For example, if a hospital is opening an open heart or transplant department, then outreach activities might be developed to attract those patients that can benefit from the specialty screening, prevention, or treatment services.

Marketing is the process of creating a product or health care service for patients. Health care facilities use the four Ps of marketing—that is, patient, product, price, and placement—to place desirable health care services or products in desirable locations at a price that benefits both patients and the health care facility. In this way, the health care facility, the patient, and the community benefit. Marketing of services does have a price tag, such as the cost of advertising campaigns on television and radio. Using printed materials, mailing information to patient residences, and advertising in journals, magazines, and newspapers are all examples of ways to educate and stimulate the public for future referrals for health care services. Once marketing strategies are implemented, most organizations attempt to measure the effectiveness or return on investment of marketing strategies. Hospitals can begin to measure the effectiveness of advertising for new cardiovascular patient care services by reviewing the number of new patient referrals to the hospital that result from the marketing and media exposure. Sometimes the measurement is not clear-cut, because results cannot always be clearly attributed to the marketing efforts.

## COMPETITIVE ANALYSIS

A competitive analysis is important, because it probes into how the competition is performing as compared to other health care organizations. A competitive analysis examines other hospitals' or practices' strengths and weaknesses, in addition to other details such as location and new or existing services and technology. Having this knowledge can influence decisions regarding the implementation of new programs, hiring of specialty staff, and purchasing of equipment. Figure 14-2 presents a competitive analysis of three different hospitals.

## REGULATORY INFLUENCES

Regulatory requirements and reimbursement rates have an effect on financial performance. Regulatory changes are influenced by several governing bodies. A government agency that has high visibility in the area of reimbursement is the Centers for Medicare & Medicaid Services (CMS), whose mission is to ensure health care security for beneficiaries. CMS (www.cms.hhs.gov) administers federal control, quality assurance, and fraud and abuse prevention for Medicare, Medicaid, and the State Children's Health Insurance Program (SCHIP). Under the aegis of the Department of Health and Human Services, it is also responsible for coordinating health care policy, planning, and legislation.

Other regulatory bodies play a role in reimbursement by ensuring that federal and state laws are adhered to through approval and accreditation. For example, the Food and Drug Administration (www.fda.gov) regulates the use of drugs, food products, and medical devices in the United States. If equipment or drugs under its jurisdiction are not approved, then organizations cannot bill for their use, by law. The Joint Commission (JC) formerly known as the Joint Commission on Accreditation of Healthcare Organizations (JCAHO) (www.jointcommission.org) accredits hospitals and ambulatory care and home health agencies and departments to ensure that organizations meet specific standards. Medicare and Medicaid will not reimburse for services unless a hospital is accredited by the JC.

Regulatory requirements may change regarding who may deliver a specific service and in what type of setting; for example, a procedure may have to be done in the hospital rather than in a practitioner's office if it is to be reimbursed by the insurance company. Medicare and Medicaid change their reimbursement rates periodically. Total and partial coverage of specific procedures can change and may not be predictable from year to year.

Managed care organizations and insurance companies typically negotiate rates on a yearly basis, which can affect hospital revenue. Consumers' willingness to pay out of pocket when a service is not covered by insurance affects revenue as well.

## STRATEGIC PLANS

Generally, hospitals have strategic plans that map out the direction for the organization over several years. Strategic plans guide the staff at all levels so that the entire organization can have a shared mission and vision, with clearly defined steps to meet the goals. The vision provides the hope, direction, and perseverance to transform an organization and to keep it moving forward while focusing on the goals.

Each department develops unit-specific plans to help the organization follow its overall strategic plan. For example, a goal may be to become the most preferred GI lab or site for inpatient hospital care in the surrounding region.

## COMPETITIVE ANALYSIS

### HOSPITAL A

Location: Rural—100 miles from metropolitan area

Affiliation: Currently negotiating with three academic hospitals

General clinical description:

- Scattered-bed approach to inpatient oncology
- Ambulatory chemotherapy clinic
- Many of the same physicians on staff at Hospital J

Radiation capability: None—refers to Hospital J

Support services: Cancer screenings offered sporadically

Miscellaneous:

- Tumor board
- Cancer committee

### HOSPITAL B

Location: Suburb of large metropolitan city

Affiliation: University hospital

General clinical description:

- Dedicated oncology inpatient unit
- Ambulatory chemotherapy department
- Comprehensive breast center
- Head and neck oncology team

Radiation therapy:

- Linear accelerator—two units
- High-dose rate
- Intraoperative radiation therapy
- Stereotactic radiosurgery

Support services:

- Home infusion and home care program
- Hospice care program
- Annual cancer awareness fair
- Support group—general cancer patients

Miscellaneous:

- Tumor registry
- Tumor board
- Committee on cancer
- Head and neck patient conferences
- Stereotactic radiosurgery conferences

### HOSPITAL C

Location: Urban city with a population of 150,000

Affiliation: For-profit corporation

- Medical oncology affiliation with University K
- Radiation therapy department affiliation with University K radiation therapy department

General clinical description:

- Dedicated inpatient medical oncology unit
- Four-bed autologous and stem cell bone marrow transplant unit (Eastern Cooperative Oncology Group Referral Center for autologous bone marrow transplants)
- Coagulation laboratory
- Therapeutic pheresis
- Pain clinic
- Oncology clinic
- Oncology rehabilitation
- Breast cancer rehabilitation program
- Ambulatory care chemotherapy unit
- Medical oncologist on staff at two hospitals

Radiation therapy:

- Linear accelerator
- Stereotactic radiosurgery
- Hyperthermia
- Brachytherapy

Support services:

- Home health and hospice program
- Cancer registry
- Cancer committee
- Physician update—quarterly cancer newsletter
- Cancer information line
- Cancer advisory council
- Cancer Survivor's Day offered annually
- Cancer screenings offered routinely
- Cancer support group—general cancer patients

Delmar/Cengage Learning

**FIGURE 14-2** Competitive analysis of three hospitals.

To meet the goal, one department may focus on patient satisfaction and room turnaround time to increase volume and decrease patient wait time for a procedure appointment. This goal is part of the organization's overall plan and takes into consideration the quality of care delivered and the cost to the organization.

## SCOPE OF SERVICE AND GOALS

During the budget-preparation phase, it is important to examine the individual nursing or hospital department or section thoroughly. Hospital systems are frequently divided into sections, departments, or units to compartmentalize them for organizational purposes. These subsections or units of an organization, commonly called **cost centers**, are created to track financial data.

Each department or cost center defines its own scope of service (Figure 14-3 and Figure 14-4 provide examples of scope of service.) The scope of service is helpful because it provides information related to the types of service and the sites at which services are offered, including the usual treatments and procedures, hours of operation, and the types of patient/customer groups.

Departmental goals may include the introduction of new technology or treatments, patient education, and creation of a special patient care environment. Staff members are generally queried to determine whether they have any proposed quality initiatives that should be included in the plans for the upcoming year. Generally, new treatments, patient education materials, and documentation tools require different types or amounts of supplies. Technically trained staff members are often needed to implement new services. Both the staff and supplies can have varying costs from the early induction phase through full implementation. Creating a new environment or "best patient" experience may require additional funding that must be identified early in the planning stages. If new services are offered that are billable to the insurance company, then a method for charging patients has to be established to ensure that the hospital can receive appropriate payment. The manager is responsible for identifying upfront the expenses associated with patient care so that they are covered by the charges. A charge is the dollar amount that the patient is responsible for paying as a result of service.

## HISTORY

Organizations typically use history, or past performance, as a baseline of experience and data to better understand activity in a department or unit. These data are used to assist in interpreting expenses associated with staff productivity and unit performance. Most often, adjustments are made to planned budgets because of the ever-changing cost of products, supplies, and buying contracts. Buying contracts are negotiated so that predetermined reduced rates can be realized when organizations purchase large quantities of supplies. For example, if a hospital purchases a large quantity of one product, the vendor may reduce the price below the list price as an incentive. If a hospital can demonstrate that a particular product is used at a certain percentage—in 60% of all procedures or departments (called penetration rate), for example—then a reduced rate may be offered.

## REAL WORLD INTERVIEW

Perhaps the most challenging part of the budget process in health care is finding an efficient and effective method to enlist the involvement and intellectual capital of our staff. Labor costs, on average, constitute 60% of a department's budget. Balancing the budget and the need to obtain staff nurse input for key budget decisions about their practice is imperative. Our organization uses a value analysis program (VAP) to ensure staff involvement in decisions about the products and services staff members use in their daily practice while using their time efficiently.

Our value analysis program ensures that, whenever quality is equivalent, cost is the greater factor in making a decision about products and services. The VAP is brought to life through our value analysis teams (VAT) for key areas, for example, medical-surgical units and the operating room. Nurses lead and participate in these teams. The teams align with our organization's commitment to thoughtful stewardship in providing the highest-quality health care to our patients. The nurses on the VATs are supported by the purchasing and finance departments with financial and utilization data. These nurses explore best practice literature and national standards such as those of the Oncology Nursing Society.

Recently, one medical-surgical VAT identified an opportunity to reduce the inventory of selected medical-surgical supplies on their unit. They looked at the amount of supplies used per day, decided on the appropriate inventory, and achieved an immediate cost savings of $4,500. They set a targeted cost savings and continue to monitor their progress through a budget dashboard that reports average daily cost of supplies. They make adjustments as needed.

Our staff feels a commitment to the process and realizes the importance of our input into decision making. The organization identifies the value of staff involvement in this process and incorporates this into our budget cycle.

**Barbara Buturusis, RN, MSN**
Executive Director, Cancer Services
Loyola University Medical Center
Maywood, Illinois

The gastrointestinal (GI) laboratory may be defined as a specialized department that performs major procedures that are both diagnostic and therapeutic in nature, such as upper endoscopies, colonoscopies, flexible sigmoidoscopies, and endoscopic retrograde cholangiopancreatography (ERCP). Conscious sedation is typically delivered to patients to provide comfort during the procedures. The gastrointestinal laboratory is operational Monday through Friday from 7 a.m. to 5 p.m. and provides after-hours service for emergent cases. Pre-procedure, intra-procedure, and post-procedure care, including full recovery, is provided on site. Services are provided to critical in-house patients at the bedside via the staff assigned to travel to inpatient units. The department employs nurses, technicians, and receptionists, who work with gastroenterologists and surgeons to provide care. The unit is equipped with 10 procedure rooms, 25 recovery bays, and a GI scope cleaning facility on site.

**FIGURE 14-3** Gastrointestinal laboratory scope of service.

A medical nursing unit provides primarily inpatient care to patients with acute or chronic medical problems, such as congestive heart failure, diabetes, pulmonary disease, cancer, and so on. The unit, equipped with 30 private beds, a full kitchen, a lounge, and conference/consultation rooms, is operational 24 hours per day, seven days per week. Patient education and support groups are held routinely in the library located directly on the unit. Team nursing is used as the model of care. Nurses, patient care technicians, and unit secretaries are employed, with a social worker and diabetes educator providing additional patient support. Patients admitted to the unit for longer than 48 hours are discussed during daily multidisciplinary rounds. The rounds include case management personnel; psychosocial counselors; and nutrition, nursing, and medical staff. Staff discusses patient problems to facilitate future care, including discharge planning.

**FIGURE 14-4** Medical nursing unit scope of care.

## CASE STUDY 14-1

The manager from an inpatient unit asks for staff input into identifying ways to decrease use of medical supply and paper items. These items have been identified as in excess of the budget by 10% to 20% during the past three months. This is the first time that the staff members have been involved in helping with cost containment. Beginning nurses, experienced nurses, and assistants have been invited to participate.

When approaching an analysis of health care supply use, what might be the first step in the process? If you were to break the staff into work groups, which members should be chosen to analyze the use of clerical supplies? How would you proceed if you were trying to determine the supply costs associated with starting an IV with continuous infusion?

Additionally, knowledge about historical volume (e.g., procedures, admissions, or patient visits and average length of stay) provides a perspective as to how a department has grown or declined over time. This information may help with anticipating future demand and capacity. The story behind a unit and its heritage related to how the department developed is equally

important because the financial numbers are tracked over time. Often, the culture and complexity of a unit unfold by interviewing staff that may have been involved in the unit during the past, including practitioners, nurses, technologists, assistants, housekeeping staff, dietary counselors, and so on. This information may provide further insight into why and how decisions were made in the past. Hence, multiple phases of data gathering are imperative to building a budget with a full knowledge base.

# BUDGET DEVELOPMENT

Once background data have been gathered, the development of the budget can follow. This includes projecting revenue and expenses.

## REVENUE

**Revenue** is income generated through a variety of means, including billable patient services, investments, and donations to the organization. Specific unit-based revenue is generated through billing for services such as X-rays, invasive diagnostic or therapeutic procedures, drug therapy, surgical procedures, physical therapy, and so on. Revenue can also be generated through the delivery of multiple services over time, such as hourly rates for chemotherapy administration or blood transfusions. The specific number and types of services and procedures have to be projected for the budget. Each type of service may have varying volume associated with it. For example, projecting the volume and type of procedures to be conducted in a gastrointestinal laboratory is based upon feedback from referring physicians and technical staff, in addition to conclusions from historical data.

Similarly, the same types of projections occur for inpatient units, including the number of patients anticipated to be admitted, along with the average length of stay (e.g., three to five days) and the projected occupancy rate. The type and amount of services and patient days can be measured. The number of patient days or the services delivered are commonly called service units or primary statistics and are measured so that productivity and efficiency can be tracked.

It is important to note that the reimbursement rates of third-party payers affect revenue and can change from year to year. Uniform rates are often used, which transfers significant financial risk to the provider. Medicare, Medicaid, managed care companies, and insurance companies dictate or negotiate rates with health care organizations that may include discounts or allowances. Payers determine what costs are allowable for procedures, visits, or services. Payment schedules vary from state to state and among plans. Additionally, the rates can change monthly, such as with the ambulatory payment classification (APC) system from Medicare, which applies to the outpatient setting. The reimbursement rates or payments received by hospitals often do not equal the actual hospital or unit charges for the services rendered. For example, there may be a fixed or flat reimbursement rate per case regardless of how long the patient stays in the hospital or how much the hospital pays for the service. If the costs exceed the reimbursement rate, then the provider absorbs the remaining costs.

Another payment classification system, called diagnosis-related groups (DRGs), is used to group inpatients into categories based upon the number of inpatient days, age, complications, and so on. Reimbursement covers room and board, tests, and therapy during a predetermined length of stay.

Some patients will not have health care insurance nor the ability to pay their bills. Therefore, the hospital may receive only a portion of the payment for services, if any.

Typically, organizations will review their payer mix to determine the percentage of patients carrying different types of health care coverage (Figure 14-5). The proportions help measure the anticipated dollars to be received for services delivered and make projections for the coming year. However, it has been more difficult recently to predict reimbursement, due to government changes. It is unclear

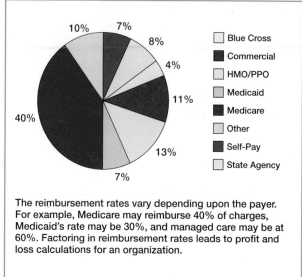

The reimbursement rates vary depending upon the payer. For example, Medicare may reimburse 40% of charges, Medicaid's rate may be 30%, and managed care may be at 60%. Factoring in reimbursement rates leads to profit and loss calculations for an organization.

**FIGURE 14-5** Inpatient payer mix.

what the future holds related to government and insurance reimbursement.

If charges for patient care are negotiated with third-party payers such as insurance companies and managed care corporations, they are preestablished and are not negotiable once established. Third-party payers often impose a penalty fee, as a disincentive, if a health care organization changes a charge under contract. The penalty often exceeds the charge amount and will usually create a loss for the organization.

Let's look at the differences in reimbursement related to a specific procedure. The charge for a central line placement is $2,500, which includes the use of the fluoroscopy equipment, supplies, nursing, and technical time. If a hospital places 200 central lines per year, then the anticipated total revenue is $500,000. The breakdown of third-party payers becomes important, because the total revenue does not mean that the hospital will be reimbursed for the full amount of $500,000. Third-party payers typically contract with health care organizations for the amount that the third-party payer is willing to pay or reimburse. Table 14-1

**TABLE 14-1   Total Charges—Central Line Placement $500,000**

| THIRD-PARTY PAYER REIMBURSEMENT | RATE (MEASURED IN PERCENT OF CHARGES) | EXPECTED REIMBURSEMENT |
| --- | --- | --- |
| Managed Care | 60% | $300,000 |
| Medicare | 40% | $200,000 |
| Medicaid | 30% | $150,000 |

**EVIDENCE** FROM THE LITERATURE

**Citation:** Hendrich, A. L., & Lee, N. (2005, July–August). Intra-unit patient transports: Time, motion and cost impact upon hospital efficiency. *Nursing Economics, 23*(4), 157–164, 147.

**Discussion:** Intrahospital transport of patients between patient care units can produce detrimental effects on patients. Inaccuracy in monitoring, patient stress, and potential nursing shift report hand-off errors have been cited as risk factors. Patients are moved to different patient care units during hospitalization primarily due to changing clinical requirements. Appropriate units are determined based upon the technical capacity, for example, of availability of oxygen in the head wall, cardiac monitoring, clinical care-giver skill level, and nursing ratio or hours per patient day. Hendrich and Lee initiated a study to examine the cost of patient transfer between patient care units, including the transfer process, time, and personnel required. Delays, disruptions, communication gaps, administrative work, and resource unavailability were factors identified that negatively affected productivity. Of the patient transfer processes, 87% was quantified as inefficient due to the previously stated reasons.

The authors recommend reevaluation of facility design and patient models of care to accommodate fluctuations in patient acuity so that patients do not need to be transferred. Lack of progressive care beds covering multiple levels of care causes bottlenecks within hospital systems and excess resource requirements.

The strength of this article was the creative analysis of the transfer process highlighting inefficiencies. Although every organization may have differences in the process, it appears that by using this valuable exercise, organizations might bring forward opportunities to rethink the cost and safety of transfers. Reducing the contributing causes for transfer delays may improve financial, clinical, and operational management and outcomes.

**Implications for Practice:** Many factors can be identified that influence operations. Using the workforce can potentially bring to light new ideas. Imagine the potential if a work group were continually analyzing patient scheduling, transport, and discharge to improve operations and decrease cost. Different team members have different perspectives. It is important to have empowered work groups address problems and challenges to create constructive measures for improvement.

demonstrates the potential reimbursement rates based upon varying third-party payers.

## EXPENSES

Expenses are determined by identifying the costs associated with the delivery of service. Expenditures are resources used by an organization to deliver services and may include supplies, labor, equipment, utilities, and miscellaneous items.

It is important to understand what it takes to deliver patient care services so that there are appropriate charges in place to pay for, or cover, the services. Expenses are commonly broken down into line items that represent specific categories that contribute to the cost of the procedure or activity, such as paper supplies, medical supplies, drugs, and so on. This breakdown helps identify where the significant expenses lie related to a service. For example, a colonoscopy may have a high medical supply cost, whereas chemotherapy administration may have a high drug cost associated with it.

## Supplies

As new procedures are introduced, or when a manager wants to ascertain the actual supply expenses associated with a procedure or activity, zero-based budgeting may be instituted. Zero-based budgeting is a process used to drill down into expenses by detailing every supply item and the quantity of items typically used. A list of supplies is developed, including large and small items, along with the itemized expense. Often, supplies are packaged in bulk and sold in quantity. Hence, the expense of the items has to be calculated and backed out of the bulk figure to accurately depict the expense.

Figure 14-6 illustrates the zero-based budgeting that may be necessary to understand all of the expenses associated with delivering a procedure. This example can be expanded further to calculate the total expense associated with the anticipated number of procedures. This calculation can be achieved by multiplying the number of anticipated procedures by the total expense per procedure, which leads to authentic projections.

| General Supplies | Quantity | Price | Drugs | Quantity | Price |
|---|---|---|---|---|---|
| 4 Chux | 4 | $ 3.60 | Fentanyl | 1 | $ 0.30 |
| Tri Pour Container | 1 | $ 0.24 | Versed | 1 | $ 2.52 |
| Sterile Water 1.000mL | 1 | $ 0.48 | **Total** | | **$ 2.82** |
| Normal Saline Vial | 2 | $ 0.12 | | | |
| Cannister/Lid | 2 | $ 3.90 | **Printed Forms** | **Quantity** | **Price** |
| Tubing | 2 | $ 0.60 | | | |
| 02 Cannula | 1 | $ 0.15 | Hospital Consent | 1 | $ 0.12 |
| Suction Catheter | 1 | $ 0.24 | Procedure Consent | 1 | $ 0.90 |
| Disposable Gowns | 2 | $ 4.56 | Nursing Form | 1 | $ 0.90 |
| Gloves | 6 | $ 0.42 | Vital Sign Sheet | 1 | $ 0.06 |
| 4X4's | 10 | $ 0.30 | Doctors Orders | 1 | $ 0.18 |
| Surgilube | 2 oz. | $ 0.18 | History/Physical | 1 | $ 0.12 |
| Photos | 2 | $ 5.16 | Discharge Instructions | 1 | $ 0.24 |
| Syringe 10cc | 2 | $ 0.18 | Education Sheet | 1 | $ 0.96 |
| Syringe 60cc | 1 | $ 0.42 | Charge Voucher | 1 | $ 0.12 |
| Emesis Basin | 1 | $ 0.12 | Procedure Education | 1 | $ 0.12 |
| Denture Cup | 1 | $ 0.18 | **Total** | | **$ 3.72** |
| Recording paper | 1 | $ 0.06 | | | |
| Alcohol Pads | 2 | $ 0.06 | **Clerical Supplies** | **Quantity** | **Price** |
| Slippers | 1 | $ 0.90 | | | |
| Mask | 2 | $ 0.30 | Patient File | 1 | $ 0.96 |
| Goggles/Face Shield | 1 | $ 1.44 | Labels | 2 | $ 0.06 |
| Cetacaine Spray | 1 | $ 0.12 | Xerox Paper | 6 | $ 0.06 |
| Bite Block | 1 | $ 2.40 | Pen | 1 | $ 0.06 |
| Patient Bag | 1 | $ 0.24 | Pencil | 1 | $ 0.06 |
| Cleaning Brush | 1 | $ 2.64 | Marker | 1 | $ 0.06 |
| | **Total** | **$ 29.01** | Highlighter | 1 | $ 0.06 |
| | | | **Total** | | **$ 1.32** |
| | | | | | |
| **IV Start** | **Quantity** | **Price** | | | |
| | | | **Grand Total** | | **$ 43.11** |
| Tourniquet | 1 | $ 0.18 | | | |
| Alcohol Wipes | 2 | $ 0.06 | | | |
| Angiocath | 1 | $ 0.06 | | | |
| IV Solution | 1 | $ 0.72 | | | |
| IV Primary Set | 1 | $ 4.80 | | | |
| Tegaderm | 1 | $ 0.18 | | | |
| Tape | 6 inches | $ 0.06 | | | |
| Band-Aid | 1 | $ 0.06 | | | |
| 4X4 | 4 | $ 0.12 | | | |
| | **Total** | **$ 6.24** | | | |

**FIGURE 14-6  Zero-based budgeting for a gastrointestinal laboratory.** *Values are for instructional purposes only (*Source:* Adapted with permission of Central DuPage Hospital, Winfield, IL).

## Critical Thinking 14-1

Staff working day to day handling patient care activities are in an optimal position to identify the best practices that impact efficiency and cost-effectiveness. Managers can learn from staff and organize processes to assist with unit-based improvement. Think back on the steps taken by a nurse during the first hour of a shift. Reflect on communication and how information is received. Examine the amount of time spent in patient care versus other activities. Create a journal of activity from different time increments during a shift. Discuss your observations with your coworkers and manager. What problems in flow of activity and gaps in communication or efficiency did you find? How can you drill down further into understanding how the unit operates and ways to increase productivity? How could you improve your team's functioning?

## TABLE 14-2  Time and Salary Expense Analysis per Procedure

| EXPENSE | PROCEDURE | TIME |
|---|---|---|
| Reception Staff, average salary per hour = $13.00 | **Pre-procedure Care** | Time (minutes) |
| | Appointment schedule | 5 |
| | Registration | 10 |
| | Escort to changing room | 5 |
| $4.33 | **Subtotal** | **20** |
| Staff Nurse, average salary per hour = $32.00 | **Direct Patient Preparation** | Time (minutes) |
| | History | 5 |
| | Patient education and consent | 15 |
| | IV start | 10 |
| $16.00 | **Subtotal** | **30** |
| Staff Nurse, average salary per hour = $32.00 | **Intra-procedure Care** | Time (minutes) |
| | Positioning | 5 |
| | Initiation of conscious sedation | 10 |
| | Procedure | 15 |
| $16.00 | **Subtotal** | **30** |
| Staff Nurse, average salary per hour = $32.00 | **Post-procedure Care** | Time (minutes) |
| | Recovery | 100 |
| | Education | 10 |
| | Discharge | 10 |
| $64.00 | **Subtotal** | **120** |
| **Grand Total Salary Expense = $100.33 per procedure** | | **Grand Total = 200 minutes per procedure** |

**Source:** Adapted with permission of Central DuPage Hospital, Winfield, IL. *Values are for instructional purposes only

## Labor

Labor is another significant expense associated with medical and nursing care. Health care services are very labor intensive. It is estimated that salaries and benefits account for 50% to 60% of operational costs. Hence, it is very important to calculate the amount of time the staff members are involved with the service. This analysis includes professional, technical, and support staff. For example, the time that it takes to schedule an appointment, register a patient, and take a patient to a procedure room or unit needs to be calculated into the overall cost of care for the patient.

In the ambulatory area, staff time is calculated relative to the delivery of a specific procedure, including preparation for the procedure, intra-procedure care, and post-procedure care. Pre-procedure preparation entails gathering of supplies, assembling equipment, and preparing the environment. Preparing the patient may involve taking a history, completing a physical, administering medication or taking specimens, placing tubes or establishing an intravenous line, and positioning the patient. Intra-procedure care is the actual care delivered after the procedure has been initiated. Post-procedure care may require activity

## Critical Thinking 14-2

When you walk onto a patient care unit, ask a staff member what key quality initiatives the unit is working on that reflect process improvement. Ask what the goals are for the unit and how staff are participating in decisions so that the goals may be achieved. Think about how these initiatives may increase productivity, increase staff or patient satisfaction, or decrease expenses. Ask the staff what impact their efforts are having.

How is the staff involved in helping the organization to meet its goals?

such as educating and discharging a patient, or extensive recovery activity requiring several hours of direct nursing care and removal of equipment and supply items. Refer to Table 14-2 for a sample time analysis.

### Staffing

The amount of staff and types of staff are often accounted for in a staffing model. The model outlines the number of staff required based upon a primary statistic such as procedures or patients. An outpatient model may focus on the number of procedure rooms that require staff. One nurse may be required to staff a gastrointestinal laboratory procedure room, and one shared technician may staff two procedure rooms.

Models may help in analyzing productivity, as illustrated in the following:

- *Scenario 1:* One nurse is assigned to a procedure room during a 4-hour period in which 8 patients are treated.
- *Scenario 2:* One nurse and a technical assistant are assigned to a procedure room for 4 hours, and 16 patients are treated.

The second scenario depicts greater productivity, because the number of procedures delivered doubled by using two staff, recognizing that the assistant staff member will cost the organization less in terms of salary expense. Because labor is one of health care's greatest operational costs, enhancing productivity will likely produce savings.

For an inpatient unit, nurses may be assigned to a fixed number of patients during all three shifts. The ratios vary depending upon the shift and patient acuity. The nurse-to-patient ratio on a medical nursing unit may be one nurse to six (1:6) patients during the day and evening shift, whereas it may be 1:8 during the night shift. The nurse-to-patient ratio may be 1:2 for all shifts on a critical care unit. Figure 14-7 and Figure 14-8 illustrate sample staffing models.

Staffing ratios and salary data are particularly important because of the cost factor. Specialty salaries fluctuate,

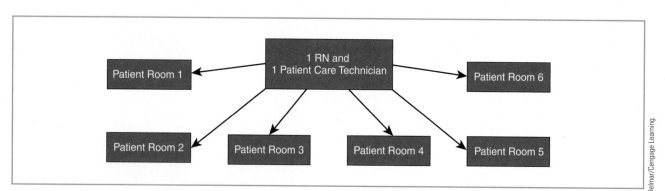

**FIGURE 14-7** Inpatient staffing model.

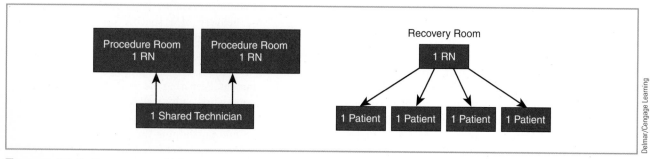

**FIGURE 14-8** Gastrointestinal laboratory staffing model.

## REAL WORLD INTERVIEW

The staff nurse plays a critical role in assisting the unit manager in planning department budgets. Without appropriate planning, department budgets are at risk of not meeting organization targets. It is important that staff provide input regarding the current environment and also discuss with the unit manager the goals and objectives for the upcoming fiscal year. Historic data related to patient volume, revenue, and expenses is a starting point. However, new programs and services need to be anticipated. In addition, practice changes, safety and quality initiatives, changes in patient populations, and acuity must all be considered in the planning process. Unit equipment requirements are another important piece that must be included in capital planning. Ideally, the staff nurse input is given on an ongoing basis to the manager so that by the time budget planning is started, the manager has received comprehensive feedback from many staff who are working different shifts.

In addition to assisting with planning, the staff nurse is responsible for managing resources on a daily basis. Resource management is done through supply management and labor management. Nurses must utilize supplies efficiently and minimize waste. This assists the organization to meet budget targets and reduces unnecessary costs for the patient. Labor costs account for a significant percentage of most nursing budgets. Overtime management is one way in which the staff nurse can assist the unit budget on a daily basis.

The staff nurse role is multidimensional. The nurse is not only a clinical care provider but is also a financial manager of the organization's use of resources on a daily basis.

**Anne K. Anderson, RN, BSN, MHSA**
Director, Patient Care Services Operations
Central DuPage Hospital
Winfield, Illinois

depending upon supply and demand. When there are shortages of certain staff, the salary tends to increase. Additionally, a health care organization may change its benefits, offering a more attractive package that includes continuing education, paid time off for education purposes, or professional membership expenses. Institutions may also look for alternative ways to supplement or deliver services during staff shortages. This means that supplemental staff—professional agency nurses, nurses from in-house registries, or patient care technicians—may be hired at a different salary rate. It is important to note whether a unit has had historical difficulty retaining or recruiting staff. Recruitment and retention—especially attracting, interviewing, hiring, and orienting staff—require dollars. For example, it has been estimated that the turnover cost per nurse, including advertising, recruitment, orientation, and time to fill the vacancy, can equate to $67,000 (Jones, 2005). The average cost to educate a nurse during a six-week orientation period is more than $5,000. Not only the salary but also benefits are frequently factored into a salary package, and they need to be included in the budget.

## Unproductive Time

Unproductive time is also calculated into a budget, because there has to be staff coverage when nurses or other staff members are not working. Unproductive time

### Critical Thinking 14-3

Calculate the added expense that a nursing unit will need to budget next year based upon the addition of five nurses (five full-time equivalents [FTE]*, taking into account a salary increase over each year). Instruction: Multiply next year's average hourly rate times the total FTE worked per year times the total paid number of nurses added per year to get the salary expense.

|  | Last Year | This Year | Next Year |
|---|---|---|---|
| Total FTE | 50 | 55 | 60 |
| Average hourly rate | $25.10 | $25.45 | $26.00 |
| Salary expense | $2,616,675.00 | $2,918,478.75 | $3,252,600.00 |

*FTE (full-time equivalent) works 2,085 hours per year.

What would the salary expense be on a clinical unit that you have recently worked on?

### TABLE 14-3 Unproductive Time

| UNPRODUCTIVE TIME | NUMBER OF HOURS | SALARY* |
|---|---|---|
| Vacation | 168 | $5,376 |
| Holiday | 56 | $1,792 |
| Sick | 40 | $1,280 |
| Personal | 24 | $768 |
| Education | 8 | $256 |
| Total | 296 | $9,472 |

*Salary dollars are based upon an average rate of $32.00 per hour. Remember to calculate the number of hours times $32.00. The salary dollars change depending upon pay rate and shifts, for example, day shift versus evening shift differential or 10- and 12-hour shift rates.

## EVIDENCE FROM THE LITERATURE

**Citation:** Needleman, J., Buerhaus, P., Mattke, S., Stewart, M., & Zelevinsky, K. (2002). Nurse-staffing levels and the quality of care in hospitals. *New England Journal of Medicine, 346*(22), 1714–1722.

**Discussion:** Administrative data from 799 hospitals in 11 states (covering 5,075,969 discharges of medical patients and 1,104,659 discharges of surgical patients) was examined to note the relation between the amount of care provided by nurses at the hospital and patients' outcomes.

The mean number of hours of nursing care per patient day was 11.4, of which 7.8 hours were provided by registered nurses, 1.2 hours by licensed practical nurses, and 2.4 hours by NAP. Among medical patients, a higher proportion of hours of care per day provided by registered nurses and a greater absolute number of hours of care per day provided by registered nurses were associated with a shorter length of stay (P = 0.01 and P < 0.001, respectively), as well as with lower rates of both urinary tract infections (P < 0.001 and P = 0.003, respectively) and upper gastrointestinal bleeding (P = 0.03 and P = 0.007, respectively). A higher proportion of hours of care provided by registered nurses was also associated with lower rates of pneumonia (P = 0.001), shock or cardiac arrest (P = 0.007), and "failure to rescue," which was defined as death from pneumonia, shock or cardiac arrest, upper gastrointestinal bleeding, sepsis, or deep venous thrombosis (P = 0.05). Among surgical patients, a higher proportion of care provided by registered nurses was associated with lower rates of urinary tract infections (P = 0.04), and a greater number of hours of care per day provided by registered nurses was associated with lower rates of "failure to rescue" (P = 0.008). No associations existed between increased levels of staffing by registered nurses and the rate of in-hospital death or between increased staffing by licensed practical nurses or nurses' aides and the rate of adverse outcomes.

**Implications for Practice:** A higher proportion of hours of nursing care provided by registered nurses and a greater number of hours of care by registered nurses per day are associated with better care for hospitalized patients and their improved outcomes.

| NET EXPENSE WORKSHEET | | | | |
|---|---|---|---|---|
| DESCRIPTION | GI LAB | MEDICAL UNIT | | |
| OPERATING REVENUE | | OPERATING REVENUE | | |
| Service Revenue | $ 24,752,800.00 | I/P Room and Board Private | $ 15,600,400.00 | |
| Deductions from Revenue | $ (17,662,544.00) | Deductions from Revenue | $ (12,750,670.00) | |
| | | | | |
| TOTAL OPERATING REVENUE | $ 7,090,256.00 | TOTAL OPERATING REVENUE | $ 2,849,730.00 | |
| | | | | |
| | | | | |
| OPERATING EXPENSES | | OPERATING EXPENSES | | |
| Salaries | $ 1,799,807.00 | Salaries | $ 2,150,000.00 | |
| Benefits | $ 440,234.00 | Benefits | $ 650,100.00 | |
| | | | | |
| Total Salaries and Benefits | $ 2,240,041.00 | Total Salaries and Benefits | $ 2,800,100.00 | |
| | | | | |
| Purchased Services | $ 104,270.00 | Purchased Services | $ 7,000.00 | |
| Supplies | $ 892,105.00 | Supplies | $ 521,421.00 | |
| Maintenance & Repairs | $ 1,600.00 | Maintenance & Repairs | $ 3,800.00 | |
| Rent, Util. & Telephone | $ 900.00 | Rent, Util. & Telephone | $ 17,071.00 | |
| Depreciation & Amortization | $ 200,000.00 | Depreciation & Amortization | $ 10,171.00 | |
| Bad Debt Expense | $ 250,000.00 | Bad Debt Expense | $ 150,000.00 | |
| Other Expenses | $ 4,558.00 | Other Expenses | $ 500.00 | |
| | | | | |
| TOTAL OPERATING EXPENSES | $ 3,693,474.00 | TOTAL OPERATING EXPENSES | $ 709,963.00 | |
| | | | | |
| OPERATING INCOME LOSS | $ 3,396,782.00 | OPERATING INCOME LOSS | $ (2,139,767.00) | |

FIGURE 14-9 **Net expense worksheet.** *Values are for instructional purposes only (*Source:* Adapted with permission of Central DuPage Hospital, Winfield, IL).

usually includes sick, vacation, personal, holiday, and education time. For example, Table 14-3 illustrates the average number of hours that a nurse at one institution may take off from work during a 12-month period. This time off may require coverage by another nurse, depending on the unit.

### Direct and Indirect Expenses

Expenses can be further broken down into direct and indirect. **Direct expenses** are those expenses directly associated with the patient, such as medical and surgical supplies and drugs. **Indirect expenses** are expenses for items such as utilities—gas, electric, and phones—that are not directly related to patient care. Other support functions frequently charged to a department that are not specifically related to patient care delivery are housekeeping, maintenance, materials management, and finance.

### Fixed and Variable Costs

**Fixed costs** are those expenses that are constant and are not related to productivity or volume. Examples of these costs are building and equipment depreciation, utilities, fringe benefits, and administrative salaries. **Variable costs** fluctuate depending upon the volume or census and types of

care required. Medical and surgical supplies, drugs, laundry, and food costs often increase with the volume. Figure 14-9 shows sample worksheets used to calculate expenses.

## BUDGET APPROVAL AND MONITORING

Once developed, budgets are submitted to administration for review and final approval. The approval process may take several months as the unit budgets are combined to determine the overall budget for the health care organization. Senior management, representing finance and operations, often makes the final decisions regarding acceptance of a budget.

The unit or department manager is responsible for controlling the budget. Budget monitoring is generally carried out on a monthly basis. The purpose of monitoring is to ensure that revenue is generated consistent with projected productivity and standards. Organizations often recognize a flexible budget that allows for adjustments if the volume or census increases or decreases. If the volume increases, it is likely that expenses will increase. If the volume decreases and expenses increase, then the manager needs to determine what actions are necessary to control or bring down costs.

| Monthly Variance Report* | | | | | | | | |
|---|---|---|---|---|---|---|---|---|
| | MTD Actual | MTD Budget | Variance | Variance from budget | YTD Actual | YTD Budget | Variance | Variance from budget |
| Volume - visits | 450 | 500 | –50 | –10% | 2905 | 2885 | 20 | 1% |
| Gross Revenue | $ 576,317.00 | $ 407,312.00 | $ 169,005.00 | 41% | $ 1,656,194 | $ 1,235,149 | $ 421,045 | 34% |
| Gross Revenue per UOS | $ 411.00 | $429.40 | $ (18.40) | –4% | $ 570.37 | $ 428.06 | $ 142.31 | 33% |
| Total Expense per UOS | $ 130.26 | $ 120.31 | $ (9.95) | –8% | $ 114.06 | $ 120.87 | $ 6.81 | 6% |
| Supply Expense per UOS | $ 13.74 | $ 23.94 | $ 10.20 | 74% | $ 12.93 | $ 23.93 | $ 11.00 | 85% |
| Total FTEs | 10.90 | 11.00 | 0.1 | 1% | 10.61 | 11 | 0.39 | 4% |

| | MTD Actual | Staffing Matrix (Fixed) Budget | Productivity | YTD Actual | Staffing Matrix (Fixed) Budget | Productivity | | |
|---|---|---|---|---|---|---|---|---|
| Paid Hours per UOS | 1.99 | 1.98 | 99.50% | 1.92 | 2 | 104.17% | | |
| Worked Hours per UOS | 1.75 | 1.79 | 102.29% | 1.7 | 1.81 | 106.47% | | |

| Finance / Action Steps |
|---|
| Volume is slightly down MTD (month to date), but still favorable to budget YTD (year to date). Total expenses are slightly above budget for this month due to changing process for ordering but these expenses are still in line YTD.  Plan: Although productivity is strong the department will continue to focus on reducing staffing when volumes are down. Anticipate additional efficiencies with implementation of electronic scheduling system. The Lab has insituted a new coding process which is anticipated to improve revenue capture. |
| **Patient Satisfaction / Action Steps** |
| Patient satisfaction overall quality of care rating for this month is at target however the quarter results are trending downward. The staff is focusing on total time spent in the department and have developed multiple tools to help patients understand how long it takes to complete a procedure. |
| **Quality Initiatives / Action Steps** |
| The staff is currently working on documentation of services and moderate sedation documentation requirements. Audits have been completed to ensure that the staff is in compliance with problem lists.  Action plans are being initiated to  educate staff so that compliance can be improved. |

\* Values are for instructional purposes only

**FIGURE 14-10** **Monthly variance report.** (Used with permission of Central DuPage Hospital, Winfield, IL).

Many organizations require managers to complete a budget variance report, which is a tool used to identify when categories are out of line and to identify the need for corrective action. Figure 14-10 illustrates a monthly variance report with finance, patient satisfaction, and quality initiative action steps.

The entire health care team is responsible for ensuring that expenses are kept within the budgeted amount and that the volume or census is maintained. The manner in which this is accomplished depends on the organization. Some institutions request that dashboards (see Figure 14-11) be developed reflecting departmental activity at a glance. Variance reports or dashboards may be posted so that all staff members have an opportunity to review them and participate in any improvement needed. Plans for corrective action based on the variance reports or dashboards are typically based upon understanding whether any particular variance resulted from revenue challenges or from supplies and labor expenses. The root cause of the variance needs to be identified so that effective action plans for improvement can be developed.

Staff can meet to discuss implementation or reinforcement of strategies that can positively affect the budget. Following are examples of such strategies:

- Analyze time efficiency of staff involved in patient care.
- Understand the process for entering patient charges.
- Educate coworkers regarding the charging process.
- Plan for supplies needed for every patient encounter and consciously eliminate unnecessary items.
- Learn how a department is reimbursed for services delivered, identifying covered and excluded expenses.
- Input charges in a timely manner.
- Discuss quality and cost differences in supplies with other staff and management.
- Evaluate staff and equipment downtime.
- Analyze cause of schedule delays, canceled cases, and extended procedure times.
- Explore new products with vendor representatives, and network with colleagues who have tried both new and modified products.
- Reduce the length of stay by troubleshooting early.
- Assist staff in organizational planning.
- Enhance productivity through rigorous process improvement.
- Post overtime and high/low productivity analysis.
- Explore how time and motion studies may increase efficiencies by identifying gaps or duplication in effort.

| Metric | Target Performance | 1-Jan | 1-Feb | 1-Mar | 1-Apr | 1-May | 1-Jun |
|---|---|---|---|---|---|---|---|
| **Efficiency** | | | | | | | |
| **GI Lab** | | | | | | | |
| FY 09 Volume | 64,904 | 5353 | 4917 | 5922 | 5798 | 6297 | 5540 |
| FY 10 Volume | 63,750 | 5540 | 5183 | 5434 | 5281 | 5670 | 5563 |
| **Medicine Unit** | | | | | | | |
| Bed request to bed Assigned (in Minutes) | 30 min | 40 | 38 | 38 | 37 | 31 | 20 |
| Bed Assigned to Pt Placement (in Minutes) | 30 min | 60 | 55 | 50 | 63 | 58 | 46 |
| **PatientSatisfaction - PRC Scores in Percentile Rank** | | | | | | | |
| **GI Lab** | | | | | | | |
| Overall Quality of Care | 85 | 99.5 | 88.5 | 92.2 | 63.9 | 92.9 | 92.2 |
| Discharge Calls - % Attempted | 90% | 81.0% | 77.0% | 79.1% | 74.8% | 75% | 67% |
| **Medicine Unit** | | | | | | | |
| Overall Quality of Care | 85 | 85.0 | 82.5 | 84.5 | 85.5 | 90 | 80.6 |
| Discharge Calls - % Attempted | 90% | 90.0% | 60.0% | 56.0% | 78.8% | 80% | 70% |
| **Regulatory Compliance & Quality** | | | | | | | |
| **GI Lab** | | | | | | | |
| Mis ID Specimens | 0 | 4 | 0 | 5 | 0 | 1 | 1 |
| Moderate Sedation Flow Sheet Documentation | 95% | 91% | 96% | 89% | 94% | 94% | 96% |
| Total Narcotic Discrepancies | 0 | 7 | 7 | 6 | 6 | 15 | 8 |
| Number of Discrepancies Resolved | 100% | 7 | 7 | 6 | 6 | 15 | 8 |
| Narcotic Counts | 100% | 100% | 90% | 86% | 80% | 82% | 100% |
| **Medicine Unit** | | | | | | | |
| PNI: (pneumonia core measures) - misses | 0 | 2 | 0 | 0 | 0 | 0 | 0 |
| AMI (acute myocardial infarction core measures) - misses | 0 | 1 | 0 | 2 | 1 | 0 | 1 |
| Stroke Tool Compliance | 100% | 75% | 83% | 44% | 86% | 88% | 100% |
| Hand Hygiene Compliance | 90% | 96% | 76% | 80% | 75% | 79% | 89% |

**FIGURE 14-11** Dashboard-Gastrointestinal laboratory and Medicine unit. (*Source:* Adapted with permission of Central DuPage Hospital, Winfield, IL) *Values are for instructional purposes only.

- Ensure that staff have the right tools and that the tools are ready when needed.
- Analyze patient supplies and review cost-per-patient encounter (e.g., chemotherapy administration, dialysis, insertion of indwelling or peripheral catheter).
- Track various steps in patient care that are time consuming or problematic for a unit (e.g., communication from front desk to recovery room, staff response to patient call lights, number of staff responding to an emergency code).
- Acquire a working knowledge of how a department/unit monitors financial and quality indicators, and participate in the development of action plans to increase patient satisfaction or to create the "best patient experience."

## KEY CONCEPTS

- Nurses play an integral role in the preparation, implementation, and evaluation of a unit or department budget.
- If nurses are not conscious of revenue and expenses, then deviation from financial performance will occur.
- Overall, organizational performance is dependent upon the insight and skills of staff members regarding patient care quality and financial outcomes.
- Hospitals use several types of budgets to help with future planning and management. These include operational, capital, and construction budgets.
- The budget preparation phase is one of data gathering related to a variety of elements that influence an organization, including demographic information, competitive analysis, regulatory influences, and strategic initiatives. Additionally, it is helpful to understand the department's scope of service, goals, and history.
- During the budget preparation phase, it is important to examine the individual nursing or hospital department or section thoroughly. Hospital systems are frequently divided into sections, departments, or units to compartmentalize them for organizational purposes. These subsections or units, commonly called cost centers, are used to track financial data.
- Organizations typically use history, or past performance, as a baseline of experience and data to better understand activity in a department or unit.
- Once background data have been gathered, the development of the budget can follow. This includes projecting revenue and expenses.
- Expenses are determined by identifying the cost associated with the delivery of service.
- Expenditures are resources used by an organization to deliver services. They may include labor, supplies, equipment, utilities, and miscellaneous items.
- Once developed, budgets are submitted to administration for review and final approval. The approval process may take several months as the unit budgets are combined to determine the overall budget for the health care organization.

## KEY TERMS

accounting
capital budget
construction budget
cost centers

dashboard
direct expenses
fixed costs

indirect expenses
operational budget
profit

revenue
variable costs
variance

## REVIEW QUESTIONS

1. An operational budget accounts for which of the following?
   A. The purchase of minor and major equipment
   B. Construction and renovation
   C. Income and expenses associated with daily activity within an organization
   D. Applications for new technology

2. Revenue can be generated through which of the following?
   A. Billable patient services
   B. Donations to service organizations
   C. Use of generic drugs
   D. Messenger and escort activities

3. Cost centers are used to do which of the following?
   A. Develop historical and demographic infor mation.
   B. Track expense line items.
   C. Plan for strategic growth and movement.
   D. Track financial data within a department or unit.

4. The purpose of monitoring a budget is to do which of the following?
   A. Keep expenses above budget.
   B. Maintain revenue above the previous year's budget.
   C. Ensure revenue is generated monthly.
   D. Generate revenue and control expenses within a projected framework.

5. Productivity can be measured by which of the following?
   A. Number of beds in a hospital
   B. Reimbursement rates for services rendered
   C. Past performance and history regarding revenue
   D. Volume of services delivered

6. Revenue and expenses are typically tracked using which of the following tools?
   A. Strategic planning
   B. Competitive analysis
   C. Operational budget
   D. Construction budget

7. Identify which of the following are indirect expenses.
   A. Salary of staff providing care
   B. Gas, electric, phones
   C. Medical supplies
   D. Monitoring equipment

8. Which of the following should be reviewed immediately after an operational budget has been put into place?
   A. Demographic information
   B. Regulatory influences
   C. Competitive analysis
   D. Budget monitoring

9. Examples of capital budget items are which of the following? Select all that apply.
   _____ A. Gauze and tape
   _____ B. Forceps
   _____ C. MRI scanner
   _____ D. Paper supplies
   _____ E. Mammogram equipment
   _____ F. Brain scanner

10. What types of indicators are typically displayed on dashboards in hospital settings? Select all that apply.
   _____ A. Patient satisfaction
   _____ B. Financial performance
   _____ C. Quality performance
   _____ D. Staff satisfaction
   _____ E. Patient outcomes

## REVIEW ACTIVITIES

1. Look around your clinical agency. Do you see any dashboard displays of quality measures? What do they reveal about your agency?

2. Using the tables in this chapter as a guideline, construct a competitive analysis of one or more of the agencies in your community. Note whether the agencies have a Web site at www.google.com.

3. Using the zero-based budgeting figure in this chapter, construct an analysis of one of the clinical procedures in your agency.

## EXPLORING THE WEB

- Go to the site for the Joint Commission. What information do you find there? www.jointcommission.org

- Review the site for the American Organization of Nurse Executives. Is the information helpful? www.aone.org

- Review these sites for helpful information. What do you find there?

   Advisory Board Company: www.advisory.com

Agency for Healthcare Research and Quality: www.ahrq.gov

American College of Healthcare Executives: www.ache.org

Centers for Medicare & Medicaid Services: www.cms.hhs.gov

Food and Drug Administration: www.fda.gov

Healthcare Financial Management Association: www.hfma.org

## REFERENCES

Advisory Board. (2009a). *Instilling cost discipline leading cost and operational excellence: The Advisory Board Company* (pp. 65–102). Washington, DC: Author.

Advisory Board. (2009b). *Weathering the Storm: Creating focus and Stability During Tumultuous Times: The Advisory Board Company* (pp. 59–68). Washington, DC: Author.

Frith, K., Anderson, F., Sewell, J. (2010). Assessing and selecting data for a nursing services dashboard. *Journal of Nursing Administration, 40*(1), 10–16.

Hendrich, A. L., & Lee, N. (2005). Intra-unit patient transports: Time, motion, and cost impact on hospital efficiency. *Nursing Economics, 23*(4), 157–164.

Jones, C. B. (2005). The costs of nurse turnover, part 2. *Journal of Nursing Administration, 35*(1), 41–49.

Needleman, J., Buerhaus, P., Mattke, S., Stewart, M., & Zelevinsky, K. (2002). Nurse-staffing levels and the quality of care in hospitals. *New England Journal of Medicine, 346*(22), 1714–1722.

## SUGGESTED READINGS

Duffield, C., Diers, D., Aisbett, C., Roche, M. (2009). Churn: Patient turnover and case mix. *Nursing Economics, 27*(3), 185–191.

Newhouse, R. (2010). Do we know how much the evidence based intervention cost? *Journal of Nursing Administration, 40*(7/8), 296–299.

O'Brien, Y., & Boat, P. (2009). Getting lean the grass roots way. *Nursing Management,* 28–33.

Joint Commission. (2010). Comprehensive accreditation manual for hospitals (CAMH): The official handbook. Oakbrook Terrace, IL: Joint Commission.

Wagner, C., Budreau, G., & Everett, L. (2005). Analyzing fluctuating unit census for timely staffing intervention. *Nursing Economics, 23*(2), 85–90.

# CHAPTER 15

# Effective Staffing

ANNE BERNAT, RN, MSN, CNAA; MARY L. FISHER, PhD, RN, CNAA, BC; BETH A. VOTTERO, PhD, RN, CNE

*Best practice staffing provides timely and effective patient care while providing a safe environment for both patients and staff, as well as promoting an atmosphere of professional nursing satisfaction.*

(CARL RAY, ET AL., 2003)

## OBJECTIVES

Upon completion of this chapter, the reader should be able to:

1. Calculate full-time equivalents (FTEs) needed to staff a typical inpatient nursing unit.

2. Analyze the impact of patient volume and work intensity on the demand for nursing care.

3. Discuss appropriate units of service used to measure nursing need by unit type.

4. Describe how informatics affects knowledge about staffing decisions.

5. Critique organizational, regulatory, staff, and patient dynamics underlying the development of a staffing plan.

6. Analyze scheduling issues that impact the matching of nursing resources to patient needs.

7. Compare and contrast models of care delivery and their impact on patient outcomes.

*You are a new nurse manager of a 30-bed medical unit that uses primary nursing as the care delivery model. You have 40 employees who work full and part time, with vacancies for eight additional full-time staff. The current schedule does not accommodate any 12-hour shifts. You have five long-term staff members who threaten to leave if they are forced to work 12-hour shifts. You have interviewed several new graduates who will come to work for you only if you offer them 12-hour shifts.*

*How can you accommodate the needs of all groups of staff?*

*What effect will the 12-hour shifts have on your care delivery model?*

The ability of a nurse to provide safe and effective care to a patient is dependent on many variables. These variables include the knowledge and experience of the staff, the severity of illness of the patients, patient dependency for daily activities, the complexity of care, the amount of nursing time available, the care delivery model, care management tools, and the organizational supports in place to facilitate care. This chapter will explore these factors and how they affect planning for staffing, scheduling staff, and patient outcomes associated with staffing factors. By the end of this chapter, you will understand how to plan staffing and measure the effectiveness of a staffing plan. You will also be able to critique the models of care delivery that are applicable to your environment and patient population.

# DETERMINATION OF STAFFING NEEDS

Historically, patient census was used to determine staffing needs. This resulted in fixed nurse-to-patient care ratios. This method of patient census staffing proved to be highly inaccurate because of the variability of patient care needs (Gran-Moravec & Hughes, 2005). The cost of nursing care and changes in reimbursement rates by Medicare-Medicaid and insurance providers has resulted in rethinking previous ways of making staffing decisions. For example, the American Nurses Association (ANA) supports taking into consideration the following variables when determining staffing: patient char-

acteristics, intensity of nursing care, context of the unit (size, layout, work flow, technology, arrangement of the unit, etc.) and staff expertise (Hyun, Bakken, Douglas, & Stone, 2008). In today's rapidly changing health care environment, many variables are considered in determining nurse staffing requirements. The effectiveness of the staffing pattern is only as good as the planning that goes into its preparation.

## CORE CONCEPTS

Gaining knowledge of the key terms—full-time equivalents (FTEs), productive time, nonproductive time, direct and indirect care, nursing workload, and units of service—is necessary for understanding staffing patterns.

### FTEs

A **full-time equivalent (FTE)** is a measure of the work commitment of an employee who works five days a week or 40 hours per week for 52 weeks a year. This amounts to 2,080 hours of work time (Figure 15-1).

A full-time employee who works 40 hours a week is referred to as a 1.0 FTE. An employee who works 36 hours (three 12-hour shifts) is considered full time for benefit purposes in many agencies but is assigned 0.9 FTE for budgeting purposes (36/40 = 0.9 FTE). A part-time employee who works five days in a two-week period is considered a 0.5 FTE. FTE calculation is used to mathematically describe how much an employee works (Figure 15-2). Understanding FTEs is essential when moving from a staffing plan to the actual number of staff required.

FTE hours are a total of all paid time. This includes worked time as well as nonworked time. Hours worked and available for patient care are designated as **productive hours**. Benefit time such as vacation, sick time, and

---

5 days per week  × 8 hours per day   = 40 hours per week
40 hours per week × 52 weeks per year = 2,080 hours per year

**FIGURE 15-1**  **Calculation of full-time equivalent hours.**

---

1.0 FTE = 40 hours or five 8-hours shifts per week
0.9 FTE = 36 hours or three 12-hour shifts per week
0.8 FTE = 32 hours or four 8-hour shifts per week
0.6 FTE = 24 hours or two 12-hour or three 8-hour shifts per week
0.4 FTE = 16 hours or two 8-hour shifts per week
0.3 FTE = 12 hours or one 12-hour shift per week
0.2 FTE = 8 hours or one 8-hour shift per week

**FIGURE 15-2**  **FTE calculation for varying levels of work commitment.**

| | |
|---|---|
| Vacation time | 15 days or 120 hours |
| Sick time | 5 days or 40 hours |
| Holiday time | 6 days or 48 hours |
| Education time | 3 days or 24 hours |
| Total nonproductive time | = 232 hours |

2,080 − 232 = 1,848 hours of productive work time available for each staff member with these benefits.

**FIGURE 15-3** Calculation of productive and nonproductive time.

| Unit Type | Unit of Service |
|---|---|
| Inpatient unit | Nursing hours per patient day (NHPPD) |
| Labor and delivery | Births |
| Operating room | Surgeries/procedures |
| Home care | Patient visits |
| Emergency services | Patient visits |

**FIGURE 15-4** Units of service—volume measures by unit type.

education time is considered **nonproductive hours.** When considering the number of FTEs needed to staff a unit, count only the productive hours available for each staff member, because this represents the amount of time required to meet patient needs. Available productive time can be easily calculated by subtracting benefit time from the time a full-time employee would work (Figure 15-3). These figures vary greatly depending on institutional policy and availability of human resource benefits. In this case, a full-time registered nurse (RN) would have 1,848 hours per year of productive time available to care for patients.

Employees who work with patients can be classified into two categories: those who provide direct care and those who provide indirect care. **Direct care** is time spent providing hands-on care to patients. **Indirect care** is time spent on activities that support patient care but are not done directly to the patient. Documentation, order entry, time consulting with people in other health care disciplines, and time spent following up on outstanding issues are good examples of indirect care. Even though RNs, licensed practical nurses (LPNs), and nursing assistive personnel (NAP) engage in indirect care activities, the majority of their time is spent providing direct care; therefore, they are classified as direct care providers. Nurse managers, clinical specialists, unit secretaries, and other support staff are considered indirect care providers, because the majority of their work is indirect in nature and supports the work of the direct care providers.

## UNITS OF SERVICE

Nursing workload is dependent on the nursing care needs of patients. Both patient volume and work intensity—that is, severity of illness, patient dependency for activities of daily living, complexity of care, and amount of time needed for care—contribute to nursing workload. **Units of service** include a variety of volume measures that are used to reflect different types of patient encounters as indicators of nursing workload (Figure 15-4). Volume measures are used in budget negotiations to project nursing needs of patients and to assure adequate resources for safe patient care.

The majority of nurses practice on inpatient units; therefore, further calculations for this chapter's examples

will be in nursing hours per patient day (NHPPD). **Nursing hours per patient day (NHPPD)** is the amount of nursing care required per patient in a 24-hour period and is usually based on midnight census and past unit needs, expected unit practice trends, national benchmarks, professional staffing standards, and budget negotiations. NHPPD reflects only productive nursing time needed. Calculation of NHPPD is displayed in Figure 15-5.

## NURSE INTENSITY

**Nurse intensity** is "a measure of the amount and complexity of nursing care needed by a patient" (Adomat & Hewison, 2004, p. 304). Nurse intensity is dependent on many factors that are difficult to measure: severity of illness, patient dependency for activities of daily living, complexity of care, and amount of time needed for care (Beglinger, 2006). It is vital that nurses measure nurse intensity, because staffing needs vary not only with the number of patients being cared for but also with the type of care provided for each of those patients (Unruh & Fottler, 2006).

Patient turnover affects nurse intensity. **Patient turnover** is a measure reflecting patient admission, transfer, and discharge, all of which entail RN-intensive procedures. As the health care industry pushes to reduce costs through shorter lengths of stay, these RN-intensive procedures related to patient turnover consume an increasing proportion of the hospital stay. As length of stay shortens, the intensity and need for NHPPD increases. What is not known at this time is the exact nature of this inverse relationship (Unruh & Fottler, 2006).

20 patients on the unit

5 staff × 3 shifts = 15 staff

15 staff each working 8 hours = 120 hours available in a 24-hour period

120 nursing hours ÷ 20 patients = 6.0 NHPPD

**FIGURE 15-5** Calculation of nursing hours per patient day (NHPPD).

# PATIENT CLASSIFICATION SYSTEMS

A **patient classification system (PCS)** is a measurement tool used to articulate the nursing workload for a specific patient or group of patients over a specific period of time. The measure of nursing workload that is generated for each patient is called the **patient acuity**. Classification data can be used to predict the amount of nursing time needed based on the patient's acuity. As a patient becomes sicker, the acuity level rises, meaning the patient requires more nursing care. As a patient acuity level decreases, the patient requires less nursing care. In most patient classification systems, each patient is classified once a day using weighted criteria that then predict the nursing care hours needed for the next 24 hours. Because patient care is dynamic, it is impossible to capture future patient care needs using a one-time measure (Gran-Moravec & Hughes, 2005). The criteria reflects care needed in bathing, mobilizing, eating, supervision, assessment, frequent observations, treatments, and so on. The ideal PCS produces a valid and reliable rating of individual patient care requirements, which are matched to the latest clinical technology and caregiver skill variables. These systems are generally applied to all inpatients in an organization. Other PCS systems exist to measure the workload associated with patient visits in the emergency department (ED) or in clinic environments, based on relative weights for visit lengths as well as complexity of care. There are two different types of classification systems: factor and prototype.

## Factor System

The factor classification system uses units of measure that equate to nursing time. Nursing tasks are assigned time or are weighted to reflect the amount of time needed to perform the task. These systems attempt to capture the cognitive functions of assessment, planning, intervention, and evaluation of patient outcomes along with written documentation processes. There are many factor systems that have been "home grown" or built for a specific organization. There are also many factor systems available for purchase on the open market. This is the most popular type of classification system because of its capability to project care needs for individual patients as well as patient groups. The time assigned or the weighted factor for different nursing activities can be changed over time to reflect the changing needs of the patients or hospital systems.

**Advantages and Disadvantages**    In the factor classification system, data are generally readily available to managers and staff for day-to-day operations. These data provide a base of information against which one can justify changes in staffing requirements. A disadvantage to this system type is the ongoing workload for the nurse in classifying patients every day. There are also documented problems with "classification creep," whereby acuity levels rise as a result of misuse

of classification criteria. These systems do not holistically capture the patient's needs for psychosocial, environmental, and health management support. And finally, these systems calculate nursing time needed based on a typical nurse. A novice nurse may take longer to perform activities than the average nurse. Recommended nursing time needed may differ from the actual time needed based on the expertise of the staff.

## Prototype System

The prototype classification system allocates nursing time to large patient groups based on an average of similar patients. For example, specific **diagnostic-related groups (DRGs)** have been used as groupings of patients to which a nursing acuity is assigned based on past organizational experience. DRGs are patient groupings established by the federal government for reimbursement purposes. DRGs are sorted by patient disease or condition. This model assumes that, on average, it will reflect the nursing care required and provided. The data are then used by hospitals in determining the cost of nursing care and negotiating contracts with payers for specific patient populations.

**Advantages and Disadvantages**    The distinct advantage of this classification system is the reduction of work for the nurse, because daily classification is not needed. A major disadvantage of this system is that DRGs do not accurately reflect patients' nursing needs, because medical diagnosis alone does not adjust for variances in patients' self-care ability and severity of illness. There is no ongoing measure of the actual nursing work required by individual patients. There are also no ongoing data to monitor the accuracy of the preassigned nursing care requirements. The prototype system is much less common than the factor system.

## Utilization of Classification System Data

Patient classification data are valuable sources of information for all levels of the organization. On a day-to-day basis, acuity data can be utilized by staff and managers in planning nurse staffing over the next 24 hours. Acuity data and NHPPD are concrete data parameters that are used to educate staff on how to adjust staffing levels. Experienced staff have the knowledge to manage staffing to acuity given the information, boundaries, and authority to do so. In many organizations, a central staffing office monitors the census and acuity on all units and deploys nursing resources to the areas in most need, using the classification system data and recommended staffing levels. The manager reviews the results of staffing over the past 24 to 48 hours to adjust staffing performance to patient requirements. At the unit level, acuity data are also essential in preparing month-end justification for variances in staff utilization. If your average acuity has risen, then there should be an expected rise in NHPPD to accommodate the increased patient needs.

At an organization level, acuity data have been used to cost out nursing services for specific patient populations and global patient types. This information is also very helpful in negotiating payment rates with third-party payers such as insurance companies to ensure that reimbursement reflects nursing costs. In most organizations, the classification or acuity data are also used in preparation of the nursing staffing budget for the upcoming fiscal year. The data can be benchmarked with other organizations to lend credence to any efforts to change nursing hours. Finally, patient acuity data and NHPPD can be used to develop a staffing pattern. Patient classification and NHPPD data provide an enormous amount of information that serves a multitude of needs.

# CONSIDERATIONS IN DEVELOPING A STAFFING PLAN

Developing a staffing plan is a science and an art. The following sections will consider other areas in addition to the acuity data and NHPPD just discussed. Each of these areas should be reviewed and the findings incorporated into development of the staffing plan.

## Benchmarking

**Benchmarking** is the continuous measurement of a process, product, or service compared to that of the toughest competitor, those considered industry leaders, or similar activities in the organization in order to find and implement ways to improve the product, process, or service (Joint Commission, 2006). Often, benchmarking data provide only comparable units of service performance and do not reflect quality of care indicators that can link quality patient care outcomes to productivity measures. In developing a staffing pattern that leads to a budget, it is important to benchmark your planned NHPPD against other organizations with similar patient populations as part of evidence-based decision making (EBD-M). Purchased patient classification systems often offer acuity and NHPPD benchmarking data from around the country as part of their system. This kind of data can be helpful in establishing a starting point for a staffing pattern or as part of justification for increasing or reducing nursing hours. Caution must be used, however, because each organization has varying levels of support in place for the nurse at the unit level. For example, a nursing unit that has dietary aides from the dietary department distribute and pick up meal trays would need less nursing time than a unit that had no external support for this activity. Practice differences such as these contribute significantly to differences in hours of care from one organization to another.

## Regulatory Requirements

Historically, few regulatory requirements have prescribed nurse staffing. This changed when the Health Services Research Administration (HRSA) released a study on over five million patient discharges. The study found a consistently strong relationship between nurse staffing and patient care outcomes (Needleman, et al., 2002). This landmark study's findings supported changes in staffing models at the national, state, and local hospital levels.

The 42 *Code of Federal Regulations* (Centers for Medicare & Medicaid Services. (2007) requires Medicare-certified hospitals to have adequate numbers of direct care providers, leaving it up to the states to determine appropriate staffing. Three methods are used to comply with the Code, i.e., placing accountability on the hospital, mandating ratios through legislation, and requiring public reporting of staffing levels. To date, 15 states have enacted legislation to address nurse staffing. Seven states require staffing committees (Connecticut, Illinois, Nevada, Ohio, Oregon, Texas, and Washington), one state (California) stipulates the minimum nurse-to-patient ratio, and five states (Illinois, New Jersey, New York, Rhode Island, Vermont) require mandatory public reporting of staffing levels. Although a start, it is not a panacea for appropriate staffing. The American Nurses Association (ANA, 2010) promotes a legislative model that empowers nurses to create staffing plans specific for their unit, patient population, and nursing skill mix. In June 2010, the ANA introduced federal regulation, the Registered Nurse Safe Staffing Bill, whereby hospital staffing committees must be comprised of at least 55% direct care nurses (ANA, 2010).

The Joint Commission (JC) surveys hospitals on the quality of care provided. The JC does not mandate staffing levels but does assess staffing effectiveness. JC standards on staffing require organizations to monitor a correlation between two clinical and two human resource indicators—for example, NHPPD against hospital-acquired pressure ulcer occurrence or NHPPD against falls. Findings must then be used to adjust staffing levels.

## Skill Mix

Skill mix is another critical element in nurse staffing. **Skill mix** is the percentage of RN staff compared to other direct care staff (LPNs). For example, in a unit that has 40 FTEs budgeted, with 20 of them being RNs and 20 FTEs of other skill types, the RN skill mix would be 50%. If the unit had 40 FTEs, with 30 of them being RNs, the RN skill mix would be 75%. The skill mix of a unit should vary according to the care that is required and the care delivery model utilized. For example, in a critical care unit, the RN skill mix will be much higher than in a nursing home where the skills of an RN are required to a much lesser degree. It is important to note that RN hours of care are more costly than those of lesser-skilled workers, but there is evidence that RNs are a very productive and efficient type of labor. It is essential to critically evaluate patient care requirements and identify who can perform necessary functions. For instance, if many patients require feeding, NAP may be most appropriate. As you consider skill mix, however, you need to clearly understand the activities in which each level of staff can engage within the scope of practice in your state. In some states, NAP may catheterize patients if they have received training and are competent.

## CASE STUDY 15-1

You are the manager of a 16-bed intensive care unit (ICU) with a nurse-to-patient ratio of one nurse to every two patients. At this time, the unit is full with 16 patients. The nurses identify four patients who have improved enough for transfer to a medical-surgical unit. The ratio of nurse to patient on the medical-surgical unit is one nurse to four patients. The hospital census (number of patients in the hospital) is high and no medical-surgical beds are available for the ICU patient transfers. Because of this, the patients will have to stay in ICU for the next two shifts.

In planning staffing for the next two shifts for your ICU unit, calculate what the budgeted NHPPD is for your unit.

What is the NHPPD for the medical-surgical unit where the nurse-to-patient ratio is 1:4?

You need to adjust staffing based on the patient information. What other factors should be considered?

What is your plan for staffing the next two shifts?

How will you communicate the plan to the staff?

---

In other states, NAP may not perform this function regardless of their training and expertise. RNs are always required for patient assessment, monitoring changes in patient status, evaluation, teaching, and patient treatments requiring judgment.

## Staff Support

Another important factor to consider in developing a staffing pattern is the supports in place for the operations of the unit or department. For instance, does your organization have a systematic process to deliver medications to the department, or do unit personnel have to pick up patient medications and narcotics? Does your organization have staff to transport patients to and from ancillary departments? The less support available to your staff, the more nursing hours have to be built into the staffing pattern to provide care to patients. Nursing areas such as critical care that have a significant amount of equipment to track and supply may benefit greatly from adding a materials coordinator. This kind of support for staff allows staff to spend their precious available time with patients rather than looking for equipment or supplies. An additional important unit-based need is secretarial support. If the unit has admissions, discharges, and transfers, it makes sense to provide unit secretarial support for the peak periods of the day. In ICUs, unit secretaries are commonly scheduled around the clock to provide support for the unit staff as well as for other disciplines.

## Historical Information

As you consider the many variables that affect staffing, it helps to ask the following questions: What has worked in the past? Were the staff able to provide the care that was needed? How many patients were cared for? What kind of patients were they? How many staff were utilized, and what kind of staff were they? This kind of information can help to identify operational issues that would not be apparent otherwise. For example, in an older part of a facility, there may not be a pneumatic delivery tube system, a system that is available in most other parts of the facility. Because it is generally available, you may overlook its absence, but the lack of a tube system means that a significant amount of time will be required to collect needed items, affecting the staffing plan you develop.

## ESTABLISHING A STAFFING PLAN

A **staffing plan** articulates how many and what kind of staff are needed by shift and day to staff a unit or department. There are basically two ways of developing a staffing plan. It can be generated by determining the required ratio of staff to patients; nursing hours and total FTEs are then calculated. It can also be generated by determining the nursing care hours needed for a specific patient or patients and then generating the FTEs and staff-to-patient ratio needed to provide that care. In most cases, you would use a combination of methods to validate your staffing plan. We will start with development of a plan from the staff-to-patient ratio.

### Inpatient Unit

An **inpatient unit** is a hospital unit that provides care to patients 24 hours a day, seven days a week. Establishing a staffing pattern for this kind of unit utilizes all the data discussed in the previous areas. Using data from all your sources, you can build a staffing plan that you believe will meet the needs of the patients, the staff, and the organization. To illustrate the concept of calculating a staffing plan, we will use a typical medical unit with 24 beds and an average daily census (ADC) of 20. **Average daily census (ADC)** is calculated by taking the total numbers of patients at census time, usually midnight,

**Scenario: A 24-bed medical unit where the ADC is 20 and NHPPD is budgeted at 8.**

*Step 1:*

Formula: Number of patients $\times$ NHPPD / care hours per day = shifts needed per 24 hours staff productive time per shift

Example: $20 \times 8 = \dfrac{160 \text{ care hours per 24 hour day}}{8} = 20$ 8-hour shifts needed per 24 hours

*Step 2:*

Allocate 20 staff to the unit by shift and skill mix

|  | % of Staff per Shift | # of Staff | RN | Tech/Unit Clerk |
|---|---|---|---|---|
| Days | 40 | 8 | 4 | 4 |
| Evenings | 35 | 7 | 4 | 3 |
| Nights | 25 | 5 | 4 | 1 |
|  |  |  | 12 | 8  (Total 20) |

*Step 3:*

Calculate FTE to cover staff days off. (Calculations are in parentheses in the matrix below.)

Formula: $\dfrac{\text{number of staff needed per shift} \times \text{Days of needed coverage}}{\text{Number of shifts each FTE works}}$

Example: $\dfrac{4 \times 7}{5} = 5.6$

|  | % of Staff per Shift | # of Staff | RN | Tech/Unit Clerk |
|---|---|---|---|---|
| Days | 40 | 8 | 4 (5.6) | 4 (5.6) |
| Evenings | 35 | 7 | 4 (5.6) | 3 (4.2) |
| Nights | 25 | 5 | 4 (5.6) | 1 (1.4) |
|  |  |  | 12 (16.8) | 8 (11.2)  = 20 (28) |

*Step 4:*

Provide coverage for benefit time off.

Formula: $\dfrac{\text{Productive hours}}{\text{budgeted nonproductive hours}}$ = percent of nonproductive hours $\times$ total FTE = additional FTE needed to cover benefits;

Productive + nonproductive FTE to cover each week = Grand Total FTE

Example: $2080/232 = 0.11 \times 28 = 3.08$;

$3.08 + 28 = 31.08$ Grand Total FTE

FIGURE 15-6  **Staffing plan template for an inpatient unit.**

over a period of time—for example, weekly, monthly, or yearly—and dividing by the number of days in the time period. Many institutions budget their staffing based on ADC and then adjust for patient census and acuity changes. Utilizing the staffing plan (Figure 15-6), plot out the number and type of staff needed during the week and on weekends for 24 hours a day for the number of patients on a clinical unit.

*Step 1:* To develop a staffing plan using NHPPD, you would start with a target NHPPD. If your target NHPPD were 8, for example, and you expected to have 20 patients on your 24-bed unit, you would multiply 8 NHPPD times 20 patients to get 160 productive hours needed every day. Dividing 160 by 8-hour shifts worked by each staff member gives you 20 staff members needed per day.

*Step 2:* Now that you know how many 8-hour shifts are needed in 24 hours, the next step is to allocate staff to the plan, based on how care is delivered on the unit. To fully understand the complexity of decisions involved in this

step, review the Models of Care Delivery section later in this chapter. Allocating FTEs cannot be separated from an intelligent understanding of how patient care is delivered and how to use the right mix of staff to accomplish that patient care. In our example, the medical unit uses team nursing to provide care.

Taking the 20 shifts that are needed for patient care in our example, the nurse manager must determine the mix of staff, the weighting of staff per shift, and whether this changes by day of the week. The census on this unit does not lower on the weekends, so we calculate the staffing for the entire week. If the unit had reduced census on specific days of the week, those days would be calculated separately. The result of step 2 is a snapshot of the staffing plan for 24 hours (see Figure 15-6).

You have now determined a staffing plan for your unit. The staffing plan calculates the number of FTEs needed per day.

*Step 3:* You must now calculate the amount of additional staff that will be needed to provide for days off. Direct caregivers will need to be replaced, but some other support staff may not need to be replaced for days off or benefited time off.

Managers typically are not replaced on days off. The formula for calculating coverage for staff days off is the number of staff needed per shift multiplied by the number of days of needed coverage, usually seven, divided by the number of shifts each FTE works per week. For each 8-hour staff member needed in the daily plan, the unit must hire 1.4 FTEs ($1 \times 7/5 = 1.4$). In a 12-hour staffing model, the manager must hire 2.3 FTEs for each 12-hour staff member needed in the daily plan ($1 \times 7/3 = 2.3$). This is true because each 12-hour staff member works only three days per week.

In our example for step 3, the calculations for seven-day coverage are in parentheses in the staffing plan. To have four RNs on day shift, for example, you need to hire 5.6 FTEs. In total, you need to hire 28 FTEs to have 20 staff working seven days per week.

*Step 4:* The next step is to provide additional FTEs for coverage for benefited time away from work. This includes vacations, educational time, orientation time, and so on. The amount of time away from work varies by organization. If every employee receives the benefits outlined in Figure 15-3, 232 benefit hours are needed per person. We then need to determine what percent of each FTE's productive time the benefits represent. In our example, 232 is 11% of 2,080. We then take our total FTEs from step 3 and multiply by 0.11 ($28 \times 0.11 = 3.08$). Thus, we need an additional 3.08 FTEs to work when other staff are taking benefit time. This figure is then added to our total from step 3 to get our grand total FTEs for the budget ($28 + 3.08 = 31.08$).

# SCHEDULING

Scheduling of staff requires attention to placing the appropriate staff on each day and shift for safe, effective patient care. There are many issues to consider as you schedule staff: the patient type and acuity, the number of patients, the experience of your staff, and the supports available to the staff. The combination of these factors should guide the number of staff scheduled on each day and shift. These factors must be reviewed on an ongoing basis as patient types and patient acuity drive different patient needs and staff expertise.

## PATIENT NEED

Patient classification systems do not tell you when the nursing activity will take place over the next 24 hours. In addition to planning for the acuity of the patients, the staffing plan must support having staff working when the work needs to be done. A good example of this would be an oncology unit in which chemotherapy and blood transfusions typically occur on the evening shift. In this scenario, staffing in the evening may need to be higher than for other shifts, to support these nurse-intensive activities. As patient types

change, so do patients' needs and staffing requirements. Adding a population of step-down patients from the ICU would likely require additional FTEs on a medical-surgical unit. Anytime patient populations change, staffing and NHPPD should be assessed. A general rule is that the higher the patient acuity, the more consistent the staffing needs are across shifts. A critical care unit is continuously monitoring patients around the clock, whereas a surgical unit has activities concentrated before and after surgeries each day, with somewhat less activity on the late evening and night shifts.

## EXPERIENCE AND SCHEDULING OF STAFF

Each nurse differs regarding knowledge base, experience level, and critical thinking skills. A novice nurse takes longer to accomplish the same task than does an experienced nurse. An experienced RN can handle more in terms of workload and acuity of patients. If your area requires special staff skills or competencies, you would also want to plan so that staff with the special skills are scheduled when the patient care need may arise. Remember, the underlying principle of good staffing is that those you serve come first. This may dictate some undesirable shifts, but your responsibility is to ensure that there are appropriate numbers and kinds of staff on hand to care for the patients you serve. Staff are plotted out across a staffing sheet (Figure 15-7).

Staff members should be scheduled for the number of days for which they are committed: five days a week for a full-time, eight-hour employee and less for part-time employees, as determined by their hiring commitment. When staff are hired, there is an agreement between the manager and the employee as to the shift, schedule, and work commitment. If the unit workload is consistent, the scheduled days should be assigned so that there are equal numbers of staff available through the week. Typically, the spread of FTEs across the 24-hour period falls within the following guidelines: days 33% to 50%, evenings 30% to 40%, and nights 20% to 33%. The spread should be based on patient need.

## Shift Variations

To attract and retain employees, organizations offer both traditional and flexible schedules to meet organizational and employee needs.

**Traditional Staffing Plans** Traditional staffing plans are generally eight-hour shifts, 7 a.m. to 3:30 p.m., 3 p.m. to 11:30 p.m., and 11 p.m. to 7:30 a.m. A full-time employee works 10 eight-hour shifts in a two-week period. The start time of eight-hour shifts may vary by organization or by nursing unit and patient need. For example, EDs are typically busiest during the evening into the night hours. An eight-hour shift for the ED may be 7 p.m. to 3 a.m. to cover the peak activity times. After you have determined what numbers of staff are necessary, it is

## REAL WORLD INTERVIEW

Given the need for staffing and financial accountability, I used spreadsheet software to improve the development of staffing patterns in our facility. We had been using a pencil and paper template for managers to use to develop staffing patterns. This manual template concentrated on the weekday staffing needs and applied an overall factor to calculate weekend and benefit time. The FTE number provided did not address orientation and education needs for any of the staff or benefit needs for the weekend staff. Although these staffing patterns were used to project the number of FTEs needed and the distribution of employees to staff the nursing unit, they were not used to drive the budgeted quota for the unit.

Using a computer software program, I developed a spreadsheet template the managers could use to accurately project FTEs needed to meet the staffing pattern. This computerized approach allows for weekday and weekend staffing to be considered independently, allowing for any differences in census or direct NHPPD. A benefit time factor, tailored to our organization's specific benefit package for each skill level, was used to calculate the number of FTEs needed to staff for benefit time. Benefit time was now calculated for weekday and weekend staffing coverage for a 24/7 operation. Additionally, an orientation and education factor is used to calculate the FTEs needed to provide coverage. For the first time, benefited time off and orientation time were built into each unit's staffing pattern. Additionally, direct and indirect NHPPD are automatically calculated as the staffing pattern is changed, and the calculated FTE needs can be compared to the current budgeted quota for variances. I also worked with Finance to use this template as the basis for a budgeted quota sheet, which is used during the budget process for determining the unit quota for the next year, a quota that now includes benefited time off and orientation time.

One of the biggest assists has been the ability of the nurse managers to use the template for what-if scenarios. When they are planning for a census or patient program change, FTE needs can be quickly calculated and compared to their current budgeted quota. This tool has become part of our business planning process.

Overall, this template has been accepted as a valid management tool, has standardized inclusion of nonproductive time into FTE budgets, and has given managers a simple tool to develop new staffing patterns. It has also helped in raising the accountability of managers, in that it helps them develop workable staffing patterns for which they can be held accountable.

**Barbara Leafer, RN, BS**
Fiscal Administrator for Patient Care
Albany, New York

---

important to attempt to schedule staff in a way that meets their needs. Some prefer to work long stretches to have several days off in a row. Others prefer to work short stretches.

**New Options in Staffing Plans**   In recent years, new options in scheduling have emerged that take into consideration both the health care worker and patient needs. Twelve-hour shifts have become very popular across the country. In many organizations, employees can work 36 hours per week and get full-time benefits. In this situation, a nurse could work three 12-hour shifts per week, have four days off, and be full time. Another popular option is weekend programs. Weekend program staff work two 12-hour shifts every weekend and are paid a rate that make the 24 hours of work equal to 40 hours of work during the week. Some of these programs include full-time benefits as well. The purpose of this kind of program is to improve weekend staffing and allow full-time staff members who usually work 26 weekends a year to work fewer weekends, for staff retention purposes.

**Impact on Patient Care**   Any time you implement a scheduling plan, it is critical to assess what the effect will be on the care of patients. For example, workweeks made up of three 12-hour shifts have in many units disrupted continuity of care. **Continuity of care** is generally defined as the follow-through in patient care that is inherent in having the same nurse return to care for patients in subsequent shifts on sequential days of the week. Disruptions in continuity of care are especially the case when 12-hour shifts are not scheduled on sequential days. To mediate this impact, 12-hour staff

| | Monday 04 | Tuesday 05 | Wednesday 06 | Thursday 07 | Friday 08 | Saturday 09 | Sunday 10 | Monday 11 | Tuesday 12 | Wednesday 13 | Thursday 14 | Friday 15 | Saturday 16 | Sunday 17 |
|---|---|---|---|---|---|---|---|---|---|---|---|---|---|---|
| **Melinda** | D | | D | D | D | D | D | | | D | D | D | | |
| **Carlos** | | 8.00 1900 | | | N | N | N | D | 8.00 1900 | | D | | | |
| **Tabitha** | 12.00 0900 | | 12.00 0900 | | D | 12.00 0900 | 12.00 0900 | 12.00 0900 | | 12.00 0900 | | N | | |
| **Maria** | D | 8.00 1100 | 13.00 2400 | E | E | E | | vac | | 8.00 1100 | E | E | E | |
| **Barbara** | | 14.00 2400 | 13.00 2400 | 13.00 2400 | D | | | | 14.00 2400 | 13.00 2400 | 13.00 2400 | | D | D |
| **Nirmala** | D | D | D | D | | E | D | | D | N | N | | E | |
| **Robert** | N | N | N | N | | | N | N | N | N | N | | | N |
| **Jacqueline** | E | E | E | | E | | E | E | E | E | E | E | E | E |
| **Irma** | D | D | D | D | D | D | D | E | D | | | D | | |
| **Sara** | E | E | E | E | 8.00 0800 | N | E | E | E | E | E | 8.00 0800 | | E |
| **Gary** | | | | | | | | E | E | | | 12.00 1500 | | |
| **Cynthia** | N | N | N | N | N | P | P | N | N | P | | P | N | |
| **Jose** | 8.00 0730 | 8.00 0730 | 8.00 0730 | 8.00 0730 | 8.00 0730 | | | 8.00 0730 | 8.00 0730 | | 8.00 0730 | 8.00 0730 | | |

The first number in a square is the number of hours scheduled, the second number is the shift start time in military time.

Standard Work Assignments

D 0700–1500
E 1500–2300
N 2300–0700
A 0700–1900
P 1900–0700

**Figure 15-7** Excerpt from the schedule for an emergency department, showing great variation in shift design.

# EVIDENCE FROM THE LITERATURE

**Citation:** Kane, R. L., Shamliyan, T. A., Mueller, C., Duval, S., & Wilt, T. J. (2007). The association of registered nurse staffing levels and patient outcomes: Systematic review and meta-analysis. *Medical Care, 45*(12), 1195–1204.

**Discussion:** A total of 2,858 studies were considered for this review, with 96 studies meeting the inclusion criteria. Findings suggested that an increase of one RN FTE per patient day was associated with a 9% reduction in ICU deaths and a 6% reduction in medical unit deaths. In addition, consistent and significant reductions in patient adverse events such as a 30% reduction in hospital-acquired pneumonia in ICU patients, a 16% reduction in failure to rescue in surgical patients, and a 49% reduction in nosocomial infections were also identified. On the other hand, one additional patient per shift resulted in an 8% increase in hospital-related mortality across the health care setting. Of interest, "... nurse-sensitive adverse events including falls, pressure ulcers, and urinary tract infections did not demonstrate a consistent association with staffing levels" (p. 1201). Causal relationships between outcomes and RN experience, competence, knowledge and education level, work environment, quality of medical care, collaboration practices, nurse retention, and job satisfaction were identified but not measured in this study.

The review found that fixed minimum RN-to-patient ratios did not always realize patient safety benefits, primarily because individual patient needs and acuity were not taken into consideration. Patient acuity–based staffing was found to be used inconsistently in the studies, largely due to the varying types and use of classification systems. The effect of public reporting of staffing on quality of care is not known at this time; there is a need for more research in this area. Pay for performance is promoted to encourage quality, but until more research is conducted, its effects are not well understood.

**Implications for Practice:** There is a strong association between RN staffing and hospital-related mortality and other outcomes. Study findings encourage health care organizations to critically review current RN staffing, mindful of all causal relationships between patient outcomes and nurse staffing.

can be paired so that the patient has the same pair of nurses every day for three days, and then the patient can be transitioned to a new pair of 12-hour staff. Units that have short patient lengths of stay may have fewer continuity problems than units with longer lengths of stay. The number of staff shift hand-off reports per day also affects continuity of care. A **hand off** occurs any time the nurse caring for a group of patients reports off to the nurse on an oncoming shift. Such shift hand-off reports are opportunities for missed communication and errors in patient care. In 8-hour shifts, there are three shift hand offs per 24 hours, whereas in 12-hour shifts, there are only two shift hand offs. This is another type of continuity of care issue that must be balanced with the number

of shifts per week that each nurse works. When implementing staffing plans, you must ensure that there are always staff members scheduled who are familiar with the patients and the events that have transpired previously.

**Financial Implications**   New staffing plans or program changes may have significant financial implications. A number of new programs are being put into place to recruit staff and encourage staff to work more hours. Weekend programs are more expensive than traditional staffing plans because of the higher rate of hourly pay, but they are a recruitment and retention tool for nursing leadership. For example, note the financial impact of a weekend option program (Figure 15-8).

Weekend staff working at $42 an hour × 24 hours = $1,008 per weekend
Regular staff working at $25 an hour × 24 hours =   $600 per weekend
Difference in cost =   $408 per weekend option FTE

Six weekend option staff members at $1,008 would cost $2,448 more than regular staff per weekend; $2,448 × 52 weekends a year would cost $127,296 more than regular staff annually.

**FIGURE 15-8**  Annual cost of a weekend option program for one nursing unit.

## REAL WORLD INTERVIEW

As a manager of an intensive care unit, I can say that self-scheduling has greatly increased my staff's satisfaction with their schedules. I think the biggest factor in the success of our process was the initial buy-in from the staff. Before implementing, staff were surveyed to assess their commitment to making the process work. I was looking for 60% to 75% staff buy-in before implementation and found greater than 70%. A second critical factor was having clear guidelines for the process. These included time lines for how and when staff can sign up for time and how time off is prioritized.

During implementation, we learned many things. One key factor was that staff needed to have confrontation and negotiation skills in order for this process to work. Inevitably there were situations when someone had to change their schedule. When confrontation and negotiation didn't take place, there were periods of short staffing and patient care needs not being met. We also learned that this is a time-consuming process. It takes about 16 hours per month for the self-scheduling committee to put the schedule together.

Another key element I found was the manager had to maintain accountability for staffing. I meet with the scheduling committee regularly and oversee the orientation of new staff to the self-scheduling process. I sign off on every schedule to ensure that the schedule maintains appropriate staffing levels at all times. I found that I needed to identify trends that may be affecting staffing and assist the staff in addressing the trends. I also work with the staff on the implementation of any new program that affects the schedule. The weekend program is a good example of this. I worked with the staff to ensure there were appropriate guidelines for staff receiving a reduced weekend commitment. And finally, the most important role I play is to be very clear about the expectations for all—the committee, the staff, and myself. This scheduling process has been one of the most positive quality of work life efforts for my staff.

**Rob Rose, BSN, MSN**
Nurse Manager, Cardiopulmonary Surgery Intensive Care Unit
Albany, New York

Reduced turnover of nurses must also be considered in evaluating the financial impact of a weekend option program.

To implement a similar program or other new programs, collaboration with the finance and human resources departments of the organization is necessary. This collaboration must be used to develop a financial analysis to measure the dollar and human resource impact of the program.

## Self-Scheduling

**Self-scheduling** is a process in which unit staff take leadership in creating and monitoring the work schedule while working within defined guidelines. Often, there is a staffing committee that is part of unit-shared governance, which is a unit model where staff manage professional practice through unit committees. Increasing staff control over scheduling is a major factor in nurse job satisfaction and retention and has been associated with reductions in sick time usage. The nurse manager retains an important role in self-scheduling through mentoring, providing open communication, and holding everyone to equal expectations.

**Boundaries of Self-Scheduling**    To implement self-scheduling, responsibilities and boundaries need to be established that clearly state expectations of staff. This is best done by a unit committee, made up of staff, that reports to the nurse manager. It is important to spell out the roles and responsibilities of all—the unit-based committee, the chairperson (if there is one), the staff, and the manager. Generic boundaries need to be established regarding fairness, fiscal responsibility, evaluation of self-scheduling, and the approval process. Table 15-1 spells out specific issues that must be addressed. During the self-scheduling process, the unit staff should be included and educated about the guidelines as they are being developed. For this process to be successful, all staff members must understand the process, their responsibilities, and the effect of their decisions on staffing. All personnel must also be committed to providing safe staffing on all shifts for their patients.

**TABLE 15-1 Issues to Be Spelled Out in Self-Scheduling Guidelines**

- *Scheduling period:* Is the scheduling period two-, four-, or six-week intervals?

- *Schedule timeline:* What are the time frames for staff to sign up for regular work commitment, special requests, overtime, and per diem workers?

- *Staffing pattern:* Will eight- or twelve-hour shifts be used? A combination?

- *Weekends:* Are staff expected to work every other weekend? If there are extra weekends available, how are they distributed?

- *Holidays:* How are they allocated?

- *Vacation time:* Are there restrictions on the amount of vacation during certain periods?

- *Unit vacation practices:* How many staff from one shift can be on vacation at any time?

- *Requests for time off:* What is the process for requesting time off?

- *Short-staffed shifts:* How are short-staffed shifts handled?

- *On call, if applicable:* How do staff get assigned to or sign up for on-call time?

- *Cancellation guidelines:* How and when do staff get canceled for scheduled time if they are not needed?

- *Sick calls:* What are the expectations for calling in sick, and how are these shifts covered?

- *Military/National Guard leave:* What kind of advance notice is required?

- *Schedule changes:* What is the process for changing one's schedule after the schedule has been approved?

- *Shifts defined:* What are the beginnings and endings of available shifts?

- *Committee time:* When does the self-scheduling committee meet and for how long?

- *Seniority:* How does it play into staffing and request decisions?

- *Staffing plan for crisis/emergency situations:* What is the plan when staffing is inadequate?

## Informatics Support for Evidence-Based Staffing

Information technology is poised to provide innovative solutions to support evidence-based nurse staffing. Hospitals currently capture data electronically from a variety of sources, i.e., admissions, discharges, transfers, bar-coding, call light timing, computerized charting of risk assessments, nurse profiles, and so forth. Some organizations even use infrared signals to identify the nurses location on the unit. When all of this data is combined, it provides meaningful information on nursing care needs, supply of qualified nurses, and quality of care outcomes (Hyun, Bakken, Douglas, & Stone, 2008).

Similar to a puzzle, pieces of data can be found throughout a hospital. Separately, they provide small glimpses of the picture. To see the entire picture, it requires a program that is able to take each piece of data and put it together to create meaning.

Standardizing how computer programs collect and represent data allows different computer programs to work together. For example, data is collected and reported from a variety of sources such as pagers, laboratory, nurse charting, glucometers, call light timing, and medication bar-code and administration, etc. The ability of each system to interact with the other and provide data on real-time care provision by nurses provides information on the actual workload of nurses.

Think of the vast amount of data continuously flowing on a clinical unit. The ability to capture and consistently gather data regarding real-time staffing needs is critical. Computer systems can pull the pieces of data together in real time and combine it with historical trends to help inform staffing decisions.

For example, a medical-surgical unit experiences multiple patient admissions, discharges, and transfers during a Monday shift. In the morning, the 30-bed unit has 26 beds occupied. Computerized staffing identifies six potential discharges and seven surgical admissions. In addition, the average number of call light signals is 67.3 per shift, with the length of time spent with the patient at 8.9 minutes per call. The computerized staffing program warns that there are four experienced nurses and one new nurse scheduled for the day shift. Another computerized warning shows that a surgeon is back from vacation, the emergency department (ED) is full, and historically Mondays have an ED admission rate of 4.2 patients. The ability to access real-time information on variables that impact staffing during a shift from a variety of sources assists in determining appropriate staffing needs.

Computerized programs that support staffing allow for automated staff scheduling, online staff bidding for shifts, and prompt notification of open shifts. Over the past decade, computerized programs have advanced to include sophisticated patient tracking through the hospital and identification of appropriate extended care placement for patient case management. The future of informatics promises to bring innovative solutions to support staffing and improve quality care outcomes.

# EVALUATION OF STAFFING EFFECTIVENESS

Many patient outcomes are driven by the available hours of care delivered and the competence of staff delivering the care. The nurse manager and the organizational nurse leader have the ongoing responsibility to monitor the effectiveness of the staffing plan. To ensure objectivity, staffing outcomes must be delineated, measured, and reviewed.

## PATIENT OUTCOMES AND NURSE STAFFING

The American Nurses Association (ANA) commissioned an integrative review of research on nurse staffing and the effects on patient outcomes (Curtin, 2003). The results show that nurse staffing has a measurable impact on patient outcomes, medical errors, nurse turnover, length of stay, and patient mortality. Additionally, study findings suggest that nurse-patient ratios are important when they are modified by the level of experience of the nurse, the characteristics of the organization, and the quality of interdisciplinary interactions. For instance, a new nurse should be assigned less-critical patients or have a lower nurse-to-patient ratio to allow for learning, whereas an experienced nurse would be expected to handle the standard nurse-to-patient ratio for the unit. Characteristics of the organization include the care delivery model (explained later in the chapter), presence of a charge nurse or team leader without assigned patients, availability of resources such as advanced practice nurses to assist with care, or an environment of autonomy over nursing practice. The quality of interdisciplinary interactions involves the presence and availability of ancillary departments (respiratory, physical therapy, etc.) as well as the utilization of hospital physician intensivists (physicians whose primary responsibility is the care of hospitalized patients) and case managers.

Other professional organizations are weighing in on staffing and patient outcomes. For example, the American Association of Critical Care Nurses (AACN) considers appropriate staffing to be one of six key measures of sustaining healthy work environments for nurses (AACN, 2005), and the Emergency Nurses Association (ENA) has issued guidelines for determining appropriate ED nurse staffing (Ray et al., 2003).

# NURSE STAFFING AND NURSE OUTCOMES

In the previous section, we reviewed outcome measures for patients directly affected by staffing. In addition to patient outcomes, nurse outcomes should also be measured. Staff's perception of the adequacy of staffing should be tracked. Kramer and Schmalenberg (2005) recommend an instrument to measure perception of adequacy of staffing (PAS) as a proxy measure of acceptable staffing levels. Nurse perception of staffing effectiveness must be monitored by hospitals seeking magnet status. Initiating such measures might lead to comparisons for benchmarking best practice in the future and linking RN staffing perception to patient outcomes (Shirey & Fisher, 2008).

There should be the ability for staff to communicate both in written and verbal form regarding staffing concerns. Nurses have the obligation to report to their supervisor their concerns regarding staffing, and every manager has the responsibility to follow up on staffing issues identified by staff. Formalizing this communication process says that you take the issues seriously and gives you data on which to act. In addition, actual staffing compared to recommended staffing should be tracked. This will identify changes in patient acuity and give you clues to other staffing issues. Medication errors is another measure that has been linked with inadequate NHPPD. When resources are scarce, data are imperative to drive needed changes. The outcomes of ineffective staffing patterns and nursing care can be devastating to both patients and staff.

## REAL WORLD INTERVIEW

We have developed a nursing practice quality scorecard. The scorecard is a tool to display data on our three organizational priorities: mission, customer orientation, and cost-effectiveness. By looking at measures in all three arenas, we can see how we are doing in these areas. We can also see if changes made in one arena positively or negatively affect the other measures. To look at nursing's mission for nursing practice, we track and trend several of the American Nurses Association national indicators. We track medication errors, patient falls, restraints, nosocomial pressure ulcers, and urinary tract infections. For customer satisfaction, we measure overall satisfaction with the nursing care provided and how well patients' pain was controlled. For cost-effectiveness, we track nursing hours per patient day. All of these measures are tracked and trended on control charts every three months. The specific data is trended, and measures that are greater than two standard deviations from the target are identified as potential points to be reviewed for identification of opportunities for improvement.

One of the areas we chose to target for improvement was medication errors. It became evident that the most prominent reason for medication errors was delayed and omitted medications. Further investigation proved that the procedures for obtaining medications were unclear and outdated. We have written new procedures to specify responsibilities of the nursing staff and the pharmacy staff. We are now monitoring our rate of medication errors to see if our changes have made any improvement in the error rate.

Another example of use of the scorecard was in the review of our pressure ulcer rate. We found that there was an increase in the incidence of pressure ulcers. In the review of causes, we found that the reporting system had been revised to include all stages of skin breakdown. Since the reporting change, we have seen an increase in the number of pressure ulcers reported. This is a positive change, as we now have accurate data on which to target our improvement efforts.

What we have learned in the development of the scorecard is that we needed to set improvement targets earlier in the process to push the search for opportunities for improvement. We also learned that many of these measures are not well defined and therefore benchmarking to other organizations is difficult. We continue to strive for further improvement and utilize the scorecard to measure our success and look for opportunities for improvement. Reviewing nursing outcome data for the entire nursing division has been a powerful tool to ensure that care provided is meeting expected outcomes, and it allows us to benchmark our outcomes to other organizations.

**Louann Villani, RN, BSN**
Nursing Quality Specialist
Albany, New York

## MODELS OF CARE DELIVERY

To ensure that nursing care is provided to patients, the work must be organized. A care delivery model organizes the work of caring for patients. Over the history of nursing, there have been many models of care delivery. Models vary regarding clinical decision making, work allocation, communication, and management, with differing social and economic forces affecting the selection (Tiedman & Lookinland, 2004). Managers have the responsibility to implement models and evaluate the outcomes in their area. Staff have the responsibility to engage in the implementation and evaluation

process. Each model has strengths and weaknesses that should be considered when deciding which to implement. Several different care delivery models are explored in the following sections.

## CASE METHOD

The case method is the oldest model for nursing care delivery. As nurse training programs began to turn out educated nurses, these nurses were found working in the homes of the sick, taking care of one individual patient. In this model of care, the nurse cares for one patient exclusively. Total patient care is the modern-day version of the case method.

## Critical Thinking 15-1

Recently, you have been able to access data on your unit's pressure ulcer rates. In researching further, you uncover that your unit's rates are significantly higher than those of other units. Your staffing has been stable and in accordance with your staffing plan. Your staff are experienced, and, in fact, you have the longest tenured staff in the hospital.

What are possible explanations for why your pressure ulcer rates are higher than those on other units? What would you do?

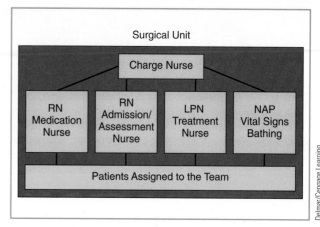

**FIGURE 15-9** Functional nursing model.

## TOTAL PATIENT CARE

In **total patient care**, the nurse is responsible for the total care for assigned patients for the shift worked. The RN has several patients for whom he or she is responsible. The nurse may have some support from LPNs or NAP, but they are not assigned to a specific group of patients.

### Advantages and Disadvantages

The advantage of total patient care and the case method for the patient is the consistency of one individual caring for patients for an entire shift. This enables the patient, nurse, and family to develop a relationship based on trust. This model provides a higher number of RN hours of care than do other models. The nurse has more opportunity to observe and monitor the progress of the patient. A disadvantage is that these models utilize a high level of RN hours to deliver care and are more costly than other models of delivery. This model works well in a specialized unit, such as hospice, where patient/family needs are unstable and require frequent RN assessment and intervention.

## FUNCTIONAL NURSING

This model of care delivery became popular during World War II when there was a significant shortage of nurses in the United States. This method allowed LPNs and NAP to take on tasks that were previously carried out by the RN in the case method. **Functional nursing** divides the nursing work into functional roles that are then assigned to one of the team members. In this model, each care provider has specific duties or tasks for which they are responsible. For instance, a typical division of labor for RNs is medication nurse, admission/assessment nurse, and so on. Decision making is usually at the level of the charge nurse (Figure 15-9).

### Advantages and Disadvantages

In this model, care can be delivered to a large number of patients. This system utilizes other types of health care workers when there is a shortage of RNs. Patients are likely to have care delivered to them in one shift by several staff members. To a patient, care may feel disjointed. A risk of this model is that patients become the sum of the tasks of care they require rather than an integrated whole. Technical rather than professional nursing care often results from a functional model of care, and communication blocks across functional roles can put the patient at peril.

## TEAM NURSING

During World War II, multilevel training programs were developed to teach auxiliary personnel how to perform simple care and technical procedures. In the military, these trained workers were called corpsmen. Outside the military, there were one-year programs developed to teach technical nursing care. On-the-job training programs were established to produce what would today be called nursing assistants. The model of team nursing was developed after the war in an effort to utilize these trained workers and to ease the shortage of nurses that most hospitals were experiencing.

**Team nursing** is a care delivery model that assigns staff to teams that then are responsible for a group of patients. A unit may be divided into two or more teams, and each team is led by an RN. The team leader supervises and coordinates all the care provided by those on the team. The team is most commonly made up of LPNs and NAP. The team leader is responsible for safely delegating specific duties to the team. The larger the team, the more the RN is stretched to safely monitor and care for the patients (Figure 15-10).

A **modular nursing** delivery system is a kind of team nursing that divides a geographic space into modules of patients, with each module cared for by a team of staff led by an RN. The modules may vary in size but center in a

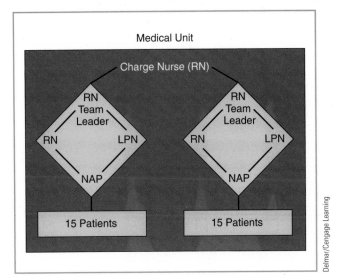

Delmar/Cengage Learning

FIGURE 15-10  **Team nursing model.**

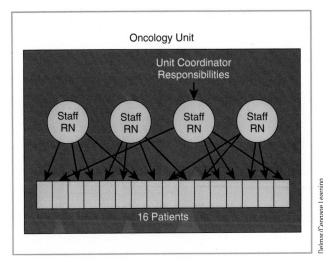

Delmar/Cengage Learning

FIGURE 15-11  **Primary nursing model.**

geographical location and may be associated with a decentralized nursing station.

## Advantages and Disadvantages

In team nursing and modular nursing, the RN is able to get work done through others, but patients often receive fragmented, depersonalized care. Communication in these models is complex. The shared responsibility and accountability can cause confusion and lack of accountability. These factors contribute to RN dissatisfaction with these models. The models require the RN to have very good delegation and supervision skills.

## PRIMARY NURSING

**Primary nursing** is a care delivery model that clearly delineates the responsibility and accountability of the RN and designates the RN as the primary provider of care to patients. The primary nurse retains 24-hour accountability for care coordination for a set of patients throughout the patients' hospital stay. Patients are assigned a primary nurse, who is responsible for developing with the patient a plan of care that is followed by other nurses caring for the patient. Nurses and patients are matched according to needs and abilities. Patients are assigned to their primary nurse regardless of unit geographic considerations. Daily care is provided by associate nurses who enact the plan of care in collaboration with the primary nurse when the primary nurse is not working (Figure 15-11).

## Advantages and Disadvantages

An advantage of this model is that patients and families are able to develop a trusting relationship with the nurse. There is defined accountability and responsibility for the nurse to develop a plan of care with the patient and family.

A holistic approach to care facilitates continuity of care rather than a shift-to-shift focus. Nurses, when they have adequate time to provide necessary care, find this model professionally rewarding, because it gives the authority for decision making to the nurse at the bedside. Disadvantages include a high cost, because there is a higher RN skill mix. The person making out the assignments needs to be knowledgeable about all the patients and the staff to ensure appropriate matching of nurse to patient. With no geographical boundaries within the unit, nursing staff may be required to travel long distances at the unit level to care for their primary patients. Nurses often perform functions that could be completed by other staff. And, finally, nurse-to-patient ratios must be realistic to ensure that enough nursing time is available to meet the patient care needs.

## PATIENT-CENTERED CARE OR PATIENT-FOCUSED CARE

**Patient-centered care** or **patient-focused care** is designed to focus on patient needs rather than staff needs. In this model, required care and services are brought to the patient. In the highest evolution of this model, all patient services are decentralized to the patient area, including radiology and pharmacy services. Staffing is based on patient needs. In this model, there is an effort to have the right person doing the right thing. Care teams are established for a group of patients. The care teams may include other disciplines such as respiratory or physical therapists. In these teams, disciplines collaborate to ensure that patients receive the care they need. Staff are kept close to the patients in decentralized work stations. For example, on a rehabilitation unit, physical therapists may be members of the care team and work at the unit level rather than in a centralized physical therapy department (Figure 15-12).

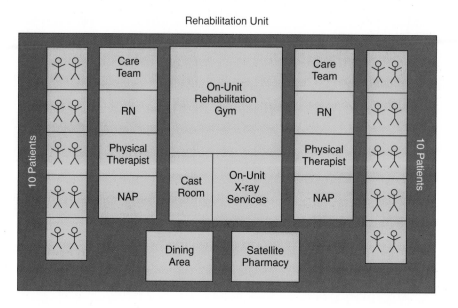

Rehabilitation Unit

FIGURE 15-12  Patient-centered care model.

Delmar/Cengage Learning

 **EVIDENCE** FROM THE LITERATURE

**Citation:** Aiken, L. H., Clarke, S. P., Cheung, R. B., Sloane, D. M., & Silber, J. H. (2003). Education levels of hospital nurses and surgical patient mortality. *Journal of the American Medical Association, 290*(12), 1617–1623.

**Discussion:** This study looked at whether the proportion of RNs with a baccalaureate or higher level of nursing education is associated with risk-adjusted mortality and failure to rescue of surgical patients in Pennsylvania. "Nurses constitute the surveillance system for early detection of complications and problems in care, and they are in the best position to initiate actions that minimize negative outcomes for patients" (p. 1617). It stands to reason that the level of education of an RN might impact the nurse's clinical judgement and ability to serve as a detector and minimizer of complications, although impact on patient outcomes has not been studied before.

Outcomes data from hospital discharges of 168 adult acute care hospitals were compared to American Hospital Association (AHA) administrative data on the hospitals and a survey of Pennsylvania nurses. There was a statistically significant relationship to the proportion of RNs with baccalaureate or higher levels of nursing education and patient mortality and failure to rescue. Each 10% increase in the RN BSN or higher proportion resulted in a 5% reduction of mortality and failure to rescue, after adjusting for both patient and hospital characteristics. Nurses' years of experience were not found to be a significant factor in the statistical models developed by the team.

**Implications for Practice:** The conventional idea that nursing experience is more important than the underlying educational preparation for the RN role is refuted by this study. These findings are additive to our knowledge that nurse-to-patient staffing ratios for medical-surgical nursing of more than 1:4 are associated with increased mortality and failure to rescue. Although further studies are needed to control for additional variables, this study strongly supports the need for leaders to encourage RNs to advance their education for the sake of patients and to set goals that increase the proportion of their nurses who hold a BSN or higher-level degree.

## REAL WORLD INTERVIEW

We consider several key elements when we make staffing decisions. First, we have developed an activity ratio that monitors the RN-sensitive tasks associated with patient movement. The formula is the number of admissions, discharges, and transfers divided by the midnight census for each unit.

Next, we look at trends in average length of stay (ALOS). For example, in one of our surgical units, we tracked a change in ALOS from seven days down to four days. This type of change dramatically impacts RN work by compressing the care into fewer days. This may necessitate a change in delivery model so the nurse can maintain appropriate surveillance of patients. Failure to rescue patients is directly related to the nurse's ability to monitor the patient closely.

The third element in staffing decisions is the level of care needed by patients. We base the intensity of care needed on frequency of patient assessments needed. If there is a significant change in the level of care, then we may need to change the RN mix or the staffing numbers. If we can keep units pure as to the level of care, then we will not be as likely to overstaff them. This is important for productivity and to assure that all of our patients receive the correct level of care. For that reason, it is important to place patients who need observation, are post-procedure, or are post-catheterization in one area, because their numbers on an inpatient unit really impact the level of care needs.

Since adopting this model of evidence-based staffing, I believe we have achieved a more rational and appropriate staffing level that assures optimal outcomes for our patients and supports our nurses' practice.

**Janet M. Bingle, MS, RN**
Chief Nursing Officer, Community Health Network
Indianapolis, Indiana

## Advantages and Disadvantages

The pros of the system are that it is most convenient for patients and expedites services to patients. But it can be extremely costly to decentralize major services in an organization. A second disadvantage is that some staff have perceived the model as a way of reducing RNs and cutting costs in hospitals. In fact, this has been true in some organizations, but many other organizations have successfully used the patient-centered model to have the right staff available for the needs of the patient population.

## PATIENT CARE REDESIGN

In the 1990s, there was significant pressure to reduce health care costs. Hospitals bore the brunt of this pressure. During this decade, patient care redesign was an initiative to redesign how patient care was delivered. The industry learned a lot about how care is delivered as it struggled with the redesign process. In addition to cost reduction, the redesign movement goals included making care more patient centered and not caregiver centered. This was accomplished by reducing the number of caregivers with which each patient had to interface and by organizing care around the patients, thus encouraging greater patient satisfaction. The concept is one of having caregivers cross-trained so they can intervene in more patient situations without having additional resources from outside the care team to assist. Examples include having the team members, such as RNs and cross-trained care technicians, draw lab specimens instead of having a phlebotomy team come to the unit, or team members doing their patients' breathing treatments instead of calling respiratory therapy to come to the unit. Team control of these functions allowed the functions to occur when patients needed them and not at the convenience of outside departments. Work flow analysis is a tool used to determine what activities are value added and how to streamline or eliminate those that do not contribute to improved patient outcomes (Capuano, Bokovoy, Halkins, & Hitchings, 2004). *Value added* refers to activities that possess the following characteristics:

- The customer is willing to pay for this activity.
- The activity must be done right the first time.
- The activity must somehow change the product or service in some desirable manner (Six Sigma, 2006).

Many of these *value added* changes in care attributed to the patient care redesign movement of the 1990s continue to this day as good practices within patient-centered care. Other aspects of redesign have not survived. Those redesign projects that failed to assist caregivers to change their roles or failed to consider unit culture in the patient care redesign were frequently not successful.

# CARE DELIVERY MANAGEMENT TOOLS

In 1983, the federal government established diagnostic-related groups (DRGs) as a payment system for hospitals. In DRGs, the national average length of stay (LOS) for a specific patient type is used to determine payment for that grouping of Medicare patients. **Length of stay (LOS)** refers to the average number of days a patient is hospitalized from day of admission to day of discharge. In DRGs, hospitals are paid the same amount for caring for a DRG patient group regardless of the actual LOS of the specific patient. This prompted initiatives in hospitals to reduce LOS and reduce hospital costs. There are further adjustments to costs based on patient co-morbidities, that is, additional conditions that add to the complexity of care needed by a patient with heart disease, diabetes, or hypertension, for example, hospitals were able to benchmark their LOS for specific patient populations against a national database published through the Medicare DRG system. As hospitals looked for opportunities to reduce costs through reduction in the LOS, clinical pathways and case management surfaced as significant strategies.

## CLINICAL PATHWAYS

Clinical pathways was a major initiative to come out of the efforts to reduce LOS and are widely used to enhance outcomes and contain costs. **Clinical pathways** are care management tools that outline the expected clinical course and outcomes for a specific patient type. Clinical pathways should be evidence based, reflecting the best knowledge to date for patient care. Typically, pathways outline the normal course of care for a patient, and for each day, expected outcomes are articulated. Patient progress is measured against expected outcomes.

 **REAL** WORLD INTERVIEW

In my role as a case manager, I work exclusively with the pediatric population at our hospital. I facilitate the care of patients while they are in the hospital and plan for their home care needs. Our main goal is to ensure that there is a safe transition between hospital and home. As an example of how this goal is achieved, we have met with the cardiac surgery patients and families prior to surgery to identify and proactively address insurance, equipment, or other issues that may arise during hospitalization and at discharge. We have found that patients, families, and staff find our function very helpful. For the staff RN, case managers take on the burden of complex discharge planning, which is enormously time consuming. Patients and families find it comforting to know there is someone who can help them plan for post-discharge.

As the pediatric case manager, I meet with the social worker and the RN staff on the pediatric unit weekly to go over each patient and their specific discharge and social work needs. We have found the work of the case managers and social workers to be complementary, and working together allows us both to have more information and to help the patient get through our system more efficiently. All of these functions help to reduce the patient's length of stay.

As case managers, we also perform some utilization management functions for our organization. We review admission charts, assessing for evidence of meeting admission criteria that the patient's insurer will accept. In addition, we review charts daily to ensure that the patient's acuity warrants continued hospitalization. On occasion, we have to inform the patient and family that the patient's stay is no longer covered by their insurance and they will be responsible for paying the remaining portion of their hospital bill. This can be a difficult situation, and we sometimes get caught between the patient and the patient's insurer. In these cases, we sometimes refer the patients back to their insurer, and sometimes we advocate for the patient to have their continued hospital stay approved.

As a case manager, I find that my diverse clinical background has enabled me to better anticipate the patient's clinical course and be proactive to support the patient's needs. This is a role that is supportive of patients, families, and staff and one I find very challenging and rewarding.

**Linda Zeoli, RN**
Pediatric Case Manager
Albany, New York

| Admission<br><br>Date/Time | Date/Time<br>Day One-Stroke Unit | Date/Time<br>Day 2-Stroke Unit | Date/Time<br>Day 3-Stroke Unit |
|---|---|---|---|
| **ASSESSMENT** | | | |
| Physical | • Adult History Assessment<br>• Adult Physical Assessment per unit protocol and PRN<br>• Neuro Checks per order | • Adults Physical Assessment per unit protocol and PRN<br>• Neuro Checks per order | • Physical Assessment per unit protocol and PRN<br>• Neuro Checks per orders |
| Psychosocial | • Communication regarding plan of care<br>• Identify support system | • Ongoing communication regarding plan of care and discharge plan | • Ongoing communication regarding plan of care and discharge plan |
| **INTERVENTIONS** | | | |
| Tests/Lab | • CBC, CMP, CPK, Lipid Profile, UA, PT, PTT<br>• PTT per Heparin Protocol<br>• CXR, EKC3<br>• CT of Brain/MRI, Carotid Doplex Echocardiogram if indicated | • Daily PT<br>• PTT per Heparin Protocol<br>• Echocardiogram if indicated<br>• Carotid Doplex Scan if indicated<br>• Metabolic Profile if indicated | • Daily PT<br>• PTT per Heparin Protocol |
| Treatments | • O2 per RT Protocol<br>• IV fluid/Saline Lock<br>• I & O<br>• Foley if indicated<br>• Bowel and Bladder Program PRN<br>• Bedside Swallow | • O2 per RT Protocol<br>• IV fluid/Saline Lock<br>• I & O<br>• D/C Foley if indicated<br>• Bowel and Bladder Program PRN<br>• Rehab Team Protocol | • D/C O2 RT protocol if indicated<br>• D/C I & O as indicated<br>• D/C IV if indicated<br>• Rehab Team Protocol |
| Medications | • Thrombolytic therapy as indicated<br>• Antiplatelet therapy as indicated<br>• Anticoagulant therapy as indicated<br>• Antianxiety therapy as indicated<br>• Antihypertensive therapy as indicated<br>• Anticonvulsive therapy as indicated<br>• Diuretic therapy as indicated | • As ordered<br>• Consider Cathartic/stool softener | • As ordered |
| Nutrition | • Check for Dysphagia, then diet as ordered<br>• Tube Feeding if indicated | • Diet as ordered<br>• Consider supplements | • Diet as ordered<br>Consider Calorie Count |
| Activity | • Turn q2 hours and PRN if indicated<br>• Up as tolerated with assist if indicated<br>• As ordered | • Turn q2 hours and PRN if indicated<br>• Up as tolerated-chair TID if indicated<br>• As ordered | • Turn q2 hours and PRN if indicated<br>• Up as tolerated-chair TID if indicated (increased length of time)<br>• As ordered |
| Teaching | • Orient to Unit/Discuss plan of care<br>• Provided Patient Information Sheet<br>• Multidisciplinary Education | • Multidisciplinary Education<br>• Referral to Health Quarters Community Education Center, if indicated | • Reinforce Education<br>• Multidisciplinary Education<br>• Smoking Cessation |
| Discharge | • Assess discharge needs | • Assess discharge needs<br>• Referrals for evaluation to appropriate level of care | • Reassess Discharge Plans<br>• Referrals for evaluation to appropriate level of care |
| **REFERRAL** | | | |
| | • Consult if indicated for concurrent conditions<br>• Rehab Team Evaluation/Treatment | • As indicated | • As indicated<br>• Complete Referrals |
| **EXPECTED OUTCOMES** | | | |
| Initial and Date Outcomes Met | _____ VS stable<br>_____ Pt/Family verbalize understanding of treatment<br>_____ Tolerates oral intake or tube feedings | _____ VS stable<br>_____ Appropriate L&O<br>_____ Neuro status stabilized or improved<br>_____ Tolerates chair (2hours)<br>_____ No nutritional compromise (75–100% meals) | _____ VS stable<br>_____ Microcoagulation levels are therapeutic<br>_____ Balanced I & O<br>_____ Increased participation in functional retraining |
| Additional Outcomes | _____ Patient/Family verbalize plan of care | _____ Patient/Family with increased knowledge of disease process and possible needs at discharge | _____ Patient/Family with increased knowledge of disease process and possible needs at discharge |

Estimated LOS: 4.5

**CEREBRAL INFARCTION PATHWAY**
La PORTE REGIONAL HEALTH SYSTEM
Page 1 of 2

Patient Label

**FIGURE 15-13** Excerpt Page 1 of 2-page Cerebral Infarction (stroke) Clinical Pathway. (*Source:* Courtesy of La Porte Regional Health System, La Porte, IN).

Any variance in outcome can then be noted and acted upon to get the patient back on track. In some facilities, pathways have practitioner orders incorporated into the pathway to facilitate care. Pathways can be comprised of multidisciplinary orders for care, including orders from nursing, medicine, and allied health professionals such as physical therapy and dietary services. Figure 15-13 is a clinical pathway from a magnet-designated hospital that is Stroke Certified by the Joint Commission.

## Advantages and Disadvantages

By articulating the normal course of care for a patient population, clinical pathways are a powerful tool for managing care. In most cases, the implementation of a clinical pathway will improve care, reduce variability, and shorten the LOS for the patient population on the pathway. Pathways also allow for data collection of variances to the pathway. The data can then be used to continuously improve hospital systems and clinical practice.

Some practitioners perceive pathways to be cookbook medicine and are reluctant to participate in their development. Practitioner participation is critical. Development of multidisciplinary pathways also requires a significant amount of work to gain consensus from the various disciplines on the expected plan of care. For patient populations that are nonstandard, pathways are less effective, because the pathway is constantly being modified to reflect the individual patient's needs.

## CASE MANAGEMENT

**Case management** is a second strategy to improve patient care and reduce hospital costs through coordination of care. Typically, a case manager is responsible for coordinating care and establishing goals from preadmission through discharge. In the typical model of case management, a nurse is assigned to a specific high-risk patient population or service, such as cardiac surgery patients. The case manager has the responsibility to work with all disciplines to facilitate care. For example, if a postsurgical hospitalized patient has not met ambulation goals according to the clinical pathway, the case manager would work with the practitioner and nurse to determine what is preventing the patient from achieving this goal. If it turns out that the patient is elderly and is slow to recover, they may agree that physical therapy would be beneficial to assist the patient in ambulating. In other models, the case management function is provided by the staff nurse at the bedside. If the patient population requires significant case management services, there needs to be enough RN time allocated for this activity. In addition to facilitating care, the case manager usually has the responsibility to monitor and improve care. In this role, the case manager collects aggregate data on patient variances from the clinical pathway. The data are shared with the responsible practitioners and other disciplines that participate in the clinical pathway and are then used to explore opportunities for improvement in the pathway or in hospital systems.

## KEY CONCEPTS

- To plan nurse staffing, you must understand and apply the concepts of full-time equivalents (FTEs) and units of service such as nursing hours per patient day (NHPPD).

- Patient classification systems predict nursing time required for groups of patients; the data can then be utilized for staffing, budgeting, and benchmarking.

- Determination of the number of FTEs needed to staff a unit requires review of patient classification data, units of service, regulatory requirements, delivery systems, skill mix, staff support, historical information, and the physical environment of the unit.

- The number of staff and patients in your staffing plan drives the amount of nursing time available for patient care.

- In developing a staffing plan, additional FTEs must be added to a nursing unit budget to provide coverage for days off and benefited time off.

- Scheduling of staff is ultimately the responsibility of the nurse manager, who must take into consideration

patient need and intensity, volume of patients, and the needs and experience of the staff.

- Whatever staffing plans are chosen, it is critical to assess the effect of staffing decisions on patient care and finances.

- Self-scheduling can increase staff morale and professional growth, but success requires clear boundaries and guidelines.

- Evaluating the outcomes of your staffing plan on patients, staff, and the organization is a critical activity that should be done daily, monthly, and annually.

- Case management and clinical pathways are care management tools that have been developed to improve patient care and reduce hospital costs and should be evidence based.

- Technology helps to gather and represent both historical trends and real-time nursing data, to inform staffing decisions.

## KEY TERMS

average daily census (ADC)
benchmarking
case management
clinical pathways
continuity of care
diagnostic-related groups (DRGs)
direct care
full-time equivalent (FTE)
functional nursing
hand offs

indirect care
inpatient unit
length of stay (LOS)
modular nursing
nonproductive hours
nurse intensity
nursing hours per patient day (NHPPD)
patient acuity
patient-centered care
patient classification system (PCS)

patient-focused care
patient turnover
primary nursing
productive hours
self-scheduling
skill mix
staffing plan
team nursing
total patient care
units of service

## REVIEW QUESTIONS

1. Patient classification systems measure nursing workload needed to care for patients. The higher the patient's acuity, the more care is required by the patient. Which of the following statements is a *weakness* of classification systems?
   A. Patient classification data are useful in predicting the required staffing for the next shift and for justifying nursing hours provided.
   B. Patient classification data can be utilized by the nurse making assignments to determine what level of care a patient requires.
   C. Classification systems typically focus on nursing tasks rather than a holistic view of a patient's needs.
   D. Aggregate patient classification data are useful in costing out nursing services and developing the nursing budget.

2. If your full-time staff members receive four weeks of vacation and 10 days of sick time per year, how many productive hours would each FTE work in that year if they utilized all of their benefited time?
   A. 2,080 productive hours
   B. 1,840 productive hours
   C. 1,920 productive hours
   D. 1,780 productive hours

3. Patient outcomes are the result of many variables, one being the model of care delivery that is utilized. From the following scenarios, select which is the *worst* fit between patient needs and the care delivery model.
   A. Cancer patients cared for in a primary nursing model
   B. Rehabilitation patients cared for in a patient-centered model
   C. Medical intensive care patients being cared for in a team nursing model
   D. Ambulatory surgery patients with a wide range of illnesses being cared for using a functional practice model

4. When calculating paid nonproductive time, the nurse manager considers
   A. overtime pay and evening and night shift differential.
   B. total hours available to work.
   C. insurance benefits and educational hours.
   D. all hours that are paid but not worked on the assigned unit.

5. The most important variable that affects staffing patterns and schedules should be which of the following?
   A. Organizational philosophy
   B. Budget allocation and restrictions
   C. Delivering safe, quality patient care
   D. Personnel policies regarding shift rotation

6. The medical-surgical unit provides 200 hours of care daily to 20 patients. Their NHPPD is which of the following?
   A. 1
   B. 10
   C. 20
   D. 200

7. Benchmarking is which of the following?
   A. A comparison of productivity data for similar nursing units
   B. A method to measure cost of care
   C. A comprehensive measure of good quality
   D. A set of written standards of care

8. Informatics supports staffing by doing which of the following?
   A. Determining the flow of information throughout the organization
   B. Organizing data from multiple sources into information for taking action
   C. Facilitating the development of an individualized care model
   D. Reinforcing staffing without considering variables

9. In developing a staffing plan for your unit, which of the following considerations would you include? Select all that apply.

_____ A. Data from census and staffing from the past quarter

_____ B. Benchmark against the organization's NHPPD from the previous year's data

_____ C. Presence of a mini-pharmacy on your unit that stocks medications

_____ D. The hiring of three new nurses for your unit

_____ E. Your hospital is located in Modesto, California.

_____ F. The scope of practice for nursing assistive personnel on your unit

10. Which of the following would you take into account when developing a staffing schedule starting next week and running for six weeks? Select all that apply.

_____ A. Susan took four vacation days last month and has three days scheduled this month.

_____ B. Tim comes out of nursing orientation next week.

_____ C. There is scheduled maintenance on the water system for rooms 201–207 next week.

_____ D. The unit secretary is scheduled for vacation and a float pool secretary will take her place.

_____ E. You have 22 nurses with over 10 years experience on the unit.

_____ F. A hospital-wide lift team will start in two weeks.

## REVIEW ACTIVITIES

1. How do you know whether the outcomes of your staffing plan are positive? What measures do you have available in your organization that indicate your staffing is adequate or inadequate?

2. You are a nurse manager of a new unit for psychiatric patients. What would you consider in planning for FTEs and staffing for this unit?

3. You are a new nurse and you have increasing concerns regarding the staffing levels on your unit. You are becoming increasingly anxious each time you go to work. What would you do?

## EXPLORING THE WEB

- To get more information on safe staffing initiatives, go to www.safestaffingsaveslives.org/. Follow related links on staffing outcomes, research articles, and evidence-based practice for staffing.

- To get more information on staffing effectiveness, go to www.jointcommission.org. Type *staffing effectiveness* into the search box, and read about staffing effectiveness standards issued by the JC.

- To get more information on staffing and quality, go to www.ahrq.gov. This government site of the Agency for Healthcare Research and Quality provides evidence-based analysis of clinical issues. Look in the section for Quality and Patient Safety. You can also search using the term *safe staffing*.

- For examples of care pathways, go to www. hospitalmedicine.org. Search for, care pathways. Scroll down and view the examples of care pathways.

## REFERENCES

Adomat, R., & Hewison, A. (2004). Assessing patient category/ dependence systems for determining the nurse/patient ratio in ICU and HDU: A review of approaches. *Journal of Nursing Management, 12*(5), 299–308.

Aiken, L. H., Clarke, S. P., Cheung, R. B., Sloane, D. M., & Silber, J. H. (2003). Education levels of hospital nurses and surgical patient mortality. *Journal of the American Medical Association, 290*(12), 1617–1623.

American Association of Critical-Care Nurses. (2005). *AACN standards for establishing and sustaining healthy work environments.* Aliso Viejo, CA: AACN.

American Nurses Association. (2010). Nurse staffing plans and ratios. Nursing World. Retrieved September 24, 2010, from http://www. nursingworld.org/mainmenucategories/ANAPoliticalPower/State/ StateLegislativeAgenda/StaffingPlansandRatios_1.aspx.

Beglinger, J. E. (2006). Quantifying patient care intensity: An evidence-based approach to determining staffing requirements. *Nursing Administration Quarterly, 30*(3), 193–202.

Capuano, T., Bokovoy, J., Halkins, D., & Hitchings, K. (2004). Work flow analysis: Eliminating nonvalue-added work. *Journal of Nursing Administration, 34*(5), 246–256.

Centers for Medicare & Medicaid Services. (2007). Conditions for coverage (CfCs) & conditions of participations (CoPs). Retrieved September 25, 2010, from http://edocket.access.gpo.gov/cfr_2007/octqtr/pdf/42cfr482.23.pdf.

Curtin, L. (2003). An integrated analysis of nurse staffing and related variables: Effects on patient outcomes. *OJIN: Journal of Issues in Nursing, 8*(3). Retrieved September 30, 2010, from http://www.nursingworld.org/MainMenuCategories/ANAMarketplace/ANAPeriodicals/OJIN/KeynotesofNote/StaffingandVariablesAnalysis.aspx.

Gran-Moravec, M. B., & Hughes, C. M. (2005). Nursing time allocation and other considerations for staffing. *Nursing & Health Science, 7*(2), 126–133.

Hyun, S., Bakken, S., Douglas, K., & Stone, P. W. (2008). Evidence-based staffing: Potential roles for informatics. *Nursing Economics, 26*(3), 151–173.

Joint Commission. (2006). Staffing standards. Retrieved October 1, 2010, from http://www.jointcommission.org.

Kramer, M., & Schmalenberg, C. (2005). Revising the essentials of magnetism tool: There is more to adequate staffing than numbers. *Journal of Nursing Administration, 35*(4), 188–198.

Needleman, J., Buerhaus, P., Mattke, S., Stewart, M., & Zelevinsky, K. (2002). Nurse-staffing levels and the quality of care in hospitals. *New England Journal of Medicine, 346*(22), 1715–1722.

Ray, C. E., Jagim, M., Agnew, J., Inglass-McKay, J., & Sheehy, S. (2003). ENA's new guidelines for determining emergency department nurse staffing. *Journal of Emergency Nursing, 29*(3), 245–253.

Shirey, M., & Fisher, M. L. (2008). Leadership agenda for change toward healthy work environments for the 21st century. *AACN* Journal. October; 28(5):66–79.

Six Sigma. (2006). Value added. Retrieved December 1, 2006, from http://www.sixsigma.com/dictionary/value-added-134.htm.

Tiedman, M. E., & Lookinland, S. (2004). Traditional models of care delivery: What have we learned? *Journal of Nursing Administration, 34*(6), 291–297.

Unruh, L. Y., & Fottler, M. D. (2006). Patient turnover and nursing staff adequacy. *Health Services Research, 41*(20), 599–612.

## SUGGESTED READINGS

Buerhaus, P. I. (2010). What is the harm in imposing mandatory hospital nurse staffing regulations? *Nursing Economics, 28*(2), 87–93.

Frith, K., & Montgomery, M. (2006). Perceptions, knowledge, and commitment of clinical staff to shared governance. *Nursing Administration Quarterly, 30*(3), 273–284.

Kane, R. L, Shamliyan, T. A., Mueller, C., Duval, S., & Wilt, T. J. (2007). The association of registered nurse staffing levels and patient outcomes: Systematic review and meta-analysis. *Medical Care, 45*(12), 1195–1204.

*Patient safety and quality: An evidence-based handbook for nurses.* (2008). AHRQ Publication No. 08-0043. Agency for Healthcare Research and Quality, Rockville, MD. http://www.ahrq.gov/qual/nurseshdbk/

Rischbieth, A. (2006). Matching nurse skill with patient acuity in the intensive care units: A risk management mandate. *Journal of Nursing Management, 14*(5), 397–404.

Unruh, L. (2007). Nurse staffing and patient, nurse, and financial outcomes. *American Journal of Nursing, 108*(1), 62–71.

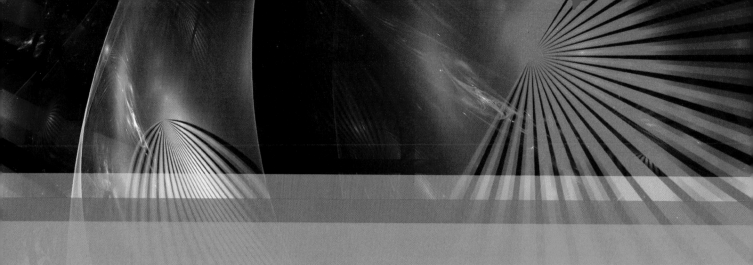

# CHAPTER 16

# Delegation of Patient Care

## MAUREEN T. MARTHALER, RN, MS; PATRICIA KELLY, RN, MSN

*Nobody can do everything, but everyone can do something.*

(GIL SCOTT-HERON)

## OBJECTIVES

Upon completion of this chapter, the reader should be able to:

1. Identify delegation, accountability, responsibility, authority, assignment, supervision, and competence.
2. Identify organizational responsibility for delegation and the chain of command.
3. Support the National Council of State Boards of Nursing Decision Tree—Delegating to Nursing Assistive Personnel.
4. Identify responsibilities of the health care team in delegation.
5. Relate the five rights of delegation.
6. Identify potential delegation barriers.
7. Outline six cultural phenomena that affect transcultural delegation.
8. Discuss direct and indirect patient care.
9. Discuss overarching principles of delegation, nurse-related principles of delegation, and organization-related principles of delegation.

Delmar/Cengage Learning

*Delegation to the appropriate personnel is an important responsibility in nursing. Inappropriate delegation can be life threatening as in the following instance.*

*A patient was admitted to 3C with the diagnosis of transient ischemic attack (TIA). She required neurological assessments to be performed at the onset of every shift and whenever necessary as indicated by a change in the patient's condition. The night nurse assessed the patient at the beginning of her shift, noting that the patient's neurologic status was fully intact. During the night, the nurse periodically checked on the patient every two hours but did not awaken the patient. A sitter was in the room with the patient. The sitter had assured the nurse that the patient was "doing fine." The sitter did not report that when the patient had been assisted to the bathroom initially, she had no difficulty. Upon assisting the patient a second time, the sitter noted that the patient was leaning to one side so badly that she could not stand and required help from two additional nursing assistive personnel. No one reported this to the nurse. When the nurse checked the patient at 6 a.m., she noted that the patient was not able to move her right side.*

*Should the nurse have checked the patient more carefully during the night?*

*What are the responsibilities of the nurse and the sitter?*

*How could delegation have been appropriately performed in this situation?*

On the National Council of State Boards of Nursing Licensure Examination (NCLEX), a student may often encounter test questions that assess the ability of the nurse to delegate care. Safeguarding patients is a number one patient care priority. To ensure that this responsibility is met, nurses are accountable under the law for care rendered by both themselves and other personnel. Multiple levels of nursing assistive personnel (NAP) give care to patients. NAP include nurse aides, nurse technicians, patient care technicians, personal care attendants, unit assistants, nursing assistants, and other nonlicensed personnel.

In support of the role of the NAP and the licensed practical nurse (LPN) in delivering patient care, the American Nurses Association (ANA) and the National Council of State Boards of Nursing (NCSBN) (2006) state, "There is a need and a place for competent, appropriately supervised nursing assistive personnel in the delivery of affordable, quality health care." However, it must be remembered that nursing assistive personnel are equipped to assist—not replace—the nurse. In order to assure that NAP can safely assist the nurse, RNs must know what aspects of nursing can be delegated and what level of supervision is required to ensure that the patient receives safe, high-quality care.

## DELEGATION

Florence Nightingale is quoted as saying, "But then again to look to all these things yourself does not mean to do them yourself. ... But can you not insure that it is done when not done by yourself?" (1859, p. 17). Nursing delegation was discussed by Nightingale in the 1800s and has continued to evolve since then. The American Nurses Association (ANA) defines **delegation** as the transfer of responsibility for the performance of a task from one individual to another while retaining accountability for the outcome. (ANA, Principles for Delegation, 2005).

In 2006, both the American Nurses Association and the National Council of State Boards of Nursing adopted papers on delegation and included them as attachments in a Joint Statement on Delegation (2006). The Joint Statement's two attachments are the American Nurses Association (ANA) Principles of Delegation and the National Council of State Boards of Nursing (NCSBN) Decision Tree—Delegation to Nursing Assistive Personnel.

Note that state nurse practice acts define the legal parameters for nursing practice (www.ncsbn.org). Research supports the link between increased professional nursing staffing and more-desired patient outcomes (Needleman et al., 2006). Most states authorize RNs to thoughtfully delegate patient care activities. Unfortunately, research also shows that NAPs are sometimes delegated tasks beyond their capabilities (Spilsbury & Meyer, 2005). The nursing profession is responsible for determining the scope of nursing practice (Joint Statement, 2006).

The nursing profession defines and supervises the education, training, and utilization of NAP involved in providing direct patient care. The RN is in charge of patient care and determines the appropriate utilization of any NAP involved in providing direct patient care. All decisions related to delegation must be based on the fundamental principles of protection of the health, safety, and welfare of the public. A task delegated to an NAP cannot be re-delegated by the NAP. When a nursing task is delegated, the task must be performed in accordance with standards of practice, policies, and procedures established by the nursing profession, the state nurse practice act, the agency, and ethical-legal standards of behavior. Standards are written and used to guide both the provision and evaluation of patient care. The nurse who delegates retains accountability for the task delegated (Delegation Concepts and Decision-Making Process, NCSBN Position Paper, 1995).

## ANA AND NCSBN

Note that the ANA and NCSBN have different constituencies. The constituency of the ANA is state nursing associations and member RNs. The constituency of NCSBN is state boards of nursing and all licensed nurses. Although for the purpose of collaboration the 2006 Joint Statement refers to registered nurse practice, NCSBN acknowledges that in many states LPN/LVNs have limited authority to delegate (Joint Statement, 2006).

## PRINCIPLES OF DELEGATION

Decisions related to delegation are based on overarching principles of delegation, nurse-related principles of delegation, and organization-related principles of delegation. Overarching principles of delegation include such elements as the idea that the nursing profession determines the scope of nursing practice. The nursing profession also takes responsibility and accountability for the provision of nursing practice (ANA, 2005b). Nursing-related principles include elements such as the idea that the RN may delegate elements of care but not the nursing process itself. RNs monitor organizational policies, procedures, and position descriptions to ensure that there is no violation of the nurse practice act, working with the state board of nursing as necessary. Chief nursing officers (CNOs) are responsible for establishing systems to assess, monitor, verify, and communicate ongoing competence requirements in areas related to delegation, both for RNs and delegates (ANA, 2005b). Organization-related principles guiding nursing delegation include such elements as the organization being accountable for documenting staff competence, developing organizational policies of delegation, and allocating resources to ensure safe staffing (ANA, 2005b).

## ACCOUNTABILITY AND RESPONSIBILITY

Nurses are legally liable for their actions and are accountable for the overall nursing care of their patients. **Accountability** is being responsible and answerable for actions or inactions of self or others in the context of delegation (NCSBN, 1995).

Licensed nurse accountability involves compliance with legal requirements as set forth in the jurisdiction's laws and rules governing nursing. The licensed nurse is also accountable for the quality of the nursing care provided; for recognizing limits, knowledge, and experience; and for planning for situations beyond the nurse's expertise (NCSBN, 2005). Licensed nurse accountability includes the preparedness and obligation to explain or justify to relevant others (including the regulatory authority) the relevant judgments, intentions, decisions, actions, and omissions, as well as the consequences of those decisions, actions, and behaviors. Nurses are accountable for following their state nurse practice act, the standards of professional practice, the policies of their health care organization, and the ethical–legal models of behavior. RNs are accountable for monitoring changes in a patient's status, noting and implementing treatment for human responses to illness, and assisting in the prevention of complications.

The RN assesses the patient; makes a nursing diagnosis; and develops, implements, and evaluates the patient's plan of care. The RN uses nursing judgment and monitors unstable patients who have unpredictable outcomes. The monitoring of other, more stable patients cared for by the LPN and NAP may involve the RN's direct, continuing presence, or the monitoring may be more intermittent. Nursing tasks that do not involve direct patient care can be reassigned more freely and carry fewer legal implications for RNs than delegation of direct nursing practice activities. The assessment, analysis, diagnosis, planning, teaching, and evaluation stages of the nursing process may not be delegated to NAP. Delegated activities usually fall within the implementation phase of the nursing process.

**Responsibility** involves reliability, dependability, and the obligation to accomplish work when an assignment is accepted. Responsibility also includes each person's obligation to perform at an acceptable level—the level to which the person has been educated. For example, an NAP is expected to provide the patient with a bed bath. He or she does not administer pain medication or perform invasive or sterile procedures. After the NAP performs the assigned duties, he or she provides feedback to the nurse about the performance of the duties and the outcome of the actions. This feedback is given to the nurse within a specified time frame. Note that feedback works two ways. It is also the RN's responsibility to follow up with ongoing supervision and evaluation of activities performed by non-nursing personnel. The nurse transfers responsibility and authority for the completion of a delegated task, but the nurse retains accountability for the delegation process. Whenever nursing activities are delegated, the RN is to follow federal regulations, state nursing practice acts, state boards of nursing rules and regulations, and the standards of the hospital.

## Critical Thinking 16-1

The Joint Commission has recommended that all patients be monitored every two hours. One way of documenting this monitoring process is to keep a record sheet on the patient's door. Every time a health care team member checks on the patient, they sign the time and their initials on the record sheet. Now that computerized charting is becoming ever more popular, how will hospitals be able to document this patient monitoring process?

# AUTHORITY

The right to delegate duties and give direction to nursing assistive personnel (NAP) places the RN in a position of authority. **Authority** is the right to act or to command the action of others; it comes with the job and is required for a nurse to take action. The person to whom a task and authority have been delegated must be free to make decisions regarding the activities involved in performing that task. Without authority, the nurse cannot function to meet the needs of patients. An understanding of the level of authority at the time the task is delegated and the level of authority that is identified by the state nurse practice act and the agency's job description prevents each party from making inaccurate assumptions about authority for delegated assignments (Kelly & Marthaler, 2011).

# ASSIGNMENT

**Assignment** is defined by both the ANA and the NCSBN (2006) as the distribution of work that each staff member is to accomplish during a given shift or work period.

The NCSBN uses the verb **assign** to describe those situations when a nurse directs an individual to do something the individual is already authorized to do (ANA & NCSBN, 2006). For example, when an RN directs another RN to assess a patient, the second RN is already authorized to assess patients in the RN scope of practice. It is necessary to ensure that the education, skill, knowledge, and judgment levels of the personnel being assigned to a task are commensurate with the assignment. During a typical shift, the care of patients varies from patients needing only occasional care to patients requiring frequent care. The charge nurse makes out the assignment sheet, taking into consideration the skill, knowledge, and judgment of the RNs, LPN/LVNs, and NAP. Assignments are given to staff who have the appropriate knowledge and skill to complete them. Assignments must always be within the legal scope of practice. Assignment sheets are used to identify patient care duties for RNs, LPN/LVNs, and NAP (Figure 16-1).

An assignment designates those activities that a staff member is responsible for performing as a condition of employment. This is consistent with the staff member's job position and description, legal scope of practice, and education and experience. *Scope of practice* refers to the parameters of the authority to practice granted to a nurse through licensure (NCSBN, 2005). Experienced RNs are expected to work with minimal supervision of their nursing practice. The RN who assigns care to another competent RN who then assumes responsibility and accountability for that patient's care does not have the same obligation to closely supervise that nurse's work as when the care is delegated to an NAP. The RN can assign patient care to the LPN/LVN or delegate to the NAP, but the RN retains accountability for the patient's care. LPN/LVNs and NAP work under the direction of the RN, but they must complete their assignments.

It may be useful to delegate NAP to work with the RN. Typical tasks for NAP include passing trays, assisting with transfers, transporting patients, and stocking supplies. The LPN may be assigned specific patients for whom to

## Critical Thinking 16-2

Note this assignment sheet excerpt. Which patient care needs can be delegated to the NAP?

| Room | Patient | Patient Care Needs |
| --- | --- | --- |
| 211 | Mr. W | Trach care 8 a.m. and prn (patient with new trach) |
| 213 | Mrs. R | Central line dressing change today (first dressing change after central line insertion) |
| 215 | Mr. L | Ambulate at 8 a.m. and 12 noon (patient is postop day 3 and has been walking) |

Unit _____

Date _____

Shift _____

Charge nurse _____

Breaks/Lunch _____

RN _____

_____

LPN _____

NAP _____

_____

Notify RN immediately if:

T    <97 or >100

P    <60 or >110

R    <12 or > 24

SBP <90 or >160

DBP <60 or >100

BS   <70 or >200

Pulse oximetry <95%

Urine output <30
cc/hour or 240cc/8 hours

Notify RN one hour prior to end of shift:

I and O

Patient goal achievement

Narcotic count _____

Glucometer check _____

Stock Pyxis _____

Pass water _____

Stock linen _____

Other _____

| Room and Initials | Patient | Staff | AM/PM Care | Weight I & O | IV | Activity | Accucheck | Tests | NPO | Comments |
|---|---|---|---|---|---|---|---|---|---|---|
| | | | | | | | | | | |
| | | | | | | | | | | |
| | | | | | | | | | | |
| | | | | | | | | | | |
| | | | | | | | | | | |
| | | | | | | | | | | |

FIGURE 16-1  Assignment sheet.

perform care, but the RN remains responsible for all nursing care. When the RN makes out the assignments for the LPN or delegates to the NAP, the RN must consider the patient ABCs, the safety and infection control needs, and the degree of supervision required. For example, in considering patient ABCs, the RN would assess the patient's level of consciousness, vital signs, physical status, changing needs, complexity, stability, multisystem involvement, and technology requirements such as the need for cardiac monitoring. The RN would also consider patient teaching and emotional support needs.

In considering the degree of supervision involved, the RN considers the level of supervision, direct or indirect, required based on the staff member's education, experience, skill level, and competence. It is useful to also use assignments to develop staff. Assigning a less experienced nurse to a more complex patient, but at the same time increasing the level of supervision, increases the less experienced nurse's skill level, competence, and confidence while maintaining safe, high-quality patient care.

Assigning full care responsibility and accountability to a new graduate or to a nurse working in an unfamiliar specialty may be unsafe at times. In these instances, the supervising nurse may have a greater responsibility and accountability to evaluate the abilities and performance of the new nurse.

Certain actions may be assigned to an LPN/LVN, in keeping with the scope of practice as designated by state regulation. If the LPN is certified in IV therapy and the policy of the state and the employing institution permits LPNs/LVNs to provide IV treatment, the RN should not have an inordinate duty to supervise the work of the LPN after the LPN's skills in this area are verified. Note that prior competency certification of the LPN may have been done through a skills day or through a competency validation that ensures that the LPN has been observed inserting a nasogastric tube or an IV line successfully three times under direct supervision of an RN in states where this practice by an LPN is allowed. The RN cannot assign responsibility and accountability for total nursing care to LPNs, but the RN can assign certain tasks to them in keeping with the state law, the job description, their knowledge base, and the demonstrated competency of these individuals. The RN is then responsible for adequate supervision of the person to whom the task is given.

## COMPETENCE

NCSBN defined **competence** as, the application of knowledge and the interpersonal, decision-making and psychomotor skills expected for the practice role, within the context of public health (NCSBN, 1996). The Canadian Nurses Association and Canadian Association of Schools of Nursing (2004) state that continuing competence is the ongoing ability of a nurse to integrate and apply the knowledge, skills, judgment and personal attributes required to practice safely and ethically in a designated role and setting. Competence is required to practice safely and ethically in a designated role and setting. Licensed nurse competence is built upon the knowledge gained in a nursing education program, orientation to specific settings, and the experiences of implementing nursing care. Nurses must know themselves first, including strengths and challenges; assess the match of their knowledge and experience with the requirements and context of a role; gain additional knowledge as needed; and maintain all skills and abilities needed to provide safe nursing care.

NAP competence is built upon formal training and assessment, orientation to specific settings and groups of patients, interpersonal and communication skills, and the experience of the NAP in assisting the nurse to provide safe nursing care.

Chief nursing officers are accountable for establishing systems to assess, monitor, verify, and communicate ongoing competence requirements in areas related to delegation (ANA & NCSBN, 2006). Written documentation of these competencies is maintained in the employee's personnel file. Most health care organizations require employees to undergo annual competency training for elements of care unique to their practice setting. Annual competency testing for RN, LPN/LVN, and NAP may include: patient safety, infection control, code blue, medication safety, IV skills, glucose testing, chain of command, HIPAA policies, and restraints.

## SUPERVISION

**Supervision** is the provision of guidance or direction, evaluation, and follow up by the licensed nurse for accomplishment of a nursing task delegated to NAP (NCSBN, 1995). ANA (2005) states that supervision is "the active process of directing, guiding, and influencing the outcome of an

### Critical Thinking 16-3

Steve, RN, is working with a new RN, Nadia. Steve tells Nadia that when she assigns patient care to another RN, that RN assumes both responsibility and accountability for the care. When Nadia delegates to NAP, she delegates responsibility but keeps the accountability for that patient's care. When Nadia asks Jill, the NAP, to give a bath, what does Nadia retain responsibility and accountability for? What is Jill responsible for?

## Critical Thinking 16-4

The RN continuously monitors unstable, complex patients who have threats to their airway, breathing, circulation, or safety. Examples of these patients might include a patient on a ventilator and an unconscious patient. The RN can delegate care of stable patients to the LPN or NAP. What are some examples of unstable patients?

individual's performance of an activity" (p. 20). Supervision can be categorized as on-site, in which the nurse is physically present or immediately available while the activity is being performed, or off-site, in which the nurse has the ability to provide direction through various means of written, verbal, and electronic communication (ANA, 2005). On-site supervision generally occurs in the acute care setting where the RN is immediately available. Off-site supervision may occur in community settings.

As a result of the rapidly increasing use of technology in patient care, some operational guidelines for supervision from the ANA are helpful. Ask yourself, who is in control of the activity? If the RN is responsible, the nurse should incorporate measures to determine whether an activity has been completed to meet the expectations. Also ask yourself, how should controls be instituted? Controls must be in place that allow the RN delegating an activity to stop the task when inappropriately done, review the measures taken, and take back control of the task (ANA, 2006).

A nurse who is supervising care will provide clear direction to the staff about what tasks are to be performed for specific patients. The supervisor nurse must identify when and how the task is to be done and what information must be collected, as well as any patient-specific information. The nurse must identify what outcomes are expected and the time frame for reporting results. The nurse will monitor staff performance to ensure compliance with established standards of practice, policy, and procedure. The supervisor nurse will obtain feedback from staff and patients, and intervene as necessary to ensure quality nursing care and appropriate documentation.

Hansten and Jackson (2009) identify three levels of supervision based on the task delegated and the education, experience, competency, and working relationship of the people involved:

- Unsupervised occurs when one RN works with another RN. Both are accountable for their own practice. When an RN is in a management position—for example,

charge nurse, nurse manager, and so on—the RN will supervise other RNs.

- Initial direction and periodic inspection occurs when an RN supervises licensed or unlicensed staff, knows the staff's training and competency level, and has a working relationship with the staff. For example, an RN who has worked with NAP for several weeks is now comfortable giving initial directions to ambulate two new postoperative patients. The RN follows up with NAP once and as needed during the shift.

- Continuous supervision occurs when the RN determines that the delegate needs frequent to continuous support and assistance. This level is required when the working relationship is new, the task is complex, or the delegate is inexperienced or has not demonstrated competency.

# RESPONSIBILITIES OF HEALTH TEAM MEMBERS

Even though the development of working skills in delegation is an outcome expectation of baccalaureate nursing program graduates (AACN, 2010), the new graduate nurse may feel overwhelmed by the amount of patient care required and the lack of time to complete the care. The new graduate may be consumed by feelings of inadequacy and failure. For instance, not knowing how to answer the phone or find a washcloth, as well as not finding time to eat lunch, can be exhausting. All of these feelings and behaviors may be a result of trying to do it all and not asking for help. New graduate nurses may quickly realize that if they do not delegate, the patient's care will not be completed in a timely and effective manner. The consequences and likely effects must be considered when delegating patient care. The AACN (2008) suggests assessment of five factors that must occur before deciding to delegate:

1. *Potential for Harm:* Determine if there is a risk for the patient in the activity delegated.
2. *Complexity of the Task:* Delegate simple tasks. These tasks often require psychomotor skills with little assessment or judgment proficiency.
3. *Amount of Problem Solving and Innovation Required:* Do not delegate simple tasks that require a creative approach, adaptation, or special attention to complete.
4. *Unpredictability of Outcome:* Avoid delegating tasks in which the outcome is not clear, causing volatility for the patient.
5. *Level of Patient Interaction:* Value time spent with the patient and the patient's family to develop trust.

Attention to these five factors will improve patient safety associated with delegation.

As health care institutions in the United States respond to shrinking budgets and nursing shortages by increasing the use of NAP, school nursing practice is changing from providing direct care to supervising activities delegated to NAP (Gordon & Barry, 2009). The RN should create an environment that encourages teaching and learning by all staff. The RN should be willing to teach and demonstrate how to perform a task rather than just telling how it should be done. The RN should strive to earn a reputation for exceptional training and mentoring, involving everyone on the health care team, including LPN/LVNs and NAP.

## NURSE MANAGER RESPONSIBILITY

The nursing manager is responsible for developing staff members' ability to delegate. Guidance in this area is necessary, because new graduates, wanting to be regarded favorably, may not ask too much of NAP. Delegation is a skill that requires practice. Graduate nurses are often sent to classes conducted by the education department in the hospital. Topics may include policies and procedures, health team members' roles, and nursing delegation to name a few. This is where the graduate nurses often learn the job descriptions of health care team members. This information is needed to determine what and to whom to delegate.

The nurse manager will determine the appropriate mix of personnel on a nursing unit. The nurse manager may have personnel with a variety of skills, knowledge, and educational levels. The acuity and needs of the patients usually determine the personnel mix. From this personnel mix, the new graduate nurse will begin to identify who can best perform assigned duties. The non-nursing duties are shifted toward clerical personnel, NAP, or housekeeping personnel to make the best use of individual skills.

## NEW GRADUATE RESPONSIBILITY

Berkow, et al. (2008) reported that the delegation proficiency of newly licensed nurses, who comprise more than 10% of a typical hospital's nursing staff, was ranked lowest among 36 competencies by frontline nurse leaders. New graduate nurses need to focus on the duties for which they are directly responsible. What duties can they delegate and to what extent? What do NAP do? What do LPNs/LVNs do? Reviewing the nurse practice act for a nurse's individual state applies to all licensed nurses, regardless of whether the nurse is a new graduate or not. The nurse practice act is the legal authority for nursing practice in each state. In the individual states, the definitions, regulations, or directives regarding delegation may be different. The state nurse practice acts determine what level of

licensed nurse is authorized to delegate (NCSBN, 2005). The RN also reviews any other applicable state or federal laws; patient needs; job descriptions and competencies of the RNs, LPN/LVN, and NAP; the agency's policies and procedures; the clinical situation; and the professional standards of nursing in preparation for delegation. Table 16-1 includes delegation suggestions for RNs.

## REGISTERED NURSE RESPONSIBILITY

The RN is responsible and accountable for the provision of nursing care. The RN is always responsible for patient assessment, diagnosis, care planning, evaluation, and teaching. Although NAP may measure vital signs, intake and output, or other patient status indicators, it is the RN who interprets this data for comprehensive assessment, nursing diagnosis, and development of the plan of care. NAP may perform simple nursing interventions related to patient hygiene, nutrition, elimination, or activities of daily living, but the RN remains responsible for the patient outcome. Having NAP perform functions outside their scope of practice is a violation of the state nursing practice act and is a threat to patient safety.

Inconsistent facility or agency expectations regarding NAP duties or tasks coupled with minimal, if any, training can lead to an unstable and, in some cases, a less qualified workforce, according to the American Nurses Association (ANA) publication *Principles for Delegation* (2005).

As the RN prepares to care for the patient, he or she should describe the health care team to the patient. For example: "Hello Mrs. Jones, my name is Luke Ellingsen. I am a registered nurse, and I will be responsible for your care until 3 p.m. today. Thelma Marks, a nursing assistant, is working with me and will be in to take your vital signs and help with your bath. Please call me if you have any questions."

## NURSING ASSISTIVE PERSONNEL (NAP) RESPONSIBILITY

The increase in numbers of NAP in all settings where patient care is provided poses a degree of risk to the patient. NAP are trained to perform duties such as bathing, feeding, toileting, and ambulating patients (Figure 16-2). NAP are also expected to document and report information related to these activities. The RN will delegate to the NAP and is liable for those delegations. According to the ANA (2005), if the RN knows or reasonably believes that the assistant has the appropriate training, orientation, and documented competencies, then the RN can reasonably expect that the NAP will function in a safe and effective manner.

## TABLE 16-1 Delegation Suggestions for RN

| DELEGATION SUGGESTIONS | EXAMPLES |
| --- | --- |
| • Be clear on the qualifications of the delegate, i.e., education, experience, and competency. Require documentation or demonstration of current competence by the delegate for each task. Clarify patient care concerns or delegation problems. Consult ANA position papers at www.nursingworld.org and your state board of nursing, as necessary. | • The charge nurse will assign a new graduate nurse a team of patients less complex than the assignment of an RN who has several years of experience. |
| • Assess what is to be delegated and identify who would best complete the assignment. | • The RN will ask a nursing assistant to pick up specific equipment, e.g., a pediatric pulse oximeter from a stock room. The assistant has worked on this unit for five years and is familiar with the type of equipment the nurse needs. |
| • Communicate the duty to be performed, and identify the time frame for completion. The expectations for personnel should be clear and concise. | • The charge nurse tells another nurse, "While I am at lunch, Mr. Jones, the patient in bed 34-2, may ask for a pain shot. Please make him comfortable and tell him that it is too early for another shot for 60 more minutes." |
| • Avoid changing tasks once they are assigned even when staff request a change. Changing duties should be considered only when the task is above the level of the personnel, as when the patient's care is in jeopardy due to a change in status. | • The NAP was delegated the task of taking vital signs on a set of patients. One of the patients is receiving a blood transfusion and is very ill. The nurse may transfer the delegated task of taking vital signs on this patient to an RN. |
| • Evaluate the effectiveness of the delegation of duties. Monitor care and check in with NAP frequently. Ask for a feedback report on the outcomes of care delivery. | • After the patient was assisted to the bathroom, the nurse asked the NAP, what amount of assistance did the patient require? Is the patient safely back in bed? |
| • Accept minor variations in the style in which the duties are performed. Individual styles are acceptable as long as the duty is performed correctly within the scope of practice and there is a good outcome. | • Both of the following nurses are successful at providing care using different and acceptable methods. One nurse assesses the assigned patients, documents care, and then passes medications. Another nurse assesses the patients while passing medications and then documents care. |

Reasons for using NAP in acute care settings include cost control; freeing RNs from duties, primarily non-nursing duties; and allowing time for RNs to complete assessments of patients and their potential responses to treatments. NAP cannot be assigned to assess or evaluate responses to treatment, because that is the role of the RN. It is more cost effective to have NAP perform non-nursing duties than to have nurses perform them. NAP can deliver supportive care; they cannot practice nursing or provide total patient care. The RN has an increased scope of liability when tasks are delegated to NAP. The RN must be aware of the job description, skills, and educational background of the NAP prior to the delegation of duties. In a school system, nurses may need to utilize NAP more than in other care settings. The student–school nurse ratio is 1:750 according to the National Association of School Nurses (2010). The Association also reports that 13 states are in noncompliance, Michigan being the least compliant with a student–school nurse ratio of 1:4,836 (Robert Wood Johnson Foundation, 2010). The NAP in this setting are the secretaries, teachers, and other individuals who may or may not be trained to perform such nursing activities as distributing medications or administering first aid to the students.

## Elements to Consider When Delegating

- Federal, state, and local regulations and guidelines for practice, including the state nurse practice act
- Nursing professional standards
- Health care agency policy, procedure, and standards

- Job description of registered nurse, licensed practical nurse/licensed vocational nurse, nursing assistive personnel
- Five Rights of delegation (NCSBN, 1997)
- Knowledge and skill of personnel

- Documented personnel competency, strengths, and weaknesses (select the right person for the right job)
- ANA Principles for Delegation, i.e., overarching principles, nurse-related principles, and organization-related principles*

RN accountable for application of the nursing process
Assessment and clinical judgment*
Nursing diagnosis
Planning care
Implementation and teaching
RN delegates as appropriate
RN retains accountability
Note that LPN/LVNs and NAP are also responsible for their actions

### RN

RNs assess, plan care, monitor, and evaluate all patients, especially complex, unstable patients with unpredictable outcomes.

Intervenes quickly to assure patients' physiologic, safety, and psychological needs.

Administer medications, including IV push and IVPBs.

Start and maintain IVs and blood transfusions.

Perform sterile or specialized procedures, for example, Foley catheter and nasogastric tube insertion, tracheostomy care, suture removal, and so on.

Educate patient and family.

Maintain infection control.

Administer cardiopulmonary resuscitation.

Interpret and report laboratory findings.

Triage patients.

Prevent adverse nurse-sensitive patient outcomes, for example, cardiac arrest, pneumonia, and so on.

Monitor patient outcomes.

### LPN/LVN

LPN/LVNs care for stable patients with predictable outcomes. They work under the direction of the RN and are responsible for their actions within their scope of practice.

Gather patient data.

Implement patient care.

Maintain infection control.

Provide and reinforce teaching from standard teaching plan.

Depending on the state and with documented competency, may do the following:**

- Administer medications.
- Perform sterile or specialized procedures, for example, Foley catheter and nasogastric tube insertion, tracheostomy care, suture removal, and so on.
- Perform blood glucose monitoring.
- Administer CPR.
- Perform venipuncture and insert peripheral IVs, change IV bags for patients receiving IV therapy, and so on.

### NAP

NAP assist the RNs and the LPNs and give technical care to stable patients with predictable outcomes and minimal potential for risk. They work under the direction of an RN and are responsible for their actions within their scope of practice.

Assist with activities of daily living.

Assist with bathing, grooming, and dressing.

Assist with toileting and bed making.

Ambulate, position, and transport.

Feed and socialize with patient.

Measure intake & output (I&O).

Document care.

Weigh patient.

Maintain infection control.

Depending on the state and with documented competency, may do the following:**

- Perform blood glucose monitoring.
- Collect specimens.
- Administer CPR.
- Take vital signs.
- Perform 12 lead EKGs.
- Perform venipuncture for blood tests.

Evaluation
RN uses clinical judgment** and is responsible for evaluation of all patient care.

*Clinical judgment means an interpretation or conclusion about a patient's needs, concerns, or health problems, and/or the decision to take action (or not), use or modify standard approaches, or improvise new approaches as deemed appropriate by the patient's response (Tanner, 2008).

**Some variation from state to state and agency.

FIGURE 16-2 **Elements to consider when delegating.** (*Source:* Adapted from Kelly & Marthaler, 2011).

## REAL WORLD INTERVIEW

I use delegation now that I have completed school. I began working as a graduate nurse immediately after graduating nursing school. Prior to graduation, I worked as a nurse technician. I feel that I do understand how it feels to be at both ends of patient care delivery. I vowed that when I became a registered nurse, I would delegate appropriately and fairly to others.

As a registered nurse, I make a point to delegate appropriately to certified nursing assistants (CNAs). I delegate duties like vital signs, changing beds, bathing patients, feeding patients, and performing an accurate intake and output. I delegate these things after giving my CNA a complete report of my patients.

I work on a medical-surgical floor where our CNAs use an automated DYNAMAP to take blood pressures, pulses, and temperatures. I will take my own manual blood pressure when I am assessing my patient if the readings from the DYNAMAP were high or low. My CNAs bring me their vital signs as soon as they are done so that I can determine what more I need to evaluate.

It is important to mention that I never delegate patient assessments or patient education. These duties are reserved for the registered nurse. I will never delegate to a CNA to watch over a patient while they take their medication. I never delegate the insertion or removal of Foleys. I do believe my CNAs take me seriously, as I do not delegate anything that I am not willing to do myself and have not done myself in the past. In essence, I do not give the impression that I am "beyond" or "better" than anyone else.

I get concerned when I see a fellow nurse walk out of the room of a patient who has just requested a bedpan and go to find a CNA to get him or her that bedpan. I would never make my patient wait to perform such a necessary and often immediate task. Like I said earlier, I have been on both ends of patient care delivery, and I know what it feels like to be unappreciated. So far, I have stuck to my promise to delegate appropriately and fairly. I truly believe my CNAs would agree.

**Shelly A. Thompson**
Registered Nurse
Dyer, Indiana

If the LPN or NAP performs poorly, the RN should tell them about mistakes privately (as much as possible), in a supportive manner with a focus on learning from mistakes. If they perform in an inappropriate, unsafe, or incompetent manner, the RN must intervene immediately and stop the unsafe activity, counsel the LPN or NAP, document the facts, and report to the nurse manager as appropriate.

## Licensed Practical/ Vocational Nurse Responsibility

Licensed practical/vocational nursing caregivers who have undergone a standardized training and competency and licensing evaluation are licensed practical nurses/licensed vocational nurses (LPNs/LVNs). LPNs/LVNs are able to perform duties and functions that NAP are not allowed to do, and they are also held to a higher standard of care and are responsible for their actions. LPN/LVNs usually care for stable patients with predictable outcomes, though they may help the RN with seriously ill patients in ICU. Common duties of the LPN include the duties of the NAP plus data collection, sterile dressing changes, colostomy irrigations, respiratory suctioning, insertion of Foley catheters, and teaching from a standard patient care plan. In some states, passing medications, including the administration and initiation of IV fluids, monitoring of IV sites, and passing of nasogastric tubes is part of the LPN role. In other states, IV therapy under the supervision of an on-site RN is allowed. The LPN does not do initial patient assessment, but after the RN has completed the patient's initial assessment and the plan of care, the LPN does the ongoing head-to-toe assessment; monitoring of vital signs, IV sites, IV fluids, and breath sounds; and so on. In some states in nursing homes, the LPN may assume the charge nurse role with an on-site supervising RN. LPNs report their findings to the RN. The RN is still primarily responsible for overall patient assessment, nursing diagnosis, planning, implementation, and evaluation of the quality of care delegated. Table 16-2 (NCSBN, 1997) lists five rights of delegation to be considered.

# DIRECT AND INDIRECT PATIENT CARE ACTIVITIES

As mentioned earlier in this chapter, **direct patient care activities** include activities such as assisting the patient with feeding, drinking, ambulating, grooming, toileting, dressing, and socializing. Direct patient care activity may also involve collecting, reporting, and documenting related to these activities. Activities delegated to NAP do not include health counseling or teaching or activities that require independent, specialized nursing knowledge, skill, or judgment.

**Indirect patient care activities** are often necessary to support the patient and the patient's environment, and only incidentally involve direct patient contact. These activities are often designated "unit routines" and assist in providing a clean, efficient, and safe patient care milieu. They typically encompass chore services, companion care, housekeeping, transporting, clerical, stocking, and maintenance tasks (ANA, 2005).

# UNDERDELEGATION AND OVERDELEGATION

Personnel in a new job role such as nurse manager, RN, or nursing graduate often underdelegate. Believing that older, more experienced staff may resent having some-

## TABLE 16-2 The Five Rights of Delegation

| | |
|---|---|
| Right Task | Does the delegated task conform to agency established policies, procedures, and standards consistent with the state nurse practice act, federal and state regulations and guidelines for practice, nursing professional standards, and the ANA Code of Ethics? |
| Right Circumstance | Does the delegated task require independent nursing management? Do the personnel have the education, experience, resources, equipment, and supervision needed to complete the task safely? |
| Right Person | Is a qualified, competent person delegating the right task to a qualified, competent person, to be performed on the right patient? Is the patient stable, with predictable outcomes? Is it legally acceptable to delegate to this patient? Do health care personnel have documented knowledge, skill, and competency to do the task? |
| Right Direction/ Communication | Does the RN communicate the task clearly with directions, specific steps of the tasks, any limitations, and expected outcomes? Are times for reporting back to the RN specified? Is staff understanding of the task clarified? Are staff encouraged to say, "I don't know how to do this and I need help," as needed? |
| Right Supervision | Is there appropriate monitoring, intervention, evaluation, and patient and staff feedback as needed? Are patient and staff outcomes monitored? Does the RN answer staff questions and problem-solve as needed? Does the staff report task completion and patient response to the RN? Does the RN provide follow-up teaching and guidance to staff as appropriate? Is there continuous quality improvement of the delegation process and patient care? Are problems, particularly any sentinel events, reported via the chain of command and as needed to the State Board of Nursing and the Joint Commission (JC)? |

**Source:** Adapted from National Council of State Boards of Nursing (NCSBN). (1997). *The five rights of delegation.*

## Critical Thinking 16-5

It was just about 6:30 p.m. on the 3 p.m. to 11 p.m. shift on 2 East. Most of the practitioners had made their rounds, so the evening was calming down. The NAP, Jill, was picking up the dinner trays from the patient's rooms. Steve, the RN, had just sat down to document his patient assessments when he heard NAP Jill yell, "I need some help in Room 2510! Mr. Olson is not breathing." As several of the nurses ran to Room 2510, the NAP ran for the emergency crash cart. The cart was wheeled into the patient's room during the overhead announcement by the operator, "CODE BLUE, Room 2510." The nurses initiated CPR. The NAP plugged the cart into the wall, turned the suction machine on, and then assisted the family out of the room and stayed with them until the nurse was able to talk with them.

How does completion of these tasks by the NAP contribute to patient care?

Does the NAP relieve the pressure on the nurse to provide acute patient care?

 **REAL** WORLD INTERVIEW

Upon evaluating delegation on several, varied nursing units, I arrive at one conclusion; we as professional nurses just do not do it well. There is the exception, of course: the individual who has developed an outstanding ability to delegate nearly all of his or her responsibilities to others in an authoritative or diplomatic manner, with the recipients either loving or hating it.

Part of the problem may lie with the job description, that black-and-white document that delineates a role in great detail right up to the final statement of "inclusive of duties as assigned." The latter statement is too vague, and the delegated task should be clearly stated somewhere in the job description.

On a nursing unit, it generally falls on the staff nurse to function in the assigning and delegating role. For this role, he or she is often criticized, most frequently behind the scenes, though occasionally criticisms are blasted right out in the open. "What do you mean I am getting the next admission; I already have two!" At best, one becomes apprehensive when assigning *anything*, from an admission to cleaning up the break room. I wonder, if that is the fate of the charge nurse, just how well would one expect the staff RN to delegate?

Perhaps our failures with delegation stem from our predominantly female, motherly gender. Moms can do it! Moms can do it all. Often, Mom finds the route of least resistance: "It's just easier to do it myself!" It is the same thing with RNs; RNs can do it, RNs can do it all.

I believe that the fine art of delegation needs to be taught more in the educational process, along with the concept of teamwork. The team is hindered when we become ineffective at delegation. The challenges in contemporary health care are tremendous, only to become even more challenging in the future. We as professional nurses would do well to acquire advanced skills in delegation, team building, and diplomacy, for these skills will become tools of survival in the very near future.

**Suzanne Kalweit, RN, MS**
Charge Nurse
Dyer, Indiana

## TABLE 16-3 Delegation Checklist

| QUESTION | YES | NO |
|---|---|---|
| Do you recognize that you retain ultimate responsibility for the outcome of delegated assignments? | —— | —— |
| Do you spend most of your time completing tasks that require an RN? | —— | —— |
| Do you trust the ability of your staff to complete job assignments successfully? | —— | —— |
| Do you allow staff sufficient time to solve their own problems before intervening with advice? | —— | —— |
| Do you clearly outline expected outcomes and hold your staff accountable for achieving these outcomes? | —— | —— |
| Do you support your staff with an appropriate level of feedback and follow-up? | —— | —— |
| Do you use delegation as a way to help staff develop new skills and provide challenging work assignments? | —— | —— |
| Does your staff know what you expect of them? | —— | —— |
| Do you take the time to carefully select the right person for the right job? | —— | —— |
| Do you feel comfortable sharing control with your staff as appropriate? | —— | —— |
| Do you clearly identify all aspects of an assignment to staff when you delegate? | —— | —— |
| Do you assign tasks to the lowest level of staff capable of completing them successfully? | —— | —— |
| Do you support your staff, even when they are learning? | —— | —— |
| Do you allow your staff reasonable freedom to achieve outcomes? | —— | —— |

**Source:** Compiled with information from Harvard ManageMentor® Delegating Tools 2004. *Delegation skills checklist*. Boston: Harvard Business School Publishing.

one new delegate to them, a new nurse may simply avoid delegation. Or new nurses may seek approval from other staff members by demonstrating their ability to complete all assigned duties without assistance. In addition, new nurses may be reluctant to delegate because they do not know or trust individuals on their team or are not clear on the scope of their duties or what they are allowed to do. New nurses can become frustrated and overwhelmed if they fail to delegate appropriately. They may also fail to establish appropriate controls with staff or fail to follow up properly; they may also fail to delegate the appropriate authority to go with certain responsibilities. Perfectionism and a refusal to allow mistakes can lead new nurses in over their head in patient care responsibilities. More-experienced staff members can help new personnel by

intervening early on, assisting in the delegation process, and clarifying responsibilities (Table 16-3).

## OVERDELEGATION

Overdelegation of duties can also place the patient at risk. The reasons for overdelegation are numerous. Personnel may feel uncomfortable performing duties that are unfamiliar to them, and they may depend too much on others. They may be unorganized or inclined to either avoid responsibility or immerse themselves in trivia. Overdelegation leads to delegating duties to personnel who are not educated for the tasks, such as LPNs and NAP. Delegating duties that are inappropriate for personnel to perform because they

## TABLE 16-4   Obstacles to Delegation

- Fear of being disliked

- Inability to give up any control of the situation

- Fear of making a mistake

- Inability to determine what to delegate and to whom

- Inadequate knowledge of delegation process

- Past experience with delegation that did not turn out well

- Poor interpersonal communication skills

- Lack of confidence to move beyond being a novice nurse

- Lack of administrative support for nurse delegating to LPN and NAP

- Tendency to isolate oneself and choose to complete all tasks alone

- Lack of confidence to delegate to staff members who were previously one's peers

- Inability to prioritize using Maslow's hierarchy of needs and the nursing process

- Thinking of oneself as the only one who can complete a task the way "it is supposed" to be done

- Inability to communicate effectively

- Inability to develop working relationships with other team members

- Lack of knowledge of the capabilities of staff, including their competency, skill, experience, level of education, job description, and so on

have been inadequately educated is dangerous and against the state nurse practice act. Overdelegating duties can overwork some personnel and underwork others, creating obstacles to delegation (Table 16-4).

# ORGANIZATIONAL RESPONSIBILITY FOR DELEGATION

Organizational responsibility for delegation includes providing sufficient resources such as adequate staffing with an appropriate staff mix; documenting competencies for all staff providing direct patient care and ensuring that the RN has access to competence information for the staff to whom the RN is delegating care; developing organizational policies on delegation with the active participation of all nurses; and acknowledging that delegation is a professional right and responsibility (ANA & NCSBN, 2006). These elements help assure nursing and medical quality. They also help clarify all health care elements within the organizational system (Table 16-5).

Organizations fulfill their responsibility to staff and patients by developing these organizational elements and maintaining a clear chain of command.

## CHAIN OF COMMAND

The RN, including the new graduate nurse, is accountable to the charge nurse and nurse manager of the unit. The nurse manager is accountable to the chief nursing officer, for example, the vice president for nursing. The chief nursing officer is accountable to the chief executive officer. The hospital's chief executive officer is accountable to the board of directors. The board of directors is accountable to the community it serves and often to another larger hospital corporation, as well as to state nursing and medical licensing boards and the Joint Commission (Figure 16-3). All are accountable for their actions to the patients and the communities that they serve.

## Delegation of the Nursing Process

Ultimately, some professional activities involving the specialized knowledge, judgment, or skill of the nursing process can never be delegated. These include patient assessment, triage, nursing diagnosis, nursing plans of care, extensive teaching or counseling, telephone advice, outcome evaluations, and patient discharges (Zimmermann, 1996). Delegated tasks are typically those tasks that occur frequently, are considered technical by nature, are considered standard and unchanging, have predictable results, and have minimal potential for risks (Westfall, 1998). As a

## TABLE 16-5 Organizational Elements Needed for Efficient Delegation

- Follow professional standards for education, licensure, and competency in all hiring decisions, orientation, and ongoing continuing education programs.

- Have clear job descriptions and ongoing licensing and credentialing policies for nursing and medical providers, LPN/LVNs, NAP, and other health care staff. The organization must ensure that all staff members are safe, competent practitioners before assigning them to patient care. Orient staff to their duties, chain of command, and the job descriptions of RN, LPN, and NAP.

- Facilitate clinical and educational specialty certification and credentialing of all health care practitioners and staff.

- Provide standards for ongoing supervision and periodic licensure/competency verification and evaluation of all staff.

- Provide access to evidence-based, professional health care standards, policies, procedures, library, Internet, and medication information, with unit availability and efficient library and Internet access.

- Facilitate regular evidence-based reviews of critical standards, policies, and procedures.

- Have clear policies and procedures for delegation and chain of command reporting lines for all staff from RN to charge nurse to nurse manager to nurse executive and, as appropriate, to risk management, the hospital ethics committee, the hospital administrator, nursing and medical practitioners, the chief of the medical staff, the board of directors, the state licensing boards for nursing and medicine, and the Joint Commission. See Figure 16-3 for illustration of one such organizational chain of command.

- Provide administrative support for supervisors and staff who delegate, assign, monitor, and evaluate patient care.

- Clarify health care provider accountability; for example, if a medical or nursing practitioner or physician assistant delegates a nursing task to an NAP, the health care provider is responsible for monitoring that care delivery. This must be spelled out in hospital policy. If the RN notes that the NAP is doing something incorrectly, the RN has a duty to intervene and to notify the ordering practitioner of the incident. The RN always has an independent responsibility to protect patient safety. Blindly relying on another nursing or health care provider is not permissible for the RN.

- Provide education and standards for regular RN evaluation of NAP and LPN/LVN, and reinforce the need for NAP and LPN/LVN accountability to the RN. RNs must delegate and supervise. They cannot abdicate this professional responsibility.

- Communicate with patients and staff and keep them informed. Treat them with kindness and respect.

- Acknowledge unfortunate patient care incidents and express concern about these events without either taking the blame, blaming others, or reacting defensively.

- Have staff avoid taking telephone and verbal orders. If necessary to maintain patient safety, however, staff should repeat the order back to the practitioner to assure clarity. Staff must document this, e.g., telephone order repeated back (TORB) or verbal order repeated back (VORB).

- Speak to staff as you would like to be spoken to. There is no need to apologize for delegation. Staff who delegate are carrying out their professional responsibility.

- Reinforce a positive use of authority, a good attitude, and dependability in staff. Include any limits on authority also. The expectations for personnel before, during, and after duty performance should be stated in a clear, pleasant, direct, and concise manner.

- Verify a delegate's understanding of any delegated tasks and have him or her repeat instructions as needed. Be clear, and welcome lots of questions until you are convinced that the delegate understands what you want done. Verify that the delegate accepts responsibility for carrying out the task correctly. Require frequent mini-reports about patients from staff, and include any specific reporting guidelines, times for interventions, and deadlines for accomplishing any tasks.

- As needed, explain to a delegate why a task needs to be done, its importance in the overall scheme of things, and any possible complications that may arise during its performance. Invite questions, and don't get defensive if your delegate pushes you for answers. Seek commitment from your delegate to complete the task according to standards and in a timely fashion.

- Provide support, and monitor task completion according to standards. Make frequent walking rounds to assess patient

*(Continues)*

**TABLE 16-5** (Continued)

outcomes. Be sure your delegate has the resources, training, and other help to get a task done.

- Try to provide for continuity of care by the same staff when possible, and consider the geography of the unit and fair, balanced work distribution among staff when assigning care.

- If delegates don't meet standards, talk with them to identify the problem. If this is not successful, inform the delegate that you will be discussing the problem with your supervisor. Document your concerns, as appropriate. Follow up with your supervisor according to your organization's policy.

- Develop a physical, mental, and verbal "no abuse" policy to be followed by all professional and nonprofessional health care staff. Follow up on any problems.

- Consider applying for magnet status for your facility. This status is awarded by the American Nurses Credentialing Center to nursing departments that have worked to improve nursing care, including the empowering of nursing decision making and delegation in clinical practice.

- Monitor patient outcomes, including nurse-sensitive outcomes, staffing ratios, and other patient, clinical, financial, and organizational quality outcomes, as well as developing ongoing clinical quality improvement practices. Benchmark with national groups.

- Maintain ongoing monitoring of incident reports, sentinel events, and other elements of risk management and performance improvement of the process and outcome of patient care.

- Develop electronic health records (EHR), including systematic, error-proof systems for medication administration that ensure the six rights of medication administration, that is, the right patient, right medication, right dose, right time, right route, and right documentation. Develop safe computerized order-entry systems and staffing systems.

- Provide documentation of routine maintenance for all patient care equipment.

- Maintain the JC Patient Safety Goals (www.jointcommission.org).

- Develop intrahospital and intra-agency safe transfer policies.

- Avoid high-risk delegation. The RN may be at risk if the delegated task can be performed only by the RN according to law, organizational policies and procedures, or professional standards of nursing practice; if the delegated task could involve substantial risk or harm to a patient; if the RN knowingly delegates a task to a person who has not had the appropriate training or orientation; or if the RN fails to adequately supervise the delegated activity and does not evaluate the delegated action by reassessing the patient (ANA, 1996).

- Monitor the Medicare "Do Not Pay" list, for example, note that Medicare will not pay for transfusions gone wrong due to human error.

professional standard for all nurses in all states, the assessment, analysis, diagnosis, planning, teaching, and evaluation stages of the nursing process may not be delegated. Delegated activities usually fall within the implementation phase of the nursing process.

# NCSBN DECISION TREE-DELEGATING TO NURSING ASSISTIVE PERSONNEL

The National Council of State Boards of Nursing (NCSBN) has developed a Decision Tree-Delegating to Nursing Assistive Personnel (Figure 16-4).

The steps of the decision tree are as follows:

- Assessment and planning
- Communication
- Surveillance and supervision
- Evaluation and feedback

# STATE BOARDS OF NURSING

Many states specify nursing tasks that may be delegated in their rules and regulations. Although the excerpts in the example in Table 16-6 are similar to the rules of different states, there is variation in rules and regulations from state to state. Check state requirements with your state board of nursing at www.ncsbn.org.

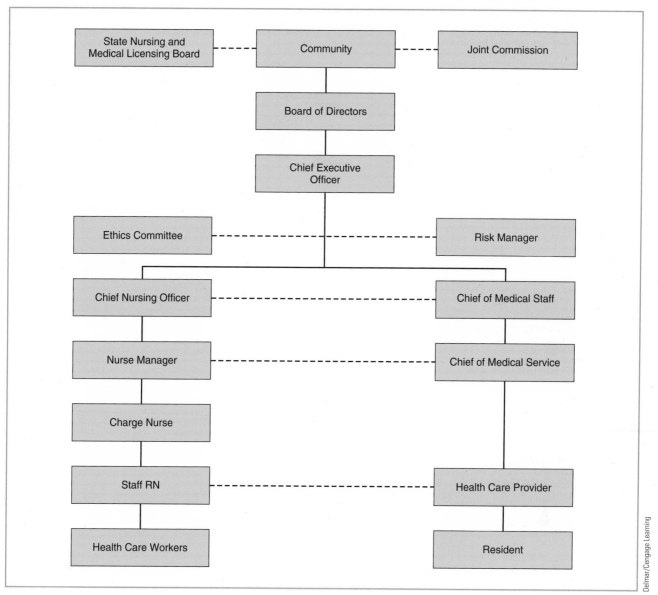

**FIGURE 16-3** Organizational chain of command.

# DELEGATION SUGGESTIONS FOR RNS

RNs concerned with appropriate delegation find it helpful to use the delegation suggestions in Table 16-7.

# TRANSCULTURAL DELEGATION

Nurses and patients come from diverse cultural backgrounds. Transcultural delegation requires that personnel perform duties with this cultural diversity taken into consideration. Poole, Davidhizar, and Giger (1995) suggest

that there are six cultural phenomena to be considered when delegating to a culturally diverse staff: communication, space, social organization, time, environmental control, and biological variations (Table 16-8).

## CULTURAL PHENOMENA

Communication, the first cultural phenomenon, is greatly affected by cultural diversity in the workforce. Communication influences how messages are sent and received (Poole, Davidhizar, & Giger, 1995). For example, if a nurse were talking to NAP in a loud voice, it could be viewed as anger. However, the nurse may be from a cultural group whose members always speak loudly—she may not be angry at all.

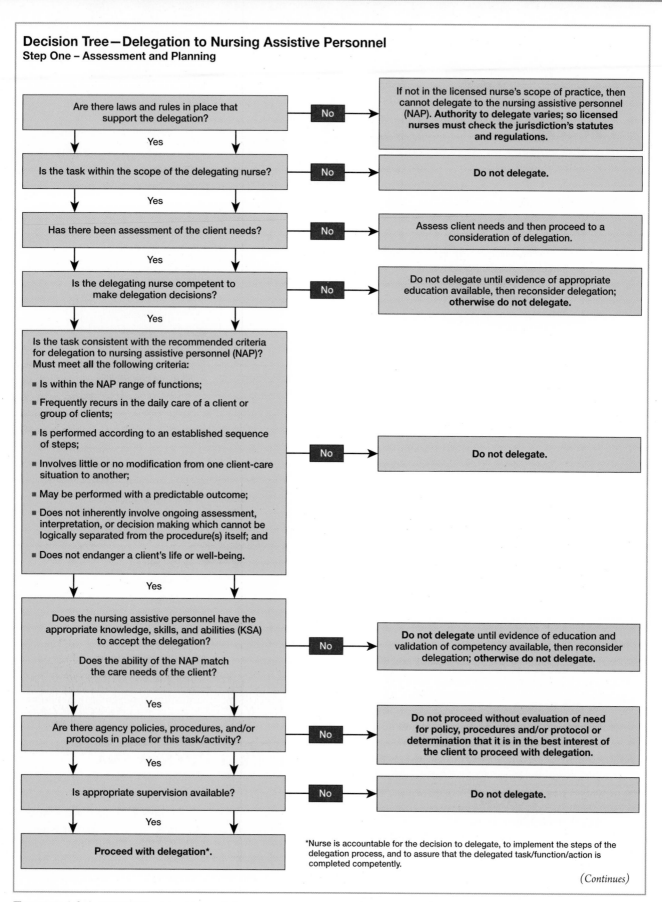

## Decision Tree—Delegation to Nursing Assistive Personnel
### Step One – Assessment and Planning

Are there laws and rules in place that support the delegation?

No → If not in the licensed nurse's scope of practice, then cannot delegate to the nursing assistive personnel (NAP). **Authority to delegate varies; so licensed nurses must check the jurisdiction's statutes and regulations.**

Yes ↓

Is the task within the scope of the delegating nurse?

No → **Do not delegate.**

Yes ↓

Has there been assessment of the client needs?

No → Assess client needs and then proceed to a consideration of delegation.

Yes ↓

Is the delegating nurse competent to make delegation decisions?

No → Do not delegate until evidence of appropriate education available, then reconsider delegation; **otherwise do not delegate.**

Yes ↓

Is the task consistent with the recommended criteria for delegation to nursing assistive personnel (NAP)? Must meet **all** the following criteria:

- Is within the NAP range of functions;
- Frequently recurs in the daily care of a client or group of clients;
- Is performed according to an established sequence of steps;
- Involves little or no modification from one client-care situation to another;
- May be performed with a predictable outcome;
- Does not inherently involve ongoing assessment, interpretation, or decision making which cannot be logically separated from the procedure(s) itself; and
- Does not endanger a client's life or well-being.

No → **Do not delegate.**

Yes ↓

Does the nursing assistive personnel have the appropriate knowledge, skills, and abilities (KSA) to accept the delegation?

Does the ability of the NAP match the care needs of the client?

No → **Do not delegate** until evidence of education and validation of competency available, then reconsider delegation; **otherwise do not delegate.**

Yes ↓

Are there agency policies, procedures, and/or protocols in place for this task/activity?

No → Do not proceed without evaluation of need for policy, procedures and/or protocol or determination that it is in the best interest of the client to proceed with delegation.

Yes ↓

Is appropriate supervision available?

No → **Do not delegate.**

Yes ↓

**Proceed with delegation*.**

*Nurse is accountable for the decision to delegate, to implement the steps of the delegation process, and to assure that the delegated task/function/action is completed competently.

*(Continues)*

**FIGURE 16-4** NCSBN Decision Tree—delegating to nursing assistive personnel. (*Source:* Reprinted and used by permission of the National Council of State Boards of Nursing, copyright 2005).

## Decision Tree—Delegation to Nursing Assistive Personnel (Continued)
### Step Two – Communication
*Communication must be a two-way process*

| The nurse: | The nursing assistive personnel: | Documentation: |
|---|---|---|
| ■ Assesses the assistant's understanding of:<br>　■ How the task is to be accomplished<br>　■ When and what information is to be reported, including:<br>　　■ Expected observations to report and record<br>　　■ Specific client concerns that would require prompt reporting.<br>■ Individualizes for the nursing assistive personnel and client situation<br>■ Addresses any unique client requirements and characteristics, and expectations<br>■ Assesses the assistant's understanding of expectations, providing clarification if needed<br>■ Communicates his or her willingness and availability to guide and support assistant<br>■ Assures appropriate accountability by verifying that the receiving person accepts the delegation and accompanying responsibility | ■ Asks questions regarding the delegation and seek clarification of expectations if needed<br>■ Informs the nurse if the assistant has not done a task/function/activity before, or has only done infrequently<br>■ Asks for additional training or supervision<br>■ Affirms understanding of expectations<br>■ Determines the communication method between the nurse and the assistive personnel<br>■ Determines the communication and plan of action in emergency situations | Timely, complete, and accurate documentation of provided care<br>■ Facilitates communication with other members of the health care team<br>■ Records the nursing care provided |

### Step Three – Surveillance and Supervision
*The purpose of surveillance and monitoring is related to nurse's responsibility for client care within the context of a client population. The nurse supervises the delegation by monitoring the performance of the task or function and assures compliance with standards of practice, policies and procedures. Frequency, level, and nature of monitoring vary with needs of client and experience of assistant.*

| The nurse considers the: | The nurse determines: | The nurse is responsible for: |
|---|---|---|
| ■ Client's health care status and stability of condition<br>■ Predictability of responses and risks<br>■ Setting where care occurs<br>■ Availability of resources and support infrastructure<br>■ Complexity of the task being performed | ■ The frequency of onsite supervision and assessment based on:<br>　■ Needs of the client<br>　■ Complexity of the delegated function/task/activity<br>　■ Proximity of nurse's location | ■ Timely intervening and follow-up on problems and concerns<br>　Examples of the need for intervening include:<br>　■ Alertness to subtle signs and symptoms (which allows nurse and assistant to be proactive, before a client's condition deteriorates significantly)<br>　■ Awareness of assistant's difficulties in completing delegated activities<br>　■ Providing adequate follow-up to problems and/or changing situations is a critical aspect of delegation |

### Step Four – Evaluation and Feedback
*Evaluation is often the forgotten step in delegation.*

*In considering the effectiveness of delegation, the nurse addresses the following questions:*
- Was the delegation successful?
  - Was the task/function/activity performed correctly?
  - Was the client's desired and/or expected outcome achieved?
  - Was the outcome optimal, satisfactory, or unsatisfactory?
  - Was communication timely and effective?
  - What went well; what was challenging?
  - Were there any problems or concerns; if so, how were they addressed?
- Is there a better way to meet the client need?
- Is there a need to adjust the overall plan of care, or should this approach be continued?
- Were there any "learning moments" for the assistant and/or the nurse?
- Was appropriate feedback provided to the assistant regarding the performance of the delegation?
- Was the assistant acknowledged for accomplishing the task/activity/function?

FIGURE 16-4 (Continued)

## TABLE 16-6  Delegation Task Examples

*Nursing Tasks That May Not Be Delegated*

- Patient assessment (physical, psychological, and social assessment, which requires professional nursing judgment, intervention, referral, or follow up). Data collection without interpretation is not assessment.

- Planning of nursing care and assessment of the patient's response

- Implementation that requires judgment

- Health teaching and health counseling other than reinforcement of what the RN has already taught

- Evaluation of the patient's response

- Medication administration

*Nursing Tasks Not Routinely Delegated*

Note that these may sometimes be delegated if the staff has received special credentialing such as education and competency testing.

- Sterile procedures

- Invasive procedures such as inserting tubes in a body cavity or instilling or inserting substances into an indwelling tube

- Care of broken skin other than minor abrasions or cuts generally classified as requiring only first aid treatment

- Intravenous therapy

*Nursing Tasks Most Commonly Delegated*

- Noninvasive and nonsterile treatments

- Collecting, reporting, and documenting data such as:

  - Vital signs, height, weight, intake and output

  - Ambulation, positioning, and turning

  - Transportation of the patient within the facility

  - Personal hygiene and elimination, including cleansing enemas

  - Feeding, cutting up food, or placing meal trays

  - Socialization activities

  - Activities of daily living

Cultural background also influences the space or physical closeness that individuals maintain between themselves. Ineffective delegation can take place when an individual's space is violated. Some delegators stand too close when speaking. Conversely, some members of a group may feel left out if they are not sitting close to the delegator. They may not feel included or important.

Social organization or social support varies in different cultures. If staff look to other staff for social support, those staff may have difficulty fulfilling any tasks delegated to them that could threaten their social organization.

Another cultural phenomenon affecting delegation is the concept of time. How often have you heard

## CASE STUDY 16-1

During your next clinical rotation, practice filling out the NCSBN Decision Tree—Delegating to Nursing Assistive Personnel in Figure 16-4 for one of your patients. Identify the patient's needs, and identify what an RN could safely delegate. What did you note?

people say, "They are on their own time schedule"? Some people tend to move slowly and are often late, whereas other people move quickly and are prompt in meeting deadlines. Poole, Davidhizar, and Giger (1995) describe different cultural groups as being either past, present, or future oriented.

Past-oriented cultures may invest time in preparation of food that is traditional even though the food can be bought already prepared in a store. Present-oriented cultures may focus on working hard for today's wages and not planning for the future. Future-oriented cultures worry about what might

## TABLE 16-7 Additional Delegation Suggestions for Nurses

- Consider prior to delegating

  - Who has the time to complete the delegated task?

  - Who is the best person for the task?

  - What is the urgency of the task?

  - Are there any time restraints?

  - Who do you want to develop their skills?

  - Who is the best person to meet the patients needs?

  - Who would enjoy completing the task?

- Be clear on the qualifications of the delegate, that is, education, experience, and competency. Require documentation or demonstration of current competence by the delegate for each task. Clarify patient care concerns or questions about the delegated task. Find ANA position statements at www.nursingworld.org and at your state board of nursing yearly to keep up to date with changes that may occur regarding delegation of nursing care.

- Speak to your delegates as you would like to be spoken to. There is no need to apologize for your need to have staff perform a task. Remember that you are carrying out your professional responsibility. Assure the delegates that you will help them when you can.

- Communicate the patient's name, room number, and the task to be performed, and identify the time frame for completion. Discuss any changes from the usual procedures that might be needed to meet special patient needs and any potential or expected changes that should be reported to the RN. The expectations for personnel before, during, and after task performance should be stated in a clear, pleasant, direct, and concise manner.

- Identify the expected patient outcome and the limits on staff authority.

- Verify the delegate's understanding of delegated tasks, and have the delegate repeat instructions as needed. Verify that the delegate accepts responsibility and accountability for carrying out the task correctly. Require regular, frequent mini-reports about patients from the health care team including NAP, LPN/LVNs, respiratory therapists, etc.

- Avoid removing tasks once assigned. This should be considered only when the task is above the level of the personnel, such as when the patient's care is in jeopardy because the patient's status has changed.

*(Continues)*

## TABLE 16-7 (Continued)

- Monitor task completion according to standards. Make frequent walking rounds to assess patient outcomes. Intervene as needed.

- Accept minor variations in the style in which the tasks are performed. Individual styles are acceptable as long as the patient standards are maintained and good outcomes are achieved.

- Try to meet staff needs for learning opportunities, and consider any health problems and work preferences of the staff as long as they don't interfere with meeting patient needs.

- If a delegate doesn't meet the standards, talk with them to identify the problem. If this is not successful, inform the delegate that you will be discussing the problem with the supervisor. Document your concerns, as appropriate. Follow up with the supervisor according to your organization's policy.

- Avoid high-risk delegation. The RN may be at risk if the delegated task can be performed only by the RN according to law, organizational policies and procedures, or professional standards of nursing practice; if the delegated task could involve substantial risk or harm to a patient: if the RN knowingly delegates a task to a person who has not had the appropriate training or orientation; or if the RN fails to adequately supervise the delegated activity and does not evaluate the delegated action by reassessing the patient (ANA, 2005).

**Source:** Adapted from Boucher (1998); Zimmermann (1996); and ANA, (2005).

# EVIDENCE FROM THE LITERATURE

**Citation:** Nursing and Midwifery Council (2008, May). Advice on delegation for registered nurses and midwives. Retrieved July 1, 2010, from www.nmc-uk.org.

**Discussion:** This article provides advice on delegation to nurses and midwives in the United Kingdom (UK). Nurses in the United States can also use this advice. The Nursing and Midwifery Council is the United Kingdom's regulatory body organization for nursing and midwifery. The primary purpose of the Council is to protect the public. This role is similar to that of the state boards of nursing in the United States. *Non-regulated personnel* is the term in the UK for "NAP." The term *registrant* is the UK term for "nurse." When a nurse or midwife has authority to delegate tasks to another, he or she will retain responsibility and accountability for that delegation. A nurse or midwife may only delegate an aspect of care to a person deemed competent to perform the task. The nurse or midwife should be sure that the person to whom care has been delegated fully understands the nature of the delegated task and what is required. Where another person, such as an employer, has the authority to delegate an aspect of care, the employer becomes accountable for that delegation. The nurse or midwife will, however, continue to carry the responsibility to intervene if he or she feels that the proposed delegation is inappropriate or unsafe. The decision whether or not to delegate an aspect of care or to transfer and/or rescind delegation is the primary responsibility of the nurse or midwife and is based on his or her professional judgment. Nurses and midwives have the right to refuse to delegate if they believe it would be unsafe to do so or if they are unable to provide or ensure adequate supervision. Those delegating care and those assuming delegated duties should do so in accordance with robust local employment practice in order to protect the public and support safe practice.

**Implications for Practice:** The decision to delegate is made either by the nurse, midwife, or by the employer. It is this decision maker who is accountable for the decision. Health care can sometimes be unpredictable. It is important that the person to whom an aspect of care is being delegated understands his or her limitations. He or she must know when not to proceed should circumstances affecting the task change. No one should feel pressured into either delegating or accepting a delegated task. When pressure is felt, advice should be sought, as appropriate, from either the nurse or the midwife's professional manager.

## TABLE 16-8  Cultural Phenomena

| PHENOMENA | EXAMPLE |
| --- | --- |
| Communication | Consider cultural communication elements such as communication volume, dialect, use of touch, context of speech, and kinesics such as gestures, stance, and eye behavior as you delegate patient care to staff. |
| Space | Consider physical closeness as you delegate patient care to staff. Some cultures prefer to stand close physically while communicating. Others prefer to maintain more physical distance between themselves and others. |
| Social Organization | When communicating with patients, consider that cultures vary in the amount of close social supports they maintain with family and others. Note that staff also vary in the amount of social support that they need from other health care staff. |
| Time | Cultures vary in their past, present, or future orientation. Note that some cultures focus on maintaining past traditions, while other cultures focus on the current activities of today. Still other cultures focus on preparing for the future. |
| Environmental Control | Note that cultures with an internal locus of control plan and take action. They don't rely on luck or fate. Cultures with an external locus of control wait for fate and luck to determine and guide their actions. |
| Biological Variations | Note that there are cultural and biological variations in attributes such as physiological strength, stamina, and susceptibility to disease. Consider these as you delegate patient care to staff. |

# CASE STUDY 16-2

You are working with an RN, NAP, an LPN/LVN, and a sitter. Which patient(s) from the list in Figure 16-5 would you give to each of them? Who would you have do the P.M. care for all patients, pass water, answer call lights, and pick up supplies? Who would give the medications and change dressings?

| ROOM/NAME | PATIENT DESCRIPTION | SPECIAL NEEDS |
| --- | --- | --- |
| | | Report vitals and outcomes to Mary, RN, at 8:30 P.M. <br><br> Report anything abnormal STAT |
| 2501/Ms. J. D. | 68-year-old female, post-op day 1, post-shoulder repair <br><br> Confused; fall risk; side rails up | Up in chair at 6 P.M. <br><br> Maintain safety <br><br> Vitals at 4 P.M. and 8 P.M. check distal pulses <br><br> Monitor level of consciousness (LOC) <br><br> Check dressings at 4 P.M. and 8 P.M. <br><br> Check voiding at 6 P.M. <br><br> Family at bedside |

*(Continues)*

## CASE STUDY 16-2 (Continued)

| 2502/Mr. D. H. | 45-year-old male diabetic, post-op day 1, amputation just below the knee; Insulin sliding scale;<br><br>Complaining of pain; restlessness; diaphoretic | Vitals and Accucheck STAT and at 4 P.M. and 9 P.M.<br><br>Up in chair 6 P.M.<br><br>Pain medication as needed<br><br>Monitor dressing<br><br>No pillow under stump |
|---|---|---|
| 2503/Mr. H. M. | 35-year-old male, history of alcohol abuse<br><br>Complaining of abdominal pain; new hematemesis of coffee-ground fluid; IV of 0.9% normal saline at 125 cc/hour; alert | Vitals Q 15 minutes<br><br>Monitor LOC, hematemesis, and possible seizures, 16 gauge IV, type and crossmatch<br><br>Possible transfer to ICU |
| 2504/Mr. J. K. | 20-year-old male college student, just admitted, threatening to commit suicide; alert and oriented | Vitals at 4 P.M. and 8 P.M.<br><br>Do not leave unattended<br><br>Maintain safety |

FIGURE 16-5  Assigment form.

(Delmar/Cengage Learning)

## TABLE 16-9

Identify which members of the health care team may do each of the following nursing activities.

| Nursing activity | RN | LPN | NAP |
|---|---|---|---|
| Administer blood to a patient | | | |
| Assess a patient going to surgery | | | |
| Develop a teaching plan for a newly diagnosed patient with diabetes | | | |
| Measure a patient's intake and output | | | |
| Provide a bath to an immobilized patient | | | |
| Give a dressing change to a patient | | | |
| Give patient report when transferring a patient from ICU to a step-down unit | | | |
| Give insulin | | | |
| Evaluate a patient's DNAR status | | | |
| Give an oral medication | | | |
| Assist a patient with ambulation | | | |
| Give an IM pain medication | | | |

# EVIDENCE FROM THE LITERATURE

**Citation:** Berkow, S., Virkstis, K., Stewart, J., & Conway, L. (2008). Assessing new graduate nurse performance. *Journal of Nursing Administration, 38*(11), 468–474.

**Discussion:** New graduate nurses comprise more than 10% of a typical hospital's nursing staff. With that being said, only 10% of hospital and health system nurse executives surveyed believe their new graduate nurses are fully prepared to provide safe and effective care. The Nursing Executive Center administered a national survey to a cross section of frontline nurse leaders on new graduate nurse proficiency across 36 nursing competencies deemed essential to safe and effective nursing practice. Based on the survey results, the most pressing and promising opportunities for improving the practice readiness of new graduate nurses were identified.

**Implications for Practice:** Most notably, all of the competencies in the category "management of responsibilities" can be found in the lowest-ranked subset. Most of the competencies in the lowest-ranked subset tend to be more "applied" competencies. New graduate nurses' greatest improvement needs seem to center on skills such as taking initiative, tracking multiple responsibilities, and delegation—all skills that are more readily taught in the clinical area as opposed to a classroom. Therefore, an evaluation of the education students receive must reflect the competencies and should be a priority.

## TABLE 16-10  Cultural Views

- What is your cultural stereotype of Hispanics? Africans? Chinese? Filipinos? Germans? Irish? Polish? Others?
- How do your grandparents, parents, family, and close friends view people from these cultures?
- Have your interactions with these cultural groups been positive? Negative? Neutral?
- How do you feel about going into a neighborhood or into the home of a person from one of these cultural groups?
- Does your cultural stereotype allow for socioeconomic differences?
- What culturally based health practices do you think characterize people from these cultures? How are these practices different from your own culturally based health practices?

**Source:** Compiled with information from Swanson, J., & Nies, M. (1997). *Community health nursing: Promoting the health of aggregates.* (2nd ed.). Philadelphia: WB Saunders.

happen in the future and prepare diligently for a potential problem, perhaps financial or health related. A nurse delegator should always be aware of duties to be completed and their deadline so that appropriate personnel can achieve their responsibilities in a timely fashion. Otherwise, people who meet deadlines in a timely fashion may be frustrated by those who do not.

Poole, Davidhizar, and Giger (1995) define environmental control as people's perception of their control over their environment. This is also called internal locus of control. Some cultures place a heavier weight on fate, luck, or chance, believing, for example, that a patient is cured from cancer based on chance. They may think the health care treatment had something to do with the cure but was not the sole cause of it. How personnel perceive their control of the environment may affect how they delegate and perform duties. Personnel with an internal locus of control are geared toward taking more personal initiative and

## CASE STUDY 16-3

A new nursing graduate, Jamilla, has been assigned to work with Abdul, an NAP, and five patients. Jamilla introduces herself to Abdul and asks him what types of patient care he usually performs. He tells Jamilla that he gives baths and takes vital signs. Jamilla asks Abdul to get all of the vital signs and give them to her written on a piece of paper. She asks Abdul if he documents them. He states that he does document them.

Later that morning, Dr. Kent is making rounds on his patients, two of whom are Jamilla's patients. He asks Jamilla for the most recent vital signs. She then asks Abdul for the vital signs on all the patients. Abdul tells her he has not taken them yet. Dr. Kent then asks Jamilla to get the vitals herself. By the time Jamilla returns with the vital signs, Dr. Kent has gone and has written orders she cannot read.

There are several factors in this delegation situation that should have been handled differently. Can you name any?

Do you think the new graduate was ready to delegate to the NAP? Why?

Were the duties delegated appropriate for the NAP?

Would review of the job description for each health care personnel identified in this case study help solve this problem in the future?

### Critical Thinking 16-7

Deborah M. Nadzam, PhD, practice leader for patient safety at the Joint Commission (2010), writes, "Fundamental is clear communication—with the patient and with each other. ... The pace of health care today necessitates better communication with each other as well. Regardless of the health care setting, we are caring for sicker patients, with fewer staff, and limited resources. Any knowledge we have about the patient should be shared with each other in writing and during hand-offs of care. ... Let's use effective methods when communicating with others to emphasize key data."

During your clinical laboratory experiences, have you observed or received hand-off reports of patient care that were clear? If yes, how do you know the communication emphasized key data? If no, what data was omitted?

not requiring assistance in decision making. They believe in taking action and not relying on fate. Personnel with an external locus of control may wait for fate and luck to determine their actions.

The last cultural phenomenon is biological variations between racial and ethnic groups. These variations include physiological differences, physical stamina, and susceptibility to disease. Such variations need to be considered. For example, it would be problematic if the care of a comatose patient who weighs more than 300 pounds and needs frequent turning were delegated to a small nurse who cannot physically handle the patient. Hospital policy must address how to meet all patient and staff needs safely. Perhaps this patient can be assigned to two nurses. A nurse who is pregnant may not be assigned to this patient because of the potential injury to the baby and the nurse. Likewise, a pregnant nurse would not be assigned to a patient with radium implants because of the risks that the radium poses to the baby and the mother. Biological variations must be considered, for the sake of both the health care providers and the patient.

## Critical Thinking 16-8

Note the following selected list of values:

| **Mainstream American Values** | **Other Cultural Values** |
| --- | --- |
| Make your own luck. | Fate and luck determine your life. |
| Like change. | Like tradition. |
| Arrive on time. | Frequently arrive late. |
| Value the individual. | Value the group. |
| Value competition. | Value cooperation. |
| Set goals for the future. | Enjoy life and just let it happen. |
| Value directness. | Value being subtle. |
| All people have a fairly equal chance to achieve status in life. | Some people will always have higher status in life. |

Which of these values do you hold? Which do members of your staff hold? How can you work to improve communication around these values and improve your working relationships?

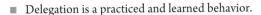

# KEY CONCEPTS

- Delegation is a practiced and learned behavior.
- The RN must have a clear definition of what constitutes the scope of practice of all personnel.
- The five rights of delegation are the right task, the right circumstance, the right person, the right direction and communication, and the right supervision and evaluation.
- Accountability is being responsible and answerable for actions or inactions of self or others.
- Responsibility involves reliability, dependability, and the obligation to accomplish work when one accepts an assignment. Responsibility also includes each person's obligation to perform at an acceptable level.
- Authority occurs when a person who has been given the right to delegate based on the state nurse practice act also has the official power from an agency to delegate.
- The RN is accountable for the delegation and performance of all nursing duties.
- There are several potential barriers to good delegation.
- Transcultural delegation is encouraged to provide a patient with optimal care.
- Supervision is the provision of guidance or direction, evaluation, and follow-up by the licensed nurse for accomplishment of a nursing task delegated to an NAP.

- Assignment is the distribution of work that each staff member is responsible for during a given shift or work period.
- To assign is to describe those situations when a nurse directs an individual to do something the individual is already authorized to do.
- Key elements must be in place in an organization for efficient nursing delegation to occur.
- Professional judgment is the intellectual (educated, informed, and experienced) process that a nurse exercises in forming an opinion and reaching a clinical decision based upon an analysis of the available evidence.
- Organizations interested in quality patient care provide staff with guidelines on how to use the chain of command.
- The NCSBN Decision Tree—Delegation to Nursing Assistive Personnel is a useful tool when developing skill in delegating patient care.
- All members of the health care team must fulfill their delegated responsibilities.
- Nursing staff provide direct and indirect patient care.

## KEY TERMS

accountability
assign
assignment
authority

clinical judgment
competence
delegation

direct patient care
activities
indirect patient care
activities

responsibility
supervision

## REVIEW QUESTIONS

1. When a nurse considers delegating a task, what five rights should be utilized?
   A. Right patient, right chart, right physician, right results, right information
   B. Right person, right patient, right task, right documentation, right time frame
   C. Right task, right circumstance, right person, right direction/communication, right supervision
   D. Right room, right time, right person, right documentation, right directions

2. The nurse has become incredibly busy with discharging two patients and expecting a new patient any second. The following are tasks that need to be completed right away. What task can the nurse delegate to an NAP to help out with managing the nurse's time with patients?
   A. Remove sutures from an incision and apply dressing to the patient's left wrist.
   B. Provide tracheostomy care to the patient.
   C. Sit with a patient who was recently diagnosed with Crohn's disease who is crying.
   D. Transport a patient to X-ray on a cardiac monitor.

3. The staff working on the unit includes four RNs, two LPNs, and an NAP for 25 patients. What assignment is the most appropriate for one of the LPNs?
   A. Gather data about a newly admitted patient.
   B. Pass medications to a group of patients.
   C. Pass water to all of the patients on the unit.
   D. Obtain a urine sample from a Foley catheter on a patient who is not assigned to the LPN.

4. What is the most appropriate task for the RN to delegate to the NAP?
   A. Silence the IV pump until the RN arrives.
   B. Notify the family of a patient who has died.
   C. Administer a soap suds enema to a patient who has requested it for constipation.
   D. Reinforce teaching to a patient who has had an above the knee amputation.

5. What parts of the nursing process cannot be delegated to a NAP? Select all that apply.
   _____ A. Assessment
   _____ B. Nursing diagnosis
   _____ C. Planning
   _____ D. Intervention
   _____ E. Evaluation

6. The charge nurse working with an RN, an LPN, and a NAP is very busy with the group of patients on the unit. One patient's intravenous line has just infiltrated, a practitioner is on the phone waiting for a nurse's response, a patient wants to be discharged, and the NAP has just reported an elevated temperature on a new surgical patient. Who should be assigned to restart the intravenous line?
   A. LPN
   B. NAP
   C. RN
   D. Charge nurse

7. A new graduate nurse is assigned a patient who is two days postoperative who has had a colostomy. The patient has an order to have a nasogastric tube inserted immediately. The new graduate has never inserted this type of tube in a patient. How should the new graduate nurse proceed in this situation?
   A. Delegate the task to a NAP.
   B. Read over the procedure, and then insert the tube.
   C. Notify the practitioner of the new graduate's inexperience.
   D. Ask an experienced RN for assistance with the procedure.

8. When a nurse considers delegating a task, which of the following are cultural phenomena that should be considered?
   _____ A. Communication
   _____ B. Space
   _____ C. Social organization
   _____ D. Environmental control
   _____ E. Biological variations
   _____ F. Right results

9. Who is responsible for educating the health care team in an organization on the policies, practices, and laws of delegation?
   A. American Nurses Association
   B. National Council of State Boards of Nurses
   C. Health care organization
   D. The state

10. When a nurse asks another nurse to observe his or her group of patients while at lunch, and one patient falls out of bed, which nurse is responsible?
   A. The nurse originally assigned to the patient who went to lunch is responsible.
   B. The nurse who was observing the group of patients is responsible.
   C. Neither nurse is responsible.
   D. The actions of both nurses will be reviewed.

## REVIEW ACTIVITIES

1. Have you had any clinical opportunities to delegate duties? Identify to whom and what you delegated. Discuss how it affected the patient and your work. What would you do differently next time?

2. Observe delegation procedures at your institution. Is transcultural delegation considered? If so, which phenomena have you observed?

3. You are caring for a new patient in Room 2510. You are trying to decide whether to delegate his care to NAP Jill or to NAP Penny. Jill is not certified and is not always easy to work with but usually does her fair share of the work. Penny is certified. She always does her fair share and is easy to work with . She is able to perform dressing changes. Use the decision grid below to decide.

4. Look at the assignment sheet for the shift your clinical day is on. Who is listed on the assignment sheet? What assignments are made on the sheet? During the shift, did you observe an RN assign another RN a job to do for them? If yes, could the NAP have done it?

5. Discuss with an NAP and an RN their preparation in regard to delegation. How much education or training has each of them received? How long ago did they receive it? What type of education or training did they receive? Is the RN familiar with the five rights of delegation and the delegation tree? How is the education of the RN and the training of the NAP different?

**Decision Grid**

|       | Certified | Easy to work with? | Do their fair share? | Other? |
|-------|-----------|--------------------|----------------------|--------|
| Jill  |           |                    |                      |        |
| Penny |           |                    |                      |        |

## EXPLORING THE WEB

- Log on to www.aacn.org
  Click on Public Policy and find the advisory team member of your state. Note what policies, if any, consider delegation of care to NAP and LPNs.
- Log on to www.nursingworld.org, the American Nurses Association (ANA) site, to view safety and quality of care issues.

- Visit http://www.iom.edu, the Institute of Medicine's Web site.
  Search for Job Delegation.

# REFERENCES

American Association of Colleges of Nursing. The essentials of baccalaureate education for professional nursing practice. Available at http://www.aacn.nche.edu/education/pdf/Baccessentials08.pdf. Accessed July 5, 2010.

American Association of Critical Care Nurses (AACN). (2008). *AACN scope and standard of acute and critical care nursing practice.* Aliso Viejo, CA: Author.

American Nurses Association (ANA). (2005). Principles for delegation. Available at http://www.safestaffingsaveslives.org/WhatisSafeStaffing/SafeStaffingPrinciples/PrinciplesforDelegatio tml.aspx#Definitions. Accessed November 14, 2010.

American Nurses Association (ANA) and National Council of State Boards of Nursing (NCSBN) (2006). *Joint statement on delegation.* Retrieved April 5, 2011, from www.ncsbn.org/Joint_statement.pdf.

Berkow, S., Virkstis, K., Stewart, J., & Conway, L. (2008). Assessing new graduate nurse performance. *Journal of Nursing Administration, 38*(11), 468–474.

Boucher, M. A. (1998). Delegation alert. *American Journal of Nursing, 98*(2), 26–32.

Center for American Nurses. (2008). Registered nurse utilization of nursing assistive personnel: Statement for adoption. Retrieved November 1, 2010, from www.centerforamericannurses.org/positions/finalassistivepersonnel.pdf.

Gordon, S. C., & Barry, C. D. (2009). Delegation guided by school nursing values: A comprehensive knowledge, trust, and empowerment. *Journal of School Nursing, 25*(5), 352–360.

Hansten, R., & Jackson, M. (2009). *Clinical delegation skills* (4th ed.). Boston: Jones and Bartlett.

Kelly, P., & Marthaler, M. (2011). *Nursing delegation, setting priorities and making assignments* (2nd ed.) Clifton Park, NY: Delmar Cengage Learning.

Marthaler, M., & Kelly, P. (2010). Delegation of nursing care. In P. Kelly, *Nursing leadership and management* (2nd ed.). Clifton Park, NY: Delmar Cengage Learning.

National Association of School Nurses. (2010). National association of school nurses announces new student-to-school nurse ratios. Silver Springs, MD. Retrieved April 5, 2011, from http://www.nasn.org/portals/0/releases/2010_02_02_ratio_ruler_release.pdf.

National Council of State Boards of Nursing (NCSBN). (1995). Delegation concepts and decision-making process, NCSBN position paper. Retrieved April 5, 2011, from https://www.ncsbn.org/323.htm.

National Council of State Boards of Nursing (NCSBN). (2005). Working with others: A position statement. Chicago: Author.

NCSBN. (1997). Five rights of delegation. Retrieved from http://www.ncsbn.org/fiverights.pdf.

NCSBN. (2005). Meeting the Ongoing Challenge of Continued Competence. Retrieved April 6, 2011, from https://www.ncsbn.org/Continued_Comp_Paper_TestingServices.pdf.

Needleman, J., Buerhaus, P. I., Stewart, M., Zelevinsky, K., & Mattke, S. (2006). Nurse staffing in hospitals: Is there a business case for quality? *Health Affairs, 25*(1), 204–211.

Nightingale, F. (1859). *Notes on nursing: What it is and what it is not.* London: Harrison & Sons.

Nursing and Midwifery Council. (2008, May). Advice on delegation for registered nurses and midwives. Retrieved July 1, 2010, from www.nmc-uk.org.

Poole, V., Davidhizar, R., & Giger, J. (1995). Delegating to a transcultural team. *Nursing-Management, 26*(8), 33–34.

Robert Wood Johnson Foundation. (2010). Majority of states fail to provide children with adequate access to health care in school. Retrieved from http://www.rwjf.org/pr/product.jsp?id=59272, retrieved April 5, 2011.

Spilsbury, K., & Meyer, J. (2005). Making claims on nursing work: Exploring the work of healthcare assistants and implications for registered nurses' roles. *Journal of Research in Nursing, 10*(1), 65–83.

Tanner, C. A. (2008). Thinking like a nurse: A research-based model of clinical judgment in nursing. *Journal of Nursing Education, 45*(6), 204–211.

The Canadian Nurses Association and Canadian Association of Schools of Nursing. (2004). Joint Position Statement, Promoting Continuing Competence for Registered Nurses. Retrieved April 6, 2011, from http://www.cna-nurses.ca/CNA/documents/pdf/publications/PS77_promoting_competence_e.pdf.

Zimmermann, P. G. (1996). Delegating to assistive personnel. *Journal of Emergency Nursing, 22*(3), 206–212.

# SUGGESTED READINGS

Beckstrand, R. L., Kirchhoff, K. T. (2005). Providing end-of-life care to patients: Critical care nurses' perceived obstacles and supportive behaviors. *American Journal of Critical Care, 14*(5), 395–403.

Bittner, N. P., & Gravlin, G. (2009). Critical thinking, delegation, and missed care in nursing practice. *Journal of Nursing Administration, 39*(3), 142–146.

Dumpel, H. (2005). Contemporary issues facing international nurses. *California Nurse, 18–22.*

Fulks, C., & Thompson, J. (2008). Charge nurses: Investing in the future. Retrieved July 21, 2010, from www.besmith.com.

Hanston, R. I. (2005), Relationship and results-oriented health care. *Journal of Nursing Administration, 35*(12), 522–524.

Hatler, C., Buckwald, L., Salas-Allison, Z., & Murphy-Taylor, C. (2009). Evaluating central venous catheter care in a pediatric intensive care unit. *American Journal of Critical Care, 18*(6), 514–520.

Institute of Medicine (IOM). (2008). Retooling for an aging America: Building the health care workforce. Retrieved on July 17, 2010, from www.iom.edu/agingamerica.

Johnson, S. H. (1996). Teaching nursing delegation: Analyzing nurse practice acts. *Journal of Continuing Education in Nursing, 27*(2), 52–58.

Kleinman, C. (2004). Leadership strategies in reducing staff nurse role conflict. *Journal of Nursing Administration, 34*(7/8), 322–324.

Lavizzo-Mourey, R., & Berwick, D. (2009). Nurses transforming care. *American Journal of Nursing, 109*(11), 3. doi:10.1097/01.NAJ.0000362008.30472.8c.

Mark, B. A., & Harless, D. W. (2007, Summer). Nurse staffing, mortality, and length of stay in for-profit and not-for-profit hospitals. *Inquiry, 44,* 167–186.

Nurse Midwifery Council. (2007). NMC issues new advice for delegation to non-regulated healthcare staff, UK. Retrieved July 12, 2010, from http://www.medicalnewstoday.com/articles/79367.php.

O'Keefe, C. (2005, Jan/Feb). State laws and regulations for dialysis: An overview. *Nephrology Nursing Journal, 32*(1), 31–37.

Pasquale, P. (2007, Winter). Expanding nurse delegation to allow insulin injections: Is it safe practice? *Washington Nurse, 37*(4), 8–9.

Ponte, P. R., Glazer, G., Dann, E., McCollum, K., Gross, A., Tyrrell, R., et al. (2007). The power of professional nursing practice: An essential element of patient and family centered care. *Online Journal of Issues in Nursing, 12.* Retrieved July 1, 2010, from http://nursingworld.org/ojin.

Quallich, S. A. (2005). A bond of trust: Delegation. *Urological Nursing, 25*(2), 120–123.

Rowe, A. R., Savigny, D., Lanata, C., & Victora, C. G. (2005). How can we achieve and maintain high-quality performance of health workers in low-resource settings? *Lancet, 366,* 1026–1035.

Tourigny, L., & Pulich, M. (2006). Delegating decision making in health care organizations. *The Health Care Manager, 25*(2), 101–113.

Whitman, M. (2005). Return and report. *American Journal of Nursing, 105*(3), 97.

# CHAPTER 17

# Organization of Patient Care

**KATHLEEN F. SELLERS, PhD, RN**

*The object of shared governance is to build a structure that supports the point of care delivery, and the patient, and sustains ownership and accountability there.*

(TIM PORTER-O'GRADY, 1995)

## OBJECTIVES

Upon completion of this chapter, the reader should be able to:

1. List the components that contribute to the organization of patient care.
2. Identify elements of strategic planning—philosophy, mission, vision.
3. Distinguish a model of nursing shared governance.
4. Differentiate among Benner's concepts of novice, advanced beginner, competent, proficient, and expert nursing practice.
5. Illustrate accountability-based nursing practice.
6. Identify measures of a unit's performance.

Delmar/Cengage Learning

*The patient care manager of an acute care surgical unit has been informed that there are plans to merge that unit with an ambulatory surgery unit that currently cares for patients requiring 24-hour observation. As a visionary with a great depth of experience, the patient care manager has been recommended to oversee the development of the new work unit. The institution believes the creation of this new unit will enhance revenue, staff productivity, and continuity of patient care. Therefore, resources are available to design and staff the new work unit in a manner that is congruent with the institution's mission, with the understanding that the investment will bring added value to the organization.*

*What are your reactions as a new nurse on this unit?*

*What unit structures and patient care processes need to be put in place?*

*What committees and/or work teams would you get involved with so that you will have a voice in creating this new work environment?*

*What care delivery system would you like to practice in?*

*How will you ensure your competency and continued professional growth on this new unit?*

Organization of patient care is the coordination of resources and clinical processes that promote patient care delivery. Coordination of resources, clinical processes, and, therefore, care delivery occurs through senior, middle, and frontline staff nurse management levels. Regardless of the level, organization of patient care management utilizes the nursing process to assess, plan, implement, and evaluate the outcomes of care for populations of patients. The organization of patient care management is akin to conducting a large orchestra. Like the conductor, the patient care manager's primary function is to lead or coordinate a team of diverse individuals with varied talents and expertise toward a common goal (MacGregor-Burns, 1979). The orchestra creates beautiful music. The patient care team provides an outcome of quality, cost-effective patient care.

Successful organization of patient care management requires governance structures, patient care delivery processes, and measures of the outcomes of care delivery.

These must be consistent with the mission and vision of the organization and are built on a philosophy of professional practice. The organization of patient care management is built on the tenets of professional nursing practice and requires a structure of shared decision making or shared governance between nursing management and clinical nursing staff. Such a framework creates an environment in which the processes of patient care delivery demand an accountability-based system where staff are able to report, explain, and justify their actions. In such an environment, the outcomes of care delivery, clinical quality, access, service, and cost can regularly be evaluated and staff can continue to grow.

# UNIT STRATEGIC PLANNING

**Strategic planning** is a process designed to achieve goals in dynamic, competitive environments through the allocation of resources.

## ASSESSMENT OF EXTERNAL AND INTERNAL ENVIRONMENT

As outlined in Figure 17-1, strategic planning involves clarifying the organization's philosophical values or what is important to the organization; identifying the mission of why the organization exists; articulating a vision statement of what the organization wants to be; and then conducting an environmental assessment, or SWOT analysis, which examines the strengths, weaknesses, opportunities, and threats of the organization. This information provides data that then drive the development of three- to five-year strategies for the organization. Tactics are then created and prioritized. Finally, goals and objectives are concretized into annual operating work plans for the organization, which can be measured. This same process is used for unit or departmental strategic planning. In developing a strategic plan, unit staff must also examine their organization's mission, vision, strategic plan, and annual operating plans. Unit strategic plans should be congruent with and support the mission and vision of the organizational system of which they are a part. Therefore, communication with the nurse executive who is responsible for the unit is a key step.

## DEVELOPMENT OF A PHILOSOPHY

A **philosophy** is a statement of beliefs based on core values—inner forces that give us purpose (Raphael, 1994). A unit's mission and vision are most authentic if they are developed based on the philosophy or core beliefs of the staff work team (Wesorick, Shiparski, Troseth, & Wyngarden, 1998). Core beliefs may be complex or they can be short statements developed from a staff brainstorming session, such as "patient centered," "partnering," "healing environment," and the like. A unit's core beliefs or values are then incorporated into the unit's mission and vision statements.

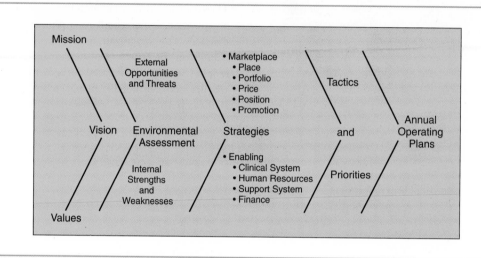

FIGURE 17-1 **Bassett Healthcare strategic planning framework.** (*Source:* Developed by Gennaro J. Vasile, PhD, FACHE).

## REAL WORLD INTERVIEW

At our academic health science center, leaders in the organization, board members who represent the community, and customer stakeholders develop the strategic plan. After the strategic plan is developed, it is published and reviewed at a centerwide management meeting. It is then reviewed in divisional meetings and presented to staff through unit staff meetings. This is where the voice of the chief nursing officer has the most impact. I essentially interpret the rationale for the corporate strategic plan to my staff and glean their reactions. I then communicate the staff's feedback to the corporate level.

Articles are also published in our organizational newsletter for all staff, describing the plan and addressing points of clarification. Each division and department then undertakes the process of developing divisional and department plans that support the strategic plan. For example, in the strategic plan a few years ago, it was articulated that our academic center would become a major cardiac center with a state-of-the-art cardiac catheterization laboratory. The department of cardiac services then included development of a state-of-the-art cardiac catheterization laboratory into its plan.

**Anne L. Bernat, RN, MSN, CNAA**
Chief Nursing Officer
Arlington, Virginia

## Mission Statement

A **mission** is a formal expression stating the primary purpose of the practice unit, i.e., its reason for being. A mission statement reflects why the unit exists and provides a clear view of what the unit is trying to accomplish. It indicates what is unique about the care that is provided (www.csuchico.edu). The mission should be a reflection of the unit's core values or philosophy.

Covey (1997) states, "An organizational mission statement—one that truly reflects the shared vision and values of everyone within that organization—creates a unity and tremendous commitment" (p. 139). For the unit mission statement to have the greatest effect, all members of the unit staff should participate in its development.

Questions to be answered by the group charged with development of the unit mission include the following:

- What do we stand for?
- What principles or values are we willing to defend?
- Who are we here to help?

Mission statements are often so broad that many nursing units adopt the nursing department's mission statement, as shown in Figure 17-2, which is based on the overall organization mission statement.

The Bassett Nursing Association is a diverse group of autonomous, professional nurses committed to patients, their families, the Bassett and larger community, and each other. We strive for nursing and patient care excellence based on a solid foundation of professional standards, multidisciplinary teamwork, ongoing education, and evidence.

**FIGURE 17-2** The Bassett Nursing Association's mission.
(*Source:* Courtesy of Connie Jastremski, MS, RN, VP Nursing and Patient Care Services, Bassett Healthcare.)

## Vision Statement

The unit vision statement reflects the organization's vision of what the organization wants to be.

Following are four elements of a vision:

1. It is written down.
2. It is written in present tense, using action words, as though it were already accomplished.
3. It covers a variety of activities and spans broad time frames.
4. It balances the needs of providers, patients, and the environment. This balance anchors the vision to reality (Wesorick et al., 1998).

An environmental assessment of strengths, weaknesses, opportunities and threats (SWOT) to the organization is useful.

## Goals and Objectives

The next step in the strategic planning process is to develop broad strategies that span the next three to five years and then develop annual goals and objectives to meet each of these strategies. A **goal** is a specific aim or target that the unit wishes to attain within the time span of one year. An **objective** is the measurable step to be taken to reach a goal. Performance measures of the goals and objectives can be included in a performance improvement plan like the plan illustrated later in this chapter.

# THE STRUCTURE OF PROFESSIONAL PRACTICE

In an organization in which professional nursing practice is valued, strategic initiatives are developed and implemented most effectively through a structure of shared governance and shared decision making between management and clinicians.

## SHARED GOVERNANCE

**Shared governance** is an organizational framework grounded in a philosophy of decentralized leadership that fosters autonomous decision making and professional nursing practice (Porter-O'Grady, Hawkins, &

Parker, 1997). Shared governance, by its name, implies the allocation of control, power, or authority (governance) among mutually (shared) interested, vested parties (Stichler, 1992).

In most health care settings, the vested parties in nursing fall into two distinct categories: (1) nurses practicing direct patient care, such as staff nurses, and (2) nurses managing or administering the provision of that care, such as managers. In shared governance, a nursing organization's management assumes the responsibility for organizational structure and resources. Management relinquishes control over issues related to clinical practice. In return, staff nurses accept the responsibility and accountability for their professional practice.

Unit-based shared governance structures are most successful if there is an organization-wide structure of nursing shared governance in place that unit-based functions can articulate with. The nursing shared governance structure is most effective if the entire health care system is supported by whole-systems shared governance. In most health care organizations, nursing shared governance was adopted first, as nursing is the largest professional work group and practices closest to the point of service delivery, the patient (Porter-O'Grady, Hawkins, & Parker, 1997). However, the principles of **whole-systems shared governance**—that is, partnership, equity, accountability, and ownership—apply to all professionals practicing in the organization. The principle of partnership connotes horizontal linkages with nursing and other staff roles that are clearly negotiated. With the principle of equity, individual staff roles are based on relationships, not titles. The contributions stemming from staff roles are understood and valued. The principle of accountability comes from within. Individuals are encouraged to report, explain, and justify their actions. This leads to the principle of ownership for the work performed.

Nursing shared governance structures are usually council models that have evolved from preexisting nursing or institutional committees. In a council structure, clearly defined accountabilities for specific elements of professional practice have been delegated to five main arenas: clinical practice, quality, education, research, and management of resources (Porter-O'Grady, et al., 1997). Figure 17-3 illustrates a shared governance model.

## Clinical Practice Council

The purpose of the clinical practice council is to establish the practice standards for the work group. Often, this council or committee is a unit-level committee that works in conjunction with the organizational committee accountable for determining policy and procedures related to clinical practice. Evidence-based practice fostered by research utilization initiatives ensures that practice standards are developed based on the state of the science of clinical practice and not merely on tradition.

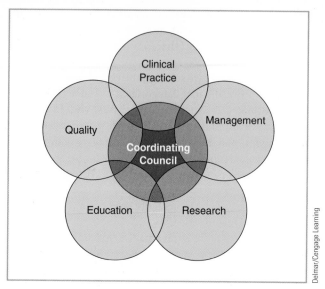

**FIGURE 17-3**  A shared governance model.

## Quality Council

The purpose of the quality council is twofold: (1) to credential staff and (2) to oversee the unit quality management initiatives. In the role of credentialing staff, this committee is responsible for interviewing potential staff and reviewing their qualifications, or credentials. It then makes recommendations regarding hiring. The quality committee also serves as the body that reviews staff credentials on an ongoing basis and makes recommendations regarding promotion.

Quality management initiatives for which the council is responsible can include review of indicators of the unit's overall clinical performance, such as medication errors, patient falls, family satisfaction, and response time in answering call lights. At times, a unit will also participate in an organizational disease management study looking at the care of a specific patient population, such as patients with diabetes.

## Education Council

The purpose of the education council is to assess the learning needs of the unit staff and develop and implement programs to meet these needs. According to Peter Senge (1990), **learning organizations** promote professional practice through the encouragement of personal mastery, an awareness of our mental models, and team learning. Personal mastery goes beyond competence and skills to include continually clarifying and deepening our personal vision and focusing our energies and developing patience to see reality objectively. Mental models are deeply ingrained assumptions, generalizations, and biases that influence how we understand the world. Mental models influence how we take action. Therefore, the more insight we have into what our mental models are, the more effective we will be at team learning. Team learning is a work group's ability to align and develop their collective talents for the purpose of attaining a shared vision (Smith, 2001).

The education council usually works closely with organizational education and training departments. Unit orientation programs and training programs related to new clinical techniques and new equipment are examples of programs sponsored by the education council.

---

## EVIDENCE FROM THE LITERATURE

**Citation:** Sellers, K., Millenbach, L., Kovach, N., & Klimek-Yingling, J. (2010). Prevalence of horizontal violence in New York State registered nurses. *Journal of the New York State Nurse*, Fall/Winter 2009–2010 (40), 20–25.

**Discussion:** Horizontal violence (HV), also known as bullying, is described as aggressive behavior towards individuals or group members by others (Hastie, 2002). It is a little-known phenomenon that recent evidence demonstrates is prevalent in the workplace of practicing registered nurses. Horizontal violence contributes to nursing turnover and undermines a culture of professional nursing practice. Examples of HV include acts of unkindness, dishonesty, and divisiveness such as gossip, verbal abuse, intimidation, sarcasm, elitist attitudes, and fault finding. It is often said that "nurses eat their young."

The Institute of Medicine (IOM) and the Joint Commission (2010) have identified HV, along with other negative actions, as contributing to patient errors. The literature has shown that HV can be addressed once it is recognized to be present in the nursing culture.

**Implications for Practice:** A three-tiered approach by the organization's administration, the frontline nurse managers who are the cultural gatekeepers of the work unit, and professional nurses themselves is necessary to change this disruptive, harmful HV behavior. Education and heightened awareness of the existence of HV as well as implementing a no-tolerance organizational policy, confronting behaviors when they occur, and encouraging staff to respond to manifestations of HV are part of the solution.

## Research Council

At the unit level, the research council advances evidence-based practice with the intent of staff incorporating research-based findings into the clinical standards of unit practice. Evidence-based practice involves staff critiquing the available research literature and then making recommendations to the practice council so that clinical policies and procedures can be based on evidence-based research findings. The research council may also coordinate research projects if advanced practice nurses are employed at the institution.

## Management Council

The purpose of the management council is to ensure that the standards of practice and governance agreed upon by unit staff are upheld and that there are adequate resources to deliver patient care. The first-line patient care manager is a standing member of this council. Other members include the assistant nurse managers and the charge or resource nurses from each shift.

## Coordinating Council

Shared governance structures also include a coordinating council. The purpose of the coordinating council is to facilitate and integrate the activities of the other councils. This council is usually composed of the first-line patient care manager and the chairpeople of the other councils. This council usually facilitates the annual review of the unit mission and vision and develops the annual operational plan (Sellers, 1996).

Unit-based shared governance structures may be less diverse. Often, some of the councils are combined into one council—for example, education and research. Or a council may contain subcommittees whose purposes are to perform very specific tasks—for example, credential and promote staff or recruit and retain staff. Unit-based structures are varied, with the primary purpose being to empower staff by fostering professional practice while meeting the needs of the work unit. A shared governance structure and shared decision making among disciplines integrates various functions and thus fosters the organization's patient care management of services (Porter-O'Grady, 2009).

# RELATIONSHIP-BASED CARE

Mary Kolorutis (2007) expounds on the constructs of shared decision making and creates a model of care delivery known as relationship-based care that places the patient and family at the center of all decision making in health care organizations. The center for organizing patient care is coordinated by a unit practice council (UPC), which reports to a larger organizational shared governance model (see Figure 17-4).

There are six major components of relationship-based care that support patients and families in a caring and healing environment:

- leadership
- teamwork
- professional nursing practice
- patient care delivery
- resource-driven practice
- outcomes measurement

Relationship-based care was a framework used by Faxton-St Luke's Healthcare in Utica, New York, to transform their organizational culture. See Real World Interview with Pat Roach.

### Critical Thinking 17-1

As a nurse practicing in a shared governance organization, you remember a decade ago when the organization decentralized, made a commitment to nursing professional practice, and implemented shared governance. Everywhere you went, people were talking about it and displaying posters and other signs of nursing's importance to the organization. That was six years ago, before managed care and all the changes and before this latest nursing shortage. Now you do not hear people talking about it so much. You wonder, does professional practice still exist? How can you tell? Is this an organization that you can be proud to work in? Are you a magnet organization?

**FIGURE 17-4 Relationship-based care model.**
(*Source:* Creative Health Care Management, Inc., Minneapolis, Minnesota).

## REAL WORLD INTERVIEW

Relationship-based care (RBC) is a model for transforming our organizational culture. I'd like to share a little bit about the six dimensions that the model is based upon. This will help create a clearer understanding of our patient care delivery system.

Note that six dimensions—i.e., leadership, teamwork, professional nursing practice, patient care delivery, resource-driven practice, and outcomes measurement—surround the patient and family in the model. This serves as a key guide to the success of our organizational cultural transformation. The entire model is surrounded by, and rests on, a caring and healing environment. This environment is promoted through the power of relationships and respecting the needs of each person in every interaction with the patient and family.

In relationship-based care, we define every employee as a leader. Leaders:

- Know the vision
- Act with purpose
- Create and sustain caring and healing environments
- Remove barriers to quality patient care
- Consistently make patients and families their highest priority
- Solve problems creatively to get results
- Model and support the changes that they desire

Healthy interdependent teamwork is one of the most statistically significant predictors of quality care. All disciplines/departments must work as a team and define and embrace a shared purpose to achieve quality. We've made a decision to roll out our RBC model as a collegial multidisciplinary team. Some may view RBC as a nursing model, but from day one we've been fortunate to have a multidisciplinary team at the table that naturally extends to the bedside and beyond.

Our support services focus on assuring that patients, families, and nurses have the resources that they need. Input from the frontline staff is solicited through the unit practice councils (UPC). UPCs are teams comprised of 20% of the staff in specific units/departments that are elected by their peers to translate agreed-upon principles into action plans. Additionally, a support practice council made up of allied health professional and support services staff meets regularly to identify opportunities to streamline processes and improve patient care delivery. In a resource-driven practice environment, the therapeutic relationship with the patient and family serves as the basis for clinical resource allocation. Our patients' needs come first. Clinical staff and managers share responsibility for resources needed to provide care.

Caring is the essence of all that a nurse delivers and is truly the cornerstone of professional nursing practice. Professional nursing practice exists to provide compassionate care delivery to patients and their families, help them heal and cope during times of stress and suffering, and help them experience a dignified, peaceful death.

Our organizational vision statement determines the quality indicators that are measured. These are:

- Quality
- People
- Service
- Growth
- Finance

As you can see, RBC serves as an excellent vehicle to help drive our success and make this a GREAT place to receive care and a GREAT place to work.

**Pat Roach, MS, RN, OCN, CNAA,**
Senior Vice President/Chief Nursing Officer
Faxton-St Luke's Healthcare
Utica, New York

# ENSURING COMPETENCE AND PROFESSIONAL STAFF DEVELOPMENT

Professional practice through the vehicle of shared governance requires competent staff. **Competence** is the ongoing ability of a nurse to integrate and apply the knowledge, skills, judgment, and personal attitudes required to practice safely and ethically in a designated role and setting (Canadian Nurses Association, 2004). Competence of professional staff can be ensured through credentialing processes developed around a clinical or career ladder staff promotion framework. A **clinical ladder** acknowledges that staff members have varying skill sets based on their education and experience. As such, depending on skills and experience, staff members may be rewarded differently and carry differing responsibilities for patient care and the governance and professional practice of the work unit.

## BENNER'S NOVICE TO EXPERT

Benner's (1984) **Novice to Expert Model** provides a framework that, when developed into a clinical or career promotion ladder, facilitates professional staff development by building on the skill sets and experience of each practitioner.

## CASE STUDY 1-1

You are a staff nurse who is a member of the credentialing committee of the quality council. A fellow peer has presented his credentials for review in hopes of being promoted to the next level on the clinical ladder. You review the packet and make the recommendation that he be promoted. However, at the credentialing committee meeting, it is revealed that the patient care manager and the individual's preceptor, another member of the committee, have not recommended promotion.

You wonder if your colleague is aware that there were concerns about his performance.

Are there guidelines and standards that you are not aware of that have not been met?

What is the next course of action for the committee?

What should your response be at this meeting?

## REAL WORLD INTERVIEW

I'm a new graduate nurse. I understand how tough it is adjusting straight out of school into a "real" job. My fellow new graduates and I did some brainstorming and came up with a couple of things to help make your transition at Bassett an easier one. Here goes!

- Ever wonder where all the Carpujets are? Just when you need to flush an IV, you can't find one anywhere. Guess what? Carpujets are located at the pharmacy! Just give them a call, and they will send you a bunch.

- How about all those tabs on the edge of the charts? It took me two months to find out what they mean on surgery! Red is for STAT orders. Blue is for medicine orders. Yellow alerts the unit clerk. Green alerts the RN.

- How to make your shift run smoother: It's always a good idea to round with the doctors. They know all kinds of information you need to know, and it's a great time to give your input. Collaboration!

- Know what team your patient is on and what doctor is on that specific team and who is coming onto the shift and who is leaving. Nothing is worse than trying to page a doctor before you realize they aren't there! That can be a little embarrassing, not to mention time consuming.

- Words of wisdom: Write things down; it helps you remember.

- Just one more tidbit: Always bring any documentation and patient medication and vital sign information with you to the phone before you page the doctor. Then, you'll have the answers for questions.

I hope that this information will be useful to you.

**Christina Denton, RN, Surgery Unit**

Bassett Healthcare
Cooperstown, New York

Benner's model acknowledges that there are tasks, competencies, and outcomes that practitioners can be expected to have acquired based on five levels of experience. Note that the Ten Year Rule states that it takes a decade of heavy labor to master any field (Ross, 2006).

Benner's Novice to Expert model is based on the Dreyfus and Dreyfus (1980) model of skill acquisition applied to nursing. There are five stages of Benner's model: novice, advanced beginner, competent, proficient, and expert. Table 17-1 discusses Benner's model and shows the

## TABLE 17-1  Benner's Model of Novice to Expert

| STAGE OF MODEL | APPLICATION TO NURSING PRACTICE |
| --- | --- |
| Novice nurses are recognized as being task oriented and focused on the rules. They need a directing, telling style from a mentor. They tend to see nursing as a list of tasks to do rather than seeing the big picture of patient care needed to meet patient care goals. After novices have mastered most tasks required to perform their ascribed roles, they move on to the phase of advanced beginner. | The novice nurse is educated in techniques associated with delegation. A nurse new to the direct patient care setting may have been educated in principles of delegation, but he or she has not used them in the clinical setting. These nurses are very task oriented and focused and are often still in orientation. They may delegate tasks clearly outlined by the hospitals; for example, they may ask the nursing assistive personnel (NAP) to pass water. They often cannot decide what else to delegate. Novices often tend to do all tasks themselves and need a directing, telling style from mentors. |
| Advanced beginner nurses demonstrate marginally acceptable independent performance. They are learning to apply newly acquired knowledge and skills to many situations. They have enough experience to grasp aspects of the situation. This nurse still focuses on the rules but is more experienced and just needs some coaching. This nurse still needs help identifying priorities. | This nurse is out of orientation, has worked for a short while on the unit, and is able to perform most nursing tasks that are required for patient care. This nurse is becoming more comfortable independently delegating simple tasks to NAP, that is, errands, assisting in positioning of patients, bathing, and taking vital signs. The nurse is often reluctant to delegate to staff whose personality is resistant to delegation. This nurse still needs coaching from a mentor. |
| Competent nurses have one to three years' experience and have developed the ability to see their actions as part of the long-range goals set for their patients. They lack the speed of the proficient nurse, but they are able to manage most aspects of clinical care. Competent nurses can cope with the contingencies of clinical nursing. They use conscious planning to help achieve efficiency and organization skills. They recognize patterns and use this knowledge to make decisions. They can personalize care for each patient. They are gaining perspective but are not yet able to select out the most important elements in the overall picture. | One to three years in the same role has allowed these nurses to develop the ability to delegate to NAP and LPNs. They have developed a higher level of ability to apply the nursing process and use nursing skills. The competent nurse is more able to assess the NAP's abilities, communicate expectations effectively, and gather clinical information from the NAP. The competent nurse is more comfortable communicating and delegating to staff, even in the presence of personality conflicts. This nurse expects that all staff must work to meet the requirements of their job description. |
| Proficient nurses usually have three to five years' experience and characteristically perceive the whole situation rather than a series of tasks. They have often been on the job several years and have been delegated total responsibility for their patients' care. They develop a plan of care and then guide the patient from point A to point B. These nurses need minimal guidance and control and only occasional support from a mentor. They draw on their past experiences and know that in a typical situation, a patient must exhibit specific behaviors to meet specific goals. They realize that if those behaviors are not demonstrated within a certain time frame, then the plan needs to change. | These nurses are often charge nurses developing plans of care for the whole unit. They can see delegation of tasks as an important part of guiding patients from point A to point B. They are able to use past experiences with patients and NAP to guide the delegation process. They may need a little occasional support from their mentors. |

*(Continues)*

## TABLE 17-1 (Continued)

| STAGE OF MODEL | APPLICATION TO NURSING PRACTICE |
|---|---|
| Expert usually have five years' experience or more and intuitively know what is going on with their patients. Their expertise is so embedded in their practice that they have been heard to say, "There is something wrong with this patient. I'm not sure what is going on, but you had better come and evaluate them." Not heeding the call derived from the intuitive sense of an expert nurse can result in a patient's condition deteriorating, with subsequent development of the nurse-sensitive outcome of cardiac arrest. These expert nurses usually seek continuing education. | This nurse intuitively knows what is going on with patients and their needs. They can quickly assess what needs to be delegated. They evaluate the situation continuously. |

**Source:** Developed with information from Benner (1984); Hersey & Blanchard (1993); and Kelly (2010).

appropriate application to nursing practice that a professional nurse would be expected to perform at each level. Proficiency in completion of these tasks contributes to readiness for promotion along a career ladder. Note that all nurses who care for patients must meet basic criteria for safe care.

The Bassett Healthcare Professional Nursing Pathway (Figure 17-5) has four stages and builds on the work of Benner (1984) regarding career ladder stages. Stage I of the Bassett Healthcare Professional Nursing Pathway is characterized as the entry/learning stage. Stage II is characterized by the individual who competently demonstrates acceptable performance, adapting to time and resource constraints. Stage III is characterized by the individual who is proficient. And stage IV is characterized by the individual who is an expert. The stages in this model are specifically defined by behaviors that are consistently exhibited or practiced over a defined period of time. Figure 17-6, the Bassett Healthcare Professional Nursing Pathway Algorithm, provides an overview of how nurses advance in this process.

## ACCOUNTABILITY-BASED CARE DELIVERY

Accountability-based care delivery is essential in today's value-driven workplace. Accountability-based care delivery systems focus on roles, their relationship to the work to be done, and the outcomes they are intended to achieve. In a professional context, accountability includes the exercise of activities inherent to a role that cannot and are not legitimately controlled outside the role. Competence in accountability-based approaches to work is evidenced not by what a person brings to the work, but instead by the results of the application of the person's

### Critical Thinking 17-2

Review the interview from Christina Denton. This was published in the newsletter of Bassett Healthcare's Nursing Association. Christina manages care as a new graduate and shares what she has learned with her peers in a public forum, the newsletter.

How might you contribute to the management and leadership of your organization as a new graduate?

## REAL WORLD INTERVIEW

We decided to update our clinical ladder at Bassett a few years ago. In doing so, the clinical ladder was transformed into the Professional Nursing Pathway, a more complex, accountable, and objective means of credentialing professional nursing staff. The Pathway continues to be based on the traditional tenets of Benner's (1986) novice to expert theory, scaled at four levels—i.e., novice, competent, proficient, and expert. (We collapsed the novice and advanced beginner level into one level.) In addition, the Professional Nursing Pathway incorporates the 14 Forces of Magnetism. The process for advancement is outlined in a document known as the Professional Development Pathway Algorithm, which is posted on the Bassett Intranet as part of the Pathway Manual. Staff have access to the Pathway Manual at all times.

Staff are expected to develop and grow a professional portfolio that showcases their competencies and accomplishments. The portfolio is reviewed at the time of the annual performance evaluation and must demonstrate the accumulation of points based on specific professional activities. A certain number of points is required to maintain a level, and additional points are required as part of promotion to the next level of the Professional Development Pathway. While all professional staff are evaluated annually, the nurse may initiate movement to the next level of the Pathway at any time with approval of the nurse's supervisor.

Additionally, we have now implemented a process called Accountability Review to audit our Pathway implementation process and to gather feedback from the nurses as end users. Every month, Credentialing Committee members interview four randomly selected nurses who have completed the annual performance appraisal and portfolio review. In a 15-minute dialogue, the nurses' experience with the Professional Development Pathway and an independent critique of the nurse's portfolio is completed to ensure that staff and their supervisors are meeting the standards of the performance appraisal process.

We have found that the Pathway has actually "pulled" the nurses into professional practice activities and has also provided a keener awareness of what professionalism is!

**Maureen Fitzgerald Murray, MS, RN, NE-BC**
Director, Professional Nursing Practice
Bassett Medical Center/Bassett Healthcare Network
Cooperstown, New York

---

skills to the work (Porter-O'Grady, 1995). Individuals who are accountable, by definition, are able to report, explain, or justify their actions (*Merriam-Webster*, 2005) (Table 17-2).

# THE PROCESS OF PROFESSIONAL PRACTICE

Ongoing professional staff development is part of the regular performance feedback that staff can expect to receive from the patient care manager or the credentialing committee. However, all patient care managers provide ongoing professional development of staff in their daily interactions on the unit by identifying projects and activities that enhance a staff member's readiness for leadership development and advancement.

## SITUATIONAL LEADERSHIP

The leadership framework developed by Hersey and Blanchard (1993), when combined with an individual's position on a clinical/career ladder, is useful to a patient care manager in discerning the best approach to take in developing the potential of staff members. **Situational leadership** maintains that there is no one best leadership style, but rather that effective leadership lies in matching the appropriate leadership style to the individual's or group's level of task-relevant readiness. Readiness is how able and motivated an individual is to perform a particular task. A

## REAL WORLD INTERVIEW

There are four levels of our clinical ladder, which is similar to Benner's Novice to Expert model. The RN novices are the new graduates and people in orientation. The experts are the clinical specialists. A lot of them have also become nurse practitioners so that the organization can receive some reimbursement for their patient care services. This is a good thing, because otherwise I'm afraid we wouldn't have these expert nurses anymore. They are the true mentors for nursing staff, especially when you are working with a very complex or difficult patient situation.

Staff nurses also mentor each other. During orientation, your preceptor guides you along the path from RN I to RN II. When you decide you'd like to advance to RN III, you can choose another mentor. RN IIIs provide much more clinical leadership for staff and for the overall unit. I decided I was ready to be promoted to that level when other staff consistently were coming to me for clinical guidance and with patient care questions. Now, as an RN III, I am the chairperson of our unit credentialing committee, which is part of the quality council of our shared governance model.

Our clinical ladder uses a portfolio as the main tool to evaluate the nurse's readiness to advance. When you are an RN I in orientation, you are first introduced to the idea of a portfolio and how to put it together. It is difficult at first, as people do not know what is expected. However, after that first time when you are promoted from an RN I to an RN II, it becomes easier. You just build on what is already in the portfolio.

A portfolio should include the following:

Licenses

Your resume

Letters of reference

Evaluations

Clinical documentation of patient care

Validations for competencies related to technical
   skills (medication administration, IV therapy)

Examples of participation in development
   of the team plan of care

Exemplars

CEU certificates

Presentations

Publications

The portfolio tells the story of your practice. When a group of people are ready for promotion, the members of the credentialing committee meet. We review the portfolios and make recommendations related to advancement. The nurse manager is a member of this committee. She always reviews the portfolio and gives us her feedback even if she is unable to attend the credentialing meeting. I enjoy reading the exemplars the best. Exemplars are mini-stories that paint the pictures of each nurse's practice, and they are all so different.

**Stacey Conley, RN, BS**
Staff Nurse
Cooperstown, New York

basic assumption of situational leadership is the idea that a leader should help followers grow in their readiness to perform new tasks as far as they are able and willing to go. This development of followers is accomplished by adjusting leadership behavior through four styles along the leadership continuum: directing, coaching, supporting, and delegating (Figure 17-7).

According to Hersey and Blanchard's (1993) leadership framework, individuals with low leadership readiness, such as novice nurses, initially require a directing, telling style on the part of the mentor. They need strong direction if they are to be successful and productive. For example, if a patient requires a new IV line to be started, the mentor should inform the novice that this needs to be done and then review

| | RN I (NOVICE) | RN II (COMPETENT) | RN III (PROFICIENT) | RN IV (EXPERT) |
|---|---|---|---|---|
| **ACADEMIC PREPARATION** | • AAS diploma required<br><br>• Nursing BS or MS or non-nursing bachelor or higher degree preferred | • AAS diploma required<br><br>• Nursing BS or MS or non-nursing bachelor or higher degree preferred | • AAS diploma required<br><br>• Nursing BS or MS or non-nursing bachelor or higher degree preferred | • Nursing BS or MS with 5 years' experience as described below required<br>or<br>• non-nursing bachelor or higher degree with 5 years' experience as described below and national certification in specialty area<br>or<br>• National certification in specialty area with 10 years' experience as described below |
| **WORK EXPERIENCE** | • New graduate, 1 year or less at time of hire | • Six months to 2 years minimum as an RN at time of hire<br><br>• licensed | • 3 years minimum as an RN, minimum 1 year experience in the specialty area | • With any bachelor's degree, 5 years minimum as an RN, 3 years' experience in the specialty area, 2 years' experience in Bassett organization<br><br>• Without bachelor's degree, 10 years' minimum experience as an RN, 3 years' experience in specialty area, 2 years' experience in Bassett organization |
| **CLINICAL PRACTICE** | • Orientation<br>• Works under preceptor<br>• Residency program<br>• Attains licensure<br>• Gaining toward competent practice | • Demonstrates accountable competent practice<br>• Works independently<br>• Coordinates care team through appropriate delegation and supervision<br>• Preceptor<br>• Developing resource responsibilities<br>• Gaining toward proficient practice | • Demonstrates proficient practice<br>• Unit leadership<br>• Resource responsibilities<br>• Problem solving<br>• Participates in setting standards<br>• Preceptor<br>• Clinical resource<br>• Gaining toward expert practice | • Demonstrates expert practice<br>• Self-directed<br>• Leads performance improvement activities<br>• Preceptor/mentor<br>• Organization-wide leadership, problem solving, and standards development |
| **PROFESSIONAL DEVELOPMENT** | • Complete orientation and residency<br>• Establish portfolio<br>• Accountable for mandatory requirements | • Uses evidence-based practice<br>• Delegates care<br>• Unit based committee or work group activities<br>• Fully successful rating on current performance appraisal | • Supports peers<br>• Develops area of specialty/resource<br>• Participates in professional practice activities<br>• Current on professional issues<br>• Fully successful rating on current performance appraisal | • Leadership in professional practice activities<br>• Promotes national standards of practice area<br>• Fully successful rating on current performance appraisal |
| **ADDITIONAL PROFESSIONAL ACHIEVEMENTS** | | • 20 minimum points in minimum of 3 categories, maximum 10 points per category.<br><br>• Graduate nurses are required to submit points at their second anniversary performance evaluation. | • 25 minimum points in minimum of 4 categories, maximum of 10 points per category | • 30 minimum points in minimum of 5 categories, maximum of 10 points per category |

**FIGURE 17-5  Bassett Healthcare Professional Nursing Pathway, Cooperstown, New York.** (*Source:* Stacey Conley, B.S., RN).

**CHECKLIST FOR AREA-BASED/UNIT-BASED CREDENTIALING PROCESS**

### DISCUSSION AND APPROVAL TO APPLY PHASE

- _____ Applicant has been functioning consistently at higher level for past 6 months.
- _____ Applicant discusses goal for advancement with Supervisor/Director.
- _____ Supervisor/Director approves of the plan to apply for promotion.
- _____ Applicant submits letter of intent to Supervisor/Director and unit-based Credentialing Committee.
- _____ Applicant and Supervisor/Director decide if a sponsor for applicant is necessary and, if so, Supervisor/Director appoints a sponsor in concert with applicant.
- _____ Applicant and Supervisor/Director agree on a time frame.

### PREPARATION PHASE

- _____ Applicant prepares portfolio as required, using policy and related documents and including verification of required points.

### PEER REVIEW PHASE

- _____ Supervisor/Director prepares a cohort of peers who have observed the nursing practice of the applicant and who agree to and are capable of engaging in a constructive, respectful, and thoughtful peer review process. Nurses are recommended for promotion based on their observed consistent performance to standards, including a brief audit of compliance with documentation standards and the preparation and submission of a professional portfolio that includes required elements. Practice must be observed over time and be evident to all members of the health care team with which nurse functions. The team's perceptions must be measurable.

**Purpose of peer review:**

- Establish a process for empowering nursing staff to utilize positive and constructive feedback to promote increased professionalism, productivity, and job satisfaction.
- Identify areas of strength as well as areas for professional/practice development.
- _____ Supervisor/Director publishes the applicant's intention and schedules peer review. Ideally, applicant is present for the peer review and discussion. In selected cases, the Supervisor/Director may facilitate a discussion after the applicant is excused.
- _____ Clinical unit or area-based credentialing committees will follow established policy and procedure for recommending a nurse for promotion on the Clinical Ladder.

### PEER REVIEW RECOMMENDATION

- _____ After peer review and discussion, peer group makes decision concerning promotion recommendation.
- _____ If promotion is not recommended, specifics of remedial plan are defined in writing below and returned to applicant.

_____

_____

_____

_____

- _____ If promotion is recommended, this checklist will be signed by Supervisor/Director and sent to Credentialing Committee Chair for review at the next meeting. The Supervisor/Director or sponsor will attend the next meeting of the Credentialing Committee and bring the applicant's professional portfolio.
- _____ Promotion to _____ is recommended for _____

Signature of Supervisor/Director _____

Date _____

FIGURE 17-6  Bassett Healthcare Professional Pathway Algorithm, Cooperstown, New York. (*Source:* Stacey Conley, B.S., RN).

---

CREDENTIALING COMMITTEE DECISION

- _____ Credentialing Committee reviews the unit-based process and compliance with policy.
- _____ If promotion is not approved, specifics of issue are defined in writing below and returned to Supervisor/Director.

_____

_____

_____

_____

- If the promotion is approved by the Central Credentialing Committee, this checklist will be signed by Credentialing Committee Chair or Co-Chair and returned to Supervisor/Director
- _____ Promotion to _____ is approved for _____

Signature of Credentialing Committee Chair/Co-Chair_____

Date _____

DIRECTOR/MANAGER/CREDENTIALING COMMITTEE ACTIONS

- _____ Supervisor/Director will then initiate the Payroll Authorization form.
- _____ Director/Manager sends letter of congratulations to nurse.
- _____ Credentialing Committee sends certificate, initiates recognition process (Bassett Works, Staff Bulletin, Nursing Matters, and Professional Practice Outcomes).

---

**FIGURE 17-6** (Continued)

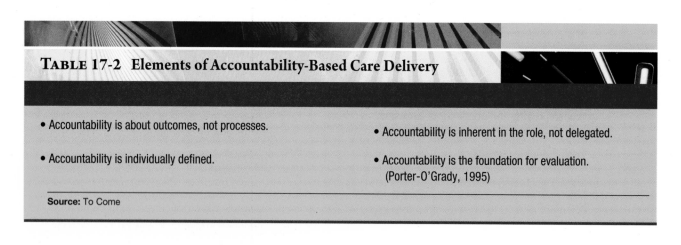

**TABLE 17-2 Elements of Accountability-Based Care Delivery**

- Accountability is about outcomes, not processes.

- Accountability is individually defined.

- Accountability is inherent in the role, not delegated.

- Accountability is the foundation for evaluation. (Porter-O'Grady, 1995)

**Source:** To Come

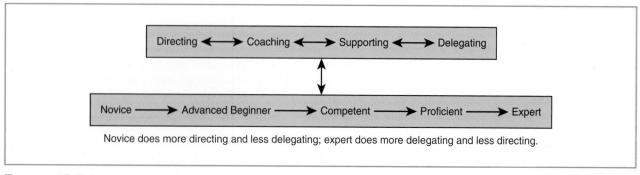

Directing ⟷ Coaching ⟷ Supporting ⟷ Delegating

Novice → Advanced Beginner → Competent → Proficient → Expert

Novice does more directing and less delegating; expert does more delegating and less directing.

**FIGURE 17-7** **Leadership continuum.** (*Source:* Compiled with information from Hersey & Blanchard (1993) and Benner (1984)).

**CASE STUDY 17-2**

You are a staff nurse working as part of the interdisciplinary orthopedic team. You notice that there are an increasing number of diabetic patients being admitted for elective total hip surgery. Because the length of stay is so short and your team has such a surgical focus in caring for patients, their underlying chronic diseases have not been a focus on the unit. However, you are aware that the larger organization is beginning to evaluate how different populations of patients, such as diabetics, are cared for across the continuum of care.

What should you do to improve care for your patients?

the steps required to start the new line using the evidenced-based clinical practice standard followed on that unit. This allows the novice to ask questions and obtain guidance. This enhances the likelihood that the task will be performed correctly the first time, improving patient satisfaction and self-confidence in the novice.

As the professional nurse grows in leadership readiness, the mentor should shift to a coaching style where staff are rewarded with increased relationship behavior, that is, positive reinforcement and socioemotional support. Then, when individuals reach higher levels of leadership readiness, as a proficient or expert nurse, the mentor should respond by decreasing control. The mentor moves first to a participatory, supportive style characterized by a high-quality working relationship with the proficient staff and a lower need to give task direction. The mentor then moves to a style of more complete delegation, communicating a sense of confidence and trust. Highly competent individuals respond best to this greater freedom.

Individuals' readiness to learn and accept new tasks may change for a variety of reasons. When patient care managers discern a change, they must readjust their style of interaction with the nurse—moving forward or backward through the leadership continuum from directing to coaching to supporting to delegating—and provide the appropriate level of support and direction to facilitate that individual's continued development, productivity, and success as a member of

the patient care team. Development of professional staff based on their innate readiness to accept new tasks and responsibilities facilitates their promotion along a continuum of novice to expert and ensures a patient care team that is able to consistently deliver accountability-based patient care.

# MEASURABLE QUALITY OUTCOMES

An important component of organizational patient care management is regular evaluation of a work unit's performance to ensure that the outcomes of care delivery are meeting the objectives of professional practice as outlined in the unit's annual operational plan. The development of process improvement measures in today's health care organizations is driven by the multiple domains of quality required by the Joint Commission (JC), formerly known as the Joint Commission on Accreditation of Healthcare Organizations (JCAHO), (www.jointcommission.org) and the National Council for Quality Assurance (NCQA) (www.ncqa.org), the credentialing organization that certifies managed care organizations.

## TRANSFORMING CARE AT THE BEDSIDE

Since to "Err is Human" was published by the Institute of Medicine (IOM, 2000), a main emphasis of performance improvement has been on patient safety. Transforming Care at the Bedside (TCAB) (www.ihi.org) is a nationwide project sponsored by the Robert Wood Johnson Foundation to look for changes in patient care processes that enhance patient safety and demonstrate improved patient care outcomes.

The TCAB process can be an invaluable process for staff to participate in to determine what variables should be measured/included in the unit-based performance improvement plan.

## UNIT-BASED PERFORMANCE IMPROVEMENT

To develop a comprehensive unit-based quality improvement program to meet the requirements of today's competitive, value-driven health care system, outcomes should be tracked from four domains: access, service, cost, and clinical quality. These outcomes reflect the unit's goals and objectives. See Figure 17-8 for the inpatient surgical unit 2010 performance improvement plan.

**REAL** WORLD INTERVIEW

We use the Transforming Care at the Bedside (TCAB) performance improvement process to improve patient care at St. Peter's Hospital. TCAB is different from other performance improvement processes. The primary characteristic that sets it apart is its focus on engaging frontline staff and unit managers. Ideas for transforming the way that care is delivered on medical-surgical units comes not from the executive suite or from a quality improvement department, but from the front-line nurses and other care team members who spend the most time with patients and their families. The TCAB performance improvement process to improve patient care includes the following:

• Brainstorming new improvement ideas with a frontline team

• Adopting the best health care practices

• Adapting improvement strategies from other industries

• Implementing and spreading successful changes

• Measuring outcomes

We started the TCAB performance improvement process at St. Peter's Hospital by doing the following:

• Planning weekly meetings and involving the leaders

• Having key staff members present from all disciplines and all roles

• Identifying a facilitator for the meetings

• Brainstorming and involving all shifts and all staff

• Encouraging ideas and solutions, not just the voicing of problems

• Posing this question to the group: "If you could create the perfect patient, family, staff experience, how would you do it?"

• Encouraging storytelling about actual good patient care delivery experiences

• Prompting all staff to engage in the TCAB process by telling their patient care stories once the weekly TCAB meetings were under way

• Limiting patient care stories to two minutes each

• Recording 107 potential new ideas of small TCAB improvements that could be made

Staff then prioritized and voted on which TCAB performance improvement projects they wanted to address first. They took the attitude of "just do it" and "try something." During weekly meetings, powerful words were heard from staff that communicated that they owned the process. These "words" were the staff's feedback about the TCAB performance improvement process and included newly brainstormed ideas, new strategies when the first plan didn't work, encouragement to each other to keep going with the change(s), and ideas on how to monitor progress. Some of these words were:

• "How might we do this?"

• "What is your prediction?"

• "What lesson did you learn?"

• "What happened that was not expected?"

• "What is the next step in this process?"

• "Small tests of change were evaluated."

• "Think small, very small"

*(Continues)*

## REAL WORLD INTERVIEW (Continued)

- "Test the outcome today, this afternoon."

- "Review results and test again tomorrow."

- "Do not wait a month, a week."

- "Adapt, adopt, abandon."

- "Spread new changes quickly."

Out of all the ideas generated, the following TCAB performance improvement projects were prioritized by staff:

- Improve equipment accessibility.

- Improve linen availability.

- Begin a thought for the day board.

- Use an RN to nursing assistive personnel report sheet.

- Use whiteboard instructions in patient rooms.

These were the staff outcomes:

- Improved staff vitality

- Staff reports of improved communication, improved teamwork, and improved staff satisfaction

**Barbara Bonificio, MS, RN, BC**
Professional Practice Specialist
St. Peter's Healthcare
Albany, New York

---

### 2010 PERFORMANCE IMPROVEMENT PLAN

As part of our organization's commitment to quality, the Surgical Unit will strive to improve performance through a cycle of planning, process design, performance measurement, assessment, and improvement. There will be ongoing assessment of important aspects of care and service and correction of identified problems. Problem identification and solution will be carried out using a systematic intra- and interdepartmental approach organized around patient flow or other key functions and in concert with the approved visions and strategies of the organization. Priorities for improvement will include high-risk, high-volume, and problem-prone procedures.

The Surgical Unit will:

- Promote the Plan-Do-Check-Act methodology for all performance improvement activities
- Provide staff education and training on integrated quality and cost improvement
- Collect data to support objective assessment of processes and contribute to problem resolution

In identifying important aspects of care and service, the Surgical Unit will select performance measures in the following operational categories:

**A. Clinical Quality**

1. Patient safety

- Patient falls
- Indicator: # of patient falls per month/# of patient days with upper control limits set by the research department based on statistical deviation

FIGURE 17-8  Inpatient surgical unit, 2010. Performance Improvement Plan. (*Source:* Courtesy of Patricia Roach, BS, RN, Bassett Healthcare).

- Medication and IV errors
- Indicator: # of patient IV/medication errors per month/# of patient days with upper control limits set by the research department based on statistical deviation

- Restraint use
- Indicator: % of compliance with policy for use of restraints and overall rate of restraint use

2. Pressure ulcer prevention
- Indicator: Rates of occurrence and quarterly tracking report

3. Surveillance, prevention, and control of infection
- Indicator: Infection control statistical report of wound- and catheter-associated infections
- Indicator: Quarterly monitoring of compliance with standards for acid fast bacilli (AFB) room use; evidence of staff validation in AFB practice

4. Employee safety
Injuries resulting from:
- Back- and lifting-related injuries
- Morbidly obese patients
- Orthopedic patients
- Indicators: # of injuries sustained by employees and any resultant worker's compensation (Human Resources quarterly report)
- 100% competency validation in lifting techniques and back injury prevention
- Respiratory fit testing
- Indicator: competency record of each employee

5. Documentation by exception
Indicators:
- 100% validation of RN/LPN staff
- Monthly chart audit (10% average daily census or 20 charts) meeting compliance with established standards

**B. Access**
- Maintenance of the 30-minute standard for bed assignment of ED admissions
- Indicator: Quarterly review of ED tracking record

**C. Service**
Patient satisfaction
Indicator: patient satisfaction survey: 90% or above response to "Would return" and "Would recommend"

**D. Cost**
- Nursing staff productivity will remain at 110% of target of 8.5 worked hours per adjusted patient day within a maximum variance range of 10%

For each of the above performance measures, this performance improvement plan will:

- Address the highest-priority improvement issues
- Require data collection according to the structure, procedure, and frequency defined
- Document a baseline for performance
- Demonstrate internal comparisons trended over time
- Demonstrate external benchmark comparisons trended over time
- Document areas identified for improvement
- Demonstrate that changes have been made to address improvement
- Demonstrate evaluation of these changes; document that improvement has occurred or, if not, that a different approach has been taken to address the issue

The Inpatient Surgical Unit will submit biannual status reports to the Bassett Improvement Council (BIC) through the Medical Surgical Quality ImprovementCouncil (MSQIC).

**Approved by:** _____   **Date:** _____

*(Chief or Vice President)*

**FIGURE 17-8** (Continued)

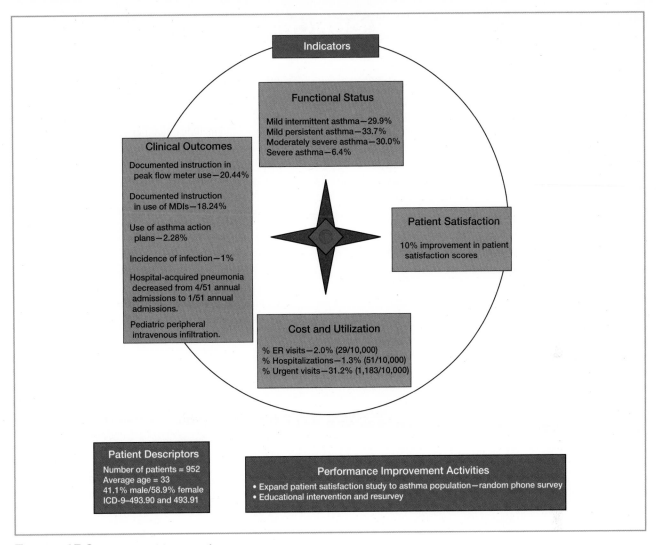

FIGURE 17-9 **Bassett Healthcare quality compass.** (*Source:* Courtesy of Kathleen F. Sellers, PhD, RN, Bassett Healthcare).

Outcomes of unit quality improvement programs can be succinctly displayed using the quality compass (Nelson, Mohr, Batalden, & Plume, 1996). The Bassett quality compass in Figure 17-9 measures quality from four domains: functional status, clinical outcomes, cost and utilization, and patient satisfaction. This quality compass depicts the outcomes of an organization-wide disease management asthma study prior to an asthma disease management intervention. The compass tells us that functionally 30% of the population has moderately severe asthma and that the majority of asthmatics have little documented teaching in use of a peak flow meter or metered-dose inhalers (MDIs), which is the current standard of care. More than 30% of the patient visits for asthma are urgent visits, indicating that a large portion of the asthma population will benefit from

the disease management intervention of increased patient teaching and development of individual specific asthma care plans. The quality compass provides a framework to guide the development of a unit-based quality improvement program and provides a tool with which to present the outcomes of quality improvement in the succinct visual format of an executive summary.

Patient care managers are the fundamental operations people in the health care system. Successful orchestration of a patient care area in today's health care system is achieved through vision-driven professional practice. Implementing this vision is achieved through a governance structure of shared decision making, an accountability-based patient care delivery system, and regular evaluation of performance based on the tenets of performance improvement.

## KEY CONCEPTS

- Successful orchestration of patient care in today's health care environment is achieved through vision-driven professional practice.

- Strategic planning is a process that is designed to achieve goals in dynamic, competitive environments through the allocation of resources.

- Shared governance is an organizational framework grounded in a philosophy of decentralized leadership that fosters autonomous decision making and professional nursing practice.

- Benner's model of novice to expert provides a framework that facilitates professional staff development.

- Situational leadership maintains that there is no one best leadership style but rather that effective leadership lies in matching the appropriate leadership style to the individual's or group's level of task-relevant readiness.

- Individuals who are accountable are, by definition, able to report, explain, or justify their actions.

- Accountability-based care delivery systems include primary nursing, relationship-based care, and case management.

- A comprehensive unit-based quality improvement program should include outcomes that are tracked from four domains: access, service, cost, and clinical quality.

- Organization of patient care management is the coordination of resources and clinical processes that promote service delivery.

- A learning organization supports the tenets of professional practice implemented through the vehicle of shared governance.

- Transforming Care at the Bedside (TCAB) is a nationwide project sponsored by the Robert Wood Johnson Foundation to look for changes in patient care processes that enhance patient safety and demonstrate improved patient care outcomes.

## KEY TERMS

clinical ladder
goal
learning organizations

mission
Novice to Expert Model
objective

philosophy
shared governance
situational leadership

strategic planning
whole-systems shared
governance

## REVIEW QUESTIONS

1. Shared governance is described as which of the following?
   A. An accountability-based care delivery system
   B. A tested framework of organizational development
   C. A competency-based career promotion system
   D. An allocation of control, power, or authority among mutually interested, vested parties

2. The five levels of a clinical promotion ladder built on Benner's theoretical framework include all but which of the following?
   A. Proficient
   B. Competent
   C. Orientee
   D. Expert

3. The principles of whole-systems shared governance include which of the following? Select all that apply.
   _____A. Partnership
   _____B. Community partnerships
   _____C. Accountability

   _____D. Ownership
   _____E. Equity

4. In developing a unit-based performance improvement plan, which of the following areas should be considered? Select all that apply.
   _____A. Service
   _____B. Cost
   _____C. Outcomes
   _____D. Clinical quality
   _____E. Access

5. A learning organization promotes professional practice through encouragement of personal mastery, team learning, and which of the following?
   A. Awareness of our mental models
   B. Taking tests
   C. Performance reviews
   D. Mandatory education

6. Transforming Care at the Bedside (TCAB) is a performance improvement process whose distinguishing factor is that it focuses on which of the following?
   A. Outcomes
   B. Clinical quality
   C. Value
   D. Ideas for performance improvement come from staff at the bedside.

7. Relationship-based care is a framework that includes the dimensions of leadership, patient care delivery, teamwork, resource-driven practice, and which of the following? Select all that apply.
   _____A. Professional nursing practice
   _____B. Organizational growth
   _____C. Patient satisfaction
   _____D. Balanced budget
   _____E. Outcomes measurement

8. In situational leadership, the development of the follower is accomplished by the leader adjusting their leadership style as the follower grows. Leadership styles include coaching, directing, delegating, and which of the following?

   A. Demonstrating
   B. Supporting
   C. Performing
   D. Evaluating

9. A professional portfolio that is reviewed periodically as a registered nurse journeys through a clinical ladder could include a copy of the nurse's resume, license, and which of the following? Select all that apply.
   _____A. Exemplars
   _____B. CEU certificates
   _____C. Minutes of a clinical practice committee meeting
   _____D. Publications
   _____E. The unit performance improvement plan

10. The strategic planning process includes values clarification defining the mission, a SWOT analysis, and which of the following?
    A. A performance improvement plan
    B. A vision statement
    C. An annual budget
    D. A patient care delivery system

## REVIEW ACTIVITIES

1. You have been asked by your organization to participate on a performance improvement team looking at the care of the diabetic patient. What areas other than clinical quality will you evaluate? Identify indicators for each area to be measured.

2. You have been practicing now for three years. This summer, you have been precepting a new graduate. He is having difficulty mastering changing a sterile dressing. You must give him feedback. You are uncertain on how to do this most effectively and wonder if you are part of the reason he is having difficulty. Review situational leadership.

At what level of readiness is this new graduate? Has your leadership style been appropriate for that level of experience and motivation?

3. You have been practicing as a new graduate for a little over a year. You are feeling more confident about your clinical practice and think you might want to expand your leadership experience. Your unit governance framework is shared governance. Review the common councils of shared governance. Given your education and experience, which council would you like to join?

## EXPLORING THE WEB

■ You have heard that your hospital is looking at implementing relationship-based care as the care delivery system to complement Shared Governance. Go to www.youtube.com and search on Relationship Based Care: The CEO Story and Relationship Based Care: Making a Difference, to hear of the impact this system has had in other organizations.

■ If you are interested in learning more about horizontal violence, go to: http://www.hvfreezone.net. This site was developed by an RN/BS completion student as part of her course work as an intervention to decrease horizon-

tal violence on her unit. The website gives information on what horizontal violence is and what are some strategies to decrease its prevalence.

■ Go to the Magnet Hospitals site, and see if there is information that would help your organization foster professional nursing practice. Striving for magnet hospital designation increases an organization's capability to recruit and retain nurses. www.ana.org.

■ Go to www.nursingsociety.org. This site provides weekly literature updates from Sigma Theta Tau International, the nursing profession's honor society. What new books and periodicals are available that may be helpful to you in your practice?

■ Go to www.vitalsmarts.com. Access your style under stress. Enhanced insight allows for improved personal

mastery and effectiveness as a patient care manager, whether you are leading others or communicating with peers.

## REFERENCES

Benner, P. (1984). *From novice to expert.* Menlo Park, CA: Addison-Wesley.

Canadian Nurses Association. (2005). Promoting continuing competence for registered nurses. Ottawa, Canada: Author.

Covey, S. R. (1997). *The seven habits of highly effective people.* New York: Simon & Schuster.

Dreyfus, S. E., & Dreyfus, H. L. (1980). *A five stage model of the mental activities involved in directed skill acquisition.* Unpublished report supported by the Air Force Office of Scientific Research (AFSC), USAF (Contract F49620-79-C-0063), University of California at Berkeley.

Hastie, C. (2002). Horizontal violence in the workplace. *Birth International.* Retrieved September 10, 2008, from http://www.birthinternational.com/articles/hastie02.html.

Hersey, R. E., & Blanchard, T. (1993). *Management of organizational behavior.* Riverside, NJ: Simon & Schuster.

Institute of Medicine. (2000). Keeping patients safe: Transforming the work environment of nurses. Retrieved August 21, 2006, from http://www.iom.edu/Default.aspx?id=16173.

Institute of Medicine (IOM). (2000). To err is human: Building a safer health system. Washington, DC: National Academy Press.

Institute for Health care Improvement. Transforming Care at the Bedside. http://www.ihi.org/IHI/Programs/StrategicInitiatives/TransformingCareAtTheBedside.htm. Accessed April 5, 2011.

Joint Commission. (2010). Comprehensive accreditation manual for hospitals. The official handbook. Chicago: Author. Retrieved August 15, 2010, from http://www.jointcommission.org.

Kelly, P. (2010). Essentials of nursing leadership and management (2nd ed.). Clifton Park, NY: Delmar Cengage Learning.

Kolorutis, M. (2007). Relationship-based care field guide. Creative Health Care Management.

MacGregor–Burns, J. (1979). *Leadership.* New York: Harper & Row.

*Merriam-Webster online dictionary.* Retrieved August 15, 2010, from http://www.merriam-webster.com/dictionary/competency.

Nelson, E., Mohr, J. J., Batalden, P. B., & Plume, S. K. (1996, April). Improving health care, part 1: The clinical value compass. *Joint Commission Journal on Quality Improvement, 22*(4), 243–258.

Porter-O'Grady, T. (1995). *The leadership revolution in health care.* Gaithersburg, MD: Aspen Publishers, Inc.

Porter-O'Grady, T. (2009). Interdisciplinary shared governance. Sudbury, MA: Jones and Bartlett Publishers.

Porter-O'Grady, T., Hawkins, M.A., and Parker, M.L. (1997). *Whole-systems shared governance: Architecture for integration.* Gaithersburg, MD: Aspen.

Raphael, D. D. (1994). Moral philosophy (2nd ed.). New York: Oxford University Press.

Ross, P. E. (2006, August). The expert mind. *The Scientific American, 8,* 64–71.

Sellers, K., Millenbach, L., Kovach, N. & Klimek-Yingling, J., (2010). Prevalence of Horizontal Violence in New York State Registered Nurses. The Journal of the New York State Nurse. Fall/Winter 2009–2010, 40, 20–25.

Sellers, K. F. (1996). The meaning of autonomous nursing practice to staff nurses in a shared governance organization: A hermeneutical analysis. Unpublished doctoral dissertation, Adelphi University, Garden City, New York.

Senge, P. (1990). *The fifth discipline.* New York: Doubleday.

Smith, M. K. (2001). Peter Senge and the learning organization: The encyclopedia of informal educaton. Retrieved March 26, 2006, from http://www.infed.org/thinkers/senge.htm.

Stichler, J. F. (1992). A conceptual basis for shared governance. In N. D. Como & B. Pocta (Eds.), *Implementing shared governance: Creating a professional organization* (pp. 1–24). St. Louis, MO: Mosby.

Transforming care at the bedside (TCAB). Available at http://www.ihi.org/IHI/Programs/StrategicInitiatives/TransformingCareAtTheBedside.htm.

Wesorick, B. (2006). The way of respect in the workplace. Retrieved August 22, 2006, from http://www.cpmnc.com.

Wesorick, B., Shiparski, L., Troseth, M., & Wyngarden, K. (1998). *Partnership council field book.* Grandville, MI: Practice Field Publishing.

# SUGGESTED READINGS

Doyle, V., & Turkie, W. (2006). Transforming the organization of care. *Nursing Management—UK, 13*(2), 18–21.

Gladwell, M. (2005). *Blink.* New York: Little, Brown and Company.

Institute of Medicine. (2000). Keeping patients safe: Transforming the work environment of nurses. Retrieved August 21, 2006, from http://www.iom.edu/Default.aspx?id=16173.

Johnson, S. (1996). Who moved my cheese? New York: Putnam.

Lavizzo-Mourey, R., & Berwick, D. (2009). Nurses transforming care. *American Journal of Nursing, 109*(11), 3.

Personality Pathways. (2006). Introduction to type. Retrieved August 22, 2006, from http://www.personalitypathways.com/ MBTI_intro.

Porter-O'Grady, T. (2003). A different age for leadership part 2: New rules, new roles. *Journal of Nursing Administration, 33*(3), 173–178.

Scott, L., & Caress, A. (2005). Shared governance and shared leadership: Meeting the challenges of implementation. *Journal of Nursing Management, 13*(1), 4–12.

Thompson, P., Navarra, M., & Anderson, N. (2005). Patient safety: The four domains of nursing leadership. *Nursing Economic$, 23*(6), 331–333.

Tracey, C., & Nicholl, H. (2006). Mentoring and networking. *Nursing Management—UK, 12*(10), 28–32.

Wesorick, B. (2006). Clinical practice model resource center. Retrieved August 22, 2006, from http://www.cpmrc.com.

# CHAPTER 18

# Time Management and Setting Patient Care Priorities

### Patsy L. Maloney, EdD, MSN, MA, RN-BC, NEA-BC

*Time is the coin of your life. It is the only coin you have, and only you can determine how it will be spent.*

(Carl Sandberg)

## OBJECTIVES

Upon completion of this chapter, the reader should be able to:

1. Apply principles of priority setting to patient care situations.
2. Apply time management strategies to the reality of delivering effective nursing care.
3. Relate time management strategies to enhance personal productivity.

## OUTCOME ORIENTATION

With the Pareto Principle in mind, it is important to recognize that more is achieved through an outcome orientation than through an emphasis on the process of task completion. Long-term goals must be determined. It is best to break long-term goals down into achievable outcomes that are the steps toward long-term goals. Long-term goals cannot be achieved overnight. Long-term goals and outcomes should be written down in a planner or in an electronic device such as a personal data assistant (PDA). Even though these goals are written, they should remain flexible. Flexibility should be built into any outcome orientation. There may come a time when the outcome is no longer realistic or should be shifted to a more realistic goal as circumstance changes (Reed & Pettigrew, 2006).

## TIME ANALYSIS

Another time management concept is analysis of time to effectively use it. Nurses cannot possibly know how to better plan time without knowing how they currently use it. When keeping track of time, it is important to consider the value of a nurse's time as well as the use of time.

### Valuing Nursing Time

Nurses often undervalue their time. Consider salary and benefits. Benefits are frequently forgotten, but they raise employer costs by 15% to 30% of salary. If a nurse is making $26.00 an hour, benefits add $3.90 to $7.25 to the hourly cost of a nurse's time. The value of nursing time in this example, excluding what the organization is paying in worker's compensation and payroll taxes, is $29.90 to $33.25 an hour. The organization has also invested in nurse recruitment, orientation, and development, which can easily exceed $20,000 per nurse. Nursing time is an expensive commodity. Keeping this in mind when considering what tasks can be delegated to personnel who receive less compensation, or when considering what tasks are busy work and do not support achieving an outcome, is invaluable.

### Use of Time

Numerous studies have shown how nurses use their time. Many studies have been done on acute care nurses, because they comprise the majority of nurses. Only 30% to 35% of nursing time is spent on direct patient care (Scharf, 1997). Twenty-five percent of a nurses' time is spent on charting and reporting, and the remainder of time is spent on admission and discharge procedures, professional communication, personal time, and providing care that could be provided by unlicensed personnel, such as transportation and housekeeping (Upenieks, 1998). Urden and Roode (1997) summarized various work sampling studies to show that RNs spend 28% to 33% of time on direct patient care, defined as activities performed in the presence of the patient or family; 42% to 45% of time on indirect care activities, which include all activities done for an individual patient but not in the patient's presence; 15% on unit-related activities, which include all unit general maintenance activities; and 13% to 20% on personal activities, which include activities that are not related to patient care or unit maintenance. Fitzgerald, Pearson, Walsh, Long, and Heinrich (2003) found a distribution similar to that of previous studies. Heindrich, Chow, Skierczynski, and Lu (2008) had similar findings except for a significant decrease in nonproductive personal time from 13%–20% to only 6%. Unit-related activities, non-nursing practice, filled 13% of the nursing time. Documentation accounted for the largest proportion of time at approximately 25%, followed by care coordination and patient care activities, each accounting for 14% (Figure 18-2).

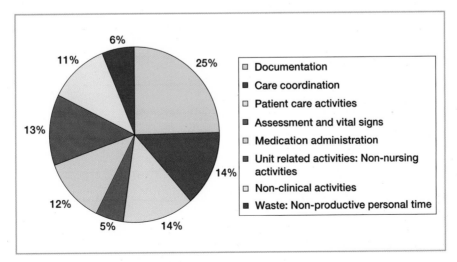

FIGURE 18-2 **Use of nursing time.** (*Source:* Compiled with information from Hendrich, A., Chow, M. P., Skierczynski, B. A., & Zhenqiang, Z. (2008). A 36-hospital time and motion study: How do medical-surgical nurses spend their time? *The Permanente Journal, 12*(3). Retrieved from http://xnet.kp.org/permanentejournal/sum08/time-study.html/clinicalTOC.html, accessed April 5, 2011).

Given such a distribution of nurses' time, shifting the use of time could have a major impact on outcomes. If non-nursing activities could be performed by non-nursing personnel instead of nurses, more time could be redirected toward essential nursing responsibilities.

How do you use your time? Memory and self-reporting of time have been found to be unreliable. Staff are often unaware of time spent socializing with colleagues, making and drinking coffee, snacking, and other nonproductive time. Self-reporting of time is not recommended for estimating the total number of activities or the average time an activity takes to complete (Barrero et al., 2008).

An **activity log** is a time management tool that can assist the nurse in determining how time is used. The activity log (Table 18-1) should be used for several days. Behavior should not be modified while keeping the log. The nurse should record every activity, from the beginning of the shift until the end, as well as periodically note feelings while doing the activities—alert, energetic, tired, bored, and so on. Review of this log will illuminate time use as well as time wasted. Analysis of the log will allow the separation of essential professional activities from activities that can be performed by someone else (Grohar-Murray & Langan, 2011; Sullivan & Decker, 2009).

## TABLE 18-1   Work Activity Log

| TIME | NAME OF ACTIVITY (MEDICATION ADMINISTRATION, VITAL SIGNS, BED-MAKING, PATIENT TRANSPORT, AND SO ON) | TIME REQUIRED AND FEELINGS (ENERGETIC, BORED, AND SO ON) | COULD BE BETTER DONE BY SOMEONE ELSE? WHO? (LPN, NURSING ASSISTANT, HOUSEKEEPER, AND SO ON) | TOWARD WHAT OUTCOME ACHIEVEMENT? (INCREASE IN PATIENT'S FUNCTIONAL STATUS, PREVENTION OF COMPLICATIONS, AND SO ON) |
|------|------|------|------|------|
| 0500 | Treadmill | 30 min – energetic | Keep for self | Fitness |
| 0530 | Shower and breakfast | 45 min – energetic | Keep for self | Health |
| 0630 | Drive to work | 10 min – alert | Keep for self | Get to work |
| 0700 | Hand-off shift report | 15 min – alert | Keep for self | Patient identification |
| 0730 | Patient rounds/planning | 15 min – alert | Keep for self | Prioritize patients |
| 0730 | | | | |
| 0800 | | | | |
| 0830 | | | | |
| 0900 | | | | |
| 0930 | | | | |
| 1000 | | | | |
| 1030 | | | | |
| 1100 | | | | |
| Etc. | | | | |

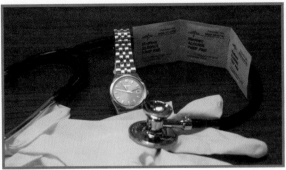

Delmar/Cengage Learning

*Inez has just completed her medical-surgical orientation as a new graduate registered nurse. This evening is her first solo shift, but she is frightened and feels like she is holding up the world on her new graduate shoulders. Although she feels that all rests on her, she is not really alone. Inez and Carole, the other RN, are responsible for 12 possible patients on this section of the unit along with one certified nursing assistant. Currently, there are 10 patients in this section, but a new admission is on the way, another patient is returning from surgery, the dinner trays are arriving, and Inez has medications to pass. Just as the dinner trays arrive, a patient's family member runs out to Inez and states that her mom is confused and incontinent and has pulled out her IV.*

*What would you do if you were Inez?*

*What would you do first?*

**M**any nurses become nurses out of idealism. They want to help people by meeting all their needs. Unfortunately, most new graduates find it impossible to meet all or even most of their patients' needs. Needs tend to be unlimited, whereas time is limited. In addition to the direct patient care responsibilities, there are shift responsibilities, charting, practitioners' orders to be transcribed or checked, medication supplies to be restocked, and reports to be given.

New graduates often go home feeling totally inadequate. They wake up remembering what they did not accomplish. One young nurse shared with tears in her eyes that once, when she answered a call bell late in her shift, the patient requested a pain medication. She went to the narcotics cabinet to get the medication but was interrupted by an emergent situation. When she arrived home, she was so exhausted that she fell asleep rapidly, only to awaken with the realization that she had not returned with her patient's medication. Her guilt was tremendous. She had gone into nursing to relieve pain, not to ignore it.

Time management allows the novice nurse to prioritize care, decide on outcomes, and perform the most important interventions first. Time management skills are important not just for nurses on the job, but for nurses in their personal lives as well. They allow nurses to make time for fun, friends, exercise, and professional development.

# GENERAL TIME MANAGEMENT CONCEPTS

**Time management** has been defined as "a set of related common-sense skills that helps you use your time in the most effective and productive way possible" (Mind Tools, 2006a, p. 1). Another way to say this is that time management allows us to achieve more with available time. Three valuable time management concepts to master are the relative effectiveness of effort (the Pareto Principle), the importance of outcome versus process orientation, and the value of analyzing how time is currently being used. It is important to analyze and manage time to achieve key outcomes effectively and efficiently.

## THE PARETO PRINCIPLE

Time management requires a shift from wasting time on the process of being busy to organizing time to achieve desired outcomes and get things done. Oftentimes, a busy frenzy of activity is reinforced with sympathy and assistance. Too often, this frenzied behavior is accomplishing very little, because it is not directed at the right outcome. The **Pareto Principle** states that 20% of focused effort results in 80% of outcome results, or conversely that 80% of unfocused effort results in 20% of outcome results (Figure 18-1).

Pareto's Principle, named after Vilfredo Pareto, was invoked by the total quality management (TQM) movement and is now reemerging as a strategy for balancing life and work through prioritization of effort (Hughes, 2006). Effective time management requires that a shift be made from doing unfocused activities that require 80% of time for achieving 20% of desired results to doing planned and focused activities that use only 20% of time or input to achieve 80% of desired outcomes. It is important to analyze how your time is being used to manage time and achieve desired outcomes.

If time management achieves more outcomes, why do so many people continue at a crazy, hurried pace? There are several possible explanations for this. They do not know about time management, they think they do not have time to plan, they do not want to stop to plan, or they love crises (Mind Tools, 2006a).

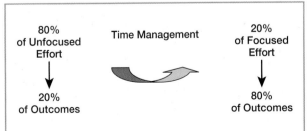

**FIGURE 18-1** The Pareto Principle.

## Critical Thinking 18-1

Four of Jose's patients were discharged today by 10:00 a.m. The nursing supervisor asked Jose to help out in the emergency room (ER). Jose agreed and was assigned to help the triage nurse. Identify the order in which patients should be seen in the ER.

Group I

- A 2-year-old boy with chest retractions
- A 1-year-old girl choking on a grape
- A 5-year-old boy with a knee laceration

How about Group II? Which patient would you see first?

Group II

- A 60-year-old female who is nonresponsive and drooling
- A 30-year-old male trauma patient who has absent breath sounds in the right side of his chest
- A 15-year-old female who cut her wrist in an attempted suicide

# PRIORITIZING USE OF TIME

To plan effective use of time, nurses must understand the big picture, decide on desired outcomes, and do first things first.

## UNDERSTAND THE BIG PICTURE

Before priorities are set, the big picture must be examined. No nurse works in isolation. Nurses should know what is expected of their coworkers, what is happening on the other shifts, and what is happening beyond the unit. If nurses know what is expected of their coworkers, they can offer assistance during coworkers' busy times and in turn receive assistance during their own busy times. If the previous shift was stressed by a crisis, a shift may not get started as smoothly (Hansten & Jackson, 2009). If areas outside the unit are overwhelmed, someone might be moved from one unit to assist on the overwhelmed unit elsewhere in the hospital. When nurses take the big picture into consideration, they are less likely to be frustrated when asked to assist others on their unit or on other units in the hospital. They can also build into their time management plan the possibility of giving and receiving assistance.

## DECIDE ON OPTIMAL DESIRED OUTCOMES

When nurses begin their shifts, they need to decide what outcomes can be achieved. Desired optimal outcomes are the best possible objectives to be achieved given the resources at hand.

As nurses decide on desired optimal outcomes, they must consider what can and should be achieved given less-than-optimal circumstances and limited resources. These circumstances could include a rough start to a busy shift; personnel late, absent, or uncooperative; and a patient crisis. It is hard for nurses to give themselves permission to do less-than-optimal work, but sometimes achievement of these outcomes is the best that can be expected. These outcomes should be achieved given less-than-optimal circumstances and limited resources.

## DO FIRST THINGS FIRST

To decide what is reasonable to accomplish, a nurse has to come to terms with the resources that are available and the outcomes that must be achieved. If someone has called in sick and no replacement is available, it might be unreasonable for a nurse to plan to reinforce teaching or discuss the home environment with a patient scheduled to leave the next day. However, there would be no question that interventions that prevent life-threatening emergencies or save a life when a life-threatening event occurs are priorities. They must be done no matter how short the staffing. It is imperative that nurses protect their patients and maintain both patient and staff safety as well as perform the activities essential to the nursing and medical care plans (Hansten & Jackson, 2009).

### First Priority: Life-Threatening or Potentially Life-Threatening Conditions

Life-threatening conditions include patients at risk to themselves or others and patients whose vital signs and level of consciousness indicate potential for respiratory or circulatory collapse (Hansten & Jackson, 2009). A patient whose condition is life threatening is the highest priority and requires monitoring until transfer or stabilization. Life-threatening conditions can occur at any time during the shift and may or may not be anticipated. Patients must be monitored to prevent the occurrence of adverse nurse-sensitive patient outcomes, such as a cardiac arrest from an inadvertent airway occlusion, etc.

A quick guide to assessing life-threatening emergencies is as simple as ABC. A stands for Airway. Is the airway open and patent or in danger of closing? This is the highest priority of care. B stands for Breathing. Is there respiratory distress? C stands for Circulation. Is there any circulatory compromise? This is a way of prioritizing actions. Although there is clearly an order of importance, ABC is

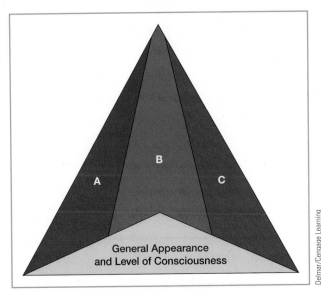

FIGURE 18-3 Quick assessment tool.

often assessed simultaneously while observing the patient's general appearance and level of consciousness (Figure 18-3 and Table 18-2). Patients with life-threatening conditions usually have an IV access line and receive continuous monitoring of their cardiac rhythm, blood pressure, pulse, respiration, and oxygen saturation level. Their temperature and urinary output and their potential for other life-threatening conditions is monitored closely as well.

## Second Priority: Activities Essential to Safety

Activities that are essential to safety are very important and include those responsibilities that ensure the availability of life-saving monitoring, medications, and equipment, and that protect patients from infections and falls. They include asking for assistance or providing assistance during two-people transfers, or turning and movement of heavy patients (Hansten & Jackson, 2009). They also include

### TABLE 18-2 Top Priority Patients with Potential Threats to Their ABCs

| PATIENT | POTENTIAL THREATS |
|---|---|
| Respiratory Patients | • Airway compromise<br>• Choking<br>• Asthma<br>• Chest trauma |
| Cardiovascular Patients | • Cardiac arrest<br>• Shock<br>• Hemorrhage |
| Neurological Patients | • Major head injury<br>• Unconscious<br>• Unresponsive<br>• Seizures |
| Other Patients | • Major trauma<br>• Traumatic amputation<br>• Major burn, especially if airway involvement<br>• Abdominal trauma<br>• Vaginal bleeding<br>• Anaphylaxis<br>• Diabetic with altered consciousness<br>• Septic shock<br>• Child or elder abuse |

**Source:** Compiled with information from the *Canadian Pediatric Triage and Acuity Scale: Implementation Guidelines for Emergency Departments.* Retrieved August 7, 2010, from www.caep.ca.

monitoring the patient for the prevention of adverse nurse-sensitive outcomes. Nurse-sensitive patient outcomes are those outcomes that improve if there is a greater quantity or quality of nursing care—e.g., pressure ulcers, falls, and intravenous infiltrations (Vanhook, 2007).

## Third Priority: Comfort, Healing, and Teaching

Activities that include comfort, healing, and teaching are essential to the plan of care and lead to outcomes that relieve symptoms and/or lead to healing. They are the activities that, if omitted, will hinder the patient's recovery. These essential activities include those that relieve symptoms—pain, nausea, and so on—and those that promote healing, such as nutrition, ambulation, positioning, medication administration, and teaching (Figure 18-4).

Covey, Merrill, and Merrill (1994) developed another way of setting priorities. Activities are classified as urgent or not urgent, as important or not important. If an activity is neither important nor urgent, then it becomes the lowest priority (Figure 18-5).

Some activities that are often thought of as important may not be. Sometimes laboratory data, vital signs, and intake and outputs are ordered to be monitored more frequently than the status of the patient indicates. Frequent monitoring of these parameters may make no significant difference in patient outcomes. When nurses begin their shifts, they should question the activities that make no difference in outcomes (Hansten & Jackson, 2009). If a practitioner orders these activities, a nurse should work to get the order changed. If there is a nursing order that does not make a difference, the nurse should change it. Nurses should give priority to the activities that they know are going to make a difference in patient outcomes.

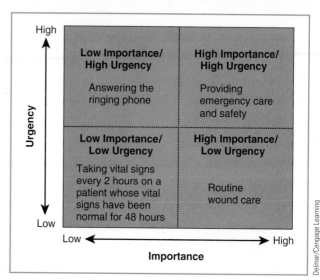

**FIGURE 18-5**　Determining priorities.

# APPLICATION OF TIME MANAGEMENT STRATEGIES TO THE DELIVERY OF CARE

After priorities are set, nurses know which are the most important activities to accomplish first. Time management strategies can be used in all areas of care delivery to maximize the effectiveness of the nurse's time and minimize lost time and efforts.

## ESTIMATE ACTIVITY TIME CONSUMPTION

Nurses need to estimate how much time each activity will take and plan accordingly. The previously discussed activity log may help estimate how much time many activities will take. Perhaps a patient tends to need more time for medication administration than do other patients, so the wise nurse will save that patient's medication administration until last. By estimating the time of activities, nurses can schedule the best time to perform activities. Nurses may notice when passing 6 p.m. medications that water pitchers are empty and juice cups dry. Scheduling the nursing assistant to fill water pitchers and pass refreshments prior to medication administration will be a prudent response to such an observation.

## CREATE AN ENVIRONMENT SUPPORTIVE OF TIME MANAGEMENT AND PATIENT CARE

Often in the frenzy of giving care, nurses forget the obvious. Where are the linens, supplies, medications, and so on located? Are there optimal locations? Is stocking things a

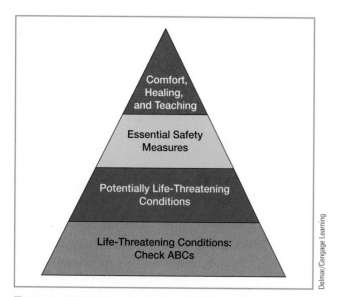

**FIGURE 18-4**　Prioritization triangle.

priority in order to make them available? Do nurses really stop and think before going to a patient's room with pain medications or for a treatment?

How many trips does one treatment take? It should take only one trip, but if a nurse hurries in and leaves something at the nurse's station, then the nurse will have to return to retrieve it. How many times do nurses count narcotics that have not been used in months? These are simple things that take time. The nurse should give consideration to all aspects of the unit environment and get together with coworkers to make a difference. Are specialty carts such as those for intravenous infusion, isolation, laboratory collections, wound care, and so on needed to become more efficient and effective?

# Utilize Shift Hand-Off Report

Prior to making a plan for the shift of duty, the end of shift hand-off report at best can lead to an efficient, effective, and safe start to the shift. At worst, it can leave the oncoming shift with inadequate or old data on which to base their plan. In 2005, the Joint Commission (JC) reported that communication breakdowns were correlated with adverse outcomes, and it included a standardized approach to shift hand-off communication as one of its patient safety goals (Riesenberg, Leisch, & Cunningham, 2010). So the day of haphazard shift hand-off reporting has ended. There are various ways to conduct an end of shift hand-off report—e.g., a face-to-face meeting, and walking rounds (Table 18-3).

Whether the report is conducted face-to-face or through walking rounds, information has to be transmitted and time must be provided for questions and answers. A procedure that allows time for both transmission of information and for questions and answers facilitates safe, effective, and efficient care. If the outgoing nurse fails to cover all pertinent points, the oncoming shift must ask for the appropriate information. See Table 18-4 for a tool for taking and giving shift hand-off reports.

## TABLE 18-3  Methods of End-of-Shift Hand-Off Report: Advantages and Disadvantages

| METHOD | ADVANTAGES | DISADVANTAGES |
|---|---|---|
| Face-to-face report | • Nurses get clarification and can ask questions.<br>• Nurse giving report has actual audience and tends to be less mechanical.<br>• Nurses are more likely to give pertinent information than they would to a tape recorder. | • It is time consuming.<br>• It is easy to get sidetracked and gossip or discuss non-patient-related business.<br>• Both oncoming and departing nurses are in report.<br>• Patients are not included in planning. |
| Walking rounds | • Provides the prior shift and incoming shift staff the opportunity to observe the patient while receiving report. Staff can address any assessment or treatment questions.<br>• Information is accurate and timely.<br>• Patient is included in the planning and evaluation of care.<br>• Accountability of outgoing care provider is promoted.<br>• Patient views the continuity of care.<br>• Incoming shift makes initial nursing rounds.<br>• Departing nurse can show assessment and treatment data directly to oncoming nurse. | • It is time consuming.<br>• There is a lack of privacy in discussing patient information. |

## TABLE 18-4   Tool for End-of-Shift Hand-Off Report

NOTES

| | |
|---|---|
| Demographics | • Room number<br>• Patient name<br>• Sex<br>• Age<br>• Practitioner |
| Diagnoses | • Primary<br>• Secondary<br>• Nursing<br>• Medical<br>• Surgery Date |
| Patient status | • Do Not Resuscitate (DNR) status<br>• Current vital signs<br>• Problem with ABCs, level of consciousness, or safety<br>• Oxygen saturation<br>• Pain score<br>• Skin condition<br>• Ambulation<br>• Fall risk<br>• Suicide risk<br>• Presence/absence of signs & symptoms of potential complications<br>• New orders/changes in treatment plan |
| Fluids/tubes/oxygen, laboratory tests and treatments | • IV fluid, rate, site<br>• Tube feedings—type of tube, solution, rate, and toleration<br>• Oxygen rate, route, other tubes (e.g., chest tube, NG tube, Foley, and so on), type of tube and type of drainage<br>• Abnormal lab and test values<br>• Labs and tests to be done on oncoming shift<br>• Treatments done on your shift, including dressing changes (times, wound description) and procedures<br>• Identify treatments to be done during next shift |
| Expected shift outcomes | • Priority outcomes for one or two nursing diagnoses<br>• Patient learning outcomes |
| Plans for discharge | • Expected date of discharge<br>• Referrals needed<br>• Progress toward self-care and readiness for home |
| Care support | • Availability of family or friends to assist in ADL/IADL (activities of daily living/instrumental activities of daily living) |
| Priority interventions | • Interventions that must be done this shift |

# FORMULATE THE PLAN FOR THE SHIFT

Having received pertinent information from the shift report, nurses can consider the big picture; decide on optimal and reasonable outcomes; and set priorities based on life-threatening conditions, safety considerations, and activities essential to comfort, healing, and teaching. The nurse can then develop an assignment sheet (Figure 18-6, Figure 18-7) after considering nursing standards, patient care routines, and several other factors that must be considered in planning for a shift (Table 18-5).

## Make Assignments

Because nurses cannot accomplish patient objectives by themselves, they must delegate and make assignments to other team members. This can be challenging and requires understanding of delegation principles, which are discussed in Chapter 16. The assignment sheet should identify who will perform the intervention. Figure 18-6 and Figure 18-7 show a completed and a blank assignment sheet, respectively. Assignments should be part of the planning process and should include delegation of nursing and non-nursing tasks to others, with specific reporting guidelines and deadlines for accomplishment of the tasks. Many factors are considered in making assignments (Table 18-6).

## Timing the Actions

Identifying the time by which an intervention should be completed is important. If the time is flexible, then that should also be noted. Assigning a completion time for the intervention makes it more likely for the intervention to be completed during the shift. This makes the outcome more likely to be accomplished (Hansten & Jackson, 2009).

# MAKE PATIENT CARE ROUNDS

If the end-of-shift hand-off report does not include walking rounds, the oncoming nurse needs to first make initial rounds on the patients at risk for life-threatening conditions or complications. As the nurse makes rounds, he or she performs rapid assessments. These assessments may vary from the information given during the shift hand-off report, so the information gathered on rounds may change the shift plan. A patient with asthma that has been calm and without respiratory distress on the previous shift may have experienced a visitor who wore perfume and/or brought bad news. As the oncoming nurse makes initial rounds and uses the quick ABC assessment, the nurse may quickly determine that the patient has suddenly developed respiratory distress. The patient may have been initially prioritized as requiring only supportive activities directed at healing, but

## REAL WORLD INTERVIEW

As an emergency room nurse, your priorities are very important. Priorities must be flexible to change with the environment. When arriving at work, prepare by checking your patients, rooms, and equipment. Review your patients' charts and receive the patient hand-off report from the nurse who is going off duty. Then prioritize patient care. Priorities change when new patients arrive or a patient's condition changes. Stay ahead of the curve by anticipating what will happen during the patient's stay. For example, if the patient's problem is nausea with vomiting, I can anticipate starting the IV and collecting lab samples before the patient sees the physician. When the patient sees the physician and orders are written, the patient can then quickly receive his or her medicine for nausea. I take care of the patient's immediate needs and anticipate his or her future needs.

**Chris Curtis, RN**
Austin, Texas

now the patient is experiencing a life-threatening reaction and requires appropriate nursing interventions as well as continuous monitoring. While assessing the patient, the nurse also checks all the patient's IV lines to make sure that the correct fluid is infusing and the infusion site is without a complication. The nurse checks all the patient's drains, tubes, and continuous treatments. The nurse also listens to the patient's concerns and desires. It is important to remember that plans are just that, plans, and have to be flexible based on ever-changing patient care needs. Times for treatments and medications may have to be changed. Often, nurses believe that the times for administering medication are inflexible, yet practitioners usually write medication orders as daily, twice a day, three times a day, or four times a day. These kinds of orders give nurses flexibility in administration times. Although unit policy dictates when these medicines are given, unit policy is under nursing control.

## EVALUATE OUTCOME ACHIEVEMENT

At the end of the shift, the nurse reexamines the shift action plan. Did the nurse achieve the outcomes? If not, why? Were there staffing problems or patient crises? What was learned from this for future shifts?

## ASSIGNMENT SHEET EXCERPT

Unit  2  South

Date October 2, 2008

Shift _____ Days

Charge nurse _____ Mary

RNs Breaks/Lunch
Steve 0900 and 1100
Lakeisha 0930 and 1130
Colleen 1000 and 1200

NAPs
Breaks/Lunch
Juan 0900 and 1100
Pat 0930 and 1130

Notify RN immediately if:
T    <97 or >100
P    <60 or >110
R    <12 or > 24
SBP <90 or >160
DBP <60 or >100
BS  <70 or >200
Pulse oximetry <95%
Urine output <30 cc/hour

Notify RN one hour prior to end of shift:
I and O
Patient goal achievement

Narcotic count Steve
Glucometer check Colleen
Stock pyxis Lakeisha
Pass water Juan
Stock linen Pat
Other Colleen attend in-service at 1300

| Room and Initials | Patient | Staff | AM/PM Care | Weight I & O | IV | Activity | Accu-check | Tests | NPO | Comments |
|---|---|---|---|---|---|---|---|---|---|---|
| 501, Mr. MM | 27-year-old with newly diagnosed AIDS, left lower lobe pneumonia | Steve, RN | Complete care | 0715 1400 | KVO | Bedrest | | Lab | | Vitals Q4H |
| 502, Mr. MG | 61-year-old with acute congestive heart failure (CHF) | Lakeisha, RN | Partial care | 0715 1400 | KVO | BRP | 1100 | Lab | Yes | Vitals Q4H |
| 503, Ms. SC | 92-year-old with new right hip fracture, in Bucks traction | Juan, NAP | Complete care | I & O at 1400 | KVO | Bedrest | | Lab and X-ray | Yes | Vitals Q4H |
| 504, Ms. NJ | 48-year-old, with new cholelithiasis | Pat, NAP | Self-care | | KVO | Up ad lib | | Ultra-sound | Yes | |
| 505, Ms. LG | 89-year-old with new onset CVA with right side paralysis | Colleen, RN | Complete care | 0715 1400 | NS @ 125 cc/hr. | Bedrest | 1100 | Lab | | Vitals Q4H |

**FIGURE 18-6** Assignment sheet excerpt.

Delmar/Cengage Learning

Unit _____

Date _____

Shift _____

Charge nurse _____

Breaks/Lunch _____

RN _____

_____

LPN _____

NAP _____

_____

Notify RN immediately if:

T    <97 or >100

P    <60 or >110

R    <12 or > 24

SBP <90 or >160

DBP <60 or >100

BS   <70 or >200

Pulse oximetry <95%

Urine output <30 cc/hour or 240cc/8 hours

Notify RN one hour prior to end of shift:

I and O

Patient goal achievement

Narcotic count _____

Glucometer check _____

Stock Pyxis _____

Pass water _____

Stock linen _____

Other _____

| Room and Initials | Patient | Staff | AM/PM Care | Weight I & O | IV | Activity | Accu-check | Tests | NPO | Comments |
|---|---|---|---|---|---|---|---|---|---|---|
| | | | | | | | | | | |
| | | | | | | | | | | |
| | | | | | | | | | | |
| | | | | | | | | | | |
| | | | | | | | | | | |
| | | | | | | | | | | |

**Figure 18-7** Assignment sheet.

## TABLE 18-5  Factors to Consider in Planning for a Shift

| CONCERN | CONSIDERATIONS |
|---|---|
| *Plan:* | |
| What is the big picture? | How many patients? Any staffing issues? Any environmental concerns? Is unit geography conducive to patient care delivery? |
| What are the desired outcomes? | If everything goes well and as expected, what does the nurse hope to accomplish? |
| | If unexpected setbacks occur, what can the nurse, staff, and patients really accomplish? |
| What are the priorities, nursing standards, and nursing routines? | Who is at greatest risk for potential life-threatening complications? Has all emergency equipment been checked? Are patients who are at a high risk for falls or suicide identified and measures taken? Who are the patients suffering from significant symptoms—airway, breathing, circulation? Are there any other complex patient or family needs? What nursing standards and patient care routines are needed by patients? |
| *Intervene:* | |
| What are the parts to be accomplished? | Monitoring? Medication administration? Treatments? Teaching? Counseling? Physical and functional care? Unit support, e.g., stocking, maintenance? |
| Who is available to do the work, and what skills and attributes do the personnel have? | RN? LPN? NAP? What are the other responsibilities of staff? What is staff's attitude and dependability? Do any patients need continuity of care by same staff? Is there a fair work distribution among staff? Do any patients need isolation or protection? Do staff need to be protected from some patients? What is the skill, education, and competency of staff, i.e., RN, LPN, NAP? |
| What can the RN do? What about the LPN? What can the NAP do? There is much variation in the titles of nursing assistive personnel (NAP), e.g., unlicensed nursing personnel, nursing assistant, nurse aide, patient care attendant, patient care aide, etc. Besides those titles, some states use the term *unlicensed assistive personnel* (UAP) (National Council of State Boards of Nursing, 2005)[*]. | RN identifies standards and patient care routines and teaches, counsels, and supervises all nursing care. LPN can do medication administration and treatments. NAP can complete physical care such as bathing, performing oral care, and obtaining vital signs. Assign and delegate accordingly. |
| When should the actions be completed by? What are the guidelines for completion? | Give a set time for completion of all tasks. |
| *Evaluate:* | |
| Are the shift outcomes achieved? | How will you check throughout the shift and at the end of the shift? Is your team in need of assistance? Has anything unexpected happened to change your plan? Has patient status changed? At the end of the shift, did you accomplish outcomes? |

[*] National Council of State Boards of Nursing (NCSBN). (2005). Working with others: A position paper. Available at https://www.ncsbn.org/Working_with_Others.pdf, accessed April 5, 2011.

## TABLE 18-6 Factors Considered in Making Assignments

- Priority of patient needs
- Geography of nursing unit
- Complexity of patient needs
- Other responsibilities of staff
- Attitude and dependability of staff
- Need for continuity of care by same staff
- Agency organizational system
- State laws, e.g., state nurse practice act
- In-service education programs
- Need for fair work distribution among staff
- Need for lunch/break times

- Need for isolation
- Need to protect staff and patients from injury
- Skill, education, and competency of staff, that is, RN, LPN, NAP
- Hospital policy and procedure
- Patient care standards and routines for surgical, medical, maternal, child, and/or mental health patients
- Environmental concerns
- Equipment checks, medication checks
- Accreditation regulations
- Needs of other units in hospital, number of staff, problems left over from earlier shifts, etc.
- Desired patient outcomes

If, at the end of a shift, the nurse did not accomplish the outcomes, the nurse might review the shift activities to see what time wasters interfered with outcome achievement. Marquis and Huston (2011) described time wasters as procrastination, inability to delegate, inability to say no, management by crisis, haste, and indecisiveness. Sullivan and Decker (2009) add interrupting telephone calls and socialization to the list. Reed and Pettigrew (2006) add complaining, perfectionism, and disorganization to the list of time wasters.

## AVOID PRIORITY TRAPS

Vaccaro (2007) states that prioritizing has several traps that nurses should avoid. These are:

- Doing whatever hits first
- Taking the path of least resistance
- Responding to the squeaky wheel
- Completing tasks by default
- Relying on misguided inspiration

A frequent trap for nurses is acting upon the "doing whatever hits first" trap. This nurse typically responds to things that happen first. For example, a nurse at the beginning of the day shift chooses to fill out the preoperative checklist for a patient going to surgery the next day rather than assess the rest of the patients first.

The second trap is the "taking the path of least resistance" trap. Nurses in this trap may make a flawed assumption that it is easier to do a task themselves, whereas the task could have been delegated, and the nurses could be completing another task that only a nurse can complete. For example, the nurse admitting a patient needs to take the vital signs, weigh the patient, get the patient settled, complete the baseline assessment, and call the practitioner for orders. Weighing the patient and getting the patient settled are tasks that can be delegated to NAP so the nurse may complete the baseline assessment of the patient and then call the practitioner for orders.

The third trap is the "responding to the squeaky wheel" trap, where the nurse feels compelled to respond to whatever need has been vocalized the loudest. In this case, the nurse may choose to respond to a family member who has come to the nursing station every half hour with some concern. To appease the family, the nurse may take time to focus on one of their many verbal concerns and overlook a more pressing patient care need elsewhere.

The fourth trap is called the "completing tasks by default" trap. This trap occurs when the nurse feels obligated to complete tasks that no one else will complete. A common example is emptying the garbage when it is full instead of asking housekeeping to complete the task.

The last trap is the "relying on misguided inspiration" trap. A classic example of this trap is when the nurse feels "inspired" to document findings in the chart and avoids tackling a higher-priority responsibility. Unfortunately, some tasks will never become inspiring. These tasks need discipline, conscientiousness, and hard work to complete them.

# CASE STUDY 18-1

You are working the day shift on a medical-surgical unit. You are responsible for six patients with the assistance of NAP and an LPN. What are the outcomes you want to achieve? Please use the criteria previously given for prioritizing, i.e., ABC, safety, etc. Which patients will you give to each care provider? Make out an assignment sheet, building on the format in Figure 18-7. Discuss your rationale.

| Patient | Priority Nursing Assessments |
|---|---|
| Ms. JD is a 68-year-old patient who is postop day 1 after a total shoulder replacement following a traumatic fall. She is confused and on multiple medications and has a history of hypertension and multiple falls. She is anxious and frightened by the "visiting spirits." | ABCs, level of consciousness, vital signs, safety, distal pulse, incision/dressing check, breath sounds. See this patient second during rounds. |
| Mr. DB is a 55-year-old patient with insulin-dependent diabetes mellitus, juvenile onset at age 12. He is postop day 2 after a right below-the-knee amputation. He complains of severe right leg pain and is restless. Mr. DB has a history of noncompliance with diet and is on sliding scale insulin administration. | ABCs, symptoms of hypoglycemia, glucoscan at 4 p.m. and 9 p.m., vital signs, safety, incision/dressing check, pain, DB teaching. See this patient third during rounds. You may need to check his glucose level STAT. |
| Mr. JK is a 35-year-old patient with a history of alcohol abuse, admitted for severe abdominal pain. He is throwing up coffee-ground-like emesis. | ABCs, level of consciousness, seizure and shock potential, hematemesis, DTs, safety, vital signs, CBC, hematocrit, type and cross-match, 16-gauge IV line for possible blood transfusion, oxygen, cardiac monitor. See this patient first during rounds. |

Now, identify the priority nursing assessments for this next group of patients.

## PRIORITY PATIENT ASSESSMENTS, GROUP II

| Patient | Priority Nursing Assessments |
|---|---|
| Ms. HM is an 85-year-old patient who was transferred from a nursing home because of dehydration. She is vomiting and has abdominal pain of unknown etiology. Intravenous hydration continues and a workup is planned. Ms. HM is alert and oriented. | |
| Mr. AB is a 72-year-old patient who is status post cerebrovascular accident. He is to be transferred to rehabilitation. He needs his belongings gathered and a nursing summary written. | |
| Ms. VG is an 82-year-old patient who is postop day 5 after an open reduction of a femur fracture. She has a history of congestive heart failure, hypertension, and takes multiple medications. Her temperature is elevated. She is confused. | |

# STRATEGIES TO ENHANCE PERSONAL PRODUCTIVITY

Time management applies not only to work but also to the nurse's personal life. Too often, nurses feel that they have no personal life because of rotating shifts, weekend work, and stressful work experiences. Proper time management will allow nurses to take control of both their work life and their personal life. Use a variation of the activity log in Table 18-1 to review how you spend your personal time. It will help to determine your most energetic time of day. Activities that take focus and creativity should be scheduled at high-energy times, and dull, repetitive tasks at low-energy times. Scheduling

## Critical Thinking 18-2

Throughout the day shift, nursing and unit staff communicate and work together to deliver quality patient care according to standards. The nursing standards and routines that the nursing staff will apply to these patients reflect the ANA Standards of Nursing Practice and include routines of care such as:

Bedfast patients—Turn Q2H, intake and output, etc.

All patients—Q4H vital signs, hygiene, fresh water, etc.

Depending on the type of patient care unit, the day shift routine might go like this:

7:00 a.m. Charge nurse reviews patient care assignments with all nurses and unit staff. Day shift takes hand-off shift report from night shift. Patient rounds and assessment begin and continue every half hour

7:30 a.m. NAP obtain vital signs (VS). Patient breakfast served.

7:45 a.m. Patient assessment, including ABCs, VS, lab work, medications, IV fluids; a.m. hygiene care begins following nursing care standards; turn and reposition patients Q2H.

8:00 a.m. Health care providers make rounds; patients sent for diagnostic tests; regular patient rounds and assessments.

8:30 a.m. VS reassessment as needed; patient rounds.

9:00 a.m. Medications given; patient rounds continue; 15-minute breaks begin for all staff.

9:30 a.m. Patient rounds continue; documentation begun.

10:00 a.m. Patient rounds; turn and reposition patients.

10:30 a.m. Patient rounds.

11:00 a.m. Patient rounds; lunch breaks start for all staff.

11:30 a.m. Patient rounds; reassessment of VS of those patients with 7:30 a.m. abnormal VS.

12:00 p.m. Patient rounds; turn and reposition patients. Patient lunch served.

12:30 p.m. Patient rounds.

13:00 p.m. Medications given; patient rounds.

13:30 p.m. Patient rounds.

14:00 p.m. Patient rounds. Intake and output reports completed; documentation completed. Turn and reposition patients.

15:00 p.m. Patient rounds. Hand-off shift report from day shift to evening shift.

Is this patient care routine similar to one you have seen on a patient care unit where you have worked? Make out a day shift routine for a unit you are familiar with. Make up an assignment sheet like the one in Figure 18-7 for the same unit.

## Critical Thinking 18-3

Nurses set priorities fast when they "first look" at a patient. As you approach your patient, get in the habit of observing the following:

FIRST LOOK
- Eye contact as you approach
- Speech
- Posture
- Level of consciousness

AIRWAY
- Airway sounds or secretions
- Nasal flare

BREATHING
- Rate, symmetry, and depth
- Positioning
- Retractions

CIRCULATION
- Color

- Flushed
- Cyanotic
- Presence of IV or oxygen
- Pain
- Vital signs (TPR and BP)
- Pulse oximetry
- Cardiac monitor

DRAINAGE
- Urine
- Blood
- Gastric
- Stool
- Sputum

Practice your "first look" the next time you approach a patient. Does this improve your assessment skills?

time for proper rest, exercise, and nutrition allows for quality time.

# CREATE MORE PERSONAL TIME

There are three major ways to create time. One is to delegate work to others or hire someone else to do work. Another is to eliminate chores or tasks that add no value. The last way is to get up earlier in the day. When you delegate a task, you cannot control when and how the task is completed. Initially, it may take more time to get others to do the chore than to just do it yourself, but this investment of time should save you time and energy in the future. If a chore is boring and mundane, it makes more sense to work an hour more at a job you enjoy to pay for someone else to do the unrewarding, boring work.

Getting up one hour earlier in the day for a year can free up 365 hours, or approximately two weeks a year, of extra time that can be used to enrich life. After several days of rising an hour earlier, an individual may feel tired and respond to the fatigue by going to bed a little earlier (Mind Tools, 2006b). This may be a good strategy for many people, especially those who are not productive in the evening and spend time doing activities that are minimally

rewarding, such as watching television. If a person does not try to get to bed earlier, however, and the end result of getting up early is fatigue, the strategy is not beneficial.

# USE DOWNTIME

During any day, there is time available that is seldom used, often referred to as "downtime." When waiting at appointments, in lines, or for others, people often have time available. Wait time can sometimes be avoided by calling ahead to verify appointments and arriving no more than five minutes early. During unavoidable waits, the time can be put to good use by having reading and writing materials handy.

Traveling or commuting time as a passenger is frequently frittered away. Listening to books on tape is a good way to catch up on reading. Many libraries have a collection of books on tape. If privacy is not an issue, phone calls can be initiated and returned during commutes when you are a passenger. Don't make phone calls while driving.

Sometimes it is important just to sit back and enjoy the scenery or the company. Time management principles aim at creating more enjoyable time, not filling every moment with chores.

## Critical Thinking 18-4

The relationship between personal lifestyle and the incidence of several diseases has been demonstrated. Many health promotion programs include the expectation that people invest in themselves. Do you invest in yourself with your daily activities to promote higher education; planned savings; healthy eating; regular exercise; deferred gratification; avoidance of smoking, tanning booths, drugs, excessive alcohol consumption; and regular physical checkups? Do you know people who seem to live only from one day to the next because their perspective of time is in the immediate and they do not seem to recognize the benefits of setting priorities and doing long-term planning?

## CONTROL UNWANTED DISTRACTIONS

Personal life is not immune from distractions that get in the way of accomplishing personal goals. These may include such distractions as visitors, unplanned phone calls, low-priority tasks, and requests for assistance (Table 18-7 and Table 18-8).

## FIND PERSONAL TIME FOR LIFELONG LEARNING

Finding time for lifelong learning is a struggle for recent graduates and even more-seasoned nurses. The complexity of health care demands an educated workforce. So as soon as one degree is completed, it is time to start thinking about the next.

### TABLE 18-7 Strategies for Avoiding Personal Time Distractions

| DISTRACTION | STRATEGIES |
|---|---|
| Casual visitors | Make your environment less inviting. Remain standing. Remove your visitor chair. Keep a pen in your hand. |
| Unplanned phone calls | Use an answering machine or voice mail. Consider a humorous message. Set a time to return calls. Don't call when driving. |
| Unwanted/low-priority jobs | Say no to jobs that have little value or in which you have little interest. Leave low-priority tasks undone. If an unwanted job must be done, pay or ask for assistance. |
| Requests for assistance | Encourage others to be more independent. Give them encouragement, but send them back to complete the job. Decisions to help should be conscious decisions, not drop-in distractions. |
| Clutter | Clear your work area of clutter, and keep it clean. Organize your work area, and take a few minutes at the end of your shift to prepare your area for the next shift. |
| Interruptions | Open your mail over the garbage can. Respond, delegate, or throw it out. Organize your papers. Keep your notebooks, calendar, and phone lists in one three-ring binder so that you have your essentials together. |
| Procrastination | Break a task down into manageable segments, and return to it again and again until it is complete. |
| Perfectionism | Become a pursuer of excellence, not a perfectionist, as you pursue your goals (Table 18-8). |
| Texting | Turn your phone off and check at scheduled intervals; let your friends and family know this. |
| Social media | Check social media no more often than once a day. It is a way of communicating but should not take over your life. |
| E-mails | E-mails should be checked at scheduled intervals and responded to appropriately. |

## TABLE 18-8    Behaviors of Perfectionists vs. Pursuers of Excellence

| PERFECTIONISTS | PURSUERS OF EXCELLENCE |
| --- | --- |
| • Hate criticism | • Welcome criticism |
| • Are devastated by failure | • Learn from failure |
| • Get depressed and give up | • Experience disappointment but keep going |
| • Reach for impossible goals | • Enjoy meeting high standards within reach |
| • Value themselves for what they do | • Value themselves for who they are |
| • Have to win to maintain high self-esteem | • Do not have to win to maintain high self-esteem |
| • Can only live with being number one | • Are pleased with knowing they did their best |
| • Remember mistakes and dwell on them | • Correct mistakes and then learn from them |

**Source:** Courtesy of White, L., Duncan, G., & Baumle, W. (2011). *Foundations of basic nursing* (3rd ed.). Clifton Park, NY: Delmar Cengage Learning.

There are ways to achieve your dreams of more education and work and still have a personal life. Flaherty (1998) offers tips for balancing school, family, and work in Table 18-9.

Returning to school is certainly a challenge, but with time management skills, the return to school can result in the accomplishment of personal outcomes, a degree, and new knowledge.

## TABLE 18-9    Personal Time Management When Returning to School

- Let your employer know that you are interested in returning to school. Most employers are supportive of additional education and will be flexible with your schedule. But they will continue to expect a competent, dedicated employee.
- Develop computer skills. By using a computer, you can e-mail professors and classmates at any time. You can do online research. You can easily incorporate constructive criticisms into papers and build on previous work. Technology is the working student's friend.
- Discover a flexible educational program. Many programs offer several classes in a row on a single day, weekend and night classes, or immersion classes for a week at a time. Some programs offer distance learning opportunities.
- Do not be surprised by the demands of school. Courses will be difficult and demanding of time. Remember that you have faced difficult demands and challenges before. Use the same techniques that helped you in the past, and develop some new ones.
- Solicit support from family and friends. They may offer emotional support, financial assistance, household assistance, child care, and so forth.
- Use all available resources at school and at work. Develop mentors and role models. Establish relationships with faculty. Discover and use academic support services such as writing centers and tutors. Read syllabi and course instructions carefully.
- Focus on the outcome. Keep the end in sight, and do not give up. Take it one course at a time. Reward yourself along the way. When a course is completed, celebrate.
- Be careful of the sacrifices. You may replace some hobbies with school. But save some time for the things that are really meaningful to you and your family.

*(Continues)*

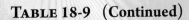

## TABLE 18-9 (Continued)

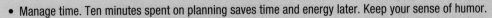

- Manage time. Ten minutes spent on planning saves time and energy later. Keep your sense of humor.
- Take care of yourself and your responsibilities. Set aside a day to take care of personal chores and errands.
- If you need a break, take one. Take time to reflect on what you are accomplishing. If you are feeling overwhelmed, take only one course or take a semester off.
- Study on the run. Taping lectures and listening to them as you commute is a great way to study on the run.

**Source:** Developed with information from Flaherty, M. (1998). The juggling act: Ten tips for balancing work, school, and family. *Nurseweek.* Retrieved on April 5, 2011 at http://www.nurseweek.com/features/98-5/juggle.html.

## KEY CONCEPTS

- General time management strategies include an outcome orientation, analysis of time cost and use, focus on priorities, and visualizing the big picture.
- Shift planning begins with developing both optimal and reasonable outcomes.
- Priority setting takes into account what is life threatening or potentially life threatening, what is essential to safety, and what is essential to the plan of care.
- The shift action plan assigns activities aimed at outcome achievement within a time frame.
- End-of-shift hand-off reports include face-to-face meetings and walking rounds.
- The shift action plan is evaluated at the end of the shift by asking if optimal or reasonable outcomes have been achieved.
- Time wasters that might interfere with outcome achievement include procrastination, inability to

delegate, inability to say no, management by crisis, haste, indecisiveness, interruptions, socialization, complaining, perfectionism, and disorganization.
- Time management applies to personal life as well as the job.
- Quality time can be achieved by analyzing time use and energy patterns.
- Delegation and getting up one hour earlier can create time.
- Additional time can be found by productive use of travel time and waiting time.
- Distractions can be controlled by making your environment less inviting, by using voice mail or an answering machine, by saying no, and by encouraging others to be independent.
- It is possible to balance work, family, and school.

## KEY TERMS

activity log
desired optimal outcomes

Pareto Principle
time management

## REVIEW QUESTIONS

1. The nurse has just finished the change of shift report. Which patient should the nurse assess first?
   A. A postoperative cholecystectomy patient who is complaining of pain but received an IM injection of morphine five minutes ago.
   B. A postoperative appendectomy patient who will be discharged in the next few hours.
   C. A patient with asthma who had difficulty breathing during the prior shift.
   D. An elderly patient with diabetes who is on the bedpan.

2. The staff RN's assignment on the 7 a.m. to 3 p.m. shift includes a newly admitted patient with pneumonia who has arrived on the unit, a new postoperative surgical patient requesting pain medication, and a patient diagnosed with nephrolithiasis who is complaining of nausea. What should the nurse do first after shift report?
   A. Assess the newly admitted pneumonia patient.
   B. Give morphine to the new postoperative patient.
   C. Set up the 9 a.m. medications.
   D. Administer Zofran (Ondansetron hydrochloride) to the patient complaining of nausea.

3. The nurse has been assigned to a medical-surgical unit on a stormy day. Three of the staff can't make it in to work, and no other staff is available. How will the nurse proceed? Select all that apply.
   _____ A. Prioritize care so that all patients get safe care.
   _____ B. Provide nursing care only to those patients to whom the nurse is regularly assigned.
   _____ C. Have the patients' families and ambulatory patients take care of the other patients.
   _____ D. Refuse the nursing assignment, as the increased number of patients makes it unsafe.
   _____ E. Quickly make rounds on all assigned patients to assess needs.
   _____ F. Decide on desired outcomes.

4. The nurse has just completed listening to morning report. Which patient will the nurse see first?
   A. The patient who has a leaking colostomy bag.
   B. The patient who is going for a bronchoscopy in two hours.
   C. The patient with a sickle cell crisis and an infiltrated IV.
   D. The patient who has been receiving a blood transfusion for the past two hours and had a recent hemoglobin of 7.2 grams/dL.

5. A new graduate RN organizing her assignment asks the charge nurse, "Of the list of patients assigned to me, who do you think I should assess first?" What is the best response the charge nurse could make?
   A. "Check the policy and procedure manual for whom to assess first."
   B. "Assess the patients in order of their room number to stay organized."
   C. "I would assess the patient who is having respiratory distress first."
   D. "See the patient who takes the most time last."

6. Of the following new patients, who should be assessed first by the nurse?
   A. A patient with a diagnosis of alcohol abuse with impending delirium tremens (DTs).

   B. A patient with a newly casted fractured fibula, complaining of pain.
   C. A patient admitted two hours ago who is scheduled for a nephrectomy in the morning.
   D. A patient diagnosed with appendicitis who has a temperature of 37.8°C (100.2°F) orally.

7. The nurse has just come on duty and finished hearing morning report. Which patient will the nurse see first?
   A. The patient who is being discharged in a few hours.
   B. The patient who requires daily dressing changes.
   C. The patient who is receiving continuous IV Heparin per pump.
   D. The patient who is scheduled for an IV Pyelogram this shift.

8. A nurse can enhance personal productivity by doing which of the following? Select all that apply.
   _____ A. Analyzing time, getting up an hour early, and delegating unwanted tasks.
   _____ B. Getting up an hour early, answering the phone, and inviting a friend in to talk.
   _____ C. Analyzing use of time, getting up early, and waiting patiently.
   _____ D. Avoiding working and going to school at the same time.
   _____ E. Controlling unwanted distractions such as texts, e-mails, and unplanned phone calls.
   _____ F. Using downtime by having reading and writing materials available.

9. You have received report on the following patients. Who would you make patient care rounds on first?
   A. Patient who is concerned that he has had no bowel movement for two days.
   B. Patient who has suffered several acute asthmatic attacks within the last 24 hours.
   C. Patient who is now comfortable but has had several episodes of breakthrough pain since yesterday.
   D. Patient who is severely allergic to peanuts who just ate potato chips fried in peanut oil.

10. Who would you make rounds on second?
    A. Patient who is concerned that he has had no bowel movement for two days.
    B. Patient who has suffered several acute asthmatic attacks within the last 24 hours.
    C. Patient who is now comfortable but has had several episodes of breakthrough pain since yesterday.
    D. Patient who is severely allergic to peanuts who just ate potato chips fried in peanut oil.

# REVIEW ACTIVITIES

1. For the next three days, complete an activity log (Table 18-1) for both your personal time and your work time. On what activities are you spending the majority of time? When is your energy level the highest? Is your energy level related to food intake?

2. Compare your use of nursing time to Figure 18-2. Are there any distractions that you can eliminate? What time management concepts might assist you in improving your time management?

3. You go to work one day and there are too many staff members on the unit. Several patients have been discharged. The nursing supervisor asks you to float to another medical-surgical unit. Note the example of the use of priority setting in caring for a group of three patients on this unit (see box).

| **PRIORITY ASSESSMENTS, GROUP I** | |
| --- | --- |
| **Patient** | **Priority Nursing Assignments** |
| Ms. JD is a 68-year-old who is postop day one after a total shoulder replacement following a traumatic fall. She is confused and on multiple medications, with a history of hypertension and multiple falls. She is anxious and frightened by the "visiting spirits." Her daughter stays with her at all times. | Vital signs, safety, distal pulse, incision/dressing check, and breath sounds. See this patient third during rounds. Safety is a prime concern with this confused patient, as well as watching for any postoperative concerns. |
| Mr. DB is a 55-year-old with insulin-dependent diabetes mellitus, juvenile onset at age 12. He is postop day two after a right below-the-knee amputation. He complains of severe right leg pain and is restless. Mr. DB has a history of noncompliance with diet and is on sliding scale insulin administration. | Vital signs, Glucoscan at 4 p.m. and 9 p.m., safety, incision/dressing check, pain, DB teaching. See this patient second during rounds. He has pain, restlessness, and a relatively new amputation. He is a diabetic and could have a postoperative complication or an insulin reaction. If in doubt, check Glucoscan. |

| | |
| --- | --- |
| Mr. JK is a 35-year-old patient with a history of alcohol abuse, admitted for severe abdominal pain. He is throwing up coffee-ground-like emesis. | Level of consciousness, seizure and shock potential, hematemesis, DTs, safety, vital signs, CBS, hematocrit, type and cross-match, 16-gauge IV line, pulse oximeter, cardiac monitor. See this patient first during rounds. He is a candidate for the development of shock. |

Now, identify the priority nursing assessments for this next group of patients back on your regular unit.

| **PRIORITY ASSESSMENTS, GROUP II** | |
| --- | --- |
| **Patient** | **Priority Nursing Assignments** |
| Mrs. Hohman, a 61-year-old with a hypertensive crisis three days ago, blood pressure decreasing daily, now 180/102. She periodically complains of headache. | |
| Mrs. Glusak, a 67-year-old transferred two hours ago from ICU, with a recent brain attack/CVA, responsive to painful stimuli, and has right-sided paralysis. Family at bedside. | |
| Mrs. Zurich, a 78-year-old with cellulitis of the right toe and a history of diabetes mellitus, needs teaching. | |

4. What are your distractions from outcome achievement? Develop a plan to minimize your distractions.

5. Use Table 18-4, Tool for End-of-Shift Hand-Off Report, to organize your patient care report.

# EXPLORING THE WEB

- If you would like to find a system for managing your time, the following Web sites offer electronic organizers:

  www.casio.com

  www.sharp-usa.com

  www.palm.com

- If you prefer a less technological time management system, the following Web sites offer nonelectronic organizers and systems for time management:

  www.daytimer.com

  www.covey.com

  www.franklin.com

- Find a free online calendar that you can access from anywhere at

  http://calendar.yahoo.com

  http://google.com

- Look at all the hints and free tools on time management at the Mind Tools Web site. Can you put any of the ideas to use?

  www.mindtools.com

- If you find time management an impossible challenge, you can find professional assistance at the Professional Organizers Web site.

  www.organizerswebring.com

- Check out this University of Michigan site for time management tips.

  www.umich.edu

- Take a look at the personal time management guide at

  www.time-management-guide.com

- Enjoy the Pickle Jar Theory

  www.alistapart.com

# REFERENCES

Barrero, L. H., Katz, J. N., Perry, M. J., Krishnan, R., Ware, J. H., & Dennerlein, J. T. (2009). Work pattern causes bias in self-reported activity duration: A randomised study of mechanisms and implications for exposure assessment and epidemiology. *Occupational Environmental Medicine, 66*(1), 38–44.

Covey, S. R., Merrill, A. R., & Merrill, R. R. (1994). *First things first: To love, to learn, to leave a legacy.* New York: Simon & Schuster.

Fitzgerald, M., Pearson, A., Walsh, K., Long, L., & Heinrich, N. (2003). Patterns of nursing: A review of nursing in a large metropolitan hospital. *Journal of Clinical Nursing, 12*(3), 326–332.

Flaherty, M. (1998). The juggling act: Ten tips for balancing work, school, and family. *Nurseweek.* Retrieved on March 5, 2006, from http://www.nurseweek.com/features/98-5/juggle.html.

Grohar-Murray, M. E., & Langan, J. C. (2011). Managing resources: Time. In *Leadership and management in nursing* (4th ed., pp. 286–296). Stanford, CT: Appleton & Lange.

Hansten, R. I., & Jackson, M. (2009). *Clinical delegation skills: A handbook for professional practice* (4th ed.). Sudbury, MA: Jones and Bartlett Publications.

Heindrich, A., Chow, M. P., Skierczynski, B. A., & Lu, Z. (2008). A 36-hospital time and motion study: How do medical-surgical nurses spend their time? *The Permanente Journal, 12*(3). Retrieved from http://xnet.kp.org/permanentejournal/sum08/time-study.html/ clinicalTOC.html, April 5, 2011.

Hughes, M. (2006). Life's 2% solution: Simple steps to achieve happiness & balance. Boston: Nicholas Brealy Publishing Company.

Marquis, B. L., & Huston, C. J. (2011). *Leadership roles and management functions in nursing: Theory and application (7th ed.).* Hagerstown, MD: Lippincott, Williams & Wilkins.

Mind Tools. (2006a). *How to achieve more with your time.* Retrieved February 27, 2006, from http://www.mindtools.com/ tmintro.html.

Mind Tools. (2006b). *How to achieve more with your time.* Retrieved February 27, 2006, from http://www.mindtools.com/ tmgetup.html.

National Council of State Boards of Nursing (NCSBN). (2005). Working with others: A position paper. Available at https://www. ncsbn.org/Working_with_Others.pdf, accessed April 5, 2011.

Reed, F. C., & Pettigrew, A. C. (2006). Self-management: Stress and time. In P. S. Yoder-Wise (ed.), *Leading and managing in nursing* (4th ed., pp. 413–430). St. Louis: Mosby.

Riesenberg, L. A., Leisch, J., & Cunningham, J. (2010). Nursing handoffs: A systematic review of the literature. *American Journal of Nursing, 110*(4), 24–34.

Scharf, L. (1997). Revising nursing documentation to meet patient outcomes. *Nursing Management, 28*(4), 38–39.

Sullivan, E. J., & Decker, P. J. (2009). *Effective leadership and management in nursing* (7th ed.). Lebanon, IN: Pearson.

Upenieks, V. B. (1998). Work sampling: Assessing nursing efficiency. *Nursing Management, 49*(4), 27–29.

Urden, L., & Roode, J. (1997). Work sampling: A decision-making tool for determining resources and work redesign. *Journal of Nursing Administration, 27*(9), 34–41.

Vaccaro, P. (2007). Five priority setting traps: Taking control of your time. *BILLD Alumni Newsletter*: A publication of the Midwestern Office of the Council of State Governments, 8(2), 1–2.

Vanhook, P. M. (2007). Cost-Utility Analysis: A Method of Quantifying the Value of Registered Nurses. The online journal of issues in nursing. Accessed at http://www.nursingworld.org/ MainMenuCategories/ANAMarketplace/ANAPeriodicals/OJIN/ TableofContents/Volume122007/No3Sept07/CostUtilityAnalysis. aspx, April 4, 2011.

White, L., Duncan, G., & Baumle, W. (2011). *Foundations of basic nursing* (3rd ed.). Clifton Park, NY: Delmar Cengage Learning.

## SUGGESTED READINGS

Barker, A. M., Sullivan, D. T., & Emery, M. J. (2006). *Leadership competencies for clinical managers.* Sudbury, MA: Jones and Bartlett Publishers.

Brafman, O., & Beckstrom, R. A. (2006). *The starfish and the spider.* New York: Penguin Group.

Carrick, L., & Yurkow, J. (2007). A nurse leader's guide to managing priorities. *American Nurse Today, 2*(7), 40–41.

Childre, D. (2008). *De-stress kit for the changing times.* Boulder Creek, CA: Institute for Heartmath.

Cohen, S. (2005). Reclaim your lost time with better organization. *Nursing Management, 36*(10), 11.

Emmett, R. (2009). *Manage your time and reduce your stress: A handbook for the overworked, overscheduled, and overwhelmed.* New York: Walker Publishing Company, Inc.

Jackson, M., Ignatavicius. D. D., & Case, B. (2006). *Conversations in critical thinking and clinical judgment.* Sudbury, MA: Jones and Bartlett Publishers.

Kaplan, R. S. (2007, January 7). What to ask the person in the mirror. *Harvard Business Review, 84,* 86–95.

Morgenstern, J. (2005). *Never check e-mail in the morning. And other unexpected strategies for making your work life work.* New York: Fireside.

Robinson, J., & Godbey, G. (2005). Time in our hands. *The Futurist, 39*(5), 18–22.

Vestal, K. (2009). Procrastination: Frustrating or fatal? *Nursing Leadership, 7*(2), 8–9.

# CHAPTER 19

# Patient and Health Care Education

PAUL HEIDENTHAL, MS; NANCY BRAATEN, RN, MS;
MARTHA DESMOND, RN, MS; SUSAN ABAFFY SHAH, RN, MS;
NANCY S. SISSON, MS, RN

*Education takes place when the learner's, not the teacher's, objectives have been achieved.*

(FLORENCE NIGHTINGALE)

## OBJECTIVES

Upon completion of this chapter, the reader should be able to:

1. Identify five major steps of a teaching methodology.
2. Relate the major learning domains.
3. Differentiate the four components of a behavioral objective.
4. Summarize the relationship between terminal and enabling objectives.
5. Construct behavioral objectives for a teaching session.
6. Create a lesson plan for a teaching session.
7. Discuss the impact of health literacy on patient health care education.

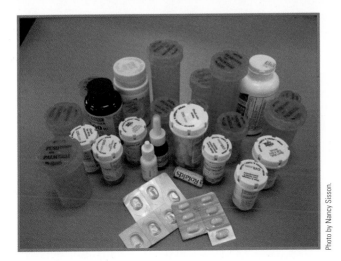

Photo by Nancy Sisson.

*Your patient, Mrs. Tulle, has arrived on the unit with multiple medication bottles from home. You need to do a medication reconciliation on her home medications. She has multiple bottles of the same pills with differing dosages.*

> *How will you determine the patient's knowledge of her medication regime?*
>
> *How will you determine her level of health literacy?*
>
> *What teaching strategies will you use to implement a discharge education plan?*

Patient and staff health care education is the communication of facts, ideas, and skills to change knowledge, attitudes, values, behaviors, and skills of patients, families, and fellow health care workers. Education is an inherent part of nursing. Teaching is a tool through which nurses bolster patients' self-care abilities by providing patients with information about specific disease processes, treatment methods, and health-promoting behaviors.

Patient education is also a legal component of the nursing process. In most states, patient education is a required function of nurses. Patient education is also mandated by several accrediting bodies, such as the Joint Commission (JC), formerly known as the Joint Commission on Accreditation of Healthcare Organizations (2010). The American Hospital Association's brochure *Patient Care Partnership: Understanding Expectations, Rights and Responsibilities* also calls for the patient's understanding of health status and treatment approaches (American Hospital Association (2006–2010).

The goal of staff education is to increase staff proficiency to function at the highest level of competency. Staff education is incorporated at all levels of nursing practice, including informally mentoring nursing assistive personnel in how to utilize safe transfer techniques in assisting a patient with a new amputation. Staff education includes distributing an evidence-based practice article or sharing an overview of a conference attended. It also includes formal ongoing education of nurses and other disciplines regarding health care issues facing nurses, patients, and their families.

## INFORMAL AND FORMAL PATIENT EDUCATION

Patient education occurs informally and formally in combination with almost all nursing interventions. Informal education can be as basic as exchanging information during a conversation with the patient, such as explaining a medication, procedure, or laboratory result. Formal education is planned, structured, and directed toward specific topics and goals. Formal education also contains evaluation, which measures the patient's success in retaining and applying information.

## INDIVIDUAL AND GROUP EDUCATION

The nurse provides education in an individual or group setting. Individual patient education frequently occurs in a clinical environment, often when an individual is facing the immediate impact of a specific health care situation. The term *patient education* may be a misnomer in many of these situations, because the nurse may be teaching the patient's family members or health care providers along with or instead of the patient. Even when others are involved, however, the focus of the education usually remains on the needs of an individual patient.

Although most patient and staff education occurs on an individual basis, nurses are increasingly becoming involved in teaching on a group level. As hospitals and other health care organizations adopt a proactive approach to health care, they are increasingly reaching out to the community, providing informational seminars and wellness classes that address common health care situations or emphasize preventive behavior.

## METHODOLOGY

Whether conducted in an individual or group setting, the educational process is more effective when it follows a structured, standardized approach. Such an approach is called a **methodology**. The methodology presented in Table 19-1 contains five major steps in the development and delivery of formal education.

### ANALYSIS

The first step in developing any educational program is to perform an analysis to define the type of education needed. The nurse should analyze three major elements:

- Context
- Learner
- Content

## TABLE 19-1 Educational Development Methodology

| ANALYSIS | DESIGN | DEVELOPMENT | IMPLEMENTATION | EVALUATION |
|----------|--------|-------------|----------------|------------|
| • Context | • Objectives | • Format | • Environment | • Learner |
| • Learner | • Sequence Content | • Strategies | • Learner | • Teaching |
| • Content | | • Media | • Presentation | |
| | | • Lesson plan | • Content | |

## REAL WORLD INTERVIEW

I had severe pain in my right foot, around the middle toe area. The diagnosis was Morton's neuroma. The nurse explained how it developed, how the doctor would remove it, and, most importantly, what the aftereffects would be. I've never had aftereffects explained to me; I never even thought to ask!

To me this represents quality care. The nurse told me what to expect. The majority of my non–health care friends and I don't even know enough to think of the questions. I really appreciated this information.

**Tessie Dybel**

Patient
Schererville, Indiana

## Context Analysis

The context consists of the situational context in which the educational need arises and the instructional context in which the education will occur.

The situational context is the situation that creates the need for education. For example, is the patient facing a particular health care procedure such as a heart operation? Has the patient expressed concerns over an existing or a potential health care condition such as diabetes? Is the patient receiving a new chemotherapy agent, or is there a practitioner initiating a new neurosurgery procedure requiring staff education?

The instructional context refers to the conditions under which education will occur. In what environment will the education be presented? Is it a health care facility, a home, a community environment, or some other setting? How will this affect the ability to provide and access resources? How will it affect the nurse's ability to effectively control the education environment? How will it affect the learner's attention and motivation?

The instructional context also includes the time for education: When will the education occur, and how long will the nurse have? Is the amount of time planned for education adequate? Is the amount of time flexible or is it fixed, such as one hour allotted at a community center for giving a course on diabetes management?

## Learner Analysis

The nurse should also conduct a learner analysis (Figure 19-1). **Learner analysis** is the process of identifying the learner's unique characteristics and needs and the ways in which these can influence the education process. Understanding the learner's particular needs and characteristics is an important consideration when developing an individualized education plan that is relevant and effective.

The learner's health literacy should be determined. **Health literacy** refers to the learner's ability to read, understand, and act on health information. Incomplete health literacy can affect anyone of any age, ethnicity, background, or education level. According to the Institute of Medicine, nearly half of all American adults (90 million people) have difficulty understanding and using health information (Partnership for Clear Health Communication of the National Patient Safety Foundation, n.d.).

Understanding the patient's unique characteristics may not guarantee success in education, but it can contribute to its effectiveness. The patient-centered and empathetic traits inherent in the nursing process can help nurses develop personal understanding and awareness of their patients. The first question facing the nurse is: Who is the learner? Is it the patient, family, nursing staff, ancillary staff, other?

In most cases, the learner will be the patient. However, in some cases, the patient may be unable or unwilling to participate in the learning process. In other instances, there may be other learners in addition to the patient, such as family members, legal guardians, or others, who make health care

Education topic _____

Target Audience (check all that apply)

_____ Patient _____ Professional staff _____ Nonprofessional staff

_____ Individual _____ Group

_____ Adult _____ Elderly _____ Adolescent _____ Child

Relationship of Group

(staff of same unit, multiple units, multidisciplinary team, family members)

_____

Purpose of Education

_____ Informational _____ Skill, procedure, technique

_____ Professional development _____ Infection control issue

_____ Introduction of new equipment _____ Quality improvement

Gender _____

Learner's primary language _____ Translation assistance needed _____

Limitations and barriers to patient/staff learning? Describe.

Cognitive _____

Physical _____

Emotional _____

Cultural/religious/lifestyle _____

Motivational _____

Language/literacy _____

Other _____

Literacy—education material at the appropriate reading level?

_____

Previous health care experience with this topic?

_____

Learning style assessment _____

_____

**FIGURE 19-1** Learner analysis for patient or staff education.

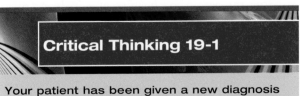

## Critical Thinking 19-1

Your patient has been given a new diagnosis of colon cancer. His brother died three months after being diagnosed with lung cancer. His physician wants to begin chemotherapy today. How do you evaluate the patient's readiness to learn about the disease and his treatment? What physical and psychosocial factors would you think about before providing education? Who else would you involve in the education?

decisions for the patient. Learners may be caregivers who, whether related or not, will be participating in the patient's care. Others may not directly participate in the care but may, with the patient's permission, require or desire information.

**Demographics** Various demographic characteristics can influence a learner's response to the health care and learning environments. These factors include the following:

- Cultural background
- Language
- Age
- Education
- Health literacy
- Health care background
- Physiological condition

## REAL WORLD INTERVIEW

When I was in the hospital with a pulmonary embolism, it became clear that I needed to go home with Lovenox shots. The nurses gave me information on Lovenox and explained why I had to take shots and not a pill. Then they started showing me how to do the shots, explaining each step. After a few practice shots, they had me do the injection. With encouragement, I was able to give the shots to myself. The nurse repeated the instructions and made sure I was comfortable. I was very nervous, but the nurse's patience with showing me each step and then letting me set the pace made it much easier for me to give my own shots at home.

**Cynthia Hooker**
Patient
Portsmouth, Virginia

The nurse's major goal is to identify characteristics that may indicate how the patient might respond to education. Some of these are physical, such as pain or physical discomfort, which will affect the patient's concentration. Others may be psychosocial, such as cultural beliefs that influence the patient's ideas about health care—for example, reliance on folk remedies. Cultural factors may also influence how a patient reacts to the nurse during an education session. Culturally influenced suspicion of or deference to health care staff may prevent the patient from effectively responding to the education.

The nurse may not be able to identify all of these characteristics prior to the education session. However, the more the nurse knows about the patient, the more the nurse can develop education that effectively incorporates the patient's unique characteristics and needs.

**Learning Styles** A **learning style** is a particular manner in which a learner responds to and processes learning. Traditional education often has expected the learner to adjust to the education style of the instructor rather than vice versa. In recent decades, this attitude has changed, and educational researchers and the educational community have come to agree on a principle that many people have always espoused: Different people learn differently.

The different ways in which people learn are still being determined and debated. Several theories of learning style exist, although they typically fall into one of

three categories: perception, information processing, and personality (Conner & Hodgins, n.d.).

**PERCEPTION** Perception theories emphasize the way in which people's senses affect learning. For example, some learners easily retain information when they can see or visualize it. Others retain best when they hear information. Still others learn best through physical action or involvement. Examples of perception theories include the Visual-Auditory-Kinesthetic (VAK) Model (Rose, 1985), shown in Table 19-2, and Gardner's Multiple Intelligences (Gardner, 1993), shown in Table 19-3.

**INFORMATION PROCESSING** Information processing theories emphasize different styles of thinking. Does the learner prefer concrete facts or abstract theory, reflective observation or direct experience? Examples are found in the Kolb Experiential Learning Style (Kolb, 1984), shown in Figure 19-2. In Kolb's theory, there are four major thinking styles: sensing and abstracting (which are opposites) and doing and watching (also opposites). An individual's style combines two of these styles, with one dominating. For example, a person whose style combines sensing and watching, with sensing dominant, is described by Kolb as a reflector personality. Such a person is likely to prefer concrete information rather than abstractions, prefers social situations to solitary ones, and judges performance by external measures rather than by personal criteria. Educationally, this person would probably learn best in an interactive group setting in which performance expectations are clearly stated and information is specific and factual.

**Personality** Personality theories emphasize how personality differences affect learning. Traits such as introversion or extroversion and preference for rational objectivity or instinctive "gut feeling" affect the individual's learning. There are examples in a theory based on the Myers-Briggs Personality Dichotomies (Figure 19-3), which in turn are based on the personality theories of Carl Jung (Briggs Myers, & Myers, 1995).

The Myers-Briggs theory suggests that personality is made up of four complementary pairs of traits: extroversion-introversion, sensing-intuition, thinking-feeling, and judging-perceiving. Within each of these pairs, there is a sliding scale, so to speak. For example, on the extroversion-introversion scale, 1 might be completely extroverted, 10 completely introverted, and 5 equally both. In most cases, one or the other of the pairs usually dominates. Through testing, an individual can be identified as one of sixteen possible types, based on the person's score in each of the four pairs. The personality type is identified by a four-letter indicator. For example, an INTJ would be an introvert-intuitive-thinking-judging personality. An ESFP would be an extrovert-sensory-feeling-perceiving personality. In

## TABLE 19-2 Rose's Visual–Auditory–Kinesthetic (VAK) Model of Learning

| STYLE | DESIGN | DESCRIPTION |
| --- | --- | --- |
| Visual learners | Learn best when they see; prefer graphic images and written text | • Use charts, graphs, pictures, diagrams, and so on.<br>• Include outlines, handouts, and other material for reading and note taking.<br>• Allow plenty of empty space in materials for learner to take notes, draw diagrams, and so on.<br>• Preview and review teaching content visually through flip charts, outlines, or other visual means.<br>• Include both textual and graphic versions of information within material. |
| Auditory learners | Learn best when they hear and say | • Present material verbally.<br>• Include verbal preview and review of material.<br>• Involve learner through verbal questions and answers, discussion.<br>• Include verbal activities such as brainstorming, discussion, and quiz show activities that require verbal response. |
| Kinesthetic learners | Learn best when they touch and move | • Use activities to get learners up and moving.<br>• Use music and color during presentation (these stimulate senses).<br>• Whenever possible, physically demonstrate learning and have learner physically practice it.<br>• Provide frequent breaks during presentation so learners can get up and move around.<br>• Provide toys, models, equipment, or other objects learners can touch.<br>• For complex tasks, have learners visualize physically performing the task. |

**Source:** Compiled from Rose, C. (1985). *Accelerated learning.* New York: Dell.

the theories of Jung and Myers-Briggs, each personality type has specific characteristics that affect the individual's approach to life experiences, including learning. Extroverts prefer to interact with people; introverts prefer to be alone. Thinking personalities prefer an objective, unwavering approach; feeling personalities are more likely to bend the rules if they think it makes people happy. Sensing people rely on their senses and prefer facts and structure to make decisions, whereas intuitive people are creative and see possibilities. Judging people like to be organized and decisive, whereas perceiving people work well in a spontaneous, flexible atmosphere.

Such personality differences can affect the way individuals approach and engage in a learning experience. Knowledge of learning styles is essential, because it directly influences the nurse's choices when it comes to selecting education strategies and media. For example, if the patient is a visual learner, using graphics and charts is a logical education choice. This choice can significantly affect the teaching process.

Knowledge of learner demographics and learning style is helpful in individual education sessions. Tailoring to specific individuals is usually not possible in group education, because members of the group will have a variety of

## TABLE 19-3 Gardner's Multiple Intelligences

| TYPE | CHARACTERISTICS | TEACHING CONSIDERATIONS |
| --- | --- | --- |
| Verbal—linguistic | Responds to rhythms and patterns of words, whether written or oral | Prefers activities that involve listening, speaking, writing |
| Logical—mathematical | Responds to reasoning, logic, and recognition of patterns and structures | Prefers activities that involve abstract symbols, formulas, numbers, problem solving |
| Musical | Responds to pitch, melody, rhythm, and tone | Prefers activities that involve audio, musical rhythms, tonal patterns, melodic sound |
| Spatial | Responds to two- and three-dimensional visual representations | Prefers using or creating graphics and models, "visualizing" abstract information |
| Bodily—kinesthetic | Responds to physical movement and activity (sports, dancing) | Prefers activities involving physical movement and gestures |
| Interpersonal | Responds to social interactions and relationships | Prefers group activities or other situations that involve human interaction or collaboration |
| Intrapersonal | Responds to personal, inner emotions to understand self and others | Prefers activities that use introspection, processing emotions, reflection |
| Naturalist | Responds to intricacies and subtleties of patterns and relationships in nature | Prefers activities that provide involvement in the natural world (plants, animals, other natural phenomena) |

**Source:** Compiled from Gardner, H. (1993). *Frames of mind.* New York: Basic Books.

characteristics. Rather than tailor group education to one style or characteristic, the nurse will need to incorporate multiple education methods that address various styles and characteristics.

## Content Analysis

After context and learner analysis, the next step in the analysis phase is content analysis.

During content analysis, the nurse begins to identify specific information that education should address. For example, the practitioner has just written an order for a medication you rarely administer on your unit. What specific information would you obtain to educate yourself and your coworkers on this medication? Could you use the same content information to educate the patient? Content information might include the following:

- Confirmation by pharmacy on the use of the medication for your unit
- Action of the medication
- Safe dose
- Side effects

- Route of administration
- Interactions with other medications or food

The nurse must also determine what information is essential to the education session. Although there may be a wealth of information available for a topic, all the available information may not be necessary information. Some of it may be irrelevant, redundant, or nonessential.

In health care, a PICO question may be used to focus evidence-based research for a patient or a staff development problem. A PICO question would include:

- P - Patient and problem. What are the characteristics of the population? What is the condition or disease that you are interested in?
- I - Intervention. What do you plan to do with this person or group?
- C - Comparison. What are the alternatives for solving the problem?
- O - Outcome. What are the relevant outcomes?

Before giving a staff in-service on renal failure, for example, the nurse educator may want to develop a PICO question

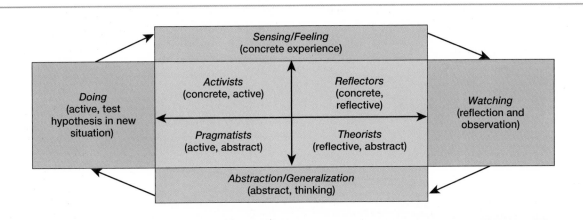

**FIGURE 19-2** **Kolb's experiential learning style.** (Compiled from Kolb, D. A. (1984). *Experiential learning.* Englewood Cliffs, NJ: Prentice-Hall/TPR).

to research the success of a support group for renal failure. For example:

- P - Patient and problem: Maintaining newly diagnosed renal failure patients at home
- I - Intervention: Involvement in a monthly support group
- C - Comparison: Patients that do not attend a support group
- O - Outcome: Decreased readmission rate over 6-month period

This is a question developed as a PICO question: "Do newly diagnosed renal failure patients that attend monthly support group meetings have a decreased readmission rate over a 6-month period when compared to a similar group of patients that do not attend a support group?" The nurse educator must analyze the evidence-based research to determine informa-

tion for the staff in-service (University of Southern California, n.d.). Using key words from the PICO statement will focus the nurse educator's search for information in the literature.

## Learning Domains

The goal of education is not just dispensation of knowledge but change in behavior. This is especially true in health care.

Learning theory suggests that learning can be classified into taxonomies, or **learning domains**, each based on the major type of learning involved.

Each taxonomy is organized into a hierarchy that progresses from simple to complex behaviors. These behaviors can be observed and measured. This means that the nurse does not just present information to a patient and hope that the patient has learned and mastered it. It also means that the nurse can document behavioral changes by measuring the

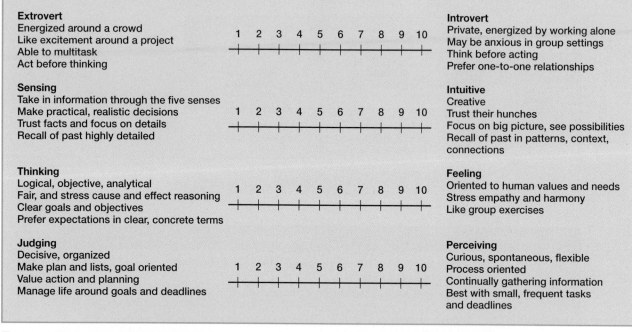

**FIGURE 19-3  Myers-Briggs personality dichotomies.** (Compiled from Briggs, I., & Myers, P. B. (1995). *Gifts differing.* Palo Alto, CA: Davies-Black; and Heidenthal, P., Braaten, H., Desmond, M., & Abaffy Shah, S. (2008). Patient and health care education. In P. Kelly, *Nursing leadership and management* (2nd ed.). Clifton Park, NY: Delmar Cengage Learning).

## REAL WORLD INTERVIEW

I work in a wound center in a small community hospital. A patient with a peripherally inserted central catheter (PICC) was due to come in for a hyperbaric chamber treatment. I would have to access the PICC prior to the patient's entry into the hyperbaric chamber. We do not usually have patients here with central lines, so I was feeling a bit apprehensive about using the PICC. I put in a call to the clinical nurse specialist who specializes in intravenous therapy to see if she could in-service the staff on my unit. She came to the unit with a policy on PICC management as well as a demonstration on how to access the PICC. She demonstrated the correct way to flush the PICC before and after disconnecting the patient from her medication. It was very helpful to have a resource person to be able to go to for this education session.

**Sabrina Mousseau, BS, RN, OCN**
Administrative Director, Medical Oncology
Albany, New York

patient's performance at appropriate levels. The nurse can also examine what types of behavior the information represents: acquiring knowledge, gaining skill, or changing attitude.

Educators have identified three domains of learning:

- cognitive (Bloom & Krathwohl, 1956)
- psychomotor (Simpson, 1971)
- affective (Krathwohl, Bloom, & Masia, 1999)

The **cognitive domain** is centered on knowledge, or what the learner knows (Table 19-4). The **psychomotor**

**domain** is centered on skill, or what the learner does (Table 19-5). The **affective domain** is centered on attitude, or what the learner feels/believes (Table 19-6).

The emerging pattern of behaviors sets the stage for the design phase of teaching development, in which the nurse translates the behaviors into objectives, develops education topics corresponding to the objectives, and arranges those topics into a structured topic sequence. This provides an organized, structured framework on which to build an effective education session.

## TABLE 19-4  Taxonomies of Learning: Cognitive Domain

| LEVEL | DESCRIPTION | ACTIONS |
| --- | --- | --- |
| Knowledge | Recalls information | Define, describe, identify, know, label, list, name, recall, recognize, select, state |
| Comprehension | Understands meaning of information, as demonstrated by the ability to restate the information in the learner's own words | Comprehend, convert, defend, distinguish, estimate, explain, generalize, give example of, interpret, paraphrase, rewrite, summarize, translate |
| Application | Uses information in a new situation | Apply, change, compute, construct, demonstrate, discover, manipulate, modify, predict, prepare, produce, relate, solve |
| Analysis | Separates information into component parts to understand the overall structure | Analyze, break down, compare, contrast, diagram, deconstruct, differentiate, discriminate, distinguish, identify, illustrate, outline, relate, select, separate |
| Synthesis | Puts information together to form a new meaning or structure | Categorize, combine, compile, compose, create, devise, design, generate, modify, organize, plan, rearrange, reconstruct, reorganize, revise, summarize |
| Evaluation | Makes judgments about the value of ideas | Appraise, compare, conclude, contrast, criticize, critique, defend, evaluate, explain, interpret, justify, support |

**Source:** Compiled from Bloom, B. S., & Krathwohl, D. (1956). *Taxonomy of educational objectives, handbook 1.* Boston: Addison-Wesley.

## TABLE 19-5  Taxonomies of Learning: Psychomotor Domain

| LEVEL | DESCRIPTION | ACTIONS |
| --- | --- | --- |
| Perception | Observes behaviors involved in a performance | Choose, describe, detect, differentiate, distinguish, identify, isolate, relate, select |
| Set | Demonstrates readiness to perform; understands steps in a task, adopts physical posture to perform the task | Begin, display, explain, move, proceed, react, show, state |
| Guided response | Imitates performance and refines it through trial and error, practice | Copy, trace, follow, react, reproduce, respond |
| Mechanism | Performs task comfortably | Assemble, calibrate, construct, dismantle, fasten, fix, manipulate, measure, mend, mix, organize |
| Complex overt response | Performs task skillfully and automatically (increased proficiency, accuracy, and coordination) | Same as for mechanism level, but adverbs/adjectives indicate increase in speed or accuracy of performance |

*(Continues)*

## TABLE 19-5  (Continued)

| LEVEL | DESCRIPTION | ACTIONS |
|---|---|---|
| Adaptation | Alters performance to adapt to new situations | Adapt, alter, change, rearrange, reorganize, revise, vary |
| Origination | Creates an original skill | Arrange, build, combine, compose, construct, create, design, initiate, make, originate |

**Source:** Compiled from Simpson, E. (1971). Educational objectives in the psychomotor domain. In M. Kapfer (Ed.), *Behavioral objectives in curriculum development.* Boston: Educational Technology.

## TABLE 19-6  Taxonomies of Learning: Affective Domain

| LEVEL | DESCRIPTION | ACTIONS |
|---|---|---|
| Receiving/attending | Displays attention, willingness to listen | Ask, choose, describe, follow, give, identify, locate, name, select, use |
| Responding | Displays active participation, willingness, motivation | Answer, assist, aid, comply, conform, discuss, help, label, perform, practice, present, read, recite, report, select, tell, write |
| Valuing | Shows acceptance, preference, and commitment for the value | Complete, demonstrate, differentiate, explain, follow, form, initiate, invite, join, justify, propose, read, report, select, share, study |
| Organization | Organizes and prioritizes values through contrasting them, resolving value conflicts, and developing a personal value system | Adhere, alter, arrange, combine, complete, defend, explain, formulate, generalize, identify, integrate, modify, order, organize, propose, relate, synthesize |
| Characterization (internalizing values) | Exhibits a value system that drives the individual's behavior | Act, discriminate, display, influence, modify, perform, practice, propose, qualify, question, revise, serve, solve, verify |

**Source:** Compiled from Krathwohl, D., Bloom, B. S., & Masia, B. B. (1999). *Taxonomy of educational objectives, handbook 2.* Boston: Addison-Wesley.

### Critical Thinking 19-2

Think about how you learn. How would you describe your own learning style? When you develop educational programs, does your learning style reflect your own learning preferences, or does it reflect the learning style of the patient? Look at a recent education session you observed. What changes would you make so that it addresses a different learning style?

## DESIGN

The purpose of the design phase is for the nurse to organize and structure the content identified in the analysis phase. The nurse accomplishes this by establishing objectives and by sequencing content.

## Establishing Behavioral Objectives

The behaviors identified in content analysis are now translated into behavioral objectives. A **behavioral objective** states a specific and measurable behavior that should result from the education session. Behavioral objectives define what the nurse will teach, thereby providing the skeleton of the education session. They also provide the basis for evaluation, because the success of learning is gauged by whether the learner achieves these objectives.

The essential component of a behavioral objective is a performance, specifically one that can be observed and measured. For example, the following is a behavioral objective: "List patient injection sites that are used to administer insulin." However, this statement leaves some elements undefined. Who is doing the listing? The person(s) who performs the behavior is the *audience*. This objective also does not state how well the person should perform the objective. Is 20% accuracy acceptable, or 50%, or 80%? The *degree* identifies how well the behavior is performed. When we add audience and degree to the objective, it reads this way: "The patient will list six injection sites that he or she can use to administer insulin, with 100% accuracy." These three elements—audience, performance, degree—should always be present in an objective. A fourth, optional component of an objective is called the condition. The condition indicates any restrictions or specific requirements involved in performing the behavior. For example: "Given an outline of the human body and a pencil, the patient will mark six injection sites where he can administer insulin with 100% accuracy." In this example, the condition is the phrase "given an outline of the human body and a pencil."

One way to remember the components of an objective is to think of the ABCD method: A = Audience, B = Behavior, C = Condition, and D = Degree (Table 19-7).

It is up to the nurse to determine how detailed objectives should be. The important consideration is that they indicate a behavior that is measurable.

**Terminal and Enabling Objectives** When developing education, the nurse often begins with a primary goal for the session—for example, the patient will demonstrate strategies for stroke management. As the nurse develops objectives, a hierarchy of objectives may evolve, much like an outline or a hierarchical organization chart. Certain primary behaviors, each supporting the session goal, become

evident. A **terminal objective** identifies a major behavior that contributes to achievement of the overall session goal. Terminal behaviors may in turn be achieved through the performance of related, secondary behaviors. An **enabling objective** identifies a secondary behavior that contributes to, or enables, achievement of terminal objectives. Table 19-8 illustrates this concept of terminal and enabling objectives.

Objectives also dictate the items for learner evaluation. Each evaluation item should correspond to an objective. If the nurse is creating objectives for performances that are not relevant or necessary, the nurse should examine those objectives and eliminate them.

## Sequencing Content

As the nurse analyzes content and develops objectives, the main topics within the education session and the order in which they appear become clearer. The nurse should then choose a specific topic-sequencing structure. In many situations, there will be only one choice, such as when teaching a procedure in the order in which the steps are performed. In other situations, no one sequence is obvious and the nurse has various options from which to choose.

Table 19-9 describes common sequencing structures.

## DEVELOPMENT

In the design phase, the nurse essentially determines what objectives education should accomplish and what information will support those objectives. During the development phase, the nurse clarifies how to teach by determining what strategies and resources to use.

Objectives should support evidenced-based practices. It is important to show how utilizing evidenced-based practice improves the quality of care. Nurses need to identify the learner's needs and target the education to achieve a positive outcome for that patient (Violana, Corjulo, Bozzo, & Diers, 2005).

**TABLE 19-7  Components of a Behaviorial Objective**

| AUDIENCE | BEHAVIOR | CONDITION | DEGREE |
|---|---|---|---|
| Who will perform the behavior? | What will the performer do? | What limitations or other conditions will be placed on the performance? | What degree of measurement will be used to determine successful performance? |
| *The patient* | *Will identify the major valves of the heart* | *Using the anatomical heart chart provided by the teacher* | *With an accuracy of 100%* |

## TABLE 19-8 Example of Session Goal, Terminal Objectives, and Enabling Objectives

*Session Goal: The patient will be able to demonstrate strategies for stroke prevention and management.*

**Terminal Objective 1:**
**Perform an accurate blood pressure reading**

Enabling Objectives:

1. Identify equipment for taking blood pressure reading

2. Demonstrate the procedure for taking blood pressure

3. State the acceptable range of blood pressure measurements

**Terminal Objective 2:**
**Develop a stroke management and prevention plan**

Enabling Objectives:

1. Identify signs and symptoms of a stroke

2. List modifiable risk factors for stroke

3. Discuss a low-salt, low-cholesterol diet plan

4. Develop an exercise/rehabilitation plan

**Terminal Objective 3:**
**Develop a Coumadin management program**

Enabling Objectives:

1. Discuss medication and precautions

2. Identify food interactions

3. Demonstrate knowledge of INR/PT laboratory tests

## TABLE 19-9 Common Sequencing Structures

| TYPE | CHARACTERISTICS |
|---|---|
| Chronological | Information is arranged in time sequence, based on the sequential occurrence of events. Typically used for teaching history. |
| Procedural (step by step) | Presents the steps in a procedure in the order in which they are performed. |
| Categorical | Information is organized into categories that are related to the primary topic. Allows the nurse to develop an arbitrary but logical structure when the information does not fit any of the other structures described here. |
| Topical | Patient is immediately placed in the middle of a topical problem or issue. Education may then address how the issue originated and the concerns surrounding its resolution. |
| Parts to whole; whole to parts | Presents the parts and then shows how they relate to the whole, or presents the whole and then talks about each part. |
| General to specific; specific to general | Similar to the preceding; presents the "big picture" first and then presents details, or presents details first and then shows how they fit into the big picture. |
| Problem to resolution | Presents a problem/situation and then presents the topics involved in its resolution. |
| Known to unknown; unknown to known | Known to unknown begins with patient's existing information/experience and uses it as a bridge to new, unknown information. Unknown to known presents unfamiliar information, later showing its connection to what the patient already knows. |

*(Continues)*

**TABLE 19-9** (Continued)

| TYPE | CHARACTERISTICS |
| --- | --- |
| Theoretical to practical; practical to theoretical | Presents the theory and then demonstrates how it is used, or shows practical applications of information and then presents the theory. |
| Simple to complex; complex to simple | Begins with information that is easier to present or easier for the patient to learn and then moves to more difficult information, or presents the most complex material first and then progresses to easier information. |

**Source:** Based on Milano, M., & Ullius, D. (1998). *Designing powerful training.* San Francisco: Jossey-Bass.

## CASE STUDY 19-1

Mr. Cabelo, a 74-year-old married man, is admitted to your unit with a new diagnosis of congestive heart failure. You are developing his education plan. His expected length of stay is three to four days. He will be adding Lasix and potassium to his home medicine regimen.

Who else can you involve in the teaching plan?

How do you determine what topics to include in the teaching?

How will you determine if your teaching is understood prior to discharge?

## Format

Modern educators have come up with various formats for teaching. One of the most common is Gagne's nine events of instruction (Gagne, 1985), described in Table 19-10. This provides an effective framework for conducting the education session.

**Repetition**   There is an old adage in adult training: Tell them what you're going to tell them; tell them what you want to tell them; tell them what you just told them.

In other words, preview information, present the information, and then review information. Previewing information allows the patient to understand what is to come and mentally prepare, much like providing a road map of an upcoming trip. It also provides a framework so that when information is presented, the patient knows where he or she is in the process. Finally, review provides the patient with an opportunity to organize the information and reinforce the learning. Therefore, repetition provides one of the most effective methods of education and should be incorporated into the education as much as possible.

**Interaction**   Learning works best when there is interaction between the learner and teacher, and when the learner is involved, whether it be nurse to patient or nurse to nurse. Interaction is stimulated using constant discussion, incorporating activities, and questioning the learner.

**Presentation, Performance, Practice**   In the case of patient education, the nurse can incorporate repetition, as well as interaction, into education through a format of presentation, performance, and practice. The nurse presents material to the patient, the nurse performs a demonstration either alone or along with the patient, and then the patient practices the performance, either independently or with nurse observation. This format allows the patient to repeat and reinforce learning while also increasing the patient's interaction with the nurse and active involvement in the learning.

**Variety**   A certain amount of variety in education maintains the learner's interest. Using different approaches and media can add variety to the education session. However, when working with patients, too much variety can break continuity and confuse the patient.

## Strategies

Although Gagne's events provide a structure for the overall education session, they do not prescribe how education will occur at the topic level. At this level, education can take various forms, commonly called strategies. Examples

**TABLE 19-10   Gagne's Nine Events of Instruction**

| EVENT | EXPLANATION |
| --- | --- |
| 1. Gain attention. | Engage the learner, and stimulate interest and motivation. |
| 2. Inform learner of objective. | Stating objectives establishes the expectations for the learner and provides an opportunity to preview the content of the education event. |
| 3. Stimulate recall of prerequisite learning. | Try to relate the content to previous learning or experience. This provides the learner with a familiar context in which to approach learning. |
| 4. Present stimulus materials. | Present the content. |
| 5. Provide learning guidance. | Perform/demonstrate the learning for the learner or in conjunction with the learner. |
| 6. Elicit performance. | Have the learner practice or demonstrate the learning. |
| 7. Provide feedback. | Help the learner refine learning by providing feedback and suggestions. |
| 8. Assess performance. | Evaluate learner performance in terms of the learning objectives. |
| 9. Enhance retention and transfer. | Review learning and encourage learner to use learning in new situations. |

**Source:** Compiled from Gagne, R. (1985). *The conditions of learning and the theory of instruction.* New York: Holt, Rinehart & Winston.

of education strategy include lecture, group discussion, fact sheets, and role play. The number and types of strategies the nurse chooses depend on several factors such as the type of content being taught, the audience's learning style, the nurse's own comfort level with various strategies, the resources available, the education environment, and so on. The nurse may use a single education strategy throughout the session or may vary strategy from topic to topic. Strategy may also be influenced by the education event; that is, it may be different during presentation and performance.

The nurse has wide latitude in selecting education strategies, but the nurse should choose strategies and be able to adapt the education methods to be effective in meeting the needs and preferences of the learner (Minnesota Department of Health, 2001).

## Media

Equally important to consider are the resources to be used during the education session. Resources consist of various media such as written, visual, audio, video, computer based, or other material, or combinations of these materials. As in selecting an education strategy, a single medium may be used throughout the event, or several may be used, depending on such factors as content type, education strategy, patient learning style, patient literacy level, teacher preference, or the limitations of what is available. The nurse has several options but should make decisions based on what would be best for the patient's learning style.

The nurse must consider two categories of media for patient education—media to be used during education and media to be provided to the patient for reference. Because the latter is likely to be used outside the presence and supervision of the nurse, it is especially important that it be appropriate for the patient.

**Media Evaluation and Selection**    When possible, the nurse should use media resources that are already available. This frees the nurse from the task of creating materials and allows for more time to be devoted to developing and managing the education event. However, the nurse must examine available materials and determine whether they are truly suitable for use in education.

The nurse should look at several factors. Is the content presented in the material appropriate for the patient's needs? Is it accurate? Is the material of suitable quality—for example, is the audio understandable, is video of appropriate visual quality, are graphics understandable, is text readable or at the appropriate reading level? Is the source reliable—for example, does information on a Web site come from a reputable and verifiable source?

Web-based education has made many advancements, and, with the increased computer knowledge of today's patients and nursing staff, this type of education can be ideal. Formal online educational programs allow nurses to learn about the most recent evidence-based practices, thus allowing them to bring this information to their patient education sessions. Many online educational programs can be accessed from home, making it more convenient for the nurse (Belcher & Vonderhaar, 2005).

**Materials Development** The nurse may use existing materials when developing an education session. In many situations, the nurse's organization will have existing materials and encourage their use. In other situations, the nurse may have greater freedom in selecting education materials.

In some situations, the nurse may find that no materials exist or existing materials are not appropriate for the content to be taught. The nurse may need to create materials, either to augment existing materials or to fill a void. Keep in mind that creating materials can be both a creative and a frustrating endeavor.

Many times, nurses are busy and unable to attend in-services and classes. Fact sheets are a good alternative. Fact sheets provide bedside nurses with up-to-date research. They are usually one page in length, with easy-to-read bullets that bring out the most important aspects nurses need to apply to their practice (Valente, 2003) (see Table 19-11).

## Lesson Plan

So far, the nurse has analyzed education content, designed behavioral objectives, organized the sequence of content, established the overall format of education, and selected appropriate education strategies and materials. All these decisions will be reflected in the lesson plan. A **lesson plan** is a document that provides the blueprint for the education session. See Table 19-12 and Table 19-13 for sample lesson plans. They provide the necessary information for the nurse or other educator to conduct the education session.

## IMPLEMENTATION

During the implementation phase, the nurse actually conducts the education. The lesson plan defines the objectives, topics, strategies, and materials needed for the session. Using this plan, the nurse begins the education session.

When the nurse begins the education session, various factors will affect the success of the session. Some of these factors are highly controllable, others less so. Some of the more influential factors are environment, patient condition, and the nurse's education and communication skills.

### TABLE 19-11 Sample Fact Sheet

#### PREVENTING VENTILATOR-ASSOCIATED PNEUMONIA (VAP)

*Ventilator-Associated Pneumonia*
- Common complication of patients in the Intensive Care Unit (ICU)
- Risk for developing VAP increases from 1% to 3% for each day a patient is intubated and on a ventilator

*Definition*
- Inflammation of lung parenchyma, caused by infectious agents

*Symptoms*
- Fever > 100.4°F
- Chest pain
- Crackles
- Mental status changes
- Purulent tracheobronchial secretions
- Decreased gas exchange
- Hypoxemia

*Diagnosis Determined*
- Chest X-ray showing infiltrate
- Leukocytosis
- Positive gram stain in sputum

*(Continues)*

## TABLE 19-11   (Continued)

### PREVENTING VENTILATOR-ASSOCIATED PNEUMONIA (VAP)

*Interventions to Prevent VAP*

- Hand washing—after removal of gloves and between patient contacts
- Semi-recumbent positioning—head of bed elevation to 35–45 degrees decreases gastrointestinal reflux and aspiration
- Sedation and neuromuscular-blocking agents (NBAs)—oversedation and use of NBAs may prolong mechanical ventilation
- Adequate oral hygiene—use of Sage Toothette for oral care
- Endotracheal (ET) tubes with subglottic suctioning—Hi-Lo Evac ET tube helps to prevent aspiration of colonized bacteria
- Lateral rotation therapy/continuous oscillation beds—should be started within 24 hours of intubation
- Ventilator suctioning—catheters should be changed and suctioning done only when necessary
- Stress ulcer prophylaxis—some agents reduce acidity, causing an environment for growth of bacteria; use is recommended when high risk of bleeding present
- Oral vs. nasal intubation—nasal intubation should be avoided whenever possible due to risk of sinusitis
- Nasogastric/enteric tubes—tube provides route for bacteria to travel from stomach to orophyranx
- Noninvasive positive pressure ventilation—eliminates need for ET tube

**Source:** Prepared by Susan Shah RN, MS. Compiled with information from Grap M, Munro C. L. (2004). Oral health and care in the intensive care unit: State of the science. *American Journal of Critical Care, 13,* 25–34; and Schleder, B. J. (2003). Taking charge of ventilator-associated pneumonia. *Nursing Management, 34,* 29–32. Reviewed by Nancy Sisson, MS, RN, August, 2010.

## TABLE 19-12   Sample Staff Lesson Plan for Stroke Management

**Setting:** 4th-floor stroke unit

**Learner:** 4th-floor unit staff

**Overall Goal:** The staff on the 4th floor will verbalize knowledge of stroke management

**Objectives:** After completion of this session, the staff will:

1. Verbalize a definition of a stroke.
2. Identify the signs and symptoms of stroke.
3. Discuss prevention and treatment of stroke.

**Topic:** Stroke management. In this lesson, the staff will receive the following information:

**Definition of stroke:** A decrease in the blood supply of the brain due to bleeding or a blood clot. This results in loss of function of the affected parts of the body.

**Signs and symptoms of stroke:** Loss of muscle tone, headaches, dizziness, tingling in the arms and legs, numbness, vision disturbances or temporary blindness in one eye, confusion, faintness, loss of consciousness, slurred speech, or inability to talk.

*(Continues)*

## TABLE 19-12 (Continued)

**Causes:** Stroke is caused by bleeding or a blood clot that causes decreased blood flow to the brain.

**Prevention:** Exercise daily, do not smoke, check blood pressure regularly, follow a recommended diet, daily aspirin may help.

**Treatment:** Thrombolytics, anticoagulants, and/or aspirin, physical therapy, speech therapy, and so on.

**Strategy:** PowerPoint presentation

**Medium:** Handouts—Stroke Management

**Evaluation:** Provide a five question true or false quiz to staff members—they must obtain a score of 80% or better.

**Evaluation tool for stroke management:** Mark the correct answer.

_____T_____ F 1. A stroke is a temporary increase in blood supply to the brain.

_____T_____ F 2. Numbness and tingling is a symptom of a stroke.

_____T_____ F 3. A stroke can be caused by full blockage of a small artery in the brain.

_____T_____ F 4. One way to avoid a stroke is to avoid smoking.

_____T_____ F 5. Two medications that may be prescribed for a stroke patient are aspirin and Coumadin.

**Source:** Compiled from www.americanheart.org (accessed in February, 2006).

## TABLE 19-13 Sample Lesson Plan

**Lesson Plan Title: Diabetic Self-Management**

| | |
|---|---|
| Patient | Juan Abado |
| Presenter | John Reilley, RN |
| Setting | Patient's hospital room |
| Brief patient/learner summary | Patient is a 50-year-old English-speaking Hispanic male, college graduate, newly diagnosed with diabetes and unfamiliar with the self-injection process. No other learners involved. No physical limitations to learning. Mild anxiety about self-injection, but high motivation. Initial interview suggests preference for learning through visual means. |
| Overall goal | The patient will understand and demonstrate diabetic self-care. |
| Objectives | After completing the session, the patient will be able to do the following: <ul><li>Perform an accurate blood sugar reading</li><ul><li>Identify equipment for taking blood sugar reading</li><li>Demonstrate the procedure for taking a blood sugar reading</li><li>State the acceptable range of blood sugar level</li></ul></ul> |

*(Continues)*

**TABLE 19-13 (Continued)**

- Develop a diabetic behavior management plan
  - Identify symptoms of hypoglycemia and hyperglycemia
  - List foods that can raise blood sugar levels
  - Discuss the ADA diet plan
  - Develop a regular exercise plan
- Administer insulin self-injection
  - Identify equipment necessary for insulin self-injection
  - Discuss insulin and method for self-injection
  - Demonstrate correct procedure for insulin self-injection

**Topic outline (excerpt)**

Time: 1 minute

Preview

In this session, we will learn about diabetes and three major elements of diabetic self-care:

- Monitoring blood sugar
- Developing appropriate diet and exercise
- Insulin self-injection

Time: 3 minutes

Objective: Identify equipment for taking a blood sugar reading

Topics:

1. Importance of taking a blood sugar reading
2. When/how often to take a blood sugar reading
3. Equipment for taking a blood sugar reading
   a. Accu-Check and similar machines
   b. Interpretation strips

Strategy: Lecture and demonstration/return demonstration

Medium: Handout—*Monitoring Blood Sugar*

Time: 10 minutes

Objective: Demonstrate the procedure for taking a blood sugar reading

Topics:

1. How to put the interpretation strip in the machine
2. How to disinfect the finger
3. How to stick the finger
4. How to apply the blood to the interpretation strip
5. How to interpret the results

Strategy: Presentation of video and discussion with patient; nurse demonstration followed by patient demonstration

Medium: Equipment—Accu-Check machine and interpretation strips; video segment—*Taking a Blood Sugar Reading;* handout—*Monitoring Blood Sugar*

(Continue with this format for each objective in the session. Each segment lists the objective, the related topics to be covered, the teaching strategy used, and the presentation medium. It is also helpful to indicate the estimated amount of time needed to conduct each segment.)

*(Continues)*

**TABLE 19-13** (Continued)

| | |
|---|---|
| Time: 3 minutes | Review |
| | (In this segment, the nurse reviews the major objectives and topics covered in the session. This is also a useful time for asking the patient whether there are any topics he is unsure of or has further questions about.) |
| | Strategies for retention/transfer |
| | (In this segment, the nurse indicates any methods for helping the patient apply the teaching to future situations.) |
| | Have patient perform own blood sugar tests four times daily during remaining time in hospital. |
| Evaluation | (This segment can take the form of a behavioral checklist, in which the nurse checks off that the patient has acceptably performed the behavior, or specific evaluation items, such as questions or other forms of evaluation. The example below contains samples of each.) |
| | Identifies equipment for taking blood sugar reading ☐ Yes ☐ No |
| | Demonstrates the procedure for taking a blood sugar reading ☐ Yes ☐ No |
| | Blood sugar readings are taken using an Accu-Check machine and _____. |
| | Before sticking the finger, you should: |
| | A. put the blood on the strip |
| | B. read the blood sugar results |
| | C. disinfect the finger |
| | D. close your eyes |

## Environment

The environment in which the education will occur can have a major impact on the effectiveness of education. Before beginning to educate, the nurse should evaluate physical environment factors such as lighting, temperature, and sound quality. Also consider the learner's privacy needs, interior, or exterior distractions, and the environment's ability to support the education session. The nurse may have to quickly adapt the education session in the case of unexpected situations.

## Learner

The nurse must also assess the learner's condition at the time of the education session and be prepared to adapt accordingly. The learner's condition can be affected by physical factors, such as discomfort or pain, and by psychological factors, such as depression or learner anxiety. Any of these factors can seriously affect the learner's ability to learn. The nurse must decide how to address these factors or whether to postpone education until a more appropriate time.

## Presentation and Content

An effective educator must bring certain qualities to the education event. Although every individual will have varying degrees of talent in each area, nurse educators must constantly be aware of these qualities and strive to exhibit them in the education session. These qualities include the following:

- *Content knowledge:* The nurse may not be an expert on a topic but must be able to demonstrate reasonable knowledge to the patient.
- *Education experience:* The nurse must demonstrate experience and professionalism.
- *Communication:* The nurse must be able to communicate clearly and at a professional level.
- *Intelligence:* The nurse must be intelligent and able to grasp the complexities of the content.
- *Adaptability:* The nurse must be able to adapt to changes in the education content and format and to unforeseen changes in the education session.
- *Patience:* The nurse must be patient and caring with learners.

# EVIDENCE FROM THE LITERATURE

**Citation:** Chang, M., Kelly, A. E. (2007). Patient education: Addressing cultural diversity and health literacy issues. *Urologic Nursing, 27*(5), 411–417.

**Discussion:** The article discusses the importance of patient education for improved health outcomes. It gives a review of the teaching and learning process, including a review of assessing learning styles and prioritizing education for the patient. How cultural influences affect the learning process is reviewed with information on how to individualize teaching strategies. An overview of health literacy and the development of educational material is included. There is a case study at the end to apply the information in practice.

**Implications for Practice:** Utilizing knowledge of cultural influences, health literacy, and the assessment process for teaching and learning will contribute to improved patient outcomes.

---

### Critical Thinking 19-3

You are caring for a patient with lung cancer. Smoking cessation has been difficult for her. Your preceptor would like you to identify key elements that would assist the patient and staff in the plan of care for this patient. Can you identify one topic that you could use to educate the patient using a fact sheet? What research literature will you use to develop your information? How can you make the fact sheet appealing to the eyes and draw the attention of the staff and patient?

---

- *Self-confidence:* The nurse must maintain a poised and professional manner when interacting with learners.
- *Self-direction:* The nurse must be able to assume initiative, identify needs, and solve problems. The nurse must be able to work independently without supervision.
- *Interactive:* The nurse must enjoy people and interacting with them and be able to work with difficult people.
- *Organization:* The nurse must be able to organize and prioritize tasks and information to work efficiently.

## Communication Skills

Although communication can be viewed as another education skill, it is more significant than all others in producing successful education. Both verbal and nonverbal communication skills are essential to effective education.

Create a welcoming and supportive environment for the patient. Offer privacy where the patient is free to ask questions and discuss concerns. Note these simple steps to increase your patient's understanding:

- Slow down the pace of your teaching.
- Avoid abbreviations and idioms. Avoid the use of "PRN," for instance; instead, use "as you need it." An idiom is a figure of speech that could be misunderstood. Avoid asking a patient if they are "feeling blue" or are "under the weather." Instead, ask the patient if they are "feeling sad."
- Use visual aids and illustrations.
- Tell patients what their test results are and also give them the normal readings, such as: "Your cholesterol level is 290. A healthy cholesterol level is less than 200. We will be reviewing information with you on how to lower your cholesterol level."
- Limit the amount of information shared at each session and repeat, repeat, repeat.
- Use a "teach back" or "show me" approach with patients to confirm their understanding. **Teach back** is a method of confirming learner knowledge where the learner can teach back the new information in their own words. This method allows the teacher to evaluate the learner's understanding. The patient should not view this as a test, but as a review of the instructions. The teach back approach includes the demonstration of a task by a patient to verify understanding.
- Be an active listener.
- Clarify and address any quizzical looks by the patient or family. Be aware that confused looks or blank stares by the patient or family may indicate a lack of understanding. If this occurs, rephrase the information, use simpler words, pause, and offer education in smaller segments.
- Remember to be respectful, caring, and sensitive to the patient and family.

Listening skills are also critical for effective education. The nurse must constantly watch and listen to the patient, looking for cues as to the patient's reaction to the education. The nurse must involve the patient in learning and look for opportunities to clarify, support, encourage, and incorporate patient responses. This further involves the patient in the learning process and increases patient motivation.

Body language says a lot about the nurse's level of interest and motivation. The nurse should maintain a professional appearance during the education event. Eye contact, use of hands, movement, and distance between nurse and patient all send messages about the nurse's attitude toward the education event.

## EVALUATION

**Evaluation** is the process of determining the effectiveness of education. The two major components of evaluation are learner evaluation and education evaluation:

- *Learner evaluation:* Did the learners learn what they were supposed to learn?
- *Education evaluation:* Was the education presented in an effective manner?

### Learner Evaluation

What to evaluate is determined by the objectives. There should be a direct correlation between the objectives established for the education and the learner evaluation that occurs during the education.

The following demonstrates an objective and the related evaluation item:

- *Objective:* Patient/caregiver able to identify four high-salt foods to avoid.
- *Evaluation item:* Patient/caregiver able to state four high-salt foods to avoid.

Learner evaluation can take many forms. Asking the learner to recall information, answer questions, perform procedures, solve relevant problems, analyze a situation, or construct a plan of action are all forms of learner evaluation. The nurse should choose evaluation events based on how effectively they reflect the associated learning objective, how *realistically* the learner can be expected to perform the evaluation, and how practically the nurse can observe and measure successful performance.

The nurse must also remember that the purpose of evaluation is to validate that the patient and staff have effectively processed and adopted the information. Many learners feel anxious at any event that has the slightest hint of "testing." The nurse can reduce learner anxiety by presenting evaluation in the context of a review of the education. Using the note method of confirming learner understanding allows the teacher to evaluate the learner's understanding and places the responsibility on the teacher. As mentioned earlier, the patient should not view this as a test, but as a review of the instructions. If the learner is having trouble with certain topics, the nurse can revisit those topics or, if conditions make that approach impractical, provide additional resources or referrals to the learner.

### Education Evaluation

Education evaluation is concerned with whether the education event itself was effectively constructed and presented. It is useful for the nurse to examine the education session and identify areas for improvement as well as areas to reinforce the effective elements.

Education evaluation can involve feedback from the nurse educator, the patient, and/or third-party observers. Measurement can be formal or informal and can involve verbal or written feedback.

Table 19-14 identifies some of the elements that can be examined in education evaluation.

### TABLE 19-14  Education Evaluation

| POSSIBLE AREAS FOR EDUCATION EVALUATION | |
| --- | --- |
| Patient learning | Did the patient learn the appropriate content, as indicated by such tools as learner evaluation results and follow-up observations? Were learner evaluation items appropriate and reliable? |
| Patient satisfaction/comfort | Was the patient satisfied with the content presented? With the effectiveness of the presenter? Did the patient feel that questions were addressed appropriately? |
| Environment | Was the environment conducive to learning? Were there distractions from inside or outside the room? Was lighting appropriate? Room temperature? |
| Design | Were the objectives appropriate? The topics? The sequence? |
| Knowledge | Did the materials and/or teacher reflect adequate knowledge of the content? |
| Organization | Was the information well organized? |

*(Continues)*

## TABLE 19-14   (Continued)

### POSSIBLE AREAS FOR EDUCATION EVALUATION

| | |
|---|---|
| Accuracy | Was the information presented accurate? |
| Relevance | Was information presented relevant to the patient's situation? Was relevant information missing? |
| Delivery | Were education strategies effective? Materials? |
| Pacing | Did education move too fast or too slow? Were demonstrations at a pace the patient could follow? Was the patient given enough time for practice? |
| Variety | Was there too much of one type of activity? Not enough variety in education methods or presentations? Too much variety, creating a sense of confusion? |
| Involvement | Was there enough patient involvement? Was there a lack or shortage of activities and/or practice time? |
| Communication | Did the nurse communicate clearly and effectively with the patient? |
| Focus | Did the materials and/or educator stay on the topics? |
| Assistance | Did the educator provide enough assistance? Did the materials provide cues or explanations to assist the user in completing activities? |

Patient and staff education is an important component of clinical practice. The staff needs to be kept up to date on the latest nursing and health care research. Evidence-based nursing care is of the utmost importance to the bedside clinical nurse.

Patient and staff education is a rewarding experience for the nurse. The nurse can provide a professional and gratifying learning situation through application of a structured approach to the design, development, and delivery of education.

## KEY CONCEPTS

- A standard education methodology contains five major phases: analysis, design, development, implementation, and evaluation.

- Analysis consists of context analysis, learner analysis, and content analysis.

- The design phase of education consists of establishing objectives and sequencing content into an organized, structured framework.

- The development phase of education consists of establishing format, selecting strategies and media, and finalizing the lesson plan.

- The implementation phase of education consists of conducting education based on the lesson plan.

- The evaluation phase of education consists of learner evaluation, which measures how well the learner learned, and the education evaluation, which measures how well the education was conducted.

- Learners have individual learning styles and respond to specific educational methods. Health care literacy must also be assessed when developing educational methods.

- Gagne's nine events of instruction provide structure for education.

- All learning can be classified under three domains: cognitive, psychomotor, and affective. Each contains a hierarchy of behaviors.

- Behavioral objectives specify the audience, behavior, condition, and degree of measurement of the education session.

- Terminal and enabling objectives clarify learning.

- The lesson plan documents the objectives, content, sequence, format, strategies, media, and evaluation methods of the teaching session.

# KEY TERMS

affective domain

behavioral objective

cognitive domain

enabling objective

evaluation

health literacy

learner analysis

learning domains

learning style

lesson plan

methodology

psychomotor domain

teach back

terminal objective

# REVIEW QUESTIONS

1. You are a new graduate on the unit, and your preceptor has asked you to demonstrate an insulin injection to your patient. The primary learning style you will utilize when the patient does his return demonstration is which of the following?
   A. Visual learning
   B. Auditory learning
   C. Kinesthetic learning
   D. Logical learning

2. Referring to Gagne's nine events of instruction (Table 19-10), a nurse has just instructed her asthmatic patient on the use of an inhaler. One event the nurse can utilize to evaluate the patient's skill is which of the following?
   A. Ask the patient to watch the nurse demonstrate the skill again.
   B. Have the patient demonstrate the use of the inhaler.
   C. Document the education session on the patient's chart.
   D. Provide the patient with a lecture on asthma.

3. The nurse has just instructed the patient on stroke management as part of discharge planning for home. The patient states, "I don't feel I am ready to go home so soon. I am scared." What should the nurse say next?
   A. Tell the patient to call the rehabilitation nurse every hour for reassurance.
   B. Ask the patient in a calm, reassuring voice to verbalize his/her concerns.
   C. Review the stroke education information again in a loud, emphatic voice.
   D. Stand over the patient without eye contact and ask the patient to verbalize his concerns.

4. The major goal of education is to spread and share knowledge as well as which of the following?
   A. Utilize group learning
   B. Change behavior
   C. Enable student learning
   D. Learn new behavior

5. Utilizing a fact sheet such as the one on VAP (ventilator-associated pneumonia), the nurse's instruction should include which of the following?
   A. A loved one should be flat in bed at all times.
   B. VAP is not a common occurrence in the ICU.

C. Hand washing is a measure to prevent VAP.
D. VAP is an inflammation of the lining of the heart.

6. Evaluating the effectiveness of an education event can be accomplished by which of the following? Select all that apply.
   _____ A. Noting a change in the learner's behavior
   _____ B. Having the patient complete a written test
   _____ C. Using the teach back method
   _____ D. Asking the patient if they have understood
   _____ E. Observing the patient do a return demonstration

7. The nurse is an important component in patient education because of which of the following?
   A. Caring attitude
   B. Knowledge and expertise
   C. Ability to comfort others
   D. Professional appearance

8. Which learning domain involves changes in the learners' feelings and beliefs?
   A. Cognitive domain
   B. Psychomotor domain
   C. Affective domain
   D. Reflective domain

9. Which of the following forms of analysis is concerned with the situation that created the need for education and with the conditions under which education will occur?
   A. Context learning
   B. Learner analysis
   C. Content learning
   D. Design analysis

10. Which of the following behaviors may indicate that your patient has low health literacy? Select all that apply.
    _____ A. Opening up pill bottles to identify the pills
    _____ B. Signing a consent form after reading all of the front page
    _____ C. Refusing to fill out a form and saying, "I'm too tired to read."
    _____ D. Responding when asked about why he takes a medication, "I take it because the doctor told me to!"
    _____ E. Becoming angry about nurses asking "too many questions."
    _____ F. Patient responds, "I'll read it later."

# REVIEW ACTIVITIES

1. Identify a patient education project that you can develop for one of your patients. Develop a lesson plan for this patient. Are your objectives in one, two, or three domains of learning?

2. Your patient is newly diagnosed with diabetes and needs instruction in checking blood glucose. Develop one objective for this education. How would you evaluate if this patient met the objective?

3. You have developed a lesson plan for nursing staff on a new computer documentation system that is to take effect on all the clinical units in your institution. Look at Table 19-14, which lists areas for education evaluation. How would you develop an evaluation tool for your lesson plan by using all or some of these areas? Are there other areas you would want to include in the evaluation?

4. Develop a staff lesson plan for staff on a unit where you have your clinical rotation. Use Table 19-12 to do this.

# EXPLORING THE WEB

- Personality types:
  Use the Web site www.humanmetrics.com to take the Jung Typology Test. This test combines the work of Carl Jung and Isabel Myers Briggs to determine basic personality types. Does the test seem to fit your personality and work ethic? Would it help to know your coworkers personality type to understand how they view the world?

- Evaluating Internet health information:
  Your patient arrives with information she has printed off the Internet. She has been reading many messages posted by the public about treatment for her disease. How do you determine the accuracy of the information on a Web site?

  Use www.medlineplus.gov. Type in a search for tutorial on evaluating Web sites. This connects to articles and to a tutorial by the National Library of Medicine on evaluating Web-based health resources.

- Health literacy:
  Explore www.npsf.org to find out more about health literacy and ways to increase your patients' understanding of health information. This Web site has a basic overview of health literacy and teach back methodology with examples. There are connections to videos about health literacy. See also www.nchealthliteracy.org and look for the tool kit.

- Try these Web sites for connections to online health care information:
  www.chcs.org
  www.medlineplus.gov

  www.webmd.com
  www.patients.uptodate.com
  www.healthline.com

  How accurate is the information these sites provide? Look at the bottom of the Web page for information on who maintains the Web site and when the information was updated. Search for the same information on each site and compare the information for level of health literacy. Determine what type of patient would benefit from information from the Web site. Would you give the Web site to a patient to look up their own information? Would you use it for professional information?

- Major United States and international health care Web sites:
  American Diabetes Association, www.diabetes.org

  American Heart Association, www.heart.org

  American Cancer Society, www.cancer.org

  American Dietetic Association, www.eatright.org

  Public Health Agency of Canada, www.phac-aspc.gc.ca

  National Health Service, United Kingdom, www.nhsdirect.nhs.uk

  How would you evaluate these sites in comparison to the sites listed previously? When would an international Web site be useful?

# REFERENCES

American Hospital Association. (2006–2010). *The Patient Care Partnership*. Retrieved August 28, 2010, from http://www.aha.org/aha/issues/Communicating-With-Patients/pt-care-partnership.html.

Belcher, J. V., & Vonderhaar, K. J. (2005). Web-delivered research-based nursing staff education for seeking magnet status. *Journal of Nursing Administration, 35*(9), 382–386.

Bloom, B. S., & Krathwohl, D. (1956). *Taxonomy of educational objectives: Handbook I: Cognitive domain.* Boston: Addison-Wesley.

Briggs Myers, I., & Myers, P. B. (1995). *Gifts differing.* Palo Alto, CA: Davies-Black.

Conner, M., & Hodgins, W. (n.d.). Putting learning styles into context. Retrieved August, 2010, from http://nwlink.com/~donclark/hrd/styles/perspective.html.

Gagne, R. M. (1985). *The conditions of learning and the theory of instruction* (4th ed.). New York: Holt, Rhinehart, & Winston.

Gardner, H. (1993). *Frames of mind: The theory of multiple intelligences* (10th anniversary ed.). New York: Basic Books.

Joint Commission. (2010). *Comprehensive accreditation manual for hospitals: The official handbook.* Chicago: Author.

Kolb, D. A. (1984). *Experimental learning: Experience as the source of learning and development.* Englewood Cliffs, NJ: Prentice-Hall/TPR.

Krathwohl, D. R., Bloom, B. S., & Masia, B. B. (1999). *Taxonomy of educational objectives: Handbook 2: Affective domain.* Boston: Addison-Wesley.

Milano, M., & Ullius, D. (1998). *Designing powerful training.* San Francisco: Jossey-Bass.

Partnership for Clear Health Communication of the National Patient Safety Foundation. (n.d.) Retrieved August, 2010, from http://www.npsf.org/askme3/PCHC/.

Rose, C. (1985). *Accelerated learning.* New York: Dell.

Simpson, E. (1971). Educational objectives in the psychomotor domain. In M. Kapfer (Ed.), *Behavioral objectives in curriculum development.* Englewood Cliffs, NJ: Educational Technology.

University of Southern California. (n.d.). Asking a good question (PICO). Retrieved August, 2010, from http://www.usc.edu/hsc/ebnet/ebframe/PICO.htm.

Valente, S. M. (2003). Research dissemination and utilization. *Journal of Nursing Care Quality, 18*(2), 114–121.

Violano, P., Corjulo, M., Bozzo, J., & Diers, D. (2005). Targeting educational initiatives. *Nursing Economic$, 23*(5), 248–252.

## SUGGESTED READINGS

Bastable, Susan, B. (2006). *Essentials of patient education.* Boston: Jones and Bartlett Publishers.

Bastable, Susan, B. (2010). *Health professional as educator: Principles of teaching and learning.* Boston: Jones and Bartlett Publishers.

Cornett, S. (2009). Assessing and addressing health literacy. *Online Journal of Issues in Nursing, 14*(3).

Cutilli, C. C. (2006). Do your patients understand? How to write effective healthcare information. *Orthopaedic Nursing, 25*(1), 39–50.

Egbert, N., & Nanna, K., (2009). Health literacy: Challenges and strategies. *Online Journal of Issues in Nursing, 14*(3).

Hamilton, S. (2005). Clinical consultation. How do we assess the learning style of our patients? *Rehabilitation Nursing, 30*(4), 129–131.

Pierce, L. L. (2010). How to choose and develop written educational materials. *Rehabilitative Nursing, 35*(3), 99–105.

Rankin, S. R., Stallings, K. D., & London, F. (2005). *Patient Education in Health and Illness* (5th ed.). Philadelphia: Lippincott Williams & Wilkins.

Ridge, R. (2005). A dynamic duo: Staff development. *Nursing Management, 36*(7), 28–35.

Rigdon, A. S. (2010). Development of patient education for older adults receiving chemotherapy. *Clinical Journal of Oncology Nursing, 14*(4), 433–441.

Scheckel, M., Emery, N., & Nosek, C. (2010). Addressing health literacy: The experience of undergraduate nursing students. *Journal of Clinical Nursing, 19*, 794–802.

Sewchuk, D. H. (2005). Experiential learning: A theoretical framework for perioperative education. *Association of Operating Room Nurses Journal, 81*(6), 1311–1316.

Wolf, M. S., Wilson, A. H., Rapp, D. N., Waite, K. R., Bocchini, M. V., Davis, T. C., et al. (2009). Literacy and learning in health care. *Pediatrics, 124*(3), 5275–5281.

# CHAPTER 20

# Managing Outcomes Using an Organizational Quality Improvement Model

MARY MCLAUGHLIN, RN, MBA; KAREN HOUSTON, RN, MS;
EDNA HARDER MATTSON, RN, BN, BA(CRS), MDE,
DOCTORAL STUDENT IN EDUCATION

*Go the extra mile, it's never crowded.*

(EXECUTIVE SPEECH WRITER NEWSLETTER)

## OBJECTIVES

Upon completion of this chapter, the reader should be able to:

1. Relate major principles of quality and quality improvement (QI), including customer identification; the need for participation at all levels; and a focus on improving the process, not criticizing individual performance.

2. Explain how quality improvement affects the patient and the organization.

3. Compare the Plan Do Study Act Cycle, the FOCUS methodology, and other methods for quality improvement.

4. Identify how data are utilized for QI (time series data, Pareto charts).

5. Outline the difference between risk management and QI.

6. Relate how the principles of QI are implemented in the organization.

University HealthSystem Consortium (UHC), a group of about 110 academic health science centers, did a benchmark study on total hip arthroplasty. Albany Medical Center, an organization that participated in that study, noted that compared to other organizations, its average length of stay (LOS) was long (the Albany Medical Center LOS was 7.07 days; the average LOS for UHC was 5.78 days). It also noticed that the percentage of patients that used a pneumatic compression device (a device to decrease the postoperative rate of deep vein thrombosis) was 85% for UHC but only 53% for Albany Medical Center. The percentage use of indwelling catheters in total hip arthroplasty patients at Albany Medical Center (indwelling catheters are associated with an increase in postoperative urinary tract infections, or UTIs) was 53%. Although this catheter use was lower than the UHC average, the team believed it could decrease the rate further. Albany Medical Center also delivered an average of three physical therapy (PT) visits postoperatively per patient, whereas UHC's average was five visits per patient. There was also an increased cost at Albany Medical Center versus the average cost at UHC.

Albany Medical Center decided to assign an interdisciplinary team the responsibility of identifying opportunities for improvement. The team began by developing a clinical pathway based on the most recent research in this area. Using data and research, the team looked for ways to improve the patient care process. The team incorporated the best practices that they found. This meant designing into the clinical pathway increased use of the pneumatic compression device, more physical therapy, earlier catheter removal, and an earlier discharge date. To prepare the patient properly for the earlier discharge, the team added preoperative home visits to the clinical pathway. The Visiting Nurses Association would make the home visit and then make a recommendation for discharge planning prior to the patient's admission. This process expedited initiation of referrals to an acute rehabilitation facility or home care following discharge, if needed. If the family home had to be rearranged to accommodate limited ambulation or stair use, these recommendations were made early in an effort to allow the family to prepare ahead of time and assure the patient's safety.

The higher costs at Albany Medical Center seemed to be related to the number of prosthetic vendors. When fewer vendors are used, the volume with each vendor is higher, allowing more competitive price negotiation among vendors. The organization worked with the surgeons on the team to decrease the number of vendors to a ratio of 0.25 vendors to surgeons.

How does attention to cost, safety, and quality improve patient care?

Traditionally, health care organizations have faced challenges related to the process of how patient care is delivered. However, "the focus is no longer just on the process of how care is delivered but on the outcomes of that care" (Thompson, 2008, p. 913). Many countries have recognized the need to change this focus from the process to the outcome of patient care delivery, placing patient safety and quality as a priority. United States federal regulations and national standards set by the Joint Commission and the National Quality Forum require the development of clinical indicators of safe and competent care within the health care organization (Governance Institute White Paper, 2009).

The Governance Institute White Paper (2009) describes health care organizations "as providing high-quality, safe care to those who seek its help, whether they are patients, residents, clients, or recipients of care" (chap. 1, p. 1). The fact that the health care organization is a system challenges all nurses to function as an interdisciplinary team. Introducing nursing students to quality improvement strategies involving all levels of the organization promotes a futuristic perspective of improving patient care rather than focusing on the limitations and errors of the past (Kyrkjebø, 2006). Preparing nursing students to be future nurse leaders is an obligation of current nursing leaders. The nursing graduates of today will provide care in a changing health care environment with increased awareness and attention to the performance of all health care staff. Reviewing the history of quality improvement is essential to appreciating the current standards developed by the Joint Commission (JC), formerly known as the Joint Commission on Accreditation of Healthcare Organizations (JCAHO).

Quality improvement is described as both a science and an art. The science of improvement is the development of new ideas, the testing of those ideas, and the implementation of change. Carey and Lloyd (2001) indicate that W. Edwards Deming, Joseph M. Juran, and Philip B. Crosby have been the gurus of continuous quality improvement and have provided important contributions to the science of quality improvement. Deming's components of appreciating a system, understanding variation, and applying knowledge and psychology are fundamental improvement principles. Quality improvement is also described as an art that taps into creative, "out of the box" ideas. It is about systematically testing evidence-based practices at all levels

## Critical Thinking 20-1

Refer back to the chapter's opening scenario. If you were a staff nurse on the orthopedic unit at Albany Medical Center, what could you do to improve the quality of care? How would you encourage the decreased use of indwelling catheters? How would you encourage the use of a pneumatic compression device?

How could you bring your ideas to other staff members without making them feel that the quality of their care was being criticized? For example, many staff members feel that catheter use is better for the patient's skin and reduces the need for assistance in ambulating the patient to the bathroom. You know that research shows that indwelling catheters and decreased ambulation increase risk of complications. How do you deal with these competing positions?

What measures would you use to ensure that while you were improving some aspects of care, you were not decreasing other critical outcome measures? For example, as you decrease catheter use, what is happening to UTI and fall rates?

of the services to improve customer care. Health care customers are patients, families, practitioners, nurses, staff, and so forth. If a quality improvement change is planned and measured, there are limitless boundaries to what can be achieved.

Organizational leadership often has a significant say in activities in which staff are involved. They can designate required resources and remove obstacles to making changes and improving care. Deming (1986), Crosby (1989), and Ransom, Joshi, and Nash (2005) all stress the importance of management and leadership commitment to the success of quality improvement. Without that commitment, successful quality improvement is jeopardized. Adapting the concepts of science and art in improving health care can create an enthusiasm for change and a passion for results.

This chapter will discuss and provide examples of the application and implementation of quality improvement principles in a health care organization. It will describe current standards developed by the Joint Commission (JC) and will review the Plan Do Study Act (PDSA) Cycle,

the FOCUS methodology, and other methods for quality improvement(QI). Data that is utilized for QI—e.g., time series data, Pareto charts, etc.—will be illustrated. The importance of customer identification and the need for quality improvement participation at all levels of the organization will be discussed, along with the need to focus on improving the health care process and avoid criticizing individual performance in order to improve quality.

# EVOLUTION OF QUALITY IMPROVEMENT INITIATIVES

Consider Berwick and Plsek's red bead example (1992):

> In a group of beads in a bag, there are 90 blue beads and 10 red beads. There are four workers whose job it is to take blue beads out of a bag. They cannot see what color bead they are taking. The supervisor watches to see how many red beads are pulled out of the bag. The first day, worker A has 1 red bead, worker B has 4 red beads, worker C has 3 red beads, and worker D has 2 red beads. The supervisor states that worker A has done a great job and worker B has done a terrible job. He tells them they have to improve. The beads go back in the bag, and they start over the next day. This day, worker A has 4 red beads, worker B has 0 red beads, worker C has 2 red beads, and worker D has 4 red beads. The supervisor praises worker B for the improvement and yells at workers A and D for not doing a good job. The truth is that these workers do not have any ability to change the number of red beads that they pull out of the bag.

This example demonstrates random variation. Using inspection in systems to reward or punish random variation results in tampering with the system rather than quality improvement. Instead of improving the process, the tampering encourages staff to look for someone to blame rather than to change the process to improve outcomes. So the question is, how much variability do we expect? This can be calculated on a time series chart and will be discussed in more detail later in the chapter.

Prior to the 1980s, the focus of improvement initiatives was on **quality assurance (QA)**, rather than on quality improvement (QI). QA began as an inspection approach to ensure that health care institutions—mainly hospitals—maintained minimum standards of patient care quality. The use of QA grew over time, as did federal and state regulatory controls. QA departments became the organizational mechanism for measuring performance against standards and reporting incidents and errors, such as mortality and morbidity rates. This approach was reactive and fixed the errors after a problem was noted.

QA's methods consisted primarily of retrospective chart audits of various patient diagnoses and procedures. The method was thought to be punitive, with its emphasis on "doing it right," and did little to sustain change or proactively identify problems before they occurred. It did, however, accomplish the task of monitoring minimum standards of performance.

## TOTAL QUALITY MANAGEMENT

Total quality management (TQM), also referred to as quality improvement (QI) and performance improvement (PI), began in the manufacturing industry with W. Edwards Deming and Joseph Juran in the 1950s. TQM, QI, and PI are terms that are frequently interchanged. For the purposes of this chapter, **quality improvement (QI)** will be referred to as a systematic process of organizationwide participation and partnership in planning and implementing continuous improvement methods to understand, meet, or exceed customer needs and expectations and improve patient outcomes. This proactive approach emphasizes "doing the right thing" for customers. It was integrated into the health care industry in the 1980s when purchasers of care and accrediting bodies began to push providers to document quality (Carey & Lloyd, 2001). Movement into QI is thought to be more of an overall management approach than a single program. Integrating concepts of quality into daily organizational operations is key to successful outcomes. Table 20-1 notes the difference in focus between QA and QI.

## GENERAL PRINCIPLES OF QUALITY IMPROVEMENT

Quality improvement principles include the following:

1. The priority is to benefit patients and all other internal and external customers.
2. Quality is achieved through the participation of everyone in the organization.
3. Improvement opportunities are developed by focusing on the work process.
4. Decisions to change or improve a system or process are made based on data.
5. Improvement of the quality of service is a continuous process.

Early QA literature focused on fixing problems, "doing it right," and having zero defects. Over time, a gap was found between theory and practice. It was determined that quality is not about being perfect. First, it is about being better, doing the right thing the first time, and being better than the competition. This increases an organization's performance during highly turbulent and financially uncertain times (Guo, 2008). The Joint Commission (2010) vision statement reflects the basis of quality improvement, i.e., "All people should always experience the safest, highest quality, best value health care across all settings" (para. 1). Second, quality is about health care professionals seeing themselves as having customers. The notion of *customer* requires major shifts

## TABLE 20-1  Difference in Focus between Quality Assurance and Quality Improvement

**Focus of Quality Assurance (doing it right)**

- Assessing or measuring performance retrospectively
- Reviewing chart audits and incident reports
- Determining whether performance conforms to standards
- Improving performance when standards are not met

**Focus of Quality Improvement (doing the right thing)**

- Meeting the needs of the customer proactively
- Building quality performance into the work process
- Assessing the work process to identify opportunities for improved performance
- Employing a scientific approach and using data for assessment and problem solving
- Improving health care performance and changing the health care system continuously as a management strategy, not just when standards are not met

in mind-set for the health care professional. The term is frequently used in business, and calling a patient a "customer" was initially thought by some to undermine the professional care provided to patients. Designing health care processes from the customer's point of view versus the professional's point of view is a challenge and requires changes in thinking and the redesign of health care processes. Health care involves work processes in which one step leads to the next step. Improving these steps in the work process is an important part of a focused attention to quality care and customer satisfaction. Customer satisfaction is rooted in the way health professionals treat their patients/customers and in the quality of their outcomes. Third, quality directs health professionals to give evidence-based care within and across disciplines (Ajjawi and Higgs, 2008). This is achieved by proactively seizing opportunities to perform better, driving for quality consistently and continuously, and not waiting for a problem to be pointed out or for pressure from a competitor to cause improvement. Improvements are sustained over time when interdisciplinary teams collaborate and decisions about change are supported by data.

The primary benefits of adopting quality improvement concepts and principles include discovering performance issues more quickly and efficiently by looking at every problem as an opportunity for improvement. QI involves staff in how the work is designed and carried out. This improves staff satisfaction and empowers staff to identify problems and implement improvement in the health care system and results in improved patient outcomes. Increasing the customer's perception that you care by designing health care work processes to meet the customer's needs, rather than the health care provider's, and decreasing unnecessary costs from waste and rework and failure to meet regulations are also quality concepts. These quality improvement concepts should be emphasized until they become work habits and part of an organization's daily operations.

## CUSTOMERS IN HEALTH CARE

A customer is anyone who receives the output of your efforts. There are internal and external customers. An internal customer is anyone who works within the organization and receives the output of another employee. Internal customers include health care staff such as practitioners, nurses, pharmacists, physical therapists, respiratory therapists, occupational therapists, pastoral caregivers, and so on. An external customer is anyone who is outside the organization and receives the output of the organization. The patients are external customers, but they are not the only external customers. Other external customers include private practitioners, insurance payers, regulators such as the Department of Health, the Joint Commission (JC), and the community you serve.

## THE MISSION OF THE JOINT COMMISSION (JC)

The mission statement of the Joint Commission (2010) is described as, "To continuously improve the health care for the public, in collaboration with other stakeholders by evaluating health care organizations and inspiring them to excel in providing safe and effective care of the highest quality and value" (para. 1). This statement shapes the JC 2010 National Patient Safety Goals. Complementing the work of the JC is the Agency for Healthcare Research and Quality (AHRQ), endorsed by the U.S. Department of Health and Human Services (Sorbero, et al., 2009). The AHRQ (2005) provides a set of readily available programs that can be downloaded without charge. For example, the AHRQ site lists references that provide valuable information such as Optimizing the Process Improvement Kit (Mark, 2010). Elixhauser, Pancholi, and Clancy (2005) advocate the use of available QI software, as it provides applications for individual institutions, state data organizations, and hospital associations at an extremely low cost.

## PARTICIPATION OF EVERYONE IN THE ORGANIZATION

QI is not the sole responsibility of the national organizations but is achieved through the participation of everyone in health care organizations at all levels. A participation and empowerment initiative must be built by first offering employees the opportunity for appropriate involvement. Standard LD.03.01.01 of the Governance Institute guide describes the role of leaders thus: "Leaders create and maintain a culture of safety and quality throughout the hospital" (2009, p. 20).

A new staff member can participate in the design and improvement of daily work practices and processes on an individual, unit, or organizational level. For example, as an individual, a nurse could change the organization of the day to spend more time with patients' families. On a unit level, a nurse could work with others on the unit to change the way in which a patient report is given, to be more time efficient. On an organizational level, a nurse could suggest that the process for notifying pharmacy about a missing medication could be improved. The nurse could participate on a team to find a solution.

The Governance Institute (2009) identifies "well-functioning teams as demonstrating certain universal characteristics:

- A shared vision and goal among members
- A shared plan among members to achieve a goal
- Clarity about each member's role
- Each member's individual competence
- Understanding other members' roles, strengths, and weaknesses
- Effective communication

- Monitoring other members' functioning
- Stepping in to back up other members as needed
- Mutual trust" (p. 4).

The goal of QI efforts and processes determines who participates on a specific team. For example, if you were trying to decrease the time a patient waits outside the radiology suite for a test, you would need to include the patient transportation staff, unit clerks, unit RNs, and radiology staff in your QI efforts. If you were trying to ensure that patients with congestive heart failure are discharged understanding the importance of weighing themselves daily, you would need to involve the patient and patient's family, the cardiac unit's RNs, the clinical dietician, the primary care practitioner, the cardiologist, the pharmacist, and the Visiting Nurses Association staff. The key in determining who participates is including the point-of-service staff: the workers on the front line who do the direct care involved in the work process you are trying to change. They are the people who have the most knowledge of the work process, so they can look for potential areas of improvement. There should be a clearly identified way for staff to suggest improvement opportunities that they see in their day-to-day work. For example, an X-ray technologist may note steps in the process of scheduling and transporting patients that create a long wait time for X-ray testing. If a mechanism for suggesting improvement exists in the radiology department, the technologist could suggest and test ideas for change.

## FOCUS ON IMPROVEMENT OF THE HEALTH CARE WORK PROCESS

Improvement opportunities are focused on the process of work that the health care team delivers. A **process** is a set of causes and conditions that repeatedly come together in a series of steps to transfer inputs into outcomes (Bandyopadhyay & Hayes, 2009). All work processes have inputs, steps, and outputs. Deming (2000) points out that "Every activity, every job is part of a process." A process in health care includes the work process or activities that constitute health care (i.e., diagnosis, treatment, rehabilitation, prevention, and patient education) and also other work processes or activities that help make the care happen (i.e., food preparation, transportation, technical support) (Donabedian, 2003). The people involved in these health care activities or work processes include all types of health care workers, such as nursing and medical practitioners, technicians, housekeeping, and so on. An example of a health care work process is illustrated by a patient who presents in the emergency department with chest pain. When it is determined that the patient has had a myocardial infarction (MI), several work processes should occur (Colligan, et al., 2010). These work processes include an appropriate set of interventions for the patient with an MI, such as pertinent blood work, the right diet, the right medication in the right time frame, and patient assessments done on a regular interval to identify complications early or before they happen. All the steps of the work process can be measured. These measurements are then reviewed, applying evidence-based principles as appropriate to improve patient care. Steps of the work process may be eliminated or changed and then standardized so that all staff use the improved work process. For instance, in the example of the patient with the MI, health care evidence shows that all patients with an MI should receive aspirin within a specific time frame. In an organization with a focus on quality improvement, the steps of aspirin administration to the patient are reviewed and changed until the measurement data shows that all patients with an MI receive aspirin in the appropriate time frame. This would mean a review of when the practitioner ordered the drug, when the order was transcribed, when the pharmacy got the aspirin to the staff, when the staff gave it to the patient, and so on. A decrease in variability of the work process leads to improved care. The care is standardized to assure the best result.

## Improvement of the System

A **system** is an independent group of items, people, or processes with a common purpose (Schmittdiel, Grumbach, & Selby, 2010). In a system, the work processes as well as the relationships among the work processes lead to the outcome. You can improve the outcome by examining these work processes and relationships. In a system, every step of a work process affects the following step. For example, if the X-ray staff members place the patient who has had a chest X-ray in the hall and call transportation to take the patient back to his room but do not monitor and consider the transportation process, they may decrease the total time a patient is in radiology but increase the patient's time in the hall waiting for transportation. You cannot improve care unless you review all the steps in the system's work process.

## A Continuous Process

Improvement of quality of service is a continuous process. Walter Shewhart, the director of Bell Laboratories in the mid-1920s, is credited with the concept of the cycle of continuous improvement. This concept suggests that products or services are designed and made based on knowledge about the customer. Those products or services are marketed to and judged by the customer. These judgments lead to improved products and services. Hence, the process of QI is continuous, because it is linked to changing customer needs and judgments.

In addition to being a continuous process, as discussed earlier in the example of the patient with an MI, quality improvement focuses on standardization of a work process. Variation in a work process increases complexity and the risk of error (Ransom, et al., 2005). Technology increasingly offers tools for standardization of work processes; however, there are challenges with technology. Ransom, et al. (2005) explain it like this: "Imagine how difficult it would be to drive a rental car if every manufacturer placed gas pedals, brake pedals, and shifts in completely different locations with varying designs." As we increase the use of technology to assist in the standardization of work processes, we need to bear in mind the need to standardize the technology itself.

## OUTCOMES MONITORING

Of all the efforts to achieve QI, monitoring of patient outcomes is the most significant criteria of safe and competent care in response to the structure and process of the delivery system. Outcomes measure actual clinical progress. Outcomes can be short term, such as the average length of stay for a patient population, or long term, such as a measure of patients' progress over time (e.g., survival rate for a transplant patient one, two, and three years after treatment). Outcomes are studied to identify potential areas of concern. This may lead to an investigation of structure and process to determine any root causes of a negative outcome. Outcomes influence ongoing decision making in the provision of health care services by each discipline. Ongoing monitoring and decision practices are best captured through data collection and interpretation, including initiatives to address the limitations.

## Improvement Based on Data

Decisions to change or improve a work system or work process are made based on data. When someone says, "The patients are waiting too long to return from radiology," it is time to look at the data. Review the waiting time data to see whether waiting times are increasing. The data clarify these issues. Using data correctly is important. Data should be used for learning, not for judging. It is critical to look at work processes rather than people for improvement opportunities. In the radiology example, if we jumped to the conclusion that someone was not doing the job correctly, we might criticize the transportation staff person who returned the patient from radiology. This would not foster improvement ideas. By not analyzing the process (patient has chest X-ray, is put in hall, clerk at desk calls transportation, transportation clerk pages transportation aide, and so on) and the relationships among the processes (waiting times between calls, transportation phone process, page system, and so on), we could miss finding where the real improvement opportunity lies. Perhaps it has nothing to do with the transportation person who

### Critical Thinking 20-2

In the opening scenario at the start of this chapter, when the original length of stay data by nursing unit was examined, one unit had a much shorter length of stay than the other. At first there was discussion about this variance, and the idea emerged of just going to the floor with the longer length of stay and fixing things there. The group members decided that rather than approaching the task from this limited perspective, they would study the work process as a whole and determine whether there were steps they could take to improve the work process. Several excellent opportunities for improvement were identified, as noted previously—for example, preoperative home evaluation, increased physical therapy involvement, and shorter Foley catheter use. All these areas contributed to the work process improvement, and the outcome was that both units ended up reducing their lengths of stay. These opportunities would have been lost had the group members used the data only to say that one unit was doing a bad job. They needed to review the work process as a whole to improve the length of stay on both units.

How could you improve care in a patient care unit that you are familiar with? What patient care work process could be improved? Who would you ask to work on the improvements with you?

returned the patient to the room. It may be that the actual root cause of the problem is a long delay in the paging system. Reviewing the wait time data is an example of examining the work process, not the people carrying out the work process.

# IMPLICATIONS FOR PATIENT CARE

The implications of quality improvement for patient care can be measured by the overall value of care. Value is a function of both quality outcomes and cost. Outcomes can be a patient's clinical or functional outcomes. For example, did the patient survive? Can the patient go back to work? Outcomes can also be measured by patient satisfaction. For example, would the patient recommend this health care facility to someone else? Cost is the cost of both direct and indirect patient care needs. Direct cost is the cost of the care of the patient, for example, cost of medications, operating room equipment, and direct patient caregiver salaries. Indirect costs are the costs of nondirect care activities, including electricity and salaries of nondirect patient caregivers such as secretaries or human resource staff.

$$Value = \frac{Quality\ of\ outcomes}{Cost}$$

In most QI efforts, as quality is improved by standardizing care delivery work processes and applying evidence-based principles, the cost of care decreases. The example in the opening scenario illustrated this. The length of stay decreased as the team found ways to standardize care and evaluate the patient prior to admission, to plan for discharge and prepare the family. A decrease in length of stay generally translates into a decrease in costs.

# METHODOLOGIES FOR QUALITY IMPROVEMENT

Implementing methods for quality improvement requires consideration of the climate of care, the data collection methodologies, and the response to the interpretation of the evidence. Gallart (2009) describes the implications of financial sustainability at a time of declining resources. The cost to address hospital-acquired infections, decubitus ulcers, and medication errors "gives incentive to identify and implement measures to protect patients from these unfortunate complications" (Gallart, 2009, p. 206). The Plan Do Study Act (PDSA) cycle and the FOCUS methodology are two examples of methods for quality improvement (Figure 20-1 and Figure 20-2).

## THE PLAN DO STUDY ACT CYCLE

The PDSA cycle starts with the following three questions (Ransom, et al., 2005):

1. What are we trying to accomplish?
2. How will we know that a change is an improvement?
3. What changes can we make that will result in improvement?

As these questions are being answered, testing needs to be done to evaluate any proposed changes. Testing is done to evaluate the effect of a proposed change and to learn about different alternatives. The goal is to increase the ability to predict the effect that one or more changes would have if they were implemented. The plan for testing should cover who will do what, when they will do it, and where they will do it.

## Critical Thinking 20-3

The third-party payer system in the United States is complex and constantly changing. Third-party payers are the organizations that pay patients' hospital bills. They include government payers, such as Medicaid and Medicare, and private payers such as health insurance companies. For example, to some payers, decreasing lengths of stay may mean decreasing payments. Other payers will pay for a patient admission using a diagnosis-related group (DRG) payment reimbursement schedule with preset fees, regardless of the length of stay. DRG payment schedules are based on groups of patients that are medically related with respect to diagnosis or condition, presence of a surgical procedure, age, and presence or absence of comorbidities or complications. So under this type of payment, if the patient is discharged in a short time, generally the facility will make money. If patients stay a long time, hospitals will lose money.

What do you think of these types of cost reimbursement? What can you do to keep health care costs down for your patient? Is there a benefit to the hospital if length of stay is decreased in the DRG system? What can an RN do to prevent delay in discharge of a patient?

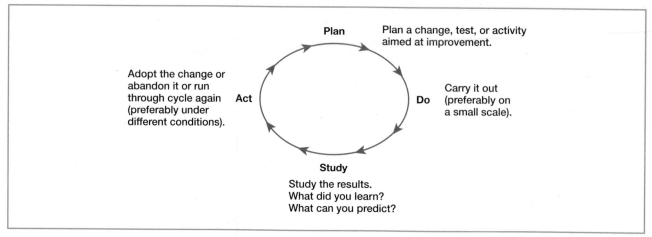

**FIGURE 20-1** **PDSA cycle.** (Courtesy of Albany Medical Center, Albany, NY).

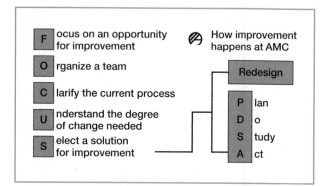

**FIGURE 20-2** **FOCUS method.** (Courtesy of Albany Medical Center, Albany, NY).

## THE FOCUS METHODOLOGY

The FOCUS methodology describes in a stepwise process how to move through the improvement process.

**F:** Focus on an improvement idea. This step asks, "What is the problem?" "What is the opportunity?" During this phase, an improvement opportunity is articulated and data are obtained to support the hypothesis that an opportunity for improvement exists.

**O:** Organize a team that knows the work process. This means identifying a group of staff members who are direct participants in the work process to be examined—the point-of-service staff. A team leader is identified who will appoint team members.

**C:** Clarify what is happening in the current work process. A flow diagram (Figure 20-3) is very helpful for this. A detailed flowchart can be analyzed in two ways to uncover possible problems—at a macro level and at the micro level.

At the macro level, scan the flowchart for any indication that the work process is broken. Red flags include the following:

- Many steps that represent quality checks or inspections for errors. When you notice too many boxes in your flow diagram describing similar steps, this could indicate rework or lack of clarity in roles.

## REAL WORLD INTERVIEW

In my job, I review a patient's chart and compare it to evidence-based guidelines from research and the literature to see if the patient's health care is being performed in the appropriate setting. I will review if the patient's care is medically necessary. If it is not, I assist the hospital case managers or physicians to move the patient to the appropriate level of care. For example, IV antibiotics can sometimes be administered at home or in another facility. When the situation at home is such that the family cannot manage it, the patient could move to a subacute facility, if available, or the patient could stay in the hospital with the hospital paid at a different rate. Documentation is critical in this type of review. An accurate clinical picture of the patient needs to be reflected in the documentation.

**Marguerite Montysko, RN**
Case Manager, Albany Medical Center
Albany, New York

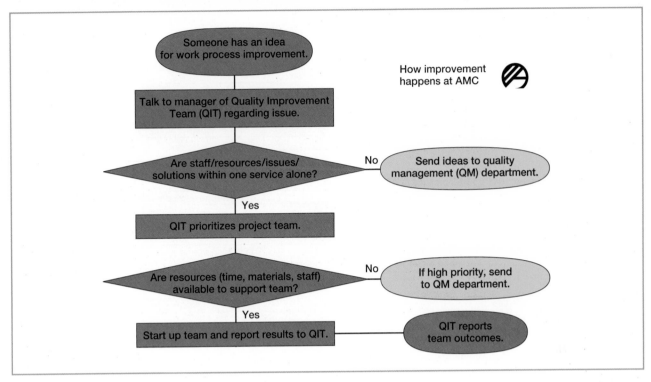

How improvement happens at AMC

**FIGURE 20-3** **Flow diagram—how improvement happens.** (Courtesy of Albany Medical Center, Albany, NY).

- Areas in the work process that are not well understood or cannot be defined. If the work process defies definition, you can be certain it is not being performed efficiently, with maximized outcomes.
- Many wait times between work processes. Wait times should always be minimized to improve efficiency of the process.
- Multiple paths that show lots of people involved in the activity or delivering the service to the customer. Too many staff involved is wasteful and confusing to the patient.

If the existing work process seems reasonable, with one or two areas needing improvement, then a micro-level analysis of your flow diagram is needed.

- Examine decision symbols (diamonds) that represent quality inspection activities. For example, in the flow diagram in Figure 20-3, the "Are resources available ... ?" diamond is a decision point. Either the resources are available or they are not. Can material, etc., be eliminated? Do some errors go undetected? Is the issue high priority? This examination will ensure limited rework and maximum clarity.
- Examine each work process in the diagram for redundancy and value. If a step in the work process is repeated or does not have any value for the customer, it should be eliminated.

- Examine work processes for waiting time areas. The work process should be changed to eliminate these wait times.
- Examine all work processes for rework loops. A step should not be repeated. Resources are always limited, especially in hospitals today.
- Check that hand offs are smooth and necessary. Hand offs are times when a work process is handed from one staff person or department to another. Hand offs always

**REAL** WORLD INTERVIEW

I felt that benchmarking was useful, because it allowed us to compare ourselves to other organizations. It allowed us to network with other similar facilities to share ideas and strategies. It allowed us to test our strategies to see if we were making any improvements.

**Karen Petronis, RN, MS**
Orthopedic Clinical Nurse Specialist
Albany, New York

**Bed Access Improvement Team**
**Phase 2 Work Plan: Transition to Daily Management and Evaluation**

| Activity | Responsible Party | 8/10 | 9/10 | 10/10 | 11/10 | 12/10 |
|---|---|---|---|---|---|---|
| **1.0 Modify the Team** | | | | | | |
| 1.1 Identify Phase 2 Tasks to Be Completed | Team | ▓ | | | | |
| 1.2 Review & Modify Team Composition/Membership | Team | ▓ | | | | |
| 1.3 Develop Work Plan | Planning Team | ▓ | | | | |
| 1.4 Review Work Plan with Team | Myers/Nolan | | ▓ | | | |
| | | | | | | |
| **2.0 Review/Modify Ideal Design** | | | | | | |
| 2.1 Identify Modifications/Opportunities for Additional Change | Team | | | ▓ | | |
| 2.2 Revise Ideal Flow Chart | Team | | | ▓ | | |
| | | | | | | |
| **3.0 Modify Structure & Supports: People/Forms Needed** | | | | | | |
| 3.1 Revise Process Management Structure   • Modify Job Descriptions—Triage Manager and Admitting Coordinator | Triage Management Subgroup | | | | ▓ | |
| 3.2 Assess Communication Needed with Nursing Units | Team | | | | | |
| **4.0 Draft/Standardize Tasks** | | | | | | |
| 4.1 Draft/Standardize Tasks | | | | | ▓ | |
| | | | | | | |
| **5.0 Transition to Daily Operations, Develop Data Collection Process, Evaluate, Monitor** | | | | | | |
| 5.1 Evaluate Bed Access Simulation   • Review ED & PACU Data   • Identify Accomplishments and Opportunities of Structure and Ideal Process | Team | | | | | ▓ |
| 5.2 Develop Plan to Transition Process and Structure to Daily Operations | Planning Team | | | | | ▓ |
| 5.3 Develop Data Collection Process | Planning Team | | | | | ▓ |
| 5.4 Evaluate Process & Structure (Milestone Meeting) | Team | | | | | ▓ |
| 5.5 Identify Subgroup of Pt Care Delivery System QIT to Monitor Progress | Team | | | | | ▓ |

FIGURE 20-4 **Gantt chart/work plan.** (Courtesy of Albany Medical Center, Albany, NY).

leave room for error (Adamski, 2007). Aviation safety is characterized by a collective sense of communication and teamwork that can be applied to patient care (Lyndon, 2006).

**U:** Understand the degree of change needed. In this stage, the team reviews what it knows and enhances its knowledge by reviewing the literature, available data, and competitive benchmarks. How are other health care organizations implementing the process?

**S:** Solution: Select a solution for improvement. The team can brainstorm and then choose the best solution. It can then use the PDSA cycle to test this solution. An implementation plan should be used to track progress and the steps required. This implementation plan can be in the form of a work plan or Gantt chart (Figure 20-4). This is a chart in the form of a table that identifies what activity is to be completed, who is responsible for it, and when is it going to be done. It outlines the steps needed to implement the change.

# OTHER IMPROVEMENT STRATEGIES

Other improvement strategies identified at the organizational level involve benchmarking, meeting regulatory requirements, and identifying the need for review of a sentinel event.

## Benchmarking

**Benchmarking** is the continual and collaborative discipline of measuring and comparing the results of key work processes with those of the best performers. Quality improvement involves clinical benchmarking and personal and professional commitment to improving safe and competent care at all levels of the organization (Ellis, 2006). Benchmarking focuses on key services or work processes, for example, length of time from the patient entering the emergency department until the time of treatment. A benchmark study will identify gaps in performance and provide options for selection of health care processes to improve, ideas for redesign of care delivery, and ideas for better ways of meeting customer expectations. There are various types of benchmarking studies, such as clinical, financial, and operational benchmarking. A clinical benchmark study will review outcomes of patient care, for example, reviewing the outcomes of care of patients with diabetes or stroke. Financial benchmarking studies examine cost/case charges and length of stay. Operational benchmarking studies review the health care systems that support care, for example, the case management system in an organization. The outcomes of clinical, financial, and operational studies are compared or benchmarked with another high-quality organization's outcomes.

## Regulatory Requirements

The JC (2010) has developed standards to guide critical activities performed by health care organizations. Preparation for an accreditation survey and the survey results will provide a wealth of information and data that can be utilized as ideas for improvement strategies. In January 2003, the first National Patient Safety Goals (NPSG) were approved by the JC. Each year, the JC publishes new goals that organizations must have in place to promote specific improvements in care related to patient safety (2010). Recognizing that system design is intrinsic to the delivery of safe, high-quality health care, the goals focus on systemwide solutions wherever possible. The NPSG 2011 goals focus on such elements as communication, patient identification, medication safety, etc., and are available at www.jointcommission.org. Specific information regarding the history and ongoing requirements can be found at the JC Web site. JC requires that specific data be collected and reported during a survey (ORYX measures).

## Sentinel Event Review

An adverse **sentinel event** is an unexpected occurrence involving death or serious physical or psychological injury to a patient. Sentinel events require immediate investigation. The process of investigating an adverse sentinel event is based on a format for systematic communication modeled after aviation and military communication hand-offs (Tamuz & Thomas, 2006). The acronym SBARR reflects the situation, background, assessment, recommendation, and response. SBARR results "in assertive communication that promotes situational awareness and timely interventions for patients" (Willingham & Eden, 2007, p. 518). An example of a sentinel event is surgery performed on the wrong side of a patient. Reviewing the surgical process and developing a system to mark the appropriate site is a change in the work process to prevent future harmful occurrences.

## Measurements

To assess and monitor outcomes, heath care organizations collect and report measures at various levels in the organization. Two examples of data collection and measuring

performance at both the strategic and operational levels in the organization are a balanced scoreboard (Schalm, 2008) and a clinical value compass. Indicators may be patient clinical or functional status, patient satisfaction measures, cost measures, or organizational performance measures. Figure 20-5 illustrates these in the form of a clinical value compass (Caldwell, 1998).

Such an approach allows those reviewing data to examine all aspects of care. For example, patient outcomes are reflected in a patient's functional status and clinical status. Patient satisfaction and cost balance this to ensure value. Data can be arranged to create a balanced scorecard in an approach that uses the organization's priorities as categories for indicators. For example, three priorities might be customer service, cost-effectiveness, and positive clinical outcomes. These priorities help sort out what should be measured to give a balanced view of whether a strategy is working. Indicators are selected based on what they have in common, so that if a change occurs in the cost-effectiveness category, it will affect the data in another category. For example, if we decrease cost for orthopedic surgery, does that affect the customer's satisfaction positively or negatively? If we decrease the length of stay for these patients, does it increase or decrease complication rates? After indicators are selected, data are tracked over time at regular intervals (every month or every quarter, for example). Figure 20-6 shows how the balanced scorecard was used on an orthopedic unit. From the control charts, you can see that the total hip pathway length of stay increased and then decreased. The satisfaction scores remained at around 90%, so even though the length of stay decreased, the satisfaction did not deteriorate. The ratio of complications went down; the average number of physical therapy visits varied and then went up. This reporting mechanism offers a balanced view. Kocakülâh and Austill (2007) support Kaplan and Norton's (1996) description of a Balanced Scorecard approach aligning customer service cost-effectiveness and clinical outcome measures. These measures are utilized to monitor organizations' priorities. The goal is to assess that changing a strategy in one area does not negatively impact another indicator. For example, as an organization increases patient volume and reduces length of stay, is there any change noted in the patient readmission rate or patient satisfaction? The Balanced Scorecard reflects the organization's mission, vision, and values (Table 20-2).

## Storyboard: How to Share Your Story

Quality improvement teams share their work with others using a storyboard. The storyboard usually demonstrates the major steps in the improvement methodology and visually outlines the progress in each step. The storyboard can be displayed in a high-traffic area of the organization to inform other staff of the QI efforts under way. Storyboarding

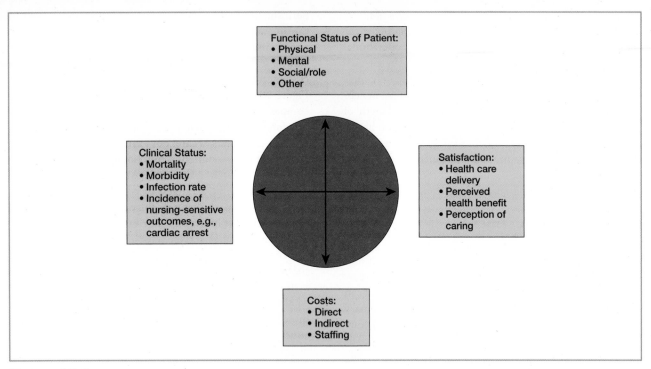

FIGURE 20-5  Clinical value compass. (Developed with information from Albany Medical Center, Albany, NY).

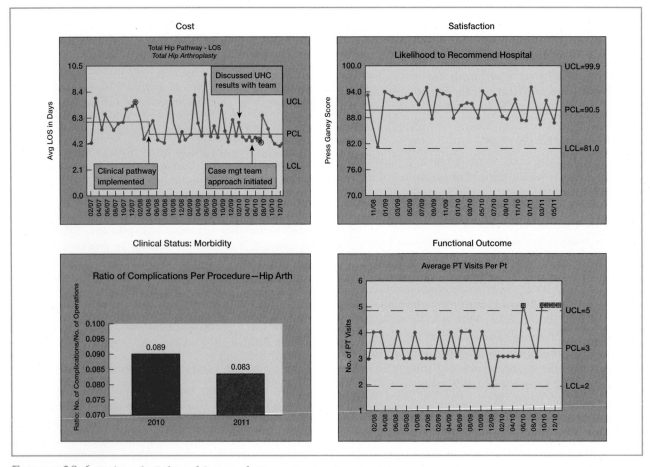

FIGURE 20-6  Orthopedic Balanced Scorecard. (Courtesy of Albany Medical Center, Albany, NY).

## TABLE 20-2   Balanced Scorecard

| | |
|---|---|
| Mission—Why we exist | Balanced Scorecard Measures |
| Values—What's important to us | • Clinical and functional |
| Vision—What we want to be | • Financial |
| Strategy—Our game plan | • Organizational |
| | • Satisfaction |

**Source:** Compiled with information from Kaplan, R. S., & Norton, D. P. (2004). *Strategy maps.* Boston: Harvard Business School Publishing.

 **EVIDENCE** FROM THE LITERATURE

**Citation:** Cossette, S., Cote, J. K., Pepin, J., Ricard, N., Di'Aoust, L. (2006). A dimensional structure of nurse-patient interactions from a caring perspective: Refinement of the Caring Nurse-Patient Interaction Scale (CNPI-Short Scale). *Journal of Advanced Nursing, 55*(2), 198–214.

**Discussion:** The development of a short version of the Caring Nurse-Patient Interaction Scale is discussed in this article. Measuring caring to assess its effect on patient health outcomes is a priority for nursing. The short version of the 70-item scale was developed based on both inductive and deductive process to assess attitudes and behaviors associated with Watson's 10 carative factors. Evidence of validity and reliability of the Short Scale is presented. The Short Scale comprises 23 items, reflecting four caring domains: humanistic care (four items), relational care (seven items), clinical care (nine items), and comforting care (three items). The items of the CNPI-Short Scale are identified in the article.

**Implications for Practice:** The CNPI-Short Scale has potential for use in clinical research settings, particularly when questionnaire length is an issue. It is a useful tool for research aimed at demonstrating that caring is indeed fundamental to nursing.

 **CASE STUDY** 20-1

Identify one outcome to measure in each of the four areas for organizational quality improvement of patient outcomes—that is, clinical status, functional status, patient satisfaction, and cost (Figure 20-5).

can be done when a QI process is complete or used during the QI process to communicate information.

## Patient Satisfaction Data

Health care facilities get feedback from patients by having them fill out a questionnaire that asks how they felt about

their health care encounter See http://www.pressganey.com/index.aspx). It is most helpful if this data can be compared or benchmarked with other organizations' data. This requires that several organizations use the same data collection tools. All patient responses are put into a database so the results can be reviewed. Another method of patient data collection is via a phone call after patient discharge. Another method for obtaining patient satisfaction information is via a focus group or post-care interview or phone call. This means talking with one or more patients after their discharge and getting feedback on their perceptions of their stay.

## INTERPRETING DATA

Several different types of charts are used to examine data in QI efforts. These include time series charts, Pareto charts, histograms, flowcharts, fishbone diagrams, pie charts, and check sheets.

## Time Series Data

Time series data (Figure 20-7) allow a QI team to see change in quality over time. A time series chart allows the user to determine whether a process is in control, meaning that the process has normal variation rather than dramatic changes that are not predictable. Although bar charts are useful, there are times in process improvement efforts when time series data display the process more clearly.

In the bar graph at the top of Figure 20-7, you can see that from year one to year two the percentage of the time that Foley catheters were in for greater than two days decreased dramatically. However, if you look at the time series chart of the same data for two different units, the actual process for each unit is quite different. Unit A had a good initial decrease after the change was implemented. However, it could not hold the gains, and the rate of Foley use has begun to creep back up. Unit B, however, made progress and has continued to decrease its rate over time. Determining next steps for these two units in this process improvement initiative would require very different strategies.

Tracking data over time allows you to see how a process is behaving. A time series graph is used for this. Graphs or charts—rather than tables of numbers—are used to display data, because graphs are faster to interpret. As you can see from the graph of the total hip arthroplasty pathway length of stay in Figure 20-6, a time series data graph contains data points at particular intervals, every two months for example. The process centerline (PCL) represents the average, and the upper control limit (UCL) and lower control limit (LCL) represent acceptable boundaries for expected performance (Wheeler, 2000). If the process were changed in some way, you would expect to see the data change at that time. The time series chart is used to look for trends, shifts, and unusual data.

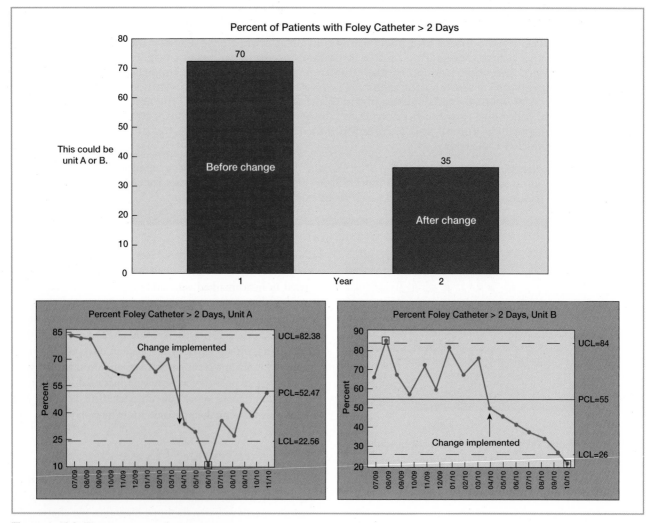

**FIGURE 20-7** **Time series versus bar charts.** (Courtesy of Albany Medical Center, Albany, NY).

## Charts: Pareto, Histogram, Flowcharts, Fishbone Diagrams, Pie Charts, and Check Sheets

In addition to time series data graphs, information can be displayed in several different ways to enhance decision making. These include Pareto diagrams, pie charts, flowcharts, and histograms. Figure 20-8 is an example of a fishbone diagram. (Fishbone diagrams are also referred to as root cause diagrams, cause-and-effect diagrams, or Ishikawa diagrams.) Note that many factors contribute to a problem. Review of a cause-and-effect chart encourages staff to look for all the causes of a problem, not just one cause. Figure 20-9 and Figure 20-10 show a flowchart, a check sheet, a Pareto chart, and a control chart. A full discussion of these tools is outside the scope of this chapter.

## ORGANIZATIONAL STRUCTURE

Most organizations today are structured to maximize QI efforts. An organization accomplishes this through an organizational structure that encourages accountability and communication and by focusing all staff on the priorities of the organization.

Figure 20-11 is an example of an organizational chart that shows a structure for quality improvement. Note that it includes members from the governance or board of directors level to staff on quality improvement teams (QITs). Within an organizational structure, it is vital that nursing leadership be represented at all levels. In the structure represented in Figure 20-11, the chief nursing officer participates at board/hospital affairs, executive management council, and center quality council. The CNO and the nursing directors colead the service quality improvement teams (QITs). Nursing and other health care staff are represented on the QITs and are vital to promoting input and ideas for change in work flow and work processes to improve patient care. Communicating priorities at all levels in the organization is key. Staff members must realize how their day-to-day work influences the accomplishment of strategic goals. Mission, vision, and value statements help accomplish this clarity of focus. The Governance Institute White Paper (2009) emphasizes that quality patient care involves teamwork, including recognition that the patient and the patient's family are members of this clinical "microsystem" (p. 4). As nursing graduates, you will be the future leaders, leading change and advancing health for all citizens (Institute of Medicine, 2010).

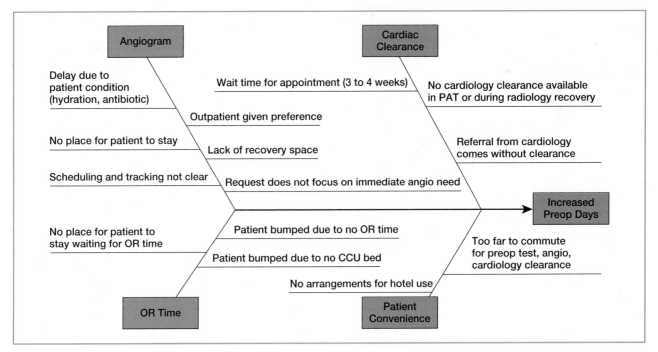

FIGURE 20-8 **Root cause/fishbone diagram.** (Courtesy of Albany Medical Center, Albany, NY).

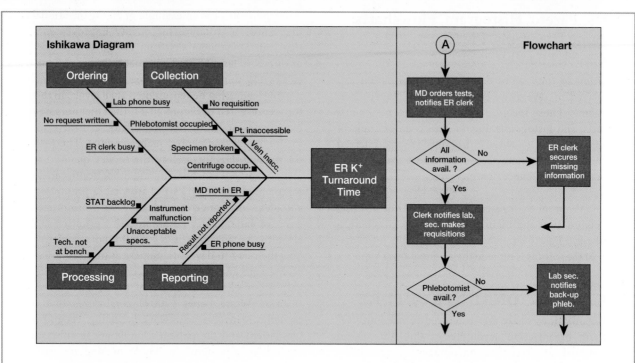

**FIGURE 20-9** **Ishikawa diagram, flowchart, and check sheet.** (Reprinted with permission from ©Clinical Laboratory Management Association, Inc. All rights reserved. Simpson, K. N., Kaluzny, A. D., & McLaughlin, C. P. (1991). Total quality and the management of laboratories. *Clinical Laboratory Management Review, 5*(6), 448–449, 452–453, 456–458, passim).

**Check Sheet**
Delays in Production of Se K$^+$ Results from 1/1/01 to 1/7/01

| Code/Delay Type | | Mon | Tue | Wed | Thur | Fri | Sat | Sun | Total |
|---|---|---|---|---|---|---|---|---|---|
| A | Request not Written by physician | I | I | | | | I | | 3 |
| B | Lab phone busy > 2 minutes | I | | I | | II | | I | 5 |
| C | Phlebotomists unavailable | III | II | III | III | II | IIII | III | 20 |
| D | Requisition not ready | II | I | I | I | I | III | II | 11 |
| E | Patient inaccessible | I | I | II | I | | II | I | 8 |
| F | Vein inaccessible | I | | II | | I | II | | 6 |
| G | Centrifuge busy | II | | I | | I | | | 4 |
| H | Specimen broken | II | | I | | | | I | 4 |
| I | STAT backlog | III | | | I | | II | I | 7 |
| J | Tech. not at bench | II | | II | | I | II | I | 8 |
| K | Unacceptable specimen | I | I | | II | | I | II | 7 |
| L | Lab. sec. unavailable to report | III | | I | | I | I | | 6 |
| M | ER phone not answered | | | I | | | II | | 3 |
| N | MD not in ER | II | | I | | II | | I | 6 |
| O | MD not answer page | I | I | II | | II | | II | 8 |
| P | Results not reported by ER sec. | II | I | II | I | III | II | II | 13 |

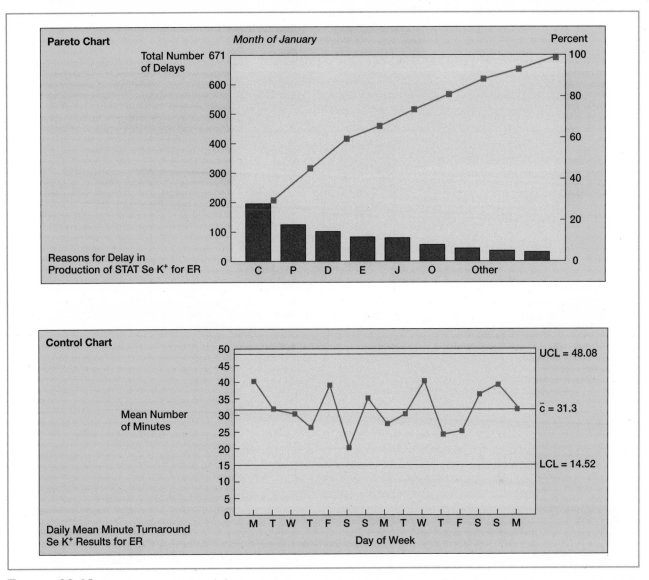

FIGURE 20-10 **Pareto chart and control chart.** (Reprinted with permission from ©Clinical Laboratory Management Association, Inc. All rights reserved. Simpson, K. N., Kaluzny, A. D., & McLaughlin, C. P. (1991). Total quality and the management of laboratories. Clinical Laboratory Management Review, 5(6), 448–449, 452–453, 456–458, passim).

 **CASE STUDY** 20-2

You have been caring for groups of patients following myocardial infarctions. You have also developed a good working relationship with the other nursing and medical staff on your unit. You believe that the care delivery on your unit could improve, thus improving patient satisfaction and clinical outcomes and decreasing the length of stay. How would you proceed? Whose support would you enlist first? Who should be involved? What quality indicators could be measured?

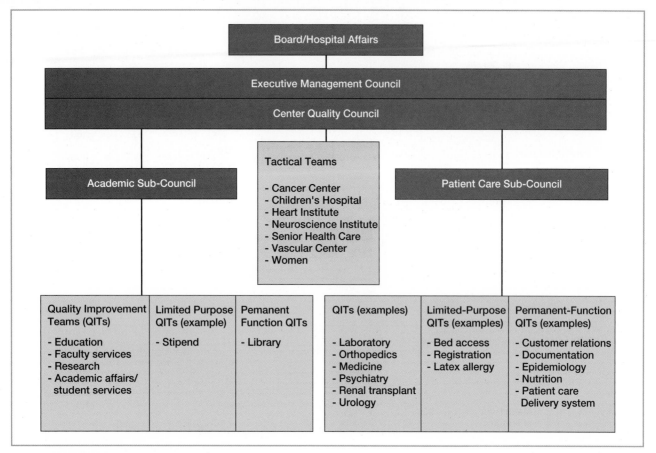

FIGURE 20-11  **Structure for quality improvement.** (Courtesy of Albany Medical Center, Albany, NY.)

## Critical Thinking 20-4

There are times when some care practitioners see clinical pathways and other standardized guidelines as cookie-cutter medicine. All health care providers like to think they give their patients the best care possible. Standardization and evidence-based clinical guidelines are meant to communicate the latest evidence as to the best practice for a given patient problem. In the absence of evidence-based practices, consensus of the team should be used to develop the clinical pathway.

A group that was developing a clinical pathway for the care of patients with acute myocardial infarction noted evidence showing that these patients should receive acetylsalicylic acid (ASA) on admission. The research in this area was very clear, and most providers believed this was being done. When a chart audit was performed to determine whether this was in fact the practice on the unit, it was discovered that only 48% of the patients were receiving ASA within eight hours of admission. The team added this to the clinical pathway. After this was implemented, 85% of the patients received ASA within the first eight hours of admission.

What clinical practices do you see on your clinical unit that are based on an evidence-based clinical pathway? How can you participate in improving the care of more patients using evidence-based clinical pathways?

## KEY CONCEPTS

- Quality improvement is a continuous process focused on maintaining regulatory compliance and improving patient care processes and outcomes.
- Patient care needs should drive improvement opportunities.
- Decisions should be driven by data.
- Improvement initiatives should be linked to the organization's mission, vision, and values.
- Organizational goals and objectives should be communicated up and down the organization.
- There should be a balance in improvement goals focused on patient clinical and functional status, cost, and patient satisfaction outcomes.

- Customers of health care are patients, nurses, doctors, the community, and so on.
- A clinical value compass identifies key outcomes that are monitored for quality improvement.
- The PDSA cycle and FOCUS method are used to improve quality in organizations.
- Everyone in an organization is part of quality improvement efforts.
- Systems for quality improvement will work to improve care in an organization.
- Benchmarks are used to monitor quality.
- Organizations are structured to improve quality.

## KEY TERMS

benchmarking
diagnosis-related group

process
quality assurance (QA)

quality improvement (QI)
sentinel event

system

## REVIEW QUESTIONS

1. The primary focus of the Joint Commission is which of the following?
   A. Remaining within the proposed budget
   B. Safe, high-quality patient care
   C. Providing employment opportunities
   D. Providing a structure for collaboration among employees

2. The current goal of quality improvement initiatives is which of the following?
   A. Developing strategies to promote ongoing and future health care practices
   B. Reviewing past breeches of clinical practices
   C. Involving the consumer in the decision-making process
   D. Improving response time to care requirements

3. Following a sentinel event, which step would be initiated first?
   A. No action required
   B. Taking corrective action on personnel
   C. Reporting the event to legislative authorities
   D. Immediate investigation

4. A process reflecting the situation, background, assessment, recommendation, and response (SBARR) results in which of the following?
   A. Procedures that promote and enhance data collection strategies

   B. Improved organizational reporting requirements to licensing bodies
   C. Communication that promotes situational awareness and timely interventions for patients
   D. Increased governmental funding

5. Which of the following describes the benchmarking process?
   A. Reviewing your own unit's data for opportunities
   B. Collecting data on an individual patient
   C. Reviewing data in the literature
   D. Comparing your data to that of other organizations to identify improvement opportunities

6. Identifying QI opportunities in the health care arena is the responsibility of which group?
   A. Administration
   B. Practitioners
   C. Patients
   D. All health care personnel

7. Standardizing a process has which of the following effects? It:
   A. removes all chance of error.
   B. removes unwanted variation from a work process, decreasing risk for error.
   C. increases complexity, because it is hard to communicate the standard to everyone.
   D. advocates for the same care for everyone, without consideration for individual differences.

8. What tool could be used to track a change in a process?
   A. Flowchart
   B. Histogram
   C. Time series chart
   D. Pie chart

9. A well-functioning team demonstrates which of the following characteristics? Select all that apply.
   _____  A. Clarity about each member's role
   _____  B. A shared plan to achieve a goal
   _____  C. Effective communication practices
   _____  D. Cost of each clinical procedure
   _____  E. Understanding the vision of the organization

10. Cost of care decreases with quality improvement initiatives such as which of the following? Select all that apply.
   _____  A. Standardizing care delivery work processes
   _____  B. Applying evidence-based principles
   _____  C. Improving patient care outcomes
   _____  D. Maintaining continuous quality improvement surveillance
   _____  E. Benchmarking with other health care organizations

## REVIEW ACTIVITIES

1. Risk management, infection control practitioners, and a benchmark study have revealed that your unit's utilization of Foley catheters is above average. Brainstorm reasons why this may be occurring. Creating a fishbone (root cause) diagram may help.

2. After you have identified the root causes for the overuse of Foley catheters, use the PDSA cycle to identify improvement strategies.

3. Think about your last clinical rotation experience. Identify one work process that you believe could be improved, and describe how you would begin improving the process. Use the FOCUS method in Figure 20-2.

## EXPLORING THE WEB

- Use these sites for potential benchmark data. University HealthSystem Consortium (UHC):
  www.uhc.edu

  Institute for Healthcare Improvement (IHI):
  www.ihi.org

- These sites are recommended for a team that is looking for evidence-based guidelines or research studies for a particular diagnosis.

  National Guideline Clearinghouse:
  www.guideline.gov

  Cochrane Library:
  www.cochrane.org

  PubMed's clinical queries:
  www.ncbi.nlm.nih.gov

  Evidence-based practice internet resources:
  http://ebm.mcmaster.ca

  www.zynxhealth.com

- The Joint Commission:
  www.jointcommission.org

- Note this AACN Web site for nursing staffing levels:
  www.aacn.nche.edu

  Health grades:
  www.healthgrades.com

## REFERENCES

Adamski, P. (2007). Implement a handoff communications approach. *Nursing Management, 38*(1), 10–12.

Ajjawi, R., & Higgs, J. (2008). Learning to reason: A journey of professional socialization. *Advances in Health Sciences Education, 13*(2), 133–150.

Bandyopadhyay, J. K., & Hayes, G. (2009). Developing a framework for the continuous improvement of patient care in United States hospitals: A process approach. *International Journal of Management, 26*(2), 179–185.

Berwick, D., & Plsek, P. (1992). *Managing medical quality videotape series.* Woodbridge, NJ: Quality Visions.

Caldwell, C. (1998). *Handbook for managing change in health care.* Milwaukee, WI: ASQ Quality Press.

Carey, R. G., & Lloyd, R. C. (2001). *Measuring quality improvement in healthcare: A guide to statistical process control applications.* Milwaukee, WI: Quality Press.

Colligan, L., Anderson, J. E., Potts, H. W. W., & Berman, J. (2010). Does the process map influence the outcome of quality improvement work? A comparison of a sequential flow diagram and a hierarchial task analysis diagram. *BMC Health Services Research, 10*(7). Available at http://www.biomedcentral.com/1472-6963/10/7.

Cossette, S., Cote, J. K., Pepin, J., Ricard, N., D'Aoust, L. (2006). A dimensional structure of nurse-patient interactions from a caring perspective: Refinement of the Caring Nurse-Patient Interaction Scale (CNPI-Short Scale). *Journal of Advanced Nursing, 55*(2), 198–214.

Crosby, P. B. (1989). *Let's talk quality.* New York: McGraw-Hill.

Deming, W. E. (1986). *Out of the crisis.* Cambridge, MA: Center for Advanced Engineering Study.

Deming, W. E. (2000). *Out of crisis.* Cambridge, MA: MIT Press.

Donabedian, A. R. (2003). *An Introduction to quality assurance in health care.* New York: Oxford University Press.

Elixhauser, A., Pancholi, M., & Clancy, C. M. (2005). Using the AHRQ quality indicators to improve health care quality. *Joint Commission Journal of Quality Patient Safety, 31*(9), 533–538.

Ellis, J. (2006). All inclusive benchmarking. *Journal of Nursing Management, 14*(5), 377–383.

Gallart, H. (2009). Best practice bundles: The new science of quality improvement. *Creative Nursing, 15*(4), 206–207.

Guo, K. L. (2008). Quality of health care in the U.S. managed care system. *International Journal of Health Care, 21*(3), 236–248.

Institute of Medicine. (1999). *To err is human.* Washington, DC: National Academy Press.

Institute of Medicine. (2010). Report: The future of nursing; leading change, advancing health. Available at http://www.iom.edu/Reports/2010/The-Future-of-Nursing-Leading-Change.

Joint Commission. (2010). *What's new on quality check.* Available at http://www.jointcommission.org/QualityCheck/06_qc_new.htm.

Joint Commission. National Patient Safety Goals (NPSG). Retrieved December 20, 2010, from http://www.jointcommission.org/assets/1/6/2011_NPSGs_HAP.pdf.

Kaplan, R. S., & Norton, D. P. (1996). *Balanced scorecard: Translating strategy into action.* Boston: Harvard Business School Publishing.

Kaplan, R. S., & Norton, D. P. (2004). *Strategy maps.* Boston: Harvard Business School Publishing.

Kocakülǎh, M. C., & Austill, A. D. (2007). Balanced scorecard application in the health care industry: A case study. *Journal of Health Care Finance, 34*(1), 72–99.

Kyrkjebø, J. M. (2006). Teaching quality improvement in the classroom and clinic: Getting it wrong and getting it right. *Journal of Nursing Education, 45*(3), 109–116.

Lyndon, A. (2006). Communication and teamwork in patient care: How much can we learn from aviation. *Journal of Obstetric, Gynecologic, and Neonatal Nursing, 35*(4), 538–546.

Mark, V. (2010). Optimizing the process improvement toolkit. *Joint Commission Journal on Quality and Patient Safety, 36*(12), 531–532.

Ransom, S. B., Joshi, M. S., & Nash, D. B. (2005). *The healthcare quality book: Vision, strategy, and tools.* Chicago: Health Administration Press.

Schmittdiel, J. A., Grumbach, K., & Selby, J. V. (2010). System-based participatory research in health care: An approach for sustainable translational research and quality improvement. *Annals of Family Medicine, 8*(3), 256–259.

Simpson, K. N., Kaluzny, A. D., & McLaughlin, C. P. (1991). Total quality and the management of laboratories. *Clinical Laboratory Management Review, 5*(6), 448–449, 452–453, 456–458.

Sorbero, M. E. , Ricci, K. A., Lovejoy, S., Haviland, A. M., Smith, L., Bradley, L. A., Hiatt, L., Farley, D. O. (2009, April). Assessment of contributions to patient safety knowledge by the Agency of Healthcare Research and Quality-funded patient safety projects. *HSR: Health Services Research, 44*(2), 646–664.

Tamuz, M., & Thomas, E. J. (2006). Classifying and interpreting threats to patient safety in hospitals; Insights from aviation. *Journal of Organizational Behavior, 27*(7), 919–940.

The Governance Institute. (2009, Winter). *Leadership in healthcare organizations, a guide to Joint Commission leadership standards.* Retrieved October 10, 2010, from http://www.jointcommission.org/standards/.

Thompson, P. A. (2008). Key challenges facing American nurse leaders. *Journal of Nursing Management, 16*(8), 912–914.

Wheeler, D. J. (2000). *Understanding variation: The key to managing chaos* (2nd ed.). Knoxville, TN: SPC Press.

Willingham, M., & Eden, T. (2007). Can you hear me yet? The importance of effective communication in healthcare. *Oncology Nursing Forum, 34*(2), 517–518.

# SUGGESTED READINGS

Albert, N. M. (2006). Evidence-based nursing care for patients with heart failure. *AACN Advanced Critical Care, 17*(2), 170–185.

Barr. J. K.,Giannotti, T. E., Sofaer, S., Duquette, C. E., Waters, W. J., & Petrillo, M.K. (2006). Using public reports of patient satisfaction for hospital quality improvement. *HSR: Health Services Research, 41*(3), 663–682.

Beyea, S. C. (2006). Surgical Care Improvement Project: An important initiative. *Association of Operating Room Nurses Journal, 83*(6), 1371–1373.

Committee on Quality of Health Care in America, Institute of Medicine. (2000). In L. T. Kohn, J. M. Corrigan, & M. S. Donaldson (Eds.). *To err is human: Building a safer health system.* Washington, DC: National Academy Press.

Fisher, E. S., & Shortell, S. M. ( 2010). Accountable care organizations, accountable for what, to whom, and how. *Journal of American Medical Association, 304*(15).

Garcia, J. L., & Wells, K. K. (2009). Knowledge-based information to improve the quality of patient care. *Journal of Healthcare Quality, 31*(1), 30–35.

Grosfeld-Nir, A., Ronen, B., & Kozlovsky, N. (2007). The Pareto managerial principle: When does it apply? *International Journal of Production Research, 45*(10), 2317–2325.

Hockenberry, M., Walden, M., & Brown, T. (2007). Creating an evidence-based practice environment. *Journal of Nursing Care Quality, 22*(3), 222–231.

Huston, C. (2008). Preparing nurse leaders for 2020. *Journal of Nursing Management, 16*(8), 905–911.

Leape, L. L., & Berwick, D. M. (2005). Five years after To Err is Human. *Journal of American Medical Association, 293*(19), 2384–2390.

Morath, J. M., Tunbull, J. E., & Leape, L. (2005). *To do harm: ensuring patient safety in health care organizations.* San Francisco: Jossey-Bass.

Mueller, J. (2007). "When doing good is just the start to being good": A possible tool to improve the organizational effectiveness of non-profit health care organizations. *Journal of Hospital Marketing & Public Relations, 17*(2), 45–60. Press GANEY Associates, Inc. (2011). Retrieved from http://www.pressganey.com/index.aspx on May 15, 2011.

Ring, N., Coull, A., Howie, C., Murphy-Black, T., & Watterson, A. (2006). Analysis of the impact of a national initiative to promote evidence-based nursing practice. *International Journal of Nursing Practice, 12*(4), 232–240.

Schalm, C. (2008). Implementing a balanced scorecard as a strategic management tool in a long-term organization. *Journal of Health Services Research & Policy, 13*(Suppl. 1), 8–14.

Shortell, S. M., & Kaluzny, A. D. (2006). *Health care management organization design and behavior* (6th ed.). Clifton Park, NY: Delmar Cengage Learning.

# CHAPTER 21

# Evidence-Based Strategies to Improve Patient Care Outcomes

MARY ANNE JADLOS, MS, ACNP-BC, CWOCN;
GLENDA B. KELMAN, PhD, ACNP-BC

*The best outcomes evaluation is likely to come from partnerships of technically proficient analysts and clinicians, each of whom is sensitive to and respectful of the contributions the other can bring.*

(ROBERT L. KANE, PROFESSOR OF PUBLIC HEALTH, UNIVERSITY OF MINNESOTA, 1997)

## OBJECTIVES

Upon completion of this chapter, the reader should be able to:

1. Discuss the use of outcomes research in evidence-based practice.
2. Describe selected evidence-based models.
3. Apply the PDSA Cycle to implement evidence-based practice in specific patient care situations.
4. Identify resources available to generate outcomes/benchmarks in clinical practice.

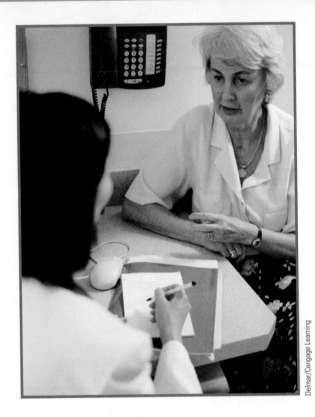

Delmar/Cengage Learning

*During shift handoff report, the staff nurse tells you about Mrs. Rose, who has recently been admitted to the inpatient hospice unit with a worsening stage three coccyx pressure ulcer.*

*Mrs. Rose is an 81-year-old female with end stage adenocarcinoma of the lung and a history of left CVA with right-sided hemiparesis. The patient has been cared for at home by her daughter, who brought her to the ER because of her deteriorating condition. According to the daughter, the patient has recently developed open skin on her bottom that is not improving. She is receiving MS Contin 15 mg around the clock, with no breakthrough medication.*

*The patient's daughter relayed that the patient has lost 15 pounds over the past month and has not taken any food or water for 2 to 3 days. Also, the patient has not walked in over a month, and her daughter has been lifting her OOB to a chair. Mrs. Rose's daughter reports that the patient has a three-month history of a pressure ulcer "on her bottom," and that over 10 days prior to admission this skin has opened up, is oozing yellow-green drainage with a foul odor, and is painful. Because of the pain, the patient's daughter has been unable to get her out of bed for a week and has found it impossible to turn her mother without her complaining of severe pain. Mrs. Rose has insisted on sleeping in a lounge chair. The patient is also becoming incontinent of urine and stool, so the daughter has found it difficult to keep a dressing on the ulcer.*

*Mrs. Rose is a widow. She has no other children, except for her daughter. The patient never smoked or used alcohol or illicit drugs. She grew up in South Carolina. The patient had worked as a housekeeper.*

*The patient has no known allergies to drugs, food, latex, or tape.*

*On exam, Mrs. Rose appears cachectic. She is very lethargic. Her eyes are open, but she is nonverbal. She is able to answer simple questions, but her speech is garbled and difficult to understand. Her respirations are shallow and irregular. Bedscale weight equals 90 pounds. Height equaled 5'4". BMI =15.0 (underweight). Mrs. Rose has no independent mobility and requires the maximum assist of two caregivers for turning and care. Physical exam of the skin is remarkable for multiple bony prominences. The occiput, shoulders, spine, hips, knees, and ankles are intact, with a very pale, slightly mottled color. The heels are intact, with pale color and slightly boggy turgor. The sacrum and coccyx have a pear-shaped wound bed with a large amount of devitalized pale yellow-gray slough and malodorous drainage. The wound measures 2.5 × 1.0 cm, and the depth is to the bone (0.5 cm). The ulcer is documented as a stage four pressure ulcer. Despite pre-medication with 2 mg of of short-acting IV morphine, Mrs. Rose grimaces with turning and direct care to the coccyx ulcer. She is able to rate her pain on admission = 10. X-ray of the sacrum and coccyx are negative for osteomyelitis. Her white blood cell count is 10,000. Her albumin is 0.8 g/dl.*

*What are you trying to accomplish in your plan of care for Mrs. Rose?*

*How will you know that a change is an improvement?*

*What changes can you make in the plan of care that will result in an improvement?*

This chapter will identify evidence-based strategies used to improve patient care outcomes. Information from nursing theory and an evidence-based practice model for improvement will be applied to a selected patient case study. By asking the simple question "Why?", you are beginning the journey of gathering data and evidence either to support your current practice or to change how you provide care and interact with patients and families to improve patient care outcomes. The focus is the patient.

## EVIDENCE-BASED PRACTICE

**Evidence-based practice (EBP)** is defined as the conscientious, explicit, and judicious use of current best evidence in making decisions about the care of individual patients (Sackett, Rosenberg, Gray, Haynes, & Richardson, 1996). EBP uses outcomes research and other current research findings to guide the development of appropriate strategies to deliver quality, cost-effective care. Outcomes research provides evidence about benefits, risks, and results of treatments so individuals can make informed decisions and choices to improve their quality of life. Research seeks to understand the end results of particular health care practices and interventions. End results may include changes in a person's ability to function and carry out routine activities of daily living. Outcomes research can also identify potentially effective strategies that can be implemented to improve the quality and value of care.

Evidence-based practice is a total process that begins with knowing what clinical questions to ask, how to find the best practice, and how to critically appraise the evidence for validity and applicability to the particular care situation. The best evidence is then applied by a clinician with expertise, based on the patient's unique values and needs. The final aspect of evidence-based practice includes evaluation of the effectiveness of care and commitment to continuous improvement of the process of patient care. Responsibilities related to achieving quality outcomes will be addressed in this chapter. They include: (1) Provide evidence-based, clinically competent care; (2) Demonstrate critical thinking, reflection, and problem-solving skills; (3) Take responsibility for quality of care and health outcomes at all levels; and (4) Contribute to continuous improvement of the health care system.

Historically, health care providers have relied primarily on biomedical parameters or measures such as laboratory and diagnostic tests to determine whether a health intervention is necessary and whether it is successful. However, these measures often do not fully reflect the multidimensional outcomes that matter most to the patients, such as quality of life, family, work, and overall level of functioning. Traditionally, outcome measures have included physical measures, such as blood pressure to assess the effectiveness of antihypertensive medications. They have also sometimes included patient satisfaction measures to assess patient satisfaction with the care or services provided.

The Institute of Medicine (IOM) has released two landmark reports on health care safety and quality, *To Err is Human* (1999) and *Crossing the Quality Chasm* (2001). These studies provided a broad agenda for the nation in addressing quality improvement in health care.

The Institute for Healthcare Improvement (IHI), in December 2004, decided to launch a national initiative, the 100,000 Lives Campaign, with a goal of saving 100,000 lives among patients in hospitals through improvements in safety and the effectiveness of care (Berwick, Calkins, McCannon, & Hackbarth, 2006). Six areas for evidence-based interventions were identified:

1. Deploy rapid response teams.
2. Deliver reliable evidence-based care for acute myocardial infarction.
3. Prevent adverse drug events through medication reconciliation.
4. Prevent central line infections.
5. Prevent surgical site infections.
6. Prevent ventilator-associated pneumonia.

All of the nation's 5,759 hospitals were invited to join and share their progress in reducing mortality. Other organizations that joined the IHI to support the campaign interventions included the Agency for Healthcare Research and Quality (AHRQ), Centers for Medicare & Medicaid Services (CMS), the Surgical Care Improvement Project, the Joint Commission (JC), and the Leapfrog Group.

## EVIDENCE-BASED STRATEGIES

Approximately 3,100 hospitals participated in the initial IHI Initiative. Based on their work on the initiative's interventions, and combined with other national and local improvement efforts, the facilities saved an estimated 122,000 lives in 18 months.

The IHI then launched a second two-year initiative in 2006, the 5 Million Lives Campaign for U.S. health care, focusing on protecting patients from five million incidents of medical harm over a two-year period from December 2006 to December 2008. The 5 Million Lives Campaign challenged American hospitals to adopt 12 changes in care that would save lives and reduce patient injuries. In addition to the six interventions from the 100,000 Lives Campaign, six new interventions targeted at harm were identified:

1. **Prevent harm from high-alert medications,** starting with a focus on anticoagulants, sedatives, narcotics, and insulin.
2. **Reduce surgical complications** by reliably implementing all of the changes in care recommended by the Surgical Care Improvement Project (SCIP) (Quality Net, 2011).
3. **Prevent pressure ulcers** by reliably using science-based guidelines for their prevention.
4. **Reduce methicillin-resistant *Staphylococcus aureus* (MRSA) infection** by reliably implementing scientifically proven infection control practices.
5. **Deliver reliable, evidenced-based care for congestive heart failure** to avoid readmissions.
6. **Get boards on board** by defining and spreading the best-known leveraged processes for hospital boards of directors, so that they can become far more effective in accelerating organizational progress toward safe care.

The IHI 5 Million Lives Campaign engaged more than 4,000 U.S. hospitals to prevent five million incidents of medical harm over a two-year period. The IHI campaign's how-to guides were developed to help organizations and providers implement evidence-based practices that were available at no cost to institutions and health care providers—for example, "How-to Guide: Improving Hand Hygiene." New standards of care were developed at hospitals throughout the United States, with other countries also adopting the standards. Another IHI Campaign legacy has been the creation of a national and international network that has helped to disseminate evidence-based practices and enable shared learning and innovation that is free and easily accessible at www.ihi.org (IHI, 2010).

IHI is currently engaged in improvement initiatives in England, Scotland, Ghana, Malawi, and South Africa. Other academic and health care organizations and programs focusing on quality improvement are listed in Table 21-1.

Other organizations focusing on evidence-based practice and quality improvement have included the Robert Wood Johnson Foundation (RWJF), which has funded the Quality and Safety Education for Nurses (QSEN) project for three phases. The overall goal through all phases of QSEN is to address the challenge of preparing future nurses with the knowledge, skills, and attitudes (KSA) necessary to continuously improve the quality and safety of the health care systems in which they work.

Six competencies were defined in Phase I of the RWJF QSEN project. They included:

- patient-centered care
- teamwork and collaboration
- evidence-based practice
- quality improvement
- safety
- informatics

In Phase II of the RWJF QSEN project, pilot nursing schools integrated the six QSEN competencies into their nursing curricula and have shared them on the QSEN Web site (Quality and Safety Education for Nurses (QSEN). (2011). Phase III continued to promote innovation in the development, evaluation, and dissemination of the competencies. Mechanisms to sustain these changes through accreditation and certification standards, as well as competency requirements, are ongoing.

## TABLE 21-1  Academic and Health Care Organizations and Programs Focusing on Quality and Safety

| ORGANIZATION | WEB SITE |
| --- | --- |
| • Academy for Healthcare Improvement | www.a4hi.org. |
| • Agency for Healthcare Research and Quality (AHRQ) Patient Safety Network | psnet.ahrq.gov |
| • American Association of Colleges of Nursing (AACN) | www.aacn.nche.edu |
| • AORN, Patient Safety First | www.aorn.org. Search for Patient Safety First |
| • Institute for Patient- and Family-Centered Care | www.ipfcc.org |
| • Institute for Healthcare Improvement (IHI) | www.ihi.org |
| • Institute for Safe Medication Practices | www.ismp.org |
| • International Nursing Association for Clinical Simulation and Learning (INACSL) | www.inacsl.org |
| • National League for Nursing (NLN) | www.nln.org |
| • National Patient Safety Foundation | www.npsf.org |
| • Robert Wood Johnson Foundation (RWJF) | www.rwjf.org |
| • Quality and Safety Education for Nurses (QSEN) | www.qsen.org |
| • Sigma Theta Tau International Honor Society of Nursing | www.nursingsociety.org |
| • Joint Commission | www.jointcommission.org |
| • North Carolina Center for Hospital Quality and Patient Safety | www.ncqualitycenter.org |
| • VA National Center for Patient Safety | www.patientsafety.gov |

## ROLE OF THE ANA

The American Nurses Association (ANA) was an active advocate of outcomes evaluation as early as 1976. Outcomes were emphasized as a measure of quality care. The 1980 ANA Social Policy Statement said that one of the four defining characteristics of nursing is the evaluation of the effects of actions in relation to phenomena. In 1986, the ANA approved policies related to the development of a classification system, including outcomes. In 1995, the ANA developed a Nursing Report Card for Acute Care Settings, which lists indicators for patient-focused outcomes, structures of care, and care processes. In 2003, the ANA revised the Social Policy Statement and included: "Nurses use their theoretical and evidence-based knowledge of these phenomena in collaborating with patients to assess, plan, implement, and evaluate care. Nursing interventions are intended to produce beneficial effects and contribute to quality outcomes. Nurses evaluate the effectiveness of their care in relation to identified outcomes and use evidence to improve care (p. 7)."

## EVOLUTION OF EBP

Evidence-based practice has evolved from a nice-to-know perspective to a need-to-know essential strategy in health care. Patients, health care providers, and payers recognize the significance of collecting data and analyzing outcomes to achieve safe, quality, cost-effective care. Outcome strategies used in EBP by nurses and members of the health care team include the creation of clinical protocols, guidelines, pathways, algorithms, and so on, which become the tools for health care interventions. A **practice guideline** is a descriptive tool or a standardized specification for care of the typical patient in the typical situation. Guidelines are developed by a formal process that incorporates the best scientific evidence of effectiveness and expert opinions. Synonyms or near synonyms include *practice parameter, preferred practice pattern, algorithm, protocol,* and *clinical standard.*

Evidence-based practice is used to guide practice interventions and is most successful when the entire organization and interdisciplinary team buy into EBP and participate and support the process. By linking the care that people receive to the outcomes they experience, EBP or outcomes research has become key to identifying and developing better strategies to monitor and improve the quality of care.

EBP is not a cookbook approach to patient care, however. The nurse and members of the health care team assess each patient and determine whether a guideline is appropriate. Many nursing theorists such as Nightingale, Peplau, Benner, and others have identified that the uniqueness of nursing is based upon the relational and integrative nature of healing, involving the person and environments. Nursing is the only discipline that views the whole person within the context of the person's environment. Nurses must not mimic medical practitioners but must focus on the art and science of nursing that comforts, cares, nurtures, heals, and builds on nursing theory to guide practice. There is a difference between evidence-based practice and evidence-based nursing practice. Ingersoll has defined **evidence-based nursing practice (EBNP)** as the conscientious, explicit, and judicious use of theory-derived, research-based information in making decisions about nursing care delivery to individuals or groups of individuals and in consideration of individual needs and preferences (2000). Evidence-based practice has a medical focus, whereas evidence-based nursing practice considers the individual's needs and preferences based on nursing theory and research.

The role of the nurse is to participate in developing a comprehensive, interdisciplinary evidence-based plan of care in conjunction with the patient and members of the health care team. This plan of care integrates the art and science of caring, not merely the medical model of the

## Critical Thinking 21-3

An elderly patient is admitted with CHF and chronic venous stasis ulcers on her legs. The doctor orders an alginate dressing to be changed every other day. The evening staff RN is giving a report to the night nurse. She states that the dressings are always wet and are saturating the bed linens. The RN questions whether the prescribed dressing regimen is the best choice for this patient.

What resources would you use to answer your clinical questions?

What health care personnel are available to you as a resource?

absence or presence of disease. Nurses must use innovation, creativity, and technology to plan care.

Benner, Hooper-Kyriakidis, Hooper, Stannard, and Eoyang (1998) identify aspects of what they refer to as the skilled know-how of managing a crisis. They state that some of this knowledge is assumed based on training, skill, and experience. Other nursing responsibilities are accepted or imposed based on necessity or a sense of moral obligation. These responsibilities might include stocking equipment at the bedside, prioritizing interventions and procedures, organizing the team and orchestrating its actions in a way that enhances its ability to function, recognizing the effect of therapies, and asking for help as appropriate. These skills are all relatively invisible in daily nursing practice. They become more visible in crisis situations. Mastery of and comfort with these skills come only with practice, practice, practice.

Nurses are expected to manage the care of acutely ill patients. Beginning nurses should be encouraged to thoughtfully acknowledge their personal abilities and limitations. There must be a blend between knowing and doing. It is important that the beginning nurse realize that gaining knowledge and skill is a gradual process. As the beginning nurse develops aspects of skilled know-how, the nurse will become aware of the ability to deliver safe patient care, minimize patient discomfort, decrease stress and anxiety, and assist in optimizing team performance. Guidelines based on EBP can help to direct care but cannot replace learning by hands-on delivery of patient care. As beginning nurses gain clinical experience and learn new theoretical knowledge, they will be able to contribute to the development and revision of EBP guidelines. Remember, EBP guidelines outline practice parameters based on evidence or research, but they do not outline how to deliver individualized patient care.

## EVIDENCE FROM THE LITERATURE

**Citation:** Orsted, H.L., Rosenthal, S., & Woodbury, G.M. (2009). Pressure ulcer awareness and prevention program: A quality improvement program through the Canadian Association of Wound Care. *Journal of Wound, Ostomy and Continence Nursing, 36(2)*, 178–183.

**Discussion:** A study was funded by the Canadian Association of Wound Care to identify the extent of pressure ulcers in Canada. Findings revealed a mean prevalence rate of 26%. Based on these findings, the Canadian Association created a continuous quality improvement program known as The Pressure Ulcer Awareness and Prevention Program. This program was pilot tested, revised and has been implemented in multiple Canadian health care facilities.

Six sites including acute care and long-term care facilities throughout Canada participated in pre-pilot (N = 1337) and postpilot (N = 1144) prevalence studies. Prepilot prevalence rates ranged from 6.4% to 33%. Postpilot prevalence rates were decreased in five of the six sites and ranged from 6% to 28%. Program outcomes reflected four areas of change: (1) awareness, (2) clinical practice, (3) policy, and (4) incidence.

**Implications for Practice:** Patients would benefit from collaborative pressure ulcer prevention programs developed across health care systems in communities. Nurses with expertise in pressure ulcer prevention and wound care can take the lead in championing collaborative community efforts which include education, clinical expertise, use of evidence-based practice, and conducting systematic prevalence and incidence studies.

# EVIDENCE-BASED MULTIDISCIPLINARY PRACTICE MODELS

Several models are used in EBP. They include the University of Colorado Hospital model and the PDSA Cycle.

## THE UNIVERSITY OF COLORADO HOSPITAL MODEL

The University of Colorado Hospital model (Figure 21-1) is an example of an evidence-based multidisciplinary practice model (Goode et al., 2000; Goode & Piedalue, 1999). This model presents a framework for thinking about how you use different sources of information to change or support your practice. The health care team or team member uses valid and current research from sources such as journals, conferences, and clinical experts as the basis for clinical decision making. The model depicts nine sources of evidence that are linked to the research core. This model provides a way for the nurse to organize information and data needed not only to care for a patient, but also to evaluate the care provided. In other words, did this patient receive the best possible care not only that this institution can offer, but that is available in this world?

The elements of the University of Colorados Practice Model can be applied to the case of Mrs. Rose in the opening scenario of this chapter (Table 21-2).

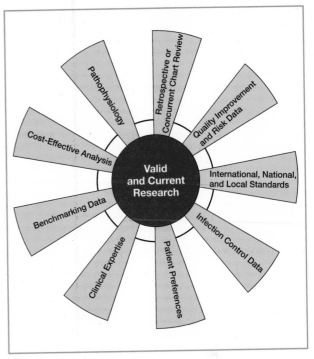

**FIGURE 21-1** University of Colorado Hospital evidence-based multidisciplinary practice model. (*Source*: By permission of University of Colorado Hospital Research Council, Denver, CO.)

**TABLE 21-2  Practice Applications to Elements of the University of Colorado Model**

| MODEL ELEMENT | APPLICATION |
|---|---|
| Benchmarking Data | Compare Mrs. Rose's comfort, palliative care, and wound care measures against institutional and national benchmarks. Review evidence-based literature. |
| Cost-Effective Analysis | Analyze cost-effectiveness of wound care regimen, including product use and nursing care hours. |
| Pathophysiology | Review pathophysiology and etiology of pain and pressure ulcer progression for the palliative care patient. |
| Retrospective/Concurrent Chart and/or Electronic Health Records (EHR) Review | Document and assess Braden Pressure Ulcer Risk Assessment and pain rating using the Numeric Pain Scale (NPS). |
| Quality Improvement and Risk Data | Review and analyze documentation regarding patient progress and outcomes and risk assessment (for example, new hospital-acquired pressure ulcer development and pain rating, as well as dosage of narcotic administration). |

*(Continues)*

**TABLE 21-2** (Continued)

| MODEL ELEMENT | APPLICATION |
| --- | --- |
| International, National, and Local Standards | Assess the effectiveness of care related to national and evidence-based guidelines such as the American Pain Society (www.ampainsoc.org/), the European Pressure Ulcer Advisory Panel (www.epuap.org/), and the National Pressure Ulcer Advisory Panel (www.npuap.org/). |
| Infection Control Data | Assess the wound for signs and symptoms of infection, review appropriate laboratory data, and institute appropriate standards or infection control and treatment. Monitor outcomes. |
| Patient Preferences | Discuss, document, and implement patient's wishes regarding advanced directives and comfort care measures. |
| Clinical Expertise | Consult acute care nurse practitioner (ACNP) and other practitioners for pressure relief, wound care, pain management, and comfort and palliative care measures. |

For example, to assess Mrs. Rose's progress, it would be important to review evidence using institutional and national benchmarks related to comfort, palliative care, and wound care. **Benchmarking** is defined as the continuous process of measuring products, services, and practices against the toughest competitors or those customers recognized as industry leaders (Joint Commission, 2009; Camp, 1994).

The wound care regime related to product use and nursing time could be analyzed for cost-effectiveness.

Pathophysiology would be analyzed by reviewing the etiology of pain and pressure ulcer progression. Findings could be discussed with the practitioners regarding implications related to the patients treatment.

A concurrent or ongoing review could be conducted using the Braden Scale for Predicting Pressure Sore Risk (Bergstrom, Braden, Laguzza, & Holman, 1987) and documenting the score daily.

Data collected for quality improvement purposes could include information on the development of hospital-acquired pressure ulcers, wound management, and pain. Pain and pressure ulcer guidelines from the American Pain Society (2005), the European Pressure Ulcer Advisory Panel (EPUAP), and the National Pressure Ulcer Advisory Panel (NPUAP) (2009) are examples of national evidence-based standards that could be used as benchmarks to assess the effectiveness of care. The nurse would also discuss with Mrs. Rose, document, and implement her wishes regarding advanced directives. Utilization of clinical expertise could include consulting the acute care nurse practitioners and other health care practitioners regarding wound, skin, and pain management initially, and on an ongoing basis.

## THE PDSA CYCLE

Another model for using EBP is the PDSA Cycle (Langley, Nolan, Nolan, Norman, & Provost, 1996). Some elements of this Cycle (Figure 21-2) were discussed in Chapter 20. The Cycle begins with these questions:

1. What are we trying to accomplish?
2. How will we know that a change is an improvement?
3. What change can we make that will result in improvement?

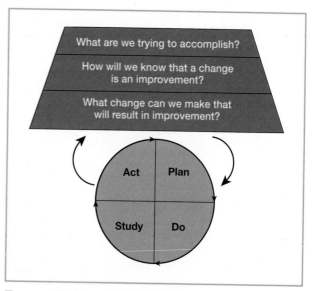

**FIGURE 21-2** The PDSA Cycle. (*Source:* From Langley, G. J., Nolan, K. M., Nolan, T. W., Norman, C. L., & Provost, L. P. 1996. *A practical approach to enhancing organizational performance* [p. 3]. San Francisco: Jossey-Bass. Reprinted by permission of Jossey-Bass, Inc., a subsidiary of John Wiley & Sons, Inc.)

These three questions provide the foundation for the Plan, Do, Study, Act (PDSA) Cycle. We will apply the PDSA Cycle to the Opening Scenario presented at the beginning of this chapter in relation to pain, pressure ulcers, and wound management.

# PAIN AND COMFORT CARE, INCLUDING WOUND MANAGEMENT, FOR THE PALLIATIVE CARE PATIENT

According to the World Health Organization, **palliative care** is defined as the active, total care of patients whose disease is not responsive to curative treatment (Fleck, 2005). The palliative care philosophy regards dying as a normal process. Palliative care interventions are directed at preventing, reducing, and relieving symptoms of disease or disorders rather than interventions intended to cure. Both the patient and family are considered the unit of care. However, end-of-life care is not synonymous with end of care giving. The focus of care can shift from aggressive curative interventions to aggressive comfort interventions (Letizia, Uebelhor, & Paddack, 2010).

In caring for a patient with a cancer diagnosis, the nurse needs to recognize that 50% to 80% of individuals with cancer do not receive adequate pain relief (American Pain Society, 2005). The American Pain Society revised and expanded their 1995 quality improvement guidelines for the treatment of acute pain and cancer pain in 2005. The revised guidelines focus on recognizing and treating pain promptly, involving patients and families in pain management planning, improving treatment patterns, reassessing and adjusting the pain management plan as needed, and continuous monitoring of processes and outcomes related to pain management.

The pain management standards in the Joint Commission *Comprehensive Accreditation Manual for Hospitals* state that the hospital assesses and manages the patient's pain (2009). The standards include conducting a comprehensive pain assessment in order to provide optimal treatment and services. The hospital is also expected to utilize methods to assess pain, consistent with the patient's age, condition, and ability to understand. Reassessment and treatment for pain is also an essential element of pain management.

Chronic, non-healing wounds are common as patients approach the end of life. Alvarez, Kalinski, Nusbaum, Hernandez, Pappous, Kyriannis, et al. (2007) noted that up to 28% of patients in long-term care and hospice programs will experience a pressure ulcer (PrU). In addition, patients with a cancer diagnosis may experience disruption in skin integrity, including PrU, related to the disease process. The transition to palliative care for patients with non-healing wounds that may include PrUs involves a focus on strategies to stabilize the wound and prevent the development of new wounds. Aspects of wound care requiring specific interventions include managing exudates, controlling odor, maximizing mobility and function, preventing infection, and controlling pain and other symptoms (Lee, Ennis, & Dunn, 2007).

## HEALING PROBABILITY ASSESSMENT TOOL

As part of providing palliative care, the program For Recognition of the Adult Immobilized Life (FRAIL) may help health care providers identify patients with wounds that are unlikely to heal and that may benefit from palliative care. The FRAIL program uses the Healing Probability Assessment Tool (www.frailcare.org) for estimating the probability for any wound to respond to aggressive local intervention. Criteria used in the tool reflect changes in a patient's status. As the number of these factors that are present increases, the greater the probability that the wound will not heal. Many factors identified in the Healing Probability Assessment Tool are similar to the Braden subscales. Additional factors include assessment of changes in oxygenation status, presence of comorbid conditions, history of falls, and history of duration of the wound (Letizia, Uebelhor, & Paddack, 2010).

The factors identified below from the Healing Probability Assessment Tool provide the basis for estimating the probability for wound healing. Although there is no specific scoring for this tool, the more items checked, the greater the probability that the wound will not heal.

- Wound(s) is over 3 months old or is a reoccurrence of a preexisting breakdown.
- Patient spends 20 or more hours daily in a dependent position (chair or bed).
- Patient is incontinent of urine.
- Patient is incontinent of feces.
- Patient has lost >5% of baseline weight, or 10 pounds, in the past 90 days.
- Patient does not eat independently.
- Patient does not walk independently.
- Patient has a history of falls within the last 90 days.
- Patient is unable/unwilling to avoid placing weight over wound(s) site(s).
- Patient's wound is associated with complications of diabetes mellitus.
- Patient's wound is associated with peripheral vascular disease (PVD).
- Patient has severe chronic obstructive pulmonary disease (COPD).
- Patient has end stage renal, liver, or heart disease.
- Patient's wound is associated with arterial disease.
- Patient has diminished range of motion (ROM) status nonresponsive to rehabilitative services.
- Patient has diminished level of mental alertness demonstrated by muted communication skills and inability to perform activities of daily living (ADLs) independently.

- Patient's wound is full thickness, with presence of tunneling.
- Patient's blood values indicate a low oxygen-carrying capacity.
- Patient's blood values indicate an exhausted or decreasing immune capacity, e.g., low lymphocyte count.
- Patient's blood values indicate below normal visceral protein levels that have not responded to nutritional support efforts (i.e., low prealbumin, transferrin, and albumin) (Letizia, Uebelhor, & Paddack, 2010).

The nurse should also consider the patient's potential feelings of embarrassment, depression, anxiety, and social isolation, which are often associated with a non-healing wound (McDonald & Lesage, 2006). An expanded view of pain management and comfort care is included in Kolcaba's application of comfort theory (1994, 2003). Kolcaba has defined **comfort** as the immediate state of being strengthened through having the human needs for relief, ease, and transcendence addressed in the context of the four dimensions of the physical, psychospiritual, sociocultural, and environmental experience (2003). When the three types of comfort—i.e., relief, ease, and transcendence—are met across the four dimensions, comfort care for the individual will be maximized (www.thecomfortline.com).

Another model, the wound associated pain (WAP) model, was developed to address the complexity of chronic wound-associated pain (Woo & Sibbald, 2009). The key components of this model include (1) patient-centered concerns, (2) wound etiology, and (3) local wound factors. The WAP model integrates principles of wound pain assessment and management related to principles of wound bed preparation.

In developing a plan of care for Mrs. Rose, consideration should be given to application of Kolcaba's comfort care model and the wound associated pain (WAP) model, with a specific focus on wound-associated pain. These models may be used to guide evidence-based practice and create an interdisciplinary culture of comfort recognized and valued by health care institutions.

In addition, Benner's research with critical care nurses in acute care settings provided a description of clinical judgment and thinking-in-action referred to as clinical grasp and clinical forethought. Benner identified nine domains or themes of practice common to complex patient care situations such as the one described in the opening scenario. One of these domains or themes of practice includes providing comfort measures. Benner explains that the skin is a point of connection and central or primary to many nursing interventions. Interventions specific to Mrs. Rose's scenario could include providing physical comfort, such as a bath or a backrub; wound assessment and wound dressing changes; minimizing and controlling wound odor; and providing care to a patient who is incontinent.

# APPLICATION OF THE PDSA CYCLE TO PAIN MANAGEMENT AND COMFORT CARE, INCLUDING WOUND MANAGEMENT

Let us revisit the scenario related to Mrs. Rose, described at the beginning of this chapter, and apply the PDSA Cycle. We start with three questions:

1. *What are we trying to accomplish?* The overall objective is to reduce or alleviate Mrs. Rose's pain rating and promote comfort, which will also include a focus on preventing wound infection, eliminating necrotic slough, controlling odor and exudates, and relieving or minimizing pain associated with dressing changes.
2. *How will we know that a change is an improvement?* Mrs. Rose's pain will be decreased or relieved as reflected by the pain rating scores, and her comfort level will be improved. In addition, Mrs. Rose's sacral pressure ulcer will show signs of decreased necrotic slough and malodorous drainage. Mrs. Rose will express less discomfort related to dressing changes and repositioning.
3. *What changes can we make that will result in improvement?* Implementation of the pain and pressure ulcer protocols will standardize the approach to patient care. Documenting Mrs. Rose's response to care will provide data to monitor and evaluate progress towards the goals. A multidisciplinary care conference may be needed to coordinate the wound and comfort care.

## Implementation of the PDSA Cycle

The PDSA Cycle can be individual or system focused. It can be used to solve a specific patient problem or to structure strategies to manage groups of patients with common problems. Based on our answers to the three questions above, we will begin the PDSA Cycle (Figure 21-2) for this patient situation.

### Planning Phase

Once the initial PDSA Cycle questions have helped staff identify what should be improved, the multidisciplinary staff would develop a plan for improvement. The plan would include routine pain assessment every shift and prior to dressing changes, as well as consultation with the nurse practitioner for management of the patient's sacral ulcer. Specific interventions will be identified to relieve pressure, manage wound exudate and odor, prevent infection, optimize mobility, and minimize pain related to dressing changes.

In the palliative care setting, wound care priorities may shift from healing the wound to stabilizing the wound, preventing new wounds, and managing symptoms with a focus on the patient's comfort and quality of life. Palliation of symptoms, odor, and exudates of chronic wounds is a realistic goal, whereas wound healing may not be a realistic goal for a palliative care patient. Wound dressings should be selected with consideration to minimizing wound trauma by gently and effectively debriding the wound of necrotic slough and minimizing the frequency of dressing changes. Dry dressings, if used, should be moistened before removal from the wound. Products that provide periwound protection to the area immediately around the wound should be considered. Non-adherent dressings may be used to minimize trauma to local tissues once the wound is clean.

The nurse can ask Mrs. Rose if the prescribed medication and comfort care measures are reducing her pain and improving her comfort. Changes to the plan of care will be made as needed. The nurse will utilize the Protocol for Assessment and Management of Pain as a guide to plan and individualize Mrs. Rose's care (Figure 21-3). The Pain Management Flow Sheet in the electronic health record (EHR) will be used to assess, document, and evaluate Mrs. Rose's response to pain intervention strategies (Figure 21-4).

The Wound Care Flow Sheet will also be initiated to document wound assessment and treatment measures (Figure 21-5). Commonly used wound care products need to be considered after completing the wound assessment. Selected examples of wound and skin care products with instructions for use are included in Figure 21-6. The acute care nurse practitioner (ACNP) and nursing staff will collaborate to review the patient care data and make recommendations to modify the plan for wound management and comfort care.

## Doing Phase

Pain management is a primary focus for individuals with non-healing wounds. Wound-associated pain is related to a number of factors, especially dressing removals. Therefore, patients should be premedicated before dressing changes with a quick-acting opioid medication, even if chronic pain is being managed with a long-acting analgesic. Topical anesthetics may be used to minimize pain associated with dressing removal (Woo & Sibbald, 2009). Both pharmacologic and non-pharmacologic approaches to reduce anxiety should be provided for patients experiencing wound-associated pain (Letizia, Uebelhor, & Paddack, 2010).

Nursing staff will utilize the Protocol for Assessment and Management of Pain (Figure 21-3), the EHR Pain Management Flow Sheet (Figure 21-4), and the Wound Care Flow Sheet (Figure 21-5) to collect data and document Mrs. Rose's care during her hospital stay. All the nurses assigned to care for Mrs. Rose were asked to complete the documentation around the clock and especially with dressing changes. Mrs. Rose will be premedicated with a short-acting opioid medication, Roxanol, for breakthrough pain and 30 minutes before dressing changes. She will continue to receive a long-acting analgesic. Data that will be monitored and documented will include the patient's pain rating, her nonverbal behavior, level of consciousness, respiratory rate, any side effects, mobility, and all non-pharmacologic and pharmacologic interventions. Mrs. Rose is receiving MS Contin, a long-acting opioid, twice daily.

## Studying Phase

Mrs. Rose's pain and wound management data was reviewed after three days. The multidisciplinary team met to review the documentation. Several areas of patient improvement were noted. However, some ongoing issues were identified. Documentation of pain assessment had been completed only 66% of the time. Staff nurses reported that they referred to the pain assessment and pain management data when giving report at the end of shift as well as when giving report to the physician and nurse practitioner. As a result, Mrs. Rose's pain management was central in discussions regarding her care, and decisions about changes in pain medication regimen were made in a timely manner. Within three days, Mrs. Rose expressed increased comfort, and her pain rating decreased from 8/10 to 3-4/10. The sacral ulcer appeared deeper, but the wound bed was cleaner with 20% less necrotic slough. Wound drainage had decreased approximately 50%. The foul odor was no longer present, and the periwound skin around the wound appeared less inflamed. The patient expressed less discomfort with the dressing changes, as evidenced by verbalization of decreased pain and increased comfort with repositioning and dressing changes. Massage therapy was initiated two days later to promote comfort. In addition, the patient was eating approximately 50% of her meals, as well as taking a daily dietary supplement.

## Acting Phase

The staff agreed to continue Mrs. Rose's current plan of care. While her pain and comfort had significantly improved, the patient still verbalized a pain level of 3-4/10. The MS Contin was increased from 15 mg orally on admission to 30 mg orally. Breakthrough pain was being treated with a short-acting opioid, Roxanol, 2.5mg every four hours as needed. It was decided to continue the same wound dressing regimen because of the presence of necrotic tissue, but to increase her short-acting opioid dose of Roxanol from 2.5 mg to 5.0 mg and continue her massage therapy. The next step in the PDSA Cycle would include the use of additional multiple PDSA cycles to improve not only Mrs. Rose's individual care, but also to improve the total care delivery system.

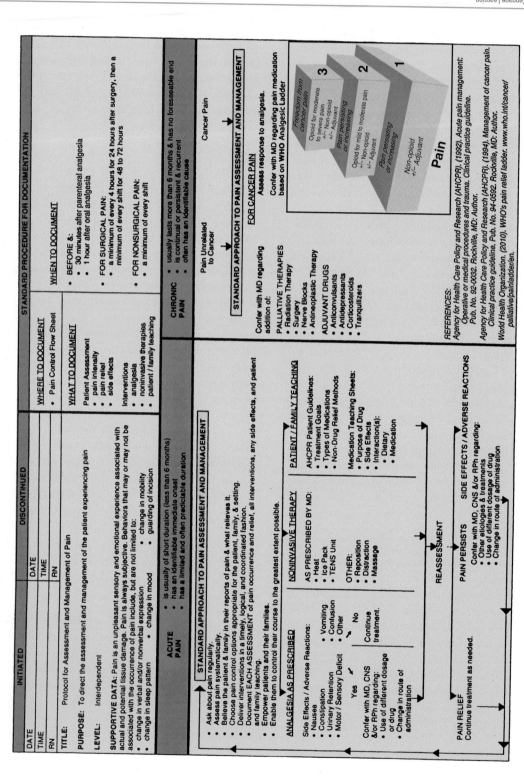

**FIGURE 21-3** Protocol for assessment and management of pain.

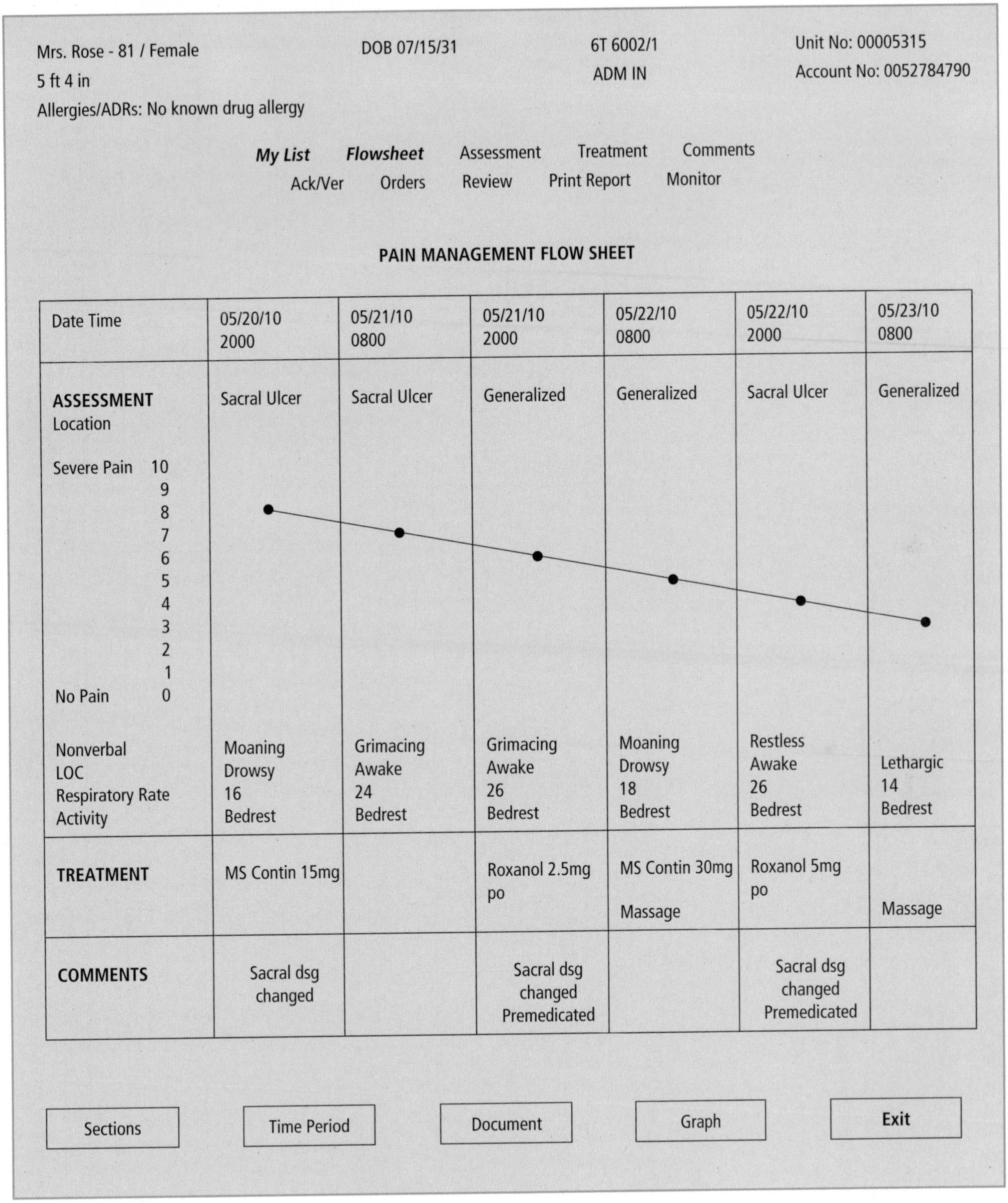

Mrs. Rose - 81 / Female

DOB 07/15/31

6T 6002/1

Unit No: 00005315

5 ft 4 in

ADM IN

Account No: 0052784790

Allergies/ADRs: No known drug allergy

*My List* **Flowsheet** Assessment Treatment Comments
Ack/Ver Orders Review Print Report Monitor

## PAIN MANAGEMENT FLOW SHEET

| Date Time | 05/20/10 2000 | 05/21/10 0800 | 05/21/10 2000 | 05/22/10 0800 | 05/22/10 2000 | 05/23/10 0800 |
|---|---|---|---|---|---|---|
| **ASSESSMENT** Location | Sacral Ulcer | Sacral Ulcer | Generalized | Generalized | Sacral Ulcer | Generalized |
| Severe Pain 10 9 8 7 6 5 4 3 2 1 No Pain 0 | | | | | | |
| Nonverbal LOC Respiratory Rate Activity | Moaning Drowsy 16 Bedrest | Grimacing Awake 24 Bedrest | Grimacing Awake 26 Bedrest | Moaning Drowsy 18 Bedrest | Restless Awake 26 Bedrest | Lethargic 14 Bedrest |
| **TREATMENT** | MS Contin 15mg | | Roxanol 2.5mg po | MS Contin 30mg Massage | Roxanol 5mg po | Massage |
| **COMMENTS** | Sacral dsg changed | | Sacral dsg changed Premedicated | | Sacral dsg changed Premedicated | |

Sections   Time Period   Document   Graph   **Exit**

**FIGURE 21-4** Electronic health record (EHR) pain management flow sheet (example).

## WOUND CARE FLOW SHEET

### DIRECTIONS:
1. Mark wound site(s) on body figure, number & circle.
2. For each site, indicate wound type / contributing factors & place ☑ of all that apply.
3. Document assessment & treatment for wounds:
 - requiring frequent dsg changes, only once a day.
 - requiring extended wear dsg, only when dsg is changed.

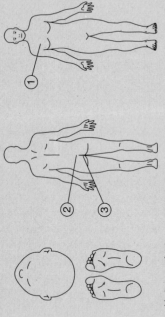

**Assessment Key:** *I = intact  P = pink  R = red  S = slough  Y = yellow  G = green  W = white  E = eschar  Gr = Gray  ** S = superficial or record in cm.
*** Type: S = serosanguinous  B = bloody  Y = yellow  G = green  BL = Blue  PU = Purple  0 = none / Amount: SC = scant  SM = small  MOD = moderate  H = heavy

**Treatment Key:** A = Aquacel  AAg = Aquacel Ag  AA = Acetic Acid  AP = Aquaphor  AZ = Accuzyme  BAC = Bacitracin  BB = Bactroban  B = Barrier Lotion  BS = Barrier Spray
DD = Duoderm  C = Cleocin Spray  DSD = Dry Sterile Dsg  F = Flexzan  GEL = Hydrogel  NS = Normal Saline  OTA = Open To Air  S = Silvadene  Other: _____

☐ Abrasion  ☐ Ecchymosis  ☐ Skin Tear  ☐ Tumor  ☐ Burn  ☐ Radiation  ☐ Surgical Wound  ☐ Venous / Arterial Insufficiency  ☐ Friction
☐ Shear  ☐ Moisture  ☐ Cellulitis  ☑ Pressure Ulcer (circle one) > Stage 1  2  3  ④ Unstageable  Deep Tissue Injury

| DATE | 5/20 | 5/21 | 5/22 | 5/23 | 5/24 | 5/25 | 5/26 | | | | |
|---|---|---|---|---|---|---|---|---|---|---|---|
| INITIALS | KR | KR | KR | BW | BW | BW | TT | | | | |
| Wound bed appearance * | S Gr | S Gr | S Gr | S Gr | S Gr | S Gr | S Gr | | | | |
| Wound edge appearance * | S Gr | S Gr | S Gr | S Gr | S | S | S | | | | |
| Size (cm.) | 2.5×1 | 2.5×1 | 2.5×1 | 2.5×1 | 2.5×1 | 2.5×1 | 2.5×1 | | | | |
| Depth ** | Bone | Bone | Bone | Bone | Bone | Bone | Bone | | | | |
| Drainage *** | G BL | G BL | G BL | G R | R SS | R SS | R SS | | | | |
| Odor (Y / N) | Y | Y | N | N | N | N | N | | | | |
| Maceration (Y / N) | N | N | N | N | N | N | N | | | | |
| Inflammation (Y / N) | Y | Y | Y | N | N | N | N | | | | |
| Undermining (Y / N) | Y | Y | Y | Y | Y | Y | Y | | | | |
| TREATMENT (use Key) | AA C | AA C | AA C | AA C | AA C | AA C | AA C | | | | |

**FIGURE 21-5** Wound care flow sheet. (*Source:* Courtesy of Northeast Health, Samaritan and Albany Memorial Hospitals, Acute Care Division, Troy and Albany, NY, with documentation.)

# WOUND & SKIN CARE PRODUCTS with INSTRUCTIONS

| Product | Goal | Application Procedure |
|---|---|---|
| **Calcium Alginate**<br>*Sorbsan / Calcicare / Tegagen* | Absorb drainage & promote granulation | *Change once a day / prn if drainage strikes through outer dressing:*<br>- place dressing into wound, filling all "dead space"<br>- cover with 4 × 4 gauze dressing & secure with tape |
| **Foam**<br>*Flexzan / Lyofoam / 3M Foam* | Absorb drainage & facilitate wound closure | *Change every 5 to 7 days / prn if non-adherent:*<br>- wipe periwound skin with "skin prep" & allow to dry<br>- place dressing<br>- wipe edges with "skin prep" or frame with tape |
| **Hydrocolloid**<br>*Duoderm / Restore / Tegasorb* | Autolytically debride necrotic tissue & facilitate wound closure | *Change every 2 to 3 days / prn if non-adherent:*<br>- remove paper backing, then apply dressing<br>- frame edges with tape |
| **Hydrofiber**<br>*Aquacel / Aquacel Ag (with silver)* | Absorb drainage, promote granulation, & decrease bacterial load | *Change once a day / prn if drainage strikes through outer dressing:*<br>- place dressing into wound, filling all "dead space"<br>- cover with 4 × 4 gauze dressing & secure with tape |
| **Hydrogel**<br>*Saf-Gel / Solosite / Tegagel* | Hydrate dry wound bed and facilitate wound closure | *Change 1 or 2 times a day:*<br>- apply coat of barrier ointment onto periwound skin<br>- apply hydrogel to wound bed<br>- cover with 4 × 4 gauze dressing & secure with tape |
| **Debridement Gel**<br>*Collagenase Santyl / Accuzyme* | Enzymatically debride devitalized tissue | *Change once a day:*<br>- apply coat of barrier ointment onto periwound skin<br>- apply debridement gel to wound bed<br>- cover with 4 × 4 gauze dressing & secure with tape |
| **Wet-to-Dry Gauze**<br>*Normal Saline 1/4 st Acetic Acid* | Mechanically debride necrotic tissue | *Change 2 or 3 times a day / prn if soiled:*<br>- moisten gauze & wring out<br>- unfold gauze & place into wound bed, with all surfaces in contact with the gauze<br>- cover with dry gauze / ABD pad dressing & secure with tape |
| **Hemostatic**<br>*Surgicel* | Stop bleeding | *Do not remove. Allow the dressing to fall off:*<br>- place the dressing onto the site of persistent bleeding<br>- cover with a dry ABD pad dressing as desired by the patient<br>- DO NOT TAPE. Mesh netting may be used to secure |
| **Topical Antibiotic**<br>*Cleocin Spray / Flagyl Spray* | Decrease bacterial load & eliminate foul odor | *Apply to secondary dressing when changed & prn:*<br>- spray antibiotic solution onto a dry gauze dressing<br>- place moist side of dressing over the wound bed<br>- cover with dry gauze / ABD pad dressing & secure with tape |
| **Topical Antimicrobial**<br>*Bacitracin / Bactroban SilvaSorb / Silvadene* | Treat local signs of infection | *Apply as prescribed by MD or NP:*<br>- apply prescribed agent onto wound bed<br>- cover with 4 × 4 gauze / ABD pad dressing & secure with tape |
| **Transparent Film**<br>*Tegaderm / OpSite* | Autolytically debride devitalized tissue | *Change once a day:*<br>- remove paper backing carefully from the dressing<br>- place over wound bed / periwound skin |

**FIGURE 21-6** **Wound and skin care products with instructions.** (*Source:* Courtesy of Northeast Health, Samaritan and Albany Memorial Hospitals, Acute Care Division, Troy and Albany, NY.)

## Multiple Uses of the PDSA Cycle

### Planning Phase

The inpatient staff across our acute care setting have continued to use the Protocol for Assessment and Management of Pain (Figure 21-3), and the EHR Pain Management Flow Sheet (Figure 21-4) as well as the Wound Care Flow Sheet (Figure 21-5). The multidisciplinary team, including nursing, was the champion in the creation and development of the EHR Pain Management Flow Sheet. Use of the EHR Pain Management Flow Sheet enhances the ability of the nurse to access patient information, document it, and evaluate patient care delivery.

### Doing Phase

The multidisciplinary committee, including staff nurses, nurse managers, nursing staff development, and medical information system (MIS) specialists, coordinated orientation and ongoing updates for the nursing staff and physicians. Once the nursing staff was comfortable with entering and accessing data from the EHR Pain Management Flow Sheet, the remaining physicians who were not initially involved in the committee were oriented to the EHR Pain Management Flow Sheet.

### Studying Phase

One of the additional advantages of using the EHR Pain Management Flow Sheet related to skin and wound care is the ability to both document a baseline Braden Scale for predicting pressure sore risk (Figure 21-7) and document the presence of pressure ulcers on admission (Bergstrom, Braden, Laguzaa, & Holman, 1987; Bergstrom, Braden, Kemp, Champagne, & Ruby, 1998). The ability to access and review this data is useful in identifying the institution's pressure ulcer prevalence and incidence of facility-acquired pressure ulcers. Another advantage of using the EHR Pain Management Flow Sheet is the opportunity to monitor and evaluate patient responses to pharmacological and non-pharmacological interventions during hospitalization. Institutional data can be used, for example, to benchmark against national standards and other institutional outcome data. Additionally, this information is useful for medical record coding and health insurance reimbursement.

### Acting Phase

In Mrs. Rose's scenario, the PDSA Cycle provides a framework to think about how to apply knowledge and increase the ability to make changes in individual patient care, ultimately resulting in improvement for all patients and the health care system. The next application of the PDSA Cycle is also related to the opening scenario.

# PRESSURE ULCER MANAGEMENT

Pressure ulcers continue to be a major health care concern in the United States and globally. The Agency for Healthcare Research and Quality (AHRQ, 2008) reported an increase in pressure ulcer–related hospitalizations of nearly 80% between 1993 and 2006. Of 503,300 pressure ulcer–related hospitalizations in 2006, pressure ulcers were the primary diagnosis in about 45,500 hospital admissions (up from 35,800 in 1993), and they were a secondary diagnosis in 457,800 hospital admissions (up from 245,600 in 1993) (AHRQ, 2006). In 2007, the Centers for Medicare & Medicaid (CMS) data reported 257,412 cases, with a cost of approximately $43,180 per hospital stay (CMS, 2008; Lyder & Ayello, 2009).

The Healthy People 2020 Initiative, which includes assessments of major risks to health and wellness related to the nation's health prevention, has also identified the need to reduce the rate of pressure ulcer–related hospitalizations among older adults. As a result, there are national efforts to reduce the prevalence and incidence of pressure ulcers. The AHRQ has developed a program for pressure ulcer prevention in nursing homes and is funding research on pressure ulcer prevention in hospitals. The Institute for Healthcare Improvement (IHI) 2020 Healthy People Initiative objectives also include the need to decrease the rate of pressure ulcer–related hospitalizations among older adults.

In 2009, the National Pressure Ulcer Advisory Panel (NPUAP) and the European Pressure Ulcer Advisory Panel (EPUAP) published evidence-based recommendations for the prevention and treatment of pressure ulcers. The recommendations outline specific measures, including establishing a risk assessment policy and comprehensive screening skin assessments as well as implementing specific strategies for the prevention and treatment of pressure ulcers. The Pressure Ulcer Scale for Healing (PUSH) tool or the Bates-Jensen Wound Assessment Tool (BWAT), formerly known as the Pressure Sore Status Tool (PSST), is recommended for use as a valid tool to document progress toward healing (Berlowitz, Ratliff, Cuddigan, Rodeheaver, & NPUAP, 2005; Harris et al., 2010).

The Braden Scale for Predicting Pressure Sore Risk (also known as the Braden Pressure Ulcer Risk Assessment) continues to be a valid and reliable tool for assessing pressure ulcer risk (Bergstrom, Braden, Laguzaa, & Holman, 1987; Bergstrom, Braden, Kemp, Champagne, & Ruby, 1998). The Braden Scale is composed of six subscales that reflect two major etiologic factors that contribute to pressure ulcer (PrU) development—i.e., the intensity and duration of pressure on skin tissue, and tissue tolerance for pressure.

## The Braden Scale for Predicting Pressure Sore Risk

| | | | | | |
|---|---|---|---|---|---|
| **SENSORY PERCEPTION** ability to respond meaningfully to pressure-related discomfort | **1 Completely Limited** Unresponsive (does not moan, flinch, or grasp) to painful stimuli, due to diminished level of consciousness or sedation OR limited ability to feel pain over most of body surface. | **2 Very Limited** Responds only to painful stimuli. Cannot communicate discomfort except by moaning or restlessness OR has a sensory impairment which limits the ability to feel pain or discomfort over 1/2 of body. | **3 Slightly Limited** Responds to verbal commands, but cannot always communicate discomfort or need to be turned OR has some sensory impairment which limits ability to feel pain or discomfort in 1 or 2 extremities. | **4 No Impairment** Responds to verbal commands. Has no sensory deficit which would limit ability to feel or voice pain or discomfort. | 2 |
| **MOISTURE** degree to which skin is exposed to moisture | **1 Completely Moist** Skin is kept moist almost constantly by perspiration, urine, etc. Dampness is detected every time patient is moved or turned. | **2 Very Moist** Skin is often, but not always moist. Linen must be changed at least once a shift. | **3 Occasionally Moist** Skin is occasionally moist, requiring an extra linen change approximately once a day. | **4 Rarely Moist** Skin is usually dry, linen only requires changing at routine intervals. | 2 |
| **ACTIVITY** degree of physical activity | **1 Bedfast** Confined to bed. | **2 Chairfast** Ability to walk severely limited or non-existent. Cannot bear own weight and/or must be assisted into chair or wheelchair. | **3 Walks Occasionally** Walks occasionally during day, but for very short distances with or without assistance. Spends majority of each shift in bed or chair. | **4 Walks Frequently** Walks outside the room at least twice a day and inside room at least once every 2 hours during waking hours. | 1 |
| **MOBILITY** ability to change and control body position | **1 Completely Immobile** Does not make even slight changes in body or extremity position without assistance. | **2 Very Limited** Makes occasional slight changes in body or extremity position but unable to make frequent or significant changes independently. | **3 Slightly Limited** Makes frequent though slight changes in body or extremity position independently. | **4 No Limitations** Makes major and frequent changes in position without assistance. | 2 |
| **NUTRITION** usual food intake pattern | **1 Very Poor** Never eats a complete meal. Rarely eats more than 1/3 of any food offered. Eats 2 servings or less of protein (meat or dairy products) per day. Takes fluids poorly. Does not take a liquid dietary supplement OR is NPO and/ or maintained on clear liquid or IV's for more than 5 days. | **2 Probably Inadequate** Rarely eats a complete meal and generally eats only about 1/2 of any food offered. Protein intake includes only 3 servings of meat or dairy products per day. Occasionally will take a dietary supplement OR receives less than optimum amount of liquid diet or tube feeding. | **3 Adequate** Eats over half of most meals. Eats a total of 4 servings of protein (meat, dairy products) each day. Occasionally will refuse a meal, but will usually take a supplement if offered OR is on a tube feeding or TPN regimen which probably meets most of nutritional needs. | **4 Excellent** Eats most of every meal. Never refuses a meal. Usually eats a total of 4 or more servings of meat and dairy products. Occasionally eats between meals. Does not require supplementation. | 2 |

*(Continues)*

FIGURE 21-7 **Braden Scale for Predicting Pressure Sore Risk.** (*Source:* Copyright Barbara Braden and Nancy Bergstrom, 1988. Reprinted with permission, documentation added.)

| FRICTION & SHEAR | 1 Problem | 2 Potential Problem | 3 No Apparent Problem | | 1 |
|---|---|---|---|---|---|
| | Requires moderate to maximum assistance in moving. Complete lifting without sliding against sheets is impossible. Frequently slides down in bed or chair, requiring frequent repositioning with maximum assistance. Spasticity, contractures or agitation leads to almost constant friction. | Moves feebly or requires minimum assistance. During a move skin probably slides to some extent against sheets, chair, restraints, or other devices. Maintains relatively good position in chair or bed most of the time but occasionally slides down. | Moves in bed and in chair independently and has sufficient muscle strength to lift up completely during move. Maintains good position in bed or chair at all times. | | |

TOTAL SCORE: **10**

9 or less = very high risk    10–12 = high risk    13–14 = moderate risk    15–18 = at risk    19–23 = not at risk

**\*\* If other major risk factors are present advance to the next level of risk**
*(advanced age, fever, poor dietary intake of protein, diastolic pressure below 60, hemodynamic instability)*

**Total Braden Score of 18 or below is considered predictive for
the development of a pressure ulcer unless preventive measures are taken.**

FIGURE 21-7  (Continued)

The Braden Scale subscales of sensory perception, activity, and mobility relate to the factors of *intensity and duration of pressure* on skin tissue. The Braden Scale subscales of moisture, nutrition, and friction and shear relate to *tissue tolerance for pressure*. The six subscales are rated from 1 (least favorable) to 4 (most favorable) except for friction and shear, which is rated from 1 to 3. Each rating is accompanied by descriptive criteria for assigning the score. When the Braden subscale scores are added together, the total Braden score ranges from 6 to 23. In general, a Braden score of 18 or below in an adult patient is considered to identify a patient to be at risk of developing PrUs.

Despite our best efforts, not all pressure ulcers are avoidable. In 2009, a panel of international experts formulated a consensus statement on Skin Changes At Life's End (SCALE). The panel proposed that, at the end of life, failure of the homeostatic mechanisms that support the skin can occur, resulting in a diminished reserve to handle insults such as minimal pressure. Ten statements were recommended based on input from 69 wound care experts. These statements said that expectations around the patient's end-of-life goals and concerns should be communicated among the members of the multidisciplinary

team and the patient's circle of care. In addition to SCALE, other factors affecting skin changes, skin breakdown, and PrUs need to be considered and addressed. Other recommendations emphasize the need for ongoing assessment of the whole body, because there may be signs and indications in the entire body that relate to skin compromise. The 5 Ps were recommended for determining appropriate intervention strategies, including (1) prevention, (2) prescription (consider treatment), (3) preservation, (4) palliation, and (5) patient preference (Sibbald, Krasner, & Lutz, 2010).

The Palliative Performance Scale Version 2 (PPSv2) (Victoria Hospice Society, 2001) was developed to assess end-of-life performance measures including ambulation, activity, evidence of disease, self-care, intake, and consciousness level. Descriptors are matched with a percentage of performance. The PPSv2 may serve as an adjunct to other existing tools such as the Braden Scale score to communicate the patient's functional level and be useful in patient care decision making.

In 2008, the Centers for Medicare & Medicaid Services (CMS) outlined financial incentives for hospitals to prevent pressure ulcers. This reinforced the need to document pressure ulcers (PrUs) existing on admission and to prevent the development of pressure ulcers during hospitalization. Acute care facilities that fail to document and prevent PrUs will not receive additional money from the CMS to cover the cost of caring for hospital-acquired Stage III or IV PrUs (Lyder & Ayello, 2009).

The prevention and treatment of pressure ulcers continues to be a priority in health care. Pressure ulcer development is considered a major risk to the health and wellness of individuals when the overall goal is to cure an illness, rehabilitate the individual, help the individual live optimally with a chronic illness, or help the patient die in comfort. Therefore, pressure ulcer risk assessment, prevention, and treatment should be priorities of patient care regardless of age, physical condition, or setting.

The IHI website (www.ihi.org) has valuable resources that individuals can download for free, including an updated annotated bibliography, tools to assist health care providers in working to prevent pressure ulcers, and measurement information forms on the process and outcome measures of preventing and treating pressure ulcers. The Wound Ostomy and Continence Nurses Society (WOCN) is also a valuable resource for current evidence-based practice and research related to pressure ulcer management. Information related to prevention and treatment of pressure ulcers is available in print and online for nurses, patients, and families. Wound product information and a global learning center for nurses is available online at www.wocn.org.

## APPLICATION OF THE PDSA CYCLE TO PRESSURE ULCER MANAGEMENT

We will now focus on the prevention and treatment of pressure ulcers in Mrs. Rose, from the opening scenario. She cannot communicate her discomfort except by moaning and facial grimacing. Her right-sided hemiparesis alters her ability to sense pain or pressure. Mrs. Rose is incontinent of urine and has loose stools, necessitating linen changes at least once a shift. She is bedbound and only makes occasional slight changes in body position. She is unable to reposition herself without assistance. Mrs. Rose's appetite is poor. She eats less than half of any food provided. She rarely finishes her dietary supplement.

Additional assessment data includes a low serum albumin of 0.8 gram/dL and an initial pain rating of 8-10/10. Mrs. Rose frequently slides down in bed, requiring frequent repositioning with maximum assistance. Based on this information, the nurse uses the Braden Scale to assess her pressure ulcer risk. See Figure 21-7 for the Braden Scale score for pressure sore risk and Mrs. Rose's actual score. Based on her Braden score of 10, Mrs. Rose is identified as being at high risk for pressure ulcer development. Additionally, in the presence of other major risk factors—including advanced age, advanced cancer, cerebrovascular accident (CVA), an existing stage four pressure ulcer, and hypoalbuminemia— Mrs. Rose would be assessed at being at very high risk for the development of additional pressure ulcers. The Skin Care Protocol was initiated (Figure 21-8). Preventive interventions were identified, implemented, and documented on the Prevention of Skin Breakdown Flow Sheet (Figure 21-9). Assessment of the coccyx pressure ulcer was documented on the Wound Care Flow Sheet (Figure 21-5).

We apply the PDSA Cycle and start with three questions:

1. *What are we trying to accomplish?* While recognizing that some skin changes, including pressure ulcers, at the end of life are unpreventable, the overall goal is to prevent or minimize further skin breakdown and provide comfort care.
2. *How will we know that a change is an improvement?* Skin integrity will be maintained and/or new or progression of existing breakdown minimized.
3. *What change can we make that will result in improvement?* The acute care nurse practitioner (ACNP) will be consulted to evaluate the patient's pressure ulcer. The ACNP will then make recommendations for wound care and additional pressure relief measures. Other members of the health care team will be consulted as appropriate, including nutrition and pastoral care. A multidisciplinary patient care conference will be conducted to coordinate comprehensive palliative care related to changes at the end of life. Specific strategies are discussed below.

## Implementation of the PDSA Cycle, the Planning Phase

Based on our answers to the three PDSA Cycle questions, we held a multidisciplinary patient care conference with the patient and her daughter, hospice staff, nursing staff, the ACNP, medical practitioners, the nutritionist, spiritual care, and the clinical resource manager. All conference members agreed that Mrs. Rose was at high risk for continued skin breakdown, as reflected by her Braden score of 10. Based on the Braden subscale scores, measures were identified to address pressure relief, friction and shear relief, and moisture relief. The Skin Care Protocol (Figure 21-8) would be implemented with a specific focus on comfort and palliative care.

### Doing Phase

The staff nurses implemented the agreed-upon plan of care. The Braden Scale for Pressure Sore Risk was used to assess the patient daily. Preventive interventions for pressure relief measures included placement of a dynamic air mattress, a turning schedule with a 30-degree angle, and elevation of the heels off surfaces, using pillows. Friction and shear measures would include maintaining the head of the bed lower than 30 degrees, using barrier lotions

TITLE: SKIN CARE PROTOCOL

Purpose: To optimize conditions for maintenance of skin integrity & the treatment of open skin for acutely ill patients.

Level: Interdependent (* requiring MD / NP order)

Supportive Data: Signs of infection. Dressing selection considerations. Pressure relief mattress selection. Standard procedures for skin breakdown. Definition of pressure ulcers. Types of partial & full thickness skin breakdown.

PROCEDURE FOR DOCUMENTATION
1. Computerized Nursing Assessment
2. Computerized Wound & Skin Assessment
3. Patient Care Notes
4. Electronic Medical Record

# PRESSURE ULCERS

## Braden Scale Score of 18 or less determines patients at risk for pressure sore development.

### Stage 1 Pressure Ulcer
**Non-blanchable erythema**

Intact skin with non-blanchable redness of a localized area, usually over a bony prominence. Darkly pigmented skin may not have visible blanching; its color may differ from the surrounding area. The area may be painful, firm, soft, warmer, or cooler as compared to adjacent tissue. May be difficult to detect in individuals with dark skin tones. May indicate "at-risk" persons.

### Stage 2 Pressure Ulcer
**Partial thickness skin loss**

Dermis presenting as: a shallow open ulcer with a red-pink wound bed, without slough; an intact or open/ruptured serum-filled or serosanguinous filled blister: shiny or dry shallow ulcer without slough or bruising. *Bruising indicates deep tissue injury.*

### Stage 3 Pressure Ulcer
**Full thickness skin loss**

Subcutaneous (SC) fat may be visible, but bone, tendon, or muscle are not exposed. Slough may be present but does not obscure the depth of the tissue loss. May include undermining or tunneling. Depth varies by anatomical location. Areas of significant adiposity can develop extremely deep Stage 3 pressure ulcers. Bone/tendon is not visible or directly palpable.

### Stage 4 Pressure Ulcer
**Full thickness tissue loss**

Full thickness tissue loss with exposed bone, tendon, or muscle. Slough or eschar may be present. Often includes undermining and tunneling. Depth varies by anatomical location. Can extend into muscle and/or supporting structures (e.g., fascia, tendon, or joint capsule), making osteomyelitis likely to occur. Exposed bone/muscle visible or directly palpable.

Bridge of the nose, ear, occiput, and malleolus do not have (adipose) subcutaneous tissue, and these ulcers can be shallow.

### Unstageable
**Unclassified full thickness skin or tissue loss - depth unknown**

Actual depth of the ulcer is completely obscured by slough (yellow, tan, gray, green, or brown) and/or eschar (tan, brown, or black) in the wound bed. Until enough slough and/or eschar are removed to expose the base of the wound, the true depth cannot be determined; but it will be a Stage 3 or 4. Stable (dry, adherent, intact, without erythema or fluctuance) eschar on the heels serves as "the body's natural (biological) cover" and should not be removed.

### Suspected Deep Tissue Injury (DTI)
**Depth unknown**

Purple or maroon localized area of discolored intact skin or blood-filled blister due to damage of underlying soft tissue from pressure and/or shear. The area may be preceded by tissue that is painful, firm, mushy, boggy, warmer, or cooler as compared to adjacent tissue. DTI may be difficult to detect in individuals with dark skin tones. Evolution may include a thin blister over a dark wound bed. The wound may further evolve and become covered by thin eschar. Evolution may be rapid, exposing additional layers of tissue even with optimal treatment.

## PARTIAL THICKNESS WOUNDS

Abrasion   Skin Tear   Superficial Burn   Tape Burn   Excoriation
Radiation Injury   Maceration   Incontinence-Associated Dermatitis

## FULL THICKNESS WOUNDS

Stage 3 Pressure Ulcer   Stage 4 Pressure Ulcer   Deep Burn   Tumor
Surgical Wound   Arterial Insufficiency   Venous Insufficiency

**Notify MD / NP if signs of infection**

purulent drainage, malodorous drainage, change in color of exudate, elevated body termperature, periwound erythema / edema / pain / fluctuance, and elevated white blood cell count

APPROVAL:


COO & CNO,
Albany Memorial & Samaritan Hospitals

Original:   8/8/89
REVIEW: _____
REVISION:
11/11/92   01/14/98   09/19/06
06/9/93   09/02/02   09/16/08
05/10/95   09/21/04
08/9/95   09/19/06

*(Continues)*

FIGURE 21-8  Skin care protocol. (*Source:* Courtesy of Northeast Health, Samaritan and Albany Memorial Hospitals, Acute Care Division, Troy and Albany, NY.)

*Patients with Braden Score of 18 or less indicates risk for pressure ulcer development.*

## STANDARD PROCEDURES FOR PREVENTION & / or WORSENING OF SKIN BREAKDOWN IN PATIENTS AT RISK FOR PRESSURE SORE DEVELOPMENT

Moisture Relief - KEEP CLEAN & DRY
- apply Barrier Cream to moist skin
- consult MD/NP re: catheter to contain urine & / or stool containment device

Pressure Relief - Heels
- HEELIFT Boot / Traction Boot
- Elevate OFF surfaces using pillows

Pressure Relief - Chair / Recliner
- air cushion (regular, gerichair, bariatric)
- limit time in chair based on care goals
- remind patient to shift own weight, if able

Pressure Relief - Bed
- air mattress (dynamic, rotational, static)
- turning schedule (30-degree angle)

Friction & Shear Relief
- HOB lower than 30 degrees, if condition permits
- apply Barrier Cream to all bony prominences
- use assistive devices to reposition/transfer patient

ICU / CCU Prone Positioner Precautions
- off-load pressure over chin, jaw, ribs, hips, knees & toes
- administer oral care to avoid skin maceration from saliva

## PATIENT SELECTION CONSIDERATIONS for PRESSURE RELIEF MATTRESS

*All ELDERLY patients AT RISK FOR PRESSURE ULCERS should be considered for pressure relief mattresses/cushions.*

Consider SizeWise Pulsate for patients with:
- Stage 3 or 4 Pressure Ulcers.
- sensory perception deficits (e.g., hemiparesis, dementia, neuropathy).
- activity deficits (e.g., bedbound - chronic or acute, breathing problems, tube feeding dependence).
- mobility deficits (e.g., CVA, s/p joint replacement surgery).

Consider SizeWise Mighty Air for obese patients with:
- complex care needs
- risk for pressure ulcer development.

Consider Static Air Mattress Overlay for patients with:
- less-complex care needs

## DRESSING SELECTION CONSIDERATIONS

Condition of wound bed.    Amount & type of wound drainage.    Location of wound.    Pain.

&

GOALS OF WOUND CARE - Healing. Elimination of infection. Prevention of wound deterioration. Comfort. Ease of patient & nursing care.

**Clean Wound Bed**
* bid or qshift Bacitracin Ointment Dsg
* daily Xeroform or Restore Foam or Tegaderm Dsg
* bid or qshift NS or 1/4st Acetic Acid wet-to-dry Dsg

**Lower Extremity Cellulitis**
* bid soap & tap water washes
If no wounds: Leave open to air
If wounds: * bid 1/4st Acetic Acid wet-to-dry Dsgs
* bid Silvadene Cream Dsgs

**Debridement**
Consult Surgery or NP

**Malodorous Wound**
* Cleocin Spray Solution with Gauze Dsg changes

**Burn or "Burn-like" Wound**
* bid Silvadene Cream Dsgs

**Devitalized Slough or Necrotic Wound Bed**
*If Wound Cx Pseudomonas &/or yellow-green foul wound drainage:*
* bid or qshift 1/4st Acetic Acid wet-to-dry Dsg

**Excoriated Skin**
* qshift/prn Aquaphor Ointment

If pain:
* qshift/prn Aquaphor / Lidocaine Ointment

FIGURE 21-8 (Continued)

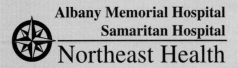

**Albany Memorial Hospital**
**Samaritan Hospital**
**Northeast Health**

# Room: 6006-1

Patient: Mrs. Rose

## PREVENTION OF SKIN BREAKDOWN
## FLOW SHEET

*Initiate if Braden Score for Pressure Ulcer Risk = 18 or less*

**Documentation Directions:**
Enter *START* date, your initials & indicate selected preventive skin care measures by placing a check mark (✔) in the spaces provided.
***Refer to the regimen daily & implement!*** Re-document ONLY if regimen *changes!*

| | | | START / CHANGE DATE | 5/20 | | | | | |
|---|---|---|---|---|---|---|---|---|---|
| | | | INITIAL | KR | | | | | COMMENTS |
| **PRESSURE RELIEF** | **Bed** | Air Mattress *Static Dynamic Rotational* | | D | | | | | |
| | | Turning Schedule (30 degree angle) | | ✔ | | | | | |
| | **Chair** | Air Cushion *Regular Bariatric* | | | | | | | |
| | | Limit sitting *based on pt's needs / goals* | | | | | | | |
| | | Remind pt to shift own weight *frequently* | | | | | | | |
| | **Heels** | HeeLift Boot / Traction Boot | | | | | | | |
| | | Elevate OFF surfaces using pillows | | ✔ | | | | | |
| **FRICTION & SHEAR RELIEF** | | HOB lower than 30 degrees | | ✔ | | | | | |
| | | Barrier lotions to skin folds / bony prominences routinely | | ✔ | | | | | |
| | | Assistive device to reposition/transfer pt | | ✔ | | | | | |
| **MOISTURE RELIEF** | | Keep CLEAN and DRY | | ✔ | | | | | |
| | | Apply Barrier Cream to moist skin | | ✔ | | | | | |
| | | Consult MD/NP re: catheter | | ✔ | | | | | |
| **PRONE POSITIONER PRECAUTIONS** | | Off-load pressure over chin, jaw, ribs, hips, knees & toes | | | | | | | |
| | | Administer oral care to avoid skin maceration from saliva | | | | | | | |

**FIGURE 21-9** **Prevention of skin breakdown flow sheet.** (*Source:* Courtesy of Northeast Health, Samaritan and Albany Memorial Hospitals, Acute Care Division, Troy and Albany, NY, with documentation.)

routinely to skin folds and bony prominences, and using assistive devices such as lift sheets to reposition the patient. Moisture relief measures would include keeping the skin clean and dry, applying barrier lotions to moist skin, and consideration of the use of collection devices for urine and stool as indicated for incontinence and comfort. All interventions were documented as illustrated on the Prevention of Skin Breakdown Flow Sheet (Figure 21-9) and on the Wound Care Flow Sheet (Figure 21-5).

The dietician was also consulted to assess Mrs. Rose's nutritional status and made dietary recommendations, which were implemented and included oral supplements. Mrs. Rose received assistance with feeding, and the head of the bed was elevated only during mealtime. The clinical resource manager met with the patient, family, and staff to further determine care options.

### Studying Phase

Five days after the proposed plan of care was implemented, Mrs. Rose's Braden score remained at 10. However there were no new signs of skin breakdown. Pressure relief measures included maintaining the dynamic mattress with heel elevation. Friction and shear and moisture relief were also continued with no modifications.

### Acting Phase

The nursing and multidisciplinary team agreed to continue the selected preventive measures. The team agreed to continue the PDSA Cycle and monitor any skin changes, as well as recommend appropriate evidence-based practice such as the Skin Changes At Life's End (SCALE) consensus document, mentioned earlier.

## REAL WORLD INTERVIEW

I was one of the nurses who cared for Mrs. Rose during her stay. She was very sick when she was admitted. Her daughter was upset that she was unable to care for her mother at home any longer. She relayed that her mother was in a great deal of pain that seemed to be related to a bedsore on the patient's bottom. Mrs. Rose was not able to talk. She did nod her head yes and no appropriately to simple questions, i.e., related to pain rating. She did indicate that she was in pain, but was unable to tell us the location. Her daughter said that the patient was taking thin liquids at home, but no solid foods. She had not had any problems with swallowing. The patient took only water at first. The nurses tried giving Mrs. Rose some soft foods, which she seemed to want. However, the patient's daughter was adamant that the patient not eat any solid foods. She allowed only Cream of Wheat, ice cream, and Ensure supplements.

Mrs. Rose was evaluated by physical and occupational therapy. However, physical and occupational therapy were not indicated based on Mrs. Rose's poor physical condition.

The nurses assessed Mrs. Rose as being at very high risk for pressure ulcer development based on her inability to call for help or verbalize discomfort, her bedbound status, her very limited mobility in bed with no ability to turn or reposition herself, urinary and fecal incontinence, and sliding down in bed from time to time. A low air loss mattress was ordered for Mrs. Rose on the day of her admission.

Our first examination of Mrs. Rose was remarkable for a necrotic coccyx ulcer that had a large amount of yellow-gray stringy slough and a large amount of purulent foul-smelling drainage. The surrounding skin had a dark purplish color. Mrs. Rose was very thin and had multiple bony prominences. I could feel bone at the base of her coccyx ulcer, but I could not actually see the bone. Mrs. Rose was incontinent of stool at the time. Stool was found on the ulcer. When we washed the perirectal area and the ulcer, Mrs. Rose grimaced and cried out in pain. Her daughter was present and was very upset about the wound and the patient's discomfort. The nurses and doctor consulted the wound care nurse practitioner(NP) to evaluate the ulcer and recommend a plan of care. MS Contin was ordered and given every 12 hours. A short-acting opioid medication was ordered to be given as needed every two to four hours.

The wound care NP saw the patient that afternoon. She was not able to stage the ulcer because of the slough, but thought that the ulcer was likely Stage IV. The patient's daughter was present. The NP asked a number of questions about the patient's condition and care at home, as well as

*(Continues)*

## REAL WORLD INTERVIEW (Continued)

specific details about the development of the coccyx ulcer. The NP reassured the patient's daughter that she did not cause the wound. She explained that because the patient had been sick for a long time with her cancer diagnosis and her recent deterioration, it was not unusual for these types of ulcers to develop. The patient's daughter understood this. She wondered if the wound was infected. The NP advised her that she did not think the wound was infected. She thought that the patient's pain was from the inflammation surrounding the ulcer and also because the ulcer extended to the bone. The NP explained that that it was likely that there were bacteria in the wound because it was frequently contaminated with stool due to the location close to the anus. The daughter asked if there was anything we could do about the ulcer and expressed particular concern about the patient's discomfort and also the odor. The NP explained that gently debriding the loose slough would help decrease the odor. Dressing changes could be started to help gently remove the remaining slough and absorb the drainage. Hopefully, the patient's pain would lessen over time as the inflammation subsided. The patient's daughter understood all the information and agreed to the proposed plan of care. An IV was started for intravenous fluids and pain medication, as needed. The NP then proceeded to gently debride a large amount of the loose, stringy slough. I held the patient on her side, and the patient did not complain of any pain during this procedure. A culture of the wound was taken. Dressing changes were then started with wet-to-dry quarter-strength Acetic Acid gauze dressings. The dressings were ordered to be changed two times a day. The patient would be premedicated with short-acting morphine 30 minutes prior to the dressing change. The NP also ordered Cleocin spray solution to be applied to the outer gauze dressing to decrease the odor from the wound. The patient's daughter was present for this care. The patient was continued on the pressure relief mattress and turned at least every two hours. We kept the head of the bed at 30 degrees or less except for times of oral intake.

The NP wrote specific orders for the wound care and pressure relief measures. After two days, the NP returned to evaluate the wound. The nurses thought that the ulcer was getting worse. The nurses relayed that the patient still had pain with dressing changes but that this seemed less. The NP agreed with the nurses that the wound did look larger, but she had expected this as the necrotic tissue was removed. The NP explained that the condition of the wound had actually improved. The odor was gone. The NP talked with the patient's daughter and other family members and relayed her impression. The daughter was relieved that our care was helping. It was decided to continue with the same wound care regimen until the slough was completely removed, and then another type of dressing would be considered. The patient was starting to eat about half of her Cream of Wheat and Ensure. Her pain had improved to the point that she was able to tolerate sitting at a 45-degree angle to eat.

The nurses, the NP, and the physicians worked as a team in caring for Mrs. Rose. Her comfort was the central goal of her care, and this improved as time went on. The NP assistance with the patient's wound care allowed for a consistent approach, and I think this consistency made a difference in improving the wound and decreasing the patient's pain. Also, the NP interaction with the patient's daughter helped to alleviate her anxiety about the wound and allowed her to understand that it was not realistic to expect the ulcer to heal. The nurses caring for Mrs. Rose were always present for these discussions, and this helped our understanding of the rationale for the wound care approach and goals.

**Jami Nazzaro**
Staff Nurse
Troy, New York

# EVIDENCE FROM THE LITERATURE

**Citation:** Woo, K. Y., & Sibbald, R. G. (2009). The improvement of wound-associated pain and healing trajectory with a comprehensive foot and leg ulcer care model. *Journal of Wound, Ostomy, and Continence Nursing 36*(2), 184–191.

**Discussion:** The purpose of this study was to validate an organized pain management approach using a Wound Associated Pain model in individuals with chronic leg and foot ulcers. The key components of the Wound Associated Pain (WAP) model include patient-centered concerns, wound etiology, and local wound factors. The study design included a prospective cohort over a four-week period, documenting the pain rating in individuals with chronic wounds. A total of 111 participants with chronic leg and foot ulcers from community and ambulatory wound care clinics in Canada were recruited. Participants (*N* = 102) were reassessed four weeks later. Pain was measured using the Numeric Rating Scale (NRS). Progress toward wound healing was assessed by estimating changes in wound surface area over time. Data was analyzed using the Statistical Package for the Social Sciences (SPSS) software program (version 15). Data was analyzed using descriptive statistics and paired *T* tests. A total of 78 leg ulcers and 96 foot ulcers were initially evaluated, and 66 leg ulcers and 85 foot ulcers were evaluated four weeks later. Sixty percent of the participants were male, and their range of age was 33 to 95 years (*M* = 66 years). At the beginning of the study, approximately half (45.6%) of the participants reported severe pain (NRS scores greater than or equal to 7). The average level of pain rating after four weeks was reduced from 6.3 to 2.8 (*P* = < .001). The average healing rate was 0.39 centimeters squared per week, and the average relative reduction in size was 59.36% (*T* = 2.31; *P* = .023). Approximately 40% of the leg ulcers closed during data collection, and approximately 30% of the foot ulcers healed.

**Implications for Practice:** The WAP model is a useful approach that focuses on patient-centered concerns, wound causation, and local wound factors. This model is useful in conducting a comprehensive patient assessment in individuals with chronic wounds and wound-related pain. Further studies across different patient populations and cultures would add in validating the reliability and validity of this model.

# CASE STUDY 21-1

Mr. Jones is a 52-year-old African American patient with metastatic rectal cancer. He was admitted to the hospital with intractable abdominal pain rated 9/10 on admission. He is alert and oriented x3 and can respond to verbal commands. He is able to verbalize and express his needs. He is incontinent of urine at least four to five times a day. Because of the intensity of his abdominal pain, he is bedbound and has been unable to get out of bed for the past week. He is reluctant to reposition himself and spends most of his time lying on his back, with the head of the bed elevated at least at 45 degrees. He needs total assistance with physical care and all other activities of daily living. The patient's appetite is very poor, and he rarely eats more than 25% of any food offered. He has refused oral dietary supplements. Using this information, calculate the Braden subscale scores and the total Braden Scale score (Figure 21-7).

## KEY CONCEPTS

■ Evidence-based practice (EBP) represents a multidisciplinary approach to the utilization of current research findings to guide the development of appropriate strategies to deliver quality, cost-effective care.

■ Outcomes research provides evidence about benefits, risks, and results of treatment so that individuals can make informed decisions and choices to improve their quality of life.

■ The American Nurses Association (ANA) has been an active advocate of outcomes evaluation.

■ EBP provides a "static" snapshot of a conclusion based on previous clinical trials about a condition or situation, but the clinician must still make a clinical judgment about an individual, considering his or her unique characteristics (such as gender, age, clinical history, socioeconomic status, support system, ethical concerns, and the illness experience).

■ The University of Colorado Hospital model is an example of a multidisciplinary EBP model for using different sources of information to change or support your practice.

■ The Model for Improvement (PDSA) can be applied to a system or an individual.

## KEY TERMS

benchmarking
comfort

evidence-based nursing
practice (EBNP)

evidence-based
practice (EBP)

palliative care
practice guideline

## REVIEW QUESTIONS

1. To participate effectively in the use of EBNP, nurses should do which of the following? Select all that apply.
   _____ A. Participate in the development, use, and evaluation of practice guidelines
   _____ B. Read and analyze outcomes of research studies
   _____ C. Involve themselves in everyday patient care and nursing practice
   _____ D. Attend professional nursing conferences on evidence based practice
   _____ E. Utilize evidence-based outcomes to provide cost-effective care

2. Why is it important for nurses to recognize and value patient-focused outcome indicators?
   A. To achieve safe, quality, cost-effective care for patients in daily practice
   B. To realize that individual nursing practice styles directly affect the rates at which patients recover
   C. To prevent development of unnecessary complications and injury
   D. All of the above

3. Which of the following are examples of national evidence-based practice guidelines?
   A. Hospital policy on how to staff a nursing unit
   B. AHRQ pressure ulcer treatment guidelines
   C. Hospital procedure on how to insert a catheter
   D. JC accreditation standards

4. Which organization released two landmark reports on health care safety and quality, entitled *To Err Is Human* and *Crossing the Quality Chasm*?
   A. The Leapfrog Group
   B. Institute of Medicine
   C. Agency for Healthcare Research and Quality
   D. Joint Commission

5. Which of the following statements are true about the PDSA Cycle? Select all that apply.
   _____ A. It provides a framework to think about how to apply evidence-based knowledge, facilitating ability to make changes in individual patient care.
   _____ B. It is a process that can be applied to a system or an individual.
   _____ C. It is a process that can be used to test hunches, theories, and ideas about patient care, resulting in actual changes that result in improvement.
   _____ D. It is a process only utilized for individual patients.
   _____ E. It is a multidisciplinary process involving all members of the health care team.

6. Which of the following is an example of a component of the planning phase of the PDSA Cycle?
   A. Revision of a current pain documentation flow sheet
   B. Implementation of a skin care protocol for bariatric patients
   C. Initiation of a multidisciplinary conference to discuss reduction in patient falls
   D. Evaluation of patient progress after implementation of a new pressure relief device

7. A source of data used to modify or adapt patient care may include which of the following?
   A. A patient's concerns, preferences, and changing clinical condition
   B. Study results of a pressure ulcer prevalence survey
   C. The Web site for the Institute for Healthcare Improvement (IHI)
   D. All of the above

8. Which organization revised its Social Policy Statement to include statements about the evaluation of the effectiveness of care in relation to identified outcomes and the use of evidence to improve care?
   A. JC
   B. IHI
   C. ANA
   D. AHRQ

9. The continuous process of measuring products, service, and practices against the toughest competitors or those customers recognized as industry leaders is referred to as which of the following?
   A. Process evaluation
   B. Benchmarking
   C. Quality improvement
   D. Cost analysis

10. The PDSA Cycle by Langley, et al., begins with what question?
    A. How will we know that change is an improvement?
    B. What change can we make that will result in an improvement?
    C. What have we learned from this process up to this point?
    D. What are we trying to accomplish?

## REVIEW ACTIVITIES

1. Review the EPUAP and NPUAP 2009 *Prevention and Treatment of Pressure Ulcers: Quick Reference Guide* section related to skin assessment. Review specific recommendations related to assessment of the elderly and implementation of evidence-based repositioning techniques and pressure relief devices. Compare these recommendations to current practice in your hospital, and review a research study or clinical practice article that would support evidence-based practice changes.

2. The University of Colorado Hospital model is one example of an evidence-based multidisciplinary practice model. It presents a framework for thinking about how you use different sources of information to change or support your practice. Select a situation from your clinical practice and apply the model. For example, if you were caring for an elderly patient admitted with a hip fracture sustained during a fall at home, what benchmarking data would you review to compare the patient's length of stay with that of other patients with fractured hips? What standards of care would be used? Are these institutional specific or do they also incorporate any specific outside organizations' guidelines?

3. Another model for achieving improvement is the PDSA Cycle. Based on what you have read, consider the three questions:
   A. What are we trying to accomplish?
   B. How will we know that a change is an improvement?
   C. What changes can we make that will result in an improvement?

   Apply the PDSA Cycle to a situation that involves patient safety.

4. Imagine that you want to initiate a pain management team in your setting. Search one of the Web sites listed in the chapter that addresses evidence-based practice related to pain management. *HINT:* Start with the IHI to review what other organizations are doing related to pain management and reducing adverse drug events. See Exploring the Web.

# EXPLORING THE WEB

Utilizing one of the Web sites listed, explore current evidence-based practice updates and/or specific clinical issues related to your area of practice (e.g., oncology, palliative care, wound care, etc.). Compare current practice guidelines utilized in your facility with current evidence-based practice guidelines, and develop recommendations to share with your institution.

## Evidence-Based Practice

- AHRQ Evidence-Based Practice: http://www.ahrq.gov
- Institute for Healthcare Improvement: http://www.ihi.org
- The Cochrane Library: http://www.cochrane.org

## Pressure Ulcers

- European Pressure Ulcer Advisory Panel: http://www.epuap.org

- National Pressure Ulcer Advisory Panel: http://www.npuap.org
- Wound Ostomy and Continence Nurses Society: http://www.wocn.org

## Pain/Comfort Care

- International Association for Hospice and Palliative Care: http://www.hospicecare.com
- International Association for the Study of Pain: http://www.iasp-pain.org
- Kolkaba, K. Comfort Theory Web site: http://www.thecomfortline.com

# REFERENCES

Agency for Healthcare Policy and Research. (2006). *Hospitalizations related to pressure ulcers among adults 18 years and older.* Rockville, MD: Author.

Agency for Healthcare Policy and Research. (2008). *Pressure ulcers increasing among hospital patients.* Rockville, MD: Author.

Alvarez, O. Kalinski, & C. Nusbaum, J. (2007). Incorporating wound healing strategies to improve palliation (symptom management) in patients with chronic wounds. *Journal Palliative Medicine, 10*(5), 1161–1189.

American Nurses Association. (2003). Social Policy Statement. Washington, DC: Author.

American Pain Society, Quality of Care Committee. (1995). Quality improvement guidelines for the treatment of acute pain and cancer pain. *Journal of the American Medical Association, 23,* 1874–1880.

American Pain Society, Quality of Care Task Force. (2005). American Pain Society recommendations for improving the quality of acute and cancer pain management. *Archives of Internal Medicine, 165,* 1574–1580.

Benner, P. E., Hooper-Kyriakidis, P., Hooper, P. L., Stannard, D., & Eoyang, T. (1998). *Clinical wisdom and interventions in critical care: A thinking-in-action approach.* Philadelphia: Saunders.

Bergstrom, N., Braden, B. J., Kemp, M., Champagne, M., Ruby, E. (1998). Predicting pressure ulcer risk: A multisite study of the predictive validity of the Braden Scale. *Nursing Research, 47,* 261–269.

Bergstrom, N., Braden, B. J., Laguzza, A., & Holman, V. (1987). The Braden Scale for Predicting Pressure Sore Risk. *Nursing Research, 36*(4), 205–210.

Berlowitz, D. R., Ratliff, C., Cuddigan, J., Rodeheaver, G. T., & the National Pressure Ulcer Advisory Panel. (2005). The PUSH Tool: A survey to determine its perceived usefulness. *Advances in Skin & Wound Care, 18*(9), 480–483.

Berwick, D. M., Calkins, D. R., McCannon, B. A., & Hackbarth, A. D. (2006). The 100,000 Lives Campaign: Setting a goal and a deadline for improving health care quality. *Journal of the American Medical Association, 295*(3), 324–327.

Centers for Medicare & Medicaid Services (CMS). (2008). Educational resources. http://www.cms.hhs.gov/HospitalAcqCond/07_EducationalResources.asp#TopOf Page. Accessed July 9, 2010.

European Pressure Ulcer Advisory Panel and National Pressure Ulcer Advisory Panel. (2009). *Prevention and treatment of pressure ulcers: Quick reference guide.* Washington DC: National Pressure Ulcer Advisory Panel.

Fleck, C. (2005) Ethical wound management for the palliative patient. *Extended Care Product News, 100*(4), 38–46.

Goode, C. J., & Piedalue, F. (1999). Evidence-based clinical practice. *Journal of Nursing Administration, 29,* 15–21.

Goode, C. J, Tanaka, D. J., Krugman, M., O'Connor, P. A., Bailey, C., & Deutchman, M. (2000). Outcomes from use of an evidence-based practice guideline. *Nursing Economic$, 18,* 202–207.

Harris, C., Bates-Jensen, B., Parslow, N., Raizman, R., Singh, M. & Ketchen, R. (2010). Bates-Jensen Wound Assessment Tool. *Journal of Wound, Ostomy, and Continence Nursing, 37*(3), 253–259.

Healthy People 2020 Initiative. http://www.healthypeople.gov/hp2020/Objectives/ViewObjective. Accessed June 30, 2010.

Healing Probability Assessment Tool. http://www.frailcare.org/images/Palliative%20Wound%20Care.pdf. Accessed November 6, 2010.

Ingersoll, G. L. (2000). Evidence-based nursing: What it is, and what it isn't. *Nursing Outlook, 48,* 151–152.

Institute for Healthcare Improvement (IHI). (2004). Protect 5 million lives from harm. www.ihi.org/IHI/Programs/Campaign/. Accessed June 22, 2010.

Institute for Healthcare Improvement. (IHI). (2010). Retrieved June 22, 2010, from http://www.ihi.org

Institute of Medicine (IOM). (1999). *To err is human.* Washington, DC: National Academies Press.

Institute of Medicine (IOM). (2001). *Crossing the quality chasm: A new health system for the 21st century.* Washington, D.C.: National Academies Press.

International Association for Hospice and Palliative Care. (2009). Assessment and research tools. http://www.hospicecare.com/resources/pain-research.htm. Accessed June 30, 2010.

Joint Commission. (2009). *Comprehensive accreditation manual for hospitals: The official handbook (CAMH).* Oakbrook Terrace, IL.

Kolcaba, K. (1994). A theory of comfort for nursing. *Journal of Advanced Nursing, 19,* 1178–1184.

Kolcaba, K. (2003). *Comfort theory and practice.* New York, NY: Springer Publishers.

Langley, G. J., Nolan, K. M., Nolan, T. W., Norman, C. L., & Provost, L. P. (1996). *The improvement guide: A practical approach to enhancing organizational performance.* San Francisco: Jossey-Bass.

Lee K., Ennis W., & Dunn, G. (2007). Surgical palliative care of advanced wounds. *American Journal of Hospice and Palliative Medicine, 24*(2), 154–160.

Letizia, M., Uebelhor, J., & Paddack, E. (2010). Providing palliative care to seriously ill patients with nonhealing wounds. *Journal of Wound, Ostomy, and Continence Nursing, 37*(3), 277–282.

Lyder, C. H, & Ayello, E. A. (2009). Annual checkup: The CMS pressure ulcer present-on-admission indicator. *Advances in Skin & Wound Care, 22,* 476–486.

McDonald A., Lesage P. (2006). Palliative management of pressure ulcers and malignant wounds in patients with advanced illness. *Journal of Palliative Medicine, 9*(2), 285–296.

Orsted, H. L., Rosenthal, S., & Woodbury, G. M. (2009). Pressure ulcer awareness and prevention program: A quality improvement program through the Canadian Association of Wound Care. *Journal of Wound, Ostomy, and Continence Nursing, 36*(2), 178–183.

Quality and Safety Education for Nurses: Quality/Safety Competencies. www.qsen.org/overview.php. Accessed June 22, 2010.

Quality and Safety Education for Nurses(QSEN). (2011). Retrieved April 15, 2011 from http://www.qsen.org/.

Quality Net. (2011). The Surgical Care Improvement Project (SCIP). Retreived April 15, 2011 from http://www.qualitynet.org/dcs/ContentServer?c=MQParents&pagename=Medqic%2FContent%2FParentShellTemplate&cid=1137346750659&parentName=TopicCat.

Sackett, D. L., Rosenberg, W. M., Gray, J. A., Haynes, R. B., & Richardson, W. S. (1996). Evidence-based medicine: What it is and what it isn't. *British Medical Journal, 312*(7023), 71–72.

Sibbald, R. G, Krasner, D. L, & Lutz, J. (2010). SCALE: Skin Changes at Life's End: Final consensus statement: October 1, 2009. *Advances in Skin & Wound Care, 23,* 225–36.

Victoria Hospice Society. (2001). *Palliative Performance Scale Version 2 (PPSv2).* Victoria, BC: Author.

Woo, K. Y., & Sibbald, R. G. (2009). The improvement of wound-associated pain and healing trajectory with a comprehensive foot and leg ulcer care model. *Journal of Wound, Ostomy, and Continence Nursing 36*(2), 184–191.

World Health Organization (2010). WHO's Pain Relief Ladder. Retrieved April 15, 2011 from http://www.who.int/cancer/palliative/painladder/en/.

# SUGGESTED READINGS

American Academy of Pain Management. AAPM facts and figures on pain. www.painmed.org/patient/facts.html. Accessed August 2, 2010.

American Professional Wound Care Association. (2010). SELECT: Evaluation and implementation of clinical practice guidelines: A guidance document from the American Professional Wound Care Association. *Advances in Skin & Wound Care, 23*(4), 161–168.

Arroyo-Novoa, C. M., Figueroa-Ramos, M. I., Miaskowski, C., Padilla, G., Stotts, N., & Puntillo, K. A. (2009). Acute wound pain: Gaining a better understanding. *Advances in Skin & Wound Care, 22*(8), 373–382.

Ayello, E. A., & Lyder, H. (2008). A new era of pressure ulcer accountability in acute care. *Advances in Skin & Wound Care, 21*(3), 134–140.

Baranoski, S. & Ayello, E.A. (2006). Wound Care Essentials. Lippincott, Williams, & Wilkins.

Bohmer, R. M. (2010). Fixing health care on the front lines. *Harvard Business Review, 88*(4), 62–69.

Carter, M. J. (2010). Evidence-based medicine: An overview of key concepts. *Ostomy Wound Management, 56*(4), 68–85.

Hiser, B., Rochette, J., Philbin, S., Lowerhouse, N., TerBurgh, C., & Pietsch, C. (2006). Implementing a pressure ulcer prevention program and enhancing the role of the CWOCN: Impact on outcomes. *Ostomy Wound Management, 52*(2), 48–59.

O'Connor, A. M., Wennberg, J. E., Legare, F., Llewellyn-Thomas, H. A., Moulton, B.W., Sepucha, K. R., et al. (2007). Toward the "tipping point": Decision aids and informed patient choice. *Health Affairs, 26*(3), 716–725.

Pieper, B., Langerno, D., & Cuddigan, J. (2009). Pressure ulcer pain: A systematic literature review and National Pressure Ulcer Advisory Panel White Paper. *Ostomy Wound Management, 55*(2), 16–31.

Sussman, C. (2008). Preventing and modulating learned wound pain. *Ostomy Wound Management, 54*(11), 38–47.

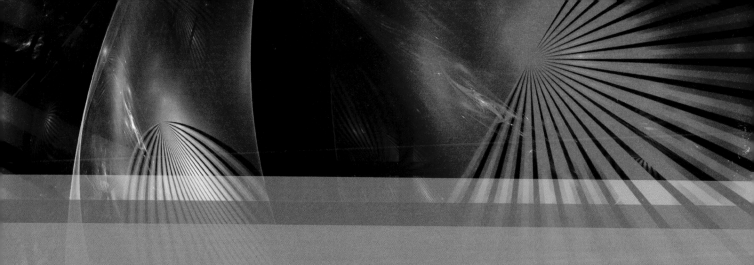

# CHAPTER 22

# Decision Making and Critical Thinking

**Sharon Little-Stoetzel, RN, MS;**
**Barbara K. Fane, MS, RN, APRN-BC**

*When you have to make a choice and don't make it, that is in itself a choice.*

(William James)

## OBJECTIVES

Upon completion of this chapter, the reader should be able to:

1. Apply decision making to clinical situations.
2. Explain how problem solving, critical thinking, reflective thinking, and intuitive thinking relate to decision making.
3. Apply decision-making tools and technology to nursing care.
4. Facilitate group decision making using various techniques.
5. Examine limitations to effective decision making.
6. Apply strategies to strengthen the nurse's role in decision making for patients.

Delmar/Cengage Learning

*You are a staff nurse in the medical intensive care unit in your hospital. The nurse manager has requested that you be a task force committee member for determining a new method of scheduling to be implemented in the unit. Currently, the manager writes the schedule after staff members have submitted their requests for days off for that time period. The manager has received consistent feedback that the staff would like to try a new method of scheduling, but staff have been unable to come to a decision as to what new method they should try. The task force consists of three RNs from the unit in addition to you. There are two nurses from night shift and two nurses from day shift. One day shift nurse has been on the unit for five years and is the chairperson of the committee.*

*What should be the first step of the task force?*

*Can the decision-making process help the group solve the situation?*

Rapid changes in the health care environment have expanded the decision-making role of the nurse. Because the health care market is more competitive than before, decision making and critical thinking by nurses now tend to be necessary skills for the agency's survival. Additionally, stringent budgets require that nurse managers and staff alike do more with less. Patient care is more complex as acuity rises. With patients being discharged from acute care institutions earlier, effective decisions regarding treatment must be made quickly. Unfortunately, many practicing nurses feel they do not have the time, access, or expertise needed to search and analyze the literature to answer clinical questions in a timely manner (Prevost, 2006).

Critical thinking is essential when making decisions and solving problems. This chapter explores the decision-making process and how it relates to critical thinking, reflective thinking, intuitive thinking, and problem solving. Application of decision-making models to clinical nursing decisions is presented. The chapter examines advantages and limitations to group decision making as well as the use of technology in decision making. Finally, it discusses limitations to effective decision making and strategies for improving the nurses role in decision making for patients.

# DECISION MAKING

In everyday practice, nurses make decisions about patient care. As the nurse gains experience in clinical practice, decision making will become somewhat automatic in certain circumstances, but other decisions will remain complex. DeLaune and Ladner (2011) define **decision making** as "considering and selecting interventions from a repertoire of actions that facilitate the achievement of a desired outcome" (p. 89). Critical, reflective, and intuitive thinking may be used during the decision-making process, as illustrated in Figure 22-1.

Although decisions are unique to different situations, the same decision-making process can be applied to most all situations. The decision-making process consists of five steps (Table 22-1):

*Step 1:* Identify the need for a decision.

*Step 2:* Determine the goal or outcome desired.

*Step 3:* Identify many alternatives, i.e., personal, financial, political, etc., along with the benefits and consequences of each alternative.

*Step 4:* Make the decision.

*Step 5:* Evaluate the decision.

Note that making clinical decisions based on best evidence, either from the research literature or clinical expertise, improves the quality of care and patients' quality of life (Evidence-Based Practice Resource Center, 2006). In the following clinical application, the decision-making process is applied to a clinical situation.

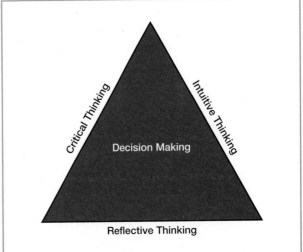

**FIGURE 22-1** Critical thinking, intuitive thinking, and reflective thinking are incorporated throughout the decision-making process.

**TABLE 22-1  The Decision-Making Process**

1. Identify the need for a decision.

2. Determine the goal or outcome desired.

3. Identify many alternatives. Remember, one choice is to do nothing.

   a. Identify the consequences of each alternative, i.e., personal, financial, political, etc.

   b. Identify the benefits of each alternative, i.e., personal, financial, political, etc.

4. Make the decision.

5. Evaluate the decision.

# CLINICAL APPLICATION

You are the night shift nurse caring for Mr. Eli. In the morning, Mr. Eli is scheduled for a permanent pacemaker insertion due to third degree heart block. Hospital policy states that no visitor may stay all night with a patient unless that patient is very critically ill. Mr. and Mrs. Eli are both requesting that Mrs. Eli stay in a chair beside Mr. Eli's bed, because both are anxious about his upcoming procedure. Use your decision-making skills to help you decide what to do.

**Step 1:** Identify the need for a decision. Should you let Mrs. Eli spend the night? Consider all the information (hospital policy, the patient's wishes, anxiety level, and so on).

**Step 2:** Determine the goal or outcome. Questions to consider include the following: Can an exception be made to hospital policy? Is the goal to alleviate Mr. and Mrs. Eli's anxiety? Will Mr. Eli's level of anxiety affect the outcome of the surgery? Will Mr. and Mrs. Eli be satisfied customers?

**Step 3:** Identify many alternatives, i.e., personal, financial, political, etc., and the benefits and consequences of each. If you enforce hospital policy, the benefits are that all patients are treated equally and the written policy supports the decision. The consequences are that Mr. and Mrs. Eli's anxiety level increases, perhaps adversely affecting the outcome of his surgery, and they will not be advocates for the health center. The other alternative is to allow Mrs. Eli to stay all night. The benefits are that Mr. and Mrs. Eli's level of anxiety will decrease and they will be satisfied customers. The consequence is that

a precedent is set that may make it difficult to enforce the existing hospital policy.

**Step 4:** Make the decision. Consider all alternatives and the benefits and consequences of each, and then implement the decision. The decision was made to let Mrs. Eli spend the night with her husband. The nurse determined that the couple's anxiety level in relation to the outcome of the surgery was the most important consideration and the goal in this scenario.

**Step 5:** Evaluate the decision. Was the goal achieved? The nurse must evaluate the decision that was implemented. The decision to allow Mrs. Eli to spend the night was a good decision. It allowed Mr. Eli to get a good night's rest and also decreased his overall anxiety. Decreasing a patient's anxiety prior to a surgery is an important nursing intervention. This decision might lead to a change in the current policy, allowing for more customer satisfaction.

From the beginning of their careers, new graduate nurses are faced with the responsibility of making decisions regarding patient care. For the beginning nurse, it is common to have more questions than answers. When nurses are faced with a difficult clinical decision, they should get others close to the situation involved. These may include other RNs on the unit or house supervisors. Experienced nurses have a wealth of information regarding clinical decision making. New nurses should tap into that valuable resource. Depending on the situation, recognize that you also have knowledge and intuition that are valuable. With more experience comes greater trust in your decision making.

# CRITICAL THINKING

What does it mean to be a critical thinker? Paul (1992) defines **critical thinking** as "thinking about your thinking while you're thinking in order to make your thinking better" (p. 7). Paul points out that there are many accurate definitions of critical thinking, and most are consistent with each other. From a nursing perspective, being a critical thinker means using a complex pattern of thought. It requires that the nurse possess a good body of knowledge and the ability to deliberately choose among evidence-based alternatives (Tanner, 2006). A good critical thinker is able to examine decisions from all sides and takes into account varying points of view. A good critical thinker does not say "We've always done it this way" and refuse to consider alternate ways. The critical thinker generates new ideas and alternatives when making decisions. The critical thinker asks "why" questions about a situation to arrive at the best decision. Critical thinking skills should be used all through the decision-making process. Four basic skills—critical reading, critical listening, critical writing, and critical speaking—are necessary for the development of critical thinking skills. These skills are part of the process of developing and using thinking for decision making. Ability in these four areas can be measured by the extent to which one achieves the universal intellectual standards listed in Table 22-2.

As you begin to apply critical thinking to nursing, use these universal intellectual standards when you are reading material from a textbook, listening to an oral presentation, writing a paper, answering test questions, or presenting ideas in oral form. Ask yourself whether the ideas are clear or unclear, precise or imprecise, accurate or inaccurate. Are they relevant or irrelevant, broad-minded or narrow-minded, logical or illogical, deep or superficial, or fair or unfair? You will improve your critical thinking skills over time and with practice.

# REFLECTIVE THINKING

Pesut and Herman (1999) describe **reflective thinking** as watching or observing ourselves as we perform a task or make a decision about a particular situation. We have two selves, the active self and the reflective self. The reflective self watches the active self as it engages in activities. The reflective self acts as observer and offers suggestions about the activities. To be a good critical thinker, you must practice reflective thinking. Reflection upon a situation or problem after a decision is made allows you to evaluate the decision. Nurse educators assist students to become better reflective thinkers through the use of clinical journals. Using journals helps students reflect on clinical activities and improve their clinical decision-making abilities.

For example, when a new nurse performs an abdominal sterile dressing change on a surgical patient, the reflective self will watch the process. The reflective self will then make suggestions as to how to make the process more efficient the next time. The reflective self might make suggestions such as having all the supplies in the room prior to the process, including an extra pair of sterile gloves, or having

**TABLE 22-2  The Spectrum of Universal Intellectual Standards**

| | |
|---|---|
| Clear | Unclear |
| Precise | Imprecise |
| Accurate | Inaccurate |
| Relevant | Irrelevant |
| Broad-minded | Narrow-minded |
| Logical | Illogical |
| Deep | Superficial |
| Fair | Unfair |

**Source:** Adapted from Paul, R. and Elder, L. (October 2010). Universal Intellectual Standards Foundation For Critical Thinking. Accessed, April 14, 2011, from http://www.criticalthinking.org/page.cfm?PageID=527&CategoryID=68.

an extra table in the room. Reflection upon a situation or problem after a decision is made allows the individual to evaluate the decision. New nurses should continue to use reflective thinking throughout their practice to build their confidence in the clinical decisions they make.

# INTUITIVE THINKING

**Intuitive thinking** is a type of discernment or insight that nurses develop that helps them to act in certain situations. Intuitive thinking, or intuition, has also been described as a gut feeling that something is wrong. When this insight is utilized, the nurse "engages the full use of self" (Dossey & Keegan, 2009). Intuitive thinking may result from unconscious assessment and analysis of data based on an individual's past experience. Nurses often make decisions about patient care based in part on intuitive thinking (Andrews, 2006; Klein, 2004; Aloi, 2006). This may seem contrary to

using the logical, evidenced-based reasoning that is so prevalent in nursing literature. Alfaro-LeFevre (2009) contends that expert thinking is usually the result of using intuition and drawing on evidence at the same time to make well-reasoned decisions. The following Real World Interview represents an example of a new nurse using intuitive thinking.

# PROBLEM SOLVING

**Problem solving** is an active process that starts with a problem and ends with a solution. Nurses address multiple needs and problems of patients on a daily basis. Some problems are uncomplicated and require one simple solution. Other problems may be complex and require more analysis by the nurse. The problem-solving process consists of the following five steps: identify the problem, gather and analyze data, generate alternatives and select an action, implement the selected action, and evaluate the action. Note the similarities to the decision-making process and to the nursing process steps of assessment, diagnosis, outcome identification, planning, implementation, and evaluation. The nursing process is applied to patient situations or problems, whereas the problem-solving process and decision-making process may be applied to a problem of any type (Table 22-3).

# DECISION-MAKING TOOLS AND TECHNOLOGY

Nurse managers are faced with complex decisions at times. Decisions related to budget are common in our current health care environment, with its emphasis on cost containment and quality maintenance. Disciplining an employee also creates a complex situation in which managers must

## REAL WORLD INTERVIEW

I was in a situation where I just didn't think my patient looked good. I decided to go ahead and start two new IV sites, just in case. The patient arrested two hours later, and we really needed those IV sites. I felt good about listening to my intuition and the decision I made.

**Cheryl Buntz, RN**
New Graduate
Independence, Missouri

## TABLE 22-3   Review of Terms

| TERM | DEFINITION |
| --- | --- |
| Decision making | Behavior exhibited when making a selection and implementing a course of action from alternatives. It may or may not involve an immediate problem. |
| Critical thinking | Thinking about your thinking while you're thinking in order to make your thinking better (Paul, 1992). |
| Reflective thinking | Watching or observing ourselves as we perform a task or make a decision about a particular situation (Pesut & Herman, 1999). |
| Intuitive thinking | An innate feeling that nurses develop that helps them to act in certain situations (Gardner, 2003). |
| Problem solving | An active process that starts with a problem and ends with a solution. |

make decisions regarding the employee's future. A decision-making grid may help to separate the multiple factors that surround a situation. Figure 22-2 illustrates the use of a decision-making grid by nurses who were told they had to reduce their workforce by two full-time equivalents (FTEs). The grid is useful in this example to visually separate the factors of cost savings, effect on job satisfaction of remaining staff, and effect on patient satisfaction. The nurse needs to determine the priorities when developing a grid.

A decision-making grid is also useful when a nurse is trying to decide between two choices. Figure 22-3 is an example of a decision grid used by a nurse deciding between working at hospital A or hospital B.

The Program Evaluation and Review Technique (PERT) is useful in determining the timing of decisions. The flowchart provides a visual picture of the sequence of tasks that must take place to complete a project. Jones and Beck (1996) provide an example of a PERT flow diagram depicting a case management project (Figure 22-4).

| Methods of Reduction | Cost Savings | Effect on Job Satisfaction | Effect on Patient Satisfaction |
|---|---|---|---|
| Lay off the two most senior full-time employees | $93,500 | Significant reduction | Significant reduction |
| Lay off the two most recently hired full-time employees | $63,200 | Significant reduction | Moderate reduction |
| Reduce by staff attrition | $78,000 | Minor reduction | Minor reduction |

Delmar/Cengage Learning

FIGURE 22-2  Sample decision-making grid.

| Elements | Importance Score (out of 10) | Likelihood Score (out of 10) | Risk (multiply scores) |
|---|---|---|---|
| If I work at hospital A: | | | |
| Learning Experience | 10 | 10 | 100 |
| Good Mentor Support | 8 | 8 | 64 |
| Financial Reward | 6 | 6 | 36 |
| Growth Potential | 8 | 8 | 64 |
| Good Location | 10 | 10 | 100 |
| Total | | | 364 |
| If I work at hospital B: | | | |
| Learning Experience | 8 | 8 | 64 |
| Good Mentor Support | 7 | 7 | 49 |
| Financial Reward | 8 | 8 | 64 |
| Growth Potential | 9 | 9 | 81 |
| Good Location | 6 | 6 | 36 |
| Total | | | 294 |

Delmar/Cengage Learning

FIGURE 22-3  Sample decision-making grid for weighing options.

The vice president for nursing plans to change all units to include case managers. She believes that this can be accomplished within a year and a half. For this to be achieved, the following activities and events have to occur.

| Activity Symbol | Activity Descriptions |
|---|---|
| A | Form a multidisciplinary advisory group |
| B | Agree upon definitions |
| C | Notify members of subcommittees |
| D | Write job descriptions |
| E | Advertise for candidates for case manager |
| F | Review qualifications of candidates |
| G | Select candidates for case manager |
| H | Review patient charts |
| I | Write patient care maps |
| J | Meet with case managers |
| K | Orient case managers |
| L | Orient unit and hospital staff |
| M | Utilize case management process |

| | Events |
|---|---|
| 1 | Project begins |
| 2 | Meeting of multidisciplinary committee |
| 3 | Formation of subcommittees |
| 4 | Subcommittee for job description meets |
| 5 | Subcommittee for patient care maps meets |
| 6 | Candidates for case managers are interviewed |
| 7 | Candidates are hired |
| 8 | Subcommittee for patient care maps meets to finalize maps |
| 9 | Orientation begins |
| 10 | Implementation begins |
| 11 | Project is evaluated |

| Expected | Time Calculations |
|---|---|
| **Activity** | **Duration** |
| A | 0.5 month |
| B | 1 month |
| C | 0.5 month |
| D | 1 month |
| E | 1 month |
| F | 2 months |
| G | 1 month |
| H | 1 month |
| I | 2 months |
| J | 1 month |
| K&L | 1 month |
| M | 3 months |

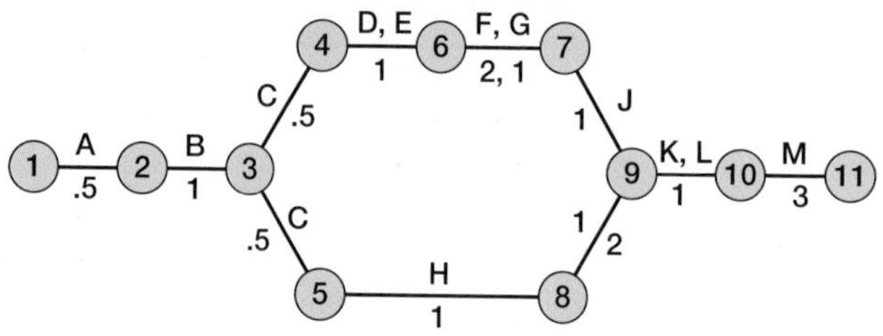

FIGURE 22-4  PERT diagram with critical path for implementation of case management.

## Critical Thinking 22-1

You have just received report from the night shift nurse in the surgical ICU. Your patient had a hip replacement the previous day and was in the ICU due to respiratory complications in the PACU. The night shift nurse reported that the patient's heart rate was starting to rise slightly during the last hour of her shift. You have been on duty for 45 minutes, and the patient's heart rate has gone from 100 beats per minute to 112 beats per minute. The patient is currently sleeping, and his blood pressure is 116/76. Use your nursing knowledge and critical thinking skills to determine possible causes of the elevation in heart rate.

## EVIDENCE FROM THE LITERATURE

**Citation:** White, A., Allen, P., Goodwin, L., Breckinridge, D., Dowell, J., & Garvy, R. (2005). Infusing PDA technology into nursing education. *Nurse Educator, 30*(4), 153–154.

**Discussion:** The article discusses the nursing education program at Duke University, where students use PDAs and software to access current drug and infectious disease information, calculations, growth charts, immunization guidelines, and Spanish- and English-language translations to improve clinical decision making. Information about this can be accessed at www.pepidedu.com.

**Implications for Practice:** Use of PDAs can improve clinical decision making.

The chart shows the amount of time taken to complete the project and the sequence of events to complete the project. An advantage of the PERT diagram is that participants can visualize a complete picture of the project, including the timing of decisions from beginning to end.

## DECISION TREE

A decision tree can be useful in making the alternatives visible. Start with the decision to be made, then diagram out the options. For example Figure 22-5 is a decision tree for choosing whether to go back to school; Figure 22-6 identifies a decision analysis tree for a patient who smokes. These decision trees help the nurse review the alternatives with the consequences and benefits of each one.

## GANTT CHART

A Gantt chart is a type of graph that shows the relationships between activities representing the phases of a project. Henry L. Gantt (1861–1919), a mechanical engineer, developed this tool for the purpose of focusing on schedule management (www.ganttchart.com). It can be useful for decision makers to illustrate a project from beginning to end. Figure 22-7 illustrates a Gantt chart used to show the progression of a nursing unit's pilot project.

## USE OF TECHNOLOGY

Nurses use technology as a support for decision making. The best source of clinical decision making and judgment is still the professional practitioner. However, computer technology has many uses to support decision making by the nurse. These include:

- Electronic health records
- Patient decision support tools, clinical and business related
- Laboratory and X-ray results reporting and viewing systems
- Computerized prescribing and order entry, including barcoding
- Population health management and information
- Patient classification staffing systems, and administrative systems
- Evidence-based knowledge and information retrieval systems
- Quality improvement data collection/data summary systems
- Documentation and care planning
- Patient monitoring, e.g., medications, vital signs, electrocardiograms, etc., including problem alerts
- Inventory control

The computer can maintain a large storehouse of information that can be useful for nurses. Data on such things as patient laboratory findings, vital signs, and medications, etc., can be monitored and used to alert nurses when abnormalities occur. Many nurses today are using PDAs and Smartphones (see Figure 22-8) to improve patient care and access information. NCLEX review questions are also available on these devices.

## GROUP DECISION MAKING

Certain situations call for group decision making. There are occasions when it is more appropriate for a group to make the decision rather than the individual nurse. Each

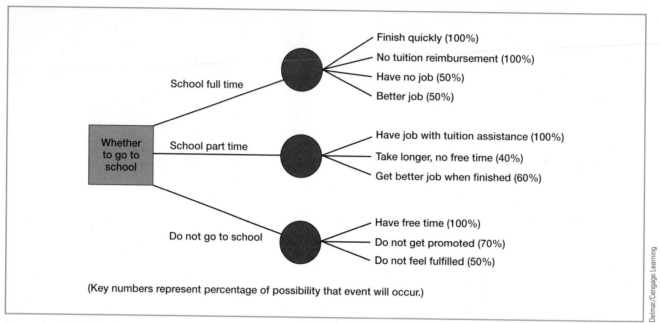

**FIGURE 22-5** Decision tree for deciding whether to go back to school.

situation is different, and an effective manager adopts the appropriate mode of decision making—group or individual. The eight questions in Table 22-4 may assist the manager in determining which mode to use.

Today's leadership and management styles include people in the decision-making process who will be most affected by the decision. Decisions affecting patient care should be made including the groups implementing the decision.

The effectiveness of groups depends greatly on the groups' members. The size of a group and the personalities of group members are important considerations when choosing participants. More ideas can be generated with groups, thus allowing for more choices. This increases the likelihood of higher-quality outcomes. Another advantage of groups is that when followers participate in the decision-making process, acceptance of the decision is more likely to occur. Additionally, groups may be used as a medium for communication.

A major disadvantage of group decision making is the time involved. Without effective leadership, groups can waste time and be nonproductive. Group decision making can be more costly and can also lead to conflict. Groups can be dominated by one person or become the battleground for a power struggle among assertive members. See Table 22-5 for a listing of the advantages and disadvantages of groups.

## TECHNIQUES OF GROUP DECISION MAKING

There are various techniques of group decision making. Nominal group technique, Delphi technique, and consensus building are different methods to facilitate group decision making.

### Nominal Group Technique

The nominal group technique was developed by Delbecq Van de Ven and Gustafson in 1971. In the first step of this technique, there is no discussion; group members write out their ideas or responses to the identified issue or question posed by the group leader. The second step involves presentation of the ideas to the group members, along with the advantages and disadvantages of each. These ideas are presented on a flip board or chart. The third phase offers an opportunity for discussion to clarify and evaluate the ideas. The fourth phase includes private voting on the ideas. The ideas receiving the highest number of votes are the solutions implemented.

**Critical Thinking 22-2**

Use Figure 22-5, and draw a decision tree. Choose a topic of your choice. Does it help to clarify your thinking?

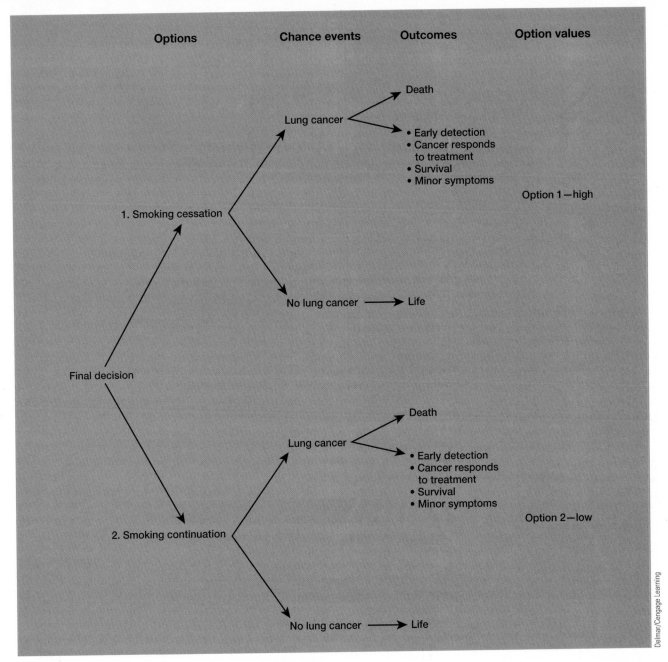

FIGURE 22-6 Decision analysis tree for a patient who smokes.

## Delphi Group Technique

Delphi technique differs from nominal technique in that group members "are not meeting face to face." Questionnaires are distributed to group members for their opinions, and these are then summarized and disseminated with the summaries to the group members. This process continues for as many times as necessary for the group members to reach consensus. An advantage of this technique is that it can involve a large number of participants and thus a greater number of ideas.

## Consensus Building

*Consensus* is defined by *Merriam-Webster's Collegiate Dictionary* (2011) as "a general agreement; the judgment arrived at by most of those concerned; group solidarity in sentiment and belief". A common misconception is that consensus means everyone agrees with the decision 100%. Contrary to this misunderstanding, **consensus** means that all group members can live with and fully support the decision regardless of whether they totally agree. Building consensus is useful with groups, because all group

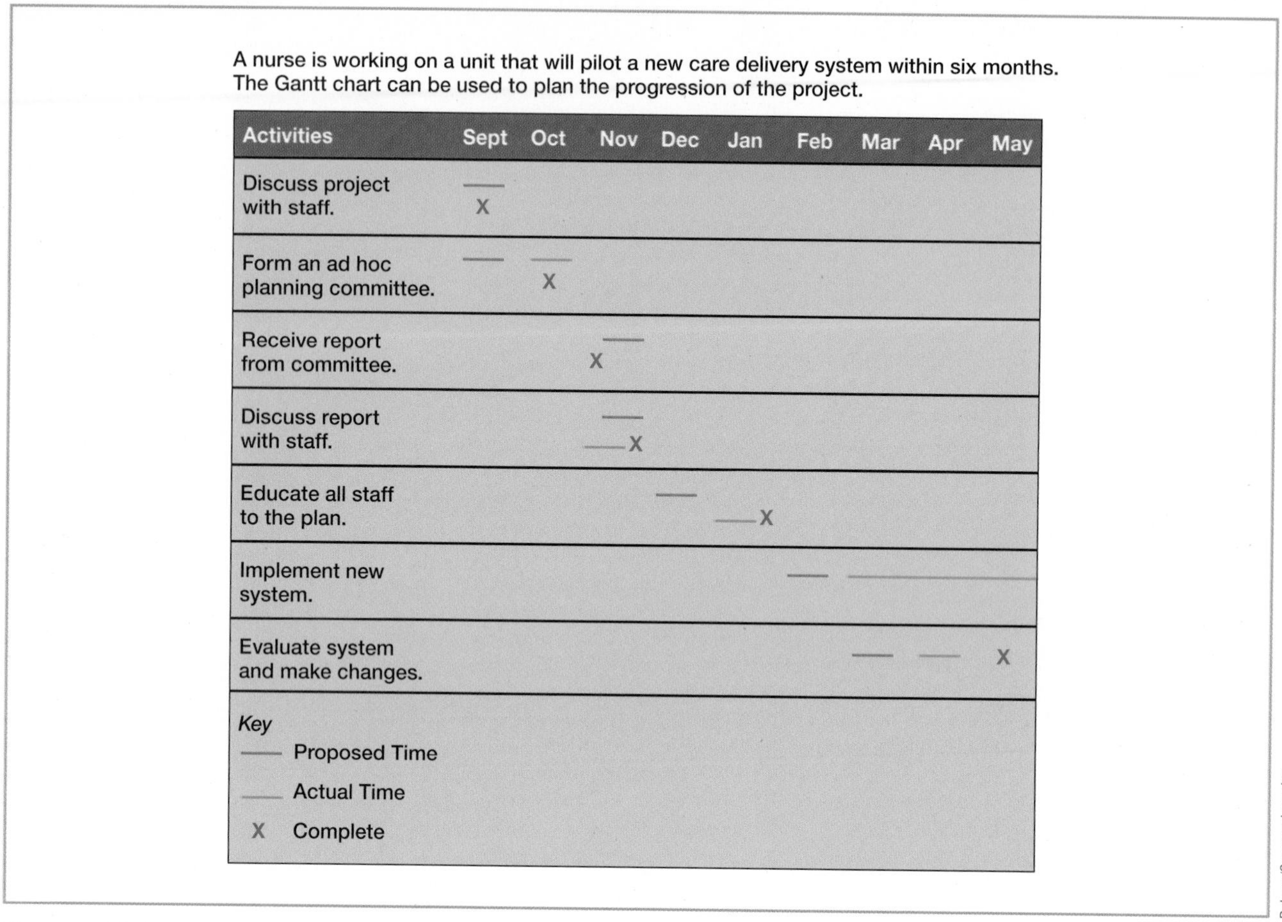

A nurse is working on a unit that will pilot a new care delivery system within six months. The Gantt chart can be used to plan the progression of the project.

| Activities | Sept | Oct | Nov | Dec | Jan | Feb | Mar | Apr | May |
|---|---|---|---|---|---|---|---|---|---|
| Discuss project with staff. | — X | | | | | | | | |
| Form an ad hoc planning committee. | — | — X | | | | | | | |
| Receive report from committee. | | | — X | | | | | | |
| Discuss report with staff. | | | — | —X | | | | | |
| Educate all staff to the plan. | | | | — | —X | | | | |
| Implement new system. | | | | | | — | — | | |
| Evaluate system and make changes. | | | | | | | — | — | X |

Key
— Proposed Time
— Actual Time
X  Complete

Delmar/Cengage Learning

FIGURE 22-7  Gantt chart: Implementation of care delivery system.

FIGURE 22-8  Nurse with a PDA. (*Source:* Courtesy PEPID, Heather Hautman.)

members participate and can realize the contributions each member makes to the decision. A disadvantage to the consensus strategy is that decision making requires more time. This strategy should be reserved for important decisions that require strong support from the participants who will implement them. Consensus decision making works well when the decisions are made under the following conditions: All members of the team are affected by the decision; implementation of the solution requires coordination among team members; and the decision is critical, requiring full commitment by team members. Although consensus can be the most time-consuming strategy, it can also be the most gratifying.

## Groupthink

In groupthink, the goal is for everyone to be in 100% agreement. Groupthink discourages questioning and divergent thinking, which can lead the group to make poor decisions. The potential for groupthink increases as the cohesiveness of the group increases; therefore, it is important for the members to recognize the symptoms of groupthink. One symptom of groupthink is that group members develop an illusion of invulnerability, believing they can do no wrong. This problem has the greatest potential to develop when the group is powerful and group members view themselves as invincible. The second symptom of

## TABLE 22-4 Individual vs. Group Decision-Making Questions

1. Does the individual nurse have all the information needed?

2. Does the group have supplementary information needed to make the best decision?

3. Does the individual nurse have all the resources available to obtain sufficient information to make the best decision?

4. Is it absolutely critical that the group accept the decision prior to implementation?

5. Will the group accept a decision I make by myself?

6. Does the course of action chosen make a difference to the organization?

7. Does the group have the best interest of the organization or patient foremost when considering the decision?

8. Will the decision cause undue conflict among the group?

**Source:** Adapted from Vroom, V., & Yetton, P. (1973). Leadership and decision-making (pp. 21–30). Pittsburgh, PA: University of Pittsburgh Press.

## TABLE 22-5 Advantages and Disadvantages of Groups

| ADVANTAGES | DISADVANTAGES |
| --- | --- |
| • Easy and inexpensive way to share information | • Individual opinions influenced by others |
| • Opportunities for face-to-face communication | • Individual identity obscured |
| • Opportunity to become connected with a social unit | • Formal and informal role and status positions evolve—hierarchies |
| • Promotion of cohesiveness and loyalty | • Dependency fostered |
| • Access to a larger resource base | • Time consuming |
| • Forum for constructive problem solving | • Inequity of time given to share individual information |
| • Support group | • Existence of nonfunctional roles |
| • Facilitation of esprit de corps | • Personality conflicts |
| • Promotion of ownership of problems and solutions | |

groupthink is stereotyping outsiders. This occurs when the group members rely on shared stereotypes—such as all Democrats are liberal or all Republicans are conservative—to justify their positions. People who challenge or disagree with the group's decisions are also stereotyped. A third symptom is that group members reassure one another that their interpretation of data and their perspective on matters are correct regardless of the evidence showing otherwise. Old assumptions are never challenged, and members ignore what they do not know or what they do not want to know.

Numerous problems can arise when groupthink is present. Groups may come to one solution to a problem early in the course of problem solving, without considering all available options (Shortell & Kaluzny, 2006). Groups may not be willing to seek expertise or be open to new information when groupthink is present. Groups may believe so strongly in their decisions that they may not develop contingency plans in case the decision is wrong (Shortell & Kaluzny, 2006). Clearly, these problems would be detrimental to any group decision-making process.

Strategies to avoid groupthink include appointing group members to roles that evaluate how group decision-making occurs. Group leaders should encourage all group members to think independently and verbalize their individual ideas. The leader should allow the group time to gather further data and reflect on data already collected. A primary

# REAL WORLD INTERVIEW

We have established a rapid response program made up of teams that stabilize a deteriorating patient and avoid cardiopulmonary arrest on medical-surgical units. In analyzing the success of this rapid response program, we look at the number of rapid response calls per month (see data below). This data reflects rapid response calls initiated by staff on all noncritical care units. Data collection is done consistently from month to month for trending and decision-making purposes.

In reviewing the rapid response data, we take into account any large variances in census or patient acuity. If the number of rapid response calls declines over several months and is not explained by changes in census, patient acuity, or patient outcomes, we assess the RN staff's knowledge of the criteria for initiating the rapid response team. We have found that reeducation of this staff at regular intervals is vital to sustaining a successful rapid response program. We also survey the RN staff initiating rapid response calls, to determine whether the relationship between the rapid response team members and the staff nurses is one of support and partnering. A lack of support from any of the rapid response team members can result in unit staff delaying or avoiding activation of the rapid response team.

Along with collecting data on the number of rapid response calls, we also monitor the incidence of cardiopulmonary arrests outside the critical care units in order to make decisions about the benefits of the rapid response program. Analyzing the events leading up to a cardiopulmonary arrest outside the critical care units can help us make decisions about whether signs of patient deterioration were identified and acted on in an appropriate and timely manner. An increase in rapid response calls and a decrease in cardiopulmonary arrests on medical-surgical units over a period of months is viewed as a positive sign that the rapid response program has been accepted by staff, is fully operational, and is saving patient lives.

**Susan Long, RN, BSN, MS**
Manager, Resource Team
Staffing Office
Central DuPage Hospital
Winfield, Illinois

Rapid Response Team Calls by Month

| Month | Jul 08 | Aug 08 | Sep 08 | Oct 08 | Nov 08 | Dec 08 | Jan 09 | Feb 09 | Mar 09 | Apr 09 | May 09 | Jun 09 | Jul 09 | Aug 09 | Sep 09 | Oct 09 | Nov 09 | Dec 09 | Jan 10 | Feb 10 | Mar 10 | Apr 10 | May 10 | Jun 10 |
|---|---|---|---|---|---|---|---|---|---|---|---|---|---|---|---|---|---|---|---|---|---|---|---|---|
| RR* | 40 | 52 | 44 | 48 | 60 | 55 | 56 | 38 | 55 | 53 | 55 | 60 | 62 | 55 | 43 | 58 | 55 | 48 | 54 | 48 | 47 | 58 | 62 | 70 |

RR* = Rapid Response Team Calls

responsibility of the managers or the group leader is to prevent groupthink from developing.

# LIMITATIONS TO EFFECTIVE DECISION MAKING

What are obstacles to effective decision making? Past life experience, including education and decision-making experience (Bakalis & Watson, 2005), values, gender (Ripley, 2005), personal biases, and preconceived ideas affect the way people view problems and situations. Restricted knowledge, limited alternatives, and the cost of alternatives may also affect a decision's quality. Incorporating

critical thinking into the decision-making process helps to prevent these factors from distorting the decision-making process. Delaune and Ladner (2011) have identified actions that may negatively affect the decision-making or problem-solving process:

- Jumping to conclusions without examining the situation thoroughly
- Failing to obtain all the necessary information
- Choosing decisions that are too broad, too complicated, or lack definition
- Failing to choose and communicate a rational solution
- Failing to intervene and evaluate the decision or solution appropriately

## EVIDENCE FROM THE LITERATURE

**Citation:** Ruth-Sahd, L., & Hendy, H. (2005). Predictors of novice nurses' use of intuition to guide patient care decisions. *Journal of Nursing Education, 44*(10), 450–458.

**Discussion:** One of the purposes of the study was to openly discuss the use of intuition in novice nurses, as they are likely to use intuition at the bedside to guide their decisions about patient care. The authors surveyed 323 novice nurses over a total of 151 nursing programs. The researchers measured the nurses' use of intuition with Miller's Willingness to Act on Intuition scale. They also questioned the nurses' personal experiences (age, gender, number of hospitalizations, self-esteem, and religiosity), interpersonal experiences (number of children and relationships with others), and professional experiences (learning methods in school, GPA, and number of months' experience), as these factors may influence the use of intuition. The results of the study suggested that older novice nurses, those with more social support, and those with more hospitalizations use intuition more often in making decisions about patient care. Additionally, the use of intuition was affected by personal and interpersonal experience as opposed to professional experience in these subjects.

**Implications for Practice:** Intuition needs to be acknowledged more openly in nursing education and clinical practice. Students should be able to discuss intuition with nurse educators, to better develop the use of their own intuition. As age was a factor associated with more frequent use of intuition, nurse educators should ensure that age groups are mixed in learning groups. Strong mentoring or support systems are also important for the new nurse, to help in guiding their decision making and use of intuition.

### Critical Thinking 22-3

The U.S. Department of Health & Human Services, (n.d.) in collaboration with the Agency for Healthcare Research and Quality (AHRQ), developed TeamSTEPPS. TeamSTEPPS is an evidence-based teamwork system aimed at optimizing patient outcomes by improving communication and teamwork skills among health care professionals. It includes a comprehensive set of ready-to-use materials and a training curriculum to successfully integrate teamwork principles into any health care system. Go to http://www.ahrq .gov/teamsteppstools. Click on Tools and Materials. Note the information there. How can you use this information to improve your ability to work on a team and make decisions?

### Critical Thinking 22-4

Use the decision-making process in Table 22-1 and ask yourself the critical thinking questions inspired by Table 22-2 when you are making decisions. Have I gathered the best, most up-to-date research and evidence to help with the decision? Is my information clear or unclear, precise or imprecise, specific or vague, accurate or inaccurate? Is it relevant or irrelevant, broad-minded or narrow-minded, logical or illogical, deep or superficial, and fair or unfair? Apply the critical thinking questions and review the research and best evidence at each step.

## STRATEGIES TO STRENGTHEN THE NURSE'S ROLE IN DECISION MAKING FOR PATIENT CARE

In today's world, patients are taking a more active role in treatment decisions. The consumers of health care are more knowledgeable and cost conscious and have more options than in previous years. Nurses must be aware of patients' rights in making decisions about their treatments, and they must assist patients in their decision making. When patients are active participants, compliance with prescribed treatments is more likely to follow. Empowering the patient in this manner ultimately promotes a more positive outcome.

## CASE STUDY 22-1

The nurse has identified a problem for the nursing unit: low patient satisfaction. Patients are dissatisfied with the long waiting periods, lack of information, and the impersonal attitude of staff. Apply the decision-making process to this problem. How would you assess the problem? How would you gather information? Who would you involve to solve the problem?

Comfort with decision making improves with experience. Early in the nurse's career, the nurse is commonly indecisive or uncomfortable with decisions. Alfaro-LeFevre (2009) has identified several strategies that help to improve critical thinking, which in turn will also help to improve decision making. Do I have all the facts that I need to make a decision? Have I explored the alternatives and the pros and cons of each alternative? At times, delaying a decision until more information is obtained may be the best approach. Asking "why," "what else," and "what if" questions will help you arrive at the best decision. When more information becomes available, decisions can be revised. Very few decisions are set in stone. Another helpful strategy for improving decision making is to anticipate questions and outcomes. For example, when calling a practitioner to report a patient's change in condition, the nurse will want to have pertinent information about the patient's vital signs and current medications readily available.

Nurses who practice strategies to promote their own critical thinking will in turn be good decision makers. A foundation for good decision making comes with experience and learning from those experiences. By turning decisions with poor outcomes into learning experiences, nurses will enhance their decision-making ability in the future.

## KEY CONCEPTS

- The ever-changing health care system calls for nurses to be effective decision makers.

- In the decision-making process, there are five steps: Step 1—identify the need for a decision; Step 2—determine the goal or outcome desired; Step 3—identify many alternatives, i.e., personal, financial, political, etc., along with the benefits and consequences of each alternative; Step 4—make the decision; Step 5—evaluate the decision.

- Critical thinking involves examining situations from every viewpoint when faced with any problem or situation. Use of the universal intellectual standards will improve a nurse's critical thinking.

- Practicing reflective thinking helps individuals become better critical thinkers.

- Nurses should recognize the importance of intuitive thinking. Recognizing their own use of intuition will help nurses develop their intuitive thinking skills.

- Improve your critical thinking skills by asking yourself whether ideas are clear or unclear; precise or imprecise; accurate or inaccurate; relevant or irrelevant; broad-minded or narrow-minded; logical or illogical; deep or superficial; or fair or unfair.

- Decision-making grids can be helpful when the nurse needs to separate multiple factors surrounding a situation during the decision-making process.

- There are situations in which a nurse needs to make individual decisions. Other decisions call for group decision making.

- Consensus building is a strategy utilized when working with a group to make a decision.

- Groupthink occurs when individuals are not allowed to express creativity, question methods, or engage in divergent thinking.

- The nurse must recognize the importance of empowering patients in making their own treatment decisions. The nurse needs to provide the patient with information and assist the patient in exploring all possible options.

- There are many strategies to improve decision making. Obtaining all the information, asking "why" and "what if" questions, and developing good habits of inquiry are a few of the strategies that will help nurses improve decision making.

## KEY CONCEPTS

consensus
critical thinking

decision making
intuitive thinking

problem solving
reflective thinking

## REVIEW QUESTIONS

1. A nurse needs to assist a patient in walking down the hall twice daily as part of the patient's postoperative activities. It is the middle of the afternoon, and the patient is asleep. The nurse would like to allow the patient to sleep, because the patient was awake a majority of the night. However, if the nurse does not ambulate the patient now, it is possible that the rest of the nurse's afternoon activities will prevent her from returning to the patient to ambulate before the end of her shift. The nurse must decide whether to ambulate the patient now. What is the next step of the decision-making process?
   A. Determine the outcome or goal that is desired.
   B. Generate alternatives and determine benefits and consequences of each.
   C. Evaluate the decision.
   D. Make the decision.

2. A nurse is asked to participate on a committee to interview and assist in making selections of staff in conjunction with the nurse manager. An applicant is interviewed, and the committee is discussing her qualifications. A committee member states, "If we hire her, then we will have all of our positions filled and no one will get any overtime." A discussion ensues about the consequences of no overtime, and the committee determines that they shouldn't hire the nurse. What dysfunctional characteristic of group decision making may be going on in this group?
   A. Consensus
   B. Groupthink
   C. Stereotyping outsiders
   D. Group believes it is invincible

3. A nursing assistant approaches the RN to tell her that her patient care assignment is too difficult of a patient load. She will be unable to complete all of her work, which is unusual for this assistant. She asks the RN if she will review the assignments to see if a more equitable assignment can be made. The RN makes rounds on the patients and reviews patient acuities as well as other nursing assistant assignments. Which step of the problem-solving process is the RN performing?
   A. Identifying the problem
   B. Selecting an action
   C. Gathering and analyzing data
   D. Generating alternatives

4. A nurse manager decides to form a task force to identify reasons and solutions for patient dissatisfaction on your unit. What are the advantages of forming this task force? Select all that apply.
   _____ A. The decisions will be made more quickly.
   _____ B. Higher-quality decisions will be made due to more solutions being generated.
   _____ C. Acceptance of the decision is more likely.
   _____ D. There is access to a larger resource base.
   _____ E. There is less likely to be conflict during the decision-making process.
   _____ F. This will help to promote ownership of the problem.

5. A new nurse is trying to set her goals for the next five years. She plans to eventually become an acute care nurse practitioner in the ICU setting. She knows she needs to become more experienced, obtain appropriate certification, go back to school, and take the practitioner exam. She would like to see a visual of the time it will take her to realistically accomplish those goals. She should use which of the following?
   A. Decision tree
   B. Gantt chart
   C. Decision grid
   D. Problem-solving process

6. A task force designed to examine solutions for low patient satisfaction in an emergency has decided to write their ideas down, present their ideas to the task force, discuss the ideas, and then vote on the ideas. This is an example of which group process?
   A. Consensus building
   B. Delphi technique
   C. Problem-solving process
   D. Nominal group technique

7. Your friend tells you that the nurse manager over the pediatrics unit, where you wanted to work, is a "terrible" nurse manager. Because of this information, you decide to pursue employment elsewhere. Making this decision might have involved which of the following?
   A. Making a decision based upon the first available information
   B. Being comfortable with the status quo
   C. Assigning inaccurate probabilities to alternatives
   D. Making a decision to justify a previous decision

8. A nurse needs to determine whether she should work overtime next week for two shifts. The nurse is having difficulty deciding, because she will miss her child's program if she works but feels the extra income might be worth it. She also believes it will be beneficial to work because the manager might give her preference for time off in the next pay period. Also, the nurse will miss a continuing education program she has been wanting to attend if she works overtime. The nurse could make the best decision by using which of the following?
   A. PERT diagram
   B. Gantt chart
   C. Group consensus
   D. Decision-making grid

9. A nurse has finished her shift and is on her way home from her job. She is mulling over the activities of the day and thinks about the teaching she completed with a patient who has diabetes. The nurse decides that the patient had some difficulty understanding the insulin injections because she (the nurse) did not explain the differences in the two types of insulin in terms the patient could understand. The nurse decides that next time she will use more nonmedical terms in her explanations. This is an example of which of the following?
   A. Critical thinking
   B. Intuitive thinking
   C. Reflective thinking
   D. Decision making

10. Today's health care consumers are taking a more active role in their treatment decisions. The nurse's confidence in critical thinking helps to empower patients in this endeavor. What are some strategies the nurse can employ that help with critical thinking and decision making? Select all that apply.
   _____ A. Base decisions on the way things were always done in the past.
   _____ B. Avoid procrastination; make decisions as you go along rather than let them accumulate.
   _____ C. Consider all those affected by the decision.
   _____ D. Make snap decisions; always "go with your gut."
   _____ E. Make only those decisions that are yours to make.
   _____ F. Trust your intuition, then also consider other ways of thinking.

# REVIEW ACTIVITIES

1. You are a nurse on a surgical unit that consists of 12 beds. Your supervisor informs you that 12 more beds will be opened for neurosurgical patients. Draw a PERT diagram to depict the sequence of tasks necessary for completion of the project.

2. The education forms are not being filled out correctly or in a timely manner on new admissions in your medical-surgical unit. Decide on your own the best action to take in this situation. Then get into a group and attempt to reach consensus on the best action to take. Compare the differences between individual and group decision making. What did you learn about developing consensus?

3. Identify a problem that you have been considering. Using the decision-making grid at the bottom of the page, rate the alternative solutions to the problem on a scale of 1 to 3 on the elements of cost, quality, importance, location, and any other elements that are important to you. Did this exercise help you in thinking through your decision?

4. Identify a current problem in health care. Use the decision-making process in a group to find a solution. Employ the nominal group technique or the Delphi technique.

| | Cost | Quality | Importance | Location | Other |
|---|---|---|---|---|---|
| Alternative A | | | | | |
| Alternative B | | | | | |
| Alternative C | | | | | |

# EXPLORING THE WEB

- Visit these critical thinking sites:
  www.criticalthinking.org
  www.insightassessment.com

- This site focuses on career excellence, decision-making, and problem-solving tools: www.mindtools.com

- Note the following site for clinical decision making, which includes software for clinical decision making: www.apache-msi.com

- Review these sites for extra information on intuitive thinking and the Delphi method
  http://en.m.wikipedia.org Search for, Delphi method
  www.typelogic.com
  www.intuitivethinking.com

- Review this site on applying artificial intelligence to clinical situations: www.medg.lcs.mit.edu

# REFERENCES

Alfaro-LeFevre, R. (2009). *Critical thinking and clinical judgment: A practical approach* (4th ed.). Philadelphia, PA: W. B. Saunders.

Bakalis, N. A., & Watson, R. (2005). Nurses' decision making in clinical practice. *Nursing Standard, 19*(23), 33–39.

DeLaune, S., & Ladner, P. (2011). *Fundamentals of nursing, standards and practice* (4th ed.). Clifton Park, NY: Delmar Cengage Learning.

Dossey, B., & Keegan, L. (2009). Holistic nursing: A handbook for practice. Sudbury, MA: Jones and Bartlett Publishers.

Jones, R. A. P., & Beck, S. E. (1996). *Decision making in nursing.* Clifton Park, NY: Delmar Cengage Learning.

Merriam-Webster's Online Dictionary (2011). Accessed April 13, 2011, at http://www.merriam-webster.com/dictionary/consensus.

Paul, R. (1992). *Critical thinking: What every person needs to survive in a rapidly changing world.* Santa Rosa, CA: Foundation for Critical Thinking.

Paul, R. and Elder, L. (October 2010). Universal Intellectual Standards. Foundation For Critical Thinking. Accessed April 25, 2011, from http://www.criticalthinking.org/page.cfm?PageID=527&CategoryID=68.

Pesut, D. J., & Herman, J. (1999). *Clinical reasoning: The art & science of critical & creative thinking.* Clifton Park, NY: Delmar Cengage Learning.

Ruth-Sahd, L., & Hendy, H. (2005). Predictors of novice nurses' use of intuition to guide patient care decisions. *Journal of Nursing Education, 44*(10), 450–458.

Shortell, S., & Kaluzny, A. (2006). *Health care management* (5th ed.). Clifton Park, NY: Delmar Cengage Learning.

Tanner, C. (2006). Thinking like a nurse: Research-based model of clinical judgement in nursing. *Journal of Nursing Education. 45*(6), 204–211.

U.S. Department of Health & Human Services. (n.d.) Agency for Healthcare Research and Quality TeamSTEPPS: National Implementation. Retrieved May 2, 2011 from http://www.ahrq.gov/teamsteppstools.

Vroom, V. H., & Yetton, P. W. (1973). *Leadership and decision-making.* Pittsburgh: University of Pittsburgh Press.

White, A., Allen, P., Goodwin, L., Breckinridge, D., Dowell, J., & Garvy, R. (2005). Infusing PDA technology into nursing education. *Nurse Educator, 30*(4), 153–154.

# SUGGESTED READINGS

Anderson, G. L., & Tredway, C. A. (2009). Transforming the nursing curriculum to promote critical thinking online. *Journal of Nursing Education, 48*(2), 111–115.

Bittner, N. P., & Gravlin, G. (2009). Critical thinking, delegation, and missed care in nursing practice. *Journal of Nursing Administration, 39*(3), 142–146.

Cholewka, P. A., & Mohr, B. (2009). Enhancing nursing informatics competencies and critical thinking skills using wireless clinical simulation laboratories. *Studies in Health Technology and Informatics, 146,* 561–563.

Evans, C. (2005). Clinical decision-making theories: Patient assessment in autonomy and extended roles. *Emergency Nurse, 13*(5), 16–20.

Hernandez, C. A. (2009). Student articulation of a nursing philosophical statement: An assignment to enhance critical thinking skills and promote learning. *Journal of Nursing Education, 48*(6), 343–349.

Lyons, E. M. (2008). Examining the effects of problem-based learning and NCLEX-RN scores on the critical thinking skills of associate degree nursing students in a southeastern community college. International *Journal of Nursing Education Scholarship, 5*(21). (Epub 2008, June 5).

McMullen, M. A., & McMullen, W. F. (2009). Examining patterns of change in the critical thinking skills of graduate nursing students. *Journal of Nursing Education, 48*(6), 310–318.

O'Callaghan, N. (2005). The use of expert practice to explore reflection. *Nursing Standard, 19*(39), 41–47.

Paul, R., Elder, L. (2008). Critical thinking concepts and tools. Santa Rosa, CA: Foundation for Critical Thinking.

Pickett, J. (2009). Critical thinking a necessary factor in nursing workload. *American Journal of Critical Care, 18*(2), 101.

Potter, P. (2005). Understanding the cognitive work of nursing in the acute care environment. *Journal of Nursing Administration, 35*(7/8), 327–335.

Romeo, E. M. (2010). Quantitative research on critical thinking and predicting nursing students' NCLEX-RN performance. *Journal of Nursing Education* (April 1), 1–9.

Wilkinson, J. (2006). Nursing process and critical thinking (4th ed.). Upper Saddle River, NJ: Prentice Hall.

Yuan, H. O., Williams, B. A., & Fan, L. (2008). A systematic review of selected evidence on developing nursing students' critical thinking through problem-based learning. *Nurse Educator Today, 28*(6), 657–63. (Epub 2008, February 11).

Zurmehly, J. (2008). The relationship of educational preparation, autonomy, and critical thinking to nursing job satisfaction. *Journal of Continuing Education in Nursing, 39*(10), 453–460.

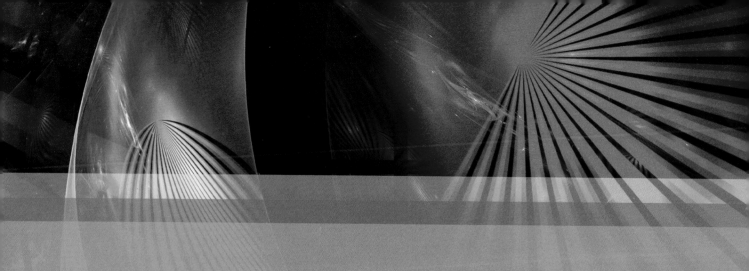

# CHAPTER 23

# Legal Aspects of Health Care

JUDITH W. MARTIN, RN, JD; SISTER KATHLEEN CAIN, OSF, JD;
CHAD S. PRIEST, RN, BSN, JD; AND SARA ANNE HOOK, MLS, MBA, JD

*Nurses in particular are held professionally accountable for their own actions and have a duty to intervene when medical care does not appear to meet the standard of care.*

(P. C. DAVIS)

## OBJECTIVES

Upon completion of this chapter, the reader should be able to:

1. Identify the sources and types of laws and regulations, and recognize their impact on nursing practice.
2. Analyze common areas of nursing practice that lead to malpractice actions, and outline actions a nurse can take to minimize these risks.
3. Relate legal protections for nursing practice.
4. Explain privacy laws related to nursing activities.
5. Analyze the nurse's role as a patient advocate and the duty to follow another practitioner's orders.

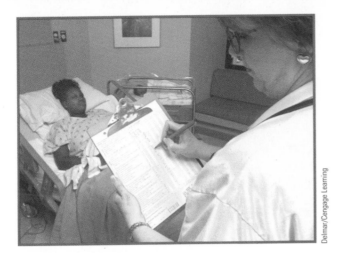

Delmar/Cengage Learning

*You are working on an obstetrical unit and have been given an order to discharge a 35-year-old female who has just had a baby. Per hospital policy, you obtain a set of vital signs before discharging her home and note her temperature of 100.9°F (38.3°C). Upon assessing the patient, she tells you that she feels a bit "chilled." You notify the practitioner of the elevated temperature and the patient's comments, but you are told to continue with the discharge.*

> *After notifying the practitioner about the elevated temperature, do you need to gather additional information about the patient's condition before you discharge her?*

> *What do you do if the patient appears to be too ill for discharge? Is there anyone else you can contact?*

> *If you discharge the patient and she develops sepsis or a serious illness, are you responsible, is the practitioner responsible, or are you both responsible?*

Law that affects the relationship between individuals is called civil law. Law that specifies the relationship between citizens and the state is called public law. This chapter reviews how laws are enacted and implemented and how the various types of law affect nursing practice.

## SOURCES OF LAW

The authority to make, implement, and interpret laws is generally granted in a constitution. A **constitution** is a set of basic laws that specifies the powers of the various segments of the government and how these segments relate to each other.

Generally, it is the role of a legislative body, both at the federal and state levels, to enact laws. Agencies under the authority of the administrative branch of the government draft the rules that implement the law. Finally, the judicial branch interprets the law as it rules in court cases. Table 23-1 gives examples of these relationships.

Also, a judicial decision may set a precedent that is used by other courts and, over time, has the force of law. This type of law is referred to as common law.

## PUBLIC LAW

**Public law** consists of constitutional law, criminal law, and administrative law and defines a citizen's relationship with government.

### CONSTITUTIONAL LAW

Several categories of public law affect the practice of nursing. For example, the nurse accommodates patients' constitutional right to practice their religion every time the nurse calls a patient's clergy as requested, follows a specific religious custom for preparation of meals, or prepares a deceased person's remains for burial.

Controversial constitutional rights that may affect the nurse's practice include the recognized constitutional right of a woman to have an abortion and an individual's right to die (see *Roe v. Wade* [1973] and *Cruzan v. Director* [1990]). Nurses may personally not believe in either of these rights and may refuse to work in areas in which they would have to assist a patient in exercising these rights. Nurses may not, however, interfere with another person's right to have an abortion or to forgo lifesaving measures.

### CRIMINAL LAW

Criminal law focuses on the actions of individuals that can intentionally do harm to others. Often, the victims of such abusive actions are the very young or the very old. These two categories of people generally cannot defend themselves against physical or emotional abuse. The nurse, in caring for patients, may notice that a vulnerable patient has unexplained bruises, fractures, or other injuries. Most states have mandatory statutes that require the nurse to report unexplained or suspicious injuries to the appropriate child or elderly protective agency. Generally, the institution in which the nurse is employed will have clear guidelines to follow in such a situation. Failure of the nurse to report the problem as required by law can result in criminal penalties.

Another aspect of criminal law affecting nursing practice is the state and federal requirement that criminal background checks be performed on specified categories of prospective employees who will work with the very young or the elderly in institutions such as schools and nursing homes. Again, this is an attempt to protect the most vulnerable citizens from mistreatment or abuse. Failure to conduct the mandated background checks can result in the institution having to defend itself for any harm done by an employee with a past criminal conviction.

**TABLE 23-1 The Three Branches of Government**

| | LEGISLATIVE BRANCH | ADMINISTRATIVE BRANCH | JUDICIAL BRANCH |
|---|---|---|---|
| **Example at federal level** | Americans with Disabilities Act (ADA) (1990) | The Equal Employment Opportunity Commission (EEOC) publishes rules specifying what employers must do to help a disabled employee. | In 1999, the U.S. Supreme Court interpreted the law to require that to be protected by the law, an individual must have an impairment that limits a major life activity and that is not corrected by medicine or appliances (e.g., glasses, blood pressure medicine). *Sutton v. United Airlines* (1999); *Murphy v. United Parcel Service, Inc.* (1999). |
| **Example at state level** | Nurse Practice Act | The state board of nursing develops rules specifying the duties of a registered nurse in that state. | Courts and juries determine whether a nurse's actions comply with the law governing the practice of nursing in a state. |

A third area in which criminal law concerns affect nursing practice is the prohibition against substance abuse. Both federal and state laws require health care agencies to keep a strict accounting of the use and distribution of regulated drugs. Nurses are routinely expected to keep narcotic records accurate and current.

Nurses' behavior when off duty can also affect their employment status. Abusing alcohol or drugs on one's own time, if discovered, can result in nurses being terminated from employment and their license to practice nursing restricted or revoked (Mantel, 1999). Frequently, boards of nursing have programs for the nurse with a drug or alcohol problem, and completion of such a program may be required before the nurse can resume practice. Additionally, health care facilities may require nurse employees to submit to random drug screens to identify those who may be using illegal substances.

## ADMINISTRATIVE LAW

Both the federal government and state governments have administrative laws that affect nursing practice. The laws pertaining to Social Security and, more specifically, Medicare, are interpreted in the *Code of Federal Regulations*, which contains the administrative rules for the federal government. These rules have specific requirements that hospitals, nursing homes, and other health care providers must adhere to if they are to qualify for payment from federal funds. Likewise, state laws are interpreted in administrative rules that specify licensing requirements for health care providers in the state.

### Federal

Administrative law deals with the protection of the rights of citizens. It extends some rights and protections beyond those granted in the federal and state constitutions. An example of this type of law, at the federal level, is the Civil Rights Act of 1964, which prohibits many forms of discrimination in the workplace. This law may necessitate that the nurse manager make some scheduling accommodations for such things as an employee's religious practices.

Another federal law that affects nursing practice is the Health Insurance Portability and Accountability Act (HIPAA), which was enacted to, among other things, safeguard certain private medical information. Under the law, disclosure of certain protected health information, such as a patient's medical diagnosis or plan of care, can result in criminal penalties. HIPAA is implicated anytime a patient's private medical information may be shared with another, whether intentionally or accidentally (Frank-Stromborg, 2004). Numerous provisions of HIPAA were expanded, clarified, and even strengthened by the American Recovery and Reinvestment Act of 2009 through Title XIII, named the Health Information Technology for Economic and Clinical Health (HITECH) Act (Committees on Energy and Commerce, Ways and Means, and Science and Technology, 2009).

Nurses in all practice areas and of all experience levels face HIPAA issues each day. For example, dry erase boards were once common features in hospital emergency rooms and inpatient units and were displayed in highly visible areas. On these boards, nurses used to list the names of all the patients on the unit, their practitioners, and perhaps their diagnoses. Today, such a practice might be a violation of HIPAA, as it would disclose the private health information of individual patients. Another example is the chart that many nurses, especially new nurses, use to organize their patient care tasks. Typically, these charts identify the patients the nurse is responsible for, their diagnoses, and what medications and

## Critical Thinking 23-1

You are a new nurse working on the OB unit of your local hospital. Your close friend George Nurse also got a job on this unit. George has a reputation as a smart, likeable, and hardworking nurse who also knows how to let loose and have a good time when work is out. However, lately George has been coming to work late and appears "out of it." You have spoken with George, who told you that he is having a difficult time at home and has not been as focused at work as he needs to be. He promised that he would try to leave his personal life at home.

For a month after your discussion with George, everything seemed fine. However, in the past two weeks, you have noticed that George has bloodshot eyes and that his speech seems slurred at times. He looks unkempt and unclean, and the narcotic count for Vicodin has been off for three separate shifts that he worked. You are concerned that George may be using drugs or alcohol and that it may be impacting his nursing care.

What action do you take? Who can you go to for help in this situation?

treatments they will need. Although these forms are invaluable tools for new nurses learning to organize their clinical practice, they contain sensitive protected health information that, if disclosed, may violate HIPAA. Therefore, nurses must be on guard to carefully destroy these forms when they are no longer needed (Adams, 2004).

As with most federal laws, the agency responsible for implementing a law has a great deal of power to draft specific rules and regulations. For example, the Occupational Safety & Health Administration (OSHA), an administrative agency, works to establish a safe workplace for employees. This includes enacting regulations concerning storage of hazardous substances, protection of employees from infection, and protection of employees from violence in the workplace. Hospitals are subject to numerous OSHA regulations designed to protect the health and safety of nurses and other health care workers. From the minute the new nurse joins the hospital staff, he or she is likely to come into contact with OSHA-mandated products or programs every day. For example, any unvaccinated nurse joining the staff of a hospital will be offered Hepatitis B vaccination pursuant to OSHA regulations. Additionally, nurses working with patients who may have tuberculosis will be issued special OSHA-approved respirators to prevent the nurse from becoming infected. Every day, nurses will utilize OSHA-mandated and approved "sharps" containers that hold used needles, and personal protective equipment such as gloves, gowns, and surgical face masks. New nurses should review hospital policies and procedures to ensure that they are using these safety devices properly.

### State

An example of a state's administrative law is its nurse practice act. Under nurse practice acts, state boards of nursing are given the authority to define the practice of nursing within certain broad parameters specified by the legislature, mandate the requisite preparation for the practice of nursing, and discipline members of the profession who deviate from the rules governing the practice of nursing. Other professions such as medicine and dentistry have similar practice acts established in state law.

An important issue to nurses is the transferability of their nursing license from one state to another. A license to practice nursing is generally valid only in the state where it is issued. In most cases, a nurse wanting to practice in a state other than where his or her license was issued must apply for a license in that state. For nurses who frequently move from one state to another, this can be a burdensome process. There is an ongoing movement to allow nurses licensed in one state to automatically receive licensure to practice in another state. The Nurse Licensure Compact, a project of the National Council of State Boards of Nursing, is an agreement among states to allow nurses licensed in other states who are parties to the agreement to practice without applying for a new license. As of the date of this writing, only 24 states have joined this agreement, meaning that about half of the states still require nurses to apply for a license in the state where they want to practice. You may check the Web to determine if your state is a member of the compact by going to www.ncsbn.org. Of course, nurses should always contact the board of nursing in any state where they intend to practice, to determine eligibility and licensure requirements.

## CIVIL LAW

Civil law governs how individuals relate to each other in everyday matters. It encompasses both contract and tort law. It does not encompass criminal law.

## CONTRACT LAW

Contract law regulates certain transactions between individuals and/or legal entities such as businesses. It also governs transactions between businesses. An agreement must contain the following elements to be recognized as a legal contract:

- Agreement between two or more legally competent individuals or parties, stating what each must or must not do
- Mutual understanding of the terms and obligations that the contract imposes on each party to the contract
- Payment or consideration given for actions taken or not taken pursuant to the agreement

The terms of the contract may be oral or written; however, a written contract may not be legally modified by an oral agreement. Another way this is often expressed is by the phrase "all of the terms of the contract are contained within the four corners of the document"; that is, if it is not written, it is not part of the agreement or contract. A contract may be expressed or implied. In an express contract, the terms of the contract are specified, usually in writing. In an implied contract, a relationship between parties is recognized, although the terms of the agreement are not clearly defined, such as the expectations one has for services from the dry cleaner or the grocer.

The nurse is usually a party to an employment contract. The employed nurse agrees to do the following:

- Adhere to the policies and procedures of the employing entity
- Fulfill the agreed-upon duties of the employer
- Respect the rights and responsibilities of other health care providers in the workplace

In return, the employer agrees to provide the nurse with the following:

- A specified amount of pay for services rendered
- Adequate assistance in providing care

### Critical Thinking 23-2

You are assigned to a medical-surgical unit, working the night shift. Your supervisor calls and says that one of the RNs assigned to the critical care unit has called in sick and you must work that unit instead of your usual assignment. You have never worked in the critical care setting before and have received no orientation to this unit. You are now asked to work there when it is short of staff.

What should you do?

- The supplies and equipment needed to fulfill his or her responsibilities
- A safe environment in which to work
- Reasonable treatment and behavior from the other health care providers with whom he or she must interact

This contract may be express or implied, depending on the practices of the employing entity. However, the clear trend is that employees at all types of companies and organizations are required to sign employment agreements. You might be asked to sign an employment agreement promising that you will not disclose trade secrets or other confidential information and that the hospital will own any materials you prepare as part of your job. You might also be asked to sign an agreement specifying whether or not you are allowed to use e-mail or the Internet for personal matters. Sometimes, what is determined to be "reasonable" by the employer is not considered "reasonable" by the nurse. For instance, after 20 years of working as a nurse on the orthopedic unit, a nurse may not view it as reasonable to be assigned to the labor and delivery unit for duty as a nurse there. It would be prudent to express any misgivings to the supervisor and to only take assignments that are in keeping with the experience one has.

## TORT LAW

Law.com (2010) defines tort as a civil wrong or wrongful act, whether intentional or accidental, from which injury occurs to another. A tort can be any of the following:

- The denial of a person's legal right
- The failure to comply with a public duty
- The failure to perform a private duty that results in harm to another

A tort can be unintentional, as occurs in malpractice or neglect, or it can be the intentional infliction of harm, such as assault and battery. In a tort suit, the nurse can be named as a defendant because of something he or she did incorrectly or because he or she failed to do something that was required. In either case, the suit is usually classified as a tort suit. Other tort charges that a nurse may face include assault and battery, false imprisonment, invasion of privacy, defamation, and fraud (see Table 23-2).

## NEGLIGENCE AND MALPRACTICE

If a nurse fails to meet the legal expectations for care—usually defined by the state's nurse practice act—the patient, if harmed by this failure, can initiate an action against the nurse for damages. The term malpractice refers to a professional's wrongful conduct in the discharge of his or her professional duties or failure to meet standards of care for the profession, which results in harm to another individual

## TABLE 23-2 Selected Torts

| TORT | DEFINITION | EXAMPLE |
|------|-----------|---------|
| Assault | Threat to touch another person in an offensive manner without that person's permission. | Nurse who threatens to give a patient a treatment against his will. |
| Battery | Touching of another person without that person's consent. | Nurse who forces a treatment against a patient's will. |
| Invasion of privacy | All patients have the right to privacy and may bring charges against any person who violates this right. | Nurse who discloses confidential information about a patient or photographs a patient without consent. |
| False imprisonment | This occurs when individuals are physically prevented, or incorrectly led to believe they are prevented, from leaving a place. | Nurse who restrains a patient who is of sound mind and is not in danger of injuring himself or others. |
| Defamation, including libel and slander | Intentionally false communication or publication, including written (libel) or verbal (slander) remarks that may cause the loss of a person's reputation. | Nurse who makes a statement that could either damage the patient's reputation or cause the patient to lose his job. |

entrusted to the professional's care. **Negligence** is the failure to provide the care a reasonable person would ordinarily provide in a similar situation.

Simply proving malpractice or negligence is not sufficient to recover damages. Proof of liability or fault requires the proof of the following four elements:

1. A duty or obligation created by law, contract, or standard practice that is owed to the complainant by the professional
2. A breach of this duty, either by omission or commission
3. Harm to the complainant (patient), which can be physical, emotional, or financial
4. Proof that the breach of duty caused the complained-of harm

A Louisiana appellate court described the plaintiff's (patient's) specific burden of proof in a negligence or malpractice claim against a nurse as follows:

*[T]he three requirements which a plaintiff must satisfy to meet its burden of proving the negligence of a nurse are: (1) the nurse must exercise the degree of skill ordinarily employed, under similar circumstances, by the members of the nursing or health care profession in good standing in the same community or locality; (2) the nurse either lacked this degree of knowledge or skill or failed to use reasonable care and diligence, along with her best judgment in the application of that skill; and (3) as a proximate result of this lack of knowledge or skill or the failure to exercise this degree of care, the plaintiff suffered injuries that would not otherwise have occurred. (Odom v. State Dept. of Health & Hospitals, [1999])*

After a plaintiff presents his or her case, the defendant nurse must refute the claims by showing that if a duty was owed it was fulfilled or by demonstrating that the breach of that duty was not the cause of the plaintiff's harm.

Proving that a duty was owed is not difficult. The person need only show that the nurse was working on the day in question and was responsible for the plaintiff's care. This can usually be accomplished by producing staffing schedules and assignment sheets.

To demonstrate a breach of duty, the courts employ a *reasonable person* standard by asking what a reasonable nurse would do in a similar situation. This is accomplished by reviewing the employing institution's policies and procedures and other evidence, including the state's nurse practice act, and hearing testimony from nurses who are accepted as expert witnesses to the standard of nursing practice in the community (Table 23-3).

The defendant nurse would employ the same methodology to refute the plaintiff's charges. The nurse would present evidence that the institution's policies and procedures were followed and that the care rendered adhered to

## Table 23-3  Selected Sources of Evidence Regarding the Standard of Care

- Evidence-based health care literature

- Nursing and medical textbooks, articles, and research

- State professional practice acts, such as nurse practice act, physician practice act

- Standards of professional associations, for example, American Nurses' Association Standards

- Equipment manufacturers manuals, for example, cardiac monitoring equipment manuals

- Written policies and procedures of a facility, such as Foley catheter insertion procedures

- Nurse, practitioner, or other health care professional expert testimony

- Professional health care accreditation agency criteria, for example, JC (Joint Commission) criteria

- Medication books, such as *Physicians' Desk Reference, American Society of Health-System Pharmacists Drug Information* Handbook, and so on.

**Note:** Not all sources are used in all states.

accepted nursing standards. To present the nurse's case, the nurse's attorney would also use expert witnesses to document that the care given fulfilled the duty owed, was the kind that would be given by a reasonable nurse in such a circumstance, and that it was not the cause of the plaintiff's harm.

It is not sufficient for a patient and/or plaintiff to show a breach of duty to prevail in a tort suit. He or she must also show that the breach of the duty caused him or her harm. Even if it is proved that a nurse made a medication error, if the error was not the cause of the plaintiff's harm, the plaintiff will not win in recovering damages from the nurse. In a recent malpractice case, a patient with sickle cell anemia died after suffering a cardiopulmonary arrest attributed to an aspiration that was witnessed by a visitor. The visitor immediately called for and obtained help. Although revived, the patient never regained consciousness and was eventually taken off life support. At trial, the plaintiff was able to prove that the nurse assigned to this patient did not follow the institution's policy of documenting frequent observations, which were mandated because the patient was receiving a blood transfusion at the time of the cardiac arrest. In reviewing the case on appeal, the appellate court noted the following:

*[T]he record contains no evidence which suggests what could have been done even if the nurse had been seated*

*at his bedside prior to the arrest. Plaintiff has failed to offer any proof that more immediate assistance would have prevented the catastrophic results of his aspiration. Based on the evidence in this record, we conclude that more frequent monitoring would have made no difference.* (Webb v. Tulane Medical Center Hospital, [1997])

Thus, even though the plaintiff successfully proved a breach of a duty, the breach was not found to be the cause of the patient's death and the nurse was not found to be guilty of negligence.

Table 23-4 represents the results of a study examining the types and frequency of nursing malpractice actions throughout the United States.

When a nurse is listed as a party in a malpractice lawsuit, the nurse's liability is determined by state laws such as the nurse practice act, the standards for the practice of nursing, and the institution's policies and procedures. Thus, if state laws mandate that a nurse must have a practitioner's order before taking action, then a practitioner's order must be present. Problems arise when the orders are verbal and later it is claimed that the nurse misunderstood and acted in error. Another pitfall is illegible writing that is then misinterpreted, resulting in harm to the patient.

The institution's policies and procedures describe the performance expected of nurses in its employ, and a

## Table 23-4  Nursing Malpractice Cases

**Treatment**

- Failure to timely treat symptoms/illness/disease in accordance with established standards/protocols/pathways

- Failure to timely implement established treatment protocols or established critical pathways

- Delay in implementing ordered, appropriate treatment

- Improper/untimely nursing technique or negligent performance of treatment resulting in injury

- Premature cessation of treatment

**Communication**

- Failure to timely report complications of pregnancy, labor or delivery to physician/licensed independent practitioner

- Failure to timely respond to patient's concerns related to the treatment plan

- Failure to timely notify physician/licensed independent practitioner of patient's condition and/or lack of response to treatment

- Failure to timely report complications of post-operative care to physician/licensed independent practitioner

- Failure to timely obtain physician/licensed independent practitioner orders to perform necessary additional treatment(s)

**Medication**

- Wrong route

- Wrong medication

- Wrong rate

- Infiltration of intravenous medication into tissue and/or sensory injury

- Wrong dose

- Medication not covered under state scope of practice

- Failure to immediately report and record the incorrect or improper administration of medication/prescription

- Wrong patient

- Wrong/delayed time of medication administration

- Missed dose

**Monitoring/Observing/Supervising**

- Abandonment of patient, including checking patient's status at appropriate intervals

- Improper/untimely nursing management of patient or medical complication

*(Continues)*

**TABLE 23-4  (Continued)**

- Improper/untimely nursing management of pre-operative, peri-operative, or post-operative treatment or complication

- Improper/untimely application of restraints, or ordering or management of physical or chemical restraints, and/or failure to remove restraints at proper increments of time

- Improper/untimely nursing management of behavioral health/mental health patient or behavioral health complication

- Improper/untimely nursing management of patients in need of physical restraints, including 1:1 supervision, timed release of restraints, comfort breaks, fluids and nourishment.

Note that "the role of the nurse in medical malpractice litigation has experienced a paradigm shift over the last several years. In the past, nurses were considered by many plaintiffs' lawyers and some judges to be mere 'functionaries' or 'custodians' who played a limited role in the care and treatment of patients." (CNA Financial Corporation. 2009. p. 35). "While nurse as custodian claims continue to be asserted, plaintiffs' lawyers have now begun to pursue claims that focus on the nurse as clinician, responsible for using professional judgment in the course of treatment." (p. 36)

"The following are examples of the new paradigm of nursing claims: (p. 36).

- Following a fall by a geriatric patient, the nurse is sued for failure to change the service plan despite increasing patient problems with gait and behavior.
- A child is born with profound brain damage and the nurse is alleged to have failed to properly interpret fetal monitoring strips.
- A lawsuit charges the nurse with failure to appreciate a patient's risk for skin breakdown and to take appropriate preventive measures.
- After a patient experiences adverse drug reactions, the family alleges that the nurse failed to properly administer and provide the correct dosage.
- A patient in the emergency department has a cardiac arrest, and a lawsuit is filed alleging that the triage nurse failed to appreciate acute cardiac symptomatology."

CNA Financial Corporation. (2009). CNA HealthPro Nurse Claims Study: An Analysis of Claims with Risk Management Recommendations 1997–2007. Hatboro, PA: Author. PDF of report available at http://www.nso.com/nursing-resources/claim-studies.jsp. Accessed April 18, 2011.

For additional statistics, see U.S. Department of Health and Human Services. The Data Bank, (no year). The NPDB (National Practitioner Data Bank) and HIPDB (Healthcare Integrity and Protection Data Bank) are information clearinghouses created by Congress to improve health care quality and reduce health care fraud and abuse in the U.S. Collectively, the NPDB and HIPDB are referred to as the *Data Bank*. http://www.npdb-hipdb.hrsa.gov/index.html. Accessed April 18, 2011.

nurse deviating from them can be liable for negligence or malpractice. Occasionally, such failure to adhere to institutional protocol can result in the employer denying the nurse a defense in a lawsuit.

Practicing nurses must also adhere to the standards of practice for the nursing profession in the community. These standards include such things as checking the six "rights" in medication administration or repositioning the bedbound patient at regular intervals.

## NEGLIGENCE AND NURSING ADVOCACY

The American Nurses Association Code of Ethics and many state nurse practice acts require nurses to serve as patient advocates (ANA Code, 2001). A patient's illness, combined with the institutional nature of hospitals, often results in patients becoming passive recipients of health care instead of active partners. Nurses are often called upon to help patients communicate their desires and needs to the health care team and to be vigilant in protecting the patient's safety and even legal rights. For example, occasionally a provider's order may appear suspect or clearly contrary to accepted medical practice. In such situations,

the nurse must exercise professional judgment and refuse to carry out the order if it would put the patient in danger. Most hospitals have policies and procedures to assist the nurse in carrying out this advocacy function. These procedures often require the nurse to take the issue up the chain of command, from the nursing manager up through the medical staff if necessary. Nurses are increasingly being held liable for negligence in failing to question potentially improper provider orders.

Nurses also serve as advocates by safeguarding patient legal interests, such as the right to make informed health care decisions. In this role, nurses frequently collaborate with other members of the health care team and provide patient education to ensure that patients understand the risks and benefits of procedures, medication regimens, or laboratory tests. Additionally, nurses may help patients express their desires regarding end-of-life decisions to the medical team. Both of these issues are discussed in detail later in this chapter.

It is not uncommon for a nurse to find conflicts between an employer's expectations and the nursing standard of care. A nurse working in a medical-surgical unit, for example, may be asked to take care of an unsafe number of patients, or a surgical nurse with no experience working

## REAL WORLD INTERVIEW

Nursing practice in today's health care environment is multifaceted and complex. Our patients and our communities expect quality care from nurses at all levels of practice. Responsibility to keep abreast of current practices is an integral component of a nurse's licensure requirements. Nursing professionals must understand their scope of practice and the legal requirements necessary to maintain their license in good standing. As a former member of a State Board of Nursing, I can assure you that this is important for all nurses, novice or experienced.

Increased emphasis is being placed on patient safety through regulatory agencies, the JC, and other national initiatives. Commitments to reducing medication errors and surgical infections and improving patient outcomes are just a couple of the ongoing issues that have become a fundamental component of nursing practice.

As a sustained focus is placed on improvement of care and patient safety, evidence-based practice must be integrated into the daily practice patterns of patient care. Proven methods and research-based policies and procedures need to dictate how patient care is administered, rather than "this is how we have always done it." Failure to adhere to best practices within your scope of practice can result in negative patient outcomes and ultimately lead to licensure problems or legal action.

Above all, staying current in skills, knowledge, and education helps nurses to meet the practice standards that are expected in today's health care environment.

**Marsha King, RN, MS, MBA, CNAA**
Chief Nursing Officer, Past President of the Indiana State Board of Nursing
Indianapolis, Indiana

in the ER may be asked to "float" to this unit. In these situations, nurses must advocate on behalf of their patients and their profession, and consider whether it is appropriate to take on such an assignment. If the nurse determines that he or she cannot safely carry out the hospital's order, the nurse must not do so. A nurse in this situation would be wise to help hospital staff maintain safety and notify the supervisor of the patient safety concerns.

## Assault and Battery

**Assault** is a threat to touch another in an offensive manner without that person's permission. **Battery** is the touching of another person without that person's consent. In the health care arena, complaints of this nature usually pertain to whether the individual consented to the treatment administered by the health care professional. Most states have laws that require patients to make informed decisions about their treatment.

Informed consent laws protect the patient's right to practice self-determination (Aveyard, 2002). The patient has the right to receive sufficient information to make an informed decision about whether to consent to or refuse a procedure. The individual performing the procedure has the responsibility to explain to the patient the nature of the procedure, benefits, alternatives, and risks and

## Critical Thinking 23-3

Review the malpractice cases in Table 23-4. Have you ever seen a similar case or one that might result in a malpractice claim? How would you handle it if you saw it?

complications. The signed consent form is used to document that this was done, and it creates a presumption that the patient has been advised of the appropriate risks.

Often, the nurse is asked to witness a patient signing a consent form for treatment. When you witness a patient's signature, you are vouching for two things: that the patient signed the paper and that the patient knows he or she is signing a consent form (Olsen-Chavarriaga, 2000). For a consent form to be legal, a patient, in most states, must be at least eighteen-years-old; be mentally competent; have the procedures, with their risks and benefits, explained in a manner he or she can understand; be aware of the available alternatives to the proposed treatment; and consent voluntarily. The nurse must also be familiar with which other people are allowed by state law to consent to medical treatment for another when that person cannot consent for himself or herself. Frequently, these include the person possessing medical power of attorney; a spouse; adult children; or other relatives, if no one is available in one of the other categories listed.

A nurse may also face a charge of battery for failing to honor an advance directive such as a medical power of attorney, durable power of attorney, or living will. Federal law requires that a hospital ask the patient, upon admission, whether he or she has a living will; if not, the hospital must ask the patient whether he or she would like to enact one. A **living will** is a written advance directive voluntarily signed by the patient that specifies the type of care he or she desires if and when he or she is in a terminal state and cannot sign a consent form or convey this information verbally. It can be a general statement, such as "no life-sustaining measures", or specific, such as "no tube feedings or respirator." Often, the patient's family has difficulty allowing health care personnel to follow the wishes expressed by the patient in a living will, and conflicts arise. These should be communicated to the hospital ethics committee, pastoral care department, risk management department, or whichever hospital department is responsible for handling such issues. If the patient verbalizes wishes regarding end-of-life care to the family, such difficult situations can sometimes be avoided, and the patient should be encouraged to do this, if possible.

The nurse should be familiar with the requirements for the implementation of a living will in the state where the nurse practices.

## Do Not Resuscitate (DNR) Orders

The attending medical practitioner may write a Do Not Resuscitate (DNR) order for an inpatient, directing the staff not to perform the usual cardiopulmonary resuscitation (CPR) in the event of a sudden cardiopulmonary arrest. The practitioner may write such an order without evidence of a living will on the medical record, and the nurse should be familiar with the institution's policies and state law regarding when and how a practitioner can write such an order in the absence of a living will. Often, a DNR order is considered a medical decision that the practitioner can make, preferably in consultation with the family, even without a living will executed by the patient. If the nurse feels such a DNR order is contrary to the patient's or family's wishes, the nurse should consult the policies and procedures of the institution. These may include going up the chain of command until the nurse is satisfied with the course of action. This may entail notifying the nursing supervisor, the medical director, the institution's chief operating officer, state regulators, or the Joint Commission (JC). Often, an institution has an ethics committee that examines such issues and makes a determination on the appropriateness of the order.

## FALSE IMPRISONMENT

**False imprisonment** occurs when individuals are incorrectly led to believe they cannot leave a place. This often occurs because the nurse misinterprets the rights granted to others by legal documents such as powers of attorney and does not allow a patient to leave a facility because the person with the power of attorney (agent) says the patient cannot leave. A **power of attorney** is a legal document executed by an individual (principal) granting another person (agent) the right to perform certain activities in the principal's name. It can be specific, such as "sell my house," or general, such as "make all decisions for me, including health care decisions." In most states, a power of attorney is voluntarily granted by the individual and does not take away the individual's right to exercise his or her own choices. Thus, if the principal (patient) disagrees with his agent's decisions, the patient's wishes are the ones that prevail. If a situation occurs in which an agent, acting on a power of attorney, disagrees with your patient regarding discharge plans, contact your supervisor for further assistance in deciding an action consistent with your patient's wishes and best interests.

The authority to make health care decisions for another may be granted in a general power of attorney document or in a specific document limited to health care decisions only, such as a health care power of attorney. The requirements for a health care power of attorney vary from state to state, as do most legal documents.

A claim of false imprisonment may be based on the inappropriate use of physical or chemical restraints. Federal law mandates that health care institutions employ the least restrictive method of ensuring patient safety. Physical or chemical restraints are to be used only if necessary to protect the patient from harm when all other methods have failed. If the nurse uses restraints on a competent person who is refusing to follow the practitioner's orders, the nurse can be charged with false imprisonment or battery. If restraints are used in an emergency situation,

**CASE STUDY 23-1**

You are working the night shift. The practitioner for one of your patients has ordered a dose of a medication to be given to a patient. The medication dose ordered is too high for this particular patient. You are unable to locate the practitioner to check the order. What would you do to ensure safe care for your patient?

the nurse must contact the practitioner immediately after application to secure an order for the restraints. Also, the nurse must check the institution's policies regarding the type and frequency of assessments required for a patient in restraints and how often it is necessary to secure a reorder for the restraints. These policies ensure the patient's safety and must be consistent with state law.

## INVASION OF PRIVACY

The nurse is required to respect the privacy of all patients. **Invasion of privacy** is defined as the intrusion into the personal life of another without just cause, which can give the person whose privacy has been invaded a right to bring a lawsuit for damages against the person or entity that intruded (Law.com Law Dictionary, 2010). The nurse may be privy to very personal information and must make every effort to keep it confidential. This often necessitates policing conversations with coworkers that have the potential for being overheard by others so that no patient information is accidentally revealed. Nurses must also remember to log off of the computer before leaving a patient's room. Be mindful that other people may be able to see both computer screens and paper records if their privacy is not assured. Sometimes the protection of a patient's privacy conflicts with the state's mandatory reporting laws for the occurrence of specified infectious diseases such as syphilis or human immunodeficiency virus (HIV). The need to protect an individual's privacy may also conflict with the state's mandatory reporting laws on suspected patient abuse, discussed previously. Other information that state or federal law may require to be revealed includes a patient's

blood alcohol level, incidences of rape, gunshot wounds, and adverse reactions to certain drugs. Failing to strictly follow reporting laws could lead to criminal liability, civil liability, or disciplinary action, including the loss of license and the suspension or termination of employment. Nurses must consult the institution's policies and confer with its risk management department to ascertain their responsibilities and course of action. The ANA Code of Ethics states that nurses must protect the patient and the public when incompetence or unethical or illegal practice compromises health care and safety. Many states have adopted this concept in their nurse practice acts, thereby creating a legal obligation to report. Nurses who observe unethical behavior in a hospital should report this as directed in the institution's policies and procedures manual or by the laws of the state.

## LEGAL PROTECTIONS IN NURSING PRACTICE

As discussed earlier in this chapter, nursing practice is guided by state nurse practice acts and agency policies and procedures. Other resources for the nurse include Good Samaritan laws, skillful communication, and risk management programs.

### GOOD SAMARITAN LAWS

**Good Samaritan laws** have been enacted to protect the health care professional from legal liability. The essential

## EVIDENCE FROM THE LITERATURE

**Citation:** Reising, D. L., & Allen, P. N. (2007). Protecting yourself from malpractice claims: Greater nursing autonomy comes at the price of increased legal exposure. *American Nurse Today, 2*(2), 39–43.

**Discussion:** This article provides a wealth of practical information on how to reduce your risk of being sued for malpractice and what to do if you are named in a malpractice lawsuit. The article defines the elements of a malpractice suit, which are based on the traditional four-factor test for negligence. According to the article, the most common malpractice claims against nurses are for failure to follow standards of care, failure to use equipment in a responsible manner, failure to communicate, failure to document, failure to assess and monitor, and failure to act as a patient advocate. The article suggests that nurses take a proactive approach, including performing only those skills that are within your practice scope, staying current in your field or specialty area, knowing your strengths and weaknesses, documenting all patient care activities and communications, and knowing how to invoke the chain of command in your facility.

**Implications for Practice:** The expansion of the nurses responsibilities, along with the public recognition that a nurse's role has evolved from a custodian to a full member of the clinical treatment team, means that a nurse is more likely to be faced with a lawsuit for malpractice.

elements of commonly enacted Good Samaritan law are as follows:

- The care is rendered in an emergency situation.
- The health care worker is rendering care without pay.
- The care provided did not recklessly or intentionally cause injury or harm to the injured party.

Note that these laws are intended to protect the volunteer who stops to render care at the scene of an accident. They would not protect emergency medical technicians (EMTs) or other health care professionals rendering care at the scene of an accident as part of their assigned duties and for which they receive pay. In performing these duties, these paid emergency personnel would be evaluated according to the standards of their professions.

## SKILLFUL COMMUNICATION

The nurse must communicate accurately and completely both verbally and in writing. In the cases detailed earlier in Table 23-4, note how many of the cases were the result of a lack of communication by the nurse. For example, the nurse may have neglected to monitor the patient and notify the practitioner of a change in the patient's status, or the nurse failed to document the assessments performed. It is essential that the nurse document everything related to the patient accurately and thoroughly. Often, a lawsuit involving medical malpractice will take several years to come to trial; by that time, the nurse may have no memory of the incident in question and must rely on the written record prepared at the time of the incident. This record is frequently in the courtroom and may be projected onto a large screen for everyone to see, as the nurse is questioned by the plaintiff's attorney. All errors are apparent, and omissions stand out by their absence, especially if they are data that should have been recorded per institutional policy. The old adage "if it isn't written, it wasn't done" will be repeated to the jury numerous times. To protect themselves when charting, nurses should use the FLAT charting acronym: F—factual, L—legible, A—accurate, T—timely:

- **F:** Charting should be *factual*—what you see, not what you think happened.
- **L:** Charting should be *legible*, with no erasures. Corrections should be made as you have been taught, with a single line drawn through the error and initialed.
- **A:** Charting should be *accurate* and complete. What color was the drainage, and how much was present? How many times, and at what times, was the practitioner notified of changes? Was the supervisor notified?
- **T:** Charting should be *timely*, completed as soon after the occurrence as possible. "Late entries" should be avoided or kept to a minimum.

## RISK MANAGEMENT PROGRAMS

**Risk management programs** in health care organizations are designed to identify and correct systemic problems that contribute to errors in patient care or to employee injury. The emphasis in risk management is on quality improvement and protection of the institution from financial liability. Institutions usually have reporting and tracking forms that record incidents that may lead to financial liability for the institution. Risk management will assist in identifying and correcting the underlying problem that may have led to an incident, such as faulty equipment, staffing concerns, or the need for better orientation for employees. After a systemic problem is identified, the risk management department may develop educational programs to address the problem.

The risk management department may also investigate and record information surrounding a patient or employee incident that may result in a lawsuit. This helps personnel remember critical factors if they are asked to testify at a later time. The nurse should notify the risk management department of all reportable incidents and complete all risk management and/or incident report forms as mandated by institutional policies and procedures. Note also that employee complaints of harassment or discrimination can expose the institution to significant liability and should promptly be reported to supervisors and the risk management, human resources, or whichever other department is specified in the institution's policies. See Table 23-5 for a checklist of actions to decrease the risk of nursing liability.

## MALPRACTICE/PROFESSIONAL LIABILITY INSURANCE

Nurses may need to carry their own malpractice insurance. Nurses often think their actions are adequately covered by the employer's liability insurance, but this is not necessarily so. If, in giving care, the nurse fails to comply with the institution's policies and procedures, the institution may deny the nurse a defense, claiming that because of the nurse's failure to follow institutional policy or because of the nurse working outside the scope of employment, the nurse was not acting as an employee at that time. Also, nurses are being named individually as defendants in malpractice suits more frequently than in the past. It is important to remember that the hospital's attorney is not the nurse's attorney. The hospital's attorney represents the hospital and its interests, which may be adverse to the nurse. Consequently, it is advantageous for the nurse to be assured of a defense independent of that of the employer. Professional liability insurance provides that assurance and covers the cost of an attorney to defend the nurse in a malpractice lawsuit.

## REAL WORLD INTERVIEW

Patient safety and risk management are synonymous terms. Health care delivery processes are inherently complex, high risk, and problem prone. In order to create an environment that promotes optimal patient outcomes, basic nursing and patient care processes and procedures must be properly designed. Well-designed nursing care processes possess the following characteristics:

- Staff-level nursing policies and procedures should be designed with input and participation by those closest to the process—that is, staff level nurses.
- Nursing care policies, processes, and procedures should be simple, practical, and written in universally understandable terms.

During a shadowing experience, I was once asked by a BSN student, "What can bedside nurses do to protect themselves and the organization from liability?" My answer was multifaceted. The single most important risk management tool is well-documented nursing care. It is a challenge in today's nursing environment to assure that an accurate record is made that includes all of the details of a patient's care. Many hospitals utilize "charting by exception," which is designed for efficiency and to capture the essence of nursing care delivery under "normal" circumstances without variation. However, in-depth narrative must be documented, with changes in patient condition or care needs, along with the patient response to our interventions.

The medical record is the only document we will have several years out if a patient care incident results in litigation. For this reason, it must tell a vivid story in complete detail about what the patient looked like, smelled like, felt like, and sounded like at accurate points in time during our care, as well as everything we did for the patient and how they responded to what we did. It is the nurse's responsibility to supplement any standard form to provide this type of information.

Communication between nurses, physicians, and other health care team members is another critical element of safe patient care and effective risk management.

Patient safety and risk management is every individual's responsibility, and everyone has a role to play.

**Tamara L. Awald, RN, BSN, MS**
Vice President of Patient Care Services/Risk Manager
Plymouth, Indiana

## TABLE 23-5 Actions to Decrease the Risk of Liability

- Communicate with your patients by keeping them informed and listening to what they say. Treat patients and their family with kindness and respect.

- Acknowledge unfortunate incidents, and express concern about these events without either taking the blame, blaming others, or reacting defensively.

- Chart and time your observations immediately, while facts are still fresh in your mind.

- Take appropriate actions to meet the patient's nursing needs. Be assertive and professional.

- Acknowledge and document the reason for any omission or deviation from agency policy, procedure, or standard.

*(Continues)*

**TABLE 23-5** (Continued)

- Maintain clinical competency and professional certifications.

- Acknowledge your limitations. If you do not know how to do something, ask for help.

- Promptly report any concern regarding the quality of care, including the lack of resources with which to provide care, to a nursing administration representative.

- Delegate patient care based on the documented skills of licensed and unlicensed personnel.

- Communicate the patient's name, room number, and expectations for staff before, during, and after duty performance in a pleasant, direct, and concise manner when delegating patient care.

- Identify realistic, attainable outcome standards for the completion of any task that is delegated. Make frequent walking rounds to assure quality patient outcomes after delegation.

- Follow evidence-based standards of care and the facility's policies and procedures for administering care and reporting incidents. Document the reason for any omission or deviation from agency policy, procedure, or standard.

- Avoid taking telephone and verbal orders. If needed, repeat the order back to the practitioner to assure clarity. Document that you did this with Verbal Order Repeated Back (VORB).

- Encourage the development of clearly written and/or computerized orders from all practitioners.

- Document the time of nursing actions and changes in conditions requiring notification of the practitioner. Include the response of the practitioner. Use the chain of command at your agency to report any concerns.

- Complete incident reports immediately after incidents occur. Discuss critical factors with the risk manager to increase your retention of the facts.

- Follow professional guidelines for safe transfer of all patients both inside and outside the agency.

## NURSE/ATTORNEY RELATIONSHIP

Despite the nurse's best intentions, a nurse may be named as a defendant in a lawsuit and need to retain the services of an attorney. LaDuke (2000) made the following suggestions for consulting and collaborating with an attorney:

1. Retain a specialist. Generalists are competent to handle many matters, but professional malpractice, professional disciplinary proceedings, and employment disputes are best handled by specialists in those areas.
2. Be attentive. Read the documents the attorney produces, and attend court proceedings to observe the attorney's performance.
3. Notify your insurance carrier as soon as you are aware of any real or potential liability issue. Inform your agent about the status of your case every few months, even if it is unchanged.

4. Keep costs sensible. Your attorney should explain initially how the fee will be computed and how you will be billed. The attorney may require you to pay a retainer fee.
5. Keep informed. The attorney should address your questions and concerns promptly. You are entitled to be kept informed about the status of your case. You are entitled to copies of all correspondence, legal briefs, and other documents.
6. Weed through writing. Your attorney needs to explain all facts and options. Examine all relevant documents, and do not hesitate to make corrections in the same way you would correct a medical record by drawing a line through the incorrect or misleading information, writing in the correction, and signing your initials after it.
7. Set your own course. Insist on a collaborative relationship with your attorney for the duration of your case.

## Critical Thinking 23-4

Go to http://www.qsen.org. Click on Quality/Safety Competencies. Then click on Teamwork and Collaboration. Scroll down and click on Nurse-Physician Communication Exercise. Scroll down again and click on Communication_among_Health_Care_Providers.doc (46Kb). Read the case. Do the role plays and answer the questions. What did you learn about Teamwork and Collaboration?

 ## REAL WORLD INTERVIEW

Most nurses are familiar with the phrase, "If it was not documented, it was not done." Insofar as this phrase is used to encourage thorough documentation, it reflects good nursing practice. Timely, accurate, and complete documentation is an excellent way to protect oneself from litigation. However, lawyers who represent plaintiffs in medical malpractice cases are aware of this "rule" and often attempt to use it against nurses in health care liability claims.

Imagine the following scenario: A patient is admitted to the hospital, and Nurse A performs an initial assessment of the patient. Nurse A notes in the patient's chart that the patient has good capillary refill. Nurse A proceeds to take the patient's vital signs, including capillary refill, hourly throughout Nurse A's eight-hour shift. The patient's capillary refill remains good, and the nurse makes no further documentation in the chart relating to the patient's capillary refill. After Nurse A's shift, Nurse B takes over the patient's care. One hour into Nurse B's shift, the patient codes and expires. The patient's family sues Nurse A. The plaintiffs' lawyer is cross-examining Nurse A.

Lawyer: "Nurse A, are you familiar with the phrase, 'If it wasn't charted, it wasn't done'?"

Nurse A: "Yes."

Lawyer: "That's a common rule in nursing practice, isn't it?"

Nurse A: "Yes."

Lawyer: "You were taught that in nursing school, weren't you?"

Nurse A: "Yes, I was."

Lawyer: "And after you documented that the patient had good capillary refill upon admission, you did not document anything relating to the patient's capillary refill for the next eight hours, did you?"

Nurse A: "Well, no."

Lawyer: "So if we use your rule, 'If it wasn't documented, it wasn't done,' we can assume you never checked the patient's capillary refills during your shift after the initial assessment, right?"

Nurse A: "No. I checked, but it hadn't changed, so I didn't chart anything..."

Do you see what just happened? Nurse A provided competent nursing care, but the lawyer made it appear as if Nurse A was negligent. A nurse involved in litigation should not blanketly agree with this documentation rule. You simply cannot document everything noted in an assessment of a patient. Moreover, most nurses would agree that patient care takes priority over charting. This rule ignores that. Bad charting looks bad. Good charting protects you. Even lapses in charting do not correlate with bad nursing care. Nurses should not lose sight of that when faced with litigation.

**Robyn D. Pozza Dollar, JD**

Austin, Texas

## KEY CONCEPTS

- Nursing practice is governed by civil, public, and administrative laws.

- Nurses are legally responsible for their actions and can be held liable for negligent care.

- Nurses have an ethical and legal obligation to advocate for patients.

- Every health care institution has its own policies and procedures designed to help nurses carry out their professional responsibilities.

- Nurses need to be familiar with their institution's policies and procedures for giving care and for reporting variances, illegal activities, or unexpected events.

- HIPAA legislation was enacted to safeguard private health care information.

- Many sources of evidence are used to identify the standard of care.

- Nursing malpractice examples include treatment problems, communication problems, medication problems, and monitoring/observing/supervising problems.

- The ANA Code of Ethics requires nurses to serve as advocates for patients.

- Legal protections in nursing practice include Good Samaritan laws, skillful communication, and risk management programs.

- Common torts in health care include negligence and malpractice, assault and battery, false imprisonment, invasion of privacy, and defamation.

- Nurses need to be familiar with their state nurse practice act.

- Good Samaritan laws exist in many states.

- Risk management programs improve the quality of care and protect the financial integrity of institutions.

## KEY TERMS

assault
battery
civil law
constitution

contract law
false imprisonment
Good Samaritan laws
invasion of privacy

living will
malpractice
negligence
power of attorney

public law
risk management
    programs
tort

## REVIEW QUESTIONS

1. You are given a written order by a provider to administer an unusually large dose of pain medicine to your patient. In this situation, which of the following is an appropriate nursing action?
   A. Administer the medication because it was ordered by a provider.
   B. Refuse to administer the medication, and move on to another patient.
   C. Speak with the provider about your concerns, and clarify whether the medication dose is accurate.
   D. Select a dose that you feel comfortable with, and administer that dose.

2. A practitioner has ordered you to discharge Mr. Jones from the hospital, despite a new temperature of 102.0°F (38.8°C). The practitioner refuses to talk with you about the patient. In this situation, which of the following is an appropriate nursing action?
   A. Administer an antipyretic medication, and discharge the patient.
   B. Discharge the patient with instructions to call 911 if he has any problems.

   C. Do not discharge the patient until you have discussed the matter with your nursing manager and are satisfied regarding patient safety.
   D. Discharge the patient, and tell the patient to take Tylenol when he gets home.

3. A practitioner has issued a Do Not Resuscitate (DNR) order for your patient, a fifty-five-year-old man with cancer. You spoke with the patient this morning, and he clearly wishes to be resuscitated in the event that he stops breathing. What is the most appropriate course of action?
   A. Ignore the patient's wishes because the practitioner ordered the DNR.
   B. Consult your hospital's policies and procedures, speak to the practitioner, and discuss the matter with your nurse manager.
   C. Attempt to talk the patient into agreeing to the DNR.
   D. Contact the medical licensing board to complain about the practitioner.

4.  The Health Insurance Portability and Accountability Act (HIPAA) protects which of the following?
    A.  A patient's right to be insured, regardless of employment status or ability to pay
    B.  The confidentiality of certain protected health information
    C.  The nurse's right to health insurance
    D.  The hospital's right to disclose protected health information

5.  Which of the following elements is not necessary for a nurse to be found negligent in a court of law?
    A.  A duty or obligation for the nurse to act in a particular way
    B.  A breach of that duty or obligation
    C.  The nurse's intention to be negligent
    D.  Physical, emotional, or financial harm to the patient

6.  Which type of law authorizes state boards to enact rules that govern the practice of nursing?
    A.  State law
    B.  Federal law
    C.  Common law
    D.  Criminal law

7.  You are a new nurse working on a medical-surgical unit. One of your patients, an elderly woman, has an advanced directive that requests that no CPR be done in the event that she stops breathing. One day she stops breathing, and someone on your unit calls a "code" and begins resuscitative efforts. You go along with the team and help to resuscitate the patient. She regains a pulse but never regains consciousness. She is now ventilator dependent, and her family is very angry with you and the staff. Which of the following is a potential legal action you will face?
    A.  Violation of patient privacy
    B.  Battery

C.  Criminal recklessness
D.  Revoked nursing license

8.  Which of the following is not an essential element of a Good Samaritan law?
    A.  The care is rendered in an emergency situation.
    B.  The health care worker is rendering care without pay.
    C.  The health care worker is concerned about the safety of the victims.
    D.  The care provided did not recklessly or intentionally cause injury or harm to the injured party.

9.  Some of the situations in which a nurse could be accused of false imprisonment are which of the following? Select all that apply.
    _____ A.  Inappropriate use of chemical restraints
    _____ B.  Inappropriate use of physical restraints
    _____ C.  Restraining a competent person
    _____ D.  Failure to follow the institutions policies regarding the type and frequency of restraints
    _____ E.  Using restraints in an emergency situation to protect the patient from harm

10. The goal of a risk management program in an institution is to do which of the following? Select all that apply.
    _____ A.  Identify and correct systemic problems in the facility
    _____ B.  Identify when better orientation is needed for staff
    _____ C.  Identify issues with equipment and staffing
    _____ D.  Punish the person responsible for the error
    _____ E.  Reduce legal and financial liability for the institution
    _____ F.  Reduce the number of errors made in patient care

## REVIEW ACTIVITIES

1.  Talk to the risk manager at the hospital in which you have your clinical assignments. Ask the risk manager how he or she handles an incident report. Is it used for improving the hospital's care in the future? If so, how?

2.  Identify the various ways in which nurses you observe in your clinical rotations discuss orders and treatments with practitioners and other nurses. How do nurses address incorrect or dangerous medication orders? Talk with nurses you encounter about how they handle these situations.

3.  Research the various companies that offer nursing malpractice insurance, and determine the cost and coverage associated with a nursing malpractice policy. Go to an Internet search engine, such as www.google.com. Search for nursing malpractice insurance. What did you find? Note the Nurses Service Organization (NSO) Web site at www.nso.com. Recent legal cases are reported there.

## EXPLORING THE WEB

- Go to this site to find malpractice information for your state: http://www.mcandl.com/states.html

- Where can you find state and federal laws regulating hospitals? www.findlaw.com

- You have a patient who is to be transferred to a nursing home for recuperation. Where can you tell the family to look to evaluate the local nursing homes regarding their adherence to the federal regulations for nursing homes? www.medicare.gov

- Note the Medical Liability Monitor at www.medicalliabilitymonitor.com Also check this site http://aspe.hhs.gov/daltcp/reports/mlupd1.htm What did you find there?

- Go to www.google.com, and type in "living wills" and "power of attorney." What did you find there?

- Go to this site to find the impact of the HITECH Act on the privacy of health information: www.hhs.gov

## REFERENCES

Adams, S. (2004). HIPAA patient confidentiality requirements. *Journal of Emergency Nursing, 30*(1), 70.

American Nurses Association. (2001). Code of Ethics. Silver Springs, MD: Author.

Aveyard, H. (2002). The requirement for informed consent prior to nursing care procedures. *Journal of Advanced Nursing, 37*(3), 243–249.

CNA Financial Corporation. (2009). CNA HealthPro Nurse Claims Study: An Analysis of Claims with Risk Management Recommendations 1997–2007. Hatboro, PA: Author. PDF of report available at http://www.nso.com/nursing-resources/claim-studies.jsp. Accessed April 18, 2011.

Committees on Energy and Commerce, Ways and Means, and Science and Technology, January 16, 2009. Health Information Technology for Economic and Clinical Health (HITECH Act). http://waysandmeans.house.gov/media/pdf/110/hit2.pdf. Accessed August 18, 2010.

*Cruzan v. Director, Missouri Department of Health,* 497 U.S. 261 (1990).

Davis, P. C. The nurse's duty to intervene: Initiating the chain of command. (2004). Emergency Care Research Institute; HRC-Risk Analysis: Risk and Quality Management Strategies 19, Suppl. A, Chain of Command.

Frank-Stromborg, M. (2004). They're real and they're here: The new federally regulated privacy rules under HIPAA. *Dermatology Nursing, 16*(1), 13–24.

LaDuke, S. (2000). What should you expect from your attorney? *Nursing Management, 31*(1), 10.

Law.com Law Dictionary. (2010). Available at http://dictionary.law.com. Accessed August 13, 2010.

Mantel, D. L. (1999). Legally speaking: Off-duty doesn't mean off *the* hook. *RN, 62*(10), 71–74.

*Murphy v. United Parcel Service,* 527 U.S. 516 (1999).

*Odom v. State Department of Health & Hospitals,* 733 So. 2d 91 (La. App. 3 1999).

Olsen-Chavarriaga, D. (2000). Informed consent: Do you know your role? *Nursing 2000, 30*(5), 60–61.

Reising, D. L., & Allen, P. N. (2007). Protecting yourself from malpractice claims: Greater nursing autonomy comes at the price of increased legal exposure. *American Nurse Today, 2*(2), 39–43.

*Roe v. Wade,* 410 U.S. 133 (1973).

*Sutton v. United Airlines,* 527 U.S. 471 (1999).

U.S. Department of Health and Human Services. The Data Bank, (no year). The NPDB (National Practitioner Data Bank) and HIPDB (Healthcare Integrity and Protection Data Bank) are referred to as *the Data Bank.* http://www.npdb-hipdb.hrsa.gov/topNavigation/aboutUs.jsp. Accessed April 18, 2011.

*Webb v. Tulane Medical Center Hospital,* 700 So. 2d 1141, 1145 (La. 1997).

## SUGGESTED READINGS

Constantino, R. E. (2007). A transdisciplinary team acting on evidence through analyses of moot malpractice cases. *DCCN—Dimensions of Critical Care Nursing, 26*(4), 150–155.

Detmer, D. E. (2010). Engineering information technology for actionable information and better health: Balancing social values through desired outcomes, complementary standards and decision support. *Studies in Health Technology & Informatics, 153,* 107–118.

Kendall, D. B. (2008). Symposium: Improving health care in America: Protecting privacy in the information age. *Harvard Law & Policy Review, 2*(2), 1–5.

Kenward, K. (2008). Discipline of nurses: A review of disciplinary data 1996–2006. *JONA's Healthcare Law, Ethics, & Regulation, 10*(3), 81–84.

Klein, K. (2010). So much to do, so little time: To accomplish the mandatory initiatives of ARRA, healthcare organizations will require significant and thoughtful planning, prioritization and execution. *Journal of Healthcare Information Management, 24*(1), 31–35.

Larson, K., & Elliott, R. (2010). The emotional impact of malpractice. *Nephrology Nursing Journal, 37*(2), 153–156.

Pastorius, D. (2007). Crime in the workplace, part 1. *Nursing Management, 38*(10), 18–27.

Pastorius, D. (2007). Crime in the workplace, part 2. *Nursing Management, 38*(10), 14–27.

Rolfsen, M. L. (2007). Medical care provided during a disaster should be immune from liability or criminal prosecution. *Journal of the Louisiana State Medical Society, 159*(4), 227–229.

Sanchez Abril, P., & Cava, A. (2008). Health privacy in a techno-social world: A cyber-patient's bill of rights. *Northwestern Journal of Technology and Intellectual Property, 6*(3), 244–276.

Weld, K. K., & Garmon Bibb, S. C. (2009). Concept analysis: Malpractice and modern-day nursing practice. *Nursing Forum, 44*(1), 2–10.

Worel, M. A., & Wirtes, D. G., Jr. (2007). Don't neglect the nurse's duty of care. *Trial, 43*(11), 50–57.

Young, A. (2009). Review: The legal duty of care for nurses and other health professionals. *Journal of Clinical Nursing, 18*(22), 3071–3078.

Zalon, M. L., Constantino, R. E., & Andrews, K. L. (2008). The right to pain treatment: A reminder for nurses. *DCCN – Dimensions of Critical Care Nursing, 27*(3), 93–103.

# CHAPTER 24

# Ethical Aspects of Health Care

**Camille B. Little, MS, RN, BSN; Joan Dorman, RN, MS, CEN**

*Moral excellence comes about as a result of habit, we become just by doing just acts, temperate by doing temperate acts, brave by doing brave acts.*

(Aristotle)

## OBJECTIVES

Upon completion of this chapter, the reader should be able to:

1. Define ethics and morality.
2. Identify historical and philosophical influences on nursing practice.
3. Devise a personal philosophy of professional nursing.
4. Analyze ethical theories, virtues, principles, and values as the basis for professional nursing practice.
5. Support participation on ethics committees in hospitals.
6. Explain values clarification.
7. Apply an Ethical Positioning System (EPS) model to an ethical dilemma.
8. Evaluate ethical issues encountered in practice, including cost containment, use of technology, and patients' rights.
9. Support ethical leadership and responsibility in professional practice and organizations.

Delmar/Cengage Learning

*In a large teaching hospital, a patient you are caring for says he does not want to go on living. He has had cancer for several years and states he is tired of being sick. When you ask him whether he has shared these feelings with his family, he says that he does not want them to think he is giving up. You report the patient's statements to the nurse on the next shift and explain how you encouraged him to talk with his family and his practitioner. That evening, the patient suddenly has a cardiac arrest and a code is called. The patient ends up on a ventilator, receives five units of blood, and is comatose. The patient did not have advance directives and Do Not Resuscitate (DNR) orders had not been signed.*

> *What are your thoughts about maintaining the patient's life in this situation?*

> *Who should make the decision about the patient's situation while he is comatose?*

Throughout its history, nursing has relied on ethical principles to serve as a guideline in determining care. Nurses are confronted with ethical dilemmas in all types of practice settings. This chapter provides an overview of the nursing profession's ethics and the increased ethical challenges faced by nurses in today's health care environment.

## DEFINITION OF ETHICS AND MORALITY

**Ethics** is the branch of philosophy that concerns the distinction of right from wrong on the basis of a body of knowledge, not just on the basis of opinions. **Morality** is behavior in accordance with custom or tradition and usually reflects personal or religious beliefs (DeLaune & Ladner, 2011). Ethics governs professional groups and provides a framework for determining the right course of action in a particular situation. For nurses, the actions they take in practice are primarily governed by the ethical principles of the profession. These principles influence practice, conduct, and relationships that nurses are held accountable for in the delivery of care. Health care ethics, also called **bioethics**, are ethics specific to health care and serve as a framework to guide behavior. An **ethical dilemma** occurs when there is a conflict between two or more ethical principles; there is no "correct" decision.

Laws, in contrast, are state and federal government rules that govern all of society. Laws mandate behavior. In some situations in health care, the distinctions between law and ethics are not clear. There are cases in which ethics and law are similar; there are others where ethics and law differ.

## HISTORICAL AND PHILOSOPHICAL INFLUENCES ON NURSING PRACTICE

Nursing practice evolved from the needs of society and has been strongly influenced by religions and women. Society created the profession of nursing for the purpose of meeting specific, perceived health needs (Burkhardt & Nathaniel, 2008). Nursing fulfilled the need to care for people with illnesses. Likewise, a strong instinct for the preservation of humanity gave people the motivation to help one another. The concern for the health of the community was evident in antiquity and continued as civilizations developed (Donahue, 2011). A complementary relationship evolved as social needs and individual motivation to care for others developed.

### RELIGIOUS INFLUENCES

Workers who were engaged in nursing, usually women, were often trained in the doctrines of the church, including unquestioning obedience, humility, and sacrificing oneself for the good of others. An individual nurse did not make independent decisions, but followed instructions given by a priest or practitioner.

### WOMEN'S INFLUENCES

Donahue states that nursing has its origin in mother's care of helpless infants and must have coexisted with this type of care from earliest times (Donahue, 2011). Mothers cared for family members when they were helpless and sick. During the Christian era, women were selected by Jesus because of the compassion they showed as they ministered to the poor and sick. Thus, Christianity greatly enhanced women's opportunities for useful social service (Donahue, 2011).

# PHILOSOPHY

**Philosophy** is the rational investigation of the truths and principles of knowledge, reality, and human conduct. Personal philosophies stem from an individual's beliefs and values. These beliefs and values, in turn, develop based upon a person's experiences in life, cultural influences, and education.

## Philosophy of Nursing

A professional nurse's personal philosophy affects that nurse's philosophy of nursing. Throughout the nursing educational process, students begin forming their philosophy of nursing. This philosophy is influenced significantly by a student's personal philosophy and experiences. One's personal philosophy should be compatible with the philosophy of the nursing department where the nurse works. This helps the nurse be an effective leader and practitioner. See Figure 24-1 for a nursing department's philosophy statement. An example of a personal nursing philosophy is the following:

> *I believe professional nursing care promotes an optimal level of wellness in body, mind, and spirit to those being served. I believe professional nurses must hold themselves to the highest standards of the profession and honor the profession's code of ethics in all aspects of practice.*

**Memorial's Professional Practice Model**
**Design for Excellence**

**Mission–**
Professional nurses and nursing teams exist to optimize the health of our patients and communities.

We believe many conditions must work together in order to nurture a professional environment. More specifically, the manner in which beliefs, practice, structure, relationships, and operations interact will determine the quality of outcomes for us as well as for those we serve.

FIGURE 24-1 Nursing philosophy. (*Source:* Adapted with permission from Memorial Health System's Professional Practice Philosophy, Springfield, IL.)

## TABLE 24-1 Selected Ethical Theories

| ETHICAL THEORY | INTERPRETATION |
| --- | --- |
| Deontology | Actions are based on moral rules and unchanging principles, such as "do unto others as you would have them do unto you." An ethical person must always follow the rules, even if doing so causes a less desirable outcome. Theory states that the motives of the actor determine the goodness or value of the act. Thus, a bad outcome is acceptable as long as the intent was good. |
| Teleology | A person must take those actions that lead to good outcomes. The theory states that the outcome of an act determines whether the act is good or of value and that achievement of a good outcome justifies using a less desirable means to attain the end. |
| Virtue ethics | Virtues such as truthfulness and trustworthiness are developed over time. A person's character must be developed so that by nature and habit, the person will be predisposed to behave virtuously. Living a virtuous life contributes both to one's own well-being and to the well-being of society. |
| Justice and equity | A "veil of ignorance" regarding who is affected by a decision should be used by decision makers, because it allows for unbiased decision making. An ethical person chooses the action that is fair to all, including those persons who are most disadvantaged. |
| Relativism | There are no universal ethical standards, such as "murder is always wrong." Ethical standards are relative to person, place, time, and culture. Whatever a person thinks is right, is right. This theory has been largely rejected. |

Many health care centers have addressed ethical concerns by developing professional practice models. The American Nurses Association Nursing Code of Ethics (ANA, 2001) and Nursing Scope and Standards of Practice (ANA, 2010) are useful in developing these models. Professional practice models are shared with employees and consumers to illustrate the institution's commitment to excellence.

## ETHICAL THEORIES

The study of ethical behavior has resulted in different theories that may apply to nursing practice and form a framework for ethical decision making. Table 24-1 describes some of these theories.

## VIRTUES

Burkhardt and Nathaniel (2008) list four virtues that are more significant than others and that are illustrative of a virtuous person: compassion, discernment, trustworthiness, and integrity. Compassion is a trait nurses have, as perceived by society. It refers to the desire to alleviate suffering. Discernment is possession of acuteness of judgment. Trustworthiness is present when trust is well-founded or deserving. Integrity may be considered firm adherence to

a code of conduct or an ethical value. Trustworthiness and integrity are traits expected in all people but are especially necessary for professional nurses. These traits form the foundation for an ethically principled discipline and have been

### Critical Thinking 24-2

New graduates should formulate a philosophy of nursing based on personal beliefs and values. Reflections on the following questions can assist in the development of a philosophy:

What do I believe about nursing practice?

Should nurses be patient advocates?

How should professionals conduct themselves when patient values differ from personal values?

How can I influence patient care based on my nursing philosophy?

Are the virtues of compassion, discernment, trustworthiness, and integrity important both personally and professionally?

## Critical Thinking 24-3

Mr. Johanssen smokes three packs of cigarettes a day and is seen in a free clinic for chronic obstructive pulmonary disease. All attempts to get him to stop smoking have failed. Mr. Johanssen tells you that smoking is the one pleasure he has in life and he does not want to give it up.

Do you respect Mr. Johanssen's wishes? Does he still have a right to the free treatment and medications? Are limits to Mr. Johanssen's treatments justified?

## ETHICS COMMITTEES

In this time of incredible technological advancement, hospitals find themselves increasingly faced with ethical dilemmas. These ethical dilemmas span the age continuum from pre-birth and birth to death and post-death. Most health care institutions have ethics committees to help deal with these ethical dilemmas. An ethics committee is comprised of an interdisciplinary group representing medicine, nursing, pastoral care, pharmacy, nutritional services, social services, quality management, legal services, and the community. On any given occasion, there may also be

endorsed throughout the profession's history (Burkhardt & Nathaniel, 2008).

## ETHICAL PRINCIPLES AND VALUES

In addition to the theories, ethical principles and values provide a basis for nurses to determine the appropriate action when faced with an ethical dilemma in the practice setting. See Table 24-2 for a summary of the major ethical principles.

## Critical Thinking 24-4

You are caring for Mr. Trout, who has been labeled a malingerer. He is in and out of the hospital frequently and always has some type of pain requiring parenteral medications. The order is to give a placebo when he asks for pain medication. When you take the placebo in, Mr. Trout asks you what the medication is.

What ethical principles can guide your actions regarding Mr. Trout? What ethical theories can guide your actions in relation to this patient? Is a nurse ethically required to disclose placebo medication?

### TABLE 24-2   Ethical Principles

| ETHICAL PRINCIPLE | DEFINITION | EXAMPLE |
|---|---|---|
| Beneficence | The duty to do good to others and to maintain a balance between benefits and harms | • Provide all patients, including the terminally ill, with caring attention.<br>• Become familiar with your local, state, and national laws regarding organ donations.<br>• Treat every patient with respect and courtesy. |
| Nonmaleficence | The principle of doing no harm | • Always work within your American Nurses Association (2010), Nursing: Scope and Standards of Practice.<br>• Never give information or perform duties you are not qualified to do.<br>• Observe all safety rules and precautions. |

*(Continues)*

## TABLE 24-2 (Continued)

| ETHICAL PRINCIPLE | DEFINITION | EXAMPLE |
|---|---|---|
| | | • Keep areas safe from hazards. |
| | | • Perform procedures according to facility protocols. Never take shortcuts. |
| | | • Ask an appropriate person about anything you are unsure of. |
| | | • Keep your education and skills up to date with competency building and life-long learning. |
| Justice | The principle of fairness that is served when an individual is given that which he or she is due, owed, deserves, or can legitimately claim | • Treat all patients fairly, regardless of economic or social background. <br> • Learn the local, state, and national laws and your facility's policies and procedures for chain-of-command handling and reporting of suspected abuse. |
| Autonomy | Respect for an individual's right to self-determination; respect for individual liberty | • Be sure that patients have consented to all treatments and procedures. <br> • Become familiar with local, state, and national laws and facility policies dealing with advance directives. <br> • Never release patient information of any kind unless there is a signed release. <br> • Do not discuss patients with anyone who is not professionally involved in their care. <br> • Protect the physical privacy of patients. |
| Fidelity | The principle of promise keeping; the duty to keep one's promise or word | • Be sure that necessary contracts have been completed. <br> • Be very careful about what you say to patients. They may only hear the "good news." Provide written information, as appropriate. |
| Respect for others | The right of people to make their own decision | • Provide all persons with information for decision making. <br> • Avoid making paternalistic decisions for others. |
| Veracity | The obligation to tell the truth | • Admit mistakes promptly. Offer to do whatever is necessary to correct them. <br> • Refuse to participate in any form of fraud. <br> • Give an "honest day's work" every day. |

FIGURE 24-2 **Ethics committee meeting.** (*Source:* Courtesy St. Catherine's Hospital Ethics Committee, July 2010.)

guests—e.g., members from a specialty area such as obstetrics or oncology. Family members may also be invited to an ethics committee meeting (Figure 24-2).

The mission of an ethics committee is to provide thoughtful and timely consultation when an ethical issue arises. This might involve an emergency meeting of some or all of the members. There is generally a written policy guiding the consultation process. Appropriate committee members, as well as family representatives, the patient's physician(s), and other care givers, may be invited. The committee needs to know the background of the case and all pertinent information. They need to know the options, along with the risks and benefits of these options, and review the ethical theories, values, and principles involved. They need to examine the possibilities for resolution and the potential outcomes of those resolutions. The committee does not make a decision. The ethics committee consults, gives guidance, and provides any resources needed for ethically sound decision making.

In the case in the opening scenario, the ethics committee may invite family members of the patient as well as nursing and medical practitioners who care for the patient to assist the ethics committee with the review.

## VALUES AND VALUES CLARIFICATION

**Values** are personal beliefs about the truth of ideals, standards, principles, objects, and behaviors that give meaning and direction to life. If you were told that you must pack a bag for a special trip but you may bring only three items from your belongings, what items would you choose? The ones selected are what you value.

### Values Clarification

Values clarification is the process of analyzing one's own values to better understand what is truly important. In their classic work—*Values and Teaching*—Raths, Harmin,

## EVIDENCE FROM THE LITERATURE

**Citation:** Hamric, A. B., & Blackhall, L. J. (2007). Nurse-physician perspectives on the care of dying patients in intensive care units: Collaboration, moral distress, and ethical climate. *Critical Care Medicine, 35*(2), 641–642.

**Discussion:** This study is designed to explore and compare the perspectives of physicians and nurses caring for dying patients in the ICU. The conclusions were that RNs experienced more moral distress and less collaboration than did physicians. RNs also perceived the ethical environment to be more negative, and they were less satisfied with the care on their units. The study found that discussions of the moral distress experienced while caring for dying ICU patients and improved collaboration between nurses and physicians might lead to improvement in RN perceptions.

**Implications for Practice:** It appears that working in an intensive care unit with healthy physician-nurse relationships and good collaboration between nurses and physicians would lessen the frustration of moral distress and poor collaboration between these groups. Both groups should work to achieve this improvement.

## Critical Thinking 24-5

Select a patient care ethical issue of your choice in class. Have the class split into two groups, one on each side of the issue. Have each group discuss the ethical issues involved. Instruct the groups to consider all pertinent ethical theories and principles; any conflicts between them; any relationship to ethical codes, legal implications, and the people involved and impacted; and any relevant sociocultural, political, or religious aspects that may influence this ethical issue. Then ask a volunteer from each group to state the group's position. Share any references that the group may have found.

What were the pros identified? What were the cons?

Did any of the group's comments alter your own position on this ethical issue?

**Source:** Developed with information from Candela, L., Michael, S. R., & Mitchell, S. (2003). Ethical debates. *Nurse Educator, 28*(1), 37–39.

and Simon (1978, p. 47) formulated a theory of values clarification and proposed a three-step process of valuing, as follows:

1. *Choosing:* Beliefs are chosen freely (that is, without coercion) from among alternatives. The choosing step involves analysis of the consequences of various alternatives.
2. *Prizing:* The beliefs that are selected are cherished (that is, prized).
3. *Acting:* The selected beliefs are demonstrated consistently through behavior.

Nurses must understand that values are individual rather than universal; therefore, nurses should not try to impose their own values on patients.

## GUIDES TO ETHICAL DECISION MAKING

Nurses have long sought guidance in the face of ethical dilemmas. The differences in knowledge and skill between novice and seasoned nurses is measurable, but the ease or torment with which these same nurses make ethical decisions is often virtually the same. For that reason, it seems that a specific ethical decision-making tool is helpful. Just as most of us rely on some sort of Global Positioning System (GPS) navigation device when traveling into "uncharted waters," nurses often find themselves looking for the route that might lead them to an ethical decision. Several authors have proposed guides for ethical decision making. Burkhardt and Nathaniel (2008) propose a five-step decision-making process to guide nurses in making ethical decisions. The five steps are as follows:

- Articulate the problem.
- Gather data, and articulate conflicting moral claims.
- Explore strategies.
- Implement the strategy chosen.
- Evaluate the outcome.

Husted and Husted (2007) discuss an ethical decision-making approach based on a Symphonological Model. Central in this model is a health care professional/patient agreement on a decision. This agreement is linked to the ethical principles of autonomy, freedom, self-assertion, beneficence, objectivity, and fidelity, and considers three contexts, the situation, knowledge, and awareness.

Jonsen, Siegler, and Winslade (2006) have published a practical approach to ethical decision making in clinical medicine. Their model is directed not only to physicians and medical students, but to hospital administrators, ethics committees, and other health care workers. Their model consists of four topics:

- Medical indications: How might this patient benefit?
- Patient preferences: Is the patient's right to choose being respected?
- Quality of life: What are the prospects for a meaningful life?
- Contextual features: What other issues need to be considered?

In examining each of these four areas, the authors ask that we consider the ethical principles inherent in the decision we ultimately make.

In 2009, this author made a presentation at Oxford University at a roundtable titled Ethics: The Convolution of Contemporary Values. At that time, the Ethical Positioning System (EPS) Model (Figure 24-3) was introduced to an audience of doctors, lawyers, theologians, journalists, philosophers, and ethicists. It was well received, and representatives from specialties other than nursing thought they might easily adapt the tool.

The Ethical Positioning System Model follows the format of the nursing process, with a few additions (Table 24-3).

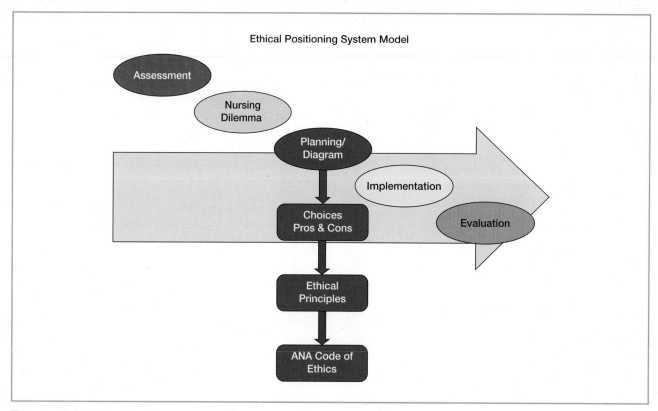

**FIGURE 24-3** **The Ethical Positioning System Model.** (*Source:* Courtesy of Joan Dorman, unpublished manuscript)

## TABLE 24-3   The Ethical Positioning System (EPS) Model

1. Assessment: In this step, the nurse gathers all available data. This includes identifying all people involved in the decision-making process and exploring everything pertinent to the context of the situation (Figure 24-4).

2. Nursing Dilemma: This is a simple statement of the problem.

3. Planning: In this step, the nurse examines all possible choices of actions and identifies pros and cons for each choice. Ethical theories and principles and the American Nurses Association Code of Ethics for Nurses with Interpretive Statements (2001) are reviewed for each choice.

4. Diagram: This is a visual image of the planning process that facilitates making a decision. Pros and cons are explored and choices are eliminated.

5. Implementation: One choice is selected and implemented.

6. Evaluation: The choice is evaluated. Did it work? What was the outcome? What was learned?

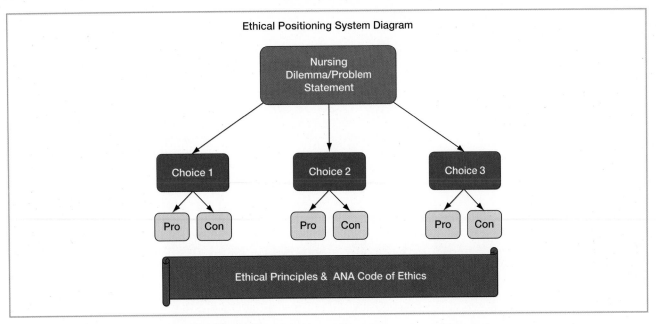

Ethical Positioning System Diagram

**FIGURE 24-4** **The Ethical Positioning System Diagram.** (*Source:* Courtesy of Joan Dorman, unpublished manuscript)

## CASE STUDY 24-1

A nursing student, Tom, was assigned to spend a clinical day in the emergency department of a hospital. Tom was sitting at the nurses' station when a patient with schizophrenia was brought into the department. The patient was shouting and acting in a threatening manner. The patient was placed in full leather restraints for her own safety and for the safety of the staff. By this time, the patient was yelling obscenities, crying, and begging to be released. The nursing staff was laughing and joking about the patient's behavior. Tom was concerned for the patient and decided to use the Ethical Positioning System model (Figure 24-3) to assess the situation, though he was not sure what he could do as a nursing student. That was the nursing dilemma.

In the planning step, step 3, Tom considered all the possible choices and made a diagram of those choices and their pros and cons. He reviewed the applicable ethical principles and the ANA Code of Ethics with Interpretive Statements (2001). His first choice was, he could do nothing. The pros to this choice were that Tom would not have to confront the nurses and would not risk being looked upon as a busybody and a know-it-all. The cons to this choice were that he would feel guilty for not acting as a patient advocate, the patient would remain distressed, and the other patients in the department would suffer, both then and in the future.

A second choice would be to approach the nurses about their nontherapeutic handling of this patient and their unprofessional behavior. The pro to this would be that Tom would feel he was actively being a patient advocate. The cons would be that the nurses might ignore him, they might criticize his interference, and the situation would not change.

A third choice is that Tom could approach the nurse manager and ask him or her to arrange for an educational program to deal with the lack of patient compassion and sensitivity. The pro of this choice would be that Tom would feel he was at least addressing the problem. However, this choice would do nothing for the immediate situation.

A fourth choice was that Tom could go sit with the patient, try to calm the patient down, and give emotional support. The pro of this choice would be that Tom would be proactive and therapeutic. The cons would be that the staff nurses might not even notice and there would be no lasting effect in their view of or approach to patients with mental health conditions.

*(Continues)*

## CASE STUDY 24-1 (Continued)

The American Nurses Association. (2001). Code of Ethics for Nurses with Interpretive Statements discusses several provisions that are applicable to this case. Some of these are respect for human dignity (1.1), primacy of the patient's interests (2.1), professional boundaries (2.4), responsibility for nursing judgment and action (4.2), moral self-respect (5.1), influence of the environment on moral virtues and values (6.1), and responsibilities to the public (8.2). The ethical principles that apply to this case are respect, justice, beneficence, nonmaleficence, and fidelity.

As Tom analyzed the diagram, he selected the fourth choice. He then took step five, implementation. Tom entered the room and spent considerable time with the patient. She calmed down and appreciated Tom's concern. Also, the other patients no longer had to listen to the patient's screams. As Tom had predicted, the staff nurses did not even notice. It appeared that there would be no lasting effect in their view of or approach to patients with mental health conditions.

In the Ethical Positioning System's final step, evaluation, Tom reviewed the process and his choice. He found that he was not completely satisfied. He believed that his choice comforted the current patient but did little to affect future patient care. What do you think? How would you have chosen from the choices listed or from additional choices? Is it always easy to make an ethical choice?

## EVIDENCE FROM THE LITERATURE

**Citation:** Hernandez, A. H. (2009). Student articulation of a nursing philosophical statement: An assignment to enhance critical thinking skills and promote learning. *Journal of Nursing Education,* *48*(6), 343–349.

**Discussion:** This article discusses a graduate or undergraduate nursing school assignment that focuses on the development of a personal philosophical statement of nursing. The finished product is a 5- to 10-page paper. The process involves class discussion and group work. The students discuss their beliefs, assumptions, and values as they relate to the four metaparadigm concepts of nursing, person, health, and environment. Although the grading criteria for the final paper is firm, students are able to express their personal thoughts and are encouraged to think critically.

**Implications for Practice:** The self-discovery process that is integral in this assignment would be beneficial for all nurses. The assignment itself could be adapted and incorporated to fit in any type of nursing program.

## AN ETHICS TEST

A practical way of improving ethical decision making is to run decisions that you are considering through an ethics test when any doubt exists. The ethics test presented here was used at the Center for Business Ethics at Bentley College (Bowditch & Buono, 2007) as part of ethical corporate training programs. Decision makers are taught to ask themselves:

- Is it right?
- Is it fair?
- Who gets hurt?
- Would you be comfortable if the details of your decision were reported on the front page of your local newspaper or through your hospital's e-mail system?
- Would you tell your child or young relative to do it?
- How does it smell? This question is based on a person's intuition and common sense.

## Critical Thinking 24-6

To better relate the study of ethics to yourself, take the following self quiz. Do you agree or disagree with the following statements?

1. I would report a nursing coworker's drug abuse.
2. I would tell the truth to a patient who asked if he was dying.
3. I see no harm in giving ordered drugs on a temporary basis to a drug-addicted patient who presents to the emergency department.
4. When applying for a nursing position, I would cover up the fact that I had been fired from a recent job.
5. If I received $100 for doing some odd jobs, I would not report it on my income tax returns.
6. I see no harm in taking home a few nursing supplies.
7. It is unacceptable to call in sick to take a day off, even if only done once or twice a year.
8. I would accept a permanent, full-time job even if I knew I wanted the job for only six months.

**Source:** C. S. Faircloth, RN, personal communication, March 13, 2003.

# ETHICAL ISSUES ENCOUNTERED IN PRACTICE

Numerous events contribute to ethical issues that professional nurses encounter in today's practice. Cost containment, technology, and patient rights have all influenced the increased numbers of ethical dilemmas nurses face.

## COST CONTAINMENT AND ISSUES RELATED TO TECHNOLOGY

Cost containment and the sophisticated technology available in health care today are related. The more technology that is developed and used, the more costly health care becomes. There are several factors that contribute to the high costs of receiving treatment in our system. ICUs are full of expensive equipment designed to breathe for patients, monitor vital signs automatically, and communicate if the heart is failing in some way. Scores of devices all have the same purpose—to let us know what needs to be done to sustain life.

Although the use of technology has brought many benefits, two significant questions arise about its use. When do we refrain from the use of technology? When do we stop using it once we have started? Who is entitled to it in our society?

# PATIENT RIGHTS

A patient's bill of rights was first adopted by the American Hospital Association (AHA), in 1973 and was revised in 1982 and 1992. This bill of rights was developed to assure that the health care system would be fair and meet patient needs. It provides patients with a guide to addressing problems with their care and encourages them to participate in staying healthy or getting well. In 2003, the American Hospital Association replaced the bill of rights with the Patient Care Partnership. The Patient Care Partnership is a booklet informing patients of what to expect during their hospital stay. It discusses their right to high-quality hospital care; a clean, safe environment; involvement in their care; protection of their privacy; help when leaving the hospital; and help with billing claims. The Patient Care Partnership can be found on the AHA Web site (www.aha.org).

The U.S. Department of Health and Human Services, in partnership with the American Hospital Association, published five steps to safer health care for patients. These steps include asking questions if you have doubts or concerns, keeping an up-to-date list of medications, obtaining results of tests or procedures, talking to your physician about which hospital best fits your needs, and making sure

## REAL WORLD INTERVIEW

One of my most difficult cases involved a man in his early forties who was in a coma, ventilator dependent, and declared brain dead. The patient was from a different culture, and when the family arrived six weeks later from the country abroad, they refused to allow him to be removed from the ventilator. His parents said they were told by the gods that their son would be well several months in the future. After two months in the hospital, the administration began to put pressure on the family to transfer the patient.

**Emily Davison, RN**
Case Manager
Pleasant Hill, Missouri

you understand what will happen if you need surgery. This was originally published in 2004 and can be found on the Agency for Healthcare Research and Quality Web site (www.ahrq.gov).

The AHA website also provides a checklist for health care administrators to assure their institution is in compliance with The Patient Care Partnership. Their suggestions include developing the following: communication in-services for employees, a language bank, an ethics committee, patient friendly billing, a statement of patient rights, a process for patient follow-up on concerns, and patient education on the use of advanced directives.

## ETHICAL LEADERSHIP AND MANAGEMENT

Nurse leaders are in the position to assure that the setting in which they practice is ethically principled and that it accepts accountability for the Code of Ethics set by the profession of nursing. The professional environment can empower nurses and can foster autonomy and an appreciation of the diversity of persons and opinions. Treating others with fairness, dignity, and respect, nurse leaders can influence the decisions made by the organization.

## ORGANIZATIONAL BENEFITS DERIVED FROM ETHICS

Health care institutions are increasingly faced with making decisions, based on the financial bottom line, that might ultimately affect the quality of care provided. Ethically and socially responsible decisions often come with a price tag. DuBrin (2009) cites several examples of companies that have allowed profit motivation and perhaps greed to lead to unethical and imprudent acts. Problems in the Enron Corporation and the banking industry represent a few recent examples of this.

Hospitals have also suffered financial losses when ethics were cast aside. There have been countless medical errors and subsequent malpractice suits that can be traced back to cutbacks in staff or other resources. Quality and safety issues in health care have increasingly come under scrutiny. The American Hospital Association, the National League of Nursing, the National Institutes of Health, and the American Association of Colleges of Nursing have all designated quality and safety initiatives as the highest priority. Additional information is available on their Web sites (www.aacn.nche.edu, www.nursingworld.org, www.nih.gov, www.aha.org, and www.nln.org).

## EVIDENCE FROM THE LITERATURE

**Citation:** Candela, L., Michael, S. R., & Mitchell, S. (2003). Ethical debates. *Nurse Educator, 28*(1), 37–39.

**Discussion:** This article discusses a format for classroom debate on ethical issues. A portion of the format is as follows:

Select an ethical issue and phrase as a question starting with the word, "Should."

Select a moderator for the group. The moderator will introduce the issue and be responsible for debate flow and adhering to time requirements.

Divide the group into two subgroups. One subgroup will argue for the question; the other subgroup will argue against the question.

Be sure to consider all pertinent ethical principles, any conflict between principles, relationship to ethical codes, legal implications, people involved and impacted, and any relevant sociocultural, political, or religious aspects that may influence this ethical issue. All major points must be grounded in the literature.

Each group member must articulate a position. Be sure to credit your sources as you speak. Be familiar with your part so that you can "talk it" versus "read it."

**Implications for Practice:** Nurses can use this format to clarify their own thinking on various ethical issues.

## CASE STUDY 24-2

As you receive a hand-off report at the start of your shift, the nurse informs you not to tell Mrs. Sun, Room 240, her diagnosis. She is being treated with chemotherapy for ovarian cancer and she is not aware of her diagnosis. The family does not want her to know, because they say she will not be able to handle it. The physician spoke with the family and decided to respect their wishes.

When you make patient care rounds on Mrs. Sun, she asks you to stop the medication. She states it is making her sick. She says she was not sick when she came into the hospital, and she wants to know what this medicine is for. What do you do? Use the Ethical Positioning System diagram to analyze this case.

Nursing dilemma: What should you tell Mrs. Sun?

Choice 1 Tell Mrs. Sun the truth, that she is being treated for ovarian cancer.

Pros: You will be telling the truth and respecting your patient's autonomy.

Cons: You do not know what her reaction will be. There is concern about her being able to handle the news. Also, you may not be able to answer questions about her options or prognosis. The physician and the family may be angry.

Choice 2 Tell Mrs. Sun that you do not know about her case and that you are unable to give her any additional information.

Pros: It lets you off the hook, and she may be pacified for the moment.

Cons: She will probably refuse treatment at this time. This is lying and is in opposition to several ethical principles and the ANA Code of Ethics.

Choice 3 Tell Mrs. Sun that you are going to look into her situation and that you will be back to speak with her. The ANA Code of Ethics (1.4, the right to self-determination; 2.3, collaboration; and 6.3 responsibility for the health care environment) and following the ethical principles of autonomy, respect, beneficence, nonmaleficence, and veracity, seem to encourage you to phone the physician and the family and let them know the current situation. Suggest to them that they reconsider their decision and give Mrs. Sun the correct information. Suggest a consultation with the ethics committee.

Pros: You are being honest with all involved. At this point, the patient may refuse treatment anyway and needs to be involved in her own decisions.

Cons: The family or physician may still not agree to discuss the diagnosis with the patient. They may blame you for the dilemma. The patient may not be able to handle hearing about her diagnosis.

What will you decide?

## CREATING AN ETHICAL WORKPLACE

Establishing an ethical and socially responsible workplace is not simply a matter of luck and common sense. Nurse managers can develop strategies and programs to enhance ethical and socially responsible attitudes. These may include the following:

1. Formal mechanisms for monitoring ethics, such as an ethics program or ethics hotline.
2. Written organizational codes of conduct (Table 24-4).
3. Widespread communication in the hospital to reinforce ethically and socially responsible behavior.
4. Leadership by example: If people throughout the firm believe that behaving ethically is "in" and behaving unethically is "out," ethical behavior will prevail.
5. Encouraging confrontation about ethical deviations: Unethical behavior may be minimized if every employee confronts anyone seen behaving unethically.
6. Training programs in ethics and social responsibility, including messages about ethics from managers, classes on ethics at colleges, and exercises in ethics (DuBrin, 2009).

## TABLE 24-4    Organizational Elements to Support an Ethical Workplace

- Follow professional standards for education, licensure, and competency in all hiring decisions, orientation, and ongoing continuing education programs.

- Have clear job descriptions and ongoing licensing and credentialing policies for nursing and medical practitioners, LPN/LVNs, nursing assistive personnel (NAP), and other health care staff. The organization must ensure that all staff are safe, competent practitioners before assigning them to patient care. Orient all staff to each other's roles and job descriptions.

- Facilitate clinical and educational specialty certification and credentialing of all practitioners and staff.

- Provide standards for ongoing supervision and periodic licensure/competency verification and evaluation of all staff.

- Provide access to professional health care standards, policies, procedures, library, and medication information, with unit availability and efficient Internet access.

- Facilitate regular evidence-based review of critical standards, policies, and procedures.

- Have clear policies and procedures for delegation and chain-of-command reporting lines for all staff from RN to charge nurse to nurse manager to nurse executive and, as appropriate, to risk management, the hospital ethics committee, the hospital administrator, medical practitioners, the chief of the medical staff, the board of directors, the state licensing board for nursing and medicine, and the Joint Commission (JC).

- Provide administrative support for supervisors and staff who delegate, assign, monitor, and evaluate patient care.

- Clarify nursing and medical practitioner accountability. For example, if a medical practitioner delegates a nursing task to NAP, the medical practitioner is responsible for monitoring that care delivery. This must be spelled out in hospital policy. If an RN notes that the NAP is doing something incorrectly, the RN has a duty to intervene and to notify the ordering practitioner of the incident. The RN always has an independent responsibility to protect patient safety. Blindly relying on another nursing or medical practitioner is not permissible for the RN.

- Provide standards for regular RN evaluation of NAP and LPN/LVN; reinforce the need for NAP and LPN/LVN accountability to RNs. RNs must delegate and supervise; they cannot abdicate this professional responsibility.

- Develop physical, mental, and verbal "No Abuse" policy, to be followed by all professional and nonprofessional health care staff.

- Consider applying for Magnet status for your facility. This status is awarded by the American Nurses Credentialing Center to nursing departments that have worked to improve nursing care, including the empowering of nursing decision making and delegation in clinical practice.

- Support the development of a strong ethics committee with an interdisciplinary team. Make it available to all staff.

- Monitor patient outcomes, including nurse-sensitive outcomes, staffing ratios, and other clinical, financial, and organizational quality indicators, and also develop ongoing clinical quality improvement practices.

- Maintain ongoing monitoring of incident reports, sentinel events, and other elements of risk management and performance improvement of the process and outcome of patient care.

- Develop systematic, error-proof systems for medication administration that ensure the six rights of medication administration—that is, the right patient, right medication, right dose, right time, right route, and right documentation. Include computerized order entry.

- Provide documentation of routine maintenance for all patient care equipment.

- Attain JC Patient Safety Goals, 2011. These can be retrieved at http://www.jointcommission.org/standards_information/npsgs.aspx (retrieved April 13, 2011).

- Develop safe intrahospital and intra-agency transfer policies.

## EVIDENCE FROM THE LITERATURE

**Citation:** Nelson, W. (2009). Ethical Uncertainty and Staff Stress. *Healthcare Management Ethics*, (July/August), 38–40.

**Discussion:** Ethical conflicts and concerns have a significant negative impact on health care organizations. Everyone is affected, from the executives to the clinicians. For staff members, managers, and even executives, moral distress greatly diminishes job satisfaction.

The article discusses how recognizing moral distress can be the first step to managing its negative impact. In this first step, staff members develop self-awareness and seek assistance. Second, an open environment is developed in which ethical concerns can be shared and discussed. Third, the need to have the members of the ethics committee available and easily accessible for consultation is recognized. Finally, the need for health care organizations to have an accessible employee assistance program to offer confidential emotional support to staff is recognized.

**Implications for Practice:** With this approach to managing ethical uncertainty in place, staff, managers, and executives of an organization are better equipped to deal with ethical dilemmas. There is support available when they are faced with moral distress.

## NURSE-PHYSICIAN RELATIONSHIPS

Nurses working in organizations often confront ethical dilemmas in working with patients and their families. To resolve these dilemmas, the nurse must often work closely with the medical practitioner. Nurses often find that medical practitioners hold beliefs different from theirs about values, communication, trust and integrity, role responsibilities, and organizational politics and economics. These beliefs affect their ethical beliefs, which in turn affect their decisions about treatment, which may lead to conflicts between nurses and practitioners. These conflicts can be limited with clear ethical guidelines and policies, established by interdisciplinary teams and overseen by an ethical administration. When an ethical issue arises, resolution might be tedious and possibly riddled with resentment without guidelines and policies.

The Gallup Organization's 2006 annual poll on professional honesty and ethical standards ranked nurses number one. Of the 21 professions tested, 6 have high ethical ratings: nurses (84%), pharmacists (73%), veterinarians (71%), medical doctors (69%), dentists (62%), and engineers (61%). For more information, see http://www.gallup.com. Search for nurse ethical rating. Click on, Nurses Top Honesty and Ethics List for 11th Year.

## ETHICAL CODES

One mark of a profession is the determination of ethical behavior for its members. Several nursing organizations have developed codes for ethical behavior. The International Council of Nurses Code for Nurses appears in Table 24-5.

The ANA has also developed a Code of Ethics for Nurses with Interpretive Statements (2001). See the reference section at the end of this chapter for a site for viewing this document online.

## THE FUTURE

Ethical issues in the future that will challenge nursing practice include the allocation of resources, advanced technologies, an aging population, and an increase in behavior-related health problems. These issues all magnify the importance of professional nurses providing leadership that emphasizes ethical behavior in all practice settings.

## Critical Thinking 24-7

The nurse is caring for a patient in the surgical ICU who is connected to a ventilator. He is on a sedation protocol with continuous IV infusion of a powerful sedative that requires constant monitoring and titration to maintain the required level of sedation. During the night shift, the nurse discovers that the medication bag is almost empty and the pharmacy, which is closed, did not send up another bag. The nurse looks up the medication and mixes the medication herself, inadvertently mixing a double-strength dose. The night charge nurse was busy supervising a cardiac arrest situation outside the ICU and was unavailable to double check how the medication was mixed.

Thirty minutes after the nurse hung the new IV bag of medication, the patient's blood pressure dropped significantly. The family was notified that their loved one had taken a turn for the worse and that they should come to the hospital immediately. The physician was notified, and she ordered a medication to raise the blood pressure. In backtracking for the cause of the hypotension, the nurse realized that she had mixed the sedative double strength, and she reduced the dose by half.

When the patient's family arrives, the patient's blood pressure had started to return to normal. They ask the nurse what happened and why their mother was on the new IV medication to raise the blood pressure.

Should the family be told about the error? If so, who should tell them? The nurse? The physician? What approach should be used? What ethical principles enter into the decision?

## TABLE 24-5 International Council of Nurses Code of Ethics for Nurses

An international code of ethics for nurses was first adopted by the International Council of Nurses (ICN) in 1953. It has been revised and reaffirmed at various times since, most recently with the review and revision completed in 2005.

### Preamble

Nurses have four fundamental responsibilities: to promote health, to prevent illness, to restore health, and to alleviate suffering. The need for nursing is universal.

Inherent in nursing is respect for human rights, including cultural rights, the right to life and choice, to dignity and to be treated with respect. Nursing care is respectful of and unrestricted by considerations of age, color, creed, culture, disability or illness, gender, sexual orientation, nationality, politics, race, or social status.

Nurses render health services to the individual, the family and the community and coordinate their services with those of related groups.

### The ICN Code

The ICN Code of Ethics for Nurses has four principal elements that outline the standards of ethical conduct.

*(Continues)*

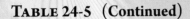

## TABLE 24-5 (Continued)

### Elements of the Code

#### 1. Nurses and people

The nurse's primary professional responsibility is to people requiring nursing care.

In providing care, the nurse promotes an environment in which the human rights, values, customs and spiritual beliefs of the individual, family and community are respected.

The nurse ensures that the individual receives sufficient information on which to base consent for care and related treatment.

The nurse holds in confidence personal information and uses judgement in sharing this information.

The nurse shares with society the responsibility for initiating and supporting action to meet the health and social needs of the public, in particular those of vulnerable populations.

The nurse also shares responsibility to sustain and protect the natural environment from depletion, pollution, degradation and destruction.

#### 2. Nurses and practice

The nurse carries personal responsibility and accountability for nursing practice, and for maintaining competence by continual learning.

The nurse maintains a standard of personal health such that the ability to provide care is not compromised.

The nurse uses judgement regarding individual competence when accepting and delegating responsibility.

The nurse at all times maintains standards of personal conduct which reflect well on the profession and enhance public confidence.

The nurse, in providing care, ensures that use of technology and scientific advances are compatible with the safety, dignity and rights of people.

#### 3. Nurses and the profession

The nurse assumes the major role in determining and implementing acceptable standards of clinical nursing practice, management, research and education.

The nurse is active in developing a core of research-based professional knowledge.

The nurse, acting through the professional organization, participates in creating and maintaining safe, equitable social and economic working conditions in nursing.

#### 4. Nurses and co-workers

The nurse sustains a cooperative relationship with co-workers in nursing and other fields.

The nurse takes appropriate action to safeguard individuals, families and communities when their health is endangered by a co-worker or any other person.

**Source:** Copyright 2006 by ICN, International Council of Nurses, 3, place Jean-Marteau, 1201 Geneva, Switzerland.

## CASE STUDY 24-3

Go to the American Hospital Association's Patient Care Partnership site at http://www.aha.org/aha/content/2003/pdf/pcp_english_030730.pdf. This document has replaced the Patient Bill of Rights. Notice what it says about the following:

- High-quality hospital care
- A clean and safe environment
- Involvement in your care
- Protection of your privacy
- Help when leaving the hospital
- Help with your billing claims

How can you use this information to help patients?

## KEY CONCEPTS

- Ethics is a branch of philosophy that concerns the distinction of right from wrong.
- Society created the profession of nursing for the purpose of meeting specific health needs.
- A person's beliefs and values will influence his or her philosophy of nursing.
- Teleology, deontology, virtue ethics, justice and equity, and relativism are examples of ethical theories.
- Ethical principles and rules include beneficence, nonmaleficence, fidelity, justice, autonomy, respect for others, and veracity.
- Ethics committees provide guidance for decision making in ethical dilemmas in the health care setting.

- Values clarification is an important step in decision making.
- The Ethical Positioning System is a helpful tool for ethical decision making.
- Numerous ethical issues face today's nurses.
- Patients' rights is an area in which nurses are held accountable.
- Nurses have the obligation to uphold the trust society places in the nursing profession.
- Nurse leaders who are dedicated to ethical principles can influence organizational ethics.

## KEY TERMS

| | | | |
|---|---|---|---|
| autonomy | ethics | morality | respect for others |
| beneficence | fidelity | nonmaleficence | values |
| bioethics | justice | philosophy | veracity |
| ethical dilemma | | | |

# REVIEW QUESTIONS

1. When the nurse is obtaining a patient's consent, the patient states that the surgeon did not give the patient information on the risks of surgery. The nurse should do which of the following?
   A. Tell the patient the risks.
   B. Report the surgeon to the ethics committee.
   C. Report the surgeon to the unit manager.
   D. Inform the surgeon that the patient is unaware of the risks.

2. The nurse notices that a coworker has been drinking and is not able to practice safely. The nurse should do which of the following?
   A. Inform the manager or shift director immediately.
   B. Warn the coworker that black coffee is in order.
   C. Discuss the situation with the other nurses working.
   D. Do nothing, but keep an eye on the nurse.

3. The nurse demonstrates nonmaleficence by doing which of the following? Select all that apply.
   _____ A. Observing the six rights of medication administration.
   _____ B. Reviewing practitioner orders for accuracy and completeness.
   _____ C. Striving to improve patient satisfaction.
   _____ D. Keeping knowledge and skill up-to-date.
   _____ E. Dressing professionally, with name badge clearly visible.

4. Which of the following represents a teleological theory?
   A. The goodness of an action is based on the intent.
   B. The end justifies the means.
   C. There are no universal ethical standards.
   D. Do unto others as you would have them do unto you.

5. The primary role of an ethics committee is which of the following?
   A. Decide what should be done when ethical dilemmas arise.
   B. Prevent the practitioner from making the wrong decision.
   C. Provide guidance for the health care team and family of the patient.
   D. Prevent ethical dilemmas from occurring.

6. Ethical dilemmas may be referred to the ethics committee by which of the following?
   A. Medical practitioners only
   B. Nursing and medical practitioners, lawyers, all health care team members, and families of patients
   C. Lawyers only
   D. Hospital administration only

7. Mrs. Jones rides the elevator to the fifth floor, where her husband is a patient. While on the elevator, Mrs. Jones hears two nurses talking about Mr. Jones. They are discussing the potential prognosis and whether Mr. Jones should be told. The nurses are violating which of the following ethical principles?
   A. Autonomy
   B. Confidentiality
   C. Beneficence
   D. Nonmaleficence

8. The nurse realizes that neglecting to inform the patient about the plan of care is a violation of which of the following? Select all that apply.
   _____ A. The Patient Care Partnership
   _____ B. The patient's right to privacy
   _____ C. The patient's right to confidentiality
   _____ D. The fifth amendment of the constitution
   _____ E. The right to self-determination

9. During morning report, the night nurse tells you that Mr. P., who is admitted for pancreatitis, is a drug addict and an alcoholic and caused all his own problems. You realize that this nurse is exhibiting a lack of which of the following?
   A. autonomy
   B. compassion
   C. discernment
   D. trustworthiness

10. A nurse caring for a cancer patient demonstrates paternalism when she cannot decide if she should give the patient discouraging lab results. This dilemma represents a conflict between which principles?
   A. Beneficence and nonmaleficence
   B. Autonomy and beneficence
   C. Justice and veracity
   D. Autonomy and justice

## REVIEW ACTIVITIES

1. An elderly woman, age 88, is admitted to the emergency department in acute respiratory distress. She does not have a living will, but her daughter has power of attorney (POA) for health care and is a health care professional. The patient has end-stage renal disease, end-stage Alzheimer's disease, and congestive heart failure. Her condition is grave. The doctors want to intubate her and place her on a ventilator. The sons agree. The daughter states that their mother would not want to be on a machine just to prolong her life.

Divide into groups and discuss the ethical theories that can be applied to this situation. Apply the Ethical Positioning System Models to this case.

2. As a hospice nurse, you are involved with pain control on a regular basis. Many of the medications prescribed for the management of pain also depress respirations.

Determine a protocol for the use of these medications, keeping in mind that the purpose of hospice is to promote comfort. Support your decisions with ethical theories and principles.

## EXPLORING THE WEB

- See what this Web site says about nursing competencies, ethics, and health care policy. www.nursingworld.org
- Go to the following site to view the American Hospital Association's Patient Care Partnership. www.aha.org

Search for, Ethics, at the following web sites:
- National League of Nursing http://www.nln.org

- National Institutes of Health
  http://www.nih.gov
- American Association of Colleges of Nursing
  http://www.aacn.nche.edu

## REFERENCES

Agency for Healthcare Research and Quality. (2004). Five steps to safer health care for patients. Accessed April 13, 2011, at http://www.ahrq.gov/consumer/5steps.htm.

American Hospital Association. Patient Care Partnership. Accessed April 13, 2011, at http://www.aha.org/aha/issues/Communicating-With-Patients/pt-care-partnership.html.

American Nurses Association. (2001). Code of Ethics for Nurses with Interpretive Statements. Accessed April 13, 2011, at http://www.nursingworld.org/MainMenuCategories/EthicsStandards/CodeofEthicsforNurses/Code-of-Ethics.aspx.

American Nurses Association. (2010). *Nursing: Scope and Standards of Practice 2nd Edition.* Accessed April 13, 2011, at http://nursesbooks.org/Main-Menu/Standards/H–N/Nursing-Scope-and-Standards-of-Practice.aspx.

Bell, J., & Breslin, J. M. (2008). Healthcare provider moral distress as a leadership challenge. *JONA's Healthcare, Law, Ethics, and Regulation, 10*(4), 94–97.

Bowditch, J. L., & Buono, A. F. (2007). A primer on organizational behavior (7th ed.). New York: Wiley.

Burkhardt, M. A., & Nathaniel, A. K. (2008). *Ethics & issues in contemporary nursing* (3rd ed.). Clifton Park, NY: Delmar Cengage Learning.

Candela, L., Michael, S. R., & Mitchell, S. (2003). Ethical debates: Enhancing critical thinking in nursing students. *Nurse Educator, 28*(1), 37–39.

DeLaune, S. C., & Ladner, P. K. (2011). *Fundamentals of nursing* (4th ed.). Clifton Park, NY: Delmar Cengage Learning.

Donahue, M. P. (2011). *Nursing, the finest art* (3rd ed.). St. Louis, MO: Mosby.

DuBrin, A. (2009). *Leadership: Research findings, practice, and skills* (6th ed.). Mason, OH: Cengage Learning.

Fry, S. T., & Johnstone, M. J. (2008). *Ethics in nursing practice* (3rd ed.). New York: Blackwell Publishing.

Hamric, A. B., & Blackhall, L. J. (2007). Nurse-physician perspectives on the care of dying patients in intensive care units: Collaboration, moral distress, and ethical climate. *Critical Care Medicine, 35*(2), 641–642.

Hernandez, A. H. (2009). Student articulation of a nursing philosophical statement: An assignment to enhance critical thinking skills and promote learning. *Journal of Nursing Education, 48*(6), 343–349.

Husted, G. L., & Husted, J. H. (2007). Ethical decision making in nursing and healthcare (4th ed.). New York: Springer Publishing Co.

International Council of Nurses. (2006). *Code for nurses.* Geneva: Author.

Jonsen, A. R., Siegler, M., & Winslade, W. J. (2006). Clinical ethics (6th ed.). New York: McGraw Hill.

Nelson, W. (2009). Ethical uncertainty and staff stress. *Healthcare Management Ethics,* (July/August), 38–40.

# SUGGESTED READINGS

Aitamaa, E., Leino-Kilpi, H., Puukka, P., & Suhonen, R. (2010). Ethical problems in nursing management: The Role of codes of ethics. *Nursing Ethics, 17*(4), 469–482.

Arries, E. (2005). Virtue ethics: An approach to moral dilemmas in nursing. *Curationis, 28*(3), 64–72.

Beidler, S. M. (2005). Ethical considerations for nurse-managed health centers. *Nursing Clinics of North America, 40*(4), 759–770.

Chen, P. W. (2009, February 6). "When doctors and nurses can't do the right thing." *New York Times.*

Faithful, S., & Hunt, G. (2005). Exploring nursing values in the development of a nurse-led service. *Nursing Ethics, 12*(5), 440–452.

Falk-Rafael, A. (2005). Speaking truth to power: Nursing's legacy and moral imperative. *Advanced Nursing Science, 28*(3), 212–223.

Ferrell, B. (2005). Ethical perspectives on pain and suffering. *Pain Management Nursing, 6*(3), 83–90.

Gallagher, A., & Wainwright, P. (2005). The ethical divide. *Nursing Standards, 20*(7), 22–25.

Gutierrez, K. M. (2005). Critical care nurses' perceptions of and responses to moral distress. *Dimensions of Critical Care Nursing, 24*(5), 229–241.

Pinch, W .J. E., & Haddad, A. M. (Eds). (2008). *Nursing and health care ethics.* Silver Spring, MD: American Nurses Association.

Tschudin, V. (2006). How nursing ethics as a subject changes: An analysis of the first eleven years of publication of the journal *Nursing Ethics. Nursing Ethics, 13*(1), 65–85.

# CHAPTER 25

# Culture, Generational Differences, and Spirituality

**Karen Luther Wikoff, RN, PhD; Sara Swett, RN, BSN, MSN**

*If we are to achieve a richer culture, rich in contrasting values, we must recognize the whole gamut of human potentialities, and so weave a less arbitrary social fabric, one in which each diverse human gift will find a fitting place.*

(Margaret Mead, 1935)

## OBJECTIVES

Upon completion of this chapter, the reader should be able to:

1. Apply cultural considerations to the role of a nurse leader.

2. Compare and contrast the demographics of the U.S. population and U.S. nurses.

3. Describe how organizational culture can influence leading a team.

4. Describe the current generations and their behaviors that influence leading and delivering patient care.

5. Identify methods of assessment and meeting spiritual needs of patients.

6. Apply knowledge of spirituality to problem solving in the nurse leader role.

Delmar/Cengage Learning

*Mr. Hernández, a recent Mexican immigrant who came over the border without any documentation, arrived in the Emergency Room bleeding profusely from his right arm following an accident with a chain saw. A pressure dressing is applied while he keeps repeating in Spanish that he is in pain. As you try to do his assessment, he grimaces in pain but refuses to answer any questions or cooperate with the nurse. You have asked another nurse to translate your questions and plans for care to Mr. Hernández, but he keeps attempting to leave.*

> *Considering his status as an undocumented worker, how will you convince him to be treated?*
>
> *How does his undocumented status affect his access to health care?*

We live in a global society rich with ever-changing cultures, traditions, religions, spiritual beliefs, and generations, all of which influence the delivery of health care. New immigrants arrive daily. Consideration of all of these differences and sensitivity to patient and health care staff diversity is necessary in all nursing roles as situations arise that call for evaluating patient and staff behavior. Though each person is first and foremost an individual, each is also a member of a cultural group. Clues to people's behaviors come from understanding the cultural, generational, and spiritual focus of the individual or group. This chapter will enable the nurse to begin the process of preparing for cultural, generational, and spiritual aspects of leadership and management with patients and health care staff. These concepts are also discussed further in Chapter 8, Personal and Interdisciplinary Communication, and Chapter 11, Effective Team Building.

## CULTURE

**Culture** refers to the integrated patterns of human behavior that include the language, thoughts, communication, actions, customs, beliefs, values, and institutions of racial, ethnic, religious, or social groups (Muñoz & Luckmann, 2004). Although people from all cultures share most human characteristics, the study of culture highlights the ways in which individuals differ and are similar to individuals in other cultures. Individuals from one culture may think, solve problems, and perceive and structure the world differently from individuals of another culture. Cultural beliefs serve as a conscious and unconscious point of reference that guides the outlook and decisions of people. Culture incorporates the experience of the past and influences the present. It transmits traditions to future members of a culture. Culture influences what we eat, the language we speak, the values we believe in, and the actions we take. Values serve important functions:

- They provide people with a set of rules by which to govern their lives.
- They serve as a basis for attitudes, beliefs, and behaviors.
- They help to guide actions and decisions.
- They give direction to people's lives and help them solve common problems.
- They influence how individuals perceive and react to other individuals.
- They help determine basic attitudes regarding personal, social, and philosophical issues.
- They reflect a person's identity and provide a basis for self-evaluation. (Muñoz & Luckmann, 2004)

Culture is not solely based on a person's ethnicity. It can be influenced heavily by spirituality, religion, socioeconomic status, acculturation, age, gender, sexual orientation, and country of origin (Mitchell & Mitchell, 2009). When interacting with members from your own culture, the rules that guide behavior are usually known. However, when interacting with members of other cultures, these rules may be unknown or not well understood.

Culture is learned and then shared. People learn about their culture from parents, teachers, religious and political leaders, and respected peers. As children grow up, they gradually internalize the values and beliefs of their culture, and they, in turn, share these values and beliefs with their children.

Normally, children learn about their culture while growing up. However, when people emigrate from their native cultures into a new culture, they often experience culture shock. Culture shock develops when the values and beliefs upheld by this new culture are radically different from the person's native culture. For successful assimilation into a new culture, immigrants must learn and internalize that culture's important values.

In addition to belonging to a major cultural group, people also belong to a variety of subcultures, or smaller groups within a culture. Each culture has its own value system and related expectations. Subcultures may be based on the following:

- Professional and occupational affiliations (e.g., nursing)
- Nationality, ethnicity, or race (a shared historical and political past)

- Age groups (adolescents, older adults)
- Gender (feminists, men's groups)
- Socioeconomic factors (income, education, occupation)
- Political viewpoints (Democrat, Republican)
- Sexual orientation (gay, lesbian, bisexual, or transgendered groups)

For example, when you studied to be a nurse, you entered a nursing subculture, and initially you probably suffered from some degree of culture shock. You had to learn a whole new value system. During your years of study, you gradually internalized the values taught by your instructors. Eventually, you became comfortable with the values and behaviors you learned in your school of nursing, and by the time you graduate, you will be assimilated into the professional nursing subculture.

Upon admission to a hospital, patients also become members of a culture. In this world filled with strange sights, unfamiliar sounds, and strangers, many patients experience culture shock. This shock intensifies for patients who are recent immigrants or who do not speak English (Muñoz & Luckmann, 2004).

# RACE AND ETHNICITY

**Race** describes a geographical or global human population distinguished by genetic traits and physical characteristics such as skin color or facial features. The 2009 U.S. Census classifications for race are White; Black or African American; Two or more races; American Indian or Alaska Native; Asian; and Native Hawaiian or Other Pacific Islander (Figure 25-1). Cultural ethnicity identifies a person or group based on a racial, tribal, linguistic, religious, national, or cultural group—for example, Jewish or Irish. Dayer-Berenson (2011) notes that as immigrants and

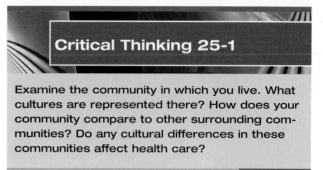

## Critical Thinking 25-1

Examine the community in which you live. What cultures are represented there? How does your community compare to other surrounding communities? Do any cultural differences in these communities affect health care?

minorities acculturate in the United States, they encounter two significant issues. The first issue is the degree of impetus, allowance, or approval to retain their identification with their original heritage and customs. The second issue is the degree of impetus, allowance, or approval to identify with the dominant American culture with which they are now surrounded. In years past, when new immigrants arrived in the United States, they sought to become acculturated to their new country by adopting the conditions, language, and customs of the United States. Acculturation has taken a generation or two in the past and often resulted in the loss of a separate cultural identify. Today's immigrant population is more likely to maintain a strong tradition of valuing their historical cultural customs and identity.

# INCREASING DIVERSITY

The ethnic and racial composition of the population of the United States has changed dramatically since the 1990s. Currently, one out of every three persons in the United States is a minority (Dayer-Berenson, 2011). Subsequently, there is increasing diversity in languages, beliefs, lifestyles, and practices among residents in rural and urban areas throughout the country. The percent of the U.S. population that is White has decreased since the 1970s. The reduction in percentage is due in part to increasing immigration from Asian and Latin American nations and in part to a higher population growth rate among Blacks (Muñoz & Luckmann, 2004). In addition to ethnic and racial diversity, health care providers must also be aware of and anticipate the care of patients with diversity of religion and spirituality, generational differences, mental and physical aptitude, as well as sexual orientation (Dayer-Berenson, 2011).

# POPULATION GROUPS

It is important to recognize that within each of the broad statistical categories of population groups, there are numerous cultural groups. These groups are characterized by variations in lifestyles, values and beliefs, health-related and illness-related practices, preferences for care, and family member patterns of interaction. For example, many cultural groups are included in the category of

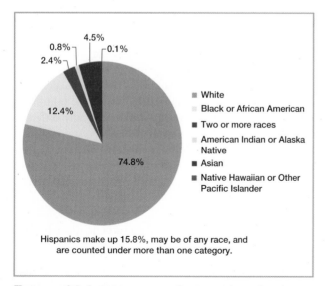

4.5%
0.8%
0.1%
2.4%
12.4%
74.8%

- White
- Black or African American
- Two or more races
- American Indian or Alaska Native
- Asian
- Native Hawaiian or Other Pacific Islander

Hispanics make up 15.8%, may be of any race, and are counted under more than one category.

FIGURE 25-1  **2009 estimate of U.S. population by race.** (*Source:* U.S. Census Bureau, 2009 American Community Survey).

*White* (sometimes also referred to as Anglo-American or Caucasian). Individuals in these groups may trace their heritage to a European nation, Australia, North America, or many other nations and regions. Among U.S. residents, a large group of 50.7 million people have a German ancestry. Another 36.9 million Americans trace their roots to Ireland, and more than 27.6 million Americans have an English ancestry (U.S. Census Bureau, 2009b).

There is also great diversity among the individuals who are included in the Hispanic population category. There are differences in their countries of origin, in the dialects spoken, and in customs and beliefs, including practices related to health and illness. Hispanics may include people from the Caribbean, Cuba, El Salvador, Guatemala, Puerto Rico, Mexico, Central and South America, and Spain (Muñoz & Luckmann, 2004).

The term *Black* is similarly very inclusive and may refer to individuals who are refugees or immigrants from African nations such as Eritrea and Kenya; those from the Caribbean, Haiti, and the Dominican Republic; or individuals who can trace their heritage through multiple generations of residency in the United States. Furthermore, the term *Native American*, too, often blurs distinctions—among members of more than 560 federally recognized Native American and Alaska Native tribes who reside in North America (for example, Cherokees, Navajo, Chippewa, Sioux, Choctaw, American

Indian/Athapaskan, Aleuts, and Eskimos). Likewise, the term *Asian* can refer to Japanese, Chinese, Indochinese, Filipino, Korean, Vietnamese, Cambodian, Laotian, Thai, Indonesian, Pakistani, Hmong, and Indian populations, each of which contains numerous, diverse subcultures. Religions practiced by this group include Buddhism, Taoism, Christianity, Confucianism, Zen Buddhism, Shintoism, and Hinduism (Muñoz & Luckmann, 2004).

## HEALTH CARE DISPARITY

Differences in health risks and health status measures that reflect the poorer health status that is found disproportionately in certain population groups are referred to as **health disparities**. These health disparities include differences in the occurrence of illness, disease, and death among minorities and other vulnerable populations in the United States (Cohen, Iton, Davis, & Rodriguez, 2010). The Office of Minority Health and Health Disparities (OMHD, 2010) notes that race and ethnicity correlate with continual, and often escalating, health disparities among residents of the United States. Although there has been significant improvement in the general health of the United States, ongoing disparities in morbidity and mortality are experienced by African Americans, Hispanics, American Indians, Alaskan Natives, Native Hawaiian, and other Pacific Islanders compared to the entire U.S. population (OMHD, 2010). For example:

- The death rate in 2006 for the black population surpassed that of the white population by 48% for cerebrovascular disease, 31% for heart disease, 21% for malignant neoplasms, 113% for diabetes, and 786% for HIV disease (CDC, 2009, p.25).
- The infant mortality rates in 2005 were highest among babies born to non-Hispanic black mothers (13.63 deaths per 1,000 live births), American Indian or Alaska Native mothers (8.06 per 1,000), and Puerto Rican mothers (8.30 per 1,000) (CDC, 2009, pp. 25–26).
- Hispanics residing in the United States have increased rates of obesity and high blood pressure comparatively to non-Hispanic whites (OMHD, 2010).
- The highest incidence of diabetes in the world is found among the Pima Indians in the United States (National Institutes of Health [NIH], 2006).
- Latinos have increased rates of ocular disease and impairment comparatively to other minority groups in the United States (NIH, 2006).

Current knowledge about the biological traits of ethnic populations does not explain the health disparities that exist among these groups compared with the non-Hispanic white population in the United States (OMHD, 2010). These health disparities are believed to arise from a complicated relationship among "genetic variations, environmental factors, and specific health behaviors" (OMHD, 2010, p.1).

### REAL WORLD INTERVIEW

The U.S. population is becoming more diverse, while access to health care for racial and ethnic minority populations has not improved. The challenge for nursing leadership is to attract a workforce that will mirror the diversity of the population and promote culturally competent clinical care. Nursing leadership in any health care work environment must recognize the uniqueness of each person and respect, protect, and advocate for the individual's right to self-determination, self-expression, confidentiality, and dignity. Nurses must embrace the belief that the relationships established while caring for the patient and family have the inherent capacity to promote health, healing, and wholeness. Caring and compassionate behaviors are the essence of nursing and therefore essential in today's work environment.

**Carol Robinson, RN, MPA, CNAA, FAAN**
Senior Associate Director, Patient Care Services
University of California Davis Health System
Sacramento, California

## REAL WORLD INTERVIEW

In the southern part of California, we have an underrepresentation of Hispanic nurses in our hospitals. To encourage and promote nursing enrollment, we have brought Hispanic nurses from Mexico to experience our health care system. They are provided with a hospital orientation of key programs in their areas of interest. They also spend time in the clinical area.

At the high school level, we have brought a group of about thirty high school students, the majority Hispanics, to the hospital. We talked to them about health care, nursing opportunities, and the need for Hispanic nursing staff.

As a nursing executive, I feel that it is important to be a role model for minority students. Recently, I participated in mock interviews for high school students. The majority of these students were Hispanic, Black, and Asian. We wanted to mentor and coach them on how to successfully interview for jobs. Many of them lack role models in their lives. Just by seeing that other people have been successful creates a high sense of inspiration and a desire to do better, because they see that it is possible.

**Pablo Velez, PhD, RN**
Chief Nursing Officer
Sharp Chula Vista Medical Center
Chula Vista, California

## SOCIOECONOMIC DISPARITIES

It is important to understand that health disparities also affect those of lower socioeconomic status (Giger, Davidhizar, Purnell, Harden, Phillips, & Strickland, 2007). Braveman, Cubbin, Egerter, Williams, & Pamuk (2010) studied patterns of socioeconomic disparities in the United States and found that there were persistent variations in health with regards to social class, including income and education. Populations with lower incomes and lower educational levels were associated with poorer health status. Socioeconomic disparities negate an individual's ability to reach his or her full health, and often economic, potential (Braveman, 2006; Braveman, Cubbin, Egerter, Williams, & Pamuk, 2010). When socioeconomic factors are controlled, the rate of health disparities decreases significantly with the exception of those of racial and ethnic minorities (Smedley, Stith, & Nelson, 2003).

## DIFFERENCES IN THE QUALITY OF HEALTH CARE

Differences in the quality of health care that an individual receives that are not justified by the underlying health condition, access-related factors, or the treatment preferences of the patient are referred to as **health care disparities** (Smedley, Stith, & Nelson, 2003; AACN, 2008). These differences are often a result of "conscious or unconscious bias, provider bias, and institutional discriminatory policies toward patients of diverse socioeconomic status, race, ethnicity, and/or gender orientation" (AACN, 2008, p.4).

## ELIMINATING HEALTH DISPARITIES

The elimination of health disparities is a significant national priority in Healthy People 2010 (U.S. Department of Health and Human Services, 2000). Providers of health care must deliver equitable services to all Americans. Without systematic and individual provider changes, America will face significant challenges in public health, public policy, ethics, social justice, and economics that will last for decades. The elimination of disparities is a multifaceted challenge that includes interventions aimed at access to care as well as factors associated with prejudice and bias. These interventions primarily lie with health care providers, health care organizations and institutions, academics, community organizations, payers, and government (Dykes & White, 2009). Several strategies for nurse leaders include:

- Increasing the diversity of the nursing staff to enhance patient and provider relations and reduce problems in cross-cultural communication (Smedley, Stith, & Nelson, 2003; Dykes & White, 2009)
- Utilizing culturally and linguistically appropriate approaches to nursing care that can be flexible to patient needs while still identifying and addressing potential barriers that are specific to the individual patient (King, Green, Tan-McGrory, Donahue, Kimbrough-Sugick, & Betancourt, 2008)
- Improving access and infrastructures at the public health and health care systems levels (Cohen, Iton, Davis, & Rodriguez, 2010)

■ Examining the organization's culture, conditions, and training needs and then advocating for changes that reduce disparities (Washington, Bowles, Saha, Horowitz, Moody-Ayers, Brown, et al., 2008)

As nurses begin to understand their own biases and stereotypes as well as identify problems in their own health care facilities and communities, they can facilitate problem-solving activities, including improving access to culturally appropriate care services and improving the cultural competency skills of staff.

## CULTURAL COMPETENCE

Nurses providing care need to ensure that the cultural needs of their patients are considered. **Culturally competent care** is a complex integration of knowledge, attitudes, and skills that enhance cross-cultural communication and appropriate and effective interactions (American Academy of Nursing (AAN), 1992). The Expert Panel on Cultural Competence of the AAN notes that cultural competence is a continual process for the nurse, wherein the personal preferences of the nurse should not interfere with the care preferences of the client (Giger, Davidhizar, Purnell, Harden, Phillips, & Strickland, 2007). A baseline for delivering culturally competent care can be developed by completing an assessment of your beliefs, your organization, and your community. Table 25-1 provides a list of items to consider when preparing to provide culturally competent care. This preparation should include education regarding cultural groups specific to your local community as well as education on methods of obtaining sensitive personal information from patients.

Many nursing schools added curricula in the 1990s to ensure that nurses have adequate preparation for cultural competence. In addition, all nurses need to commit themselves to enhancing their knowledge of different cultures by reading about various cultures and their traditional health practices, seeing films about different cultures, talking with patients and coworkers about their cultural backgrounds, or acting as a participant observer in a cultural setting such as an ethnic community celebration or gathering. Nurses also have opportunities to participate in continuing education for cultural competency. For example, the Office of Minority Health, U.S. Department of Health and Human Services (2007), offers a free online continuing education program titled "Culturally Competent Nursing Care: A Cornerstone of Caring." You can also develop cultural competence by assessing your level of skill, practicing various techniques to improve your skills, evaluating your new skills, and deciding what improvements you still need to make. Note that every new person you meet who is from another culture will help to broaden your appreciation of different cultures and improve your cultural competence.

A new nurse leader is often asked to work with people who are different from the nurse because of their ethnicity, race, culture, religion, or age. This can result in workplace difficulties with communication and the development of problems. To manage and understand these potential difficulties, cultural nursing theories and conceptual models offer some direction. Table 25-2 reviews several of these:

### Critical Thinking 25-2

You discovered what seems to be the perfect job, based on your family scheduling needs, as a clinic nurse for a nearby Native American tribe. What research will you do about this tribe? How do you prepare for the interview? What questions will you ask to enhance your understanding of their culture?

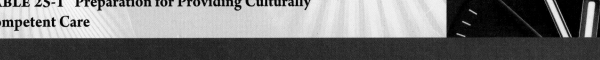

**TABLE 25-1 Preparation for Providing Culturally Competent Care**

- Identify cultural groups in your community.
- Assess your own feelings, values, and beliefs to increase self-awareness and decrease ethnocentric bias about working with different cultural groups.
- Review common cultural barriers often encountered in working with various cultural groups.
- Review any challenges you have had working with patients and families from diverse cultures.
- Develop culturally competent approaches to these challenges.
- Review organizational polices and procedures regarding cultural diversity.
- Attend community and organizational educational offerings to improve your knowledge.

## TABLE 25-2   Cultural Nursing Theories and Models

| THEORIST | CHARACTERISTICS | BASIC TENETS | IMPLICATIONS FOR NURSING PRACTICE |
|---|---|---|---|
| *Leininger (2006)* Culture Care Theory | Formal approach to the study and practice of comparative holistic cultural care and health and illness patterns while respecting differences. | All cultures have caring behaviors that may vary from culture to culture. Each culture identifies what it values for care. An understanding of each culture is important for providing care and meeting patient needs. Need to evaluate world view, social structure, language, ethnohistory, environmental context, and folk and professional systems. | Initial theory was created in the 1960s and is frequently updated and edited. Seeks to provide culturally congruent nursing care that is meaningful to the recipient. |
| *Purnell (2008)* Purnell Model for Cultural Competence | An organizing framework for assessing culture across disciplines. States that cultural competence is a process, not an endpoint. | The model is represented by a circle with rims moving from the global society to the community, to the family, and to the individual. The inner circle has 12 pie slices representing 12 domains of culture—that is, overview/heritage, communication, family roles and organization, workforce issues, biocultural ecology, high-risk behaviors, nutrition, pregnancy, death rituals, spirituality, health care practices, and health care practitioners. | This grand theory offers direction for practice. Analyzing the patient belief system in each of the 12 domains gives direction to the nursing process. |
| *Campinha-Bacote (2003)* Process of Cultural Competence in the Delivery of Health Care Service | Cultural competence is a process where in the nurse works to effectively meet needs within a cultural context. | *Cultural desire* leads the nurse to engage in developing cultural competence. *Cultural awareness* is a self-examination and exploration of one's own cultural background. *Cultural knowledge* is learning about diverse groups and understanding beliefs and values specific to the patient. *Cultural skill* is the ability to collect data and complete a cultural assessment. *Cultural encounters* occur when the nurse is engaged in interactions with diverse patients, meeting their needs in a culturally sensitive manner. | This process reflects the need for the nurse to have an awareness of patient needs and then to actively seek to meet those needs. The model serves as a guide for becoming culturally competent. |
| *Giger and Davidhizar (2008)* Transcultural Assessment Model | Method for completing and evaluating the outcomes of cultural assessment. | Has five concepts that focus on transcultural nursing and the provision of culturally diverse nursing care, provision of culturally competent care, identification of the cultural uniqueness of individuals, development of culturally sensitive environments, and development of culturally specific illness and wellness behaviors. | Consider communication, space, social organization, time, environmental control, and biological variations when assessing culture. |

# CASE STUDY 25-1

Following are different levels of response that you might have toward a person.

**Levels of Response:**

- *Greet:* I feel I can greet this person warmly and welcome him or her sincerely.
- *Accept:* I feel I can honestly accept this person as he or she is and be comfortable enough to listen to his or her problems.
- *Help:* I feel I would genuinely try to help this person with his or her problems as they might be related to or arise from the label or stereotype given to him or her.
- *Background:* I feel I have the background of knowledge and experience to be able to help this person.
- *Advocate:* I feel I could honestly be an advocate for this person.

The following is a list of individuals. Read down the list and place a check mark by anyone you would not greet or would hesitate to greet. Then move to response level 2, Accept, and follow the same procedure for all five response levels. Try to respond honestly, not as you think might be socially or professionally desirable. Your answers are only for your personal use in clarifying your initial reactions to different people. How did you do?

| Individual | 1<br>Greet | 2<br>Accept | 3<br>Help | 4<br>Background | 5<br>Advocate |
|---|---|---|---|---|---|
| 1. White Anglo-Saxon American (WASP) | | | | | |
| 2. Child abuser | | | | | |
| 3. Jewish | | | | | |
| 4. Iranian | | | | | |
| 5. Hispanic | | | | | |
| 6. Prescription drug abuser | | | | | |
| 7. Catholic | | | | | |
| 8. Homeless adult | | | | | |
| 9. Native American | | | | | |
| 10. Prostitute | | | | | |
| 11. Jehovah's Witness | | | | | |
| 12. Person with Cerebral Palsy | | | | | |
| 13. Asian | | | | | |
| 14. Person who is Gay/Lesbian/Bisexual/<br>  Transgendered | | | | | |
| 15. Muslim | | | | | |
| 16. Person with HIV | | | | | |
| 17. Chinese | | | | | |
| 18. African American | | | | | |
| 19. Unmarried expectant teen | | | | | |
| 20. Person with Schizophrenia | | | | | |
| 21. Amputee | | | | | |
| 22. Pacific Islander | | | | | |
| 23. Filipino | | | | | |
| 24. Alcoholic | | | | | |
| 25. Amish | | | | | |
| 26. Protestant | | | | | |
| 27. Asian | | | | | |
| 28. Obese | | | | | |
| 29. Rural Appalachian | | | | | |
| 30. Alaskan Native | | | | | |

**Source:** Compiled with information from Muñoz & Luckmann, 2004.

**EVIDENCE** FROM THE LITERATURE

**Citation:** DeRosa, N., & Kochurka, K. (2006, October). Implement culturally competent healthcare in your workplace. *Nursing Management,* 18–26.

**Discussion:** This article discusses selected verbal and nonverbal communication patterns of culture. Even when two people speak the same language, communication may be hindered by different values or beliefs. Nonverbal differences or ethnic dialects can also block mutual understanding. Communication differences include the following:

- *Conversational style:* Silence may show respect or acknowledgement. In some cultures, a direct "no" is considered rude, and silence may mean "no." A loud voice or repeating a statement may mean anger or simply emphasis, enthusiasm, or a request for help.
- *Personal space:* Beliefs about personal space vary. Someone may be viewed as aggressive for standing "too close," or as "distant" for backing off when approached.
- *Eye contact:* In some cultures, direct eye contact may be a sign of respect. In other cultures, direct eye contact may be seen as a sign of disrespect.
- *Subject matter and conversation length:* Even appropriate subject matter, when it's appropriate to discuss, and how long the discussion should last, vary from culture to culture. Some cultures value communication that's subtle and circumspect; forthright discussion is considered rude. In some cultures, it's acceptable to discuss topics such as sexuality and death, whereas in other cultures, these topics are taboo.

The article also discusses elements of a cultural assessment, including nutrition, medications, pain, and psychological and primary language assessment. It also explores how to approach the patient with educational needs.

**Implications for Practice:** Be aware of these cultural differences when working with patients and staff from other cultures. Remember that not everyone sees things the way you do.

Leininger's Transcultural Nursing (2006), Purnell's Model for Cultural Competence (2008), Campinha-Bacote's Process of Cultural Competence in the Delivery of Health Care Service (2003), and Giger and Davidhizar's Transcultural Assessment Model (2008). Each of these models offers the nurse insight into examining the cultural beliefs and needs of culturally diverse patients. Choosing a model to direct your practice offers a path for implementation of the nursing process. In lieu of choosing a model, the nurse must be aware of his or her personal cultural values, be sensitive to and accepting of the influence of culture, have a commitment to continual learning about different cultures, and provide care that is in harmony with the cultural beliefs and lifestyle of patients.

Although it is not realistic to learn every nuance of every culture, it is essential to learn as much as possible about the cultures represented in your community. Table 25-3 provides an overview of cultural norms, health care beliefs, and religious beliefs seen often in America. The content in this table provides information for beginning a dialogue between the nurse and the patient. Note that the norms and beliefs may not be valid for all individuals within a cultural group.

# ORGANIZATIONAL CULTURE

Before leaving the concept of culture, it is important to discuss organizational culture. In an organizational culture, the norms of behavior and beliefs influence the outcomes for the organization. **Organizational culture** is the system of shared values and beliefs that actively influences the behavior of organization members. The term *shared values* is important, because it implies that many people are guided by the same values and that they interpret them in the same way. Values develop over time and reflect an organization's history and traditions. The culture of an organization consists of such things as being helpful and supportive toward new members.

Five dimensions of organizational culture are of major significance in influencing organizational culture (Ott, 1989):

- *Values:* Values are the foundation of any organizational culture. The organization's philosophy is expressed through values, and values guide behavior on a day-to-day basis.
- *Relative diversity:* The existence of an organizational culture assumes some degree of similarity. Nevertheless, organizations differ in how much deviation can be tolerated.

## TABLE 25-3 Cultural Norms, Health Care Beliefs, and Religious Beliefs*

| CULTURAL GROUP | CULTURAL NORMS | HEALTH CARE BELIEFS | RELIGIOUS BELIEFS |
|---|---|---|---|
| Hispanic | • Maintaining eye contact is valued.<br>• A pat on the back or arm is considered friendly.<br>• Treating others with respect is valued.<br>• Cakes and sweets may be a regular part of the diet.<br>• Children are highly valued and loved.<br>• May have different perception of time; for example, may have a problem being on time for appointments. | • Fatalistic and may view illness as a punishment from God.<br>• View health as the ability to rise in the morning and go to work.<br>• May or may not follow medical advice.<br>• May consult a folk healer, for example, curandero.<br>• Will use Western medications but will stop when they feel they can no longer afford it. | • Are often Roman Catholic but may be member of other Christian group; may light candles, attend mass, pray to God, Jesus, the Virgin Mary, and saints.<br>• Traditional men view religion as a preoccupation of women.<br>• May have statues of saints at home. |
| Muslim | • Have modest lifestyles for both men and women. Women wear loose-fitting clothing that includes a head covering called a *hijab.*<br>• Immediate and extended family needs are very important, often above the needs of the individual. | • Often have a fatalistic view of health.<br>• Disease is often viewed as "God's will," a test of an individual's conviction or as retribution for transgressions.<br>• Due to issues of modesty, having a health care provider and interpreter of the same sex is generally preferred. | • Pray daily. While praying, often face southeast towards Mecca.<br>• Have a restrictive diet similar to that of Orthodox Judaism.<br>• Believe the Koran to be the book of divine guidance and direction for humanity and consider the text in its original Arabic to be the literal word of God (Rahman, 2009). |
| Black and African American | • Have tradition of involving many in raising children.<br>• Many households are headed by women.<br>• May be frank and direct in speech.<br>• Unrelated persons often live in the home.<br>• High incidence of poverty.<br>• Oriented to the present. | • Often distrust or have discomfort with majority group and health care system.<br>• May be private about their health and may not want family members present during care or treatment.<br>• May try self-care first and use all forms of pharmacological and some nonpharmacological alternatives and complimentary medicines prior to seeking care.<br>• View health as being in harmony with nature, and view illness as disharmony.<br>• Some have fatalistic attitude about illness. | • Are heavily involved in church religious groups.<br>• Black minister is strong influence in community.<br>• May use faith healers or herbalists.<br>• Are active in singing and praying.<br>• Illness is between the individual and God; illness may be viewed as punishment from God.<br>• May see illness as the will of God. |

*(Continues)*

**TABLE 25-3   (Continued)**

| CULTURAL GROUP | CULTURAL NORMS | HEALTH CARE BELIEFS | RELIGIOUS BELIEFS |
| --- | --- | --- | --- |
| Asians | • Work hard, have respect for elders and nature, have esteem for self-control and loyalty to all family and extended family.<br>• Are traditionally patriarchal.<br>• Have respect for elders.<br>• May not consider shaking hands to be polite.<br>• Submissive to authority.<br>• Pride and honor are extremely important. | • Prefer a same-sex health care practitioner.<br>• Expect health care to include an injection or prescription.<br>• May not make important decisions without checking with an astrologer or almanac for a lucky day. | • Have broad group of practices from Christianity to Buddhism, Taoism, ancestor worship, Muslim, and many others, depending on the geographic area.<br>• Prayer and offerings are dominant in many groups.<br>• May use faith healers or herbalists. |
| Pacific Islanders | • May ascribe to a holistic world view—interconnectedness of family, environment, self, and spiritual world.<br>• Family and community play an important role and often live in close proximity or tightly knit communities.<br>• Interpersonal and social behavior is based on mutual respect and sharing. | • Often distrust Western style of health care. Rarely respond positively to health education and treatment based on scare tactics.<br>• Stoic; do not complain.<br>• May use Western medication but choose over-the-counter drugs for minor ailments.<br>• Massage is a method to achieve harmony. | • Have deeply rooted spiritual connections.<br>• Hold belief in unity, balance, and harmony.<br>• Use traditional healers.<br>• Some may be Christian. |
| Native Americans | • Family and tribal affiliations are part of daily life.<br>• May have extended family structure and live with relatives from both sides of the family.<br>• Have a holistic view of life and health.<br>• Often suffer from poverty, poor nutrition, and inadequate access to health care.<br>• Avoid eye contact.<br>• Elders often assume leadership role.<br>• Share goods with others.<br>• Cooperate with others.<br>• Work for good of the group. | • Physical illness may be due to violation of a taboo or being out of harmony.<br>• Skeptical regarding the benefit and habit-forming properties of medications.<br>• Holistic orientation to health.<br>• May wait to see Western practitioner until seen by a healer.<br>• Oriented to the present.<br>• Accept nature rather than try to control nature. | • Religion or spiritual affiliation is based on personal choice.<br>• May have Christian beliefs and traditional beliefs.<br>• Have spiritual orientation.<br>• May fear witchcraft as cause of illness and use a medicine bag received from a healer, which should be kept with the patient at all times.<br>• May carry object at all times to guard against witchcraft. |

*The information presented is from multiple sources and is meant to serve as a starting point to understanding. All people are individuals and these norms and beliefs may not be valid for all within a cultural group identified in the table.

**Source:** Compiled with information from *The provider's guide to quality and culture.* Retrieved October 21, 2010, from http://erc.msh.org/ mainpage. cfm?file=1.0htm&module=provider&language=English; Lipson, J. G., & Dibble, S. L. (2005). *Culture and clinical care.* San Francisco: UCSF Nursing Press; and Muñoz, C., & Luckmann, J. (2004). *Transcultural Communication in Nursing.* Edition 2. Clifton Park, NY: Delmar Cengage Learning.

**TABLE 25-4  Questions to Ask When Assessing Organizational Culture**

- Are the organization's values consistent with your values?

- Is the organization or the department centralized or decentralized?

- What is the formal chain of command?

- What is the informal chain of command?

- Do individuals participate in changing policies or procedures?

- What are the rules about how things should be done?

- Where does one take new ideas or suggestions?

- Are risk and change encouraged?

- How are individuals rewarded for quality improvement, or are all rewards oriented toward the group?

- How does the team work together?

■ *Resource allocation and reward:* The allocation of money and other resources has a critical influence on culture. The investment of resources sends a message to people about what is valued in the organization.

■ *Degree of change:* A fast-paced, dynamic organization has a culture different from that of a slow-paced, stable one. Top-level managers, by the energy or lethargy of their stance, send messages about how much they welcome innovation.

■ *Strength of the culture:* The strength of a culture, or how much influence it exerts, is partially a byproduct of the other dimensions. A strong culture guides employees in many everyday actions. It determines, for example, whether employees will inconvenience themselves to satisfy a patient. If the culture is not so strong, employees are more likely to follow their own whims, that is, they may decide to please patients only when convenient for them.

Each organization will have embedded in its environment the dos and don'ts that are specific to its workplace. When beginning in a new organization, observe and ask questions to learn about the organization and culture and how decisions are made (Table 25-4).

# ORGANIZATIONAL SOCIALIZATION

As a new member of a team or work group, it is important to be socialized into the organization. In most organizations, this socialization begins as part of the new employee's

orientation process. This allows the organization to promote the organization's values to the new employee from the beginning. The individual responsible for a new hire's orientation is often also responsible for enhancing the socialization process. When there is a good fit of ethics, values, and behaviors between the preceptor and a new individual, the socialization process goes smoothly and often occurs rapidly. To ensure socialization that meets the organizational goals, it is important to monitor the orientation process. Frequent evaluation by both the preceptor and the new hire are critical to the success of the new employee and the organization.

Socialization is beneficial to the organization when employees are a good fit—that is, the employee has a high commitment to the organization, little intention to leave, a high level of job satisfaction, and little work-related stress. Things to consider when entering a new work environment in any culture are the organizational behavior style for greetings, titles, punctuality, body language, and dress (Table 25-5).

These organizational styles are important to the workplace milieu. Additionally, it is important for employees to avoid assumptions about how other individuals think, act, or speak. New employees should consider these organizational behavioral styles and work to ensure organizational cohesiveness (Table 25-6).

Most organizations have a workplace culture that has a strong mission and vision, and its members work tirelessly to ensure its success. However, there will be organizations that can only be described as toxic. In these organizations,

## TABLE 25-5 United States' Organizational Behavioral Styles

| CONCEPTS | THINGS TO CONSIDER |
|---|---|
| Greetings | • Americans usually acknowledge each other with a smile, nod of the head, and/or verbal greeting such as "Hello" or "Hi."<br>• When greeting someone in a business situation, a firm handshake is appropriate, such as when greeting a manager of nursing or human resources. |
| Titles | • When introducing yourself or others, give your first name followed by your last name.<br>• Use the appropriate title the first time you address an individual, such as Mrs., Dr., Ms., or Mr. Wait to be directed to call them by their first name. |
| Time | • Punctuality is highly respected in nursing, so be on time to interview appointments and work.<br>• Know where you are going, and plan to be on time. |
| Body language | • Use of direct eye contact is expected in all work situations and when working with patients. However, some patients may not respond to direct eye contact, depending on their culture.<br>• In conversation, keep a distance of approximately one arm length from the speaker; closer proximity is often considered rude. |
| Dress | • When in doubt, use a professional business attire, i.e., a professional suit with white blouse or dress shirt for meetings and interviews, a professional nursing uniform for working with patients, etc.<br>• For your work environment, ask what traditional dress is in a particular work area or nursing unit before starting a new job or purchasing new uniforms.<br>• Even in areas where daily dress is casual, business situations should be considered formal. |

## TABLE 25-6 Workplace Behavior Guidelines

• To be successful in health care, work to adapt to your organization's culture.

• People from other cultures often may not think and behave the way that you do. What might be normal behavior in your culture may be inappropriate in another culture and vice versa.

• Communication requires listening and clarifying meaning to ensure understanding. It is often a good idea to rephrase what you have heard to test your understanding.

• When seeking clarification, go to the source of the communication. Do not ask other coworkers to clarify work requests that originated from a third party or from your leadership personnel.

• Observe for cultural differences in the workplace, and attempt to understand and accommodate those differences.

• In health care, holidays and weekends are considered part of the work requirements for nurses. If you need time off to celebrate a cultural or religious holiday, make arrangements early with your manager. Also, realize that health care is a 24/7 business.

## Critical Thinking 25-3

In today's diverse workplace, you will work closely with RNs, nursing students, medical practitioners, and ancillary personnel who are from different cultures and who speak English as a second language. How do you feel about working with practitioners who are from foreign countries, or from different racial or ethnic groups? Take a minute to answer the following questions. You do not need to share your answers with anyone, so be honest with yourself.

| | Agree | Neutral | Disagree |
|---|---|---|---|
| I would rather work with an American nurse than a foreign nurse. | | | |
| I find it frustrating to work with medical and nursing practitioners who are not proficient in English. | | | |
| If I thought that a medical or nursing practitioner was not fulfilling duties because of cultural or language problems, I would hesitate to report the person for fear that I would be considered prejudiced. | | | |
| I enjoy working with a skilled foreign nurse. I feel that I can learn a lot from this person. | | | |
| I like to attend classes and informal meetings where I can learn more about how nurses from other countries are educated. | | | |
| I do not feel prepared to work with or supervise an unlicensed assistive worker who has some problems with understanding English. | | | |

*Source:* Compiled with information from Muñoz & Luckmann, 2004.

## Critical Thinking 25-4

You are hired to work on a 10-bed nursing unit. During your orientation, you discover that you are the only individual of your cultural group. What actions will you take to enhance your socialization into the group? What actions should the group take?

the staff and/or leadership is dysfunctional. Instead of problem solving, the goal of these organizations is fault-finding and placing blame. Staff may observe excessive control on the part of the leader or a unit or worker in a constant state of crisis. This dysfunctional toxic environment may be a unit within a hospital or the entire organization. Choosing your workplace environment wisely will make your work life more satisfying.

# WORKING WITH STAFF FROM DIFFERENT CULTURES

Staff nurses from different cultures may have different perceptions of staff responsibilities to each other, different perceptions of the nurse's role in patient care, a different locus of control, a different time orientation, and a different language.

## Different Perceptions of Staff Responsibilities

Cultural values deeply influence what a person feels is most important, that is, the welfare of the individual or the welfare of the group. Individualism emphasizes the importance of individual rights and rewards. Collectivism emphasizes the importance of group decisions and places the rights of the group as a whole above the rights of any individual in the group. For example, some nurses tend to accept difficult assignments without complaint. They also may be more willing to do what other nurses might consider demeaning, for example, cleaning cabinets. Because nurses from some

## Critical Thinking 25-5

When you are asked to work with people from a different culture, it is natural to have some concerns. How do you feel about this? Are you worried that you will not be able to communicate clearly with people from other cultures? Try to respond as honestly as possible to the following statements. The answers are for your use only.

| | Agree | Neutral | Disagree |
|---|---|---|---|
| People are the same. I don't behave any differently toward people from a cultural background that differs from mine. | | | |
| I always know what to say to someone from a different cultural background. | | | |
| I look forward to working with people from a different cultural background. | | | |
| I can learn something when I work with people from diverse cultural backgrounds. | | | |
| I always introduce myself to new people who I work with. | | | |
| I prefer to work with staff from my own cultural group because it is easier. | | | |
| Responsible adults prepare for the future and strive to influence events in their lives. | | | |
| Intelligent, efficient people use their time well and are always punctual. | | | |
| It is disrespectful to address people by their first name unless they give you permission to do so. | | | |
| It is rude and intrusive to obtain information by asking direct questions. | | | |

*Source:* Compiled with information from Muñoz & Luckmann, 2004.

cultures value the group, they believe that a nurse's individual duties have less value than the combined work of all the nurses on the unit. Maintaining face and ensuring harmony are Eastern cultural values that may be more important to some nurses than upsetting a supervisor to get an easier assignment. An emphasis on minimizing conflict and maintaining harmony, teamwork, and commitment to group loyalty typifies most Asian cultures.

In contrast, nurses educated in Western culture often place value on individualism and independence. Individualistic cultures are commonly achievement oriented. Personality characteristics such as assertiveness and competitiveness aid in achievement endeavors (Andrews, 2008). Western nurses may also complain to the supervisor if they feel assignments are unfair or involve menial work. This assertive behavior is consistent with ingrained

values of equitable work distribution. In a work setting, most employees will tend towards either individualistic or collectivist values. Nurse leaders need to recognize these values in their employees to aid in understanding workplace behaviors (Andrews, 2008).

## Different Perceptions of the Nurse's Role

Mattson (2009) notes that nurses in many countries outside the United States usually focus their time on patient care. They are astonished by the amount of documentation as well as the cost-effective focus that is required in the United States. For example, in socialized health care, no evidence needs to be provided to an insurance company. In Asian countries, where families are very involved in patient care, it would be considered rude for the nurse to assist in bathing or feeding a patient. However, in American health care facilities, if the nurse does not ensure that these services are done, staff members could view the nurse as a "slacker" that is not completing his or her job duties (Mattson, 2009).

## Differences in Dress

The vast majority of health care agencies have a policy statement regarding clothing and accessories worn by employees in the institution. This is often referred to as a "dress code". Nurse leaders need to review these policy statements regularly from a cultural standpoint. The dress code may need to be revised to accommodate traditional dress such as saris worn by Hindu, turbans worn by Sikh men, and head coverings such as the hijab or veil worn by Muslim women. Accommodations may also need to be made for Native Americans, African Americans, and others who traditionally put jewelry and other items in their hair (Andrews, 2008).

## Differences in Locus of Control

Locus of control refers to the degree of control that individuals feel they have over events. People who feel in control of their environment have an internal locus of control. People who believe that luck, fate, or chance controls their lives have an external locus of control.

Health care providers who are trained in the United States often have an internal locus of control. American medical and nursing practitioners and nurses feel that it is their duty to diagnose disorders, plan interventions, carry out procedures, and do everything possible to save the patient's life.

Conversely, health care providers from cultures that promote an external locus of control may have a more fatalistic attitude toward their patients and thus feel that they cannot control matters. This may be revealed when staff from differing cultures are late, take excessive break periods, or fail to complete work on time. Differences are often correlated with the cultural values and significance of work; with spiritual beliefs and practices; as well as with issues in language and communication.

## Differences in Time Orientation

Cultural groups are either past-, present-, or future- oriented. Americans generally value the future over the present. Southern blacks and Puerto Ricans value the present over the future. Southern Appalachians, traditional Chinese Americans, and Mexican Americans also value the present more (Muñoz & Luckmann, 2004).

The ways in which different cultural groups value time can create challenges in the health care workplace. This may be revealed when staff from differing cultures are late, take excessive break periods or fail to complete work on time (Andrews, 2008).

These differences are often correlated with the cultural value and significance of work, spiritual beliefs and practices, as well as issues in language and communication. As a nurse leader, it is essential to give clear work expectations regarding shift schedules, promptness, break schedules and time allocations for job related duties (Andrews, 2008).

## Educational Differences

Foreign nurses are educated differently from American nurses. Generally, nursing education outside of the United States is less theory oriented, focusing primarily on the development of clinical skills. Also, there is less emphasis on meeting the psychosocial needs of patients.

Another cultural difference in the education of nurses revolves around who provides the majority of care—the nurse, the patient's family, or the patient. Some nurses may feel it's the nurse's duty to give patients complete physical care. Other nurses, educated outside of the United States, may have been taught that it is the family's duty to bathe the patient and provide personal care. In contrast, nurses educated in the United States are taught that patients should perform self-care, whenever possible. In keeping with the American values of independence and self-reliance, American nurses encourage patients to be ambulatory, active, and self-sufficient as soon as possible.

Even when nurses are from other English-speaking countries, they are educated differently from American nurses. English nurses, taught under the system of socialized medicine, may find it difficult to adjust to the concept that health care in the United States is a business and that practitioners are in private practice. These nurses may not be familiar with practitioner referral services, charging patients for supplies, or the use of extreme measures to prolong the lives of terminally ill patients.

## Communication Differences

A preponderance of conflicts in the culturally diverse health care setting stem from problems related to verbal and nonverbal cross-cultural communication (Andrews, 2008). Today, large health care centers in the United States may be primarily staffed by nurses and practitioners for whom English is a second language. For example, in urban medical centers on the East Coast, it is not unusual to hear a Filipino nurse and a Haitian nurse attempting to communicate with a resident practitioner who has been educated in India. Unless these caregivers take the time to clarify their communications, serious errors may result. The potential for miscommunication exists (especially over the telephone) unless words are clarified by a coworker or practitioner (Muñoz & Luckmann, 2004).

Even when the nurse leader is working with staff from the same cultural background, it requires astute leadership skills to decide whether to speak with a staff member face-to-face, send an e-mail, or contact the person via telephone about a particular issue. The nurse leader exercises substantial judgment when making choices about successful communication methods with staff and patients from differing cultural backgrounds, including the timing of the communication, voice tone and pitch, and the location for face-to-face exchanges (Andrews, 2008).

Language differences are also a source of friction between American nurses and foreign nurses. When frictions escalate, foreign nurses may form cliques on a unit. By speaking in their native language and excluding English-speaking nurses, foreign nurses may feel unified, even though they are alienating themselves even further from the rest of the nursing staff (Mattson, 2009). English-speaking nurses may believe that they are being talked about by the foreign nurses and thus demand that personnel speak only English on the unit. Foreign nurses, denied the right to speak their own language at work, can feel even more threatened and angry, a feeling that presents further obstacles to communication (Muñoz & Luckmann, 2005).

## IMPROVING COMMUNICATION ON THE TEAM

If you are assigned to work with a nurse or staff member who is from a different culture and who speaks English as a second language, try these techniques to facilitate communication:

- Recognize that your coworker probably has an educational background in nursing that is very different from your own.
- Acknowledge that the coworker's value system and perception of what constitutes good patient care may differ from your own.
- Try to assess your coworker's level of understanding of verbal and written communication. For example, ask a coworker to explain a practitioner's order to you in his or her own words. It also helps to assess a patient with the coworker and note what terms the person uses to describe the patient's signs and symptoms.
- Avoid the use of slang terms and regional expressions. For example, Chinese, Japanese, and Filipino nurses may not understand such expressions as "piggybacking," "doing a double," or "rigging" something to work.
- Provide your coworker with resources, such as written procedures and protocols, that may help to reinforce your verbal communication.
- Remember to praise your coworker's competency in technical skills. Inspiring self-confidence in a foreign nurse will make it easier for that person to ask for assistance when needed.
- Appreciate the knowledge that you can gain by working alongside a skilled nurse from another culture. Observe how foreign nurses relate to patients who are from their culture.

- When offering constructive criticism, try to use *I* statements instead of *you* statements. For example: "I think that it's very important to address the patient's emotional state when you chart" is better than "You never seem to chart anything about the patient's emotional state."

## Communication with Others

Sometimes you may need to work with a foreign practitioner who is difficult to understand because of language differences or a strong accent. In this case, do not take verbal orders, particularly over the telephone. Even when an order is written, take the time to clarify the order with the practitioner. Because patients may also find it difficult to understand a foreign practitioner, you may need to listen carefully and then explain the practitioner's remarks to the patient (Muñoz & Luckmann, 2004).

Another group of health care workers who may have difficulty understanding and speaking English are nursing assistive personnel (NAP) (Walton & Waszkiewiez, 1997). If you are called upon to supervise an unlicensed worker who speaks English as a second language, follow these cautions:

- Delegate appropriate tasks to an unlicensed worker. Match assignments to the worker's level of understanding and skill.
- Do not stop at just delegating an assignment or giving instructions. Instead, make sure that the worker understands your instructions.
- Restate your instructions in clear, concrete terms, and give a demonstration of a procedure if necessary.
- To reduce miscommunication, check for understanding by asking the worker to repeat instructions or do a return demonstration.
- If you are still not satisfied that the communication between the two of you is accurate and effective, repeat your directions and request a repeat demonstration.
- Establish a time frame for the worker to complete assigned tasks. For example, "I want you to feed Mr. González before you get Mr. Jones out of bed."
- Observe how the worker communicates with patients and performs duties.
- Give workers clear feedback concerning their communication skills and performance of duties. If the worker has performed a procedure incorrectly, offer suggestions for improvement. Demonstrate the procedure as it should be done, and ask for a return demonstration.
- If, despite your best efforts, the worker is still unable to perform because of language difficulties, ask your supervisor to work with the person. You do not want to be held responsible for a worker with whom you cannot communicate (Muñoz & Luckmann, 2004).

## MANAGERIAL RESPONSIBILITY

Jamieson and O'Mara (1991) have laid out a broad, six-step program for nurse managers to follow to actively manage a diverse nursing staff:

1. Determine which cultural groups are represented on staff.
2. Understand the organization's values and goals.
3. Decide on what is best for the future of the organization.
4. Analyze present conditions within the organization.
5. Plan ways to reach the desired future state, and decide how to manage transitions.
6. Evaluate the results.

Nurse managers might consider using the following approaches to diminish tensions between staff members and improve communication:

- Plan informal meetings for nurses to discuss their cultural values. For example, it may benefit Asian nurses to share with American-born nurses their cultural values concerning respect for authority.
- Provide cultural workshops, and ask knowledgeable individuals to present information about the values, behaviors, and communication patterns of the different cultural groups that are represented on staff.
- Provide classes in English as a second language for foreign nurses who do not speak fluent English or who have difficulty pronouncing words.
- Establish a program for orienting foreign nurses to the hospital or agency (Jein & Harris, 1989). The orientation program should be designed to help newcomers adjust to the new work environment. It is helpful to assign each new nurse to a preceptor who will assist in the orientation process. If possible, the preceptor should be a member of the nurse's cultural group. For maximum benefits, the nurse manager needs to interview each new nurse every week to find out how that person is adapting to the new hospital culture (Williams & Rodgers, 1993).
- Confer with specialists in transcultural communication; also hire experts to identify potential areas of conflict and resolve conflicts peacefully before they erupt into legal battles (Muñoz & Luckmann, 2004).

# GENERATIONAL PERCEPTIONS

Generational perceptions and different values and beliefs are created as each generation deals with the experiences of their lives as these are altered by the changing times. A **generation** is a group that shares birth years as well as a common connection or bond based upon significant and influential life events (Calhoun & Strasser, 2005). A generation is approximately fifteen to twenty years in length and has a different value system from the preceding generation and later generations. Like culture, we take our generational differences with us into patient care and the work environment. We often assume that those around us are like us and think like us. This is not always true.

The nursing workforce is more age diverse today than ever before in history (Manion, 2009). Four distinct generations make up the current workforce population. These generations are the Veteran or Silent Generation, born between 1922 and 1945; the Baby Boomers, born between 1946 and 1964; Generation X (Gen Xers), born between 1965 and 1979; and the newest generation to hit the workforce, the Generation Y or Millenial Generation, born between 1980 and 2000. The generation that follows Generation Y (the Millennials) is called the New Silent Generation. This generation began in 2001 and is not in the workforce population yet.

The Veteran or Silent Generation was raised in difficult times, including the Great Depression and World War II (Manion, 2009). They value consistency, hard work, conformity, and organizational fidelity (Manion, 2009). The Veteran or Silent Generation was followed by the Baby Boomers, who came of age during a time of much available education and economic expansion. They work for the challenge of work and career advancement (Calhoun & Strasser, 2005). Baby Boomers have been characterized as workaholic, strong-willed individuals who are working for material gain, promotions, recognition, job security, and corner offices. Baby Boomers are the largest generation. They have had a dramatic financial impact on the present and are anticipated to impact the future dramatically as they continue to retire.

The Gen Xers are often called latch-key kids, as their parents were often away working. They learned to be self-reliant and independent. Subsequently, they are used to doing things themselves, and this formative dynamic translates into staff who respond well to goal setting (Paterson, 2010). Gen Xers often observed their parents going through multiple changes in their work organizations, such as downsizing and rightsizing. Therefore, they tend to be skeptical, independent workers who seek a balance between work and leisure. Gen Xers want a work environment that is technologically current, has competent leadership, and provides a mentor or coach for a boss.

Gen Yers, or Millenials, grew up with the end of the Cold War. The Cold War was a struggle for global supremacy from approximately 1945 to 1991 that pitted the Western world, primarily the United States and its allies, against the Communist World, primarily the Soviet Union and its allies (Black, A., Hopkins, J., et. al. 2003). Gen Yers, or Millenials also grew up with the Internet and a speak-your-mind philosophy. Their lives have been characterized by demanding schedules filled with structure and activities (Manion, 2009). This generation is just beginning to make its mark on the workforce. What is known is that they are focusing on early retirement. The Millenials are considered a true "global generation" where diversity is the cultural norm (Manion, 2009, p.13). They are a very hi-tech generation, growing up with advanced computer technology including instantaneous and continuous communication through cell phones and text messaging (Manion, 2009).

Nurse leaders will be working with a mixed generation staff and need to successfully communicate to a group of nurses with highly diverse and often conflicting preferences and expectations (Paterson, 2010).

The generational diversity in the nursing workforce is an incredible challenge for the nurse leader, as differences can lead to disagreement and tension in the work setting. To address this challenge, the International Council of Nurses and the International Centre for Human Resources in Nursing recommend four key approaches developed by Manion (2009):

- Creating an organizational environment that connects with all staff, thereby increasing retention rates
- Continually exploring the workplace culture and conditions and maintaining the ability to be flexible in intervention and response mechanisms
- Working successfully with problems that arise due to generational differences
- Utilizing effective approaches that are specific to the applicable staff generation

For example, if the nurse leader needs to work on motivation in the multigenerational nursing workforce, consider the following strategies (Manion, 2009):

- *Veteran or Silent Generation*: Consider sending personal letters to them or use them as a preceptor for new nurses.

---

### Critical Thinking 25-6

When working on a nursing team, consider the values, goals, and outcomes of the members and generations on the team. Make a list of how your generation is different from the one before it and the generation after your own. How can you improve your ability to work together?

---

- *Baby Boomer Generation*: Give public acknowledgement and reward for good performance.
- *Generation X*: Make opportunities known for growth and development, and give as much autonomy and independence in practice as possible.
- *Generation Y or Millennial Generation*: Emphasize the importance and significance of their work.

## SPIRITUALITY

In nearly every culture, spirituality is a component of healing, yet many nurses seldom assess the concept of spirituality or work to help patients meet their spiritual needs. In the workplace, spiritual considerations are often ignored unless an individual pushes the issue.

During the first 60 years of the twentieth century, there was considerable attention paid by nursing to religious issues. Then, during the cultural change of the 1960s and 1970s, little was written on religion in the nursing setting. In the mid-1980s, interest developed in the holistic component

---

## EVIDENCE FROM THE LITERATURE

**Citation:** Grossman, C. L. (2010, October 7). "How America sees God; Analysts cast our views in four ways." *USA Today*, 1A.

**Discussion:** This article reports on a book by Baylor University sociologists, P. Froese and C. Bader. They conducted religion surveys in 2006 and 2008 on 1,648 and 1,721 Americans, respectively, along with more than 200 interviews of Americans. Participants were asked to react to 12 suggestive images, including a "wrathful old man slamming the Earth, a loving father's embrace, an accusatory face, or a starry universe" (p. 1). The research findings defined four ways in which Americans see God—that is, authoritative (28%), benevolent (22%), critical (21%), and distant (5%). The first wave of survey results found that 91.8% of those surveyed said they believed in God, a higher power, or a cosmic force. The other respondents said they were atheists, did not answer, or weren't sure. The book also discusses how the four views of God affect views of morality, science, and money.

**Implications for Practice:** Many Americans believe in God and may want assistance with their spiritual needs during times of illness and stress. Nurses can play a role in helping patients obtain the assistance they need.

of spirituality. Each succeeding year thereafter, there has been an increasing amount of research and thought on spirituality. For many years, the focus was on defining the term *spirituality*. The researchers found the term difficult to quantify and finally concluded that spirituality is a complex and enigmatic concept (Martsolf & Mickley, 1998; McSherry & Draper, 1998). Foley, Wagner, and Waskel (1998) described spirituality as a multifaceted concept specific to the spiritually lived experience of an individual. This description gives the nurse little direction in assessing or intervening when there is evidence of spiritual need or distress.

## SPIRITUAL ASSESSMENT

Providing holistic patient care includes assessing the spiritual needs of the patient in addition to the biophysical and psychosocial needs. To provide spiritual care, an understanding of the patient's beliefs can be used to plan appropriate interventions. In contemporary nursing, spirituality has become an important area of assessment during hospitalization. However, asking the questions, "What is your religion?" or "What religious needs can we meet during your hospitalization?" leaves the nurse with only descriptive labels. To meet the spiritual needs of patients, nurses need to understand more than these labels. The national health care accrediting body, the Joint Commission, recently focused on the assessment of spirituality, requiring that patients be asked some questions regarding their spiritual needs. The Joint Commission (2011) provides examples of spiritual assessment components that may be utilized by health care providers, including: the use of prayer; philosophy of life; the meaning of dying to the patient; and how faith has helped with the coping abilities of the patient (The Joint Commission, 2011). Patients may be asked if they would like to see their spiritual leader or advisor, with the understanding that not all patients will want to meet with a clergy member. Nurses need to be familiar with and use various resources available for providing spiritual support to patients. Many inpatient health care facilities have professional chaplains

that can assist the health care team in multiple ways, such as the determination of appropriate spiritual interventions for individual patients; visitation of the patient and family; facilitation of prayer; provision of referrals to support groups; as well as aid in end-of-life issues (Nance, Ramsey, & Leachman, 2009). Additionally, professional chaplains can also help meet the spiritual needs of organizational staff. Professional chaplains are increasingly becoming part of the multidisciplinary health care team along with physicians, nurses, and pharmacists (Carson & Koenig, 2008).

There are a number of research tools used for measuring spirituality, including:

It is important for the nurse to conduct a spiritual assessment to determine the specific spiritual concerns that the patient may have. There are a number of simple spiritual assessment frameworks that can help guide spiritual assessments. Several are described here:

- Carson and Koenig's (2008) spiritual assessment involves five broad questions that the nurse can utilize while caring for the patient to respond to both verbal and nonverbal patient indicators that communicate the presence of a spiritual need. The questions assess: how any spiritual beliefs may conflict with medical care; involvement and support from any religious communities (e.g., churches, mosques, synagogues); feelings of comfort or stress associated with the spiritual beliefs; how spiritual beliefs may impact health care decision making; and the need for additional referrals.
- Jenkins, Wikoff, Amankwaa, and Trent (2009, p. 35) developed a mnemonic assessment tool for nurses titled "FAITH." The tool assesses five areas:
  - **F**amily, friends, or the patient's support network
  - **A**ffiliations, e.g., religious denomination, church membership, clubs
  - **I**llness consequences for the patient's life, e.g., employment, education, and parental role consequences
  - **T**reatment concerns of the patient
  - **H**ow the nurse can assist with meeting the spiritual needs of the patient, e.g., prayer, facilitating clergy visitation, etc.
- Parsian and Dunning's (2009) spirituality questionnaire evaluates self-awareness, e.g., attitude, self-confidence, compassion; the significance of spiritual beliefs in one's life; spiritual practices, e.g., meditation, involvement in ecological programs; and spiritual needs, e.g., the importance of relationship maintenance, searching for life purpose.

Which spiritual assessment framework the individual nurse decides to utilize should be based on patient needs, the environment in which care is being provided, and the individual nurse's ability to implement the assessment during care. Regardless of the assessment tool chosen, assessing the spiritual needs of patients is an important component of providing holistic health care.

### REAL WORLD INTERVIEW

Only when you meet the spiritual needs of your patient can you say that you have treated the patient as a whole. Spiritual realities exist whether one believes in them or not. In fact, nurses consider all world views when directing or providing care.

**Celeste Lynn Hagen Proctor, MS, RN**
Roseville, California

# SPIRITUAL DISTRESS

**Spiritual distress** is a North American Nursing Diagnosis Association (NANDA) term used to identify when an individual has an impaired ability to integrate meaning and purpose in life through the individual's connectedness with self, others, art, music, literature, nature, or a power greater than themself (Ackley & Ladwig, 2011). To decrease or eliminate spiritual distress, it is expected that an individual will connect with the elements he or she considers important to arrive at meaning and purpose in life. These elements may include meditation; prayer; participating in religious services or rituals; communing with nature, plants, and animals; sharing of self; and caring for self and others.

Once spiritual needs have been identified, the nurse must develop interventions to help patients. Nursing interventions for spiritual needs could include the nurse requesting a visit from a patient's spiritual leader or offering the patient uninterrupted quiet time to allow the patient time for personal prayer or reading of spiritual and religious material. Most religious groups from Christians to Jews to Muslims to Buddhists communicate spiritually through prayer or meditation. It is also important for the nurse to serve as a patient advocate when religious or spiritual needs arise (Carlson & Koenig, 2008).

# BARRIERS TO SPIRITUAL CARE

Barriers to providing spiritual care are often more common for nurses than barriers to providing any other type of care or service. One of the barriers that nurses face may be those of a personal nature where the nurse believes that spiritual needs are private and not the purview of the nurse and, therefore, not a nursing responsibility (McEwen, 2005). The nurse may be uncomfortable or embarrassed by his or her own spirituality and find it upsetting to deal with the issues that bring up spiritual distress, such as death and dying or suffering and grief. A lack of knowledge regarding the specific beliefs of a patient's religion and how to facilitate those spiritual needs are other potential barriers to quality nursing care delivery. Finally, there is often insufficient nursing time or privacy to allow patients to discuss their spiritual beliefs.

These barriers to spiritual care sometimes form an apparently legitimate reason for nurses to avoid the necessary interaction to determine the patient's spiritual needs. Spiritual needs are often ignored, forgotten, and never dealt with. As a result, patients experience even more distress and suffering. This distress can be avoided by supporting and spending time with the patient. Other nursing interventions to help patients are included in Table 25-7.

# CHAMPIONING SPIRITUALITY

The nurse leader who champions spirituality for all staff ensures that this component of holistic nursing is not forgotten or marginalized. For example, there may be an employee requesting a Saturday or Sunday off every week to attend religious services. When nurses are needed to work every other weekend, this can pose a problem for the nurse leader. A possible solution is to pair two nurses, where one works every Saturday and the other works every Sunday. Both nurses are then able to meet their spiritual needs without a negative impact on staffing the nursing unit.

## REAL WORLD INTERVIEW

It is extremely important to recognize the problems that caregivers experience as they make decisions and struggle through their loved ones' stages of Alzheimer's disease. These problems include stress, guilt, and religious, spiritual, and emotional turmoil. In northern Indiana, there have been over 36,500 patients diagnosed with Alzheimer's disease. Seven out of ten patients live at home. Seventy-five percent of them are cared for by family and friends. Those family members and friends need to take care of themselves also, as they are vital to their loved ones. They must maintain some normalcy in their own lives and stay mentally and physically healthy for their own well-being as well as for the well-being of their loved ones. To assist them, Alzheimer's Services of Northern Indiana (ASNI), a nonprofit organization, shares information, at no cost to the families, about respite home care, adult day care centers, specialized Alzheimer's units in nursing homes, local support groups, and family outreach programs. We offer a 24-hour toll-free HELP line (1-888-303-0180). ASNI's mission statement, which we take quite seriously, is "Bringing hope and help to Alzheimer's families in northern Indiana."

**Leona Bachan, Community Relations Coordinator**
Alzheimer's Services of Northern Indiana
Munster, Indiana

# EVIDENCE FROM THE LITERATURE

**Citation:** Tzeng, H. M., & Yin, C. Y. (2006). Learning to respect a patient's spiritual needs concerning an unknown infectious disease. *Nursing Ethics, 13*(1), 17–28.

**Discussion:** While the location of this article is Taiwan, there are parallels for nursing practice anywhere a new disease might develop. The first readily transmissible disease of the twenty-first century was Severe Acute Respiratory Syndrome (SARS). It occurred in 29 countries. In a Taiwanese hospital, there was a substantial outbreak of SARS. All staff and patients were quarantined within the hospital. Despite their medical knowledge in a time of rapid technological development, the Taiwanese community began to see the disease from the framework of a religious concept called a taboo. A taboo in Taiwan acts as a part of folk religion that serves as a cultural and psychological remedy for society. It is thought that this religious concept of the taboo worked to unite the community to overcome the crisis. To illustrate the shift to folk belief by patients during a time of crisis, a diary written by a medical doctor who developed a case of SARS is shared in the article. The story the medical doctor tells is of his own increasing reliance on prayer to the gods, seeking divine advice, and asking others to pray for him during his episode of SARS. While his traditional medical beliefs were important, he also prayed to Jehovah, the Virgin Mary, the Goddess of Mercy, the Buddha, and Bodhisattva. He discusses his sense of powerlessness in the article and his sense of not knowing what he could do to aid his recovery during his illness.

**Implications for Practice:** In times of crisis, it can be expected that many will return to their cultural religious heritage for reassurance regarding healing and a return to wellness. Nurses need to be willing to respect personal beliefs that may be foreign to them and even against known scientific data. These personal beliefs may help the patient shift toward or away from survival.

## TABLE 25-7 Spiritual Nursing Interventions

- Open a dialogue with the patient regarding the purpose and meaning of life.

- Allow the patient to describe his or her spiritual life.

- Ask the patient if prayer plays a role in his or her life. If comfortable, you can offer to pray with the patient.

- Offer to seek the spiritual or religious leader of the patient's choice, such as their pastor, priest, imam, rabbi, and so on.

- Be physically present, and listen to the patient. Seek a quiet, noninterrupted environment.

- Use therapeutic touch by holding the patient's hand or gently touching the patient's arm.

- Seek an answer to how you may provide support to the individual patient.

- Support patient-directed spiritual activities, such as receiving sacraments, anointing meditation, and so on.

- Focus on spiritual relationships and how you might provide support for patients with spiritual needs.

**Source:** Compiled with information from Ackley, B. J., & Ladwig, G. B. (2011). *Nursing diagnosis handbook: A guide to planning care* (9th ed.). St. Louis, MO: Mosby.

## Critical Thinking 25-7

How does your own spirituality influence the way you provide nursing care? Are you comfortable with your spirituality?

There are religious holidays and celebrations that have spiritual and cultural significance. An understanding and empathetic approach to vacation requests will ensure contented staff and minimize turnover. It is also important to consider important markers of life such as weddings, births, and deaths. There are many cultural and spiritual overtones to these life events that need to be respected by nursing leadership practices. Sensitivity to the spiritual practices of the staff will enable the nurse manager to provide a compassionate and caring leadership.

## DEVELOPING SPIRITUAL LEADERSHIP

Spiritual leadership involves using values and beliefs as the basis for dealing with all staff and patients within an organization. It is using compassion, caring, and nurturing to create an environment that reflects the values and beliefs of the leaders, patients, and staff. Spiritual leaders develop trust and connect with their staff on both a personal and professional level. This connection is then the basis for change and growth within the organization. The ability to connect and build relationships among staff and leadership, inspire others, and spot problems early on results in a cohesive and positive workplace.

## EVIDENCE FROM THE LITERATURE

**Citation:** Villagomeza, L. R. (2006). Mending broken hearts: The role of spirituality in cardiac illness: A research synthesis, 1991–2004. *Holistic Nursing Practice, 20*(4), 169–186.

**Discussion:** This research synthesis analyzed research on spirituality in cardiac illness from 1991 to 2004 to identify progress, gaps, and priorities for research. Articles were retrieved from PubMed and CINAHL. Twenty-six studies met the inclusion criteria. Moody's Research Analysis Tool, Version 2004, was used to analyze studies. Lack of a conceptual model and a universal definition of spirituality are major knowledge gaps identified. A proposed conceptual model is presented.

The proposed conceptual model of spirituality has seven distinct but overlapping constructs: a sense of connectedness, a value system, a sense of self-transcendence, a sense of inner strength and energy, a sense of inner peace and harmony, a sense of purpose and meaning in life, and faith and a religious belief system.

**Implications for Practice:** Spirituality—the dimension of human life that been regarded as the central artery of a heart that permeates, energizes, and enlivens all other dimensions—may serve as a buffer for the stressful physical, emotional, and psychologic events associated with heart disease and illness. Spirituality is the manner by which human beings make sense of life events and establish the meaning of their existence amid potentially life-threatening illnesses. Nurses concerned with quality patient care can help strengthen this spirituality buffer.

## KEY CONCEPTS

- Culture affects both the nursing staff and the patient.
- Cultural competence is an important component of nursing.
- Organizations that nurses work in have distinct cultures.
- Each culture has its own values.
- Health care disparities continue to exist for racial and ethnic minorities.

- Nurses are moving to increase the number of various population groups in nursing.
- Four distinct generations make up the current workforce population, i.e., the Veteran or Silent Generation, the Baby Boomers, Generation X, and the Generation Y or Millennial Generation. The generation that follows Generation Y is called the New Silent Generation.

This generation began in 2001 and is not in the work force population yet. Each generation has different values, goals, and expected outcomes from its work and life experiences.

■ Spirituality may include religion and reflects one's values and beliefs.

■ Spiritual assessment is a requisite of holistic nursing.

■ Many barriers can interfere with meeting a patient's spiritual needs.

# KEY TERMS

culture
culturally competent care

generation
health care disparity

health disparity
organizational culture

race
spiritual distress

# REVIEW QUESTIONS

1. Stephanie, the Gen X night shift charge nurse, is requesting more time off than any other charge nurse. What reason for this best represents her generation? She does which of the following?
   A. Prefers to work the day shift and is hoping for a schedule change
   B. Believes that there are other nurses just as capable to fill her role
   C. Wants to increase her leisure time to balance with work
   D. Seeks to be rewarded for time spent at work

2. Nurses may fail to meet spiritual needs because of which of the following?
   A. Lack of compassion
   B. Lack of knowledge regarding the patient's illness
   C. Lack of understanding of the patient's spiritual beliefs
   D. Lack of chaplains in the care area

3. During orientation to work on a new unit, the nurse experiences a sense of isolation from his preceptors. Which of the following actions will best increase his socialization into the preceptor group? The nurse should do which of the following?
   A. Ask as many questions as he can
   B. Request that his orientation be increased for two more weeks
   C. Study the differences between his values and his preceptor group's values
   D. Arrive late for duty frequently

4. An organization's workplace culture reflects which of the following?
   A. The cultural affiliations of the staff
   B. The religion practiced by the hospital's leadership group
   C. The culture of its preceptors
   D. The values and beliefs of the organization

5. The Veteran or Silent Generation may do which of the following?
   A. Value working as one's duty
   B. Have a speak-your-mind philosophy
   C. Value working primarily for the challenge
   D. Use the Internet daily as a part of their life

6. Strategies to improve cultural competency skills include which of the following? Select all that apply.
   _____ A. Participation in continuing education offerings about culturally congruent care
   _____ B. Development of culturally competent approaches to care
   _____ C. Talking with patients about their cultural views on health care
   _____ D. Assessing your own skill level and seeking improvements
   _____ E. Talking with staff and colleagues about their cultural views on health
   _____ F. Identifying and reducing ethnocentric bias

7. Causes of health care disparities include which of the following? Select all that apply.
   _____ A. Patient health behaviors
   _____ B. Conscious bias
   _____ C. Institutional discrimination
   _____ D. Unconscious bias
   _____ E. Culturally specific provider rapport
   _____ F. Biological variation

8. After shift report, the nurse is preparing to enter the room of patient, Guadalupe González. While reviewing the patient's chart, the nurse notes that the patient is listed as of "Hispanic/Latino origin." What is the best way for the nurse to determine which culture Mrs. González identifies with?
   A. Ask the patient which culture she identifies with
   B. Determine her culture by her hair color and skin complexion
   C. Determine her culture by her accent
   D. Determine her culture by her body shape and size

9. A spiritual assessment includes which of the following? Select all that apply.
   _____ A. Prayer needs
   _____ B. Affiliations with religious communities
   _____ C. Medicinal regimes
   _____ D. Coping mechanisms
   _____ E. Pastoral care visits

10. A practicing Muslim nurse requests to wear her hijab while on duty. Which of the following actions would be most appropriate?
    A. Decline the request
    B. Advocate for modification of the organization's dress code
    C. Review the organization's dress policy
    D. Make the accommodation

## REVIEW ACTIVITIES

1. You are called to problem-solve a situation between a patient and a staff nurse. The patient is refusing to allow the nurse to take care of him because the nurse is wearing a headscarf. How will you handle the situation?

2. You live in a small community with little diversity. A patient that you are scheduled to take care of is visiting from a large city. This patient is from India and speaks only Hindi, a language unfamiliar to the staff. When you walk in the room, there are no visitors and no one to translate. How will you approach your care in a culturally competent manner?

3. You are hiring several new graduate nurses to work on a unit that has had little turnover in the past 10 years. Many of the current staff are from the Veteran and Baby Boomers generations and have been in this organization for the entire length of their career. How will you go about integrating the Generation Xers and the Millenial Generation of nurses into this group?

4. You are taking care of a trauma patient in the ER. The patient has burns over 70% of his body. The likelihood of the patient's survival is unknown. There is a lot of noise and distraction in the ER as well as many family members within earshot. The patient asks you to pray for him and his family. How will you respond?

5. As the nurse manager, you are asked to intervene with a patient and his family. The nursing staff are concerned that the adult children of the patient do not allow the patient to do anything for himself. The patient always has someone at his bedside who does all the care and refuses to allow the father to do anything, including feeding himself. The family's behavior is out of a cultural belief of caring for and respecting their elders. As the patient has had a recent cerebral vascular accident (CVA), you know that self-care is a requisite to healing. How will you intervene in a culturally competent way?

## EXPLORING THE WEB

- This Web site is devoted to developing cultural awareness and diversity in health care. It includes a Listserv. www.diversityrx.org
- Sigma Theta Tau International Honor Society of Nursing position paper on global diversity. What does the nursing honor society believe about diversity? www.nursingsociety.org
- This site contains information about cultural beliefs and medical issues pertinent to the health care of recent immigrants to the United States. Who are the most recent immigrants, and what are their cultural needs? www.ethnomed.org
- Visit the the Office of Minority Health Web site at https://ccnm.thinkculturalhealth.hhs.gov. Complete the online training titled "Culturally Competent Nursing Care: A Cornerstone of Caring."
- The Provider's Guide to Quality and Culture: http://erc.msh.org

- Canadian Nurses Association: www.cna-nurses.ca
- National Alaskan Native American Indian Nurses Association (NANAINA): www.nanainanurses.org
- National Black Nurses Association, Inc.: www.nbna.org
- Transcultural Nursing Society: www.tcns.org
- U.S. Census Bureau: www.census.gov
- Islamic Information Center of American (IICA): www.iica.org
- U.S. Citizenship and Immigration Services: www.uscis.gov
- American-Arab Antidiscrimination Committee: www.adc.org
- American Civil Liberties Union (ACLU): www.aclu.org
- American Indian Heritage Foundation: www.indians.org
- American Jewish Community: www.ajc.org
- Anti-Defamation League: www.adl.org

- Asian and Pacific Islander Partnership for Health: www.apiph.org
- National Association for the Advancement of Colored People: www.naacp.org
- Urban League: www.nul.org

- International Council of Nurses: www.icn.ch
- Global Health Council: www.globalhealth.org
- World Health Organization: www.who.org
- United States Committee for Refugees and Immigrants: www.refugees.org

## REFERENCES

Ackley, B. J., & Ladwig, G. B. (2011). *Nursing diagnosis handbook:* An evidenced-based guide to planning care (9th ed.). St. Louis, MO: Mosby.

American Academy of Nursing. (1992). AAN expert panel report: Culturally competent health care. *Nursing Outlook, 40*(6), 277–283.

American Academy of Nursing. (2008). Cultural competency in baccalaureate nursing education. Retrieved November 1, 2010, from http://www.aacn.nche.edu/education/pdf/competency.pdf.

Andrews, M. (2008). Cultural diversity in the health care workforce. In M. Andrews & J. Boyle, *Transcultural concepts in nursing care* (5th ed., pp. 297–354). Philadelphia: Lippincott, Williams and Wilkins.

Black, A., Hopkins, J. et. al. (2003). The Eleanor Roosevelt Papers. "Cold War." *Teaching Eleanor Roosevelt,* ed. by (Hyde Park, New York: Eleanor Roosevelt National Historic Site, 2003). Accessed April 14, 2011 from http://www.nps.gov/archive/elro/glossary/cold-war.htm.

Braveman, P. (2006). Health disparities and health equity: Concepts and Measurement. *Annual Review of Public Health, 27*(1), 167–194.

Braveman, P. A., Cubbin, C., Egerter, S., Williams, D. R., & Pamuk, E. (2010). Socioeconomic disparities in health in the United States: What the patterns tell us. *American Journal of Public Health, 100*(S1), S186–196.

Calhoun, S. K., & Strasser, P. B. (2005). Generations at work. *AAOHN, 53*(11), 469–471.

Campinha-Bacote, J. (2003, January 31). Many faces: Addressing diversity in health care. *Online Journal of Issues in Nursing, 8*(1), manuscript. Retrieved April 18, 2006, from http://nursingworld.org/ojin/topic20/tpc20_2.htm

Carson, V., & Koenig, H. (Eds.). (2008). *Spiritual Dimensions of Nursing Practice* (Revised ed.). West Conshohocken, PA: Templeton Foundation Press.

Centers for Disease Control and Prevention. (2009). *Health, United States, 2009.* Retrieved November 2, 2010, from http://www.cdc.gov/nchs/data/hus/hus09.pdf#highlights.

Cohen, L., Iton, A., Davis, R., & Rodriguez, S. (2010). A time of opportunity: Local solutions to reduce inequities in health and safety. Institute of Medicine. Retrieved on November 4, 2010, from http://www.iom.edu/~/media/Files/Activity%20Files/SelectPops/HealthDisparities/Commissioned_local_disp.pdf.

Dayer-Berenson, L. (2011). *Cultural competencies for nurses: Impact on health and illness.* Boston: Jones and Bartlett Publishers.

DeRosa, N., & Kochurka, K. (2006, October). Implement culturally competent healthcare in your workplace. *Nursing Management,* 18–26.

Dykes, D., & White, A. (2009). Getting to equal: Strategies to understand and eliminate general and orthopaedic healthcare disparities. *Clinical Orthopaedics and Related Research®, 467*(10), 2598–2605.

Foley, L., Wagner, J., & Waskel. S. (1998). Spirituality in the lives of older women. *Journal of Women and Aging, 10*(2), 85–91.

Giger, J., & Davidhizar, T. (2008). Transcultural nursing: Assessment and intervention (5th ed.). St. Louis, MO: Mosby Year Book.

Giger, J., Davidhizar, R. E., Purnell, L., Harden, J. T., Phillips, J., & Strickland, O. (2007). American Academy of Nursing Expert Panel Report. *Journal of Transcultural Nursing, 18*(2), 95–102.

Grossman, C. L. (2010, October 7). "How America sees God; Analysts cast our views in four ways." *USA Today,* 1A.

Jamieson, D., & O'Mara, J. (1991). *Managing workforce 2000: Gaining the diversity advantage.* San Francisco: Jossey-Bass.

Jein, R. F., & Harris, B. L. (1989). Cross-cultural conflict: The American nurse manager and a culturally mixed staff. *Journal of the New York State Nurses Association, 20*(2), 16–19.

Jenkins, M. L., Wikoff, K., Amankwaa, L., & Trent, B. (2009). Nursing the spirit. *Nursing Management, 40*(8), 29–36.

Joint Commission. (2003). Comprehensive accreditation manual for hospitals. Chicago: Author.

Joint Commission. (2008). Provision of care, treatment, and services: Spiritual assessment. Retrieved October 29, 2010, from http://www.jointcommission.org/AccreditationPrograms/HomeCare/Standards/09_FAQs/PC/Spiritual_Assessment.htm.

King, R. K., Green, A. R., Tan-McGrory, A., Donahue, E. J., Kimbrough-Sugick, J., & Betancourt, J. R. (2008). A Plan for Action: Key perspectives from the Racial/Ethnic Disparities Strategy Forum. *Milbank Quarterly, 86*(2), 241–272.

Leininger, M. (2006). Culture care diversity and universality theory and evolution of the ethnonursing method. In M. M. Leininger & M. R. McFarland (Eds.), *Culture care diversity and universality: A worldwide nursing theory.* (2nd ed., pp. 1–42). Sudbury, MA: Jones and Bartlett.

Lipson, J. G., & Dibble, S. L. (2005). *Culture and clinical care.* San Francisco: UCSF Nursing Press.

Manion, J. (2009). Managing the multi-generational nursing workforce: Managerial and policy implications. Retrieved October 26, 2010, from International Centre for Human Resources in Nursing: http://www.ichrn.com/publications/policyresearch/Multigen_Nsg_Wkforce-EN.pdf.

Martsolf, D. S., & Mickley, J. R. (1998). The concept of spirituality in nursing theories: Differing world-views and extent of focus. *Journal of Advanced Nursing, 27,* 294–303.

Mattson, S. (2009). A culturally diverse staff population: Challenges and opportunities for nurses. *Journal of Perinatal & Neonatal Nursing, 23*(3), 258–26.

McEwen, M. (2005, July–August). Spiritual nursing care. *Holistic Nursing Practice,* 161–168.

McSherry, W., & Draper, P. (1998). The debates emerging from the literature surrounding the concept of spirituality as applied to nursing. *Journal of Advanced Nursing, 27*(4), 683–691.

Mead, M. (2001). *Sex and temperament in three primitive societies.* New York: Harper Collins Publishers.

Muñoz, C., & Luckmann, J. (2004). *Transcultural communication in Nursing. 2nd Ed.* Clifton Park, NY: Delmar Cengage Learning.

Nance, M. S., Ramsey, K. E., & Leachman, J. A. (2009). Chaplaincy care pathways and clinical pastoral education. *Journal of Pastoral Care Counsel, 63*(1–2).

National Institutes of Health (2006). Fact Sheet: Health disparities. Retrieved November 2, 2010, from http://www.nih.gov/about/researchresultsforthepublic/HealthDisparities.pdf.

Office of Minority Health, U.S. Department of Health and Human Services, (2007). Culturally Competent Nursing Care: A Cornerstone of Caring. Accessed April 14, 2011 at https://ccnm.thinkculturalhealth.hhs.gov.

Office of Minority Health and Health Disparities. (2010). About minority health. Retrieved October 26, 2010, from http://www.cdc.gov/omhd/AMH/AMH.htm.

Office of Minority Health. (2010). Culturally competent nursing care: A cornerstone of caring. Retrieved October 18, 2010, from http://minorityhealth.hhs.gov/templates/content.aspx?ID=5036&lvl=2&lvlID=12.

Ott, J. (1989). *The organizational culture perspective* (pp. 20–48). Chicago: Dorsey Press.

Parsian, N. & Dunning, T. (2009). Developing and validating a questionnaire to measure spirituality: A psychometric process. *Global Journal of Health Science, 1*(1), 2–11.

Paterson, T. Generational considerations in providing critical care education. *Critical Care Nursing Quarterly, 33*(1), 67–74.

Purnell, L. (2008). The Purnell Model for Cultural Competence. In L. Purnell, & B. Paulanka (Eds.), *Transcultural health care: A culturally competent approach* (3rd ed.). Philadelphia: F.A. Davis.

Rahman, F. (2009). *Major Themes of the Qur'an* (Second ed.). University of Chicago Press.

Smedley, B., Stith, A., & Nelson, A. (2003). *Unequal treatment: Confronting racial and ethnic disparities in health care*: Washington, DC: Institute of Medicine.

The Joint Commission. (2011). Spiritual Assessment. Retreived April 14, 2011, at http://www.jointcommission.org/standards_information/jcfaqdetails.aspx?StandardsFaqId=290&ProgramId=1.

The Provider's Guide to Quality & Culture. Retrieved October 21, 2010, from http://erc.msh.org/mainpage.cfm?file=1.0.htm&module=provider&language=English.

Tzeng, H. M., & Yin, C. Y. (2006). Learning to respect a patient's spiritual needs concerning an unknown infectious disease. *Nursing Ethics, 13*(1), 17–28.

U.S. Census Bureau. (2009). American Community Survey. 2009 estimate of U.S. population by race.

U.S. Department of Health and Human Services (2000). *Healthy People 2010: Understanding and improving health* (2nd ed.). Washington DC: U.S. Government Printing Office.

United States Census Bureau. (2009a). ACS demographic and housing estimates: 2009. American Community Survey, 2009. Retrieved October 18, 2010, from http://factfinder.census.gov/servlet/ADPTable?_bm=y&-geo_id=01000US&-qr_name=ACS_2009_1YR_G00_DP5&-ds_name=&-_lang=en&-redoLog=false&-format=.

United States Census Bureau. (2009b). Selected social characteristics in the United States: 2009. American Community Survey, 2009. Retrieved October 18, 2010, from http://factfinder.census.gov/servlet/ADPTable?_bm=y&-geo_id=01000US&-ds_name=ACS_2009_1YR_G00_&-_lang=en&-_caller=geoselect&-format=.

United States Census Bureau. (2010). Census 2010: Explore the form. Retrieved October 18, 2010, from http://2010.census.gov/2010census/how/interactive-form.php.

Villagomeza, L. R. (2006). Mending broken hearts: The role of spirituality in cardiac illness: A research synthesis, 1991–2004. *Holistic Nursing Practice, 20*(4), 169–186.

Walton, J. C., & Waszkiewiez, M. (1997). Managing unlicensed assistive personnel: Tips for improving quality outcomes. *Medsurg Nursing, 6*(1), 124–128.

Washington, D., Bowles, J., Saha, S., Horowitz, C., Moody-Ayers, S., Brown, A., et al. (2008). Transforming clinical practice to eliminate racial–ethnic disparities in healthcare. *Journal of General Internal Medicine, 23*(5), 685–691.

Williams, J., & Rodgers, S. (1993). The multicultural workplace: Preparing preceptors. *Journal of Continuing Education in Nursing, 24*(3), 101–104.

# SUGGESTED READINGS

Cuellar, N. (2006). *Conversations in complementary and alternative medicine.* Boston: Jones and Bartlett Publishers.

D'Avanzo, C. (2008). *Pocket guide to cultural health assessment* (4th ed.). St. Louis, MO: Mosby.

Dossey, B. M., Keegan, L., & Guzzetta, G. E. (2005). *Holistic nursing: A handbook for practice* (4th ed.). Boston: Jones & Bartlett.

Douglas, M., & Pacquiao, D. (Eds.). (2010). *Core curriculum for transcultural nursing and health care* (Vol. 21). Los Angeles: Sage.

Hunt, B. (2007). Managing equality and cultural diversity in the health workforce. *Journal of Clinical Nursing, 16*(12), 2252–2259.

Kawi, J., & Xu, Y. (2009). Facilitators and barriers to adjustment of international nurses: An integrative review. *International Nursing Review, 56*(2), 174–183.

Keepnews, D. M., Brewer, C. S., Kovner, C. T., & Shin, J. H. (2010). Generational differences among newly licensed registered nurses. *Nursing Outlook, 58*(3), 155–163.

McClung, E., Grossoehme, D. H., & Jacobson, A. F. (2006). Collaborating with chaplains to meet spiritual needs. *Medsurg Nursing, 15*(3), 147–156.

Noble, A., Rom, M., Newsome-Wicks, M., Engelhardt, K., & Woloski-Wruble, A. (2009). Jewish laws, customs, and practice in labor, delivery, and postpartum care. *Journal of Transcultural Nursing, 20*(3), 323–333.

Salimbene, S. (2005). *What language does your patient hurt in? A practical guide to culturally competent patient care.* Amherst, MA: Diversity Resources, Inc.

Villagomeza, L. R. (2005, November–December). Spiritual distress in adult cancer patients. *Holistic Nursing Practice*, 285–294.

Washington, D. Moving to understanding and change. *Policy, Politics, & Nursing Practice, 11*(2), 158–163.

# UNIT 5
## Leadership and Management of Self and the Future

# CHAPTER 26

# Collective Bargaining

JANICE TAZBIR, RN, MS, CCRN

*You will have much opposition to encounter. But great works do not prosper without great opposition.*

(FLORENCE NIGHTINGALE, 1864, CITED IN ULRICH, 1992)

## OBJECTIVES

Upon completion of this chapter, the reader should be able to:

1. Relate the history of collective bargaining and associated legislation.
2. Identify collective action models and associated terminology.
3. Outline the steps of whistle-blowing.
4. Identify the process of unionization.
5. Identify collective bargaining agents.
6. Summarize professionalism and unionization.
7. Relate the process of managing in a union environment.
8. Analyze pros and cons of collective bargaining in the workplace.

Delmar/Cengage Learning

*You are a new nurse on an orthopedic unit. You walk into a discussion between two nurses. Juanita, a registered nurse with 10 years of experience, states, "I'm tired of low pay and work assignments that are unsafe." Peggy, a registered nurse with five years of experience, says, "Have you brought your complaints to management?" Juanita replies, "Of course. I point out unsafe situations and the lack of raises, but no one cares." Peggy says, "I bet we would have better success with these issues if we nurses came together as a group."*

*What are your thoughts about this situation?*

*What are some of the choices the nurses have?*

Historically, nurses have been perceived by some as hardworking, submissive staff who do what they are told. The scope of nursing has changed so drastically that today nurses cannot afford to have a submissive image and do only what they are told. Patients, their illnesses, and their families are more complex than ever. Nurses are educated to advocate for their patients and themselves. Clinical situations arise in which nurses must voice their opinions and stand up for what is best for patients. To be effective in today's world, nurses must understand the tools available to deal with problems.

**Collective action**, or simply acting as a group with a single voice, is one method of dealing with problems. **Collective bargaining** is the practice of employees, as a collective group, bargaining with management in reference to wages, work practices, and other benefits. This chapter discusses different types of collective action models as they may function in the health care environment, and also includes information concerning unionization as well as professionalism within the context of unionization.

# HISTORY OF COLLECTIVE BARGAINING AND COLLECTIVE BARGAINING LEGISLATION IN AMERICA

Collective bargaining and unionization have existed in the United States since the 1790s. Traditionally, people who formed and joined unions were highly skilled craftspeople. People found that by working collectively, they could set wages and standards for their trades. The Erdman Act, passed in 1898, was the first federal legislation to deal with collective bargaining. Since then, numerous legislative acts have been passed to ensure the rights of employees (Table 26-1). The rights many workers have today came from the struggles of others with the fortitude to stand up for what they believed was right.

### TABLE 26-1 Summary of Selected Legislation Affecting the Workplace

| YEAR AND TITLE OF LEGISLATION | SUMMARY |
| --- | --- |
| 1898: Erdman Act | Outlawed discrimination by employers against union activities |
| 1935: National Labor Relations Act (Wagner Act) | Gave private employees the right to organize unions to demand better wages and safer work environments |
| 1938: Fair Labor Standards Act | Set minimum wage and maximum hours that can be worked before overtime is paid |
| 1947: Taft-Hartley Act | Returned some rights to management; somewhat equalized balance between unions and management |

*(Continues)*

**TABLE 26-1** (Continued)

| YEAR AND TITLE OF LEGISLATION | SUMMARY |
| --- | --- |
| 1962: Kennedy Executive Order 10988 | Amended National Labor Relations Act to allow public employees to join unions |
| 1964: Civil Rights Act | Set equal employment standards such as equal pay for equal work |
| 1965: Executive Order 11246 | Set affirmative action guidelines |
| 1967: Age Discrimination Act | Protects against forced retirement |
| 1973: Rehabilitation Act | Protects rights of disabled people |
| 1973: Vietnam Veterans Act | Provides reemployment rights |
| 1974: Taft-Hartley Amendments to the Wagner Act | Allows nonprofit organizations to join unions |
| 1986: False Claims Act | Allows whistle-blowing without fear of retribution |
| 1993: Family Medical Leave Act | Allows up to 12 weeks of job-protected leave based on medical reasons |
| 2004: California State Hospital RN to Patient Staffing Law | Legislates minimum staffing rations in acute care setting |
| 2008: American with Disabilities Act with ADA Amendment Act of 2008 | Prohibits discrimination and assures equal opportunities in the workplace for those with disabilities |

# COLLECTIVE ACTION MODELS

The main focus of most nursing collective bargaining units is to give nurses a voice (Pittman, 2007). Many nurses belong to numerous collectives, including specialty nursing organizations, church organizations, special interest clubs, community groups, and so on. The reason most people belong to these organizations is to better themselves and their communities or to promote and support the special interests of a group. Two types of nursing collective action are discussed in this chapter: workplace advocacy and collective bargaining. Shared governance, another type of collective action, is discussed in Chapter 17.

## WORKPLACE ADVOCACY

**Workplace advocacy** refers to activities nurses undertake to address problems in their everyday workplace setting. This type of collective action is probably the most common in nursing. An activity that falls under workplace advocacy is forming a committee to address problems, devising alternatives to achieve optimal care, and inventing new ways to implement change.

An example of an issue that would be addressed by workplace advocacy is patient advocacy. Patient advocacy is preserving and protecting the wishes of patients (Beyea, 2005). Patients rely on nurses to do this. Often, in the workplace, nurses are too busy to serve as a patient advocate, which causes the nurses and patients distress.

**Critical Thinking 26-1**

Go to, http://www.qsen.org/. Click on Quality/Safety Competencies. Then click on Safety. Scroll down and click on Patient Safety Teaching Case–Hyperkalemia. Scroll down again and click on Patient_Safety_Scenario_-_Hyperkalemia.doc. Read the case. What could have been done differently in this case to avoid a patient care delivery problem and lawsuit?

## Other Forms of Workplace Advocacy

Nurses in many hospitals serve on professional practice councils. These councils are often part of a shared governance organization within the hospital that works to improve patient care and the environment for staff.

Note that a supportive management will view workplace advocacy as a way to strengthen staff and promote teamwork. If the management is authoritative, however, workplace advocacy may not be encouraged, because it may be perceived as a threat to management and its policies.

## COLLECTIVE BARGAINING

In collective bargaining, the group is bargaining with management for what the group desires. If the group cannot achieve its desires through informal collective bargaining with management, the group may decide to use a collective bargaining agent to form a union.

## Factors Influencing Nurses to Unionize

Unions can help nurses by providing educational opportunities and creating a safer workplace (Pittman, 2007). When nurses feel powerless, they initiate attempts to unionize. Other motivations to unionize include job stress and physical demands. Nurses are also motivated to join unions when they feel the need to communicate concerns and complaints to management without fear of losing their jobs. Some nurses believe that they need a collective voice so that management will hear them and changes will be instituted.

Issues that are commonly the subject of collective bargaining include low wages, work environment, job dissatisfaction, nurse turnover, and workload issues (Porter, Kolcaba, McNulty, & Fitzpatrick, 2010). Many nurse managers believe that it is best to deal quickly and effectively with issues that arise, and to avoid collective bargaining, because of the increase in costs to the hospital that results from collective bargaining and the limitations it places on managers. A new nurse or a nurse changing his or her place of employment should identify clinical decision-making mechanisms within an institution; and if there is a union, the contract should be reviewed (Curtis, 2008).

## Unions

A **union** is a formal and legal group that works through a collective bargaining agent to formally present desires to management within the legal context of the National Labor Relations Board (NLRB).

Table 26-2 lists some collective action terminology. This is useful in understanding the collective bargaining process.

## WHISTLE-BLOWING

As patient advocates, nurses protect patients from known harm. Nurses are often aware of health care fraud in the form of people violating laws or endangering public health or safety. However, some nurses who are aware of health care fraud do nothing because of fear of retribution. Fraud costs the federal government and ultimately costs the taxpayer.

**Whistle-blowing** is the act in which an individual discloses information regarding a violation of a law, rule,

 **EVIDENCE** FROM THE LITERATURE

**Citation:** Porter, C. A., Kolcaba, K., McNulty, S. R., & Fitzpatrick, J. J. (2010). The effect of nursing labor management partnership on nurse turnover and satisfaction. *Journal of Nursing Administration, 40*(2), 205–210.

**Discussion:** This study was performed in a large magnet hospital that measured nurse turnover and satisfaction before and after the implementation of a nursing labor-management partnership. There was a significant decrease in nurse turnover (9.9% versus 6.8%) and improved nurse satisfaction after the partnership. Administration often views working with a unionized environment as difficult. This study indicates that communication and mutual goal setting helps both administration and nurses and that a nursing labor-management partnership can be successful in the union environment.

**Implications for Practice:** Communication and mutual goal setting improves nurses' job satisfaction and decreases nurse turnover rates. Job satisfaction and lower turnover rates help the nurses working in that environment and helps administration as well. Working together, staff nurses and administration can create an improved work setting.

## Critical Thinking 26-2

You are caring for Mr. Archie Payne, a 65-year-old man who was admitted for congestive heart failure. He is a retired steelworker from an area steel mill. He states, "I worked in that mill for 30 years, and I am thankful for the union. Because of the union, my medical costs are covered for the rest of my life. The union served me well. Do nurses have unions or groups that help them get what they want?"

How will you respond to Mr. Payne? Name a collective group to which you belong. What are collective groups able to get done as a whole? Are they more effective and stronger than you are as an individual in their interest areas? What are the downsides of belonging to a collective group?

### TABLE 26-2 Collective Bargaining Terminology

| TERM | DEFINITION |
| --- | --- |
| Agency shop | Synonymous with "open shop." Employees are not required to join the union but may join it. |
| Arbitration | Last step in a dispute. Indicates a nonpartial third party will be involved and may make the final decision. Arbitration may be voluntary or imposed by the government. |
| Collective bargaining | The practice of employees, as a collective, bargaining with management in reference to wages, work practices, and other benefits. |
| Collective bargaining agent | An agent that works with employees to formalize collective bargaining through unionization. |
| Contract | A set of guidelines and rules voted and agreed upon by union members that guides their work practices, wages, and other benefits. |
| Dispute | A disagreement between management and the union. A dispute may go through (1) mediation and conciliation, (2) arbitration, and possibly (3) a strike. A dispute may be settled at any stage. |
| Employee at will | An employee working without a contract. The employee agrees to work under given rules and may be terminated if the employee breaks any rules imposed by management. |
| Fact finding | Fact finding is used in labor management disputes that involve government-owned companies. It is the process in which claims of labor and management are reviewed. In the private sector, fact finding is usually performed by a board of inquiry. |
| Grievance | A grievance occurs when a union member believes that management has failed to meet the terms of the contract or labor agreement and communicates this to management. |
| Grievance proceedings | A formal process in which a union member believes that management has failed to meet the terms of a contract. The steps usually include (1) communication of the grievance to management, (2) mediation with a union representative and a member of management, and possibly (3) arbitration. The dispute may be settled at any step. |
| Lockout | The closing of a place of business by management in the course of a labor dispute to attempt to force employees to accept management terms. |
| Mediation and conciliation | A step in the grievance process in which a nonpartial third party meets with management and the union to assist them in reaching an agreement. In this step, the third party has no actual power in decision making. |

*(Continues)*

**TABLE 26-2** (Continued)

| TERM | DEFINITION |
|---|---|
| National Labor Relations Board (NLRB) | The National Labor Relations Board was formed to implement the Wagner Act. The two major functions of the board include (1) determining and implementing the free democratic choice of employees as to whether they choose to be or choose not to be in a union and (2) preventing and remedying unfair labor practices by employers or unions. |
| Professional | A person who has knowledge from formal studies and has autonomy. |
| Self-expression | "The expressing of any views, argument, or opinion, or the dissemination thereof, whether in written, printed, graphic or visual form[,] if such expression contains no threat or reprisal or force or promise of benefit" (National Labor Relations Act, 1994). |
| Strike | An act in which union members withhold the supply of labor for the purpose of forcing management to accept union terms. |
| Supervisor | A person with the authority to (1) impart corrective action and (2) delegate to an employee. |
| Union | A formal and legal group that brings forth desires to management through a collective bargaining agent and within the context of the National Labor Relations Board. |
| Union dues | Money required of all union employees to support the union and its functions. |
| Union shop | A place of employment in which all employees are required to join the union and pay dues. *Union shop* is synonymous with the term *closed shop*. |
| Whistle-blowing | Whistle-blowing is the act by which an individual discloses information regarding a violation of a law, rule, or regulation, or a substantial and specific danger to public health or safety. |

### Critical Thinking 26-3

You are a nurse working at an institution in which there is limited flexibility in the scheduling. You want to institute self-scheduling, with the staff nurses responsible for making and maintaining the schedule. Make a plan to present this idea to the manager. How will you elicit the support of other nurses? Now put yourself in the role of the manager. How would you respond to this request?

or regulation, or a substantial and specific danger to public health or safety. The government has recouped billions from whistle-blowing—$1.12 billion in 2009 in health care fraud alone (Solnik, 2009). Health care fraud can range from filing false claims to performing unnecessary procedures. As patient advocates, nurses have an ethical and moral duty to protect their patients. In 1986, the False Claims Act was modified to encourage whistle-blowers to come forward. Whistle-blowing claims are brought in *qui tam* lawsuits (OSHA, 2010), which anyone can file on both the government's behalf and their own behalf. If the government believes an individual has a case of fraud, the government will pay all expenses for the lawsuit, and the individual will be entitled to 15% to 25% of the government's recovery. The name of the person filing the suit will not be divulged if the government does not consider the matter to involve health care fraud, thereby protecting the person from any retribution from the employer. The employer will not know who blew the whistle. If nurses are aware of fraud in their practice setting, the proper steps for them to take include the following:

- File a *qui tam* lawsuit in secret with the court.
- Do not let the agency or hospital know you filed a lawsuit.

## REAL WORLD INTERVIEW

When I started as a nurse at this institution 27 years ago, you could choose if you wanted to be in the union or not. Initially, I chose not to join. I realized my mistake in the 1980s, when the union became mandatory for all nurses. Even though I was here longer than other nurses, the ones that were in the union before me were considered to have more seniority in the eyes of the union. I can see that unions can be good for issues like seniority. As for other issues, it really doesn't matter to me because I have a fair manager, so I don't have problems. My advice to younger nurses is to choose a hospital with a union, because it will protect you.

**Brenetta Ireland, RN BSN CNRN**

Chicago, Illinois

■ Serve a copy of the complaint to the Department of Justice with a written disclosure of all the information you have concerning the fraud.

■ If the government decides to go forward with the lawsuit, the government will bear responsibility for litigating the lawsuit and will pay for it.

# PROCESS OF UNIONIZATION

The process of choosing a collective bargaining unit and negotiating a contract may take three months to three years. A **collective bargaining agent** is an agent that works with employees to formalize collective bargaining through unionization. The American Federation of Labor and Congress of Industrial Organizations (AFL-CIO) is a federation of unions that organize through collective bargaining units. The steps to organize are outlined through a state nurses' association in Table 26-3.

## MANAGERS' ROLE

RNs have the legal authority to participate in collective bargaining in the majority of health care facilities in the country.

**TABLE 26-3  Steps in Organizing a Collective Bargaining Unit**

- Bring together a group of nurses supportive of collective bargaining.

- Arrange a meeting with a representative of the state nurses' association to discuss organizing.

- Assess the feasibility of an organizing campaign at your facility.

- Conduct the necessary research, such as what are the needs and/or complaints of the employees, to develop a plan of action.

- Establish an organizing committee and subcommittees to facilitate organizing.

- Begin the process of obtaining union authorization cards from the National Labor Relations Board to legally vote on a collective bargaining agent.

- Schedule an informal meeting for nurses eligible for the collective bargaining unit.

- Keep the lines of communication open with nurses.

- Seek voluntary recognition from the employer.

- Move toward formal organization of the unit.

- Seek certification by the NLRB as the exclusive bargaining agent of the unit.

- Initiate contract negotiations.

**Source:** Compiled with information from Minnesota Nurses Association. (2010). Steps to Organizing. Retrieved April 21, 2011 from http://www.mnnurses.org/action/organize/steps-organizing.

## CASE STUDY 26-1

You are a nurse working in a cardiac catheterization unit. You notice that a certain practitioner routinely performs cardiac catheterizations on patients who are in their early forties, have no cardiac risk factors or cardiac history, and are on Medicaid. The catheterizations are always negative for disease. You love your job but are troubled by this practice. You are fearful that patients will have complications. You ask the practitioner why these procedures are performed on patients who do not appear to need this testing. The response is, "You don't worry about what I do; these procedures keep us all employed with healthy paychecks." You discuss this with your nursing manager and the chief nursing executive, who both say, "Just do your job and let the practitioner decide what is best for your patients."

You decide that whistle-blowing is your next action. What is your first step? Should you notify management of your whistle-blowing? What policies exist in your agency to guide the nurse when the nurse finds unprofessional activities?

Over the years, there has been debate over the composition of collective bargaining units in the health care industry. In 1989, the NLRB deemed eight collective bargaining units, including one for RNs, appropriate in the hospital setting. Some other collective bargaining units in the hospital include licensed practical nurses (LPNs), secretaries, and housekeepers. Managers who work in a union setting may have up to eight different contracts for various employees. Unionization may result in increased costs for the hospital and may limit the authority of its managers. Table 26-4 lists some ways managers can respond to a call for collective bargaining (National Labor Relations Board, 2010).

### EMPLOYEES' ROLE

Nurses desiring to choose a collective bargaining agent must be sure they carefully follow the laws pertaining to

unionization. It is important to carefully choose the collective bargaining agent, such as National Nurses United. It is useful to find out about the former success of the union, details of how nurses will be supported, and where and how the union dues are spent. Spend time talking to other nurses in union settings to see how their contract is structured and if collective bargaining has helped with the issues that led them to unionize in the first place. Table 26-5 lists some suggested activities for the nurse during the process of unionization (National Labor Relations Board, 2010).

### STRIKING

Many nurses are morally opposed to unions, because they believe if they are members of a union they may be forced to strike. In reality, a collective bargaining agent cannot make the decision to strike. The decision to strike is made

### TABLE 26-4 A Manager's Role During Initiation of Unionization

- Know the law; make sure the rights of the nurses as well as the rights of management are clearly understood.

- Act clearly within the law, no matter what the organization delegates to you as manager.

- Find out the reasons the nurses want collective action.

- Discuss and deal with the nurses and the problems directly and effectively.

- Distribute lists of cons of unionization such as paying dues.

- Distribute examples of unions that did not help with patient care issues.

**Source:** National Labor Relations Board (2010). Employer/union rights and obligations. Retrieved April 17, 2010 from http://www.nlrb.gov/rights-we-protect/employerunion-rights-obligations.

## TABLE 26-5  A Nurse's Role During Initiation of Unionization

- Know your legal rights and the rights of the manager.

- Act clearly within the law at all times.

- If a manager acts unlawfully, such as firing an employee for organizing, report it to the NLRB.

- Keep all nurses informed with regular meetings held close to the hospital.

- Set meeting times conveniently around shift changes, and assist with child care during meetings.

**Source:** National Labor Relations Board (2010). Employer/union rights and obligations. Retrieved April 17, 2010 from http://www.nlrb.gov/rights-we-protect/employerunion-rights-obligations.

only if the majority of union members decide to do so. Most nursing collective bargaining agents insert in the contract a no strike clause, stating that striking is not an option for its members. The union members decide upon the no strike clause. Provisions set forth in the 1974 Taft-Hartley Amendments to the Wagner Act guarantee the continuation of adequate patient care by requiring the union to provide contract expiration notice and advance strike notice, making mediation mandatory, and giving the hospital or agency the option of establishing a board of inquiry prior to work stoppage.

In June 2010, more than 12,000 nurses went on strike in Minnesota to ensure safe staffing at all times (National Nurses United, 2010). Minnesota nurses have had strikes in the past. The issue of safe staff to patient ratios is being addressed by many unions throughout the nation.

In 1995, the California Nurses Association separated from the ANA. The California Nurses Association was responsible for mandating through law safe patient to nurse staffing ratios in the state of California. In 2009, the California Nurses Association, the National Nurses Organizing Committee, the Massachusetts Nurses Association and United American

## REAL WORLD INTERVIEW

I graduated from a diploma nursing program in 1962. I worked for 5 years and then was home for 15 years raising my children. I wanted to return to nursing and took a refresher course. It was very hard. So much had changed. I made it, though. I think that with the shortage of nurses now, hospitals would be smart to try to make it easier for nurses who have left nursing to return by offering reasonable, supportive refresher courses. I stayed at the hospital I went back to and later retired with just short of 19 years of service. When I retired, I was shocked to find out what my pension was going to be. It was $425 a month—this after almost 19 years of service. If it wasn't for my husband's pension, who, with a high school education, gets almost 10 times what I get, I would never have been able to retire. My husband worked through a union. I understand that teachers who work through unions often get 75% of their salary when they retire. Some nurses who are single or divorced would like to retire but simply can't afford to do so. You keep hearing about the poor pay for teachers, but, while I agree it should be better, at least they can afford to retire. Who thinks about nurses? It seems to me that more and more of the doctors' work is being given to the nurses, and yet a survey I read said that the gap between the doctors' and nurses' pay is greater than what it was at the end of World War II. New nurses should start thinking about retirement benefits when they look for their first job. I know my 40 years as a nurse went fast.

**Gerri Kane, RN**
Retired Staff Nurse
Cedar Lake, Indiana

## Critical Thinking 26-4

You are a new nurse in a unionized hospital. You are just completing orientation. You are asked to float to another unit with an entirely different patient population than you have ever cared for. You are afraid of not being able to care for patients safely. How will you respond?

nurses formed National Nurses United (National Nurses United, 2010). National Nurses United is now the largest union and professional association of registered nurses with over 150,000 members (National Nurses United, 2010). It is a professional organization and a union. Its goal is for nurses to improve practice and have a greater voice in decisions that affect patient safety.

# COLLECTIVE BARGAINING AGENTS

Various organizations act as collective bargaining agents for millions of workers including nurses, other health care workers, laborers and specially trained workers such as actors and fire fighters. Some of these organizations are the Teamsters Union, the General Service Employees Union, the National Union of Hospital and Health Care Employees, the Service Employees International Union, the United Autoworkers of America, and the United Steelworkers of America. The two largest and most recognized national professional nursing collective bargaining agents include the American Nurses Association (ANA) and National Nurses United (NNU). The ANA was part of collective bargaining at the state level through United American Nurses (UAN) until 2003 (Hackman, 2008). Since 2003, the ANA has disaffiliated itself from the UAN (Hackman, 2008). In 1995, the California Nurses Association separated from the ANA's California Nurses Association and joined the existing National Nurses Organizing Committee (NNOC). The NNOC was responsible for mandating through law safe patient to nurse staffing ratios in the state of California. In 2009, the California Nurses Association, the NNOC, the Massachusetts Nurses Association, and UAN (formerly with the ANA) formed National Nurses United (National Nurses United, 2010). Beginning in 2009, the NNU has unionized nurses through collective bargaining at the state and national level (National Nurses United, 2010). NNU is the only exclusive nurses union affiliated with the American Federation of Labor and Congress of Industrial Organizations (AFL-CIO). NNU represents more than 160,000 nurses

through collective bargaining in fifteen states (National Nurses United, 2010). National Nurses United is now the largest exclusively nursing union that is also a professional association of registered nurses (National Nurses United, 2010). Its goal is for nurses to improve practice and have a greater voice in decisions that affect patient safety. NNU is considered to be a more aggressive union that has initiated and supported striking for nurses' rights.

## AMERICAN NURSES ASSOCIATION (ANA)

The American Nurses Association (ANA) is a full-service professional organization representing the nation's entire RN population whether they are members or not. The ANA is not a collective bargaining agent; The ANA represents 3.1 million RNs in the United States through its 54 constituent state and territorial associations (American Nurses Association, 2010). The ANA's mission is "Nurses advancing our profession to improve health for all" (American Nurses Association, 2010). The ANA Code of Ethics states, "While seeking to assure just economic and general welfare for nurses, collective bargaining, nonetheless, seeks to keep the interests of both nurses and patients in balance" (ANA, 2010). The ANA represents the interests of nurses in healthy work environments and in many other areas as well. The ANA advances the nursing profession by fostering high standards for nursing practice and lobbies Congress and regulatory agencies on health care issues affecting nurses and the general public. The ANA initiates many policies involving healthcare reform. It also publishes its positions on issues ranging from whistle-blowing to patients' rights. The ANA recently launched a major campaign to mobilize nurses to address the staffing crisis, to educate and gain support from the public, and to develop and implement initiatives designed to resolve the staffing crisis. ANA has created the National Database of Nursing Quality Indicators (NDNQI) that helps show, with the use of data, the link between good nursing care and healthier patients. American Nurses Credentialing Center, a subsidiary of the ANA, created the Magnet Recognition Program to recognize health care organizations that provide the very best in nursing care. Since 1994, more than 372 institutions have received this award (American Nurses Association, 2010). In 2003, the American Nurses Association created a new workplace advocacy structure known today as the Center for American Nurses. The ANA feels workplace advocacy is an alternative to unionization and promotes collaboration and communication among nurses. Key ANA programs include conflict competency training and consultation; workshops and publications on lateral violence and bullying; nurse investor's education project; and the offering of a host of resources through its affiliate, the Institute for Nursing Research and Education (Scott, 2008).

## EVIDENCE FROM THE LITERATURE

**Citation:** Mays, D., Janzen, S. K., & Quigley, P. A. (2009). Administration and union partnership: One magnet hospital's story. *Nursing Administration Quarterly, 33*(2), 105–109.

**Discussion:** This is an article about one hospital that achieved magnet status by working with the union. They implemented the ANA Quality Indicators, used the ANA staffing principles, looked at nurses' satisfaction, and created a shared governance model. The RN satisfaction survey found that nurses wanted more involvement in organizational decision making. To achieve this goal, the shared governance model was created and implemented with input from staff, union leaders, and hospital administration.

**Implications for Practice:** When administration, unions, and nurses work toward common goals, outcomes can be creative and positive for all. The group was able to use standards from the American Nurses Association to help shape and create common language and achieve success.

## REAL WORLD INTERVIEW

I believe it is professional to be in a union, because you have more opportunities to stand up for your patients and your own nursing practice. Having worked in both a union and a nonunion environment, I think being in a union allows you to speak your mind without fear of losing job security. They can't dismiss you for just any reason. There are grievance procedures. In a nonunion environment, if they don't like you or what you say, they can punish you. But I've also seen the downside of unions. An example is when a contract comes out. The more-senior union nursing staff wants to hold out from agreeing on a contract that does not address all of our concerns, while the junior union nursing staff wants to agree on the first contract that is presented. Holding out for what you want is why there is arbitration. The junior nurses don't realize the power of the bargaining unit in nursing. I think most nurses don't realize what we as nurses can accomplish if we stick together.

**Susan Zielinski, RN**
Staff Nurse
Chicago, Illinois

# PROFESSIONALISM AND UNIONIZATION

Requirements for a vocation to be considered a profession include: (1) a long period of specialized education, (2) a service orientation, and (3) the ability to be autonomous (Jacox, 1980). Jacox (1980) defines autonomy as a characteristic of a profession in which the members are self-regulating and have control of their functions in the work situation. Nurses agree that specialized education and a service orientation are necessary to become a nurse, but many nurses disagree on the concept of autonomy. This disagreement is the central argument that divides nurses with regard to whether it is professional to be part of a union.

Many nurses believe that for nursing to be considered a profession, nurses must exercise autonomy and like most professionals, work out issues themselves. Many argue that this cannot be done without unionization. The debate about whether it is professional to be a part of a nurses' union continues.

## DEFINITION OF SUPERVISOR

Much discussion in nursing unions has revolved around the definition of a supervisor. The National Labor Relations Act (1994), in Title 29 of the United States Code, defines a supervisor as "any individual having authority, in the interest of the employer, to hire, transfer, suspend, lay off, recall, promote, discharge, assign, reward, or discipline

## REAL WORLD INTERVIEW

Being a manager in a union environment doesn't affect me much with my staff, as they are pretty reasonable. Recently, when our hospital changed unions, it was a bit difficult because there were representatives from two unions going around to staff. Each of these union representatives was trying to win the vote by saying their union would have more power against management. Many things in the union agreement are reasonable and make my job easier, because the rules are written out—e.g., seniority rules. I can see why many nurses want a union environment, because of what could happen if you have an unfair manager. Other things in the union aren't that fair, however, such as raises. If I have a nurse that is very productive and sits on many committees, I can't give him or her more of a raise than I do a nurse that does only patient care and not one thing else. This doesn't provide any initiative for staff to do more.

**Peggy Zemansky, RN**
Patient Care Manager
Chicago, Illinois

other employees, or the responsibility to direct them, or to adjust their grievances, or effectively, to recommend such action, if in connection with the foregoing, the exercise of such authority is not of a merely routine or clerical nature, but requires the use of independent judgment."

Using this definition, conceivably every nurse may be considered a supervisor—if not to another RN, then of LPNs, nursing assistive personnel (NAP), and other unlicensed personnel. The larger issue for discussion is, if all nurses are supervisors by definition, can they legally be in a union? Nursing unions do not allow nursing managers or supervisors to unionize. Only nurses defined as employees can unionize. The ambiguity of the terms *employees* and *supervisors* has caused legal disputes (Fine, 2006). Dependent on clarification from the legal system, nurses may not always have the privilege to unionize. This very definition of supervisor has not allowed many other professionals to join unions because, by definition of their roles, they are supervisors.

## PHYSICIAN UNIONIZATION

As health maintenance organizations (HMOs) and other health care groups change the face of health care, they are changing the role of medical practitioners. These practitioners are considered employees in some settings, instead of supervisors, and now, like nurses, have the ability to join unions. The recent loss of medical practitioner autonomy and lowered wages have prompted many medical practitioners to join unions (Fine, 2006). Similar to what has occurred in nursing, medical practitioner discontent has led to unionization. Approximately 50,000 medical practitioners in the country are already unionized (Fine, 2006). The Service Employees International Union is the largest collective bargaining agent for medical practitioners. The American Medical Association

(AMA) supports medical practitioners engaging in collective action with employers but does not favor them formally joining unions (Fine, 2006).

## UNIONIZATION OF UNIVERSITY PROFESSORS

The unionization of kindergarten through twelfth-grade teachers is established in this country. Now, though, the number of professors at higher education institutions who are choosing to unionize is increasing. As with nurses, wages and work environment have been reasons stimulating university professors to join unions. As the average age of university faculty increases and fewer people show interest in teaching, unions may be able to protect professors from becoming overburdened and financially reward those who enter teaching at the university level.

## MANAGING IN A UNION ENVIRONMENT

Managers must work with the union to manage within the rules and context of contract agreements. In some ways, managing after a union is in place is less difficult because of the explicit language in most union contracts. Corrective actions, rules concerning allowed absences, and so on are agreed upon, voted on, and written in the contract.

## GRIEVANCE

When a union member believes that management has failed to meet the terms of the contract or labor agreement and communicates this to management, this process is called a **grievance**. All union contracts specify grievance procedures for union members. These usually start with an employee who believes there has been wrongdoing on the part of management. Next, the member talks with a union representative, who helps the employee judge whether the act or condition actually justifies a complaint. The union representative uses knowledge of the contract, knowledge of the NLRB, and judgment to assist the employee. Next, the union member and the union representative meet with the manager to voice the grievance. At this step, the conflict may be resolved. If the conflict is not resolved, the next step may be to appeal management's decision and mediate with a higher-level manager. Grievance procedures may differ from union to union.

# PROS AND CONS OF COLLECTIVE BARGAINING

The decision to support or not to support collective bargaining in the form of a union is a personal one. Table 26-6 summarizes a number of pros and cons of collective bargaining.

Nurses practicing in the United States have the luxury of many laws to protect individuals in the workplace. If nurses prefer a particular collective action model, they can find that model in action in numerous work settings. Nurses have the ability to choose where they practice and under which model they practice.

### TABLE 26-6 Pros and Cons of Unionization

| PROS | CONS |
|---|---|
| The contract guides standards. | There is reduced allowance for individuality. |
| Members are able to be a part of the decision-making process. | Other union members may outvote your decisions. |
| All union members and management must conform to the terms of the contract without exception. | All union members and management must conform to the terms of the contract without exception. |
| A process can be instituted to question a manager's authority if a member feels something was done unjustly. | Disputes are not handled with an individual and management only; there is less room for personal judgment. |
| More people are involved in the process. | Union dues must be paid even if individuals do not support unionization. |
| Union dues are required to make the union work for you. | Employee may not agree with the collective voice. |
| Unions give a collective voice to employees. | Unions may be perceived by some as not professional. |
| Employees are able to voice concerns to management without fear of losing job security. | |

## REAL WORLD INTERVIEW

As a new graduate, I researched employee work environments before I applied to any organization. In doing so, I determined that I wanted to be part of a union to help guarantee safe patient-nurse ratios. Patient safety (patient:nurse ratio) was at the top of my list. This in turn would decrease stress levels in my own personal work environment and increase employee satisfaction. Also, having union support with a grievance policy and disciplinary actions enables me as an employee to feel like I have someone to turn to if I need help. Lastly, I chose to work in a facility with a union because unions provide help with all other contract negotiations.

**Michael Pankowski, RN, BSN**
Chicago, Illinois

# KEY CONCEPTS

- Collective bargaining has existed in the United States since 1790.

- The Wagner Act of 1935 gave private employees the legal right to form unions. Since then, numerous legislative acts have been passed to protect employees from unfair work practices.

- Workplace advocacy is a collective action model that is more informal and encompasses the everyday creativity and problem solving that occur in nursing.

- Collective bargaining through unionization is a collective action model that is formal and legally based. It uses a written contract to guide nursing and workplace issues.

- Nurses may be aware of fraud and be fearful to report it. *Qui tam* lawsuits allow people to discreetly expose health care fraud.

- Nurses who are unhappy in the workplace because of issues such as wages and unsafe staffing often attempt to unionize to rectify workplace problems. Nurses who are not managers have the legal right to unionize. There are specific steps that can be taken to unionize. Employees and managers must be aware of what steps to take during the initiation of unionization.

- The American Nurses Association is a full-service professional organization that represents the nation's entire registered nurse population. The ANA strives for excellence with initiatives such as magnet status certification and the National Database of Nursing Quality Indicators. The ANA is politically active and lobbies on issues affecting nursing and the general public.

- Some other professionals who do not have a tradition of unionization are opting to unionize. Medical practitioners and university professors are joining unions for the same reasons that some nurses have chosen to join unions.

# KEY TERMS

American Nurses Association (ANA)

collective action

collective bargaining

collective bargaining agent

grievance

union

whistle-blowing

workplace advocacy

# REVIEW QUESTIONS

1. Which statements concerning unions are true? Select all that apply.
   - _____ A. Unions work through a collective bargaining agent.
   - _____ B. Unions represent only hourly employees.
   - _____ C. Unions represent only salaried employees.
   - _____ D. Unions formally present a group's desires to management.
   - _____ E. Nurse managers are part of the union.
   - _____ F. All union agreements support striking.

2. A manager observes a paper on the unit that states there will be a meeting in the hospital cafeteria to discuss the nurses' rights to organize and choose a collective bargaining agent. Which response by the nurse manager is most appropriate?
   - A. Explain to the nurses that the meeting should take place off hospital property.
   - B. Tell them they will be fired if they attend the meeting.
   - C. Ask if you could join them in the cafeteria.
   - D. Explain that nurses cannot join unions because they are supervisors.

3. A staff nurse tells a coworker, "I don't want any part of a union. Unions restrict your individuality, other union members may outvote what I want, they cost too much, and management can still fire you for no reason." Which of those comments by the nurse is not true of a union environment?
   - A. Other union members may outvote what you want.
   - B. Unions restrict individuality.
   - C. Some feel they cost too much.
   - D. Management can still fire you for no reason.

4. Which are the correct steps when nurses feel they have witnessed health care fraud? Select all that apply.
   - _____ A. Serve a copy of the complaint to the Department of Justice.
   - _____ B. Do not tell your employer you have filed a suit.
   - _____ C. File a *qui tam* lawsuit in secret.
   - _____ D. Verify to the best of your knowledge that the action witnessed is health care fraud.
   - _____ E. Hire a lawyer to litigate the lawsuit.

5. In which situation does the employee have the right to grieve an action by the manager?
   A. Making the employee work their scheduled weekend
   B. Talking to the nurse in private to discuss a comment made by a patient about the nurse
   C. Changing scheduled work days after the schedule has been put out, without consent or knowledge of the nurse
   D. Refusing to grant the vacation request of a nurse with one year seniority in order to grant the request of a nurse that has 15 years' seniority

6. Which is correct concerning collective bargaining? Select all that apply.
   _____ A. Collective bargaining is formal and only occurs through unionization.
   _____ B. Collective bargaining agents represent the interests of the nurses.
   _____ C. Collective bargaining is done by a group acting with a single voice.
   _____ D. Workplace advocacy is a type of collective action.
   _____ E. Collective bargaining status can be attained with the NLRB.
   _____ F. Hospitals with magnet status cannot have collective bargaining.

7. Workplace advocacy is best defined as
   A. a management-defined solution for the workplace.
   B. holding managers and nurses accountable.
   C. a formal structure that is voted on.
   D. activities nurses undertake to address problems in the workplace.

8. Common reasons nurses unionize include all but which of the following?
   A. patient care issues.
   B. wages.
   C. staffing issues.
   D. being content in the workplace.

9. Which legislation gave unions the right to organize?
   A. National Labor Relations Act (1935)
   B. Fair Labor Standards Act (1938)
   C. Taft-Hartley Act (1947)
   D. Executive Order 11246 (1965)

10. Which are large collective bargaining agents that nurses commonly belong to? Select all that apply.
    _____ A. National Nurses Union
    _____ B. Service Employees International Union
    _____ C. International Nurses Union
    _____ D. California Nurses Association
    _____ E. Healthcare Union of Ohio

## REVIEW ACTIVITIES

1. You are a new graduate nurse and have begun working on a medical unit. The nurse manager explains to you that the unit uses workplace advocacy. What is workplace advocacy? How will it affect your functioning as an RN on the unit?

2. You are hired in a hospital that is a union shop. How does unionization differ from other collective action models such as workplace advocacy? Give three examples of how unionization differs from workplace advocacy.

3. You are a graduate nurse, and you have found out that you passed the NCLEX examination. As an RN, you are represented by the ANA. What is the mission of the ANA? Is the ANA active in politics?

## EXPLORING THE WEB

- What site would you recommend to someone inquiring about collective bargaining?
  www.nursingworld.org
  Search for collective bargaining.

- Go to the site for the American Nurses Association and find your state nurses' association. What did you learn about your state nurses' association?
  www.nursingworld.org

- What site would you access to find out the history of collective bargaining?
  www.nlrb.gov

- Visit the American Nurses Credentialing Center Web site and see what magnet status is all about. Would magnet status have an impact on your decision of where to be employed?
  www.nursingworld.org

- Go to the National Nurses United home page and see which national nursing issues are being addressed.
  www.nationalnursesunited.org

# REFERENCES

American Federation of Labor and Congress of Industrial Organizations (AFL-CIO), (n.d.). Unionize the workplace. Retrieved May 7, 2010 from http://www.aflcio.org.

American Nurses Association. (2006).Who we are: ANA's statement of purpose. Retrieved May 7, 2010 from http://nursingworld.org/FunctionalMenuCategories/AboutANA.aspx.

American Nurses Association. (2010). Code of ethics for nurses. Retrieved May 22, 2010 from http://nursingworld.org.

Curtis C. P. (2008). Finding the job that's right for you. *American Journal of Nursing Career Guide 2008*, 13–15.

Fine, S. (2006). Emergence of unionization a threat or salvation for physicians? *The Osteopathic Family Physician News*. Retrieved from www.acofp.org/member_publications/print/ busmar_02.html.

Hackman, D. (2008). ANA and UAN part ways (American Nurses Association, United American Nurses ). Georgia Nursing. Retrieved April 19, 2011 fromhttp://goliath.ecnext.com/coms2/gi_0199-9596517/ANA-and-UAN-part-ways.html.

Jacox, A. (1980). Collective action: The basis for professionalism. *Supervisor Nurse*, 11(9), 22–24.

Mays, D., Janzen, S. A., Quigley, P. A. (2009). Administration and union partnership one magnet hospital's story. *Nursing Administration Quarterly*, 33(2), 105–109.

Minnesota Nurses Association. (2010). Steps to Organizing. Retrieved April 21, 2011 from http://www.mnnurses.org/action/organize/steps-organizing.

National Labor Relations Act. (1935). Retrieved January, 2007, from www.nlrb.gov.

National Labor Relations Board (2010). Employer/union rights and obligations. Retrieved April 17, 2010 from http://www.nlrb.gov/rights-we-protect/employerunion-rights-obligations.

National Nurses United. (2010). National Nurses United salutes the Minnesota Nurses Association for historic patient care strike of 2010. Retrieved May 7, 2010 from http://www.nationalnursesunited.org.

Occupational Safety and Health Administration. (2010). The whistle-blower protection program. Retrieved May 14, 2010 from http://www.osha.gov/dep/oia/whistleblower/index.html.

Occupational Safety & Health Administration. (2010). Whistleblower.gov offers quick worker access to whistleblowing protection information. Retrieved May 21, 2010 from http://www.osha.gov.

Pittman, J. (2007). Registered nurse satisfaction and collective bargaining unit membership status. *The Journal of Nursing Administration*, 37 (10), 471–476.

Porter, C. A., Kolcaba, K., McNulty, S. R., Fitzpatrick, J. (2010). The effect of nursing labor management partnership on nurse turnover and satisfaction. *The Journal of Nursing Administration*, 40 (2), 205–210.

Scott, D. (2008). The Center for American Nurses-celebrating five years of workforce advocacy. Nurses First, 1 (1), p 5–7.

Solnik, C. (2009). Government whistleblowing pays off. Long Island Business News. Retrieved May 14, 2010 from http://libn.com/blog/2009/07/07/government-whistle-blowing-pays-off.

Ulrich, B. (1992). *Leadership and management according to Florence Nightingale*. Norwalk, CT: Appleton & Lange.

# SUGGESTED READINGS

Armalegos, J., & Berney, J. (2005). 30 years of collective bargaining autonomy, voice in practice. *Michigan Nurse*, 78(2), 6–8.

Bunn, W. B., Weinstein, L. M., & Peskin, S. R. (2006). Unions and collective bargaining. *Managed Care*. October; 15(10 Suppl 9):16–18.

Converso, A., Martin, S. L., Markle-Elder, S. (2007). Is your hospital safe? *American Journal of Nursing*, 107(2), 37–39.

Lawson, L. D., Miles, K. S., Vallish, R. O., & Jenkins, S. A.(2011). Recognizing nursing professional growth and development in a collective bargaining environment. *Journal of Nursing Administration*. May; 41(5):197–200.

Pittman, J. (2007). Registered nurse job satisfaction and collective bargaining unit membership status. *Journal of Nursing Administration*. October; 37(10):471–6.

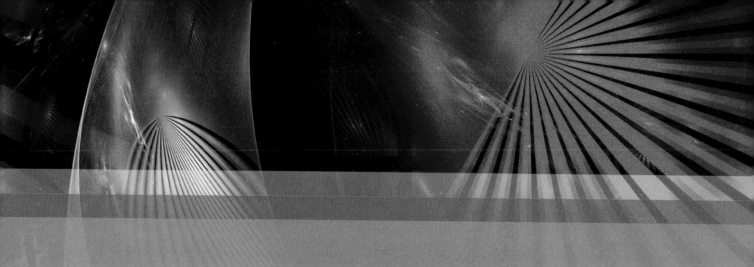

# CHAPTER 27

# Career Planning

EDNA HARDER MATTSON, RN, BN, BA(CRS), MDE;
KARIN POLIFKO-HARRIS, PhD, RN, CNAA

*Do what is right—and the future will take care of itself.*

(KATHERINE HARRIS, 2002)

## OBJECTIVES

Upon completion of this chapter, the reader should be able to:

1. Outline the process of career planning.
2. Recognize the importance of clarifying your values as a tool for the development of a fulfilling career.
3. Identify characteristics of goal setting.
4. Initiate and conduct an effective job search.
5. Prepare an outstanding cover letter and resume.
6. Describe the factors that contribute to a productive interview.
7. Relate appropriate follow-up to a job interview.

*Your family, friends, and fellow students are frequently asking you where you are going to work after graduation. You know that you need to get into the workforce to apply the skills you have learned. In addition, your student loans are coming due. You feel the pressure of securing gainful employment and appreciate the importance of being successful in your first nursing job.*

*Where will you begin your job search?*

*How can you plan for a successful career as a registered nurse?*

You are ready to consider your future in nursing at this point in your educational program. By now, you have experienced several clinical rotations and know that the nursing shortage will affect your job search. There may be other considerations, such as the location of the job and your financial resources. **Career planning** is an ongoing process that involves a personal and professional self-assessment, setting goals, searching for job opportunities, researching potential employers, preparing a cover letter and resume, and participating in an interview, including follow-up. You began the strategic planning process for your career with your selection of nursing. Your vision was to be a nurse. Your plans to achieve this vision included completion of the nursing program entrance and course requirements. Now, as you complete the final year of the nursing program, your academic and clinical preparation has placed you in an admirable position. You begin the strategic planning process again, this time to attain the career you want as an entry-level nurse and begin to plan for the future (Schoessler & Waldo, 2006).

The focus of this chapter is to illustrate the process of career planning. Applying the career-planning process requires adapting your job search to your personal space and place and the workforce at that time. Carry out every step in the career-planning process with a view toward marketing yourself even if rejected. Hopefully, you will leave the impression with the nursing department interviewers that you could meet their employment needs in the future, if not now. Career planning implies thriving both now and in the future rather than surviving just today.

In today's nursing world, nursing management is seeking nurses who are willing to enter into partnerships with them. The health care industry recognizes that nurses have a unique capacity to influence the health status of its patients as well as improve the nursing work environment (Holden, 2006). Forging respectful partnerships between the employer and employee promotes quality health care for all patients. Also, as nursing graduates share their job expectations, they are taking greater control of their careers. The career-planning process encourages the health care industry to adapt current human resource practices to meet the needs of the health care agency, the patients, and the new graduate (Arvidsson, Skarsater, Oijervall, & Friglund, 2008).

This chapter presents career planning as a linear process. However, in reality, career planning is a repetitive strategic planning process that occurs throughout the life span of a nurse. This process requires developing a vision, creating goals, and making plans to achieve them. Strategic career planning has many similarities to the nursing process and involves assessment, planning, implementation, and evaluation. Strategic career planning requires the following:

- Assessing and clarifying your values, interests, and the job market
- Determining your vision and goals
- Planning and implementing a job search, including sending a cover letter and resume to employers, participating in a job interview, and following up with a potential employer regarding the outcome of your application and interview
- Performing ongoing evaluation to assure alignment with your strategic planning vision and goals

Throughout this chapter, we will follow three case scenarios. With each exercise, you will integrate the concepts of strategic planning for your career (Table 27-1).

## STRATEGIC PLANNING PROCESS

A significant aspect of strategic planning for your career is to assess and clarify your vision and goals, taking into consideration your values, interest, and the job market. Career planning consultants describe value clarification as self-assessment. It is suggested that this assessment may be an indication of your resilience in nursing (Hodges, Keeley, & Troyan, 2008). Resilience in nursing describes the ability to recover from or adjust to a misfortune or significant change. Resilience is especially required if you need to demonstrate your improvement from a previous job, especially if you had received corrective action or discharge from that job. Within the past decade, nursing has undergone rapid changes, and

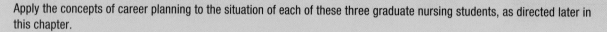

## TABLE 27-1  Career-Planning Scenarios

Apply the concepts of career planning to the situation of each of these three graduate nursing students, as directed later in this chapter.

### Student A

Julie, age 21, finished high school with a B average. She had changed her mind frequently during high school regarding a career goal and had finally applied to nursing. Julie was pleasantly surprised by how much she enjoyed nursing. Her decision as to what to do after graduation is now looming. She had very little difficulty with the academics of the nursing program, and at times found her social life interfering with being at the top of her class. The possibility of going on to study at the Masters level led her to review her priorities this year.

### Student B

James, age 39, entered nursing following a nine-year career as an emergency ambulance first responder with the fire and ambulance services. He is married and has two children, ages seven and nine. Although emergency work is interesting, he recognizes that life goes on for patients and their families after an emergency. He wants to be part of helping patients with their care after the emergency. This means sacrificing study time to allow time for his studies. He works evening and night shifts during the weekend. He recognizes that this is a necessary trade-off. His wife works as a legal assistant, which often means long and uncertain hours. James is pulling a B average, and at times is helped by his experience as a first responder.

### Student C

Jane is 28 and comes from a low-income family. Her immediate family is supportive of her aspirations to become a nurse. Jane has worked for 10 years as a nursing assistant on a pediatric unit. She enjoys working with the pediatric patients but doesn't feel fulfilled in her work.

Jane did not initially complete high school. She returned to school to complete the nursing entrance requirements. She found science courses particularly challenging. Her family is very proud of the effort she is making to upgrade her education. Jane continues to apply herself in nursing school. Despite feeling overwhelmed at times by the amount of reading demanded by her nursing courses, she maintains an A average.

these changes have required many decisions for those considering entering nursing and staying with it (Jackson, Firtko & Edenborough, 2007).

Some of these changes include dealing with significant diseases such as severe adult respiratory syndrome (SARS) and coping with a changing male and female, multigenerational, multicultural nursing workforce in times of nursing shortages. The Registered Nurses Association of Ontario (RNAO) submitted a report on the nursing experience with SARS in Ontario, Canada, to an independent commission. This report illustrates the value of the nursing workforce during a challenging time. The report honored nurses for their resilience while exposing their vulnerability in challenging nursing situations (RNAO, 2003).

Many changes have occurred in the nursing workforce, testing the resilience of nurses. Fifty years ago, nursing was primarily a female occupation, with the students entering following completion of high school. Childrearing frequently prevented continuation of nursing employment while the children were at home. At that time, nurses who returned to their nursing careers after their children were older did so with a goal of lifelong employment and retirement pensions. Nurses today are both female and male. They are often raising a family and working simultaneously. Current research is indicating that many nurses are now leaving nursing in five years or less (Hodges, Keeley, & Troyan, 2008). These changing values are shaping the future of nursing and affecting career planning.

Within the current nursing workforce, several generations of nurses working together may have different values. The older nurse may view the recently graduated nurse as placing his or her own interests before that of the organization and seeming to lack commitment. For example, working overtime may be seen by the older, long-term nurse as an expectation, whereas the younger, newly graduated nurse may resist working overtime and do so only if the remuneration is acceptable or time is given for some desired activity. With all attrition rates. Career planners suggest setting short-term career goals for one to three years, intermediate career goals for three to five years, and long-term career goals for six to twenty years. Keep in mind that goals are often reviewed as part of your ongoing performance review during employment. Employers expect you to take responsibility for your ongoing career development and goal setting. Your SMART career planning goals may look like Table 27-2.

The shortage of nurses has also created the need for unprecedented migration of nurses from other countries,

these changes, it is important to clarify your values and seek employment that is a good fit for both you and a health care agency and that prepares you for your vision of the future.

## DETERMINING YOUR GOALS

Determining your goals using the SMART acronym for goal setting is useful. SMART stands for Specific (S), Measurable (M), Achievable (A), Realistic (R), and Timely (T) goal setting. Being SMART will help you describe specifically what you want to accomplish with strategic planning for your career. For example, you may want to work in a specialty patient care unit after graduation. You may also want to include goals for continuing your education. Visit www.allnursingschools.com and search all programs for listings of bachelor's, master's, nurse practitioner, and doctoral nursing programs. Approved schools of nursing are often listed on state Web sites (e.g., www.nursing.illinois.gov) along with valuable in-depth data for students, such as each school's NCLEX pass rate for the past five years, graduation data, and

## TABLE 27-2 Smart Career Planning Goals

| SMART | EXAMPLE |
| --- | --- |
| **S**pecific | Employment as an RN in an emergency department (ED) |
| **M**easurable | Function independently full-time |
| **A**chievable | Employment at hospital that allows recently graduated RNs to work in ED |
| **R**ealistic | Presence of other new graduates that were able to achieve goal |
| **T**imely | Achieve goal within two years |

resulting in a multicultural workforce. The face of nursing has changed and will continue to change (Callister, 2006). Increasingly, internationally educated nurses mirror the patient population. Generic nursing programs are also witnessing an increase in their culturally diverse student populations (Myers & Dreachslin, 2007).

Setting strategic planning goals sounds like it is part of a business plan or venture. Although you may resist this approach at first, consider that your career is your business, and your business is nursing care. Putting your goals in writing permits you to analyze the current situation and make the necessary changes to achieve your goals. Your employer will appreciate your being proactive in goal setting and including such things as joining committees, volunteering for assignments, etc.

## PLANNING AND IMPLEMENTING A JOB SEARCH

You can use several methods to search for a job. The first is networking through family, friends, and your acquaintances. These persons may be able to identify job opportunities you have not noticed. Do you want to work in a large city teaching hospital? A smaller private community hospital?

The next method is to look at positions advertised through newspapers, bulletin boards, professional associations, or online job listings (Kluemper, 2009). Check job boards specific to health care such as the following:

- *www.careercity.com*
- *www.healthcareerweb.com*
- *www.rn.com*
- *www.healthcareersinteraction.com*
- *www.aone.org* (This site requires membership to use.)

Also access information about any specific health care organizations in which you are interested. You can search for these by name and location using a search engine, such as www.google.com. Note the specific job descriptions of any posted nursing positions. Study the job descriptions carefully so that you can tailor your cover letter, resume, and interview to meet the needs of the organization. Accessing the job description online electronically is proving to be a distinct advantage for both nurses and human resources personnel. It saves time for the human resources personnel when triaging multiple applications. It gives applicants the ability to use key words from the advertised job description on their cover letter and resume to highlight their ability to meet any important organizational requirements. Having access to information about the organization and the job description alerts the applicant not only to information about the health care organization but also to the variety of skills the new nurse will be expected to master. If the job description suggests that the position involves floating to different departments, then it may not be the most appropriate job for an entry-level nurse initially. Gaining experience in different acute care settings is helpful later, however, in preparing for advanced nursing roles and community health settings.

It may be helpful to apply for a position that is not advertised at a health care agency that is of interest to you. This may be very effective, because the health care agency you would like to work for may have many hidden job opportunities. Your skills and aptitudes may be just what they are looking for. Even if you do not receive a positive response immediately, it is common for prospective employers to keep desirable resumes on file for future reference.

Another method of searching for a job is attending a job fair. This allows you to be exposed to many opportunities in various health care agencies in a limited time. Always review any organizations carefully that you are considering for employment. Is it a magnet hospital? Is it a hospital that you would be proud to be associated with? In 1993, the American Nurses Credentialing Center (ANCC) established the Magnet Services Recognition Program. The ANCC Magnet Program has certified many hospitals in the United States and is expanding internationally (see Chapter 3). Review the Hallmarks of the Professional Nursing Practice Environment, available at www.aacn.nche.edu.

Consider looking for a one-year nursing residency program such as the one implemented through a partnership between the American Association of Colleges of Nursing (AACN) and the University HealthSystem Consortium (UHC). In addition to developing clinical judgment and leadership skills for new nurses at the point of care, the goal of the residency program is to strengthen the new nurse's commitment to practice in the inpatient setting by making the first critical year a positive working and learning experience (Robert Wood Johnson Foundation. 2010).

Determine if there is a multistate licensure compact in place in any states where you are interested in working. This allows a nurse to have one license (in his or her state of residency) and to practice in other states (both physically and electronically), subject to each state's practice law and regulation. Under this mutual state recognition, a nurse may practice across participating state lines unless otherwise restricted. View guidelines of the multistate licensure compact at www.ncsbn.org. Find the nurse licensure compact map and click on it.

Finally, call to make an appointment with the nurse recruiter at an agency that you are interested in. Have your resume in front of you when you call. Be professional and do not call when you can't focus on the recruiter and his or her questions—for example, because you are driving, cooking, or doing something else. Respect the recruiter's time. When contacted for an interview, display knowledge about the organization. Link your personal experience, skills, and ability to the organization's job needs and mission and values. This tells the interviewer that you have done your homework and your interest is serious. It is often helpful to begin a job tracking file (Figure 27-1).

## Preparation of a Cover Letter and Resume

Your cover letter and resume are a form of marketing strategy. You are marketing and advertising yourself to a potential employer. Develop your opening sentences carefully. Highlight your accomplishments and contributions, not just your tasks and responsibilities. This is important, as it demonstrates what you've actually accomplished, not just what tasks you have performed day to day. For instance, instead of noting that you served on a process improvement committee, add "which resulted in a 5% reduction in patient wait times and a 5% reduction in supply costs." Consider your cover letter to be a brief commercial about yourself. It is a brief opportunity to catch the attention of the nurse recruiter. Address your cover letter to a person rather than to a company. It should fit on one page and use dynamic language. Limit repeating information contained in your resume. Do not indent paragraphs. Sign your name in blue or black ink. See Figure 27-2 for an example.

### Critical Thinking 27-3

Identify job search methods that may be used by each of our three student nurses from Table 27-1.

How will their strategic planning goals and vision and values focus their job search?

Where should they begin their job search?

| AGENCY AND REFERRAL SOURCE | TELEPHONE NUMBER | CONTACT NAME | RESUME SENT/ DATE | THANK-YOU LETTER | FOLLOW-UP |
|---|---|---|---|---|---|
| | | | | | |
| | | | | | |
| | | | | | |
| | | | | | |
| | | | | | |
| | | | | | |
| | | | | | |
| | | | | | |
| | | | | | |
| | | | | | |
| | | | | | |

FIGURE 27-1 **Tracker for job leads.** (*Source:* Courtesy of Karen Polifko-Harris, 2003.)

James Mattern
214 Christie Avenue
Gladstone, OH 43523
(604) 775–3424

April 11, 2011

Ms. Eileen Carter, BSN, RN
Director of Human Resources
Concordia Hospital
100 Seaside Drive
Austin, NJ 12356

Dear Ms. Carter:

I am requesting the opportunity to discuss my career plans with you. I will be graduating on June 30, 2011, from the University of Ohio with a Baccalaureate of Science Degree in Nursing. I will take my NCLEX-RN on July 30, 2011.

I have served as an ambulance attendant for 9 years. This employment has provided me with the skills to handle emergency calls, including mass disasters such as airline crashes and hotel and apartment fires. I have also performed many tasks of varying priorities within many fire and police departments. I feel that these skills, combined with my newly acquired nursing skills, would be an asset to your emergency department.

I would appreciate the opportunity to discuss employment possibilities with you. I will call you next week to schedule an appointment. In the meantime, I can be contacted at (604) 775–3424 or at James123@school.edu.

Thank you for your time and consideration of my resume.

Sincerely,

*James Mattern*

**James Mattern**

FIGURE 27-2 Cover letter.

Delmar/Cengage Learning

Remember that your first opportunity to market yourself is a well-written cover letter and resume. It gives the nurse manager or human resources personnel the opportunity to see an example of your work. Both your cover letter and your resume highlight your credentials and skills. Be brief and specific. Double-check all information; pay attention to the layout of the cover letter and resume. Keep margins, indentations, dates, and places consistent. Use only approved abbreviations, e.g., IL for Illinois. Proofreading is essential. Be sure all dates are accurate. Have someone else read your cover letter and resume. Do not rely only on the spell-checker to find a word that is spelled incorrectly or a sentence that has a grammar error. Your written communication skills will be evident in your cover letter and resume. It is important to not have any typographical, spelling, or grammar errors, as this may suggest to the reader that you may have a problem in nursing documentation skills. Use white, off-white, or ivory top-quality 8 1/2 × 11-inch paper with matching envelopes. Print only with a laser printer, and, if sending the same resume to more than one potential employer, print multiple originals instead of making photocopies. Any sloppiness in your cover letter and resume indicates a lack of attention to detail. A prospective employer may question whether your performance in nursing would also be sloppy. Use action verbs in your documents (Table 27-3). Action verbs are powerful. Highlight your skills of communication, time, and resource management; team support and leadership; as well as your organizational, analytical, and technical skills. Never bad-mouth a former employer, and avoid any humor or sarcasm in your documents. Your nursing curricula, including both the academic and clinical components, has prepared you to meet employment needs (Ervin, Bickes, & Schim, 2006).

## Critical Thinking 27-4

You are helping some of the new nurses in Table 27-1 prepare their cover letters and resumes.

What job application method is illustrated in James's cover letter?

How would you strengthen Julie's position?

**TABLE 27-3  Action Verbs**

| COMMUNICATION | TIME AND RESOURCE MANAGEMENT | TEAM SUPPORT AND LEADERSHIP | ORGANIZATIONAL SKILLS | ANALYTICAL AND TECHNICAL SKILLS |
|---|---|---|---|---|
| address | adapt | demonstrate | arrange | analyze |
| arrange | advocate | design | classify | apply |
| clarify | collaborate | eliminate | compile | assess |
| debate | conceive | explore | distribute | critique |
| develop | coordinate | generate | generate | detect |
| document | delegate | innovate | incorporate | examine |
| illustrate | encourage | institute | order | exercise |
| introduce | expedite | manage | organize | identify |
| present | facilitate | master | process | implement |
| read | modify | motivate | revise | inspect |
| relate | prevent | negotiate | schedule | investigate |
| report | refer | oversee | select | perform |
| summarize | resolve | promote | supply | practice |
| teach | simplify | respect | update | research |
| translate | support | | verify | solve |
| write | volunteer | | | utilize |
| | | | | validate |

## Resume

It is customary to write resumes in either a chronological or a functional style. The choice of resume style is dependent upon the message you want to convey. The chronological style lists jobs in reverse chronological order. It illustrates your employment history and is good for applicants with little or no gaps in work history in the same field in which they are seeking employment. (Be prepared to discuss any gaps in employment in your interview.) This style of resume may also serve to highlight a progression of your work experiences from a position of lesser to greater responsibility. A functional style of resume gives the applicant the opportunity to illustrate experience in multiple careers or to dramat-ically change a career focus. It emphasizes skills and abilities rather than a sequence of job experiences. A resume contains educational status, including any certifications, clinical rotations, personal attributes, societal contributions, and related work experience. In the body of your resume, double-space between sections. See Figure 27-3 for an example of a functional resume. Figure 27-4 is an example of a chronological resume. Note that elements of the two resume styles may be combined in one resume.

Itemize your educational qualifications on your resume, including the name of your academic institution as well as any certifications (such as cardiopulmonary resuscitation [CPR]) and dates obtained. List additional

**Julie Martin**
111 Norberry Place
Maryland, NY 06701
(609) 323–4562
ljm@uscotia.net

## Objective

An entry-level staff nurse position on a medical-surgical patient care unit.

## Education

Bachelor of Nursing (June 2011)
University of Scotia

- GPA 3.7
- Dean's List, 2011
- Computer Informatics certificate, June 2010

## Certifications

- CPR, June 2010
- Diabetic Monitoring Devices, June 2010

## Clinical Rotations

Senior Practicum                                                    Winter Session, 2010
Calgary Mayo Care Center

- Determined patient need priorities and provided nursing interventions
- Engaged in self-evaluation of clinical performance
- Identified hospital and community resources available to patients
- Communicated with disaster team leader during disaster drill

Maternal Child Nursing                                            Fall Session, 2009
Nightingale Health Center

- Conducted a physical assessment of an infant at 4 hours of age and daily
  until discharge
- Assessed postpartum mothers, including breasts, fundus, lochia discharge,
  and perineum at 1 hour, 4 hours, and daily

Community Health                                                   Winter Session, 2009
Riverdale Community Health Center

- Conducted well-baby clinics, including physical assessment using the Rourke
  assessment form
- Offered contraceptive information
- Provided vaccinations for children under the supervision of the community health nurse
- Participated in a diabetic clinic focusing on accurate blood glucose monitoring

Societal/Professional Contributions                               2007–present

- Taught swimming to children ages 6–10 at YMCA
- Organized a health fair for seniors in an assisted living complex
- Provided regular reading opportunities to a child experiencing developmental delays
- Tutored first-year nursing students in anatomy and physiology
- Seek additional learning opportunities
- Take responsibility for personal learning needs
- Demonstrate conflict resolution skills

FIGURE 27-3 Functional resume.

**Caitlin O'Malley**
2424 Sailing Avenue
Cherry Hill, NJ 08080
(609) 444–2212 (home)
Cat24@excite.net

**Objective**

An entry-level position as a pediatric registered nurse

**Education**

Bachelor of Science in Nursing, May 2010
University of Pennsylvania, Philadelphia, PA

- Maintained 3.66 GPA, Dean's List
- Senior class president, junior class advocate
- 220-hour preceptorship on the oncology unit at Children's Hospital of Philadelphia
- Computer Informatics certificate, June 2009

**Experience**

Patient Care Assistant, Labor and Delivery
St. Mary's Medical Center, Philadelphia, PA
(August 2006–present)

- Assist in preparation of the operating room
- Provide basic patient care monitoring, including vital signs, phlebotomy, glucose screening
- Prepare and stock patient rooms
- Monitor fetal heart rate and progression of labor

Life Guard and Camp Counselor
Camp Perry, Point Pleasant, NJ
(Summers 2006–2010)

- Supervise waterfront for 150 campers along with three additional lifeguards
- Perform basic camp counselor duties, including direct supervision of campers ages 9 to 14

**Certification**

Certified as a Basic Life Support Provider, 2005–present

**Professional Organizations**

Nursing Student Association, University of Pennsylvania Chapter
National Student Nurses Association
American Red Cross, blood drive volunteer
Philadelphia Free Clinic, registration volunteer

FIGURE 27-4 **Resume—Chronological.** (*Source:* Courtesy of Karen Polifko-Harris, 2003.)

education you have taken to enhance your knowledge base, such as online courses or computer technology training. Demonstration of a strong knowledge of drug therapy is significant, including knowledge of commonly used drugs (Luk, 2008). Include a list of your clinical rotations, with specific competencies you have achieved. If your college has a senior clinical practicum experience, highlight the skills you have mastered. For example, note such items as "administered intravenous medications using a patient-controlled analgesic pump for three patients, gave antibiotic medications via the burretrol, identified normal and abnormal laboratory results, notified the practitioner of health care problems, and contributed significant data in an interdisciplinary team conference."

Nursing employers are looking for staff with employability skills, just like any other employer. Include personal attributes (such as the following) on your resume that position you as a continuous learner and consistent performer:

- Pay attention to detail
- Take responsibility for your own learning
- Seek out learning opportunities
- Provide a safe and comfortable environment for patients experiencing dementia
- Demonstrate resilience in resolving conflict
- Work well with a team
- Interact with others
- Accept constructive feedback
- Demonstrate reliability in attendance and punctuality
- Perform therapeutic nursing interventions.

For applicants with limited formal work experience, recruiters may consider societal and professional contributions as significant for the entry-level nurse. It can suggest that the individual is motivated and self-directed. Employers are looking for workers with employability skills that can be transferred between settings. For example, participating in the organization of a health fair for seniors and conducting a session on the need for regular foot care for the diabetic patient demonstrates your interest in the promotion of wellness as well as in caring for patients with an illness.

Skills needed for employment are commonly recognized as good communication, teamwork, adaptability, problem solving, and positive attitudes and behaviors. Mastering these skills will assist your transition into the workplace setting, whether you are working on your own or as part of a team. There are many opportunities within the nursing workplace that may promote your marketability and advance your career. Taking advantage of these opportunities helps improve critical thinking skills in beginning nurses (Forneris & Peden-McAlpine, 2007).

Many nursing licensing bodies require regular submission of continuing education units (CEUs) to maintain licensure (College of Registered Nurses of Alberta, 2007). National nursing associations provide position statements that outline nurses' responsibility to demonstrate professional competence throughout their careers (American Nurses Association, 2008). Documentation of your formal professional development and practice should be initiated at the outset of your first position. Keep a record of your orientation material and any in-service education sessions attended. It is helpful to have the in-service education presenter or your immediate supervisor verify your attendance. These records, along with documentation of your nursing education, will become a portable portfolio of your career (Byrne, Schroeter, Carter, & Mower, 2009; Williams & Jordan, 2007). An excellent example of structuring and maintaining a portfolio is illustrated by Byrne, Schroeter, Carter, & Mower, 2009. Your portfolio may include such items as CEUs, a patient care plan demonstrating the nursing process, and a clinical pathway for a specific patient population.

## Electronic Internet Job Searches and the Resume Distribution Process

Performing an electronic Internet job search and submitting an application to the human resources department will provide you with an opportunity to become familiar with the vision, mission, and services of the organization in which you are interested. Sending cover letters and resumes via the Internet is an acceptable practice. It does require some additional considerations in terms of both safeguards and catching the attention of the reader (Honaman, 2009). The resume and cover letter should be created as Microsoft Word documents. Check with the human resources department of the organization that you are interested in to identify if they require use of any special version of Microsoft Word. This should assure that your resume and cover letter receive prompt attention. Do not use a PDF file format or a PowerPoint slide format for your resume and cover letter, as they may take more time for the human resources department to read.

Note that many agencies now scan resumes electronically and search them for key nouns associated with a particular job opening. Those resumes containing the most key word nouns are selected and then ranked. In constructing your resume, be sure to:

- Keep it simple. Use a plain typeface such as Helvetica for headings and Times Roman for body text, and 10- to 14-point type.
- Avoid fancy highlighting. Use boldface or ALL CAPS for emphasis. Avoid fancy fonts, italics, underlining, slashes, dashes, parentheses, and ruled lines.
- Avoid a two-column format. Multiple columns can be jumbled by scanners that read across the page.

It is safest to initially send the e-mail to yourself or to a mentor to determine how it will appear to the reader. Send the resume as an attachment to your cover letter. The human resources personnel or nurse manager can reproduce it readily to circulate it to other managers or members of the interviewing committee.

Catching the attention of human resources personnel is vital. Rather than sending your resume and cover letter to an organization's e-mail address, do some research and try to identify a person to whom the cover letter should be addressed. Call the human resources department to identify the appropriate person. Then, when you send the e-mail, instead of entering *Resume* in the subject line of your e-mail, enter *Resume for nursing position with 9 years EMT experience*, or *Resume for entry-level RN seeking pediatric nursing position*, as appropriate. Human resources personnel are more likely to read this resume quickly. If you really want to stand out, follow up with the human resources department by sending them a hard copy of your cover letter and resume. If you receive a call from someone who indicates that they have read your cover letter and resume, verify the caller's position and the organization before revealing any personal information (Honaman, 2009).

## Preparation for the Interview

Congratulations for securing an interview! Sometimes it may be a phone interview. Minimize interruptions—e.g., a barking dog, TV—and focus on the interview. It is acceptable to set up another time to talk to the interviewer if you have been caught driving, etc. Minimize any distractions, and know that it is okay to ask to set an appointment for the call so that you can focus on the conversation.

In preparation for the interview, learn more about the agency and the possible questions they may ask you or that you should ask (Hart, 2006). You will be wise to ask for a copy of the job description beforehand if it was not available to you earlier. Go online again and find the agency's Web site. Review what you find there. Talk to others who work there. Familiarize yourself with the Web site and the job description, as this will demonstrate your interest in the position. It will also give you an opportunity to prepare appropriate interview questions. For example, if the job description requires the nurse to demonstrate the use of medical equipment, you can clarify what type of medical equipment is used in the unit.

Arrive shortly before the interview to demonstrate your time management skills. Smile and shake the interviewer's hand. Be prepared to complete an application even if you have already sent a resume. Maintain good eye contact, and greet the interviewer formally. Don't sit down before the interviewer sits unless directed to do so. Maintain good posture when seated. Don't chew gum or play with your hair or pen during the interview. Prepare a folder that contains a description of the organization and its services, extra copies of your resume, questions you have researched and are prepared to ask, and blank paper as well as a pen and any documents that may be helpful. Note that your ability to meet the job requirements will be assessed as part of the interview process. The nurse manager or representatives of the human resources department will verify your license, assess your competency, review your employ-

ment references, and complete background and criminal checks, as appropriate. They will assess your ability to meet any health requirements or any other job requirements of a nursing position. Your ability to fit in with the agency's culture as well as the patient care unit's culture will be assessed. Your communication skills, maturity, dependability, learning and nursing skills, as well as your ability to delegate, use initiative, use judgment, and be loyal and dedicated to your work are all items that may be assessed. The nursing representatives and the human resources representatives will usually try to offer you a competitive salary or hourly rate within approved budget guidelines, and they will assure the completion of any required organizational and governmental paperwork. Some organizations may have multiple persons interview you for some positions, to assess such things as your ability to work on a team, stay calm, keep focused on questions, establish rapport with the group, etc. For entry-level position interviews, it is customary to have only the nursing manager or the nurse manager and a nurse recruiter or human resources person present during the interview. In some situations, other staff nurses are included in interviews for new unit staff.

You will want to assess such items as whether the organization offers a nurse internship program or nurse residency program for new graduates, what the program consists of, who serves as preceptors for the program and their backgrounds, and what the salary is during the internship or residency. Note that internship or residency programs may vary from organization to organization in content, length of program, preceptor requirements, salary during internship, and so on.

Rehearse an interview scenario with a trusted colleague or by video. Types of interviews can vary from one-to-one interviews, panel interviews, telephone interviews, and follow-up interviews, all with varying types of questions involving hypothetical case scenarios. You are applying for an entry-level position, and therefore the questions will be directed at your nursing care knowledge. For example, if you are applying for a nursing position on a general medical unit, be ready to give the nursing interventions for a patient experiencing chest pain or hypoglycemia. You may also be asked to recall a difficult nursing situation and describe your behavior in that situation. For example, you may be asked, "If you are faced with a demanding patient who has been waiting for a long time to have his dressing changed, what would you do?" To respond, use the STAR acronym and include each component. Describe Specifically (S) what happened; the Task (T), problem, or issue; the Action (A) you took; and the Result (R) of the action. What the interviewer is looking for is what you learned from the situation and how you would handle a similar situation in the future (Table 27-4). When you are asked a question, avoid rambling. Think about the question and answer it. Actively participate in the interview, but don't monopolize the conversation; let the interviewer lead.

### Critical Thinking 27-5

Prepare a cover letter and resume for Jane from Table 27-1.

How can Jane capture the attention of a human resource person who has had an extremely busy day and is reading his or her e-mail subject lines at the end of a workday?

How should Jane develop her cover letter and resume so that she is interviewed for a job on a pediatric patient care unit?

## TABLE 27-4 STAR Interviews

| STAR | EXAMPLE |
| --- | --- |
| **S**pecifics | A patient was overdue for his dressing change. He became angry and demanded that I come now to change his dressing. |
| **T**ask | I was busy with other high-priority patients. I was having trouble getting to this dressing change. |
| **A**ction | I called my charge nurse and asked for help. The charge nurse was able to change the patient's dressing, talk with him, and help him to relax. I stopped in to tell the patient I was sorry for the delay. |
| **R**esult | I asked the charge nurse to review assignments for future care of this type of patient who has extensive dressing change needs. I also resolved to examine the way I prioritize my patients at the beginning of a shift to determine the best way to meet patient care needs. I resolved to change my future patients' dressings early in the shift before it gets busy. |

Interviews that ask about your behavior are designed to provide the employer with information about how you have handled both negative and positive experiences in the past. Employers are seeking employees who are able to reflect on their past performance and learn from it. In this information age, nursing employers are recognizing the need to transform work sites into learning sites (Holden, 2006).

During the introductory phase of the interview, the employer should outline the job and the conditions of employment. If the job and conditions do not reflect your understanding of the position, be sure to clarify by asking questions at this time.

The working phase of the interview will begin with the employer asking you questions regarding your cover letter and resume. Many of the questions during the interview will refer to the job description. Familiarize yourself with the legal and illegal questions that may be asked. See the Web site www.hospitalsoup.com. Search for, "inappropriate interview questions." Legally acceptable questions include your reason for applying, your career goals, any problems you foresee, and your strengths and weaknesses. Illegal questions include asking if you are a citizen of the country, your age, cultural heritage, membership in social organizations, family characteristics, and medical history (Canadian Human Rights Commission, 2010).

Rather than refusing to answer an illegal question, which may be seen as being uncooperative or confrontational, respond as if it is a legally acceptable question. For example, should the interviewer ask how many children you are caring for at home, respond by indicating that you are able to handle the demands and hours of the job for which you are applying. Responding in this manner may signal to the interviewer your ability to serve as a team player without compromising the legal or ethical issues of the job requirements.

Highlight specific personal and professional accomplishments, as these reflect your ability; however, be careful not to inflate them as this can raise doubts concerning your truthfulness and accuracy. If you give the interviewers reason to question your veracity, you may lose the job opportunity. Note that everything on your resume and application is subject to verification. Respond in a calm, problem-solving fashion to all questions. See Table 27-5 for interview questions you may be asked.

Avoid any discussion of how bad your last employer or faculty was or how incompetent you think your coworkers or classmates are. Keep the entire interview process as positive as possible. Avoid any discussions of personal problems. If an employer has a choice between you and the person who lost their last job because they kept calling in sick over child care or personal problems, they're going to pick you every time.

### Dressing for the Interview

Dress appropriately for the position by wearing professionally acceptable, comfortable, and neatly pressed clothing. For women, this may be a solid-color conservative suit with a modest, coordinated blouse; medium-heeled polished shoes; limited jewelry; and a neat, professional hairstyle. Skirt length should be long enough so you can sit down comfortably. Use light makeup and no perfume, and have neat, short, manicured nails. Avoid visible tattoos, nose rings, extra earrings, or wild hair or clothing colors or styles.

## TABLE 27-5  Interview Questions

| QUESTION | POTENTIAL RESPONSE |
|---|---|
| Tell me about yourself. | Do not go into a long list, but have two to three traits that are solid (for example, "I am a positive person and look for new learning experiences."). |
| Why do you want to work here? | Describe several attributes of the work environment, the staff, or the patients (for example, "I enjoyed my rotation on 5 West—the staff worked as a team, and I am looking for that type of support in my first position."). Comment on any attractive organizational strengths you saw on the organization's Web site. |
| What do you want to be doing in five years? | Identify a long-term goal and your plans to achieve it with progressive responsibilities and achievements. |
| What are your qualifications? | Discuss experiences that you have had that qualify you for the new position. |
| What are your strengths? | This is a favorite question. Look at the job description. What qualities do you have that are required? Are you able to work under stress, are you organized, are you eager to learn new skills, do you enjoy new challenges? |
| What would your references say? | You may want to ask your references this question. Would they say you are easily distracted or focused? A team player or solo player? A problem solver or one who ignores problems? |
| Are you interested in more schooling? | Most who have just graduated may want to say no, but an employer wants someone who is interested in lifelong learning, especially in the nursing profession. |
| What has been your biggest success? | Think of a success ahead of time that may fit with the organization. It does not have to be in nursing. |
| What has been your greatest failure? | Again, think ahead, but this time make sure you can state what you learned from the negative experience. After all, to fail is to learn, so state what you would do differently next time and why. |
| Why do you want to leave your current job? | For an RN, you can say that you are seeking new responsibilities, experiences, and challenges. Give an example of a new experience you are looking for. |

**Source:** Compiled with information from Polifko-Harris, 2003.

For men, appropriate dress may be a solid-color dark blue, gray, muted pinstripe, or very muted brown conservative suit with a white long-sleeve shirt and conservative tie. Use a conservative stripe or paisley tie that complements your suit—silk or good-quality blends only. Wear dark socks with professionally polished leather dress shoes—brown, cordovan, or black only. Wear limited or no jewelry, and have a neat, professional hairstyle. Limit aftershave, and have neatly trimmed nails. Both women and men should avoid body-piercing jewelry, and cover tattoos. Avoid food, and don't chew gum or use a cell phone or iPod during the interview. Turn off your cell phone before you enter the office. Use a breath mint before you enter the setting for the interview.

## TERMINATION PHASE OF THE INTERVIEW

Terminating an interview is important. The employer will close an interview by asking if you have any questions (Table 27-6). If the interviewer does not raise the issue of salary, it is appropriate to ask about the anticipated salary expectations and any differential for advanced nursing degrees or academic preparation. Note the regular salary surveys done by some nursing journals, for example, *Nursing* (October 2006). It is also appropriate to ask about the type of health, dental, vacation, holiday time, sick time, continuing education, and educational reimbursement benefits offered. As an entry-level nurse, the range of

## TABLE 27-6  Sample Questions to Ask During an Interview

- How can I prepare myself to work on this unit and do a good job?

- May I have a copy of the job description and performance appraisal form? How often will I be evaluated?

- Is there a clinical ladder program that identifies clinical, managerial, and educational levels for promotion?

- Who is my preceptor?

- What shift will I be scheduled to work? Will I rotate shifts? Are special requests for time off honored?

- What holidays and weekends am I scheduled to work?

- What type of orientation or internship will I receive? How long is it? Does it address how to work well with other practitioners?

- Is this a magnet hospital? Do you monitor nurse-sensitive outcomes?

**Source:** Compiled with information from Polifko-Harris, 2003, and Kelly, 2010.

a salary is often standard. However, this should not deter one from asking about the possibility of position advancements. Try to defer specific salary discussions until a job offer has been made. If feeling pressured to make a decision, you may suggest that you would like to have a few days to make sure you feel comfortable that the position is the right fit for you in terms of your career goals (Betterton, Lewis, & McElroy, 2007).

Expect to be quite exhausted by the end of the interview. However, take time to review your notes, seek clarification for any concerns, and conclude the interview by asking when you can expect to hear from the employer. Asking when you might hear back from the employer indicates that you are actively seeking employment and suggests that, if they are serious about hiring you, they may want to offer a position.

### Obtaining References

Seek permission to use your references prior to your interview, to avoid delay and illustrate that you are not hesitant to provide references. Seeking permission to submit a person's name for a reference alerts that individual that he or she can expect a call and from whom and for what type of job. This will prevent any hesitancy on the part of the reference in agreeing to provide information to a potential employer while trying to recall who you are, especially if there has been a delay since you had contact with him or her. It is wise to build your network of contacts and maintain contact with them over time.

Your past employment history will guide you as to whom to list as references. If you have had past health care

### Critical Thinking 27-6

For many nurse applicants, the most anxiety-producing aspect of searching for a job is the interview. Assume you are seeking employment. Practice answering the questions in Table 27-5. Have a friend interview you using the questions, or even just practice answering them out loud on your own. What questions would represent your level of knowledge and skill? What characteristics would you emphasize when you respond to the questions?

employers, as Jane and James have (Table 27-1), you will be wise to list those employers and your manager. You should also include any character references, including at least one nursing professor. For graduates who have not had many work-life experiences, references from volunteer service and high school contributions are helpful.

### Following Up Your Interview

Within 24 hours, follow up your interview with a simple thank-you note (Figure 27-5).

If you sensed that the interview did not go well, reflect objectively upon the event with a colleague or friend, avoiding excessive negative talk about yourself or the interviewers.

James Mattern
214 Christie Avenue
Gladstone, OH 43523
(604) 555–1212

April 26, 2011

Ms. Eileen Carter, BSN, RN
Director of Human Resources
Concordia Hospital
100 Seaside Drive
Austin, NJ 12356

Dear Ms. Carter:

Thank you for the time you spent with me as I interviewed for a position as a registered nurse at Concordia Hospital. I enjoyed meeting the emergency department nurse manager and several of the staff nurses yesterday and was especially impressed with the sense of professionalism among the staff. I believe I would be a good fit for the position in your emergency department. I worked there for a year during my nursing education program and both enjoyed and profited from working with the high-caliber staff.

I have requested that my transcripts be sent directly to your office, and I will have three of my instructors complete the reference forms you gave me. I look forward to hearing from you soon about my second interview and will contact you in two weeks as directed.

Sincerely,

*James Mattern*

James Mattern

FIGURE 27-5 **Interview follow-up letter.** (*Source:* Compiled with information from Polifko-Harris, 2003.)

Recognize the confidentiality of the interview for both the employer and yourself. Consider it a good learning experience. If you are not the successful applicant, ask what areas were weak and how you might address them. For example, you may not be considered for a community health nursing position because employers may require you to have two to three years of previous acute medical and surgical nursing experience. By asking for this information, you are demonstrating your interest in addressing your weaknesses and learning from the interview. If the employer is reluctant to spend time with you, as an entry-level nurse, to answer these questions, this employer may not have been the best fit for you. If you have not heard back from the employer in the time indicated at the interview, follow up with a phone call. This will demonstrate your continued interest in the position and your willingness to learn about your weaknesses for future interviews. Don't just say, "I am following up. Have you decided?" Instead say, "I am really interested in your position and here is why…"

If the job is offered to you, suggest a follow-up meeting to clarify any questions that need further explanations, such as salary. Even within unionized health care agencies, there is negotiation room for where the employer may want to place you on the pay scale. It is important to know your value, know the average salary paid for similar positions with other agencies, and to clearly communicate your expectations.

Should you find the job offer unacceptable, clearly state the reason. You will leave the door open should the employer return with a counteroffer. In addition, be sure to thank them for the opportunity to discuss your career goals and plans.

## Critical Thinking 27-7

To help pinpoint the most important areas where graduate nurses need help adapting to their new clinical role, the Advisory Board's Nursing Executive Center commissioned a survey that asked participants to rank new graduate nurses' competency in 36 areas that are considered essential to safe and effective practice (Mosby's Nursing Suite, 2010). These 36 areas are organized under the categories of Clinical Knowledge, Technical Skills, Critical Thinking, Communication, Professionalism, and Management of Responsibilities. Go to the website in the references and note the 36 areas. How are you doing on the 36 competencies? How can you improve your skills in these areas?

## EVIDENCE FROM THE LITERATURE

**Citation:** Holden, J. (2006). How can we improve the nursing work environment? *American Journal of Maternal/Child Nursing, 31*(1), 34–38.

**Discussion:** The mother of our profession, Florence Nightingale, was the first to set the most important patient care goal, which was to do the sick no harm. The article highlights the need for organizations to transform the environment of nurses into learning organizations. When using the framework of a learning organization, health care organizations achieve their goals just as business and industry have done. The article articulates the principles of a learning organization to include systems thinking, personal mastery, team learning, mental models, and shared vision.

Through systems thinking, the processes that exist within an organization are viewed as interrelated. Activities or work processes done in one part of an organization affect the entire organization. Personal mastery relates to the development of the person as an evolving, growing individual and professional. This can best be realized in an environment where not only individual learning but also team learning is actively promoted. Having a mental model suggests that individuals develop a shared mental vision of a healing environment where individuals are given permission and freedom to address health care deficiencies and promote quality improvement. Having a shared vision of such environments promotes a common goal with a stronger sense of commitment by the entire team. The author relates making a successful and integrated change to a new work environment through the well-known steps of the nursing process: assessment, planning, implementation, and evaluation.

Implications for Practice: When planning your career in nursing, consideration of a potential employer's working environment is important in selecting with whom and for whom you want to work. Before accepting a nursing position, determine the congruence between your values and goals and the operation of the organization. As mastery over your own destination is important, select a learning organization that will promote your ultimate success.

## EVALUATION

Your academic and clinical preparation has placed you in an admirable position to apply the standards of practice expected of the entry-level nurse, in accordance with the licensing and professional associations. Carry out every step in your strategic planning for your career with a view toward marketing yourself even if you are rejected. You will leave the impression with the interviewers that you may meet their employment needs in the future. Career planning means thriving rather than surviving.

In today's nursing world, nursing management is seeking nurses who are willing to enter into partnerships and communicate and collaborate with other nurses, health care providers, etc. The health care industry recognizes that nurses have a unique capacity to influence the health status of its patients as well as improve the nursing work environment (Holden, 2006). Forging respectful partnerships between the employer and employee promotes quality health care for all our citizens. Sharing the mutual goal of providing safe and competent care will be satisfying for both you and your employer. Your consistent application of employability skills will result in a highly marketable reference for any future endeavor. Seek to build relationships early in your career with one or more experienced career mentors.

## KEY CONCEPTS

- Career planning for the professional nurse necessitates taking control of the strategic planning, vision, and goal-setting process.

- Marketing strategies designed to demonstrate your academic preparation and the critical thinking skills that you honed during your education are key features in your success.

- Values clarification enhances organizational fit and success for both employer and employee.

- Establishing short-term, intermediate, and long-term goals will shape your job search.

- Consider your cover letter and resume as a commercial for marketing yourself.

- Preparation for a job interview requires knowledge of the agency, the job description, and standards of nursing practice.
- Practice your job interview with a trusted colleague.
- Do a job search on the Internet to locate positions that you are interested in.

- Recognizing personal strengths and weaknesses is helpful when working collaboratively as a member of the interdisciplinary health team.
- Follow up your interview with a thank-you note.

## KEY TERM

career planning

## REVIEW QUESTIONS

1. You are being considered for an entry-level nursing position on a busy medical nursing unit. During the interview, the nursing manager poses the following case scenario: During the night shift, one of the patients on the unit becomes unstable. It is difficult to care for him and all the other patients in your assignment. How should you handle the situation to provide a safe practice environment?
   A. Ask the NAP to watch the unstable patient while you care for the other patients.
   B. Inform the nursing supervisor of the patient's condition and the patient assignment, and request assistance.
   C. Prepare a detailed documentation report to give to the nursing manager in the morning.
   D. Place the patient near the desk so that anyone passing by can check on the patient.

2. During an interview for a position in the emergency department (ED), James was asked how he would handle the following situation: The ED is very busy, and a patient arrives with his wife. Some of the nurses refer to him as a "frequent flyer" and tell you to put him in the end room until you have time to care for him. What would be your initial response?
   A. After smelling alcohol on his breath, you realize that he may be intoxicated and comply with the nurses' direction.
   B. You place him on a stretcher and tell his wife to watch him so that he doesn't fall off.
   C. You assess him more thoroughly and take the appropriate nursing action.
   D. You ask a nurse what was meant by the term "frequent flyer," because you had never heard it before.

3. During a job interview, the nurse recruiter presents the following hypothetical question: You are working the evening shift and receive a sick call for the night shift. The unit has been very busy on the evening shift. How would you handle the situation?

   A. After an hour, call the nurse back to determine if her health has improved.
   B. Fill out a heavy workload form, and submit it to the union representative.
   C. Assess the unit's patients conditions, and notify the nursing supervisor of the sick call.
   D. Inform the nursing manager that you think the nurse is abusing her sick time.

4. The employer is requesting applicants to take an assessment test that includes the following case scenario: You are making early morning rounds and discover that a patient was restrained to the side rails of the bed. What is your initial response?
   A. Assess the patient's condition.
   B. Check the policy of the hospital with respect to the use of restraints.
   C. Contact the nursing manager to witness the presence of the restraint.
   D. Check the practitioner's order for the use of restraints.

5. During orientation to the workplace, the staff educator provides information that is not familiar to you, and you are not sure what to do. What would be the best approach to address your lack of knowledge?
   A. Research evidence-based material.
   B. Ask one of the other participants.
   C. Ask for clarification from the staff educator.
   D. Contact a former nursing faculty member.

6. During your job interview, you are asked to consider the following scenario: A patient on a rehabilitation unit requests to see his dog. You realize that the hospital policy does not permit animals on the unit. What would you do?
   A. Inform the patient that the hospital policy does not permit animals in the unit.
   B. Suggest that the family sneak the dog into the unit using the stairwell.

C. Explore approaches with the hospital's management that could meet the patient's need.

D. Forward a research document to administration for their consideration that addresses the importance of pet therapy.

7. You are assisting a nurse who is performing a sterile procedure. Accidentally, the nurse contaminates the sterile field and does not appear to notice. What would be the most appropriate action?

A. Ignore the accidental contamination.

B. Indicate that you will obtain a different sterile tray immediately.

C. Speak to the nurse about the break in technique after the procedure

D. Complete an adverse event form.

8. During your nursing employment interview, the nurse recruiter asks if you can provide references. Your best response would be which of the following?

A. Contact a high school teacher and a nursing teacher to provide references.

B. Give the recruiter the names of possible references.

C. Provide the names of references who have consented to provide a reference.

D. Provide the name of your former employer at a fast food chain.

9. You have received a letter offering you a nursing position. However, you have some concerns about the standards of practice of the organization. Which of the following concerns may you want to consider regarding the practice of a registered nurse before accepting the offer? Select all that apply.

_____ A. Authority of registered nurses.

_____ B. Progression of salary increments.

_____ C. Responsibility of entry-level nurses.

_____ D. Accountability of entry-level nurses.

_____ E. Perceived image of the organization.

10. What factors should be considered in delegating a nursing assignment to others? Select all that apply.

_____ A. Job description of the individual.

_____ B. Confirmation of delegate's understanding of the assignment.

_____ C. The high school education of the individual.

_____ D. The complexity of the patient's care needs.

_____ E. Previous experience with the assigned task.

## REVIEW ACTIVITIES

1. You have been hired as a career planner by James and Jane. Your services require you to prepare a cover letter and resume for them.

   What method of job application is illustrated in James's cover letter?

   How would you position Jane so that the human resource person would want to continue reading her cover letter or return to it?

2. Send an e-mail for Julie and Jane to a human resource manager. How would you begin to capture the attention of a human resource person who has had an extremely busy day and is reading his or her e-mail at the end of the workday?

   How would you position Jane so that the human resource personnel would consider reading her e-mail?

   What employment skills does Jane bring to the job?

## EXPLORING THE WEB

- For U.S. statistics, review this reference from the U.S. Department of Labor, Bureau of Labor Statistics: www.bls.gov

- For the Canadian labor market, review this reference from Statistics Canada: www.statcan.gc.ca

- Browse this site: career-advice.monster.com

  Click on Profile and Resume.

- Check these job boards specific to health care www.healthcareerweb.com www.medjobs.com

- Review these general Web sites for nursing issues: www.nursingworld.org www.nursingcenter.com www.nurseweek.com www.ncsbn.org www.rwjf.org www.aacn.nche.edu

- There are many Web sites specific to nursing employment opportunities. Try some of these: www.americanmobile.com www.rnwanted.com

www.healthopps.com
www.healthcareersinteraction.com
www.healthcaretraveler.com

■ For online journal articles:
www.medscape.com
www.google.com

■ For nursing articles: www.nursingcenter.com (click on Library of Nursing Journals)
www.nursingmanagment.com

## REFERENCES

American Nurses Association. (2011). ANA Position Statements. Retrieved April 25 from http://www.nursingworld.org/search.aspx?searchMode=1 &SearchPhrase=ANA+Position+Statements&SearchWithin=2.

Arvidsson, B., Skarsater, I., Oijervall, J. & Friglund, B. (2008).Process-oriented group supervision during nursing education: nurses' conception 1 year after their nursing degree. *Journal of Nursing Management, 16*(7), 868–875.

Betterton-Lewis, D. G. & McElroy, M. (2007). Salary negotiation: How and when to talk money. *American Medical Writers Association Journal, 22*(3), 127–129.

Blythe, J., Baumann, A., Zeytinoglu, I. U. Denton, M., Skhtar-Danesh, N., Davis, S., Kolotylo, C. (2008). Nursing generations in the contemporary workplace. *Public Personnel Management, 37*(2), 137–159.

Byrne, M., Schroeter, K., Carter, S. & Mower, J. (2009). The professional portfolio: an evidence-based assessment method. *The Journal of Continuing Education in Nursing, 40*(12), 545–552.

Callister, L. C. (2006). Global health and nursing: It's a small, small world. *American Journal of Maternal/Child Nursing, 31*(1), 63–63.

Canadian Human Rights Commission. (2010). Guide to screening and selection in employment. Retrieved April 20, 2011 from http://www.chic-ccdp.ca?pdf/screen.pdf.

College of Registered Nurses of Alberta. (2007). Interpretive document: documenting your continuing competence activities. Retrieved April 25 from http://www.nurses.ab.ca/carna/index.aspx?WebStructureID=2227.

Ervin, E. E., Bickes, J. T., & Schim, S. M. (2006). Environments of care: A curriculum model for preparing a new generation of nurses. *Journal of Nursing Education, 45*(2), 75–80.

Forneris, S. G. & Peden-McAlpine, C. (2007). Evaluation of reflective learning to improve critical thinking in novice nurses. *Journal of Advanced Nursing, 57*(4), 410–421.

Harris, K. (2002). Center of the storm, practicing principled leadership in times of crisis. Nashville, TN: Thomas Nelson, Inc.

Hart, K. (2006). Student extra: The employment interview: Tips for success selecting an employer for the perfect fit. *American Journal of Nursing, 106*(4), 72AAA–72CCC.

Hodges, H. F., Keeley, A. C. & Troyan, P. J. (2008). Professional resilience in baccalaureate-prepared acute care nurses: first steps. *Nursing Education Perspectives, 29*(2), 80–89.

Holden, J. (2006). How can we improve the nursing work environment? *The American Journal of Maternal/Child Nursing, 31*(1), 34–38.

Honaman, J. C. (2009). Managing the electronic job search. *Healthcare Executive, 24*(3), 68–70.

Jackson, D., Firtko, A. & Edenborough, M. (2007). Personal resilience as a strategy for surviving and thriving in the face of workplace diversity: a literature review. *Journal of Advanced Nursing, 60*(1), 1–9.

Kelly–Heidenthal, P. (2004). Essentials of nursing leadership and management. Clifton Park, NY: Delmar Cengage Learning.

Kelly, P. (2010). *Essentials of nursing leadership and management* (2nd ed.). Clifton Park, NY: Delmar Cengage Learning.

Kluemper, D. H. (2009). Future employment selection methods: evaluating social networking web sites. *Journal of Managerial Psychology, 24*(6), 567–580.

Lenburg, C. B., Klein, C., Abdur-Rhaman, V., Spencer, T. & Boyer, S. (2009). THE COPA MODEL: A comprehensive framework designed to promote quality care and competence for patient safety. *Nursing Education Perspectives, 10*(5), 312–317.

Luk, L. A. (2008). Nursing management of medication errors. *Nursing Ethics, 15*(1), 28–39.

Myers, V.L. & Dreachslin, J. L. (2007). Recruitment and retention of a diverse workforce: challenges and opportunities. *Journal of Healthcare Management, 52*(5), 290–298.

Polifko-Harris, K. (2003). Career Planning. In P. Kelly-Heidenthal (Ed.), *Nursing Leadership and Management.* Clifton Park, NY: Delmar Cengage Learning.

Registered Nurses Association of Ontario. (2003). Report on the nursing experience with SARS in Ontario. RNAO: Public Hearing, September 29. Retrieved March 31, 2006, from http://www.rnao.org/html/PDF/SARS_Unmasked.pdf.

Robert Wood Johnson Foundation. (2010). Recent Research About Nursing. Retrieved April 20, 2011 from http://www.rwjf.org/pr/product.jsp?id=71349.

Schoessler, M. & Waldo, M. (2006). The first 18 months in practice, a development transition model for the newly graduated nurse. *Journal of Nurses in Staff Development. 22*(2), 47–52.

Williams, M. & Jordan, K. (2007). The nursing professional portfolio: a pathway to career development. *Journal for Nurses in Staff Development, 23*(3), 125–131.

## SUGGESTED READINGS

Atkinson, P. A., Martin, C. R., & Rankin, J. (2009). Resilience revisited. *Journal of Psychiatric and Mental Health Nursing, 16*(2), 137–145.

Bagnardi, M., & Perkel, L. K. (2005). The learning achievement program, fostering student cultural diversity. *Nurse Educator, 30*(1), 17–20.

Barrick, M. R., Shaffer, J. A., & DeGrassi, S. W. (2009). What you see may not be what you get: Relationships among self-presentation tactics and ratings of interview and job performance. *Journal of Applied Psychology, 94*(6), 1394–1411.

Brewer, C. S., Zayas, L. E., Kahn, L. S., & Sienkiewicz, M. J. (2006). Nursing recruitment and retention in New York State: A qualitative workforce needs assessment. *Policy, Politics & Nursing Practice, 7*(1), 54–63.

Carson, K. D., Carson, P. P., Fontenot, G., & Burdin, J. J. (2005). Structured interview questions for selecting productive, emotionally mature, and helpful employees. *Health Care Manager, 24*(3), 209–215.

Hackett, R. D., Lapierre, L. M., & Gardiner, H. P. (2004). A review of Canadian human rights cases involving the employment interview. *Canadian Journal of Administrative Sciences, 21*(3), 215–228.

Hankin, H. (2005). *The new workforce.* New York: AMACOM.

Heilmann, P. A. K. (2010). Employer brand image in a health care organization. *Management Research Review, 33*(2), 134–144.

Hill, S. (2006). Inspiring the next generation. *Nursing Times, 102*(21),18–19.

Hoye, G. V. (2008). Nursing recruitment: Relationship between perceived employer image and nursing employees' recommendations. *Journal of Advanced Administration, 63*(4), 366–375.

Johnson, S. A., & Romanello, M. L. (2005). Generational diversity, teaching and learning approaches. *Nurse Educator, 30*(5), 212–215.

Kidder, M., & Cornelius, P. B. (2006). Licensure is not synonymous with professionalism: It's time to stop the hypocrisy. *Nurse Educator, 31*(1), 15–19.

Lake, E. T., & Friese, C. R. (2006). Variations in nursing practice environments, relation to staffing and hospital characteristics. *Nursing Research, 55*(1), 1–9.

Langford, B. (2005). *The etiquette edge, the unspoken rules for business success.* New York: AMACOM.

Laschinger, H. K. S., Purdy, N., Cho, J., & Almost, J. (2006). Antecedents and consequences of nurse managers' perceptions of organizational support. *Nursing Economic$, 24*(1), 1–29.

Maxwell, M. (2005). It's not just black and white: How diverse is your workplace? *Nursing Economic$, 23*(3), 139–140.

McIntyre, M., Thomlinson, E., & McDonald, C. (2005). *Realities of Canadian nursing: Professional, practice, and power issues* (2nd ed.). New York: Lippincott Williams & Wilkins.

Morath, J. M., & Turnbull, J. E. (2005). *To do no harm.* San Francisco: Jossey-Bass.

Poster, E., Adams, P., Clay, C., Garcia, B. R., Hallman, A., Jackson, B., et al. (2005). The Texas model of differentiated entry-level competencies of graduates of nursing programs. *Nursing Education Perspectives, 26*(1), 18–23.

Roberts, K. Lockhart, R., & Sportsman, S. (2009). A competency transcript to assess and personalize new graduate competency. *Journal of Nursing Administration, 39*(1), 19–25.

Thrasher, C., & Staples, E. (2005). Expanding community health nursing practice, primary health care nurse practitioners. In L. L. Stamler & L. Yiu (Eds.), *Community health nursing, a Canadian perspective* (pp. 333–336). Toronto: Pearson Prentice Hall.

Wagner, S. E. (2006). Staff retention: From "satisfied" to "engaged." *Nursing Management, 37*(3), 25–29.

White, K. M. (2005). Policy highlight: Staffing plans and ratios. What is the latest U.S. perspective? *Nursing Management, 37*(4), 18–24.

Xu, Y., Zaikina-Montgomery, H. Shen, J. (2010). Characteristics of internationally educated nurses in the United States: an update from the 2004 National Sample Survey of Registered Nurses. *Nursing Economic$,* Jan-Feb;28(1):19–25, 43.

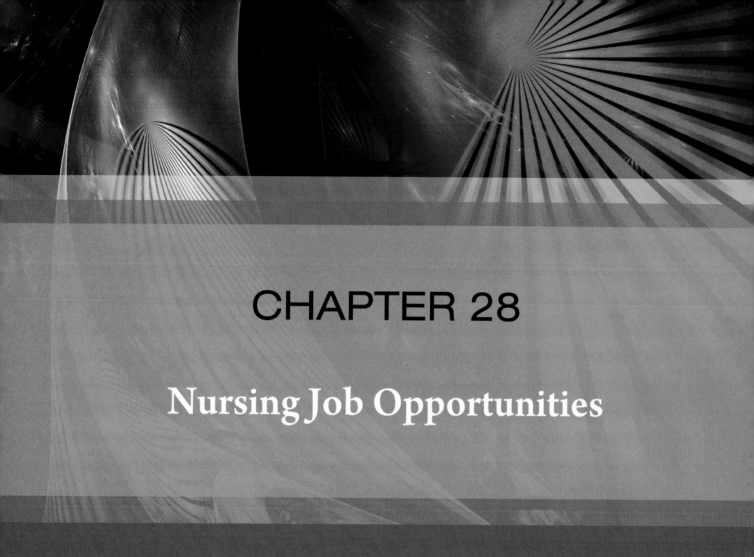

# CHAPTER 28

# Nursing Job Opportunities

## STEPHEN JONES, MS, RN, CPNP, ET

*Luck is a matter of*
*preparation meeting*
*opportunity.*

(OPRAH WINFREY)

## OBJECTIVES

Upon completion of this chapter, the reader should be able to:

1. Discuss the many nursing opportunities available upon graduation.
2. Identify advanced nursing practice and other nontraditional nursing roles.
3. Identify various opportunities for certification.
4. Describe hospital-based and nonhospital-based nursing practice.
5. Identify directions to take to ensure the future of nursing.

Delmar/Cengage Learning

*You are within four months of graduation. Your thoughts are focused on moving to a nice city and getting a good job in a health care setting that will stimulate and educate you. You also know individuals who have been nurses for a while. They are doing seemingly incredible activities and interventions. Many are able to write prescriptions, and some even have flexible hours. There are numerous questions that you, the graduating nurse, need to consider:*

> *What do you see yourself doing one year, five years, and ten years from now?*
>
> *What are your real-life situations and circumstances regarding family, ability to relocate, hours to work?*
>
> *Do you have financial constraints?*
>
> *What are your strengths and weaknesses?*

Health care facilities have been experiencing nursing shortages since the mid-1980s. In the early 1990s, there was a temporary ease in the nursing shortage. This easing, however, was short-lived; at the turn of the twenty-first century, the nursing demand far exceeded the supply of registered professional nurses. As of 2005, one in ten nursing jobs were unfilled, and according to the U.S. Bureau of Labor and Statistics, the nursing shortage will reach one million by 2012 (Erickson et al., 2005). That said, according to a study by the U.S. Bureau of Labor Statistics, registered nursing is the top occupation in terms of the largest job growth from 2008 to 2018, with estimates of more than 580,000 new RN jobs being created (American Association of Colleges of Nursing, 2011b). Further, this study projects the rising complexity of acute care and notes that demands for RNs could increase by 36% by 2020. Nurses apply their holistic knowledge to help individuals manage the changes brought on by illness or disease, educate them about preventive health care, and, in general, attempt to improve the quality of health care. Many factors have contributed to the nursing shortages, especially since the mid-1980s. In particular, the number of chronically ill and frail elderly patients has increased, and the demand for nurses who can care for them far exceeds the supply. In addition, the various restructuring efforts in our nation's health care system have made nursing jobs less attractive. This nursing shortage is expected not only to continue, but to worsen by the year 2020, as mentioned. This shortage is different from those of the past in several ways. Factors contributing to the most recent shortage include an aging nursing population, a declining number of nursing students in the academic pipeline, and the need for more accommodating working conditions for nurses.

Statistical data from the American Association of Colleges of Nursing (AACN) and the National League for Nursing (NLN) (National League for Nursing, 2011) reveal the latest data on nursing as a profession:

- Nursing is the United States' largest health care profession, with more than 3.1 million registered nurses (RNs) nationwide.
- Registered nurses comprise one of the largest segments of the U.S. workforce as a whole and are among the highest-paid large occupations.
- Nurses comprise the largest single component of hospital staff, are the primary providers of hospital patient care, and deliver most of the nation's long-term care.
- Most health care services involve some form of care by nurses. From 1980 to 2004, the percentage of employed RNs working in hospitals decreased from 66% to 56.2% as health care moved to sites beyond the hospital and nurses increased their numbers in a wide range of other settings.
- There are over four times as many RNs in the United States as physicians.

Further data reveals the following statistics regarding schools of nursing and the types of degrees conferred. The 2008–2009 academic year indicated that the capacity of the nation's nursing education programs continued to fall short of demand as the weakened economy nearly halted the expansion of programs. Almost 40% of all qualified applicants to basic RN programs were turned away during this time period. The average age of instructors was also increasing (American Association of Colleges of Nursing, 2011a). Forty-eight percent of nurse educators were age 55 and over, and fully one-half of nurse faculty said they expected to retire within the next 10 years. 2008 statistics showed that the diploma was the highest educational credential for 13.9% of RNs, while a bachelor's degree was the highest educational credential for 36.8% of RNs. The associate's degree was the highest educational credential for 36.1% of RNs. Advanced degrees, e.g., master's degree or doctorate degree, were held by 13.2% of registered nurses. The average age of RNs was 46 years (American Association of Colleges of Nursing, 2011b).

Throughout its history, nursing has always responded to changes in society's health care needs. From the days of Florence Nightingale and her service in the mid-1800s to

Lillian Wald and public health nursing in the early 1900s to Martha Rogers in the 1930s and 1940s suggesting that nurses prepare themselves to handle the health issues that arise from space travel, nursing has met society's changing health care needs by expanding the roles of nurses. For decades, nurses in acute care settings have been taking on advanced and expanded roles on evening, night, and weekend shifts.

Many of the changes in nursing have evolved naturally and logically. Positions such as staff nurse, nurse manager, director of nursing, and case manager, and organizations such as visiting nurse associations and public health organizations, are a direct reflection of these evolutions. Nursing has established itself in hospitals, ambulatory clinics, practitioners' offices, and community and school settings. An issue confronting nursing, however, has been identifying what the appropriate entry-level educational degree should be and what a nurse really is, because of the many levels of educational preparation. In December 1965, the American Nurses Association (ANA) House of Delegates (HOD) adopted a motion that the ANA continue to work toward baccalaureate education as the educational foundation for professional nursing practice. By 1985, the ANA HOD agreed to urge state nursing associations to establish the BS degree with a major in nursing as the minimum educational requirement for licensure and to retain the title of registered nurse (RN) (American Nurses Association [ANA], 2000). It has been 45 years since the American Nurses Association (ANA) established this position. During that time period, only one state, North Dakota, was successful at getting the entry into practice proposal fully implemented, and even there it was rescinded in 2003 (Smith, 2009). To date, there are still a variety of paths an individual can take to become an RN. These include a two-year associate degree, a three-year diploma, and a four-year baccalaureate degree. Many four-year baccalaureate schools of nursing have changed the traditional four-year pathway to a degree and are admitting students who already have a non-nursing baccalaureate degree or an associate's degree. In deciding which education option to pursue, students should consider their future, that is, where they want to be in five to ten years. If advanced practice is a desired goal, then a baccalaureate education is the required first step toward that goal.

A major advancement within nursing has been the emergence of several areas of advanced nursing practice: certified registered nurse anesthetist (CRNA), clinical nurse specialist (CNS), nurse practitioner (NP), and certified nurse-midwife (CNM). Historically, the oldest American advanced nursing practice role is that of nurse anesthetist, which began in the mid-1800s, with a professional nursing association later established in 1931 (Murphy-Ende, 2002). Nurse midwives were the next role to develop, incorporating as a national group in 1955. Clinical nurse specialists (CNSs), which began in psychiatric nursing in the early 1940s, developed specialties and certification by the mid-1970s. The newest, and perhaps most visible, role due to its significant presence in primary care, is the nurse practitioner (NP) role, with its inaugural program beginning in 1965. Currently, there are a variety of patient populations served by the CNS and NP roles. Nurses within these roles assume expanded functions and additional responsibilities, often crossing over into what had been viewed as the role of medicine. The twentieth century was a unique time in nursing's history. The profession struggled to assume its rightful role in the health care

## EVIDENCE FROM THE LITERATURE

**Citation:** Khomeiran, R., Yekta, Z. P., Kiger, A. M., & Ahmadi, F. (2006). Professional competence: Factors described by nurses as influencing their development. *International Nursing Review, 53*(1), 66–72.

**Discussion:** This study was done to explore and define factors that may influence competence development. This oftentimes controversial issue in health care settings affects many aspects of the nursing profession, including education, practice, and management. The study was qualitative, with data obtained via tape-recorded, semistructured interviews. Interviews were then transcribed verbatim and analyzed according to the qualitative methodology of content analysis. Six descriptive categories were identified from the data: experience, opportunities, environment, personal characteristics, motivation, and theoretical knowledge.

**Implications for Practice:** The findings suggest that the six identified factors influencing the process of developing professional competence in nursing extend across both personal and extra-personal domains. Further understanding of and research into these areas may enhance the ability of nursing leaders, including educators, to enable student and qualified nurses to pursue effective competency development pathways to prepare themselves for a higher standard of care delivery.

## EVIDENCE FROM THE LITERATURE

**Citation:** Woodburn, J. D., Smith, K. L., & Nelson, G. D. (2007). Quality of care in the retail health care setting using national clinical guidelines for acute pharyngitis. *American Journal of Medical Quality, 22*(6), 457–62.

**Discussion:** Rates of adherence to an acute pharyngitis practice guideline in the retail clinic setting were measured as an indicator of clinical quality. An analysis of 57,331 patient visits for the evaluation of acute pharyngitis was conducted. In 39,530 patients with a negative rapid strep test result, nurse practitioner and physician assistant staff adhered to guidelines in 99.05% of cases by withholding unnecessary antibiotics. Of 13,471 patients with a positive rapid strep test result, 99.75% received an appropriate antibiotic prescription. The combined guideline adherence rate for groups with positive and negative rapid strep test results was 99.15%. Strep cultures were performed on 99.1% of patients with a negative rapid strep test result, and 96.2% of patients with a positive culture were treated with an antibiotic. Finally, 0.95% of patients with a negative rapid strep test result were provided an antibiotic outside clinical guidelines; however, approximately half of these prescriptions (n = 190) were supported by documentation of clinical concerns for which an antibiotic was a reasonable choice.

**Implications for Practice:** Nurse practitioners and physician assistants deliver health care according to guidelines. As health care needs continue to grow in the general population, these practitioners are an important element of health care delivery for patients with primary health care needs.

---

arena, with its members performing traditional nursing roles, yet it significantly upgraded its educational, clinical, research, and managerial focus. This chapter will examine emerging opportunities for both new graduates who would prefer not to assume advanced or expanded nursing roles and those who are contemplating these roles.

## EMERGING OPPORTUNITIES

Within the traditional hospital setting, the levels of nursing hierarchy are established: vice president/director of nursing, nurse managers, and staff nurses. This hierarchy allow for promotion and advancement. Since the mid-1980s, however, the trend has been to flatten the levels of nursing management—that is, have fewer managers and additional clinical nurses, including bedside nurses, nurses with expanded roles, and advanced practice nurses.

## CERTIFICATION

Professional certification in nursing is a measure of distinctive nursing practice. The rise in consumerism in the face of a compelling nursing shortage and the profession's movement to elevate nursing as a career option have given prominence to the value of certification in nursing (Shirey, 2005). Certification in nursing represents an example of professional credentialing and is a voluntary process undertaken by practicing nurses. It is a marker of the knowledge and experience of a professional RN and is more than just

a symbolic title, whether at the BSN or advanced degree level. It is also a means of validating a nurse's knowledge in a specific practice area (Watts, 2010). The American Board of Nursing Specialists (ABNS) defines the certification process as formal recognition of the specialized knowledge, skills, and experience demonstrated by the achievement of standards identified by a nursing specialty to promote optimal health outcomes (Stromberg et al., 2005).

Certification attests to the education, skills, and training of a registered nurse, and this recognition is achieved through testing for standard knowledge by examination (Hittle, 2010). Certification should not be confused with credentialing and/or privileging, which is the process by which a physician, advanced practice nurse, or other health care provider obtains authorization from a health care organization to practice in a particular health care setting. The credentialing and/or privileging process involves an objective evaluation of a subject's current licensure, training or experience, competence, and ability to provide particular health care services.

Information about specialty nursing certification is available (American Nurses Credentialing Center, 2011). Table 28-1.

The Accreditation Board for Specialty Nursing Certification (ABSNC), formerly the ABNS Accreditation Council, is the only accrediting body specifically for nursing certification (American Board of Nursing Specialties, 2009). ABSNC accreditation is a peer-review mechanism

## Critical Thinking 28-1

Many nurses are certified in their area of specialty. You are thinking about taking the certification examination but are not sure you really want to spend the time required to prepare for it. Some reflections are useful as to what certification might do for you:

What is the reason you want to take this examination? Will becoming certified make a difference in your job? Will becoming certified allow you to further progress in your position? Is certification required for licensure?

 **EVIDENCE** FROM THE LITERATURE

**Citation:** Marrelli, T. (2006). Nursing in flux. *American Journal of Nursing, 106*(1), 19–20.

**Discussion:** The author states, "When I compare career trajectories with nurses in various fields, I realize that my many shifts in roles and interests are not unusual. The sheer range of opportunity for nurses often beckons us toward change, whether to a new workplace, a new specialty, or another area within the current system. Nursing today offers complex variations on all the old pros and cons. There are more opportunities than ever, more work setting choices, more specialties and certification areas, and a wide range of geographic locations limited only by time and the availability of flights."

**Implications for Practice:** Nurses must share information about nursing opportunities with potential nurse candidates. This will help nurses in the future prepare for careers they are interested in.

that allows nursing certification organizations to obtain accreditation by demonstrating compliance with the highest quality standards available in the industry. Several of the larger certifying organizations include the ANA; American Association of Nurse Anesthetists (AANA); National Association of Pediatric Nurse Associates and Practitioners (NAPNAP); Association of Women's Health, Obstetric and Neonatal Nurses (AWHONN); and The Association of Critical Care Nurses (ACCN).

## TRAVELING NURSE

As demand for nursing has increased, supply has often been very low, and hospitals are frequently understaffed. One option to fill the nursing shortage is the traveling nurse. These nurses usually work in three-month assignments on the same unit. They travel to various locations throughout the United States. The financial charge by the traveling nurse company to the employing hospital for a traveler is usually very high. The traveling nurse's salary however, is not as high and is similar to that of fellow employees.

The benefits to the health care institution of using traveling nurses include having a nurse with a variety of experiences providing continuity of care for a three-month period. These traveling nurses often need only the basic hospital and unit orientation because they come with skills applicable to their area of practice. Traveling nurses need to be aware of differing nursing methodologies and licensure requirements from state to state and will require a license for each state in which they practice unless there is a multistate compact for licensure in the new state in which they want to practice. Traveling nurses should also ensure that their contract stipulates clearly what their assignment is and what the institution and agency expect of them. Check this carefully before signing a contact. Most travelers exhibit flexibility, adaptability, assertiveness, strong organizational and interpersonal skills, confidence, independence, and the ability to learn new skills and techniques.

If traveling is in your blood, adventure also lies outside the United States, as many foreign countries actively recruit American nurses, especially Middle East countries such as Saudi Arabia and Kuwait. This is an opportunity to see other

**TABLE 28-1   Certifications Available from American Nurses Credentialing Center**

| Nurse Practitioners (NP) | Specialties |
|---|---|
| • Acute Care NP | • Ambulatory Care Nursing |
| • Adult NP | • Cardiac Rehabilitation Nursing |
| • Adult Psychiatric & Mental Health NP | • Cardiac Vascular Nursing |
| • Diabetes Management—Advanced | • Case Management Nursing |
| • Family NP | • College Health Nursing |
| • Family Psych & Mental Health NP | • Community Health Nursing |
| • Gerontological NP | • General Nursing Practice |
| • Pediatric NP | • Gerontological Nursing |
| • School NP | • High-Risk Perinatal Nursing |
| **Clinical Nurse Specialists** | • Home Health Nursing |
| • Adult Health CNS | • Informatics Nursing |
| • Adult Psychiatric & Mental Health CNS | • Medical-Surgical Nursing |
| • Child Adolescent Psych & Mental Health CNS | • Nurse Executive |
| • CNS Core Exam | • Nursing Professional Development |
| • Diabetes Management—Advanced | • Pain Management |
| • Gerontological CNS | • Pediatric Nursing |
| • Home Health CNS | • Perinatal Nursing |
| • Pediatric CNS | • Psychiatric & Mental Health Nursing |
| • Public/Community Health CNS | • School Nursing |
| **Other Advanced-Level** | |
| • Diabetes Management—Advanced | |
| • Forensic Nursing—Advanced | |
| • Nurse Executive—Advanced | |
| • Public Health Nursing—Advanced | |

**Source:** *American Nurses Credentialing Center (2011). ANCC Nurse Certification.* Retrieved April 18, 2011 from http://www.nursecredentialing.org/CERTIFICATION.ASPX.

areas of the world, work with different cultures, and learn other interventions. Many of the traveling nurse companies advertise in nursing journals as well as over the Internet.

# FLIGHT NURSING

In numerous tertiary care centers, nurses are functioning in the role of flight nurse for both helicopter and fixed-wing transports. Over the years, numerous television shows and movies have portrayed these nurses, in their jumpsuits and helmets, landing on the helipad on the top of the hospital. Flight nursing actually started in 1933 with the emergency flight corps of the armed services and was present in both the Korean and Vietnam wars. The concept of an air ambulance was initiated in Denver in 1972. One needs numerous advanced technical skills to practice flight nursing, such as patient intubation, EKG interpretation, intravenous (IV) and chest tube insertion, medication administration, sedation, and central line placement.

The vast majority of flight teams consist of an RN and a respiratory therapist. Although team members follow established health care protocols, they still make many independent decisions regarding crisis intervention. These nurses are truly functioning in an expanded role, providing care to infants, children, and adult patients while performing a variety of therapeutic interventions. Flight nurses also provide education to outlying communities and volunteer in emergency situations.

Although there are no true national standards for becoming a flight nurse, most of the opportunities available nationally require the following:

- Two to three years critical care experience
- Advanced Cardiac Life Support certificate (ACLS)
- Pediatric Advanced Life Support certificate (PALS)
- Neonatal Resuscitation Program (NRP)
- Graduation from a nationally recognized trauma program, such as Prehospital Trauma Life Support (PHTLS), Basic Trauma Life Support (BTLS), Trauma Nurse Core Course (TNCC), and/or Transport Nurse Advanced Trauma Course (TNATC).

Certifications such as Critical Care Registered Nurse (CCRN), Certified Emergency Nurse (CEN), or Certified Flight Registered Nurse (CFRN) may also be required.

Most of these courses and certificates are nationally known and offered by a variety of providers. The certification exams are offered and given by their governing bodies, and most exams are offered online, for example, by AACN.

If you are interested in flight nursing, it is important to note that a broad-based background is immensely helpful for this position. Large, busy medical centers will afford you a broad range of experiences, such as adult and pediatric critical care, high-risk obstetrics, emergency department care, and so on.

Many flight nurses have a baccalaureate degree (BS) and many of the certifications listed previously. Some flight nurses may have additional critical care certifications such as those of critical care registered nurse (CCRN) or certified emergency nurse (CEN). In addition to the advanced certifications, education, and clinical skills, nurses qualifying for flight nursing need a strong critical care/emergency room background with demonstrated clinical skills in a variety of areas. An extensive training program is also provided by the hospital. These training programs include clinical rotations, ambulance and flight observation/ride time, and classroom training. Upon becoming a flight nurse, it is essential that nurses maintain continuing education and certification credits and attend courses offered by agencies such as the U.S. Department of Transportation, the National Flight Nurses Association, and the Emergency Nurses Association, in addition to ASTNA (Air and Surface Transport Nurse Association).

## HEALTH CARE SALES/ PHARMACEUTICAL REPRESENTATIVES

Some nurses have left clinical staff nursing and have become representatives for companies that work in conjunction with traditional health care institutions. These include companies involved in pharmaceutical sales, durable

**REAL** WORLD INTERVIEW

Becoming a flight nurse after being a pediatric intensive care nurse for seven and a half years was a huge role change, especially because of the autonomy I now have. I am responsible for performing an assessment on a patient and then delivering care based on my findings. The care a flight nurse delivers is based on diagnosis-specific standards of care. Many times the patient diagnosis is known, such as when a patient is being transferred from a hospital, but it is the flight nurse's responsibility to ensure that the previous care initiated is still appropriate for the patient. The physical environment is also very different from that of my previous role. Many of the patients I encounter are located in a very small community hospital with limited resources. When I have a flight mission at an accident scene, the patient may be in an ambulance, trapped in a vehicle, on the ground, or in their house. Once the patient is initially assessed and stabilized, they are moved to the confined and noisy environment of the helicopter. I think that the biggest change from staff nurse to flight nurse is the limited resources available while on a call. Although physician consult is only a radio or phone call away, I am expected to make autonomous clinical decisions in order to save the patient valuable time in receiving lifesaving therapies. At times, I must rely solely on my own experience and that of my partner.

**Allison Goodell, BSN, RN, NREMT-P**
Flight Nurse
Albany, New York

medical equipment, home care, and insurance coverage, as well as health maintenance organizations (HMOs). These opportunities afford the nurse a perspective into corporate America and the workings of the world of business. There are frequently many salary enhancements, or perks, given with these positions, including having a company car, going on business trips, attending national and regional meetings and conferences, participating in profit sharing and stock option plans, and meeting a variety of health care personnel. It can be difficult, however, especially with pharmaceutical sales, to meet sales quotas, provide for your customers, and survive the ups and downs of the business world.

RNs are usually seen as desirable employees for these positions because of their knowledge of pharmacology, technology, and health care systems. They understand how items are supposed to work and are excellent problem solvers when events are not going well.

## CASE MANAGER

The Case Management Society of America defines **case management** as a "collaborative process of assessment, planning, facilitation and advocacy for options and services to meet an individual's health needs through communication and the use of available resources to promote quality, cost-effective outcomes" (Case Management Society of America, 2011). Case managers "actively participate with their clients to identify and facilitate options and services, providing and coordinating comprehensive care to meet patient/client health needs, with the goal of decreasing fragmentation and duplication of care, and enhancing quality, cost-effective clinical outcomes" (McCullough, 2009, p.124).

The Case Management Society of America indicates that case management serves as a means for achieving client wellness and autonomy through advocacy, communication, education, identification of service resources, and service facilitation. The case manager helps identify appropriate providers and facilities throughout the continuum of services while ensuring that available resources are being used in a timely and cost-effective manner in order to obtain optimum value for both the client and the reimbursement source (Case Management Society of America, 2011). In accomplishing these goals, case management integrates both patient and provider satisfaction, taking into consideration cost factors. The hope is that this role will optimize the patient's self-care and decrease length of stay while also decreasing fragmentation of care by ensuring seamless and coordinated care. The case management specialty of health care was created, in large part, because of the influence of managed care, third-payer payers, and reimbursement systems (Thomas, 2009). Care coordination, access to and delivery of care, resource use, and financial reimbursement are key aspects of the role fulfilled by case managers to organize care and reduce costs. In the mid-1980s, when hospitals began to feel the financial impact of a changing reimbursement system, internal types of case management were developed to ensure that hospitals were appropriately reimbursed for their services (Yamamoto & Lucey, 2005). During the 1990s and into the twenty-first century, the growth of managed care has prompted the next level of case managers, in which not only discharge planning but also utilization review is incorporated into the role. In performing the role, the nurse case manager should have expert clinical skills and knowledge of the health care system, health care finances, and legal issues, as well as be an effective

## REAL WORLD INTERVIEW

Nurse **case management** is no longer just discharge planning. It is a nursing practice field that encompasses utilization review, quality management, and the coordination of timely discharges. It is a collaborative profession that works closely with multiple disciplines including social work, nursing, physicians, risk management, and quality management. It also works closely with patients and their families, community-based vendors, health care facilities, and insurance companies. Nurse case managers (NCM) utilize standardized, nationally accepted criteria as one tool to review patient charts following a patient's hospitalization, for appropriateness of admission; documentation of diagnoses, procedures, and treatments; and readiness for discharge. The NCM follows a caseload of patients, contributing to each patients efficient "flow" through their hospitalization and promoting optimal patient outcomes and safe and timely discharges. Nurse case management has proven itself to be a valuable nursing field in which the case managment nurse acts as a patient advocate by contributing towards optimal patient outcomes in a fiscally complex and changing environment.

**Katherine L. Jones, MS RN**
Nurse Case Manager
Albany, New York

communicator. Within the case management model, the nurse case manager will utilize various tools—such as practice guidelines, critical pathways, variance analysis, protocols, and outcome measurement tools—to achieve quality and cost-effective outcomes. Nurse case managers must provide care that focuses on outcome achievement and assist in arranging, coordinating, and monitoring patient care services. Patient goals include improving access to health care services, providing services that meet the patient's needs, and facilitating support for informal caregivers. System goals include coordinating service delivery systems so that services are accessed more easily, preventing unnecessary use of services and containing costs.

## PUBLIC HEALTH NURSING

Public health nurses integrate community involvement and knowledge about the entire population with personal, clinical understandings of the health and illness experiences of individuals and families within the population (American Public Health Association, 2011a). According to the American Public Health Association, "the primary focus of public health nursing is improving the health of the community as a whole rather than just that of an individual or family" (American Public Health Association, 2011b).

The work of a public health nurse is essentially primary prevention; i.e., preventing disease, injury, disability, and premature death. Public health nurses work collaboratively as a team with other public health professionals such as health educators, epidemiologists, physicians, advanced practice nurses, nutritionists, etc.

To become a public health nurse usually requires a Bachelor Degree in Nursing from an accredited four-year college. Some communities will employ public health nurses with an Associate Degree in Nursing. Public health nurses are found in a variety of settings, including schools and the workplace, the latter called occupational health nurses. Public health nurses also often work for government agencies, non-profit groups, and community health centers.

## NURSE ENTREPRENEUR

Many nurses are leaving the bedside for the world of entrepreneurship in a variety of consultative, educational, and technical areas. With this risk-taking move, nurses individuals quickly learn that success is based on high-quality work, patient satisfaction, and establishing and building effective relationships.

The concept of entrepreneurship is not a new one. The term is an interpretation of a French word that means "to undertake" (Simpson, 1998). The *Merriam-Webster Collegiate Dictionary* describes an entrepreneur as "one who organizes, manages, and assumes the risks of a business or enterprise" (*Merriam-Webster Collegiate Dictionary*, 2011). Nurses have always been independent thinkers and somewhat entrepreneurial. At the start of the twentieth century, many nurses functioned independently and contracted directly with the patient or family to provide care.

Bedside hospital nursing, as we know it today, was in its infancy, and it was not until the 1930s that nurses moved into the hospital setting and became employees. Nurse entrepreneurs plan, organize, finance, and operate their own businesses (Leong, 2005). Some of the major characteristics and attributes of nurse entrepreneurs include the following:

- Are visionary, self-motivated, and risk-takers
- Have common sense
- Are good decision makers and problem solvers
- Are self-confident, assertive, autonomous, and creative
- Are responsive to a perceived need
- Are market driven, with good financial foresight
- Recognize the possibility of success as well as the possibility of failure (Leong, 2005; Faugier, 2005)

For Web sites that provide information on nurse entrepreneurship, see Table 28-2.

As with any job, there are benefits and drawbacks to becoming a nurse entrepreneur. The benefits include job satisfaction, flexibility in choosing opportunities, and being able to do what you want to do. Some of the downsides of entrepreneurship include enduring tough competition, riding the highs and lows of the market, finding the right product or service to sell, and providing for your own health insurance. It is important to decide what type of product or service you want to provide and to develop a solid business plan. Expect to use your personal savings to cover initial start-up expenses, and plan to develop marketing strategies for spreading the word about your business.

Table 28-3 and Table 28-4 provide additional information about establishing a business plan. As Manthey indicated in 1999, "The process of deciding how to package my experience in such a way as to sell it was one of my most important learning experiences as a consultant" (p. 82).

## WOUND, OSTOMY, CONTINENCE NURSE SPECIALIST

The field of enterostomal therapy was initiated in 1958 at the Cleveland Clinic, with the first enterostomal therapists (ETs) being non-nurses. The first nursing training program was started in 1961, and in 1972 new standards for the schools of enterostomal therapy were established. The year 1976 marked a significant change when the governing body of the International Association of Enterostomal Therapists determined that only RNs would be admitted to enterostomal therapy educational programs. After this, the scope of practice was expanded beyond just caring for ostomies to include skin care, management of draining wounds and fistulas, pressure sores, and incontinence. At this point, the entry requirements into an enterostomal therapy nursing education program (ETNEP) changed to a bachelor's degree with a major in nursing.

Nurses with this training and education practice both in hospitals and community-based settings such as visiting nurse associations, public health, nursing homes, and long-term care facilities. Over the past 20 years, these specialists have truly become the clinical experts in managing patients with ostomies, alterations in skin integrity, and wounds.

## Table 28-2  Web Sites with Information and Guidance on Nurse Entrepreneurship

- www.nurse-entrepreneur-network.com

- www.nurse.com

- www.nnba.net (National Nurses in Business Association)

- www.nursingentrepreneurs.com (nurse entrepreneur)

- www.nursebiz.com (travel nursing)

- www.urmc.rochester.edu (University of Rochester Center for Nursing Entrepreneurship)

- www.independentrncontractor.com

- www.sba.gov

- www.bbb.org (Better Business Bureau of USA and Canada)

## TABLE 28-3  Process of Establishing a Business Plan

| NURSING PROCESS | BUSINESS PROCESS |
| --- | --- |
| Assess | • Develop an idea/concept: short term and long term.<br>• Perform a market survey and feasibility study: determine consumer, clientele, location, and business forecast. (Check Web sites in Table 28-2 for information.)<br>• Identify resources available: financial, technology, and business support based on your designed product or service to market. |
| Plan | • Develop market strategies and financial plan based on market survey and feasibility studies.<br>• Develop product information: literature, brochures, pamphlets.<br>• Develop advertising/public relation methods and material.<br>• Schedule appropriate time to deliver product information and services. |
| Implement | • Implement business concepts: direct and indirect methods with follow-up (mailings, telephone, Internet).<br>• Perform services/deliver products or service. |
| Evaluate | • Perform periodic assessment of business plan: monthly, biannually, and annually.<br>• Identify strengths and weaknesses, and implement changes. |

## TABLE 28-4  Elements of a Business Plan

| ELEMENT | EXPLANATION |
| --- | --- |
| Resources | • Financial: required capital, personal savings, loans, investors<br>• Technology: computer with encryption capabilities, phone, fax, cell phone, beeper, car, credit card provider (usually coordinated through a bank)<br>• Business support services: Better Business Bureau, Chamber of Commerce, personal contacts, accountant/financial planner |
| Expenses | • Labor: self, employees, benefits, wages, health care, retirement<br>• Supplies: office supplies (paper, envelopes, stamps, business cards, and so on), technology (computer, phone, fax, car, telephone, and so on)<br>• Fees: professional services, equipment repair/maintenance, purchased services and/or products<br>• General and administrative: utilities, leases/rentals/mortgages, phone services, taxes, depreciation, continuing education/tuition, travel, health insurance, post office box |
| Revenue | • Direct result of services and/or products provided and sold<br>• Fair market value |

## REAL WORLD INTERVIEW

The wound, ostomy, and continence nurse (WOCN) specialty has grown tremendously over the past 50 years. It was started as a lay specialty by those affected by ostomy surgery. WOCN is a complex, interdisciplinary, evidence-based area of specialty nursing practice that has evolved into the care of the person with wound, ostomy, and/or continence needs. The WOCN functions in an independent nursing role as a specialist who can care for people across the life span and a continuum of care settings. WOCN's practice areas include clinical practice, advocacy, consultation, education, research, and administration. Many WOCN hold positions that encompass all of these areas. WOCN is a specialty area of practice that is stimulating, is challenging, and allows the nurse to have an impact on both the individual and the health care system.

**Jody Scardillo, MS RN ANP-BC, CWOCN**
Clinical Nurse Specialist
Albany, New York

# CLINICAL RESEARCH NURSE (CRN)

This is a relatively new, emerging role and one that allows the nurse a unique opportunity to offer novel treatment approaches to patients who may have otherwise reached a dead end in their treatment course (Posten & Buescher, 2010). In this role, the CRN coordinates the day-to-day management of a research trial, though the primary research investigator still has ultimate responsibility for all study activities. The CRN is employed by the health care site, which may be a hospital, university setting, private health care practice, etc.

# NURSING INFORMATICS (NI)

The term **nursing informatics** first appeared in the literature in the 1980's, though nurses have contributed to the purchase, design, and implementation of information technology well before then. Perhaps one of the first definitions of the nursing informatics role was that it is a ". . . combination of nursing, information, and computer sciences to manage and process data into information and knowledge for use in nursing practice" (Graves & Corcoran, 1989 in Murphy, 2010). The early pioneers in NI most likely got into it accidentally.

Numerous definitions of nursing informatics have been developed by the American Nurses Association over the years, but a recent definition (2008) by them states that NI is a specialty that integrates nursing science, computer science, and information science to manage and communicate data, information, knowledge, and wisdom in nursing practice.

In 2009, the Nursing Informatics Special Interest Group of the International Medical Informatics Association fine-tuned the NI definition by stating that "Nursing Informatics science and practice integrates nursing, its information and knowledge and their management, with information and communication technologies to promote the health of people, families and communities world wide (Murphy, 2010).

# LEGAL NURSE CONSULTANT

Over the past two decades, our society has become inundated with commercials, newspapers advertisements, and Web-based pop-ups regarding individual or class action lawsuits. Litigation is increasing, along with medical malpractice lawsuits. One response to this has been the creation and development of the legal nurse consultant, especially by Vickie Milazzo, MSN, RN, JD (Vickie Milazzo Institute, n.d.). She advanced this nursing role in the 1980s and 1990s and established a certification process. The legal nurse consultant is an individual who utilizes knowledge and clinical expertise to provide services such as expert opinions or advice on health, illness, and injury-related issues (Elliott & Larson, 2010). According to the American Association of Legal Nurse Consultants (AALNC), a legal nurse consultant evaluates, analyzes, and provides informed opinions on health care delivery and the resulting outcomes by identifying applicable standards of care, conducting literature reviews, interpreting medical records, attending depositions and trials, and/or developing written reports (America Association of Legal Nurse Consultants, 2011).

Nurses choosing this type of nursing must be currently licensed, have credentials that equal or exceed the credentials of the nurse defendant, and not have a professional or private relationship with anyone involved in the legal case. There are essentially two types of legal nurse consultants: testifying legal nurse consultants who are expected to testify at depositions and/or trials, and consulting legal nurse consultants who do not testify. Both types identify current standards of care and any deviations from them as well as review, interpret, and analyze medical records.

# ADVANCED PRACTICE NURSING

The concept of advanced practice nursing originated in the mid- to late nineteenth century, with the creation of the role of nurse anesthetist. The concept further developed in the twentieth century with nurse-midwifery (1925 with the Frontier Nursing Service in rural Kentucky), clinical nurse specialists (1955 at Rutgers University), and finally, nurse practitioners (1965 at University of Colorado with Loretta Ford).

Why is there a need for advanced practice nurses (APNs)? The last quarter of the twentieth century taught that detection, prevention, promotion, early intervention, and education are not only cost effective but also rational. APNs are ideally suited to deliver this type of health care. Many Americans have been disenfranchised from advances in the health sciences, namely the poor, minority groups, the uninsured or underinsured, and individuals suffering from chronic poor health. This segment of our society is especially vulnerable and one that APNs can certainly assist.

Definitions of advanced practice nursing have been presented since the early 1990s. The AACN has provided a position statement that describes the APN as "an umbrella term appropriate for a licensed registered nurse prepared at the graduate degree level . . . with specialized knowledge and skills that are applied within a broad range of patient populations in a variety of practice settings" (McCabe & Burman, 2006; Walden & Wright, 2005). Numerous states allow independent practice for NPs, while the remaining states require that NPs obtain a collaborating agreement with a physician. Nurse practitioners are first licensed registered nurses who have obtained specialized and advanced education. In time, they become licensed nurse practitioners who provide primary and/or specialty nursing care in a variety of ambulatory, acute, and long-term care settings. Masters, post-masters, and/or doctoral preparation is required for entry-level NP practice (American Academy of Nurse Practitioners, 2011). Advanced practice nursing builds on the foundation of professional nursing practice and responds to the health care needs of the country. Nurses in these roles are expected to be leaders in the professional practice of nursing (Koerksen, 2010). The future for APNs will be built on their ability to be politically astute and savvy and take an active role in their own destiny, giving evidence of and demonstrating their worth.

This evidence can be provided through research studies as well as through the APN's day-to-day contributions. Are the APN's contributions unique and valuable, and can this be shown to others? Consider the following:

- *Certified registered nurse anesthetist (CRNA)*: In evaluating anesthesia services in a chronic low-back pain clinic, the CRNA should clearly document quality of service and patient outcomes.
- *Certified nurse-midwife (CNM)*: The CNM's ability to better meet patient needs, or to provide services to groups of patients at a lower cost than those provided by practitioners, should be measureable.
- *Clinical nurse specialist (CNS)*: In a hospital setting, the CNS must be able to identify how performance contributes to the patient-focused mission and goals of the organization. Does the CNS's practice reduce length of stay, improve patient outcomes, or enhance the efficiency of staff nurses?
- *Nurse practitioner (NP)*: In both inpatient and outpatient settings, the NP needs to document both the quantity and quality of services provided to patients and the NP's ability to reduce hospitalization rates.

## CERTIFIED REGISTERED NURSE ANESTHETIST (CRNA)

The first recorded nurse administering anesthesia was Sister Mary Bernard, a Catholic nun, in 1877. Alice Magaw, however, is considered the mother of anesthesia for her outstanding contributions to the field. She was also the first nurse anesthetist to publish articles and perform research on the practice of anesthesia. Between 1909 and 1914, four formal educational programs for nurse anesthetists were established. Currently, there are more than 90 approved programs, a list of which may be viewed at www.aana.com (American Association of Nurse Anesthetists).

The **certified registered nurse anesthetist** (CRNA) is an APN specialty requiring the graduate to obtain a master's degree. This individual takes care of the patient's anesthesia needs before, during, and after surgery or other procedures alone or in conjunction with other health care professionals. Today, CRNAs enjoy a high degree of autonomy and professional respect. CRNAs provide anesthetics to patients in every practice setting and for every type of surgery or procedure. They are the sole anesthesia providers in two-thirds of all rural hospitals, as well as the main provider of anesthesia to expectant mothers and to men and women serving in the U.S. Armed Forces. CRNA programs require a bachelor's of science in nursing as well as at least one year of acute care nursing for entry into a program.

## CERTIFIED NURSE-MIDWIFE (CNM)

Midwifery has existed from the beginning of humankind, and throughout history has played a vital and integral role in a community's growth. In the 1920s, Mary Breckenridge (a British midwife), along with other British midwives, worked with the Frontier Nursing Services in Kentucky to provide services for the rural population. The first American midwifery program was established in 1932, with the curriculum adapted and modified from the British model. In 1955, the American College of Nurse-Midwives (ACNM) was formed, with its main goals focused on setting standards for practice and education.

Currently, there are numerous midwifery educational programs. Since 1980, CNMs have been allowed to practice in a variety of settings, including hospitals, homes, and birthing centers, providing care for women throughout the childbearing cycle as well as postpartum. Nurse-midwives and collaborating practitioners agree upon the protocols and procedures. Health education, including the teaching of self-care skills and preparation for childbirth and child rearing, comprises a large part of clinical activities in addition to the actual delivering of babies. The American Academy of Nurse-Midwives asserts that the CNM can provide "services to women and their babies in the areas of prenatal care, labor and delivery management, postpartum care, normal newborn care, well-women gynecology, and family planning (American Academy of Nurse-Midwives, 2011).

## CLINICAL NURSE SPECIALIST (CNS)

By the turn of the twentieth century, the greatest percentage of APNs were the CNSs, whose origins can be traced back to 1938. Reiter first coined the term *nurse clinician* in 1943 to designate a nurse with advanced clinical competence and recommended that such clinicians get their preparation in graduate nursing education programs. Educators first developed the clinical nurse specialist (CNS) role because of their concern for improving nursing care. They believed that nursing care improvement was dependent upon increasing expertise at the bedside, giving both direct and indirect care, and incorporating role modeling and consultation. The CNS role has evolved to include many specialties. The first CNS master's program in psychiatry was started by Peplau in 1955 at Rutgers University.

The impressive development of the psychiatric CNS role helped to initiate the other CNS specialty programs. Following the passage and enactment of the Nurse Training Act in 1965, clinical specialization in graduate education increased tremendously. Graduates would provide a high level of specialized nursing care, as well as serve as change agents in hospital settings.

The CNS is primarily a hospital-based APN, serving as a clinical expert in evidence-based nursing practice within a specialty area. In this capacity, the CNS uses clinical expertise to influence patients, nurses, and nursing practice, as well as the organization/system, with a focus on providing quality, cost-effective care (Darmody, 2005). The CNS role was developed with the intent to enhance overall quality of patient care. Initially, there were three spheres of influence that described CNS practice (patient, nurse, and organization). The next step was the historic development of five separate CNS role components of expert clinical practice, consultation, education, clinical leadership, and research (University of Rochester, Strong Memorial Hospital, 2011). These CNS role components are often used to provide a template for CNS practice (Mayo et al., 2010). A good portion of time is spent in the hospital, in both staff and patient/family education, as well as in developing protocols, standards, and pathways that will guide nursing practice. Within the role, the CNS focus can be broad, encompassing adult, pediatric, or obstetric patients, or narrow, including areas such as oncology, the cardiopulmonary system, or the pulmonary system. It may also focus on patient care problems, e.g., wounds, or a type of unit, e.g., intensive care unit (Wetzel & Kalman, 2009). The specific activities of the CNS are dynamic as they respond to specific organizational needs and challenges. A study by Wyers, Grove, and Pastorino in 1985 listed essential CNS competencies that are still valid today:

- Developing an in-depth knowledge base
- Demonstrating clinical expertise in a selected area of clinical practice
- Serving as a role model
- Serving as a practitioner/teacher, consultant, researcher

"When thinking about Graduate Education, consider the CNS role if your goal is to impact practice and care of patients beyond a one-on-one encounter" (Wetzel & Kalman, 2009, p. 30).

## NURSE PRACTITIONER (NP)

The last decade of the twentieth century witnessed a large increase in the number of nurse practitioner programs and graduates (Ford, 1997). This was driven partially by the changing health care system, hospital downsizing, an increase in ambulatory care, and constraints on managed care. In addition, the American Medical Association (AMA) called for a decrease in admissions to medical school. Simultaneously, the federal government practically eliminated the graduate medical education (GME) funds given to hospitals with residency programs. Many of these circumstances are not unlike those that existed when Loretta Ford, a doctorally prepared RN, and Henry Silver, a medical practitioner, started the first nurse practitioner (NP) training program at the University of Colorado in 1965 (Hoekelman, 1998).

The NP role has had a tremendous evolution from its humble beginnings. This graduate-level-educated RN is still primarily practicing within primary care; however, the role has expanded greatly. These NPs are practicing throughout the continuum of health care in a multitude of settings and patient populations. Within 10 years of Dr. Ford's first NP program, 65 other NP programs were created. The second NP program was started in 1966 at Duke, the birthplace of the first physician's assistant program. By 1999, NPs in all states could be directly reimbursed by Medicare and could write prescriptions (Ventura & Grandinetti, 1999).

Although NPs have become fully integrated into the clinical delivery system, they have often been legally and financially dependent on medical practitioners for their

## REAL WORLD INTERVIEW

As NPs face the new millennium, it is advisable to listen to the wisdom of the famous author on China, Pearl Buck, who said, "To understand today, you have to search yesterday." Further, to envision the future, think "outside the box" creatively, constructively, and globally. Unfortunately, most people hate change; so do professionals. By their very nature, professionals can become myopic, territorial, and conservative. Some that are very resistant to change become arrogant, self-important, and greedy. Nursing must face the future differently. Tomorrow's practitioners will face globalization, not only of economics but of every field of human endeavor. Demographics, technological advances, transportation, and communication will expand beyond imagination and at lightning speed. Health information will no longer belong exclusively to the health professions. The Internet will see to that. The challenge for NPs is to be proactive rather than reactive in creating a social, cultural, political, and physical environment in which to successfully live, work, and thrive as responsible members of the new society and as advocates for our patients and their families. So, thoroughly examine the past; keep the enduring human values of caring, compassion, and courage in nursing; listen to your best teachers, the patients; and create your own future accordingly.

**Loretta Ford, EdD, RN, FAAN**
Founder, Nurse Practitioner Program
Rochester, New York

jobs. While some states allow NPs to work independently, most states allow NPs to work collaboratively with a medical practitioner. This differs from the case of the physician assistant, who works under the direct supervision of the physician. NPs are competent in delivering primary care that is satisfactory, acceptable to patients, and cost effective. The keys to the success of the NP role have been the autonomous yet collaborative nature of the practice; accountability as a direct provider of health care services; emphasis on clinical decision making as a foundational clinical skill; the focus on health and healthy lifestyles as a foundation of practice; and the cost-effective, accessible nature of the practice. These basic attributes of NP practice hold true regardless of setting or specialty focus (Ford, 1997).

Although many NPs have practiced in primary nonhospital, nonacute care settings, the neonatal nurse practitioner (NNP) has been hospital based for many years. Recently, a new role, the acute care nurse practitioner (ACNP), has been created. These positions also have a collaborative rather than subordinate relationship with medical practitioners (Kleinpell, 2005). As with all NPs, the ACNP is blending nursing and medicine by taking a holistic patient management approach while using collaborative treatment protocols frequently involving procedures previously done only by medical practitioners, such as lumbar punctures, chest tube insertion, writing medication prescriptions, and so on.

Although the performance of such roles bodes well for nursing in general and NPs specifically, there is also some discussion about NPs taking on too much within the health care system; the concern is that "if nurses take on an increasing amount of technical and medical work, then characteristics highly valued in the nursing profession may be threatened. It is clear that while NPs provide autonomous practice and competent patient management, they also must protect their holistic, caring nursing role (Kleinpell, 2005).

## CLINICAL NURSE SPECIALIST/ NURSE PRACTITIONER (CNS/ NP): A COMBINED ROLE

Since the mid-1980s, there has been frequent discussion and published data on the controversial issue of merging the advanced practice CNS and NP roles. There is no doubt that both these APNs are key providers meeting the health care needs of many Americans. Historically, the major differences between the CNS and NP were the setting and focus of their practice. This has changed, and, in today's health care arena, CNS and NP roles have evolved in response to the changing needs of patients as well as the changing health care system. Undoubtedly, many aspects of the roles are similar, and both roles have served their patients and each other well. The CNS role has been credited with creating an advanced level of nursing with an eye toward theory-based practice. NP's have been credited with the movement of nursing beyond traditional roles and increasing the public's awareness of advanced practice nursing (Stacey et al., 2002). In 1990, the ANA Council of Clinical Nurse Specialists and the Council of Primary Health Care Nurse Practitioners merged, which sent a powerful message regarding their positions on the issue. Despite this, and while there are some dual-role educational programs and individuals functioning as both a CNS and NP, many education programs and clinical practice areas are separate.

The most significant differences between the CNS and NP roles today are that the CNS provides expert nursing care for a specialized client population, playing the leading role in the development of clinical guidelines and protocols, evidence-based care, and expert consultation to facilitate change, while the NP provides direct care in the treatment and management of health conditions (Livingstone, 2009).

Another method of differentiating between the two roles considers the CNS as being focused primarily on quality of care issues at a system level rather than focusing on issues as an individual practitioner and the NP as offering advanced practice nursing as a cost-effective alternative to physician care, especially in underserved populations (Cukr, 1996 in Elsom et al., 2006).

It might be said that the role of the APN is emerging with both depth and breadth. Both the NP and CNS possess the skills and knowledge that promote the application of these roles in a variety of settings, in those settings that currently exist and those yet to be created. Research skills and change agent skills allow these APNs to function at multiple and varied levels within the health care system. Through these roles, nursing can resume its mission to provide health care to all individuals and reshape health care with an eye toward wellness and prevention, as described originally in Nursing's Agenda for Health Care Reform (1992) and more recently in the national health promotion and disease prevention initiative available at www.healthypeople.gov, Healthy People 2010 (Stacey et al., 2020).

There are both advantages and disadvantages to merging the roles (Table 28-5). Some predict that eventually the two roles will merge as a result of supply and marketplace demands. Additional educational programs for each, however, could still be available for CNSs not desiring to spend the majority of their time in clinical practice, such as CNS/consultant or CNS/educator.

### TABLE 28-5 CNS/NP Merger: Advantages and Disadvantages

| ADVANTAGES | DISADVANTAGES |
| --- | --- |
| • Many similarities: education, clinical settings | • Although the CNS is hospital based, NPs who are primarily community based in ambulatory and primary care settings are increasingly being utilized within the hospital as patient care providers. |
| • Expanding and overlapping practice settings | • Legal issue of trying to include CNS in existing advanced practice legislation |
| • Increased power in numbers | • Increased length of graduate education programs |
| • Cost savings to universities and education programs | • Continued blurring and role confusion |
| • Increased marketability | |

## REAL WORLD INTERVIEW

Currently, I am functioning as a CNS/PNP and enterostomal therapist. When I made the decision to pursue graduate school in order to become an advanced practice nurse, I carefully examined many different schools of nursing and came to the decision that if I wanted to be a nurse practitioner, then I should go to the school with the individual that started the entire field. For me, that was the University of Rochester, where Dr. Loretta Ford had moved in the mid-1970s from Denver. The knowledge and skills I obtained throughout my graduate education—including clinical rotations, course work, and finally my master's thesis—were supportive of where I wanted to be in the years to come as I combined the two roles both educationally and clinically. I have also had the wonderful opportunity to practice what I preach in the pediatric clinical setting where I have been working since the 1980s.

**Stephen Jones, MS, RN, CPNP, ET**
Averill Park, New York

## CASE STUDY 28-1

It is March of your senior year at college. You have been offered a position as a staff nurse at a hospital where you really wanted to work. For the past several weeks during your last clinical rotation, however, you have had the opportunity to work with an APN. You start to think that perhaps this is what you would like to do, and you know of a few colleges that will take graduates right into their master's program. You begin to check educational opportunities.

Is this something you could do? Should do?

In coming to an answer, consider the following:

• Have the clinical rotations you had as a student prepared you for an advanced practice role?

• What resources are available to you that would provide information in guiding your decision?

• Would spending time working as a staff nurse better prepare you to be an APN?

## DIRECTION FOR THE FUTURE

In examining the evolution of nursing and the direction in which it is heading, the ANA believes that more effective utilization of RNs to provide primary health care services is part of the solution to the cost and accessibility problems in health care today.

The twenty-first century holds a great deal of promise, but there are many questions to be answered regarding the cost, accessibility, and quality of the health care system in the United States. Nursing has a wonderful opportunity to be a leader in the changing health care delivery system. Few other professions provide as many options. There is a world of emerging opportunities as you prepare to enter the workforce. Take your time, research the possibilities, organize your plan, and live your dream. As Sophia Palmer said in 1897 at the first convention of the American Society of Superintendents of Training Schools for Nursing, "Organization is the power of the day. Without it, nothing great is accomplished" (ANA, 2001).

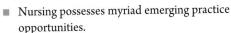

## KEY CONCEPTS

■ Nursing possesses myriad emerging practice opportunities.

■ Certification is a process readily available to any RN.

■ Case management is a growing subspecialty within the hospital setting.

■ Entrepreneurship and other nontraditional positions such as travel nursing have proven very successful for nurses.

■ Nursing practice has become increasingly specialized.

■ Numerous types of APNs are now practicing in both the hospital and community settings.

■ Although the CNS is primarily hospital based, NPs practice both within the hospital and in community-based settings.

■ For any RN, a plan to develop professional opportunities is essential.

 KEY TERMS

certified registered nurse anesthetist

case management

nursing informatics

# REVIEW QUESTIONS

1. Traveling nurses should do which of the following?
   A. Carry a notarized copy of their home state license.
   B. Become a Basic Life Support (BLS) instructor.
   C. Obtain state licenses for each state in which they will be practicing.
   D. Work for only one travel company.

2. The hospital-based case manager's primary focus is which of the following?
   A. Providing care and services that focus on outcome
   B. Making sure all the patient's expenses are taken care of
   C. Guiding medical and nursing protocols
   D. Ensuring that each patient has a primary nurse

3. To be a successful nurse entrepreneur, it is imperative that the individual nurse do which of the following?
   A. Obtain a sizable loan from a bank to help with start-up costs.
   B. Attain credit card approval.
   C. Understand how the stock market functions.
   D. Develop a solid business plan.

4. Which of the following is the best method of determining the effectiveness of an APN's practice?
   A. Patient satisfaction guide
   B. Fewer hospital admissions
   C. Improved patient outcomes
   D. Number of research studies

5. APNs provide both nursing and health care services. APNs can best fulfill this mission by doing which of the following?
   A. Performing and then publishing research
   B. Responding to patients' health care needs
   C. Maintaining as many certifications as possible
   D. Managing patients in both the hospital and outpatient settings

6. The CNS/NP role is multidimensional and provides a unique opportunity to practice in a variety of settings. Select all that apply to the CNS/NP role.
   _____ A. The CNS/NP role is mostly primary care based.
   _____ B. The CNS/NP role often has an unclear overlap in responsibilities.
   _____ C. Both roles are vested in utilizing research to advance care.

   _____ D. The CNS role is more geared toward systems management and enhancement.
   _____ E. The CNS/NP role requires ACLS.

7. What are some factors that may influence the development of clinical competence? Select all that apply.
   _____ A. Management preparation
   _____ B. Experience, motivation, and theoretical knowledge
   _____ C. Working within the intensive care setting
   _____ D. Opportunities, environment
   _____ E. Personal characteristics
   _____ F. Motivation and initiative

8. Which of the following is true about certification?
   A. Mandatory in most states
   B. Achieved through an examination that attests to the education and training of a registered nurse
   C. Usually covered when you apply for your state license
   D. One of the first processes you will complete after passing the NCLEX

9. Which of the following is correct concerning a public health nurse?
   A. You would interface with pharmaceutical representatives and companies routinely.
   B. You would work primarily with injured patients in a home environment.
   C. You would be required to work with ventilatory dependent patients.
   D. The work is essentially primary prevention, i.e., preventing disease, injury, disability, and premature death.

10. It is important that nursing not only encourages and welcomes new graduates in the field, but also implements strategies to retain nurses already employed. From research, it is known that the major predictor of intent to leave a nursing job is job dissatisfaction. A significant strategy that will enhance job satisfaction and the desire to continue in nursing is which of the following?
    A. Encourage certification at the earliest opportunity
    B. Maintain a 40 hour work week
    C. Work at several institutions within the first several years of employment
    D. Practice in a setting that creates and maintains a positive work environment

# REVIEW ACTIVITIES

1. Finally, you have graduated and moved to the city of your choice and are working at the health care facility of your choice. You are starting to apply all the knowledge and skills that you gained at school. You are around all levels and types of mentors and role models and are witnessing firsthand the activities of new and experienced staff nurses, as well as those of APNs.

What are some of your initial thoughts on where you will be in one, three, five, or ten years from now? Discuss how you will determine your progress.

2. Review categories such as travel nursing, job hunting, and licensure and certification at *Nursing's* (the journal) Web site, located at www.nursing2010.com. (This will change with the specific year this journal is being published). What did you see there?

3. Review nursing issues/programs, information, services, and certification at ANA's Web site (www.nursingworld.org). What did you see there?

4. Peruse and discover hot topics, various resources, and so on at Modern Healthcare's Web site (www.modernhealthcare.com). What did you see there?

5. For information on salaries, practice issues, and so on, log on to NLN's site (www.nln.org). What did you see there?

6. Review the RN salary survey at RN-123.com (2011). How will this information help you plan your future?

7. Review the surgeon's salaries in Scarborough JE, Bennett KM, Schroeder RA, Swedish TB, Jacobs DO, & Kuo PC. in the Annals of Surgery, September; 250(3):432–9. Note how surgeons are compensated.

# EXPLORING THE WEB

Which sites should you visit regularly, both for your personal and professional growth?

**General Interest and Nursing Issues:**

- Centers for Disease Control and Prevention:
  www.cdc.gov

- National Institutes of Health:
  www.nih.gov

- National League for Nursing:
  www.nln.org

- American Nurses Association (ANA):
  www.nursingworld.org

- ANA certification listing:
  www.nursingworld.org

- Discover Nursing:
  www.discovernursing.com

- Center for Nursing:
  www.nursingcenter.com

- National Council of State Boards of Nursing:
  www.ncsbn.org

- General nursing interest site:
  www.allnurses.com

- Health care information:
  www.medscape.com www.docguide.com

- General answer site
  www.answers.com

- Information on a wide variety of health careers.
  www.explorehealthcareers.org

**Specialty Issues:**

- American Association of Nurse Anesthetists:
  www.aana.com

- Flight nursing:
  www.flightweb.com

- Small Business Administration:
  www.sbaonline.sba.gov

- Service Corps of Retired Executives:
  www.score.org

- Traveling nurses:
  www.healthcareers-online.com

- American Academy of Nurse Practitioners:
  www.aanp.org

- Nurse Practitioner Central:
  www.npcentral.net

- American Association of College of Nursing (AACN):
  www.aacn.nche.edu

# REFERENCES

America Association of Legal Nurse Consultants (2011). What is a LNC. Retrieved April 19, 2011, from http://www.aalnc.org/about/Whatis.cfm.

American Academy of Nurse-Midwives (2011). Becoming a midwife. Retrieved April 19, 2011, from http://www.midwife.org/Become-a-Midwife RN-123.com (2011). RN's salaries. Retrieved April 19, 2011 from www.rn-123.com.

American Academy of Nurse Practitioners (2011). NP Scope of Practice. Retrieved April 19, 2011 from http://aanp.org/NR/rdonlyres/FCA07860-3DA1-46F9-80E6-E93A0972FB0D/0/2010ScopeOfPractice.pdf.

American Association of Colleges of Nursing (2011a). Nursing Faculty Shortage. Retrieved on November 5, 1020 from www.aacn.nche.edu/media/factsheets/facultyshortage.htm.

American Association of Colleges of Nursing (2011b). Nursing Shortage. Retrieved April 19, 2011 from http://www.aacn.nche.edu/Media/FactSheets/NursingShortage.htm.

American Association of Colleges of Nursing (2011c). Nursing Fact Sheet. Retrieved April 19, 2011 from http://www.aacn.nche.edu/Media/FactSheets/nursfact.htm.

American Board of Nursing Specialties (2009). AMERICAN BOARD OF NURSING SPECIALTIES. Retrieved April 25, 2011 from http://www.nursingcertification.org.

American College of Nurse-Midwives (2011). Retrieved April 19, 2011 from http://midwife.org/become_midwife.cfm.

American Nurses Association [ANA]. (2000). Press release. Retrieved February 25, 2000, from www.ana.org.

American Nurses Association [ANA]. (2001). Where we come from. Retrieved February 28, 2002, from http://www.nursingworld.org/centenn/index.htm.

American Nurses Association. [ANA]. (2008). Nursing informatics: Scope and standards of practice. Silver Springs, MD: Accessed April 20, 2011 at http://www.nursebooks.org.

American Nurses Credentialing Center (2011). ANCC Nurse Certification. Retrieved April 18, 2011 from http://www.nursecredentialing.org/CERTIFICATION.ASPX.

American Public Health Association (2011a). Public health nursing. Retrieved April 19, 2011 from http://www.apha.org/membergroups/sections/aphasections/phn/about.

American Public Health Association (2011b). Public health nursing. Retrieved April 19, 2011 from http://www.answers.com/topic/public-health-nursing, 2010.

Case Management Society of America (2011). What is a Case Manager. Retrieved April 19, 2011 from (www.cmsa.org/Home/CMSA/WhatisaCaseManager/tabid/224/Default.aspx).

Cukr, P. L. Viva la difference! The nation needs both types of advanced practice nurse: Clinical nurse specialists and nurse practitioners. *Online Journal of Issues in Nursing.* Accessed April 20, 2011 from http://www.nursingworld.org/ojin/tpc1/tpc1_4.htm.

Dias, M., et al. (2010, June). The consultation component of the clinical nurse specialist. Canadian Journal of Nursing Research, 42(2), 92–104.

Doerksen, K. (2010). What are the professional development and mentorship needs of advanced practice nurses? *Journal of Professional Nursing, 26*(3), 141–51.

Elango, B., et al. (2007). Barriers to nurse entrepreneurship: A study of the process model of entrepreneurship. Journal of the American Academy of Nurse Practitioners, 19, 198–203.

Elliott, R., & Larson, K. (2010). Legal nurse consultant: A role for nephrology nurses. *Nephrology Nursing Journal, 37*(3), 297–300.

Elsom, S. et al. (2006). The clinical nurse specialist and nurse practitioner roles: Room for both or take your pick? *Australian Journal of Advanced Nursing, 24*(2), 56–60.

Faugier, J. (2005). Developing a new generation of nurse entrepreneurs. *Nursing Standard, 19*(30), 49–53.

Ford, L. C. (1997). Advanced practice nursing. A deviant comes of age … the NP in acute care. *Heart & Lung: Journal of Acute & Critical Care, 26*(2), 87–91.

Graves, J., & Corcoran, S. (1989.) The study of nursing informatics. *Image: Journal of Nursing Scholarship, 21*(4), 227–231.

Hittle, K. (2010). Understanding certification, licensure and credentialing: A guide for the new nurse practitioner. *Journal of Pediatric Healthcare, 24*, 203–206.

Hoekelman, C. R. (1998). A program to increase health care for children: The pediatric nurse practitioner program by Henry K. Silver, MD, Loretta C. Ford, EdD, and Susan G. Stearly, MS, *Pediatrics, 102*(1 Pt. 2), 245–247.

Khomeiran, R., Yekta, Z. P., Kiger, A. M., & Ahmadi, F. (2006). Professional competence: Factors described by nurses as influencing their development. *International Nursing Review. 53*(1), 66–72.

Kleinpell, R. M. (2005). Acute care nurse practitioner: Results of a 5-year longitudinal study. *American Journal of Critical Care, 14*(3), 211–221.

Leong, S. (2005). Clinical nurse specialist entrepreneurship. *Internet Journal of Advanced Nursing Practice, 7*(1).

Livingstone, P. (2009). What is the difference between a CNS and the registered nurse practitioner. Newsbulletin, April/May. 11(2), 7.

Manthey, M. (1999). Financial management for entrepreneurs. *Nursing Administration Quarterly, 23*(4), 81–85.

Marrelli, T. (2006, January). Nursing in flux. *American Journal of Nursing, 106*(1), 19–20.

Mayo, A., et al. (2009). Clinical nurse specialist practice patterns. *Clinical Nurse Specialist, 24*(2), 60–69.

McCabe, S., & Burman, M. (2006). A tale of two APNs: Addressing blurred practice boundaries in APN practice. *Perspectives in Psychiatric Care, 42*(1), 3–12.

McCullough, L. (2009). The case manager: An essential link in quality care. Creative Nursing, 15(3), 124–126.

*Merriam-Webster Collegiate Dictionary* (2011). Retrieved April 19, 2011 from http://www.merriam-webster.com/dictionary/entrepreneur.

Murphy-Ende, K. (2002). Advanced practice nursing: reflections on the past, issues for the future. *Oncology Nursing Forums, 29*(1), 106–112.

Murphy, J. (2010). Nursing Informatics: The intersection of nursing, computer and information sciences. *Nursing Economic$, 28*(3), 204–207.

National League for Nursing (2011). Nurse Educator Shortage Fact sheet. Retrieved April 19, 2011 from http://www.nln.org/governmentaffairs/pdf/NurseFacultyShortage.pdf.

Poston, R., & Buecsher, C. (2010). The essential role of the clinical research nurse (CRN). *Urologic Nursing, 30*(1), 55–59.

Scarborough JE, Bennett KM, Schroeder RA, Swedish TB, Jacobs DO, Kuo PC. ( 2009). Will the clinicians support the researchers and teachers? Results of a salary satisfaction survey of 947 academic surgeons. Annals of Surgery. September; 250(3):432–9.

Schwarz, T. (2006). What is past is prologue? *American Journal of Nursing, 106*(1), 13.

Smith, T. (2009). A policy perspective on the entry into practice issue. *OJIN: The Online Journal of Issues in Nursing, 15*(1). Available at http://www.nursingworld.org/MainMenuCategories/ANAMarketplace/ANAPeriodicals/OJIN/TableofContents/Vol152010/No1Jan2010/Articles-Previous-Topic/Policy-and-Entry-into-Practice.aspx.

Smith, T. (2009). A policy perspective on the entry into practice issue. *OJIN: The Online Journal of Issues in Nursing, 15*(1). Retrieve April 20, 2011 from http://www.nursingworld.org/MainMenuCategories/ANAMarketplace/ANAPeriodicals/OJIN/TableofContents/Vol152010/No1Jan2010/Articles-Previous-Topic/Policy-and-Entry-into-Practice.aspx.

Stacey, R., All, A., & Gresham, D. (2002). Role preservation of the clinical nurse specialist and the nurse practitioner. *Internet Journal of Advanced Nursing Practice, 5*(2).

Stromberg, M. F., Niebuhn, B., Prevost, S., Fabrey, L., Muenzen, P., Spence, C., et al. (2005, May). More than a title. *Nursing Management, 36*(5), 36–46.

Thomas, P. (2009). Case management delivery models. *Journal of Nursing Administration, 39*(1), 30–37.

University of Rochester. Strong Memorial Hospital (2011). Nursing practice model. Retrieved April 19, 2011 from http://www.urmc.rochester.edu/strong-nursing/about/professional-practice-model.cfm

Vickie Milazzo Institute. (n.d.). retrieved April 20, 2011 at http://www.legalnurse.com/.

Walden, M., & Wright, K. (2005). Advanced practice registered nurse update: Issues of evolving complexity . . . stay informed (guest editorial). *Journal of Wound, Ostomy & Continence Nursing, 32*(1), 1–2.

Watts, M. (2010). Certification and clinical ladder as the impetus for professional development. *Critical Care Nursing Quarterly, 33*(1), 52–59.

Wetzel, C., & Kalman, M. (2010). Critical care clinical nurse specialist. Is this the role for you? *Dimensions of Critical Care Nursing, 29*(1), 29–32.

Woodburn, J. D., Smith, K. L., & Nelson, G. D. (2007). Quality of care in the retail health care setting using national clinical guidelines for acute pharyngitis. *American Journal of Medical Quality, 22*(6), 457–62.

Wyers, M. E., Grove, S. K., & Pastorino, C. (1985). Clinical nurse specialist: In search of the right role. *Nursing Health Care, 6*(4), 202–207.

Yamamoto, L., & Lucey, C. (2005). Case management "within the walls." A glimpse into the future. *Critical Care Nursing Quarterly, 28*(2), 162–178.

# SUGGESTED READINGS

Adams, M. H., & Crow, C. S. (2005). Development of a nurse case management service: A proposed business plan for rural hospitals. *Lippincott's Case Manager, 19*(30), 148–158.

*American Journal of Nursing.* (2006, January supp.). Your guide to certification. *106,* 50–63.

Cummings, G., & McLennan, M. (2005). Advanced practice nursing: Leadership to effect policy change. *Journal of Nursing Administration, 35*(2), 61–69.

Dias, M., CHAMBERS-EVANS, J AND REIDY, M. (2010, June). The consultation component of the clinical nurse specialist. Canadian Journal of Nursing Research, 42(2), 92–104.

Elango, B., Phil, M; Hunter, G; Winchell, M. (2007). Barriers to nurse entrepreneurship: A study of the process model of entrepreneurship. Journal of the American Academy of Nurse Practitioners, 19, 198–203.

Ford, L. C. (1995). Nurse practitioners: Myths and misconceptions. *The Pulse, 32*(4), 9–10.

Fulton, J. S. (2005). Calling blended role programs to account. *Clinical Nurse Specialist, 19*(5), 221–222.

Grainer, P., & Bolan, C. (2006). Perceptions of nursing as a career choice of students in the baccalaureate nursing program. *Nurse Education Today, 26*(1), 38–44.

Hader R. (2005). Salary survey 2005. *Nursing Management, 36*(7), 18–27.

Hudspeth, R. (2008). Understanding clinical nurse specialist regulation by the boards of nursing. *Clinical Nurse Specialist, 23*(5), 270–75.

Kleinpell, R. M., & Hravnak, M. M. (2005). Strategies for success in the acute care nurse practitioner role. *Critical Care Nursing Clinics of North America, 17*(2), 177–181.

Lofmark, A., Smide, B., & Wikbald, K. (2006). Competence of newly graduated nurses: A comparison of the perceptions of qualified nurses and students. *Journal of Advanced Nursing, 53*(6), 721–728.

Nelson, R. (2005). AJN reports: Is there a doctor in the house? A new vision for advanced practice nursing. *American Journal of Nursing, 105*(5), 28–29.

Pellico, L. (2006, January). We'll leave the light on for you. *American Journal of Nursing, 106*(1), 32–33.

Robbins, C. L., & Birmingham, J. (2005). Issues and interventions: The social worker and nurse roles in case management. *Lippincott's Case Management, 10*(3), 120–127.

Smolenski, M. C. (2005). Cedentialing, certification and competence: Issues for new and seasoned nurse practitioners (Fellows column). *Journal of the American Academy of Nurse Practitioners, 17*(6), 201–204.

Steefel, L. (2005). New doctoral degree aims to advance nursing practice. *Nursing Spectrum: New York & New Jersey edition, 17*(5), 14–15.

Szanton, S., et al. (2010). Taking charge of the challenge: Factors to consider in taking your first practitioner job. Journal of the American Academy of Nurse Practitioners, 22, 356–60.

Tuite, P., & George, E. (2010). The role of the clinical nurse specialist in facilitating evidence-based practice within a university setting. Critical Care Nursing Quarterly, 33(2), 117–125.

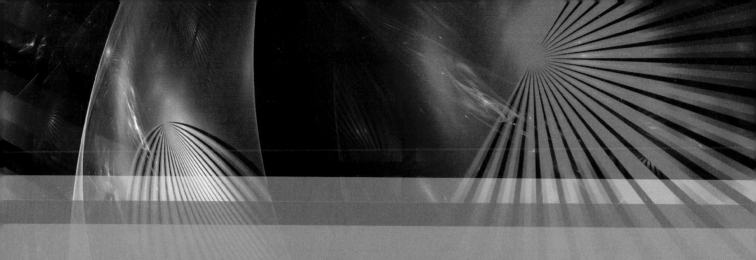

# CHAPTER 29

# Your First Job

**Lyn LaBarre, MS, RN; Miki Magnino-Rabig, PhD, RN;
Erin C. Soucy, MSN, RN**

*Practice nursing with love,
faith, and passion....*

(Miki Magnino-Rabig, 2006)

## OBJECTIVES

Upon completion of this chapter, the reader should be able to:

1. Identify key elements to consider in choosing a nursing position.
2. Compare and contrast typical components of health care orientation.
3. Explain types of performance feedback.
4. Compare and contrast organizational responses to performance.
5. Relate specific strategies to enhance the beginning nursing manager role.
6. Integrate material presented to develop mechanisms to enhance professional growth.

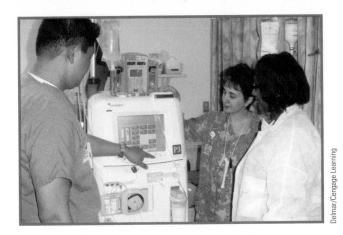

Delmar/Cengage Learning

*Congratulations! You have just completed your nursing educational requirements, and graduation is one week away. You have decided to stay in this geographic area and have received three job offers: a 12-hour night position in the surgical intensive care unit of a regional teaching hospital, a rotating shift 8-hour position on a community hospital's medical-surgical floor, and a public health nursing position with your county's health department.*

*Which position should you accept?*

*What factors will help you decide which is the best fit for you?*

Graduation brings the transition from the role of student to that of RN. A nurse's first job is an opportunity to solidify skills learned in school. It is also the time to establish relationships with mentors and to set down a foundation for future professional growth. This chapter will discuss important considerations regarding your first job.

# CHOOSING A POSITION

According to a recent study conducted by the Council on Physician and Nursing Supply, 30,000 additional nurses will be needed annually in the future, which is about a 30% increase over the number of graduates currently completing nursing school (2008). In the current job market, new nurses in some states are in the enviable position of having broad choices for their first job, while those in other states are having to relocate to find a position or must alter their expectations of where they will work. Some hospitals recruit new nurses to specialty areas such as obstetrics, psychiatry, and critical care, as well as to traditional medical-surgical units. In some states, some community health organizations may also be willing to hire recent graduates.

A year-long University HealthSystem Consortium (UHC) and American Association of Colleges of Nursing (AACN) Nurse Residency Program (NRP) has been developed by the UHC, an alliance of 107 academic medical centers and 232 of their affiliated hospitals (The University HealthSystem Consortium, 2007). The NRP has helped members achieve an astounding 4.4% turnover rate among first-year nurses. This figure compares with a national turnover rate of 27.1%, according to the PricewaterhouseCoopers Health Research Institute. The NRP programs emphasize critical thinking skills that promote patient safety. Hospitals are complex institutions providing a broad spectrum of services to the public they serve. New nurses are expected to have a great deal of knowledge and are challenged to apply that knowledge in real-life settings. The NRP offers nurse graduates the tools they need to become successful professionals. NRP is a year-long orientation and mentoring program that has proven beneficial to new graduates and to the hospitals employing them. Hospitals report less turnover, better job satisfaction, improved patient outcomes, and a decrease in the number of nurses who leave the profession.

## PATIENT TYPES

One of the most important considerations in selecting a job is choosing the best fit for you in a patient care environment. New nurses who start their career on a general medical or surgical unit typically manage patients with a variety of diagnoses. They learn diverse technical and assessment skills. These nurses develop a working knowledge of many common medications and patient teaching scenarios. In contrast, nurses who choose an entry position on a specialty unit focus on patients with specific diseases, body system disorders, or age groups. Specialty units that hire new graduates may include cardiology, obstetrics, or neonatology units. Nurses working in community health often have roles similar to those of medical and surgical nurses or may have more-specialized roles such as those of community health neonatal nurses.

## WORK ENVIRONMENT

Another facet to consider in choosing a first position is the opportunity to develop organizational skills. In a critical care or specialty area, nurses need to develop the ability to prioritize and plan care for limited numbers of patients who need highly specialized assessments and technical care. In contrast, the nurse whose first job is on a medical floor must organize care for a diverse and much larger group of patients. A new nurse in the community may work alone for most of the day, seeing patients one at a time. Even though each area requires specific skills, the organizational skills for all these nurses include effective time management.

Another consideration in weighing possible positions is the available schedule. Many health care organizations now offer a variety of schedules. Twelve-hour shifts are particularly popular, because a full-time nurse can work as few as three shifts per week. For the patient, this means fewer changes in nursing personnel within a day but less continuity throughout the hospital stay. For the new nurse, a 12-hour shift can be a long work day, but it allows increased flexibility in personal time. Some organizations offer rotations between daytime and other shifts. Others award the more popular day shifts by seniority.

When choosing a position, it is important to find out about the process for changing to a different schedule after hiring. Some hospitals restrict new nurses from changing positions for a set period of time. Following are some other questions to ask about your schedule: What will my weekend commitment be? How many holidays will I be expected to work each year? Does the health care organization use a self-scheduling system (in which nurses select their own schedule according to unit guidelines), or is time assigned? If time is assigned, how much notice will be required to have a certain day off?

Pay is an obvious element in choosing your first job, but the best-paying job offer is not necessarily the wisest choice, even from a financial point of view. Health care organizations in the same geographic area tend to offer competitive salaries at the start of employment. It is important to ask about an employer's salary policy. Does the hospital you are considering give a raise after you have passed your NCLEX, or would any potential raise be held until your first-year anniversary? Are nurses paid extra for having a BSN degree? What are the differentials, if any, for weekend work or working night or evening shifts? Be sure that nurses are paid for orientation shifts and required courses. When comparing offers between two health care organizations, ask how many hours are paid for a typical workweek. Some employers pay for a full 8- or 12-hour shift by allowing for a 30-minute overlap at change of shifts. Other organizations do not expect staff nurses to overlap, resulting in a shorter shift. Thus, at some facilities a typical pay week includes 40 hours, whereas at others, nurses routinely work 37.5 hours per week.

Finally, work environments in health care organizations can vary tremendously. Are you more comfortable in a smaller hospital setting in which it is relatively easy to find your way around and everyone knows each other? Or do you prefer the more complex, perhaps less personal setting of a large teaching facility? Do you enjoy working with resident practitioners in a teaching hospital, or will you be more satisfied interacting with community-based, private attending practitioners?

# ORIENTATION TO YOUR NEW JOB

Orientation fosters a smooth transition from graduate to practicing nurse. At its completion, a new nurse should be able to demonstrate competency in the basic skills needed for safe patient care.

# ORIENTATION CONSIDERATIONS

Different health care organizations also can have very different approaches to new employee orientation and education. Because orientation is a key component of the transition between being a student nurse and becoming a first-time manager of patients, it is important to establish what the organization offers during orientation. Consider the following questions:

- How long should I expect to be in orientation?
- Is it tailored to my learning needs, or is it the same for all incoming nurses?
- Does it all occur at the beginning of my new position, or will it be offered in stages?
- What ongoing education will be available to me?
- Will I be paid for time in education programs?
- In case of short staffing, will I be pulled from orientation?

Many new nurses feel pressured to find just the right setting for their first job, particularly if they have a long-term goal of working in a subspecialty. The focus in the first job needs to be on refining assessment and technical skills and learning to be organized in the delivery of nursing care. These skills, coupled with a positive work record regarding attendance, flexibility, and attitude, will ensure the new nurse of many future opportunities.

# GENERAL ORIENTATION

Many health care organizations divide nursing orientation into general sections and unit-specific sections. General orientation includes information and skills measurement, which all nurses new to the facility need, regardless of their eventual unit assignment. Two examples of information discussed at general orientation are validation of CPR competency and an introduction to policies regarding medication administration. General orientation also typically includes explanations of human resource policies and opportunities to hear from representatives of various departments within the organization. Recent concerns about patient safety have expanded orientation to include information about the Joint Commission (JC), formerly the Joint Commission on Accreditation of Health Care Organizations (JCAHO), and National Patient Safety Goals.

Some organizations offer sections of general orientation as written materials or on videotape. This allows a more flexible orientation schedule. It is particularly beneficial for new employees who are available to attend orientation only outside daytime hours. Figure 29-1 is a sample schedule for the first week of general orientation at one medical center.

Most facilities offer general orientation first, followed by unit-specific orientation. In this case, new nurses may not actually spend a shift on their unit for two weeks after

RN Orientation Template
Week One

| Monday Perdiems/Weekend Staff attend May 21 | Tuesday Perdiems/Weekend attend May 22 0745 meet in main lobby | Wednesday Perdiems/Weekend attend May 23 | Thursday Perdiems/Weekend attend May 24 | Friday May 25 |
|---|---|---|---|---|
| Human Resource/ Safety Education | 8:00–11:30 Intro/Tour Nursing at AMC Education Opportunities | 08:00–11:00 Documentation Standards of Care, protocols, I/O, graphic, Clinical Pathways Unit Day Prep *(ED exempt: modules)* | 08:00–09:30 Modules | 07:30–10:30 SMS Computer/P Building– if needed |
| *Remember to sign in on your unit if you want to get paid for the days you attend orientation.* | 11:30–12:00 Tina Raggio-Project Learn | | 09:30–11:00 Epidemiology *Carolyn Scott* | 10:30–12:00 Independent Activities on Unit of Hire or Modules |
| | 12:00–1:00 Lunch | 11:00–11:30 Restraints | 11:00–11:30 Lunch | 12:00–1:00 Meet with Director Main 4 Office Lunch Provided |
| | 1:00–2:00 Delegation/Assigning *Donna Harat* | 11:30–12:00 Back Video Nurse Scheduling | 11:30–2:30 SMS Computer Class | 1:00–2:00 Pastoral Care Room U477 |
| | 2:00–4:30 Modules | 12:00–1:00 Lunch | 3:00–4:00 Math Calculation Class (**optional-check with Educator**) | 2:00–4:00 Modules |
| | **some orientees may need to attend the SMS Computer class in the P building from 11:30–2:30–check with Educator** | 1:00–4:30 Skills Lab Afternoon Emergency Care and Mock Code (ACLS or PALS exempt and does not apply to NICU), IV/Phlebotomy Skills, Accucheck, PCA Pump | 4:00–4:30 Planning for next week/core orientation/orientee assessment forms | |

**Required Modules:**

| | | |
|---|---|---|
| Age Specific ☐ | IV Therapy ☐ | Blood and Blood Products ☐ | RN Medication ☐ | Patient Rights ☐ |
| Peds or Adult Emergency Care ☐ | Latex Allergy ☐ | Patient Classification ☐ | Patient Safety ☐ |
| Order Transcription (not for ED, PACU) ☐ | Pain Management ☐ | Documentation ☐ | CPR: (see handout) ☐ |

Dept of Education- #262-3705

**FIGURE 29-1** Registered professional nurse general orientation schedule template—week one. (*Source:* Courtesy of Albany Medical Center, Albany, NY.)

starting work. Other nurse educators plan orientation so that nurses go to their home unit very early, reserving some of the general content for later in the orientation schedule. The facility works to get information to the new graduates in a meaningful way.

Most organizations tailor their general orientation to individual learners. Thus, an experienced nurse may opt to challenge particular orientation classes by successfully completing the demonstration or written test or demonstrating competency in some other fashion.

## UNIT-SPECIFIC ORIENTATION

Unit-based orientation, whether it follows the general orientation or is interspersed throughout, focuses on the specific competencies a new nurse needs to care for the diagnoses and ages of patients typical to the assigned unit. These competencies include technical skills as well as beginning mastery of unit-specific processes. Some of the content covered may include topics such as what paperwork is necessary for new admissions and how to get an IV pump for medications.

Most organizations have developed unit-specific competency tools that list those skills orientees need to demonstrate. These lists provide a useful road map with which to plan a learner-specific orientation. Competency tools

are often used on an annual basis by all nurses as a mechanism to demonstrate competence with certain clinical skills. Figure 29-2 is an excerpt from an emergency department's unit-based orientation tool.

## Identifying Your Own Learning Needs

New graduates begin orientation with varied clinical experiences and competencies. Often, beginning nurses are asked to self-rate their level of knowledge or experience with various patient care skills. It is important for new nurses to identify their own learning needs and ask themselves: What skills do I need to develop to do a good job in this position? New graduates have much to learn, and orientation is the ideal time for the new nurse to observe coworkers and establish learning priorities. One way to do this is to ask questions of the preceptor or nurse educator. This provides feedback and molds the orientation to the learner's needs.

In addition, plan to do your own self-study to prepare for your new patients. Be sure to review the patient care for the top 10 nursing and medical diagnoses seen on the unit where you will be working. Review the common medications, lab tests, diagnostic procedures, and treatments done on your unit. This review will help prepare you to deliver quality patient care delivery and meet your responsibilities in a professional fashion.

---

### Critical Thinking 29-1

You are responsible for being the nurse you want to be. To do this, set your goals and monitor and evaluate them regularly. Gather data on the following indicators of being a professional nurse and add to the list:

- Monitor data so that I am up to date on evidence-based care for my patients.
- Monitor data that my patients are satisfied, pain free, and feel cared about.
- Monitor data that my patients are complication free and have no nurse-sensitive outcomes.
- Offer professional nursing service to my patients and my community.
- Give and receive professional respect to health care team.
- Speak up about the important role that nurses play in preventing patient complications.
- Network with other professionals.
- Participate in professional committees at work.
- Communicate assertively with the health care team.
- Receive professional salary and benefits.
- Take good care of myself and work for professional and personal balance.
- Continue my education, for example, certification, formal education, continuing education, and so on.
- Join my professional organization.
- Dress like a professional.
- Communicate pride in being a nurse.

| Name | Preceptor | Unit/Dept. | Emergency Dept. | Date: |
|---|---|---|---|---|

At the completion of orientation, the RPN will perform technical nursing skills specific to the age and characteristics of the patients served, consistent with the Standards of Nursing Practice.

| Self-Evaluation Scale 1 2 3 | RPN Technical Skill Checklist | Method of Validation/Code | Date Met/ Initials |
|---|---|---|---|
| | 1. Cardiovascular<br>   A. Initiate IV therapy<br>      1. Adult, non-trauma<br>      2. Trauma patient<br>      3. Pediatric<br>      4. Newborn<br>      5. Phlebotomy percutaneous approach<br><br>   B. Blood sampling<br>      1. Arterial line<br>      2. Blood sampling: port-a-cath<br>      3. Triple lumen/trauma cath/central line<br><br>   C. Central venous line management: securing/dressing/<br>      caps/tubing<br>      1. Trauma catheter/triple lumen<br>      2. Implanted device, external access (i.e., Hickman)<br>      3. PICC line<br>      4. Port-a-cath<br><br>   D. Infusion pumps<br>      1. IV pumps<br>      2. Syringe pumps<br>      3. Programmable pediatric pump<br>      4. Patient-controlled analgesia<br><br>   E. Spacelab bedside and central monitors<br>      1. Cardiac rhythm interpretation<br><br>   F. Defibrillator operation<br>      1. Zoll<br>      2. Physiocontrol 10 and 9<br><br>   G. External transcutaneous pacer Zoll<br><br>   H. Transvenous pacer pack: emergent<br><br>   I. Transvenous pacer pack: urgent<br>      1. Pulse generator<br>      2. Ushkow's lead<br><br>   J. Blood products administration<br><br>   K. Level I blood warmer and rapid infuser<br><br>   L. Spun hematocrit<br><br>   M. Utilization of doppler for vascular assessment<br><br>2. Gastrointestinal<br>   A. Tubes & drains<br>      1. Salem sump/Nasogastric tube (age-appropriate size)<br>         a. Measuring<br>         b. Cetacaine administration<br>         c. Securing<br>      2. Gastric decontamination/lavage (Code Blue) | | |

FIGURE 29-2 **Emergency department competency-based orientation tool sample page, excerpt.** (*Source:* Courtesy of Albany Medical Center, Albany, NY.)

## Socialization

Socialization to the new workplace is another important part of orientation. Preceptors can play a key role in introducing the new nurse to coworkers and other members of the health team, both on and off the unit. This helps the new nurse identify relationships within the unit and between the unit and the larger health care organization. In practice areas where staff are infrequently together, such as a home health agency, socialization can be difficult. Some nurse managers may arrange a luncheon or coffee hour to introduce new staff members to the work group.

## Working with Patients

When you begin to work with patients during your orientation, you soon begin to realize that you are a "real" nurse now. Patients expect you to have the answer. This can be a little intimidating at first. The first time you experience an emergency by yourself, it can be unnerving to realize that you are the nurse in charge of your patient. Your nursing education and hospital orientation should have prepared you for this moment. You can gain confidence by keeping your knowledge base up to date and by looking and acting professional. The more experience you have, the easier this will become.

Work to put your patients at ease, and demonstrate a sense of caring to them. Work to become a nursing expert in your patient's eyes, and relay to the patient that you possess a body of nursing care knowledge. Demonstrate to your patients that they can trust you, and they will want to continue their relationship with you. Never be afraid to ask other, more experienced nurses for advice or help if you are unsure of a situation; not asking for help is where you can get into trouble.

## Working with Physicians

Sometimes, new graduates are intimidated by the practitioners with whom they work. Cardillo (2001) gives several tips on working with practitioners. She suggests that it is useful to establish rapport and introduce yourself to the practitioners you work with. Do not be intimidated. You and the practitioner are both on the health care team to meet the patient's goals. Both you and the practitioner are important to your patient's welfare; one could not function well without the other.

Nurses should be assertive but sincere when calling practitioners. Never fear calling practitioners because they may "yell" at you. If they do so, it's their issue, not yours. Remember that you are the patient's advocate. If you do not understand something, ask questions. Many practitioners love to teach. Be honest and up front. Tell the practitioner if something is new to you.

Give due respect to the practitioners you work with, and expect the same from them. If the practitioner is out of line, you might say, "I don't appreciate being spoken to in that way" or "I would appreciate being spoken to in a civil tone of voice, and I promise to do the same with you" or something similar.

Nurses should always seek clarification from the practitioner if an order is unclear and repeat the order to clarify it. If an order is inappropriate or incorrect, rather than saying, "This order does not seem appropriate for this patient," which would likely put the practitioner on the defensive, try "Teach me something, Dr. Jones; I've never seen a dose of Lopressor that high. Can you explain the therapeutic dynamics to me?" or "Dr. Smith, I can't figure out why you ordered a brain scan on this patient. Can you help me out here?" This approach usually results in the practitioner either reevaluating an order or changing it. If the practitioner does not change an order that you think is inappropriate, let your supervisor know and follow the guidelines of the agency that you work for regarding what to do when an inappropriate order is given. Never give any medication if you have a doubt; clear the doubt first and foremost. *Remember, diplomacy often works wonders, and it is your license on the line.*

## Learning Styles

Educators have long realized that people learn in different ways, based in part on their previous experiences. At each stage of the learning process, individuals have different learning styles and need different interventions from their preceptors or leaders. There are more than 20 different learning styles in the literature. New graduates orienting to the clinical area need a preceptor who gives specific directions. They need details and demonstrations of skills. New nurses benefit if the preceptor or educator breaks tasks down into components so that they can readily see the proper order or priority of items. As nurses become more experienced, they do well with a teaching style that emphasizes collaboration and relates the new material to the learner's frame of reference. Taking a learning style inventory can help bring awareness to the orientation process. It can be difficult to match a teaching style to a learning style.

New graduates may find it helpful to develop learning objectives that can be shared with a preceptor and, later, with a nurse manager. The new graduate and his or her preceptor or manager may choose to review and update the learning objectives each week. This will help give direction and purpose to the new graduate.

## Preceptors

Ideally, the nurse manager assigns each new orientee to a preceptor who understands the need to match the teaching style to the new nurse's learning needs. In many organizations, the learner follows the preceptor's schedule so that orientation is consistent. A successful preceptor is clinically experienced, enjoys teaching, and is committed to the role. However, if you find it difficult to work with your assigned preceptor, make this known early in the orientation process. The nursing manager or educator should be

## REAL WORLD INTERVIEW

I just finished my unit-based orientation a few weeks ago. Because I'm working in a critical care area, my general orientation included a critical care course, so I wasn't on my unit too much at first. I had two preceptors most of the time—one while I was on the day shift so I could attend classes some days, and the second when I moved to my regular night hours. My preceptors were great—they supported me, taught me new technical skills, and helped me figure out the order to do things. The idea of coming off orientation was scary at first, but I was able to work my schedule so I was on duty the same shifts as my preceptor the first few nights. This gave me the security of knowing I would have a resource available when I needed it.

**Jennifer Holscher, RN**
Albany, New York

notified and the situation discussed and resolved. Good preceptors are familiar with the organization's policies and procedures, willing to share knowledge with their orientees, and able to model behaviors for their orientees.

In some larger organizations, one preceptor is assigned to a group of new nurses. Together, the several orientees work with the preceptor to master core competencies before being assigned to their home unit. For example, several new graduates hired for medical or surgical floors may all be assigned temporarily to one unit, with one preceptor. This has the advantage of providing peer supports to the new nurses and may be more efficient and less expensive than a traditional one-on-one relationship.

In 1974, Kramer described "Reality Shock" and discussed the difficulties some new graduates have in adjusting to the work environment. Kramer identified a conflict between new graduates' expectations and the reality of their first nursing position. A skilled preceptor can assist new nurses through this transition by offering them opportunities to validate their impressions. The support of other new nurses in a similar situation, such as those participating in the same core orientation, is particularly helpful. Note that all nurses may experience reality shock throughout their career whenever they enter a new career area (Brunt, 2005).

## PERFORMANCE FEEDBACK

"So, how am I doing?" Everyone wants feedback about their performance, particularly when they are in a new position. Some preceptors and managers recognize new employees for their progress, but in many cases the new nurse needs to solicit their feedback. A concrete mechanism to measure your performance is through the objective assessment materials provided by nurse educators. New nurses must successfully pass the written and technical portions of orientation. If the organization has a competency-based orientation tool, the new nurse must meet its performance criteria. New nurses may find it helpful to develop and review learning objectives. A new nurse can ask for feedback from preceptors and supervisors regarding progress toward meeting these learning objectives.

## PRECEPTOR ASSESSMENT

New graduates should meet at regular intervals with their preceptor and manager to review progress. This evaluation time is important to make certain that new nurses are being assigned to clinical experiences that match their learning needs. It also provides a chance to ensure a smooth interpersonal relationship between orientee and preceptor. At these meetings, the preceptor, manager, and learner should set goals for the next interval. For example, "By the end of next week, the orientee will have progressed to an independent patient assessment, completed a patient admission, and increased the workload to a four-patient assignment."

At each of these sessions, it is important for the new nurse to solicit feedback. Ask, "How do you think I'm doing? Am I at the level you would expect? What should I focus on next?" Answers to questions such as these allow the orientee to measure progress. Many new graduates take negative evaluation personally. It is important that the preceptor identify that it is the skill or behavior that is inappropriate, and not the nurse as a person.

## FORMAL PERFORMANCE EVALUATION

To maintain accreditation, health care organizations are required to administer performance evaluations for each employee at regular intervals. Individual facilities set their own policies identifying the process and time frames. For many, annual evaluations are the norm. Evaluation forms

## EVIDENCE FROM THE LITERATURE

**Citation:** Snodgrass, S. G. (2001, June 10). Wish you were a star? Become one! *Chicago Tribune,* D1.

**Discussion:** The most logical way to predict your future is to create it, so if you want to be a star, start by becoming a top performer now. Companies are drawn to those who use up-to-date skills and leadership to produce measurable results. Organizations seek such people out. Surprisingly, few people understand this. You can begin to position yourself now with exceptional performance.

Start by delivering more than you promise, and consistently outperform yourself. Exceed expectations on a regular basis, seek more responsibility, value teamwork and diversity, provide leadership, and always go beyond the call of duty. Communicate effectively, and know how to network with others. Be resourceful, comfortable with ambiguity, and open to saying, "I don't know, but I'll find out." In addition, take initiative and persevere until you reach quantifiable results. Finally, assume some personal risk by thinking outside the box and exploring bold, new solutions to challenges. Provide yourself with a margin of confidence through lifelong learning. Be open, be flexible, and adapt to new ideas. Spend time with those who challenge your thinking.

You should also be creative, seek innovative solutions, and supplement your past experience with a fresh perspective. Learn how to put your ideas into action and be persistent, because achieving results takes time. In addition, do your homework. Understand the business agenda and close any gaps between what you are and what you could be. In other words, define your goals, then create and implement a personal development plan. Finally, demonstrate respect for others, and apply the Golden Rule. Achieving great results with great behavior enables your star to rise. You can begin the process right now with these specific steps:

- See the big picture. Know why your job was created, how it relates to your organization, and what opportunities it contains. You can positively influence outcomes through performance and achievement.

- Invest in your organization; make decisions as if you owned the company. Determine which actions promise the most significant impact, and then pursue them with zeal.

- Push your comfort zone by seeking challenges, finding the positive in negative situations, taking action, and learning from the past.

- Make time for people; understand the culture, values, and beliefs of the organization; keep things in perspective; and have fun.

- Inspire those around you to exceed expectations; also, convey a sense of urgency, and consistently drive issues to closure.

After you do all this, how do you ensure that you will be noticed? Ask how your company identifies and rewards top performers. Inquire as to whether there is a high-potential category. You should pursue an environment in which the best are recognized and valued. It should be an organization that provides career growth, lifelong learning, and development opportunities.

You also want meaningful work, an opportunity to contribute, and an environment that prizes new ideas and fresh perspectives. In addition, you deserve honest feedback and the opportunity to provide the same in return. Finally, seek an organization that energizes and empowers you, encourages your good health, respects your point of view, and honors your performance. Many such organizations abound.

**Implications for Practice:** Although a business professional wrote this article, the advice rings true for nurses as well.

are usually standardized but they also allow for individual performance feedback from the nurse manager.

What should nurses expect from their first formal evaluation? The individual and the nurse manager meet to review progress since either the previous evaluation or date of hiring. The evaluation should be objective, based on the nurse's performance as measured against the job description. See Figure 29-3 for an example of a job description.

Most performance evaluations use some sort of checklist, reflecting whether the individual being evaluated meets

ALBANY MEDICAL CENTER HOSPITAL PATIENT CARE SERVICES
Job Description

JOB TITLE: REGISTERED PROFESSIONAL NURSE

| | |
|---|---|
| Exempt (Y/N): No | JOB CODE: |
| SALARY LEVEL: N25.1–4 | DOT CODE: |
| SHIFT: | DIVISION: PATIENT CARE SVC |
| LOCATION: NURSING UNITS | DEPARTMENT: |
| EMPLOYEE NAME: | SUPERVISOR: NURSE MANAGER |
| PREPARED BY: AMY BALUCH | DATE: 03/22/2011 |
| APPROVED BY: | DATE: |

**SUMMARY:** The registered professional nurse utilizes the nursing process to diagnose and treat human responses to actual or potential health problems. The New York State Nurse Practice Act and A.N.A. Code for Nurses with Interpretive Statements guide the practice of the registered professional nurse. The primary responsibilities of the registered professional nurse as leader of the patient care team is coordination of patient care through the continuum, education, and advocacy.

**ESSENTIAL DUTIES AND RESPONSIBILITIES** include the following. Other duties may be assigned.

— Performs an ongoing and systematic assessment, focusing on physiologic, psychologic, and cognitive status.

— Develops a goal-directed plan of care that is standards based. Involves patient and/or significant other (S.O.) and health care team members in patient care planning.

— Implements care through utilization and adherence to established standards that define the structure, process, and desired patient outcomes of the nursing process.

— Evaluates effectiveness of care in progressing patients toward desired outcomes. Revises plan of care based on evaluation of outcomes.

— Demonstrates competency in knowledge base, skill level, and psychomotor skills.

— Demonstrates applied knowledge base in areas of structure standards, standards of care, protocols, and patient care resources/references. Practices in compliance with state and federal regulations.

— Demonstrates knowledge of Patient Bill of Rights by incorporating it into their practice.

— Demonstrates ability to identify, plan, implement, and evaluate patient/S.O. education needs.

— Participates in development and attainment of unit and service patient care goals.

— Organizes and coordinates delivery of patient care in an efficient and cost-effective manner.

— Documents the nursing process in a timely, accurate, and complete manner, following established guidelines.

— Utilizes standards in applying the nursing process for the delivery of patient care.

— Participates in unit and service quality management activities.

— Demonstrates self-directed learning and participation in continuing education to meet own professional development.

— Participates in team development activities for unit and service.

— Demonstrates responsibility and accountability for professional standards and for own professional practice.

— Supports research and its implications for practice.

— Adheres to unit and human resource policies.

— Establishes and maintains direct, honest, and open professional relationships with all health care team members, patients, and significant others.

— Seeks guidance and direction for successful performance of self and team, to meet patient care outcomes.

— Incorporates into practice an awareness of legal and risk management issues and their implications.

*(Continues)*

FIGURE 29-3 **Albany Medical Center hospital patient care services job description for registered professional nurses.** (*Source:* Courtesy of Albany Medical Center, Albany, NY.)

**QUALIFICATION REQUIREMENTS:** To perform this job successfully, an individual must be able to perform each essential duty satisfactorily. The requirements listed below are representative of the knowledge, skill, and/or ability required. Reasonable accommodations may be made to enable individuals with disabilities to perform the essential functions.

**EDUCATION and/or EXPERIENCE:** Graduate of an approved program in professional nursing. Must hold current New York State registration or possess a limited permit to practice in the State of New York.

**LANGUAGE SKILLS:** Ability to read and interpret documents such as safety rules and procedure manuals. Ability to document patient care on established forms. Ability to speak effectively to patients, family members, and other employees of organization.

**MATHEMATICAL SKILLS:** Ability to add, subtract, multiply, and divide in all units of measure, using whole numbers, common fractions, and decimals. Ability to compute rate, ratio, and percent.

**REASONING ABILITY:** Ability to identify problems, collect data, establish facts, and draw valid conclusions.

**PHYSICAL DEMANDS:** The physical demands described here are representative of those that must be met by an employee to successfully perform the essential functions of this job. Reasonable accommodations may be made to enable individuals with disabilities to perform the essential functions.

While performing the duties of this job, the employee is regularly required to stand; walk; use hands to probe, handle, or feel objects, tools, or controls; reach with hands and arms; and speak or hear. The employee is occasionally required to sit or stoop, kneel, or crouch.

The employee must regularly lift and/or move up to 100 pounds and frequently lift and/or move more than 100 pounds. Specific vision abilities required by this job include close vision, distance vision, peripheral vision, depth perception, and the ability to adjust focus.

**WORK ENVIRONMENT:** The work environment characteristics described here are representative of those an employee encounters while performing the essential functions of this job. Reasonable accommodations may be made to enable individuals with disabilities to perform the essential functions.

While performing the duties of this job, the employee is regularly exposed to bloodborne pathogens.

The noise level in the work environment is usually moderate.

**FIGURE 29-3** (Continued)

standards, exceeds standards, or falls below the organization's standards.

Formal performance evaluation between the manager and staff member serves several purposes. The evaluation is used to ensure competence in the skills required for safe patient care. It is also an opportunity to recognize the nurse's accomplishments in the evaluation period, which can be a real morale boost. This is the ideal time for the manager to enhance future performance by coaching, setting goals, and identifying learning needs. At the end of the performance evaluation, both the manager and the staff member should have a clear understanding of what needs to happen in the next year for that nurse to grow and continue to be successful. Feedback is most useful when it identifies actual examples of good and poor performance and gives suggestions for change.

## 360-Degree Feedback

Some health care organizations have moved to an evaluation program known as **360-degree feedback**. In this system, an individual is assessed by a variety of people to provide a broader perspective. For example, a nurse may complete a self-assessment and submit a packet detailing the year's progress. This may include documentation of in-services completed in the assessment period, samples of charting, and details related to committee work. The appraisal process also includes peer reviews, evaluation by the nurse's immediate supervisor, and patient interviews.

With 360-degree feedback, the individual can potentially receive a broader, more balanced assessment. To be consistent and objective, nurses who are asked to evaluate their peers need orientation to the process and the specific tool being used. Overall, 360-degree feedback can be time-consuming to complete, yet it provides valuable information.

## Setting Goals

A key component of any performance appraisal is the opportunity to set goals for the coming year. Goals that are measurable and clearly articulated are more likely to be met. These should be developed jointly by the nurse being evaluated and the nurse's manager.

A sample performance goals outline might look like the following:

By the next scheduled performance assessment, nurse Joanne Johnson will do the following:

- Successfully complete the advanced pediatric assessment course
- Assume the primary nurse role for patients with an anticipated length of stay of greater than three days
- Become an active participant on a unit-based or hospital-wide committee

# ORGANIZATIONAL RESPONSES TO PERFORMANCE

Many health care organizations have a merit-based compensation structure that is tied to performance evaluations. Employees' pay raises are matched to their performance. But most health care organizations are looking for other ways—in addition to money—to create job satisfaction. Recognition is an important way in which health care organizations can motivate employee performance.

## EMPLOYEE RECOGNITION PROGRAMS

Many health care organizations have developed formal recognition activities. These may take the form of surprising an employee of the month with balloons and a plaque, bringing in a national speaker for a celebration of Nurses Day, or presenting recognition pins for years of service. One popular recognition involves selecting an employee from each unit to attend a quarterly luncheon with the organization's administrator. At the luncheon, employees are recognized individually for their contributions to patient care, based on the narratives submitted by the nominating individuals. Figure 29-4 and Figure 29-5 are sample forms used in such a program.

## CORRECTIVE ACTION PROGRAMS

Sometimes, appraisal feedback indicates the need for significant performance improvement. Most health care organizations have a prescribed corrective action program.

One of the first steps in helping employees improve their performance is identifying whether their poor performance is developmental or related to a failure to follow policies or procedures. For example, a nurse may be having difficulty completing assignments in an appropriate time frame. It is unlikely that the nurse's problem is related to a lack of understanding of the rules. Instead, the manager needs to coach the employee, assisting the nurse with whatever support will help the nurse improve. It may be that the nurse needs remedial work in some particular technical skill or feedback specifically directed to organizing a patient assignment, either of which can affect the nurse's ability to complete the shift on time.

Another category of corrective action is disciplinary corrective action. In this case, an employee receives feedback for failing to follow the organization's policies. Excessive absenteeism is an example. As with the previous example, the goal is to assist the nurse to improve performance. Most organizations have a series of progressive steps for corrective action in cases in which employee performance does not improve. For example, a manager may begin by providing a verbal warning to an employee whose attendance is minimally acceptable. If the nurse's attendance problem continues, the nurse may receive a written warning. Without improvement, this could proceed to a suspension, final warning, and eventually termination.

In **progressive discipline**, the manager and employee's mutual goal is to take steps to correct performance to bring it back to an acceptable level. In a union environment, the employee may have the right to union representation after a verbal reprimand. It offers a stepwise process with opportunities for continued feedback and clarification of expectations. In any event, the corrective action applied by the manager must be fair. Employees should be forewarned of the consequences of violating an institution's policies so that there are no surprises. The corrective action should be consistent and impartial—each person is treated the same each time the rule is broken. Figure 29-6 is a sample corrective action documentation tool.

# THE NURSE AS A FIRST-TIME MANAGER OF A SMALL PATIENT GROUP

The RN responsible for direct care of a small group of patients is functioning as a first-line nurse manager. This nurse is responsible for linking each patient to the resources that the patient needs. This often involves supervising other licensed LPNs and NAP involved in direct patient care.

**Success Stories Nomination Form**

Name: _____

Position: _____ Unit/Dept.:_____

Reason for Selection:

_____

_____

_____

_____

_____

_____

_____

_____

_____

_____

Submitted by:   Name: _____

Unit/Ext.: _____

Please return complete form to:
Marketing and Retention Committee
M4 Mailbox / MC 73

FIGURE 29-4  **Success stories nomination form.** (*Source:* Courtesy of Albany Medical Center, Albany, NY.)

# RELATING TO OTHER DISCIPLINES

Given the complexity of health care organizations in the United States, the successful interconnection among departments is a potential source of tremendous strength. The RN who understands the functions of the respiratory therapists, pharmacists, dieticians, social workers, case managers, and vendors for durable medical equipment will be able to efficiently incorporate their contributions in planning for effective patient care. In many settings, diagnosis-specific care plans or clinical practice guidelines articulate the anticipated relationships among disciplines. For example, a care plan for a patient admitted for a CVA

43 New Scotland Avenue, Albany, New York 12208-3478

ALBANY MEDICAL CENTER

October 8, 2011

Dear Managers,

The Marketing and Retention Committee, along with Mary Nolan, has been sponsoring Success Stories luncheons to recognize staff. This December 14, 2011, we will extend this luncheon to include a new staff member, of less than a year on your unit, to accompany the staff member who has been chosen by you, another staff member, or the previous Success Story candidate.

Please take this opportunity to submit the name of a staff member whom you feel has a positive impact on your department and helps make it a successful one. The individual chosen will receive an invitation to a luncheon with Mary Nolan.

Submissions must be returned to Carole West, Marketing and Retention mailbox on M4, MC73, by November 16, 2011.

Thank you, and we look forward to honoring your Success Story. If you have any questions, please feel free to call me in the Emergency Department, 262-3131. Your staff member will be honored from noon to 2:00 p.m. on December 14, 2011.

Sincerely,

Carole West, RN
Success Luncheon Chairman

**FIGURE 29-5** Invitation to Success Stories luncheon. (*Source:* Courtesy of Albany Medical Center, Albany, NY.)

may include consultation with physical therapy on day two and an evaluation for home care needs on day three. It is important for nurses to develop strong relationships with representatives of the many other disciplines whose practices interface with the nursing role.

## DELEGATION TO TEAM MEMBERS

In the current health care environment, lengths of stay (LOS) are shorter despite increased patient acuity and complexity. The nurse who is responsible for a group of patients also needs to work with other nurses and licensed and unlicensed personnel to provide safe patient care. This usually involves delegating specific responsibilities to other staff, including some staff who may be either from an older or a younger generation with different beliefs and working styles. For example, a Generation X charge nurse may be working with baby boomer nursing assistive personnel (NAP) and a Millennial Generation LPN. Communication among staff of all generations is critical in

ALBANY MEDICAL CENTER

## Corrective Action Notice

Employee's Name: ______________________ Job Title: ______________________

Division: ( ) Center ( ) College ( ) Hospital Department: ______________________

**PART I: CORRECTIVE ACTION HISTORY**

| Date of Corrective Action | Reason for Corrective Action | Level of Corrective Action Applied |
|---|---|---|
| 1. / / | ______ | ______ |
| 2. / / | ______ | ______ |
| 3. / / | ______ | ______ |
| 4. / / | ______ | ______ |
| 5. / / | ______ | ______ |
| ( ) | Check here if no previous corrective action issued | |

**PART II: CURRENT OFFENSE REQUIRING CORRECTIVE ACTION**

Date of Offense: ____/____/____

Level of Corrective Action Being Applied: ( ) Written Warning ( ) Final Warning

Category of Offense: ( ) Job Performance ( ) Absenteeism/Tardiness ( ) Misconduct ( ) Other

Description of Offense: ______________________

Expected Improvement and Plan for Correction: ______________________

Suspension without Pay (pending investigation)

Description of Incident: ______________________

______________________ ____/____/____ Follow-Up Date: ____/____/____

Manager's Signature Date

The offense(s) described above is in violation of Albany Medical Center's policies governing the conduct and/or performance standards of it's employees. The reason for and level of corrective action being issued has been fully explained to me, and I understand that I must correct my job performance and/or conduct immediately. My job performance and/or conduct must remain at an acceptable level following improvement, or further action up to and including discharge will be taken.

______________________ ____/____/____

Employee's Signature Date

Original Copy: Human Resources (Hospital Paid Staff) College Personnel Services (College Paid Staff) Department Copy: Manager Employee Copy

H-170

**FIGURE 29-6 Corrective action notice documentation tool.** (*Source:* Courtesy of Albany Medical Center, Albany, NY.)

assuring safe patient care and good working relationships. A nurse who delegates effectively assigns routine tasks to a coworker, freeing the RN for more-complex planning or care. As a starting point, an RN needs to refer to the nurse practice act for the state in which the nurse practices. This document limits or defines the responsibilities that may be delegated. For example, in some states, LPNs may draw blood but may not start IV lines. This can also vary from hospital to hospital. The nurse needs to follow all such policies in decisions about delegation.

The RN must also consider the skills and knowledge of coworkers in order to delegate well. This allows the nurse to give teammates the opportunity to work within their competencies. The RN needs to match the coworkers skills with the delegated task, while remembering which skills cannot be delegated, such as ongoing evaluation and supervision of tasks. The RN may ask the LPN what materials he or she usually uses for a particular wound dressing. At the same time, the RN needs to ensure that the LPN knows what must be reported back, in this case a saturated dressing. The RN must continue to evaluate the ongoing needs of the patient.

It is easy to fall into the trap of overdelegating or underdelegating, particularly for new nurses. Some nurses are hesitant to delegate activities to others, because they are afraid their teammates will resent being asked to do a specific task. They worry they will be seen as lazy or lacking ability. Or they may hesitate to delegate out of a belief that they can do the task better or faster themselves.

Other nurses delegate more care than is appropriate or safe. Nurses who overdelegate may do so because they are poor time managers or because they personally lack the skill required. It may be that they failed to first assess their patients or are unfamiliar with their coworkers' competencies.

Performance feedback is a crucial element of delegation. It is important to openly recognize team members' contributions to safe patient care. In an instance in which the nurse is not satisfied with the outcome of a delegated task, it is equally important to discuss the assignment with the coworker individually. Perhaps the nurse's directions were unclear, misunderstood, or failed to include an important time frame. Taking the time to provide feedback demonstrates the respect and value a nurse places on his or her teammates' contributions.

## Levels of Authority

Sometimes when one delegates an assignment to a team member, that person questions the parameters of the assignment. If, for example, the RN in charge of a group of patients delegates a patient's ostomy teaching to another RN on the team, that RN may hear that assignment several ways. For example, the RN may think, "I need to do the ostomy appliance change while the patient's wife is here." Or "I need to

assess what the patient has learned already and report back to the RN." Or, finally, "I need to develop a teaching plan with the patient and begin to implement it today."

These possibilities demonstrate the importance of delegating clearly and specifically. It is important to remember that even when a task is delegated, the person delegating the task is still responsible to assure that the task is completed correctly. Communication is key to effective delegation. The person to whom the task is assigned must clearly understand the expectations. In the above scenario, the RN who is delegating should explicitly tell the second RN what the expected outcome and time frame are for the delegated task. For example, the delegating RN might state, "Please construct a teaching plan for this patient's ostomy care today. Be sure that the patient and his wife will be able to perform effective ostomy care at home." See Chapter 16 for more discussion of delegation.

## The Charge Role

Many hospital-based nurses, especially those who work evening or night shifts, rapidly progress from being assigned responsibility for a small group of patients to being assigned to the charge role for the shift. Particularly on medical and surgical floors, the charge nurse continues to care for a group of patients but also may coordinate care for the rest of the unit. This nurse may be responsible for assigning the workload of the nursing staff for the shift.

First-time charge nurses often have high expectations for their own performance, and they can easily become stressed in the new role. It is helpful to recognize that the nursing process—assessing, diagnosing, planning, implementing, and evaluating care—requires organizational and priority-setting skills that directly apply to the charge nurse role. It is a matter of perceiving and delegating patient care needs from the perspective of the unit as a whole. The new charge nurse must let go of the need to be perfect. Instead, the nurse should concentrate on staying organized and focused on what is best for the patients. It is also important to recognize and utilize the available resources for problem solving, such as coworkers or the facility supervisor.

### Articulating Expectations

As a first-time charge nurse, it is important to build relationships with other staff members as well as coworkers from other disciplines. One way to develop these relationships is by sharing expectations. This may be as simple as sitting down over coffee and agreeing to certain behaviors, such as "We will maintain a patient focus, as evidenced by answering call lights quickly." Some performance expectations may be more generic, applying to relationships more than specific patient care items, but they still need to be clearly spelled out. For example, "If you disagree with me, you will talk to me about it before you discuss our disagreement with others." These specific expectations help establish a level of

## REAL WORLD INTERVIEW

I remember my first job in nursing about 40 years ago. It was as a staff nurse on a general medical-surgical unit. I had a wonderful preceptor, Ed Fuss, RN. He worked with me and helped me until gradually I could assume my full role as a staff nurse on the unit. It took a while. It was very stressful to care for the 42 patients on that unit. After my orientation period, I would sometimes be the only RN, though I would have several licensed practical nurses and nurse aids working with me. I had to quickly learn appropriate delegation techniques.

Nursing has been a great career for me. I have worked as a nurse in Indiana, Illinois, New York, Oklahoma, and Wisconsin. Besides other nursing positions that I have held, I often work as a per diem agency nurse in various emergency departments. I find I can move quickly into the culture of a new unit by being friendly, helping others on the unit, and keeping my nursing skills up to date. I notice that people in other professions often complain about being concerned that they will lose their job. In nursing, I have not had to worry about job layoffs. I always have been able to get an interesting nursing job doing something I like.

I have had nurse friends who have worked as nurse practitioners, nurse chaplains, traveling nurses, seminar teachers, missionary nurses, nurse lawyers, nurse managers, informatics nurses, nurses in a homeless shelter, nurses on a cruise boat, etc. There are all kinds of nursing opportunities. I also like stopping at the scene of an accident and knowing I can help. It has been my privilege to be a nurse. How many other professions can say they save lives for a living?!

**Patricia Kelly, RN, MSN**
Chicago, Illinois

## CASE STUDY 29-1

You are the charge nurse for the evening shift on a general surgical floor. A patient care associate (PCA) working on your shift brings you a complaint. He says that the nurse he is assigned to work with this evening is not doing her share. She is sitting at the desk visiting with the unit secretary while he answers all the call lights. She has also assigned him the task of setting up a traction bed, which he has done only once in orientation two months ago. As the charge nurse, how would you respond to the PCA? Would you talk to the nurse he is working with? What is the priority issue in this situation?

trust and prevent the need for mind reading. They open the doorway for clear communication, so that when a problem develops, it is easier to approach the individual involved.

## STRATEGIES FOR PROFESSIONAL GROWTH

New nurses are more likely to stay in their positions if they are challenged and have opportunities for professional growth. Some health care facilities have a wealth of available educational opportunities. Others, particularly smaller organizations, may require the nurse looking for experiences to be more creative. The best place to start is with the experts on and around the nursing unit. Suppose you have developed an interest in learning more about cardiac arrhythmias. If your hospital offers an EKG interpretation course, great! Sign up! If not, there are lots of other ways to grow in this area. Talk to your nurse manager and educator about your interest. Ask them about classes, or ask them to refer you to experts in your geographic area. Speak to the cardiologist making rounds on your unit. Ask for the name of an interventional cardiologist so you can observe a cardiac catheterization. Ask to spend a day shadowing in a coronary care unit. Do not limit your search to nurses and practitioners. Often, other health disciplines overlap with nursing's interests, and you may be able to tap into opportunities with another discipline.

Not all nurses have the motivation or time for a lot of formal professional growth activities. What is important is to stay challenged. Find a particular skill or interest in your position, and expand it. What do you like best of all the things you did today? Working with the patient's family? Teaching the new diabetic? Starting that IV? Whatever it is, look for opportunities to become your unit's expert at it.

## CROSS-TRAINING

Given today's shortage of nurses, there is increased floating and offering nurses cross-training to new areas. Cross-training is another opportunity for individual growth. Although some organizations have strict guidelines to limit the practice of floating nurses, other health care facilities expect nurses to routinely float to either a related unit or an area particularly in need of assistance. It is useful for all staff to recall that a hospital is a business and must have staff available to care for patients at all times.

It is important for nurses in their first job to be articulate about their competencies for a new patient population if they are asked to float. They need to be sure the manager assigning them is aware of their experience level. Nurses should not accept total responsibility for an area or population in which they have not achieved competency. It may be more appropriate to assign an inexperienced nurse to specific tasks to help on the unit rather than asking that nurse to take a typical patient assignment on an unfamiliar unit.

One way to minimize the stress of being asked to float to a different unit is to volunteer ahead of time to cross-train to the new area. This has many advantages. It allows the nurse the opportunity to experience working with different ages or types of patients. Besides learning new skills, it gives a nurse the chance to see how the other half lives. For example, a nurse who has worked only on a medical floor may regard accepting an admission from the emergency department (ED) as something to be worked into the shift, based on other patient care needs. After cross-training to the ED and seeing admitted patients waiting on stretchers in the hall, that nurse may have a new appreciation of the need to negotiate for timely acceptance of ED admissions to the floor.

Cross-training has some long-term benefits as well. If a new nurse is considering a career in a specialty, spending some time cross-training to that population can help the nurse decide whether he or she wants to pursue that field. Cross-training is also beneficial to the nurse seeking a new position. When a nurse is applying for a new job, experience in more than one clinical area enhances a resume and makes the individual a stronger candidate.

Some health care organizations reward nurses who volunteer to cross-train so they can safely float to different areas. These rewards may be monetary. Other institutions offer nonmonetary incentives, such as reduced weekend or holiday commitments, as rewards for cross-training or floating.

## IDENTIFYING A MENTOR

Developing a mentoring relationship with a more experienced, successful nurse is another mechanism for professional growth. A mentor coaches a novice nurse and helps the novice develop skills and career direction. A mentor may introduce the younger nurse to professional networking opportunities. A good person to assist the new nurse in a workplace ethical dilemma may well be his or her mentor.

How does a new nurse find a mentor? First, the new nurse needs to communicate a willingness to learn and grow. A newer nurse usually needs to seek out a prospective mentor rather than wait to be approached by one. An ideal mentor is an experienced nurse who is willing to support and counsel other nurses when asked. This may lead to a formal structured relationship or a more informal role-modeling association.

Nurses who have been successful preceptors are often potential mentors, because they are committed to helping another nurse learn and grow. Even though the preceptor role is more narrow and defined, the role can easily be expanded to a more informal mentoring relationship.

The Internet is a newer mentoring resource. Nurses can develop relationships through special-interest chat rooms or by e-mailing experts in other geographic areas. There are forums for questions and answers, often on specific patient populations, disease processes, or operational issues. Want to get some expert advice on a particular patient problem? Spend some time on the Internet.

## DEVELOPING PROFESSIONAL GOALS

After a new nurse has mastered the skills for day-to-day nursing care, what is next? How does a nurse measure professional growth? For many nurses, the answer to these questions is a clinical ladder.

### Clinical Ladder

A **clinical ladder** is a program established by some health care organizations to encourage nurses to earn promotions and gain recognition and increased pay by meeting specific requirements. Although the criteria may vary, most programs have three or four distinct levels. Some also offer the nurse the opportunity to seek promotion in a specific track, within a clinical, educational, or managerial focus. Thus, it is possible for a new nurse to choose a clinical nursing track and move through the organization's promotional levels by meeting those requirements.

## REAL WORLD INTERVIEW

Here are a few lessons I learned as a new graduate:

1. You will manage to get every single type of body fluid on you at one time or another (blood, trach gunk, fistula juice, stomach residual, stool, urine). Bring a pair of backup scrubs to work.

2. You learn from your mistakes. I was taking care of a patient on an insulin drip during the night shift. She was up all night long, and her daughter didn't think too highly of the care she was getting. When the patient and her daughter finally fell asleep, I skipped her 3 a.m. Accucheck. Her 4 a.m. Accucheck was 27. Needless to say, I have never skipped an Accucheck since.

3. If a preceptor shows you something, don't say that they are doing it wrong and then pull out a policy book. Your preceptor will hate you for life.

4. Never pass up an opportunity to learn something, even if you think you have seen it before. Maybe someone will teach you a new and better way to do it.

5. You will think you are ready for your first really sick patient, but you are not. The senior staff will help you through it. I still replay my first really sick patient in my head and think back to all the things I wish I had done differently. Luckily, no one else dwells on it. I am my worst critic!

6. Find a person besides your preceptor whom you admire, maybe for their nursing skill, their personality, or their way of making everything look easy. Ask them if they will mentor you. It doesn't need to be a super-serious conversation. I made mine a joke and asked a nurse if she would be my Nighttime Sensei. She accepted, and to this day she has my back when things get crazy.

7. Always do the little things. I was a patient care tech before I was a nurse. I always make sure my rooms are stocked, everything is put away, my patient is clean, the patient's meds are in the drawer, etc. Little things like this can really help out your next coworker. There is nothing worse than walking into a patient's room and it looks like a bomb went off.

8. Always help out your coworkers. I work in an ICU, and there is a real sense of teamwork. As a new graduate, I always felt like I never had enough time to do my work, but I always made time to help the other nurses turn their patients, do a bath, move patients to a chair, etc. That stuff doesn't take that long, and your coworkers really appreciate it. Plus, next time you need help with something, easy or not, they will be there for you.

9. I think my first month off of orientation I cried in the shower at least once a week. Some of it was about the patients I took care of, some was about working with not-nice people, some of it was just because I needed to cry. I always tried to keep my emotions out of my workplace. Some people are very emotional at work, and it makes others uncomfortable. I am not saying that you can never cry at work or with a patient's family, but if you are crying during every shift, your coworkers will start to think that you can't handle your job.

10. I live by the mottos "never show fear" and "do it right." Always walk into the patient's room with confidence. If you are doing something for the first time, run through it with someone experienced before going into the patient's room. You are taking care of someone's mom, dad, or child, and they are trusting you with their lives. Don't give them a reason to lose that trust.

**Erin Mahoney, BSN, RN**
Loyola University Hospital
Maywood, Illinois

## REAL WORLD INTERVIEW

There is a national nursing shortage, most acutely realized in nursing specialty areas such as critical care, operating room, and pediatrics. Staffing and scheduling practices directly affect nursing personnel costs, patient care outcomes, and recruitment and retention of nurses. At a Midwest university hospital, a specialty cluster-nursing program was implemented to respond to the nursing shortage. The specialty cluster-nursing program consisted of grouping together several inpatient units and related specialty clinics with similar patient care requirements. Nurses hired into the cardiac, oncology, pediatrics, and trauma clusters are offered a special fellowship orientation program.

Nurses working the cardiac cluster can work on the cardiac medical intensive care unit, cardiac medical step-down unit, cardiac surgery intensive care unit, and cardiac surgery step-down unit. Nurses working the medical cluster can work on pulmonary, geriatric, psychiatry, and general medicine inpatient units. Nurses working the neuroscience cluster can work on orthopedics, rehabilitation, and neurology inpatient units. Nurses working the oncology cluster can work on the pediatric oncology clinic, adult oncology unit, or the adult hematology/oncology clinic or adult surgical unit. Nurses working the pediatric cluster can work on the infant/toddler, adolescent, hematology, and pediatric intensive care inpatient units. Nurses working the surgical cluster can work on general surgery, plastics and otolaryngology, transplant, peripheral vascular, and security inpatient units. Finally, nurses working the trauma cluster can work on the trauma life support intensive care unit, Emergency Department, burn unit, and general surgery inpatient unit.

This program creates opportunities for nurses to develop expertise and specialty knowledge in patient populations that cross multiple units. Nurses skilled in the specialty cluster are preassigned to one or more settings within the cluster based on projected staffing requirements. Staffing adjustments are made for each scheduling period for patient acuity changes, changes in patient volume, extended leaves, and sick leaves. Scheduling specialty cluster staff prior to the beginning of a work schedule minimizes floating of unit-based staff on a shift-by-shift basis. The cluster program with the fellowship orientation has become an effective nurse retention strategy.

**Patricia Dianne Padjen, RN, MBA, MS, EdD**
Manager, EMS Program
Madison, Wisconsin

For example, to be promoted from a new graduate level to a Level II RN, the nurse may be required to complete a specialty course such as Advanced Cardiac Life Support (ACLS) or EKG interpretation, join a unit-based or hospital-based committee, and finish the preceptor course. Besides offering opportunity for promotion, these programs offer an objective way to measure a nurse's achievements. Clinical ladders can be time-consuming to complete, yet they provide valuable information.

## Specialty Certifications

Many health care organizations encourage their staff to become certified. Nearly all nursing specialties now offer board certification exams to validate expert knowledge of that particular discipline. Emergency nurses may sit for the Certified Emergency Exam and the Advanced Cardiac Life Support Exam. Nurses who specialize in critical care may take the critical care certifying exam to earn their

### Critical Thinking 29-3

You have just begun interviewing for your first nursing job.

What type of nursing recognition programs would appeal to you? For you, would monetary or scheduling rewards be more of an incentive to cross-train on another unit? What are some measurable professional goals for your first year as an RN?

CCRN. Successfully passing a certification exam is another measure of professional growth and offers the benefit of national recognition of one's credentials.

## KEY CONCEPTS

- When choosing a first nursing position, it is important to contemplate the differences in developmental opportunities between specialty and general medical-surgical units. Environmental, scheduling, and orientation options are also important considerations.

- Organizational orientation is both general and unit based. Orientation is a time for developing strong relationships with preceptors and members of other disciplines, as well as for mastering competencies needed for safe patient care.

- Nurses receive performance feedback both informally and as part of periodic evaluations. This input is valuable in developing personal goals.

- Health care organizations have mechanisms to recognize employee contributions. Many of these programs reward success, both monetarily and through recognition programs. Corrective action programs can be used to coach an employee who is having performance problems and to foster change in an employee who is failing to follow policies.

- Given the increasing complexity of health care today, it is crucial for the first-time nurse to develop strong relationships with team members and representatives of other health care disciplines. The new nurse needs to delegate appropriately and identify specific levels of authority with coworkers. Relationships with coworkers are enhanced when staff members mutually agree to performance expectations.

- Professional growth is important for job satisfaction. Organizational opportunities for growth include clinical ladders and developing mentoring relationships. Cross-training is another means to expand experiences and can be helpful in defining future career plans.

## KEY TERMS

clinical ladder          progressive discipline          self-scheduling          360-degree feedback

## REVIEW QUESTIONS

1. General orientation includes which of the following?
   A. Information all nurses new to a facility need
   B. Mastery of unit-specific processes
   C. Patient care for a specific diagnostic group of patients
   D. Patient care for a specific age group of patients

2. Which of the following is usually NOT necessary to do in your first job in nursing?
   A. Learning to be organized
   B. Developing a good attendance record
   C. Refining your assessment skills
   D. Completing written performance evaluations of the Nursing Assistive Personnel that report to you

3. Preceptors who work with new nursing graduates should have all but which of the following characteristics?
   A. Be clinically experienced
   B. Enjoy teaching
   C. A committment to the preceptor role
   D. The ability to float to specialty units

4. The corrective action process usually contains all but which of the following?
   A. A verbal warning
   B. A written warning
   C. A final warning
   D. A transfer to another unit

5. Setting goals is most important for a new nurse because of which of the following?
   A. It gives the manager a tool to use to find out what the new nurse is doing wrong.
   B. It gives the preceptor a tool to use to find out what the new nurse is doing wrong.
   C. It can help give the new nurse direction and help managers and preceptors objectively evaluate the new nurse.
   D. New nurses must have a list of goals so they know what to do.

6. An appropriate reason to cross-train nurses to another patient care unit is which of the following? Select all that apply.
   _____ A. To expand nursing skills
   _____ B. To learn about new patient populations and their needs
   _____ C. To improve a nurse's marketability
   _____ D. To keep nurses only minimally competent in each area
   _____ E. To discover interest in other specialty units

7. New nurses should develop rapport with other practitioners such as physicians so that they can accomplish which of the following?
   A. Become effective patient advocates.
   B. Become friends with new colleagues.
   C. Become better patient educators.
   D. Learn more about the role of registered nurse.

8. It is important to build self-confidence and expertise as a nurse because of which of the following? Select all that apply.
   _____ A. Patients expect nurses caring for them to be knowledgeable.
   _____ B. Patients will be more at ease if their nurse is confident.
   _____ C. Nurses gain greater job satisfaction when they are confident in their skills and knowledge.

   _____ D. Colleagues have more respect for confident nurses.
   _____ E. Managers may have more respect for confident nurses.

9. Nurses consider which of the following when selecting a new position? Select all that apply.
   _____ A. Pay scale
   _____ B. Expected nursing role
   _____ C. Patient unit
   _____ D. Small or large hospital setting
   _____ E. Scheduling and benefits

10. Which of the following governs a nurse's practice?
    A. Trends in nursing
    B. State nurse practice acts
    C. Patient preferences
    D. Physicians who direct nursing care

## REVIEW ACTIVITIES

1. You will be graduating from your nursing program in three months. Identify several possible employment opportunities in your desired locale. Prepare examples of questions you will ask as part of choosing a position. What factors are most important to you?

2. You have been working as a new graduate for a year and have done well. Your nurse manager asks you to be the relief charge nurse on your unit for the 3 P.M. to 11 P.M. shift. What type of orientation will you need for this position? How can you work with a mentor to do well in this position?

3. You are interested in the concept of 360-degree feedback. Who could you ask to give you feedback on your clinical performance to achieve 360-degree feedback?

4. Review a recent nurse salary survey in a nursing journal (e.g., Mee, 2006: Bacon, 2010; Litchfield, 2010; Rollett, 2010). How do nursing salaries in your area compare?

## EXPLORING THE WEB

- There are many Web sites specific to nursing employment opportunities. Try some of these:
  www.rnwanted.com
  www.healthopps.com
  www.healthcareers-online.com
  www.nursingcenter.com
  www.healthjobsusa.com
- If you have a specialty area in mind, it is worth the time to explore the Web for more details. How will this help you as you prepare for interviews?

- Look up the Association of Pediatric/Hematology Oncology Nurses: www.apon.org

  Association of Rehabilitation Nurses: www.rehabnurse.org

  Association of Women's Health, Obstetric and Neonatal Nurses: www.awhonn.org
- If you are interested in trauma nursing, try www.ena.org/

## REFERENCES

Bacon, D. (2010). Results of the 2010 AORN Salary and Compensation Survey. *AORN Journal.* December; 92(6):614–30.

Brunt, B. A. (2005). Models, measurement, and strategies in developing critical thinking skills. *Journal of Continuing Education in Nursing,* 36(6), 255–262.

Cardillo, D. W. (2001). *Your first year as a nurse.* Roseville, CA: Prima.

Council on Physician and Nurse Supply. (2008). *Finding solutions to the healthcare shortage.* Philadelphia: Author.

Kramer, M. (1974). *Reality shock: Why nurses leave nursing.* St. Louis, MO: Mosby.

Litchfield, S.M. (2010). Salary negotiations and occupational health nurses. American Association of Occupational *Health Nurses Journal*. May;*58*(5):174–6. doi: 10.3928/08910162-20100428-03.

Mee, C. L. (2006). Nursing 2006 salary survey. *Nursing, 36*(10), 46–51.

Rollet, J. (2010). Compensation for new grads. What salary should you expect? *Advance for Nurse Practitioners*. April;*18*(4):40–2.

University HealthSystem Consortium. (2007). UHC nurse residency program reduces turnover, offers case study for addressing nationwide nursing retention problems. Retrieved August 7, 2010, from https://www.uhc.edu/docs/003743961_UHC_NRPPressRelease2010.pdf.

# SUGGESTED READINGS

[No authors listed]. (2010). 2009 salary survey results. Quality is a hospitalwide responsibility now more than ever. *Hospital Peer Review. January;35*(1):supplement 1–4.

Albaugh, J. A. (2005). Resolving the nursing shortage: Nursing job satisfaction on the rise. *Urology Nursing, 25*(4), 293–284.

Amos, M. A., Hu, J., Herrick, C. A. (2005). The impact of team building on communication and job satisfaction of nursing staff. *Journal of Nursing Staff, Development, 21*(1), 10–18.

Andrews, D. R., & Dziegielewski, S. F. (2005). The nurse manager: Job satisfaction, the nursing shortage and retention. *Journal of Nursing Management, 13*(4), 286–295.

Hansen, R. S. (2010). 10 tips for successful career planning. Retrieved August 7, 2010, from http://www.quintcareers.com/career_planning_tips.html.

Hegney, D., Plank, A., & Parker, V. (2006). Extrinsic and intrinsic work values: Their impact on job satisfaction in nursing. *Journal of Nursing Management, 14*(4), 271–281.

Murrells, T., Clinton, M., & Robinson, S. (2005). Job satisfaction in nursing: Validation of a new instrument for the UK. *Journal of Nursing Management, 13*(4), 296–311.

Orlovsky, C. (2008). How to avoid new nurse burnout. Retrieved August 7, 2010, from http://www.nursezone.com/recent-graduates/recent-graduates-featured-articles/How-to-Avoid-New-Nurse-Burnout_20086.aspx.

Tremayne, P., Moriarty, A., & Harrison, P. (2005). Starring role. Preparing for job interviews is often an ad hoc affair for nursing students. *Nursing Standards, 19*(27), 80.

University HealthSystem Consortium. (2011). The University HealthSystem Consortium (UHC) and American Association of Colleges of Nursing (AACN) Nurse Residency program. Retrieved April 20, 2011, from https://www.uhc.edu/cps/rde/xchg/wwwuhc/hs.xsl/16807.htm.

University HealthSystem Consortium/American Association of Colleges of Nurses. (2008). Executive summary of the post-baccalaureate residency program. Washington, DC: American Association of Colleges of Nursing. Retrieved August 7, 2010, from http://www.aacn.nche.edu/education/pdf/NurseResidencyProgramExecSumm.pdf.

Vogel, L. (2010). Nursing degree: Opens doors beyond bags, beds and bedpans. Canadian Medical Association Journal, 182(2), 131–132.

Wu, L., & Norman, I. J. (2006). An investigation of job satisfaction, organizational commitment and role conflict and ambiguity in a sample of Chinese undergraduate nursing students. *Nursing Education Today, 26*(4), 304–314.

# CHAPTER 30

# Healthy Living: Balancing Personal and Professional Needs

**MARY ELAINE KOREN, RN, PhD;**
**CAROLYN CHRISTIE-McAULIFFE, PhD, FNP**

*TO NURSE*

*To Care*

*To Solace*

*To Touch*

*To Feel*

*To Hurt*

*To Need*

*To Heal others,*

*As well as ourselves.*

(CAROL BATTAGLIA, 1996)

## OBJECTIVES

Upon completion of this chapter, the reader should be able to:

1. Generate a personal definition of health.
2. Apply the six concepts of physical, intellectual, emotional, professional, social, and spiritual health to your life.
3. Devise strategies to maintain physical, intellectual, emotional, professional, social, and spiritual health.
4. Summarize occupational health hazards that are present in the nursing work setting.
5. Devise methods of personal financial planning.

Delmar/Cengage Learning

*You get up early to work the day shift. On your drive to work, you grab a cup of coffee and a doughnut to sustain you through the morning. It is one of those busy days. The phone is ringing off the hook, patients are not stable, family members are demanding, practitioners are slow to answer your page, and the laboratory delivers misinformation. It is now noon, and there is no time for lunch. You run down to the vending machines for a Coke and pea-nut butter crackers to keep you going until the end of your shift. Five o'clock rolls around and you have worked two hours over-time and are exhausted. You are already late for your community meeting this evening. You race through a fast-food restaurant for a hamburger, fries, and a milkshake. When you finally arrive home late in the evening, you reward yourself with cookies and a bowl of ice cream and fall into bed. You have given so much throughout the day. There has been little time for yourself and for good nutrition.*

*What factors contributed to your busy day and poor eating habits in this scenario?*

*What recommendations would you have to decrease your stress and improve your nutrition in this situation?*

*Can eating become a crutch for your daily stress?*

*How can you model the behaviors you teach your patients?*

**N**ursing is a caring profession. Nurses spend their days helping others, many times at the expense of themselves. But who is there to care for the nurse at the end of the shift? If there is nothing left for them, they will not be able to give to their patients. Those they care for also look to them to model healthy living. If they do not try to live by the standards set for their patients, they will lose a certain amount of credibility in their patients' eyes. How, then, can they balance the demands of work with their personal needs? The first step is to gather information. This chapter provides an over-view of good health. Many strategies for healthy living are discussed based on six organizing concepts. This chapter also discusses financial planning and occupational health hazards for nurses.

# DEFINITION OF HEALTH

Patients and nurses alike strive to maintain good health. But what does health mean? The World Health Organization (2006) describes **health** as a "state of complete physical, social, and mental well-being, and not merely the absence of disease or infirmity. Florence Nightingale described health as "being well and using every power the individual possesses to the fullest extent" (Nightingale, 1969 [1860], p. 334). Health is a resource for everyday life, not the object of living. It is a positive concept emphasizing social and personal resources as well as physical capabilities."

Pender, Murdaugh, and Parsons (2005) view health as multidimensional and consisting of biophysical, spiritual, environmental, and cultural aspects. Health changes over the course of a lifetime, and gender differences are often evident. Health can also be affected by the environment and culture we live in. Roy and Andrews (1999) define health as "a state or a process of being and becoming an integrated and whole person" (p. 31). Not only is health a complex concept, but it is also dynamic and in a constant state of change.

Health is holistic and multidimensional. All the parts of the whole must be in balance and work together to pro-duce the end result of good health.

# HEALTHY PEOPLE—2020 AND BEYOND

Since 1979, U.S. scientists and public health officials have set goals and objectives aimed at improving the health of the nation in realistic 10-year increments through docu-ments entitled *Healthy People*. Input from a cross section of society and health care professionals helped to outline specific initiatives to be taken in this endeavor.

*Healthy People 2010* was developed with two overall goals for the nation: (1) to encourage those of all ages to increase their life expectancy as well as improve their over-all quality of life and (2) to eliminate disparities among various pockets of the population. In similar fashion, the development of goals and objectives for 2020 is based on discussions with society, health care professionals, and sci-entists, focused on health promotion and disease preven-tion for every person in our country. The goals address the environmental factors that contribute to our collective health and illness by placing particular emphasis on the determinants of health. Health determinants are the range of personal, social, economic, and environmental factors that determine the health status of individuals or popula-tions. Health determinants are embedded in our social and physical environments. Social health determinants include family, community, income, education, sex, race/ethnicity, geographic location, and access to health care, among others. Physical health determinants include our natural

**TABLE 30-1 Proposed Healthy People 2020 Vision, Mission, and Goals**

*Vision*

A society in which all people live long, healthy lives.

*Mission*

To improve policy and practice by:

Increasing public awareness and understanding of the underlying causes of health, disease, and disability;

Providing nationwide priorities and measurable objectives and goals;

Catalyzing action using the best available evidence;

Identifying critical research and data collection needs.

*Goals*

Achieve health equity, eliminate disparities, and improve the health of all groups.

Eliminate preventable disease, disability, injury, and premature death.

Create social and physical environments that promote good health for all.

Promote health development and healthy behaviors across every stage of life.

**Source:** U.S. Department of Health and Human Services. (2008). HealthyPeople.gov. Third Meeting: June 5 and 6, 2008. Secretary's Advisory Committee on National Health Promotion and Disease Prevention Objectives for 2020. Retrieved April 20, 2011 from http://healthypeople.gov/2020/about/Advisory/FACA3Minutes.aspx.

### Critical Thinking 30-1

Maintaining good health is not only important to a nurse's overall well-being, but it will also affect the quality of care provided to patients. Only when you feel good are you able to deliver optimal patient care. Nurses also serve as role models for patients.

How, then, can nurses, when under tremendous stress on a daily basis, strive toward good health? As you think about your own health, how would you define health? What inhibits you from engaging in healthy behaviors?

environment and the environments that we build. In 2010, a vision, mission, and four overarching goals were tentatively established for *Healthy People 2020* (Table 30-1).

## AREAS OF HEALTH

Health is a complex and dynamic state of being. A healthy person must balance various aspects in life to achieve and maintain good health. When one area of life is affected, general health is also affected. There is overlap among the different areas, but for purposes of discussion in this chapter, health has been divided into the following six areas: physical health, intellectual health, emotional health, professional health, social health, and spiritual health (Figure 30-1).

Our bodies are dynamic and ever changing. Each area of health is constantly adjusting to outside stimuli and attempting to bring balance in life. For example, when someone is tired, fatigue can slow mental acuity. It is easy to become short tempered, which in turn can affect our relationships with others. By making a conscious decision to sleep longer, we can potentially influence not only our

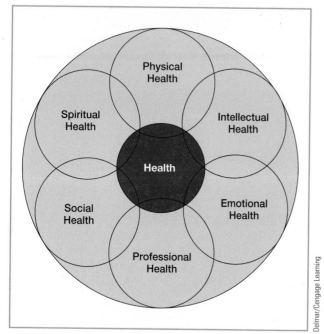

FIGURE 30-1   Areas of health.

physical well-being but also our ability to think clearly, our emotional state, and how we relate to others. The remainder of this chapter will explore in more depth each of the various areas of health.

# PHYSICAL HEALTH

**Physical health**, the first area of health, encompasses nutrition as well as exercise, coupled with a balanced amount of sleep. Physical health also includes preventive health behaviors such as avoiding smoking and having annual breast and prostate examinations and other screening procedures that detect health problems early.

The first step in maintaining good physical health is to do an assessment. Table 30-2 provides a self-assessment of physical health. This tool is designed to assess trends only.

## NUTRITION

Maintaining good nutrition—one aspect of physical health—helps us feel better and perform tasks at a higher level. The U.S. Department of Agriculture (2011) has Dietary Guidelines that describe a healthy diet as one that emphasizes fruits, vegetables, whole grains, and fat-free or low-fat milk and milk products; includes lean meats, poultry, fish, beans, eggs, and nuts; and is low in saturated fats, trans fats, cholesterol, salt (sodium), and added sugars, at www.mypyramid.gov.

## Calculation of Body Mass Index

A method for assessing body weight in relation to height is to calculate the body mass index (BMI). Note the calculator

| TABLE 30-2   Physical Health Assessment | | |
|---|---|---|
| PHYSICAL HEALTH BEHAVIOR | YES | NO |
| 1. I eat three balanced meals a day. | ____ | ____ |
| 2. I exercise 30 minutes every day. | ____ | ____ |
| 3. I enjoy exercising. | ____ | ____ |
| 4. I exercise with a friend. | ____ | ____ |
| 5. I rarely eat between meals. | ____ | ____ |
| 6. I sleep 8 hours a night. | ____ | ____ |
| 7. I wake up refreshed most mornings. | ____ | ____ |
| 8. I sleep soundly without waking up during the night. | ____ | ____ |
| 9. I don't smoke. | ____ | ____ |
| 10. I implement recommended health screenings for my age category, such as breast exam, prostate exam, dental cleaning, and so on. | ____ | ____ |
| 11. I avoid risky health behaviors, for example, drugs, tanning booths, unprotected sex, and so on. | ____ | ____ |

Delmar/Cengage Learning

for this at the National Heart, Lung, and Blood Institute web site, www.nhlbisupport.com/bmi. The National Institute of Health (n.d.) has established the following guidelines for interpretation of the BMI:

- 18.5 to 24.9 is optimal health
- 25 to 29.9 is overweight
- 30 and above is obese
- Below 18.5 is considered underweight

## BENEFITS OF EXERCISE

Following nutritional guidelines is not enough to maintain physical health. Daily exercise is another essential ingredient for healthy living. Exercise provides many benefits. It can improve cardiovascular function by lowering cholesterol and blood pressure and strengthening heart muscle. Exercise can boost the immune response to disease. Weight-bearing exercises are especially helpful for calcium uptake in bones. Exercise also improves flexibility and endurance and decreases fat deposition. Exercise can also make you feel better mentally. With exercise usually comes fewer depressive thoughts, less anxiety, an increase in self-confidence, and increased mental acuity (World Health Organization, 2006). The Centers for Disease Control. (2010) has specific physical activity guidelines (Figure 30-2). If the intent is to maintain weight, prevent further weight gain, and/or decrease the incidence of chronic disease, about 60 minutes of daily exercise is suggested. You may also be interested in specific exercises for a particular sport.

## Practical Exercise Suggestions

There are many different types of exercise from which to choose. Find one that you are passionate about and one that you truly enjoy. Finding friends who participate in the same activity can help you keep the commitment. When you decide, try to engage in the activity daily for 30 minutes at a time. Start slowly, and gradually build the intensity and duration.

You can set up an exercise program using DVDs or reference books. There are many ways to get exercise. Following is a list of some suggestions:

- Walking
- Jogging
- Cycling
- Swimming
- Rowing
- Yoga and tai chi
- Golf and tennis
- Skiing
- Team sports, such as volleyball or basketball
- Dancing

If you are having trouble getting started, you can spend time at a health spa. There are many spas throughout the country, tailored to various budgets. Visit www.spafinder.com for ideas.

## SLEEP

Sleep is another aspect of physical health. It is not uncommon for nurses to sleep less than eight hours per night. Nurses who work nights may find it especially difficult to sleep for an uninterrupted block of time. Nurses who are constantly changing shifts are more susceptible to sleep deprivation. It is estimated that it can take from four to six weeks to change sleeping patterns. Despite this, nurses may work multiple shifts within a week. Still other nurses work 10- and 12-hour shifts and do not have a lot of time between shifts before they are back at work again. If it is necessary to swing to a different shift, it is best to rotate from days to evenings to nights. People generally adapt better if shift rotation is done clockwise. Practitioners, family members, and critically ill patients place heavy demands on nurses. Many nurses find it a challenge to leave thoughts of patients and the day's activities at work. This also contributes to insomnia.

## Assessment of Sleep Deprivation

How do you know if you are sleep deprived? Assess your sleepiness index by answering eight simple questions using the Epworth Sleepiness Scale at the following Web site: http://provigil.com. Test your knowledge regarding sleep deprivation with the tools, quizzes, and sleep diary at the National Sleep Foundation Web site, www.sleepfoundation.org. Click on Tools and Quizzes.

## Deleterious Effects of Sleep Deprivation

Current research indicates that there are many negative effects of sleep deprivation. Sleep-deprived individuals become petulant and find it difficult to remember or to concentrate on the simplest tasks. Lockley, Barger, Ayas, Rothschild, Czeisler, & Landrigan (2007) found that patient care errors were significantly increased when nurses' hours of work were extended and/or shift work was necessary.

> Adults need at least 2 hours and 30 minutes (150 minutes) of moderate-intensity aerobic activity (i.e., brisk walking) every week and muscle-strengthening activities on 2 or more days a week that work all major muscle groups (legs, hips, back, abdomen, chest, shoulders, and arms).

FIGURE 30-2 **Physical activity needs of adults.** (*Source:* Compiled with information from the Centers for Disease Control. (2010). Physical Activity for Everyone. Retrieved April 28, 2011, from http://www.cdc.gov/physicalactivity/everyone/guidelines/adults.html).

## Critical Thinking 30-2

Keep a diary for one week of all that you eat, the type and amount of exercise you do, and how many hours of sleep you get each night. See Figure 30-3 for a sample activity diary. At the end of the week, assess to see how well you have taken care of yourself. Is this a typical week? Do you need to make any changes? Were there any surprises? You can also record for several weeks and compare the outcomes.

| | Breakfast | Lunch | Dinner | Snacks | Exercise Type/ Duration | Hours of Sleep |
|---|---|---|---|---|---|---|
| Monday | | | | | | |
| Tuesday | | | | | | |
| Wednesday | | | | | | |
| Thursday | | | | | | |
| Friday | | | | | | |
| Saturday | | | | | | |
| Sunday | | | | | | |

Delmar/Cengage Learning

FIGURE 30-3  Sample activity diary.

Patient safety is the utmost concern. The drive home after work for a nurse can be equally dangerous. For each extra shift worked during one month, there is a 9.1% increased risk of a motor vehicle accident during the commute home from work. Nurses who work rotating shifts are at the greatest risk of fatigue compared to those committed to one shift. Nurses who rotate to evenings or nights are almost twice as likely to nod off while driving home (Frank, 2005).

There is no magic formula to guarantee a good night's sleep, but the following suggestions may improve the quality of your sleep life:

- Make sleep a priority. Make a conscious decision to obtain adequate sleep every night.
- Do not use caffeine as a stimulant to stay awake or alcohol as a tool to fall asleep.
- Try drinking warm milk or decaffeinated tea at bedtime. Establish a routine before bed that is repeated nightly.
- Reserve your bed only for sleeping. Watching television or doing paperwork in bed can cause sleepiness early and interfere with sleep patterns.
- If thoughts of work prevent you from falling asleep, write your worries on a piece of paper. Leave your thoughts on paper, and make a plan to deal with worries the next day while awake.

Other steps that nurses can take to prevent fatigue and help to reduce errors include the following:

- Use evidence-based fatigue guidelines.
- Take an uninterrupted 15-minute break every four hours. Take a short walk or climb a set of stairs.
- Develop a system with coworkers to monitor fatigue levels and double-check important tasks.
- Join the safety committee at work.
- Rotate shift work clockwise, when possible—that is, days, evenings, nights.
- Build in a sanctioned short nap, especially on the night shift.
- Share a ride home or use public transportation (Frank, 2005; Scott, Wei-Ting, Rogers, Nysse, Dean, & Dinges, 2007).

Adequate sleep, good nutrition, and proper exercise all go hand in hand. When you are tired, you may eat more to compensate for the lack of sleep. Overeating can lead to unnecessary weight gain. And the weight gain and fatigue can lead to a lack of exercise. Proper balance among the three is critical.

# INTELLECTUAL HEALTH

Intellectual health is the second area of health and encompasses those activities that maintain intellectual curiosity. Intellectual health consists of the knowledge you accumulate and the ability to think. Intellectually *healthy people* are able to clearly process information and make sound decisions. They learn from experience, are flexible, and remain open to new ideas. For purposes of this chapter, the term *intellectual health* also includes personal financial planning.

The first step in maintaining intellectual health is assessment. Table 30-3 contains an assessment tool for intellectual health; this tool is designed to assess trends only.

## INTELLECTUAL ACUITY

Just as it is important to find a type of exercise you are passionate about, so it is important to find some activity outside nursing that is of interest. The list is endless—antique shopping, reading, painting, sewing, photography, and so on. Develop a new hobby. Keep your mind sharp by staying abreast of developments within your interest area. Another way to maintain intellectual acuity is to establish and maintain a financial portfolio.

## PERSONAL FINANCIAL PLANNING

The first step in personal financial planning is to identify your annual salary. In 2009, the average salary for all categories of registered nurses was $66,530. The top 10% earned over $93,000, while the lowest 10% earned less than $44,000 (Bureau of Labor Statistics, 2009).

Next, begin to think about the percentage of your salary you want to save; most experts recommend 10% to 15% (Table 30-4). There is no better time than now to invest in your future. Now is the time to begin saving for such things as a home, your children's education, and even retirement, no matter what your age.

Note that a nurse who is making $50,000 annually will make $1,800,000 in a 30-year working career. If this nurse invests $200 per month at 12% interest for the 30 years, the nurse will have more than $1,000,000 in a retirement account at age 65. Check the Web resources in this chapter for investing sites to review. Savings for retirement are three pronged: (1) Social Security funds, (2) retirement funds, and (3) additional personal savings.

## Social Security

Social Security is automatically taken out of every paycheck by your employer. The benefits from Social Security will not cover all retirement expenses, especially with today's projected life span and inflation. You will need other retirement money. You should annually check the accuracy of your Social Security account by reviewing the information sent to you by the Social Security Administration.

## Retirement Funds

The most common retirement funds are the 401K or 403b plans. The primary difference between the two is that the 403b is a plan offered by a nonprofit organization and the

### TABLE 30-3 Intellectual Health Assessment

| INTELLECTUAL HEALTH BEHAVIOR | YES | NO |
|---|---|---|
| 1. I read at least one book a month. | ___ | ___ |
| 2. I have a hobby I enjoy. | ___ | ___ |
| 3. I belong to a club or organization. | ___ | ___ |
| 4. I save 10% of my income in a 401K, 403b, or a Roth IRA savings plan for retirement. | ___ | ___ |
| 5. I have invested in a mutual bond fund. | ___ | ___ |
| 6. I have invested in a mutual stock fund. | ___ | ___ |
| 7. I know how much money I have invested in Social Security. | ___ | ___ |
| 8. I have a money market account. | ___ | ___ |
| 9. I plan to buy a home soon. | ___ | ___ |

401K is offered by a for-profit organization. For purposes of this discussion, the term 403b will be used.

Both the employee and employer contribute money to a 403b. This is a great way to save, because many health care institutions will match the funds that you contribute. The maximum annual contribution by law is $10,500 or 20% of earned salary. After your money is put into the fund, it is tax sheltered, meaning that you do not pay any taxes on the amount contributed until it is withdrawn. For example, if you earn $58,000 per year and contribute $5,800 to the 403b, you will be taxed on only $52,200 of income. If the money is withdrawn before you reach the age of 59 1/2, you will pay a 10% federal penalty. This is an incentive to keep the money in the account until retirement; the plan should be considered a long-term investment.

Each retirement plan offers a number of investment options such as stock funds or bond funds. Some plans severely limit the number of investment options. To learn more about the risks and benefits of various funding options, consult *Morningstar* or *Consumer Reports*. (See Exploring the Web at the end of this chapter.)

**Individual Retirement Account (IRA)** Another type of retirement fund is the individual retirement account (IRA). This fund may or may not be tax deductible, depending upon your income level. There are two kinds of IRAs: the traditional IRA and the Roth. Funds placed in a Roth IRA account grow tax free and are tax free when withdrawn at retirement. Money from a Roth IRA can be withdrawn early, tax free, if used for the purchase of your first home.

## Personal Savings Vehicles

After investing in retirement funds, you also have a few more options for investment. You can open a **money market account**. This is similar to a bank checking account, although it often requires a larger minimum amount of money to open the account. The interest rate is higher than that of a passbook savings account or a traditional bank checking account, and you have check-writing privileges. This is a place for money that you may need to access quickly.

You also have the option to invest in stock mutual funds, bond mutual funds, individual stocks, or individual bonds outside your retirement account. You have the option of investing in individual stocks and bonds, but such an investment requires more research. You can start by reviewing Valueline at your local library (see Exploring the Web at the end of this chapter for Valueline). Review Kiplinger's 25 Favorite Funds (Kiplinger Washington Editors Inc., 2011). There are many suggestions for a diversified portfolio of mutual funds for your consideration at the Kiplinger web site, listed in the references. The key to successful investment is to diversify, meaning to spread your money around in many different types of investment options, including stocks, bonds, mutual funds, and so on.

## Buying Real Estate

In the past, owning your own home has been a smart investment. When making a decision to purchase a home, obtain as much information as possible. Research properties in the area where you want to buy. Investigate the school system, tax base, typical list price, and average time homes are on the market. Drive by and examine the neighborhood. Is this where you would want to live? Next, plan how to finance the property. Work with a real estate agent.

## How to Educate Yourself

There are many ways to learn more about investments. Try taking a course on personal finance at your junior college. You can also go to the Internet for advice (see Exploring the Web at the end of this chapter).

Another option is to hire a financial planner; however, this can become expensive. Check Fidelity for free advice at www.fidelity.com. The last suggestion is to read. Many books on the best-sellers list discuss personal finance. Suze Orman (2010) and Jean Chatzky (2008) are both excellent authors on this subject. Many people find them easy to understand and very relevant. Good financial magazines include *Money*, *Kiplinger's*, and *Consumer Reports*; read an article periodically—they become more interesting as you learn.

**TABLE 30-4    Time Value of Money**

A nurse who invests $200 per month at a 6% annual rate of return beginning at age 20 will earn $527,031 by the age of 65. However, even if the nurse increases this amount to $300 per month but does not begin investing until the age of 30, the nurse will only earn $414,077 by age 65 (Fiscal Agents, 2006).

## REAL WORLD INTERVIEW

Buying your first home is probably the smartest and biggest investment you will ever make. If you are currently renting, buying a home is a good start to a secure financial future. While renting has its pluses—that is, no long term commitment and no lawn care or maintenance—it does have its drawbacks. For example, the rent you are paying every month goes into your landlord's pocket and does not benefit you in any way. When your lease is up and you move on, you have nothing to show for your hard-earned money.

Let us say that the rent you have been paying is $600 a month. For the same monthly payment, you may be able to qualify for a $100,000 mortgage. Of course there will be other expenses such as utilities, maintenance, insurance, and taxes, but every penny you spend towards your mortgage will build financial security for you.

Once you are a homeowner, there are many advantages. You can write off all of your house taxes and interest on April 15th. This can reduce the tax you owe or give you a bigger tax refund. Another advantage to owning your home is that year after year your property may go up in value. This can build more financial security for you. Make sure you always choose a reputable realtor who knows the market and has your best interests in mind. This will be the first step to gathering information that you can use to find your dream home.

**Pamela Gwozdz, Realtor**

Next Chicago Realty
Chicago, Illinois

## EMOTIONAL HEALTH

Emotional health is the third area of health. Our emotions express how we are feeling about an event. Emotions can be intense, and each emotion evokes a strong response. Our challenge as human beings is to acknowledge the emotion and then respond appropriately. It is important to have balance between our thought processes and the emotion we are feeling; otherwise, disharmony occurs. Emotions are what make us human. Truly, emotions are one of our greatest gifts and add spice to our lives (Dossey, Keegan, & Guzzetta, 2005).

### EMOTIONAL INTELLIGENCE

Emotional intelligence is the ability to recognize your own feelings and the feelings of those around you and manage your emotions in a positive manner. Emotional intelligence requires a mix of practice and skill. Emotional intelligence includes these five basic emotional and social competencies:

- *Self-awareness:* Knowing what you are feeling in the moment and using these preferences to guide your decision making; having a realistic assessment of your abilities and a well-grounded sense of self-confidence.
- *Self-regulation:* Handling your emotions so that they facilitate rather than interfere with the task at hand,

being conscientious and delaying gratification to pursue goals, and recovering well from emotional distress.
- *Motivation:* Using your deepest preferences to move and guide you toward your goals, to help you take initiative and strive to improve, and to persevere in the face of setbacks and frustrations.
- *Empathy:* Sensing what people are feeling, being able to take their perspective, and cultivating rapport and attunement with a broad diversity of people.
- *Social skills:* Handling emotion well in relationships and accurately reading social situations and networks, interacting smoothly; using appropriate skills to persuade and lead, negotiate, and settle disputes for cooperation and teamwork (Goleman, 1998).

Take a minute to assess your emotional health (Table 30-5). This tool is designed to assess trends only. The first step toward emotional health is acknowledgment of feelings.

### ANGER

Anger is a common emotion that we all may feel at one time or another. As a matter of fact, anger is pervasive in our society today. There is road rage, airplane rage, outraged customers, and rage at sporting events. The causes for the anger are numerous. Some of the anger may be due to rapid changes in society related to high technology; a lack of privacy, because

**TABLE 30-5 EMOTIONAL HEALTH ASSESSMENT**

| EMOTIONAL HEALTH BEHAVIOR | YES | NO |
|---|---|---|
| 1. I don't often get angry. | ____ | ____ |
| 2. When I do get angry, I keep my anger under control. | ____ | ____ |
| 3. I laugh every day. | ____ | ____ |
| 4. My friends make me laugh. | ____ | ____ |
| 5. I reward myself for something every day. | ____ | ____ |
| 6. I am aware of my emotions and am confident in my decision making. | ____ | ____ |
| 7. I can handle my emotions appropriately. | ____ | ____ |
| 8. I am motivated to constantly improve. | ____ | ____ |
| 9. I am empathetic to others. | ____ | ____ |
| 10. I can handle social situations. | ____ | ____ |

we are accessible to work at all hours through cell phones, BlackBerries, and beepers; a sense of entitlement; a lack of family connection; and overcrowding.

There seems to be a spillover of anger into the nursing profession. Many sources for this anger are evident when one considers the increasing complexity of patient care. Add to this, economic problems, staffing shortages, high turnover of experienced staff, and demeaning treatment by other health care staff, and one begins to see some of the causes of anger.

## Ways to Cope with Anger

An effective way of coping with anger is to prevent its development. Teaching resilience as part of nursing education is one approach to this goal. Resilience is the ability to cope with and adapt to adversity, which is a desirous quality to have in the stress-filled environment of health care (McAllister & McKinnon, 2009). Evidence is growing that the skill of resilience can be learned.

Many strategies exist for coping with anger. Sometimes, anger and frustration are caused by sensory overload or overcommitment. It is important to have time for yourself. If nurses are to be effective caretakers, they must first care for themselves. Saying no, be it to a supervisor, friend, or family member, may at times be necessary. Learn to say no.

## HUMOR

Laughter is the best medicine. Laughter has many benefits, such as helping to boost the immune system, promoting relaxation, and decreasing blood pressure, heart rate, and respiratory rate (Holistic Online, 2006). But best of all, laughter is contagious and it is free. Laughter is a critical stress reliever (Figure 30-4).

## Ways to Make Yourself Laugh

Set a goal for yourself—not to let a day go by without a hearty laugh. Surround yourself with people who can joke about life. Read a humorous book; watch a funny television program or movie. See how many people you can get to smile in a day. Treat others at work to a little laughter. It is a great way to engender trust and teamwork. You can also learn to appreciate the humor at work. Visit "Ivy Push," the acting name of Hob Osterlund, MS, RN, CHTP, who has created DVDs of comical nursing situations. You can obtain more information about her shows at www.ivypush.com.

## STRESS MANAGEMENT

Many things can be done to relieve the emotional stress of life. See Table 30-6 for a few suggestions.

If these suggestions do not help, there are other options. You can seek out professional counseling. And,

FIGURE 30-4 **Laughter is an effective stress reliever.**

yes, sometimes nurses need a little extra help. You would be the first to call a counselor for a patient, so why not help yourself? There are also employee assistance programs that can help. Ask your human resource department at work.

A system of cognitive-behavioral self-help techniques for helping with stress and changing attitudes is available through Recovery International. People who practice the living skills detailed in the Recovery International Method learn to change their thoughts and behaviors and manage their stress better. See the Recovery International web site at http://www.lowselfhelpsystems.org/. There are Recovery meetings available in many areas of the United States, Canada, and the world.

## AVOIDING AUTOMATIC NEGATIVE THOUGHT DISTORTIONS

Research on thinking processes has shown that people make mistakes in the way they perceive information and think about the world around them. When people are depressed, their automatic thoughts can be loaded with distorted negative thinking. If you can recognize these automatic negative thoughts, (Table 30-7), you can begin to turn life in a more positive direction.

## PROFESSIONAL HEALTH

Professional health is the fourth element of health. You are professionally healthy when you are satisfied with your career choice and have continual opportunities for growth. A professionally healthy individual is goal directed and seeks

## TABLE 30-6 Stress Relief Suggestions

| | | |
|---|---|---|
| Meditate | Do relaxation exercises | Be polite to all |
| Think peaceful thoughts | Do something different for lunch | Take a walk |
| See things as others might | Give yourself a pat on the back | Read |
| Forgive your mistakes | Join a support group | Join a club |
| Do not procrastinate | Talk about your worries | Sing a song |
| Set realistic goals | Be affectionate | Forgive and forget |
| Do a good deed | View problems as a challenge | Listen to music |
| Vary your routine | Get/give a massage | Take a hot bath |
| Appreciate what you have | Say a prayer | Call an old friend |
| Focus on the positive | Expect to be successful | Let go of the need to be perfect |

# EVIDENCE FROM THE LITERATURE

**Citation:** Cooper, J. R. M., Walker, J. T., Winters, K., Williams, P. R., Askew, R., Robinson, J. C. (2009). Nursing students' perceptions of bullying behaviours by classmates. *Issues in Educational Research, 19*(3), 212–226.

**Discussion:** The purpose of this study was to explore types, sources, and frequency of bullying behaviors that nursing students experience while in nursing school. Bullying behaviors explored included: yelling or shouting in rage; inappropriate, nasty, rude, or hostile behavior; belittling or humiliating behavior; spreading of malicious rumors or gossip; cursing or swearing; negative or disparaging remarks about becoming a nurse; assignment, task, work, or rotation responsibilities made for punishment rather than educational purposes; a bad grade given as a punishment; hostility after or failure to acknowledge significant clinical, research, or academic accomplishment; actual threats of physical or verbal aggression; being ignored or physically isolated; and unmanageable workloads or unrealistic deadlines. The study also evaluated responses utilized by nursing students to cope with these bullying behaviors. These were: did nothing; put up barriers; pretended not to see the behavior; reported the behavior to a superior/authority; went to a doctor; perceived the behavior as a joke; demonstrated similar behavior; shouted or snapped at the bully; warned the bully not to do it again; spoke directly to the bully; and increased the use of unhealthy coping behaviors such as smoking, overeating, and increased alcohol consumption. Six hundred thirty-six participants completed the Bullying in Nursing Education Questionnaire (BNEQ). All respondents reported at least one encounter with bullying behavior. Fifty-six percent reported that the most frequent source of bullying behaviors was school of nursing (SON) classmates. Cursing and swearing, inappropriate behaviors, and belittling or humiliating behaviors by classmates were the most frequently reported bullying behaviors. The most frequently reported behavior used in response to bullying was "did nothing."

**Implications for Practice:** This study illustrates that bullying by nursing classmates occurs frequently. Recommendations to address bullying include adoption of "zero tolerance" policies and education and training for students, faculty, and health care agency employees.

## TABLE 30-7  Automatic Negative Thoughts

| AUTOMATIC NEGATIVE THOUGHT | EXAMPLE |
| --- | --- |
| All-or-nothing thinking: seeing things only in absolutes | If I leave this area of nursing, no one will respect me. |
| Overgeneralization: interpreting every small setback as a never-ending pattern of defeat | Everyone here is so smart; I'm a real loser. |
| Dwelling on negatives: ignoring multiple positive experiences | I made a mistake. I'm not good enough to be a nurse. |
| Jumping to conclusions: assuming that others are reacting negatively without definite evidence | I don't know why I study. Everyone thinks I'm going to fail the NCLEX-RN test anyway. |
| Pessimism: automatically predicting that things will turn out badly | It's only a matter of time before everything falls apart for me. |
| Reasoning from feeling: thinking that if one feels bad, one must be bad | My head hurts because I'm a bad person. |
| Obligations: living life around a succession of too many "shoulds," "shouldn'ts," "musts," "oughts," and "have-tos" | I should volunteer for that committee because my nursing director wants me to. It will impress my boss. |

**Source:** Compiled with information from Burns, David D. (1999). *Feeling Good: The New Mood Therapy.* Avon Books: New York, NY.

## TABLE 30-8 Professional Health Assessment

| PROFESSIONAL HEALTH BEHAVIOR | YES | NO |
|---|---|---|
| 1. I have professional goals, including certification and/or additional nursing education. | ____ | ____ |
| 2. I have a mentor. | ____ | ____ |
| 3. I have attended at least three educational workshops in the past year. | ____ | ____ |
| 4. I subscribe to three nursing journals. | ____ | ____ |
| 5. I belong to at least one professional organization. | ____ | ____ |
| 6. I use appropriate personal protective equipment. | ____ | ____ |
| 7. I never recap a needle. | ____ | ____ |
| 8. I use good body mechanics when transferring patients. | ____ | ____ |
| 9. I follow standards of care when handling gaseous waste, disinfectants, and chemotherapy. | ____ | ____ |
| 10. I follow standards of care in dealing with radiation equipment and environmental hazards. | ____ | ____ |
| 11. I don't abuse alcohol and/or drugs. | ____ | ____ |
| 12. I protect myself from workplace violence. | ____ | ____ |

every opportunity to obtain knowledge and new learning experiences. You can assess your professional health by using Table 30-8. This tool is designed to assess trends only.

## WAYS TO MAINTAIN PROFESSIONAL HEALTH

There are numerous ways to continue to advance your career. Ask yourself where you want to be 5, 10, and 15 years from now. If you do not have an overall plan, work can become very monotonous.

It is important to seek out others within the health care field. Find a more experienced nurse you can relate to or who can act as a mentor. This nurse can provide guidance and support when problems arise. Network with other nurses and health care professionals. You can learn an enormous amount from others. Keep in touch with some of your favorite faculty members. They will enjoy hearing from you and can also offer useful advice. Join at least one professional organization. Attend as many workshops and professional meetings as possible. Many of these presentations may be paid for by your employer. After attending professional conferences, do not forget to apply for continuing education units (CEUs). You might want to return to school for an advanced degree. Refer to

Table 30-6 for ways to deal with the stress of school, work, and family commitments.

Read as often as possible. If questions arise at work, come home and look up the information in your books or on the Internet, or go to the health care library. Subscribe to at least three professional journals related to your specialty area.

## OCCUPATIONAL HAZARDS COMMON AMONG NURSES

An important aspect of professional health is avoidance of occupational hazards. The U.S. Department of Labor Statistics (2006) reports that health care and social assistance workers rank second highest in percentage of nonfatal workplace injuries. The survey reports the number of new work-related illness cases that are recognized, diagnosed, and reported during the year. Some conditions, for example, long-term latent illnesses caused by exposure to carcinogens, are often difficult to relate to the workplace and are not adequately recognized and reported. These long-term latent illnesses are believed to be understated in the survey's illness measures. In contrast, the overwhelming majority of the reported new illnesses are easier to directly relate to workplace activity—for example, contact dermatitis, carpal tunnel syndrome, or back injuries.

The cumulative weight lifted by a nurse providing direct patient care in a typical 8-hour workday is estimated to be 1.8 tons. Unfortunately, nurses accept back pain as part of their job, with 52% to 63% of nurses reporting musculoskeletal pain that lasts for more than 14 days; in 67% of cases, pain was a problem for at least 6 months (Nelson, Collins, Knibbe, Cookson, Castro, & Whipple, 2007).

The newest approach to this serious occupational problem has been the use of safe patient handling policies—at the facility, state, and national levels. More information about developing a safe patient handling policy is available at Patient Safety Center of Inquiry (Tampa, FL), Veterans Health Administration and Department of Defense (2005). Ergonomics for the Prevention of Musculoskeletal Disorders is available at U.S. Department of Labor, Occupational Safety and Health Administration (2009).

According to research conducted by Liberty Mutual Group (2007), the top 10 causes of disabling injury for all professions were:

| | |
|---|---|
| 1. Overexertion | 24% |
| 2. Falls on the same level | 14.6% |
| 3. Falls to a lower level | 11.7% |
| 4. Bodily reactions | 10.2% |
| 5. Struck by an object | 9.0% |
| 6. Highway incident | 4.7% |
| 7. Caught or compressed | 3.9% |
| 8. Repetitive motion | 3.8% |
| 9. Struck against object | 3.8% |
| 10. Assault/violent act | 1.1% |

There are numerous suggestions for safeguarding against various hazards in the workplace. For purposes of discussion, occupational hazards can be divided into four major categories: (1) infectious agents, (2) environmental agents, (3) physical agents, and (4) chemical agents. See Table 30-9.

## TABLE 30-9  Safeguards for Occupational Hazards

### Infectious Agents

- Do not recap needles.

- Use needle-free intravascular access devices.

- Place needle disposal containers near point of use of needles.

- Use personal protective equipment.

- Report all needle stick injuries.

- Wash hands before and after each patient contact.

### Physical Agents

- Follow standards of care for dealing with radiation/laser equipment.

- Assess work area for amount of noise.

- Eliminate excessive noise in the workplace.

- Implement good body mechanics.

- Follow ergonomic safety guidelines from OSHA (Dept. of Government Affairs, 2006b).

### Environmental Agents

- Develop a violence reduction plan.

- Rotate shifts clockwise—day to evening to night (Rogers, 1997).

- Assess for dangerous chemicals in your workplace.

- Determine whether OSHA standards are in place.

- Be aware of bioterrorist alert plans.

### Chemical Agents

- Utilize effective ventilation systems.

- Develop standards of care for handling hazardous agents.

- Protect pregnant nurses from handling chemotherapy during the first trimester.

- Use appropriate nonlatex barrier protections.

- Develop policies and procedures to ensure safety from latex allergies.

## Infectious Agents

Infectious agents can be transferred through direct contact with an infected patient or through exposure to infected body fluids. The major infectious agents for health care workers are HIV, herpes, tuberculosis, and hepatitis. A recent survey of 58 hospitals found that 44% of the needle-stick injuries were sustained by nurses compared to 15% by medical practitioners (IHCWSC, 2006). Nurses have the largest amount of direct patient contact and are at the greatest risk for exposure to blood-borne pathogens. The chance of a seroconversion, which occurs when a serological test for antibodies changes from a negative reading to a positive reading, varies according to the disease exposure. The estimated risk for HIV infection after needle exposure is from 0.29% to 0.56%. The risk for hepatitis B ranges between 6% and 30% for nonimmunized personnel, whereas hepatitis C carries an estimated risk of 6% (Lee, Botteman, Xanthakos, & Nicklasson, 2005).

Although the majority of infectious agents are transmitted through blood, herpes simplex virus can be transmitted by direct contact with an infected lesion. Hepatitis A is transmitted primarily via diarrhea as a result of poor hygienic practices among health care workers. The overall incidence of tuberculosis (TB) is gradually increasing, and there are certain regional differences. For example, in 2004, the rate of cases per 100,000 population differed by states: 1.0 in Wyoming, 7.1 in New York, 8.3 in California, and 14.6 in the District of Columbia. The overall rate for TB in the United States in 2004 was 4.9 per 100,000 people, and in 2000 the rate was only 3.5 per 100,000 people (CDC, 2006).

## Environmental Agents

Another group of occupational hazards is environmental agents. These include all those agents within the hospital that may lead to injury. The most prevalent include violence, shift work, air quality, mold and fungus, and bioterrorism. Nurses are at risk for workplace violence. A job in a health care facility is considered one of the most dangerous in the United States. In a recent study, nearly half a million nurses per year reported being victims of some type of violence in the workplace (Department of Government Affairs, 2006).

Poor air quality in the workplace is yet another environmental risk that may lead to symptoms such as shortness of breath, eye and nose irritation, headaches, contact dermatitis, joint pain, memory problems, and reproductive difficulties. Glutaraldehyde, a chemical used to disinfect many commonly used instruments, can emit a hazardous gas. Lasers used in operating rooms can emit hazardous gaseous material in the form of either laser plume or chemical by-products of laser smoke. It is essential that nurses be protected from the by-products of lasers by high-efficiency smoke evacuators. Nurses should always be aware of and follow safety guidelines associated with the technology and equipment they work with and around. Air quality can also be influenced by mold and fungus, which are often found in carpeting and in ceiling tiles. The presence of mold and fungus can lead to asthma and other respiratory problems.

Most hospitals have some type of bioterrorist alert plan. Nurses play an active role in dealing with any type of bioterrorist disaster.

## Physical Agents

Physical agents are another occupational hazard and include radiation, noise, and back strain. Radiation exposure is common among health care workers. Radiation is used for both diagnostic and treatment interventions. Persons exposed to excessive amounts of radiation are at risk for cancer. Nurses can protect themselves from the effects of radiation by following agency guidelines and wearing a dosimeter that measures the amount of radiation exposure. Pregnant nurses working with radiation should declare their pregnancy to their employer as soon as possible (Duke University, 2006). Laser treatment carries the risk of eye and skin injury if the instruments are not handled properly.

Noise is another physical hazard and can lead to hearing loss. Excessive noises that occur over a long period of time can lead to irritability and inability to concentrate. High levels of noise are deleterious for both nurses and patients. Special care units are especially noisy, with alarms, ventilators, suction equipment, monitors, call lights, and so on.

## Chemical Agents

Another occupational hazard is chemical agents such as anesthetic agents, antineoplastic drugs, disinfectants, latex gloves, hazardous agents, and drug and alcohol abuse. Questions still remain as to the negative effects of anesthesia on health care workers. Nurses and other health care professionals working in the operating room should continuously protect themselves by ensuring proper ventilation and appropriate disposition of waste products.

Hazardous drugs are any medications that nurses may be exposed to, either during preparation or administration of medication, that pose a potential health risk. These drugs require special handling due to inherent toxicities (Nixon & Schulmeister, 2009). Strict adherence to the agency guidelines are essential when handling these hazardous drugs.

Disinfectants are chemical agents that are of concern to nurses. Besides the gaseous effects of glutaraldehyde, which were discussed earlier, glutaraldehyde can also cause skin problems. Ethylene oxide, a chemical commonly used to sterilize surgical equipment, has been reported to have carcinogenic effects.

## EVIDENCE FROM THE LITERATURE

**Citation:** Tabone, S. (2005). Safe patient handling. Texas Nursing, 3, 10–11.

**Discussion:** The profession of nursing continues to demonstrate one of the highest rates of back injuries in comparison to other professions. Nurses also have high compensation claims, which cost approximately $20 billion per year. Sadly, over 50% of nurses report the existence of chronic back pain. Up to 12% of nurses who leave the profession do so because of back injury.

Though the Occupational Safety and Health Administration (OSHA) published standards in the year 2000 to help prevent work-related back injury, hospitals and nursing homes were slow to adopt the suggested changes in practice. In 2004, OSHA issued mandatory guidelines that minimized lifting and injury when moving patients. The implementation of the guidelines has decreased injuries among health care professionals in these settings by up to 80%. As an added benefit, some institutions also saw an increase in staff retention, with a subsequent decrease in the need for training and other administrative expenditures. These findings have led to a new program offered through the American Nurses Association (ANA) to encourage further adoption of OSHA guidelines. States such as Florida and Texas have considered proposing legislation to augment efforts by OSHA and the ANA. Note, however, that legislation and specialized equipment can only work if the institution embraces a comprehensive approach to this issue, including:

- A change in culture with a commitment of management and staff to use ergonomics workplace design to maximize productivity and reduce staff fatigue and discomfort
- The provision of training about prevention of injury and the use of specialized equipment
- The offering of comprehensive occupational management of musculoskeletal disorders
- Conducting ongoing evaluation to assess how to continue to improve

**Implications for Practice:** Prevention of musculoskeletal injuries is possible without sacrificing quality of patient care. Guidelines have been developed by OSHA and endorsed by the ANA. When the use of these OSHA guidelines is combined with a comprehensive commitment to this issue, a significant impact can be made in reducing injury to nurses.

Latex glove exposure is another type of chemical concern for nurses. It has been reported that approximately 8% to 13% of health care workers are allergic to latex (AANA, 2006). Reactions to latex range from a simple contact dermatitis to a more systemic reaction to an anaphylactic crisis.

Yet another concern for nurses is the potential for personal misuse of drugs and/or alcohol. The ANA estimates that 6% to 8% of nurses may have a substance-abuse problem. This is similar to the national average for drug abuse. Nurses should be aware of the potential for addiction and abuse.

## SOCIAL HEALTH

Social health is another significant area of health. The essence of social health is interacting with other people. Having the ability to relate to others is essential for life. Few can survive completely alone. We relate to people at various levels—some we know intimately, and others are mere acquaintances. These relationships occur within the immediate and extended family; at work; and within

the local, national, and international community. These relationships give meaning to our lives. There are times when relationships cause distress and pain and other times when they bring great joy. We strive toward harmony in all relationships. It is human nature to seek out others and grow in relationships (Dossey, et al., 2005). Take a minute to assess your status with relationships by taking the self-assessment in Table 30-10. This tool is designed to assess trends only.

## IMPACT OF SOCIAL RELATIONSHIPS

The positive effects of social support on health outcomes have been documented in the literature (Breedlove, 2005). If these interactions are frequent, that only adds to good health. In other words, the more you see your friends, the healthier you become. The variety of those relationships may also keep you healthy. The greater the diversity of the relationships—such as professional, family, neighborhood, or church relationships—the more likely you are to remain healthy.

## CASE STUDY 30-1

You are a nurse admitting Mrs. Zakima, an 84-year-old female. You received the following report from the emergency room:

Her vital signs are, BP 140/70, RR 30, HR 80. The patient has bilateral rales, 2+ pitting pedal edema, weighs 350 pounds, and her height is 5' 5". Mrs. Zakima is a transfer from a nursing home. She is being admitted for symptoms related to congestive heart failure. She has a past history of hypertension, diabetes, rheumatoid arthritis, dementia, and MRSA. Mrs. Zakima is being admitted to your unit. In preparing the room that she will be admitted to, what do you need to keep in mind with regards to noise, safety, and universal precautions?

What safety precautions would you put in place to protect yourself, that is, ergonomics, universal precautions, and so forth?

After reviewing her chart, you find she is taking the following medications: methotrexate, zestril, toprol, and lasix. What other precautions should you take with this patient?

**TABLE 30-10  Social Health Assessment**

| SOCIAL HEALTH BEHAVIOR | YES | NO |
|---|---|---|
| 1. I go out with my friends at least once a week. | ___ | ___ |
| 2. I see my family at least twice a month. | ___ | ___ |
| 3. I have one friend I can confide in and feel close to. | ___ | ___ |
| 4. I have a wide diversity of professional, family, neighborhood, and church friends. | ___ | ___ |
| 5. I have other friends besides nurses. | ___ | ___ |
| 6. I do volunteer work. | ___ | ___ |

Try to interact with friends in person as often as possible. Cell phones and e-mail are helpful forms of communication, but nothing is more effective than face-to-face conversation. Be careful, though, to stay away from negative relationships and people who do not treat you well. Sometimes it is difficult to end a friendship, but if the relationship is destructive, it may be in your best interest to end it.

Another way to build relationships is through rituals. For example, celebrate Christmas with friends by seeing *The Nutcracker* or play tennis with a friend once a week. But do not forget about your friends of the past. Phone a friend from the past or contact a family member you have argued with.

Finally, another way to establish friendships is through volunteerism. For example, join the American Red Cross as a disaster volunteer or pursue another activity that you find rewarding. It is also a good way to be aware of community issues.

## SPIRITUAL HEALTH

Spiritual health is the last area of health. *Spirituality* is an elusive term that is difficult to define. It can be viewed as the essence of being, that which gives meaning and direction in life, and the principles of good living. It may also be your relationship with a higher being, other individuals, or your environment (Dossey, et al., 2005). See Table 30-11.

Nurses function in a fast-paced and stressful environment, which often leads to job dissatisfaction, burnout, and potentially suboptimal patient care. Dossey and Keegan (2009) describe how nurses can gain insight into their

## REAL WORLD INTERVIEW

I try to integrate my personal and professional life by organizing and prioritizing at the beginning of each day. Each night I outline with my family who needs to be where and how my husband and I will get the children to their various activities. In this way, I know my family will be taken care of during the day while I am at work. My family is my first priority. I view my social life in much the same manner. What needs to be accomplished today, and how will I go about achieving this?

I organize my professional life every morning before I officially begin my day. I ask myself what the issues are that most need my attention today. Each Friday is "turnaround time." I assess the events of the week and determine all the good things that I did and those things that I could improve upon. I make sure I reward myself for the positives. I encourage all my managers to implement this weekly method of evaluation.

I have also developed a competency-based checklist for my managers. This is a tool that I use to continually assess and guide improvement of the managers. The tool outlines communication, leadership, decision making, and interpersonal skills, just to name a few. I may begin by going over the assessment areas weekly and gradually taper the time needed to teach and assess the managers. Once the manager has mastered an area, there is no longer a need to use the tool. It is my goal for the managers to exhibit competency in all areas.

Another way that I try to keep balance in my life is to work hard and play hard. I make sure that I have fun on my weekends. It is important that I have humor in my life on a daily basis. My staff can laugh at me and I at them. This is healthy and a great stress reliever. Family and friends are very important to me. I try to incorporate family and friends into my daily life.

**Corinne Haviley, RN, MS**
Director of Medicine Nursing
Northwestern Memorial Hospital
Chicago, Illinois

## TABLE 30-11   Spiritual Health Assessment

| SPIRITUAL HEALTH BEHAVIOR | YES | NO |
|---|---|---|
| 1. I pray or meditate every day. | ____ | ____ |
| 2. I believe in a higher power. | ____ | ____ |
| 3. I regularly attend meetings at a place of worship. | ____ | ____ |
| 4. I spend time each day in quiet contemplation. | ____ | ____ |
| 5. I read spiritual books. | ____ | ____ |
| 6. I listen to spiritual audiotapes. | ____ | ____ |
| 7. I am at peace. | ____ | ____ |
| 8. I feel confident when confronted with new situations. | ____ | ____ |
| 9. I maintain a daily journal. | ____ | ____ |

own spirituality by exploring ways to nurture themselves through ritual, rest, play, and expressions of creativity.

## RELIGION AND HEALTH

Koenig (1999), a leader in spirituality research, found that older adults who are religiously active (based on church attendance) are more physically fit and live longer than those who are less religious. In the past decade, there have been numerous studies documenting the positive effects of religiosity and spirituality on psychological and physical health (Berry, 2005). It appears that those who are more religious or spiritual have a greater chance of adjusting to

life and maintaining good health. See Suggested Readings for other resources on spirituality.

## DECISION MAKING

Every day, we make decisions. Some days, we may decide to care for others and put their needs before our own. But on other days, we need to care for and nurture ourselves. The decision is always up to us. These decisions are often complicated and difficult to make. The goal of this chapter has been to provide you with some thoughts that will guide your decisions concerning your health.

### Critical Thinking 30-3

There is evidence suggesting that people can improve their happiness. This evidence suggests that everyone can benefit, to varying degrees, from the various habits and practices identified by Positive psychology (Seligman & Csikszentmihalyi, 2000). Take a few minutes each evening to sit and think about the day's activities. Make sure you are comfortable and will not be interrupted by outside stimuli. Ask yourself, what are the 3 things that made me happiest today? Write them down in your Journal. No matter how busy you become, do not forget to save this time for yourself. The busier you become, the more important it is to take time to reflect and relax. Note that some happiness-boosting habits include expressing gratitude and appreciation and practicing altruistic selfless concern for the welfare of others. Getting better exercise and a healthier diet have also been proven to have strong effects on mood. Be good to you!

### REAL WORLD INTERVIEW

I have learned a lot in the nine months that I have practiced nursing. I used to say yes to overtime each time when asked. I thought I needed the experience and that I owed it to the staff. I have since learned to say no. I have also learned that I need to take care of myself. I used to come home and eat a lot and fall into bed. I gained weight and felt awful. Now I either go for a walk or to the health club after work. I volunteer at my church youth organization part-time whenever I can. I have also developed my own routine in delivering patient care. I am better organized and leave on time. I have gained more confidence in my clinical decisions. For example, I had a patient who I thought should not be extubated from the ventilator. I tried to voice my opinion to the resident on call but was overridden by the attending physician. The charge nurse was also supportive of keeping the patient intubated. As it turned out, the patient was extubated for a half hour and then reintubated. I felt bad for the unnecessary procedure for the patient, but it did boost my confidence. I know now that in the future I will be even more assertive with physicians concerning the welfare of my patients.

When I am scared in a clinical situation, I try not to let fear paralyze me. I try to think everything through logically. I have found that the better I understand what I am doing, the better nurse I become. I'm always trying to learn new things. I've been to about five seminars this year, and I subscribe to three nursing journals. I want to learn as much as I can.

**Nayiri M. Birazian, RN**
New Staff Nurse
Maywood, Illinois

## CASE STUDY 30-2

You have been working for a home health agency for about one year now. One of your best friends at work, another nurse, asks your advice about how to maintain a healthier lifestyle. She states that she is 10 pounds overweight and does not really exercise much. She says she feels like all she does is work and come home and go to sleep. She says she does not have fun in life anymore and that college life was great compared to all the stress at work.

What advice would you give your friend about how to stay healthy? What would your first priority be in offering suggestions? What other major areas of consideration would you explore with your friend?

## KEY CONCEPTS

- Nurses can care for themselves by maintaining a healthy lifestyle.

- Health is not just the absence of disease, but a balance among physical, intellectual, professional, social, and spiritual well-being.

- Health is multidimensional, with areas that interact and overlap. If one dimension is altered, all other areas are impacted.

- Physical health includes good nutrition, proper exercise, and adequate sleep.

- It is important to strengthen your intellectual health and keep current in both personal and professional areas.

- An important piece of intellectual health is adequate financial planning. Now is the time to begin saving.

- Emotional health includes laughter and control of anger.

- Maintaining many different types of relationships helps keep you healthy.

- Spiritual well-being provides direction and meaning in life.

- To stay healthy, you must make a conscious decision to maintain each of the six areas of health: physical, intellectual, emotional, professional, social, and spiritual health. The choice is yours.

## KEY TERMS

health                money market account                physical health                resilience

## REVIEW QUESTIONS

1. You are trying to maintain a healthy diet. Which of the following would you include in your daily consumption? Select all that apply.
   _____ A. Peanut butter
   _____ B. Four glasses of fruit juice
   _____ C. Green leafy vegetables
   _____ D. White bread
   _____ E. Nuts
   _____ F. Whole milk

2. You have just calculated your BMI to be 28. How much exercise would be appropriate for you?
   A. 30 minutes three days a week
   B. 30 minutes every day
   C. 60 minutes three days a week
   D. 60 minutes daily

3. You work in the operating room and notice that your hands have become very itchy and have small papules. What would be the most appropriate first action to take?
   A. Wear only nonlatex gloves in the future
   B. Ignore the situation
   C. Make a conscious effort to apply hand cream after washing your hands
   D. Apply a corticosteroid cream

4. The CNS for your unit has just passed you in the cafeteria. You are generally on very good terms with her. She does not acknowledge your presence. The most probable explanation for this is which of the following?
   A. You must have done something wrong with one of your patients yesterday.
   B. She has something else on her mind and didn't see you.
   C. She doesn't want to acknowledge you in front of other nursing leaders because you are only a staff nurse.
   D. She doesn't want to acknowledge you because you are one of the worst nurses on your unit.

5. Lack of attention to spiritual health can lead to which of the following? Select all that apply.
   _____ A. Lack of sleep
   _____ B. Job dissatisfaction
   _____ C. Burnout
   _____ D. Increased sense of well-being
   _____ E. Suboptimal patient care
   _____ F. Shorter life span

6. Social support is essential to health because it does which of the following?

A. Facilitates weight loss
B. Provides an opportunity to relate to others
C. Improves intellect
D. Improves creative thinking

7. Research has shown that sleep deprivation can lead to a significant increase in which of the following?
   A. Salary
   B. Job satisfaction
   C. Spirituality
   D. Patient care errors

8. Retirement financial planning includes all but which of the following?
   A. Social Security
   B. Playing the lottery
   C. Retirement funds
   D. Personal savings

9. Examples of activities promoting intellectual health include all but which of the following?
   A. Drinking coffee
   B. Reading books
   C. Engaging in a hobby
   D. Becoming a member of an organization

10. Establishing professional goals is an example of which of the following?
   A. Aggression
   B. Physical health
   C. *Healthy People 2020* indicators
   D. Professional health

## REVIEW ACTIVITIES

1. Try doing a short relaxation exercise. Take a deep breath in and let it out. Take slow, deep breaths that originate from the diaphragm. Tighten the muscles in your right arm for 30 seconds and release. Your arm should feel totally limp and relaxed. Do the same with your left arm. Tighten the muscles in your right leg for 30 seconds and release. Repeat with the left leg. Pull in your stomach muscles for 30 seconds and release. Tighten the muscles in your buttocks and release. Continue to breathe in and out deeply. You can practice this brief exercise anytime or anywhere. If you are having a particularly hectic day, take a minute to do a relaxation exercise. You can vary the exercise by flexing any group of muscles you want.

2. Your best friend is getting married next month, and you are the maid of honor. You have been invited to a shower to be held out of town in honor of your friend this next weekend. You have already purchased a nonrefundable airline ticket to attend the shower. You work in a very small ICU. You have been working 10- and 12-hour shifts and are near exhaustion. Your head nurse calls you two days before you are to leave for the shower and asks you to work the weekend. One of the staff has been involved in a serious car accident, and there is no one else to work. What would you do?

   If you work this weekend, you would disappoint your best friend and lose the money for your ticket. You are already exhausted, and you do not know how effective you would be at work. You know you need the break.

   If you do not work, you would be letting the rest of the staff down. They have been there for you, and now it is your turn to help them. You find it very hard to say no. You are a young nurse and you feel you should "pay your dues." How could you relax when you know that you are needed elsewhere?

3. You are interviewing for your first job. You are attending the human resource session of orientation. Which investment benefits would you be most concerned about and why?

# EXPLORING THE WEB

- Calculate your BMI and determine your life expectancy and health risks at www.healthstatus.com

- Try one of the following sites to retrieve information on dietary supplements, nutrition, and alternative medicine:
  www.mypyramid.gov
  www.nutritionsite.com

- Attend a meeting of The Recovery International Method to deal with feelings of stress and anxiety, available at www.lowselfhelpsystems.org

- Vanguard: www.vanguard.com

  Click on What We Have to Offer, and then click on the Overview tab, where you will find basic retirement planning information. Note especially the index funds.

- Note Web sites for retirement information: Charles Schwab (www.schwab.com; click on Planning and Retirement). Kiplinger (http://www.kiplinger.com; search for Kiplinger 25 Fund Portfolios).

- Fidelity Investments: www.fidelity.com Click on Planning and Retirement, and follow the different retirement options.

- Valueline: www.valueline.com

- Morningstar: www.morningstar.comClick on Real Life Finance.

- Consumer Reports: www.online.consumerreports.org

## Resources for Violence Prevention

- The ANA's "Workplace Violence: Can You Close the Door?"

Call (800) 274-4ANA
www.nursingworld.org

- Guidelines for Preventing Workplace Violence for Healthcare and Social Service Workers, U.S. Department of Labor, OSHA 3148-1996, available online: www.osha.gov

## Resources for Needlesticks

General resource links:

- Safer Needle Devices: Protecting Health Care Workers: www.osha.gov

- American Nurses Association: www.nursingworld.org

- Centers for Disease Control and Prevention: www.cdc.gov

## Resources for Latex Allergy

- ANA's position paper on latex allergy: www.nursingworld.org or call (800) 274-4ANA

- OSHA: www.osha.gov

## Resource for Back Strain

- Occupational Safety and Health Administration's ergonomics information: www.osha.gov
  Call (202) 693-1999.

# REFERENCES

American Association of Nurse Anesthetists. (2006). *AANA latex glove protocol.* Retrieved January 30, 2006, from http://www.aana.com/crna/prof/latex.asp.

Berry, D. (2005). Methodological pitfalls in the study of religiosity and spirituality. *Western Journal of Nursing Research, 27*(5), 628–647.

Breedlove, G. (2005). Perceptions of social support from pregnant and parenting teens using community-based doulas. *Journal of Perinatal Education, 14*(3),15–22.

Bureau of Labor Statistics. (2009). Occupational employment and wages, May 2009, 29-1111 Registered Nurses; Retrieved April 26, 2011, from www.bls.gov/oes/current/oes291111.htm.

Burns, D. D. (1999). Feeling Good: The New Mood Therapy. Avon Books: New York, NY.

Centers for Disease Control and Prevention. (2006). Guidelines for preventing the transmission of Mycobacterium tuberculosis in health-care settings. Retrieved January 20, 2006, from http://www.cdc.gov.

Centers for Disease Control. (2010). Physical Activity for Everyone. Retrieved April 28, 2011, from http://www.cdc.gov/physicalactivity/everyone/guidelines/adults.html.

Dawson, D., & Reid, K. (1997). Fatigue, alcohol and performance impairment. *Nature, 338*(6639), 235.

Department of Government Affairs. (2006). *Health care worker safety.* Retrieved January 30, 2006, from http://www.anapoliticalpower.org.

Dossey, B. M., Keegan, L., & Guzzetta, C. E. (2005). Holistic nursing: A handbook for practice. Sudbury, MA: Jones and Bartlett Publishers.

Dossey, B., & Keegan, L. (2009). Holistic nursing: A handbook for practice. Jones and Bartlett Publishers. Boston: MA.

Duke University Medical Center. (2006). *Radiation safety considerations for nurses at Duke.* Retrieved January 30, 2006, from http://www.safety.duke.edu/RadSafety/nurses/default.asp.

Frank, M. B. (2005). Practicing under the influence of fatigue (PUIF): A wake-up call for patients and providers. *National Association of Neonatal Nurses, 5*(2), 55–61.

Goleman, D. (1998). *Working with emotional intelligence.* New York, NY. Bantam Doubleday Dell Publishing Group.

Holistic Online. (2006). *Therapeutic benefits of laughter.* Retrieved January 28, 2006, from http://www.holistic-online.com/ Humor_Therapy/humor_therapy_benefits.htm.

International Health Care Worker Safety Center (IHCWSC). (2006). Uniform needlestick and sharp injury report 58 hospital, 2001. University of Virginia. Retrieved January 28, 2006, from http://www.healthsystem.virginia.edu/internet/epinet/soi01.cfm.

Kiplinger Washington Editors Inc. (2011). Kiplinger's 25 Favorite Funds. Retrieved April 20, 2011 at http://www.kiplinger.com/investing/funds/kip25/tables/index.php.

Koenig, H. G. (1999). The healing power of faith. *Annals of Long-Term Care, 7*(10), 381–384.

Lee, J. M., Botteman, M. F., Xanthakos, N., & Nicklasson, L. (2005). Needlestick injuries in the United States: Epidemiologic, economic and quality of life issues. *American Association of Occupational Health Nurses Journal, 53*(3), 117–134.

Liberty Mutual Research Institute for Safety. (2009). The most disabling workplace injuries cost industry an estimated $53 billion. Retrieved April 20, 2011 from www.libertymutual.com/researchinstitute.

Lockley, S., Barger, L. K., Ayas, N. T, Rothschild, J. M, Czeisler, C. A., & Landrigan, C. P. (2007). Effects of health care provider work hours and sleep deprivation on safety and performance. *Joint Commission Journal on Quality and Patient Safety, 33*(1), 7–18.

McAllister, M., & McKinnon, J. (2009). The importance of teaching and learning resilience in the health disciplines: A critical review of the literature. *Nurse Educator Today, 29*(4), 371–379.

Money Management Editor. (2006). Tips and more to help you make the most of your Registered Retirement Savings Plan (RRSP). Retrieved April 20, 2011 from www.fiscalagents.com/newsletter/4tentips.shtml.

National Institute of Health. (n.d.). *Body mass index calculator.* Retrieved February 2, 2002, from http://www.nhlbi.nih.gov/ guidelines/obesity/bmi_tbl.htm.

Nelson, A.L.,Collins, J., Knibbe, H., Cookson, K., Castro, A.B., Whipple, K.L. (2007). Safer patient handling. Nursing Management, 38 (3): 26–33.

Nightingale, F. (1969). *Notes on nursing.* New York: Dover. (Original work published 1860.)

Nixon, S., & Schulmeister, L. (2009). Safe handling of hazardous drugs: Are you protected? *Clinical Journal of Oncology Nursing, 13*(4), 433–439.

Occupational Safety & Health Administration. (2003). Guidelines for nursing homes: Ergonomics for the prevention of musculoskeletal disorders, OSHA 3182. Available at http://www.osha.gov/ergonomics/guidelines/nursinghome/final_nh_guidelines.html.

Orman, S. (2010). Suze Orman's action plan: New rules for new times. New York: Spiegel and Grau. *Random House, Inc.*

Patient Safety Center of Inquiry (Tampa, FL), Veterans Health Administration and Department of Defense(2005). Patient Care Ergonomics Resource Guide: Safe Patient Handling and Movement. Retrieved April 26, 2011 from www.visn8.va.gov/visn8/patientsafetycenter/resguide/ErgoGuidePtOne.pdf.

Pender, N. J., Murdaugh, C. L., & Parsons, M. A. (2005). *Health promotion in nursing practice.* Upper Saddle River, NJ: Pearson Prentice Hall.

Rogers, B. (1997). Health hazards in nursing and health care: An overview. *American Journal of Infection Control, 25*(3), 248–261.

Roy, C., & Andrews, H. A. (1999). *The Roy Adaptation Model.* Stamford, CT: Appleton & Lange.

Scott, L. D., Wei-Ting, H., Rogers, A. E, Nysse, T., Dean, G. E., & Dinges, D. F. (2007). The relationship between nurse work schedules, sleep duration and drowsy driving. *Sleep, 30*(12), 1801–1807.

Seligman, M., & Csikszentmihalyi, M. (2000). "Positive Psychology: An Introduction". *American Psychologist 55*(1): 5–14.

U.S. Department of Health and Human Services.(2008). HealthyPeople. gov. Third Meeting: June 5 and 6, 2008. Secretary's Advisory Committee on National Health Promotion and Disease Prevention Objectives for 2020. Retrieved April 20, 2011 from http://healthypeople.gov/2020/about/Advisory/FACA3Minutes.aspx.

U.S. Department of Labor, Occupational Safety and Health Administration, OSHA 3182. (2009). Guidelines for Nursing Homes, Ergonomics for the Prevention of Musculoskeletal Disorders. Retrieved April 26, 2011 from http://www.osha.gov/ergonomics/guidelines/nursinghome/final_nh_guidelines.html.

United States Department of Agriculture (2011). MyPyramid.gov. Steps to a Healthier You. Retrieved April 20, 2011 from http://www.mypyramid.gov.

World Health Organization. (2006). *Benefits of physical activity.* Retrieved December 30, 2005, from http://www.who.int/ moveforhealth/advocacy/information_sheets/benefits.en.index.html.

## SUGGESTED READINGS

Blocks, M. (2005). Practical solutions for safe handling. *Nursing 2005, 35*(10), 44–45.

Boston Women's Health Book Collective. (2005). *Our bodies ourselves.* New York: Simon & Schuster Incorporated.

Frankl, V. (1984). *Man's search for meaning.* New York: Simon & Schuster, Inc.

Gladwell, M. (2005). *Blink.* New York: Little Brown & Company.

Greenblatt, E. (2009). Restore yourself: The antidote for professional exhaustion. Execu-Care Press. Los Angeles, CA.

Hunt, L. (2005). Sit-down comedy. Meet Ivy Push, nursing's funny girl. *American Journal of Nursing, 105*(7), 110–111.

Moreo, J. (2007). *You are more than enough.* New York: Stephens Press, LLC.

Ramen, R. N. (2000). *My grandfather's blessings: Stories of strength, refuge, and belonging.* New York: Berkley Publishing Company.

Ruiz, D. M. (1997). *The four agreements: A practical guide to personal freedom, a Toltec wisdom book.* San Rafael, CA: Amber Allen Publishing.

Weil, A. (2009). *Why one's health matters: A vision of medicine that can transform our future.* New York: Penquin Group.

Wright, L. A. (2005). *Spirituality, suffering, and illness.* Philadelphia: F. A. Davis Company.

# CHAPTER 31

# NCLEX Preparation and Professionalism

PATRICIA KELLY, RN, MSN

*For us who nurse, our nursing is a thing, which, unless in it we are making progress every year, every month, every week, take my word for it, we are going back.*

(FLORENCE NIGHTINGALE, 1872)

## OBJECTIVES

Upon completion of this chapter, the reader should be able to:

1. Outline preparation for the NCLEX.
2. Relate factors associated with NCLEX-RN performance.
3. Outline components of organizing a review to prepare for the NCLEX.
4. Identify elements of a profession.
5. Analyze commitment to professional development.

*Delmar/Cengage Learning*

*Anwar will be graduating from his nursing education program in two months. He plans to focus his current efforts on preparing to take the NCLEX-RN Licensure Examination. He knows that three areas of examination preparation are having the knowledge, being adept at testing, and controlling test anxiety.*

*How should he prepare for the examination?*

*Where should he focus?*

*How can he decrease his test anxiety?*

A new graduate from an educational program that prepares RNs will take the **NCLEX**, the national nursing licensure examination prepared under the supervision of the National Council of State Boards of Nursing. The NCLEX is taken after graduation and prior to practice as an RN. It is wise to schedule the exam date soon after graduation. The examination is given across the United States at professional testing centers. Graduates submit their credentials to the state board of nursing in the state in which licensure is desired. After the state board accepts the graduate's credentials, the graduate can schedule the examination. The examination ensures a basic level of safe registered nursing practice to the public and is essential for working as a professional RN. RNs are the single largest health care profession in the United States. There are close to 3 million RNs holding licenses to practice and more than 1.3 million RNs working in hospital settings. There are approximately 600,000 to 900,000 LPN/LVNs, with approximately 39% working in hospitals. There are 1.5 million nurse aides in the United States, with approximately 27% working in hospitals (Kurtzman & Jennings, 2008).

Gallup has asked Americans to rate the honesty and ethical standards of professions since 1976, and annually since 1991. Gallup first asked Americans to rate nurses in 1999, and nursing has topped the list since then in all but one year, 2001, when firefighters finished first, following reports of their heroism after the 9/11 terror attacks (Gallup, 2010).

The examination follows a test plan formulated on four categories of client needs that RNs commonly encounter. Integrated processes include the nursing process, caring, communication and documentation, and teaching/learning. These are integrated throughout the four major categories of client needs (NCSBN, 2010). See Table 31-1. This chapter discusses preparation for the NCLEX. It also discusses components of professionalism needed by RNs.

### TABLE 31-1 NCLEX Test Plan

| CLIENT NEEDS TESTED | PERCENT OF TEST QUESTIONS |
|---|---|
| *Safe, effective care environment:* | |
| Management of care | 16–22% |
| Safety and infection control | 8–14% |
| *Physiologic integrity:* | |
| Basic care and comfort | 6–12% |
| Pharmacological and parenteral therapies | 13–19% |
| Reduction of risk potential | 10–16% |
| Physiological adaptation | 11–17% |
| *Psychosocial integrity:* | 6–12% |
| | 6–12% |
| *Health promotion and maintenance:* | |

**Source:** The National Council of State Boards of Nursing. (2010a). NCLEX-RN Examination Test Plan for the National Council Licensure Examination for Registered Nurses. Retrieved April 20, 2011, from https://www.ncsbn.org/2010_NCLEX_RN_TestPlan.pdf.

# NCLEX EXAMINATION

Candidates receive between 75 and 265 questions on the NCLEX examination during the testing session. Of these questions, 15 questions are being piloted to determine their psychometric value and validity for use in future NCLEX examinations. Students cannot determine whether they have passed or failed the NCLEX examination from the number of questions they receive during their session. Candidates are allowed up to 6 hours to complete the NCLEX, which means that they could take the entire 6 hours to complete as few as 75 test items. However, because candidates do not know how many test items they will be required to answer, they should progress through the exam as though they will have to answer all 265 items. This means they should allow an average of 1 minute per question. A 10-minute break is mandatory after 2 hours of testing. An optional 10-minute break may be taken after another 90 minutes of testing.

Each test question is presented to the student. If the student answers the question correctly, a slightly more difficult item will follow, and the level of difficulty will increase with each item until the candidate misses an item. If the student misses an item, a slightly less difficult item will follow, and the level of difficulty will decrease with each item until the student has answered an item correctly. This process continues until the student has achieved a definite passing or definite failing score. The least number of questions a student can take to complete the exam is 75. Fifteen of these questions will be pilot questions, and they will not count toward the student's score. The other 60 questions will determine the student's score on the NCLEX.

## HOW THE EXAMINATION IS CONSTRUCTED

The National Council of State Boards of Nursing (NCSBN) is the central organization for the independent member boards of nursing, which includes those of the 50 states, the District of Columbia, Guam, and the Virgin Islands. The member boards are divided into four regional areas, which supervise the selection of test item writers, representing educators and clinicians, whose names are suggested by the individual state boards of nursing. This provides for regional representation in the testing of nursing practice. All test items are validated in at least two approved nursing textbooks or references.

The National Council contracts with a professional testing service to supervise writing and validation of test items by the item writers. This professional service works closely with the Examination Committee of the National Council in the test development process. The National Council and the state boards are responsible for the administration and security of the exam. The test is a computer examination known as CAT (Computerized Adaptive Testing). It is taken on a computer, utilizing state-of-the-art technology. Students are identified by their fingerprints and palm vein recognition. Palm vein recognition works by scanning the student's hand veins and creating a digital template that represents the vein pattern. This is unique to each individual. Pearson VUE began utilizing the new palm vein technology in October 2009 and will continue to use fingerprints as well (The National Council of State Boards of Nursing, 2011a).

## TEST QUESTION FORMATS AND SAMPLES

There are several formats for questions. Questions may be four-option, single-answer questions or alternate style questions. An alternate style question is an examination question that uses a format other than a standard, four-option, single choice question to assess candidate ability. An alternate item question may include a(n):

- Spot questions that ask a candidate to identify one or more areas on a picture or graphic
- Chart/exhibit question where candidates are presented with a problem and need to read the information in the chart/exhibit to answer the problem
- Ordered Response question that requires a candidate to rank order or move options to provide the correct answer
- Audio item question where the candidate is presented an audio clip and uses headphones to listen and select the option that applies
- Graphic question that presents the candidate with graphics instead of text for the answer options and they will be required to select the correct graphic answer.

Any questions, including standard single choice questions, may include multimedia, charts, tables or graphic images (The National Council of State Boards of Nursing. 2010b).

## Sample Questions

Some of the formats for test questions are illustrated here.

*Test Question 1*—Fill in the blanks

A man underwent an exploratory laparoscopy yesterday. He is on strict intake and output. Calculate his intake and output for an 8-hour period.

| Intake | Output |
|---|---|
| IV-0.9% NS at 125 mL/hr | Foley urine output 850 mL |
| PO-1 ounce ice chips | NG tube-200 mL |
| NG irrigant-NS 15 mL Q 2 H | |
| Intake _____ | Output _____ |

*Test Answer 1*

Intake = 1,090 mL; Output = 1,050 mL

125 mL/hr (125 mL × 8 hr) is 1,000 mL.

1 ounce of icechips is 30 mL;

NG irrigant 15 mL q 2 hr (15 mL × 4) is 60 mL for a total of 1,090 mL.

Output is 850 mL urine and 200 mL of nasogastricdrainage for a total of 1050 mL (Stein, 2005).

*Test Question 2*—Question that requires more than one response

The nurse is assessing a patient with left-sided heart failure. Which of the following does the nurse expect to find? Select all that apply.

\_\_\_\_\_A. Dyspnea
\_\_\_\_\_B. Crackles
\_\_\_\_\_C. Neck vein distention
\_\_\_\_\_D. Hepatomegaly
\_\_\_\_\_E. Tachycardia

*Test Answer 2*

A, B, and E are correct. The patient with left-sided heart failure will have symptoms of lung congestion, i.e., dyspnea and crackles. He will also have tachycardia. Neck vein distention and hepatomegaly are associated with right-sided heart failure.

*Test Question 3*—Single-answer, multiple-choice questions

The following patients are on a medical-surgical nursing care unit:

359-1 Mr. A, 59, had an exploratory laparoscopy with permanent colostomy 2 days ago. He has an IV, PCA, indwelling urinary catheter, and nasogastric drainage to low wall suction. He is receiving IV push Morphine.

359-2 Mr. B, 86, suffered a cerebrovascular accident 1 week ago. He has a private sitter because he is confused. He has left-sided paralysis.

360-1 Mrs. C, 35, is a 23-hour admit for a myelogram. She is ready for discharge and needs her discharge teaching reinforced.

360-2 Miss D, 29, has severe asthma. She is experiencing some respiratory difficulty. She is on IV fluids, nebulizer albuterol, and oral steroids.

361-1 Mrs. E, 75, a newly diagnosed diabetic, requires reinforcement about insulin administration.

361-2 Mrs. F, 95, was admitted from a nursing home with dehydration and hypokalemia. She will be receiving KCl 10 mEq IV in 50 mL D5W × 2.

These patients will be assigned to one LPN and one RN. Which is the best assignment?

1. RN: 359-1, 359-2, 360-1
   LPN: 360-2, 361-1, 361-2

2. RN: 359-1, 360-2, 361-2
   LPN: 359-2, 360-1, 361-1
3. RN: 360-2, 361-1, 361-2
   LPN: 359-1, 359-2, 360-1
4. RN: 360-1, 360-2, 361-1
   LPN: 359-1, 359-2, 361-2 (Stein, 2005)

*Test Answer 3*

The RN should assume care for the patient in 359-1 because he is receiving recurring IV push medication and requires teaching about his colostomy; 360-2 because of the acute problems with asthma requiring very close respiratory assessment, nebulizer albuterol, and steroids; 361-2 as she will be receiving IV potassium and must be observed closely for signs and symptoms of fluid and electrolyte imbalances. The LPN should assume care for 359-2, 360-1, and 361-1. The LPN can teach the patient going home from a myelogram using a standard teaching protocol and can reinforce postmyelogram diabetic teaching.

The LPN may not be able to start the IV that the patient in 360-2 requires. The patient in 361-2 requires an IV and is more appropriately assigned to the RN.

*Test Question 4*—Fill in the blanks

A medication is ordered to be given at 6 mL/hour. The solution is 20,000 units/500 mL. How many units per hour will that deliver to the patient?

*Test Answer 4*

$$\frac{20,000 \text{ units}}{500 \text{ mL}} = \frac{40 \text{ units}}{1 \text{ mL}} = 40 \times 6 = 240 \text{ units per hour}$$

*Test Question 5*—Identify the height of the fundus at 22 weeks on this picture.

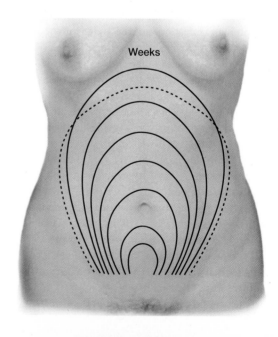

*Test Answer 5*—The fundus is located at this site at 22 weeks.

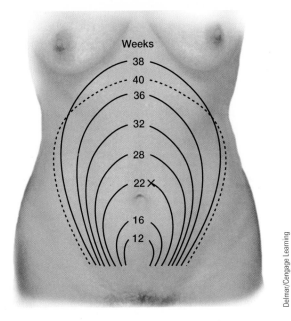

*Test Question 6*—Arrange responses in the correct order

Arrange these steps in the correct order to insert a Foley catheter.

1. Check integrity of Foley balloon.
2. Attach drainage bag to bed.
3. Send urine specimen to laboratory.
4. Cleanse meatus.
5. Spread the labia.
6. Insert the foley.

*Test Answer 6.* (Note that you will click on these responses, drag them into a new adjacent list of responses, and then drop the responses into the correct order). The correct order is, 1, 5, 4, 6, 2, 3.

1. Check integrity of Foley balloon.
5. Spread the labia.

4. Cleanse meatus.
6. Insert the Foley.
2. Attach drainage bag to bed.
3. Send urine specimen to laboratory.

## ORIENTATION TO THE EXAMINATION

Each candidate is oriented to the computer before the examination starts. Note there is a NCLEX-RN licensure examination tutorial available http://www.pearsonvue.com/nclex/. Because the exam is geared to the candidate's skill level, each candidate will have a unique computer adaptive test (CAT). Each exam will include some experimental questions. The experimental questions are interspersed throughout the examination so the candidate will answer all questions with equal effort. The time period allowed for each candidate to complete the exam includes time for the exam instructions explaining how to use the mouse, the spacebar, and the Enter key, as well as samples representing each type of question in the exam and rest breaks.

The NCLEX-RN is scored by computer, and a pass/fail grade is reported. A criterion-referenced approach is used to set the passing score. This provides for the candidate's test performance to be compared with a consistent standard criteria. After administration of the CAT, candidates are notified of their success or failure by the board of nursing of the state in which they took the examination. Successful candidates are notified that they have passed. Unsuccessful candidates are provided with a diagnostic profile that describes their overall performance on a scale from low to high and their performance on the questions testing their abilities to meet client needs (Stein, 2005).

## POSSIBLE PREDICTORS OF NCLEX SUCCESS

Several factors have been identified in research studies as being associated with performance on the NCLEX examination. Some of these factors are identified in Table 31-2.

---

### TABLE 31-2  Possible Predictors of NCLEX Success

- HESI Exit Exam,* ATI Exit Exam**
- Verbal SAT score
- ACT score
- High school rank and grade point average (GPA)
- Undergraduate nursing program GPA

- GPA in science and nursing theory courses
- Competency in American English language
- Reasonable family responsibilities or demands
- Absence of emotional distress
- Critical thinking competency

\* Evolve Learning System. (2011). HESI Frequently Asked Questions. Retrieved April 20, 2011, from https://evolve.elsevier.com/staticPages/hesi-faq.html.

\*\* Assessment Technologies Institute (2011). About ATI. Retrieved April 20, 2011, from http://www.atitesting.com/About.aspx.

## EVIDENCE FROM THE LITERATURE

**Citation:** DiBartolo, M. C., & Seldomridge, L.A. (2005). A review of intervention studies to promote NCLEX-RN success. *Nurse Educator, 30*(4), 166–171.

**Discussion:** This article summarizes numerous intervention studies done to enhance NCLEX-RN success. Some of the interventions were study sessions, test-taking strategies, support groups, study plans, NCLEX review courses, NLN Diagnostic Readiness testing to develop a report card of weakness, computer testing, remediation, etc.

**Implications for Practice:** Though NCLEX-RN success improved with the interventions studied, the researchers were limited in their ability to attribute success specifically to the interventions. More study is needed to clarify this important area of research.

## NCLEX-RN REVIEW BOOKS AND COURSES

NCLEX-RN review books often include nursing content, sample test questions, or both. They frequently include computer software disks with test questions. The test questions may be arranged in the NCLEX book by clinical content area, or they may be presented in one or more comprehensive examinations covering all areas of the NCLEX. Listings of these review books are available at www.amazon.com. It is helpful to use several of these books and computer software when preparing for the NCLEX. Focus on NCLEX review books developed in the past three years.

NCLEX review courses are also available. Brochures advertising these programs are often sent to schools and are available in many sites nationwide. The quality of these programs can vary, and students may want to ask former nursing graduates and faculty for recommendations. Access to an online NCLEX-RN review course and a test question of the week is available at the NCSBN Learning Extension site (National Council of State Boards of Nursing, Inc. 2011c).

## REAL WORLD INTERVIEW

My best advice to anyone preparing for the NCLEX is to take lots of practice tests. I answered close to 1,500 questions in preparation, and I feel it did me a world of good. I kept my nursing textbooks handy, and when I ran into something I didn't know, I looked it up.

**Amanda Meadows, RN, BSN**
Chicago, Illinois

## EXIT EXAMINATIONS

Many nursing programs administer an exit examination to students at the completion of their nursing program. Two of these exit exams are the Assessment Technologies Institute (ATI) Exit Exam and the Evolve Testing & Remediation/Health Education Systems, Inc. (HESI) Exit Exam. New graduates will want to review their performance on any of these exit exams, because these results will help identify their educational weaknesses and help focus their review sessions.

## KNOWLEDGE, ANXIETY MANAGEMENT, AND TEST-TAKING SKILL

Successful test performance requires nursing knowledge, anxiety management, and test-taking skill. Knowledge of the test content is the first critical element. Students gain knowledge of nursing as the result of a course of nursing study. Nursing students most commonly attend either a two-year associate degree nursing program, a three-year diploma nursing program, or a four-year baccalaureate nursing program to gain the knowledge needed to satisfactorily complete the NCLEX-RN examination.

Anxiety management techniques such as exercise, visualization, relaxation, maintaining PERMA (Seligman, 2011), and avoiding automatic negative thoughts (Beck, Emery, & Greenberg, 2005) are the second element of successful test performance. Plan to walk or exercise daily for 30 minutes. There is a relaxation audio CD-ROM on test anxiety prevention by Rosenthal (2004) and a book on test-taking strategies by Kesselman-Turkel and Peterson (2007). Plan to use one of them or any of the following to control your anxiety:

- Guided imagery, which requires using your imagination to create a relaxing sensory scene on which to concentrate

■ Breathing exercises
■ Relaxation exercises (Stein, 2005)

## Maintaining PERMA and Avoiding Automatic Negative Thoughts

In order to do well generally in life, it is important to maintain PERMA (Seligman, 2011) and avoid automatic negative thoughts (Beck, Emery, & Greenberg, 2005). To do well in life, we need positive emotions (P); more engagement (E) in our work, friendships, and love; better human relationships (R); more meaning and purpose (M) in life; and we want to be moving towards our goals – achievement (A). Maintaining PERMA in your life should also help you do well in preparing for NCLEX.

It is also important to avoid automatic negative thoughts as you prepare for the NCLEX-RN. Automatic negative thoughts include the following cognitive thinking distortions:

1. All-or-nothing thinking: You see things in black-and-white categories. If your performance falls short of perfect, you see yourself as a total failure.
2. Overgeneralization: You see a single negative event as a never-ending pattern of defeat.
3. Mental Filter: You pick out a single negative detail and dwell on it exclusively so that your vision of all reality becomes darkened, like the drop of ink that discolors the entire beaker of water.
4. Disqualifying the positive: You reject positive experiences by insisting that they don't count for some reason or other. In this way you can maintain a negative belief that is contradicted by your everyday experiences.
5. Jumping to conclusions: You make a negative interpretation though there are no definite facts that convincingly support that conclusion.
   a. Mind reading: You arbitrarily conclude that someone is reacting negatively to you and you don't bother to check this out.
   b. The fortune-teller error: You anticipate that things will turn out badly, and you feel convinced that your prediction is an already-established fact.
6. Magnification (catastrophizing) or minimization: You exaggerate the importance of things (such as your goof-up or someone else's achievement), or you inappropriately shrink things until they appear tiny (your own desirable qualities or the other person's imperfections). This is also called the "binocular trick."
7. Emotional reasoning: You assume that your negative emotions necessarily reflect the way things really are: "I feel it, therefore it must be true."
8. Should statements: You try to motivate yourself with shoulds and shouldn'ts, as if you had to be punished before you could be expected to do anything. "Musts" and "oughts" are also offenders. The emotional consequence is guilt. When you direct should statements toward others, you feel anger, frustration and resentment.
9. Labeling and mislabeling: This is an extreme form of overgeneralization. Instead of describing your error, you attach a negative label to yourself: "I'm a loser." When someone else's behavior rubs you the wrong way, you attach a negative label to him: "He's a damned louse." Mislabeling involves describing an event with language that is highly colored and emotionally loaded.
10. Personalization: You see yourself as the cause of some negative external event that in fact you were not primarily responsible for.
11. Self-worth: You make an arbitrary decision that in order to accept yourself as worthy, okay, or to simply feel good about yourself, you have to perform in a certain way—usually, most, or all the time (HealthyMind.com, 2004).

## Test-Taking Skill

Test-taking skill is the final critical element needed for successful test completion. Strategies to improve test-taking skills include practicing 60 test questions daily from different NCLEX-RN review books and CDs. Note the NCSBN Learning Extension review courses at National Council of State Boards of Nursing, Inc. (2011c). Practice test questions until performance is satisfactory on all areas of the NCLEX exam. Note that successful students often practice 60 questions daily in their knowledge weakness areas until their performance improves. Sixty questions daily for 30 days exposes students to 1,800 questions (60 × 30 = 1800). Use the results of an exit exam or comprehensive exam to guide you in your selection of test questions.

## ARKO STRATEGIES AND ABC-SAFE-COMFORT-TEACHING TIPS

When you are presented with a difficult test question, use these test-taking ARKO strategies and ABC-Safe-Comfort-Teaching tips to improve your test performance. Use ARKO Strategies as follows:

■ A Is the question stem asking for you to take Action or take no Action?
■ R Reword the question.
■ K Identify any Key words in the question stem.
■ O Option elimination.

Apply this ARKO strategy to the test question below:

What should the nurse do first for a patient with a spinal cord injury who complains of a headache?

A. Insert a Foley catheter.
B. Assess the patient's pupils.
C. Take the patient's blood pressure.
D. Administer a beta adrenergic blocker medication.

Applying the ARKO strategy to the question above:

A   Stem asks for nurse to take Action—i.e., What should the nurse do first?

R   Reword the question as follows: What is a priority nursing action for a patient with a spinal cord injury who has a headache?

K   Key words are "first," "spinal cord injury," and "headache."

O   Option elimination.

Note the following:

Option A may be useful if the blood pressure is elevated, as a plugged catheter can trigger autonomic dysreflexia associated with a spinal cord injury.

Option B does not give us useful information about this patient with a spinal cord injury.

Option C is useful to assess blood pressure for signs of autonomic dysreflexia.

Option D will reduce blood pressure, but stem does not say blood pressure is elevated.

The correct answer is C.

## ABC-Safe-Comfort-Teaching Tips

As you review a test question, it is helpful to review Maslow's Hierarchy of Needs and use the ABC-Safe-Comfort-Teaching tips. Assess your patient's physiological needs, i.e., their ABCs, then assess their Safety. After ABC-Safe is secured, then assess your patient's Comfort. Finally, after ABC-Safe-Comfort is secured, assess your patient's Teaching needs. Remember ABC-Safe-Comfort-Teaching when prioritizing your nursing actions:

A – Airway
B – Breathing
C – Circulation
Safety
Comfort
Teaching

Apply the ABC-Safe-Comfort-Teaching tips to this test question:

The nurse is unable to obtain a pedal pulse on Doppler examination of the cold painful leg of a patient who has just been admitted with a fractured femur. What is the priority intervention for this patient?

A.  Give morphine, as ordered.
B.  Teach the patient cast care.
C.  Notify the medical practitioner.
D.  Comfort the patient and keep the leg elevated.

Apply ABC-Safe-Comfort-Teaching tips to the question's options:

A.  Comfort can be given with the pain medication, but this is an emergency. Call the medical

practitioner. Comfort care is done after ABC & Safety are assured.

B.  Teaching is done after ABC-Safe-Comfort is assured.

C.  Patient has an absent pulse on Doppler examination and a cold, painful leg. This is an emergency! Patient's arterial circulation cannot be occluded long before there is permanent damage to tissues.

D.  Patient is not safe or comforted if there is no arterial circulation to leg. Comfort care is done after ABC & Safety are assured.

The answer is C.

Recall that NCLEX often wants you to take all nursing actions before calling the medical practitioner. When necessary, in an emergency, however, do not hesitate to call the medical practitioner. Always recall that Maslow's Hierarchy of Needs directs us to monitor our patients' ABCs and then keep them Safe. After this is done, we can Comfort and Teach our patients.

## ORGANIZING YOUR REVIEW

In preparing for the NCLEX, identify your strengths and weaknesses. If you have taken an exit exam, note any content strength and weakness areas. Look carefully at the elements of the NCLEX test plan (refer to Table 31-1). Complete a NCLEX-RN learning needs analysis (Table 31-3), and establish a schedule that permits you to cover completely all the material to be reviewed. Note any nursing program course or clinical content areas in which you scored below a grade of B. Purchase one or more of the NCLEX review books. It is useful to review questions developed by different authors. Review content in the review books in any of your weak content areas. Take a comprehensive exam in a review book or with the computer software, and analyze your performance. Try to answer as many questions correctly as you can. As you study, be sure to actually practice taking the examinations. Do not just jump ahead to the answer section until you have completed the examination.

Next, after you have completed the comprehensive examination, review the answers and rationales for any weak content areas, score the test, and then take another comprehensive exam. Repeat this process until you are doing well in all clinical content areas and in all areas of the NCLEX examination plan. Completing the examination in this way may improve your examination performance. Use methods of memory improvement that will work for you. Mnemonic devices, where a letter represents the first letter of each item in a sequence, are a useful means of recalling information as you study (Table 31-4).

Mental imagery is the technique of forming pictures in your mind to help you remember details of a sequence of events, such as the administration of an injection. Try practicing self-recitation to improve your study

## TABLE 31-3  NCLEX-RN Learning Needs Analysis Needs Analysis

Anxiety level (circle)        1        2        3        4        5        6        7        8        9        10

Weak content areas identified on NCLEX Test Plan in Table 31-1, exit exam, or another comprehensive exam:

_____

_____

Nursing courses below B: _____

Predictors from Table 31-2: _____

_____

Weak content areas identified in any common patient conditions:

- Mental health—for example, schizophrenia, bipolar disorders, anxiety, personality disorders, eating disorders, abuse, suicide, eating disorders, and so on. _____

- Women's health—for example, antepartum care, intrapartum care, postpartum care, newborn care, and so on.
_____

- Adult health—for example, cancer, myocardial infarction, diabetes, pneumonia, HIV, hepatitis, cholecystectomy, lobectomy, nephrectomy, cardiac arrest and major cardiac arrhythmias, thyroidectomy, shock, CVA, appendectomy, and so on.
_____

- Children's health—for example, leukemia, cardiovascular surgery, fractures, cancer, tonsillectomy, asthma, Wilms' tumor, diabetes, cleft palate, and so on. _____

Weak content areas identified in any of the following:

- Therapeutic communication
- Growth and development (developmental milestones and toys)
- Management, delegation, referrals, and priority setting
- Medications
- Defense mechanisms
- Immunization Schedules
- Diagnostic tests and laboratory values
- Food choices and various types of diets (Table 31-8)
- Isolation techniques, i.e., standard, airborne, droplet, contact

Organize Your Study

Your study schedule could look like the following:

Day 1:  Practice 60 adult health test questions. Score the test, analyze your performance, and review test question rationales and content weaknesses. Practice deep breathing, relaxation exercises, and positive thinking as needed. Continue this process until you are doing well in adult health questions.

Day 2:  Practice 60 women's health test questions. Repeat  this process until you are doing well in women's health questions.

Day 3:  Practice 60 children's health test questions. Repeat this process until you are doing well in children's health questions.

Day 4:  Practice 60 mental health test questions. Repeat this process until you are doing well in mental health questions.

Day 5:  Continue with content review and test question practice in all weak content areas. Practice deep breathing, relaxation exercises, and positive thinking. Continue this process until you are doing well in all areas of the exam. Study when you are most alert.

## TABLE 31-4 Mnemonic Device

It is "OK" to have your blood tested while on anticoagulants. This is a memory device to assist you in remembering the antidote for Coumadin overdose, that is, the antidote for Oral Coumadin is Vitamin K (OK). Remembering this can help you eliminate the antidote for the other anticoagulant, Intravenous Heparin. Heparin's antidote is Protamine Sulfate.

habits. Reciting to yourself the material being learned will promote retention of the information being studied. Concentrate on the information you identified in your self-needs analysis.

Do a general review of the common patient diseases, medications, diagnostic tests, and nursing procedures in each major nursing content area, as well as defense mechanisms, communication tips, food choices in various diets, and growth and development.

Organize the material so that you will be able to review all the need-to-know information within the allotted study time period. Your schedule should allow you to complete your review so that you can close your books and do something relaxing on the night before the examination.

## Nutrition, Sleep, and Wardrobe

You will function best if you are well nourished. Plan to eat three well-balanced meals a day for at least three days prior to the examination. Be careful when choosing the food you consume within 24 hours of the examination. Avoid foods that will make you thirsty or cause intestinal distress. Minimize the potential of a full bladder midway through the examination by limiting the amount of fluids you drink and by allowing sufficient time at the test site to use the bathroom before entering the room.

Plan to allow sufficient time in your schedule the week before the examination to provide yourself with the sleep you need to function effectively for at least three days prior to the examination. Plan your wardrobe ahead of time. Shoes and clothes that fit you comfortably will not distract your thought processes during the examination. Include a comfortable sweater. Your clothes for the test day should be ready to wear by the night before the examination. If you wear glasses or contact lenses, take along an extra pair of glasses. If you are taking medications on a regular basis, continue to do so during this period of time. Introduction of new medications should be avoided until after completion of the examination (Stein, 2005).

## CASE STUDY 31-1

Analyze your NCLEX-RN learning needs based on Table 31-3. Use Table 31-5 to identify your best time to practice NCLEX question and review content. Complete the schedule for the next week. When is your best time to study? Review the question of the week and note the NCLEX review course at NCSBN (National Council of State Boards of Nursing, Inc. 2011c). Note the availability of an NCLEX-RN licensure examination tutorial also (Pearson Education, Inc. 2011).

## WHEN TO STUDY

Identify your personal best time. Are you a day person? Are you a night person? Study when you are fresh. Arrange to study one or more hours daily. Use Table 31-5 to organize your study.

Students who use this technique should increase their confidence in their ability to do well on the NCLEX.

## SOME FINAL TIPS

Table 31-6 has some final tips on reviewing for the NCLEX. Table 31-7 is a medication study guide to aid you in your NCLEX preparation. Table 31-8 has common food sources for various nutrients and diets. Use these guides as a starting point in your preparation for the NCLEX.

## PROFESSIONALISM

After you successfully pass your NCLEX-RN, you will join the profession of nursing. Experts in the social sciences are considered the authorities on what makes an occupation

**TABLE 31-5  Organizing Your NCLEX Study**

|        | M | T | W | R | F | S | S |
|--------|---|---|---|---|---|---|---|
| 8–9    |   |   |   |   |   |   |   |
| 9–10   |   |   |   |   |   |   |   |
| 10–11  |   |   |   |   |   |   |   |
| 11–12  |   |   |   |   |   |   |   |
| 12–1   |   |   |   |   |   |   |   |
| 1–2    |   |   |   |   |   |   |   |
| 2–3    |   |   |   |   |   |   |   |
| 3–4    |   |   |   |   |   |   |   |
| 4–5    |   |   |   |   |   |   |   |
| 5–6    |   |   |   |   |   |   |   |
| 6–7    |   |   |   |   |   |   |   |
| 7–8    |   |   |   |   |   |   |   |
| 8–9    |   |   |   |   |   |   |   |
| 9–10   |   |   |   |   |   |   |   |

## REAL WORLD INTERVIEW

I believe some students deny their anxiety about taking the NCLEX-RN by saying it is not a big deal. If they fail it, they can always retake it. However, it *is* a big deal—you have given years of your life preparing to become a registered nurse, and you want to be successful the first time you take the licensing exam. My advice to you is to use your HESI Exit Exam score sheets to identify your weaknesses. Study carefully all subject content areas for which you received a score of less than 850, particularly if the content area is medical-surgical or management, each of which makes up a large part of the licensing exam.

**Susan Morrison, RN, PhD, FAAN**
Nurse Consultant
Houston, Texas

# EVIDENCE FROM THE LITERATURE

**Citation:** Gordon, S., & Nelson, S. (2006). *Moving beyond the virtue script in nursing in the complexities of care.* Ithaca, NY: ILR Press.

**Discussion:** The author discusses the concept that images of hearts and angels and the emphasis on caring and health in nursing trivializes what is in fact highly skilled knowledge work that addresses the important role of nurses in the care of the sick. The author states that hospital administrators, politicians, and the public won't expend the effort to decode nursing. Nurses must articulate a vivid picture of how nurses as a profession protect their patients. Nursing is facing a crisis because potential recruits don't understand that nurses prevent fatal complications and assess and monitor their patients using knowledge-based scientific care. An emphasis on caring devalues the nursing role, attracting the wrong recruits and driving practitioners from the role. Many nurses are tough-minded professionals who value their technical and practical expertise. They are often dismissed as uncaring. The author asks, isn't using technology and knowledge to prevent complications an act of caring?

**Implications for Practice:** Nurses must monitor nursing-sensitive indicators and document the lifesaving role nurses play with their patients using knowledge-based scientific practice.

## TABLE 31-6  Selected NCLEX Tips

- Remember Maslow. Physical needs are met first—for example, airway, breathing, circulation (ABCs) threats
  - Airway
    - Altered level of consciousness (LOC)
    - Unconscious
    - Foreign object in airway
  - Breathing
    - Asthma
    - Suicide threat
  - Circulation
    - Cardiac arrest
    - Shock
- Safety needs are met second—for example, safety and infection control threats
  - Confusion
  - Tuberculosis
  - Isolate infectious patients from noninfectious patients
- Needs for psychological and physical comfort and teaching are met after physical (ABC) and safety needs are met (ABC-Safe-Comfort-Teaching). Don't choose a test question

answer that gives the patient psychological or physical comfort or meets teaching needs until the patient's physical ABC and safety needs have been met.

- Remember the nursing process: Assess your patient first, then plan, implement, and evaluate.
- Keep all your patients safe—airway open; side rails up; IV access line in place on unstable patient; monitor vital signs, pulse oximeter, cardiac rhythm, urine output, as needed.
- Know delegation guidelines for RNs, LPNs, and NAP. Observe the five rights of delegation, that is, the right task, the right circumstance, the right person, the right direction/communication, and the right supervision. See Chapter 16.
  - The RN assures quality care of all patients, especially complex patients. RNs delegate care of stable patients with predictable outcomes.
  - The RN uses patient care data such as vital signs, collected either by the nurse or others, to make clinical judgments. The RN continuously monitors and evaluates patient care and gives care involving standard, unchanging procedures to LPNs and NAP.
  - The RN never delegates patient assessment, teaching, evaluation (ATE), or judgment.

*(Continues)*

## TABLE 31-6 (Continued)

- The RN makes appropriate referrals to community resources.
- LPNs can perform medication administration (includes IV meds in some states), sterile dressings, Foley insertions, and so on.
- NAPs can perform basic care—for example, vital sign measurement, bathing, transferring, ambulating, communicating with patients, and stocking supplies.
- In some states, with documented competency, LPNs can insert IVs, pass nasogastric tubes, and so on.
- In some states, with documented competency, NAP can perform venipuncture, do blood glucose tests, insert Foley catheters, and so on.
- When delegating care, don't mix the care of a patient with an infection with a patient who has decreased immunity—for example, a patient with HIV, diabetes, steroids, the very young, the very old, and so on.
- In answering test questions, do the following:
  - When choosing priorities, choose the first answer you would do if you were alone and could only do one thing at a time. Don't think that one RN will do one thing and another RN will do another thing.
  - Assume that you have the practitioner's order for any possible choices. The NCLEX is usually looking for the correct nursing action, not the medical action.
  - Assume that you have perfect staffing, plenty of time, and all the necessary equipment for any possible test question choices. Choose the answer that indicates the best nursing care possible.
  - Assume that you are able to give perfect care "by the book." Don't let your personal clinical experience direct you to choose a test answer that is less than high-quality care.
  - Remember to care for the patient first and then check the equipment.

- Know the most common adult, maternal-child, and mental health care disorders. For each disorder, know the medications, laboratory and diagnostic tools, procedures, and treatments commonly used.
- Know common medications (see Table 31-7).
- Know common laboratory norms—for example, sodium, potassium, blood sugar, complete blood count, hematocrit, prothrombin time, partial thromboplastin time, international normalized ratio (INR), arterial blood gas (ABG), cardiac enzymes, digoxin level, dilantin level, lithium level, blood urea nitrogen (BUN), creatinine, and the specific gravity of urine.
- Know communication techniques—look for answers that give patients support and allow the patient to keep talking and verbalize concerns and problems. Be a comforting nurse, not a therapist. Avoid advice.
- Know common food sources for various nutrients and diets (Table 31-8).
- Know the defense mechanisms.
- Know growth and development, such as the developmental tasks for each childhood stage, toys for each childhood stage, and so on.
- Know immunization schedules (Centers for Disease Control and Prevention 2011a).
- Prepare mentally with the following:
  - Anxiety control and relaxation techniques
  - Regular exercise
  - Maintaining PERMA (Seligman, 2011)
  - Avoiding automatic negative thoughts (Beck, Emery, & Greenberg, 2005).
  - Visualize your name with RN next to it on your name tag.
- Remember—you graduated from an accredited nursing program. You can do it!

a profession. Although there is some variation in actual criteria, there is general agreement in several areas:

- Professional status is achieved when an occupation involves a unique practice that carries individual responsibility and is based upon theoretical knowledge.
- The privilege to practice is granted only after the individual has completed a standardized program of highly specialized education and has demonstrated an ability to meet the standards for practice.

- The body of specialized knowledge is continually developed and evaluated through research.
- The members are self-organizing and collectively assume the responsibility of establishing standards for education in practice. They continually evaluate the quality of services provided to protect the individual members and the public.

There is a trend in recent years to call every occupation a profession. Have you heard of "professional baseball play-

**TABLE 31-7   Medication Study Guide Excerpt**

**General Tips**

1. Drowsiness and changes in vital signs are a side effect of many medications given for their analgesic, antiemetic, antiseizure, tranquilizer, sedative/hypnotic, antihistamine, or antianxiety effects.

2. Note that if drowsiness is a side effect of the medications listed in the table below, consider the need to monitor airway, level of consciousness (LOC), blood pressure (BP), pulse (P), respirations (R), pulse oximetry, and cardiac monitor. Use side rails/fall precautions; avoid driving.

3. Note that if a drug is a cardiac drug, consider the need to monitor airway, level of consciousness (LOC), blood pressure (BP), pulse (P), respirations (R), pulse oximetry, and cardiac monitor. May need to have an IV line in place to safeguard patient and use side rails/fall precautions; may need to avoid driving and alcohol.

4. Many medications cause renal, liver, heart, neurological, and bone marrow side effects. Monitor laboratory results that reflect these organs' functions. Check allergies.

5. A Haldol, Ativan, and Cogentin mixture is used PRN for chemical restraints. Be sure to try nonrestrictive approaches first.

| Category | Root/Prefix/ Suffix | Some Examples | Nursing Implications |
|---|---|---|---|
| Thrombolytic | ase plase | Streptokinase (Streptase) | |
| Benzodiazepines Tranquilizer Hypnotic Antianxiety Antiseizures Anesthetic | azepam aze | Diazepam (Valium); Lorazepam (Ativan); Clonazepam (Klonopin); Alprazolam (Xanax); Midazolam (Versed) | |
| Antifungal | azole | Fluconazole (Diflucan) Micronazole (Monistat) | |
| Antiarrhythmic | caine | Lidocaine | |
| Calcium | cal | Calcium gluconate | |
| Anti-infective | ceph cef | Cephalexin (Keflex) Ceftriaxone (Rocephin) | |
| Anti-infective | cillin | Amoxicillin Penicillin | |
| NSAID | cox | Celecoxib (Celebrex) Valdecoxib (Bextra) | |
| Antibiotic | cycline | Tetracycline (Sumycin) Doxycycline (Vibramycin) | |

*(Continues)*

**TABLE 31-7** (Continued)

| Category | Root/Prefix/Suffix | Some Examples | Nursing Implications |
|---|---|---|---|
| Calcium channel blocker Antihypertensive | dipine | Amlodipine (Norvasc) Nifedipine (Adalat, Procardia) | |
| Bone resorption inhibitor | dronate | Olendronate (Fosamax) Pamidronate (Aredia) Ibandronate (Boniva) Risedronate (Actonel) | |
| Antibiotic | floxacin | Ciprofloxacin (Cipro) Levofloxacin (Levaquin) Moxifloxacin (Avelox) | |
| Diuretic | ide | Lasix (Furosemide) | |
| Anti-infectives | mycin cin | Gentamycin (Garamycin) Vancocin (Vancomycin) | |
| Vasodilator | nitr | Nitroglycerin (Nitrostat) | |
| Beta adrenergic blocker Antihypertensive | olol | Metoprolol (Lopressor) Propanolol (Inderal) Atenolol (Tenormin) | |
| Steroids Anti-inflammatory | one | Decadron (Dexamethasone) Prednisone (Deltasone) Methylprednisolone (Solu-Cortef, Depo-Medrol) | |
| Anticoagulant | parin | Heparin Enoxaparin (Lovenox) Dalteparin (Fragmin) | |
| Bronchodilator | phylline | Aminophylline | |
| Proton pump inhibitors | prazole | Esomeprazole (Nexium) Pantoprazole (Protonix) Rabeprazole (Aciphex) Lansoprazole (Prevacid) | |
| Antihypertensive | pres | Clonidine (Catapres) Hydralazine (Apresoline) | |

*(Continues)*

**TABLE 31-7** (Continued)

| | | |
|---|---|---|
| Angiotensin converting agent (ACE Inhibitor) Antihypertensive | pril | Lisinopril (Zestril) Enalapril (Vasotec) Ramipril (Altace) Captopril (Capoten) |
| Angiotensin receptor blocking agents Antihypertensive | sartan | Telmisartan (Micardis) Irbesartan (Avapro) Losartan (Cozaar) |
| Histamine-2 Blockers | tidine | Ranitidine (Zantac) Famotidine (Pepcid) |
| Anti-hyperlipidemic | vastatin | Lovastatin (Mevacor) Atorvastatin (Lipitor) Pravastatin (Pravachol) Simvastatin (Zocor) |
| Antiviral | vir | Zidovudine (Retrovir) Acyclovir (Zovirax) Valacyclovir (Valtrex) |
| Phenothiazine Antiemetic Antipsychotic Antianxiety | zine | Promethazine (Phenergan) Prochlorperazine (Compazine) Hydroxyzine (Vistaril) |

**SELECTED OTHER MEDICATIONS**

| Antipsychotics | Antidepressants | Mood Stabilizers | Cardiac Drugs |
|---|---|---|---|
| Zyprexa | Celexa | Lithium | Amiodarone (Cordarone); |
| Seroquel | Cymbalta | Depakote | Digoxin; |
| Haldol | Lexapro | Tegretol | Diltiazem; |
| Prolixin | Zoloft | | Epinephrine; |
| Abilify | Paxil | | Nitroglycerin; |
| Resperdal | Prozac | | Norpace (Disopy-ramide); |
| | Remeron | | Rythmol (Propa-fenone); |
| | Venlafaxine | | Tambocor (Flecainide); |
| | | | Betapace (Sotalol) |

*(Continues)*

## TABLE 31-7 (Continued)

### OTHER COMMON MEDICATIONS

Acetaminophen, Albuterol, Aspirin, Atropine, Benadryl, Chemotherapy (guidelines), Codeine, Cogentin, Coumadin, Ethambutol, Heparin, Ibuprofen, Insulin, Isoniazid, Lactulose, Magnesium Sulfate, Morphine, Neurontin, Oxycodone, Pitocin, Potassium, Pyrazinamide, Rhogam, Rifampin, Synthyroid, Terbutaline, Zyloprim, Miotic eye drops, Mydriatic eye drops, etc.

*Learn to recognize medications by the Root/Prefix/Suffix
*Learn the generic and trade names, action, nursing implications, and side effects of all the medications in this table.
**Source:** Unpublished manuscript, Patricia Kelly, RN, MSN, and Gabriel Hernandez, RPh.

## TABLE 31-8  Common Food Sources for Various Nutrients and Diets

| Nutrient: | Common Sources: |
|---|---|
| Iron | Iron in animal foods (heme iron) is best absorbed when eaten in a meal containing Vitamin C: Liver/liver sausage, beef, chicken, eggs, pork |
| | Non-heme iron sources (Less efficiently absorbed than heme iron): Dark green leafy vegetables, fortified breakfast cereals, kidney beans and other legumes, whole grain breads/cereals |
| Potassium | Bananas, cantaloupe, oranges/orange juice, potatoes, prune juice |
| Vitamin C (antioxidant) | Citrus fruits (e.g., oranges, grapefruit), strawberries, sweet peppers, tomatoes |
| Vitamin D | Exposure to sunshine! Cod liver oil, fatty fish (e.g., tuna, salmon/canned salmon with bones), most milk, cheese, yogurt (check label) |
| Vitamin E (antioxidant) | Nuts, salad dressings, vegetable oils, wheat germ, whole grain breads/cereals/starches |
| Calcium | Milk, cheese, yogurt, dark leafy green vegetables, fortified cereal, fortified orange juice |
| **Therapeutic Diets:** | **Foods to limit/restrict on diet:** |
| Bland/Soft | Caffeine, fatty/fried foods, fresh fruit, raw vegetables, spicey foods, whole grain breads/cereals/pasta Eat smaller more frequent meals. |
| Diabetic | Foods/beverages containing regular sugar (e.g., cakes, candy, cookies, ice cream, pies, soda, jams/jellies, syrup) Plan meals in advance for consistent carbohydrate intake daily. |
| Gluten Free | Foods containing white flour, wheat flour (e.g., breads, cereals, noodles/pasta), barley, oats, rye Look for "Gluten Free" on the package label. |
| Low Fat/Low Cholesterol | Bacon, fried foods, gravies/sauces made with butter, whole milk/whole-milk cheese/cream, sausage Remove visible fat from meat; remove skin from poultry. |

*(Continues)*

**TABLE 31-8** (Continued)

| Low Sodium | Bacon, corned beef, ham, hot dogs, sausage, canned soups, snack foods (e.g., potato chips, pretzels), canned vegetables |
| | Do not add salt in cooking or at table. |
| Low Protein | Beef, chicken, eggs, fish, pork, turkey, dairy products, legumes |
| | Restriction varies. In general, 1 oz. meat = 7 grams protein. Plan meals in advance and monitor correct portion sizes. |
| **Selected Diet needs:** | **Food choices:** |
| High Calorie | Avocados, butter, cheese, cream, mayonnaise, oil-based salad dressings, olives, peanut butter |
| High Fiber | Fresh fruit (with peel or skin, when applicable), legumes, nuts, peas, raw vegetables, whole grain breads/cereals/starches |
| High Protein | Beef, chicken, dairy products, eggs, fish, legumes, pork, turkey |

**Source:** Developed by: Georgia Hammerli, RD, LDN, Registered Dietitian, Arlington Heights, Illinois.

ers" and "professional automobile mechanics"? There has been a tendency to confuse professionalism and profession. The term *professionalism* generally refers to an individual's commitment and dedication to the occupation. It often also refers to the attitude, appearance, and conduct of the individual. Whether an occupation is a profession requires more analysis. Figure 31-1 refers to some of the classic studies about the characteristics of a profession

**Flexner, 1915**

- Professional activity is based on intellectual action along with personal responsibility
- The practice of a profession is based on knowledge, not routine activities
- There is practical application rather than just theorizing
- There are techniques that can be taught
- A profession is organized internally
- A profession is motivated by altruism, with members working in some sense for the good of society

**Pavalko, 1971**

- Work based on systematic body of theory and abstract knowledge
- Work has social value
- Length of education required for specialization
- Service to public

- Autonomy
- Commitment to profession
- Group identity and subculture
- Existence of a code of ethics

**Manthey, 2002**

- An identifiable body of knowledge that can best be transmitted via formal education
- Autonomy of decision making
- Peer review of practice
- Identification with a professional organization as the standard setter and arbiter of practice

**Public Law 93-360 on Collective Bargaining**

- Predominantly intellectual work
- Varied work requirements
- Requires discretion and judgment
- Results cannot be standardized over time
- Requires advanced instruction and study

**FIGURE 31-1** **Characteristics of a profession. Compiled with information from:** (*Source:* Mitchell, G. M., & Grippando, P. R. (1994). *Nursing perspectives and issues.* Clifton Park, NY: Delmar Cengage Learning; and Manthey, M. [2002]. *The practice of primary nursing: Relationship-based, resource-driven care delivery.* 2nd ed. Minneapolis, MN: Creative Healthcare Management Inc.)

**Professional Values:**

| | | |
|---|---|---|
| Caring | Freedom | Justice |
| Altruism | Aesthetics | Truth |
| Equality | Human dignity | Ethical |
| Nonjudgmental | | |

**Professional Behaviors and Attributes:**

| | |
|---|---|
| Appearance | Stress management |
| Time-management skills | Self-evaluation |
| Self-discipline | Initiative |
| Maintenance of licensure/certification | Motivation |
| Participation in institutional/community activities | Creativity |
| Participation in continuing education | Effective communication |
| Political awareness | |
| Reading professional journals | |
| Participation in nursing research | |

FIGURE 31-2 **Professional values, behaviors, and attributes.** (*Source:* Courtesy Mitchell, G. M., & Grippando, P. R. (1994). *Nursing perspective and issues.* Clifton Park, NY: Delmar Cengage Learning.)

(Mitchell & Grippando, 1994). Figure 31-2 lists some of the various values, behaviors, and attributes that may be exhibited by a "professional" (Mitchell & Grippando, 1994). Table 31-9 identifies many of the nursing theorists that have contributed to the development of the profession of modern nursing.

## THE FUTURE

As nursing goes forward into the future, the answer to whether nurses are professionals will continue to be developed by you, practitioners of the nursing profession, and

society as a whole (Table 31-10 and Table 31-11). Some of the answers depend on the development of nursing theory and the move to increase nursing academic credentials to a minimum of a baccalaureate degree. This degree is increasingly viewed as minimum professional preparation.

## NURSE-SENSITIVE OUTCOMES

The existence of professional nursing benefits patients who develop lower rates of nurse-sensitive outcomes. Nursing-sensitive indicators reflect the structure, process, and outcomes of nursing care. Nursing structure indicators

### Critical Thinking 31-1

You may use a Web site to contact your members of Congress/Senate at www.house.gov and www. senate.gov. You may also contact them through the U.S. Capitol at (202) 224-3121. How could you use these contact sites to improve care for your patients or advocate for an important health care issue or the professional role of the nurse?

### Case Study 31-2

Note the professional behaviors and attributes characteristic of a professional shown in Figure 31-2. Which professional behaviors and attributes do you have? Make a one-year and a five-year plan in any areas you feel that you need to work on. Use Table 31-10. Review your list with a professional mentor.

## TABLE 31-9  Selected Nursing Theorists and Their Models

| THEORIST | MODEL |
| --- | --- |
| Nightingale (1859) | Environmental Theory |
| Orem (1971) | Self-Care Deficit Theory |
| Peplau (1952) | Interpersonal Process |
| Roy (1976) | Adaptation Model |
| Henderson (1955) | Basic Needs |
| Paterson & Zderad (1976) | Humanistic Nursing |
| Levine (1969) | Conservation Theory |
| Neuman (1972/1995) | Systems Model |
| Rogers (1970) | Science of Unitary Beings |
| Watson (1979/1989) | Human Caring Theory |
| King (1971) | Goal Attainment Theory |
| Parse (1981/1995) | Human Becoming Theory |
| Erickson | Modeling and Role-Modeling Theory |
| Leininger | Transcultural Nursing |
| Orlando (Pelletier) | Nursing Process Theory |
| Kolcaba | Comfort Theory |
| Benner-Caring | Clinical Wisdom and Ethics in Nursing Practice |
| Eriksson | Theory of Caritative Caring |
| Boykin & Schoenhofer | Nursing as Caring |
| Mishel | Uncertainty in Illness Theory |

Compiled with information from Nursing Theories (2011). Retrieved May 18, 2011 from http://currentnursing.com/nursing_theory/introduction.html.

### Critical Thinking 31-2

Review how the nursing theories of Nightingale, King, Orem, Benner, Roy, and others influence your practice of nursing as a professional (Nursing Theories, 2011). Pick one of the theories, and comment on how your practice of nursing may fit with the theory.

measure the supply, skill level, and education/certification of nursing staff. Nursing process indicators measure aspects of nursing care such as assessment, intervention, and RN job satisfaction. Outcome indicators that are determined to be nursing sensitive are those that improve if there is a greater quantity or quality of nursing care (e.g., pressure ulcers, falls, and intravenous infiltrations) (American Nurses Association, 2010). Nursing-sensitive outcomes include lower death rates of patients from one of the following life-threatening complications: pneumonia, shock, cardiac arrest, urinary tract infection, gastrointestinal bleeding, sepsis, deep vein thrombosis, and "failure to rescue" (Needleman, Buerhaus, Mattke, Stewart, & Zelevinsky, 2002; Simpson, 2005).

## TABLE 31-10  Plan for Professionalism

| Behavior/Attribute | One-Year Goals | Five-Year-Goals |
|---|---|---|
| Appearance* | | |
| Time-management skills | | |
| Self-discipline | | |
| Licensure/certification | | |
| Institutional/community activity participation | | |
| Continuing education | | |
| Political awareness | | |
| Professional journals | | |
| Nursing research participation | | |
| Stress management | | |
| Exercise** | | |
| Initiative | | |
| Motivation | | |
| Creativity | | |
| Effective communication | | |

*U.S. Department of Health and Human Services. National Heart, Lung, and Blood Institute. Calculate Your Body Mass Index. Retrieved April 23, 2011, from http://www.nhlbisupport.com/bmi/bminojs.htm.

**Centers for Disease Control and Prevention. (2011b). Physical Activity for Everyone. Retrieved April 23, 2011 from http://www.cdc.gov/physicalactivity/everyone/guidelines/adults.html.

## TABLE 31-11  Developing a Professional Style

1. Assess your current education and experience.

2. As you start your new nursing role, review the following on your unit:

   Most common
   - medical diagnoses
   - nursing diagnoses

(Continues)

**TABLE 31-11** (Continued)

- medications and IV solutions
- diagnostic tests
- laboratory tests
- nursing and medical interventions and treatments

3. Set goals for any additional education and experience that you may need specific to the patient care unit you are working on and the community that you serve.

4. Review your own job description and the role and the job description of nursing and other health care and medical staff you work with.

5. Identify the names and contact information of all nursing, medical, and health care staff you work with.

6. Discuss delegation with your preceptor, and observe how the preceptor delegates to others.

7. Observe the impact of delegation on both the delegator and the person delegated to.

8. Remember the Golden Rule: do unto others as you would want them to do unto you.

9. Recognize that, under the law, the RN holds the responsibility and accountability for nursing care.

10. Communicate assertively. Work at being direct, open, and honest in your new role.

11. Exercise your power with kindness to all. Participate in professional committees at work.

12. Hold others accountable for their responsibilities as spelled out in their job description.

13. Be open to performance improvement feedback about your personal delegation style.

14. Modify your communication approach to fit the needs of patients, staff, and yourself.

15. Monitor your care, to assure that your patients receive high-quality, evidence-based care.

16. Monitor literature so you are up to date on evidence-based care for patients.

17. Monitor data that show that your patients are satisfied; are pain-free; feel cared about; are complication-free; and have no nurse-sensitive outcomes.

18. Offer professional nursing service to your patients and community.

19. Speak up about the important role that nurses play in preventing patient complications.

20. Network with other professionals and join your professional organization.

21. Participate in professional committees at work.

22. Receive professional salary and benefits.

23. Take good care of yourself and strive for professional and personal balance. Maintain a healthy diet and exercise program.

*(Continues)*

## TABLE 31-11 (Continued)

24. Dress like a professional. This tends to convey a higher level of knowledge and a sincere interest in advancement. A disheveled appearance may give the impression of a disinterested, marginal performer.

25. Communicate pride in being a nurse. Expect to be treated like a professional!

26. Refuse to be part of horizontal violence—e.g., bullying, gossiping, criticism, innuendo, scapegoating, undermining, intimidation, witholding information, insubordination, ostracizing, and physical, verbal, and passive aggression—toward other nurses and health care workers

## Critical Thinking 31-3

Have you ever witnessed a "failure to rescue" or witnessed a nurse "rescuing" a patient? Would the patient have lived if the nurse was not present? How will you rescue your patients?

## EVIDENCE FROM THE LITERATURE

**Citation:** Wu, Y., Larrabee, J. H., & Putman, H. P. (2006). Caring behaviors inventory. *Nursing Research, 55*(1), 18–25.

**Discussion:** A short instrument for patient use is presented to measure the process of nurse caring. Elements of the patient instrument include patient assessment of the knowledge and skill, respectfulness, connectedness of the nurse, and so forth. Some of the areas presented to be assessed by the nurse include responding quickly to the patient's call; knowing how to give shots, IVs, and so on; attentively listening to the patient; and spending time with the patient. All 24 elements of the Caring Behaviors Inventory are presented.

**Implications for Practice:** Nurses interested in measuring the impact of their patient care can use this instrument to do so and thereby improve their practice.

## KEY CONCEPTS

- The NCLEX-RN Test Plan reviews patient needs in the following areas:
  safe, effective care environment; physiologic integrity; psychosocial integrity; and health promotion and maintenance.
- There are several new types of questions on the NCLEX-RN.

- Possible predictors of NCLEX-RN success are identified in Table 31-2.

- Nursing exit examinations are given by many schools of nursing.

- It is useful to have a plan to review any NCLEX-RN weaknesses.

- NCLEX-RN tips can help you prepare for the examination.
- A medication guide and a food guide are useful to study for the NCLEX.
- There are many characteristics of a profession.
- There are many professional values, behaviors, and attributes.
- Nurses regularly take nursing action to prevent the incidence of nurse-sensitive outcomes in their patients.
- Several nursing theorists have identified a theory of nursing.

## KEY TERM

NCLEX

## REVIEW QUESTIONS

1. Which of the following isolation measures would the nurse institute when a patient has tuberculosis? Select all that apply.
   _____ A. Use gloves when there is risk of exposure to blood or body fluids.
   _____ B. Use gloves at all times when caring for patient.
   _____ C. Place patient in a private, negative airflow pressure room.
   _____ D. Use a mask at all times while in the patient's room.
   _____ E. Place a mask on the patient when transporting him or her out of the room.

2. The nurse is caring for a patient with Alzheimer's disease. Which of the following factors are associated with this disease? Select all that apply.
   _____ A. Reversible disease
   _____ B. Amyloid plaques
   _____ C. Acute onset
   _____ D. Personality changes
   _____ E. Impaired memory

3. A three-year-old girl is admitted with epiglottitis. She is drooling, sitting upright, unable to swallow, and looking panicky. Which of the following equipment is most important for the nurse to place at the patient's bedside?
   A. Suction
   B. Croup tent
   C. Tracheotomy set
   D. Padded bedsides for seizure precautions

4. A nurse has just drawn arterial blood gases on a patient. Which of the following is important for the nurse to do?
   A. Apply pressure to the puncture site for 5 minutes.
   B. Shake the vial of blood before transporting it to the lab.
   C. Keep the patient on bed rest for 1 hour.
   D. Encourage the patient to cough and deep-breathe.

5. The nurse is administering an intramuscular injection to a 1-year-old patient. Which of the following sites is most appropriate for the nurse to select?
   A. Ventral forearm
   B. Ventral gluteal
   C. Vastus lateralis
   D. Dorsal gluteal

6. A low-sodium, low-fat diet has been prescribed for a patient with hypertension. The nurse knows that the patient understands his diet when he selects which of the following menus?
   A. Steak, baked potato, peas, and hot coffee
   B. Macaroni and cheese casserole, tossed salad with dressing, and hot chocolate
   C. Baked chicken, steamed broccoli and cauliflower, steamed rice, and hot tea
   D. Fried chicken, mashed potatoes, green beans, and milk

7. The nurse is caring for a patient who is having a panic attack. Which symptom will the patient be least likely to exhibit?
   A. Choking
   B. Chest pain
   C. Bradycardia
   D. Fear of going crazy

8. The nurse is caring for a patient with angina. Nitroglycerin is prescribed for what purpose?
   A. To assist smooth muscles to contract
   B. To increase venous return to the heart
   C. To reduce preload and afterload
   D. To slow and strengthen the heart rate

9. Which special precaution must the nurse take when assisting a patient with self-monitoring of blood glucose?
   A. Wear gloves when performing the test.
   B. Rinse the lancet between uses.
   C. Recalibrate the glucometer before each use.
   D. Give the patient a machine for his use only.

10. The nurse is caring for a postoperative patient who develops a wound dehiscence. Which of the following will the nurse do when this occurs?
    A. Approximate the wound edges with tape.
    B. Irrigate the wound with sterile saline.
    C. Cover the wound with sterile, moist saline dressings.
    D. Hold the abdominal contents in place with a sterile, gloved hand.

# REVIEW ACTIVITIES

1. Set up a group to study for the NCLEX with several of your friends. Arrange to meet to discuss how your NCLEX review is going. Have each member of the group buy a NCLEX review book from a different publisher. Practice answering 60 questions for one hour daily. Don't mark your answers in the review book. Share your review books with each other to increase your exposure to various authors' test questions.

2. Review Flexner's characteristics of a profession in Figure 31-1. Is nursing a profession?

3. Review the medication guide in Table 31-7. Become comfortable with the root/prefix/suffix for different categories to increase your ability to recognize medications. Add nursing implications to the Table as you study.

# EXPLORING THE WEB

- Search for NCLEX review books at www.amazon.com How many different books published in the past three years did you see on the topic of NCLEX review?

- Go to www.ncsbn.org Click on NCLEX Examinations. Explore the site.

- Go to www.learningext.com Note: NCSBN's review for the NCLEX is offered through this NCSBN learning extension. This self-paced, online review features NCLEX-style questions, interactive exercises, topic-specific course exams, and a diagnostic pretest that can help you develop a personal study plan. Visit this site every Monday to see its new NCLEX-RN sample test question.

# REFERENCES

American Nurses Association (2010). Nursing-Sensitive Indicators. Retrieved October 28, 2010, from http://www.nursingworld. org/MainMenuCategories/ThePracticeofProfessionalNursing/ PatientSafetyQuality/Research-Measurement/The-National-Database/Nursing-Sensitive-Indicators_1.aspx.

Assessment Technologies Institute (2011). About ATI. Retrieved April 20, 2011, from http://www.atitesting.com/About.aspx.

Beck, A.T., Emery, G., and Greenberg, R.L. (2005). Anxiety Disorders And Phobias: A Cognitive Perspective. Basic Books.

Centers for Disease Control and Prevention. (2011a). Adult Immunization Schedule. Retrieved April 23, 2011 from http://www. cdc.gov/vaccines/recs/schedules/adult-schedule.htm.

Centers for Disease Control and Prevention. (2011b). Physical Activity for Everyone. Retrieved April 23, 2011 from http://www.cdc.gov/ physicalactivity/everyone/guidelines/adults.html.

DiBartolo, M. C., & Seldomridge, L.A. (2005). A review of intervention studies to promote NCLEX-RN success. Nurse Educator, 30(4), 166–171.

Evolve Learning System. (2011). HESI Frequently Asked Questions. Retrieved April 20, 2011, from https://evolve.elsevier.com/ staticPages/hesi-faq.html.

Gallup. (2010). Nurses Top Honesty and Ethics List for 11th Year. Retrieved April 22, 2011, from http://www.gallup.com/poll/145043/ Nurses-Top-Honesty-Ethics-List-11-Year.aspx.

Gordon, S., & Nelson, S. (2006). Moving beyond the virtue script in nursing in the complexities of care. Ithaca, NY: ILR Press.

HealthyMind.com. (2004). Cognitive Distortions. Retrieved April 23, 2011 from http://www.healthymind.com/s-distortions.html

Kesselman-Turkel, J. & Peterson, F. (2007). Test-Taking Strategies. University of Wisconsin Press. Madison, WI.

Manthey, M. (2002). The Practice of Primary Nursing. Creative Healthcare Management. Minneapolis, MN.

Mitchell, G. M., & Grippando, P. R. (1994). Nursing perspectives and issues. Clifton Park, NY: Delmar Cengage Learning.

National Council of State Boards of Nursing, Inc. (2011c). NCSBN Learning Extension. Weekly "NCLEX Style" Question and online NCLEX Review course. Retrieved April 20, 2011, from http:// learningext.com/pages/home.

Needleman, J., Buerhaus, P., Mattke, S., Stewart, M., & Zelevinsky, K. (2002). Nurse-staffing levels and the quality of care in hospitals. New England Journal of Medicine, 346(22), 1715–1722.

Nibert, A., Young A., & Adamson, C. (2002). Predicting NCLEX success with the HESI Exit Exam: Fourth annual validity study. Computers, Informatics, Nursing, 20(6), 261–267.

Nursing Theories (2011). Retrieved May 18, 2011 from http://current-nursing.com/nursing_theory/introduction.html.

Pearson Education, Inc (2011). Online Tutorial for NCLEX® Examinations. Retrieved April 20, 2011, from http://www.pearsonvue. com/nclex.

Rosenthal, H. G., (2004). Test anxiety prevention. London, UK: Routledge Pub.

Seligman, M. E. P. (2011). Flourish: A New Understanding of Happiness and Well-Being - and How To Achieve Them. Simon & Schuster.

Simpson, K. R. (2005). Failure to rescue: Implications for evaluating quality of care during labor and birth. Journal of Perinatal Neonatal Nursing, 19(1), 24–36.

Stein, A. M. (2005). NCLEX-RN review (5th ed.). Clifton Park, NY: Delmar Cengage Learning.

The National Council of State Boards of Nursing. (2010a). NCLEX-RN Examination Test Plan for the National Council Licensure Examination for Registered Nurses. Retrieved April 20, 2011, from https://www.ncsbn.org/2010_NCLEX_RN_TestPlan.pdf.

The National Council of State Boards of Nursing. (2010b). Alternate Item Formats Frequently Asked Questions. Retrieved April 20, 2011, from https://www.ncsbn.org/2334.htm

The National Council of State Boards of Nursing. (2011a). NCLEX Administration Frequently Asked Questions. Retrieved

April 20, 2011, from https://www.ncsbn.org/2325.htm#Why_do_you_need_palm_vein_reading_if_you_have_fingerprints.

U.S. Department of Health and Human Services. National Heart, Lung, and Blood Institute. Calculate Your Body Mass Index. Retrieved April 23, 2011 from http://www.nhlbisupport.com/bmi/bminojs.htm.

Wu, Y., Larrabee, J. H., & Putnam, H. P. (2006). Caring behaviors inventory. *Nursing Research, 55*(1), 18–25.

## SUGGESTED READINGS

Adamson, C., & Britt, R. (2009). Repeat testing with the HESI Exit Exam-sixth validity study. *Computer Informatics in Nursing, 27*(6), 393–397.

Aucoin, J. W., & Treas, L. (2005). Assumptions and realities of the NCLEX-RN. *Nursing Education Perspective, 26*(5), 268–271.

Bondmass, M. D., Moonie, S., & Kowalski, S. (2008). Comparing NET and ERI standardized exam scores between baccalaureate graduates who pass or fail the NCLEX-RN. *International Journal of Nursing Education Scholarship, 5,* Article16. Epub. 2008, April 1.

Bonis, S., Taft, L., & Wendler, M. C. (2007). Strategies to promote success on the NCLEX-RN: An evidence-based approach using the ACE Star model of knowledge transformation. *Nursing Education Perspectives, 28*(2), 82–87.

Davenport, N. C. (2007). A comprehensive approach to NCLEX-RN success. *Nursing Education Perspectives, 28*(1), 30–33.

Dorsey, D. (2006). Most nurses understand that the purpose of the National Nursing Licensure Examination (NCLEX) is to assess a candidate's ability to provide safe, effective nursing care upon entry into practice in the United States and its territories. *Nursing Outlook, 54*(5), 261–262.

Downey, T. A. (2008). Predictive NCLEX success with the HESI Exit Examination: Fourth annual validity study. *Computer Informatics in Nursing, 26*(5 Suppl.), 35S–36S.

Frith, K. H., Sewell, J. P., & Clark, D. J. (2005). Best practices in NCLEX-RN readiness preparation for baccalaureate student success. *Computer Informatics in Nursing, 23*(6), 322–329.

Giddens, J. F. (2009). Changing paradigms and challenging assumptions: Redefining quality and NCLEX-RN pass rates. *Journal of Nursing Education, 48*(3), 123–124.

Giddens, J., & Gloeckner, G. W. (2005). The relationship of critical thinking to performance on the NCLEX-RN. *Journal of Nursing Education, 44*(2), 85–90.

Greenspan, V. C., Springer, P., & Ray, K. A. (2009). Tri-nodal model for NCLEX-RN success. *Nurse Educator, 34*(3), 101–102.

Hahn, J. (2010). Integrating professionalism and political awareness into the curriculum. *Nurse Educator, 35*(3), 110–113.

Hart, K. (2005, Nov.). What do men in nursing really think? *Nursing,* 46–48.

Henderson, D., Sealover, P., Sharrer, V., Fusner, S., Jones, S., Sweet, S., et al. (2006). Nursing EDGE: Evaluating delegation guidelines in education. *International Journal of Nursing Education Scholarship, 3,* Article 15. Epub. 16.

Herrman, J. W., Johnson, A. N. (2009). From beta-blockers to boot camp: Preparing students for the NCLEX-RN. *Nursing Education Perspectives, 30*(6), 384–388.

Higgins, B. (2005). Strategies for lowering attrition rates and raising NCLEX-RN pass rates. *Journal of Nursing Education, 44*(12), 541–547.

Jacobs, P., & Koehn, M. L. (2006). Implementing a standardized testing program: Preparing students for the NCLEX-RN. *Journal of Professional Nursing, 22*(6), 373–379.

Johnson, A. N. (2009). NCLEX-RN success with boot camp. *Nursing Education Perspectives, 30*(5), 328–329.

Kurtzman, E. T., & Jennings, B. M. (2008). Capturing the imagination of nurse executives in tracking the quality of nursing care. *Nursing Administration Quarterly, 32*(3), 235–246.

Lewis, C. (2005). *Predictive accuracy of the HESI Exit Exam on NCLEX-RN pass rates and effects of progression policies on nursing student exit exam scores.* Unpublished doctoral dissertation, Texas Woman's University, Denton, TX.

Lyons, E. M. (2008). Examining the effects of problem-based learning and NCLEX-RN scores on the critical thinking skills of associate degree nursing students in a Southeastern Community College. *International Journal of Nursing Education Scholarship, 5,* Article21. Epub. 5.

Major, D. A. (2005). OSCE's—seven years on the bandwagon: The progress of an objective structured clinical evaluation programme. *Nursing Education Today, 25*(6), 442–454.

Manojlovich, M., & Talsma, A. (2007). Identifying nursing processes to reduce failure to rescue. *Journal of Nursing Administration, 37*(11), 504–509.

McDowell, B. M. (2008). KATTS: A framework for maximizing NCLEX-RN performance. *Journal of Nursing Education, 47*(4), 183–186.

McGann, E., & Thompson, J. M. (2008). Factors related to academic success in at-risk senior nursing students. *International Journal of Nursing Education Scholarship, 5,* Article 19. Epub 15.

Miller, J. C. (2005). Tips on taking the NCLEX-RN. *Imprint, 52* (1), 28–32.

Morin, K. H. (2006). Use of the HESI Exit Examination in schools of nursing. Commentary from the perspective of an expert in academic policy. *Journal of Nursing Education, 45*(8), 308–309.

Morrison, S. (2005). Improving NCLEX-RN pass rates through internal and external curriculum evaluation. In M. Oermann & K. Heinrich (Eds.), *Annual review of nursing education: Strategies for teaching assessment and program planning* (Vol. 3, pp. 77–94). New York: Springer Publishing Co.

Morrison, S., Free, K. W., & Newman, M. (2008). Do progression and remediation policies improve NCLEX-RN pass rates? *Computer Informatics in Nursing, 26*(5 Suppl.), 67S–69S.

Morton, A. M. (2008). Improving NCLEX scores with structured learning assistance. *Computer Informatics in Nursing, 26*(5 Suppl.), 89S–91S.

NCLEX practice questions. (2005). *Nursing, 35*(8), 68–70.

Newman, M., Britt, R. B., & Lauchner, K. A. (2008). Computer predictive accuracy of the HESI Exit Exam: A follow-up study. *Informatics in Nursing, 26*(5 Suppl.), 16S–20S.

Newton, S. E., & Moor, G. (2009). Use of aptitude to understand bachelor of science in nursing student attrition and readiness for the National Council Licensure Examination-Registered Nurse. *Journal of Professional Nursing, 25*(5), 273–278.

Nibert, A. (2005). Benchmarking for progression: Implications for students, faculty, and administrators. In L. Caputi (Ed.), *Teaching nursing: The art and science* (Vol. 3, pp. 314–335). Chicago: College of DuPage Press.

Nibert, A. T., Adamson, C., Young, A., Lauchner, K. A., Britt, R. B., & Hinds, M. N. (2006). Choosing a theoretical framework to guide HESI Exit Examination research. *Journal of Nursing Education, 45*(8), 303–307.

Norton, C. K., Relf, M. V., Cox, C. W., Farley, J., Lachat, M., Tucker, M., & Murray, J. (2006). Ensuring NCLEX-RN success for first-time test-takers. *Journal of Professional Nursing, 22*(5), 322–326.

O'Neill, T. R., Marks, C. M., & Reynolds, M. (2005). Reevaluating the NCLEX-RN passing standard. *Journal of Nursing Measures, 13*(2), 147–165.

Oermann, M. H., Saewert, K. J., Charasika, M., & Yarbrough, S. S. (2009). Assessment and grading practices in schools of nursing: National survey findings part I. *Nursing Education Perspectives, 30*(5), 274–278.

Romeo. E. M. (2010). Quantitative research on critical thinking and predicting nursing students' NCLEX-RN performance. *Journal of Nursing Education, 1*, 1–9.

Schwarz, K. A. (2005). Making the grade: Help staff pass the NCLEX-RN. *Nursing Management, 36*(3), 38–44.

Sifford, S., & McDaniel, D. M. (2007). Results of a remediation program for students at risk for failure on the NCLEX exam. *Nursing Education Perspectives, 28*(1), 34–36.

Sitzman, K. L. (2007). Diversity and the NCLEX-RN: A double-loop approach. *Journal of Transcultural Nursing, 18*(3), 271–276.

Spector, N., & Alexander, M. (2006). Exit exams from a regulatory perspective. *Journal of Nursing Education, 45*(8), 291–292.

Spurlock, D. R., & Hunt, L. A. (2008). A study of the usefulness of the HESI Exit Exam in predicting NCLEX-RN failure. *Journal of Nursing Education, 47*(4), 157–166.

Sutherland, J. A., Hamilton, M. J., & Goodman, N. (2007). Affirming At-Risk Minorities for Success (ARMS): Retention, graduation, and success on the NCLEX-RN. *Journal of Nursing Education, 46*(8), 347–353.

Uyehara, J., Magnussen, L., Itano, J., & Zhang, S. (2007). Facilitating program and NCLEX-RN success in a generic BSN program. *Nursing Forum, 42*(1), 31–38.

Wendt, A., & Harmes, J. C. (2009). Developing and evaluating innovative items for the NCLEX: Part 2, item characteristics and cognitive processing. *Nurse Educator, 34*(3), 109–913.

Woo, A., Wendt, A., & Liu, W. (2009). NCLEX pass rates: An investigation into the effect of lag time and retake attempts. *JONA's Healthcare Law Ethics Regulation, 11*(1), 23–26.

Wood, R. M. (2005). Student computer competence and the NCLEX-RN examination: Strategies for success. *Computer, Informatics, Nursing, 23*(5), 241–243.

Wray, K., Whitehead, T., Setter, R., & Treas, L. (2006). Nursing use of NCLEX preparation strategies in a hospital orientation program for graduate nurses. *Administration Quarterly, 30*(2), 162–177.

# ABBREVIATIONS

| | | | | |
|---|---|---|---|---|
| AACN | American Association of Critical-Care Nurses | | CCRN | Critical Care Registered Nurse |
| AACN | American Association of Colleges of Nursing | | CCU | Coronary Care Unit |
| AAHP | American Association of Health Plans | | CDC | Centers for Disease Control and Prevention |
| AAN | American Academy of Nursing | | CEO | Chief Executive Officer |
| AANA | American Association of Nurse Anesthetists | | CEU | Continuing Education Unit |
| AARP | American Association of Retired Persons | | CFO | Chief Financial Officer |
| ACLS | Advanced Cardiac Life Support | | CHF | Congestive Heart Failure |
| ACNP | Acute Care Nurse Practitioner | | CINAHL | Cumulative Index to Nursing and Allied Health Literature |
| ACS | American Cancer Society | | | |
| ADA | American Dietetic Association | | CIS | Clinical Information System |
| ADL | Activity of Daily Living | | CMP | Comprehensive Metabolic Panel |
| ADN | Associate Degree in Nursing | | CMS | Centers for Medicare and Medicaid Services |
| AHA | American Hospital Association | | CN3 | Clinical Nurse 3 |
| AHRQ | Agency for Healthcare Research and Quality | | CNA | Canadian Nurses Association |
| AIDS | Acquired Immune Deficiency Syndrome | | CNM | Certified Nurse Midwife |
| AMA | American Medical Association | | CNS | Clinical Nurse Specialist |
| ANA | American Nurses Association | | CNS/NP | Clinical Nurse Specialist/Nurse Practitioner |
| ANCC | American Nurses Credentialing Center | | | |
| AONE | American Organization of Nurse Executives | | COBRA | Consolidated Omnibus Budget Reconciliation Act |
| APHA | American Public Health Association | | | |
| AWHONN | Association of Women's Health, Obstetric, and Neonatal Nurses | | CON | Certificate of Need |
| | | | COPC | Community-Oriented Primary Care |
| BLS | Basic Life Support | | CPR | Cardiopulmonary Resuscitation |
| BMI | Body Mass Index | | CQI | Continuous Quality Improvement |
| BSN | Bachelor of Science in Nursing | | CRNA | Certified Registered Nurse Anesthetist |
| BTIPA | Brooks' Theory of Intrapersonal Awareness | | CU | Consumers Union |
| CAMH | Comprehensive Accreditation Manual for Hospitals | | CVA | Cerebrovascular Accident |
| | | | DM | Disease Management |
| CARING | Capital Area Roundtable on Informatics in Nursing | | DNR | Do Not Resuscitate |
| | | | DRG | Diagnosis-Related Group |
| CCQHC | Consumer Coalition for Quality Health Care | | EBC | Evidence-Based Care |

| | |
|---|---|
| EBM | Evidence-Based Medicine |
| EBN | Evidence-Based Nursing |
| EBP | Evidence-Based Practice |
| EHR | Electronic Health Record |
| EMTALA | Emergency Medical Treatment and Active Labor Act |
| ENA | Emergency Nurses Association |
| ENIAC | Electronic Numerical Integrator and Computer |
| ERCP | Endoscopic Retrograde Cholangiopancreatography |
| ERG | Existence-Relatedness-Growth Theory |
| ERISA | Employee Retirement Income Security Act |
| ET | Enterostomal Therapy |
| FDA | Food and Drug Administration |
| FTE | Full-Time Equivalent |
| GI LAB | Gastrointestinal Laboratory |
| HCFA | Health Care Financing Administration |
| HIMSS | Health Information and Management Systems Society |
| HIPAA | Health Insurance Portability and Accountability Act |
| HIV | Human Immunodeficiency Virus |
| HMO | Health Maintenance Organization |
| HRSA | Health Resources and Services Administration |
| HSA | Health Savings Account |
| IADL | Instrumental Activity of Daily Living |
| ICN | International Council of Nurses |
| ICU | Intensive Care Unit |
| IHS | Indian Health Service |
| IOM | Institute of Medicine |
| IRA | Individual Retirement Account |
| JBIEBNM | Joanna Briggs Institute for Evidence- Based Nursing & Midwifery |
| JC | Joint Commission |
| LOS | Length of Stay |
| LPN/LVN | Licensed Practical Nurse/Licensed Vocational Nurse |
| MBNQA | Malcolm Baldrige National Quality Award |
| MBTI | Myers-Briggs Type Indicator |
| MDI | Metered-Dose Inhaler |
| MEDLARS | Medical Literature Analysis and Retrieval System |
| MeSH | Medical Subject Headings |
| MIS | Medical Information System |
| MRI | Medical Records Institute |
| MS-HUG | Microsoft Healthcare Users Group |
| MSN | Master's Degree in Nursing |
| NANDA | North American Nursing Diagnosis Association |
| NANN | National Association of Neonatal Nurses |
| NAP | Nursing Assistive Personnel |
| NAPNAP | National Association of Pediatric Nurses and Practitioners |
| NAPQ | Nosek-Androwich Profit: Quality Matrix |
| NCLEX | National Council of State Boards of Nursing Licensure Examination |
| NCQA | National Committee on Quality Assurance |
| NCSBN | National Council of State Boards of Nursing |
| NGC | National Guideline Clearinghouse |
| NHPPD | Nursing Hours Per Patient Day |
| NIH | National Institutes of Health |
| NLRB | National Labor Relations Board |
| NLM | National Library of Medicine |
| NLN | National League for Nursing |
| NNP | Neonatal Nurse Practitioner |
| NQF | National Quality Forum |
| NP | Nurse Practitioner |
| NRP | Neonatal Resuscitation Program |
| NWIG-AMIA | Nursing Working Informatics Group of the American Medical Informatics Association |
| OB | Organizational Behavior |
| OR | Operating Room |
| OSHA | Occupational Safety and Health Administration |
| P4P | Pay For Performance |
| PALS | Pediatric Advanced Life Support |
| PC | Personal Computer |
| PCA | Patient Care Associate |
| PCS | Patient Classification System |
| PDSA | Plan Do Study Act |
| PERT | Program Evaluation and Review Technique |
| P-F-A | Purpose-Focus-Approach |
| PHN | Public Health Nurse |
| PI | Performance Improvement |
| POD | Postoperative Day |
| POS | Point of Service |
| POSDCORB | Planning, Organizing, Supervising, Directing, Coordinating, Reporting, and Budgeting |
| PPO | Preferred Provider Organization |
| PPS | Prospective Payment System |
| QA | Quality Assurance |
| QI | Quality Improvement |
| QSEN | Quality and Safety Education for Nurses |
| RVU | Relative Value Unit |
| SBARR | Situation-Background-Assessment-Recommendation-Response |
| SCHIP | State Children's Health Insurance Program |
| SPAN | Staff Planning and Action Network |

| | | | |
|---|---|---|---|
| **SWOT** | Strengths, Weaknesses, Opportunities, Threats | **UTI** | Urinary Tract Infection |
| **TB** | Tuberculosis | **VA** | Veterans Affairs |
| **TEFRA** | Tax Equity and Fiscal Responsibility Act | **VHA** | Veterans Health Administration |
| **TQI** | Total Quality Improvement | **VAK** | Visual, Auditory, Kinesthetic |
| **TQM** | Total Quality Management | **VR** | Virtual Reality |
| **UC** | Ubiquitous Computing | **WHO** | World Health Organization |
| **URL** | Universal Resource Locator | **WOC NURSE** | Wound, Ostomy, Continence Nurse |
| **USDHHS** | United States Department of Health and Human Services | **WWW** | World Wide Web |

# GLOSSARY

**360-degree feedback**   System in which an individual is assessed by a variety of people to provide a broader perspective.

**absenteeism**   The rate of employee absences from work.

**accountability**   Being responsible and answerable for actions or inactions of self or others in the context of delegation.

**accounting**   Activity that nurse managers engage in to record and report financial transactions and data; it assists with budget documentation.

**acculturated**   Becoming similiar to people in a new country by adopting their conditions, customs, and language.

**activities of daily living**   Activities related to toileting, bathing, grooming, dressing, feeding, mobility, and verbal and written personal communication.

**activity log**   Time-management technique to assist in determining how time is used by periodically recording activities.

**administrative law**   Body of law created by administrative agencies in the form of rules, regulations, orders, and decisions to protect the rights of citizens.

**administrative principles**   General principles of management that are relevant to any organization.

**affective domain**   Learning domain centered on attitudes, or what the learner feels and believes.

**assault**   Threat to touch another in an offensive manner without that person's permission.

**assign**   Act of a nurse directing an individual to do something the individual is already authorized to do (ANA and NCSBN, 2006)

**assignment**   Distribution of work that each staff member is to accomplish on a given shift or work period.

**authority**   The right to act or to command the action of others; it comes with the job and is required for a nurse to take action.

**autocratic leadership**   Centralized decision-making style with the leader making decisions and using power to command and control others.

**autonomy**   An individual's right to self-determination; individual liberty.

**balanced scorecard**   Device used to monitor customer perspectives; financial perspectives; internal processes and human resources; and learning and growth (Kaplan & Norton, 2004) for strategic management and as a way to examine performance throughout the organization.

**battery**   Touching of another person without that person's consent.

**behavioral objective**   Statement of specific and measurable behavior that should result from a teaching session.

**benchmark**   A quantitative or qualitative standard or point of reference used in measuring or judging quality or value.

**benchmarking**   The continuous measurement of a process, product, or service compared to those of the toughest competitor, to those considered industry leaders, or to similar activities in the organization in order to find and implement ways to improve it (The Joint Commission, 2009).

**beneficence**   The duty to do good to others and to maintain a balance between benefits and harms.

**bioethics**   Ethics specific to health care; serves as a framework to guide behavior in ethical dilemmas.

**bureaucratic organization**   Hierarchy with clear superior-subordinate communication and relationships, based on positional authority, in which orders from the top are transmitted down through the organization via a clear chain of command.

**capital budget**   Accounts for the purchase of major new or replacement equipment.

**career planning**   Ongoing process that involves a personal and professional self-assessment, setting goals, searching for job opportunities, researching potential employers, preparing a cover letter and resume, and participating in an interview including follow-up.

**case management**   Collaborative process of assessment, planning, facilitation, and advocacy for options and services to meet an individual's health needs through communication and available resources to promote quality, cost-effective outcomes (McCullough, 2009).

**change**   Making something different from what it was.

**change agent**   One who is responsible for implementation of a change project.

**civil law**   That body of law that governs how individuals relate to each other in everyday matters.

**clinical information system**   Comprehensive, integrated information system that concentrates on patient-related and clinical-practice-related data.

**clinical judgment**   Interpretation or conclusion about a patient's needs, concerns, or health problems, and/or the decision to take action (or not), use or modify standard approaches, or improvise new ones as deemed appropriate by the patient's response (Tanner, 2008).

**clinical ladder**   A promotional model that acknowledges that staff members have varying skill sets based on their education and experience. As such, depending on skills and experience, staff members may be rewarded differently and carry differing responsibilities for patient care and the governance and professional practice of the work unit.

**clinical pathway**   Care management tool that outlines the expected clinical course and outcomes for a specific patient type.

**coercive power**   Influence in the form of ability to administer punishment or take disciplinary actions against others to influence them to change their behaviors.

**cognitive domain**   Learning domain centered on knowledge, or what the learner knows.

**communication**   An interactive process that occurs when a person (the sender) sends a verbal or nonverbal message to another person (the receiver) and receives feedback.

**comfort**   Immediate state of being strengthened through having the human needs for relief, ease, and transcendence addressed in the context of the four dimensions of the physical, psychospiritual, sociocultural, and environmental experience.

**competence**   Ongoing ability (of a nurse) to integrate and apply the knowledge, skills, judgment, and personal attitudes required to practice safely and ethically in a designated role and setting (Canadian Nurses Association, 2004).

**conflict**   Disagreement about something of importance to the people involved.

**connection power**   Extent to which nurses are connected with others having power.

**consensus**   All group members can live with and fully support the decision, regardless of whether they totally agree.

**consideration**   Activities that focus on the employee and emphasize relating and getting along with people.

**constitution**   A set of basic laws that specifies the powers of the various segments of the government and how these segments relate to each other.

**construction budget**   Developed when renovation or new structures are planned; it typically includes items such as labor, materials, building permits, inspections, and equipment.

**contingency theory**   Style that acknowledges that other factors in the environment influence outcomes as much as leadership style and that leader effectiveness is contingent upon or depends upon something other than the leader's behavior.

**contract law**   Rules that regulate certain transactions between individuals and/or legal entities such as businesses. Also governs transactions between businesses.

**co-payment**   Fixed health care fee paid by the patient to the health care provider at the time of service; this amount is paid in addition to the money the health care provider will receive from the insurance company.

**cost shifting**   Process of assigning financial charges from one cost center to another cost center.

**cost center**   Subsection or unit of an organization created to track financial data.

**critical thinking**   Thinking about your thinking while you're thinking in order to make your thinking better.

**culture**   A broad term that includes the beliefs, customs, and patterns of behavior and institutions of a particular group of people.

**cultural competence**   Providing culturally sensitive care through behaviors, attitudes, and policies that are congruent within health care.

**dashboard**   Documentation tool providing a snapshot image of pertinent information and activity reflecting a point in time.

**decision making**   Considering and selecting interventions from a repertoire of actions that facilitate the achievement of a desired outcome.

**deductible**   A predetermined out-of-pocket fee paid by a patient for health care services before reimbursement through health insurance begins to be paid.

**delegation**   Transfer of responsibility for the performance of a task from one individual to another while retaining accountability for the outcome.

**democratic leadership**   Style in which participation is encouraged and authority is delegated to others.

**desired optimal outcomes**   Best possible objectives to be achieved given the resources at hand.

**diagnostic-related groups**   Patient groupings, approximately 500 Medicare groups, established for hospital reimbursement purposes; these groupings are identified by patient diagnosis or condition, surgical procedure, age, comorbidity, or complications, and are expected to use similar hospital resources and have similar needs and outcomes.

**direct care**   Time spent providing hands-on care to patients.

**direct patient care activities**   Patient care activities that include patient contact, such as bathing, providing medications, and so forth.

**direct expenses**   Expenses that are directly associated with patient care (for example, medical and surgical supplies and drugs).

**disease management**   System of coordinated health care interventions and communications for populations with conditions in which patient self-care is significant.

**employee-centered leadership**   Style with a focus on the human needs of subordinates.

**enabling objective**   Objective that identifies secondary behaviors that contribute to, or enable, achievement of terminal objectives.

**episodic care unit**   Unit that sees patients for defined episodes of care; examples include dialysis or ambulatory care units.

**ethical dilemma**   A conflict between two or more ethical principles for which there is no correct decision.

**ethics**   The doctrine that the general welfare of society is the proper goal of an individual's actions rather than egoism; the branch of philosophy that concerns the distinction between right from wrong on the basis of a body of knowledge, not just on the basis of opinions.

**ethnicity**   Identifies a person or group based on religious or national cultural group.

**ethnocentrism**   Belief that one's own culture or ethnic group is better than all other groups.

**evaluation**   Process of determining the success of teaching; it can measure the patient's learning and the teaching's effectiveness.

**evidence-based care (EBC)**   Recognized by nursing, medicine, health care institutions, and health policy makers as care based on state-of-the-art science reports. It is a process approach to collecting, reviewing, interpreting, critiquing, and evaluating research and other relevant literature for direct application to patient care.

**evidence-based nursing (EBN)**   Integration of the best evidence available, nursing expertise, and the values and preferences of the individuals, families, and communities who are served (Sigma Theta Tau International, 2005).

**evidence-based practice (EBP)**   The conscientious, explicit, and judicious use of current best evidence in making decisions about the care of individual patients; means integrating individual clinical expertise with the best available external clinical evidence from systematic research.

**expert power**   Power derived from the knowledge and skills nurses possess.

**external forces**   Influences originating outside the organization, for example, the labor force and the economy.

**false imprisonment**   Occurs when people are incorrectly led to believe they cannot leave a place.

**fidelity**   The principle of promise keeping; the duty to keep one's promise or word.

**fixed costs**   Expenses that are constant and are not related to productivity or volume.

**focus groups**   Small groups of individuals selected because of a common characteristic (for example, a specific patient population, patients in day surgery, new diabetics, and so on) who are invited to meet in a group and respond to questions about a topic in which they are expected to have interest or expertise.

**formal leadership**   When a person is in a position of authority or in a sanctioned role within an organization that connotes influence.

**full-time equivalent**   Measure of the work commitment of an employee who works five days a week or forty hours per week for fifty-two weeks per year.

**functional health status**   Ability to care for oneself and meet one's human needs.

**functional nursing**   Care delivery model that divides the nursing work into functional roles that are then assigned to one of the team members.

**gap**   The space between where the organization is and where it wants to be.

**gap analysis**   An assessment of the differences between expected magnet requirements and the organization's current performance on those requirements.

**generation**   A group that shares birth years as well as a common connection or bond based upon significant and influential life events.

**goal**   Specific aim or target that the unit wishes to attain within the time span of one year.

**Good Samaritan laws**   Laws that have been enacted to protect the health care professional from legal liability for actions rendered in an emergency when the professional is giving service without pay.

**grapevine**   An informal communication channel where information moves quickly and is often inaccurate.

**gross domestic product (GDP)**   Economic measure of a country's national income and output within a year; reflects the market value of goods and services produced within the country.

**groupthink**   Phenomonon that occurs when the desire for harmony and consensus overrides members' rational efforts to appraise the situation.

**Hawthorne effect**   Term coined to reflect the findings of a research study that demonstrated that change in employee behavior occurs as a result of being observed.

**health**   State of complete physical, social, and mental well-being, and not merely the absence of disease or infirmity.

**health assets**   Health-promoting attributes of individuals, families, communities, and systems.

**health care disparity**   Difference in the quality of health care that an individual receives that is not justified by the underlying health condition, access related factors, or treatment preferences of the patient.

**health care systems disparities**   Differences in health care system access and quality of care for different racial, ethnic, and socioeconomic population groups that persist across settings, clinical areas, age, gender, geography, health needs, and disabilities.

**health care transparency**   Ability to discover information about health care costs, medical errors, or practice preferences, preferably before receiving the service; also known as truth in reporting.

**health determinants**   Variables that include biological, psychosocial, environmental (physical and social), and health systems factors or etiologies that may cause changes in the health status of individuals, families, groups, populations, and communities. Health determinants may be assets (positive factors) or risks (negative factors).

**health disparities**   Differences in health risks and health status measures that reflect the poorer health status that is found disproportionately in certain population groups.

**health information system**   Automated or manual system that uses people, machines, and/or methods to collect, process, transmit, and disseminate data about health care in general.

**health literacy**   Learner's ability to read, understand, and act on health information.

**health-related quality of life**   Those aspects of life that are influenced either positively or negatively by one's health status and health risk factors.

**health risk factors**   Modifiable and nonmodifiable variables that increase or decrease the probability of illness or death.

**health status**   Level of health of an individual, family, group, population, or community; the sum of existing health risk factors, level of wellness, existing diseases, functional health status, and quality of life.

**high quality-of-work-life environments**   A type of work environment in which the quality of the human experience in the workplace meets and surpasses employee expectations.

**high-performance organizations**   An organization that operates in a way that brings out the best in people and produces sustainable high performance over time.

**HIPAA privacy rule**   Rule which protects all individually identifiable health information held or transmitted by a covered entity or its business associate, in any form or media, whether electronic, paper, or oral.

**horizontal violence**   Aggressive behavior towards individuals or group members by others (Hastie, 2002); also known as bullying.

**hospital information system**   Comprehensive, integrated information system designed to manage the administrative and financial information in a hospital.

**indirect care**   Time spent on activities that support patient care but are not done directly to the patient.

**indirect expenses**   Expenses that refer to such items as utilities, gas, electric, and phones, that are not directly related to patient care.

**indirect patient care activities**   Activities which are often necessary to support the patients and their environment, and only incidentally involve direct patient contact.

**informal leader**   Individual who demonstrates leadership outside the scope of a formal leadership role or as a member of a group, rather than as the head or leader of the group.

**information power**   Nurses who influence others with the information they provide to the group are using information power.

**initiating structure**   Style that involves an emphasis on the work to be done, a focus on the task and production.

**innovation**   "Process for inventing something new or improving on that which already exists" (Blakeney, Carleton, & Coakley, 2009, p. 2).

**inpatient unit**   Hospital unit that provides care to patients twenty-four hours a day, seven days a week.

**instrumental activities of daily living**   Activities related to food preparation and shopping; cleaning; laundry; home maintenance; verbal, written, and electronic communications; financial management; and transportation, as well as activities to meet social and support needs, manage health care needs, access community services and resources, and meet spiritual needs.

**intellectual capital**   An individual's knowledge, skills, and abilities that have value and portability in a knowledge economy.

**interpersonal communication**   Concerned with communication between individuals.

**intrapersonal communication**   Self-talk.

**intuitive thinking**   A type of discernment or insight that nurses develop that helps them to act in certain situations.

**invasion of privacy**   Intrusion into the personal life of another, without just cause, which can give the person whose privacy has been invaded a right to bring a lawsuit for damages against the person or entity that intruded.

**job-centered leaders**   Style that focuses on schedules, cost, and efficiency with less attention to developing work groups and high-performance goals.

**job satisfaction**   How organizational members feel about their job.

**justice**   The principle of fairness that is served when an individual is given that which he or she is due, owed, deserves, or can legitimately claim.

**knowledge workers**   Health care professionals who are well educated and technologically savvy and see themself as owning their intellectual capital.

**laissez-faire leadership**   Passive and permissive style in which the leader defers decision making.

**leader-member relations** Feelings and attitudes of followers regarding acceptance, trust, and credibility of the leader.

**leadership** Process of influence whereby the leader influences others toward goal achievement.

**learner analysis** Process of identifying the learner's unique characteristics and needs.

**learning domains** Taxonomies, or classifications, of learning.

**learning organization** Learning organizations promote professional practice through the encouragement of personal mastery, an awareness of mental models, and team learning.

**learning style** Particular manner in which an individual responds to and processes learning.

**legitimate power** Power derived from the position a nurse holds in a group; it indicates the nurse's degree of authority.

**lesson plan** Document that provides the blueprint for the teaching session; it lists the objectives, topics, format, strategies, materials, and evaluation used in the teaching session.

**living will** Document voluntarily signed by patients that specifies the type of care they desire if and when they are in a terminal state and cannot sign a consent form or convey this information verbally.

**magnet hospitals** High-quality health care organizations that have met the rigorous nursing excellence requirements as determined by the American Nurses Credentialing Center (ANCC) and that are a supportive and collegial practice setting that incorporates principles of organizational behavior to achieve positive individual, group, and organizational outcomes.

**maintenance or hygiene factors** Elements such as salary, job security, working conditions, status, quality of supervision, and relationships with others that prevent job dissatisfaction (Herzberg, 1968).

**malpractice** Professional's wrongful conduct in discharge of professional duties or failure to meet standards of care for the profession, which results in harm to another individual entrusted to the professional's care.

**management** Process of coordinating actions and allocating resources to achieve organizational goals.

**management process** Function of planning, organizing, coordinating, and controlling.

**margin** Profit.

**marginalization** Separation of a group away from the mainstream because of religious or cultural beliefs.

**marketing** Process of creating a product or health care service for patients which uses the four Ps of marketing: Patient, Product, Price, and Placement, to place desirable health care services or products in desirable locations at a price that benefits both patients and the health care facility.

**meta-analysis** Systematic review that uses quantitative methods to summarize the results of multiple studies. It often produces a summary statistic that represents the effects of an intervention across multiple studies and, therefore, is more precise than individual findings from any one study used in the review (Ciliska, Cullum, & Marks, 2001).

**methodology** Structured, standardized approach for developing teaching.

**mission** A formal expression stating the primary purpose of the practice unit, i.e., its reason for being.

**mission statement** A formal expression of the purpose or reason for existence of the organization.

**modular nursing** Care delivery model that is a kind of team nursing that divides a geographical space into modules of patients, with each module having a team of staff led by an RN to care for them.

**money market account** Similar to a bank checking account though it often requires a larger minimum amount of money to open the account and often has a higher interest rate for money.

**morality** Behavior in accordance with custom or tradition; usually reflects personal or religious beliefs.

**motivation** Whatever influences our choices and creates direction, intensity, and persistence in our behavior.

**motivation factors** Elements such as achievement, recognition, responsibility, advancement, and the opportunity for development that contribute to job satisfaction (Herzberg, 1968).

**NCLEX** The national nursing licensure examination prepared under the supervision of the National Council of State Boards of Nursing.

**negligence** Failure to provide the care a reasonable person would ordinarily provide in a similar situation.

**nonmaleficence** The principle of doing no harm.

**nonproductive hours** Paid time not devoted to patient care; includes benefit time such as vacation, sick time, and education time.

**novice to expert model** Framework that when developed into a clinical or career promotion ladder, facilitates professional staff development by building on the skill sets and experience of each practitioner. Benner's model acknowledges that there are tasks, competencies, and outcomes that practitioners can be expected to have acquired based on five levels of experience: novice, advanced beginner, competent, proficient, and expert.

**nursing assistive personnel (NAP)** Unlicensed personnel to whom nursing tasks are delegated and who work in structured nursing organizations.

**nursing informatics** Specialty that integrates nursing science, computer science, and information science to manage and communicate data, information, knowledge, and wisdom in nursing practice.

**nursing-sensitive indicators** Measures that reflect the outcome of nursing action.

**objective**    Measurable step that must be taken to reach a goal.

**operational budget**    Account for the income and expenses associated with day-to-day activity within a department or organization.

**open systems**    Entities that must interact with the environment to survive.

**organization**    A coordinated and deliberately structured social entity consisting of two or more individuals functioning on a relatively continuous basis to achieve a predetermined set of goals.

**organizational behavior**    The study of human behavior in organizations.

**organizational change**    Planned change in an organization to generally improve efficiency.

**organizational culture**    Mix of deep underlying assumptions, beliefs, and values that are shared by members of an organization and typically operate unconsciously.

**organizational commitment**    How committed or loyal employees feel to the goals of the organization.

**organizational effectiveness**    An organization's sustainable high performance in accomplishing its mission and objectives.

**palliative care**    Active, total care of patients whose disease is not responsive to curative treatment.

**Pareto principle**    Principle, developed by Pareto, a 19th century economist, which states that 20% of focused effort results in 80% of outcome results, or conversely that 80% of unfocused effort results in 20% of results.

**patient acuity**    Measure of nursing workload that is generated for each patient.

**patient-centered care**    Care delivery model in which care and services are brought to the patient.

**patient classification system (PCS)**    System for distinguishing among different patients based on their acuity, functional ability, or need for nursing care in order to predict staffing needs and the cost of nursing care.

**patient-focused care**    A model of differentiated nursing practice that emphasizes quality, cost, and value.

**personal change**    Alteration made voluntarily for one's own reasons, usually for self-improvement.

**philosophy**    Statement of beliefs based on core values and rational investigations of the truths and principles of knowledge, reality, and human conduct.

**philosophy of an organization**    A value statement of the principles and beliefs that direct an organization's behavior.

**physical health**    Encompasses nutrition and exercise coupled with a balanced amount of rest; health preventive behaviors such as avoiding smoking; and health screening behaviors that detect health problems early such as an annual Pap smear.

**political voice**    An increase in the number of voices supporting or opposing an issue.

**politics**    Process by which people use a variety of methods to achieve their goals.

**population-based health care practice**    Development, provision, and evaluation of multidisciplinary health care services to population groups experiencing increased health risks or disparities, in partnership with health care consumers and the community in order to improve the health of the community and its diverse population groups.

**population-based nursing practice**    Practice of nursing in which the focus of care is to improve the health status of vulnerable or at-risk population groups within the community by employing health promotion and disease prevention interventions across the health continuum.

**position power**    Degree of formal authority and influence associated with the leader.

**power**    Ability to create, get, and use resources to achieve one's goals.

**power of attorney**    Legal document executed by an individual (principal) granting another person (agent) the right to perform certain activities in the principal's name.

**practice guideline**    Descriptive tool or standardized specifications for care of the typical patient in the typical situation; these guidelines are developed by a formal process that incorporates the best scientific evidence of effectiveness and expert opinion. Synonyms or near synonyms include practice parameter, preferred practice pattern, algorithm, protocol, and clinical standard.

**primary nursing**    Care delivery model that clearly delineates the responsibility and accountability of the RN and places the RN as the primary provider of nursing care to patients.

**problem solving**    Active process which starts with a problem and ends with a solution.

**process**    Set of causes and conditions that repeatedly come together in a series of steps to transfer inputs into outcomes.

**productive hours**    Hours worked and available for patient care.

**productivity**    Quantity and quality of output an employee generates for an organization.

**professional change**    Alteration made in position or job such as obtaining education or credentials.

**professional judgment**    Intellectual (educated, informed) process that a nurse exercises in forming an opinion and reaching a clinical decision based upon an analysis of the available evidence.

**profit**    Determined by the relationship of income to expenses.

**progressive discipline**    System in which the manager and employee's mutual goal is to take steps to correct performance in order to bring it back to an acceptable level; it offers a stepwise process with opportunities for continued feedback and clarification of expectations.

**protective factors**   Patient strengths and resources that patients can use to combat health threats that compromise core human functions.

**psychomotor domain**   Learning domain centered on skills, or what the learner does.

**public law**   General classification of law, consisting generally of constitutional, administrative, and criminal law. Public law defines a citizen's relationship with government.

**quality assurance**   Inspection approach to ensure that minimum standards of patient care quality are maintained in health care institutions.

**quality improvement**   Systematic process of organization-wide participation and partnership in planning and implementing continuous improvement methods to understand and meet or exceed customer needs and expectations.

**quality of life**   Level of satisfaction one has with the actual conditions of one's life, including satisfaction with socioeconomic status, education, occupation, home, family life, recreation, and the ability to enjoy life, freedom, and independence.

**race**   Geographical or global human population distinguished by genetic traits and physical characteristics such as skin color or facial features.

**referent power**   Power derived from how much others respect and like any individual, group, or organization.

**reflective thinking**   Watching or observing ourselves as we perform a task or make a decision about a certain situation.

**resilience**   The ability to cope and adapt to adversity which is a desirous quality to have in the stress filled environment of health care (McAllister & McKinnon, 2009).

**resources**   People, money, facilities, technology, and rights to properties, services, and technologies.

**respect for others**   Acknowledgement of the right of people to make their own decisions.

**responsibility**   Reliability, dependability, and the obligation to accomplish work when one accepts an assignment.

**resume**   Brief summary of your background, training, and experience as well as your qualifications for a position.

**revenue**   Income generated through a variety of means (for example, billable patient services, investments, and donations to the organization).

**reward power**   The ability to reward others to influence them to change their behavior.

**Risk-management program**   Program in health care organization that is designed to identify and correct systemic problems that contribute to errors in patient care or to employee injury.

**SBARR technique**   (Situation, Background, Assessment, Recommendation, Response) Approach designed to improve communication among health care personnel and improve patient safety.

**self-scheduling**   Process by which staff on a unit collectively decide and implement the monthly work schedule.

**sentinel event**   Unexpected occurrence involving death or serious physical or psychological injury to a patient.

**shared governance**   Situation where nurses and managers work together to define their roles and expected outcomes, holding everyone accountable for their role and expected outcomes.

**situational leadership**   A framework that maintains that there is no one best leadership style, but rather that effective leadership lies in matching the appropriate leadership style to the individual's or group's level of motivation and task-relevant readiness.

**skill mix**   Percentage of RN staff compared to other direct care staff (LPNs and NAP).

**sources of power**   Combination of conscious and unconscious factors that allow an individual to influence others to do as the individual wants.

**spiritual distress**   A NANDA nursing diagnosis where an individual has an impaired ability to integrate meaning and purpose in life through the individual's connectedness with self, others, art, music, literature, nature, or a power greater than oneself.

**staffing plan**   Plan that articulates how many and what kind of staff are needed, by shift and day, to staff a unit or department.

**stakeholder**   Provider, employer, customer, patient, or payer who may have an interest in, and seek to influence, the decisions and actions of an organization, for example, competitors, suppliers, government, and regulatory agencies.

**stakeholder assessment**   A systematic consideration of all potential stakeholders to ensure that the needs of each of these stakeholders are incorporated in the planning phase.

**strategic plan**   The sum total or outcome of the processes by which an organization engages in environmental analysis, goal formulation, and strategy development with the purpose of organizational growth and renewal.

**strategic planning**   A process that is designed to achieve goals in dynamic, competitive environments through the allocation of resources.

**substitutes for leadership**   Variables that may influence or have an effect on followers to the same extent as the leader's behavior.

**supervision**   Provision of guidance or direction, oversight evaluation, and followup by the licensed nurse for accomplishment of a nursing task delegated to NAP.

**SWOT analysis**   A tool that is frequently used to conduct environmental assessments. SWOT stands for Strengths, Weaknesses, Opportunities, and Threats.

**system**   Interdependent group of items, people, or processes with a common purpose.

**systematic review**    Summary of evidence on a particular topic that uses a rigorous process for retrieving, critically appraising, and synthesizing studies in order to answer a question about a burning clinical question (Melnyk & Fineout-Overholt, 2005).

**task structure**    Involves the degree that work is defined, with specific procedures, explicit directions and goals.

**taxonomy**    System that orders principles into a grouping or classification.

**team**    Small number of people with complementary skills who are committed to a common purpose, performance goals, and approach for which they are mutually accountable.

**team nursing**    Care delivery model that assigns staff to teams that then are responsible for a group of patients.

**terminal objective**    Objective that identifies major behaviors that contribute to achievement of the overall session goal.

**Theory X**    View that in bureaucratic organizations, employees prefer security, direction, and minimal responsibility; coercion, threats, or punishment are necessary because people do not like the work to be done.

**Theory Y**    View that in the context of the right conditions, people enjoy their work, they can show self-control and discipline, are able to contribute creatively and are motivated by ties to the group, the organization, and the work itself; belief that people are intrinsically motivated by their work.

**Theory Z**    View of collective decision making and a focus on long-term employment that involves less direct supervision.

**time management**    Set of related common-sense skills that helps individuals use time in the most effective and productive way possible.

**tort**    Civil wrong resulting in injury to another.

**transformational leader**    Leader who is committed to a vision that empowers others.

**turnover**    Number of employees who resign divided by the total number of employees during the same time period.

**values**    Personal beliefs about the truth of ideals, standards, principles, objects, and behaviors that give meaning and direction to life.

**variable costs**    Costs that vary with volume and that will increase or decrease depending on the number of patients.

**variance**    Difference between what was budgeted and the actual cost.

**veracity**    The obligation to tell the truth.

**vision statement**    Statement which tells us how the people of the organization plan to actualize the mission.

**voting block**    Group that represents the same political position or perspective.

**whole systems shared governance**    When the entire organization adopts an organizational structure based on the principles of partnership, equity, accountability, and ownership.

# INDEX

## C

# D

## E

# H

# M